Technical Topics Scrapbook 1995 - 99

Pat Hawker, G3VA

Diagrams drawn by
Bob Ryan

Radio Society of Great Britain

Published by the Radio Society of Great Britain, Cranborne Road, Potters Bar, Herts EN6 3JE.

First published in book form 2000

ISBN 1 872309 61 5

Cover design: TYGA, Potters Bar
Production: Mike Dennison
Reproduction: JJ Typographics, Southend on Sea

Printed in Great Britain by Nuffield Press, Abingdon

Contents

Preface

This is the third quinquennial volume of the *Technical Topics Scrapbook* series. It contains the complete collected pages of sixty monthly *Technical Topics* columns as originally published from January 1995 to December 1999 in *RadCom*, the journal of the Radio Society of Great Britain. As for the two earlier *Scrapbook* volumes (1985-89 and 1990-94) it continues the policy, established over the 40-plus years that I have been writing and compiling *Technical Topics*, of bringing together a miscellaneous collection of new and established circuit ideas, HF/VHF antennas, power supply systems, hints and tips, new technology and comments felt to be of practical value or of interest to those who run, or wish to run, an amateur radio station on a modest budget and are experimentally minded..

For those new readers who may not have previously encountered this long-lasting column, it began in April 1958 and has continued ever since. For the first ten years it appeared six times a year but it has been a monthly column since January 1968. Early material appeared in digested and rearranged form in 1965 under the title *Technical Topics for the Radio Amateur* followed by a further six editions as *Amateur Radio Techniques,* the final seventh edition appearing in the first of several reprints in 1980. This has now been superseded by the three *Scrapbooks* each containing some 300 pages of material as they appeared in *RadCom.*

Introducing this new column in 1958, I wrote: "We cannot promise that this new feature will solve all difficulties. All we can hope to do is to survey ideas from the technical press, provide a few hints and tips that have come to notice; with perhaps an occasional comment thrown in for good measure." Ever since, despite the near-revolutionary changes in technology, the original formula has been followed other than the additional bonus that many readers worldwide have submitted their own technical ideas, observations and comments, making *TT* a unique forum for experimental Amateur Radio.

Technical Topics must have broken many records in having been compiled without breaks by a single author for more than 40 years. It has always concentrated on providing in understandable language with a minimum of mathematics, information that has real or potential practical value for everyday Amateur Radio operation, relevant and/or interesting to the majority of both professional and non-professional experimenters, and all those who remain fascinated by the marvels of radio communication. Latest developments are intermingled with the occasional look back to ideas and technology used in the earlier days of this remarkable hobby, often showing that half-forgotten antennas and techniques still deserve a fresh look even in this increasingly computer-dominated age of information technology. Simplicity and KISS ("keep it simple, stupid!") approaches have been a prime aim, but every effort is made to introduce amateurs to advances in practical aspects of basic radio communications receivers, transmitters, frequency control, simple wire antennas, the effects of the sunspot cycles on the propagation of HF signals, anomalous VHF propagation intermingled with the provision of electric power supplies for domestic, field and mobile operation, including the latest developments in batteries and fuel cells.

The final quinquennium of the 20th Century has seen the role of the Amateur as Experimenter more important than ever as professional and military research funds have increasingly been directed at broadband digital satellite communications and personal UHF systems. The phenomenal growth of the Internet and short distance personal/mobile communications at UHF is an expression of the universal importance of the need to communicate - indeed the modern compact handheld units owe much to the pioneering work of amateurs that led first to military and clandestine units, and then to the simple super-regenerative transceivers that at one time dominated the Citizens Band market. As long ago as October 1987, *TT* noted that "Already telecommunications people are thinking beyond the concepts of mobile cellular radio and the cordless telephone to an era of universal digital portable radio communications".

Optical fibres and space satellites have not spelt the end of HF and VHF as an ideal medium for amateur experimentation and enjoyment. But the competition has led to renewed efforts to improve digital communications on HF with new modes such as PSK31 and the resurgence of Hellschreiber in electronic computer-software controlled systems. New families of antennas based on fractals have emerged. New surface-mounted chips such as FST3235 have been adopted by radio amateurs for use in H-mode super-linear mixers (described first in *TT* items included in this volume). New, improved frequency stabilisers have come from the late Klaus Spaargarten, PA0KSB to whose work I paid tribute in the December 1999 *TT* but also including a "fast" version devised by Peter Lawton, G7IXH (December 1997) suitable for use with variable-frequency oscillators up to at least 30MHz.

The release of narrow bands of frequencies in the LF spectrum, first for a limited period at 73kHz and more recently at 136kHz, with an effective radiated power of up to one watt has proved a great incentive for experimentally minded amateurs. LF transmitters using relatively low-cost MOSFET or bipolar semiconductors or still available RF or audio power valves can readily be used. A number of ideas and circuit diagrams for LF will be found in this volume. There is also a comprehensive review of the advantages and disadvantages of thermionic valves (tubes) compared to semiconductors in audio and RF practice in the November 1998 pages of this volume. The age-old problem of dry joints gets an airing in the July 1998 pages, and there are many tips for re-activating some older receivers and transceivers.

TT has always aimed at providing a launching pad for controversial experimental ideas and designs. This means inevitably that claimed results may not always prove reproducable. Examples in this volume include two examples drawn originally from professional presentations: (1) The belief that the relatively good results that can be achieved with compact tuned loop antennas might be explained by their acting as short folded dipoles. It has since become clear that although there is significant dipole-mode radiation from small transmitting loops it is far from being the dominant mode and that despite the excellent results that can be achieved the radiation-efficiency of small transmitting loops is usually low. (2) The suggestion by American academics that extremely compact "meander" antennas with a span on 14MHz of only one-metre (see August 1999) can have an efficiency and bandwidth virtually equal to that of a conventional half-wave dipole has proved impossible to validate, although there is no doubt that meander dipoles with a size of one-half normal span can be highly efficient. Another suggestion that proved to have a basic flaw was the idea that the two halves of a folded dipole could be fed in push-push (March 1996). But against these interesting if doubtful ideas can be set virtually hundreds of useful ideas and reproduceable antennas and circuit-designs that are adding to the fund of Amateur Radio technology.

I have found that most Radio Amateurs, whether old-timers or newcomers, share my interest in the way radio has developed during the past Century. This has moved from an era of spark transmitters and cat-whisker receivers, the development of the thermionic valve and long-distance low-power transmission in the 1920s, the significant role played by radio amateurs in the UK and in Europe in the wartime clandestine radio links that required new ideas for equipment that could be carried in an attache case, to the post-war development of the transistor and the subsequent implementation of the integrated circuit.

The development of the high-performance communications receiver from the 1930s onwards has emphasised that good and still usable receivers such as the valved Racal RA17 and the

Hammarlund Super-Pro or the German WW2 Koeln E52 with its interesting but flawed variable bandwidth crystal filter - all of which feature in the following pages - can still provide results comparable to the latest "bells and whistles" transceivers, despite the increasing use of digital signal processing which in the better designs provide excellent selective filtering.

The origins of electricity and radio go back a long way. 1999 marked the bicentennial of the development by Volta of the first source of current electricity - the silver-zinc battery - leading to the later discoveries of Faraday, Hertz, Lodge and Marconi on which our hobby depends (see the March 1999 pages). Among the historical miscellany noted in the June 1999 pages was the discovery by an American amateur that the very first person - the Grandfather of Amateur Radio - to suggest in the January 1898 issue of *The Model Engineer* that amateurs should experiment with wireless telegraphy and told them how to get started was a London electrical engineer.named Leslie Miller. Again, much interest was aroused in items determining just when the last telegrams were sent from ships using spark transmitters (March and May 1999).

The deposit of surviving files of the Special Operations Executive in the Public Record Office has brought to light a number of fresh details about wartime clandestine radio including the remarkable story of two Norwegian brothers whose skill in successfully concealing their suitcase radios almost beggars belief: For details turn to the December 1996 pages. Wartime clandestine equipment and operations including those of SIS can be found in a number of items.

In short, I hope and believe that this book will be of practical value and make interesting reading to all those who remain fascinated by the marvels of radio communication, now already heading into its second century of exciting developments.

Writing *TT* has always been part chore, part pleasure. It has proved to me, as I hope it will to you the reader that "it's fun finding out". A bonus has been that it has brought me in touch with large numbers of enthusiastic experimental amateurs and professionals to whom RF engineering has continued to be a source of fascination and wonder. Even if, in the writing, it has left me less time to spend on the air with Morse key or microphone!

Grateful acknowledgement is made to all the many radio-amateurs, professional engineers and scientists worldwide and to the many amateur-radio and professional engineering publications from which so much of the material has been drawn. It is my policy always, wherever possible, to acknowledge fully the source of all abstracts and ideas and to credit the originators. Acknowledgement is also made to Mike Dennison, G3XDV, the RSGB's Publications Manager, who conceived the idea of providing these *TT Scrapbooks* and persuaded the Society to publish them, and to his editorial staff during 1995 to 1999 who had the onerous task of producing the material, including, from 1997, Editor Steve White, G3ZVW, Technical Editor (until Novermber 1996) Peter Dodd, G3LDO, Technical Illustrator Bob Ryan 2E1EKS and Designer Suzanne Dunnett and her predecessors. For their patience and forebearance I am most grateful.

Pat Hawker, G3VA

London, March 2000

40 YEARS OF TECHNICAL TOPICS

BY GORDON L ADAMS, G3LEQ*

An Appreciation of the Work of Pat Hawker, G3VA

I DON'T KNOW about you, but my wife keeps telling me to clear out all my old radio magazines. Apparently she is worried about being left with them, in the event that I might not survive my sixth sunspot cycle! Recently I decided to venture into the loft, and started tugging at some of the piles of aged journals. The first heap tumbled over, leaving a selection of orange and green covered *RSGB Bulletins* at my feet.

Not being particularly inspired with the notion of throwing my hoard away, I started to thumb through the copies dated from November 1957 to July 1958. Strangely enough, the first copy contained a number of technical contributions from various radio amateurs who were trying to explain how the G5RV dipole antenna might work. The last copy contained an article by Louis Varney, G5RV, explaining how he came to design the aerial in the first place! Intrigued by this apparent coincidence, I picked up the April 1958 edition, and came across a column called *Technical Topics*, by Pat Hawker.

HUMBLE BEGINNINGS

TECHNICAL TOPICS was introduced as an attempt "to keep abreast of current technical progress and practice in the amateur radio field". It continued, "we cannot promise that this new feature will solve all your difficulties. All we can hope to do is survey a few ideas from the technical press; a few hints and tips that have come to our notice; with perhaps an occasional comment thrown in for good measure." Not only had I stumbled across the very first *Technical Topics*, but I was reading Pat Hawker's original modest aims for what has become a highly regarded institution. Yes, Pat has been writing for *The Bull*, later *Radio Communication*, and now *RadCom* for exactly 40 years!

I expect G3VA is known to all RSGB members through his *Technical Topics* column, but how many of us know just what a prolific reader and writer he is? Over the years he has written for *Electronics Weekly*, *Radio Bygones*, *Wireless World*, *Morsum Magnificat*, and the (house journal) *IBA Newsletter*.

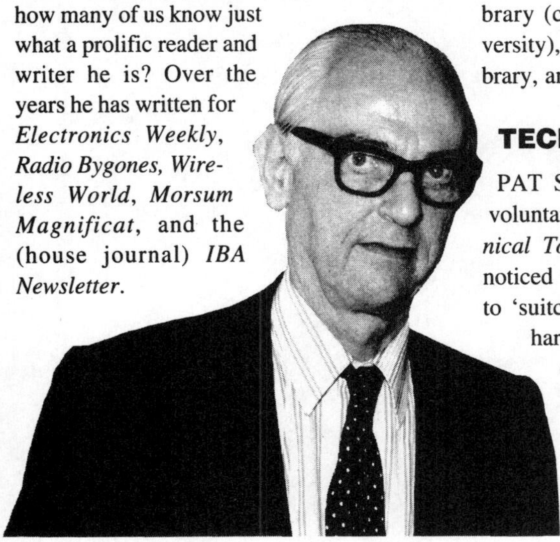

I imagine Pat as a veritable magpie, gathering together fascinating material on radio communication matters from innumerable sources. He lives not far from the Vintage Wireless Museum in West Dulwich, where, to this day, parts of the 600kW medium wave war-time transmitter codenamed 'Aspidistra' are still to be seen. Pat's research sources include the Science Museum Library (closely connected with London University), the National Science Reference Library, and IEE Library.

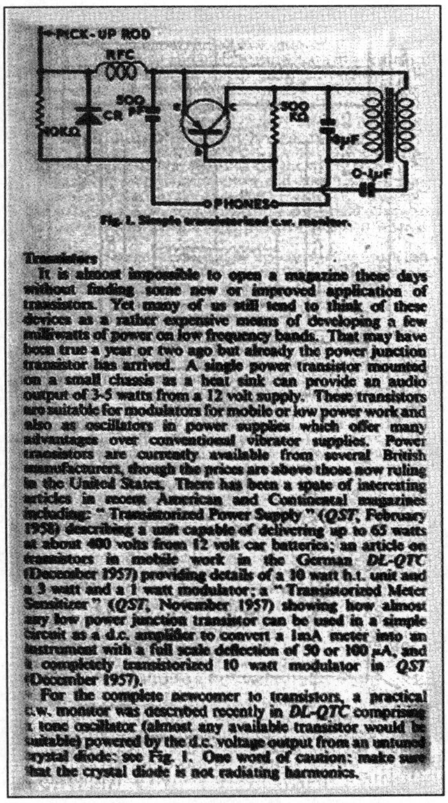

Fig. 1. Simple transistorized c.w. monitor.

Transistors

It is almost impossible to open a magazine these days without finding some new or improved application of transistors. Yet many of us still tend to think of these devices as a rather expensive means of developing a few milliwatts of power on low frequency bands. That may have been true a year or two ago but already the power junction transistor has arrived. A single power transistor mounted on a small chassis as a heat sink can provide an audio output of 3-5 watts from a 12 volt supply. These transistors are suitable for modulators for mobile or low power work and also as oscillators in power supplies which offer many advantages over conventional vibrator supplies. Power transistors are currently available from several British manufacturers, though the prices are above those now ruling in the United States. There has been a spate of interesting articles in recent American and Continental magazines including: "Transistorized Power Supply" (*QST*, February 1958) describing a unit capable of delivering up to 65 watts at about 400 volts from 12 volt car batteries; an article on transistors in mobile work in the German *DL-QTC* (December 1957) providing details of a 10 watt h.t. unit and a 3 watt and a 1 watt modulator; a "Transistorized Meter Sensitizer" (*QST*, November 1957) showing how almost any low power junction transistor can be used in a simple circuit as a d.c. amplifier to convert a 1mA meter into an instrument with a full scale deflection of 50 or 100 µA, and a completely transistorized 10 watt modulator in *QST* (December 1957).

For the complete newcomer to transistors, a practical c.w. monitor was described recently in *DL-QTC* comprising a tone oscillator (almost any available transistor would be suitable) powered by the d.c. voltage output from an untuned crystal diode: see Fig. 1. One word of caution: make sure that the crystal diode is not radiating harmonics.

From the RSGB archive, the first circuit diagram ever to appear in *TT*, along with its associated text.

TECHNICAL CAREER

PAT STARTED WORLD War Two as a voluntary interceptor, but those who read *Technical Topics* from start to finish will have noticed that he occasionally makes reference to 'suitcase transceivers'. From this it is not hard to determine that he was in 'special communications' during the last few years of WWII. Pressed on this point he admits to having been 'in the racket' - to use a somewhat esoteric expression - to which I can only reply that I was with 'the friends' too, albeit during the Cold War years. Pat recently attended a reunion of ex-Radio Security Service personnel.

For some 20 years Pat worked for the IBA, retiring from that body in 1987. He had an office at the IBA's HQ in Knightsbridge, piled high with technical books. Pat is a little cautious about his age, but, suffice to say, he is about the same age as the BBC, who celebrated a very important anniversary last year.

During a recent telephone discussion with Pat, I asked him if he had ever possessed a 'G7 plus two letter callsign'. Pat revealed that he might have worked a G7 whilst using a Belgian callsign from Germany, at the end of World War Two, but he wouldn't be any more specific. Let us all hope he considers writing his memoirs sometime.

One thing I know Pat wrote was the draft for a booklet on the 'History of the Radio Society of Great Britain', but sadly this was never published. However, one of the editions of the *RSGB Bulletin* on top of the next pile in my loft had a gold and red cover, and was dated July 1963. It contained some 12 pages which turned out to be based upon Pat's original, much longer manuscript of RSGB history.

THE TT YEARS

I SAT DOWN to get stuck into this, when the XYL's voice floated up from downstairs. "How are you getting on with clearing out all those old magazines?" "I'm having a slight problem moving all our suitcases and Christmas decorations," I replied.

The second ever *TT*, in May 1958, ended up with closing paragraphs entitled 'Here and There', and mentioned amongst other things a fully transistorised electronic Morse key. OK, so one of them was an OC16, but this was 1958!

I have only met Pat Hawker once, and that was on a London to Manchester train, when he started-up a conversation with me because he saw I was reading a copy of *RadCom*.

If you are a newcomer you can still benefit from Pat's wide experience, either by seeking-out a used copy of *Amateur Radio Techniques*, or, better still, you can purchase a copy of *Technical Topics Scrapbook 1985-1989* from the RSGB bookshop. Looking through some of the publications Pat has been responsible for, it's truly amazing how he has kept us all so well informed for 40 years.

Well, I suppose I shall have to throw away some of the old copies of *The Bull* now. On second thoughts, aren't people living longer these days? Perhaps I'll last until my seventh sunspot cycle, in which case those old *Bulls* can gather dust a while longer.

More power to your pen, Pat!

[Pat assures me he is not planning to stop writing TT at this time - *Ed*] ♦

* *2 Ash Grove, Knutsford, Cheshire WA16 8BB.*

Pat Hawker's Technical Topics

PAT HAWKER, G3VA
London 37 / SE22 8SS

LOW-NOISE OSCILLATORS

RECENT DISCUSSION IN *TT* (September and November 1994) on crystal and LC oscillators concentrated on the achievement of good frequency stability with only a brief mention of another factor that in recent years has become of increasing importance: the reduction of oscillator sideband noise. In September it was noted that the phase noise floor of a precision crystal oscillator can be reduced at the expense of the long term ageing rate. With LC oscillators, achievement of the lowest phase noise similarly requires contradiction of some of the classic requirements for high stability including the lowest possible output from the oscillator device.

But why is low phase noise important? As F N H Robinson noted in his book *Noise and Fluctuations in Electronic Devices and Circuits* (Clarendon Press, Oxford, 1974): "The treatment of oscillators in electronic texts is generally rather brief and mainly concerned with questions such as power output, efficiency and frequency stability (p172)" and "In general, except in nuclear resonance spectrometers and where an oscillator is required as a local oscillator in a superheterodyne system, noise in oscillators is not of much practical consequence. We do see, however, that as a general rule, whenever a few millivolts of signal are required, it is better to obtain it by attenuating a larger signal from a more powerful oscillator than to attempt to design a low-level oscillator (p184)."

Yet only a few years later we find Professor Mike Underwood, G3LHZ, in discussing 'Oscillator Noise Limitations' (IERE Conference Publication No 40, 1978) pointing out that "The noise sidebands of the oscillators in a radio system ultimately limit the interference generated on or received from adjacent channels. A new approach is outlined for deriving the fundamental limiting factors for oscillator noise performance. These are shown to be circuit Q, amplifier dynamic range and phase shift, and varactor RF voltage handling capability."

One of the first engineers to underline the importance of oscillator noise (in microwave systems) was W A Edson 'Vacuum Tube Oscillators' published in 1953, and first brought to the notice of *TT* readers by Walter Schreurer, K1YZW (see *ART7*. pp170-171). Edson wrote: "It is well known that the small voltages within solid conductors and the corresponding random emission of electrons within vacuum tubes set a lower limit on the magnitude of electrical signals which may be amplified and detected. . .It is not so commonly realised that noise voltages also affect the operation of oscillators. It is true that in most oscillator applications the effects of noise are quite small; but in some cases, for example in microwave oscillators used in superheterodyne receivers, the noise sidebands seriously restrict the choice of IF."

He showed that the output power from an oscillator spreads out over a band of frequencies in the type of curve one associates with high-Q tuned circuits: **Fig 1**. The noise output represents a form of jitter in the oscillator, and if fed into the mixer means that the receiver will have some (though much reduced) response to signals over a range corresponding to the spread of oscillator noise power: **Fig 2**. The purity of the output increases with in-

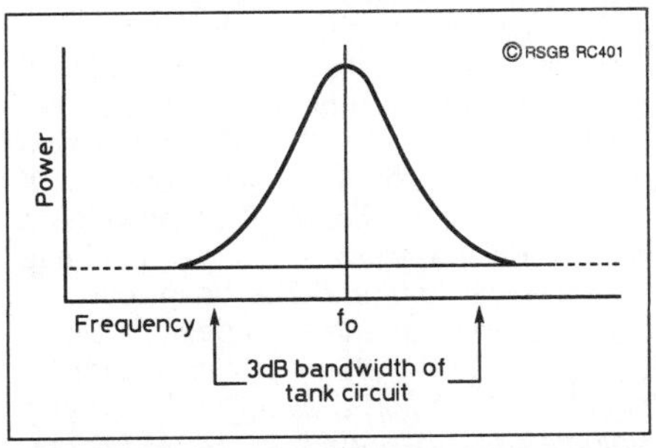

Fig 1: Noise sidebands around a carrier (fo). At the frequency corresponding to the 3dB points of the tank circuit, the noise spectrum begins to rise out of the noise plateau (-174dBm/Hz degraded by the amplifier noise figure) at the rate of 6d/B octave. That is to say the noise power per Hertz of bandwidth increases by four times each time the offset from the carrier is halved. A low noise oscillator requires a strong carrier output and maximum circuit Q. A carrier plus one sideband is mathematically equivalent to the simultaneous amplitude and angle modulation.

crease of oscillator output and selectivity, and with decrease of frequency and inherent noise of the valve or semiconductor device.

In HF and VHF superhet practice, appreciable local oscillator output is barely likely to spread significantly to the signal frequency although care should be taken that oscillator floor noise does not enter the IF channel. However, any significant jitter causes the receiver to respond to loud signals close to the desired signal frequency (Fig 2) due to reciprocal mixing (**Fig 3**). Similarly, if the transmitter oscillator has a corresponding spread of output it will be amplified in the main transmitter stages and there will be radiation in adjacent channels. The output may appear

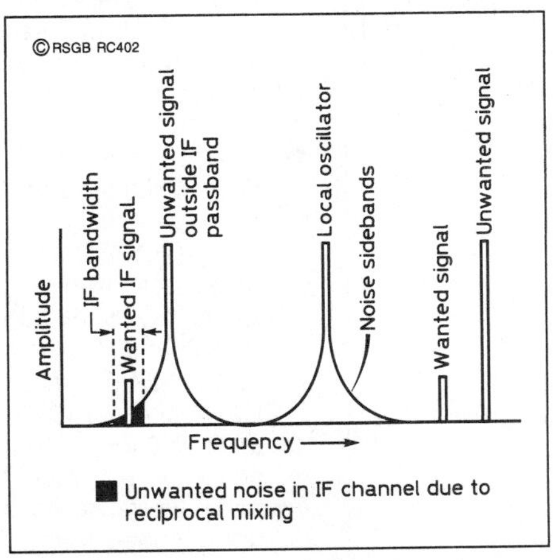

Fig 2: The mechanism whereby reciprocal mixing degrades the received SNR of weak signals in the presence of strong signals due to the noise sidebands of the local oscillator.

stable and apparently 'clean': **Fig 4** , when checked by normal techniques. At HF, oscillator noise falls off rapidly and these effects often only become apparent when trying to receive a weak signal alongside a very strong one. However oscillator noise effects in both receivers and transmitters tend to be much more significant where a PLL-type frequency synthesiser is used: **Fig 5**.

One of the problems in assessing the practical importance of oscillator noise in radio systems is the difficulty of measuring close-in noise without state-of-the-art high-grade professional instruments, seldom found even in otherwise well equipped laboratories. Lord Kelvin, that 19th century colossus of physics, once wrote: "When you can measure what you are speaking about, and express it in numbers, you know something about it; but when you cannot express it in numbers, your knowledge is of a meagre and unsatisfactory kind: it may be the beginning of knowledge, but you have scarcely, in your thoughts, advanced to the stage of science."

Colin Horrabin, G3SBI, in his quest for a super-linear receiver that could be duplicated by others, has devoted considerable effort to devising a phase-noise measurement set-up which enables him to assess the effect of component and other changes etc, on the phase noise of oscillators (**Fig 6**). He has continued to develop the super low-noise oscillator described in *TT*, July 1994. There is, unfortunately, a practical limit to the improvement in adjacent channel reception that can result from super-linear receivers, at least until the noise sidebands generated in transmitters is reduced to similarly low levels.

The July notes from G3SBI covered the development of a low phase-noise two-coil oscillator using a grounded-gate FET. The most promising version of this circuit used stripline inductors against a groundplane. A further development uses spiral printed circuit inductors to reduce the physical size and achieves a similar basic performance. The July *TT* provides a good example of the design approach to a good low-phase-noise oscillator.

The original work was done with only a high-quality spectrum analyser available to make phase-noise measurements. Subsequently, his phase noise measuring system shows that with an oscillator carrier frequency of 72MHz a noise performance of the order of -153dBc/Hz occurs at about 20kHz. The aim is to develop a frequency synthesiser using

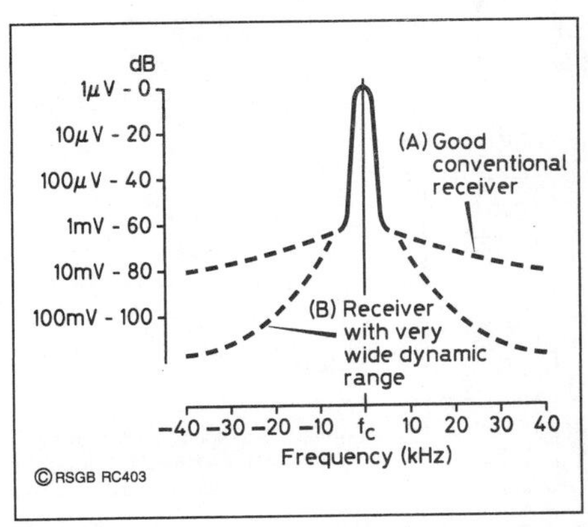

Fig 3: How reciprocal mixing due to oscillator noise can modify the overall selectivity curve of an otherwise very good receiver.

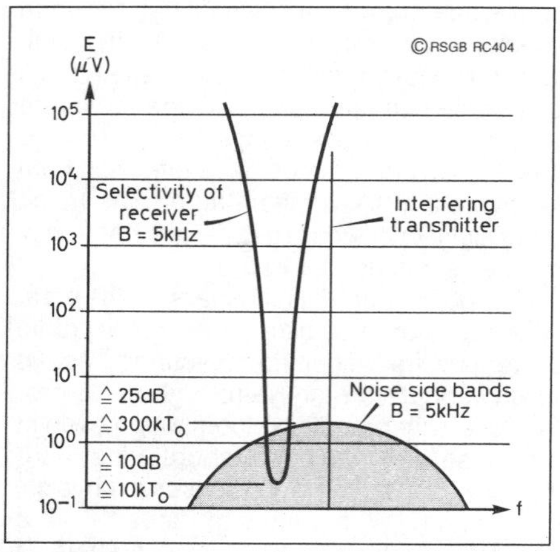

Fig 4: Showing how oscillator noise radiation from a powerful adjacent channel transmission can limit weak signal reception even in the absence of any cross-modulation in a receiver of wide dynamic range.

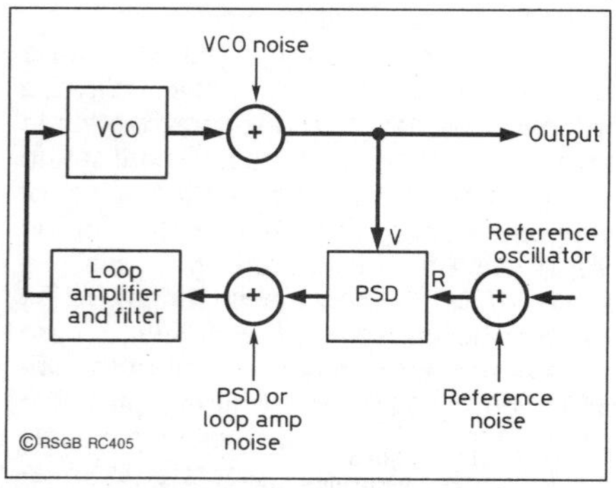

Fig 5: A phase-locked-loop oscillator arrangement is prone to be far more 'noisy' than a straightforward crystal or LC oscillator. The three main sources are the voltage-controlled-oscillator noise, the phase-sensitive-detector noise, and the reference-oscillator noise.

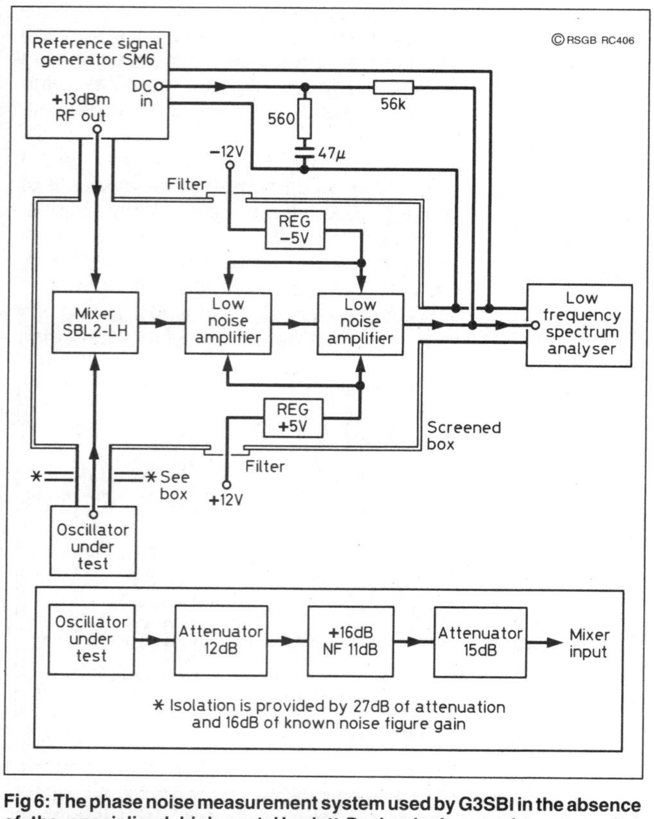

Fig 6: The phase noise measurement system used by G3SBI in the absence of the specialised high-cost Hewlett-Packard phase-noise measuring system.

direct digital synthesis (DDS) chips operating between 60 and 72MHz (to be divided by sixteen for the proposed 9MHz tunable-IF double-conversion receiver). Space limitations mean that a detailed description of this project must be deferred until another month.

In order first to provide background notes on the significance of oscillator noise, it is useful to note that the voltage regulator used on the VCO had a significant effect on the noise performance achieved. An LM317T was used to deliver 9.5V but subsequent RC bypassing was necessary to improve phase-noise performance further out from carrier. The use of the TL431 shunt regulator gave good far-out performance but the series regulator gave significantly better close-in performance (ie at less than 5kHz from carrier).

In *Proc IEEE, 1966*, pp329-330, W A Leeson provided 'A simple model of feedback of oscillator noise spectrum'. This provides a linear model that shows the origins of phase noise in oscillators and has been found to agree well with experimental data. Dr Ulrich Rohde, KA2WEU/DJ2LR provided a check list in his 1983 book *Digital PLL Frequency synthesisers - theory and design* (Prentice-Hall). The check list was based on his interpretation of the Leeson model on how to minimise phase noise in oscillators. Much of KA2WEU's text on oscillator phase noise in his book has been republished in ARRL's *QEX* as 'All about phase noise in oscillators' (Part 1, December 1993; Part 2, presumably January 1994). This is a highly mathematical treatment but general points from his check list include:

- Maximize the unloaded Q.
- Maximize the reactive energy by means of a high RF voltage across the resonator and obtain a low LC ratio. The limits are set by breakdown voltages of the active devices and the tuning diodes and the forward-bias condition of the tuning diodes.
- Avoid saturation at all cost, and try either to have limiting or AGC without degradation of Q. Isolate the tuned circuit from the limiter or AGC circuit. Use anti-parallel tuning diode connections to avoid forward bias.
- Choose an active device with the lowest noise figure. Currently (ie 1983), the best bipolar transistor is the Simians BFT66 and the lowest noise FETs are U310 and 2N5397 up to 500MHz. The noise figure of interest is the noise figure obtained at the actual impedance at which the device is operated. Using FETs rather than bipolar transistors, it is preferable to deal with the equivalent noise voltage and noise currents rather than the noise figure. . . .
- Phase perturbation can be minimized by using high-impedance devices such as FETs, where the signal-to-noise ratio of the signal voltage relative to the equivalent noise voltage can be made very high. This also indicates that in the case of a limiter, the limited voltage should be as high as possible.
- Choose an active device with low flicker noise. The effect of flicker noise can be reduced by RF feedback. An unbypassed emitter resistor of 10 to 30Ω in a bipolar circuit can improve the flicker noise by as much as 40dB.

KA2WEU notes that for low-noise LC oscillators, the highest quality signal generators use air-variable tuning capacitors rather than tuning diodes. He also stresses that while it is relatively easy to calculate oscillators and understand how they work, this does not necessarily optimise their design. He points to the Rohde & Schwarz SMDU signal generator as currently state of the art but to develop such a circuit from design equations is not possible. "This circuit is a result of many years of experience and research and looks fairly simple. The grounded-gate FET circuit provides the best performance because it fulfils the important requirements of the basic equation that allows an estimate of the noise performance of an oscillator."

G3SBI has drawn attention to a new software package that runs on a PC and can predict phase noise performance of oscillators offered as an analysis tool in Compact Software's Microwave Harmonica and claimed to have been successfully validated up to 39GHz but clearly practical experience still counts. Such matters as screening and filtering from 50Hz fields etc become increasingly important where extremely good noise performance is required to provide wide dynamic range.

Geoff Bagley, G3FHL writes: "In the quest for better stability and/or low drift, amateurs have often forgotten the noise output of an oscillator. The output of an oscillator is an approximation to a sine wave. It is really a very, very narrow-band of noise. The spectrum of this noise falls off at various rates, very steeply close to the carrier, perhaps as the inverse fourth power of the frequency offset, then the inverse third power, then the inverse second power, etc, until we reach the zeroth powers, ie the flat noise pedestal on which our 'sinusoid' sits. The quality and usefulness of an oscillator therefore depends not only upon the freedom from drift, but on the width and on the level of the output noise.

"Thus the power level of an oscillator is significant, and so is the noise figure at the input of the active device. According to notes made some time ago, a bipolar transistor is shot-noise limited (in the junction), whilst an FET is thermal-noise limited (in the channel). [Junction intermediate-power-FETs such as the U310, BF247A or 2N4856A have been favoured - G3VA]

G3FHL, who brought to my notice the 1974 book by F N Robinson, writes: "Another topic is the means whereby the level of the oscillation is defined. When you switch on from cold, some small noises or disturbance are amplified and recirculated round the closed loop of the oscillator until it builds up to our desired output level. Something must be done at this point to restrict the loop gain exactly to unity, so that the signal does not go on growing for ever. This can be done either by AGC or by reaching some form of non-linear limiting level.

"Robinson distinguished between two types of oscillator limiter: (1) A soft limiter with a gentle non-linearity with small curvature, and (2) a suddenly abrupt hard-limiter which chops the tips of the sinusoid to the limiting level of oscillation. According to Robinson the second type of limiter is superior in reducing the close-in noise, such as flicker-noise. He advocates for optimum results a high-level active device with a low noise figure and abrupt hard limiting used with any of the stable oscillator circuits such as the Vackar, preferably not using a dual-gate MOSFET."

SIMPLE VALVE TRANSMITTER PRACTICE

THE SEPTEMBER *TT* item on 'Project 6L6 - 1984 Style' showed that there is still a place for one or two-stage transmitters on the lower-frequency bands that can be built in a couple of evenings from junk box or components salvaged from discarded valve receivers, etc. Valves such as the 807 and to a lesser extent the original 6L6 or post-war derivatives such as the EL84 and the higher power PL504 TV line-scan valves are still available if not found in the junk box. More difficult for the more recent addicts are high-voltage mains transformers and high-voltage, high-current capacitors (electrolytic or RF types) but even these can be found at rallies etc.

But for those who have gained experience only with solid-state black-box rigs the circuit techniques may present something of a puzzle. And even for those who ceased to use valves a decade or more ago, it may be necessary to brush up on such matters as handling high-voltage supplies, tuning pi-network output circuits, neutralising amplifiers with tuned grid and tuned anode circuits, etc etc.

In connection with the 'Project 6L6' transmitter, I must admit that it was not until the circuit was published that I noticed with regret that KH6B had omitted the important safety feature of connecting an RF choke directly across the output, as a safety precaution if the one nanofarad anode coupling capacitor failed and thus put the 350V HT supply directly on to the antenna. Furthermore the coupling capacitor should have been specified as at least 1kV rating since it has to cope with the peak RF voltage added to DC voltage, as well as passing appreciable RF current.

Mike Hall, G3USC (with professional expertise in the field of quartz crystals) makes some pertinent comments on this circuit germane also to valve amplifiers and oscillators in general. He writes: "Can I draw your attention to the circuit of the September *TT*, Fig 11 (the KH6B 'Project 6L6'). It is often forgotten that a blocking capacitor has no effect whatever to remove DC bias from another component which also is a capacitor. The applied voltage simply divides between the two in inverse relationship to their value.

"As an example, one comes across circuits of valve power amplifiers in which the HT feed is to the output end of the pi-network tank circuit and blocking capacitors have been inserted in series with the tune and load capacitors in the vain hope they will be removing DC bias from them. This can only be effective, of course, if they are shunted with a resistor to keep them discharged. Even in Fig 11 (September), if the pi-tank is AC coupled to its load, as could be the case with certain types of antenna tuner, a large proportion of the HT voltage will be applied across the variable capacitors upon which the AC signal will also be riding (a 1nF blocking capacitor seems a bit lowish for use on Top Band). It is usual, but not shown, to keep the variable

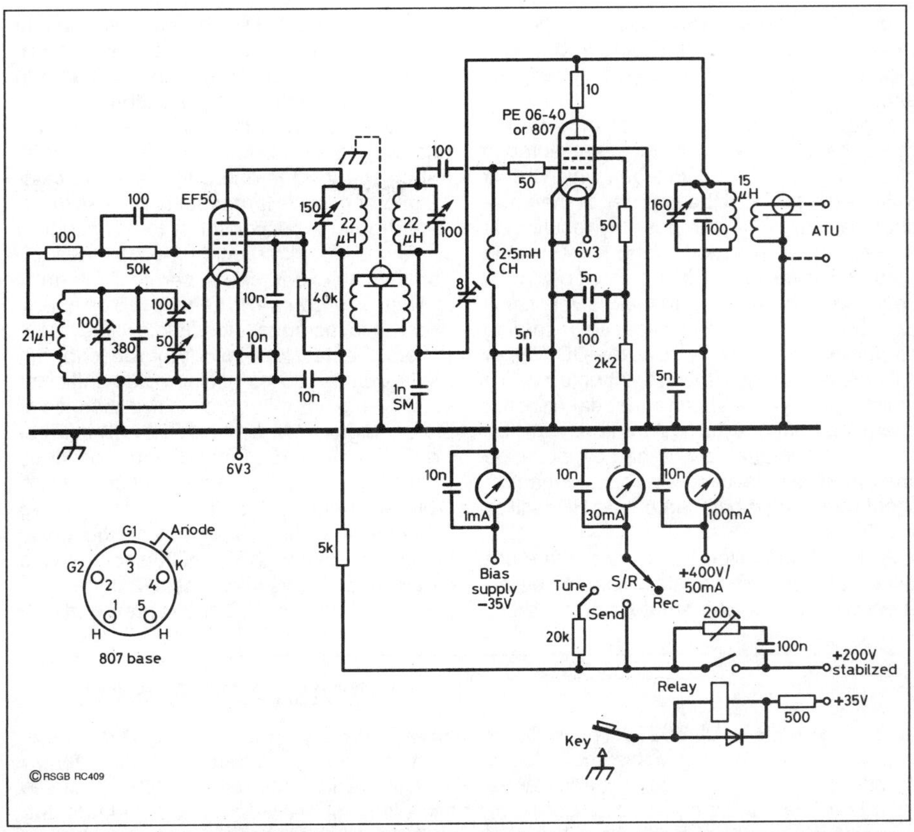

Fig 7: ECO-PA valve transmitter of PA3ZFK. Note that the keying relay is a high-speed type (Philips S50). The oscillator coil is wound on a ceramic former with, diameter 38mm, 26 turns, 0.65mm copper wire, length 40mm to provide 20-22uH inductance. Taps 3.5 and 13 turns from the earthy end. Bandfilter coils on ceramic formers diam 36mm, 30 turns, litz wire 50 x 0.07mm strands, length 35mm; link coupling 5 turns litz wire over 'cold' ends. PA tank coil ceramic former, 51mm dia, 18 turns, 3mm silvered wire, length 50mm, L 15uH, output link coil 2.5 turns insulated wire over the 'cold' ends (used with ATU).

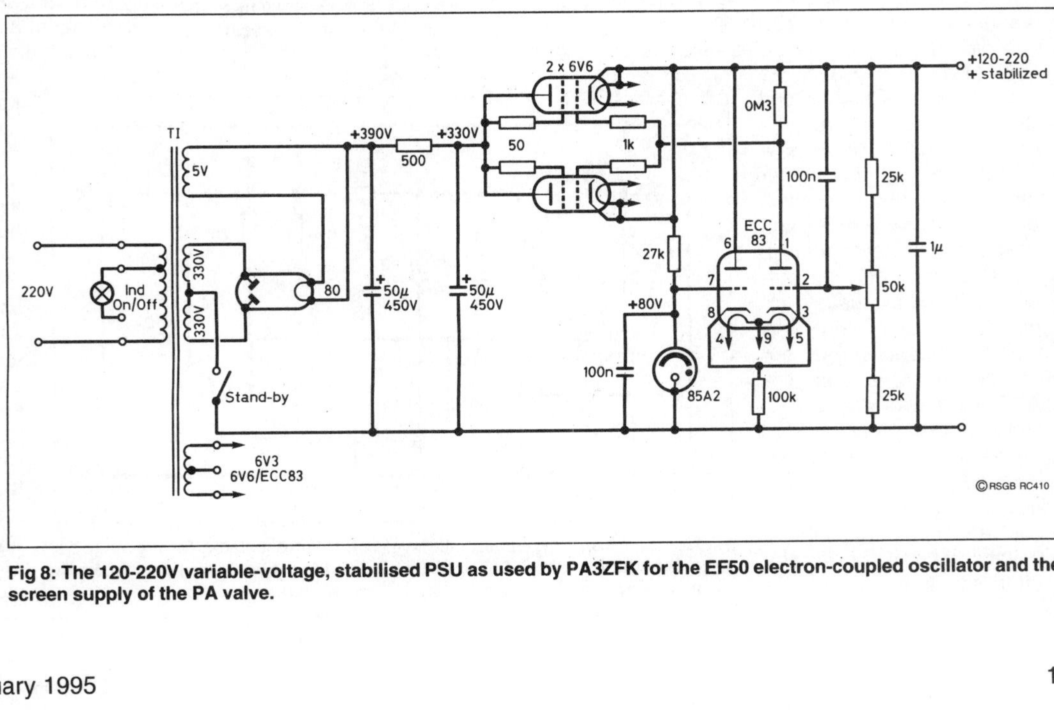

Fig 8: The 120-220V variable-voltage, stabilised PSU as used by PA3ZFK for the EF50 electron-coupled oscillator and the screen supply of the PA valve.

capacitors discharged by fitting a high-current RF choke across the output; this also blows the HT fuse if the series blocker goes short circuit."

"That is by the way. But I also wanted to make the point that crystals are subjected to stress by high applied voltages which affect their frequency and the linearity of their mechanical vibration, both possibly resulting in unpredictable operation. Without a resistor in parallel with the crystal in the circuit of Fig 11, because at DC the crystal is only a very small capacitor, the 5n blocking capacitor is serving no purpose at all. To remove the DC bias of the crystal, I suggest that it be shunted with a resistor of, say, ten times its impedance at the operating frequency or, put another way, ten times the reactance of the load capacitance presented by the circuit: a one megaohm should serve for most Pierce-type HF oscillators."

A crystal-controlled transmitter with a selection of just a few crystals can still be an effective set-up for an amateur prepared to put out CQ calls, but some degree of variable frequency control represents a great convenience and likely to bring more contacts in allowing an operator to reply to other people's CQ calls. Where it is only a question of a relatively small shift to avoid a busy channel, the crystal-VXO is one answer. A ceramic-resonator permits larger shifts but has lower Q and greater susceptibility to temperature variations. An older technique was the vari-gap crystal holder (see separate *TT* item).

There is of course no need to rely on quartz or ceramic resonators for simple valve transmitters: the classic LC oscillators described in the November *TT* can still provide sufficient stability and flexibility even in the absence of buffer stages. P M Bakker, PA3FZK in *Electron*, November 1994, pp607-610 describes 'Een CW-zender voor 80m met buizen' (A CW-transmitter for 3.5MHz with valves). **Fig 7** shows the circuit diagram of the basic transmitter, although PA3FZX also provides details of a stabilised, variable-voltage PSU (120-220V) for the VFO and screen-grid (g2) of the PA (**Fig 8**); a variable negative bias voltage PSU (-20 to -300V); a variable (0 to 35V) relay PSU for the high-speed keying relay etc; and a 400V PSU for the PA.

While this design is by no means the simplest possible it is clearly intended that the performance as a 3.5MHz CW transmitter should meet modern standards, and the stability is stated to be within 250Hz per hour after a ten-minute warm-up period. Interestingly the ECO VFO uses the Lampkin technique (*TT*, November 1994, p80, Fig 2) of tapping the grid of the EF50 down the Hartley-type oscillator coil. This resonant circuit tunes 1750 to 1800kHz with the second harmonic of 3500 to 3600kHz taken from the anode circuit with link coupling to the neutralised PA. The PA valve (PE06-40) is the Philips/Mullard equivalent of the ubiquitous 807.

VARI-GAP CRYSTAL UNITS

GERALD STANCEY, G3MCK was interested in the September *TT* item on quartz resona-

OSCILLATOR AGC

THE GENERAL QUESTION of oscillator AGC as a means of achieving a level, stable output over a wide frequency range rather than primarily for low oscillator noise is currently of great interest to Jack Hardcastle, G3IIR (who some years ago did so much to promote the use of ladder crystal-filters). He writes: "Back in the 1950s, I constructed a valve HF signal generator which used AGC to maintain a constant output from 100kHz to 20MHz but by about 1992 I felt the performance had fallen off so much that it needed replacing. While looking for ideas I came across an IEE conference paper 'VCO with negative dynamic feedback' by E Efstathiou & Z Odrzygozdz of the Warsaw University of Technology (IEE Conference Publication No 303, April 1989) and the circuit in *Amateur Radio Techniques* which you show again as Fig 10 in the September *TT*.

The method of using negative feedback to achieve level control, described in the Polish paper, is shown in **Fig 9;** but I wonder whether the BF961 has protective diodes on the input gates (like the BF981) which may in fact be rectifying the RF). My circuit (**Fig 10**) also uses a dual-gate MOSFET but includes a DC amplifier to give tighter amplitude control. This oscillated happily between 3.5 and 21MHz using appropriate L and C. Without AGC the output exhibited the usual amplitude variation as it was tuned across its range, but this became level once the AGC loop was connected. Unfortunately pressure from professional electronics then diverted me and the signal generator project was shelved. Thus the circuit was never optimised nor was any effort made to extend the range HF or LF, so much remains to be done.

"As I have now taken early retirement, I have revived my interest in this project. At the same time I substituted varicap diodes for the variable capacitor and discovered some unexpected problems caused by their tendency to act as rectifiers when the RF voltage exceeds the DC tuning bias. When this happens it restricts the tuning range at the LF end where the DC bias is at its lowest.

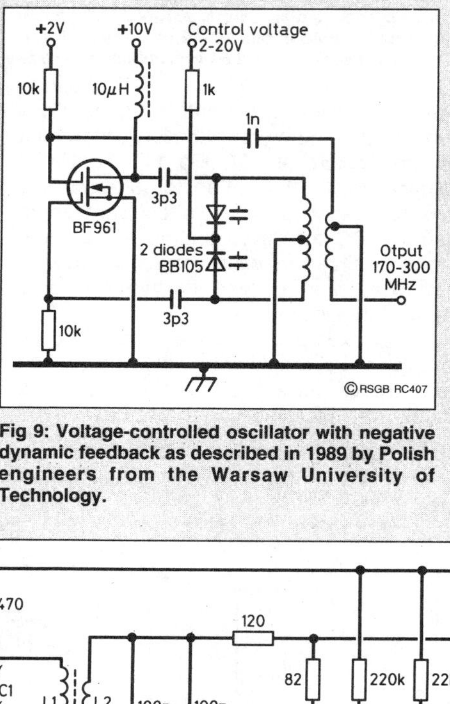

Fig 9: Voltage-controlled oscillator with negative dynamic feedback as described in 1989 by Polish engineers from the Warsaw University of Technology.

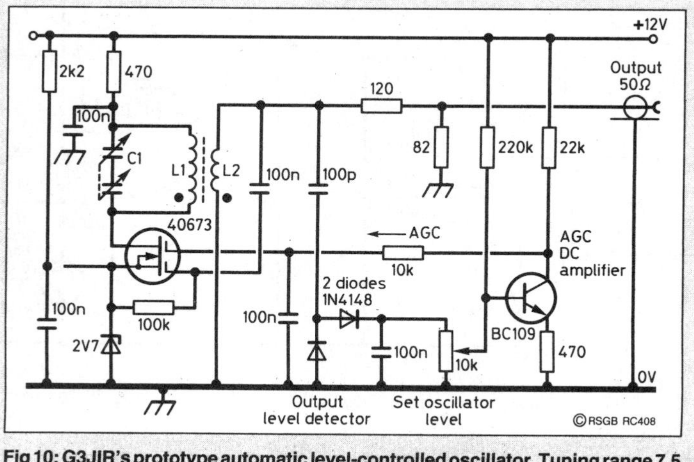

Fig 10: G3JIR's prototype automatic level-controlled oscillator. Tuning range 7.5 to 21MHz with C1 300 +300pF, L1 21 turns, L2 4 turns both wound on 0.25inch diameter slug-tuned former.

I have never seen this effect reported before, possibly because the tuning diodes have been used in the gate or base circuit of the oscillator where the RF voltage is much lower. However this entails using a shunt-fed collector circuit and the RF in the collector circuit will cause stray resonance problems at some point within the range of a wide-band signal generator (or receiver, etc). My aim was to circumvent this by using the series-fed arrangement with the tuned circuit in the collector. By regulating the amplitude of the oscillation it is possible to avoid exceeding the varicap bias and achieve the fullest tuning range.

"Using AGC makes oscillators much more 'designable' allowing variations in inductor Q and crystal activity to be readily compensated. Besides the obvious application in frequency standards and VFOs it could also be applied to GDOs, allowing the oscillator to be kept at its most sensitive level. A number of published designs have used dual-gate MOSFETs and would readily lend themselves to modification to include AGC.

"Someone (possibly Benjamin Franklin) once said "nothing is so powerful as an idea whose time has arrived". Perhaps this could be said of oscillator AGC. In keeping with this I have been planning to apply AGC to a Colpitts VFO but in view of the September *TT* comments I think I will change my mind and apply it to a Tesla (Vackar) oscillator instead."

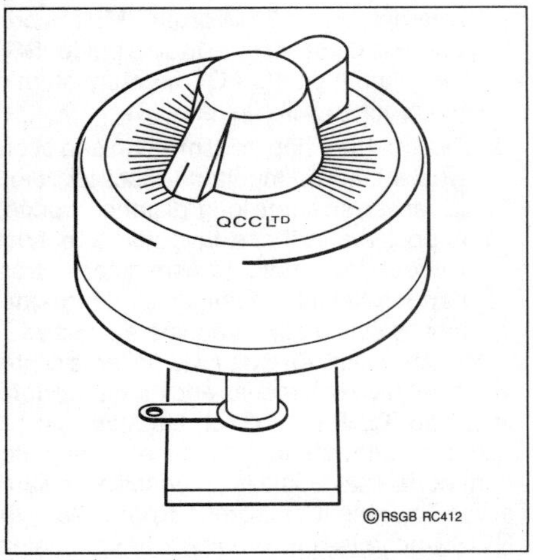

Fig 11: The 1938 QCC vari-gap crystal holder unit provided a useful frequency shift simply by turning the knob. Air-gap holders were soon superseded by plated crystals which provided higher activity.

tors but remains puzzled as to why the vari-gap crystal holders of the late 1930s and early 1940s were not more popular with radio amateurs and soon disappeared from the market. He writes: "Some years ago I acquired two of these vari-gap crystals made by QCC of New Malden. They give about 4.5kHz shift on 3.5MHz and 9kHz on 7MHz."

[An advert in the May 1938 *T&R Bulletin* (pre-war name of *RadCom*) for "A new QCC variable air gap crystal holder. Type AG" (**Fig 11**) states: The gap adjustment is controlled by knob as shown. Used with a 3.5Mc crystal it gives a frequency change of over 10kc on the fundamental frequency . . . . This invaluable feature often means the difference between a 100% QSO and a lost contact. Price including base 27/6d (£1.38p). With a 3.5Mc AT cut crystal 52/6d (£2.68p). With 7Mc LD cut crystal (shift about 7kc) 57/6d. - G3VA]

G3MCK continues: "I am puzzled as to why these were not used more by amateurs as even in those days of not very selective receivers the shifts would seem to have been most useful. In terms of cost the 1940 *ARRL Handbook* gives prices of American vari-gap crystal units as $5.50 compared with $3.00 for normally mounted crystals. This seems to be good value for money. One of the advertisements for vari-gap crystals also refers to cooling and the advantages of the large exposed metal case top! This may well be true as one of the designs in the 1940 handbook is for a 60-watt single-stage crystal oscillator transmitter using a 6L6. The design refers to short valve life but not to possible crystal damage."

I suspect the main reason for the disappearance of the vari-gap units was the development of the centre-plated crystal which rapidly displaced the earlier air-gap mounting. It was found that the activity of the plated crystals was much higher, although it was recognised that the elimination of the gap usually made the coupling between crystal and circuit much closer than was formerly the case. Whatever the mounting it is necessary to remember that any quartz oscillator unit uses material which can break readily whether shocked mechanically or electrically.

The original form of mounting (with the crystal directly between two plates larger than the round crystal plate) in an open rather than enclosed holder, often led to crystal puncturing or other damage from electrical stress. Quite early on air-gap holders were developed to minimise this problem, and it was relatively simple to arrange for mechanical variation of this air-gap. All crystals can be damaged by excessive RF current, and - particularly for such circuits as the power tritet arrangement it was common practice to wire a small 60mA fuse-bulb in series with the crystal to show whether the crystal current was excessive.

In the post-war period, crystal manufacturers pointed out that the centre-plated crystals had proved superior to gap-mounting. They also mentioned that difficulties could arise when a plated and wired unit was substituted in a circuit design primarily intended for the older types of gap-mounting. This is important for those building 'replica' valve transmitters based on pre-WW2 designs which may result in crystal currents that can damage modern plated crystals.

HERE & THERE

REG MOORES, G3GZT, provides a useful technique for winding coils for traps, loading coils, ATUs and PA tank coils. He points out that it is not easy to wind coils with the turns securely fixed so that they do not slip etc, especially when using ceramic and paxolin formers. G3GZI provides an easy solution to this problem: "The answer is simple. Obtain a coil of self-adhesive copper strip, as sold in 'hobby' shops, as used for making 'Terrariums' etc selecting a suitable width as required (typically about 5mm or 0.25inch)".

This can be wound (securely) on the former, the ends being terminated with a soldered on tag and small nut and bolt. A further advantage is that it is easy to solder any 'taps' to the strip at the required points. [I must confess I had to look up 'terrariums' in my dictionary which gives two definitions - 1) an enclosure for keeping small animals, 2) a glass container, often a globe, in which plants are grown - but I recall that self-adhesive copper strip has featured before in *TT* - G3VA].

CORRECTIONS

IN THE caption for the Vackar basic oscillator circuit (Fig 5(c) of *TT* November 1994, p61, the ratios C2:CV and C1:Cx should both be about 6:1 and not 1:6. The text information is correct. In the G3PDM diagram, Fig 9, p62, the FET oscillator should be type 2N3819 (or equivalent) and not the bipolar 2N3794 shown. Incidentally, Dr A J Smith, G4OEP suggests that the stability of the G3PDM Vackar VFO (Fig 7) can be further improved by adding standard source-bias components (say 560ohms to 1K in parallel with 100nF) since this reduces the power dissipation in the FET, stabilises the operating point, and consequently reduces the warm-up drift by a factor, in his experience, of about 5. ♦

WEAK SIGNAL LIMITATIONS

G3IPV BELIEVES THERE is a simple explanation for this phenomenon based on his experience while experimenting with signal-frequency RF filters. He writes:

"It appears that intermodulation occurs between noise and signals in the upper and lower frequency halves of a bandpass filter and this normally happens in the mixer stage of the receiver or in any non-linearity occurring in amplifying devices.

"When this intermodulation occurs a peak of noise appears at the centre of the receiver passband although this is not normally observed by the radio operator as the peak of the noise is wider than the IF passband, but can be observed when the gain control is reduced as the signals become clearer.

"This phenomenon can be considered as a transitional state between an amplifier and an oscillator. Its effect on degrading weak-signal reception can be reduced by decreasing the RF bandwidth between antenna and the first RF stage of a receiver and ensuring that positive feedback in the RF stages is kept to the minimum possible. Some improvement can also be achieved by placing a gain control between mixer and the preceding RF amplifier or by inserting additional filtering at this point to reduce the sideband intermodulation noise produced in any earlier stages."

A typical signal-frequency filter used by G3IPV is shown in **Fig 12**. Coupling capacitors should be kept to the minimum value required to reduce feedback in the RF stage and increase selectivity.

Although not often mentioned in connection with HF reception, it has, in fact, long been recognised that positive (regenerative) feedback in a pre-amplifier has the effect of increasing the noise factor above that which can be achieved with an entirely stable amplifier. But whether this effect, or other causes of non-linearity, is a complete explanation of the observation that it is easier to read a signal only just above the noise level by reducing receiver gain still seems open to some doubt. Noise is in effect a series of 'spiky transients' with peak levels well above the apparent 'noise floor level' and this could show up also the non-linearity of the human auditory system.

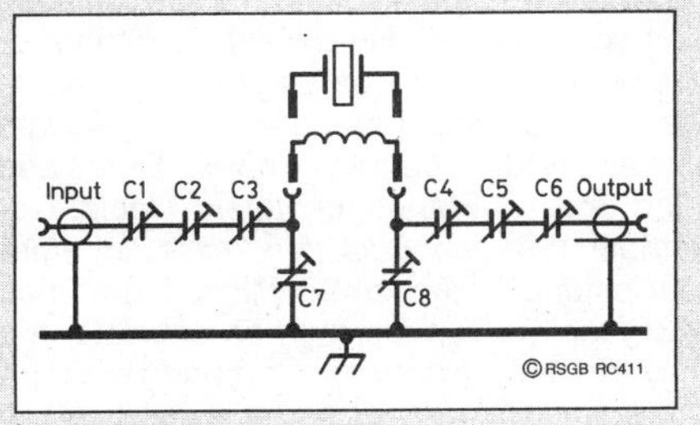

Fig 12: Basic front-end single-section signal-frequency filter (crystal/L-C version) as used by G3IPV in either single section form, or with repeated similar filter sections as required to increase selectivity. C1, C2, C3, C4, C5 and C6 coupling capacitors air-spaced vanes 0.5 to 5pF. C7, C8 tuning capacitors 150pF air-spaced vanes, keep equal in value when adjusting. Metal enclosure 8 by 6 by 3 inches.

Pat Hawker's Technical Topics

PAT HAWKER, G3VA
London 37/SE22 8SS

NEAR-TO-EARTH ANTENNAS

ON A NUMBER OF OCCASIONS attention has been drawn to the value of NVIS (Near Vertical Incidence Skywave) propagation for medium-distance contacts on such bands as 3.5, 7 and 10MHz as a means of largely overcoming the problem of the skip ('dead') zone. By directing the bulk of radiation at a high vertical angle at frequencies around the critical frequency, reliable working at distances up to a few hundred miles, without a skip zone, can usually be achieved. For military applications, this requirement has encouraged the use of compact transmitting loops or very low dipoles not more than about 10ft above ground (*TT* noted many years ago the marketing by Racal of a low dipole for this application). Conventional vehicle whip antennas, on the other hand, have a pronounced null in the vertical direction.

Traditionally, amateurs plan their antennas to provide maximum low-angle radiation and so enhance long-distance working, or use vertically-polarized antennas to maximise ground-wave propagation. As William McLeod, VK3MI puts it in 'Low Radiators and High Ground Planes' (*Amateur Radio,* November 1994, pp10-14): the accepted amateur criteria for horizontal HF radiators has traditionally been 'as high as possible' not only to take advantage of ground reflection but also to clear obstructions, particularly metal conductors and sizeable buildings. Yet, on the lower HF bands, for most suburban and portable locations these conditions of height and space are virtually impossible. Furthermore, in the real world, the ground reflector is anything but perfectly conducting and should be regarded as a lossy dielectric.

VK3MI points out that for 7MHz a height of 10m is a bare quarter-wave above ground; on 3.5MHz only an eighth-wave. This raises the question whether in practice it is worth striving even for this height. What sort of performance can be expected from horizontal antennas only a metre or two above ground? VK3MI summarises the factors involved with low horizontal radiators as follows:

- For low practical heights the radiation resistance at the centre of a resonant dipole remains within the 2:1 VSWR range for the usual coaxial cable feeder so matching procedures are minimal, more so when an electrical half-wave of cable is used to transfer the centre impedance directly to the transmitter: see **Fig 1.**
- Whereas the resonant length of a dipole remote from ground is mainly determined by the length-to-diameter ratio of the conductor, when the ground becomes an increasing part of the dielectric the length is determined by the height to diameter ratio: see Tables 1 and 2. Due to the wide spread of dielectric constant no simple formula can determine this ratio.
- The losses increase as height decreases towards ground level but do not become prohibitive until very low levels are reached; for a 7MHz dipole above common clay this can be as low as λ/40 (1m). [In desert conditions, an antenna can be laid directly on the sand, or even buried a few inches below the surface and yet still radiate reasonably well - G3VA].
- The 'cone' of radiation directed vertically, then reflected back from the ionosphere, can produce non-direction communication with no 'skip distance' to some 400-500km. This is NVIS transmission and is the mode supporting most of those semi-local nets on the 3.5MHz and 7MHz bands. There is usually some fading but for SSB reception the long AGC time-constant of the receiver will alleviate this.
- Two or three hop transmission can occur where the intermediate reflection points fall at sea so some long-distance working is possible in these favoured directions without low-angle transmission lobes. Land reflection points include greater losses which soon become excessive.

VK3MI in his Amateur Radio article provides detailed test results and tabulated data, including **Tables 1** and **2**. He sums up his results as showing that: "In general, the resonant horizontal dipole is an effective radiator at very low height from ground, particularly for NVIS transmission, or where a 'concealed' antenna is required. Losses increase seriously below λ/30 and the high impedance ends of the elements should have at least this amount of separation from ground or metallic earthed objects, towers and poles - but λ/30 is only 1.5m for the 7MHz band.

"Kevlar, Black Dacron, Polypropylene Baler Twine and Nylon Rope are all suitable insulating supports with far less end effects than the single egg-shaped strain insulator wired back to a steel tower which has been commonly used. Supports of this nature have been measured with 6 to 15pF coupling to the earthed object and Table 2 can be used to estimate the end effects of this type of support.

"With the elements double-insulated inside the popular 12mm polypropylene garden irrigation piping erected at 1.5m on the post side of a suburban wooden fence a very effective concealed radiator should result. For portable use a couple of 4m bamboo poles for end supports and a saggy dipole radiator require no apology as to effectiveness for NVIS transmission but directivity, if any, depends on local obstructions and reflectors."

Apart from considerations of possible RF

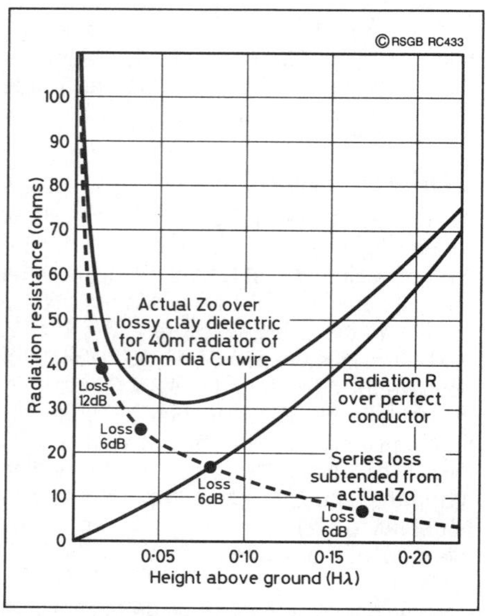

Fig 1: Radiation resistance of a half-wave dipole less than a quarter-wave above ground.

Shape	Height (metres) End	Centre	End	Freq. kHz	Impedance Ω	'E'factor %	Measured C = pF
	Horizontal dipole - 8m94 + 8m94 - nominal 33μH						
Level	2	2	2	7765	31.5	92.5	38
Saggy	2	1	2	7810	32	93	37
Level	1	1	1	7705	35.5	91.75	39
Droopy	0.05	1	0.05	6977	47	83	43
Level	0.05	0.05	0.05	4710	116	56	110

©RSGB RC477

Table 1: Effects of shape, close to clay ground for Dipole of 1mm dia. PVC covered wire - (hot, dry weather - green grass).

Level horizontal dipole - 8m94 + 8m94 - nominal 33μH

Height	Freq. kHz	Impedance Ω	'E'factor %	Measured C = pF	Resonant C = pF
2m0	7767	31R5	92.5	38	12.5
1m5	7745	32R	92.3	37	=
1m	7727	35R5	92	38	=
0m5	7550	45	90	40	=
0m2	7135	61	83	46	15
0m1	6400	70	76	87	=
0m05 (50mm)	4710	116	56	110	35

©RSGB RC478

Table 2: Effects of height above clay ground for Dipole of 1mm dia. PVC insulated copper wire - (hot, dry weather - green grass).

health hazards with high-power operation if the radiator is close to a living area (most current guidelines stipulate a minimum safe height of 30-35ft for high-power HF operation) there are other safety aspects that need to be considered. VK3MI writes: "Safety is an important consideration for both low radiators and for elevated ground planes. One part is physical in that any wires below 3m can be regarded as a trap for man and beast, including horses and wandering cattle. Even in daylight a thin wire can disappear against some backgrounds and at night is a very serious hazard. Therefore a protective, non-metallic guard-rail or fence is necessary, not just a coloured streamer tied in the middle of the hazard.

"Another aspect of safety is electrical as even at low power a nasty sting and RF burn can occur which, for non-technical people or for climbing children, can produce an emotional reaction far in excess of the initial injury. At medium power, around 100 watts, these effects can become severe and for higher powers the effect of corona and irradiation must also be considered. The use of unprotected low installations is not recommended for high powers and even for low power the radiator should be double insulated by enclosure in plastic pipe or conduit."

In his article, VK3MI also describes how by erecting an antenna above the double-pitched metal roof of a building about 20m long and some 6 to 8m wide, it is possible to obtain low-angle radiation: a system he dubs a 'Woolshed Reflector'. He also provides information on elevated ground-plane antennas and vehicle whip antennas.

SHARPER-RESONANCE STRIP-LINE FILTER

THE JANUARY 1994 *TT* (p40) included the experiences of P R Kemble, G3UYK in using a 144MHz strip-line filter (as originally described in *Radio Communication Handbook*) in overcoming the problem of the bursts of 153MHz data (paging) interference to 144MHz mobile reception, particularly in town centres, emphasising the relatively poor selectivity of VHF front-ends.

This item proved of interest to Uwe F W Keonneker, DL80BF of the Institute of Space Technology and Reactor Technology in Braunschweig, Germany. His problem was the reverse situation to that of G3UYK. Interference from a local 144.675MHz amateur radio digipeater to reception of BREMSAT-signals on 137.800MHz +/- Doppler-shift. DL80BF built the strip-line filter as described in the January *TT* and found it no problem to detune the filter to 137.8MHz. Unfortunately the -3dB bandwidth proved to be 15MHz so that attenuation of the digipeater signals was insufficient. However, the insertion-loss was less than a dB despite the use of less than optimum materials for the strip lines (copper- and zinc-plated sheets of metal). This underlined that this filter-design would be beneficial for broad bandwidth and low insertion losses but unable to overcome the digipeater interference to their satellite receiver.

With the help of Ullrich Wintzer, DL7FZ, the filter was modified to give a sharper resonance though at some increase in insertion-loss. The modified filter design is shown in **Fig 2** and its characteristics shown in **Fig 3**. DL8OBF writes:

"Only two strip-lines are used with looser coupling, reducing the bandwidth but resulting in a rather higher insertion-loss of roughly 1.8dB. We have chosen galvanic coupling to the coax as this achieves better far-off rejection. For mechanical reason I used the more easily-available 'radial' capacitors with the value unchanged at 50pF. With the dimensions shown it is no problem to detune the filter to the amateur-band nor to use it for satellite reception. After cascading this filter with the pre-amplifier and a further commercially-made filter we completely eliminated the interference caused by the 144MHz digipeater.

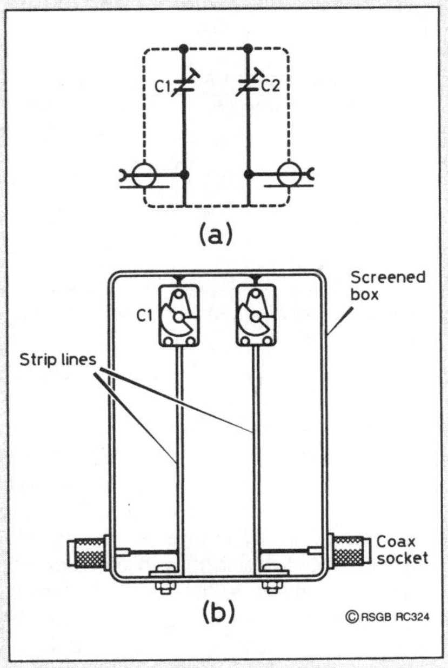

Fig 2: Modified strip-line filter with sharper resonance than the design shown in *TT* January 1994.

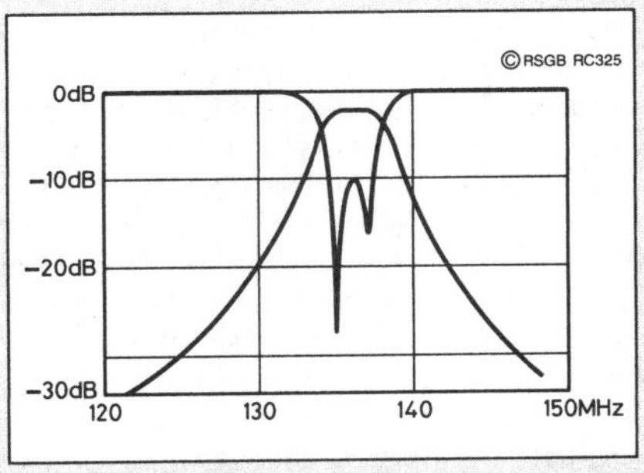

Fig 3: Return-loss and insertion-loss of the modified strip-line filter.

WHO NEEDS INTEGRATION?

TT, MAY 1991 told the 'Sad story of an electronic hobbyist' based on the account of Robert W Lucky in *IEEE Spectrum*. He explained how he had found that electronics home-construction, a hobby he had followed since youthful building of progressively more complicated radio receivers, hi-fi amplifiers, home-designed computers and writing software programs in the golden age of home-construction, had gradually been overtaken by 'keep-your-hands-off' commercial packages. He questioned whether there was any connection with the steady drop in enrolments in electronic engineering. I pointed out that amateur radio was one of the few remaining areas in which to some degree home-construction was still attractive and worthwhile - even if the golden age has lost much of its glitter.

Pierre Mosrin, F2WW draws attention to an article 'Integration: who needs it?' by an Edinburgh University lecturer, Alistair Armitage, in *Physics World* (November 1994, p80). Like Dr Lucky, his interest in electronics had been sparked off early on - in his case the result of "the long hours I spent as a youth mucking about with an electronics construction kit."

Building a succession of projects, with never enough components for more than one at a time, he found that "most of the fun came in building and debugging the circuit". Admittedly he learned "surprisingly little theory, despite the simple explanations included in the manual. Theoretical rigour (yawn) came later during my student days, and when I began teaching electronics. But I was left with an impression that tinkering with electronics was fun. This led me on to the hobby electronics magazines and the joys of our local hobby electronics store [but not apparently to Amateur Radio] . . . . Now, some 25 years later, electronics has moved on. The hobby shop has closed down due to a decline in the market. Why on earth should this be, when electronics is now seemingly all-pervasive . .

"The answer is simple: integration has killed the fun of tinkering with electronics. A glance at an RS catalogue illustrates my point. Why spend a rainy afternoon soldering transistors when you can buy a 'monolithic integrated audio amplifier' for the princely sum of 53p. turn on three pages and there is the ZN414Z receiver. this disgusting 10-transistor device provides a complete RF amplifier, detector and AGC circuit on one chip for 90p. What's even worse, the dratted thing only has three leads, labelled i/p o/p and gnd. I defy anyone to get any technical satisfaction out of building a circuit with that . . !"

But Alistair Armitage does find some positive benefits in integration: "The most important seems to be in digital circuitry and computing - an area only made accessible to the hobbyist through the advent of integration. and perhaps there will also be a renaissance in hobby electronics when we can design our own chips using home computers. Will we take our designs into the local high street silicon foundry for processing, in the way that we now get our photographs developed . . . We might then get back to the stage of mucking around with circuits, useful and useless, just for the hell of it. . . .While waiting, pass me my soldering iron - I've still a few transistors to stick into this Walkman I'm building."

TESTING A SCIENTIFIC CALCULATOR

A BBC TV programme recently drew attention to the errors that can arise when using certain computers; such chip errors may in some circumstances be very significant. An

WATERPROOFING DIPOLE TEES

THE PROBLEM OF providing effective waterproofing of the connection between a coaxial-cable feeder and an antenna element has been raised a number of times in *TT* and elsewhere but it seems worth emphasising that this is a perennial problem. As Dr J A Share, G3OKA puts it: "One of the curses of the British climate is the rain and the resulting tendency of rainwater finding its way into small openings and ruining any electrical contact with which it comes into contact. Over the past thirty years countless lengths of coax cable have been destroyed by rain water penetrating dipole-centre connections." [A major problem is that, particularly with air-spaced and semi-air-spaced cables, a single entry-point can ruin many metres of costly cable - G3VA].

G3OKA continues: "Some six years ago a simple waterproofing idea was noted in the *ARRL Handbook* in connection with a lightweight portable dipole design. This used a plumbers plastic Tee, three rubber bungs, DIY sealant, waterproof adhesive and some scrap fibreglass circuit board. Since adopting this practice and despite the rigours of six winters and summers, I have experienced no repetition of water penetration and the design (**Fig 4**) seems worthy of greater exploitation.

"The Plumbers Tee can be purchased from DIY stores (22mm ideal size) and suitable rubber bungs are available from stores selling home wine-making materials. A tube of black waterproof glue is recommended but clear Bostick would no doubt suffice. Superglue can be used but this makes it impossible subsequently to take the connection apart. Only a small amount of silicone sealant is required (left over from some other DIY project?). Since the full strain of the antenna is taken on the centre insulator, 3.2mm fibreglass PCB is recommended, but alternatively two pieces of standard 1.6mm board could be superglued together, but in this case it is essential to roughen the surfaces to be joined, otherwise the pieces will not adhere.

"Hold the Tee in a vice, force in the bungs and drill the holes with a hand drill, using sharp bits somewhat smaller than the required hole sizes. If done slowly and carefully the bungs will be a tight fit onto the coax/antenna-wire. Cut the centre insulator so that it is about 6mm less than the distance between the inside ends of the bungs; drill holes for the connections; and remove the unwanted copper from the centre of the insulator using a rough file or by tinning the entire area and then lifting off the copper using a sharp knife while the solder is molten.

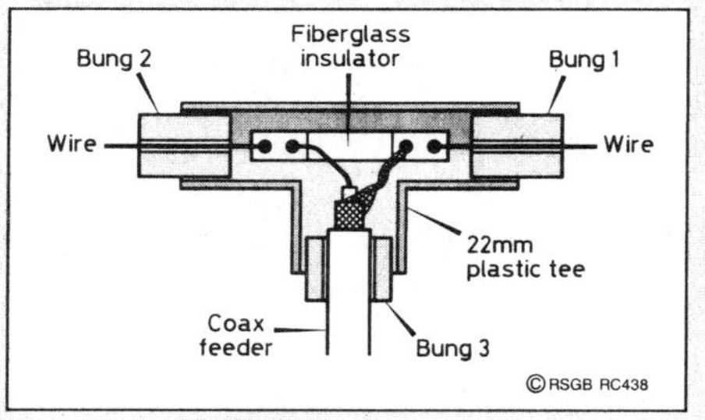

Fig 4: Detail drawing of the waterproof dipole Centre Tee.

"Fit bung (3) onto the coax and feed the coax into the tee so that it exists at one of the wire holes; feed the antenna wire through bung (1) and then through the Tee. Feed the second antenna wire through bung (2) and assemble the centre insulator. Ensure the solder joints are of high quality because it is difficult to resolder once the centre is fully assembled. It takes a little dexterity to push the insulator back into the tee, rather akin to a ship in a bottle.

"Apply Waterproof Bostick to the antenna wires close to the centre insulator and also to the inside of the Tee where bungs are going to be fitted. Slide the bungs into place and leave until the adhesive is cured. Fill the Tee-piece with the silicone sealant leaving space to fit the final bung. Curing time varies according to the sealant but at least 24 hours should be allowed. The final step is to coat the coax with more waterproof Bostick and glue the last bung into position.

"In practice, the whole assembly is light yet strong, and has withstood the wind-loading on a W3DZZ trap-dipole at 40ft in a location only a few hundred metres from the Irish Sea for the past six years without failure or water penetration.

anonymous *TT* reader has sent along a rigorous 'Engineer's test for any scientific calculator. Put calculator readout into degree mode and then press buttons as follows: 29°; sin; cos; tan; x^2; log x; 1/x; 1/x; 10^x; $\sqrt{x}$; arc tan; arc cos; and finally arc sin. The resulting readout should be approximately 29° of arc +/- 0.01.

A NEW LOOK AT THE MULTEE ANTENNA

IN CONNECTION WITH the quarter-wave folded dipole antenna, I included in the November 1944 *TT* (p62) an outline of the two-band Multee antenna originally devised by W6BCX and which has appeared for many years in editions of the *Radio Handbook*. However the dimensions given in Fig 11 were those given in *TT*, May 1965 and appear to have been taken from an article in *73* or *CQ*, although the rather different dimensions given in the text for a 1.8/3.5MHz version are as given in the *Radio Handbook*. The version reproduced in *TT* suggested open-wire line and not 300ohm ribbon with its lower velocity ratio for both the vertical and horizontal sections whereas recent editions of the *Radio Handbook* (including the 1992 edition) suggest that 300ohm open-wire TV-type line can be used. **Fig 5** shows the Handbook dimensions.

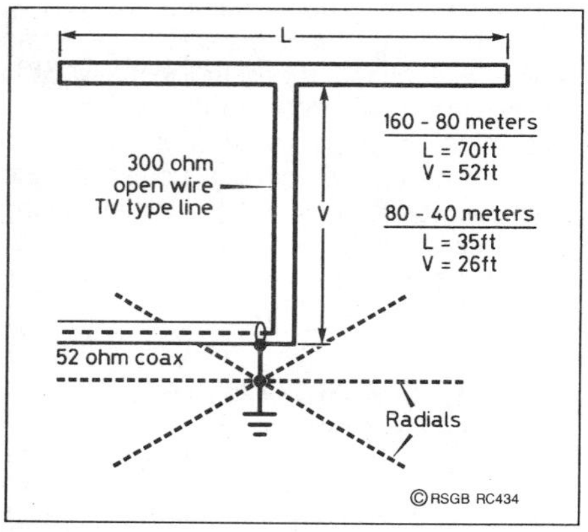

Fig 5: The Multee two-band antenna with dimensions as shown in *The Radio Handbook*.

Dr John S (Jack) Belrose, VE2CV has sent some pertinent comments on this antenna. He writes:

"I wonder whether anyone has fabricated a Multee - in my view it is certainly not a resonant antenna, but it does have interesting radiation characteristics. I can computer-model antennas, including folded antennas, provided that open wire is used.

"The purpose of my letter is to question the dimensions given in *TT*. You have suggested making the antenna out of 300W twin lead. [Actually the November *TT* referred to 300ohm *open wire* - G3VA]. This is not a good suggestion, since whatever the frequency (low-band or high-band) there is a large standing wave on the antenna. Also re-dimensioning the antenna raises the question of dimensioning the length of a radiator (which is independent of velocity factor) and dimensioning a transmission line (where the velocity factor is important).

"A folded dipole is a (sort of) complex radiator, since it carries both transmission line currents (out of phase currents) and radiating currents (the in-phase currents) on the two conductors. The velocity factors for these two modes are quite different when using 300Ω ribbon.

Let us consider a 3.75MHz half-wave folded dipole made with 300Ω twin-lead. The resonant dipole length for a half-wave radiator is about 123.2ft (37.55m), [half-wavelength times antenna factor] with an arm length 61.6ft or 18.77m. Whereas for the transmission line mode the length for a quarter wavelength is 53.86ft (16.43m quarter-wavelength times velocity factor). Therefore, for the antenna to function properly (better impedance match) you have to include shorting straps at the shorter dimension locations: see **Fig 6** for a folded dipole made from 300ohm ribbon as described in many editions of the *Radio Handbook*.

[The need for such shorting straps with folded-dipoles using 300ohm ribbon was described many years ago in *TT*, and an example appears in *ART7*, p265, derived from *The Radio Handbook* and which, incidentally, was

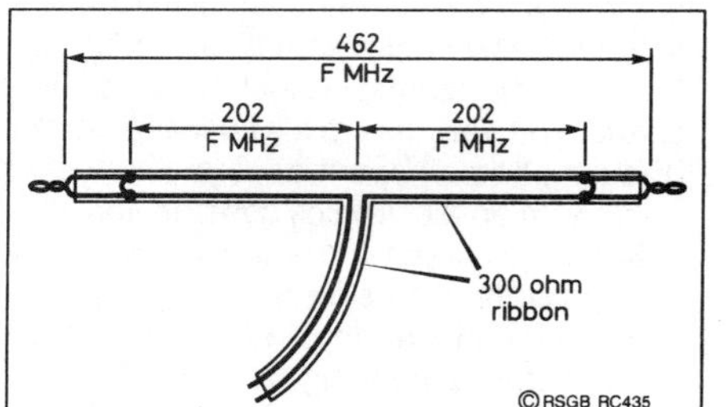

Fig 6: Single-band folded dipole formed from 300ohm ribbon feeder with shorting straps to compensate for the velocity factor of the cable as suggested for many years in *The Radio Handbook*.

used as a 14MHz antenna for many years at G3VA].

VE2CV continues: "All this raises the question: how to dimension the Multee when changing from open-wire line to 300ohm (ribbon or ladder) twin-lead? This begs the question concerning the original dimensions (dimension L2 is not a high-band half-wave times velocity factor). Furthermore, you cannot consider that the two-conductor transmission line forming he vertical element of the antenna acts like a linear (impedance) transformer, when it is fed in an unbalanced way (ie fed on one leg only).

"I have rather carefully modelled the 160/80-metre Multee, having the original dimensions for open-wire line, using a version of NEC-2. I modelled the antenna using 600ohm open-wire line (neither MININEC nor NEC can model transmission lines, particularly shorted (folded) lines, if the spacing between conductors is too small). I have assumed that the antenna is used with four insulated radials, 15.24m (50ft) long, as in the November *TT*. The height of the radials for my model is one metre, but this is not a critical dimension. I have assumed average ground characteristics (σ= 3mS/m, ε = 13).

The antenna is in the X-Z plane, the radials in the 45° planes. The input impedance, according to NEC-2 is 16 - j247Ω at 3.75MHz and 72 -j1059Ω at 1.9MHz (remember that the impedance of the open wire line for the model is 600ohm, not 300Ω). However it is clear that the antenna is not resonant in either the 1.8MHz or 3.5MHz bands.

"**Figs 7** and **8** show the computed radiation patterns. Indeed, the radiation characteristics of the antenna are as described, viz dominantly vertically polarized in the 1.8MHz band; for the 3.5MHz band the polarization is horizontal in the plane orthogonal to the horizontal element, and vertically polarized in the plane of the antenna. The asymmetry of the pattern in the plane of the antenna is interesting. The antenna's impedance will depend on the characteristic impedance of the transmission line used, the dimensions of the antenna, the length and height of the insulated radials, and to some extent on the characteristics of the ground beneath the antenna. A low-loss coaxial feeder must be used together with a good antenna system tuning unit (ASTU)."

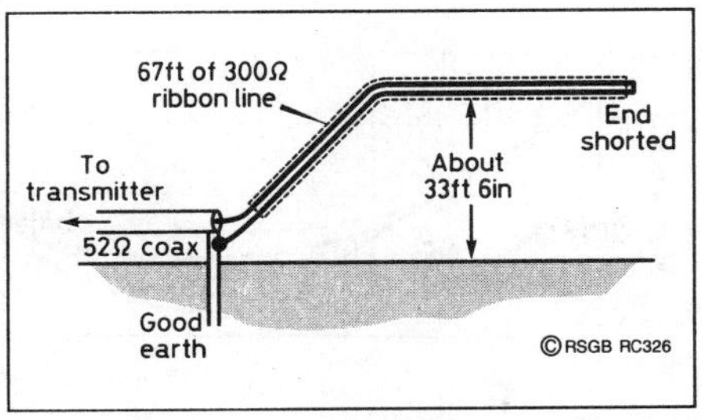

Fig 9: λ/4 ground Marconi antenna for 3.5MHz using 57ft of 300- or 450-ohm ribbon line. A good earth system (preferably including radials or elevated counterpoise) is needed for good performance.

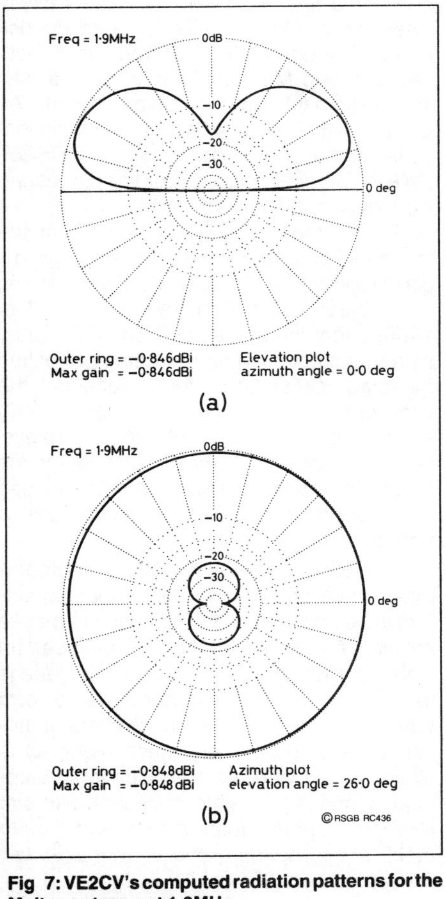

Fig 7: VE2CV's computed radiation patterns for the Multee antenna at 1.9MHz.

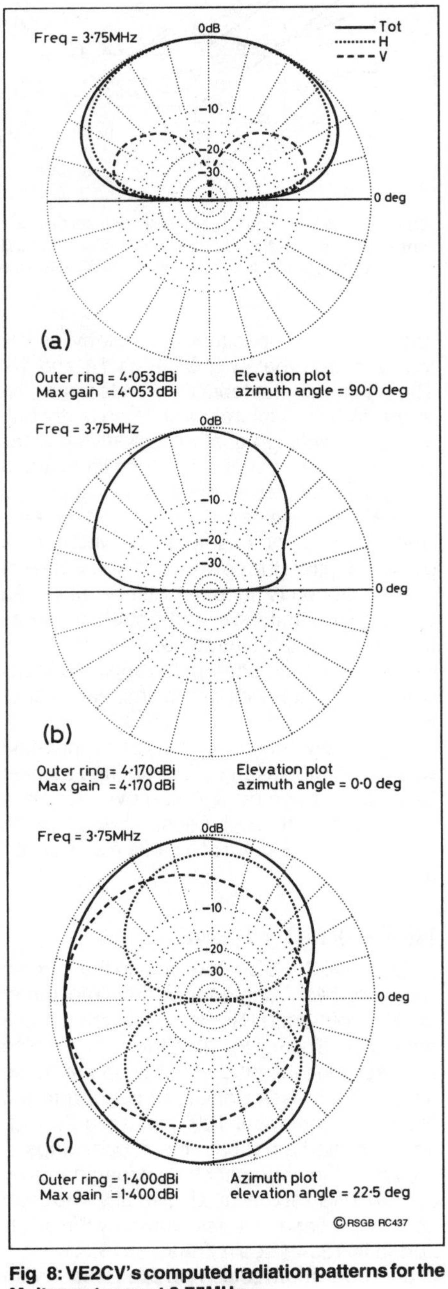

Fig 8: VE2CV's computed radiation patterns for the Multee antenna at 3.75MHz.

QUARTER WAVE MARCONI ANTENNA

AN OLD BUT STILL useful compact antenna idea for the lower-frequency bands is revived in *CQ*, June 1994. This is the λ/4 grounded Marconi antenna fed from coaxial cable using 300Ω or 450Ω ribbon feeder (or open line): **Fig 9**. For 3.5MHz some 67ft of cable is used and the length can be halved for 7MHz or doubled for 1.8MHz (preferable at double the height of the horizontal span). A good earth or radials system is essential. At low power, it is believed that a length of coaxial cable could be used short-circuited at the far end.

'CORROSION' BETWEEN DISSIMILAR PLASTICS

BILL COOPER, G4CIA, RAISES the question of whether 'corrosion' can occur between dissimilar plastics. This followed the discovery that after keeping a short power lead coiled up and held by a strong elastic band to the clear plastic cover of his computer chess set for a period of several months the power lead had etched several concentric grooves into the cover, deep enough not to be removable by polishing. He points out that this could have significant implications since it is not unusual for dissimilar plastics to be in close contact in expensive electronic equipment.

Since my knowledge of plastics technology is virtually non-existent, it seemed worth seeking expert opinion from Dr Dick Biddulph, G8DPS who writes: "The report from G4CIA of 'corrosion' between flexible PVC cable and a 'hard plastic clear cover' (probably polystyrene) is almost certainly migration of the plasticiser from the PVC which then attacks the polystyrene. Flexible PVC contains about 50% of one of several types of plasticiser some of which migrate faster than others. This can be felt with some cables as they become sticky with age.

"This is only likely to occur when flexible PVC is one component and the other is a simple thermoplastic such as polystyrene or acrylic plastic such as Perspex. It is very unlikely to affect thermosetting plastics such as phenolics (Bakelite), epoxys (Araldite) or polyurethanes (solder through wire insulation)."

THE 'OS-CON' ELECTROLYTIC CAPACITOR

EUGENE TRUNDLE in *Television* (December 1994, pp98-99) draws attention to a new form of electrolytic capacitor to add to the present ranges using aluminium and tantalum as the positive electrode in foil or solid form. This is the Os-CoN (Organic Semiconductor) capacitor, introduced by Sanyo, which has an electrolyte in solid form rather than the gel or solution of semiconducting manganese dioxide used in conventional types (**Fig 10**).

The electrolyte is based on the organic semiconductor TCNQ, a complex salt in a fine-powdered black crystalline form termed N-n-butyl isoquinolinium which in production is melted, inserted, immersed and cooled in carefully-controlled conditions. This new type of electrolytic capacitor would seem to be of particular interest for amateur radio equipment and switched-mode PSUs since its main feature is its good high-frequency characteristic, approaching that of a film capacitor and much better than that of a conventional electrolytic. Eugene Trundle states that "at frequencies above 100kHz a 47uF os-con capacitor is superior to a special low-impedance 1000uF aluminium capacitor whose physical size is twenty times greater." The temperature and frequency characteristics of an os-con are claimed to be such that it can replace three components in a noise/ripple filtering circuit: **Fig 11.**

Thus it would appear that the os-con eliminates the need to parallel a ceramic capacitor across an electrolytic in HF circuitry - a practice which, as shown in *TT* (May 1993, p58), may prove self-defeating. With the os-con, apparently, the increase in impedance above 1MHz, as shown in **Fig 12 and 13**, is caused by the inductance of the lead-out wires, which at HF should be kept as short as possible.

Fig 19, from the *Television* article, shows the construction of an os-con capacitor. There is an aluminium case with aluminium oxide as dielectric. But instead of a porous or vented seal the os-con has an impervious resin seal, since there is no electrolyte evaporation when in solid form. Sanyo are marketing the capacitors in tubular and surface-mounting forms. For a given type (capacitance value and working voltage) they are roughly equivalent in size and price to solid tantalum electrolytic capacitors, that is slightly bigger and rather more expensive than aluminium types; they are currently available with capacitance values from 0.1μF to 220μF and voltage ratings from 6.3V to 25V. The temperature coefficient of the os-con is superior to that of aluminium

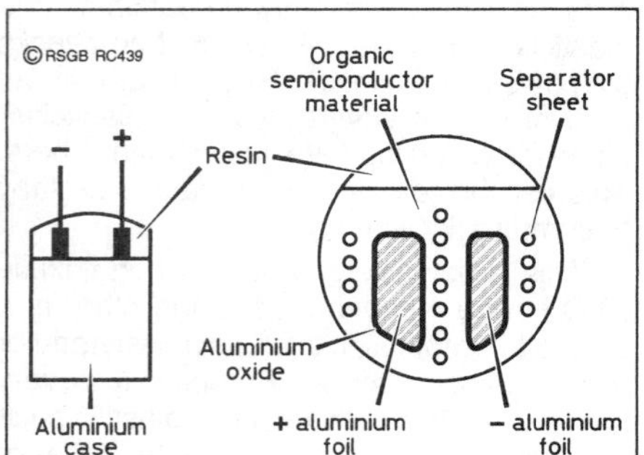

Fig 10: Construction of the OS-CON electrolytic capacitor introduced by Sanyo. (Source *Television*, December 1994).

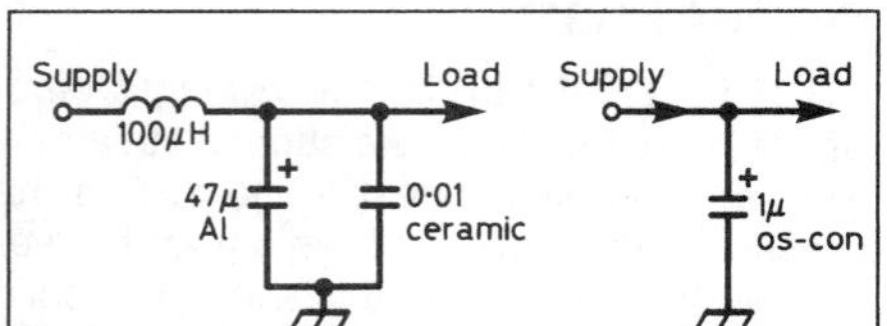

Fig 11: The temperature and frequency characteristics of an os-con capacitor enable it to replace three components in a typical noise/ripple filtering circuit. It is claimed that these two configurations both provide similar filtering over a wide temperature and frequency range. (Source *Television*).

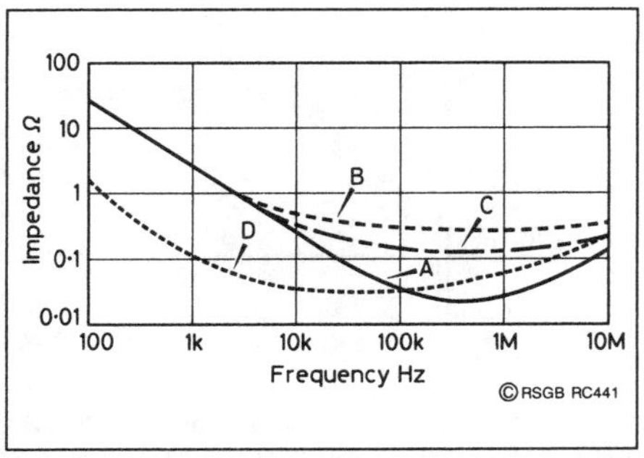

Fig 12: Frequency characteristics of different types of electrolytic capacitors at 25°C. A os-con (47μF, 16V); B low-impedance aluminium electrolytic (47μF, 16V); C tantalum electrolytic (47μF, 16V); D low-impedance aluminium electrolytic (1000μF, 16V). (Source *Television*).

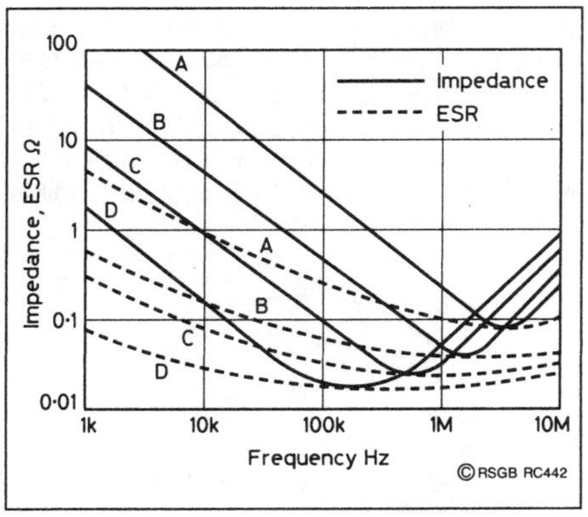

Fig 13: Os-con impedance and ESR ratings over the range 1kHz to 10MHz (at 25°C). A 0.47μF, 25V; B 4.7μF, 25V; C22μF, 6.3V; D 150μF, 16V. (Source *Television*).

and tantalum capacitors particularly at low temperatures (below 0°C): **Figs 14** and **15.** The ageing characteristics are also very different. With no electrolyte solution to dry out, there is a very gradual and relatively small capacitance loss over a very long time, with a life-span expected to extend to centuries at normal working temperatures. It is also claimed that, for example in switch-mode power supplies, the permissible ripple current with an os-con is about four times that of an aluminium electrolytic and about ten times that of a solid tantalum type. It would clearly be inadvisable to attempt to replace a faulty os-con capacitor with a conventional electrolytic.

Apparently the basic crystal formulation was first synthesised by DuPont and studied for capacitor use by Sprague over 30 years ago but production only became possible with Sanyo's development of a melting immersion method.

HERE AND THERE

IVAN JAMES, G5IJ hopes that in the flurry of radio pioneering centennials, we do not forget the work of Admiral Sir Henry Jackson, president of the Society in 1922, the year in which the decision was made to change the name from the Wireless Society of London to the RSGB. In December 1895, the then Captain Jackson read a paper in the *Proceedings of the Royal Society* describing experiments by Jagadis Bose at Calcutta University who used equipment based on that shown by Sir Oliver Lodge in 1894 (see *TT* August 1994).

Jackson soon began to experiment with

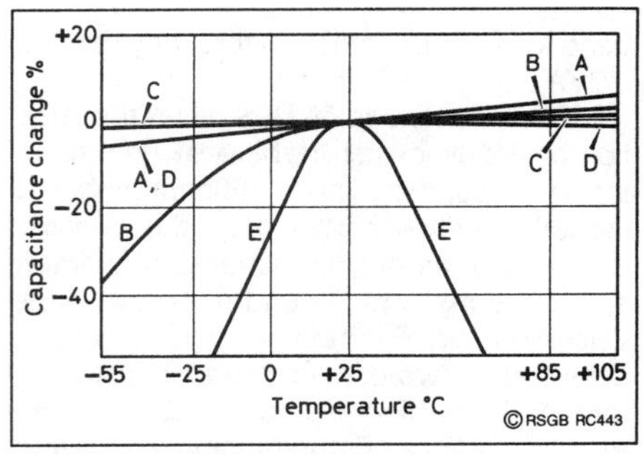

Fig 14: Temperature/capacitance characteristics of different types of capacitor. A os-con; B aluminium electrolytic; C tantalum electrolytic; D Mylar film capacitor; E ceramic capacitor. (Source *Television*).

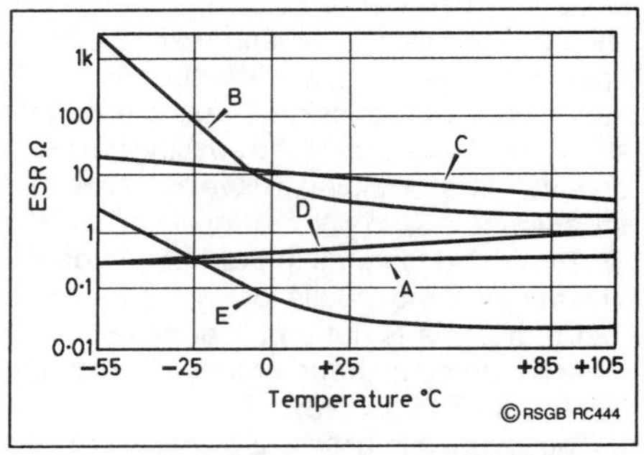

Fig 15: ESR/temperature characteristics of different types of capacitors. A os-con; B aluminium electrolytic; C tantalum electrolytic; D Mylar film capacitor; E ceramic capacitor. (Source *Television*)

wireless telegraphy on the deck of HMS Defiance, then used as the Torpedo School in the harbour at Devonport. This was in early 1896, some months before the arrival in the UK of Marconi in March 1896, by which time Jackson had succeeded in sending and receiving messages over a distance of nearly one-and-a-half miles. The full story is told in *The Origins of Maritime Radio.* by R F Pocock and G R M Garratt (G5CS) from which both Jackson and Marconi (with whom he was soon in contact) emerge with great credit. As G5IJ puts it: "Let us hope that 1996 does not pass without some appreciation (by RSGB, RNARS or local societies) of the pioneering work of Admiral Jackson (1855-1929)".

Gordon Mather, G3GKA notes that the 'cohering' of lightly packed copper filings in a glass tube caused by 'oscillatory' current was the subject of a British Patent (No 156 of 1866) under the names of C and S S Varley as a means of providing lightning protection for telegraph stations, thus preceding the coherers of Oriesti (1884), Branley (1890) and Lodge (1894), although not for 'wireless' communications. (Source *'Science for the Citizen'* by Lancelot Hogben, p742). The use of a coherer for radio was indisputably a Lodge contributiion.

A practical application of 'high-temperature' (liquid nitrogen cooled) superconductors is being tested in the USA by the mobile cellular radio company Ameritech based on work by Illinois Superconductor. The idea is to use superconducting materials to form bandpass VHF/UHF bandpass filters at mobile base stations. Very much higher-Q is claimed than with more conventional techniques, making possible 'brick-wall' filters at UHF. See report 'Superconductors move into mobiles' by Charles Arthur in *New Scientist*, 24 September 1994, p17. ♦

THE ART OF QRP

ALTHOUGH PERSONALLY not a low-power devotee, I would readily acknowledge that the hobby owes much to the substantial band of amateurs who spend much of their time using powers ranging from about 5W down to a few milliwatts. They have not only tended to keep alive the use of homebrew equipment with emphasis on KISS techniques but also maintain interest in achieving the high levels of efficiency with simple wire antennas that once marked the UK scene in pre-war days when so many of us were restricted to DC-input powers of less than 10 watts.

David (Doc) Wescombe-Down, VK4CM/VK5HP in Amateur Radio (January 1995, pp10-11, who during the past 20 years has made more than ten thousand contacts using 5 watts input (or less) on CW, mostly 14MHz, with a variety of home-brew, kit and factory rigs, underlines that each time you reduce transmitter output by four-times (6dB) the received signal falls by only one S-point. For example, if, with 256 watts output, you put in an S8 signal, then (assuming the recommended S-meter calibration of 6dB per S-point) with 4 watts the signal reduces only to S5 or with 250mW to S3. By careful choice of band, propagation conditions, and a certain degree of low-cunning in operating techniques, a 250mW rig with an efficient antenna can provide acceptable HF contacts; while with 4 watts there will be many occasions when a distant QRP station should seem as strong as a QRO rig with a less efficient antenna.

VK4CMY stresses that "antenna efficiency makes or breaks any amateur radio station but especially a QRP station!". He advises that, while any antenna that works on the desired frequency is suitable, a number of constructional practices need to be 'spot on'. Near enough is not good enough. Good practice includes:

- conductive paste used in all telescoping joints in elements/radiators;
- resonant radiators;
- open wire line (300Ω TV ladder line or 600Ω air-spaced lines) are superior to coaxial cable [note this would also apply to 450Ω ladder line - G3VA]
- an antenna site clear of obstruction of at least a half-wavelength in all directions at the desired frequency;
- elevated antenna - the higher the better.
- soldered joints, including radial wires:
- avoid conductive supports for inverted vees and other wire antennas - they play havoc with absorption and patterns of radiation; and
- use a low-loss ATU when possible.

In general, VK4CMY and most other low-power enthusiasts recommend a

Pat Hawker's Technical Topics

PAT HAWKER, G3VA
London 37/SE22 8SS

variable-frequency rig for serious DX work. Crystal-control although providing the simplest approach, largely restricts operation to CQ-induced contacts, except possibly on the recognised QRP frequencies. While the simple QRP transceiver is good fun, there are often advantages in using a separate high-performance receiver. Send a little slower than usual on CW and speak clearly on SSB using standard phonetics: give the receiving station all the help you can. It is worth indicating that you are a QRP station as this often results in greater attention being given to your calls - but don't abuse this practice when using say 15 or 20 watts!

Many designs for simple low-power rigs have been published, for example in Sprat,

Band (MHz)	28	21	14
C1(pF)	330	470	680
C2(pF)	100	150	220
L1	3turns 5/8th in long	4turns 5/8th in long	5.5turns 5/8th in long
L2	7turns 5/8th in long	10turns 1 in long	12turns 1 in long

C1 and C2 are mica or ceramic. L1 and L2 are 'air-wound' from Nr 14 wire, with an inner diameter of 3/8th inch.

Table 1: Output filter component values

but my eye was caught by the experimental 0.5 watt transmitter (*QST*, November 1994, 'Hints & Kinks' p84) with which Lew Smith, N7KSB reports working all continents and over 30 countries using just a roof-mounted ground-plane antenna. The rig is suitable for 14, 21 or 28MHz (or with suitable crystals etc also 18 and 24MHz). RF output about 0.5W on 14 and 21MHz reducing slightly to about 0.47W on 28MHz.

This transmitter is based on a 74HC240 octal inverting buffer IC with one section working as crystal oscillator, four sections as a parallel-connected amplifier and three sections unused: **Fig 1**.

N7KSB also notes that the 74HC240 can alternatively be used to drive a power MOSFET IRF15 power amplifier to provide up to 15 watts RF. The output stages in 74HC— devices are designed to have equal value pull-up and pull-down transistors. This minimises even order harmonics, simplifying the output filtering: see **Table 1.**

Among the notes supplied by N7KSB is that since the 74HC240 dissipates 0.5W on 14MHz and some 0.9W on 28MHz it must be heat-sunk; however epoxying the IC to the ground-plane in dead-bug or 'ugly' construction provides adequate cooling. L1 is a low-inductance coil and is sensitive to lead length "so if you build this transmitter with plug-in filters (Lew Smith uses phono plugs and jacks for this purpose), you may need to remove one or two turns from L1 to compensate for the extra lead length involved in wiring the coil plugs and sockets".

Table 1 gives suggested output filter component values. Note that the logic chips have built-in input and output buffers. The extra gain provided by the extra stage makes it harder to get rid of key clicks. The click-filter shown uses an unusually large (33 millisec) time constant.

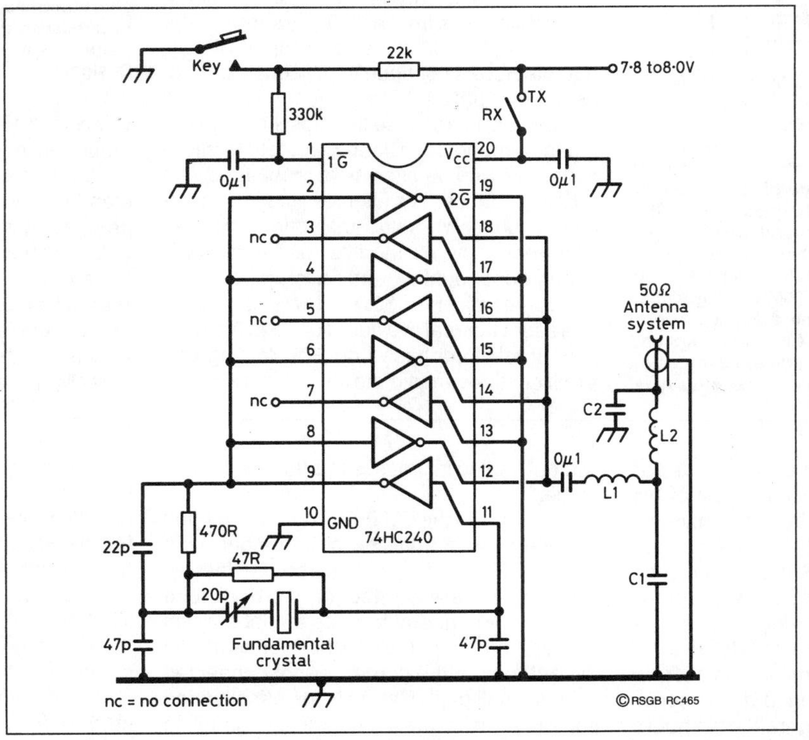

Fig 1: The experimental 0.5W CW (code) transmitter described in *QST* by Lew Smith, N7KSB and based on a 74HC240 high-speed CMOS octal buffer integrated circuit. One section serves as a crystal oscillator, four sections form the amplifier, and three sections are unused. It can put out 0.51W on 14MHz and 21MHz and 0.47W on 28MHz.

SLIM JIM & ELNEC

TT, DECEMBER 1994, pp62-63 and Fig 3(b) gave the dimensions of the original G2BCX 'Slim Jim' 144MHz antenna but noted that in 1986 Greek university authors had made a computer analysis of the design and found that these dimensions resulted in a 50Ω resistive match at 155MHz rather than 145MHz and with maximum radiation at an undesirable 22° above the horizon; modified dimensions for a 101.8MHz broadcast antenna were given.

This resulted in an interesting letter from Bernard Spencer, G3SMW which, inter alia, shows that despite the many benefits that have stemmed from the development of user-friendly versions of 'method of moments' (NEC) antenna evaluation software, some care is still needed in setting up the 'model' for

PITFALLS OF FILTER TESTING

OVER A QUARTER-CENTURY AGO, two Collins Radio engineers drew attention to the fact that the multi-section low-pass filter imposed between (valve) transmitters and the antenna were presented with attenuation characteristics that, if realised, would virtually stop any VHF harmonics from reaching the antenna: "Yet, increasingly, there have been doubts whether the high theoretical attenuation figures of these filters are always (or even often) being achieved in practice".

As still noted in *ART7* (p193) the Collins engineers (Richard Weinreich, K0UVU and R W Carroll) suggested that it is possible to increase rather than to suppress harmonics with conventional LC filters. Basically, this paradox is because such filters are designed to be driven from a purely resistive source and loaded into a resistive termination. Now, even in a correctly set up filter system, the typical transmitter output impedance tends to be resistive only at the output frequency, and

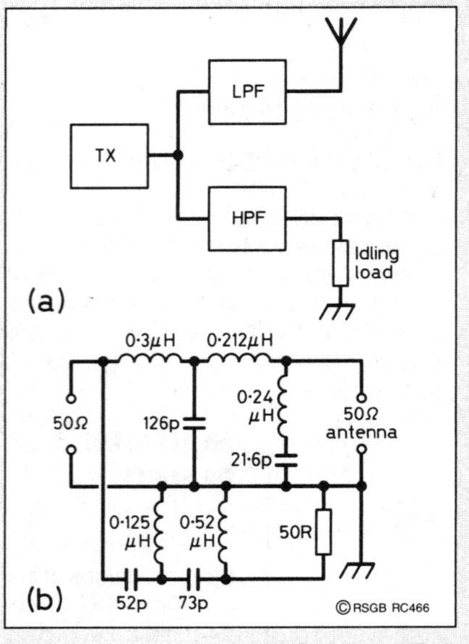

Fig 2: The absorptive type of low-pass filter proposed by Collins Radio Engineers in *QST*, November 1968. (a) General principle. (b) Representative four-pole filter with design cut-off at 40MHz and with a rejection peak in the American TV channel 2 (54-60MHz using double-clad printed board having a capacitance of 10pF per square inch for screening and capacitors. The original article gave design formulae for two to six pole filters of this type.

is likely to be highly reactive at some at least of the harmonic frequencies. The transmitter reactance at these harmonics may partially, or even wholly, cancel the filter input reactance, and thus permit VHF power to reach the antenna elements."

They presented a practical solution to this 'unhappy chapter in filtering' in the form of filters which achieve filtering by absorption of the harmonic energy, dissipating it in a resistor, rather than by attempting to reflect the harmonics back into the transmitter: **Fig 2**. This type of filter uses a cross-over filter, much as in audio loudspeaker practice, to separate the unwanted VHF harmonics from the wanted HF energy and then matching the VHF energy to an idling resistive load where it can do no harm.

Absorptive filters turn up from time to time although the technique has never been widely used. With the coming of solid-state power amplifiers, additional low-pass filters are usually built-in to transceivers, and the problem of VHF harmonics has been alleviated in the UK by the change to UHF television.

But the problem of 'over-optimistic' characteristics has not gone away, and is found to a marked degree in the insertion loss and attenuation data presented for power-line filters. John Greenwell, G3AEZ draws attention to an article 'the pitfalls of EMC filter testing; beware of half measures' by Gareth Nutt of Schaffner (*Electronic Product Design*, December 1994, pp45-47). this is introduced as follows:

"The use of powerline-filter insertion loss and attenuation curve data (as published by filter manufacturers) can be a frustrating experience. Many experienced EMC engineers consider these curves, generally prepared using data taken from a 50Ω test set-up, to be of extremely limited value. In spite of this, manufacturers continue to publish 50Ω data, as popular 50Ω equipment, connectors and test cables make these the most easily taken measurements . . . Clearly, in real-life situations, a powerline-filter is not terminated with 50Ω impedance. The filter termination is usually an unknown value that often changes with frequency. As the filter performance is largely dependent on termination impedance, the curves given in 50Ω set-ups can never represent the real live situation." **Fig 3** shows how predictions based on 50Ω testing can significantly overestimate the attenuation provided in the case of a typical switched-mode power supply.

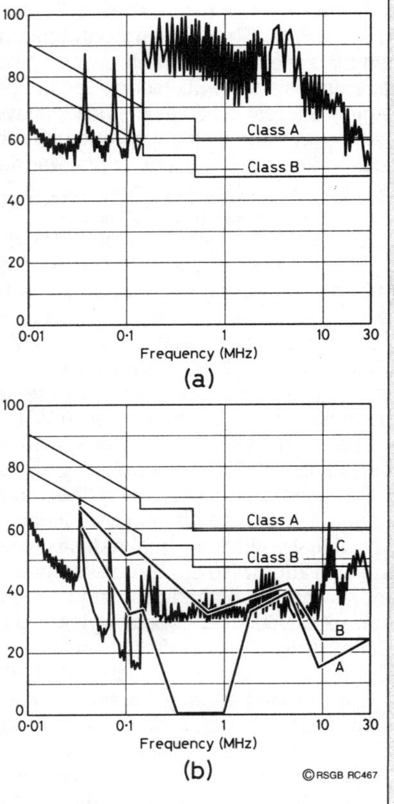

Fig 3: (a) Noise output of a typical switch-mode powder supply unit. (b) Noise level predictions and the measured results after fitting filter. (A) Noise level predicted using 50Ω test. (B) Noise level predicted using 0.1/100Ω curves. (C) Real system noise level measured after fitting filter, underlining the over-optimism of published data based on 50Ω test set-up. (Source Schaffer EMC in Electronic Product Design).

Gareth Nutt makes a strong plea for standardised, acceptable test methods to be used by manufacturers for published data. He shows that powerline filter curves based on 0.1/100Ω test systems are more realistic although even then he notes that at frequencies above 10MHz coupling effects can bypass the filter and give an unrealistic shape to the attenuation curves.

analysis. While professional antennas can usually be erected in the clear, most of our antennas are surrounded by conductive objects which affect radiation patterns etc. G3SMW writes: "I put the Slim Jim configuration of your Fig 3(b) in my ELNEC program, and sure enough got a main-lobe elevation of some 22°, and a best match on 50Ω at about 155MHz. I also got a similar elevation at 145MHz for the same antenna dimensions.

"While the Fig 3(b) configuration (without a feeder) is the obvious way to put this antenna into ELNEC, I was puzzled by the main-lobe elevation, having over the years made and tested Slim Jims and other vertical antennas without finding in practice this large elevation angle.

"So I then, in the ELNEC program, dangled a piece of wire from the bottom right-hand corner of Fig 3(b) to simulate a coaxial feeder. This had a marked effect on the main-lobe elevation. In general it made the lobe much nearer the horizontal, whether connected to the antenna at the corner, or not connected (with a small gap). Naturally the effect varied somewhat with the length of the dangling wire, being greatest for a half wavelength. With this, and for some other lengths, the main lobe given by ELNEC was within about 5° above or below the horizontal.

"This was not an exhaustive investigation, but perhaps shows that the G2BCX Slim Jim usually proves satisfactory in practice when fed as usual by vertical coaxial feeder, although admittedly this does have any bearing on the matter of the best match being at 155MHz is that it may be advisable to scale the dimensions from the Greek antenna as suggested in *TT*.

"From both my amateur and professional studies of the nasty VRPs of many vertically polarised VHF antennas, I believe that these do not arise simply from currents on the outside of the coax braid due to its connection to the antenna but also (as shown by my

ELNEC run with a gap) by currents induced on the feeder from the dipole element above. This implies that no single choke-balun or gap would prevent them. I hope to investigate this further, as this represents a nasty problem with many vertical antennas."

EARLY OVER-THE-HORIZON VHF

WHILE THE ROLE OF radio amateurs in the opening of the 'short waves' (200 metres and down) in the 1920s as a long-distance medium for international communications and broadcasting is still widely recognised, much less seems to be known about the early development of 'over-the-horizon' VHF propagation by means of tropospheric scatter and ducting in which amateurs played a less dramatic, but still useful role.

Much of the credit goes to Marconi and his work in Italy, Sardinia and on board Elettra in the Mediterranean. But hard on his heels was the British Post Office which in the early 1930s began a series of experiments leading to a fully operational good-quality radio-telephone link between the UK and the Channel Islands (Guernsey to Chaldon, a path length of some 85 miles of which some 36 miles was outside the optical range) during 1937-39. Very large antenna arrays were used and it is clear that the reliability of the circuit depended, at least in part, on troposcatter propagation.

A close observer of the Post Office progress was teenager Christopher Henn-Collins, former GU5ZC who, a few years later, as a Lt-Colonel in the Royal Corps of Signals, was responsible for the first practical use of OTH radio-relay. This was in North Africa in 1943 when he was in charge of the Radio Division of AFHQ in Algiers. He has recently published an account of how these OTH links were the result of his knowledge of the early Channel Island links combined with one of those 'happy accidents' that change the course of history: 'Over the Horizon Radio and Beyond - or Lost Opportunities', *The Journal of the Royal Signals Institution,* Autumn 1994.

This is a fascinating story of how the first military RTTY VHF radio-relay system came into being, born out of a need to prevent a serious breach of wartime signals-security and fathered by GU5ZC's interest in long distance VHF communications, first aroused by the pre-war GPO links to their station near his home in the Channel Islands. But sadly the story also emphasises how British conservatism allowed the post-war commercial exploitation of the troposcatter mode to pass largely to the USA.

To quote a few key paragraphs: "My interest in OTH radio was reawakened in 1943, not just as some esoteric planning exercise but this time with hardware. If you check the bibliography, and the source of the information is American, it is likely to credit the first successful OTH links to the work of the US National Bureau of Standards in the late 1940s; if, on the other hand, the source is British, it is likely to be credited to GPO/Telecoms trials in the early 1950s. However the earliest references I have traced were in US Amateur Radio literature (*QST*, September 1945) which credits our work in North Africa.

"From 1942 I was in charge of the Radio Division of AFHQ and early in 1943 the Americans shipped a number of high power (250 watts output) police VHF radio sets, to provide General Eisenhower with personal voice communication from anywhere in the 'Theatre' back to AFHQ (Algiers). Brigadier Willie Scott, who was then the Deputy Chief Signals Officer, took me aside and warned me that because of the total lack of security with such voice equipment the Americans must not be allowed to use it for its intended prime purpose as the implications could be appalling . . . I was faced with a dilemma . . . It was up to me to divert this new equipment to some acceptable alternative use - otherwise I was for the chop!

"It took little prompting to come up with the idea of using the equipment to provide a radio teleprinter link forward from Algiers, using 'Apparatus 2 Tone Telegraph' (which could work on a two-tone basis to provide mark and space conditions) to modulate the transmitters. The first link was to Constantine with a path length of about 130km, much the same as the prewar UK-Guernsey GPO link, and it worked well. Soon we added a couple of more links in cascade and were through to 18 Army Group HQ, near Tunis with an automatic RTTY 600km link which turned out to have excellent reliability and which often worked as a direct link without the intermediate relays.

"The success of this project was duly reported to Washington and to London but with very different responses. Soon Dr Beveridge, representing RCA, and a senior engineer from Bell Labs came to Algiers and on their return to the States, the US Army Signal Corps accelerated the manufacture of VHF equipment to take advantage of this new mode of radio communication. Within six months radio-relay systems were arriving in Algiers. By contrast, a news letter came from the British Director of Signals containing a post-script that Colonel Henn-Collins was to be thanked for his description of a radio link!

"About this time, my US No 2, Lt-Col E C Page, suggested that I should patent the idea after the war as there would be money in it, he also asked me if he could have the US rights. I remember telling him that as a regular officer I would have to assign any patent to the Crown leaving me with a lot of extra work and not a penny in royalties, but I had no objection to him exploiting it in the United States . . . Subsequently he set up Page Communication Engineers and made a fortune as prime contractor to the US Department of Defence (as the main supplier of troposcatter and long-haul VHF radio relay systems)."

After the war, the GPO believed that the success of OTH techniques was being greatly

HF CONVERTER FOR AM BROADCAST RECEIVERS

THE 'CIRCUIT IDEAS' FEATURE of the December 1994 issue of *Electronics World + Wireless World* includes a simple HF converter intended for use with medium-wave AM car radios (better screened than most domestic models with ferrite rod antennas). The converter as described by Peter Parker, Bentley Australia comprises a dual-gate MOSFET frequency converter with a separate crystal controlled oscillator: **Fig 4**. By using just two low-cost crystals and making use of oscillator-high and oscillator-low mixing, he achieves continuous coverage from 3.4 to 7.6MHz in four bands, tuned on the car radio.

Good frequency stability and easy tuning are among the benefits and he points out that if a BFO is added to the car radio, it is possible to receive SSB/CW. Inductors L are wound on ferrite toroids (14 turns on 9mm toroid, Philips 432202097180 4C6 ferrite) and the converter should be housed in a small metal box to avoid direct pick up of local broadcast signals. The car radio must have continuous tuning.

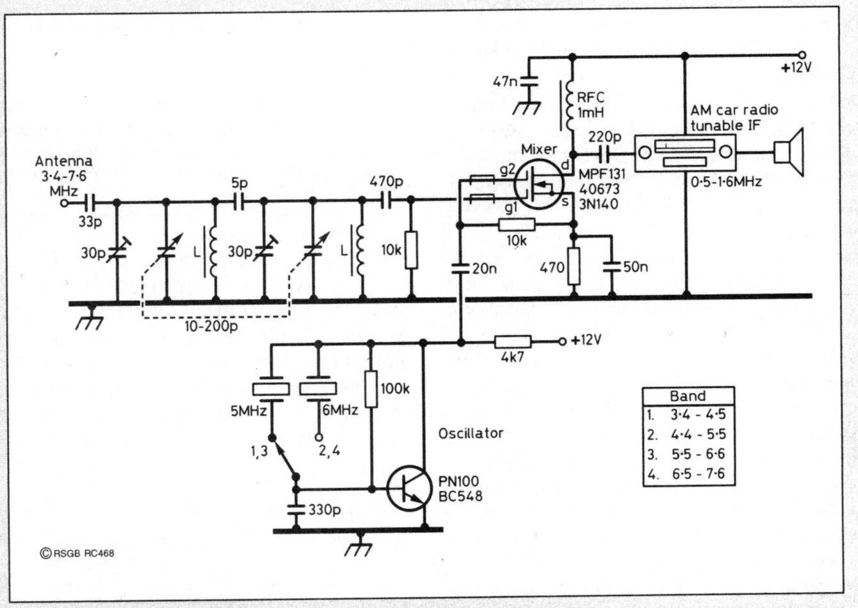

Fig 4: Band-pass tuned front-end converter suitable for feeding medium-waves AM car radio, etc. (Source EW + WW).

oversold; they considered them unreliable and of limited range. It was not until years later that the GPO re-discovered many of the techniques they had originally used in the 1930s when oil platforms in the Noth Sea became important.

I recall that about 1960 I added a section 'VHF Scatter Communications' to an edition of the *Radio & Television Engineers' Reference Book* with a bibliography that included credits to H G Booker & W E Gordon (*Proc IRE*, 1950, p401) who organised the American NBS post-war study (in which America radio amateurs participated) and to Eric Megaw (G6MU) of GEC Research (*Nature*, 1950, p1100). Dr R L Smith-Rose (RSGB President 1959) with a Miss A C Strickland published 'A study of propagation over the USW radio link between Guernsey and England on wavelengths of 5 and 8 metres (60 and 37.5MHz)' (*JIEE*, Vol 90, Part III, 1943). This noted that the first suggestion that VHF propagation over land was influenced by weather (ie troposphere) was made in 1935 by Ross Hull, the Australian radio amateur who became Technical Editor of *QST* and was later electrocuted working on TV equipment. Marconi's original work in the 1930s can be found in 'Guglielmo Marconi and communication beyond the horizon: a short historical note' by G A Isted (IEE Symposium on long-distance propagation above 30MHz, January 1958).

MEDIUM OR MESSAGE?

A NOTE IN THE November *RadCom* (p7) showed that in the year to April 1994, the total number of UK Amateur Radio licences increased by some 6% from 59242 to 63033 but the number of Full A licences actually dropped by some 600 to 31817. Even more striking was that the number of Novice A licences is only about one-tenth of the Novice B licences. Does this signify that HF operation is no longer the prime aim of newcomers, or simply the belief that the Morse barrier, even with a simple 5 word per minute practical test, may eventually disappear so that it is no longer worth the learning effort involved? For those of us who believe that CW on HF is fundamental to amateur radio this seems an unhappy state of affairs. But there is no doubt that the hobby is changing, throwing up an exciting forest of new challenges but seeing some of its traditional trees under threat.

David Sumner, K1ZZ, Secretary of ARRL and Publisher of *QST*, in an editorial (*QST*, November 1994, p9) points to the growing diversity within amateur radio as both a great strength and our principal weakness: "Sometimes it seems as if we're 50 separate interest groups marching in 50 different directions, passing up opportunities for mutual support when it's needed and bumping into one another even when collisions could be avoided . . . We must work together better . . . To do that we have to understand one another - perhaps even ourselves - a little better . . . What is more important to you, the medium or the message? Is your interest in Amateur Radio fed primarily by a sense of wonder at the magic of radio communication? Or is it derived from what that magic allows you to accomplish? . . . All too often, we amateurs denigrate one another for not being 'real hams' or for engaging in what seems to the observer to be pointless activity. Perhaps some of that intolerance is rooted in a lack of understanding what the other fellow values in Amateur Radio. Perhaps it springs from a lack of knowledge about an aspect of Amateur Radio we haven't personally experienced. Maybe we've simply forgotten how ignorant we were when we started, and how much we learned because others were willing to help us along."

Frank Merritt, VE7FPM in 'Challenges face amateur radio - New technology offers unlimited opportunities - go for it!" (the *Canadian Amateur Radio Magazine*, October 1994, pp18-19) believes "There is a crisis in amateur radio. It has been coming for a number of years and now we are in the middle of it . . . In the Chinese language, there is no difference in symbol between the word-concept 'challenge' and the word-concept 'crisis' . . . For many years, it was relatively easy to explain to a non-amateur what amateur radio was and what we do. There used to be a few HF bands and we had either code (CW) or voice (then AM - amplitude modulation) operation . . . That is the way it was and it was relatively uncomplicated.

"We have undergone a metamorphosis. The picture of amateur radio today is much more complex, a complexity that makes it more difficult to explain to a non-ham what amateur radio is all about."

VE7FPM traces the move away from AM to SSB, the emergence of VHF-FM in +/- 30kHz

LOW-DISTORTION LC OSCILLATOR

JOHN CRONK, GW3MEO has drawn attention to two designs appearing in the Maxim Engineering Journal (Volume 16) issued by Maxim Integrated Products (UK) Ltd of Pangbourne near Reading (Tel: 01734 845 2550), the UK representatives of the Californian IC manufacturers. One design, using two MAX436, provides an allpass network with 90° phase shift and by using two of these networks with different corner frequencies it is possible to obtain a wideband 90° shift (for example from 180kHz to 740kHz. All-pass networks can be used for phasing-type SSB modulation and demodulation.

The other design, based on a MAX436 single wideband transconductance amplifier (WTA) is for a negative-resistance LC oscillator which, for the 9.3MHz oscillator shown in **Fig 5**, has a claimed third harmonic output below -40dB (THD less than 1%). It is pointed out that negative resistance is easily synthesised with a WTA: "Connect the WTA's positive input to its output and its negative input to ground. Then, a positive voltage applied to the output causes current to flow out of the amplifier, in proportion to the applied voltage, the circuit acting like a resistor whose current flows in the opposite direction; hence the negative value . . . By itself, the combination of tank circuit and regenerative element (negative resistance) simply drives the output amplitude to saturation. To achieve steady oscillation the circuit needs an amplitude limiter. R4 serves that purpose; it appears (in parallel with R3) only when the amplitude is sufficient to turn on one of the diodes D1 or D2 . . . The oscillator, whose tank circuit consists of a mica capacitor [C4] and air core inductor [L1], has an output of 9.3MHz. You can trim the output frequency to any reasonable value, but above 10MHz the layout should include short connections and a ground plane."

The design note does not include any comments on the stability of the oscillator but the high C/L ratio suggests this should be good despite the use of an IC which presumably has a rather high temperature coefficient; the note does however include details of how the component values are calculated.

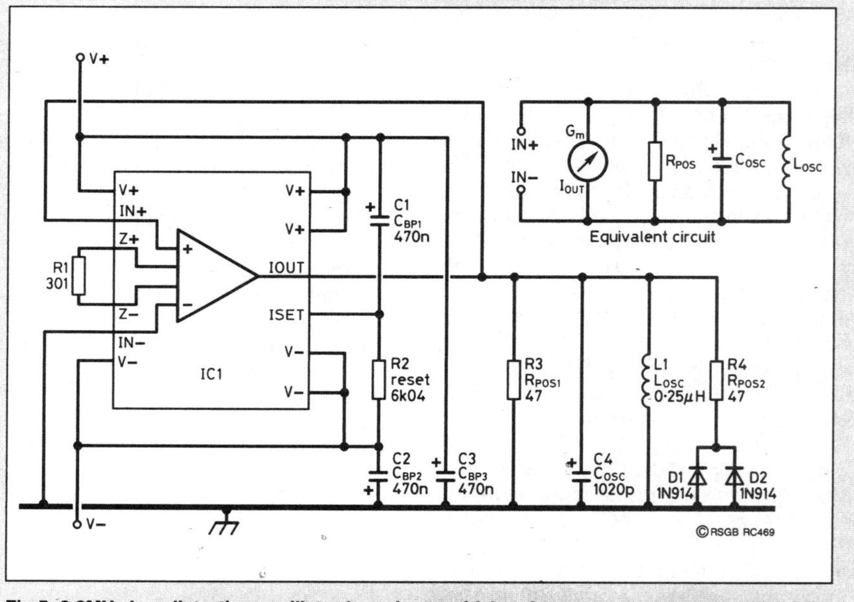

Fig 5: 9.3MHz low-distortion oscillator based on a wideband transconductance amplifier (IC1) whose negative resistance counters losses in the L1/C4 tank circuit. Total harmonic distortion less than 0.1%. (Source Maxim Engineering Journal).

systems which eliminated the problem of drift, and the steady growth of digital systems from mechanical RTTY in which the only electrical element was a selector magnet to the all-electronic PACKET (first proved in Canada by Dr John DeMercado) and more recently PACTOR and the radically new CLOVER developed by Ray Petit.

VE7FPM believes that CLOVER may in time become the dominant amateur data system and that: "We now have computer-based systems that provide hitherto unheard-of possibilities for radio communications. The biggest challenge today is not operating our systems, both digital and non-digital, but to communicate our relevancy in the modern world . . . This is where we need to sharpen up! The crisis is a challenge for us to communicate to others just what we are doing and its charm.

"Originally, virtually every amateur had to be proficient in the technology of the day. . . this has to some degree become irrelevant . . . Increased reliability puts a lot of us more firmly in the business of communicating rather than continually patching up our equipment . . . Our computers provide data-handling capabilities that were not dreamed of in years gone by. To be sure not every operator is interested in digital techniques. CW operation is still with us and giving genuine pleasure . . . There are more SSB stations than ever . . . Diversity has become the key . . . We find individuals operating in advanced areas such as moonbounce or meteor scatter . . . Modern challenges are certainly with us but so are the old ones."

Frank Merritt, VE7FPM concludes that while the future direction of amateur radio is only partially clear, a number of new vistas and pathways are evolving: "What is your choice? Go after it!"

I have to confess that, personally, I still find pleasure in keeping old, fallible but technically-understandable, pre-computer HF equipment on the air with the aid of manual Morse keys and little more than a multimeter and soldering iron. My experience of reading text off a TV screen is largely confined to a word-processor VDU and broadcast teletext. But undoubtedly K1ZZ and VE7FPM raise serious questions. The more we diversify our interests, the less we as a body will have in common, the more likely to fall out over frequencies and band-planning, and to engage in bruising mode wars. The magic of communicating by radio nowadays contributes just the provision of a link between computers to form super-highways in pale imitation of Internet! This may lead to less comprehension not only by the public but also by newcomers unaware of the traditional role of a hobby founded on technical curiosity and the wonder of unreliable but strangely fascinating radio propagation. Diversity is good and necessary, and it would be a sad day if we all opted out of state-of-the-art technology and up-to-date communication techniques but this still needs to be based on a common heritage and mutual understanding.

HERE & THERE

RICHARD FORMATO in *EW + WW* (December 1994 pp 1026-1028) shows how the bandwidth of VHF/UHF monopole antennas on ground planes can be maximised by choice of suitable length to diameter ratios. He gives some simple, easy-to-remember rules for computing maximum monopole bandwidth (50Ω characteristic impedance, VSWR less than 2.5:1: (1) Maximum bandwidth occurs when the ratio of monopole length to diameter is five; (2) maximum bandwidth is about 50% of the frequency at which the monopole is a quarter-wave long; (3) frequency of minimum VSWR is about 1.3% less than the quarter-wave frequency; (4) approximately two thirds of the bandwidth is above the quarter-wave frequency, and about a third is below; (5) VSWR minimum is a near-perfect 1.009:1.

Daisuke Yamazaki *et al* of the Sony Corporation in IEEE Trans on Consumer Electronics, August 1994 report that "For the first time in the world, a complete single-chip AM-stereo/FM-stereo radio IC which can be used for a portable radio has been developed." 2100 elements are integrated in a chip size 3.36mm by 4.00mm operating from a supply voltage of 2-12V with supply current AM 14.8mA, FM 12.2mA.

R C Arnold, G8ZDU adds to the saga of 'solid-state firsts' (*TT*, August 1994) by sending a newspaper cutting (unfortunately not dated) which was sent to him in 1953 while serving in the RAF in Egypt. This was headed 'Valveless Radios in Contact' and continued: "Two-way contact between radio sets using transistors instead of valves has been set up by amateur radio operators in Leicester. The communication was made by Mr Cyril L Wright [presumably G3CCA] and other members of the Leicester Radio Society, who received signals from a station at Maids Moreton, Bucks. The transistor is a tiny unit which consumes little power, has no filament, needs no attention after being installed and will run for well over 100 years. Mass-production of them is not expected before 1956 or 1957."

Jack Althouse, K6NY (Palomar Engineers) remains convinced that the health dangers of RF radiation (*TT*, October 1994) and low-frequency power fields have been widely exaggerated. He encloses a letter (Radio World, August 10, 1994) from a broadcast chief engineer who spent many years working in close proximity to high power broadcast antennas: "Of the hundreds of technical people that I have met or worked with in all of these years, only one man died of mysterious illnesses such as cancer. This was undeniably due to his daily ingestion of many packs of cigarettes, not his lifelong pursuit as a broadcast chief engineer . . . An organization has grown up to develop a very financially rewarding career for all concerned, with the worst yet to come. Surely there must be a way to halt this costly fiasco. Produce the corpus delicti or get off the pot".

K6NY also enclosed an article by Professor W R Bennett of Yale University 'Power Lines are Homely, Not Hazardous' (The Wall Street Journal, date unknown) and graphs from his book 'Health and Low-Frequency Electromagnetic Fields' (Yale University, 1994) which underlines that RMS magnetic fields from domestic appliance are many magnitudes lower than from electric railways [and tube trains]: **Figs 6** and **7**. Professor Bennett writes: "If fields of two milligauss really are a serious threat in Denver, Los Angeles and Sweden, then commuters on East Coast electric trains - where the fields at power line frequencies can be hundreds of times larger - ought to be dying like flies. So much for the electromagnetic hoax. To grasp how much damage it's caused, consider what useful developments in cancer research, in education, or in any other field, could have been accomplished with the $23 billion already squandered on this scare. That ought to be more shocking to people than any electromagnetic field." ♦

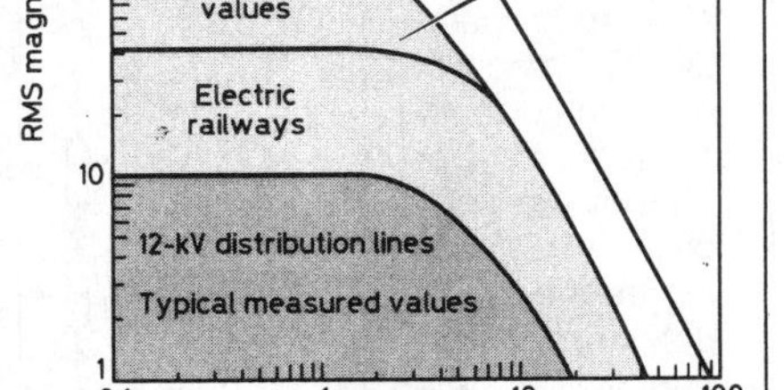

Fig 6: RMS magnetic fields at head level for electric railroads and American three-phase 12kV urban distribution lines. Note that the solid curves represent the calculated worst-case limits; the shaded curves represent the typical range in measured values obtained with a three-coil magnetometer at 60Hz. (Source 'Health and Low-Frequency Electromagnetic Fields' Yale University, 1994).

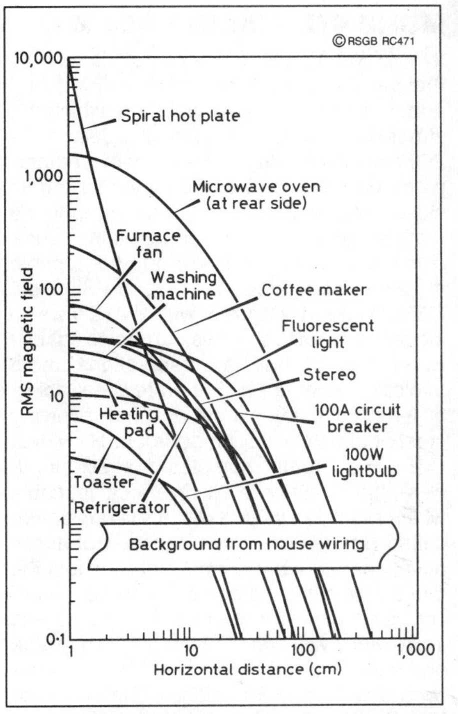

Fig 7: Measured magnetic fields from common household appliances. Note that horizontal distance is in centimetres compared with metres in the previous diagram. Calculated values are shown for the spiral hot plate because the spatial resolution of the meter available was inadequate to determine such rapidly increasing fields at small distances. (Source 'Health and Low-Frequency Electromagnetic Fields, Yale University, 1994).

MORE ON SEALING COAX

IN THE MARCH 93 *TT*, I included comments that the *QST* 'Hints & Kinks' editor WJ1Z, had added to the W0KKG item, in which he stressed the need to seal the SO-239 and PL-259 socket and plug to make them weatherproof. However he warned against using, in closed compartments, caulking compounds that liberate acetic acid vapour as they cure, since this can severely corrode susceptible metals such as steel wool. Gerald Stancey, G3MCK, heartily agrees with WJ1Z's warning on the need to seal coaxial cables against water ingress, but has reservations about avoiding sealants which liberate acetic acid – at least for conventional use not involving steel wool in a closed compartment. He writes:

"I have rcently completed a six-month weathering test on a piece of coaxial cable which has been sealed with a silicone sealer called Hermetite. The seal reeked of acetic acid (vinegar) while curing. After the test the cable was cut open and showed no sign of corrosion or water ingress. Perhaps I was lucky with my choice of sealer but perhaps the problem occurred in the past but no longer occurs with modern silicone sealers." The March 1993 *TT* also included some comments by G2HCG and GW3TMP that have not passed entirely unchallenged. J K Todd, G2KV, notes that: "G2HCG doesn't like the impedance discontinuity of the PL259 plug. But I suggest that it is adequately satisfactory on HF and not too bad on VHF. It is about 2cm long and therefore only 0.02-wavelength long at 300MHz. Put this on a Smith Chart and the resultant effect of the discontinuity is surely negligible."

GW3TMP in advocating the use of his Ferromagnetics choke balun for G5RV-type multiband dipoles may have given the impression that such a device is always advisable. A number of writers, apart from G0GSF, have shown that provided the element and feeder are erected symmetrically there will normally be very little outer-braid RF current flowing on the coaxial cable, in spite of the junction of unbalanced coax to the balanced 300ohm line. This of course is usually also the case with any symmetrical dipole. Baluns are likely to be necessary only where the element is not symmetrical and at 90° to the first downward part of the transmission line. The answer would seem to be to check for outer-braid current before fitting a balun.

Pat Hawker's Technical Topics

PAT HAWKER, G3VA

London 37/SE22 8SS

COAXIAL CABLES AND CONNECTORS AT VHF/UHF.

IVO CHLADEK, ZS6AXT, in Part 3 of his 'VHF/UHF/Microwave Primer' (*Radio-ZS*, June 1994, pp4-5) has some pertinent comments on the selection of coaxial cables and connectors for use above 50MHz. He writes:

- For the frequencies above 50MHz and cable runs longer than 10m, one should not consider thin coaxial cables such as RG58 or similar, since their losses are too high. Some makes of thicker coaxial cables such as RG213 have very poor outer braiding and these must also be avoided. The most suitable is 0.50inch Heliax-type cable for long runs, but this cable and connectors for it are rather expensive.
- Buying secondhand RG213 or similar cables is rather dangerous - even a minute amount of humidity in the cable will permanently damage it and increase its losses. Heliax-type cable with a foam dielectric is effectively sealed and humidity cannot penetrate it. 30m of good RG213 cable has a loss of over 2dB at 144MHz but losses can be much greater for cable in poor condition.
- Another problem is ageing of the cable when exposed to UV (ultra-violet, sunlight) radiation. Cables used outside for a few years will tend to become yellowish in colour and such cables will have high losses.

ZS6AXT notes that there is a large selection of coaxial connectors on the market: "The UHF or PL259 series is suitable for the 50 and 144MHz bands but with reservations. First, it is quite difficult to fit this type of connector and unless done correctly may result in excessive losses. Second, these connectors are not watertight and must be well sealed when used outside. Thirdly many of these connectors use inferior insulating materials and that increases losses.

"The BNC-type of connector is suitable only for the thinner cables such as RG58 and for use inside the shack. They tend to be noisy due to rather poor insulation, avoid them in the front end of a receiver.

"For the thinner cables and interconnections between equipment modules, the best choice is the SMA-type of connectors or the cheaper SMB-type. Both are quite expensive but for frequencies above 432MHz they are virtually the only choice. They are suitable only for the thinner cables.

"Probably the best choice is the N-type which is suitable for use up to the GHz region. They are fairly easy to work with and they are

A SIMPLE 'HF STARTER RADIO'

ALTHOUGH VERY SIMPLE 'straight' receivers cannot be recommended for serious amateur use, they can provide a low-cost and eminently satisfying project for anyone, young or old, who has never experienced the fun of listening to HF broadcast stations and amateur SSB/CW transmissions on a home-built receiver.

Bill Orr, W6SAI, in his column in *CQ* (December 1994, pp102-103) provides the circuit diagram of a "Nifty shortwave receiver for the beginner" originally designed by Charles Kitchen of Analog Devices in the 'Design Ideas' of *Electronic Design News* (EDN) with a component cost of less than about £20: see **Fig 1.**

With the coil shown, the tuning range is about 5 to 10MHz, covering several broadcast bands and the 7 and 10.1MHz amateur bands, but adding or subtracting turns from L1 will change the tuning range as required. TR1 can be regarded as either a regenerative RF amplifier or more correctly as a signal-frequency Q-Multiplier, with D1 acting as the signal detector. Regeneration is controlled by RV1. With TR1 carefully adjusted to the 'just-oscillating' condition, CW and SSB signals can be received. It is claimed that the low power of TR1, with the complete receiver drawing less than about 10mA from a 9V battery, means that although the oscillator is coupled directly to the antenna (through C1) it is most unlikely to interfere with other local receivers. C1 removes 'dead spots' which could be caused by directly coupling a resonant antenna to the resonant circuit but with a very short whip antenna C1 can be omitted. If required, a volume control can be added by making R3 (2K) a potentiometer and then connecting C3 to the variable arm. R4/C4 form a low-pass filter to improve sound quality and prevent circuit instability. D2, D3, D4 provide voltage regulation for TR1 and minimise drift. L1 can be wound on a plastic film can or a pill bottle of one-inch diameter.

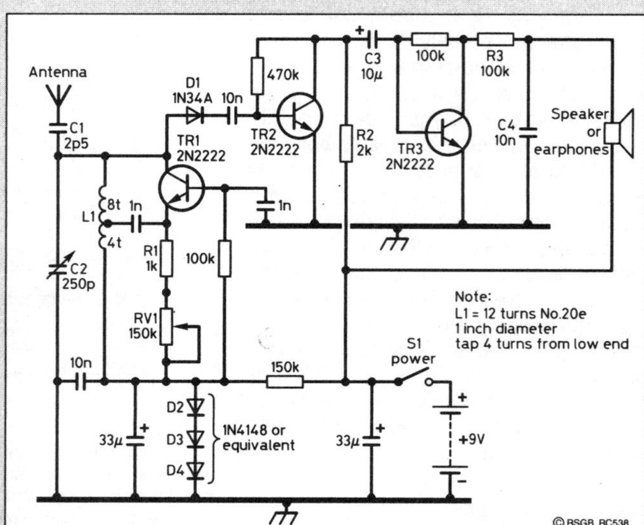

Fig 1: Inexpensive shortwave receiver for 'first steps' on HF and home construction.

watertight. A few makers use inferior insulating material. This should be PTFE (Teflon) which is quite easy to recognise. They are expensive but good value for money. The normal type of N connector can be used for 0.50inch Heliax-type cables with a small modification. It is a good idea to lubricate the thread on those connectors which are often plugged in and out with a small drop of oil. This will greatly prolong the life of the connector.

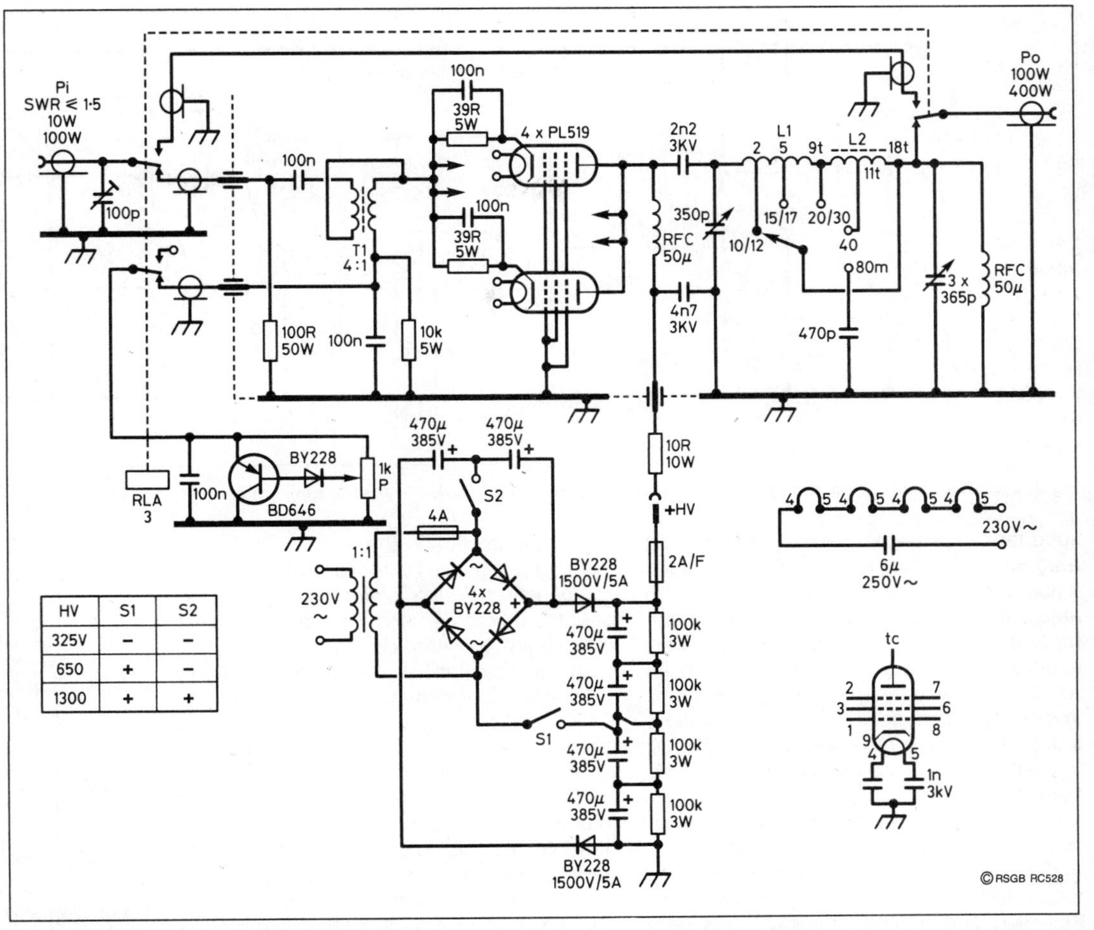

Fig 2: Annotated circuit diagram of PA0FRI's grounded-grid 400W 'Finear' linear amplifier based on four PL519 valves including quadrupler-type PSU. The four heaters of the PL519 valves are series-connected with a 'wattless' voltage-dropping capacitor. Note that the high voltages involved are potentially lethal. L1 – 9t 2.5cmm dia, 3mm wire, 6cm long; L2 – 11t+7t 2mm dia wire, T200-2 toroid; T1 - 9t bifiliar, 1mm dia wire on 5cm ferrite rod, 1cm dia; TC – set for minimum SWR on 29MHz; P – set for quiescent current 80-100mA; 100R – 2x 50Ω/25W non inductive, TO220 type; BD646 – power PNP darlington.

FRINEAR 400W GROUNDED-GRID LINEAR

IN RECENT YEARS, Frits Geerligs, PA0FRI, has presented through *Technical Topics* a series of passive-grid linear amplifiers based on a still-available colour-TV line output (sweep) valve, type PL519. Among the ones featured were a 400W linear using four PL519 valves (*TT*, August 1992, p39 and reprinted in the new 6th edition of the *Radio Communication Handbook,* (p5.32-3) and an earlier, rather more complex three PL519 version (*TT*, February 1990, p30).

PA0FRI has now sent along details of a new grounded-grid amplifier which is based on four fan-cooled PL519 valves used in association with a novel power supply unit which provides switchable output voltages. He writes: "After experimenting over the years with transmitting valves in various modes, I have finally come up with the basic (KISS) arrangement shown in **Fig 2** and developed for use with the current 'standard' range of 100W solid-state transceivers. When properly assembled and installed it will provide a comfortable 350W at 28MHz, rising to 450W at 3.5MHz, including the WARC bands. Signal reports confirm a clean signal with excellent audio quality; in fact, equal to that of the transceiver driving it, in my case a Kenwood TS-50."

The main points of difference from the arrangements used in the earlier Frinlear amplifiers are the input circuit and the voltage-quadrupler PSU.

Input circuit: The broadband input circuit is through a bifiliar transformer connected to give a 4:1 impedance step-down from the 50Ω line from the driver. The cathode input impedance is approximately 25Ω and this is thus transformed to 100Ω. A 100Ω, 50W resistor in parallel with the transformer on the input side thus provides a reasonably close match to the transceiver. This non-resonant (dummy-load) input circuit makes the linear much easier to drive from a 100W transceiver and prevents overdriving. The measured input-VSWR from 3.5 to 30MHz is less than or equal to 1.5, with an average SWR of 1.2:1.

The 100Ω, 50W swamping load is formed by two 50Ω, 25W, non-inductive T0-220-type resistors in series. These are bolted to a small heatsink which in turn is placed close to the computer-fan used for cooling the PL519 valves. The 100pF trimmer across the input socket provides low VSWR at 29MHz.

Voltage quadrupler PSU: The special feature of this arrangement (as can be seen more clearly from **Fig 3**) is that by means of two manually-operated switches, the unit provides rectification, doubling and quadrupling of the AC-voltage delivered from a 1:1 mains isolating transformer (note this must be of suitable rating to handle the wattage and heavy current involved in a voltage quadrupling circuit - G3VA). PA0FRI has not seen this particular arrangement published previously and believes that he may be the originator of this type of flexible PSU.

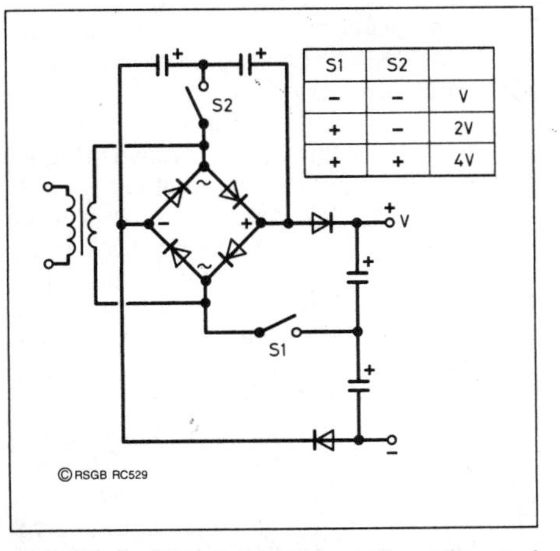

Fig 3: The flexible power supply configuration used by PA0FRI could have general application at high and also low voltages, provided suitably high-value capacitances are used to ensure reasonable voltage regulation.

'THIRD WORLD' DIRECT-CONVERSION RECEIVER

JAN-MARTIN NOEDING, LA8AK has felt rather frustrated at seeing a number of simple receiver and transceiver designs using ICs and other components not readily available in many countries, particularly in Third World countries where direct-conversion receivers and simple equipment can do much to stimulate interest in amateur radio.

He therefore set out to design a receiver using readily available and inexpensive transistors and components often found in surplus PCBs and other local sources, rather than aiming at a minimum component count.

He writes: "It was not my intention to use few components. For example, more than the usual number of bypass capacitors do not necessarily increase the cost to any great degree but can improve performance and stability. Most direct-conversion receivers have the problem that it is very difficult, but

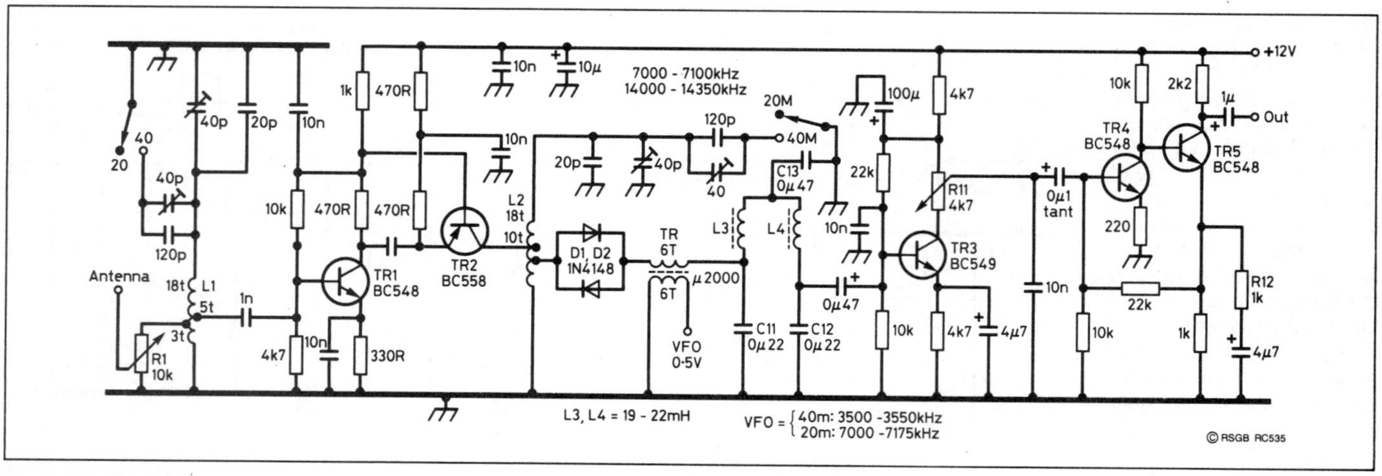

Fig 4: LA8AK's 'Third-World' direct-conversion receiver strip using RA3AAE-type harmonic mixer.

important, to prevent the VFO from radiating a signal through the mains supply. As a result a hum-modulated signal comes back via the antenna, causing a steady 100Hz hum in the receiver output. This problem can be largely avoided by using a subharmonic mixer (in which the oscillator outputs at half the signal frequency) such as the RA3AAE bilateral anti-parallel diode mixer (ART7, pp131-132 or *Technical Topics Scrapbook 1985-89*, pp46-47).

"I therefore designed a 14MHz receiver using the RA3AEE mixer. The first version had no RF amplifier but sensitivity was none too promising with too much gain needed in the audio section and with noise problems arising from the first audio stage needed to amplify signals of less than 1µV. This version used a low-noise audio pre-amplifier and this remains for the later version together with an added RF amplifier.

"14MHz is an interesting band but not too well suited for experimenters late in the evening when the 7 and 3.5MHz bands are much better and provide many strong signals. After testing the 14MHz receiver with signal generator and some on-air tests, I changed to 7MHz which presents totally different problems.

"On 14MHz post-mixer selectivity is not too critical and an active AF filter may be sufficient for providing the main selectivity. But on 7MHz some neighbouring broadcast stations can provide signals strong enough to intermodulate in the AF stages. It is thus virtually essential to use an LC filter cutting off at about 4kHz or so ahead of any AF amplifiers; standard 88mH toroid coils (or 77mH European pupin coil cores) are very suitable, using only the single winding and leaving about 20mH to be used. Experience shows that single-coil filters do not have sufficient harmonic attenuation in the range 10-100kHz; a two-coil filter improves performance (C11 = C12 = 220pF) but you really need some instruments to measure such a filter when it is to be used for high or low audio frequencies.

"This presents a problem when the main intention is to use only readily available components. However lower-inductance coils will have some effect; the use of 2.2mH RF chokes may increase the cut-off frequency to about 12kHz and such a filter is still quite useful in reducing some audio intermodulation.

"The first receiver used only audio gain control but it was later found that RF gain control was much more important on 7MHz and AF gain control was then not really necessary. The AF gain is rather high and R12 is used to reduce it. As my high-impedance headphones were missing, I cheated by using an available audio amplifier.

"Oscillator drive for the mixer is fairly critical and should be roughly 0.5V RMS. Too high or too low a drive level will dramatically reduce the sensitivity but fortunately the target level has a range of about 10dB.

"It proved possible to hear extremely weak 0.1µV CW signals on both 14 and 7MHz when sensitivity was checked with a signal generator. The need to prevent RF from getting into the AF amplifier must be strongly emphasised; otherwise some strange cross-over effects may be caused."

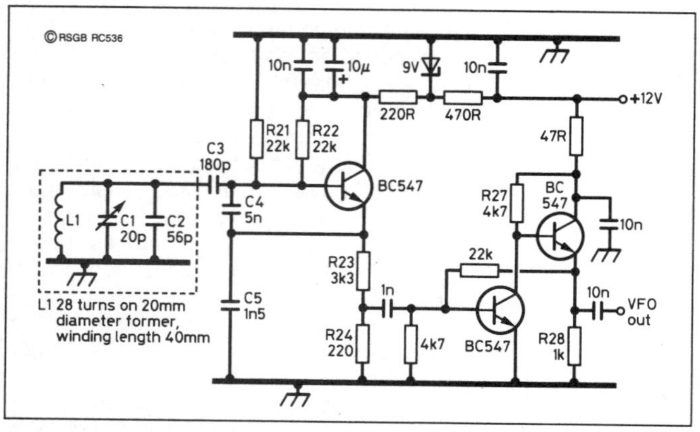

Fig 5: 'Synthetic rock' oscillator/doubler as used by LA8AK for the 'Third World' direct-conversion receiver.

The basic receiver strip is shown in **Fig 4**, and is suitable for use on 7MHz with a VFO output on 3.5MHz or on 14MHz with a VFO output on 7MHz. For the VFO of **Fig 5** LA8AK used a low-cost version of the long-established 'W3JHR synthetic-rock' design (see *TT*, November 1994, Fig 6) but using three BC547 bipolar transistors with the oscillator on 3.5MHz, with a different buffer and with a 3.5/7MHz frequency-multiplier-driver (2 x BC238 = BC547). However, he feels that, for European/American constructors, with access to a wider range of components, it would be a better idea to use as the basic oscillator stage a ceramic-resonator VXO (**Fig 6**) with a 3.579MHz ceramic resonator (see *TT*, May 1994, pp54-55).

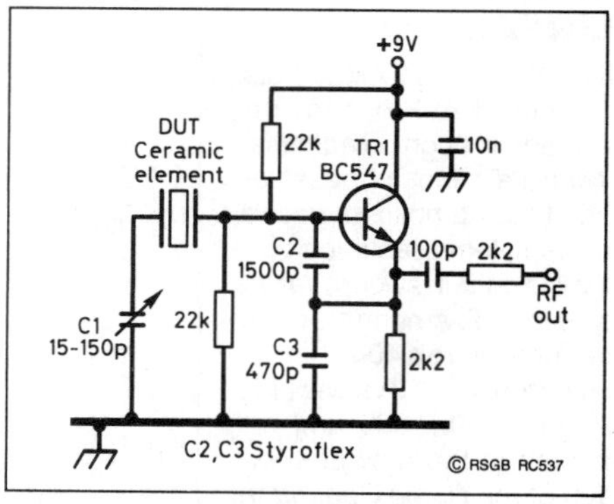

Fig 6: Alternative ceramic-resonator VXO could be used as substitute for oscillator stage of the VFO/doubler.

Component notes on the VFO of **Fig 5**: LA8AK used silvered mica capacitors and 20mm diameter 40mm on plexiglass coil former with 28 turns and a winding length of 25mm. Frequency stability was within 250Hz during the first hour of operation on 3.5MHz.

Component notes for receiver strip of **Fig 4**: The coils are made on 15mm plexiglass formers, 18 turns of 0.2mm enam. copper, 15mm long, tapped as shown. Coils are dipped with a 47 to 50pF capacitor in parallel at 14MHz. Adjustment is by spreading or closing up the turns (it might be easier to use coils with adjustable cores but this is not really necessary). The ferrite bead (u = 1000 to 2000) used for VFO injection was of the type found on heaters in VHF equipment but is not critical); two by six turns (0.2mm enam copper wire) is about the maximum possible with the small hole in these beads. L3 and L4 are 19 to 22mH RF chokes. C11 and C12 are 22OnF, C11 is 470nF.

RUSSIAN POWER VALVES

IT HAS FOR LONG been evident that manufacture of those 'old-fashioned' (but still relevant for high power amplifiers) RF power valves is increasingly being phased out in many countries but continues in some eastern European countries and in China etc. Some time ago, attention was drawn in *TT* to the rather wide range of manufacturing skills found in RF power valves stemming from China as observed by Bill Orr, W6SAI. Some factories were turning out good quality, close tolerance valves but others less so.

A 'New Products' item in *QST* (November 1994, p104) draws attention to a range of Svetlana Electron Devices power valves including a new, relatively low-cost, 4CX800A ceramic/metal transmitting tetrode which used as a linear amplifier is rated at up to 750W PEP output (a pair giving 1500W PEP) and which is being manufactured at a Svetlana valve factory in St Petersburg, Russia.

The air-cooled 4CX800A is a high-performance valve intended for grounded-grid or grounded-cathode service with a passive (untuned) 50Ω input circuit and capable of good linearity at relatively low anode voltages. The SK1A socket, with built-in screen bypass capacitor, increases the maximum frequency rating to 250MHz.

Interestingly enough, *QST* gives the address for data enquiries etc as George Badger, W6TC, Svetlana Electron Devices Inc, 3000 Alpine road, Portola Valley, CA 94028, USA (tel 415-233-0429). W6TC is well-known in the UK not only for his detailed work on coaxial-cable baluns but also for his many visits to Europe on behalf of Varian-Eimac.

THE T-NETWORK ANTENNA TUNER

THE CLASSIC PI-NETWORK or LC/CL two-component matching networks can be used as the basis of an antenna tuning unit (ATU). These are theoretically capable of matching any transmitter to any antenna impedance (resistive or reactive). However, in practice the matching range is dependent on the component values. For the widest step-up and step-down transformations, the high-voltage variable capacitors need to have low minimum and very large maximum capacitance values - a significant disadvantage these days. The pi-network and the standard LC configurations do however possess the advantage that they not only transform impedance but also form a low-pass filter and so provide additional harmonic and higher-frequency spuriae attenuation: see **Fig 7.**

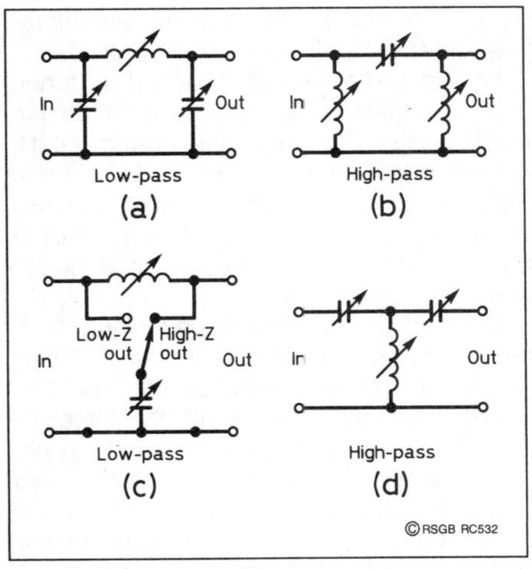

Fig 7: Some basic ATU configurations. (a) Pi-network in conventional 'low-pass' form. (b) Inverse pi-network with components interchanged but providing high-pass filter. (c) Switched LC network providing either step-up or step-down of the impedance with a degree of low-pass filtering. (d) T-network (high-pass).

But modern solid-state transceivers include built-in low-pass filtering tailored to the individual bands, with the result that there is far less requirement for the harmonic attenuation previously provided by the ATU. This has opened the way to much greater use of the T-network which can provide an acceptably wide range of impedance transformations without a requirement for large-value variable capacitors. The disadvantage that they form a high-pass rather than a low-pass filter is no longer regarded as a real disadvantage. Many of the current ATUs on the amateur market now utilise the T-network: see **Fig 7(d).**

Andrew S Griffith, W4ULD, provides a useful article on 'Getting the most out of your T-network antenna tuner' (*QST*, January 1995, pp44-47). He describes "how to adjust this popular tuning circuit so that it transfers maximum power to your antenna without going snap, crackle and pop."

He shows that the T-network of **Fig 8** with variable capacitors with a range of 20-240pF and a roller-coaster inductance variable from 0.1 to 35µH when used with a transmitter designed to feed 50Ω line can match purely resistive loads of about 10Ω to 3000Ω from 1.8 to 21MHz. On 24 and 28MHz the range narrows to about 10 to 1500+ since C_{in} and C_{out} cannot be adjusted to less than 20pF. With reactive loads, the matching range narrows. However, W4ULD points out: "Even with reactance present, very few cases should occur in which the antenna cannot be matched with the proper tuning technique."

He provides detailed tuning procedures for both roller-coaster tuners and tapped-inductor tuners but warns that while power loss in a T-network is often less than 0.3dB "it may be considerably higher. For a given impedance transformation, minimum loss occurs when C_{out} is as high as possible when a match has been achieved. The loss in a T-network with the component values shown in Fig 8 can approach 2dB when matching load impedances lower than 20Ω at 1.8, 3.5 and 7MHz. Under these conditions, component heating and/or arcing may occur, and the tuner's power-handling capability may have to be derated. With the proper tuning techniques, however, an acceptable impedance transformation - as indicated by a 1:1 SWR - should be obtainable under most circumstances. Tips include:

- To achieve the highest possible efficiency at a given impedance transformation, tune the network with the highest output capacitance that allows a match.
- When matching loads of less than 25Ω on 3.5 and 1.8MHz, you may have to reduce output power to reduce tuner heating or to keep it from arcing.

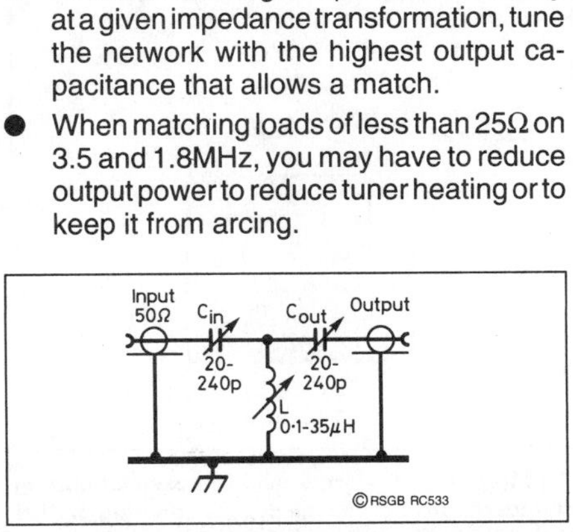

Fig 8: The T-network ATU as described by W4UMD.

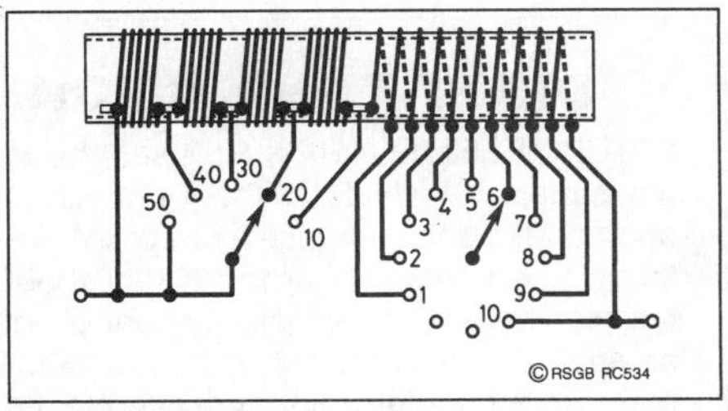

Fig 9: A variable-inductance ATU coil described by Hector Cole, G3OHK in *TT* (May 1989) using two switches and just 14 taps to permit selection of from one to 50 turns of a 50-turn coil and which can be quickly reset to any number of turns previously found suitable without the turns counters required for roller coaster coils.

- When operating high power, do not feed short (less than 0.3λ) loaded dipoles with a feed-line that is a multiple of 0.50λ (electrical) long. Such antennas may have feedpoint impedances of 5 to 9Ω, and the tuner will see this very low load impedance.
- When operating high power, do not operate a 1.8MHz dipole on 3.5MHz or a 3.5MHz dipole on 7MHz with a coax feed-line that is an odd multiple of 0.25λ (electrical) long. The antenna's high feedpoint impedance will be transformed to 1.5 to 2Ω at the tuner. To add insult to injury, the feed-line loss will be excessive - over 6dB and so wasting 75% of the transmitter's output power as heat.

The tuning procedure recommended by W4ULD for roller-inductor tuners is given below, but it is advisable to practice with low power into a dummy load fed via a length of coaxial cable, preferably with a variable capacitor of about 100pF in series with the centre conductor of the coax at the dummy antenna and to provide practice at matching reactive loads.

1. Set C_{out} at maximum capacitance and leave it there.
2. Set C_{in} to about half scale.
3. Adjust roller inductor for an SWR dip (this may be barely noticeable).
4. Slightly increase or decrease C_{in} and readjust the inductor for a dip.
5a. If the SWR is lower than it was in (3), slightly vary C_{in} in the same direction as in (4).
5b. If the SWR is higher than before, adjust C_{in} in the opposite direction to (4). Alternatively, inch C_{in} in the step (4) direction and redip the SWR with the inductor until an SWR of 1:1 is obtained.
6. When you have almost reached the match point, the SWR may start to go up as C_{in} is adjusted, but make the change anyway and redip with the inductance.
7. Continue to adjust C_{in} in the same direction until adjusting the inductor produces a higher SWR than before. Inch the capacitor back to the previous setting.
8. If you cannot obtain a 1:1 SWR reduce C_{out} and repeat the process, beginning at step (2). If you cannot acceptably minimise the SWR at some setting of C_{out}, the antenna impedance is out of range of the tuner.

Clearly, in operational circumstances and when the correct settings of the ATU have not previously been determined, it will be helpful to fellow operators to use a 'quiet tuning'

QUIET-TUNING FOR 100W TRANSCEIVERS

FRITS GEERLIGS, PA0FRI, uses the arrangement shown in **Fig 10** to permit the adjustment of an ATU without radiating more than a tiny fraction of the transceiver output. This does not give a direct SWR reading but makes it easy for zero tune-up of the ATU without radiating more than about 50mW. With the silicon diode in parallel with the meter, the usual sensitivity-control potentiometer can be omitted. The only critical component is the 50Ω non-inductive (dummy load) resistor formed from two 100Ω, 25W resistors in parallel. For further information on quiet tuning see 'Simple quiet tuning and matching of antennas' by Professor Mike Underhill, G3LHZ, *RadCom*, May 1981, pp420-422.

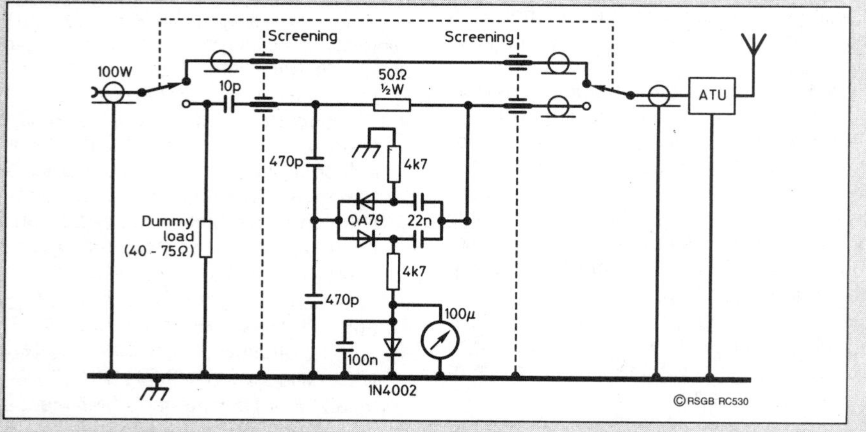

Fig 10: PA0FRI's quiet-tuning unit for use with 100W transceivers. Note that the OA79 diode is a germanium type (the germanium-type 1N34 is suitable.

device such as that described by PA0FRI or the article by G3LHZ in *RadCom*, May 1981, pp420-422. **Fig 9** shows a useful substitute for a high-cost roller-coaster.

A point emerging from W4ULD's article is the significant power loss (even with a 1:1 SWR) that can be incurred with ATUs when the component values are restricted, often detected by heating of the coil. It has been noted before in *TT* that, particularly on the lower frequency bands where very low feedline impedances may be presented to the ATU, more than half the transmitter output power (3dB) may easily be lost in an otherwise efficient ATU, usually the result of insufficient maximum capacitances.

CHEQUERBOARD CONSTRUCTION JIG

COLIN WALKER, G3VTS, reiterates the view often expressed in *TT* that for one-off construction of prototypes or operational circuits the home-etched PCB offers few advantages and several disadvantages.

He writes: "Many amateurs often have the need to construct simple items quickly for use in the shack but cannot be bothered to design and etch a PCB particularly if the components to be used come from the junk box.

"Perforated (Veroboard) board or 'dead bug' techniques are suitable but if you have odd pieces of single or double-sided board available it is much easier to cut the surface in chequerboard fashion to a size that suits the project.

"In this connection, a simple jig to cut through the copper or piece of PCB quickly and accurately is a useful tool. Such a jig is shown in **Fig 11(a).** No dimensions are given for the jig as it can be made to suit particular requirements and the material available. I used chipboard and aluminium for the saw guides.

"When complete, the jig is held in a vice and after marking the board with a pencil to show where to cut, it becomes a simple operation to cut through the copper with a fine tooth hacksaw and then clean up with 'Scotchbrite' to remove any burrs and clean the board ready for soldering: see **Fig 11(b).** The board can be easily held by hand and the guides at each side hold the hacksaw-blade square. With care it is possible to cut close enough to solder the pins of an IC onto adjacent lands but I find it easier to cut larger lands and only solder the four corner pins down, then use short wires or components for the remaining pins. With double-sided board, small holes can be drilled through to ground pads as necessary."

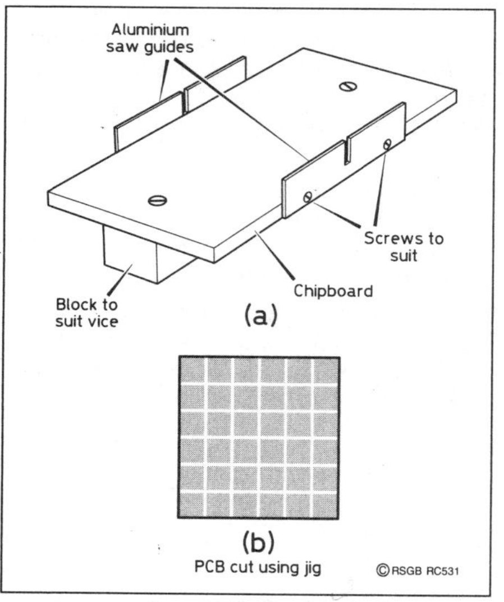

Fig 11(a): G3VTS's home-made jig used for cutting chequerboard slots in the copper covering of PCB material. (b) Typical appearance of the board after cutting, providing square or rectangular lands to which component leads can be soldered.

RF SWITCHING & TUNING DIODES

TT FEBRUARY 1993 REPORTED briefly an important article by Dr Ulrich Rohde, KA2WEU/DJ2LR, which was published simultaneously in English and German (*QST* and *CQ-DL*, November 1992) on "Recent advances in shortwave receiver design". He subsequently published a series of three articles (*QST* May, June and July 1994) on "Key components of modern receiver design" and a recent follow-up "Key components of modern receiver design: a second look" (*QST*, December 1994). In these articles he stressed that for receivers intended to have a very wide dynamic range, the intermodulation distortion that arises from the use of unsuitable RF switching and tuning diodes imposes an important limitation. He has recommended the use (or substitution) of such special-purpose RF diodes as the Hewlett-Packard HP5082-3081 PIN diodes.

Dr Rohde's articles encouraged Tom Thomson, W0IVJ, to investigate how bad in practice are the more distortion-prone switching diodes and how good are those designed for low distortion ('Exploring intermodulation distortion in RF switching and tuning diodes', *QST*, December 1994). He carried out laboratory tests on four types of diodes: The IN4153 generic PN switching diode; the Motorola MPN 3700 PIN diode intended for RF switching; the BAT-17 Siemens PIN switching diode; and the low-cost 1N4007 which is a generic 1kV-PIV rectifier diode with a PIN structure but not intended for RF switching.

He has tabulated results in terms of diode switch insertion loss (dB) at 10MHz with 0, 5, 10 and 20mA bias currents; and similarly the second- and third-order intercept points (IP2, IP3 and dBm). He draws the following conclusions: "RF-specified PIN diodes are the devices of choice for low-distortion switching at HF and above, for bandpass filter selection and IC switching in a narrow-band pre-selector. Although the presence of a PIN structure in the 1N4007 makes it seem attractive as a low-cost alternative to RF-specified PIN diodes, its insertion-loss performance when unbiased and reverse-biased - and its IMD performance when unbiased - is demonstratively inferior to RF-specified PIN diodes.

He adds: "The manually switched and tuned front-end filters of the 1960s and 1970s had much to offer in terms of second-order IMD, but we need not retrogress to those techniques to achieve improved IP2 and IP3 performance today. More attention paid to front-end filtering in general can produce the improvement we need."

Dr Rohde in commenting on W0IVJ's finding, notes that many amateurs had reported difficulty in obtaining HP5082-3081 diodes. He recognises that even with the Motorola MPN3700 with a US price of less than £1, replacing all 20-plus filter-switching diodes can be expensive. Nevertheless he recommends changing all the diodes between the antenna and the first mixer, which includes the diodes on both sides of the bandpass filters of a transceiver but not the transmit/receive switching diodes which typically are already high-quality PIN types. He also adds some notes on Japanese switching diodes which might be used "to replace the 'bad' diodes seen in the past". ♦

Pat Hawker's

Technical Topics

PAT HAWKER, G3VA
London 37/SE22 8SS

AMATEURS AND THE VE DAY AND VJ DAY ANNIVERSARIES

AS WELL AS representing '100 years of Radio', 1995 is also the 50th anniversary of the end of the war in Europe (May 1945) and in the Far East (August 1945). While some may feel it is time to think more of the future and less of the past, for those of us who were in any way involved in the events of 1939-45, it is difficult not to regard those years as memorable in many ways. Indeed, it was a time of technical development in radio communication, radar and electronic warfare. Much of our present technology can be traced back to the 1930s and 1940s.

I must confess that I missed the original VE day celebrations, having spent the entire day (8 May) travelling by ship and train from London to Brussels, en route back to Eindhoven where the clandestine Dutch Inland Radio Service - in which Dutch amateurs served and too often died - was in course of being wound up. On loan to the Dutch Intelligence Bureau (BI), I had been privileged to act as a control station operator from 1 January 1945. An unidentified Dutch operator at the Apeldoorn station of the Inland Service, is shown in the photo below. This station used Philips 'Luxor' diathermy equipment adapted as a 3.0MHz power oscillator with a three-valve regenerative 'straight' receiver. This station was raided by the Germans on 10 January, 1945. Two of the operators, one of them PA0MB and the other a professional Radio Officer, were both shot in March 1945, as were many others of that Dutch service. Technical details are shown in **Fig 1 (over page)**. At the Alkmaar station, which survived, the Luxor 'transmitter' was modified by PA0XY/PA0ZY into a two-stage MOPA

Most of the 5000 or so pre-war British amateurs used their technical and/or operating skills in aid of the Allied cause, as did many in uniform or as civilians who aimed to become post-war amateurs. Among these were the Y-services (Army, Navy and Air Force intercept services), the Radio Security Service (special intercept service by Voluntary Interceptors or those specially enlisted in SCU3/4 etc), engineers, technicians and operators for the Services, including SOE, and for MI6. Amateurs used their expertise in virtually all branches of the Armed Services and in the civilian radio/radar industry.

Some were required to keep their activities secret for many years; and some have never openly described their wartime work. For example, only after the recent death of Ronnie Ree, G2RX, has it become possible to reveal that he was one of the small group of amateurs who played a vital role in the highly important 'Double Cross' operations. He spent most of the war monitoring the sending of, or himself sending, 'chicken feed' information to the Abwehr by such double agents as Tate, Snow, Celery and Zigzag (the British safebreaker who became, in effect, a triple agent), culminating in the vital Fortitude deception.

It was largely this operation that misled Hitler and the German High Command into keeping elite troops in the Pas de Calais area for several weeks instead of rushing them to counter the vulnerable Normandy landings - and so undoubtedly saving literally many thousands, if not tens of thousands, of British, Canadian, American, French and Polish lives.

G2RX was in fact the anonymous author of Appendix 3 'Technical problems affecting radio communications by the Double-Cross Agents' in Volume 4 of *British Intelligence in the Second World War* (see my Review in *Radcom*, August 1990). In the post-war years he became a senior member of the Security Service (MI5) as principal technical adviser but remained an active amateur for over 50 years until his recent death aged 77 years.

A full obituary appeared in *The Daily Telegraph* (brought to my notice by Ray Herbert, G2KU) but perhaps I can add one story that shows the difficulty of catching out a fully trained Russian GRU radio-agent. G2RX was one of the two MI5 officers who, following the return to England of Alexander Foote, interviewed 'Sonya' who beyond doubt was the most successful clandestine radio-agent active in the UK in wartime and in the immediate post-war period. She had trained Foote to take over her Swiss radio network before coming to England early in 1941 as the wife of an RAF serviceman.

Foote had disclosed to M15 Sonya's past involvement in clandestine radio but there was no proof that she had operated in the UK (for example, sending atom secrets given her by Karl Fuchs). She maintained to G2RX and his colleague her innocence of any illegal activities in the UK and later returned to East Germany, her record as a radio-agent unblemished.

Among others who 'worked' the Abwehr control stations during wartime was Stan Riese, G5SR (post-war with GCHQ) who died some time ago, and also (in the Middle East) two still very well known amateurs: Ken Ellis, G5KW, and Roly Shears, G8KW, who later set up KW Electronics.

A pre-war amateur whose secret work for SOE has never, as far as I am aware, been mentioned in any amateur radio publication is Douglas Brewer, G4LJ, Croix de Guerre. He was one of the Tunbridge Wells group of amateurs who joined the Civilian Wireless Reserve and were called up on 20 May, 1939 when conscription was introduced. Early in 1942, as a fluent French speaker (through his mother's family) he was interviewed in London and then underwent training for SOE at Wanborough Manor, Arisaig, Ringway and Beaulieu. Despite being a skilled radio operator he was, after an internal struggle within SOE, chosen for industrial sabotage. He went on three secret missions to Occupied France, during one of which he delivered and for a time operated a B-2 transmitter-receiver. His story deserves far more space than is available in *TT*.

Another member of the Tunbridge Wells 'Early Birds' was the late Bert Allen, G2UJ, much respected for his VHF column in the *RSGB Bulletin*, who was awarded an MBE for his undercover work in Sweden in 1944 where he was involved with the interception of signals relating to the V2 trials - a story that he never disclosed after the war even to other local amateurs.

Following my remarks on the HRO and AR88 receivers in the December *TT*, I received several comments pointing out that another receiver that played an important role. This was the Hallicrafters S27 receiver, which was used for interception; and for such purposes as an RAF (83 Group for the Tactical Air Force) over-the-horizon duplex phone/rtty radio relay system. S27 receivers were used with 5kW vehicle mounted transmitters and rhombic antennas and proved capable of providing a direct link between Dover and Brussels on about 34MHz (recently brought to my notice by Dennis Willis of Royston who was in charge of one of these units).

The S27 and successors S36 and S37 were high-performance three-band tuneable VHF receivers.

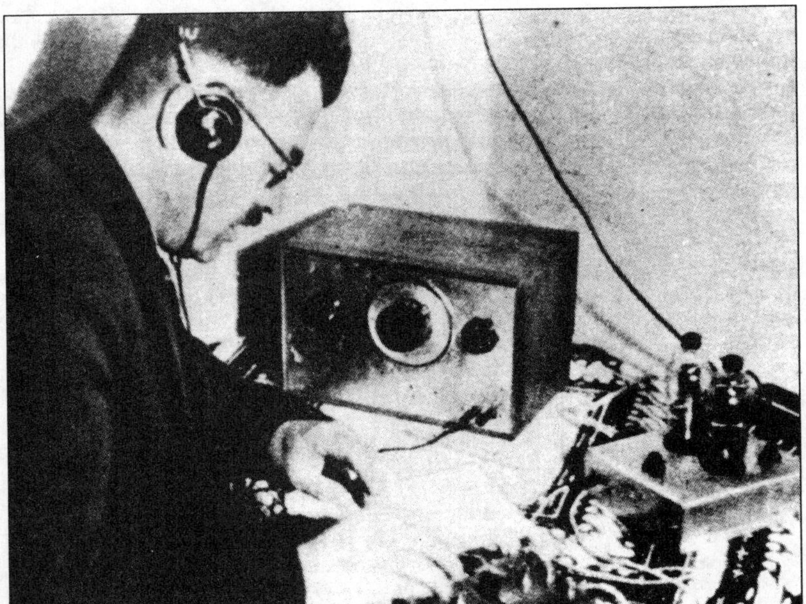

An ill-fated clandestine station (G6) of the Dutch Inland Radio Service at Apeldoorn. It was raided in January 1945 - one of a number that suffered a similar fate around the turn of the year - and two of the operators were shot in March 1945. (*Courtesy* Dick Rollema, PAOSE).

The S27 covered 27-145MHz (S27B 36-165mhz) suitable for AM/FM/CW reception with an IF of 5250kHz. The front-end used three of the famous acorn valves (956 RF, 954 Mixer and 955 Oscillator) with a VR150 voltage regulator and a push-pull audio output stage using two 6V6 valves. A very advanced model for 1940 which was much used and appreciated by the Y-Services for tactical VHF interception of Luftwaffe etc R/T signals.

When UK amateur radio resumed in January 1946 (initially on 1.8 and 28MHz only), it soon became obvious that many changes were coming into the hobby largely as the result of the hot-house of wartime developments. For instance, 'netting' with VFOs was in; crystal control on HF with the need to search the entire band for replies on its way out. Beam tetrodes such as the 807 were in; power with triodes largely out. The superhet communications receiver was in; while the classic 1-V-1 regenerative receiver was on its way to the back shelf. There was the wartime arrival in Europe of Armstrong's frequency modulation, both wideband and narrowband. There was an upsurge in the use of teleprinters on radio links along with the availability of klystrons, opening the way to amateur use of microwaves. Crystal control for VHF with superhet converters replaced the pre-war super-regenerative portable 56MHz transceivers. Small-signal valves included miniature types with battery types having low-

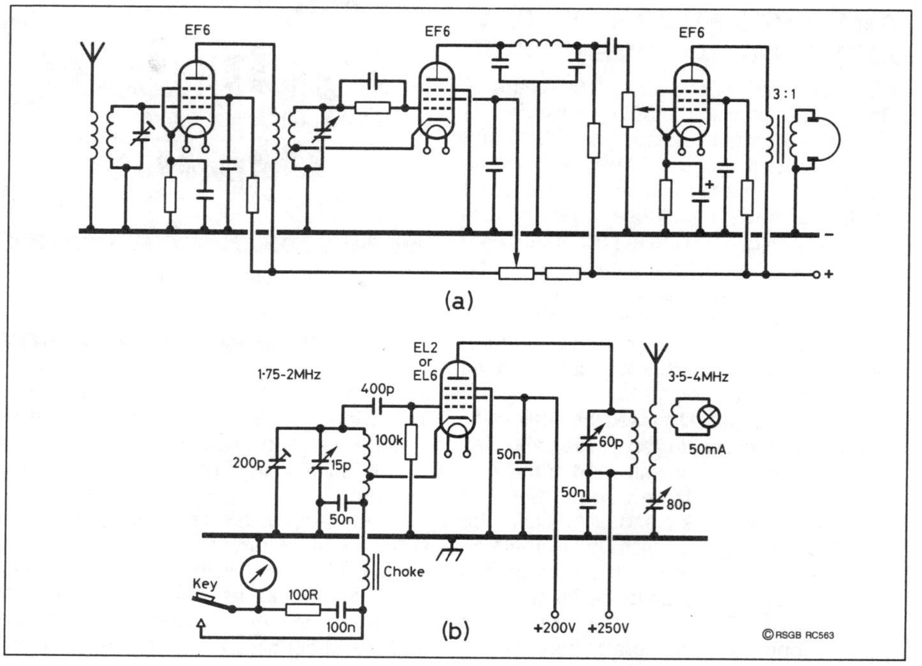

Fig 1: Typical equipment used by the Dutch Inland Radio Service from September 1944 for passing cipher traffic to Eindhoven from the Occupied provinces of Holland. (a) three-valve 'straight' receiver. (b) Single-stage tuneable transmitter improvised from medical diathermy equipment. The single stage master-oscillator inevitably produced rather rough, 'chirpy' notes even at 3MHz. This system is not recommended for modern amateur radio! For the most successful and prolific station at Alkmaar (later at Zaandam, now called Phoenix) Jan Zandbergen, PAOZY, modified the transmitter, adding a separate oscillator stage and using a PEO6/40 power amplifier. The Alkmaar operator was usually Jack Verhagen, a professional marine radio officer. (source PAOSE).

AC SUPPLY MONITORING OF VOLTS AND HERTZ

WITH THE SEASON of field day operation looming ahead, the sound of petrol-electric generators will soon be once again disturbing the rural peace. Where these are DC generators used to charge vehicle batteries there are few problems. But uncertainty on either voltage or frequency or both can arise with 230V or 115V AC generators used directly to power AC mains equipment. A slow-running generator delivering say 30Hz can seriously damage equipment. Again, for home supplies in residential areas, the voltage delivered towards the end of the spur can, at times, fall well below 240V (or even the nominal 230V) affecting transmitter efficiency and the filament/heater voltages for high-cost RF power valves, unless the line-voltage is monitored and a variac or similar device is available to adjust the voltage.

In an article entitled 'A 240 volt AC Line Monitor' (*Amateur Radio,* March 1995, p10-11) Ken Taylor, VK3KAV, describes how to construct a metering device (**Fig 2**) which can usefully be employed both in the field and in the shack. But he warns: "The unit operates directly from the AC mains. It has no protective isolation such as a transformer. All parts of the circuit must be assumed to be operating at mains potential, even with the unit switched off. This can happen if the active and neutral leads are reversed, an all too frequent occurrence."

The author recommends using a fully insulated chassis for this project such as a plastic 'Zippy' box with the meters mounted on the lid. Remember AC supplies can be lethal.

Both meters are readily available 1mA FSD meters. The voltage indicating meter is supplied via a bridge rectifier from the AC input, through a 390k resistor and a 50k trimpot. The frequency indicating portion uses a pulse counting detector, the output of which is supplied to its meter via a 4.7k resistor and a 5k trimpot.

Two back-to-back 15V zener diodes form pulses at the peaks of the voltage waveform. As the polarity of the line voltage reverses so does the action of the two zener diodes, and 5V is developed across the pair in opposite polarity. The resultant voltage at point X is almost a square wave with a frequency the same as that of the line input and an amplitude of 30V peak-to-peak. C1 and R3 differentiate this signal forming positive and negative pulses at point Y. These pulses are rectified by anode bridge rectifier. With an increase in frequency the number of rectified pulses per second increases, causing more current to flow through M2; a lower frequency has the opposite effect.

R1 33k should be rated at least 2W but may be synthesised from two paralleled 68k 1W or better from four 150k 1W resistors if the unit is to be left switched on for long periods.

When completed and carefully checked as regards insulation, circuit faults etc, it can be calibrated with the aid of an AC meter of known accuracy so that 250V represents full-scale deflection of M1 (use a well insulated screwdriver when adjusting the trimpots). Then adjust RV2 so that M2 reads 5 on the scale. If a motor-powered alternator is available, note the rate of change on the meters by slowing the motor down so that the motor reads 4 on the scale (40Hz) and speeding it up so that the meter reads 6 on the scale (60Hz).

VK3KAV makes no claim for originality as he came across a similar device in an ARRL publication.

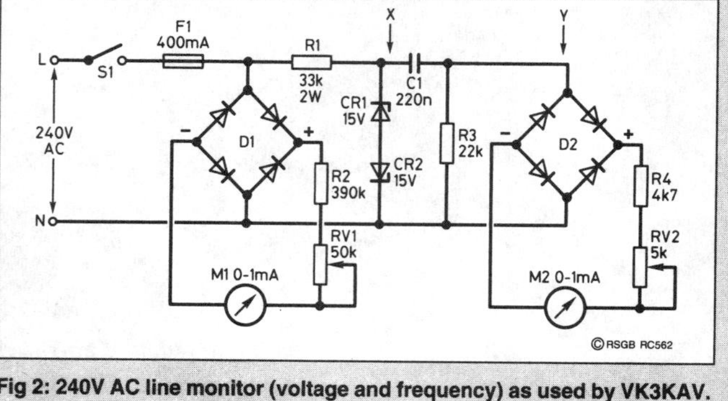

Fig 2: 240V AC line monitor (voltage and frequency) as used by VK3KAV. Note that all components may be at the AC line voltage since there is no isolation. Full precautions should be taken when using or adjusting this device. RV1 is 50k trimpot, RV2 is 5k trimpot. ZD1 and ZD2 are 15V Zener diodes. D1, D2 eight 1N4002 diodes. C1 0.22uF 100VW capacitor. Suitable enclosure is a 50 by 90 150mm plastic Zippy box.

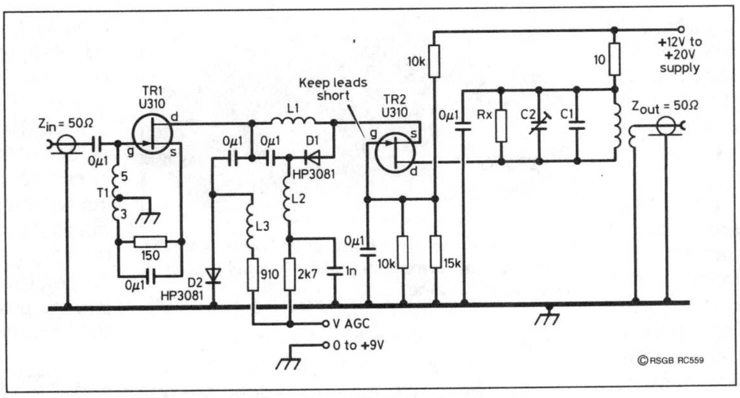

Fig 3: G3SBI's low-noise, AGC-controlled cascode IF amplifier.

consumption filaments (initially 1.4V, 50mA but soon 25mA). Miniaturisation of components also began, which in the final years of the war had led to pocket receivers and transmitters for clandestine links.

CW was the dominant DX mode and the Yagi close-spaced rotary beam soon became the ambition of many. Coaxial feeder cable put new life into dipole antennas. Meanwhile, surplus ex-Services equipment increasingly replaced home-construction in the period when currency restrictions froze out new American imports. Nobody even imagined the day would come when Japanese firms would dominate the world's amateur radio market!

LOW NOISE AGC-CONTROLLED IF AMPLIFIER

SEVERAL TIMES IN the past *TT* has emphasised, when discussing modern approaches to the design of receivers with super-linear front-ends, the importance of the IF amplifier that immediately follows a mixer having conversion loss rather than gain. For example, in *TT*, February 1993 I stated: "It should be noted that with any mixer operating directly on the incoming RF signals without pre-mixer amplification, the IF amplifier that follows the mixer, either directly or after a roofing filter, must have a low noise figure and a high intercept figure, with a diplexer arrangement often used to achieve constant input impedance over a broad band of frequencies."

A simplified diagram was included of a grounded-gate FET amplifier capable of providing a 2dB noise-figure, 9dB gain in a 50Ω system for a 45, 70 or 100MHz up-converted first IF stage was reproduced from the IEE's *Radio Receivers* book edited by Dr William Gosling.

Colin Horrabin, G3SBI, recently sent details of a 9MHz cascode-FET amplifier that he proposes to use immediately behind the roofing filters in any receiver using his super-linear H-mode mixer (see *TT*, October 1993, the new edition of *Radio Communication Handbook* or *Communications Quarterly*, Fall 1994). He points out that this could also be a useful approach for anyone constructing high-performance receivers because of the low noise-figure, reasonable gain and intercept point, and the 50Ω input and output impedance which remains constant over the 45dB AGC range.

	16dB gain	13dB gain
Noise figure	0.6dB	0.6dB
Input impedance	50Ω	50Ω
Output impedance	50Ω	50Ω
Third order Intercept		
Vs = +12V (max gain)	23dBm	26dBm
Vs = +20V (max gain)	28dBm	30dBm
Input for 1dB compression		
Vs=+12V (max gain)	OdBm	+3dBm
(max AGC)	+7dBm	+11dBm
Vs=20V (max gain)	+5dBm	+8dBm
(max AGC)	+11dBm	+14dBm

Amplifier input and output impedance's are 50Ω regardless of AGC controlled gain. Gain range 45dB.

Table 1: Performance of 9MHz AGC-controlled Amplifier.

The circuit diagram of the 9MHz version of this amplifier is given in **Fig 3**. Note that the AGC amplifier must be capable of sinking the current through D1 at 0V (ie maximum gain). The warning to keep leads short in the drain circuit of the second U310 FET arises from G3SBI's experience in finding the initial IP3 measurements were poor due to this stage oscillating at around 400MHz. Performance is given in **Table 1**.

Component notes: The two U310 are Siliconix low-noise JFETs. C1 82pF ceramic, C2 60pF ceramic trimmer (Cirkit), all other capacitors monolithic ceramic (RS Components). Resistors 1/8th watt metal film (RS Components). D1, D2 HP3081 PIN diodes (Farnell). T1 5 + 3 turns of 0.224mm diameter Bicelflux enamel on Fairite Balun core 28-43002402 (Cirkit). T2 (primary) 2.81uH, 31 turns of 0.314mm Bicelflux enamel on Micrometals toroid T37-6 (Cirkit). T2 (secondary) (1) for 16dB gain 3 turns, Rx 8k2; (2) for 13dB gain 4 turns, Rx 3k9. Note that (1) and (2) could be relay switched for use with an SSB or CW filter (loss 10dB or 3dB). L1, L2 and L3 7 turns 0.314mm enamel on balun core 28-43002402 (Cirkit).

ELECTRICITY SUPPLIES

LAST YEAR THERE was mention in *TT* (February 1994, p53-54) of the change from 1 January 1995 of the UK 50Hz mains supply to a nominal 230V AC with a tolerance of -6%, +10-%. I noted: "Since these tolerances cover the present 240V, I remain uncertain whether a voltage change will actually be introduced next January."

This seems to have been an accurate supposition, at least in the London area. With my last electricity account there was enclosed a leaflet - *Standardisation of the supply voltage in the European Union* - which confirms that although the nominal voltage is now 230V, in practice nothing will change.

The leaflet states that "the electricity supply voltage throughout the European Union is to be standardised from 1 January 1995. The voltage at which we presently supply electricity falls within both the existing range and the new range and will not need to be changed . . . . We are presently required to supply you with electricity at between 225.6 and 254.4 volts; after standardisation, the range will be between 216.2 and 253 volts.

Our declared (sic) voltage will be 230 volts . . . . Most equipment currently shows a voltage rating of 240 volts, or 220/240 volts. Over the next few years, the voltage rating shown on new electrical equipment will change, and will show either 230 volts, or possibly 230/240 volts."

So, in typical British compromise, it seems that although officially we are now on 230V supplies, an AC voltmeter will continue to indicate 240V except when the local load brings it down possibly to under 220V. What price European harmonisation!

MORE ON THE 144MHZ RIBBON J-POLE ANTENNA

THE DECEMBER 1994 *TT* item on J-Pole and Slim-Jim antennas continues to attract comment and suggestions. Dr John Belrose, VE2CV, has been using various versions of the ribbon J-antenna since about 1979. He helped to develop this versatile and useful antenna which makes a very effective roll-up and put-in-your-pocket antenna, particularly for hand-held rigs.

"It can be used as a back-up antenna if a standard whip or rubber duck antenna is broken or mislaid. It can provide improved performance over a vehicular mounted half-wave whip if it is elevated above ground level, by suspending it from the branch of a tree. The gain over a rubber-duck antenna can be greater than 7dB."

However, VE2CV takes issue with W1ICP (whose article I quoted in *TT*) and also with J Reynante, KD6GLF, whose article appeared in *QST*, September 1994. Regarding the comments from W1ICP which appeared in *TT*, VE2CV points out that while a ribbon J-Pole antenna can certainly be slipped inside a PVC pipe to form a more permanent base station antenna it then requires re-dimensioning by a significant factor.

KD6GLF proposed its use as a dual-band (146/435MHz) antenna. VE2CV comments that while ribbon J-antenna dimensions for two metres may have a low or acceptable SWR on the 70cm band, the radiator length for the UHF band is 1½-wavelength so that the major lobe is no longer directed at the horizon; in fact the lobe towards the horizon is a minor lobe: see **Fig 4**.

VE2CV writes: "The J-antenna is a half-wave length radiator, fed and matched by a shorted stub one quarter-wavelength long. The feeder transmission line is tapped onto the stub at an appropriate point to match its impedance. If coaxial cable is used to feed the antenna, a balanced-to-unbalanced (current balun) transformer should be used to minimise the current flow on the outside of the cable shield, which contributes to the radiated field.

"A flexible 2-metre J-antenna can be made out of 300Ω twinlead: see the photo (overpage). The length of the radiating part of the antenna is about 0.95 times a free-space half-wavelength (where 0.95 is the experimentally determined antenna factor). The length of the quarter-wave stub is about 0.83 times a free space quarter wavelength (the factor 0.83 corresponds to the velocity factor for Beldon Type 8230 twinlead). The tap point for a 144MHz J-antenna was determined experimentally to be about 0.0136 times the wavelength.

"In most designs, including my early versions, no balun was used and current flow on the outside of the cable shield affected the radiating properties and impedance (hence dimensions experimentally determined) of the antenna. The effect was noticeable particularly on receive. A variation of the received signal strength could be observed as one ran one's hand along the coaxial cable when receiving a distant station.

"The improved design of **Fig 5** uses a choke balun. This is formed by coiling up a few turns of the feedline which, in effect, inserts an RF choke in the outer conductor (shield) of the coaxial feedline. For the frequency range 144-148MHz, six turns of the feed line coiled into a 2.5cm diameter coil provides an inductor with more than enough impedance to minimise feed-line radiation. It was found by experiment that the optimum location for this stub is one quarter of a wavelength from the tap point. The addition of this coaxial choke coil not only isolates the antenna from its feeder, it broadens the bandwidth, and increases the effective gain of the antenna. The SWR versus frequency response is shown in **Fig 6** (note that for the smaller European two-metre band - 144-146MHz - it would be better but not essential to aim for minimum SWR at 145 rather than 146MHz - G3VA).

"Since the antenna is not self-supporting it can be hung from a tree branch, from the ceiling of a hotel room or suspended from a non-metallic mast or rod. But it should not lie against the supporting device, even if this is non-conducting. It should certainly not be slipped inside a PVC pipe without re-dimensioning it."

PARAMETRIC UP-CONVERTERS

B J MITCHELL, G3HJK, has reminded me of a form of up-converter mixer which came into use for professional/military HF communication receivers in the late 1960s before the development of high-performance quad-FET mixers. These had extremely good strong-signal performance as well as providing conversion gain rather than conversion loss. This was the parametric up-converter using varactor diodes. It is suitable for use with 'single-span' frequency synthesisers and provides a VHF IF of the order of 70 to 150MHz or so. Although never widely used for amateur equipment, G3HJK points out that there was a paramp up-converter in a Datong HF converter intended for use with 144MHz receivers (although in such a set-up the overall dynamic range would inevitably be limited by that of the VHF receiver).

VE2CV's two-metre Ribbon-J antenna showing the current balun formed by coiling a small length of the coaxial feedline.

Looking back to the late 1960s, I found that I had mentioned the paramp up-converter in *TT*, January and February 1968. This resulted in a detailed appraisal (*TT*, April 1968) of the paramp up-converter by Walter Schreuer, K1YZW/G3DCU, who developed and patented this system with Robert Sproul while at National of Malden, Mass(US patent No 3,063,011 of November 1962, filed July 6, 1959).

These items were included in some early editions of *ART* (eg *ART3* of 1970) but, in view of the problem of oscillator noise with VHF sources which limited reciprocal mixing performance, was dropped from later editions. K1YZW explained that for amateur receivers (not then often using frequency synthesis) there were significant advantages in using a 9MHz rather than a VHF first IF. Most of the professional receivers with up-conversion adopted a first IF of 112MHz as in **Fig 7(a)**, since the higher the frequency, the higher the gain of a paramp mixer.

It may be worth looking again at the paramp up-converter now that most high performance receivers use frequency synthesis - increasingly with low noise direct digital synthesis (DDS) - and aim at general-coverage reception (ruling out a first IF within the tuning range). However, one would presumably run into the problem that the dynamic range of a VHF crystal roofing filter would be less than with an HF filter.

K1YZW provided a simplified circuit of the paramp up-converter developed at National (**Fig 7(b)**). He explained that the paramp up-converter could be considered as a cross between a balanced modulator and a coupled pair of circuits. For proper match and maximum power gain, the total capacitance swing (the varactors are considered to be in parallel) should be:

$$\Delta C = \sqrt{\frac{C1 \times C2}{Q1 \times Q2}}$$

Q1 is the Q of the signal circuit when loaded by the antenna only, and Q2 that of the IF circuit loaded by the filter and subsequent amplifier. C1 and C2 are the corresponding tuning capacitances. For simplicity, Fig 7 (b) omits the blocking capacitors and bias feed resistors for the varactors, each of which must be reverse biased. If the circuit was designed for approximate match at all frequencies, the gain at the low frequency end would be excessive.

However, fairly constant gain can be obtained by matching near the high end of the tuning range, purposely mis-matching at lower frequencies and using capacitive tuning. The varactors should have a large inverse voltage rating (100 is typical) and be subjected to a peak-to-peak swing almost as large. The larger these voltages, the stronger the signal which can be handled linearly.

K1YZW added: "The basic idea is to obtain an HF receiver which is virtually free from the usually disastrous effects of strong off-channel interfering signals, such as blocking, cross-modulation etc. If the desired signal together with the strong interference is converted linearly (in amplitude) to a fixed frequency, the latter can then be filtered out.

A conventional (or time variable resistance) mixer would require excessive local oscillator power which must always be large compared to the signal plus interference power, and is quite noisy. The paramp con-

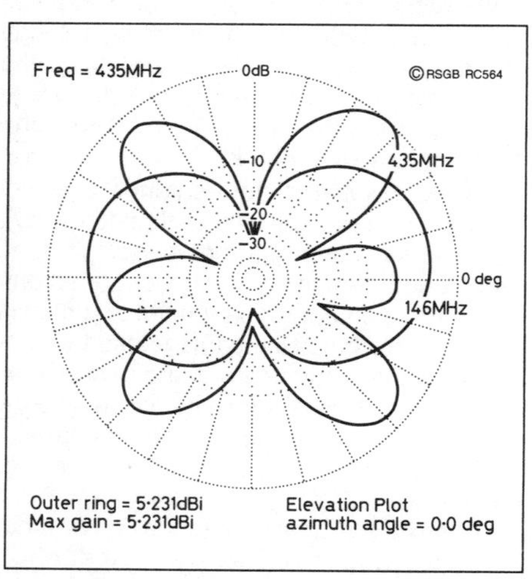

Fig 4: Vertical radiation pattern of a 2-metre Ribbon-J antenna as found with ELNEC by VE2CV at 146 and 435MHz. This shows that while such an antenna could have an acceptable SWR at 435MHz, the horizontal lobe is a minor lobe.

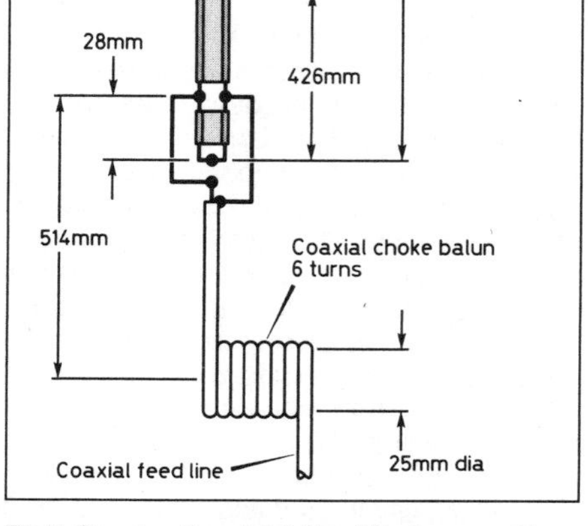

Fig 5: Constructional details of the improved two-metre Ribbon-J antenna as developed by VE2CV.

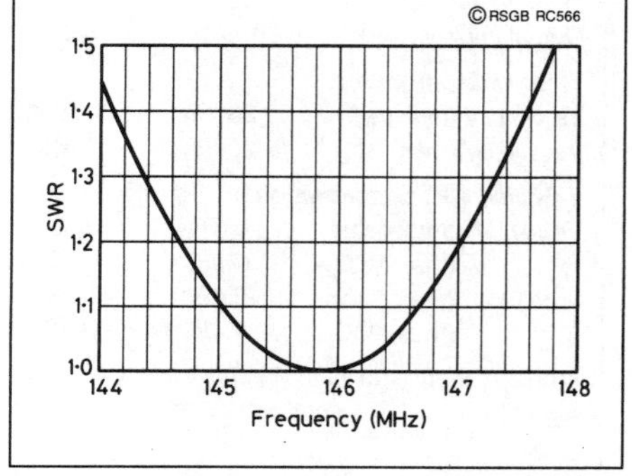

Fig 6. SWR performance of the VE2CV Ribbon-J antenna over the North American two-metre band (144-148MHz).

AGC AUDIO PREAMP FOR DSP FILTER OR TNC

DAVE MILLER, NZ9E in the February *QST* (p58) presented 'An AGC Audio Preamp for All Seasons' with the sub-title 'Make sure your audio filter or TNC is getting exactly what it needs'. He stated: "For the past year, I've been using a pair of the W9GR multiprogram DSPs in my own station set-up and I often wonder how I managed without them. To get the best performance, however, you must maintain the receive audio between certain levels. W9GR kindly provided an input-level LED bar-graph to help you set the correct level. That assistance notwithstanding, it can still be tricky to keep the audio level into the DSP low enough to prevent overload on unusually strong signals, yet high enough to hear the weakest of the weak ones.

"This isn't a problem confined to DSP audio filters. Packet and multimode TNCs also require a reasonably stable audio input. Many modern HF transceivers provide a fixed audio level output . . . . this is fine but the level is still at the mercy of the AGC and AVC circuitry in the radio. And what if your transceiver doesn't offer a fixed output? . . . . you have to tap the audio at the external speaker hack and constantly adjust the volume control as you tune from one signal to another.

"My solution is simplicity itself. It involves the addition of a 'brick wall' audio AGC amplifier between the radio and the DSP audio filter (or TNC). With the AGC preamp in the line, any audio reaching the DSP will not exceed a predetermined level (since it) effectively boosts weak signals and puts a cap on strong ones."

The circuit configuration described by NZ9NE, based on a design used by CW Electronics some years ago, is shown in **Fig 8**. It uses an easily obtained 741 op amp set for an internal gain of about 200. A portion of the output signal is rectified by the 1N4148 diodes, then filtered and fed to the gate of the FET input shunting circuit. As the output rises more of the input signal is bypassed keeping the output level constant. The arrangements offers a 100:1 limiting action, with the output level set from less than unity all the way up to nearly the gain of the amplifier, making the circuit useable for other applications.

The hang time can be adjusted by changing the value of C3 although NZ9E found the 10uF shown about right. If used with a W9GR filter it needs to be slow enough so that the background noise doesn't rise during short pauses in normal SSB conversations, yet fast enough so that sudden audio bursts do not numb the output of the receiver. The preamp's internal gain can be adjusted by replacing R8 with a 100K pot (with a fixed 1K resistor in series with it).

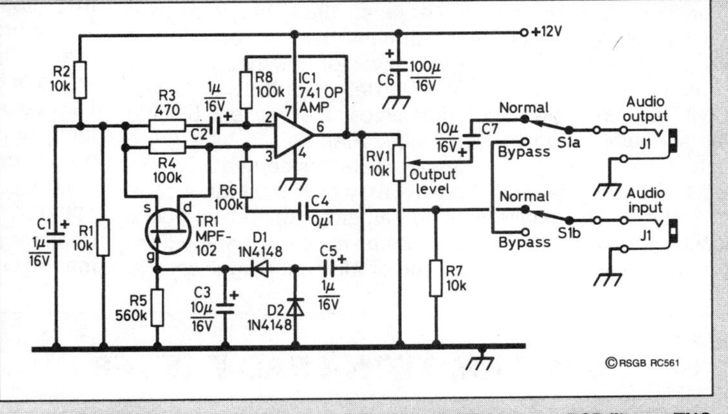

Fig 8: Audio AGC preamplifier as used to provide a level input to a DSP filter or TNC etc. C1, C2 and C3 are tantalum capacitors. C4 100nF mylar capacitor. TR1 N-channel FET (MPF102 or similar). Resistors 1/4 watt rating. RV1 10K linear-taper potentiometer. S1 DPDT switch.

verter, consisting of a varying capacitor, needs only volt-amps, not watts, and being reactive, is almost noiseless.

"The non-inverting 'up-converter' is the only circuit configuration which is stable with changing source (antenna) impedance and also provides some gain. The latter will be larger the higher the output frequency of the 'first IF'. All up-converting receivers suffer from the disadvantage of strong signals at submultiples of the 1st IF causing spurious responses. This is particularly serious in a receiver with a very large linear (or dynamic) range, so that a very high 1st IF is again desirable.

"Another factor in its favour is its help in minimising local oscillator radiation from the antenna. Against the high IF is the need for the highly selective (ie crystal) filter with low insertion loss and low noise amplifier. As a compromise, in the first receiver design 112MHz was chosen, being over three times the highest signal frequency of 30MHz and crystal filters at this frequency have been marketed . . . . I have obtained a useful linear signal input range of about 136dB. In more exact terms, the minimum useful input (10dB above noise) was 0.35uV emf behind 50Ω, with a detector bandwidth of 3kHz and maximum input of 2V which caused a 1dB departure from linearity. The minimum level represents a noise figure of 6dB, which is quite respectable . . . . Theoretically there is no limit to the dynamic range which can be obtained with a single signal . . . . However the real value of a large dynamic range is in combating the effects of large amplitude off-channel interference, and here we run into a severe limitation which is the noise power associated with the local oscillator . . . .

"From the noise point of view the high first IF is a disadvantage since this makes for high local oscillator frequencies and consequently higher oscillator noise power . . . . Because of the limitation represented by the 'purity' of signals, other recent devices are almost as useful, especially for amateur receivers. The latter have a considerable advantage in not being continuous coverage, so that a first IF of 9 or 35MHz may be used . . . ."

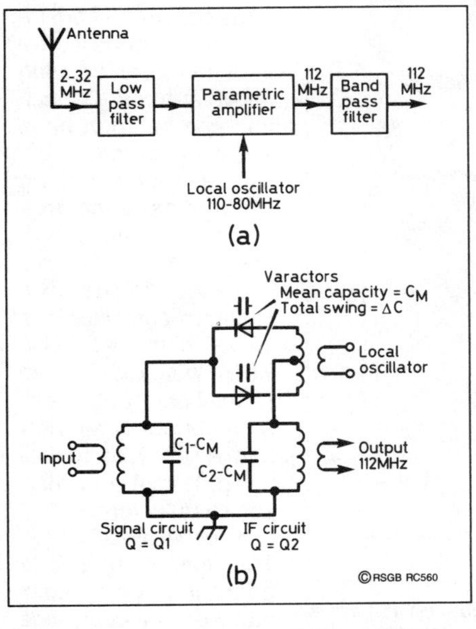

Fig 7: The parametric up-converter mixer. (a) How it was used in the front-end of high-performance HF communications receivers. (b) Simplified circuit diagram of a paramp up-converter as developed by K1ZYW/G3DCU and National engineers.

TECHNICAL TOPICS CORRECTIONS

TT, February 1995, p68. Fig 1, shows the same "loss" of 6db at three different heights. This should be 3db at 0.08 wavelength and 1db loss at a height of 0.17 wavelength.

TT, April, Fig 1. Several errors in the circuit diagram for the 'HF Starter Radio'.

R3 shown as 100k should be only 200ohms. The earthy end of RV1 should be connected to the chassis line and not as shown. The text reference to R3 should be R2. Although the values of RV1 and the unnumbered fixed capacitor connected between D2 and the 33uF capacitor are both shown as 150k in the CQ source, it seems likely that these values should be much lower, possibly both both 1.5k or 15k. Apologies for these errors.

Also the orientation of two of the diodes in the full wave rectifiers is incorrect in Figs 2 and 3. The diodes should be orientated as shown below. ♦

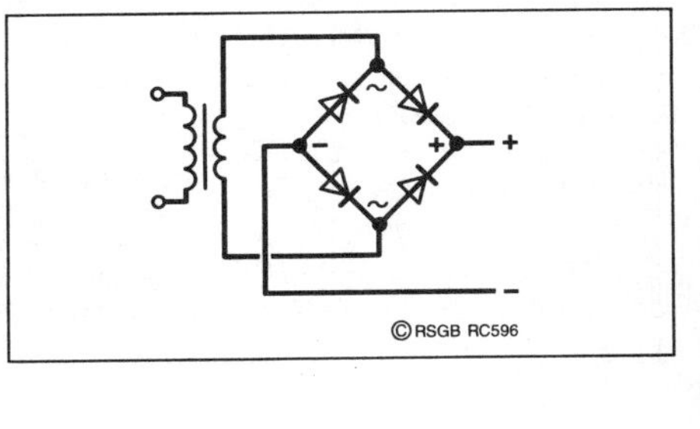

Pat Hawker's Technical Topics

PAT HAWKER, G3VA
London 37/SE22 8SS

MORSE KEYS - THEIR STORY AND THEIR USERS

AS MARK TWAIN might have put it "the death of Morse has been greatly exaggerated". Certainly interest in the design and use of Morse keys in the early years of telegraphy is being kept refreshingly alive by the publishers of *Morsum Magnificat*. Their latest venture is the publication of the first of a series of compact (A5 format) books based on 'The Best of MM'. Volume 1 is the 60-page *The Story of the Key* by the late Louise Ramsey Moreau, W3WRE, together with a listing of 'American Telegraph Instrument Makers 1837-1900'. [See review p57, May - *Ed*]

The Story of the Key led to me re-reading a very different hardback, 264 + viii page book, published some years ago entitled *The American Telegrapher: A Social History 1860-1900* by Edwin Gabler (Rutgers University Press, 1988). The book describes how, in the years following the American Civil War, an entirely 'new white-collar employee' and semi-middle-class profession of telegraphist (largely recruited from the working classes) came into being.

Members of this profession soon earned the reputation of being irresponsible, foot-loose and dissolute, largely at the mercy of the ruthless, near monopoly of Western Union and leading to the Great Telegraphists Nationwide Strike of 1883. Yet from their ranks emerged many notable figures - not least the great inventor Thomas Edison - many highly-skilled telegraphists and the first electrical communications network binding together the United States.

For amateurs more concerned with the perennial controversy over the Morse code requirement for amateur operation below 30MHz, it is worth pointing out that the February 1995 issue of *Morsum Magnificat* (No 38) contains a detailed account 'Morse Code and Amateur radio - The IARU Clarifies its Position' by Tony Smith, G4FAI, which provides a detailed summary of the 26-page document issued by the IARU on the work of the IARU CW Ad Hoc Committee. For those who believe that the Code has passed its sell-by-date this could prove an illuminating read.

CLOSELY-COUPLED RESONATORS FORM MULTIBAND ANTENNA

IN THE NOVEMBER 1994 issue of *RF Design*, editor Gary Breed, K9AY, provides design and construction information for an HF or VHF antenna that operates effectively on two, three, four or more different frequencies. The approach permits multi-band operation without the use of reactive decoupling networks or tuned stubs or traps etc. It also has the advantage of providing control over the feedpoint resistance and reactance at each frequency.

He has a patent pending (application July 1994) for "a method for constructing multiple-frequency dipole or monopole antenna elements using closely-coupled resonators" and

300 OHM TWIN-LEAD FEEDER

'BILL' MCLEOD, VK3MI, has noted recent references to 300 ohm ribbon feeder cables in *TT* and in several overseas journals, including WIA's *Amateur Radio*, both for antenna elements and for matching sections. He warns of a problem that was discovered the hard way some 50 years ago but seems often to be overlooked in recent discussions. With the standard form of exposed ribbon cable the impedance characteristics have proven far from stable in wet weather see **Fig 1(a), (b)**. In the UK a more stable form of tubular cable was marketed for a number of years, **Fig 1(c)** but is apparently no longer available.

VK3MI writes: "When a film of moisture covers the web between the two conductors the nominal impedance nose-dives as the capacitive reactances increases dramatically, with the 'K' factor of the water now forming a large part of the dielectric. This factor is about 80 (paragraph 3.3 in the ARRL Antenna Book) and compared to the 'K' factor for polythene (2.3) forming the web itself it does not require much water in the electrostatic field to cause vast changes!"

As a rough test it will be found that the 16pF or so of a 1-metre length rises to 60pF when only half immersed in a bucket of water! Enclosing the cable in a plastic pipe (as in the J-Pole antenna, *TT*, December 1994) will keep it dry although, as noted by VE2CV, this may itself change the characteristics of the matching section etc.

VK3MI adds that slotted web ('ladder') twinlead still has about half the web to give a large change when wet; the only current variety of twin lead for non-critical outdoor use appears to be the 'super low-loss foam enclosed TV twin lead Archer brand, Tandy Catalogue number 15-1174' which excludes water from the immediate field between the conductors, though fringing could still have some effect. However, measurements cannot be directly transposed when using published designs as the velocity constant of the foam filled cable is about 77% indicating that there are more solids in the foam dielectric than the 82% velocity factor of the common webbed variety.

"For similar reasons antenna loading cells and traps should always be enclosed to keep them dry; or, alternatively, be self-supporting with spaced turns separated from insulating supports to prevent moisture bridging between turns. Capacitors can be short open stubs of coaxial cable with sealed ends, eg RG58 has a capacitance of some 100pF per metre."

In the 1950s, the change of characteristics of 300 ohm feeder line when wet, led Telcom (The Telegraph Construction and Maintenance Co Ltd) to develop, patent and market a tubular form of 300 ohm feeder: See Fig 1(c). This had more stable characteristics under bad-weather conditions with an attenuation at 100MHz of about 1.4dB/100ft.

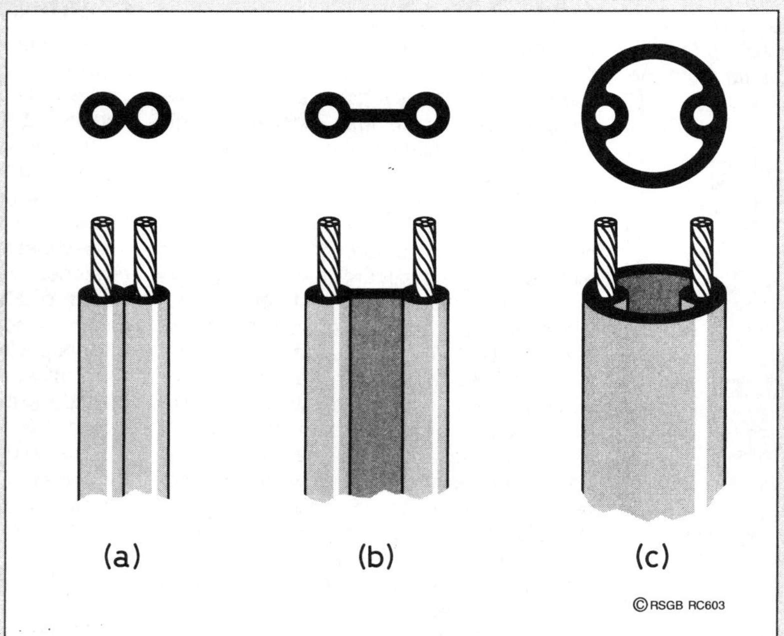

Fig 1: Low-loss twin feeder cables as marketed in the 1950s by Telcon. (a) 150 ohms nominal impedance with attenuation at 50MHz 2.1dB/100ft. (b) 300 ohm flat ribbon-type twin with attenuation at 50MHz 1.0dB/100ft. (c) 300 ohm twin tubular feeder with stable characteristics in varying weather conditions. Attenuation at 50MHz 0.92dB/100ft and power rating at 100MHz of 550 watts.

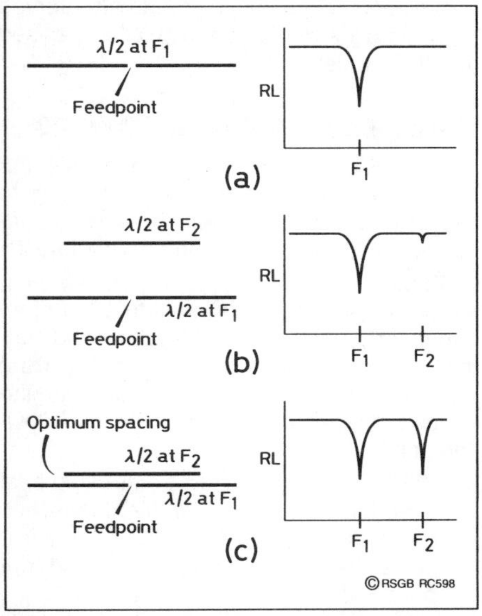

Fig 2: Principles of the close-coupled multiband antenna. (a) Shows a simple half-wave dipole element and its typical return-loss sweep; (b) shows the effect of an additional conductor, resonant at a higher frequency (F2), placed in the vicinity of the F1 driven dipole; and (c) shows how an effective antenna for both F1 and F2 emerges when the spacing of the second conductor is such that coupling is optimum. (Source *RF Design*)

is investigating applications including VHF/UHF cellular and mobile bands, HF broadcasting and amateur radio. However, it seems doubtful whether practical use of this technique is original. For instance, C B Sibley, G0LMC, reminded me that in the days of VHF television in the UK, one of the main antenna manufacturers patented and marketed a combined Band I and Band III antenna. This appeared to follow the same basic principle in which only the lowest frequency resonant dipole element is fed directly with the other dipole element being closely spaced from this element.

Similarly, Bill Orr, W6SAI, in reporting on K9AY's article (*CQ*, February, 1995) suggests that the principle was the subject of some six US patents issued between 1946 and 1950 and, as in the UK, was used for TV antennas and some log-periodic designs and is described as the 'Open Sleeve Dipole' in the 17th edition of *The ARRL Antenna Book*. But, without entering too deeply into any arguments about patents, we should all be grateful to K9AY for bringing this system more fully into the public domain and, especially, for providing computer analysis and practical details of an antenna for the 10, 18 and 24MHz bands.

K9AY introduces the Coupled-Resonator (C-R) principle as follows: "It is well known that conductors in close proximity exhibit strong mutual coupling. A design technique called the C-R principle has been developed which uses this coupling to great advantage. The C-R principle defines the conditions for optimum coupling, creating a system with multiple resonant frequencies, driven at a single feedpoint. Such a multiple-resonant structure consists of a driven dipole or monopole at the lowest frequency of operation, with additional resonant conductors surrounding it, placed at the appropriate distances.

"**Fig 2** demonstrates the C-R principle in its simplest form, a two-frequency system. A half-wave driven dipole is resonant at F1 and driven at the centre. A typical return loss sweep for such a dipole is depicted in (a). In (b) an additional conductor, half-wave resonant at an arbitrarily-chosen higher frequency, F2, is placed nearby. Some degree of coupling will exist between this conductor and the driven dipole, and the return loss sweep of the dipole shows a 'bump' at the resonant frequency of the second conductor.

"The main premise of the coupled-resonator principle is that there is an optimum spacing distance where the coupling results in a

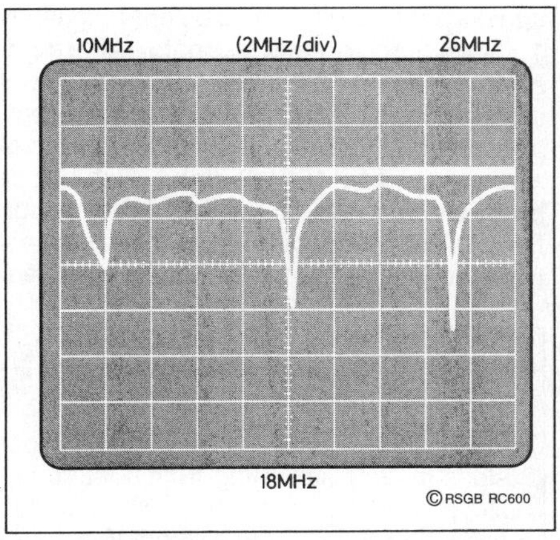

Fig 4: Return loss sweep of the triband antenna. The scales are 2MHz/division horizontal (centred on 18MHz) and 10dB/div vertical.

matched condition at F2 as in Fig 2 (c). The return loss remains good at F1 and, therefore, the system is matched at both frequencies."

K9AY shows that the same principle can be applied to monopole as well as dipole elements and can be expanded to three, four, five or more frequencies by adding additional resonators and placing them radially around the fed dipole or monopole. A practical upper limit is reached when the complexity of multiple interactions obscures the desired coupling, but systems up to seven frequencies have been successfully modelled.

He points out that the variables involved in the design of C-R antennas are: conductor diameter, conductor spacing, feedpoint impedance, and the ratio of frequencies. The feedpoint impedance at each additional frequency can be controlled by adjustment of resonator spacing and length.

The five-page *RF Design* article provides information on the design equations, radiation characteristics, advantages and limitations and also details of a practical C-R antenna. K9AY states: "Various antennas were constructed to verify the accuracy of the computer models, and to assure that the concept was valid. The first versions of these antennas were designed for HF amateur radio bands, where they could be evaluated 'on the air' and compared with other antennas of known performance . . . this arrangement allowed extensive experimentation in conjunction with enjoyment of the hobby.

"**Fig 3** shows the dimensions of a three-frequency dipole constructed from No 12 AWG wire. The driven dipole is resonant at 10.1MHz, with additional resonators for 18.1 and 24.9MHz . . . Choosing 50 ohms as the design impedance for F2 and F3, the required

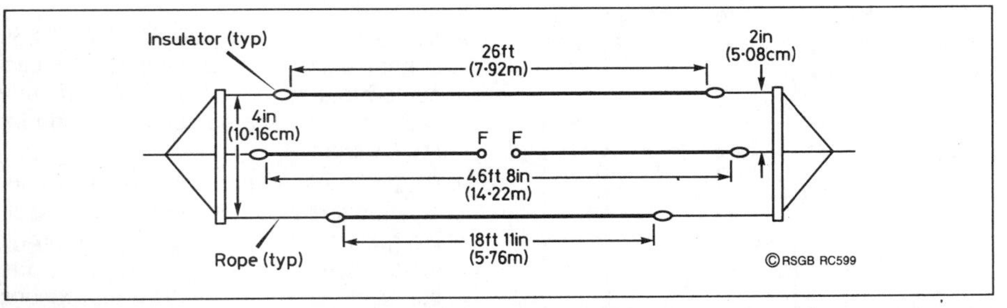

Fig 3. K9AY's experimental tri-band antenna for the 10, 18 and 24MHz band using the close-coupled (open-sleeve) approach. Dimensions are for an antenna made from No 12 AWG wire at a height of 45ft. Insulated spacers are required to keep the wires in alignment. Extra closely coupled wires could be added radially around the driven element for 14 and 21MHz but there will be a practical limit to the number of bands due to the growing complexity of interactions and structure.

DIGITAL VOICE COMING?

YOU MAY BE interested in the extract below taken from a section *Overview of Mobile & Personal Communication* by Andrew J Viterbi (QUALCOMM Inc.) in the book *Modern Radio Science 1993* (Oxford University Press for URSI, 1993):

"Based partly on Shannon's source coding rate-distortion theory, but equally on psychoacoustic and psychovisual principles, in the eighties and early nineties major improvements have occurred in speech and video digital source compression, with the result that digital bit rates for high quality speech and video are now lower than the sampling rate. This means not only that the transmission bandwidth of digital can now be lower than that of single-sideband analogue modulation, but also that the transmission quality is better and more robust because requirements for accurately transmitting a single bit per dimensions are obviously easier to meet than for transmitting an analogue value per dimension.

"These significant advances in source compression, particularly the so-called code-excited linear predictive (CELP) speech coder, have made digital mobile and personal telephony more accessible . . . "

spacing was determined to be approximately 1.75in (4.5cm) for each resonator. After modelling the design in ELNEC the spacing was increased to 2.0in (5cm), primarily to compensate for installation above real ground at a design height of 45ft (13.7m) . . . Return loss is greater than 20dB at the three design frequencies, exceeding 30dB at the highest two frequencies (those added by coupled-resonators): see **Fig 4**. 30dB corresponds to a VSWR of 1.06:1. Bandwidth tends to be reduced slightly at the higher frequency resonances.

"On-air performance of the three-band antenna proved to be indistinguishable from that of separate dipoles for each band. The radiated performance, the feedpoint impedance, and the variations in impedance with installation height above ground also served to confirm the validity of the MININEC-based ELNEC model."

200-WATT AC 50/60HZ INVERTER

OVER THE YEARS, several inverter designs, mostly stemming from *Electronics Australia,* have been noted in *TT*. Heralded as rugged and efficient, a single-IC square-wave inverter for use in automobiles and boats - designed by M S Nagaraj of the ISRO Satellite Centre, Bangalore, India - has been awarded a prize in the joint Wireless World/International Rectifier design competition and appears in the April, 1995 issue of *EW + WW*. (pp 346 and 348): **Fig 5**.

It is based around an IR2151 self-oscillating half-bridge driver which can supply complementary square waves with 50% duty cycle, at a frequency immune to supply voltage variations and with under-voltage lockout. Although the system works as a push-pull square wave amplifier with the output delivered from a centre-tapped double-wound transformer, the final output waveform is not shown and would presumably depend to some degree on the characteristics of the output transformer. Previous inverters described in *TT* have had stepped output waveforms possibly making them more suitable for the operation of transceivers etc primarily intended for use on AC mains with near sine-wave waveforms.

M S Nagaraj points out that lockout in the event of low battery voltage is a very important feature. "Without it, under low battery voltage conditions, the power mosfets would not conduct fully and would dissipate power. This in turn reduces inverter efficiency and could result in damage. In this unit, when battery voltage falls below 8V, both power mosfets are switched off."

He adds: "Although the dead time of about 1µsec generated by the IC is sufficient to prevent the cross conduction of the mosfets, it is not sufficient to allow the transformer ringing signals to be dissipated in the appropriate snubber circuits. Fortunately, dead time can be increased to the desired value by adding a few inexpensive components to the basic inverter circuit.

"Supply voltages Vcc and VB, and hence the outputs HO and LO are limited to 9.1V by components R1, C1 and D1. Oscillator frequency is set by R2 and C2. When HO goes high, TR1 holds the gate of power mosfet TR3 low for a dead period of about 300µsec, determined by R3 and C3. Diode D2 discharged the gate capacitance of TR3 instantaneously when the HO output goes low. Although the turn-on time of TR3 increases to about 30µsec, it is a very small part of the period of the inverter output.

"TR2, C4, D5, R4, D4, and R8 provide an equal dead period to the low side output LO. R5, C5, D6 and R6, C6, D7 provide the snubber action for TR3, TR4."

It is noted that this circuit configuration is not limited to 50/60Hz inverters. High-frequency inverters with ferritecore transformers (with proper dead times) could form elements of electronic ignition systems and DC-to-DC converters, and by changing R1, the inverter could be powered by batteries of other voltages.

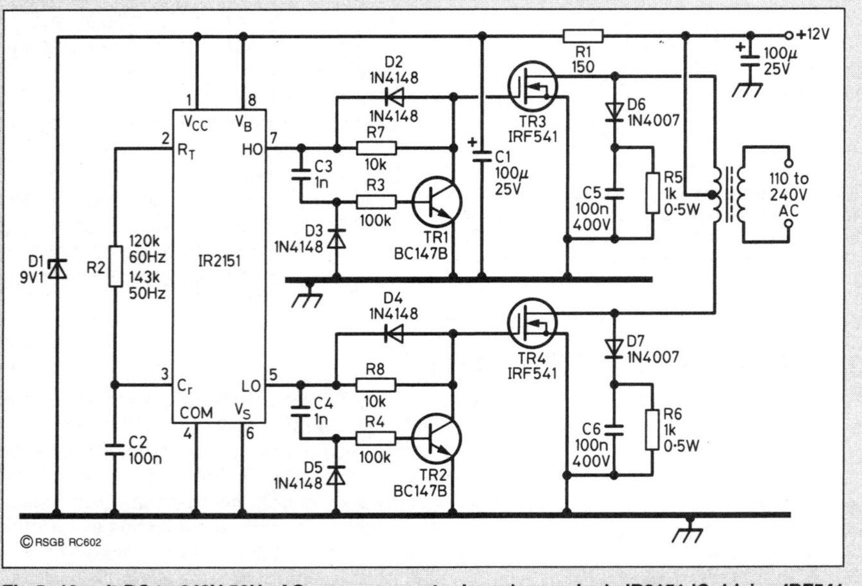

Fig 5: 12-volt DC to 240V 50Hz AC power converter based on a single IR2151 IC driving IRF541 mosfet switches. Most of the complexity represents the discrete transistor circuitry required to extend the IR2151's dead time. (Source *EW + WW*)

CABLES AND CONNECTORS

THE APRIL *TT* notes on coaxial cables and the choice of connectors and the necessity to seal them effectively against moisture ingress resulted in several pertinent comments.

Paul Gaskin, G8AYY, a long-standing microwave enthusiast, believes that the comments on selecting suitable cables and connectors for VHF/UHF, stemming from ZS6AXT, are rather confused and require some further elucidation. He lists the following points:

- RG58C/U is a small diameter cable best used for mobile whips or interconnection between equipment at VHF. It has a low power-handling capacity and a tinned copper braid which increases its losses.
- RG213/U is a MIL-C17D cable and has an adequate braid. Some RG8/U type cables made for the CB market do have a very poor braid. The plasticiser used in some older coaxial cables used to attack the copper braid.
- Andrew Heliax cable of the LDF 4-50 type is not very flexible and is expensive. It would be better to use RG213 with rotatable antennas and telescopic towers.
- RG213/U cable has a solid polyethylene dielectric and a non-contaminating PVC outer sheath. It is weather resistant and not easily damaged unless water is allowed to enter the ends in which case the copper braid could become corroded.
- I do not understand how the polyethylene dielectric of RG213 could be affected by UV light as it is enclosed in the braid and PVC sheath. It is more likely that deterioration occurs because of high temperatures. The standard black PVC sheath will give the maximum absorption of heat from the Sun!
- It is best to use silver-plated UHF or PL259 connectors with PTFE insulation. Coaxial cable braid cannot be soldered properly with the nickel-plated versions made for the CB market.
- BNC connectors normally have PTFE insulation and the cable entries have sealing gaskets. They have a *bayonet* fixing which may lead to misalignment of the contacts with wear. In critical applications it would be better to use TNC connectors which have a *threaded* fixing which ensures proper contact alignment.
- SMA *sub-miniature* connectors are not normally used on amateur equipment except at 10GHz. N-type or TNC connectors can be used on lower frequency bands!

In connection with G8AAY's comments on the braid used on such cables as the RG213/U, it is worth noting that this matter has been tackled in *TT* on various occasions. While cables manufactured to American MIL-SPEC standards do have virtually full coverage of the outer copper braid, the braiding on typical cables intended for civilian applications such as television downleads has been progressively reduced in order to neutralise the rising cost of copper (see *TT*, November 1988 and page 271 of *Technical Topics Scrapbook*

1985-1989). Robin Addie, G8LT, mentions that some years ago he obtained a large length of RG214-U cable which, unlike the RG-213, is *double-screened* and has proved excellent for HF/VHF/UHF even in 120ft runs to his tower.

After his wartime service as an Engineering Officer at SCU3, Hanslope Park, G8LT had a professional concern for many years with connectors of all types, including coaxial connectors. He contributed a fascinating paper, *Connectors Since the '20s,* for the Diamond Jubilee of the Institution of Electronic and Radio Engineers (published in *JIERE*, October/December 1986, pp317-324) and the following are some brief extracts from his Section 7 on 'Coaxial connectors':

"When C S Franklin developed his concentric line feeder system for the Marconi beam short-wave stations, he drew attention to the mechanical precision with which the branched and jointed feeders had to be made so that the phase at the feed point could be controlled accurately, thus ensuring that different sections of the antenna array were all driven in phase. Adjustment was achieved by minor changes in feeder lengths. The same principle applies where coaxial connectors are concerned . . .

"Cable impedances have varied over the years and a figure of around 75 ohms used to be the norm, whereas today 50 ohms is regarded as standard both for cable and coaxial connectors. The latter have now become one of the most highly developed components available to the radio engineer. The average family accepts the small coaxial plug (originally by Belling & Lee) as part of daily life, even if so many installers still 'forget' to tip solder the centre pin!

"The development of polythene as a dielectric for coaxial cables gave impetus to a part of radio engineering that had remained static since Franklin's day. Wartime radar with its high pulse powers and antenna arrays requiring careful phasing brought into being several forms of coaxial connector. Many were actually moulded onto the cables using small polythene moulding presses. Only later did PTFE form the internal insulation material within the connectors themselves. It has the advantage of being able to withstand the heat of soldering whilst being produced to precise dimensions.

"Anyone involved with signals and similar matters during the war will recall those magic type numbers PL259 and SO239, which designated a coaxial plug and socket pair made by Amphenol in the USA and which featured in so much equipment that came to the forces in Europe and the Far East . . . This UHF type has survived the passage of years and has been kept not only by the original makers but copied by the Japanese and others . . . being relatively inexpensive, there are versions for both commercial and professional uses. An assortment of adapters, from straight-through to right-angle are available but it must be remembered that two different screw threads are in use with are inter-matable.

"This UHF type is of non-constant impedance and, with advances in the uses of coaxial connectors, types having low VSWR and tailored to the cables of the day were more in demand . . . Two series of constant impedance connectors were introduced, namely types 'C' and 'N' for both 75 and 50 ohm cables, differing mainly in the type of securing mechanism, the difference being one of mechanics . . . more recently the 'N' type has tended to supersede the 'C' type.

"These are but two out of ten major series of RF connectors, and three of these are subdivided into families - standard, miniature and subminiature. Standard connectors are used with cables like RG8, RG213, RG214 etc. Cables appear as either flexible, semi-rigid, or rigid, and in diameters around 0.5-in for standard types, down to 0.250in and 0.14-in for semi-rigid cables . . . A theme common to all is that the use of a connector, of whichever design, shall not disturb the characteristic impedance of the line in which it is used.

ADVANCES IN LITHIUM RECHARGEABLE BATTERIES

WHILE THE SEARCH for improved large capacity batteries is being stimulated by the work on electric powered 'environmentally-friendly' vehicles, the growth of portable telephone systems, camcorders, laptop computers etc is fuelling the development of improved lightweight rechargeable batteries. It has long been evident that the limited capacity/weight ratio of nicad batteries is a practical disadvantage when using amateur-band handheld transceivers in the field, particularly when these are used at full power in the transmit mode.

Rechargeable lithium batteries began to appear on the market a few years ago and were noted in *TT,* but cost has remained high with the number of charge/discharge cycles usually rather low (about 200 maximum). Recently, the American firm Ultralife Batteries announced what it claimed to be "the World's first rechargeable lithium-based battery made entirely from lightweight solid materials". The battery, which was developed in the UK by Dowty Batteries of Abingdon (acquired by Ultralife last year), appears to overcome many of the previous problems encountered with such cells. **Fig 6** shows the new form of cell which is termed a 'shuttlecock' or 'rocking chair' battery because it relies on lithium ions passing to and fro as it is charged and discharged. Similar forms of battery are also under development by Japanese firm Yuasa and Danish firm Danionics. Ultralife is setting up a production plant in Newark, New York and later in Europe, with the first product designed for use in mobile phones due to be supplied to a communications company later this year.

The new battery, unveiled recently in London and described by staff-writer Andy Coghlan in the *New Scientist* (1 April 1995), can apparently be recharged more than 1000 times with only a tiny loss of capacity. Because it is based on lithium, it stores significantly more power weight-for-weight than nicad batteries at a cost, when in mass production, roughly the same per watt-hour (about one dollar watt-hour). The structure permits shaping to suit specific requirements; one prototype is a fraction of a millimetre thick, but has an area of about an A4 sheet of paper; another is the size of a credit card. Single cells can be stacked and sealed to form batteries shaped like small matchboxes.

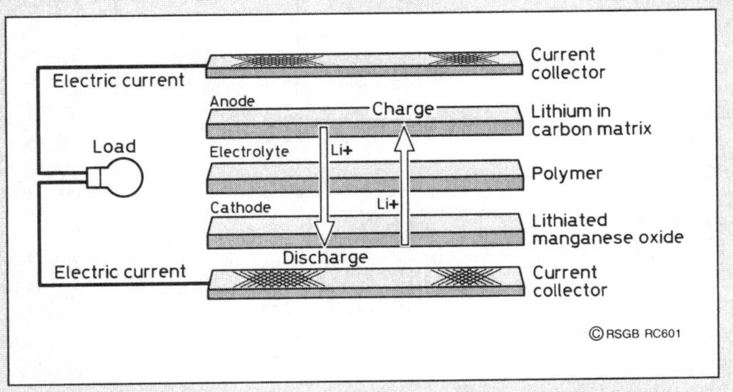

Fig 6: The solid-state lightweight rechargeable lithium-based cell as developed by Dowty Batteries, UK and due to be manufactured in the USA by Ultralife Batteries for use in handheld, portable equipment with an energy storage some three-times that of nicad batteries of equivalent weight. The battery contains lithium-based electrodes and a solid polymer electrolyte whose composition has not been disclosed. (Source *New Scientist*)

Andy Coghlan reports: "When the battery is discharging, lithium ions migrate from where they are stored in a flat, carbon anode . . . The ions cross a layer of polymer electrolyte and combine with the cathode material on the other side to form lithiated manganese oxide. Sheets of metallic mesh on the top and bottom of the cell act as current collectors. When the battery is recharged, this process is reversed. The cell is sealed in laminated aluminium foil to keep moisture out . . . heart of the battery is the plastic electrolyte, made by blending a lithium salt, a polymer and plasticisers to make a rubber mixture. The electrolyte has to perform two conflicting functions. It has to be an electronic insulator to divide the electrodes and stop the cell shorting out and it has to conduct lithium ions."

Ultralife claims that its lithium-based batteries should be able to undercut other forms of lithium-based batteries, which contain relatively expensive cathode materials such as lithiated cobalt or nickel oxide much more costly than the lithiated manganese oxide material of the new battery. Production is expected to begin this year.

The US government is supporting a large billion-dollar programme to develop large capacity lithium batteries for possible use in electric cars.

On this point alone hangs much of the complexity found in all modern coaxial connectors.

"Of course they have to do their fundamental job of good mechanical and electrical connection . . . Let us examine how this is achieved: Taking for example, 'C' and 'N' types and the stages of assembly that have to be followed in putting a plug or socket on to a cable end . . . Cable end preparation comes first. The protective sheath, the screening, the insulating dielectric and, finally, the centre conductor must each be exposed and cut accurately to fit the piece parts, both metal and insulating, that will be offered to them in the assembly process. The final work must give protection against the entry of moisture and/or dirt, and be able properly to mate with its counterpart.

"All makers supply detailed cutback dimensions for cables, and many offer reducers where there is a change in cable diameter, as well as adaptors to enable one type to intermate with another. Backend parts support the cable outer, while the screen is gripped and pressed into the metal shell, with the whole firmly anchored by an expanding grommet driven from the rear by a threaded locking ring. Before this is done, the internal PTFE insulation is put in place and the centre contact pin soldered or crimped, as appropriate. The resulting assembly should now be ready for use.

"Type C, which is of medium size, is weatherproof with a low VSWR up to 10GHz. It is characterised by a two stud bayonet locking ring, with a milled surface, which enables quick connect/disconnect . . . Type N is also a medium size weatherproof connector, where the bayonet locking ring is replaced by a threaded collar with a milled surface for finger tightening.

"The continued expansion of activity in the microwave region has brought new types of connectors. Amphenol's 'SMA' and 'SSMA' types are good examples. The former is a subminiature semi-precision (3mm) unit, with good performance up to 18GHz and, with semi-rigid cable, to 26.5GHz . . . Stainless steel gold plated bodies are used, catering for cables from 0.14-in to 0.25-in. At these frequencies, any connector range must include an ability to interface with stripline. A high performance Amphenol version gives a very low VSWR up to 40GHz. For many antenna applications, precise phasing is very important, and therefore features have been incorporated in the SMA type, for example, when used with semi-rigid cable, for phase adjustment in the connector itself by means of a nut which, when turned, brings about small changes in cable length and hence phase . . . Demands for an even smaller type have resulted in the SSMA, which is smaller than the SMA and designed for rigid and semi-rigid cables of 0.085-in diameter.

"We must also mention the workhorse of the instrument makers, namely the BNC. As the letters imply it is a 'C' type in miniature, having a two-stud bayonet lock and make to take small diameter flexible cables. It is used below 4GHz but special versions are available up to that limit. Its quick connect/disconnect feature renders it ideal for input/output in measuring and test equipment . . . "

On the topic of sealing coaxial cables (*TT*, April 1995) Dr R H Biddulph, G8DPS, comments: "I use RTV silicone rubber tube 744 made by General Electric of the USA. It cures liberating *alcohol* (Ethanol) which is non-corrosive and smells nicer than acetic acid! It is available from Industrial Silicones and Lubricants Ltd, International Centre, Spindle Way, Crawley, West Sussex RH10 1TZ. It costs about £8 for 310ml and would be a good buy for a club since it has a reasonable shelf life if sealed after use with aluminium foil."

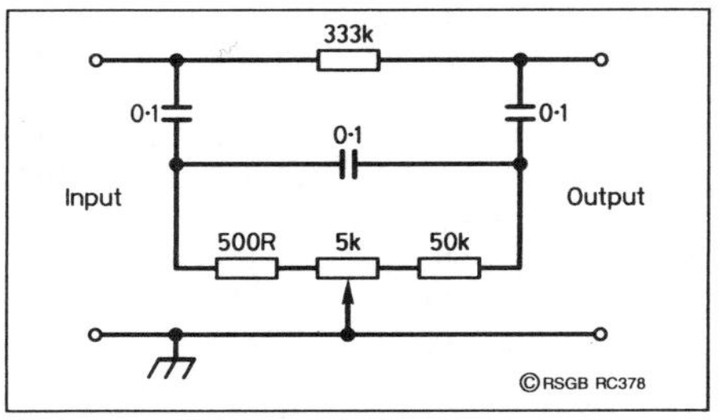

Fig 7: 60Hz notch filter (tuneable 40-120Hz) as noted by DA1PE.

A 40-120HZ TUNEABLE NOTCH FILTER

PETER COLE, DA1PE/G3JFS (Comms Branch (Tels Division) HQ UKSC (Ger), BFPO 140) while scanning some notes he made about 15-20 years ago, came across the filter circuit shown in **Fig 7**. He writes: I've no idea of its origins, but it certainly works. It is clearly some form of bridge network and I would very much like to know (a) how the circuit is derived; and (b) the design equations. Perhaps one of your mathematically-minded readers could help."

It appears to be an R-C (resistance-capacitance) form of the Bridged-T network but I willingly leave it to others to explain the mathematics to G3JFS!

HERE AND THERE

IT SEEMS HIGHLY likely that before long amateurs will begin to use digital speech systems based on the new techniques that permit communications quality speech to be transmitted at relatively low bit rates compared with the standard 64kbit/s of pulse code modulation. A recent book on this subject is *Digital Speech - Coding for low bit Rate Communications Systems* by Ahmet Kondoz of the University of Surrey (John Wiley and Sons, 1994). In his preface, the author writes: "Speech has remained the most desirable medium of communication between humans. Nevertheless, analogue telecommunication of speech is a cumbersome and inflexible process when transmission power and spectral utilisation, the foremost resources in any communication system, are considered. Digital transmission of speech is more versatile, providing the opportunity of achieving lower costs, consistent quality, security and spectral efficiency in the systems that exploit it . . . . The advent of faster and more reliable DSP chips has made possible the easy real-time implementation of highly complex algorithms. The sophistication is also exploited in the implementation of more effective echo control, equalisation and forward error control systems."

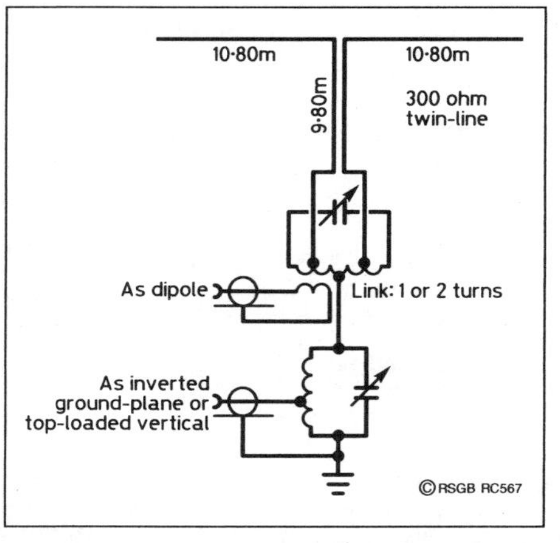

Fig 8: How an inverted Ground Plane (Vertical-T) antenna can be arranged to function alternatively as a horizontal dipole by using two matching networks which may be permanently connected. The coaxial feedline is then plugged into the appropriate socket. Dimensions shown are for 7MHz but could be scaled for other bands.

With DSP chips becoming available for speech bit rate compression to 16Kbit/s and less, intended for use in low-cost mobile and handheld telecommunication systems, it should also be possible to reduce the VHF spectrum needs for amateur VHF/UHF systems. It will, of course, be necessary for everyone to agree on which one or more standards should be used for amateur digital speech. Ahmet Kondoz explains it is also necessary to remember that "as bit rate falls, acceptable speech quality can only be maintained by (a) employing very complex algorithms which are difficult to implement in real time even with the new fast processors with their associated high cost and power consumption, or (b) incurring excessive delay which might induce echo control problems elsewhere in the system." However, a degree of time delay is entirely acceptable for simplex speech.

It has been noted several times in *TT*, that the inverted ground plane (vertical-T) antenna can readily be adapted to provide either horizontal or vertical polarization. 'Reflecties door PAOSE' (*Electron*, March 1995, p91) draws attention to an interesting way of providing alternative feed inputs for such an arrangement stemming from Ingo Huettie, DJ6YC: **Fig 8**. It would be feasible to provide remote switching by means of a coaxial relay or (for HF) a less expensive relay. With the dimensions shown, the system is intended for 7MHz.

Most of us tend to associate 'acid rain' with forestry problems in Norway and elsewhere, but as, Richard Genaille, W4UW, points out in 'Acid Rain and Your Antenna' (*CQ*, February 1995, p13-20) acid rain damages antenna systems. One of the solutions proposed by W4UW is to enclose insulators etc in empty, plastic soft-drink bottles to protect them from the acid rain.

Motorola have announced a second generation of chips specifically intended to form the heart of dual conversion narrowband FM receivers or as the lower IF of a triple-conversion system. Applications of the new MC13135 and MC13136 ICs include cordless telephones, short range data links, walkie-talkies, low-cost land mobile, amateur radio receivers, baby monitors and scanners. It is claimed that these chips offer significant advantages over the first genera-

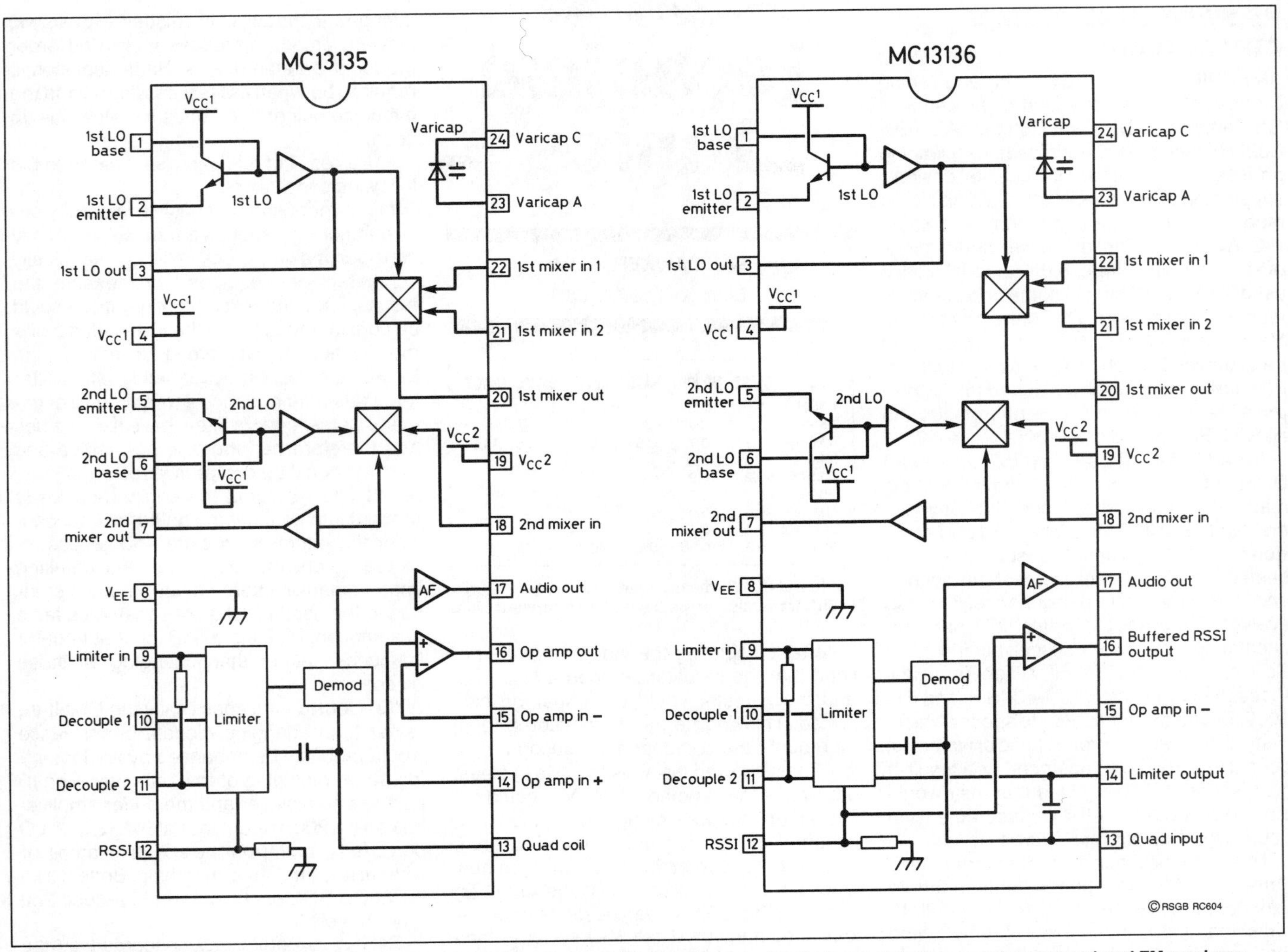

Fig 9: Pin connections for Motorola's 'second generation' MC13135 and MC13136 chips forming the heart of dual conversion, narrowband FM receivers.

tion in signal handling and first oscillator operation, etc. Both devices include a Colpitts oscillator, VCO tuning diode, low-noise first and second mixer and LO, high-gain limiting IF and RSSI. The MC13135 is designed for use with an LC quadrature detector and has an uncommitted op-amp that can be used either for an RSSI buffer or a data comparator. The '36 can be used with either a ceramic discriminator or an LC quad coil and the op-amp is internally connected for a voltage-buffered RSSI output. **Fig 9** shows the pin connections.

Irwin Math, WA2NDM, (*CQ*, March 1995) shows how a Harris HA7210 chip can form "a minimum component, ultra-simple crystal oscillator" for use between 10kHz and 10MHz. Apart from the IC it requires only the crystal, an 0.1uF bypass capacitor and jumper wires for different frequency ranges: **Fig 10**.

Harry Leeming, G3LLL, of Holdings Amateur Electronics noted the item on RF switching diodes in April's *TT* which advocated the use of PIN-type Schottky diodes because of their improved dynamic range but remains unconvinced that such diodes can be relied upon not to deteriorate and become noisy in service. He writes: "Take the FT757 as an example. These have a reputation of being noisy on receive; indeed on some samples if you switch in the RF preamp the noise comes up more than the signal. Check the dozen diodes (ISS97 Schottky barrier types) around the input to the band-pass filters and the transmit/receive switch and up to half of them can sometimes be found to be leaky. Despite being 'yesterday's technology' I replace them with 1N4148 or similar and the receiver becomes as good as new and usually *stays* that way. I wonder if anyone can suggest modest priced diodes that are better than the 1N4148 in this application and will stand-up in service? I confess to being unable to detect any difference in the performance of a new FT757 and one that has had 1N4148s fitted."

CORRECTION, MAY *TT* P59

UNFORTUNATELY, the names of both 'Double-cross' amateurs ('Amateurs and the VE Day and VJ Day Anniversaries, *TT*, May) were printed incorrectly. They were Ronnie Reed, G2RX and Stan Reisen, G5SR. ♦

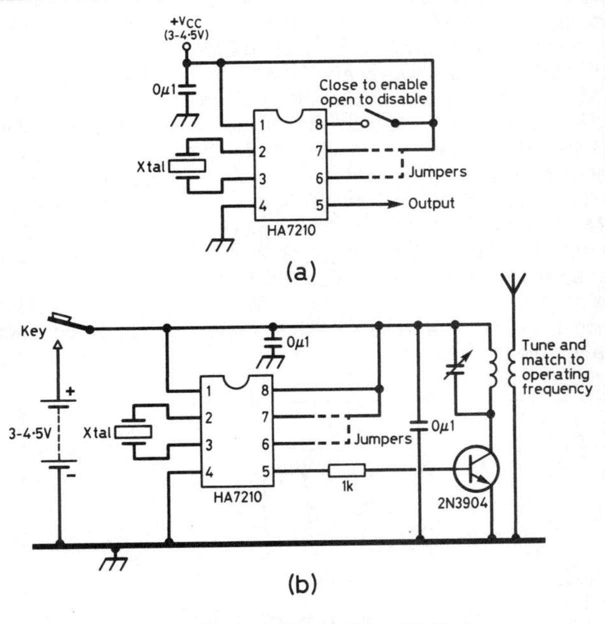

Enable	Pin 6	Pin 7	Frequency range
Closed	Jump	Jump	10kHz - 100kHz
Closed	Jump	Open	100kHz - 1MHz
Closed	Open	Jump	1MHz - 5MHz
Closed	Open	Open	5MHz - 10MHz
Open			High impedance

©RSGB RC605

Fig 10: (a) Minimum component, ultra-simple IC crystal oscillator based on the HA7210 chip. (b) Milliwatt QRP CW transmitter using the HA7210 chip with 2N3904 or similar transistor.

Pat Hawker's Technical Topics

PAT HAWKER, G3VA
London 37/SE22 8SS

RF SWITCHING DIODES CONTROVERSY

AN APRIL *TT* item ('RF Switching and Tuning Diodes', p63) drew attention to the series of QST articles by Dr Ulrich Rohde, KA2WEU/DJ2LR - recognised world-wide as a leading professional expert in HF receiver design - supplemented in a separate *QST* article by measurements made by Tom Thomson, WOIWJ. These highlighted the shortcomings of some general purpose PN and PIN diodes used for RF switching in some popular amateur HF transceivers. Dr Rohde pointed out that the second order IMD performance could be improved in such cases by substitution of PIN diodes specifically designed for RF purposes, and in particular recommended the Hewlett Packard HP5082-3081.

In part three of his article (*QST*, July 1994) Dr Rohde gave results of measurements made on unmodified and modified transceivers - an ICOM IC-765, a Yaesu FT890 and a Kenwood TS-50. These measurements suggested significant improvements in second and third order IMD performance. He also evaluated the second-order IMD performance of several other transceivers including Collins KWM-390, TS950SDX, Ten-Tec OMNI VI (second order intercept +43dBm) and FT-1000. But he did not appear to specify which, if any, of this second group would or would not benefit from diode replacement. As a result of his findings, ARRL decided that they would include second-order IMD measurements in future *QST* Equipment Reviews.

This is highlighted in a letter from Dave Farn, G4HRY, who reported the unfortunate experience of G4KPT who replaced all 40 switching diodes in his Omni VI only to find that sensitivity had suffered. As a result, G4HRY has now replaced the original diodes and believes that "the validity of the original articles is brought into doubt". G4HRY, however, was not able to check on second-order IMD performance before or after modification.

While I am sure that Dr Rohde could provide a convincing reply to his criticisms, G4HRY does make some comments that deserve to be aired. He writes:

"G4KPT read the *QST* articles and as the OMNI VI was specifically mentioned, decided to replace all of the switching diodes in the transceiver front end with HP5082-3081 types. He obtained 40 diodes at a cost of about £1 each and did the modification. This is where the trouble started.

Freq	BA482	HP3081	BAT85	1N4148	BA439	IN4007
1.8MHz	-0.6	-0.9				
3.5MHz	-0.4	-0.75				
7MHz	-0.2	-0.7	-0.6	-0.7	-0.55	-0.5
10MHz	-0.3	-0.8				
14MHz	-0.2	-0.9				
18MHz	-0.3	-0.8				
21MHz	-0.3	-0.9				
24.5MHz	-0.3	-0.9				
28MHz	-0.3	-0.98	-0.6	-0.7	-0.55	-0.5

Table 1: G4HRY's measurements of diode loss (dB) in 50Ω transmission path with 10mA forward bias.

"After completing the modification he noticed that the receiver seemed a little deaf and the S-meter could no longer be calibrated. Thinking he had introduced a fault, he brought the set to me for a second opinion. Tests showed the sensitivity was at least 5dB worse than another OMNI VI. I could not find a hard fault with the rig and decided that, as it had worked well before modification, it was probably something to do with the new diodes. To prove this I built the small jig shown in **Fig 1**. This enabled diode through-loss to be measured in a 50Ω system which can be equated to diode RF resistance. The jig was used to measure the BA482 types and then the HP PIN diodes. Out of interest, I took a quick look at a variety of other general purpose diodes and this indicated that in respect of through-loss, the original diodes selected by Ten-Tec were a good choice: see **Table 1**.

"The receive RF path of the OMNI VI includes 5 diodes before the 1st RF amplifier. The first two isolate the transmitter from the receiver input, the next two select the appropriate bandpass filter. The final diode in the chain feeds the input of the RF amplifier in the Tx/Rx switching. The RF amplifier has 54 parallel-connected FETs with an input impedance of about 22Ω. The losses at this point would therefore be higher than those measured in 50Ω. Changing the diodes to HP 3091 types had introduced about 4dB additional loss which was clearly not acceptable. G4PKT had also changed diodes on the 1st IF board between the IF roofing filter and the narrow IF filters; this accounted for further losses.

"When all diodes were restored to the original BA482 type normal operation was restored - a task taking days to achieve. I realise that the original purpose of the exercise was to reduce 2nd order effects ascribed to the switching diodes. My existing test equipment introduces more 2nd order products than the diodes. Better isolation is required between the test oscillators and the hybrid combiner and I hope to follow this up soon.

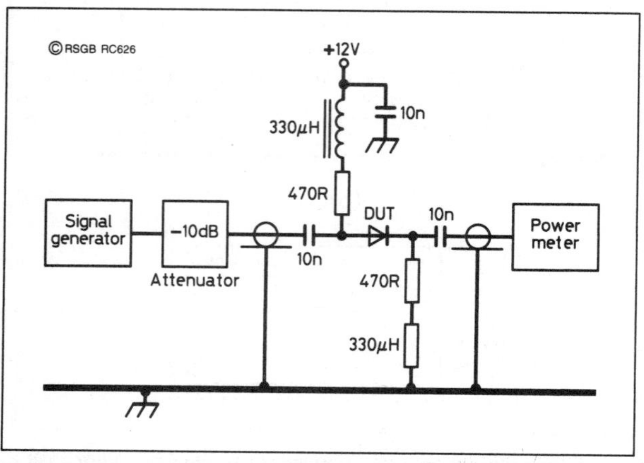

Fig 1: Test set up used by G4HRY to measure the insertion loss of various diodes.

"As a result of this exercise, I came to the following conclusions:

(1) Owners should consider carefully before attacking expensive transceivers. Only consider making modification if technically competent and equipped to measure the results. The actual circuit configuration should be considered to judge the likely effects of a modification. It may prove to induce high losses and will almost certainly effect filter termination impedance's required realignment. Some modern filters have fixed values and therefore performance on receive and transmit could be compromised.

"(2) The validity of the original articles is brought into doubt. If the author did not consider the effect of an extra 4dB of loss inserted before the first mixer, the resulting improvement in intermod performance credited to the use of PIN diodes may be a false assumption. Building a 4dB input attenuator is much cheaper than changing all those diodes!

"(3) Considering specifically the OMNI VI, rather than changing diodes, performance would probably be enhanced by implementing better matching of the 1st mixer. The IF port has no diplexer and the buffer amplifier has only a 20dB return loss at 9MHz. The LO port is fed directly from the LO power amplifier without any attempt at matching. Better filtering at the RF signal input would reduce 2nd order effects."

G4HRY is particularly concerned by the unquestioning faith often put in published articles, including *QST* and *RadCom*. He urges others to follow his own philosophy and become professional sceptics!

In his three-part article, Dr Rohde noted that second-order IMD products change 2dB for every decibel of input-signal change, and appear at frequencies that result from the

Diode Type	Bias conditions per diode				
	Reverse	0mA	5mA	10mA	20mA
1N4153	75	75	2	1	0.5
MPN3700	70	55	0.1	0.1	0.1
BAR17	75	70	0.3	0.1	0.1
1N4007	35	20	0.1	0.1	0.1

Table 2(a): WOIVJ's Diode Switch insertion loss (dB) at 10MHz

Diode Type	Bias conditions per diode				
	Reverse	0mA	5mA	10mA	20mA
1N4153	>80	>80	18	30	42
MPN3700	>80	80	66	70	72
BAR17	>80	>80	60	70	75
1N4007	>80	40	>80	>80	>80

Table 2(b): diode switch second-order intercept point (IP2) dBm

Diode Type	Bias conditions per diode				
	Reverse	0mA	5mA	10mA	20mA
1N4153	>50	>50	20	30	37
MPN3700	>50	47	>50	>50	>50
BAR17	>50	50	>50	>50	>50
1n4007	>50	35	>50	>50	>50

Table 2(c): diode switch third-order intercept points (IP3), dBm

simple addition and subtraction of input-signal frequencies. His introductory notes on switching diodes were as follows:

"The receiver sections of amateur MF/HF transceivers generally use diode-switched front-end filtering. The switching diodes used have low junction capacitance and can typically handle medium DC levels (10 to 100mA). These characteristics are important because we want these diodes to contribute minimal loss when turned on and leak very little RF when turned off.

"The two-tone, third-order MD dynamic-range testing routinely done to amateur transceivers seems to point up no weakness in these switching diodes. In real life, however, a huge number of signals simultaneously appear at a transceiver's antenna connector. Periodically, their voltages all sum in phase producing, for short durations, enough voltage to change the bias of the diode at the input of the filter in use. This causes intermodulation distortion - generally ,second-order IMD. This is ironic for two reasons: First, this diode-generated IMD generates exactly the interference the filters switched by the diodes are supposed to prevent! Second, amateur radio equipment reviews have long let second-order front-end IMD go unmeasured because we have long assumed that our radios front-end filtering reduces this IMD to a non problem. Later, I will present measurement results that prove that second-order IMD is a very real problem today. (The test jig used by WOIVJ is shown in **Fig 2** with some of the results in **Table 2** - *G3VA*).

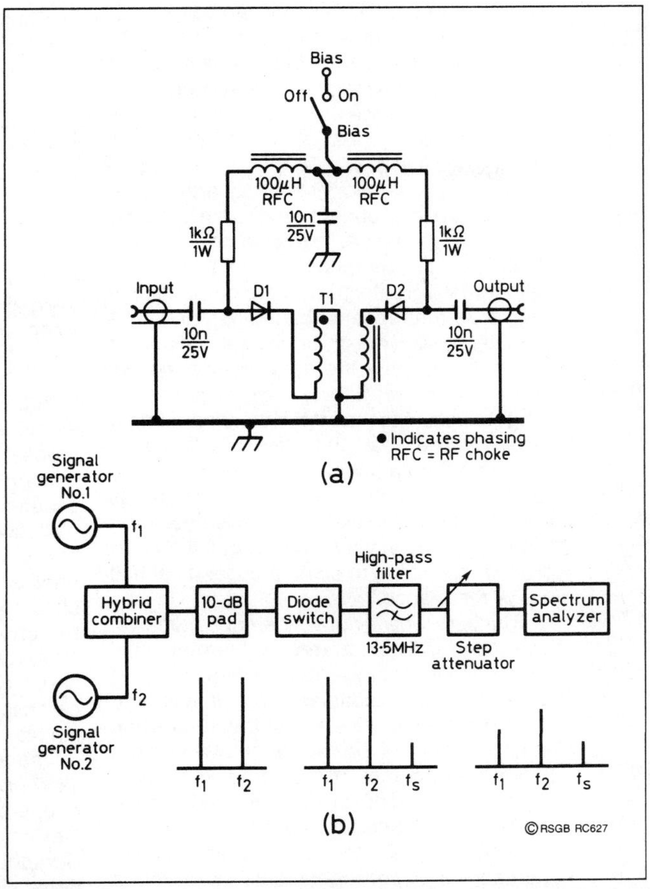

Fig 2: (a) The diode switch used by WOIVJ for his tests. D1 and D2, the diodes under test, included PN and PIN (power-rectifier and RF-specified) types. Capacitors are disc or monolithic ceramics. T1 consists of 11 bifiliar turns of Nr28 enamelled wire on an FT-37-43 ferrite toroidal form; the inductance of each winding is about 50uH. (b) Set up for measuring the diode switch's second - and third-order intercept points.

"The best way to avoid switching-diode IMD is to switch the filters with relays instead of diodes, and military and commercial gear generally take this approach. Relays are costly, however. A less expensive work-around that is acceptably good for amateur radio equipment is to use diodes - PIN diodes - designed for this application. The two best-known US manufacturers of PIN diodes for this type of low-frequency application are Hewlett-Packard and Alphas Industries. The best diode for the shortwave range is the HP 8052-3081 . . ."

Harry Leeming, G3LLL, was also concerned at the idea of using RF PIN diodes. He writes: "It is all very well testing equipment when new, but how well do the modern Schottky (hot-carrier) diodes stand up in service? Take, for example, the FT75. These have a reputation of being noisy on receive. Indeed, on many samples if you switch in the RF preamp, the noise comes up more than the signal. Check the dozen diodes around the input to the band pass filters and the Tx/Rx switching and up to half of them are likely to be found leaky. (They are ISS97 Schottky-barrier diodes).

"Despite their being 'yesterday's technology' I replace these diodes with IN4148 or similar. The receiver then becomes as good as new, and usually stays that way. I wonder if anyone can suggest modest-priced diodes that are better than the 1N4148 and will stand up in service? Meanwhile, I am unable to detect any difference between a new FT757 and one that has had 1N4148s fitted".

My own feeling is that the experiences of both G4HRY and G3LLL highlight an increasingly serious problem involving modern technology. Without the most advanced (and expensive) laboratory test equipment, it is extremely difficult to evaluate fully the performance of equipment. With equipment which is new, or has been in service for some months, it is hard to assess how important these laboratory measurements are likely to prove in normal operational use on the amateur bands. In the case of HF receivers/transceivers, the 'old technology' of variable-capacitor tuned RF filtering with mechanical wavechange switching did have significant advantages over current broadband 'low-pass' or even sub-octave bandpass filtering. However, the 'old technology' was not without its own problems and costs.

L, T & OTHER MATCHING NETWORKS

RECENT ITEMS IN *TT* and elsewhere have drawn attention to the power losses that can be involved in otherwise effective ATU matching networks. It was therefore interesting to read a three-part article, 'An L of a Network' by Graham Thornton, VK3IY, (*Amateur Radio*, March, April and May 1995) and subtitled 'all you wanted to know about the L-Match, but were afraid to ask'. This is introduced as follows:

"The simplest solution to the problem of transforming a complex load impedance into a pure resistance is an L network. It consists of just two components, a variable capacitor and an adjustable inductor. Since load resistance and load reactance are independent variables, it is not possible to have any practical ATU with less than two adjustments. The L-Match (**Figs 3** and **4**) may be seen as the progenitor of all other ATUs. Correctly applied, it can deal with a wide range of impedances: this may come as a surprise, considering its simplicity. It is par-

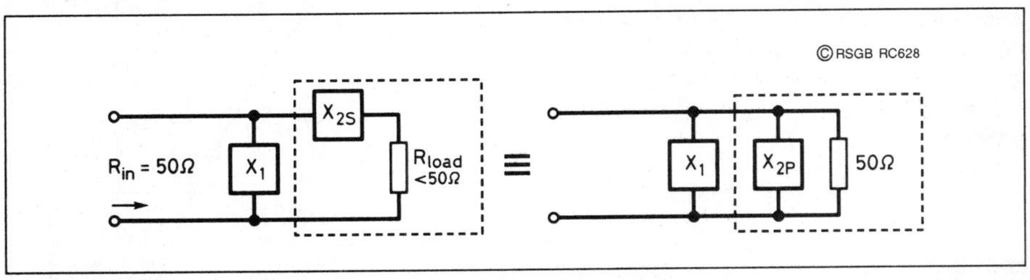

Fig 3: Practical circuit (left) and parallel equivalent circuit for the low-resistance-output case of an L network. Note that, for resonance, X1 and X2P must be conjugate reactors; ie same magnitude but opposite sign. If the concepts implicit in this and the following diagrams can be grasped, VK3IY considers that your understanding of the topic, and ATUs in general, will take a giant leap forward!

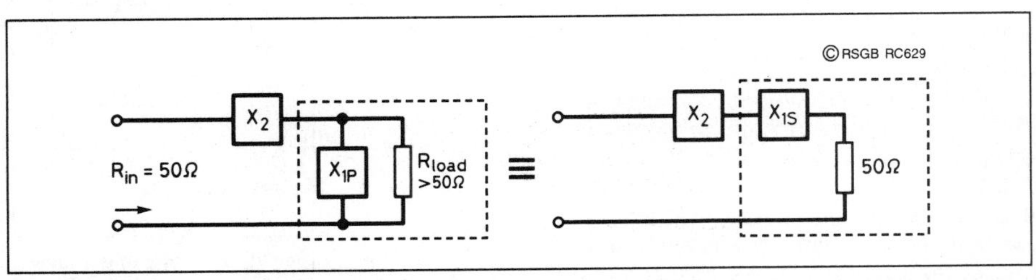

Fig 4: Practical circuit (left) and series equivalent circuit for the high resistance output case. Note that, for resonance, X2 and X1S must be conjugate reactors.

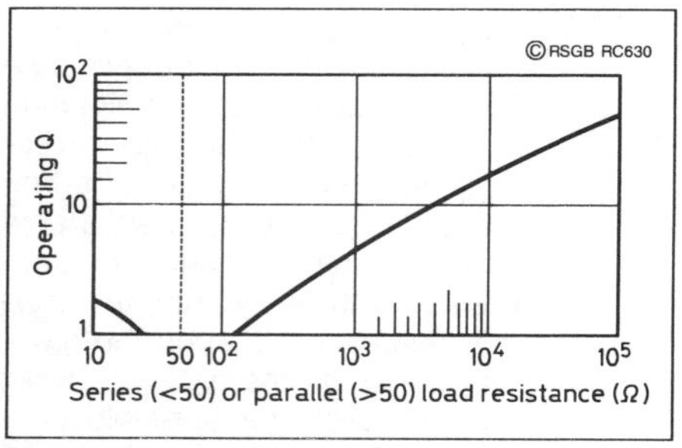

Fig 5: Network systems operating Q.

ticularly useful for unbalanced loads such as random wires, yacht backstays and mobile whips.

"The design, if done from first principles, involves quite a lot of mathematical manipulation which, at best, is tedious and, for some amateurs, utterly forbidding. This article seeks to simplify matters by using a graphical approach. After all, we are designing an adjustable device and high precision is not justified. Ball park answers are all we need. The same approach is used to investigate operating Q, insertion loss and the peak voltage developed across the capacitor. As we shall see, each load problem has at least two solutions; sometimes four . . ."

Part 3 discusses power rating, system operating Q, insertion loss and a general round-up. In view of current interest, the following extract is confined largely to VK31Y's comments on insertion loss. He writes:

"There is a price to pay for everything and some energy is wasted within the network itself. This treatment assumes that the loss is confined to the inductor. We can draw some consolation from the fact that low operating Q implies low losses.

"In general, the insertion loss depends both on the unloaded Q of the inductor and on the operating Q. The greater the former quantity, the lower the loss will be. The converse applies to the latter. If no means of measuring Q is to hand, we need to indulge in some 'guestimation' . . . If the inductor is of considerable diameter, not too long in proportion and substantially air-spaced we can take a stab at a Q of 150. If it's wound on a somewhat lossy former, we can claim a Q of about 50. For cases in between, let's say 100 (these perhaps err on the conservative case).

"The efficiency of the circuit may be expressed as:

$$\eta = \frac{Q_{UNLOADED} - Q_{LOADED}}{Q_{LOADED}}$$

and insertion loss as:

$$10\mathrm{Log}_{10}\eta \ \mathrm{dB}$$

"**Fig 5** shows a plot of insertion loss for the three values of Q. This shows that for most practical situations, the loss (being less than 1dB) is insignificant. It also means (providing the network operating Q is reasonably low: see **Fig 6**) we don't have to be too fussy about the design of the inductor, with respect to efficiency."

In his concluding summary, VK3IY stresses that the L-match can cover a wide range of load impedances with high efficiency and has one of the greatest virtues possible in any equipment - simplicity. Lack of harmonic rejection is the one deficiency. He notes that more complex ATUs, if they can be reduced to equivalent L networks, may be handled by the procedures given for the L-network. Since his treatment assumes some knowledge of the load impedance that is to be matched, he provides a thumb-nail sketch of the values likely to be met in practice:

"For antennas shorter than a quarter-wave, the resistive component varies from about 5Ω to 36Ω at resonance. Its reactance may be as high as 1000Ω capacitive (the thinner the element the greater the reactance), diminishing to zero as resonance is approached. A counterpoise, if mostly horizontal and not too high, would have the same reactance but zero resistance. Simply add its reactance (algebraically) to that of the antenna. Longer antennas have an inductive reactance increasing from a few hundred ohms to some 2000Ω or so for a thin antenna, and perhaps 500Ω or so for a thick one. Resistance over this range may vary from about 40Ω just above resonance to some thousands at half wave resonance."

The item 'The T-network antenna tuner' in the April *TT*, illustrated a 'vernier-type' switched inductor as an alternative to a more expensive roller-coaster inductor (incidentally in Fig 9 the rotor arm of the 12-position switch should of course be shown as connected to contact '10'). Some alternative means of providing an adjustable inductance have been reported by readers.

Jesper Fogh Bang, 0Z1XB, writes: "I have made a number of experiments with antenna tuners over the years. One of my favourites is the Z-match but the network tuner is also very useful. I do however, find the conventionally switched inductor too coarse in its use at certain settings and a roller coaster inductor is rather expensive. During my experiments I have found that a variable capacitor in series with a coil acts smoothly as a roller coaster inductor: **Fig 7**. It may be that losses increase

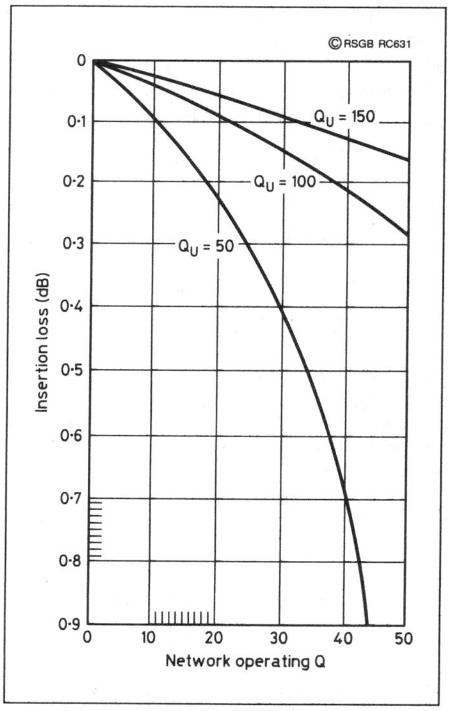

Fig 6: Insertion loss for three values of unloaded, showing that the inductance Q has insignificant effect at low network operating Q but becomes increasingly important at higher operating Q.

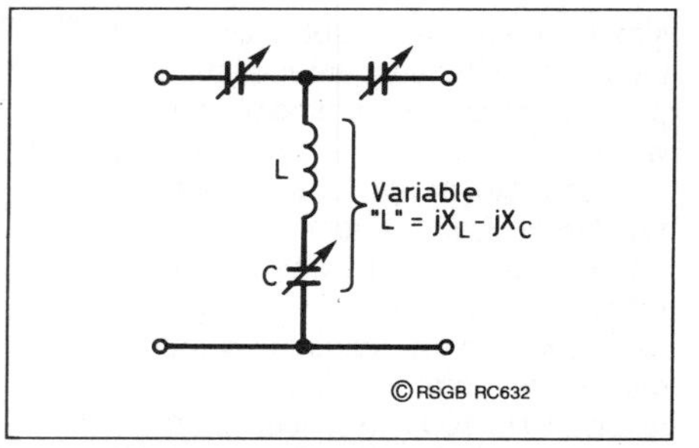

Fig 7: OZ1XB finds that a variable capacitor in series with a coil acts smoothly as a roller coaster inductor.

with this setup, but you sometimes have to sacrifice something for convenience (and money)!"

John Roscoe, G4QK, writes: "The roller-coaster is expensive, and switches are - well, switches. What's wrong with the good old 'variometer'. Twiddling one coil inside another with a range of L1 + L2 to L1 - L2 should be enough and there are no contacts to worry about. They were used in both British and American (and German) rigs in WW2; and I came back from Egypt with a nice ceramic one 'liberated' from a crashed Italian plane.

There were in fact several *TT* items during 1987 on variometers including an illustration of one of the excellent Lorenz units used in German FuG10 aircraft transmitters (*TT Scrapbook*, 1985-89, p202). Undoubtedly these were extremely good devices for antenna matching networks. But not to be found in many junk boxes, and it would probably not be easy to make one as good as the WW2 devices. Having written that, someone may come up with an easy-to-build variometer design!

CUTTING THE COSTS OF AMATEUR RADIO

THOSE OF US WHO entered the hobby of Amateur Radio as schoolboys in the days when pocket money was doled out in pence rather than pounds were able to get on the air by using components salvaged from thrown-out broadcast receivers plus a few additional short-wave components marketed by Eddystone, Raymart, Formo etc. Quartz crystals (about ten shillings [50p] including the frequency calibration certificates required by the Post Office licensing authority), mains transformers and power triodes such as the Taylor T-20 required careful planning of birthday and Christmas presents! But the total cost of entry for amplitude-modulation and CW could be kept to under about £10.

Then came the era of war-surplus equipment from which could be salvaged high-grade components, valves and metal cabinets that must have cost the Government many pounds but often available for a few shillings or at most a few pounds. It was not until the rapid spread of SSB, in the 1950s and early 1960s, that the shortcomings of receivers and transmitters designed for AM/CW began to drive up costs.

But in the late 1960s, HF SSB/CW factory-built receivers and transceivers, many Japanese built but marketed under such names as Sommerkamp and Lafayette, were beginning to appear at relatively attractive prices. For example, in 1967 the 'Lafayette' HA-500 amateur-band receiver with two mechanical filters, product detector and crystal calibrator cost some 42 guineas (for the younger generation a guinea was 21 shillings, ie £1.05p), a Sommerkamp (Yaesu) FT-100 150-watt PEP transceiver £180, the British firm of KW Electronics ((Rowley Shears', G8KW, company) offered the KW2000A complete with PSU for £220, while Heathkit had the single-band HWI2 (3.5MHz) and HW32 (14MHz) 200-watt PEP-input SSB kits for £53.10s (assembled £68).

With a wire antenna slung over a tree-branch, costing shillings rather than pounds, and with 'deferred terms' usually available an effective HF station could he put together without negotiating a second mortgage! The schoolboy could still find plenty of salvaged components for recycling, or acquire, at reasonable cost, the separate transmitters and receivers discarded by others in the rush to transceivers, and then to solid-state with bells and whistles at prices beginning to break through the £1000 barrier.

Costs continued to climb not only because of rapid inflation but also because of the extra cost of building in the extra WARC HF bands (10, 18 and 24MHz), international authorisation of 1.8MHz, frequency synthesizers with memories, digital read-out, automatic ATUs etc etc. While it can be argued, fairly, that in terms of real-money costs, HF transceivers remain excellent value for money, current prices certainly impose a formidable 'cost of entry' barrier for the newcomer to HF. The only alternative is for the operator to be confined to QRP CW operation from simple equipment assembled from the many available kits. But, even as a dedicated CW enthusiast, I have to admit that the attraction for most newcomers is SSB at reasonably high power. Enthusiasm for QRP CW tends to come later.

With the addition of I44MHz to Class B licences (originally introduced in 1964 for 420MHz and above), the majority entered amateur radio via VHF. HF began to be regarded by newcomers as a high-cost no-go area. But with the aid of surface-mount technology, more bands and modes are being crammed into VHF/UHF hand-helds. The result is that entry costs even into VHF-only are no longer within the reach of many would-be amateurs.

In the October issue of *Practical Wireless*, Tex Swann, G1TEX has an article "VHF Operation – It needn't cost an arm and a leg" in which he suggests some ways of getting going above HF cheaply, including the use of second-hand, surplus, home-brew and modification of commercial equipment. He introduces these ideas as: "I've heard it said that amateur radio is becoming less of a hobby for two reasons. The first is that electronics is seen with 'chuck-a-chip-at-it' as the answer to all problems. The second comes partly from the first; users are becoming sated with bells and whistles. This year's model has a tambourine and triangle; but beware the engineers are working on a full orchestra in next year's model! With £2500 to spend anyone can start a hobby. But merely throwing money at an interest doesn't maintain the pastime. A hobby needs to spur us on to other things, not just deeper into the bank manager's profits."

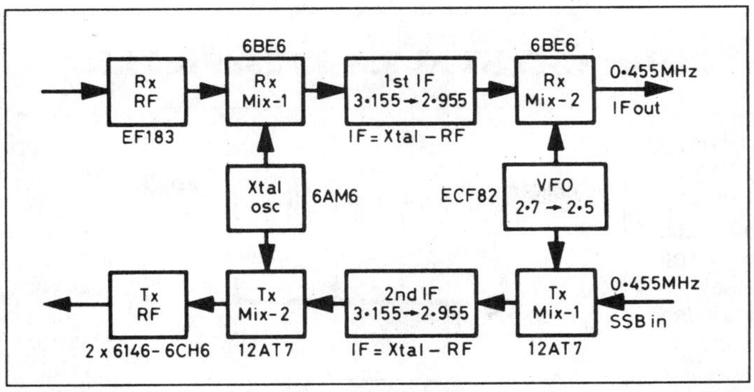

Fig 8: The basic mixing system of the 1960s KW2000 series of HF transceivers that still provide a low-cost entry into 100 watt HF SSB/CW operation.

Stephen Dyke, G3ROZ, has similar beliefs about HF. He writes: "To be a radio amateur you first pass the RAE and (if you like HF) the Morse test; then visit the local 'emporium', come away with a load of expensive kit courtesy of the finance company, install it and start collecting cards for DXCC. Right? Wrong! That hobby is *not* 'amateur radio', it is 'radio DX-ing' and always has been. This is not to say radio DXing is not fun, or that those who concentrate solely on it are somehow 'lesser men' .... they are not .... but it is a different hobby requiring different skills and different equipment."

G3ROZ has put on record a detailed description of a project designed to put a newcomer or Class B licensee firmly on the road to 'amateur radio' not with a QRP rig but with 100 watts SSB/CW on all amateur frequencies (including WARC bands) from 1.8 to 29MHz for a total outlay of under £250 and, if followed careful, results in a transceiver which should be readily disposable for more than it cost if or when the owner later succumbs to the appeal of radio DXing with bells and whistles all-solid-state gear.

G3ROZ writes: "How is it done? In the 1960s, KW Electronics produced the '2000' series of SSB/HF valved transceivers which sold widely in the UK and elsewhere. The 2000, 2000A and 2000B versions all covered eleven 200kHz segments within the (then) six HF bands. This gave all of the 1.8MHz, 3.5MHz, 7MHz and 14MHz bands, plus 400kHz on 21MHz (two segments, 21,000-21,200 and 21,300-21,500kHz) and 600kHz on 28MHz (28,000-28,200, 28,400-28,600 and 28,600-28,800kHz).

"These transceivers are wonderful from an amateur radio point of view. Virtually all of the components were standard and nearly all (apart from the Japanese-made mechanical filter) are still to be found in the RS (Radiospares) and Maplin catalogues, including wavechange switch wafers. The design is basic, parts are accessible and available. They can be aligned or repaired by an amateur with the testgear found in the average shack. They are everything an amateur's transceiver should be .... everything the current offerings from the Japanese firms are not!

As the owner of an unmodified KW2000A, I would not wax quite as lyrical about these transceivers as G3ROZ, but they certainly represent a useful means of getting on HF at relatively low cost. Mine, in nearly new condition cost me £I40 about ten years ago and has proved extremely good value for money, despite one or two annoying features, such as the trouble-prone tuning drive (changed on the KW2000B).

G3ROZ states: "Buy a KW2000A or B (two 6146 output valves) if you want 100 watts of RF; the KW2000 (single 6146) if satisfied with 50 watts. They are frequently offered in Members Ads [see pp 66, 67 *–Ed*] or elsewhere for about £160 in good order, or as low as about £40 in grim as-seen, not-working condition. Don't pay more than £160 even if a few owners who read these notes decide they should ask more .... these rigs are now around 30 years old and will usually need work to get them back to their original specification. Newcomers may value advice from an experienced amateur when contemplating purchase."

G3ROZ's description of how to correct the effects of old age and then to modify the equipment to cover the three WARC bands, to complete coverage of 21MHz, and, if required, all or at least more of the 28MHz band occupies some 20 pages of typescript, ten diagrams and several tables, 33 sheets in all, and clearly cannot be squeezed into *TT*. **Fig 8** shows the basic mixing processes of the KW2000 series.

In brief, to extend frequency coverage, he adds an external crystal-controlled oscillator with eleven switch positions. The main bandswitch on the modified rig covers the 200kHz segments immediately above: 1.8, 3.5, 3.7, 7.0, 10.0, 14.0, 14.2, 18.0, 21.0, 24.8 and 28.0MHz. Then, with this bandswitch turned to the 21.0 or 28.0 positions, the external oscillator gives 21.2, 21.4, 28.2, 28.4, 28.6, 28.8MHz segments (as implemented by G3ROZ) but extendable to 29.0, 29.2, 29.4, 29.6MHz positions for those prepared to purchase the four additional crystals at about £5.50p each. G3ROZ recommends McKnight Fordahl Ltd, from whom he has had excellent service, as a source of crystals.

It is perhaps worth pointing out, that the transceiver can be modified to cover all HF CW-only frequencies without an external oscillator, by fitting just three new crystals and three new anode resonating coils for the existing crystal oscillator in place of phone segments. But be warned that the audio-tone keying system of these transceivers results in a distinctive, although not really unpleasant, note (modifications for carrier keying were described some years ago in *Short-Wave Magazine*).

From Fig 8, it can be seen that the output of the crystal oscillator for any band is the lower segment edge frequency plus 3.155MHz; however the crystal itself is half the output frequency on all bands above and including 14MHz; thus 13.155MHz is required for 10MHz but 10.5775MHz for 18.0MHz and 13.9775MHz for 24.8MHz. G3ROZ (13 Abbey Grove, Sandy, Beds SGI9 IQP) may be prepared to supply photocopies of the 33 A4 sheets of his detailed blow-by-blow description for those prepared to pay the copying costs etc.

The ease and low cost of repair of these old

COHN CRYSTAL LADDER FILTERS

A NUMBER OF MULTI-POLE high-performance SSB and CW crystal ladder filters were described in *TT*, July 1994, pp56-67 and January 1994, pp38-39. These used the Chebyshev approach optimised to provide the best stopband attenuation (steepest skirt response), together with the design coefficients needed to determine the value of the coupling capacitors. There are, however, other possible approaches. These include Butterworth bandpass filters designed to provide optimum flatness at the centre frequency, or the compromise Cohn ('minimum-loss') filter which is optimised to exhibit minimum insertion less when built with practical resonators, while preserving a reasonably good shape factor. Cohn filters, with LC tuned circuits, have been used in the front-ends of a number of receiver projects, including the G3PDM high-performance, hybrid (valve/semiconductor) receiver of the 1970s.

Some years ago, Wes Hayward, W7ZOI, in 'Designing and Building Simple Crystal Filters' (*QST*, July 1987, pp24-29), showed how simple and inexpensive crystal filters that perform well at a fraction of the cost of commercial crystal lattice filters could be implemented using the Cohn approach, without involving capacitor coefficients. All capacitors are of equal value and all crystals have the same resonant frequency.

Although for purists this approach may seem a compromise, one notes that Cohn ladder crystal filters now figure frequently in home-brew receiver projects. The Cohn filters are more symmetrical, making them very attractive for narrow-band CW filters but not ruling out their use as SSB filters. As W7ZOI put it: "The Cohn filter, crystal or otherwise, is a rather simple circuit. This becomes more apparent when we view the filter using coupled-resonator methods. All normalised coupling coefficients are equal. Moreover, the normalised end-section loaded-Q factor is the reciprocal of the coupling coefficient.

The practical simplification becomes apparent if we examine the generalised crystal filter shown in **Fig 9**. All capacitors are of equal value. The shunt capacitors are coupling elements while the series capacitors in the filter end sections are included to properly tune the circuit.

"The 1987 *QST* article included practical designs based on three (**Fig 10**), four and six (**Fig 11**) crystals implemented either by W7ZOI Bruce Williams, WA6IVC ARRL. They used low-cost American NTSC colour-burst 3.579mhz crystals or rather more expensive 4.000Mhz crystals. Terminating impedances for practical CW filters at these frequencies may be as low as 50Ω and 200 - to 500Ω for SSB filters. **Fig 12** shows one way in which a filter can be incorporated in a receiver so that terminations can be arranged to achieve the proper filter shapes.

Michael O'Beirne, G8MOB notes early professional use of a two-crystal (Chebyshev) ladder filter in the Racal RA1772 (a 1972 design) receiver. A 1.4MHz filter is placed at the input to the 1.4MHz amplifier which feeds the product detectors, and used to select the 7th harmonic from the 200kHz square wave spectrum.

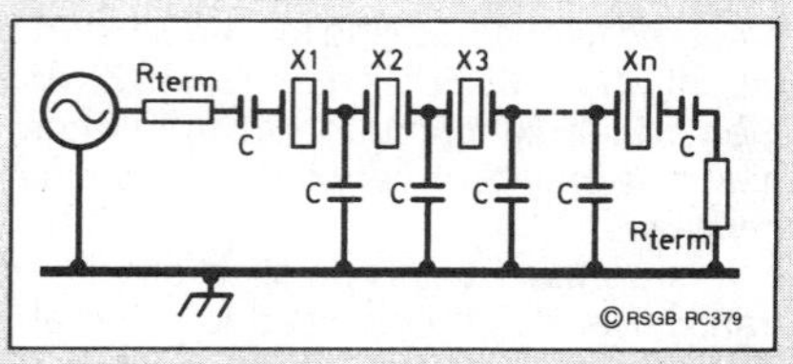

Fig 9: Generalised Cohn crystal ladder filter suitable for empirical construction. Note all capacitors are of equal value. All crystals have the same resonant frequency.

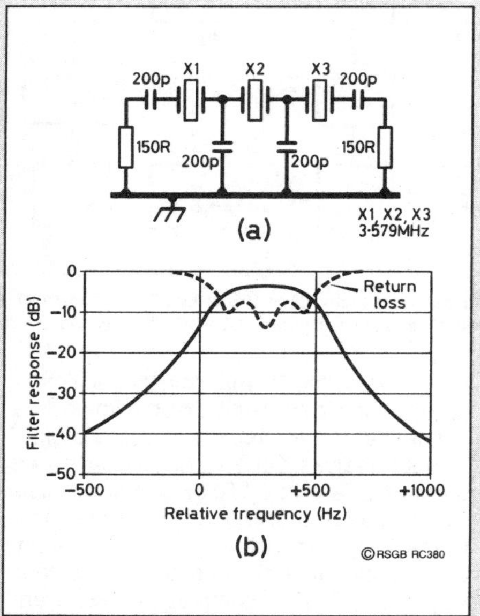

Fig 10. (a) Simple Cohn CW filter using three 3.579 NTSC colour-burst crystals. (b) Frequency response. Dotted input return loss indicates the quality of the impedance match.

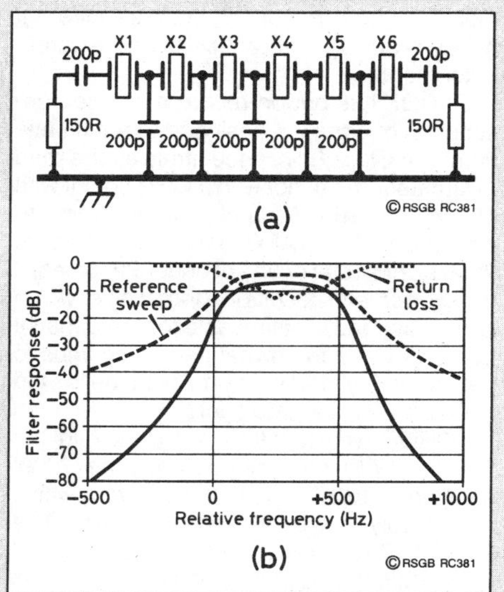

Fig 11: (a) Cohn filter with six 3.579MHz crystals. (b) Frequency response. The reference sweep shows the response of the three-crystal filter and this indicates the improved characteristics at the expense of some extra insertion loss.

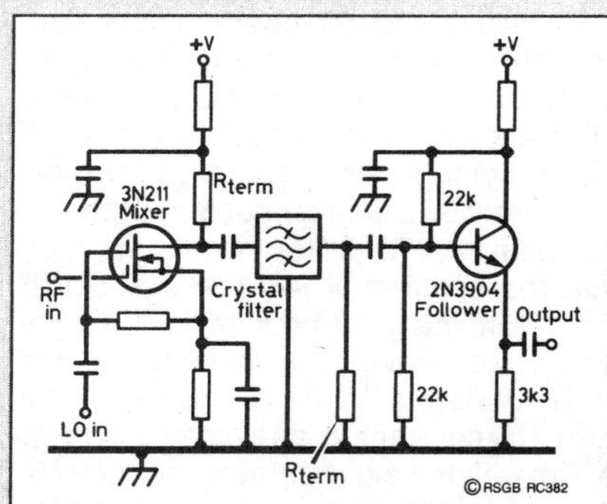

Fig 12: Showing one method of connecting a ladder filter, with correct terminations, in a receiver

valve equipments was brought home to me recently when a fault developed in the receiver section of my KW2000A (I tend to use it as a receiver-only) with no signals but a curious random clicking. With a little patience this was traced to the 6BE6 first mixer having gone 'soft' and ionized, a very unusual fault for a small-signal valve. Although my junk box contains no 6BE6s, these (and all the other valve-types used in the KW2000-series) are still readily available from such firms as Langrex Supplies Ltd of Croydon. A new valve plugged in restored the receiver to full life at a total cost of £1.50 for the valve, plus £1 p&p and 44p VAT. Not a bad repair bill for ten years use! The few other faults have been repaired at no cost with junk-box components. ♦

PAT HAWKER, G3VA
London 37/SE22 8SS

VEHICLE ANTENNAS FOR HF NVIS

THOUGH COVERING all types of communications systems, the recently published book *Naval Shipboard Communications Systems* by John C Kim and Eugen I Muehldorf of TRW Inc (Prentice Hall PTR, 1995) points out that "over the past few years HF radios staged a comeback as an alternative means of shipboard long-haul communications, since HF communications feature portability, versatility of location, and long-distance connectivity. In addition, users became aware of problems associated with satellite communications due to high costs and congested satellite channels. Newly automated and micro-processor-controlled (HF) radios now offer capabilities previously not available for automatic HF link establishment (ALE)".

A standard form of US Navy HF antenna is the 10-metre Fibreglass whip antenna but such antennas have a null in the vertical direction, and for near vertical incidence skywave (NVIS) medium-distance operation a form of this antenna, known as a 'Towel Bar' antenna, comprises a whip bent over to run 14in above a metal deck.

As mentioned previously in *TT*, NVIS HF systems, which minimise the 'dead zone' (skip zone) by maximising radiation vertically upwards, are of increasing interest for all forms of military and some civilian communications systems. A paper 'Novel antenna configurations for vehicle-borne NVIS applications' by K P Murray and Dr Brian Austin (G0GSF), University of Liverpool, was presented last April at the IEE's Antennas and Propagation Conference (ICAP95) in Eindhoven and is published in *IEE Conference Publication No 407* (Vol 1, pp415-418).

This again emphasises that "HF (3-30MHz) propagation still has an important role to play in certain communication scenarios. Of particular importance is the so-called Near Vertical Incidence Skywave (NVIS) mode of propagation where the ionosphere is utilised for relatively short-range communication . . . Of particular importance is the ability to communicate beyond the line of sight (BLOS) but within the normal HF skip-zone associated with the usual vertical whip antenna. The only way to achieve this, without the use of artificial relay facilities, is skywave propagation with a high angle of incidence to the ionosphere. However, this presents particularly demanding constraints on the antenna designer, especially when the terminals are mounted in vehicles.

"Firstly, in order to ensure signal return, frequencies close to the prevailing critical frequency must be employed. The electrical size of the vehicle is generally below the first fundamental resonance. Hence, the input impedance characteristic is generally categorised by a small resistive component with a relatively large reactance. Secondly, the usual vertical whip antenna is inherently unsuitable because of the pronounced null (greater than 20dB) in its radiation pattern towards the zenith."

The paper analyses two types of tilted whips and a vertically mounted (magnetic) loop using the Moment Method (NEC). In the first simulation, a typical 4.8m whip antenna was mounted at the rear corner of the vehicle roof and was tilted forward and backwards along its longest dimension. The best performance was seen with the whip tilted backwards away from the vehicle at a substantial angle. At tilt angles less than 50° from the vertical, performance remains similar to that of the vertical whip. This introduces severe practical constraints in terms of the antenna's robustness. (And for civilian vehicles would hardly be popular with other road users! - G3VA).

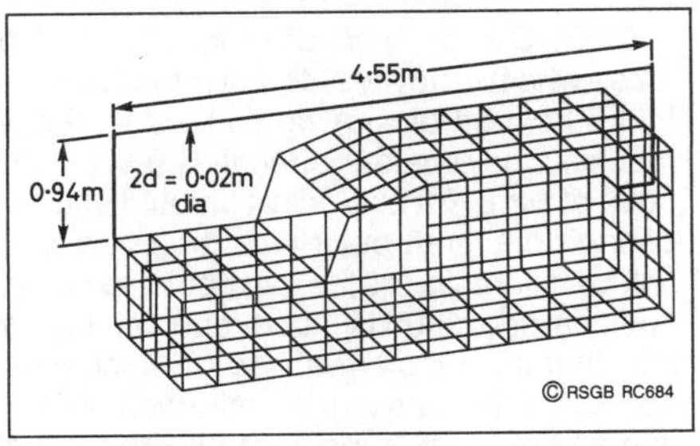

Fig 1: The vehicle wire grid model with loop antenna used for computer simulation by the Moments Method at the University of Liverpool.

The authors conclude that for NVIS systems "a better antenna was found to be the vertically mounted loop (simulated as in **Fig 1**) . . . A variety of configurations is possible using either single and multiple feed-points. On the basis of robustness, the percentage of radiated power at elevations greater than 45° and the power lost due to coupling into lossy ground, the loop is clearly a superior antenna to the whip. A possible disadvantage, though not considered here, is the lower loop input resistance and the associated feed-system efficiency".

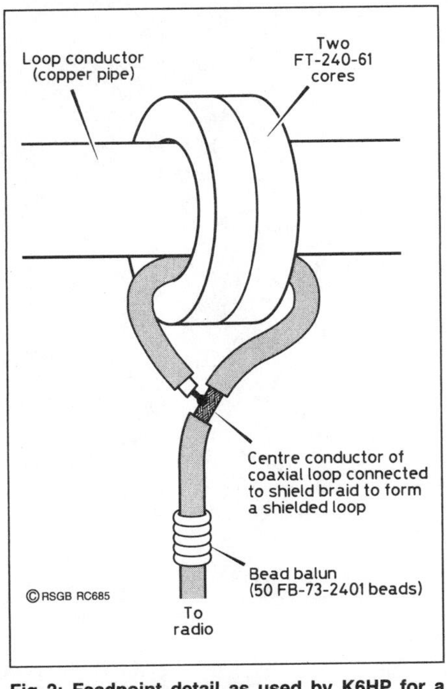

Fig 2: Feedpoint detail as used by K6HP for a mobile (magnetic) loop antenna on the 1.8, 3.5 and 7MHz bands.

This ICAP95 paper is concerned with computer simulation rather than practical hardware and one wonders how the preferred multiple feedpoints could be implemented.

Ken Hirschberg, K6HPX, in the 'Technical Correspondence' section of *QST* (March 1995, pp72-73) reports on his use of a ferrite toroid core as a current transformer to feed a compact transmitting loop (**Fig 2**). He views this as a preferred alternative to coupling by a one-turn link (transformer), a tap, or by a capacitive divider that can also be used to resonate the loop.

He writes: "The central feedpoint of the loop (directly opposite the resonating capacitor) is passed through the toroid, making the loop itself a one-turn secondary of the transformer. In this particular set-up, the coaxial feed line was used as the primary by passing it through the core, then soldering (terminating) its centre conductor to its shield through a small hole cut into the cable jacket (ie a Faraday link - *G3VA*). This forms a shielded link, which serves to reduce the capacitive coupling between the feed line and the loop.

"Based on limited experimentation, I have discovered that (at least at the moderate power level of 100 watts) this means of coupling is physically smaller, does not have to be adjusted for use on different bands, and can be made rigid and weatherproof (eg by encapsulation) for mobile operation. The toroid transformer provided the same apparent coupling as that observed with an ordinary link or balun, each of which encompassed an area about 20% of the size of the main loop's area. For these experiments, the loop I used was an 11 by 4.5ft rectangle made from 1in diameter copper water pipe, with a coaxial capacitor or vacuum variable capacitor; the toroid was made of two Amidon FT-240-61 cores.

"I operated the loop on 1.8, 3.5 and 7MHz bands with an SWR of near 1:1 easily obtained at resonance without the use of an antenna tuner. The antenna - designed for mobile NVIS use - exhibited the expected high-Q properties, and a quite narrow bandwidth. The loop is mounted in the vertical plane, with its feed just above the metal roof of the van. On-the-air results are outstanding, with S9+ reports on 3.5MHz over western North America in the early morning".

KITES AND BALLOONS - THE REGULATIONS

PAT PAINTING, G3OUC, noted the *Product News* item in June's *RadCom* on Allsopp Helikites. He is concerned at the possibility of amateurs using kites, balloons or these new kite/balloons as antenna supports without being aware of the civil aviation legislation governing the flying of large kites or tethered balloons.

He writes: "I have used kites for many years to support antennas for 1.8MHz portable operation with much success. The use of balloons has never been attempted due to the danger of line breakage allowing the wire to drag over power lines etc. With a kite this problem is not so serious since the kite will descend when the line breaks."

G3OUC has sent photocopies of letters he received in March this year from the Airspace Utilisation Section of the National Air Traffic Services (NATS) setting out the current regulations pertaining to both kites and balloons, since he feels that many amateurs who use or may use such means of supporting antennas are unaware of them. Flying of both kites (if more than 2kg in weight) and captive balloons is governed by Air Navigation Order 1989 and the Rules of the Air Regulations 1991 in which both large kites and captive balloons are classified as 'aircraft'.

For example, Article 75(1) of the Order requires that, if kites (of more than 2kg in weight) or captive balloons are to be flown above 60 metres above ground level, within 60 metres of any vessel, vehicle or structure, or within 5km of an aerodrome at any height, then written permission is necessary from the Civil Aviation Authority and if granted its validity will be subject to certain conditions. Again, Article 50 requires that a person shall not recklessly or negligently cause or permit an aircraft to endanger any person or property. Rule 14 describes the need during night-time flying at heights exceeding 60m to display lights. A copy of a CAA "Application for Permission to Fly Above 60 metres or within 5km of an aerodrome" requires completion of some 10 sections and a signed declaration "that I have checked the above information and that to the best of my knowledge it is correct and that I am aware of my obligations under the Air Navigation Order 1989 as the operator of a kite".

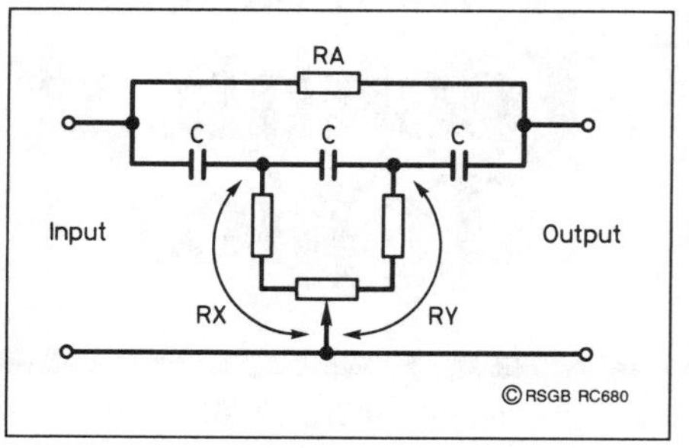

Fig 3: The tuneable notch filter redrawn to indicate how the filter frequency can be calculated.

G3OUC adds: "I am often taken to task by those who tell me my antenna is too short and that 300-400ft of wire would produce better results. One amateur often flies a kite using 500ft of wire. Not only is this amount of wire a danger to aircraft but it puts the operator at risk from lightning as the gap between ground and charged cloud is much reduced. It is my fear that if a problem arises in which injury occurs the 'powers that be' may well see fit to ban all antennas supported by kites or balloons or kite/balloons. Yet, if people are careful, there is little risk when employing this interesting means of radiating a good signal on 1.8 or 3.5MHz".

UNDERSTANDING THE TUNEABLE NOTCH FILTER

JUNE'S *TT* (p71, Fig 7) provided the circuit diagram of a '40-120Hz tuneable notch filter' recalled by Peter Cole, DA1PE/G3JFS, from notes he made some 20 years ago. He confessed that he had no idea of its origins, "but it certainly works", and he appealed for information on the derivation of the circuit and the design equations.

Bob Pearson, G4FHU, has responded. He writes: "I have worked out the required relationships as follows based on the circuit as redrawn in **Fig 3**.

Notch frequency: $f = \dfrac{1}{2\pi C\sqrt{3 \times RX \times RY}}$

Resistance $RA = 6(RX + RY)$

"Substituting C = 100nF, RX = 3000 ohms, RY = 52,500 ohms which correspond to the *TT* circuit when the 5kΩ potentiometer is centre, then f becomes 73.218Hz and RA becomes 333kΩ. To check my algebra I used a PSPICE circuit simulation and obtained the same result.

"The circuit as redrawn in Fig 3 shows that the filter is related to, although not identical with, the arrangement used long ago by Dippy in what, strangely, has ever since been known as the 'phase shift oscillator' (when, in fact, virtually all oscillators use a phase shift network).

"The simplest mathematical procedure for deriving this kind of solution is to note that if the network is to produce a true null at the output, then *at the frequency of the null*, one can solve as if the circuit has a short circuit at the output and find the condition for the cur-

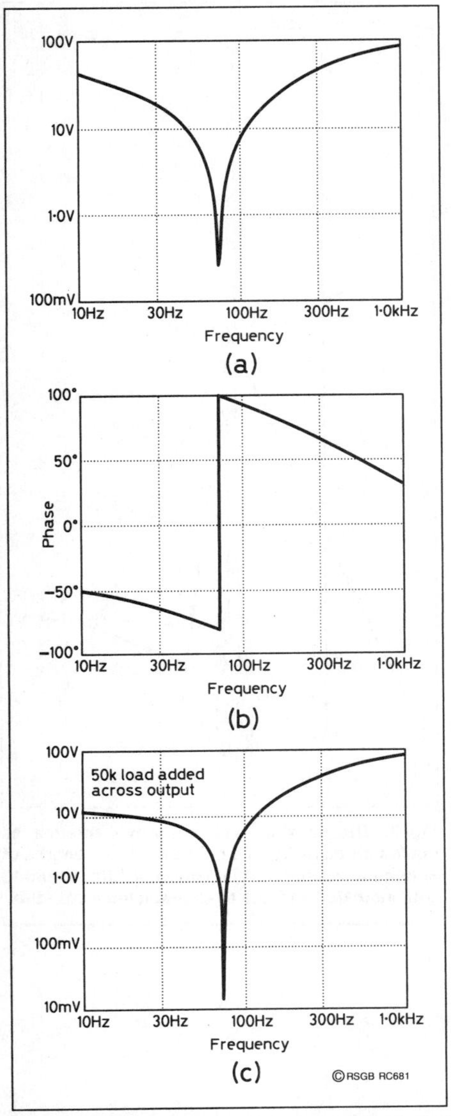

Fig 4: (a) Amplitude response of unterminated filter. (b) Phase shift swings virtually instantaneously at the notch frequency. (c) Amplitude response when filter is loaded by 50kΩ.

rent at the output to be zero. This is simpler than producing a general expression for the output at all frequencies.

Nevertheless, it still involves a certain amount of algebra with complex numbers to be inserted and sorted out at the end. For the circuit shown, the tuning range would be 55.4Hz to 175Hz. (Rather than the 40-120Hz suggested but covering 60Hz as claimed for the given component values - *G3VA*).

"If the output is loaded by a further resistance, it does not change the notch frequency but does reduce the response at frequencies below the notch frequency but at the same time sharpens the notch. **Fig 4 (a)** shows the amplitude response unloaded **Fig 4(b)** shows the phase shift which swings virtually instantaneously at the notch frequency. **Fig 4(c)** shows the amplitude response when loaded by 50kΩ. In each case the simulation input voltage was 100V and the output plotted on logarithmic scales. The output voltage theoretically goes to zero at the notch frequency but the calculating and plotting limitations mask this. In practice, of course, parasitic effects would also make the result less than perfect".

SIMPLE SPEECH PROCESSOR

REG MOORES, G3GZR, writes: "**Fig 5** shows a simple, very cheap but effective, speech processor which I developed many years ago for mobile operation. In my case, it was used with a miniature electret microphone but the circuit can be used in the shack with a dynamic microphone providing excellent quality. However, when used with a dynamic (moving coil) microphone the voltage feed shown in Fig 5 for the electret microphone (components marked ⋆) is not required.

"With relatively few components it can be made small enough to be incorporated into most rigs; if used with a stand microphone it can be mounted in the base. It is based around a 1458 (dual 741 op-amps) IC, with the first op-amp providing the required microphone gain; the second op-amp sets the processed audio signal delivered to the transceiver. The variable 50kΩ resistor sets the microphone gain; the 4.7kΩ potentiometer sets the output audio level.

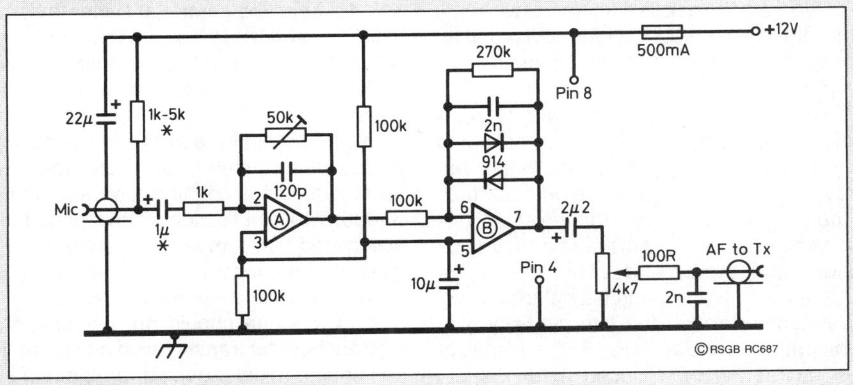

Fig 5: G3GZT's 'KISS' speech processor.

It would presumably be possible to use this filter configuration as a sharply peaked CW filter by placing it in the feedback loop of an op-amp or similar amplifier.

COAXIAL TRAVELLING WAVE ANTENNA?

SEVERAL ARTICLES IN the Australian *Amateur Radio* since 1992 have drawn attention to a form of antenna based on a patent granted in 1927 to a Romanian inventor, Emil Geles. This appears to have interesting features which, if achievable in practice, could offer significant advantages as a broadband 'travelling wave' antenna, for example, as a vertical antenna without the usual significant RF ohmic earth losses. It needs to be used with a suitable antenna tuner or balun providing balanced output from a coaxial cable feedline. However, although successfully exploited by several Australian amateurs, it still calls for a degree of experimentation.

The latest article, by Leo Weller, VK3YX (*Amateur Radio*, June 1995, pp 4-6) acknowledges that his work was initiated by an article written by George Thatcher, VK2EHN, in the September 1993 issue, and after re-reading the earlier note by Clive Cook, VK4CC, which drew attention to the UK patent issued in 1927 to Emil Geles of Bucharest. Although VK4CC did not give the Patent Number, it was possible to trace the full patent specification at the National Science Reference Library in High Holborn, London. It proved to be UK patent No 260,005 (UK Application Date October 19, 1926, accepted March 24, 1927).

In summary, this covered a closed-circuit antenna comprising a central wire surrounded by a tubular conductive (metal) structure (**Fig 6**).

"The central wire may be completely separated from, or connected directly to, the top of the tube; or it may be capacity coupled with the tube as shown in **Fig 6 (b).** The outer tube screens all radiation from the inner wire. The system functions practically as an ordinary open aerial with small ohmic losses. No earth or counterpoise is required."

In more detail, the patent states: "This invention relates to an improved aerial for the transmission of electromagnetic waves such as used in wireless telegraphy or telephony.

"It is stated in various text books on wireless telegraph that the radiation of a closed circuit aerial is nil. This is not strictly true and the proof is that in certain cases use is made of frame aerials for transmitting purposes. On

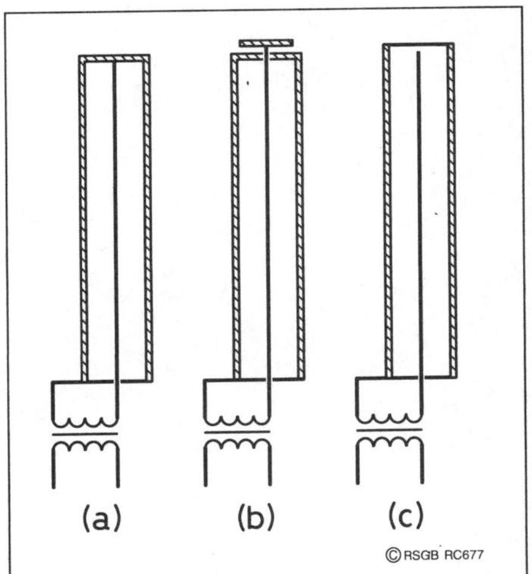

Fig 6: Coaxial closed-circuit antenna principle from UK Patent 260,005 of 1927. (a) Internal wire connected at top to metal tubes as with VK3YX's antenna. (b) Wire coupled capacitive to top of tube. (c) Wire left isolated from tube.

the other hand, it has been shown, and can be verified experimentally, that the radiation of an aerial increases with the surface of the aerial . . . It is also known that the penetration of high frequency currents either from the outside to the inside or vice versa through the walls of a metallic screen is very small.

"Aerials for transmitting and receiving purposes are broadly known in which a vertical wire or inner tube is enclosed in an outer vertical tube from which the said wire or inner tube is insulated and connected to earth or a counter-poise. The invention consists of a transmitting aerial constituted by a vertical tube wherein is enclosed a vertical wire with which the said vertical tube forms a radiating closed circuit without earth connection or counter-poise . . ."

"The field produced by the wire is completely stopped by the tube which acts as a screen. Consequently, it can no longer destroy the field produced by the outer surface of the tube which is free to radiate exactly as an open aerial. The aerial may be excited by connecting the lower ends of the tube and of the inner wire to the transmitting apparatus, for instance through the intermediacy of a transformer as shown in Fig 6. The losses of such an aerial are reduced to the ohmic losses which are exceedingly small and to the losses produced by eddy currents".

It was claimed that the main advantage of such antennas is that no earth connection or counter-poise is employed and that, consequently, neither a large earth area nor a great number of poles which are required for a high power transmitting antenna is required. It is claimed that this would be of special advantage "for use on aeroplanes".

It does appear that a coaxial antenna of this type, together with a black box tuner, was marketed in Australia some years ago (CTW Antenna Coupler), the CTW standing for 'coaxial travelling wave'. If, in fact, this structure does represent a true form of travelling wave antenna (ie an antenna with constant current along the radiating element rather than a resonant standing wave), then clearly this would represent a further valuable feature. Experimental systems by both VK2EHN and VK3YX seem to bear this out, with VK3YX warning that the antenna should not be of a length resonant in any amateur band.

In effect, it appears to be a low impedance closed-circuit antenna of a form rather like an extended transmitting loop antenna but without the complication of the high-Q tuning associated with compact loop antennas. Provided that the losses in the tuner/balun can be kept low, there seems no reason why this type of antenna should not be reasonably efficient both for transmission and reception, and should avoid the major problem of conventional monopoles of the high RF ohmic ground losses and the inconvenience of elevated ground-plane radials.

VK3YX's practical realisation of this form of antenna is outlined in **Fig 7**, with details in

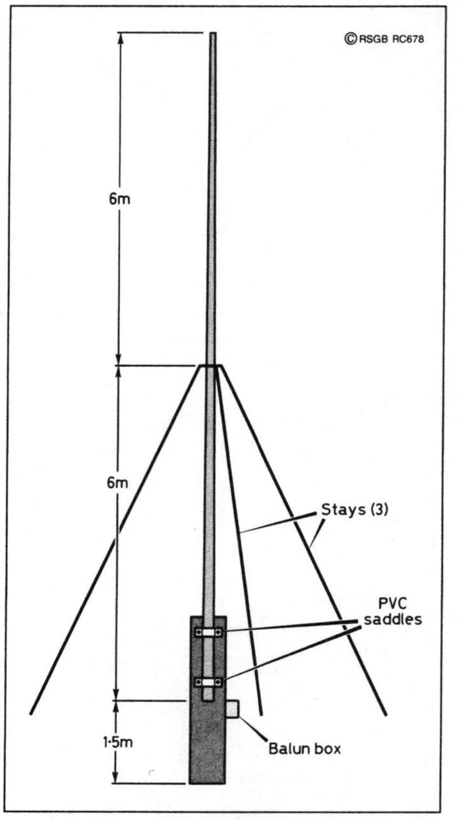

Fig 7: The coaxial travelling wave antenna as implemented by VK3YX based on three lengths of aluminium tubing of total length of 12m, mounted with insulated saddles to an insulated mast stub.

Fig 8 (a), (b) and **(c).** His antenna is constructed from three equal lengths of imperial aluminium tubing of total length 12m. As the three lengths are of different diameters (1.75/1.5/1.25in), its characteristic impedance as a 'coaxial cable' is not constant, but this does not seem important. In the centre of the tubing, kept in place by triangular spacers as in Fig 8(a), is a PVC covered wire of 5mm diameter, or 47/03.

VK3YX emphasises that "with so little information available it is an experimenter's delight. Nevertheless, looking through my notes, more than 80 different baluns have been wound with an endless number of readings. Although this was time consuming the result is most gratifying".

He concludes: "It is unlikely that, in the first experiment, maximum performance can be achieved. Much more work will be needed to arrive at fully understanding this system, and the real design considerations. I would like to collect data from other experimenters by letter, telephone or amateur radio regarding their experience with this project". (His address is 46 Peperell Avenue, Syndal, VIC 3150, Australia).

VK3YX lists a number of observations, stressing that "with an antenna so different from all accepted theories and practices, one does not know what to expect and every observation is a discovery for which one tries to find an explanation: (1) I noticed no reverse TV interference, which has plagued me for years. (2) No TVI and I could remove all suppressers, braid-breakers and the like from my own TV set. (3) No earth wire or radials . . . every attempt to use them reduced receive sensitivity. (4) Easy and smooth tuning with a Z match antenna tuner . . . SWR of 1.1 right across the band, even on 21MHz. (5) Good signal to noise ratio which compared favourably with a 3.5MHz dipole with open feeders. (6) Sensitivity on receive and field strength on transmit increases step by step on each higher frequency amateur band. (7) No high voltage spots. Even at 50 watts, a sensitive neon bulb did not glow. Touching the antenna with the hand pushes the SWR meter off the scale. (8) High current. I measured 0.5A at 5 watts on all bands, except the 7MHz band, including 1.8MHz and all WARC bands. (9) The 7MHz band needs more power to achieve 0.5A and tuning is different. This does not affect performance. (10) Three turns of heavy, insulated wire around the tubing and short circuited reduced receive sensitivity. This is the reason for the PVC saddles and clamps . . ."

It seems important to note that the RF feed to this form of 'closed-circuit' antenna must be balanced. There is also an interesting possibility that, at least for low power, it might be possible simply to use a length of conventional coaxial cable - short circuited at the far end - rather than the home made variety with a wire enclosed within a metal tube used by the Australian experimenters. I tried this on receive, using the ATUs that I use for my large looped wire antenna mentioned in June's *TT* and certainly it seemed to receive signals quite well. It seems as though the 1927 idea is worth reviving and developing further, whether or not it forms a genuine 'travelling wave' antenna. It could presumably be used as a horizontal or sloping antenna.

It should be stressed that this form of

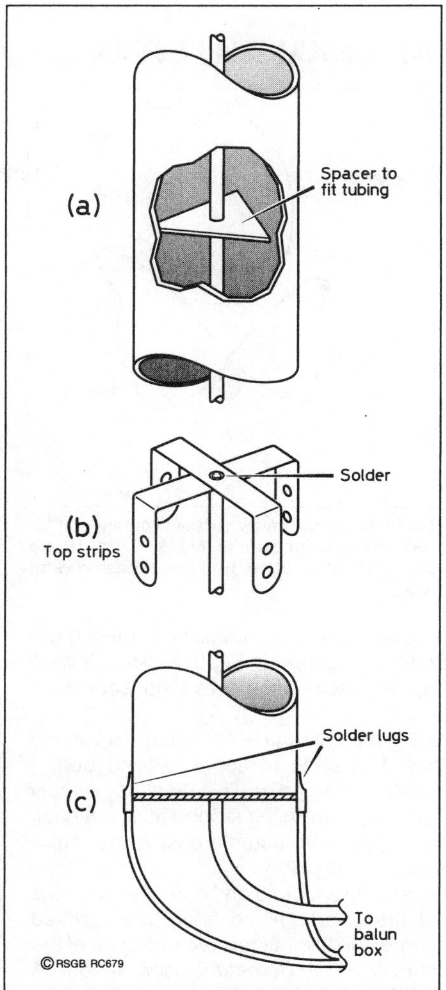

Fig 8: Constructional details of VK3YX's antenna. (a) Triangular perspex spacers 2mm thick and spaced 30cm apart are used to keep the PVC covered wire centred in the tubes, held in place by suitable plastic to plastic glue. (b) The centre wire is connected to the outside on the top of the tubing, via two crossed copper strips. After soldering, the strips are bent down over the outside of the tubing and secured with 8 self-tapping screws. (c) On the base of the antenna, on the outside, are two solder lugs 180° apart. These are fastened with 6mm copper screws, washers and nuts. Note that half way up the mast is a PVC clamp holding three 6mm polyester stays.

untuned closed-circuit (loop) antenna requires a low RF ohmic resistance approach akin to that of a compact (magnetic) tuned transmitting loop. Otherwise, there is the likelihood of low transmit efficiency due to RF power losses in both the antenna structure and the tuner or balun transformer needed to provide a low-impedance balanced-output. It must be noted that there will be high RF currents flowing even at relatively low power. As with a conventional frame or loop antenna, the very low radiation resistance may be of minor consequence on receive but could greatly reduce efficiency on transmit. This is presented as an interesting experimental project rather than a fully-proven design. Good performance on transmission cannot be guaranteed!

DAYTIME SKYWAVE PROPAGATION ON 1.8MHZ

P HOBSON, G4HOJ, reported in *TT* (August 1994, p48-49) the results of investigations he had made into daylight propagation of 1.8MHz signals. He showed that on this, the only amateur MF band, "contrary to simplified theory, skywave propagation is not limited to night-time".

For example, he found that over a 126-mile path during a September day with "daylight" existing for some 12 - 13 hours, only 4 - 5 hours showed low levels of skywave enhancement. Even then signals still peaked at regular intervals (*TT* August, 1994 Fig 11). In general, he noted that from a quiet site, skywave enhanced signals are useful on almost every day with, occasionally, high signal levels even at noon. He argued that his results proved, to his satisfaction, that 1.8MHz completely out-performs the alternative bands, particularly for mobile working, when short/medium distance working is the objective.

In an introductory note, I pointed out that in the 1970s when I was involved with answering listeners' complaints about interference with the reception of the early Independent Local Radio medium-wave stations, it became obvious that the D-layer which absorbs MF sky-waves from distant high-power stations during 'daylight' does not reform immediately at dawn and (particularly in the winter months) begins to lose its absorptive characteristics some time before dusk, so that interference from distant stations could be quite severe during much of a 'winter day'.

It was therefore interesting to come across a paper 'LF/MF skywave propagation at daytime' by John C H Wang of the American Federal Communications Commission (*IEEE Transactions on Broadcasting*, March 1995) replying to an ITU-R question on this topic which is of current concern to broadcasting service planners etc. The FCC writer notes that interference from the daytime skywave signal of one broadcasting station to the groundwave signal of a co-channel station located several hundred kilometres away has been observed in certain parts of the world. He stated: "It is extremely difficult to collect skywave data during the day. First of all, signals are very weak. Secondly, groundwave signals may be strong enough to mask skywave reception. Furthermore, co-channel interference from stations near the monitoring site may also be a problem . . . Night time data has been studied extensively and the literature is fairly rich . . . Daytime data, however, has not been studied systematically".

The paper presents daytime measurements of more than 30 MF/LF paths both in the USA and in Europe. It discusses (1) the effects of seasonal variation; (2) effects of latitude; and (3) effects of solar activity.

Among the points made are that daytime field strengths display a consistent seasonal variation pattern with maximum occurring in winter months and minimum in summer months, both on MF and LF. The winter-to-summer ratio is typically 10 to 20dB.

Daytime field strengths display much less dependence on latitude than during the night when low-latitude paths may be some 11dB stronger than high-latitude paths. As might be expected from amateur radio experience, skywave field strengths tend to be maximum in the year immediately after a sunspot minimum in both night time and daytime.

NEW FORM OF HELICAL-BEAM ANTENNA

DR JOHN D KRAUS, W8JK, has long been recognised by amateurs and professionals as one of the world's leading antenna specialists. He is highly regarded both for the many designs he has pioneered - including the 8JK which in 1936 was the first array to make use of close-spaced driven elements; the first folded dipoles; the first corner reflector; the first axial-mode helical antennas - and for his classic textbook *Antennas*, first published in 1950.

In *IEEE Antennas & Propagation* Magazine (April 1995, p45) he contributes a short article, 'A Helical-Beam Antenna Without a Ground Plane', in which he writes: "The helical-beam antenna (also known as axial-mode helix), which I devised in 1946, was fed via a ground plane. This arrangement has proven very satisfactory for many applications. Now, almost 50 years later, I present a version in which the ground plane is replaced by one or more loops. This design is well suited for pole mounting, as shown in **Fig 9**. It has less wind resistance, and presents a much-cleaner appearance, while providing equivalent performance to a helix with ground plane.

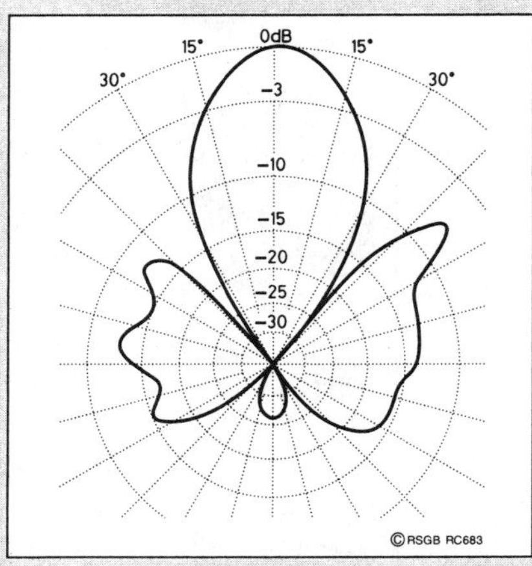

Fig 10: Radiation pattern of the experimental 10-turn antenna shown in Fig 9 at its centre frequency of 1680MHz. Half-power beamdwidth is 34°.

"The inner conductor of the coaxial line feeding the antenna is connected to the end of the helix, and the outer conductor to the base of an adjacent loop. A second (buffer) loop may be situated 1/3 to 1/2 wavelength from the feed point. This loop may or may not be continuous. A third loop, one wavelength or less from the feedpoint, is optional. The helix and all loops are approximately one wavelength in circumference at the centre frequency of operation. A typical pattern is shown in **Fig 10**".

Dr Kraus stresses that the new version retains the broadband and non-critical nature of the ground-plane helix to a marked degree. In the 'Microwaves' chapter of the current *Radio Communication Handbook* (p9.57), it is noted, in describing an axial-helix antenna (with ground-plane) for 1.3GHz, that such antennas "are useful both as test antennas and also as a general-purpose, medium-gain antennas of wide bandwidth. An eight-turn antenna will have about 10dB gain over an approximately 2:1 frequency range".

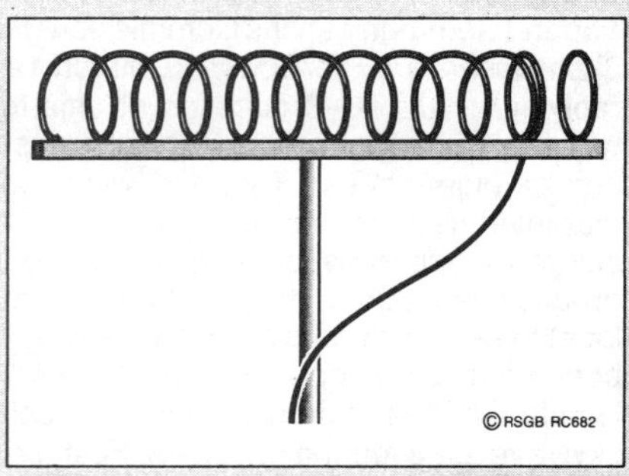

Fig 9: A ten-turn pole-mounted helical-beam antenna, using loops instead of the conventional ground plane. The turn spacing is 0.27-wavelength, and the gain is about 15dBi.

The new form of 10-turn, pole-mounted antenna with a centre frequency of 1680MHz and a turn spacing of 0.27-wavelength, as described by Dr Kraus, is stated to have a gain of about 15dBi.

The American MF broadcasting band has for some years extended to 1705kHz so that results in this region of the broadcast band are close to what can be expected in the 1.8MHz band. It would appear from measurements made in Germany and Japan that the annual median value of skywave field strength at noon is about 45dB lower than the corresponding value six hours after sunset (SS+6). The difference could be slightly more than this at 1.8MHz although, for the entire band 535-1705kHz, John Wang suggests that a figure of 42.5dB would be reasonable.

However, he notes that short-term fluctuations of daytime field strengths at the higher frequencies are usually much larger than those at lower frequencies. "Consequently, the difference between the upper-decile and the median values vary with frequency. At 1500kHz, for example, the upper-decile value is typically 10 to 15dB greater than the median value . . . On 1% of the days of the year, field strengths are expected to be 20dB higher than the median value."

It is interesting that this paper complements rather than detracts from the findings reported last year by G4HOJ and should give a further stimulus to the use of 1.8MHz in daytime as we approach and pass the coming sunspot minimum.

HERE & THERE

TT (*RADCOM*, April 1995, pp61-62) and *Product News* (*RadCom*, June 1995, p47) drew attention to the range of high power RF valves, including popular 3CX, 4CX and 5CX types now being manufactured in Russia and marketed in the USA by Svetlana Electron Devices Inc. David Porter, G4UYX, who works at the BBC Woofferton HF broadcasting site points out that the exclusive UK distributor of Svetlana products is Force Associates, 3 Twyford Business Units, Station Road, Twyford, Berkshire RG10 9TU (tel 01734 320711). The range includes some 40 types, including very high power broadcast types, and also several sockets including the SK2A/4CX250B socket.

A press release from Ampsys Electronics Ltd (Ampsys Laboratories, University of Paisley, High Street, Paisley, tel 0141 848 3444) announces the introduction of the company's 'FM 201 Demodulator' based on an 'Amplitude-Locked Loop' added to a normal FM demodulator for use in PMR, Cellular and Radiopaging FM systems (and presumably applicable to amateur FM transceivers). This apparently enables two stations of equal strength, and using the same channel, to be received and understood "without the head-on crash effect and mutual destruction" as in conventional practice but heard simply as cross-talk similar to a hard-wired telephone link. No further technical details were provided.

Arguments about the possibility of a cancer risk from low-level electromagnetic fields show few signs of fading away. A letter from Jeffrey D Saffer and Sarah J Thurston of the American Pacific Northwest Laboratory, in the 'Scientific Correspondence' section of *Nature* (May 4, 1995, p22), notes that the issue of health hazards associated with weak, extremely low-frequency electromagnetic fields as produced by power lines and household appliances continues to be debated.

They state: "Epidemiological studies . . . have not clarified the issue. Thus, researchers have looked to laboratory studies on EMF effects to determine whether there is biological plausibility for increased cancer risk. Unfortunately, those laboratory studies themselves remain equivocal.

"Kobayashi *et al* have now raised an important concern, suggesting that magnetite particles may be responsible for the apparent biological effects of weak EMF fields. Investigators reporting these effects have not controlled for such particles.

"Nevertheless, we do not consider it necessary to invoke this alternative explanation for much of the bioelectromagnetics literature. Before considering the role of potential confounders to explain the effects of EMFs on biological systems, it is of foremost importance to determine whether there is indeed an effect to explain - that is do the data reported support the conclusions drawn by the investigators?

"Almost all the reported biological effects of EMFs have been at the limit of resolution for the assays used. Given this situation, one would expect that every useful control would be included. But, on the contrary, even the most highly cited reports of EMF biological effects . . . lack essential controls"

Later the writers conclude: "Even though our data question the entire notion that changes in gene expression can be induced by magnetic fields, the results do not necessarily imply that field effects on gene expression or other cellular processes would not occur under different conditions. They do, however, demonstrate the importance of essential controls in this research. Critical evaluation of much of the bioelectromagnetics literature uncovers similar shortcomings . . ." ♦

Pat Hawker's Technical Topics

PAT HAWKER, G3VA
London 37/SE22 8SS

SOLAR FLARES - PREDICTING EFFECTS?

IT APPEARS THAT an improved model has been created into the way that solar flares interact with the Earth's magnetic field and affect the ionosphere. The model stems from Ashok Kumar and David Rust of Johns Hopkins University, Baltimore, USA, as reported by Kurt Kleiner in the *New Scientist* (17 June 1995, p21, 'A twist in the tale of magnetic storms').

It is suggested that Kumar is able to make accurate predictions about the temperature, shape, strength and orientation of the storms originating on the surface of the Sun, where filaments of twisted magnetic fields are generated. But although most of these filaments, carrying trapped hydrogen gas, spread out into space, as shown in **Fig 1**, only a few come close enough to Earth to interact with its magnetic field, affecting radio communication and sometims even damaging electrical power transformers and space satellites.

Because of the frequent 'disturbances' to HF propagation, it seems worth reviewing briefly some of the fundamentals of solar flares as outlined in a recent survey paper 'The effect of solar and geomagnetic activity on ionospheric propagation' by Jed A Sutton (University of Maryland) resulting from work while assigned in 1994 to Voice of America (*IEEE Transactions on Broadcasting*), March 1995, pp28-33).

This points out that the solar flare is the phenomenon which causes the most direct disturbances in the ionosphere, representing an explosive release of energy and particles within a relatively small region of the solar atmosphere. Radio communication may be affected immediately after the flare or this may not occur until one or two days after the onset of the flare: see **Fig 2**.

By convention, solar flares are divided into three classes: C, M and X depending on the amount of X-ray energy flux associated with it. A C class flare is the least poweful and does not immediately affect the ionosphere, although the particles from it may disturb the ionosphere several hours later. The flux of M class or X class (the most powerful) flares is sufficient to disturb the ionosphere immediately following a flare as well as producing delayed effects from solar particle radiation.

Electromagnetic radiation from an active flare - ultraviolet, X-ray, visible light and radio noise - all reach the Earth's ionosphere with the same delay of 8.3 minutes so that disturbances to the ionosphere from an X flare may begin at the same time as the flare is observed visual. This may result in ionospheric disturbances to HF and noise bursts on VHF and UHF. Another 'instant' effect of a major solar flare is the 'sudden ionospheric disturbance (SID)' also known as a 'short wave fade (SWF)', resulting from a large increase in the absorption of the D-layer. An SID may block out virtually all sky-wave signals over a large part of the HF spectrum and produce a severe black out of HF signals (particularly on the lower frequencies). Since SIDs and SWFs are caused by intense bursts of X-rays, they occur only on the daylight side of the globe. During an intense SID, an operator may span through many MegaHertz without hearing a signal!

Particle radiation, made up mostly of protons, causes the ionosphere, and hence HF signals, to weaken or disappear entirely on some paths and may result in multiple delayed effects including polar cap absorption (PCA), magnetic storms, visible auroras, and ionospheric storms. PCA results from an increase in enhanced ionisation of the D-region. Magnetic storms cause a fluctuation in the Earth's geomagnetic field which in turn causes ionospheric storms which limit ionospheric propagation. Magnetic storms may also result in auroras which may or may not be visible as far south as the UK but provide changes in the conductivity of the air and result in the reflection of radio signals up into the UHF region.

The energetic stream of charged particles, mainly electronics and protons, are carried through the solar wind towards Earth, increasing the velocity and composition of the solar wind. These take one or two days to reach the Earth's ionosphere where they may cause similar effects to the faster-moving high energy particles noted above. It may take several days (and nights) for the ionospheric layers to return to normal.

Ionospheric storms cause the lowest usable frequency (LUF) to rise and the maximum usable frequency (MUF) to fall, narrowing the spread of frequencies on which communication may be established. A band which may be wide open on undisturbed days may be devoid of sky-wave signals during disturbed days, or received only very weakly. Thus, from one day to the next, the MUF may vary by some 15% regardless of the mean sunspot level. A primary means of defining the disturbed days is the A-index (**Table 1**) which is related to the commonly used K-index as shown in **Table 2**.

As noted by Jacobs and Cohen: "With few exceptions, the higher the value of solar flux and the lower the level of magnetic activity, the better will be ionospheric propagation conditions on the HF bands."

The variation of conditions from day to day shows a strong correlation with solar and geomagnetic activity with the MUF dropping significantly on disturbed days. Unfortunately, it has always been difficult to tell how strong the magnetic storm will be and how it will be orientated in relation to the Earth's magnetic field - information critical to predicting how severe the effects will be.

To return to the recent work at Johns Hopkins University,

Level	A-index	Potential for impact
Quiet	0 - 7	Low
Unsettled	8 - 15	Low
Active	16-29	Moderate
Minor storm	30-49	Moderate
Major storm	50-99	High
Severe storm	>100	Very high

Table 1: Levels of Geomagnetic Activity

k-index	0	1	2	3	4	5	6	7	8	9
A-index	0	3	7	15	27	48	80	140	240	400

Table 2: Conversion from k-index to the A-index.

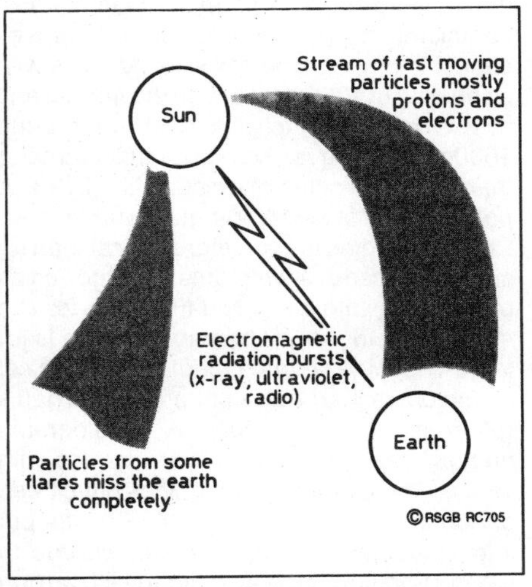

Fig 1: Solar flares emit streams of fast moving particles, mostly protons and electrons that later impact on the Earth's magnetosphere and result in geomagnetic storms, but the emissions from the flares may miss the Earth completely so that they produce no delayed effects on the ionosphere. The electromagnetic radiation from flares reach the Earth's ionosphere with the speed of light and if severe can cause the complete black out of HF signals

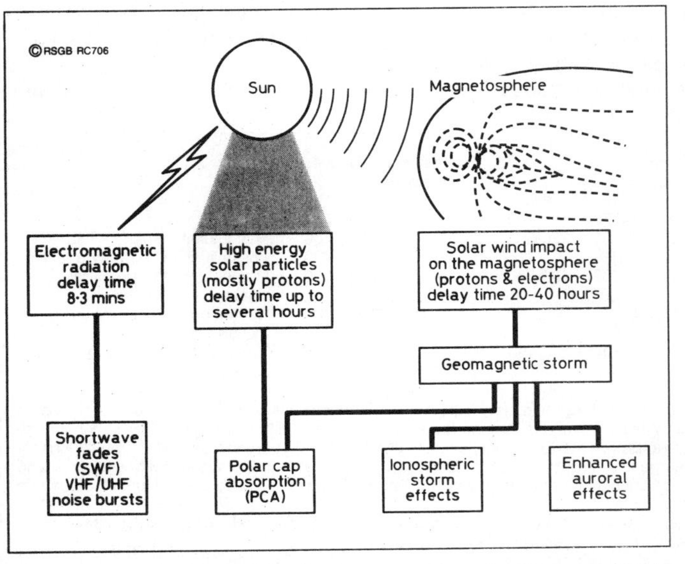

Fig 2: Effects on the ionosphere that are caused by solar flares and related activity on the Sun.

the new model tackles the problem by assuming that the magnetic storms conserve their original 'twist' in space. When filaments originate on the surface of the Sun they form helices; those formed on the Sun's northern hemisphere twist to the right. By assuming that the direction of twist stays the same as the storm travels through space, Kumar is able to make more accurate predictions. It is claimed that such predictions could be used to warn when to take steps to protect satellites in space and power transformers on the ground. They could also help warn astronauts working in deep space, outside the protection of the Earth's magnetosphere plus, of course, better information on likely affects on radio communication.

During May, June and July, 1995, there was plenty of confirmation of the theory that Sporadic E occurs most frequently and most pronounced during the mid-months of low sunspot years. While such conditions are welcome to VHF operators on 50, 70 and occasionally 144MHz, it seems seldom mentioned that the effects are also pronounced on the bands from 14 to 28MHz, bringing in strong European and inter-G/GM/GW signals on these bands even in sunspot minimum years, but also tending to blanket out completely or severely weakening the DX signals that would normally be expected on 14 and 18MHz. Sporadic E is thus a mixed blessing for HF operators.

SEEKING BETTER BATTERIES

JUNE'S *TT*, p70, reported how the search for improved batteries has encouraged the current development of new types of lithium rechargeable 'rocking chair' polymer cells which could be in production later this year. However, this is only one of several recent developments in better batteries, some of which - such as nickel-metal hydride rechargeable batteries - are already available and gaining an increasing share of the market formerly dominated by nickel cadmium (NiCad) batteries. At the same time, new developments in one of the oldest forms of cell - the zinc-air cell - promise to minimise the problem that once these cells are put into use there is a rapid rate of self-discharge.

Two useful survey articles in the American publications of the IEEE discuss the widening range of rechargeable cells available for use in such applications as hand-held transceivers and consumer appliances. One of the pressing reasons for this spurt of research interest in a mature and 'unfashionable' branch of electrochemical products is the laptop computer which often imposes a load of some 10W on the battery, with the result that a typical NiCad battery provides only about two hours of operational use without recharging (take advertising claims of about double this figure with a pinch of salt).

Michael J Riezen-man, a senior editor of *IEEE Spectrum* (May 1995, pp51-56) writes of "the Search for Better Batteries" stressing that "to handle small, power-hungry electronic systems, researchers are exploring air and lithium electrodes and solid electrolytes". An invited paper 'Batteries for Low Power Electronics' by Robert A Powers appears in *Proceedings of the IEEE* (April 1995, pp687-693). Both provide information on the new generations of batteries at a thoroughly practical level, and the following notes are based on them. See **Table 3**.

It is made clear that the small rechargeable battery market is experiencing a near 20% growth rate fuelled by the explosion in cellular phones, portable computers, camcorders and entertainment devices all of which require more power than can be provided economically by primary cells. The most rapid growth is for nickel-metal hydride and lithium-ion types, with a promising future for lithium-polymer batters when these reach the market. However, there have also been useful developments in most types.

Several new constructions have appeared for small sealed lead acid cells. Small cells about the size and shape of a pack of chewing gum in 1-2Ah capacities are being marketed. These offer about 200 cycles of life at discharge times as short as an hour (1C), with larger sizes being offered to power notebook computers. They offer the attraction of relatively low initial cost, low self-discharge, and can cope with relatively heavy loads. It is stated that small cylindrical sealed lead-acid cells are under development which can be charged and discharged in as little as 10 minutes.

The most common type of small rechargeable battery for electronics remains the NiCad battery, with over 1000-million produced worldwide in 1993. Major improvements have been made recently in energy density resulting from both better electrodes and better packaging. A small 'AA' NiCad cell can now have a capacity of 800-850mAh, compared to 500mAh only a few years ago and some have passed the 1000mAh mark. NiCad cells are capable of being charged and discharged in 10 minutes or less (ie a rate of 10C rather than the usual 1/10th C) and remain attractive for power tools. A prime objection to nicad cells is the bad environmental reputation of cadmium with the result that some manufacturers have initiated the collection and recycling of expended nicad batteries.

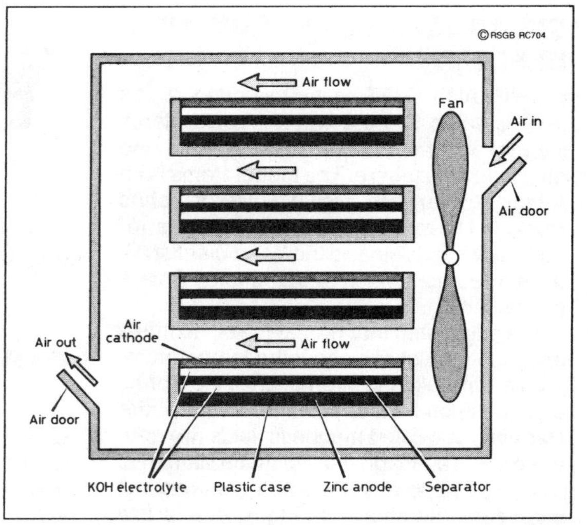

Fig 3: Air-management schemes are regarded as critical for the successful operation of modern zinc-air batteries. When no current is being drawn, the air doors seal the battery from the ambient air and so greatly reduce the self-discharge rate - the old bugbear of this type of battery.

To eliminate cadmium, the nickel-metal hydride cell was first marketed around 1991 and now accounts for more than 10% of the nicad market. Metal hydride is the result of incorporating hydrogen ions and electrons or hydrogen into a metal or alloy. In addition to overcoming the environmental problem of cadmium, the hydride electrode increases capacity for the same cell size. AA cells with a capacity of 1200mAh are on the market and this is expected to rise soon to around 1600mAh. However, self-discharge is greater than for nicad cells; rate capability (2C) and performance at low temperatures are poorer. Proper charging is even more critical than for standard nicad cells with 'smart charger' chips built directly into some Ni-MH batteries. The main thrust for Ni-MH battery research is for large-capacity batteries for electric vehicles.

Lithium rechargeable batteries have had a rather chequered history. A considerable number based on solid lithium anodes with various liquid organic electrolytes have been announced but only one (Li-MoS2) was put into production and later withdrawn due to safety problems. However, for the lithium ion battery, the solid lithium anode is replaced by a carbon material which stores electrons and lithium ions on charge and gives them up on discharge. Similarly, the cathode consists of an oxide which can store electrons and lithium ions on discharge and give them up on charge. Thus there is no lithium metal. Up to 1200 recharging cycles can be achieved with none

Cell type	Nominal Voltage (v)	Energy density (Wh/L)	Power density (W/L)	Self discharge (% per month)	Cycle life	Comments
Lead-acid	2.0	70	<400	4-8	250-500	Lowest cost
Nickel-cadmium	1.2	60-100	220-360	10-20	300-700	May exhibit memory effect
Nickel-metal hydride	1.2	220	475	30	300-600	Possible slight memory effect
Lithium-ion	3.6	260	400-500	5-10	500-1000	Intrinsically safe; no metallic lithium
Lithium-polymer	3.0	150-350	>350	<1	200-1000	Not yet on market contains metallic lithium
Zinc-air	1.2	204	190	<5	<200	Requires air manager to limit self discharge

NOTES

Cycle lives are strongly dependent on how the battery is treated. Lithium-ion cells with petroleum coke anodes drop from about 4.0V to about 3.0V fairly linearly as they go from fully charged to fully discharged. Figures for lithium-polymer cells are predicted (and pre-date the announcement of Ultralife cells). Power density figure for lithium-polymer cells apply at full charge and will drop as cell is discharged. Ni-MH data apply to 2.9Ah 4/3A cells of the kind made by Duracell and Toshiba. Zinc air data are for a cell that Zinc-Air Power is about to use in a prototype battery for electric vehicles.

It should be emphasised that these figures give at best only a rough idea of how the technologies compare.

(Source: abridged from a table in *IEEE Spectrum*, May 1995, p56).

Table 3: Comparison of some rechargeable batteries.

of the safety problems associated with high area lithium deposits. Operating voltage is 3.6V per cell so that one lithium-ion cell can replace three nicad or NI-MH cells. Strict control of charge and discharge is required for both safety and long cycle life, preferably with 'smart charger' chips in control.

Rechargeable lithium-ion cells using lithium cobalt oxide as the cathode were developed in Japan and marketed in 1991.

Initial energy density was 180Wh/L but this has risen to about 330Wh/L depending on cell size and construction, with a target of some 360Wh/L. Apart from the large production capacity in Japan, all major battery companies in Europe and North America intend to introduce such batteries. A potential problem is the availability and cost of cobalt with the result that many firms are researching lithiated oxides of nickel and manganese which also reduce environmental problems.

Rechargeable lithium polymer electrolyte cells using a solid polymer as electrolyte are also attracting considerable research effort, offering high speed production using web equipment, thin flexible cells, safety, shelf life and energy density. One of the major difficulties is the low conductivity of the electrolytes when at room temperature; even for very thin layers a temperature of some 60°C may be required to achieve a reasonable current. In some cases an organic solvent is added to the polymer but this reduces cycle life and introduces the safety problems of liquid electrolytes.

The answer appears to be in combining the electrodes of the lithium ion cell with highly conducting polymers and this appears to be the approach adopted in the Ultralife Battery described in June's *TT* (p70). Performance has still to be assessed in practice since no batteries of this type are yet on the market.

Other forms of rechargeable cells listed by Robet Powers include:

- Solid lithium anode cells using a solid lithium anode and a manganese dioxide cathode with a patented electrolyte which becomes nonconductive if the cell is abused by overcharge, cell reversal or overheating (not yet on the market).
- Secondary Zn-Manganese dioxide cells. Rechargeable alkaline manganese dioxide cells were marketed in the 1960s but proved unable to recover from deep discharge and had poor cycle life. These problems appear to have been overcome in a range of units marketed since 1993 in the USA under the 'Renewal' trademark with which special 'smart chargers' must be used; cycle life is put at 25 cycles by which time capacity is halved. Construction is similar to standard alkaline-manganese primary cells and cost is relatively low but power density lower than nicad although self-discharge is also lower. Initial capacity is better than nicad although less than for primary alkaline cells.
- Secondary Zinc-Air cells. A rechargeable zinc-air battery has been announced by one company intended for portable computers based on a patented 'bifunctional' air electrode and an air management system. Energy density at 170Wh/L similar to Ni-MH batteries but energy per unit weight more than twice as much (about 155Wh/kg). Claimed cycle life of 50-plus cycles with 100 as the objective. Zinc-air primary cells have been around for very many years and the rechargeable version is basically similar. Both have a zinc negative electrode, a potassium hydroxide electrolyte, and a carbon positive electrode which is exposed to the air. The difference is that the oxygen electrode in the rechargeable cell allows the reaction to be reversed. Not only can it absorb oxygen from the air and convert it into oxygen ions but also evolve oxygen while the battery is being charged. Air management schemes (**Fig 3**) are needed to cope for the increase in volume that occurs when the zinc is oxidized and also to restrict or completely block the air flow when little or no energy is being drawn, vastly reducing the self-discharge which occurs with the older forms of primary zinc-air cells. (see **Table 3**).

SEALING COAXIAL CABLES

DAVID APPLETON, ZL2DA, was surprised to read recently in both *TT* and *In Practice* that some amateurs still recommend or use glues and sealing compounds to waterproof coaxial connectors.

He writes: "Equally surprising to see are antenna designs that use open-ended coaxial connections, joined directly to antenna elements. Admittedly, RTV silicone sealer is certainly valuable in some applications, such as the waterproofing of coaxial cable terminations on mobile antenna bases. In this application, I have seen no signs of corrosion problems with hundreds of mobile antennas, throughout the past 20 years.

"But surely, anything other than temporary, experimental antennas warrant a well-engineered cable-to-antenna junction. Good quality coaxial cable is not cheap and, as many amateurs have discovered, water gaining access at the antenna junction soon appears at a lower point, even in the shack. Air-spaced cable functions quite well as a miniature drainpipe, with the wet braid soon decaying and the cable rapidly becoming severely degraded.

"While it is not too difficult to craft a weatherproof and mechanically reliable dipole centre-piece (see, for example, February's *TT*, p70 - G3VA) or Yagi connection box, using modern plastic mouldings or boxes, some

SAW VCOS ABOVE 120MHZ

THE USE OF a surface-acoustic-wave filter to provide a VHF voltage-controlled-oscillator is described by Di Paolo Franco (Ericsson Fatme, Rome) in an item in *Electronics World + Wireless World* (July 1995, p580), reprinted from EDN's Designer's Companion. It is pointed out that SAW filters are now available from US, European, and Japanese manufacturers in an increasing range of frequencies and styles at a cost in the region of £20. Di Paolo Franco states: "They permit the direct implementation of VHF and UHF high-stability oscillators without the doublers or triplers needed with crystals, and with a wider pulling range."

Fig 4 shows an VCO based on a Siemens 180° phase-shift SAW filter with a free running frequency of 140MHz. A VCO suitable for use with a 0° phase-shift filter is also described. The operating frequency is solely dependent on the centre frequency of the SAW filter and with suitable filters can be as high as 1000MHz.

Drawing about 20mA from a 5V supply, the oscillator has a pull range of nearly 500ppm when the control voltage of the BBY31 varactor diode is varied by 4V. It provides an output of about 600mV pk-pk output into a 50Ω load. It is claimed that all harmonics are at least 25dB below carrier and has a temperature variation of about 100ppm in the free-running mode. While the stability is rather less than with a crystal oscillator, SAW filters are available with centre frequencies starting at 120MHz to more than 1GHz. The SAW filter is the only expensive component.

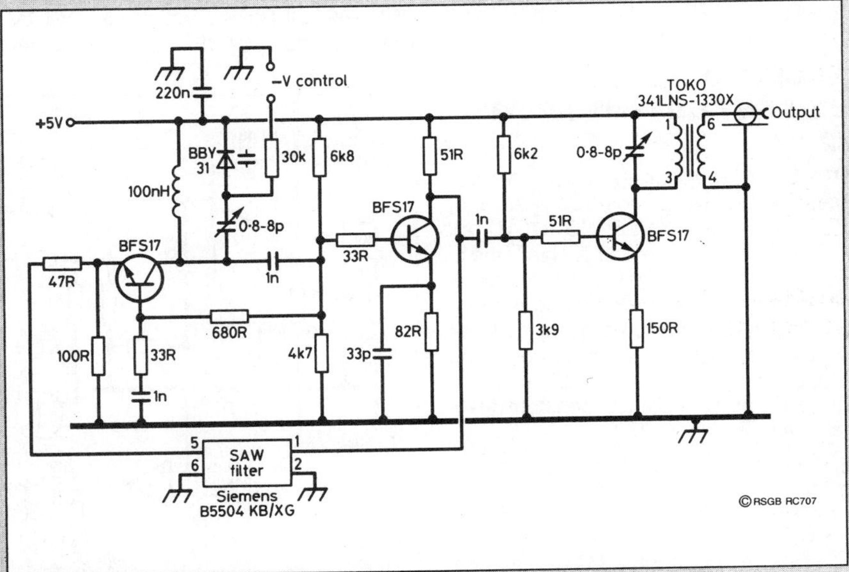

Fig 4: Voltage-controlled oscillator based on a 180° phase shift SAW filter with a free running frequency of 140MHz. SAW filters with frequencies up to about 1GHz are available.

may prefer to purchase such items. For example, SMC has, for many years, retailed reliable and relatively low-cost dipole units that have an S0239 connector moulded into them.... I have used these in both a tropical environment and in high-UV light zones, without problems.

"Self-amalgamating, butyl-rubber insulation tape has, for more than 20 years, eliminated any need for sealing coaxial connectors with glues, etc. If applied as directed, such tape makes a completely water-proof and UV-light resistant moulding that can easily be removed with a sharp knife. This tape is clean and easy to use, provides an instantly effective joint sealer and remains flexible. Protected by such tape, coaxial connectors remain in new condition indefinitely and can be re-used, if required.

"There is nothing more annoying than antenna failure during a very wet day, especially if a gale happens to be blowing, and that is just when shoddily made antenna connections will fail."

LOW-DISTORTION, CONSTANT OUTPUT AF PRE-AMP

IN MAY'S *TT* I revisited the parametric up-conversion mixer which, some 25 years ago, Walter Schreuer, K1YZW/G3DCU (ex VK2AWU), had described in *TT*. This came to his notice and writing from Ipswich, Mass, he admitted to being pleased at being remembered. One of the other items in May's *TT* that interested him was the AC audio pre-amp developed by Dave Miller, N9ZE, for such applications as providing a constant level signal to a DSP filter or TNS.

He writes: "Many years ago in the late 1970s when developing the VOMAX speech processor, I had need for a constant output pre-amp with extremely low distortion. This was needed to make the device idiot-proof and ensure constant 12-15 dB of compression on SSB. Any distortion produced in the pre-amp in such units can, in some cases, be amplified by the same 12-15dB. The schematic of the AGC section of the VOMAX unit is shown in **Fig 5**.

The device uses a P-type FET to control the gain of the gain of IC2 (one section of the quad op-amp type LM324). As is well known the distortion caused by non-linearity of a FET can be greatly reduced by negative feedback. The two 470k resistors at the gate of the E177 FET achieve this nicely. Manual gain control is easy; the green LED should light frequently, the red one only occasionally. Distortion is low at inputs 10dB above the red LED threshold; however, background noise may be excessive. If the indicators are not needed, omit op-amps 4, 5 and 6 and the associated components".

The VOMAX SBP-4 low distortion, SSB compatible speech processor was designed to overcome the problems associated with applying considerable compression to an SSB signal, increasing the 'talk power' by as much as 10 to 12dB. It seems to have adopted the technique - as used for the professional LINCOMPEX system developed by Post Office Research and Bell in the 1960s and also by the original Dolby noise reduction system - of dividing the audio spectrum into four separate channels and processing each channel independently. It was, like a clipper, an instantaneously acting peak limiter, yet its distortion was very low.

As outlined in **Fig 6**, the original audio frequency spectrum is amplified, level detected, AGC controlled and passed through four branching pre-limiter active band-pass filters. The split band signals are limited to reduce the voice peaks by 16dB, fed through the post-limiter band-pass filters to remove the distortion products produced by limiting or clipping. The signals are combined in the phase equalizer section and connected to an exciter via a 3-wire (PTT) cable. This approach was claimed to provide very sharp cut off below 400Hz and above 2400Hz at an increase 6-12 dB of 'talk power' depending on the nature of the difficulty, ie noise, fading, interference, etc of the communications link.

MODERN RECEIVER DESIGN

THE 1994 SERIES of articles in *QST* by Dr Ulrich Rohde, KA2WEU, continue to attract comment. Henry Rech, writing from Victoria, Australia to the Technical Correspondence feature of *QST* (June 1995 pp73-74) endorses the view that the second-order intermodulation distortion (IMD) of a receiver is an important but often neglected characteristic of the signal-handling capabilities of modern receivers. In particular, he stresses the need for better pre-mixer RF selectivity than is usually provided in current receivers too often based on low-pass, octave or sub-octave filters.

He writes: "I believe the trend to broadband, sub-octave front-end filters in particular - and wideband filters in general - to be the most deleterious development in receiver design in many years... It is sometimes difficult to believe that receiver manufacturers in general have succumbed to this trend and do not have the design skills to implement satisfactory narrowband-filter tracking schemes, mechanical or electrical. Perhaps their consideration is the usual one of cost against efficacy. It appears to me that front end performance is (now) almost subsidiary to the

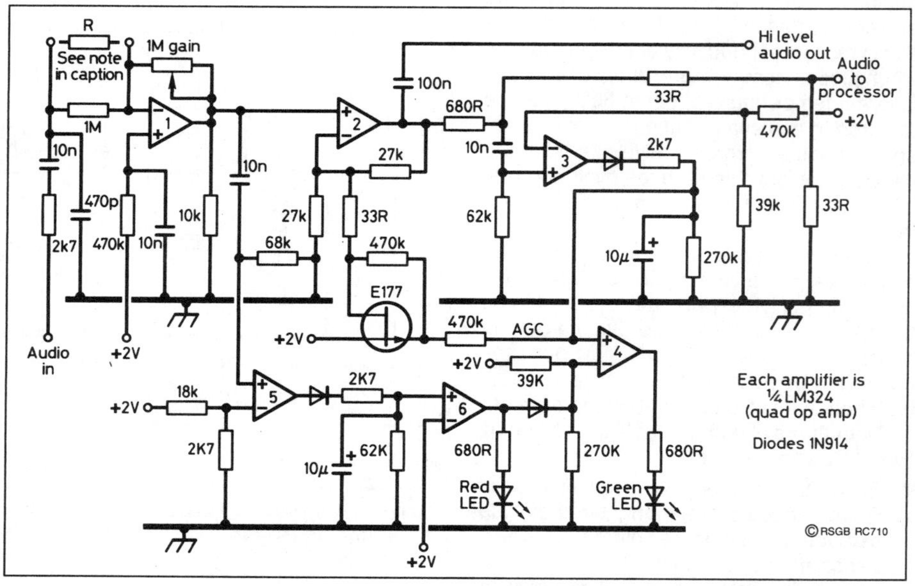

Fig 5: The audio pre-amplifier with AGC and low distortion that formed a part of the VOMAX model SBP-4 SSB speech processor developed by Walter Schreuer in the late 1970s. For high-Z microphones omit resistor R. For low-Z types R should be 120k. Input for Green LED to turn on 12mV pk (for low-Z input 2mV pk). For Red LED to turn on 90mV pk (for low-z input 15mV pk). Power for SBP-4 10-16V DC to ground, 40mA. Ripple less than 50mV p-p.

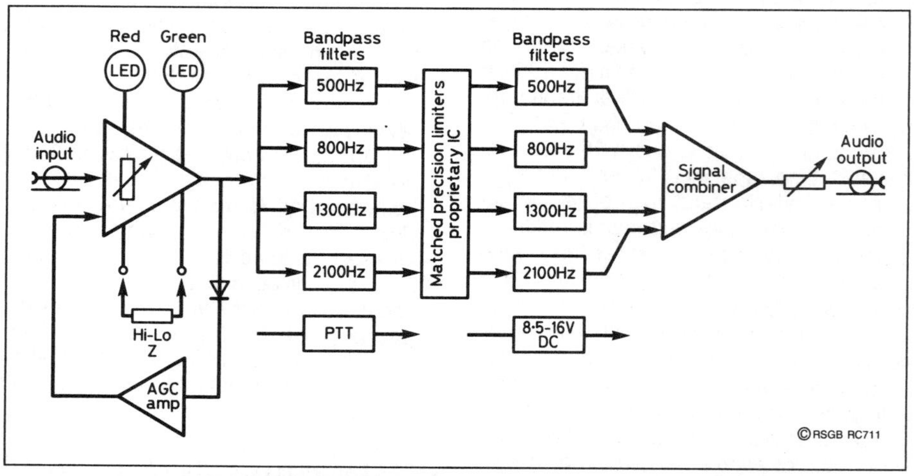

Fig 6: Functional block diagram of the VOMAX unit Model SBP-4 with split band processing designed to cause less IMD and distortion than RF type speech processors, providing a talk-power gain of some 10dB with low distortion.

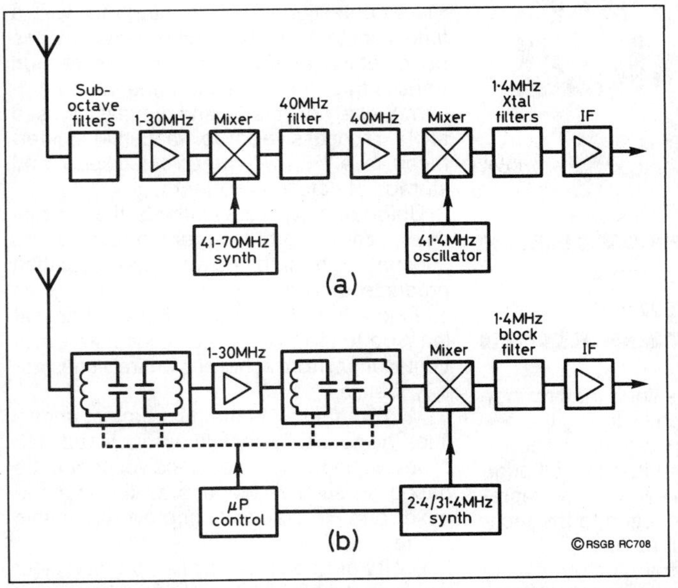

Fig 7: (a) Representative outline of the front-end of typical professional communications receivers of the 1980s using a frequency synthesiser and up-conversion to a first IF of 40MHz. (b) The 'ideal' single-conversion front-end proposed by R A Barr in 1981-2 and implemented in a Rediffusion (Redifon) receiver of that era. It used four ganged-tuned RF circuits to provide good pre-mixer selectivity and is 'ganged' to the frequency-synthesiser by means of microprocessor control.

development of rear-end 'bells and whistles' eg, DSP magic boxes. Front-end narrowband filters would go a long way to ameliorating the ills of modern receivers."

He reminds us that while it is generally claimed that sub-octave filters mitigate the problem of second-order IMD, this is correct (only) to the extent that they have very steep skirts; since otherwise they may not be sufficiently selective to reduce troublesome out-of-band signals. "Remember, too, that second-order IMD begins to appear at lower input levels than third-order IMD."

Henry Rech also points out that wideband front-end filters have a deleterious impact not only on the large-signal-handling capacity of receivers but also on the noise performance which comes about by the mixing of strong in-band, extra received-channel signals with the noise 'sidebands' of untuned RF amplifier and mixer stages, such that the mixed products fall within the wanted channel, increasing the noisiness of receivers - an important consideration on the higher frequency bands. This, he stresses, is in addition to, but very much related to, reciprocal mixing, which is a phenomenon related to the local oscillator sideband noise.

He is not happy with the present practice of describing receiver performance in terms of dynamic range, intercept points and MDS (minimum detectable signal). He writes: "I have always found it curious that a fictitious number (the intercept point) is so used. It might be better to characterise the second and third-order IMD performance by reporting the input-signal level at which the IMD products appear (at some defined level) above the noise floor. This would reinforce the notion that the intercept point is fictitious, and - more relevantly - reinforce the notion that the post-compression-point characteristic of a given receiver is more important than the intercept point. This is because receivers being compared may have similar intercept points, but completely different compression characteristics. Similarly dynamic range is an ambiguous calculation. Being the arithmetic difference between the MDS and the blocking level, of itself it does not allow one to tell whether the particular receiver is either particularly good at weak-signal reception, strong-signal reception or both. Two receivers with similar dynamic range may be totally different receivers."

While it is probably too late to change the present methods of specifying receiver performance, the points made by Henry Rech about the benefit of having narrowband RF tuned filters that are tracked with the local oscillator echo comments made a number of times in *TT*, dating from the report (*TT*, October 1981) of a classic paper presented by R A Barrs of Rediffusion Radio Systems at a 1982 IERE conference ('A reappraisal of HF receiver selectivity', *The Radio and Electronic Engineer*, vol 52 no 7 pp315-320, July 1982) in which he advocated a front end selectivity on the signals applied to the mixer of the order of -37.5dB at five per cent off-tune, and -20dB at 2.5per cent. **Fig 7** shows the differences between the front-ends of a typical modern professional receiver, with up-conversion and sub-octave mixers, and the approach advocated in 1981-2 by R A Barr using four mechanically tuned RF circuits ganged to the frequency synthesiser by means of microprocessor control.

He noted that the configuration shown in **Fig 7 (b)** achieved and bettered the pre-mixer selectivity specification given above using four tuned circuits each with a working Q of 30 to 40. He admitted that such an arrangement, requiring a matching accuracy of the tuned circuits of about one per cent over the tuning range is difficult, and very expensive, to achieve. This is one reason why so much effort has been put into improving the performance of packaged doubly-balanced mixers of which G3SBI H-mode mixer (*TT*, October 1993, correction November 1993) has unquestionably advanced the state of the art. However, as noted in the early 1980s, premixer selectivity can be added to existing receivers with an external, low-gain or no-gain pre-amplifier provided that any devices used are capable of handling strong signals. A possible arrangement would be to use a Cohn tuneable three-pole minimum loss filter of the type suggest by William Sabine in 1970 for use as a roofing filter in double conversion receivers; **Fig 8**.

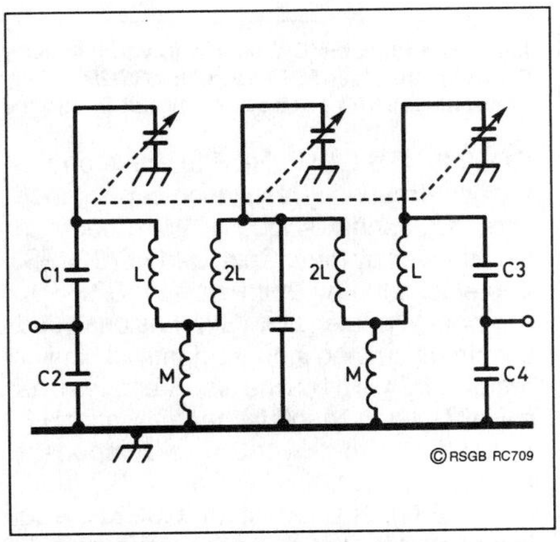

Fig 8: The Cohn tuneable three-pole minimum-loss RF or IF bandpass filter can provide a high degree of selectivity before a first or second mixer. Ratios C1/C2 and C3/C4 govern input and output impedance matching.

In July's *TT*, I quoted Dr Rohde's suggestion that "the best way to avoid switching-diode IMD is to switch the filters with relays instead of diodes, and military and commercial gear generally take this approach".

Harry Leeming, G3LLL, strongly disagrees, at least as far as amateurs are concerned. He writes: "Relay switching is the ultimate example of excellent new performance but poor long-term reliability.

"In the receiver RF chain, the average transceiver has one antenna change over relay. On average it goes intermittent and needs the contacts cleaned or replacement once every five or ten years. Fit 12 and the chances are that at least one would need attention every year!

"A prime example of this is the FT102. When this rig first appeared I thought it was the best thing since sliced bread, but with six relays in the RF signal switching chain, the trouble with intermittent problems has been horrendous (particularly when operated by smokers). On average, for smokers, a new set of relays is needed about every two years; for non-smokers about 2-4 years.

"Professional equipment gets heavy use and is only expected to have a lifetime of a few years and may have expensive non-oxidising relay-contact materials such as gold. Amateur equipment is built to a cost, used intermittently and expected to last some 20 years.

"Again, for relays that switch DC or the volts of the transmit signal there will be few problems of oxidation since this will be broken down. It is a different matter with the microvolts of RF signal which need absolutely clean contacts. Putting loads of relatively inexpensive relays into the RF chain is virtually certain to be a major cause of unreliable operation after a few years. If you want to avoid switching diodes, how about a good rotary wave-change switch with rotary self-cleaning contacts. New ideas are not always the best!

TT FEEDBACK

July *TT*, Fig 6. The insertion loss scale should be 0 to 9dB *not* 0 to 0.9dB as shown. Note this greatly increases the power loss in ATU networks! The efficiency formula should be:

$$\eta = \frac{Q_{UNLOADED} - Q_{LOADED}}{Q_{UNLOADED}}$$

July *TT*. G3ROZ's 32-page details for modifying the KW2000 series can be obtained from G3VA (QTHR) and not from G3ROZ. £2 covers the cost of copying and postage. ♦

COLLOQUIUM ON RADIO SCIENCE & ANTENNAS

I RECENTLY ATTENDED the Joint URSI (International Union of Radio Science) 12th National Radio Science and QMW 1995 Antenna Symposium at Queen Mary & Westfield College of the University of London. This was a busy three-day conference at which more than 40 short presentations and a number of invited papers were given.

These mostly represented reports on current postgraduate research projects at British universities but there were also papers from Germany and the Ukraine. While many of the papers were largely mathematical, way beyond my comprehension, some covered topics of interest to radio amateurs.

Professor Peter Clarricoats (son of John Clarricoats, G6CL, general secretary of the RSGB and editor of the *RSGB Bulletin* from the 1930s to 1964), in one of the keynote addresses, emphasised that university research faces a period of change as a result of the decline of the former Soviet Union as a military force.

One result is that there is now less funding of blue sky research: "You have to see an application if you are going to get funds - in the foreseeable future support for research in electromagnetics will be dependent on the expectation of a successful application. But because of the growth of Information Technology the future prospects remain quite bright, although the reduction in support arising from the decline in defence spending is unlikely to be fully compensated."

A subject of increasing interest to industry is more accurate calibration of the standard dipole antennas used for EMC testing. Dr Martin Alexander of the National Physical Laboratory explained how Europe's largest metal ground plane, measuring 30m by 60m, has been set up at NPL's antenna range. The range has been used to calibrate accurately over 40 different models of dipole antennas over the range 20MHz to 1GHz with uncertainties of less than +/- 0.7dB (+/- 0.2dB if required). It has been discovered that some EMC antennas can show uncertainties, according to the way they are mounted, as much as +/- 10dB (typically due to poorly balanced baluns).

David Palmer, radio amateur and mature student at the Centre for Satellite Engineering Research, University of Surrey, the department headed by Prof Martin Sweeting, G3YJO, outlined the proposed UoSAT Topside Sounder Satellite. This third-generation topside sounder for probing the upper side of the F2 ionospheric layer will require only 10W of electrical power during normal operation, compared to about 400W peak of earlier topside sounders which used antennas of up to 73m from tip-to-tip.

The new sounder is to be launched in cooperation with the Russians on an inexpensive low earth orbit microsatellite. It will feature an advanced receiver using DSP techniques as well as an automatic tuning unit which can be programmed to follow the rapid frequency sweep of the sounder. Although in the 1960s, top-side sounding was seen as a major advance in probing the ionosphere, it proved difficult to reconcile the topside measurements with those made by ground-based sounders. It is refreshing to see how the work

Pat Hawker's

Technical Topics

PAT HAWKER, G3VA
London 37/SE22 8SS

under Prof Sweeting is making a major contribution to satellite research including the setting up of a commercial operation. One hopes that it will not be forgotten that this all originated in conjunction with AMSAT and represents an important contribution to the technology based on amateur radio.

One of the few undergraduate projects to be reported was that of a group of four students at Liverpool University under the guidance of Dr Brian Austin, G0GSF. They were given the task of building a 151MHz phase-switched interferometer to detect noise radiation from the Sun.

This required construction of a low-noise receiver with good dynamic range, a solid-state-switched 180° phasing line and down-converter to a 51MHz which could be mounted close to the twin 6-element Yagi antennas. Signal detection is via an integrator with a time constant of 200s, determined by the period of the interference fringes as the Sun transits through the lobes of the interferometer. Signal conversion and processing used an 8-bit successive approximation ADC interfaced to a PC with graphical display and storage of data in 24h blocks.

Unfortunately for the students, the minimal solar activity in the current phase of the sunspot cycle combined with intermodulation products of two very strong local signals spaced a few MHz above 151MHz has - at least up to the time of the colloquium - prevented them from achieving entirely satisfactory results.

My own feeling is that this project underlines the necessity on VHF as much as on HF of developing super-linear receiver front-ends based on such techniques as the H-mode mixer and RF stages of improved dynamic range.

But it was good to learn of a student project involving the practical skills of soldering and construction as well as the design of high-performance RF hardware!

Another paper that I found of great interest was 'Large Bearing Errors at HF' presented by Dr E Warrington of the University of Leicester. This added to a paper 'Some Recent Measurements of the Direction of Arrival of HF Signals Propagated Over a Path Tangential to the Mid-latitude Trough' presented at the 1995 IEE Conference on Antennas &

THAT TUNEABLE NOTCH (OR PEAK) FILTER

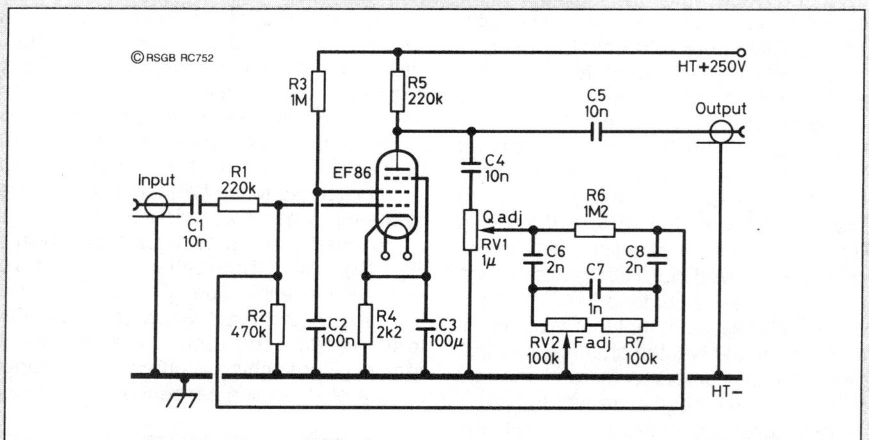

Fig 1: The tuneable CW filter with variable selectivity based on the Hall RC filter as described by G3GKS in the May 1966 issue of the *RSGB Bulletin*. Tuneable about 540 to 3600Hz (highest Q at lower frequencies). VR1 controls the circuit Q, and the frequency of the peak is set by VR2.

COMMENTS CONTINUE to arrive on the variable-frequency resistance-capacitance filter (*TT*, June & August 1995) brought to my attention by Peter Cole, DA1PE/G3JFS, and elucidated by Bob Pearson, G4FHU.

Roger Wheeler, G3MGW, has analysed the circuit arriving independently at similar results to G4FHU but he also identifies it as a 'Hall Network' quoting the reference: H P Hall, 'RC Networks with Single Component Frequency Control' (*IRE Trans CT-2 1955, pp283-284*). R C Christian, G3GKS, also identifies P Hall as the originator giving as a reference 'Single-Component-Controlled RC Null Circuits' *(General Radio Experimenter*, July 1961).

Interestingly, G3GKS recalls that he described in the May 1966 issue of the *RSGB Bulletin* (as *RadCom* was then named) 'An Audio CW Filter - a versatile design featuring variable frequency and selectivity'.

His CW filter was based on the Hall circuit and offered a tuning range of about 540Hz to 3600Hz (highest Q at lowest frequency). As shown in **Fig 1**, this was based on an EF86 valve, but the basic idea could be adapted for use with an op-amp IC device. Since wrting the above notes a high-performance CW-type filter design, based on this type of RC-filter, developed in 1964, has arrived from DrTom Seed, ZL3QQ.

Propagation and published in *IEE Conference Publication No 407*. The mid-latitude trough is an F-region electron density depletion region a few degrees wide in latitude on the equatorward edge of the auroral oval in which the critical frequencies drop by a factor 2-4 or more and the altitude of the electron density maximum rises by over 100km.

Signals over the paths between the UK and Canada, and some paths within Canada have been shown at times to deviate greatly from the Great Circle.

During the last few years, an experiment has been undertaken in conjunction with Canadian researchers accurately to measure the bearings and signal characteristics of HF transmissions from CFH Halifax; Iqaluit, Baffin Island; and Wainwright, Alberta. The transmissions on approximately 6, 10 and 18MHz have been measured at Ottawa and Alert in Canada and Cheltenham in the UK where the long-distance D/F system has an instrumental accuracy on the arrival bearings of approximately 0.1°.

It has been shown that deviations in excess of over 50° can occur at high-altitude sites at the beginning and end of propagation at 10 and 15MHz on roughly 50 per cent of days. The magnitude and frequency of occurrence are largest at the end of the day's 'opening', ie as the MUF of the path drops below the transmitted frequency. A broad maximum in the frequency of occurrence takes place during the spring with a minimum in the summer months.

All three of the frequencies, roughly 6, 10 and 18MHz, exhibit large deviations but not necessarily of the same magnitude or even at the same time intervals. The magnitude of the bearing swings and the time intervals at which they occur are strongly dependent on geophysical conditions, with the geomagnetic index Ap found to provide a good indication of the magnitude and times of the occurrence of these large bearing errors.

Many years ago in *TT*, I mentioned that Roly Shears, G8KW, while acting as one of the link stations with the special station G7AE, Point Barrow, Alaska during the 1968 British Trans-Arctic Expedition, found that his rotary beam antenna for optimum signals often needed to be swung well away from the Great Circle path.

The present measurements not only confirm the very wide deviations of the direction of signal arrival but also strongly suggest that they are due to the ionospheric tilts that occur when the Mid-latitude Trough opens and closes. The Trough opening time varies from day to day with changes in geomagnetic activity.

The closing time is essentially constant and is determined by the sunrise when photo-ionisation rapidly fills the depleted region of the Trough. The Canada-UK path tends to open around 0930UTC or around 1120UTC indicating a strong dependence on the solar-zenith angle.

The path deviations tend to be towards the South. It all adds up to the fact that it is wrong to think of the ionospheric layers as smooth mirrors in the sky.

Indeed, the layers, in practice, are complicated by troughs, blobs, travelling ionospheric disturbances (TIDs), patches and tilts; all of which are phenomena that can greatly affect amateur HF operation.

The objective of this particular investigation is to provide D/F operators with information from which they could determine the probable errors in the bearings that they obtain on accurate direction-finders.

The callsign of the Halifax transmitter, CFH, has apparently become known to D/F operators, unaccustomed to wide deviations from the Great Circle path as 'Can't Find Halifax'!

It was evident from many of the short presentations that computer simulation and progressing is now the dominant design tool throughout the field of antennas. However, as emphasised in an invited paper from Dr Pat Foster (Microwave and Antenna Systems, Malvern), the Method of Moments with versions of NEC (as now commonly used by both professionals and amateurs at HF) runs into practical difficulties as the frequency is increased and has to give away to systems based on the Geometrical Theory of Diffraction (GTD/UTD code).

The reason for this is that the Method of Moments depends on modelling the surface of the structure using wire segments or elemental surfaces over which the current can be assumed to be nearly constant. This means that as the frequency increases the number of segments rises rapidly, requiring not only more preparation time but also, even with the largest mainframe computers, vastly more computing time.

Even when the largest CRAY computers are used there is a practical limit of about 30,000 segments representing a structure of seven wavelengths.

On the other hand, the GTD/UTD system has a lower frequency limit depending on each item of a structure whereas the MoM limitations are of the entire structure. Dr Foster pointed out that although hybrid methods can be used whereby MoM is used for the antenna and its local structure and the computed results fed into a GTD/UTD code to model the parts of the structures which are further away.

Such a mixture has to be treated with care and some physical insight into the problems in order to avoid creating yet more problems. The best results using this technique, Dr Foster suggested, are obtained when the antenna and the local structure are physically divorced from the rest of the structure.

However, she emphasised that there is no agreed methodology for using hybrid methods and therefore there is no perfect solution.

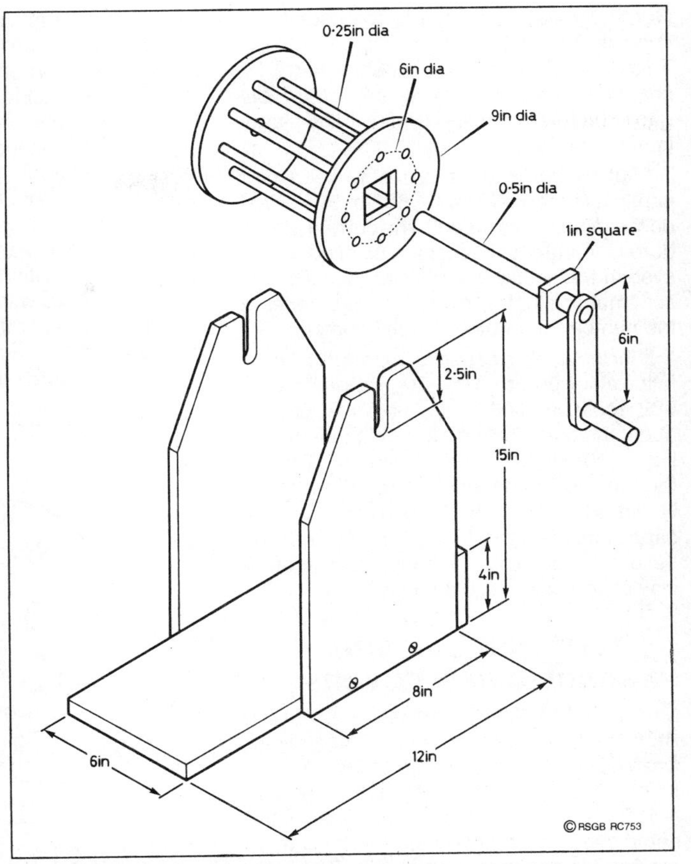

Fig 2: G3YXJ's drum and drum winder makes it simple to wind up antennas and feeder cables without kinks or twist

EASING THE WIND-UP!

DANNY BEGG, G3YXJ, has always found it a tedious task to wind up the soft wire antennas and feeder cables used at special events or portable stations without creating kinks or twists.

He writes: "Probably the best method is to wind the wire on a drum, but winding wire on a drum by hand is a slow process. **Fig 2** shows an easy-to-construct frame which will hold a drum and also enable the user to wind up wire or coaxial cable quickly and easily.

"The prototype was constructed from 1/2 inch chipboard salvaged from old, redundant kitchen units but any available timber of similar thickness could be used.

"The dimensions shown in Fig 2 are also not mandatory, but the best plate should be wide enough to accommodate a large foot, used to steady the frame whilst winding. Clearly, the drum must be of a width to fit freely between the sides of the frame.

"A number of identical drums can be made on which to wind different antennas or lengths of coax. The handle/drum combination can be fitted from either side to suit right or left handed users."

G3YXJ adds that the device is being used by the Braintree Radio Club. To receive a report on its performance, contact GB20AY, the club's special event station on the air most weekends and Monday club evenings to celebrate their 20th anniversary.

CAPACITOR CONFUSION

JAY CRASWELL, WB0VNE, in the Technical Correspondence columns of *QST* (July 1995, p76) writes: "Disc capacitors have undergone changes in the way they're identified. In 'The

Good Old Days' you'd find, for instance, a capacitor marked .0068 to indicate 0.0068µF. This worked too well and everything had to be changed. In many schematics 0.0068µF began to be referred to as 6.8nF with nF standing for nanofarads.

"But by now you've probably seen disc capacitors marked just 15, 682 or 821. What do those numbers mean? The current procedure is to mark capacitors in picofarads in a system that is like our resistor-identification scheme. The last number of the digit series is the number of required zeros - sometimes!

"For instance, a capacitor marked 821 is 82 followed by one zero, ie 820pF. One labelled 682 would be 6800pF, that is 6.8nF but what about the one labelled 15. You might expect it to be 100,000pF ie 100nF or 0.1µF. Wrong, this is a 15pF component. An 0.1µF capacitor would be marked 104! So remember that capacitors marked with digits between 1 and 99 are indicating picofarads. Just another way of confusing us old-timers!"

1.3GHZ CONICAL AND PARABOLIC REFLECTORS

MANY YEARS AGO, I noticed in *TT* (still in *ART7* pp316-7) that for the UHF and microwave operator the ultimate in high-gain antennas has for long been the parabolic dish reflector, the larger the better. But for the home constructor a true paraboloid is not an easy shape to come to terms with. The main difficulty is that the paraboloid is a doubly-curved surface: if constructed from flat sheets of material, part of the material needs to be either stretched or compressed or both.

An effective alternative was brought to attention by Fred Brown, W6HPH (then also G5AWI). He showed in a 1966 issue of the *VHF'er* and later in *TT* that conical reflectors, constructed very easily from flat sheets of material (hardware cloth in his case) could provide an extremely useful substitute. A conical reflector can give considerably more gain than would a corner reflector antenna yet it is just as easy to build.

A shallow cone has the advantage of being a singly-curved surface and can be made from flat circular sheet by removing or overlapping a segment of the material. Performance depends on the fact it is not really vital for the surface of a parabolic reflector to be a true paraboloid. W6HPH pointed out that it is usually accepted that there can be departures of up to one-sixteenth of the wavelength at any point on the surface without suffering any significant deterioration of gain and directivity.

While one-sixteenth of a microwavelength may not sound much, in practice it permits quite drastic changes in the overall shape. Up to a certain size, in terms of wavelength, it can be two shallow cones, one inside the other: **Fig 3**. A two-cone reflector can be satisfactory up to a diameter of 13.86-wavelengths in certain conditions. This means that for 1296MHz one could build a reflector of some 10ft diameter, or even 31.5ft at 432MHz without having to worry about double-curved surfaces, and yet obtain virtually as much gain and directivity as with a true paraboloid.

A single conical reflector is satisfactory up to 3.46-wavelengths. W6HPH provided details of a 1296MHz single conical reflector antennas with a gain of about 16dB: **Fig 4**. He claimed this could be constructed in a few hours for a cost (in 1966) of about $1.50. The cone has a radius of 17in. Both the cone and the small 4.5in square reflector for the folded dipole element were made from 0.5in mesh hardware cloth.

I was reminded of W6HPH's suggestions by an article 'High-Performance Antenna for 24cm' by John Cronk, GW3MEO (*Electronics World + Wireless World,* August 1995, pp 699-700). He introduced his design as having fewer critical dimensions than most other configurations, and capable of being easily customised. It comprises a lightweight parabolic-shaped plane reflector which is illuminated with a horizontally polarised dipole and reflector feed.

Gain is given as about 15dBd (ie about 17dBi) representing a power increase of almost 32 times. Bandwidth is more than ample even for amateur-television. Wind resistance is lower than for expanded aluminium mesh.

The mesh reflector was fashioned from a 3ft by 2ft sheet of wire mesh called Handy mesh obtained from a local hardware store. The mesh consists of half-inch squares formed by 22swg tinned wire. The sheet was cut in half to form a strip 60in long with the overlapped section strengthened by binding some of the coincident wires with 22swg tinned copper wire and then soldering. The mesh is then fixed to an aluminium square former which is shaped as a parabola.

For full details on building GW3MEO's antenna see the original article in *EW + WW*. The design includes a balun feed etc and is considerably more complex than the W6HPH design. But it seems worthwhile to show how he obtains a parabola profile, without resorting to mathematics, by using a pin and string: **Fig 5**.

GW3MEO explains this as follows: "First draw line AB, and then at its centre draw line PFX, at right angles to AB. Next either draw line Xy parallel to AB, or use a long rule or tape parallel to AB, this must be marked off with regular divisions - X1, X2 etc. Now fix one end of a piece of string at point F using a pin. Take the string around another pin at P1, and then up to the point X1, and mark this length with a knot. Now plot the curve by moving the knot to X2, and keeping the string at right angles to XY, prick a mark at P2, and so on, until half the curve is marked out. Repeat for the other half

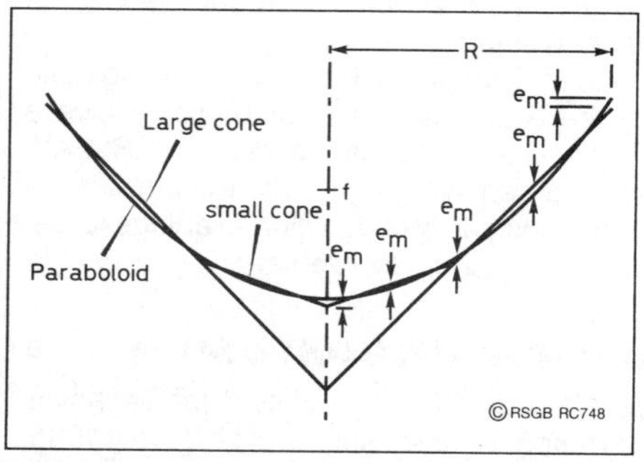

Fig 3: Principle of the polyconic reflector showing the small departures (m) from the true paraboloid.

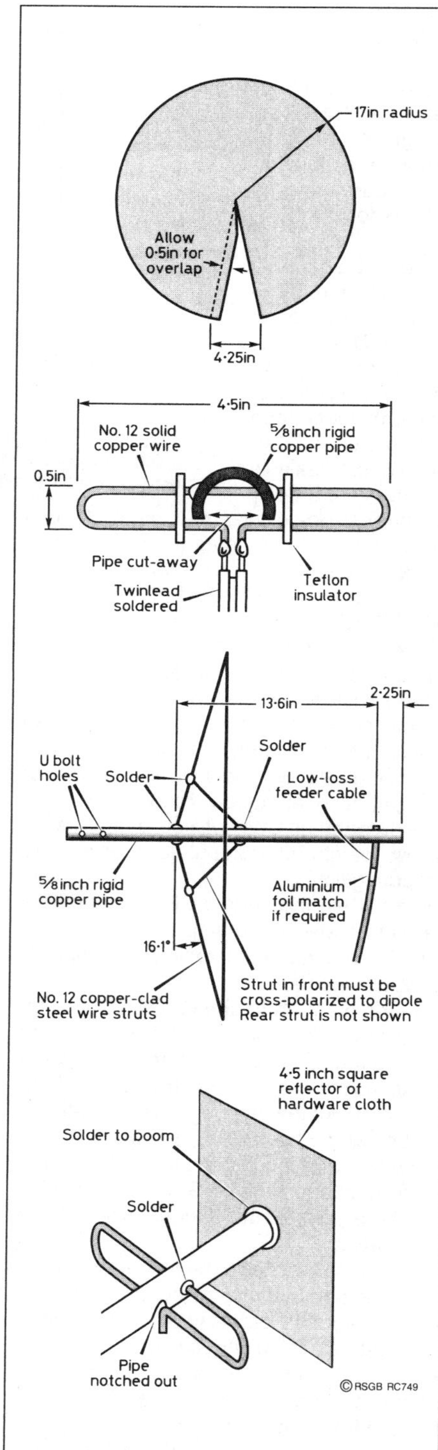

Fig 4: Details of the 1296MHz single conical reflector antenna as designed by W6HPH providing about 16dB gain.

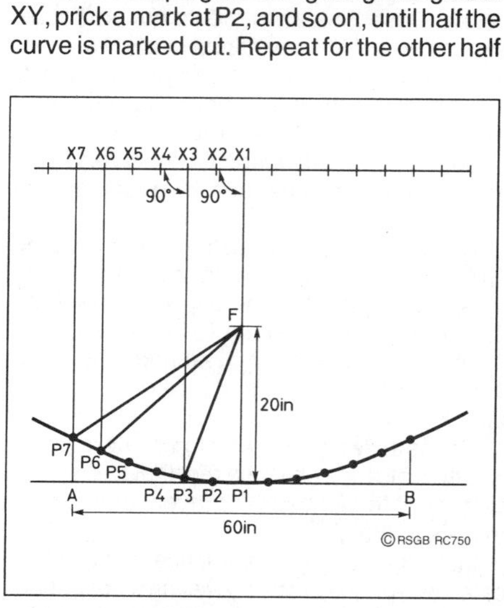

Fig 5: How GW3MEO produces his antenna's parabolic profile drawn without resorting to mathematics by using a pin and string.

of the curve. Draw a line smoothly through the pin pricks to show the shape of the reflector surface."

The curve is later transferred to a homemade bending device using hardwood blocks and a vice to avoid crumpling the tubular reflector former which comprises 60in of 0.5in square aluminium tube stock, as commonly used for TV yagi antenna booms: **Fig 6**.

Since this is a planar reflector with a parabolic profile, the problem of the double curve is avoided. GW3MEO gives the acceptable tolerance of one-tenth of a wavelength (2.5cm at 1.3GHz), a looser tolerance than W6HPH's one-sixteenth wavelength. It would be interesting to compare the two designs in terms of performance, cost and time of construction.

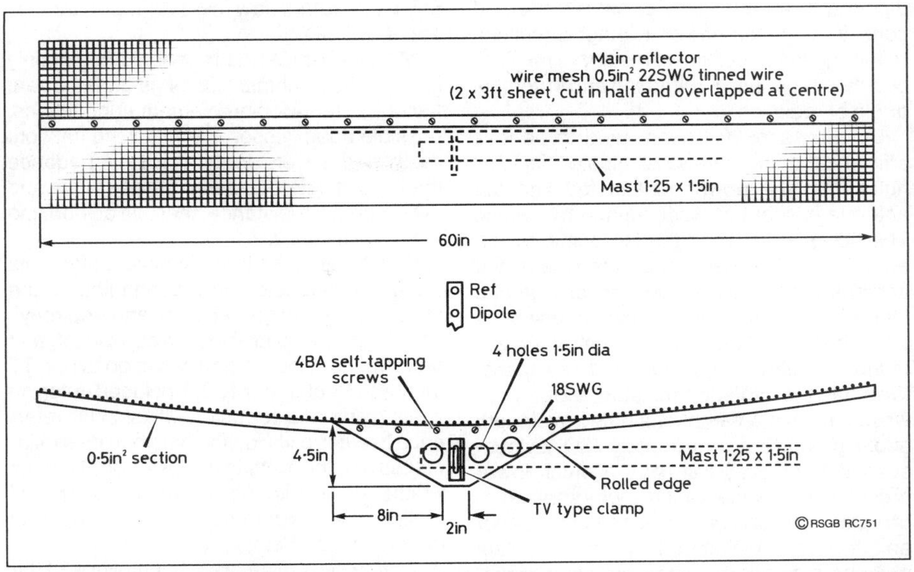

Fig 6: Reflector detail of the GW3MEO antenna. Mesh is a commercially-available type comprising half-inch squares formed from 22SWG tinned wire.

HIGH EFFICIENCY MOSFET POWER AMPLIFIERS

AN ARTICLE 'High efficiency power amplifiers for 13.56MHz ISM and HF communications' by Ken Dierberger, Frederick H Raab, Bobby McDonald and Lee Max (*RF Design*, May 1995, pp28, 30, 32 and 34-36) describes the design of a 400 watt class-C amplifier and a 250 watt class-D amplifier using recently introduced power MOSFET devices from Advanced Power Technology (APT). Both amplifiers provide high-gain and high efficiency for the 13.56MHz industrial, scientific and medical (ISM) applications but, it is claimed, may also be used for HF communications. The class D amplifier is specified as providing 250 watts or more output over 1.8 to 13.56MHz and the values of frequency-dependent components are given for 1.8, 3.5, 7, 10 and 12MHz. It should be appreciated that Class C and D amplifiers cannot be used for SSB unless, for example, in conjunction with cartesian-loop techniques (see *TT*, June 1994, p54). It would also probably be necessary for amateur applications (ie CW) to use additional harmonic filtering.

The introductory note states: "The boom in RF equipment operating in the UHF region has overshadowed the growth in lower frequency ISM systems. As a result, ISM applications have had to use high-cost RF power devices that are optimised for higher frequencies or use devices that are intended mainly for switching power supply requirements at lower frequencies. These latter devices, although low cost, are packaged in a standard common-drain configuration, requiring insulated mounting with added shunt capacitance, and which have significant inductance in the source connection, which combine to limit RF performance.

"The APT devices used in the amplifiers described here are provided in a common-source configuration, eliminating the need for an insulator between case and heat sink. The additional internal Be0 insulator required to achieve this configuration adds some thermal resistance, but less than would be added by the usual mounting insulator. The isolated die also permits the interchange of gate and drain leads, making 'left-hand' and 'right-hand' devices possible".

The 400 watt class C amplifier (**Fig 7**) has a 19dB gain with 4-5-wattsdrive; it operates from a 100Vline with an efficiency of some 75%. It uses ARF442/ARF443 300V RF power MOSFET devices in TO-247 plastic packages.

The more complex Class D switching amplifier, designed by Frederick Raab (who holds an amateur licence) uses ARF440/ARF441 devices in push-pull, driven to act as switch generating square-wave voltage. This design has a much larger component count, including two Schmitt trigger/limiter ICs, four MOSFET drivers, as well as the ARF440/ARF441 MOSFET power devices.

It is claimed that "the use of devices that are both lower in cost than VHF-capable RF power MOSFETs, and provided in standard plastic package, results in designs that can be manufactured at costs that are comparable to vacuum tube techniques". For more information contact Ken Dierberger, Applications Engineering Manager, Advanced Power Technology Inc, 405 SW Columbia Street, Bend, OR 97702, USA.

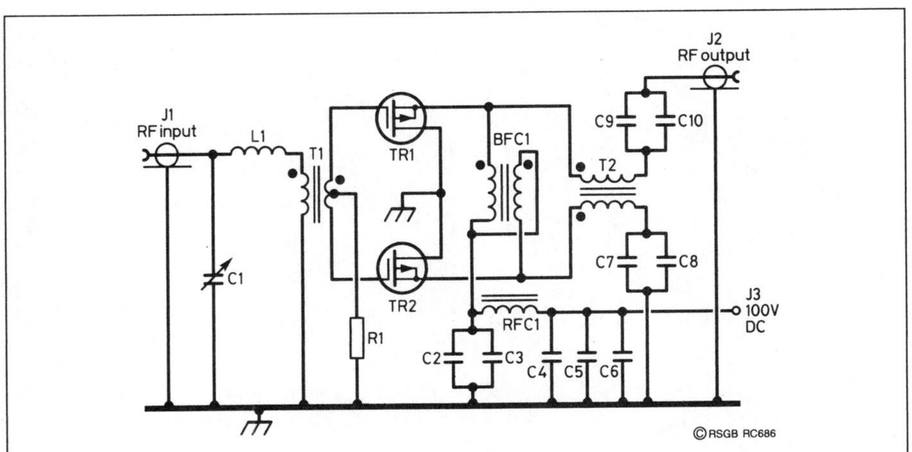

Fig 7: The 400 watt class-C amplifier suitable for 13.56MHz ISM systems but which could be adapted for CW use on HF bands up to the 14MHz band, using the relatively low-cost power MOSFETs introduced by Advanced Power Technology Inc. Component information: C1 75-480pF mica compression trimmer. C2 to C8 0.01µF 200V (CK06). C9 0.1µF 100V (CK06). C10 10µF 100v electrolytic. R1 10K, 5%, ¼W carbon. TR1, TR2 ARF442, ARF443. L1 0.5µH (7 turns, No 18 AWG, int dia 0.438-in. BFC1 Balanced DC feed choke, 7 turns, No22 stranded PTE insulated twisted pair on Indiana general F624-19-Q1 toroid, permeability 125. RFC1 2 turns No 18 stranded PTFE on a Fair-rite 2677006301 bead (µ 2000. T1 4:1 (z) conventional transformer, 2:1 turns, 22 stranded on a Fair-Rite 2843000202 balun core, µ= 850. T2 1:1 (z) transmission line transformer, 4 turns (approx 22in) of mini 50-ohm PTFE coax, OD 0.095in on a two-hole core made from two Fair-Rite 2643102002 cores, µ = 850. PCB 0.062in G10 epoxy glass.

ANTENNA SYSTEM EFFICIENCY

LEW MCCOY, W1ICP, in 'Antenna Efficiency - What is it?' (*CQ*, April 1995, pp 14, 17, 18) reminds us that antenna efficiency is not a question of gain, radiation patters or directivity but is nevertheless an important factor that is too often overlooked.

He stresses that the most efficient antenna, which can radiate about 97% of the power fed to it, is the basic half-wave resonant dipole. Compare this with some mobile and aircraft antennas which at about 2MHz may radiate only some 10mW for every 100 watts of transmitter power.

W1ICP is primarily concerned with the RF ohmic losses in antenna elements, but it should not be forgotten that wasted power may also be a feature of the ATU matching network and the transmission line (feeder). As a simple illustration, he points out that a beam antenna with a 3dBd gain but an efficiency of about 50%. This will radiate in the favoured direction only the same power as a dipole with an efficiency of near 100% and considerably less power in other directions (although the directivity of the beam may still be advantageous for reception and in vertical angle of radiation).

But why should a beam tend to be less efficient than a half-wave dipole? W1ICP notes that the power that is fed into the antenna is spent or dissipated in two ways. The power used up in the RF ohmic resistance is spent in heating up the wire in the antenna, or the traps, bad connections, ground losses, etc. Only the remaining power is spent in the "radiation resistance" of the antenna - in other words, launched into space. Whereas the radiation resistance of a half-wave dipole at a height of a half-wavelength above ground is approximately 73Ω, a close-spaced Yagi array will have a much lower radiation resistance (which is why some form of matching is required between a coax feeder and the driven element). Thus in a Yagi antenna the ratio between radiation resistance and RF ohmic resistance will normally be lower than for a dipole. To quote W1ICP:

"Let's look at beams. We must first consider the impedance. Many newcomers get hung up when they assume that the impedance of all beams is of the order of 50Ω. This is because the antenna is fed with 50Ω cable and produces a match of nearly 1:1 SWR. But remember that nearly all Yagi-type beams have some sort of 'matching' device in order to transform the real impedance up to 50Ω... A monoband three-element beam using close-spaced elements, say one-tenth spacing, will have the rather low overall impedance of about 4 to 5Ω. . .

"A beam with 5-10Ω impedance is certainly going to have ohmic losses as an important factor . . . telescoping element connections, possible boom losses the matching network itself, and so on. With a 10Ω impedance made up of, say, 5Ω radiation-resistance and 5Ω RF ohmic resistance, then we are looking at 50% efficiency."

W1ICP stresses that aluminium elements are good conductors but, depending on the atmosphere, can get corroded and 'scummy'. If this happens, ohmic losses can rocket, and performance and efficiency can go to pot. *TT* (July and September 1993) included notes on fighting antenna corrosion with specific reference to electrolytic action between dissimilar metals and the 'fretting corrosion' that occurs whenever there is a small repetitive or cyclical motion at a metal-to-metal joint or seam, in the presence of moisture.

W1ICP lives in an area that has one of the few remaining really clean atmospheres - 6400ft up in the Rockies - yet he still had joint corrosion of the aluminium elements of his log-periodic type antenna. The elements and joints had developed a scum covering which certainly would increase ohmic losses. He advises: "If your beam has been up a long time, take it down and clean and polish all connections. It will pay off. There is no magic in antennas - just good common sense."

Salty and industrial atmospheres can result very quickly in deterioration of sliding joints in antenna elements. Moisture ingress into coaxial feeder cables and baluns significantly increases losses and lower overall efficiency. Loading coils used with electrically short elements need to be of excellent high-Q construction if they are not to seriously reduce efficiency.

The need for extreme precautions to reduce losses in small transmitting 'magnetic-loop' antennas is well recognised, but serious doubts remain as to the overall efficiency of small loop systems since the losses in the matching networks are seldom taken into account.

A few years ago, Tony Henk, G4XVF, in a two-part article 'Loop Antennas - Facts not Fiction' (*RadCom*, September, October 1991) showed clearly that the overall efficiency of small transmitting loop antennas, in terms of the amount of power actually radiated compared with the output power of the transmitter, is usually quite low, with more energy being lost in the matching components, including the capacitors, than in the actual loop element.

The radiation resistance of a 1.8MHz mobile whip, like that of a small loop antenna, will be only a small fraction of an ohm and there will usually be significant ground losses in using the vehicle as a ground plane. Again, high-Q construction of the loading coil is necessary, although the overall efficiency will be very low.

W1ICP notes that, on the other hand, quad-type (wire) elements with one-wave or longer elements have relatively high impedance, largely comprising radiation resistance, resulting in efficiencies in the order of 90% or more; a significant advantage over the typical tri-band Yagi.

For many years, I have used at G3VA a long-wire antenna (of indeterminate length, initially about 42m, partly indoors) fed against quarter-wave counterpoises to overcome the problem of the long earth lead to an upstairs shack. While this performed satisfactorily, I recently attached and brought back into the shack sufficient wire to form an irregular narrow loop fed from a couple of balanced pi-network tuners (one for 14 to 28MHz, the other covering 3.5 and 7MHz). This overcomes the question of the ground losses of unbalanced antennas and seems to work at least as well as and probably rather better than the long-wire version.

BACKGROUND TO COUPLED-RESONATOR ANTENNAS

TT, JUNE 1995, pp67-69 described the construction of multiband antennas using several closely-coupled dipole elements each cut for a particular band based on a detailed article by Gary Breed, K9AY, in *RF Design* (November 1994).

I noted that similar ideas had been patented and used some years ago for combined Band I/Band III television antennas and added: "Without entering too deeply into any arguments about patents, we should all be grateful to K9AY for bringing this system more fully into the public domain..... and for providing practical details of an antenna for the 10, 18 and 24MHz bands."

Shortly after publication I received a letter from K9AY who is the Editor of *RF Design* explaining the background to his designs and to his patent application.

He wrote: "One of the most remarkable insights gained while investigating this type of antenna is how few people know about it. It has both advantages and drawbacks, as does any multiband technique, but if you don't know it's there, you can't give it consideration!

"The main purpose of this letter is to clarify what portion of my work I consider original. I make no claim to have 'discovered' the principle of one conductor imposing its resonance on another.

"This concept was discovered in the 1940s, and has appeared in many products, including the TV receiving antennas noted by G0LMC. The *RF Design* article was not meant to describe my patent application, but rather to introduce the couple-resonator principle to the readers. By using the simplest examples of two and three frequency antenna configurations, I meant only to illustrate the principle on a territorial level, not to suggest that I was the sole originator of that work.

"I also intentionally avoided using the term 'sleeve' or 'open sleeve' to describe this type of antenna, although these terms have been used by others. Past antenna developments using these names are clearly members of the coupled-resonator antenna family. But I wanted to emphasise the generality of the concept and chose to use a more broadly descriptive label.

"The work I have done that has been accepted as original is the mathematical characterisation of the principle. I have derived the equation given in the *RF Design* article, which establishes the required spacing distance between the driven conductor and an additional coupled-resonator.

"To recap, the basic relationship is:

$$\frac{\text{Log}(d)}{\text{Log}(D/4)} = 0.54 \quad \text{or} \quad d = 10^{[0.54 \log(D/4)]}$$

where d is the required spacing in wavelengths at the coupled-resonator frequency, and D is the element diameter (assuming elements of equal diameter) also in wavelengths at the coupled-resonator frequency. This relationship is the situation where the impedance seen at the coupled-resonator frequency is equal to that of a dipole in free space (approximately 72Ω). It ignores the effect of the impedance that the driven dipole has at this frequency. I have developed a correction factor for obtaining other impedances, and another factor for small frequency differences, where the impedance of the driven element becomes a significant contributor. The driven element also has a significant effect when it is 3/2-wave resonant.

"These factors are noted in the *RF Design* article. My patent application covers a family of antenna elements obtained by using the design equation.

"Finally, readers attempting to duplicate my 10/18/24MHz example antenna must remember that the required spacing is a function of conductor diameter - the 2in (5.07cm) spacing is valid only for No 12 AWG wire! Use the design equation or an antenna modelling program to determine spacings for other wire sizes.

"I am pleased to have received many constructive comments from antenna experts, and lots of interest from the amateur community. Although I didn't discover this antenna type, I'll gladly take credit for 'uncovering' it once again, and adding to our understanding of its behaviour! Thanks again for introducing it to *RadCom* readers." ♦

MORE ON METAL-OXIDE VARISTORS

IN *TT* (October 1984, p63) attention was drawn to N2EMT's warning that care is needed when using metal-oxide varistors (MOVs or VDRs) to remove spikes from mains supplies. He specifically warned against using MOVs of insufficient voltage rating and recommended a voltage rating of some 1.5 times the nominal mains voltage. This would imply the use on 240V mains of MOVs rated at 360V AC, which seems rather excessive; 275V AC MOVs are generally used to remove spikes from UK mains supplies.

N2EMI pointed out that if an MOV turns on and stays on, the least hazardous outcome would be a blown fuse. But there was the possibility that an overheated MOV package could either burst into flames and/or expel hot material, possibly igniting nearby components or materials.

David Cutter, G3UNA, of David Cutter Engineering writes: "I designed a small 275 MOV into the power supply circuit of a hot drinks vending machine as part of a 'bomb proof' design to prevent spikes getting into the microprocessor circuits: these machines are located in the most electrically noisy places anyone could imagine.

"In my first production batch of 150, I had four failures and these were all catastrophic, ie the MOV exploded, depositing metal film onto nearby components. I thought I'd taken enough precautions regarding series resistance upstream of the MOV by putting it after a PTC and a 28mH (yes mH) double inductor.

"Now, since I moved the MOV further downstream after another smaller PTC used to protect the mains transformer (a BS safety requirement) I have had no failures in a couple of thousand installations. The revised arrangement (**Fig 8**) offers a better choice from a suppression point of view, but was not done originally because the PTC was then built into the transformer.

"The lesson is: if you want to use small (ie cheap) MOVs at close to the supply voltage, make sure they have plenty of resistance between them and the mains; in my case about 24Ω in the complete return circuit using a 20-joule MOV (7mm diameter). It is possible to use a fireproof (ceramic ie concrete look-alike) resistor if you do not have other circuit elements that add up, as in my case.

"Incidentally, no further failures occurred in the original batch of 150!"

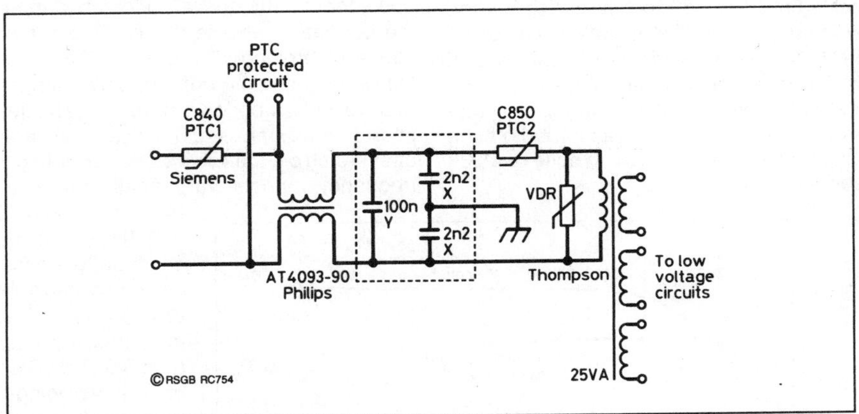

Fig 8: G3UNA's arrangement to increase safety factor when using MOVs with a voltage rating near that of the supply mains.

TEMPERATURE EFFECTS ON LC OSCILLATORS

AN ITEM ON 'Single-valve Transmitters' which I prepared for the July 1995 *TT*, but which has so far had to be held over on account of space limitations, included criticism by John Roscoe, G4QK, of the Hartley cathode-tap ECO in the rather over-complex valve transmitter by PA3ZFK noticed in the January 1995 *TT*. This surprised me as for many years I made good use of this type of VFO, but G4QK seemed convinced that this circuit contributed to the long warm-up frequency drift of the classic HRO when switched on from cold. He suggested that a possible explanation was that the heater/cathode insulation of most receiving valves was determined more by price than by choice of material with good dielectric properties.

In correspondence I mentioned this to Dick Rollema, PA0SE, who has been busy restoring his HRO (see *TT*, December 1994) and had commented on the warm-up drift (there was no voltage regulator tube in this 1934 design).

In a recent progress report on his HRO PA0SE writes: "For my National HRO receiver I have now constructed a loudspeaker cabinet out of a wooden box that once contained bottles of Bordeaux wine. My wife painted it black and gave it a finish that more or less (less than more) looks like the black crackle finish of the receiver. I also put the power supply in the loudspeaker enclosure. The PSU is not an original National unit and is built on a copper chassis! It had a conventional capacitor-input filter but when the standby switch on the receiver was turned off, the voltage rose to nearly 400V in spite of a 33k bleeder resistor.

Pat Hawker's Technical Topics

PAT HAWKER, G3VA
London 37/SE22 8SS

"This seemed excessive and I have changed it to a choke input filter by disconnecting the input capacitor and putting it in parallel with the output one (which now totals 100μF). The voltage under load is now 205V, still more than adequate, but unloaded (apart from the bleeder) rises only to 270V . . . smoothing with the single LC ripple filter is so good that only with a turned down volume control and my ear near the speaker can a slight hum be heard.

"For further protection of the receiver, I have connected a pilot lamp in series with the centre-tap of the transformer, with a fuse and double-pole on/off switch on the mains (primary) side.

"I do not believe, as G4QK apparently does, that the heater/cathode impedance of the oscillator valve in the HRO is a major cause of prolonged frequency drift. It certainly causes some drift but only during the relatively short period before the valve temperature has stabilised. I noticed that, when, after switch on, the valve starts oscillating the frequency changes very rapidly downwards by some 3kHz on 3.5MHz. But this ceases after a minute or two and then a slow drift starts. I have measured the drift on both 3.5 and 28MHz after a warming-up period of 5 minutes: see **Fig 1 (a)** and **(b)**.

"Measurements were also taken after inserting a cold coil pack into the receiver after this had been switched on for more than 24 hours. This shows, as I had expected, that most of the drift results from the coil packs and probably from the coils themselves.

"Though not impressive by current standards, my feeling is that for a 1934 design, the results are pretty good".

PA0SE's finding that coils are a major cause of drift is a further endorsement of the views put forward in a book first published in 1939 *Theory and Design of Valve Oscillators* by Dr H A Thomas (Chapman and Hall) which stresses the important contribution to frequency drift made by the effect of temperature on the inductance of the tuned circuit. While temperature changes on the maintaining amplifier (ie valves in those pre-transistor days) have some effect, "variations of the parameters associated with the oscillation circuit are responsible for far greater aberrations of frequency" - the effects of temperature on inductances are more important than on most capacitors (although even air-dielectric capacitors can contribute to frequency changes with temperature) unless these have a temperature coefficient specifically intended to compensate for the overall changes.

Dr Thomas provided detailed data on the temperature coefficients of various types of formers and conductors (recommending flat wire, ie copper tape and heat-shrunk windings on ceramic formers). He also showed how complex temperature-compensated coils and capacitors could be constructed. Characteristics of coil formers are important and it would be interesting to know how modern ferromagnetic cores compare with some of the ceramic and PTFE formers used in the days before toroid cores became popular. It is important to realise that although semiconductors reach thermal stability much quicker than valves, the stability of transistor LC oscillators is not necessarily better, and may be worse. For optimum stability an LC VFO should be left running continuously to avoid the effects of heat-cycling on inductances and capacitances.

Following the *TT* item on 'Stable LC Oscillators' (November 1994), R McEwan Reid, G4GTO, pointed out that it omitted the part played in the 1930s by W H F Griffiths of H W Sullivan Ltd. He developed for laboratory use a wide-range LC oscillator which had a long-term frequency accuracy of +/- 100ppm and a short period stability better than 1ppm. The

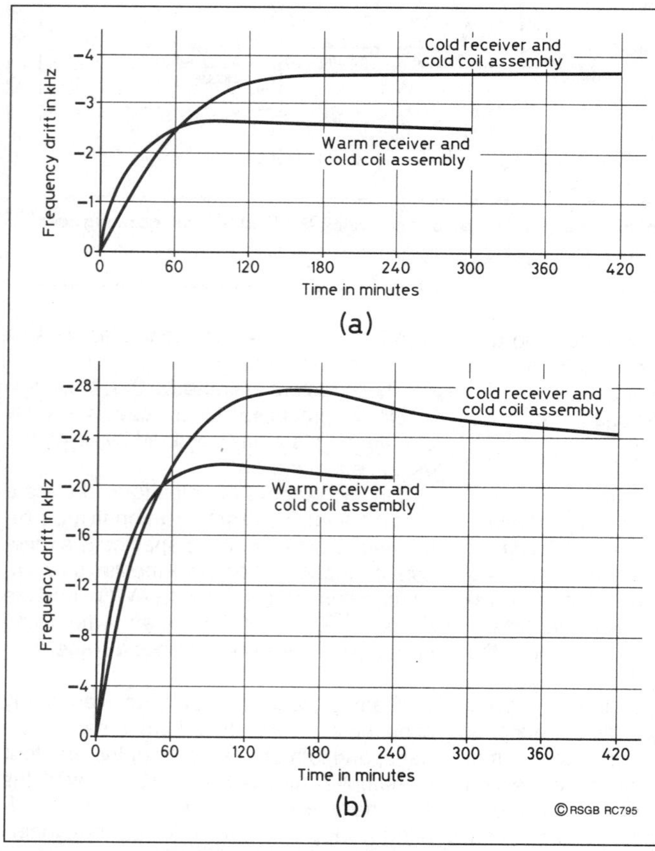

Fig 1: Measurements of the frequency drift of a 1930s HRO receiver made by PA0SE. (a) Receiver tuned to 3650kHz. (b) Receiver tuned to 28,500kHz.

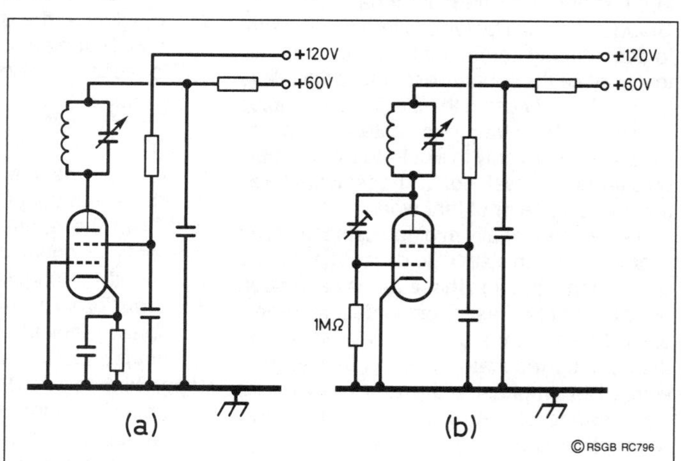

Fig 2: (a) Basic form of the Dynatron negative resistance oscillator. Note that the HT applied to the screen must be much greater than that applied to the anode. (b) Modified form for operation at frequencies higher than about 15MHz.

range of 1000kHz to 15MHz was covered by nine temperature-compensated inductors and a variable temperature-compensated air capacitor all of which had temperature coefficients (L or C) of less than 5ppm (covered by many British Patents of that period). In the mid-1930s the drive was an ACSG screened-grid (tetrode) valve as a dynatron oscillator. Later, when this valve became unavailable, considerable effort was made to use an EF50 pentode valve as a transitron oscillator without impairing the stability. The familiar red aluminium screen of the EF50 had to be removed to reduce the warm-up drift and valve capacitance effects.

Since the inductor was only partly screened, it was unsuitable as a drive for a transmitter. Under normal conditions, in the London area, the oscillator (wavemeter) could be 'pulled-in', over a few Hertz, by the BBC Droitwich long-wave transmitter on 200kHz.

G4GTO believes the fact that both the dynatron and transitron negative-resistance oscillators used 'two-terminal' inductors contributed much to the high degree of frequency stability achieved.

According to Dr Thomas's book, the upper frequency limit of oscillation with the basic dynatron oscillator **Fig 2(a)** was about 15MHz but F M Colebrook developed a modified form **Fig 2(b)** that maintained oscillation up to the VHF region (eg 54MHz). It would appear that the presence of the extra capacitance between anode and control-grid leads to an apparent decrease of negative resistance and consequently an extension of the operating frequency range.

ANOTHER LOOK AT RF SWITCHING DIODES

DAVE FAR, G4HRY, in 'RF Switching Diodes Controversy' (*TT*, July 1995, pp67-78) questioned the wisdom of wholesale substitution of HP 5082-3081 RF PIN diodes for the switching diodes fitted in the front-ends of typical modern HF transceivers. This item interested a number of readers concerned with the development of high-performance receivers including Colin Horrabin, G3SBI. He felt that the performance data provided by G4HRY and earlier by Tom Thomson, W0IVJ, in *QST*, December 1994 (summarised in *TT*, April 1995 with W0IVJ's measurement data in *TT*, July 1995, p67) deserved further investigation.

G3SBI writes: "**Tables 1** and **2** show some IP3 and RF isolation measurements on various diodes popularly used for low-level RF switching. This work was carried out by two vacation students - Mike Smith and Alex Macdonald - using test equipment in the RF laboratory of the SERC at Daresbury.

"The original test circuit (**Fig 3**) was designed to simulate different bandpass filters being switched in and switched out, permitting insertion loss and the isolation in the off condition also to be measured. Note that diodes in the off condition see about 12V reverse bias. In practice we abandoned the bandpass filters in order to widen the range of frequency measurements. However, Fig 3 is shown to provide a possible diode switching scheme for ladder filters. The insertion loss measurements in Table 2 are therefore the total for two transformers and two diodes.

"From the tabulated results, it can be seen that an important factor with some diodes is the sensitivity of insertion loss to forward current. The HP 3081 diode, severely criticised by G4HYR, shows a marked reduction of loss and hence improved performance as the forward current is increased from 6 to 10mA. As noted by G4HRY the Siemens BA482 (as used in the Ten Tec Omni) shows the lowest insertion loss (this would seem also to confirm G4HRY's view that it is inadvisable to replace BA482 diodes with HP3081 diodes - *G3VA*).

"Clearly anyone contemplating changing the RF switching diodes used in his transceiver must first estimate or measure the 'on' current used in the particular model concerned.

"All the measurements shown in Tables 1 and 2 were made in 50Ω systems. Some transceivers use 200Ω design impedances so that the insertion loss of the switching diode is of less consequence, but in this case the signal voltages are higher so that the IP3 intercept point may be lower.

"We could not achieve the sort of third order intercept figures reported by W0IVJ but it became clear that the IP3 performance of diodes is very much affected by the frequency. One is tempted to suggest that the designer of the 1N4007 diode must have been a radio amateur since its best performance seems to be around 7MHz where we measured an IP3 of +50dB. In fact, it is surprising how well the 1N4007 performs.

DIODE TYPE	FREQUENCY 1.815MHz		FREQUENCY 7.015MHz		FREQUENCY 28.015MHz	
	(dBm)	(dB)	(dBm)	(dB)	(dBm)	(dB)
BYD11M (@6 & 10mA)	38.0	1.16	45.0	1.23	35.5	1.93
IN4007 (@6 & 10mA)	38.0	0.25	42.0	0.97	37.0	2.0
HP3081 (@6mA)	38.0	3.82	37.0	3.9	36.0	3.36
HP3081 (@10mA)	35.5	2.68	41.0	2.5	35.5	2.84
BA482 (@14mA)	25.0	0.51	38.0	0.81	38.0	2.32
BA482 (@6mA)	18.5	3.49	23.5	0.71	36.0	2.0
IN4148 (@6mA)	21.2	6.61	19.0	7.47	17.3	5.83
IN4148 (@10mA)	12.5	3.61	15.0	4.1	16.5	2.84

Table 1: Third order intercept (dBm) and Test circuit insertion loss (dB)

DIODE TYPE	FREQUENCY 1.815MHz	FREQUENCY 7.015MHz	FREQUENCY 28.015MHz
BYD11m	73.9	53.9	50.0
1N4007	63.0	43.9	35.3
HP3081 @6mA	80.0	63.5	57.0
HP3081 @10mA	81.9	63.9	56.8
BA482 @14mA	80.0	65.0	57.0
BA482 @6mA	84.0	61.8	52.4

Table 2. Test circuit off isolation (dB).

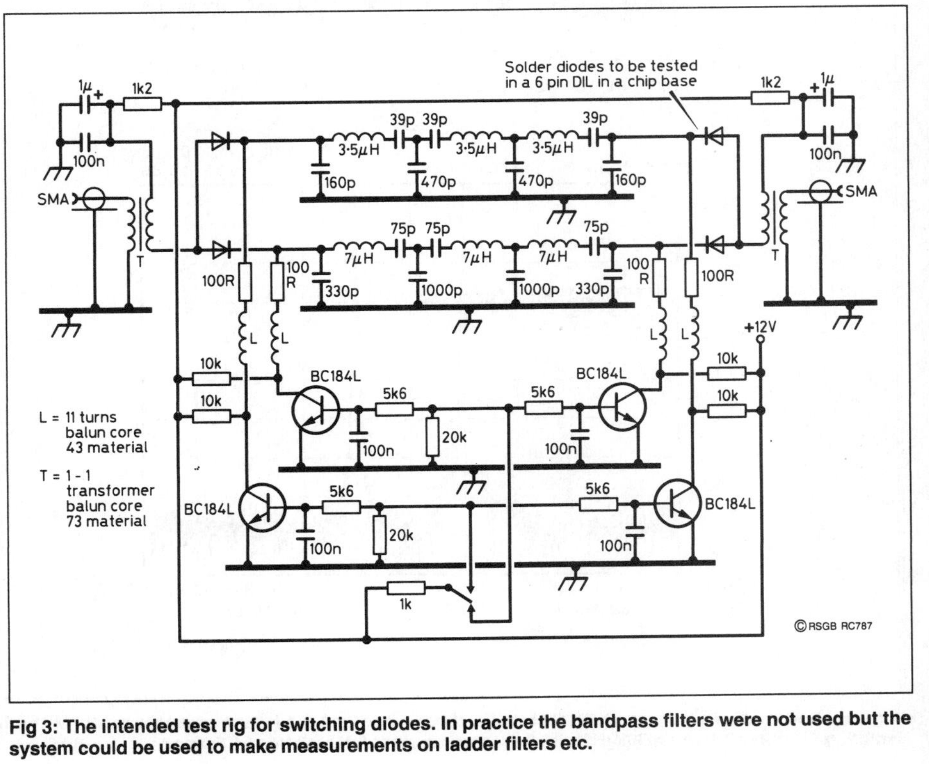

Fig 3: The intended test rig for switching diodes. In practice the bandpass filters were not used but the system could be used to make measurements on ladder filters etc.

"The 1N4007 performance encouraged us to obtain some BYD11M diodes from RS Components (29p each). These are rated as 1000V PIV, 0.5A rectifier diodes and are the same physical size as normal signal diodes. Performance as switching diodes was good: relatively low insertion loss, good IP3 intercept point, and good degree of 'off-isolation' at 50MHz, even with a fairly low 'on' current: see **Fig 4 (a)** and **(b)**.

"It should be appreciated that a very high degree of 'off-isolation' between different bandpass filters is less important in up-conversion receivers (50dB is probably adequate) since the image frequency will be in the VHF region and will be largely taken out by low-pass filters before the mixer. However, in the case of a receiver with a 9MHz IF and a 5MHz local oscillator, the image will fall in the 3.5 and 14MHz bands. In this case, isolation between the filters should be greater than 90dB so that two diode switches in series with a shunt diode would be needed to achieve this degree of isolation. It would be better to use double-pole relays, one bandpass filter to ground. It may then be necessary to use the technique of 'DC-wetting' ie arranging to pass a few mA of direct current through the relay contacts, to improve long-term contact reliability."

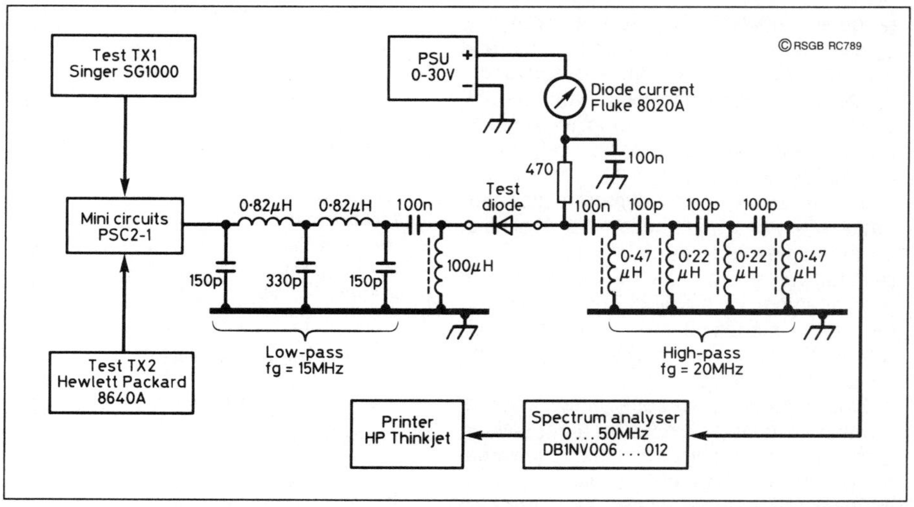

Fig 5: Test rig used by DB1NV for the measurement of intermodulation characteristics of switching diodes.

B J Mitchell, G3HJK, commenting on the strictures by G3LLL on the long-term reliability problems of RF switching by relays (*TT*, September, p68) also draws attention to the better reliability that can be achieved by DC-wetting, a long-established Post Office dodge, both in minimising oxidation and in reducing migration of contact material. This can be achieved with the aid of suitable blocking capacitors and resistors to feed DC across the contacts without changing the biasing of the active devices. G3HJK uses this technique with his FT102 and, despite being a pipe smoker, has not had to replace any of the six RF switching relays over years of use.

Another source of information on the 'Intermodulation properties of switching diodes' is an article which appeared in *VHF Communications* (Vol 26, Spring, 1/1994, pp12-18). Dr Ing Jochen Jirmann, DB1NV, used the measuring rig shown in **Fig 5** to measure intermodulation characteristics of a selection of switching diodes most of which have not been investigated by either W0IVJ or by G3SBI etc at SERC. DB1NV measured both IP2 and IP3. For the IP2 data he used test frequencies of 12 and 15MHz measuring IP2 at 27MHz while varying the diode DC from 2mA to 20mA: **Fig 6(a).** For IP3 he used test frequencies of 6 and 15.5MHz with the IM product evaluated at 25MHz with diode currents of 2mA and 5mA only: **Fig 6 (b).**

The conclusions drawn by DB1NV tend to differ in some respects from those of G3SBI. He wrote:

- Good repeatable intermodulation figures can be obtained only through the use of 'correct' PIN diodes, but they have their price. Miniature relays are even better, but more expensive and bigger.
- Universal diodes misused as HF switches can yield very good results (ISS53) or catastrophically poor results (1N4148). Moreover, it cannot be calculated what effect variations in the manufacturing parameters will have (different production lines, different production methods).
- The relatively good cut-off results obtained in practice from apparatus fitted with tuner switching diodes is not consistent with the poor measurement results from the BA244.
- The existing test rig of Fig 5 needs to be improved or re-constructed in order to check whether sufficient DC is flowing through the diodes. An attempt should be made to set a value of about 20mA . . . It can be concluded from the results that the main cause of intermodulation interference in current amateur HF receivers/transceivers should be sought in the area of the HF input switching diodes.

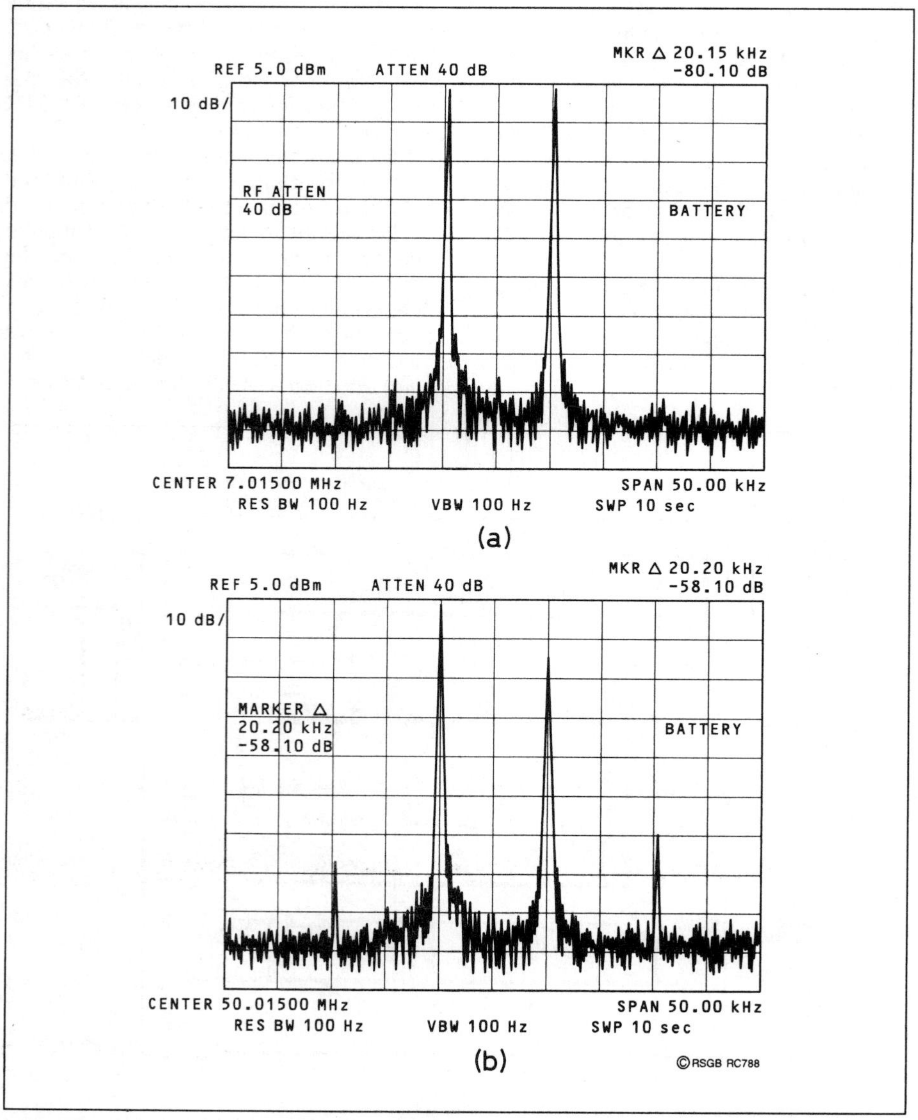

Fig 4: (a) Spectrum of a low-cost BYD11M rectifier diode with 10mA diode current measured at 7MHz (off isolation 53.9dB). BYD11M at 50MHz with 6mA diode current (off isolation 54.5dB).

ANOTHER LOOK AT COAXIAL ANTENNAS

THE AUGUST 1995 *TT* item 'Coaxial Travelling Wave Antenna?', based on the implementation by Australian amateurs of the 1927 patent of Emil Geles, has attracted comment from a number of readers. Several share my own view that it is difficult to see how this structure can provide for transmission an efficient form of non-resonant broadband travelling wave antenna, despite its apparently successful use as such by VK3YX and VK2EHN, especially when powered from a truly balanced feed. The point is made that there must be common-mode current on the outer surface of the tubular mast.

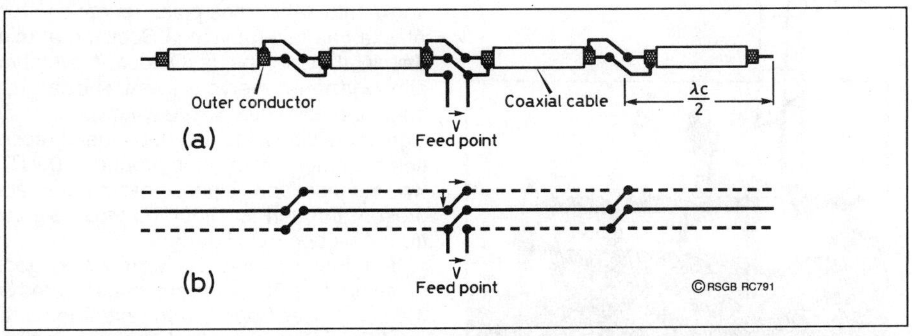

Fig 7: Basic geometry of the 'coco' (coaxial collinear) antenna formed from a series of electrical quarter-wave sections of cable incorporating cable velocity factor and transposing the inner and outer connections.

Nic Hamilton, G4XTG, pointed out: "Unless the radius of the tube is large, the field produced by the (inner) wire is completely stopped by the opposite field produced by the balancing current flowing inside the tube. So, as the patent says, the radiation is due to the current flow on the outside of the tube, which is not balanced by the current on the wire. So far, so good. But where does this unbalanced current come from? The balun (a current type) is there to make certain that the currents at its terminals are equal. If this in fact happened, there would be no radiation.

"The unbalanced current on the outside of the tube must be provided by the balun. This may be why VK3YX found it necessary to experiment with some 80 different baluns. It would seem that the antenna will only work well if the balun doesn't! The unbalanced current flows from the outer of the feeder coax, through the balun, and on to the outside of the tube.

"I believe that a better system for this type of antenna would be to dispense with the balun, and to connect the outer of the feed coax to the inner wire of the 'radiating' element, and the inner of the feed coax to the tube of the 'radiating' element. This type of feed is well known; the most recent reference is 'A Modular Coaxial Collinear Antenna' by B Lagoun and L Bertel (IEE Conference on HF Radio Systems and Techniques, July 1994, *IEE Conference Publication No 392*, pp234-238)". (This type of coaxial collinear antenna has been noted several times in *TT*, for example the Coco array in *TT*, September 1987, pp662-3 or *Technical Topics Scrapbook*, 1985-89, pp191-2, see **Fig 7** - *G3VA*).

G4TXG points out that there will be a practical difficulty with this type of feed; the efficiency of the antenna will be higher with the result that the bandwidth will be reduced. It is likely that TVI will return with a vengeance, as it is more or less bound to do with an antenna that uses the outer of the feeder as a radiator.

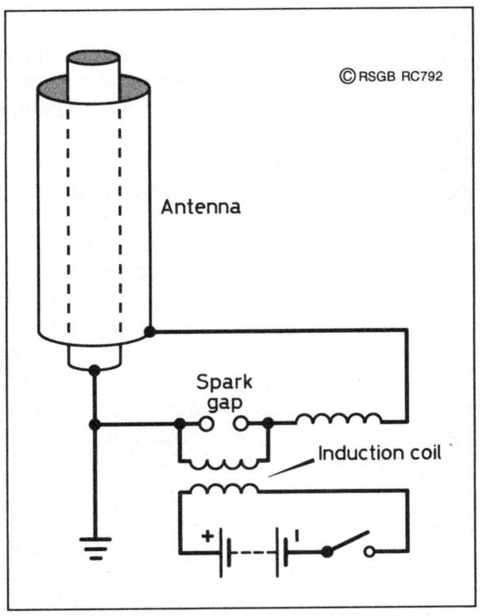

Fig 8: Basic form of Marconi's 1901 concentric cylinder antenna used as a spark transmitting antenna.

DB1NV also investigated the IM products resulting from saturated ferromagnetic cores, as used in both the aperiodic case (RF chokes) and for tuned (resonant) circuits. He found, for example, that Amidon ring cores of various sizes and intended for HF applications were practically free from intermodulation effects under normal conditions. He did, however, provide some design tips "some of which are not new, but which have probably fallen into oblivion in Japan". He wrote:

(1) Input filters effectively resistant to IM can be produced only using sufficiently large iron powder ring cores as inductances. They offer the best compromise between the space requirement and the level controllability.

(2) In compact rigs, rod core chokes, such as the Siemens MCC, can be considered as alternatives.

(3) Chokes in the filter structure, eg on the operating voltage feed, are largely non-critical provided they do not resonate.

DB1NV recalled the band-pass filters using ring core coils that were publicised many years ago by VE3TP. He commented: "These were not exactly cheap to construct but . . . solved every receiver IM problem so far . . . (proving) that it is possible to produce receiver input components which can meet today's requirements in relation to sensitivity and high-level signal strength." Dr Rohde's reply to G4HRY's criticisms will be published next month.

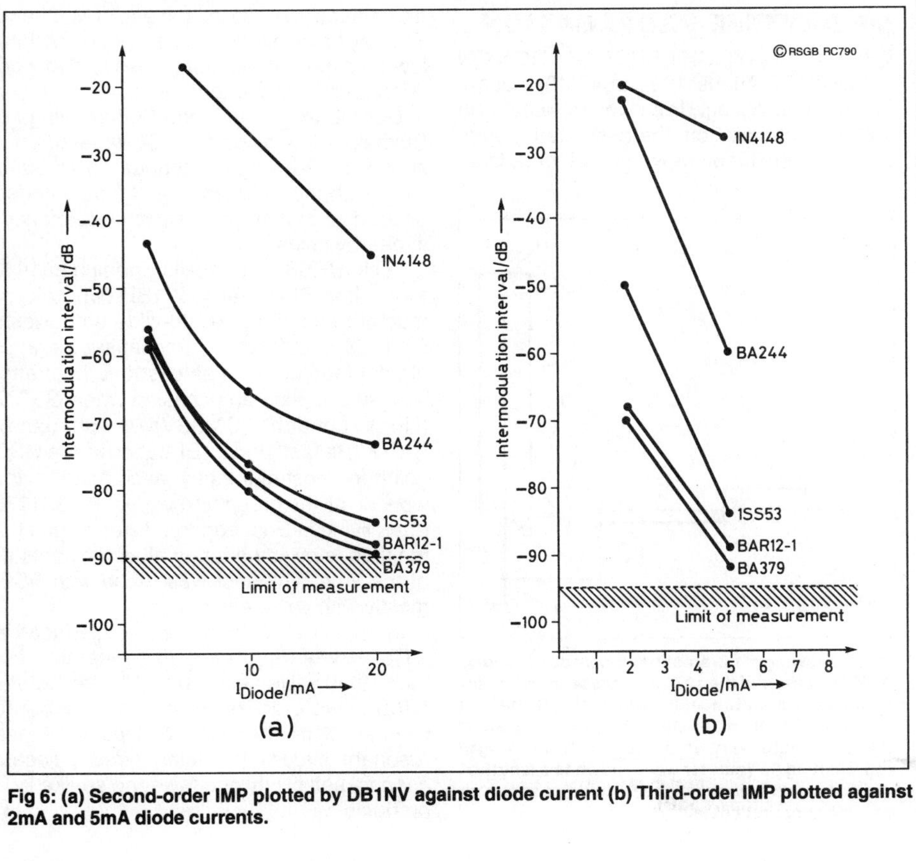

Fig 6: (a) Second-order IMP plotted by DB1NV against diode current (b) Third-order IMP plotted against 2mA and 5mA diode currents.

He also believes that the logical conclusion with this type of antenna would be to reconfigure it as a quarter-wave vertical with a good ground-plane or elevated radials. He is convinced that in the arrangement used by VK3YX it does not work in the manner intended by its Romanian inventor.

Dr Brian Austin, G0GSF, is similarly puzzled at the description of the system as a broadband travelling-wave independent of frequency. But he also reminds us that the basic idea of using a coaxial (ie concentric cylinder) antenna dates back much further than 1927. W J Baker in *A History of the Marconi Company* writes that Marconi took out patent No 5387 on 21 March, 1900 for a transmitting antenna consisting of two concentric cylinders, the outside one forming the radiator and the inside one earthed: **Fig 8**. The receiving antenna was similar. It is claimed that this "provided a fair degree of syntony"; that is to say it helped to "tune" the spark transmitter fairly sharply with an antenna of limited height.

G0GSF, who has been researching the use of radio communication during the Boer War, also draws attention (**Fig 9**) to Marconi's mobile wireless station of 1901 - intended for army use - with what appears to be a hinged concentric cylinder antenna mounted above a steam Thornycroft vehicle (the antenna lay flat on the roof of the vehicle during travel). The equipment was rejected by the army and transferred to the navy.

To bring the story of coaxial antennas up to the 1970s, John Pegler, OBE, G3ENI, while working in the Ministry of Defence (Navy), tried some experiments on a basically similar form of antenna which was officially recorded in October 1976 to protect the government's interest (under Section 4692) of the Patents Act 1949.

This comprised a coaxial structure a quarter-wave or less long exhibiting an inductive reactance at the base, which is matched to 50Ω by means of the capacitive potentiometer comprising two capacitors. Like the early Marconi antenna this was fed against earth, using the inner of the coax feeder connected to the outer tubing as described above by G4TXG.

G3ENI writes: "From time to time, I have used versions on 7, 3.5 and 1.8MHz and have conducted tests with G3ZUN who currently uses one on 3.5MHz. An unusual feature is the current distribution on the outer. Using a toroid transformer probe the measured current is almost uniform rising to a maximum at the end which seems to indicate a tendency to be a travelling wave."

G3ENI's draft patent specification of 1976 describes the system (**Fig 10**) as a "radio aerial and impedance matching system" with the object of providing a short antenna system of robust construction for reception and transmission of radio signals. "Another object of this invention is to provide an easily adjusted feeder to aerial impedance matching device which makes use of the inherent characteristics of the aerial in its operation. A further object is to provide a low-loss high-efficiency feeder to aerial impedance matching device which avoids the losses in coils often used for this purpose. An embodiment includes an arrangement for exciting with radio frequency energy a part of the structure of a ship, aircraft, armoured vehicle or other metallic structures and to provide a simple means of tuning the aerial to resonance over several decades of frequency.

Fig 9: The Marconi mobile of 1901 intended for (but not used by) the army during the Boer War. Apparently it was rejected by the army and transferred to the navy.

"Basically the invention consists of an aerial which is substantially shorter than 90° of electrical length and which consists of an inner rod or wire of high conductivity metal which is coaxially surrounded by a tube of high conductivity metal, and separated by a dielectric. In the case when the aerial is erected vertically, the inner rod and outer tube are electrically connected at their top ends . . ." A further explanation of the coax antenna has come from G4LU and will appear later.

MF DAYTIME PROPAGATION

THE ITEM 'DAYLIGHT PROPAGATION ON 1.8MHZ' (*TT*, August 1994, pp48-49) brought in comments not only from those interested in 1.8MHz but also from the dedicated groups that comb the medium-wave (AM) broadcast band for long-distance or unusual stations. Steve Whitt, G8DKL who is the editor of *Medium Wave News* (journal of the Medium Wave Circle) writes:

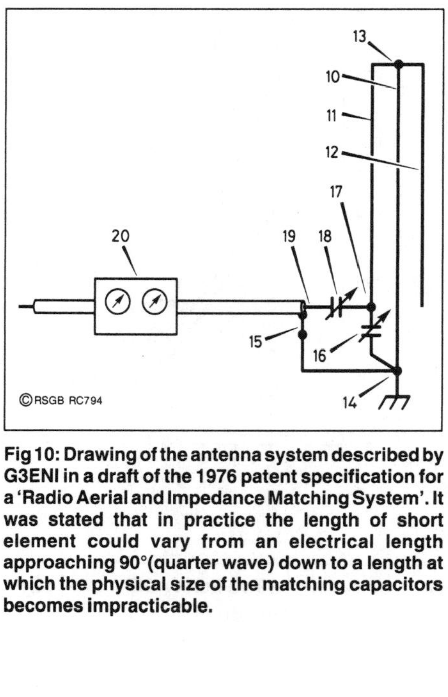

Fig 10: Drawing of the antenna system described by G3ENI in a draft of the 1976 patent specification for a 'Radio Aerial and Impedance Matching System'. It was stated that in practice the length of short element could vary from an electrical length approaching 90°(quarter wave) down to a length at which the physical size of the matching capacitors becomes impracticable.

"I am a very keen MW-DXer using Beverage antennas and have conducted numerous MW DX-peditions. You may be interested to know that MW trans-polar reception from Alaska has taken place in Scotland at midday in the months of October/November, although not an everyday event. The path has been approximately 40% daylight but with no sign of the signals fading out. Indeed reception becomes established around 0900UTC when transatlantic signals start to fade and present through to 1330UTC when signals from Asia began to interfere.

"For these events we were using good receivers plus Beverage antennas located on the coast near Cape Wrath (extreme northwest Scotland) so we had sea-gain and low interference and noise.

"The most striking aspect noted during DX-peditions is the way MW signals fade about two hours after sunrise and can be chased down into receiver noise rather than atmospheric noise. Also, on many occasions, signals begin to come back up again between three and four hours after sunrise – this has been observed a number of times although we have never measured exact signal strengths. We have heard trans-polar signals on frequencies as low as 660kHz through to 1170kHz but this appears to be limited only by suitable signal sources.

"A measure of MF groundwave daylight propagation is shown by the distance at which some of the recent restricted service licence (RSL) stations have been heard. RSL stations are authorised to use 1W and are intended to provide a maximum range of some 3 miles. Many have been heard at distances up to 20 miles and I recently heard one 85km away on 1413kHz limited only by co-channel interference despite the high groundwave attenuation towards the top end of MW band. The key to such reception is a good antenna, low ambient atmospheric noise and low co-channel electrical interference."

G8KDL who lives in the Suffolk village of Buxhall, a few miles from Stowmarket, has space for Beverage antennas up to 500m long, with the main source of local interference some daytime pulses from electric fences in nearby fields.

Mick, BRS31976 recalls listening in 1968 to the late Stew Perry, W1BB (who did so much to encourage world-wide enthusiasm for 1.8MHz DX) whose transatlantic signals did not fade out until at least one hour after sunrise and also an occasion when G3TZZ (North London), GM3SVK/A (Shetland), GI6TK (Belfast) and an EI station kept a QSO going for most of the day, each hearing and working all the group all the time. BRS31976 (then living in East London) listened most of the time, similarly hearing all participants all of the time. This test was made with SSB, mostly with valve equipment.

Indeed, the last time this topic surfaced in *TT*, a few years ago, there was general agreement that valve receivers still offer useful advantages for MF reception, probably due to the use of front-ends with capacitor-tuned resonant circuits providing good adjacent channel selectivity and fewer noise-like IMD products. ♦

Pat Hawker's Technical Topics

PAT HAWKER, G3VA
London 37/SE22 8SS

100 YEARS OF RADIO

THE RECENT IEE three-day conference succeeded in holding the interest of the nearly 200 delegates from the first day - largely devoted to the beginnings of radio with emphasis on the work of Hertz, Lodge, Marconi, Fessenden and the early application of radio to maritime and even military (Boer War) communications - to the third day with its reviews of the current use of space satellites, cellular radio, spread spectrum techniques, and the start of digital radio broadcasting.

In between, we were treated to a description and a special demonstration of the oldest working transmitter in the world - the 200kW Alexanderson alternator transmitter, SAQ, at Grimeton, near Varberg, Sweden, which can still put out a potent signal on 17.2kHz. This was received at the BBC monitoring station at Caversham with the signals sent along digital landline to Savoy Place. We also had reminiscences of the home-construction era of broadcast receivers; a review of developments in Germany 1920 - 45 by Dutch amateur Arthur Bauer, PA0AOB; radio propagation research; the pioneering FM work of Howard Armstrong; pulse-code-modulation (PCM) devised by Alec Reeves.

Antennas and propagation - from ELF to microwaves - were also well covered, with R C Kirby, formerly of the ITU Radio Bureau (and incidentally an amateur of 1938 vintage), considering the history and trends in spectrum regulation.

It was interesting to note how, after 94 years, the 'reception' of those three-dot signals by Marconi at Signal Hill, Newfoundland - the world's first attempt at long-distance radio - can still spark off controversy and continued study and experimentation. Canadian Dr John Belrose, VE2CV, is among those who retain doubts that signals from the spark transmitter tuned only by the fan antenna (**Fig 1**) at Poldu could have reached Newfoundland strong enough to be detected. J C B MacKeand, WA3ZKZ, (an Englishman living in the USA) believes from his studies that the signals made it, though not necessarily on the fundamental frequency imposed by the fan antenna.

The very first 'DX-pedition'. Marconi (centre) with G S Kemp and P Paget at Signal Hill, New Foundland in December 1901. In the background is the kite used to carry the receiving antenna. The basket on which they are sitting contained the balloons which they had intended using but which were blown away when they attempted to fly them. Scientists and engineers still argue whether the dots heard by Marconi and his assistants were really signals transmitted from Poldhu.

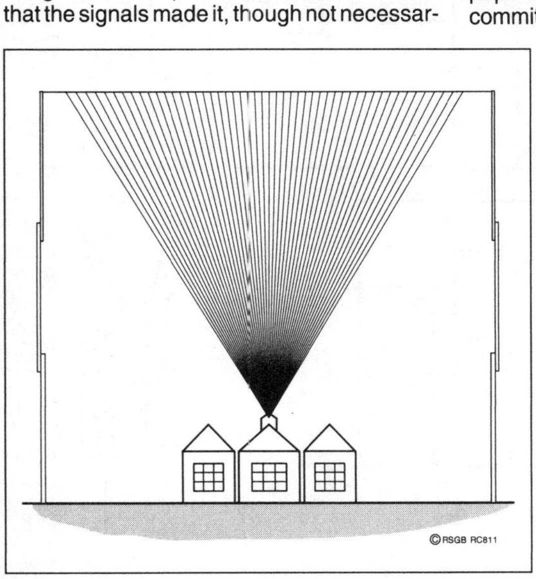

Fig 1: The emergency fan antenna used at Poldhu, Cornwall in December 1901 after the original, more complex and higher antenna had been destroyed by gales. Two wooden masts, 170ft high, were hastily erected and the fanwise antenna consisting of 60 bare copper 7/22 wires was strung from a triactic stretched between them. The aerial was not excited by an induction coil, but by a 25kW, 45Hz alternator charging high voltage capacitors through a transformer.

Amateur radio was not forgotten; a considerable proportion of the 'professional' delegates including many of those delivering papers and five members of the organising committee hold, or had held, amateur licences. A paper 'Amateur Radio: past, present and future' written by Joe Kasser, K3/G3ZCZ, was presented by Richard Lambear, G3RWL. The paper on 'Developing digital radio communications using small satellites - the UoSAT programme' describing work at the University of Surrey gave full credit to its origins in the AMSAT programme.

My personal regret, although in the circumstances inevitable, was that there were relatively few delegates at the threshold of their careers in radio communication. The rewarding 250-page IEE Conference Publication (No 411) is priced (at almost £50) well beyond student and amateur budgets. I hope that many will seek it out in the university and other libraries that hold IEE conference publications.

A 'poster presentation' by Dr Peter Excell and D. Excell of the University of Bradford explored 'What use is the history of radio?' This stressed that radio is a particularly impressive achievement of human, scientific and technological investigations which have made it a major force in the shaping of the modern way of life. This makes its history of use in education, technological forecasting. Museums with radio artefacts should be encouraged and should seek to establish and show the historical context of radio relics.

Examples of some of the subjects covered at the conference are as follows:

"Quartz crystals have been used in oscillators continuously since 1923, and are likely to remain the major form of stable low-noise frequency sources for many years to come" - Professor Mike Underhill, G3LHZ, conference. (I hope to refer to this excellent paper on another occasion - *G3VA*)

"Noise has been described as 'The ubiquitous, unwanted, insistent, unwelcome gatecrasher of electronic systems'. Whatever the description, noise adversely affects a radio system's performance. It can be naturally occurring or man-made and sometimes intentionally introduced as jamming. Good engineering practice based on 100 years of research and development can reduce noise and its effects to manageable levels, but there are some fundamental limits. 'Know your enemy' has always been a good maxim . . ." - David Evans, consultant.

"There does not, at present, seem to be any alternative to electromagnetic radiation for communication between places not connected by wire or fibres. Neither do there seem to be any major propagation modes which are not known at present, at least in general outline. Thus propagation research will continue to be devoted to more precise characterisation and modelling of the medium. This should yield better prediction methods, with less statistical uncertainty, which may be applied to give greater utilisation of the spectrum... There is no doubt that the small community of propagation researchers throughout the world will have the important work to do for many years to come". - Professor Les Barclay, OBE, G3HTF.

"Already (communications receivers) are available which digitise the signal at some point and perform, for example, filtering in software.... The approach is to retain a substantial degree of analogue filtering, digitise the baseband signal and use the digital signal processor to perform tasks which are more efficiently done in the digital rather than the analogue domain. Squelch, noise reduction, speech compression, audio AGC and data demodulation are typical examples. Efforts are being made by receiver manufacturers to digitise the signal as near to the antenna as possible, thus eliminating analogue circuitry. The ultimate goal is the antenna signal directly feeding an analogue-to-digital converter with all signal processing and conversion to baseband being performed in software. All that then remains is too digitise the operator!" - R T Sutton, Eddystone Radio.

"Early antennas were designed empirically and any theoretical explanation was produced after a working antenna had been developed. This was largely the case up to the mid - 1920s. From that time onwards electromagnetic theory was used to aid the development and design process. Design methods can be divided into 'before' and 'after' computers were developed" - Professor A D Olver.

The associated exhibition was similarly well balanced in covering many aspects of pioneering and current equipment and components, and clearly interested delegates between sessions.

MORE ON COAXIAL ANTENNAS

COMMENTS HAVE CONTINUED to come in about August's *TT* item 'Coaxial travelling wave antenna?' (see also November's *TT*). Stan Brown, G4LU, writes: "I would suggest that the so-called 'coaxial travelling wave antenna' is anything but that. It seems to me to be just a coaxial stub almost in parallel with the feed point of a vertical antenna, *viz* the outer surface of the tube. I say 'almost in parallel' because the coaxial stub is connected across the whole of the transformer secondary winding whereas the vertical antenna is connected between one end and the secondary's virtual earth point. Taking the simplest form of Fig 6(a) (August), the reactance of the short-circuited coaxial stub will be its characteristic impedance times the tangent of its electrical length expressed as an angle. The reactance of the vertical antenna will be opposite in sign and its magnitude will be its characteristic impedance divided by the tangent of the same angle. The stub characteristic impedance will be, due to its larger diameter, somewhat lower than that of a wire of the same length. When referred to the primary of the transformer there will be a cancellation of reactances which in effect will enhance the bandwidth of the antenna over that obtained from a simple vertical.

"However, the tangent of an angle and its inverse have a 'see-saw' relationship. When one is infinite, the other is zero and vice-versa; but in mid-range, when the angle is about 45° (or its odd multiples) they are about equal. Thus cancellation of the reactances will take place only over a small range of angles of mid-value; but at the extremes, when the angle is a multiple of 90°, the reactances will diverge considerably in magnitude. Hence the statement that the antenna will not work if it is resonant (nor anti-resonant). At these latter points the SWR will go haywire so one has to trade off the inferior performance of a short vertical antenna against the improved bandwidth over what one would normally obtain."

G4LU continues: "The radiation resistance of the antenna will be that of a short vertical and lower than one could expect with a longer wire antenna. Consequently, the transformer must be designed to suit the lower value or alternatively designed to be compatible with the ATU in use.

"It is interesting to note, from the dimensions quoted, that the tangent of the angle corresponding with the electrical length is between 0.5 and 1.6 for the 1.8, 3.5, 14 and 28MHz bands; whereas it is above 4.5 at 7MHz, which probably explains the quoted performance difference on that band. The other variants in Fig 6 (August) are modified versions of the stub and provide different values of reactance at its input, but the same analysis will apply.

"Since the antenna is essentially a vertical antenna, albeit with reactance compensation at its feed point (which adds nothing to its radiation properties), I cannot see why its performance should not be enhanced if it were mounted above an earth plane (ie earth mat or radials)."

HIGH PERFORMANCE, HIGH GAIN IF AMPLIFIER

AN ITEM, 'Low Noise AGC Controlled IF Amplifier' in *TT*, May 1995, p61 emphasised that modern approaches to receivers with super-linear front-ends and with DSP at the 'rear-end' require new thinking in most sections of the receiver. In May's *RadCom* a 9MHz IF amplifier design by Colin Horrabin, G3SBI was described, intended for use after

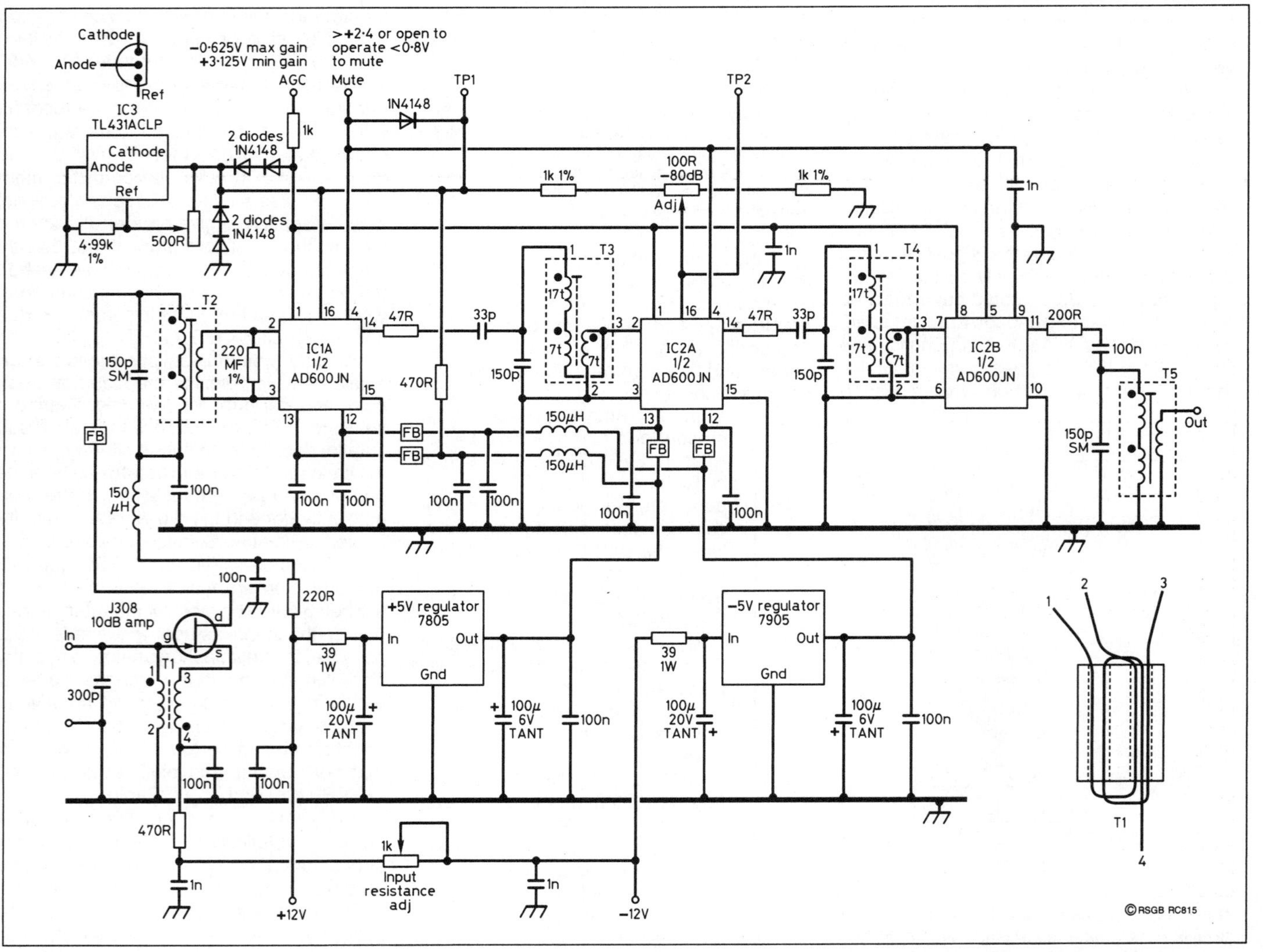

Fig 2: The high-gain, low-noise, AGC-controlled state-of-the-art 4.434MHz IF amplifier as developed by Bill Carver, K6DLG/7. Input and output impedance 50-Ω. Gain 95dB. AGC range 120dB. Max input 0dBm. Max output - 25dBm. Noise figure 1dB. Power: +12V and -12V DC at 100mA. Transformers: T1, T2, (see sketch) 1-2, 4t Nr 28. 3-4 3.5t Nr 28. Core Amidon BN-61-202. T2, T5, 15t trifilar Nr 30, slug tuned form Amidon L43-6, primary 6-9.5µH (set at 7.2µH). IC1B not used. Ground pin 6, 7, 8, 9, 10 with pin 11 not connected. Test point 1. Set to 2.596V with '-40dB pot' and refine for smooth gain/voltage around 40dB gain-reduction point (AGC +1.875V. Test Point 2. Set to 1.298V with '-80dB pot'. Refine for smooth gain/voltage around 80dB gain-reduction point. (AGC 0.625V).

his state-of-the-art H-mode mixer as presented in *TT* in October 1993 (correction November 1993) and in the current edition of *Radio Communication Handbook*.

The H-mode mixer is currently arousing considerable interest in the USA and one result has been that Bill Carver, K6DLG/7, has developed an alternative low-noise IF amplifier with exceptional characteristics including a 1dB noise figure, gain of 100dB, maximum input of 0dBM, maximum output of -20dBm and an AGC range of 120dB. K6DLG has sent details of his design (**Fig 2 and 3**) to G3SBI together with permission for its publication in *TT*.

G3SBI writes: "This IF amplifier intended for use with an H-mode mixer front-end is based around the Analog Devices AD600 devices; although the interstage networks are centred at 4.4MHz it would be useful with modified transformers for any IF up to about 21MHz.

"I expressed some concern to K6DLG that with so much gain at a single frequency, there could be a problem from BFO leakage to the front-end. In my case I may convert from say 9MHz to 6MHz for final detection, alternatively a final IF of 100kHz could be used but if this were done it might be better to use 2-pole crystal ladder filters as the interstage coupling network. Otherwise there might still be a BFO leakage problem.

"Bill is a gifted radio engineer although he now teaches mathematics. He has produced a most attractive IF amplifier which has already been successfully duplicated by some of his radio friends. It should be most useful to constructors of advanced receivers."

Subsequently, Bill Carver, K6DLG/7 sent me additional notes on this state-of-the-art IF amplifier. He is planning later to publish a fairly detailed article describing the complete 'back end' of a receiver, including this IF amplifier, in *Communications Quarterly*.

He writes: "First some philosophical issues - *Broadband Noise*: This amplifier has high gain so even though the noise figure is low due to the JFET with gate-source feedback, the broadband output noise is considerable. When it operates a product detector that noise is translated to audio, folded over around the BFO frequency (there goes 3dB!), but band limited. The result is acceptable, a minimum detectable signal (MDS) in the region of 0.02μV.

"However, an AGC detector is *not* a linear translation. Band limiting the output of a DC rectifier does no good. I find the broadband noise is about 100mV RMS and this is far too large. Therefore, if AGC is used, a post-amplifier bandpass filter is mandatory. It also offers the opportunity of preventing the loss of 3dB when the noise spectrum is folded at the BFO frequency.

"I successfully patched in the 2-pole 'roofing' crystal ladder filter that I have following the H-mode mixer. It is 3kHz wide at -3dB, with the holder capacitance 'neutralised' to maintain RMS, but I would like it lower, so I could reduce the AGC threshold. I have no interest in increased time delay through any filter, the bane of AGC stability, but intend to experiment 'soon' with a 3-pole filter.

"*Gain, output level etc*: We usually have only a general idea of what the final gain of an IF amplifier will be. Furthermore, we have even less idea of how that gain is distributed or which stage will overload first.

"The AD600 changes that. The gain and gain distribution depend only upon the accuracy of the coupling networks. In this amplifier the J308 delivers 10dB of gain from the input to pin 2 of ICIA. IC1A delivers 0-40dB of gain to T3. T3 has 10dB of loss to the input on IC2A. Similarly T4 has 10dB of loss, IC2B has 0 - 40dB of gain and T5, with the 200Ω resistor, has 12dB of loss.

"When a finished 3-stage plus JFET amplifier delivers only 95dB of gain we have extra loss in the interstage networks. If we are to avoid overload at 0dBm input, the output must be limited to (120-gain) dBm. In the event of say 95dB gain, the nominal output level should be limited to -25dBm if distortion is to be avoided over the whole 120dB gain range of the AD600.

"Where the overall gain differs from 100dB, I've found the culprit is usually a 33pF capacitor on the low side of its tolerance or (once) out of tolerance. A few pF in parallel will bring the gain up.

"The earliest amplifier had no separate reference voltage for the AGC section of the AD600. It used divided-down supply voltage. Since this determines the point at which each chip's gain-control section operates this proved far too casual an approach. I've added

IMPROVED STOPBAND OF CRYSTAL LADDER FILTERS

STEIN TORP, LA7MI, has found a relatively simple way to increase greatly the stopband attenuation. One result would be to provide an effective roofing filter in a double-conversion receiver, but the technique could also be applied to narrow-band filters using three or more crystals.

LA7MI writes: "My experiments were carried out with fifth harmonic 92.8611MHz (HC18/U) crystals. At their fundamental frequency (18.556MHz) they were free of spurious responses and I first constructed a simple ladder filter at that frequency: **Fig 4**. The stop-band attenuation was 35dB, which is about usual for such a filter.

"To build a better filter (**Fig 5**) I first modified the matching network. L1 was replaced by a 51pF polystyrene capacitor; C1 was replaced by an inductance wound on an Amidon T37-2 toroid. By using a bifiliar winding and a 2-4pF ceramic trimmer the internal capacitance of the crystal (approximately 3.8pF) can be neutralised. After careful screening, the stopband attenuation is now close to 90dB which is the limit of measurement with the HP8505A Network Analyser.

"The passband is more symmetrical compared to the original unneutralised filter. Insertion loss is only 1.5dB."

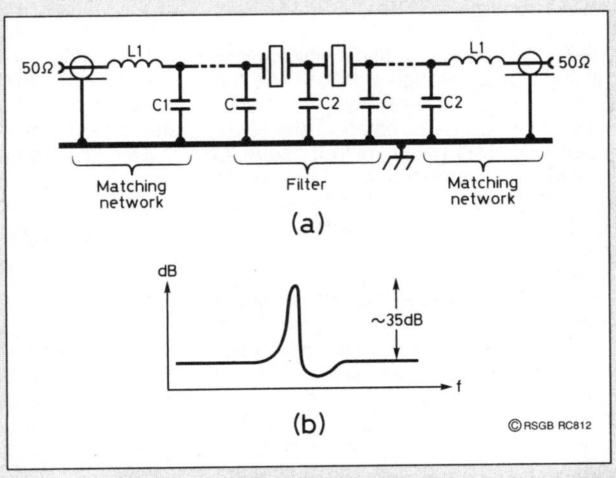

Fig 4: (a) LA7MI's prototype two-crystal ladder filter. (b) Approximate shape of the response curve showing a stopband attenuation of about 35dB.

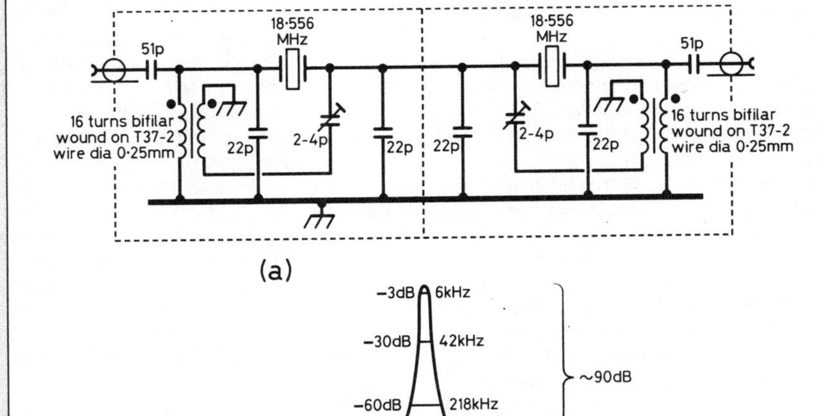

Fig 5: (a) LA7MI's improved two-crystal 18.5MHz crystal with the crystal capacitance neutralised. (b) Response curve showing a stop-band attenuation of almost 90dB. Such a filter would make an excellent low-cost roofing filter.

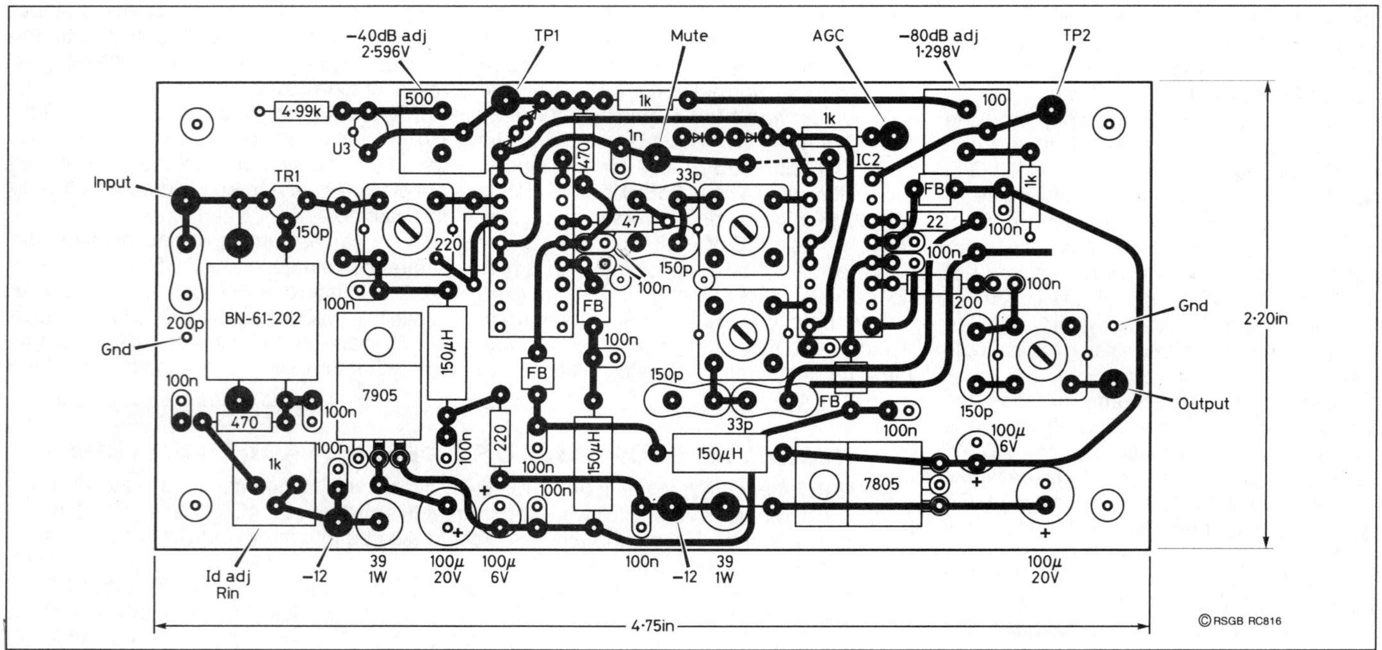

Fig 3: Printed circuit board details for K6DLG's IF amplifier.

the TL341 to provide a nominal 2.6V reference which is divided to 1.3V for the second 40dB stage.

"The Tl431 is a *shunt* regulator which works very well to clamp, through signal diodes, the AGC and mute pins, as follows: The AGC pins of the AD600 must be limited to the +/- 5V supplies. Two 1N4148 diodes from the AGC pin to ground limit negative excursions to about -1.4V, and two more from the AGC pin to the TL431's +2.5V reference limit positive excursions to about 3.9V. If used, the *mute* pin needs protection. It is hoped to incorporate this in a future design. G3SBI's concern with IMD led me to block the DC path from pin 11 of IC2B in order to prevent the possibility of a few mA of DC being bled to ground through T5.

"*Transformers*: T1, for those not familiar with dot convention, Fig 2 shows the direction of windings. The transformer is resonated with a 300pF capacitor, but the value depends upon the precise permeability of the core and the IF frequency, but it is very low Q. A return loss bridge could be used to find experimentally the capacitance required for any particular situation.

"T3, T4: W4ZCB has pointed out that the caption notes on the windings of T3 and T4 may be unclear. I used one length of Nr 28 gauge wire. A short bit is folded back on itself and twisted to form a bifilar pair hanging on the end of the single Nr 28 strand. The loose, short wire where the twisted pair turns back into a single wire is the dot end of the secondary and is soldered to pin 3. The seven turns are wound on the former, the extra length of bifilar wire trimmed off, and an ohmmeter used to find which one is the other end of the pin 3 wire. This other end is soldered to pin 2. The remaining wire of the bifilar pair is soldered to pin 3. Then the remaining long single wire is wound for 17 more turns, continuing the same direction as the first seven turns are going from pin 2. This is soldered to pin 1. Note that although the winding are the same, for board layout reasons the windings of T3 and T4 are not wired the same way to the header.

"*ACG adjustments*: Adjustment of thresholds to align the gain 'handoff' between stages seems to be uncritical. However, in the interests of taking every possible measure a simple test generator is being built to optimise the exceptional AGC performance. The AD600 is so well defined that it is possible to design the circuits and write a procedure for such an adjustment, but this will wait until the test circuits are built and the procedure verified."

SWITCHING DIODES: DJ2LR/KA2WEU'S REPLY

THE ITEM 'RF Switching Diodes Controversy' *TT*, July 1995, included G4HRY's criticism of the advice given by Dr Ulrich Rohde, DJ2LR/KA2WEU/4, in his excellent articles in *QST* that the second order IMD performance of several popular amateur HF transceivers could be improved by judicious substitution of PIN diodes, such as the Hewlett Packard HP5082-3081, specifically intended as RF switching diodes. I pointed out that the criticisms were based solely on RF losses and that G4HRY had not made any IMD measurements. I added that I was sure that Dr Rohde could provide a convincing reply. However, in view of his experiences with G4KPT's Omni VI I felt it would be right to include G4HRY's view that it was unwise to put unquestioning faith in published articles, including even those in *QST* and *RadCom*.

The detailed measurements provided by G3SBI (*TT*, November) and those published by DB1NV in *VHF Communications* showed clearly that there is a wide difference between different diodes used for RF switching both in insertion loss and in IMD performance and that IMD is significantly affected both by the forward current through the diode and by frequency.

As a result of an unfortunate delay, the November item was written before I received a fax sent by Dr Rohde on July 18th. This, in a slightly abridged form, reads: "I feel really concerned and sorry about G4KPT and the results of his experiments. As a matter of record, I would like to point out that intentionally I had not changed any of the diodes myself, but had the authorised service departments of AES, Milwaukee replace the diodes in the Yaesu FT890; ICOM changed the diodes in two IC765s; and Kenwood made the same changes in a TS50. The itemised ICOM repair bill shows 0.12uV for 12dB SINAD. I also had the other companies involved validate that following the diode changes, the receivers were within specifications.

"This validates my statement that this was a repeatable effort and the changes were not done at the expense of performance in any respect. It is also a matter of record that the HP5082-3081 diodes were used in the production of the Collins KWM380, one of which I still own and whose noise figure is on target with 0.3μV without a pre-amplifier and whose 2nd order IMD is superior to other diode applications. This should remove any doubts as to the correctness of my *QST* article.

"I have had no experience in modifying an Omni VI nor did I do any measurements or modifications with it. The ARRL edited in the Omni VI because it is a popular US-made transceiver and there had been some discussion as to whether or not the European version had different diodes or relays. Before fingers are pointed at specific diodes, I would like to examine the circuit diagram because there can be no need to change all 40 diodes. As an experiment, I may want to supply one set of more modern diodes.

"Everyone who has contacted me as the result of the *QST* articles had been advised not to use the HP 3081 (for reasons of cost and availability) but rather to use a Siemens BAR17 diode or MI204 diode, which is available through ICOM dealers/repair centres. Those diodes are much less expensive and more readily available.

"To the best of my knowledge, the companies who changed the diodes in the equipments involved did not change the diode bias. It is questionable why any one should wish to change diodes in the IF section; similarly diodes in the transmit/receive switches should not be touched.

"Finally, there is no question that relays provide the best of all worlds as far as IMD characteristics are concerned, but not necessarily the best solution in terms of space and costs. I have just tested a soon-to-be-released transceiver which uses PIN-type diodes and exhibits superb IMD characteristics while maintaining a good noise figure.

"As to multi-tone functionality, once 2nd and 3rd order IMD tests have been done, one can predict the higher-order IMD effects, especially since they are based on diode characteristics and this type of test is a legitimate test to evaluate receivers.

"Hopefully, your readers will not deduce from this experiment that *QST* or other reputable magazines publish articles which are technically incorrect."

In a subsequent letter, dated September 19, 1995, Dr Rohde confirms that he has run into a lot of people who have modified their RF switching diodes and have been extremely happy with the results. Further, after refining his test set-up he finds the improvement is now slightly more dramatic than outlined in his *QST* article.

In regard to Dr Rohde's endorsement of the technical accuracy of articles, I would enter a caveat. While most writers strive for complete accuracy, the mechanics and Murphy's Law of publication make it difficult to avoid some errors, particularly in columns produced to a tight deadline. Many years ago, I stressed that I regard *Technical Topics* as a forum for new ideas, not all of which are likely to prove repeatable or even strictly accurate. No guarantees can be given on experimental ideas still under development! I welcome comments from sceptical readers or those spotting printing errors etc. Fortunately, there is good evidence that the vast majority of *TT* items do work as intended, and often provide useful additions to amateur lore!

TIPS AND TOPICS

MAC MCMULLIN, W6DSY, in the 'Hints & Kinks' section of *QST* (September 1995, p49) writes: "Rather than threading a 100ft or so of reluctant antenna wire through holes in 30 or 40 spreaders (two holes in each) and repeating the process on the other side of the ladder, I prefer to take a couple of inches of scrap wire, push it through a hole in the spreader, and twist the ends around the antenna wire as shown in **Fig 6**. But then I'm lazy."

Bruce Carter, GW8AAG, offers the following tips: (1) While flicking through some old electronics catalogues before scrapping them I noticed several pages showing full-size illustrations of meter scales. I kept these as they may later be used to paste over existing meter dials when these are being amended.

(2) Many amateurs store small components in screw-top glass jars (even through breakage can prove disastrous). While the labels on the jars can be floated off when immersed in hot, soapy water, small areas of heat-sensitive adhesive remains. This will spread but not dissolve. Complete removal of the adhesive can be achieved by placing the dry jar in the freezer or freezing compartment of a refrigerator and leaving it there for three or four hours, then taking it out of doors and shaving off the adhesive with a knife using a 'pencil sharpening' motion. The adhesive will shatter and drop off. Afterwards, the jar should be left for an hour or two to regain ambient temperature. If the jar is washed out too soon, even cold water is likely to shatter the glass.

Mike Whitaker, G3IGW, as GM3IGW, has for many years chased 1.8MHz DX during the 'CQ contests' from a coastal site in south-west Scotland. Initially, he used hydrogen-filled ex-WD meteorology balloons but later kites. He writes: "Until we knew better we just let out as much wire as we dare! Later we understood better that the lowest angle of radiation is from base-fed wires of from five-eighths to three-quarter wavelengths; anything longer reduces the amount of low-angle radiation. Optimum vertical heights mid-band are 176ft +/- 16ft on 3.5MHz 338ft +/- 30ft on 1.8MHz. Remember that long vertical wires pick up a lot of static! Wear gloves and arrange a static-leakage path (note the regulations for heights above 60m - *TT*, August 1995, pp64-65 - *G3VA).*

Lightning is a particularly severe problem in South Africa. 'What is lightning?' compiled by J C Greenway, ZS5JER, part 3, *Radio-ZS* March 1995 provides an exhaustive list of some 15 'safety rules', of which the following seem most applicable to amateurs operating in the great outdoors during thunderstorms: Don't work on fences, telephone or power lines, pipe lines or structural steel fabrication. Don't use metal objects like fishing rods, golf clubs (or metal antenna supports). Golfers wearing cleated shoes form particularly good lightning rods. Don't handle flammable materials in open containers. Get out of the water or off small boats. Stay in your car which offers excellent lightning protection. Seek shelter in buildings. If no buildings or cars are available your best protection is a cave, ditch, canyon, or under head-high clumps of trees in open forest glades. Where there is no shelter, avoid the highest object in your area. If only isolated trees are nearby, your best protection is to crouch in the open, twice as far away from isolated trees as the trees are high. Avoid hilltops, open spaces, wire fences, (metal) clothes lines, exposed sheds and any electrically conductive elevated objects.

HERE & THERE

"IF YOU HAVEN'T SPUN the Sperry PW epicyclic dial of an HRO receiver then you haven't lived" - ZS4NP/VE3NPK in *Radio ZS*. It is clear that there are still a significant number of amateurs, old timers and newcomers, using valve transceivers produced in the 1960s and 1970s. By modern standards, most suffer from a significant degree of frequency drift, up to about 2 - 3kHz over a couple of hours. For those who find this unacceptable, one solution is to add 'huff and puff' frequency stabilisation as devised many years ago by PAOKSB and described on several occasions in *TT*. Following his WARC-bands modifications for the KW2000-series of transceivers, Steve Dyke, G3ROZ, produced a detailed article on adding a huff and puff stabiliser. While this was specifically concerned with a KW2000, the circuit board could be adapted for other models of that period. The typescript runs to 10 sheets plus 8 sheets of diagrams and tables. Providing I am not swamped with requests, photocopies (including postage) will be supplied to members seriously interested for £1.50p (G3VA QTHR). G3ROX estimated the total cost of constructing the stabiliser as just under £70 (£20 if you have a good junk box).

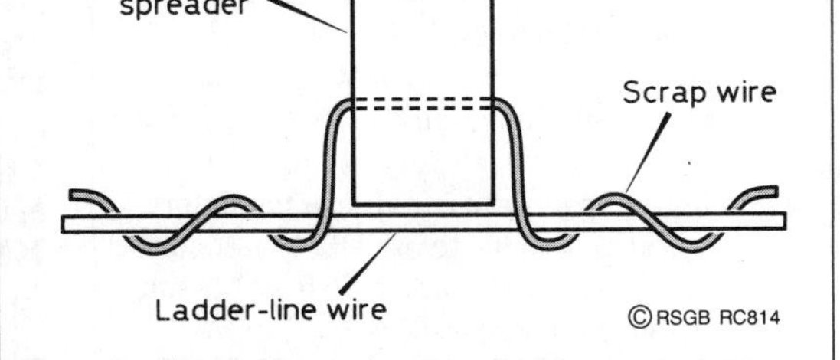

Fig 6: How W6DSY overcomes the tedious task of stringing open-wire feeders on spreaders. He uses twisted pieces of scrap wire holding the feeder wire in place. He recommends soldering the holding wires to the feeder wires if you encounter difficulty with 'rusty joint' noise or harmonics.

TT, November 1994 in 'Here & There' noted that G30UC had drawn attention to the development by Trevor Baylis of a radio receiver powered from a clockwork motor, giving about 20 minutes operation from a single winding. It was seen as of value in countries where the cost of batteries and absence of mains supplies limits the value of radio broadcasting. G30UC felt that here was a challenge for the amateur making the first "clockwork-driven" QSO.

I have not heard of anyone yet taking up the challenge, but Alan McCulloch, G4HTL/ZS6KU, who works for the Liberty Life Association of Africa Ltd, forwards an article 'Wind Up the Radio' from a recent issue of a South African financial journal indicating that South Africa is "rushing ahead with the clockwork invention that could change the direction of wars" with the International Red Cross interested in using the radio in Chechneya and in similar situations where effective communication may greatly assist humanitarian efects. Liberty Life provided the funds for this development as a UK university project. It indicates that a three-band radio covering short wave, medium wave and FM can be wound up in 40 seconds and then plays for 45 minutes. G4HTL/ZS3KU comments: "One wonders when the day will be reached when this idea could be applied to at least amateur-radio receivers, but I fear it will be a long time before the HF transmitters could use it."

Nature (11th May 1995) notes that 'super-conducting superwires' seem possible with the development at the Los Alamos National Laboratory of thick films of Y-123 material deposited on flexible metal tapes 5cm long by 2cm wide, an embodiment that could in principle be scaled up to kilometre lengths. The tapes are capable of carrying 100A when cooled to 77K (ie by liquid nitrogen) a major improvement on the current-carrying capacity of earlier 'high temperature superconductors'. The development could open the way for the use of superconductors in the practical distribution of electricity.

Scientific America (July 1995) has a long article entitled 'Plastics Get Wired' which suggests that by tailoring the electrical properties of conducting polymers, researchers hope to render electronics a bit more organic. It is admitted, however, that so far copper conducts 100,000 times as much current as the most suitable polymer - polyaniline - and costs half as much. Early applications may be for LEDs or even entirely plastic transistors (though again so far much slower than silicon). ♦

Pat Hawker's Technical Topics

PAT HAWKER, G3VA
London 37/SE22 8SS

ADVANCES IN MININEC

AN OUTLINE OF the history of *MININEC Professional for Windows* has been given by John Rockway and James Logan - who wrote the original MININEC software in 1980 - in an article 'Advances in MININEC' (*IEEE Antennas and Propagation Magazine*, August 1995, pp7-12).

This emphasises the differences between MININEC and the original Numerical Electromagnetics Code (NEC). To quote from their introduction: "Because of the similarity in names, it has often been stated that MININEC is a personal computer (PC) version of its big brother, NEC. But this could not be farther from the truth. There are significant differences between these two codes. Both codes use the method of moments to solve for currents on electrically thin wires. However, each code starts with a different version of the integral formulation for the currents and fields for wires, and follows significantly different algorithms for implementation of the method of moments. NEC calculates for the currents on segments. MININEC calculates based on currents at the nodes defining the segments. It is the intuitiveness of this node-based construct, and the relative ease of defining a problem, that has resulted in the thousands of users of MININEC . . .

"The original MININEC was written by John Rockway, with a little prodding and support from Jim Logan. One day in 1980 they were discussing the methods available to amateur radio enthusiasts for antenna design. The point was made that most amateurs did not have access to large, mainframe computers and highly developed antenna-modelling computer programs such as NEC. PCs had not been on the market for very long and they were then relatively expensive, and quite limited in performance and were then generally regarded as mere novelties or toys . . . it was no small feat in trimming down NEC to run on a PC with some 16K memory and 8 bit word length. There was no FORTRAN and would have to be written in BASIC.

"The first public release of MININEC was in 1982. The code was 550 lines of BASIC and would run on an APPLE II computer. It could compute the current distribution, impedance and far field pattern of an arbitrarily oriented set of wires in free space or over a perfectly conducting ground plane. Lumped-impedance loads were allowed at segment junctions, except for segments intersecting with the ground planes. Wires intersecting the ground plane were restricted to right angles. The problem size was limited to 10 wires and 50 currents (or 70 segments with junctions) . . ."

In 1984 came MININEC2. MININEC3 was released in 1986 with which a math coprocessor and a BASIC compiler could solve antenna problems up to 50 wires and 50 current unknowns. In 1988 came the MININEC SYSTEM. Many others have also attempted to improve on MININEC: "Most notable are the innovative user interfaces and graphics displays offered by Roy Lewallen, [W7EL, ELNEC etc] in 1991 and Brian Beezley, [K6STI, MN-PRO etc] in 1992."

The latest effort by the Rockway-Logan team is *MININEC Professional for Windows* (available as a commercial product from EM Scientific, Inc 2533 N Carson Street, Suite 2107 Carson City, NV 89706 USA Tel: (702) 888-9449. Fax: (702) 883 2384. This uses triangular testing functions rather than pulses resulting in greater accuracy . . . End cap corrections have been added to improve the modelling of Yagi antennas; treatment of bent wires has also been improved so that square loops eg quads, can be accurately modelled. **Fig 3** shows the result for the computation of a V-dipole with the input resistance computed as a function of the acute angle between the two wires. A Fresnel-reflection-coefficient approximation improves the calculation of currents in the vicinity of real ground and can also be applied as a correction to radiation patterns.

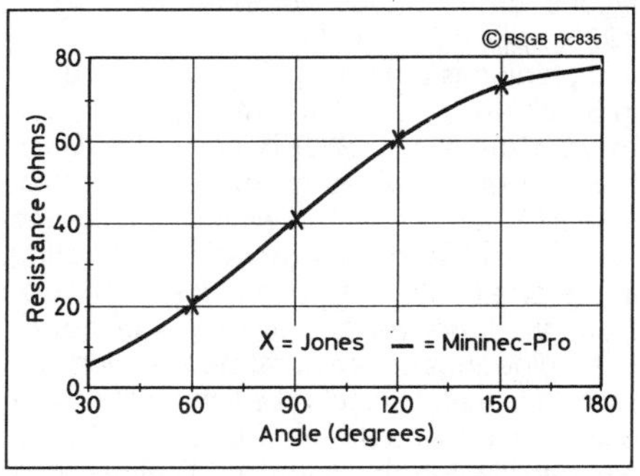

Fig 3: The feed resistance of a V dipole compared with the analysis of J E Jones in *IEEE Trans Ant & Prop*, 1976. The input resistance is computed as a function of the acute angle between the two wires. Six segments were used for each wire, a total of 12 segments. An increase in segmentation appears to improve the agreement for angles approaching 180°. In MININEC-PRO it is not necessary to employ a tapered-segmentation scheme to model wires with acute angles.

The user interface to *MININEC Professional* is through Microsoft *Windows*. Input data screens provide spreadsheet-like tabular entry, in individual customised windows. Output products are displayed in both tabular and graphic forms; input and output data screens are fully interfaced to Windows printer drivers, word processors and spread sheets.

The authors provide details of the problems that can be solved, the constructs that can be modelled and the capabilities of the integrated graphics. They claim that MININEC has progressed from a novelty to a real engineering tool.

3.3V POWER SUPPLY STANDARD

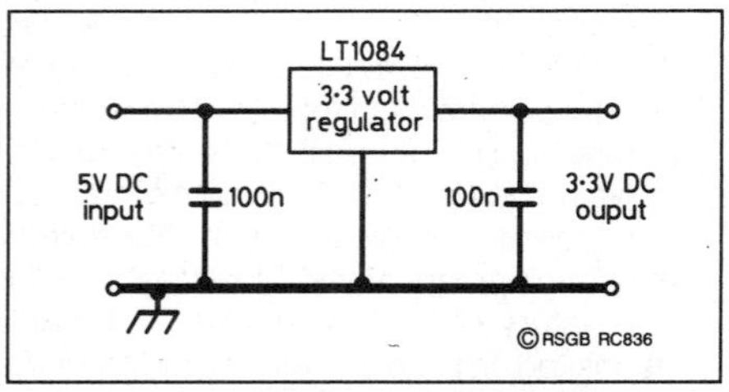

Fig 1: Use of a 3.3V linear regulator to power the new generation of 3.3V IC devices. (*CQ*)

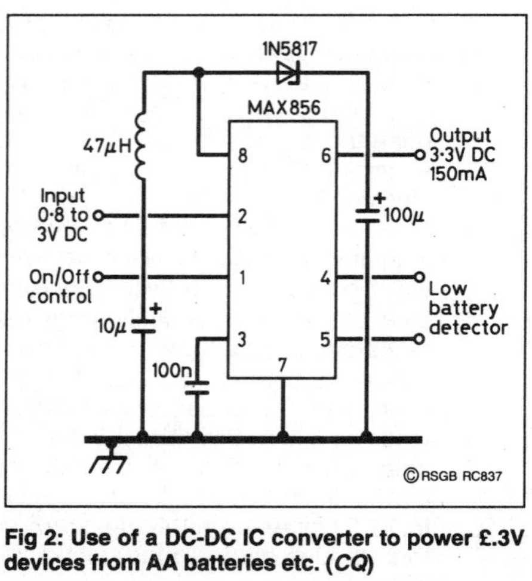

Fig 2: Use of a DC-DC IC converter to power £.3V devices from AA batteries etc. (*CQ*)

IRWIN MATH, WA2NDM, (*CQ*, September 1995 pp82 - 83) draws attention to the recent introduction of the 3.3V power supply standard for integrated circuit devices with the long standing 5V standard for digital logic on its way out. Main incentive for this change is the need to reduce power consumption in battery operated lap-top computers, portable cordless telephones etc. The 3.3V devices have a tolerance of only +/- 0.15V compared to the +/- 0.25V of the 5V devices. As WA2NDM points out this makes the choice of regulators a bit more critical.

There are several suitable three terminal regulators available including the Linear Technology LT1084CT-3.3 (5A maximum) and LT1086CT-3.3 (1.5A maximum) and equivalents from other manufacturers already on the market: **Fig 1**. Apart from such linear regulators, it is also possible to use DC-to-DC converters such as the MAX856 from Maxim Integrated Circuits: **Fig 2**.

WA2NDM lists a number of 3.3V ICs from Texas Instruments and Maxim but devices to the 3.3V standard are being introduced by most IC manufacturers.

LOW COST-HIGH PERFORMANCE, HF SWITCHING-MIXER

GIANCARLO MODA, I7SWX, has drawn attention to an article 'A Low-cost, High Performance Mixer for HF Applications' by South

African P J Coetzee (based on an idea by J M Grundlingh) in the first issue of 'Engineer's Notebook' published as part of *RF Design* (June 1995). As a result of reading this article, I7SWX has tested the idea first using standard CMOS and then the suggested high-speed devices.

P J Coetzee points out that the performance of an HF receiver is determined largely by the first mixer: "For strong signal handling capabilities the double balanced diode ring mixer is nearly universally used in high performance receivers. For applications requiring very high dynamic range from the receiver, diode ring mixers driven with up to 0.5W (+27dBm) local-oscillator power are used. Power consumption and LO leakage at these LO levels become a problem [for some applications - *G3VA*]. This article describes a high performance, low cost double-balanced mixer suitable for input frequencies up to and above 20MHz with low local oscillator and DC power requirements."

He provides test measurements, see **Table 1**, of conversion loss and two-tone tests at various frequencies, and also for a 7MHz direct conversion receiver using this mixer based on the 74HC4066 device with the 180° phase shift of the LO oscillator achieved with a 74HCO4 hex inverter, which also converts the sine wave LO to the square wave required to control the switches in the 74HC4066. One of the RF transformers required by diode ring mixers is thus replaced by an inexpensive CMOS IC.

He points out that the CD4066 quad CMOS switch has been used as a chopper modulator for low frequency applications for many years. With the development of high speed CMOS it is now possible to use these devices up to more than 20MHz. By replacing the diodes in a mixer ring with 74HC4066 CMOS switches, the performance of the mixer can be improved by the linear transfer function and low 'on' resistance of the switches at the lower HF frequencies.

The basic arrangement of such a mixer is shown in **Fig 4**. The RF input signal couples into the mixer with aid of an RF transformer which could be a Mini-Circuit T4-1 or a home-made unit. Two switches are used in parallel to reduce the 'on' resistance, with Vcc/2 DC bias applied via the transformer. As for all diode ring mixers, it is important to terminate the IF port in a wideband 50Ω load to achieve full performance.

For the test measurements of insertion loss, signal handling capability and LO leakage at the RF port, the mixer was driven by a voltage of sufficient amplitude to switch the biased high-impedance gate. For the tests, the output level driving the LO port was set at +7dBm, which corresponds to 4V p-p signal into 50Ω. As with all CMOS logic, the DC power requirement of this mixer increases with frequency.

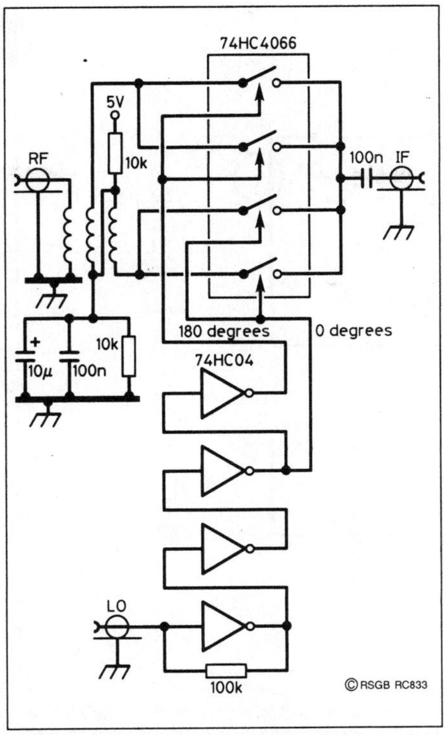

Fig 4: Basic configuration of the low cost, high performance switching mixer using 74HC4066 mixer and 74HC04 squarer. (*RF Design*)

P J Coetzee also constructed a direct-conversion receiver covering the 7MHz amateur band in order to evaluate the noise figure and dynamic range for down conversion to baseband applications. He notes that it is difficult to measure accurately the 3rd order intercept point since the spurious outputs and wideband noise at high output levels from a synthesised signal generator can easily cause faulty measurements; ideally the test signals should be generated by high performance crystal-controlled oscillators.

He also points out that this mixer is not suitable for transmitting applications without adequate filtering because of the presence of harmonics from the square wave LO signal at the IF port. However, this can be utilised for harmonic down conversion (with some reduction of performance compared to fundamental operation). This makes it possible to build a multiband receiver without having to switch the LO frequency.

Finally, he contemplates the exciting possibility of using an even higher-speed 74AC4066 should such a device become available.

I7SWX first tested the idea as the second mixer of his Yaesu FT729A transceiver which has an 10.8MHz input and 455kHz IF output using available MC4066 (maximum rated clock frequency 10MHz) and 74SO3 devices with the existing 8V receiver bus. Subsequently he was able to obtain the higher speed devices from the Italian branch of RS Components. He implements this as shown in **Fig 5**. He reports that the 74HC04 is switching beautifully at 11.2MHz and easily driving the 74HC4066 mixer.

He writes: "The complete circuit is made on a PCB measuring 20 x 35 mm which just fitted in place of the CW filter using 'ugly' (ie dead bug) construction and connected in place of the original JFET (2SK125) mixer and using the +8V receiver bus. No drilling or circuit track cutting on the FT726 was needed. The two JFET post mixer amplifier gives the same output as the original JFET mixer, including the conversion loss of the 4066.

17SWX does not have the test gear needed to check the IMD values but he is well satisfied with the results, although accepting that the overall performance on 144MHz and 50MHz is probably limited by the first mixers (Mosfet 3SK51) in the 144 and 50MHz converters and he intends in due course to replace these (first 50MHz then 144MHz).

THE MORSE REQUIREMENT

I AM OLD ENOUGH to have come to amateur radio via the pre-war 'Licence to Establish Wireless Telegraphy Station for Experiments in Wireless Telegraphy' which required the passing of a 12WPM Morse test, but no technical examination other than an ability to dream up some possible experiments involving an on-air transmitter. In the 57 years since then, I have not only made good use of Morse but continue to believe that a Morse requirement remains an excellent means of sorting out those who are prepared to go to some trouble to obtain an amateur HF licence and so demonstrate their commitment to the hobby whether or not they subsequently use the code.

Les Mitchell, G3BHK, who must have obtained his licence during the few post-war years when the first year had to be spent using Morse-only but who in recent years has packed away his key, has recently commented on the present debate over the New Zealand wish to delete RR2723 from the ITU Radio Regulations.

Before coming to G3BHK's comments it seems worth reprinting some remarks by that highly

a) Conversion Test Results

F/LO (MHz)	F/IF (MHz)	F/RF (MHz)	Insertion Loss (dB)	Pwr I/p (dBm)	P/LO at RF port (dBm)
5.5	9	3.5	7	+15	-32
5.5	9	14.5	6.5	+15	-32
30.7	10.7	20.0	6.5	+15	-20
40.7	10.7	30.0	7.5	+ 9	-17

b) Two Tone Test Results

Tone 1 (MHz)	Tone 2 (MHz)	3rd order products (MHz)		3rd order intercept point (dBm)
3.6	3.7	3.5	3.8	+29
14.6	14.7	14.5	14.8	+31
20.1	20.2	20.0	20.3	+29
30.1	30.2	30.0	30.3	+19

c) 7MHz direct-conversion receiver

FRx) (MHz)	RxB/W (kHz)	MDS (dBm)	Tone 1 (MHz)	Tone2 (MHz)	Dynamic range (dB)
7.02	2.4	-128	7.04	7.06	105

Table 1: Test Performance of the 74HC4066 switching mixer.

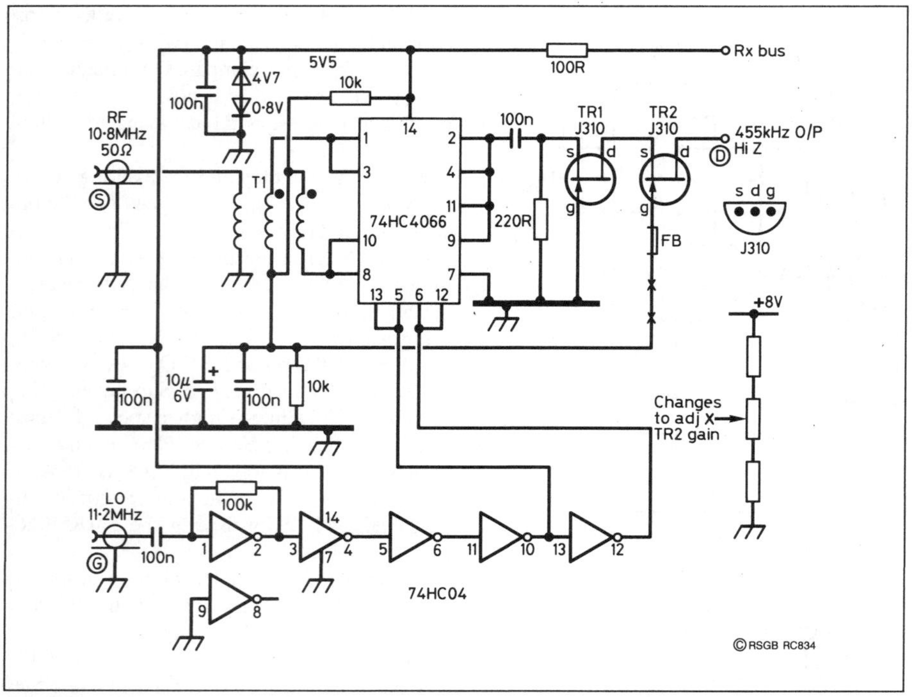

Fig 5: I7SWX's implementation of the 74HC4066 mixer to provide a second mixer (10.8MHz to 455kHz) for his FT726A 144/50MHz transceiver. S source; G gate, D drain connections to the original mixer JFET (2SK125). T1 home made broadband transformer, 3 x 5 turns magnet wire twisted. TR1, TR2 J310 FETs but J309 can be used with slightly lower gain.

respected old timer Tom Clarkson, ZL2AZ, who used to represent New Zealand at many of the WARCs. In 1972 he wrote: "CW should be recognised as a worthy participant and all plans should provide for its welfare and growth. Established amateurs who concentrate on telephony and other modes should respect the claims for CW, and newcomers to our ranks should be encouraged to discover its special character - to find out why so many are drawn to work on the CW bands."

G3BHK endorses the closing remarks by Kirsti Jenkins-Smith, VK9NL, in an article 'The Simple Things in Life' in the Summer 1995 issue of the *RNARS Newsletter*: "While commercial and defence interests move on to stake their faith in high technology, it may be up to us, radio amateurs, to preserve the art of communicating, using less vulnerable, complicated and expensive means. Detection, recognition and processing of Morse by ear is a very complex psycho-acoustic process that enables us to communicate over great distances and in adverse conditions with the most simple of man-made equipment . . . Remember, when all else fails, *manual Morse can get through*.

"If the supporters of dropping CW from the Class A licences win, how long will it be before they begin to claim that, since the vast majority of amateurs now use commercially constructed equipment [designed to meet an increasing number of EMC and other European Union specifications - *G3VA*] why do we need complicated technical examinations? A simple practical demonstration of the ability to operate a modern transceiver would surely suffice. Why not just have the technical examinations for those who wish to build and use their own equipment?"

That would then not only bring about the slow death of Morse but also the virtual elimination of home-built equipment. As G3BHK puts it, the hobby on all bands would soon become a glorified form of CB, with little or no long-term commitment to what has traditionally been a lifelong hobby. Even now, the turnover in licensees appears to be increasing, as more people drop out of the hobby from lack of interest or financial or accommodation reasons than by becoming Silent Keys.

G3BHK also sent along a photocopy of a 1922 report by a sub-committee of the 'Radio Research Board' of the then Department of Scientific and Industrial Research (DSIR) set up to consider the future of 'radio telephony'. To summarise the conclusions of their report: the development of Radio Telephony for long ranges is (in 1922) in an extremely elementary stage. "We see no line of development which would be likely to lead to its establishment on a commercial basis within a measurable period . . . the power required for Radio Telephony would be much greater than that required for satisfactory communication by Radio Telegraphy over the same distance . . .the position is more hopeful for ranges of the order of 200 miles . . .we cannot, however, recommend the use of radio telephony as a substitute for any other means of telegraphic communication except in those cases where the special requirements can be met in no other economic way; for example, the broadcasting of . . . general information where one costly transmitting station supplies a great number of simple inexpensive receiving stations."

Chairman of the sub-committee was Admiral Sir Henry Jackson, who exactly 100 years ago pioneered maritime radio at Devonport and who, in 1922, was President of the Wireless Society of London (which shortly afterwards became the RSGB)!

Those who believe that the future lies solely in digital forms of speech and data should look carefully at Figs 5 to 15 in the October 1995 *Eurotek* column (pp 70 - 72) which show the spectrum occupancy of various forms of well adjusted and badly adjusted AmTOR and packet transmissions and compare these to Fig 5 showing the narrow bandwidth of a typical CW signal.

Dick Biddulph, G8DPS, wants me to join him in attempting to ban 'CW' - not the Morse Code, but the use of the abbreviation for 'Continuous wave' (originally used to distinguish modes other than Spark) as an abbreviation for Morse Code. He notes that ARRL and some other Americans tend to call Morse transmission 'Code' rather than 'CW'. While sympathetic to G8DPS's view, I feel that attempting to change editorially what is virtually worldwide 'common-usage' is usually doomed to failure. Among amateurs, but not professional engineers, CW is universally accepted as Morse code transmission, and I feel that any attempt to 'Ban CW!' will be misunderstood!

ELEVATED RADIALS VERSUS BURIED EARTHS

THE AVAILABILITY OF computer programs based on the Method of Moments formulated by R P Harrington in the late 1960s - including progressively 'easier to use' and more accurate MININEC software developed with Amateur Radio applications in mind - is unquestionably playing an ever greater role in the design of antenna systems. While early programsignored the presence of real earth, this is no longer the case, and there are now even programs that can calculate elevation patterns over sloping ground: See 'The effects of local terrain on HF launch angles' by R Dean Straw, N6BV, of ARRL staff (*QEX*, July 1995, pp3-15) which introduces the YTAD (Yagi Terrain Analysis, with Diffraction) program.

N6BV states that YTAD.ZIP is available, free, on Internet by anonymous FTP from ftp.cs.buffalo.educ in the /pub/ham-radio/qex directory. He adds: "What I'd like in return is validation data for the results computed by YTAD for your QTH. I don't expect you to hire a helicopter and a team of dedicated scientists to generate validation data, but for those of you with access to stacked antennas, the methodology shown in Fig 15 (of the *QEX* article) may be useful . . . using a well calibrated receiver S-meter." For anyone interested is using YTAD it would be necessary to read N6BV's *QEX* article.

For those of us who remain computer illiterates, it is satisfying to note that while programs such as ELNEC and the later EZNEC are proving powerful design tools, many of the most valuable antenna system ideas have stemmed from old-fashioned back of envelope ideas or measurements based on trial and error, even if later honed by the MININEC programs.

An excellent example of an important development of this type is the current interest in the use on the lower HF bands of elevated radials with electrically-short vertical antennas rather than striving for the 120-buried radials that for so long were held up by broadcasters and amateurs as a necessity for an efficient system on MF and lower HF bands.

The 120-buried radial system stemmed from the classic 1930s MF work of Dr George Brown and his RCA colleagues R F Lewis and J Epstein (*Proc IRE*, June 1937, pp753-787). 120 radials were subsequently adopted by FCC as a standard requirement for MF broadcast channels. The RCA work, however, was specifically concerned with 'buried earths' although it seems to have for many years been assumed that their results ruled out the use of multiple elevated radials (counterpoises and ground screens). Few amateurs could hope to implement a 120-buried radial system, but 1.8 and 3.5MHz enthusiasts were encouraged post-war to bury as many wires as possible.

George Brown, it needs to be stressed, was not averse to elevated, artificial ground screens. He was, in fact, the inventor of the HF/VHF artificial ground-plane antenna. He once confessed to me over lunch that his original version used only two radials, but to satisfy early police-radio customers this was increased to four radials without significantly increasing radiation efficiency.

It was not until the early 1980s that the buried earth screen concept was challenged by three retired professional engineers who were also radio amateurs: Arch Doty, K8CFU, John Frey, W3ESU, and Harry Mills, K4HU. K8CFU kindly sent me a copy of a paper he had presented in the US to an IEEE meeting entitled 'Characteristics of the Counterpoise and the Elevated Ground Screen' based on literally *thousands* of measurements. I was delighted to include some notes on this work in *TT* (February 1983) and a longer version of their paper 'Efficient Ground Systems for Vertical Antennas', appeared in *QST*, (February 1983 pp20-25).

As I wrote in *TT* (February 1983): "A privately-funded research project of such a magnitude, undertaken by three elderly engineers with an aggregate age of 203 years, setting out and succeeding in showing that a good deal of what has appeared in the professional antenna literature during the past 60 years on this subject is in error, or at least seriously flawed, is remarkable. That the project succeeds in pinpointing how the errors occurred and in resurrecting the importance of the counterpoise is even more remarkable!"

I continued: "This ambitious project has unearthed (if I dare use that word) several important new facts; in particular it shows that the currents flowing in buried radials of the conventional type are not, as usually postulated, uniform, but depend upon ground conductivity in the immediate vicinity of the individual wires. It indicates that there can be unexpectedly large variations in surface ground conductivity within the area covered by the ground system of a typical vertical antenna." The *QST* article showed average conductivity varying with seasonal rainfall etc by a factor of over 2:1 in a matter of months: see **Fig 6**. Ground conductivity along different radials may vary by a factor of as much as 7.5:1 within a distance of 10ft at the site in North Carolina.

In *TT* I wrote: "But the tests further show that the distribution of return currents collected by wires elevated by 6 to 8ft is distinctly different from that found in buried radials. Lossy ground between the antenna and the radials reduces their efficiency in providing return currents. All the initial indications are that elevated ground systems (whether counterpoise or elevated but earthed ground screens), used with electrically-short vertical antennas *require substantially fewer radials than needed by a buried wire system of equal efficiency*'. A conclusion calculated to upset conventional thinking of the past 40 years! This does not mean that buried radials do not make good earth systems, but it does suggest that elevated wires are more efficient (and much cheaper and simpler to install)." **Fig 7** shows some typical elevated ground screens as used by K8CFU.

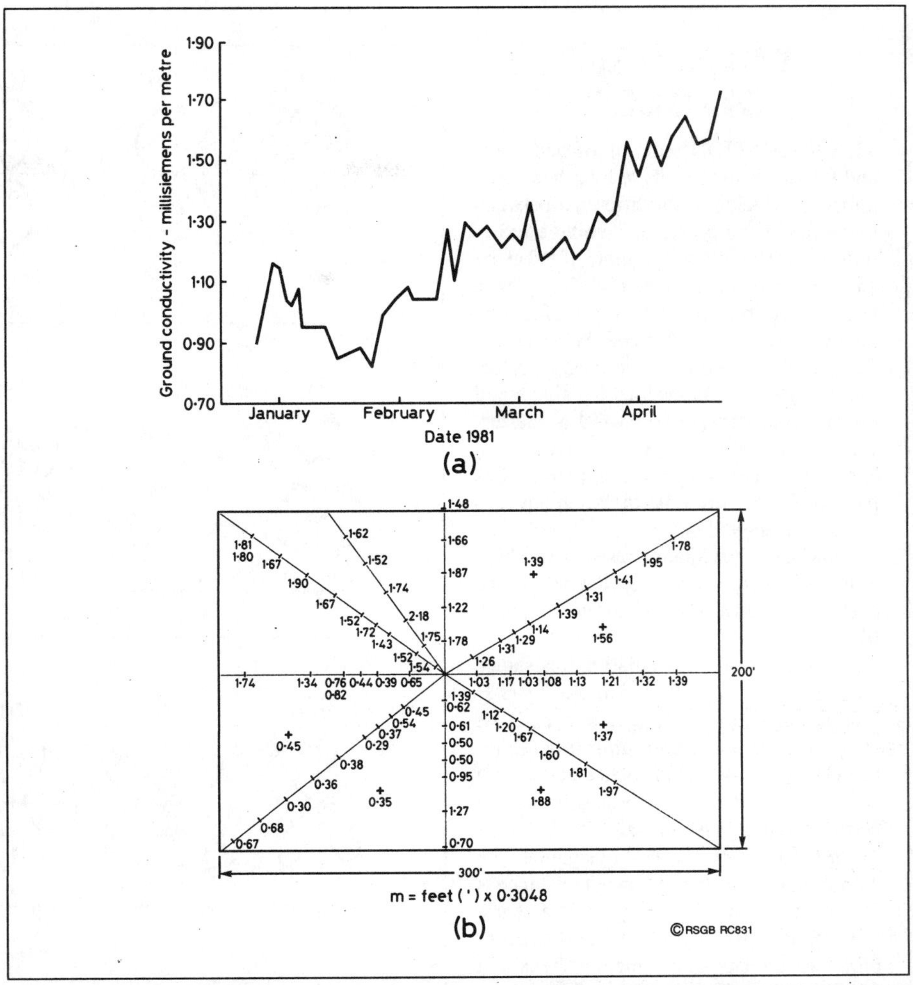

Fig 6: a) Ground conductivity at one fixed location in K8CFU's test site at Fletcher, North Carolina. Measurements were taken daily during the first four months of 1981; (b) Ground conductivity along different radials at the test site. Note there are differences of as much as 7.5:1 within a distance of 10ft. (*QST*)

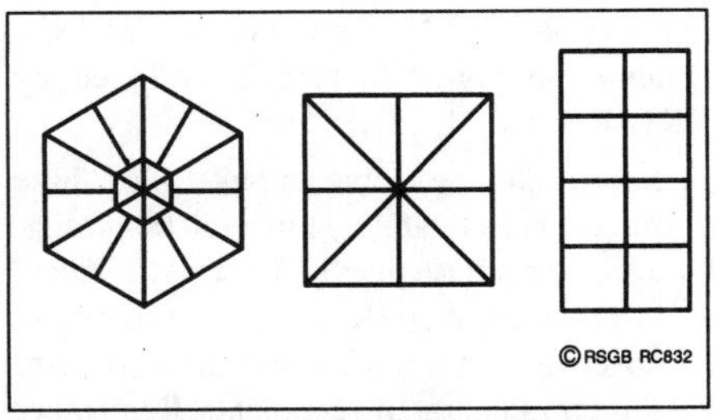

Fig 7: Typical multi-wire counterpoise or elevated ground screen mounted below a vertical antenna as investigated by Arch Doty, K8CFU, and his colleagues. They showed clearly that such systems are more efficient than buried systems of comparable size. (*TT*, Feb 83)

This outstanding amateur-professional research did not immediately make much impression on professional (or amateur) users of vertical antennas. But in the 1990s, a number of articles, based on NEC-type calculations, have been appearing - for example 'Using Elevated Radials with Ground Mounted Towers' by Al Christian, KB8I (*IEE Trans on Broadcasting*, September 1991, pp77-82, see a summary in *TT*, February 1992, p38). This predicted that "it is possible to use four elevated radials together with a conventional ground-mounted, base-insulated tower for MF broadcasting purposes, and achieve the same level of performance which is normally obtained from a classic 120-buried-radial system".

I noted than that "such an approach would clearly be most suitable for use by amateurs on the 1.8, 3.5 and 7MHz bands for verticals; on the higher frequency bands it is already conventionally adopted in the form of the popular ground-plane antenna (GPA)".

Regretfully, as far as I am aware, none of the subsequent articles on elevated ground screens has acknowledged the breakthrough investigation by Arch Doty and his colleagues, and to my mind computer simulations take more for granted (for example uniform ground conductivity) than was found during the field measurements made by K8CFU *et al.*

FULL-WAVE DIPOLES

THE RADIATION patterns of half-wave and full-wave dipoles have long been well understood including the difference between those fed at the centre and voltage fed at both ends. The full wave dipole, if fed at the ends or centre, needs to be high-impedance voltage fed but can be current fed one quarter wave from either end. When centre fed the full wave dipole becomes, in effect two collinear half-waves with a sharpened up radiation pattern; when fed a quarter-wave from one end or voltage fed at one end, it provides the clover leaf four lobes pattern. **Fig 8** shows the radiation patterns of these arrangements.

But what is the radiation resistance (which will also be the feed impedance) of the current-fed arrangement shown in Fig 8 (d)?

An 'Analysis of the Radiation Resistance and Gain of a Full Wave Dipole' by S H Idris and C M Hadzer appeared in the *IEEE Antenna & Propagation Magazine* (October 1994 pp45-47) This notes that "the full wave dipole antenna is very seldom used in communication systems, and yet this type of antenna offers a very high gain". The authors show that the radiation resistance is 194Ω (in free space). The gain of a centre fed collinear pair Fig 8 (c) is 3.94dB compared to an isotropic antenna (ie about 1.8dBd) with a 42° lobe to the -3dB points. Most of us would consider 1.8dBd a useful rather than a very high gain but it is equal to a 50% increase in transmitter power.

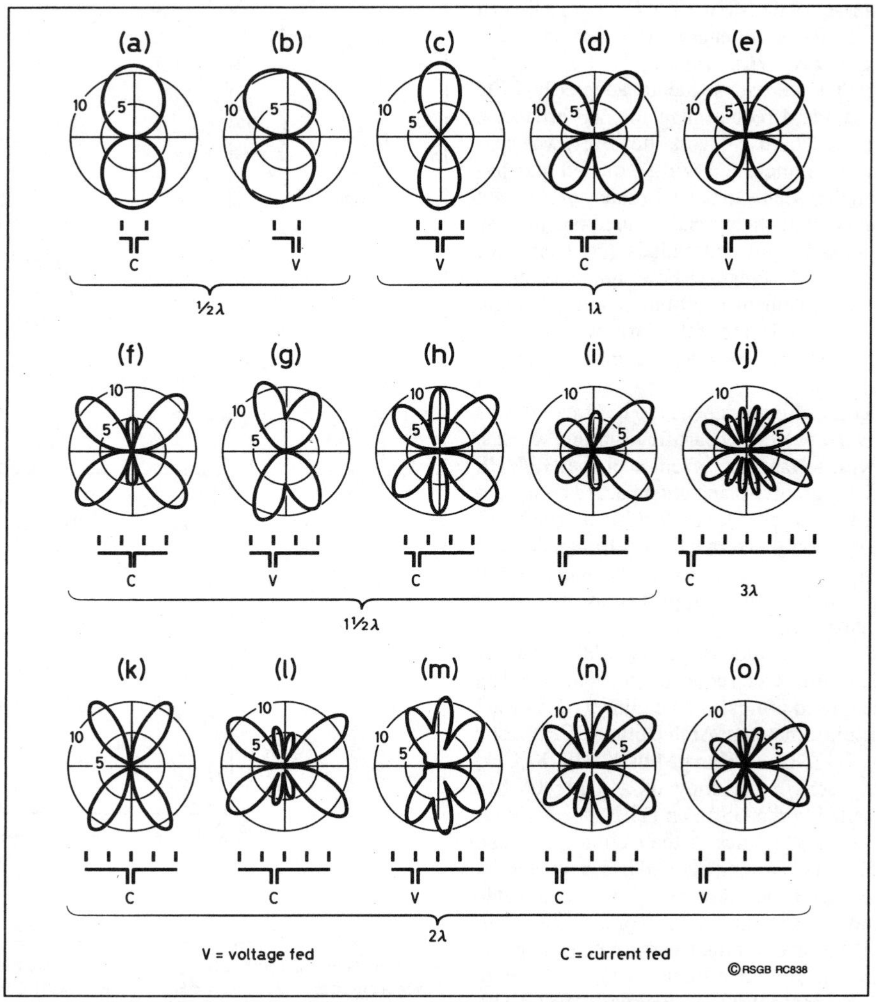

Fig 8: Horizontal radiation patterns of various dipole antennas showing the effect of the current and voltage feedpoints. Originally based on model antenna range measurements by D C Cleckner.

Nevertheless, a useful addition to the amateur literature based on computer simulations is a recent article 'Short Vertical Antennas and Ground Systems' by Ralph Holland, VK1BRH (*Amateur Radio,* October 1995 pp9-13). He presents a number of graphs based on the NEC-81 code which can model antennas in proximity to lossy ground. NEC-81 is a PC version of NEC-2.

His introduction states: "There has been a number of articles discussing the merits or otherwise of various types of ground systems. The analysis of such systems is complicated by the facts that practical ground system parameters are difficult to quantify, antennas are situated in less than ideal locations, and literature that may provide insight into what is happening often does not present the information in a practical or applicable form. Often the reader is left to extrapolate beyond the bounds of presented data and arrive at the incorrect conclusion. Some folklore has been generated as a result of these types of difficulty; one such lore is 'the higher the better' which is an over-generalisation if said without qualification."

While I have an uneasy feeling that his simulations leave some questions hanging in the air, he provides some conclusions and points to remember that seem to be essentially correct: "It is obvious that the simulated antenna efficiency declines dramatically when the groundplane displacement (that is height above ground) approaches zero. With three or more radials the radial length is more important to the overall efficiency than the number of radials, with a few exceptions.

"An interesting side-effect is that under certain conditions a vertical antenna with one radial was substantially more efficient than one with any other number; such a hybrid antenna has been studied and applied in marine and land-based systems: see for example 'HF Antennas for all Locations' by Les Moxon, G6XN, (2nd edition page 183). Among the points to remember listed by VK1BRH are:

1) Short radiating elements have a low base resistance, a relatively low radiation resistance and consequently lower efficiency; nevertheless short elements can still be quite efficient. The shorter radiators require shorter radials at lower heights. Remember that short radiators have high RF voltages at the base.
2) There is an optimal height for an elevated groundplane; typically this is about 0.05-λ. Do not place radials on or below the ground unless there is space for some 120 radials and the site has good ground conductivity; (with poor ground conductivity it is useful regularly to water the ground! (Preferable with salty water if this does not harm the garden).
3) Efficiency curves are tolerant in regard to the height of the groundplane; displacements can be as little as 0.005-λ and in some cases efficiency peaks around 0.01-λ. [Note that safety requirements mean that elevated garden radials should be at least 7 - 8ft (over 2m) high - *G3VA*]
4) Large numbers of radials are not required for elevated groundplane systems. Three of four is sufficient. Doubling the number does not double the efficiency.
5) Substantially better results can be obtained by siting an antenna near swamps, lakes or in close proximity to the sea.

VK1BRH emphasises that simulated results are only as good as the computer program. However he considers that NEC-81 is good at modelling linear antennas close to lossy ground. It has been validated numerous times but, even so, it has failed even more times due to inappropriate use. ♦

THE ENVELOPING ELECTRONIC SMOG

THERE CAN BE few amateurs, excepting possibly those who live in remote sites miles from their nearest neighbours, who are not experiencing a steady rise in man-made electrical interference, in the form of both discrete rough signals and broadband noise. Personally, I find that some days the effective noise floor on say 21MHz is several S-points above that on the ever fewer days when the band is limited only by atmospheric noise. Traditional weak-signal contacts are becoming less and less possible with incoming signals lost in the pervasive electronic smog that surrounds most urban and suburban environments.

On my bookshelves is a 40-year old book *Electrical Interference* by A P Hale of Belling & Lee Ltd (Heywood, 1956) which dealt with what was then "the wide range of higher-frequency 'man-made' noise normally the result of operating electrical machinery". For several decades, one struggled with ignition noise from passing vehicles, clicks from thermostats, buzzes from shavers and similar small motors, etc. The preferred solution was to 'suppress at source' after tracing the appliance concerned.

Interference was most often conducted along the supply wires to the receiver, with the signals sometimes originating one or two kilometres away. Mains filtering could often cope with such problems, although the interference could be re-radiated from the supply mains prior to the filter and reach the receiver through the antenna socket.

The invasion of homes by DIY power tools and a multitude of microprocessor-controlled devices with their harmonic generating clocks has brought about a situation which largely defines traditional remedies and which threatens not only amateur radio but also the increasing use of 'wireless' (cordless) telecommunications.

Gary Breed, K9AY, in an editorial in *RF Design* (October 1995) writes: "There is a potential environmental disaster lurking in the shadows beyond the wild success of wireless communication - interference. The rapidly growing use of electrical and electronic equipment of all types is causing a corresponding increase in the noise level in urban areas . . . some interference sources are obvious . . . others are unseen and often unexpected. Any common electrical apparatus can be a big problem - kitchen appliances, power tools, doorbell and furnace transformers, water bed heaters and aquarium heaters all can create an arc . . . as consumers we buy and use more electrical devices than ever, so this kind of noise will continue to increase.

"The amount of electronic equipment in our homes and offices is increasing even faster than electrical appliances. We have computers on our desks, and computer circuits in places we don't expect. Microwave ovens have microprocessor controllers, as do television sets, cable-TV boxes, VCRs and video games, CD players and other audio equipment. Every one of these devices has digital signals that create RF energy, with at least a little being radiated (or coupled into the mains supply - G3VA).

Pat Hawker's Technical Topics

PAT HAWKER, G3VA
London 37/SE22 8SS

Even if initially well designed, some of these units may develop problems, be tampered with, or simply be put close to other equipment that picks up their unwanted signals . . .

"A typical household can easily have 10 or more 'wireless' devices - cordless phones, garage-door opener, baby monitor, intercom, extension speakers, radio-controlled toys, security system, etc.

"Some of these sources may be benign but, taken as a whole, they can create a substantial sweep of RF energy . . . Can you guess the effects of large numbers of new paging transmitters, cellular base stations, microcells and PCS base stations?

"The challenge (for the professionals) is to create products that resist interference and create as little unwanted energy as possible. Is there motivation to do this? Or will things come crashing down when we finally get to the point where our wireless devices choke to death in the electronic smog? Think about it!"

In Europe, the EMC Directive of the European Union provides motivation, but it would be rash to believe that this will clear away the existing problem although it should slow down its growth. For amateur radio, we are concerned with levels of electrical noise often way below those likely to interfere with personal and mobile phones, etc.

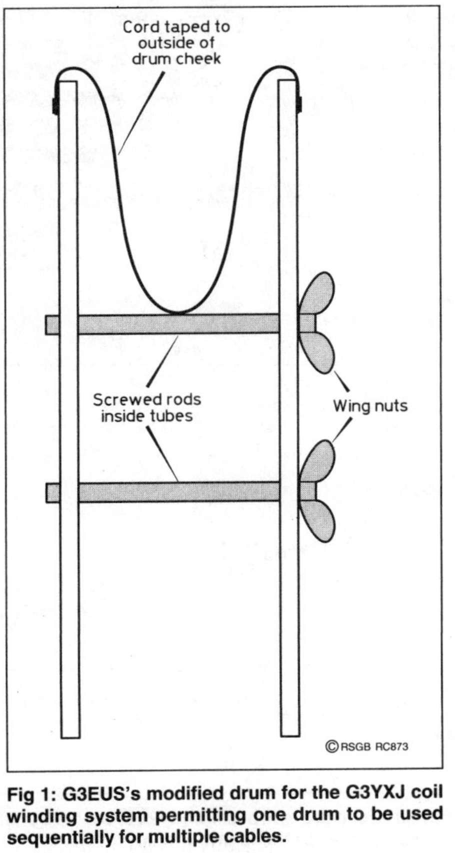

Fig 1: G3EUS's modified drum for the G3YXJ coil winding system permitting one drum to be used sequentially for multiple cables.

What can be done to alleviate the existing problems? The traditional remedies still hold good: additional screening and filtering of source where possible such as mains filtering in the shack. Horizontally polarised receiving antennas rather than vertical polarisation; even better, separate loop-type (ferrite-rod or small screened loops) receiving antennas rather than using the transmitting antenna for receiving. Antennas erected as far from houses as possible, with screened feeder (coaxial cable) possibly buried in a sealed conduit and possibly with an RF choke-type 'balun' at the receiver/transmitter as well as at the antenna element. But it is unlikely that any of these measures will produce a completely clear electronic environment.

Nulling techniques within the receiver can cope well with interference arriving from a single direction. Noise limiters and digital signal processing can reduce the impact of electrical interference to some degree. As I noted in the December's *TT*, p69, David Evans has pointed out that "good engineering practice based on 100 years of research and development can reduce noise and its effects to manageable levels, but there are some fundamental limits. 'Know your enemy' has always been a good maxim".

MULTIPLE-CABLE WINDER

J G FITZGERALD, G3EUS, was most interested in the DIY cable winding system of Danny Begg, G3YXJ (*TT*, October 1995, p74) but suggests a modification that permits cables to be wound and secured using only one drum.

He writes: "The cheeks of the drum are made as before, but the dowels are replaced by threaded rods inside plastic or metal tubes. The rods are held in place by nuts and washers, preferably using wing nuts for one set to ease rapid dismantling. When the drum is loaded into the winding frame two collars are provided against the drum to prevent the nuts fouling the frame. Similarly, the square 'clutch' is modified to accommodate the spacers, or dispensed with.

"The empty drum is placed on the stand and a piece of cord, about two feet long, is lightly attached between the edges of the cheeks with adhesive tape so that it falls in line with one of the rods. Another piece of cord is fitted 180° opposite on the drum. These cords will be trapped near the spacer rods when the cable is wound on the drum.

"The drum is wound with the cable in the usual manner, and then securely tied with the cords pulled free from the sticky tape. If desired, three cords can be used, spaced about 120° around the periphery of the drum cheek.

"The drum (**Fig 1**) is then taken off the stand, and one set of nuts and one drum cheek removed. The neatly packaged and secure cable can now be taken from the drum, and the cheek and nuts replaced in readiness for the next cable.

"If required, the assembly can be used to uncoil cables without fear of kinking, by reversing the above procedure, simply by cutting the strings once the cable is in position on the frame. The beauty of the modification is that one drum, though rather more difficult to make, can be used for winding many cables, as may be required by Field Day groups etc.

"It might also be possible to modify a wooden drum of the type often discarded by electrical suppliers."

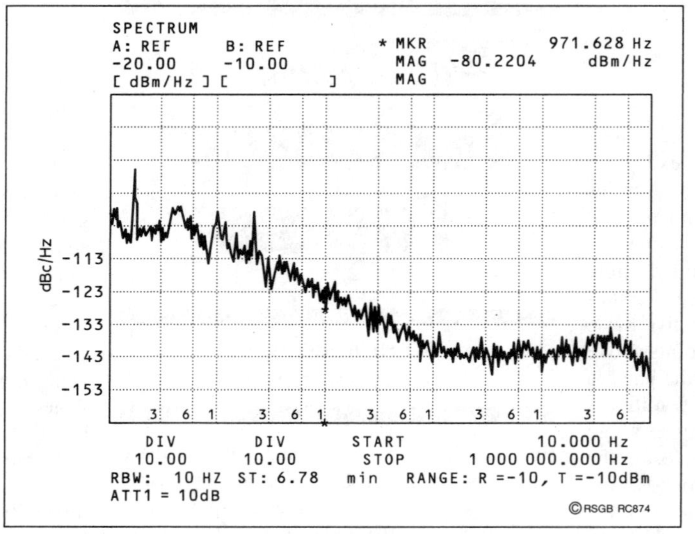

Fig 2: Phase noise spectrum of the G3JIR Vacker VFO as measured at SERC.

OSCILLATORS AND SYNTHESISERS

AT THE "100 YEARS of Radio" Conference professor Mike Underhill, G3LHZ, in his paper, 'Frequency Control in Radio - from Quartz to Direct Digital Synthesis' stated that: "For a technique which is so fundamental to the operation of radio, it is perhaps surprising that what little of its history that has been openly recorded is not readily accessible . . . The problem is compounded by the undoubted commercial importance of this area . . . The favoured technique is to keep all new advances hidden at least until the next product using such an advance is ready for release to the market . . . The skill of commercial patenting is revealing just enough to be able to claim ownership (while) obscuring the important supporting knowledge necessary to make the idea work in practice . . . The Armed Forces in a number of countries have all separately wished to have better frequency control than their allies, enemies or potential enemies. The tendency then is that a leading edge technique will not even be patented but will be kept a secret so that competitive edge in the military sense is maintained."

Fortunately, development for cost-conscious civil rather than military applications is becoming more important, and the commercial incentives for semiconductor and component manufacturers mean that more is known of the leading edge technology than was the case a few years ago, when the ramifications of the 'cold war' threw a cover over radio communication technology.

One result is that amateurs are becoming more aware of the desirability, second only to frequency stability, of achieving extremely low phase noise oscillators, except for simple receivers and QRP transmitters.

G3LHZ emphasised that: "The importance of the phase noise of an oscillator was not at first appreciated. For a transmitter the phase noise superimposes noise sideband on the transmitted signal components . . . Only if at too high a level will the demodulated received signal appear noisy . . . much more importantly, oscillator phase noise spreads the transmitter signal into adjacent channels . . . this is particularly problematic if a receiver very near the transmitter is trying to receive a weak signal on an adjacent channel ('near-far' effect).

"What was originally even less understood was that phase noise on the oscillator of a superhet *receiver* causes exactly the same effect. A strong adjacent-channel unwanted incoming signal mixes with the noise sideband of the receiver oscillator giving an output at the same IF frequency as the wanted signal. Given strong enough adjacent channel interference, the wanted signal is swamped . . ."

It is important to note that this swamping occurs regardless of the degree of adjacent channel selectivity provided by IF crystal filters or DSP. This important problem was discussed in *TT*, January 1995 (see Fig 4 of that month). The January *TT* pointed out that "with LC oscillators, achievement of the lowest phase noise requires contradiction of some of the classic requirements for high stability including the lowest possible output from the oscillator device".

A further contradiction was described by G3LHZ: "Until the importance of low phase noise was realised the rule for the design of a good oscillator was that the LC tuned circuit should be as loosely coupled as possible to the oscillator feedback amplifier stage. However, an oscillator design for low phase noise requires that the loss resistance should be power matched to the input noise resistance of the oscillator, if oscillator efficiency is of no importance; or arranged to be about twice the actual input resistance of the amplifier if maximum efficiency and low phase noise is desired. Power matching implies tight coupling to the LC circuit."

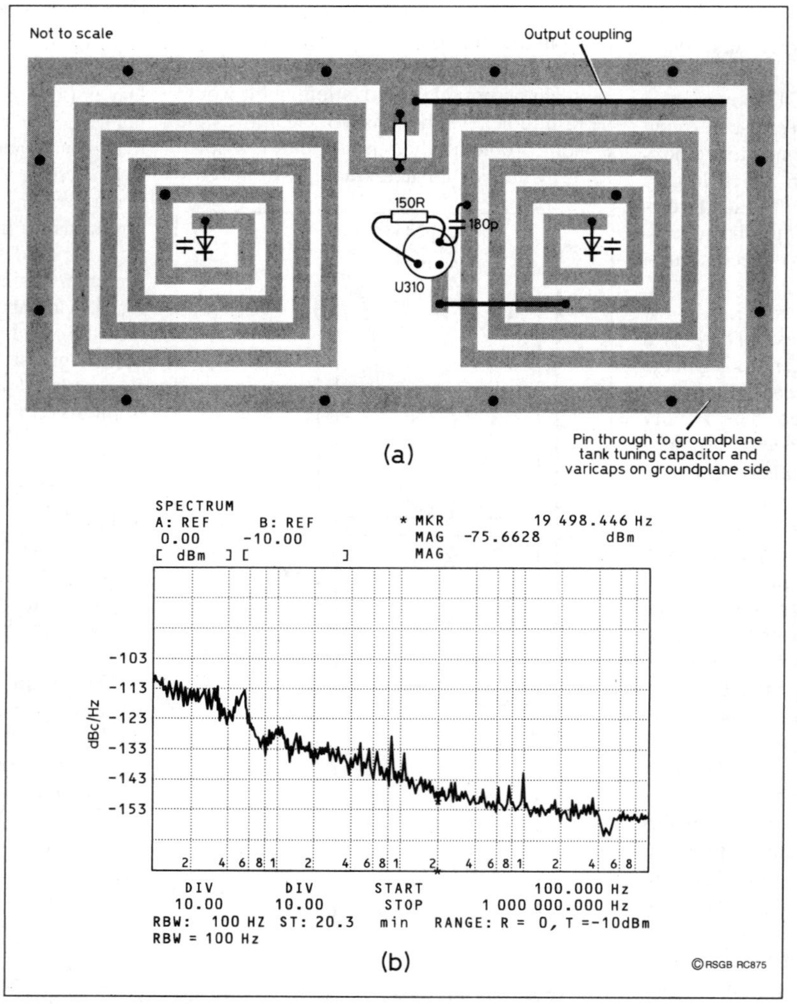

Fig 3: (a) Modified printed circuit two-coil oscillator (not to scale). Double-sided PCB forms groundplane on the reverse side. (b) Noise plot of basic oscillator.

What could well be a practical example of this contradiction between the requirements for optimum stability and optimum phase noise seems to have arisen in the excellent Vacker VFO with automatic level control presented by Jack Hardcastle, G3JIR, in *RadCom* (October 1995, pp45-46). This tuned from 4.00MHz to 4.35MHz with an output of +7dBm and a worst case drift of 28Hz for a 5° C rise in temperature - a truly excellent stability specification for an LC oscillator.

G3JIR, however, took the VFO along to the SERC Laboratory at which Colin Horrabin, G3SBI, works. Together they plotted the noise spectrum of the VFO using the high-performance facilities set up by G3SBI at SERC: **Fig 2**. G3SBI admits to being rather disappointed with the results: "I would have expected the graph to have hit the -148dBc/Hz noise floor of the HP8640 reference oscillators much closer in. In fact, the HP8640 noise floor was not reached, and a noise shelf at -143dBc/Hz was the limit. Even traditional analogue oscillators are seemingly not an open and closed book when it comes to phase noise!"

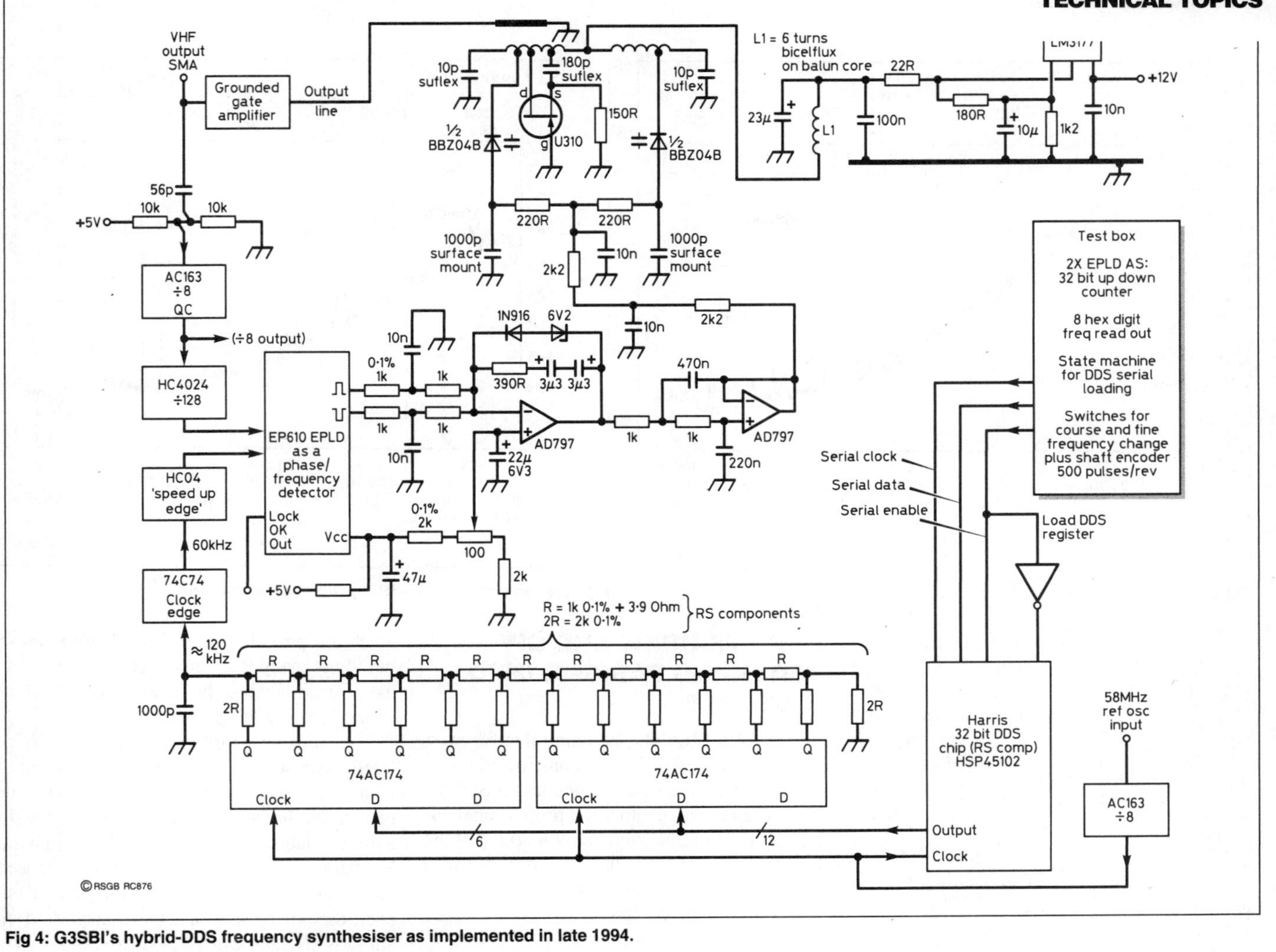

Fig 4: G3SBI's hybrid-DDS frequency synthesiser as implemented in late 1994.

While the phase noise figures of the VFO are good by conventional standards, they fall a bit short of those required for a super-linear, state-of-the art receiver and G3SBI has already achieved better noise performance from his design for a DDS synthesiser (see below). But he also reports: "While on the subject of phase noise measurement, I have discovered a little trick that may be useful to experimenters. If the signal to be characterised is a square wave, the test oscillator can be run at, say, eight times the frequency and locked in by the phase noise bridge. The result is that it is almost as if the test oscillator is divided by eight and the effective phase noise of the reference is improved. Using this method with an SMG signal generator it reduced the phase noise floor to the system limit of -160dBc/Hz and had better close-in phase noise performance. I checked the results against the HP8640 and they were similar close-in."

Despite its low phase noise characteristics, it should not be assumed that direct digital synthesis (DDS) is problem-free. It requires careful design. As G3LHZ put it at the 100 Years of Radio Conference: "A DDS with a quartz crystal as its clock source automatically has excellent phase noise. Furthermore, because there are no feedback loops in the DDS the frequency of the output can be made to change instantaneously whenever a new frequency is commanded... There are, however, two major deficiencies in current DDS synthesisers, and both these are because IC technology is still not sufficiently advanced.

"First the DDS suffers from a high level of spurious output signals, deriving from inadequacies of the digital-to-analogue converter (DAC) . . . A 10-bit DAC gives spurious components which can be as bad as -60dBc (-60dB relative to the carrier). Fortunately for a given output frequency, each time the sampling frequency is doubled the spurious power level drops by eight times or 9dB. This means that the spurious level can be reduced by making the sampling frequency much higher than the output frequency... With current technology a ten-bit DAC working at a few hundred megahertz is about the state-of-the-art limit, and so the -90dBc limit can only be achieved at a few MHz.

"A second deficiency is because, as the sampling frequency increases, the power required both for the digital waveform computing circuitry and the output DAC increases approximately in proportion. State-of-the-art DDS designs consume around a watt of power for outputs of only a few tens of MHz and with barely good enough spurious performance. It could be at least another five to ten years before the present type of DDS circuits could supplant the PLL circuitry at present used in cellular radio handsets. On the other hand, in fixed base station applications, the low phase noise with less worry about power consumption will see DDS find increasing application.

"In the meantime, the DDS can be well combined into a multi-loop hybrid PLL arrangement where, with careful design, the best of both worlds can be achieved... small step size, low phase noise, and faster switching, while the PLL keeps power consumption low even at GHz frequencies. Such a combination can also be arranged so that the PLL filters the worst of the DDS spurious signals so that DDS with lower performance (and requiring less power) can be used."

Late in 1994, G3SBI developed a low-noise frequency synthesiser based on DDS techniques as part of his overall intention to develop a super-linear receiver capable of taking full advantage of his H-mode-type mixer, etc. He wrote: "A note in *TT*, July 1994, covered the development of a low phase-noise, two-coil oscillator using a grounded-gate FET. The most promising version used stripline inductors against a groundplane. A further development of the basic circuit uses spiral printed-circuit inductors (**Fig 3(a)**) to reduce the physical size. It has similar basic performance.

"At the time of the original work, only a high-quality spectrum analyser was available to make phase-noise measurements. Subsequently, a phase noise measuring system was constructed (Fig 6 of *TT*, January 1995, p69) and the performance of the two coil oscillator is shown in **Fig 3(b)**. The carrier frequency was 72MHz and -153dBc/Hz occurs at about 20kHz from carrier.

"A frequency synthesiser (**Fig 4**) was designed to operate in the 60 to 72MHz region. This would be divided by 16 for a proposed tuneable IF receiver to provide a 3.956 to 4.456 MHz VFO with a performance near to -153dBc/Hz at 1500Hz from carrier.

"The initial design used a single helical resonator of Q 800 as the oscillator tank, but phase noise plots proved very disappointing with only -130dBc/Hz at 20kHz. The two-coil strip-line oscillator was soldered end onto the

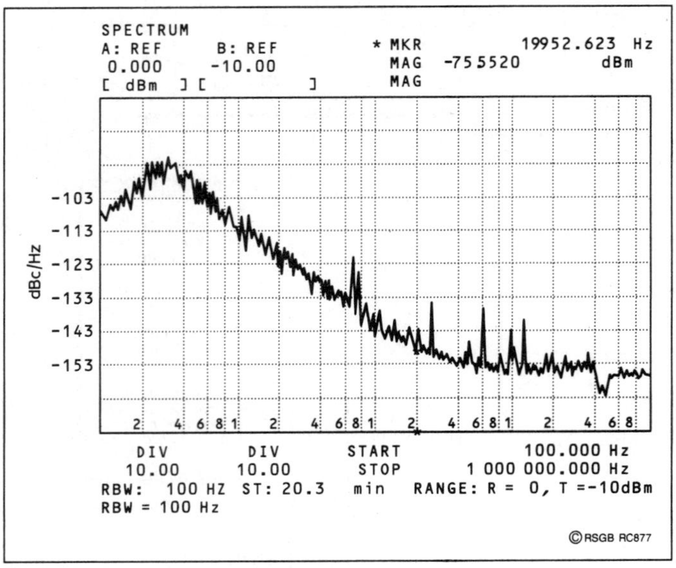

Fig 5: Noise plot of the G3SBI synthesiser at 63.3MHz.

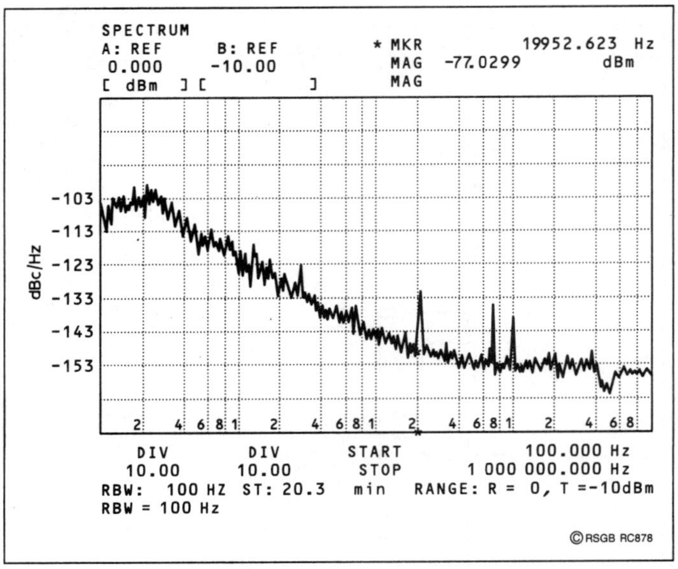

Fig 6: Noise plot of the G3SBI synthesiser at 71.3MHz.

earthplane and wired into the circuit. It achieved -150dBc at 20kHz at the high end of the range (**Fig 5**) and about -140dBc/Hz at the low end (**Fig 6**), presumably due to varicap diode conduction as the DC control voltage dropped. Despite trying various biasing and capacitive division techniques there was no significant improvement at the low frequency end of the range.

"The varicap diodes were SMV 1200 - 20 (from Cirkit) and were chosen for their high-Q at 50MHz although some BB204B dual varicaps were used for comparison and produced excellent results. At low frequencies the phase noise is higher than the basic oscillator, but above 10kHz it drops to the basic noise profile.

"The voltage regulator used on the VCO was very important in achieving the low phase noise. An LM317T delivered 9.5V with RC bypassing following the regulator necessary to improve phase noise further out from carrier. The use of a TL431 shunt regulator gave good far-out performance but the LM317T series regulator gave significantly better close-in results (less than 5kHz). The phase detector is in EPLD (see **Fig 7**). It could be made in HC or AC74 series equivalents. The EPLD version can see time changes of less than 200pS.

"Although phase noise is good, the main problem in achieving a 120dB dynamic range is reference frequency breakthrough spurs limiting true dynamic range to about 100dB. There are five poles of low-pass filtering effective on the VCO control line and pick up of the spurs has been shown to be due to radiation from the divider logic. The answer would be to construct the divider and the direct digital synthesiser (DDS) system together with the digital phase detector in the separate screened box with the VCO loop amplifier and LP filter outside; and also to screen separately the two-coil oscillator unit.

"A receiver used to listen to the VCO divided by eight output sounds fine; even thumping the bench has no audible effect on the carrier. The basic design could be used for a 5 to 5.5MHz unit by using a DDS output reference frequency which allowed a VCO tuning range of 80 to 88MHz. Tuning resolution at 5MHz can be a fraction of a Hz."

IS HOME-BREW AND DIY SERVICING PASSING INTO HISTORY?

LET US BE FRANK and admit that for most amateurs the days of home-constructed transceivers or receivers have faded into nostalgia. The exceptions are either the professional engineers in our ranks who enjoy working at the leading edges of the technology or those who find satisfaction in low-power operation with relatively simple equipment.

For the rest of us, it is either a matter of keeping older, often-valved equipment on the air or becoming content with building ancillary equipment (less and less of which is needed with modern all-embracing black boxes) or contenting ourselves with antennas (it still sounds strange to hear amateurs proudly proclaiming that their antenna is 'home-built' when at one time it would have been more unusual for them to have been purchased!)

For many years the large Eastern Bloc of

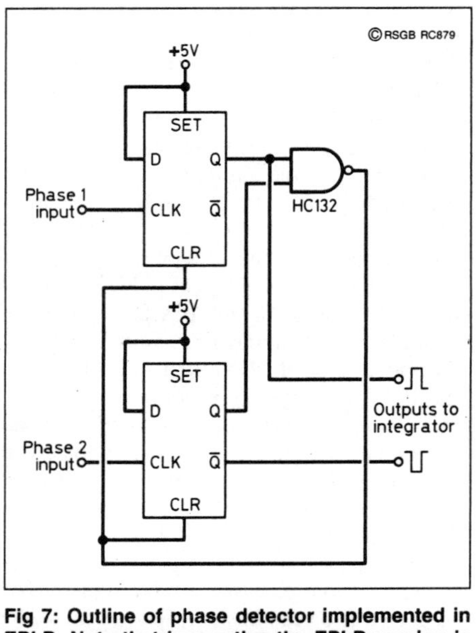

Fig 7: Outline of phase detector implemented in EPLD. Note that in practice the EPLD version is more complex than this. Polarity sense can be reversed easily and there is out-of-lock detection. The basic circuit acts as a phase/frequency detector and gives +/- 360-deg lock range.

amateurs provided a bastion of home-brew, but today many Russian etc amateurs seem to have succumbed to the undoubted attractions of the Japanese black boxes.

The modern factory-built transceiver, HF or VHF, with its tiny surface mounted components is no longer something that can even be serviced at home without an extremely well-equipped laboratory. When the almost inevitable fault (even though these are well spaced out) occurs there is often no alternative but pack the rig up and arrange transport to an authorised dealer with reliable after-sales service. And if it is a very recent model, it may turn out that even the dealer is still waiting for the essential service manual!

Watson Peat, GM3AVA, has pointed out that as the reliability of components and solid-state devices has steadily improved, a continuing problem is the traditional one of 'dry' soldered joints developing after the equipment has been working well for some months. Oxidation, heat-cycles and the like can affect any less than perfect soldered joint and result in an intermittent or continuous high-resistance or open connection. It used to be possible to poke around with a knitting needle or similar insulated probe until the offending joint was found, and then quickly restore things with a fresh dab of solder. Today, with the closely packed boards and multitude of dip- or wave-soldered joints, not to mention the complex circuitry in which it is far from easy to associate a particular fault with a particular section of the equipment, even a single dry-joint may involve sending it to a repair depot. Since, even for the professional, tracing a poor joint may take several hours, costs can quickly mount up despite the fact that nothing has to be replaced.

At one time, manufacturers of military radios etc tended to use crimping rather than soldering for many joints, but I cannot see anyone trying to crimp a surface-mounted component! Maybe there is still room for someone to come up with a good alternative to 60/40 solder. And soon, since this type of fault is now one of the main causes of equipment malfunction!

LA8AK'S CERAMIC-RESONATOR VFO

THE APRIL 1995 *TT* (pp50-51) provided details of a 'Third World' direct-conversion receiver by Jan-Martin Noeding, LA8AK, designed around low-cost components readily available in places where choice is limited to a few standard types. This used an RA3AAE-type harmonic mixer with a 'W3JHR synthetic-rock' VFO but it was suggested that a ceramic-resonator VXO might prove a useful substitute.

Recently, LA8AK has developed a fully-fledged ceramic resonator VXO which would be suitable for this receiver, although intended for a low-power portable transceiver (possibly solar powered) for the 7 or 10.1MHz bands he is building for use while mountaineering.

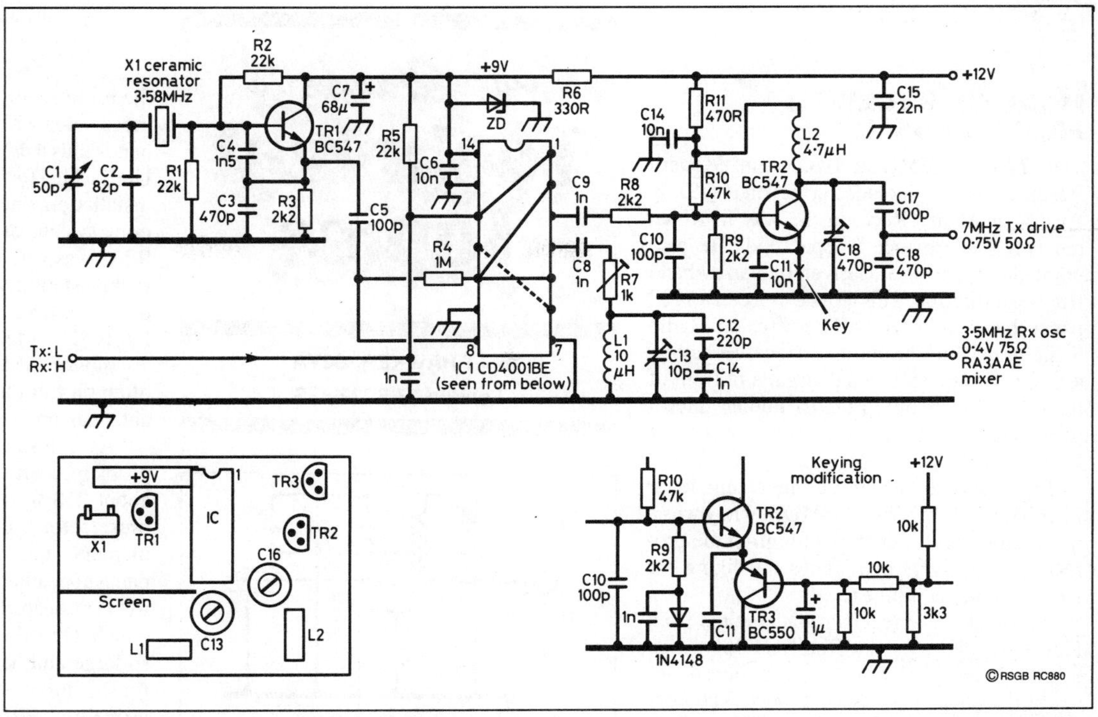

Fig 8: LA8AK's 3.5MHz ceramic-resonator VXO for 7MHz low-power direct-conversion transceiver using a receiver based on the RA3AAE harmonic mixer and frequency-doubler transmitter section. VXO is suitable for the 'Third World' DC receiver described in the April 1995 *TT*.

He writes: "A ceramic VXO based on the Clapp circuit has the drawback that the RF output varies significantly with frequency. For use as a local oscillator, it is necessary to add extra gain and apply some form of limiting or AGC. CMOS inverters are suitable for this purpose. As shown in **Fig 8**, a CD4001 (quad nor-gates) is used as the input amplifier and logic switch. To save current, the receiver oscillator drive is present only during receive, and the transmitter drive is switched off during receive. With the 7MHz DC receiver using the RA3AAE twin-diode harmonic mixer, the local oscillator is in the 3.5MHz band, and the transmitter drive is doubled to 7MHz.

"Since a square-wave signal is not recommended for driving frequency doubler stages, a simple low-pass RC filter (R8, R9, C10 100 nanoseconds) leaves 0.7V RF at the base of the TR2 and 0.75 in 50Ω (+10dBm) at the RF output. The output may be increased by 3dB by using a lower value for R11. The simple single-tuned circuit is not generally regarded as sufficient for proper filtering against spurious signals, but is shown for functional simplicity. Used for a 1W portable transmitter, it may possibly not be too important. A CMOS low-drop voltage regulator will be used later to supply the +9V to the VFO circuit. A Morse key may be connected in the emitter circuit of TR2, but an improved 5mS-delay is shown. With this modification (R11 220Ω) the RF output is 9.9V RMS. To vary the RF output drive for the transmitter, the DC collector voltage for TR2 can be varied.

"Instead of winding coils for this project, use was made of standard value inductors, intended for use as RF chokes. The sketch shows how components can be mounted in 'dead-bug' fashion on a board measuring 40x55mm. Total current consumption during receive is 5mA and with the key down it is about 20mA."

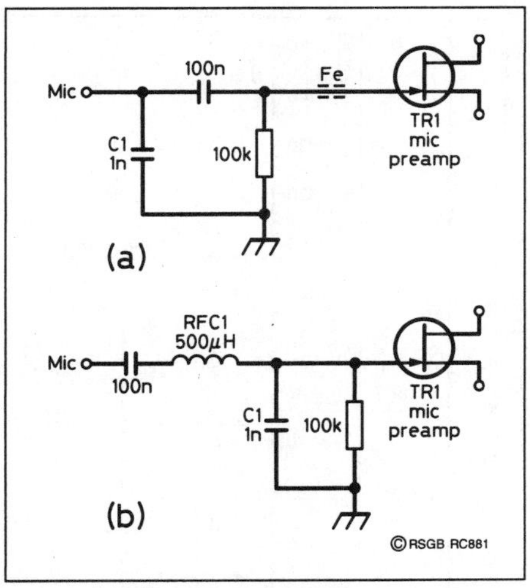

Fig 9: Basic RFI suppression devices for use with audio equipment such as an electret amplifier-microphone. C1 and Z1 (or RFC1) are added to the preamplifier to help prevent the effects of stray RF energy. RFC1 for protection against HF may be any value form 220 to 500μH. Z1 is a No 43 mix miniature ferrite bead for VHF and UHF suppression.

RFI PROTECTION OF AMPLIFIER MICROPHONES

DOUG DEMAW, W1FB, in his article 'Curing Amateur Radio Station RFI' (*CQ*, November 1995, pp58-59) notes that a number of solid-state transceivers are designed for use with amplified desk microphones. Unfortunately, some makers cut costs by not including RFI suppression in the circuit. Such unprotected desk microphones can cause distortion when the transceiver operates above about 50 watts RF: "The mere act of touching the microphone while transmitting may set up feedback conditions that produce howls and squeals in the transmitted audio."

W1FB points out that the cure for RFI feedback in microphones is simple and inexpensive, requiring only the addition of disc ceramic capacitors and a miniature RF choke. **Fig 9** shows typical RFI protection of preamplifiers for amplified electret microphones. Although some may use bipolar transistors rather than FETs, the cure is the same.

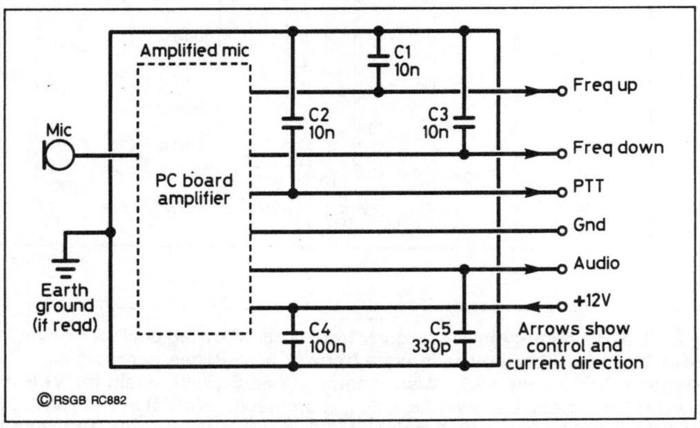

Fig 10: Block diagram of a typical amplified microphone. C1-C5 are added inside the case to suppress RFI effects.

When using a miniature ferrite bead (Z1) this should be placed as close to the transistor as practicable. It is usually effective at VHF or UHF, but for HF it is usually necessary to use a miniature 200-500μH RF choke. C1 used to bypass RF energy should not be larger in value than 0.0033μF or it may impair the high-frequency response of the audio.

Fig 10 shows the generic arrangement of an amplified microphone. Most have metal frames that are common to the amplifier circuit earth, and the added bypass capacitors C1-C5 may be earthed to the case. ♦

Pat Hawker's

Technical Topics

PAT HAWKER, G3VA
London 37/SE22 8SS

MORE ON RECEIVER FRONT ENDS

THE *TT* (SEPTEMBER 1995) item 'Modern Receiver Design' included criticism in *QST* by Australian Henry Rech on the way in which receiver performance is specified. He also pointed out the benefits given by narrowband front-end filtering. This led me to recall a 1982 paper by R A Barrs of Rediffusion Radio Systems in which he outlined a receiver having six signal-frequency tuned circuits in front of the mixer (September Fig 7(b)) under microprocessor control from the frequency synthesiser.

This resulted in a most interesting letter from Michael O'Beirne, G8MOB, who owns a most impressive collection of high-grade professional receivers and shares my liking for large, chunky equipment.

He writes: "The receiver that resulted from R A Barrs' paper was the Redifon R500 which appears to have been purchased in some quantity by the Royal Navy as the R500N. A number have appeared recently on the surplus market at about £450 - £500. I have one complete with the huge handbook giving a general description and the full specification which in practice is much bettered.

"Unfortunately from an amateur viewpoint, tuning is entirely by key pad with no tuning knob. Fine for channellised use by the professionals or for SWL listening to specific stations, but fairly useless for casual listening. There is a delay of several seconds while the servo motors controlling the capacitors and range switches in the front-end module are in action.

"I also use mine as a somewhat expensive Q5-er tuned to 100kHz or 455kHz or whatever the IF of the front receiver may be (though the sideband inversions get complicated and with some amateur receivers there are no off-set crystal filters so that much mental arithmetic is needed when shifting sidebands!)

"The front end module is a work of art. Since the variable capacitors do not have and exact linear law, the microprocessor system includes pre-set corrections, stored in part of the EAROM, and it adds or subtracts pulses to the stepper motor from the calculated value at a number of frequencies in each band. Effectively, hand calibration with provision to change the corrections.

"The mixer is a switching ring-mixer using four BSV81 FETs, not unlike that used in the Racal RA1772. Earlier Redifon receivers were usually based on diode-ring mixers."

G8MOB recognises the value of additional pre-mixer RF selectivity and draws attention to the presence on the surplus market of a cheap preselector which covers 2 - 27MHz in four bands plus a direct link: "This is the 'Acceptor Unit' which formed part of the D11/R234 radio installation, once a popular radio communications system of the Royal Signals. The D11/R234 was usually mounted either as a single mobile station in the old Austin one-ton trucks or as multiple stations in a container on the Bedford four ton truck. The station could work all modes AM/SSB/ISB/CW/FSK but most often on full-duplex FSK with, usually, close antennas on closely-spaced frequencies.

"As a result, blocking was an inevitable problem, particularly with a full output of about 250 watts. Hence the need for a good preselector. **Fig 1** shows roughly (from memory) the circuit diagram and the front panel appearance. The size is 19in wide by 7in by 12in deep. Housed originally in a Creeth case but mostly now sold without as the case is so large and heavy, with dark green hammer finish; chrome handles, black knobs and white lettering. Fitted with an excellent 10:1 Eddystone slow-motion drive on the tuning knob.

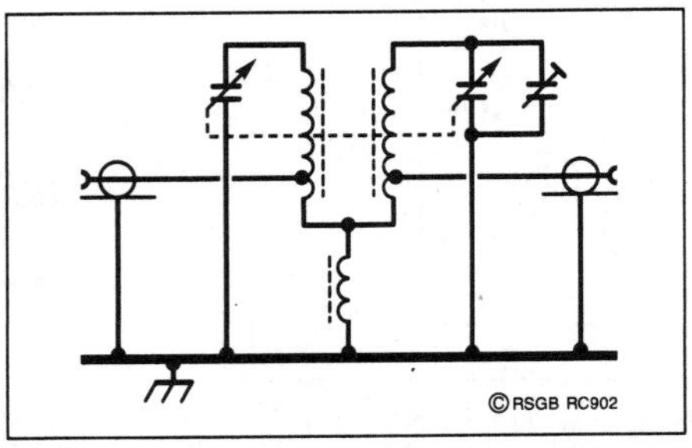

Fig 1: (a) Sketch (from memory) of the Acceptor Unit of the D11/R234 military radio now appearing on the surplus market. (b) Circuit diagram (although the in/out couplings may be link coupled rather than tappings).

"In use, the first modification is to fix 2 x BNC connectors to the rear chassis and reroute the coaxes. Input and output impedances are 75Ω nominal. On tune the signal attenuation is about 3dB. At 10% off tune this drops to around 37dB, providing good additional RF selectivity.

"The 'Acceptor Unit' is solidly constructed and easy to work on, with plenty of space to add, say, an attenuator network or antenna input switching. Mine cost £6 but dealers are now charging a lot more since discovering what they can do. By modern amateur standards it is a huge item - but I like chunky equipment and it lives in a rack alongside an RA17, RA1771, RA1792, R500 and Marconi 2541, Eddystone 958/7, 1650 and more! Not forgetting the lovely old AR88."

A useful nomogram for understanding receiver specifications, etc, appears in an article 'Das Dezibel in der Nachrichtentechnik' by Prof Dr Hans-Hellmuth Cuno, DL2CH, (CQ-DL, 10/95, pp738 - 739): see **Fig 2**. It should be noted that the prefix f (fermto) or femtowatt is 10 to the power of minus 15, and a (atto) or attowatt is 10 to the power of minus 18.

Don Nappin, G3MLS, in the course of a long letter, comments on the topics of receiver specification and pre-mixer selectivity, and the following notes are based directly on his letter. On the question of receiver evaluation he draws attention to an article by Robert Watson of the Watkins-Johnson Company (*Microwaves and RF*, De-

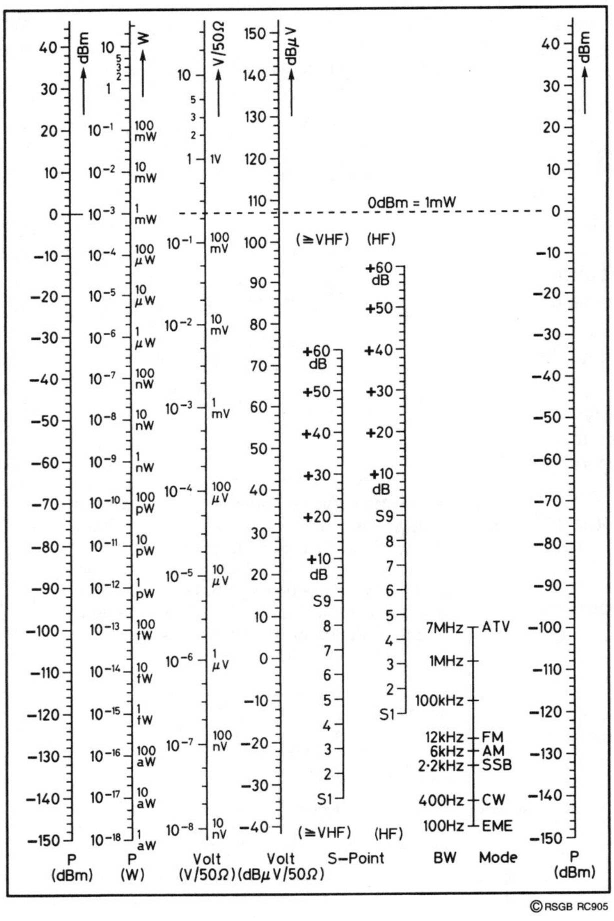

Fig 2: DL2CH's nomogram for receiver front ends, showing decibel equivalents for power in watts dBm; power in watts; power in voltages across 50Ω; power in voltages dBuV across 50Ω . Also recommended S-point levels for VHF and HF receivers; and mode bandwidths. Calibration marks: P(dBM) 2dB; P(W) 1,2,3,5; V/50Ω 1,2,3,5; V(dBuV)/50 2dB; S-points HF S9 = 50uV; VHF S9 = 5uV. BW is bandwidth for a noise voltage across 50Ω at 25°C. Prefixes m - milli, n - nano, p - pico; f - femto; a - atto (see text).

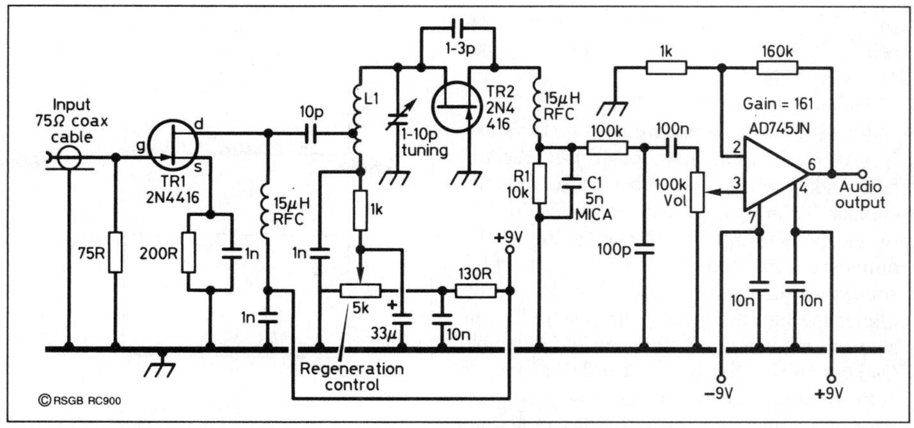

Fig 3: The low-cost VHF super-regenerative receiver designed by Charles Kitchen, N1TEV.

cember 1986, pp113 - 122 and January 1987, pp99 - 108) which recommends a method of testing known as the noise power ratio figure of merit (NPRFOM). This uses a test signal comprising broadband noise, together with a notch filter which removes the noise at the measurement frequency. The basic technique is to set the receiver to produce a standard output from the noise outside the notch, and also to produce the same output from the notched-out noise, when the noise has been increased by 40dB. The notch is presumed to be >50dB deep and the formula takes into account the receiver noise figure and the notional noise floor of - 174dBm/Hz.

In these articles, figures are quoted for measurements taken on VHF communication receivers. A valved design is shown to be inferior to a later design with high dynamic range mixers and amplifiers. This second receiver was tested with only a low-pass filter on the input, with sub-octave filtering, and with a tuneable preselector: the NPRFOM values at 39MHz were respectively: 51 (valved receiver); 65, 81 and 87dB. This clearly shows the advantage to be gained from good signal frequency filtering, even where the tuned preselector is said to have a 10% bandwidth. No doubt a narrower input bandwidth would have given a still better figure of merit.

The preselector recommended in RA Barrs' paper had a 3% bandwidth, reflecting his choice of a capacitor with only 1% ganging accuracy with the consequent choice of relatively low-Q tuned circuits (but see above the notes on ganging correction in the R500-G3VA). B M Sosin (Marconi Communication Systems) also considered these matters (*Radio & Electronic Engineer*, July 1971, pp321-9) in discussing the design of the H2900 communications receiver. He similarly concludes that tuned input circuitry is required but settles on two tuned circuits, giving a 12% passband at -30dB.

The opposite point of view was put forward by R F E Winn of Racal (*Electronics & Power*, 13 June 1974, pp440 - 444) who shows that whilst a poor receiver is improved by the use of a substantial amount of antenna attenuation, a good one, both in terms of intermodulation and selectivity, benefits less by its use. This accords with Barrs, who holds that a good receiver - he again mentions 90dB imp - should not need an input attenuator.

G3MLS believes that home construction of high-grade preselectors should be able, if thought necessary, to improve on Barrs' 10%: "Mullard used to have a range of precision ganged capacitors, which claimed 0.7% accuracy and, with many other makes, featured slotted end plates, permitting individual adjustment to a higher degree of accuracy, perhaps 0.25%. This would allow the use of circuits with a loaded Q of 100, giving about 0.44% bandwidth at -3dB from four tuned circuits. This would represent 132kHk at 30MHz, though only 8kHz at 1.8MHz, where some overcoupling to give a bandpass characteristic would be advantageous."

G3MLS owns a Standard Radio B46 receiver, dating from the 1940s, which has a roller-coaster tuned local oscillator, with three signal-frequency circuits, tuned by a three-gang capacitor. Separate tuning controls are provided for the oscillator and the RF tuning. This approach should be acceptable for amateur operation.

He notes there are other problems arising from the use of a preselector: "If one requires a loaded Q of 100, then the unloaded Q must be much higher to avoid undue loss. A Q of 200 would give 3dB loss. Clearly, a balance has to be struck between these factors, and this may by why Barrs chose a loaded Q of 30 - 40. This, with an unloaded Q of 100 would give a loss of 1.5 - 2.2dB. A higher Q would give less loss, but would require physically larger coils....

"Whilst it might not be thought that small losses are very important, they come to the fore in considering the overall noise figure. For a receiver with a preselector with four tuned circuits, each with 2dB loss, feeding a passive mixer with 6dB or more loss, perhaps a roofing filter with a further 2dB loss, the losses amount to 16dB. Even if the IF amplifier were noiseless, the receiver noise figure would be over 16dB, a figure acceptable only up to about 10 - 15MHz."

G3MLS considers a number of other factors and sums up: "It seems that the super receiver should strive to achieve high linearity, low noise and good front-end selectivity, but should not need an input attenuator."

A LOW-COST VHF RECEIVER

SUPER-REGENERATION, invented by the great Howard Armstrong and first disclosed in *Proceedings*, Radio Club of America, June 1922, has never really lived up to its early promise - yet has never entirely faded away. It lives on for low-cost receiver applications such as garage-door openers and in some CB handsets etc. Amateurs made use of super-regen for the early 56MHz transceivers; in wartime they went into aircraft identification friend or foe (IFF) systems and the clandestine S-phone and American Joan-Eleanor hand sets used by OSS agents dropped into Germany in the final months of the war. They were also used in some standard Army sets.

The last two years have seen a number of articles devoted to VHF receivers based on super-regeneration including a two part constructional article by Brian Adkinson in *Short Wave Magazine*, May and June 1995, and an in-depth article by Charles Kitchen, N1TEV, of Analog Devices in *Communications Quarterly,* Fall 1994. Charles Kitchen also presented a simple (but award winning) design in

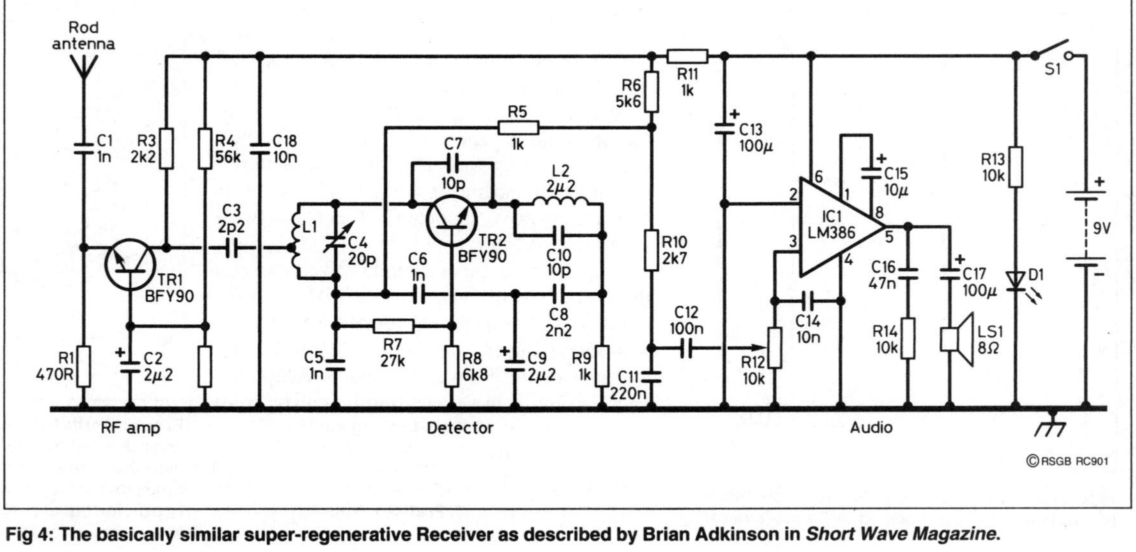

Fig 4: The basically similar super-regenerative Receiver as described by Brian Adkinson in *Short Wave Magazine*.

the *EDN* (July 1995) and more recently reprinted by Bill Orr, W6SAI, in *CQ*, December 1995): see **Fig 3**.

This uses two FETs (2N4516 RF stage, 2M4516 super-regenerative detector) and one IC (AD745JN as audio amplifier). It receives both wideband and narrowband FM, and AM signals. Tuning range can be anywhere between about 88 and 210MHz depending on the number of turns and size of L1. The tap on L1 should be determined experimentally: the higher the tap, the greater the sensitivity but also the greater the loading on the detector. Oscillation is controlled by the 2pF capacitor between source and drain and may need variation to compensate for variations in layout. Output drives a low-noise op-amp with a low-pass filter removing the quench voltage from the audio output of the detector. Remember that stray capacitances or multiple earth paths can inhibit oscillation. L1 must be located clear of other objects, particularly chassis ground, and the bottom and sides of a metal enclosure etc.

The typical loud inter-station noise of a super-regen receiver is the result of the extremely high-gain of this mode of detection which can provide a sensitivity of the order of 1μV. N1TEV's receiver draws less than 10mA, and the cost of the components in the USA is given as less than $20. With super-regen detectors, it is essential to include an RF stage, not to increase sensitivity but to prevent the radiation of interference sprayed over a wide frequency band.

The *Short Wave Magazine* design uses a basically similar configuration (**Fig 4**) but uses BFY90 bipolar transistors and LM386 IC audio amplifier. This was intended for use over the VHF band from about 105MHz to 185MHz but this can be changed by altering the dimensions of L1 or the value of the tuning capacitor C4.

G5IJ PUSH-PUSH-FED TWIN LINE ANTENNA

COMMENTS CONTINUE to arrive on the coaxial antenna in its various forms (*TT*, August, November & December 1995) including interesting letters from Kenneth Parker, G3PKR, and also J A Mullin, DA1MU, who encloses copies of 'Unipole Antennas - Theory and Practical Applications' by Ron Nott, K5YNR, etc from the *ARRL Antenna Compendium Vol 2*.

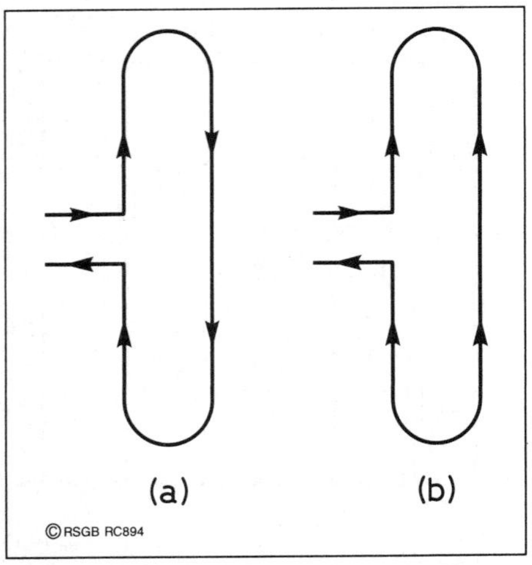

Fig 5: The current modes on a folded dipole antenna. (a) Transmission line mode. (b) Antenna (radiating) mode.

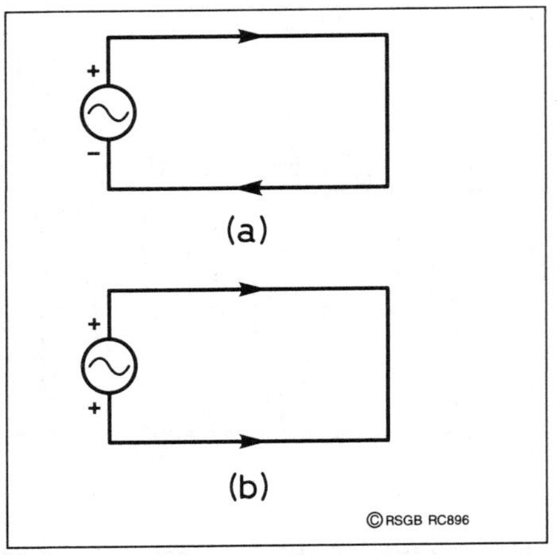

Fig 6: Mode excitation and current for a voltage V applied to the terminals of a folded dipole. Superimposition of these modes gives the complete folded dipole model. (a) Transmission line mode; (b) Antenna mode.

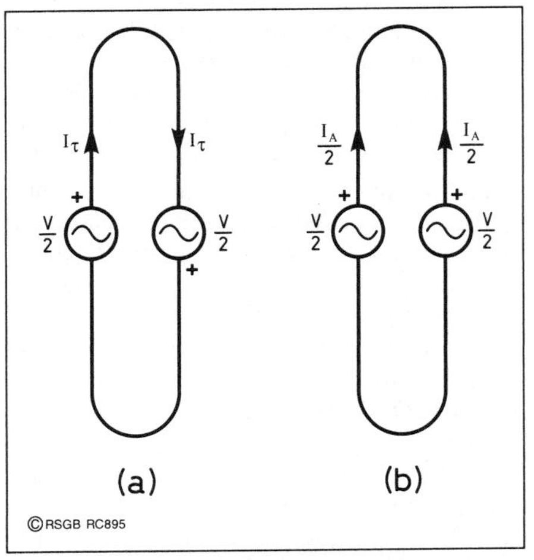

Fig 7: With half of a folded dipole (folded monopole) a voltage generator drives the system as a transmission-line mode. (b) Radiation mode calls for push-push drive (in effect turning the element into a fat element).

However, this month it seems time to move on to what could to be an interesting development in the implementation of 'folded monopole' and transmission-line antennas. This seems to offer the possibility of providing a novel and practical form of a multiband, twin-wire antenna fed in 'push-push' mode at one end from 50Ω coaxial cable via a toroidal transformer when using a simple ATU between the transmitter and the coax feeder cable. It would appear to offer reduced dependence on ground losses in the earth image and on higher bands and may just possibly be acting as a travelling-wave 'launcher' rather than a twin-wire form of Marconi antenna.

The new push-push antenna drive system is the brainchild of Ivan James, G5IJ, who for many years before his retirement was an eminent television engineer with EMI, following wartime membership of the radar development team led by Alan Dower Blumlein. The team worked on the AI Mk VI and later on H2S (during which Blumlein and some other members of the EMI team lost their lives when the field trial aircraft crashed in South Wales in 1942).

First, I have a confession to make. When G5IJ originally sent me details of his push-push antenna driving system, I began to prepare an item for *TT* but found it difficult to conceive how the G5IJ system could work efficiently as a multiband transmitting antenna independently of earth radials or counterpoise or terminating resistors. I wrote to him of the problem I was having in coming to grips with his material. He promptly offered to lend me the antenna to try. For various reasons, the tests had to be postponed for several weeks, but finally I have been able very briefly to try it out on 3.5, 7 and 14MHz bands (and check that it would load satisfactorily on 21MHz although the band was virtually dead). It functioned on all three bands, particularly well on 7MHz with strong lobe(s) into Eastern Europe. At my location, the twin wire runs roughly from west to east with the free end pointing eastwards. I must stress that while these tests show that the drive system is certainly capable of delivering RF power into the twin wires from the coax line, it would need many more checks than I have been able to make to determine exactly how the antenna functions on the various bands and the radiation patterns etc.

G5IJ has provided further information and explanations that answer some of the points I raised. But I should perhaps emphasise that it seems unlikely the antenna really functions as a broadband travelling wave antenna on the higher frequency bands. All one can say for sure is that it puts out a potent signal with a degree of directivity on 7MHz. I tend to think of it as a form of half a fan dipole or a conical broadband monipole. I now have no doubts that push-push drive of twin wires works - and works well - although I am still uncertain of the part played by the earth on bands in which the wires are not one half-wave long or multiples thereof. I can see no reason why his antenna should not be duplicated to provide practical antenna system while still offering plenty of scope for further experimentation and development. To my mind, the main advantage is end feeding multiband twin wires from a length

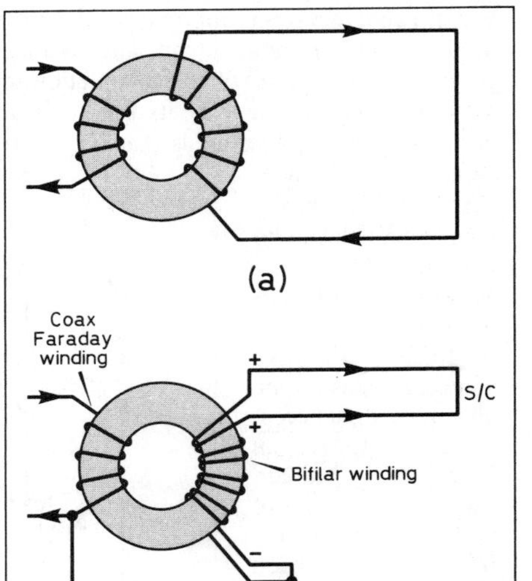

Fig 8: Substituting a transformer-coupled arrangement for case (a) can be met by a single-wound secondary winding. For case (b) the antenna mode can be met by using a bifiliar winding arranged to provide equal outputs to the two wires. (b) is the system used for the G5IJ antenna.

of low-impedance coax able via an ATU in the shack - a configuration that suits many practical locations, but not one to be used without a good ATV or with unprotected solid-state PAs!

To return to G5IJ's letter: "When I read the *TT* items in the August and November [1995] issues of *RadCom*, I was reminded of some experiments I did in 1988 and have been refreshing my memory in regard to monopole antenna and small gardens.

"I wanted to operate on 3.5MHz from my upstairs bedroom, the window of which is 72ft from the back fence. If I were to use a Marconi quarter-wave antenna with a 66ft wire a problem would be the long earth lead. The grounded folded monopole antenna using 67ft of 300Ω ribbon line, which has appeared in all editions of *Amateur Radio Techniques* (eg, 7th edition, p266) stemming from an early edition of *The Radio Handbook* showed promise, but I did not feel that the feed arrangement was optimum. I wondered whether a closely wound toroidal transformer would be a better method with the object of raising the radiation resistance and reducing earth currents.

"But first I felt it was necessary to understand how the basic folded dipole operates. An excellent exposition of the theory appears in *Antenna Theory and Design* by Stutzman & Thiele (John Wiley & Sons, 1981, pp205 - 212).

"To quote the book: 'The folded dipole antenna is basically a half-wave dipole parallel with another dipole, the latter being endfed from the main dipole, raising the drive impedance from the usual 75Ω to 300Ω. As shown in **Figs 5** and **6** there are two current modes in a folded dipole: (a) the transmission line mode; and (b) the antenna (or radiating) mode'.

"If a folded dipole is cut in half it can be seen that to form a folded monopole a generator is required to voltage drive the system: **Fig 7 (a)** shows the transmission line mode; **Fig 7 (b)** shows the radiating mode. With a transformer-coupled arrangement this becomes: **Fig 8(a)** and for the radiating mode **Fig 8(b).** It is important to note that the short-circuited end (s/c) can be open-circuited, if required. Also note that the transmitting (drive) terminals can be short-circuited without virtually affecting the operation, since both terminals (+ +) are at equal potential at any instant in time.

"Put in an alternative way, the transmission line mode needs a push-pull generator and the radiating mode favours a push-push generator.

"To the best of my knowledge all previous proposals for feeding twin, multiple, or coaxial systems (such as the Emil Geles 1927 patent or the earlier Marconi Patent No 260,005) or the arrangement shown in **Fig 9** from ART use push-pull drive, balanced or unbalanced to ground. That must be why they may be relatively inefficient as non-resonant radiators or else confined to resonant operation with the standing wave-patterns producing narrow-band effects.

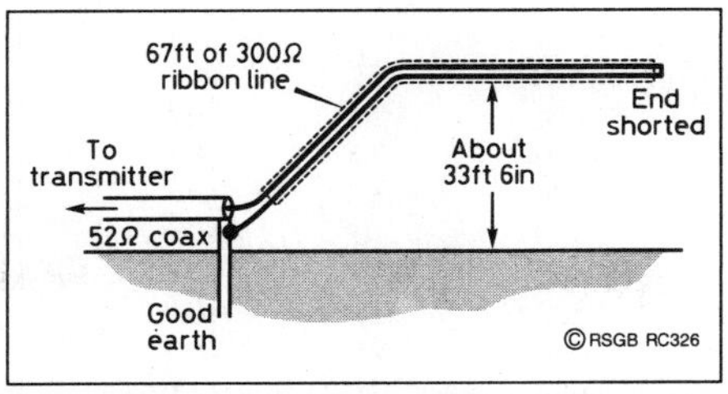

Fig 9: Quarter-wave grounded Marconi antenna for 3.5MHz using 67ft of 300Ω or 450Ω ribbon line. A good earth system (preferably including multiple radials or elevated counterpoise or ground screen) is needed for good performance.

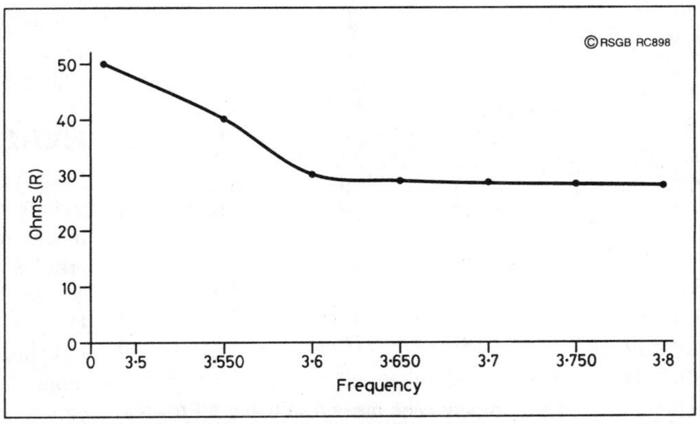

Fig 10: Measurements made by G5IJ in November 1995 using his original BM-1 transmitter on 3.5 - 3.8 MHz. Antenna 20m of heavy-duty slotted 300Ω line 6 to 15ft above ground. Bifilar winding on Amidon T200/2 (120μH per 100 turns) Output 50Ω, third winding using 50Ω coaxial cable arranged as faraday screen.

"My early experiments in 1988 were carried out in the upstairs bedroom with the slotted 300Ω cable coming through the upper opening in the window. Because maximum radiation was thus taking place within the room, the system was very prone to line time-base interference, etc, and also caused RFI problems with audio equipment. As a result I gave up these experiments and ended up with a half-size G5RV and EL40/EL40X traps. I began to write a patent specification covering the bifilar push-push monopole driving system but did not pursue this.

"However, I did make measurements of input resistance versus frequency with a Palomar Engineering R-X bridge which showed how flat the curve is over the European 3.5MHz band (3.5 to 3.8MHz) compared with the usual 100kHz or so bandwidth of a standard dipole. The measurements applied to a horizontal 300Ω slotted feeder, 19m in length and probably about 20ft high.

"After reading the *TT* items on coaxial transmission line antennas, my interest in the push-push system revived and just recently I have plotted a set of measurements (**Fig 10**) using the original toroid transformer (BM1) and using about 20m of (UR43) see **Fig 11** Faraday-screened coupling coil. Unfortunately, the height of the horizontal twin-wire element varied from 10ft to 6ft to 15ft. From 3.8MHz to 3.6MHz the resistance averaged 25 to 30Ω and rose to 50Ω at 3.5MHz.

The toroid Amidon T200/2 core was wound with bifilar winding (40 turns with inductance about 18uH at 3.78MHz). The 20swg length of wire was folded into two parallel lengths and then twisted loosely into about 2in twists in a hand drill, and finally wound around the toroid as evenly as possible. The coaxial winding was wired as a Faraday screen (to reduce capacitance efforts). It was a 1.5m length of UR43 which gave 18 turns in the toroid. Tests with a trifilar closely coupled set of windings shows that little capacitance effects should be apparent since all three windings at each toroidal turn have equal potentials.

"Incidentally, the closely-coupled bifilar winding was used by Alan Blumlein in his first patent in bridge measurements; and led to the so-called 'in-situ bridge' and the wartime capacitance altimeter.

"The suppression of the transmission line mode by the use of the bifilar circuit seems to produce remarkable results compared with the standard push-pull arrangements, and I feel the message should be spread to stimulate further experimental verification. My rough experiments so far indicate not only that the SWR ratios over the bands remain reasonable, but also operation at multiple frequency bands, and even sub-frequency bands, is effective.

"With a Lowe HF225 receiver and with the system connected to the 50Ω input socket, signals are effectively received all the way from 30kHz to 90MHz. With the increasing use of 'black boxes' and smaller gardens, the scheme should encourage antenna experimentation at relatively little expense. There seems endless possibilities for the optimisation of the transformer and its applications."

G5IJ prepared the above notes towards the end of November 1995 but subsequently (early December) sent a series of measurements made with his R-X noise bridge with a new BM3 torodial transformer: see **Table 1**. BM3 comprises an Amidon T200/2 toroid (120uH per 100 turns). Bifilar winding 27 turns (8.5uH) 4.40m length of UR43 50Ω coax used as Faraday screen taking 1.30m of the cable leaving about 3m to reach outside the window. The inner end of coax is soldered to braid of coax and also connected to common end of the bifilar winding. The antenna was 20m of 300Ω heavy duty twin slotted feeder (Westlake) weight 1lb 4oz (575 grams) stranded wire only 6 - 10ft high.

Although as shown in Table 1, it was originally impossible to get a balance on 24.800MHz (R = 50, XL => 7.5); by connecting a 350pF mica capacitor across the coax at entry to the noise bridge (to cancel

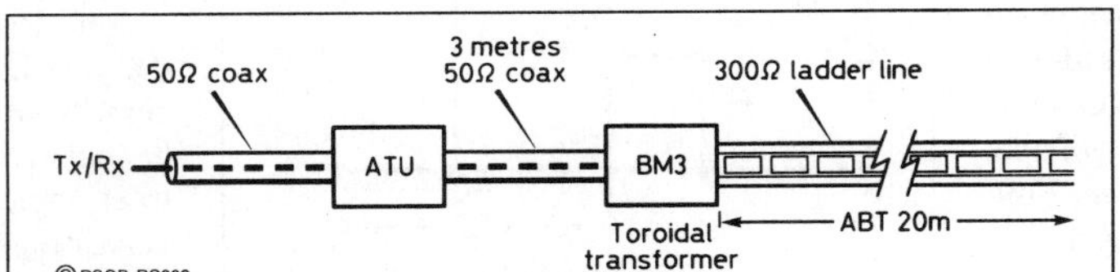

Fig 11: The G5IJ twin-wire antenna system as tried at G3VA from upstairs shack with part of the ladder line and the toroidal transformer indoors.

out the inductive reactance of the antenna) it proved possible to get a balance of R = 18Ω, Xc 0.8Ω on 24.930MHz. During these tests the toroid was inside the window with the 3m of feeder cable.

To calculate the knob in actual ohms positive or negative reactance is rather difficult, depending on frequency (see *QST*, May 1988). The Xc (XL) capacitor has a maximum capacitance of 140pF and is divided into 7 divisions from the 0 (in 20pF sectors). It would seem, at least with the particular dimensions of this specific antenna, that the coax feeder must be fed through an ATU to balance out the reactances. Whether this could be overcome by a little tweaking of lengths etc is one of the many still unknowns. G5IJ hopes that these notes will encourage others to explore the possibilities offered by push-push drive systems but the results are not guaranteed.

Freq(MHz)	R	X *
1.800	20	+0.5
1.850	20	+0.8
1.900	37	+0.7
1.950	40	+0.6
2.000	40	+0.6
3.500	87	+0.4
3.600	75	+0.8
3.650	73	+1.0
3.700	65	+1.2
3.750	60	+1.8
3.800	65	+2.2
7,050	18	- 0.6
10.100	23	+0.2
14.000	23	+2.2
14.150	21	+2.4
14.250	20	+3.0
14.350	22	+3.6
18.100	55	+0.1
21.000	20	- 0.6
21.450	20	- 0.2
21.150	20	- 0.7
28.000	75	- 3.0
28.500	87	- 3.0
29.000	70	- 3.8
29.180	50	- 3.9
29.250	45	- 3.8
29.500	26	- 3.5

Table 1: Impedances measured using a Palomar Engineering R-X noise bridge. In the reactance column + indicates inductive reactance and - capacitive reactance. The reactance values are divisions on the dial and not ohms, see text.

SINGLE SIDEBAND BIBLE

PAUL MOORE, G0OJA, draws attention to a book which he recently reviewed for his company library: *Single Sideband Systems and Circuits* edited by William E Sabin (W0IYH) and Edgar O Schoenike, both now retired members of the Collins Divisions of Rockwell Corporation (2nd edition, 1995, 652 pages, McGraw-Hill, Inc, US price $75). This massive tome has also been reviewed by Doug DeMaw, W1FB, in *CQ*, December 1995.

G0OJA comments: "Not a book for the faint-hearted but one no serious and technically-competent SSB home-brewer ought to be without."

This new edition of a book first published in 1966 (a copy of which is in the IEE Library in London), like its 594 + xii page predecessor, contains 17 chapters, many written by a number of Rockwell-Collins engineers. It is accompanied by a 3.5in computer diskette which has a number of public-domain programs for use with IBM-compatible PCs.

The book continues the succession on books stemming from Collins on HF SSB which the Collins company did so much to popularise both for amateurs and professional engineers. W1FB, while continuing to cherish his first edition, welcomes the publication of the new expanded edition which he feels belongs in the libraries of broadcast engineers, RF design engineers, and technically inclined radio amateurs. It is a pity that the high cost of professional text books seems likely to limit the number of home bookshelves on which this excellent book will be found.

TRANSCEIVERS AND BATTERY VOLTAGES

JOE WONOSKI, N1KHB, in *QST* (December 1995, p79) emphasises the importance when operating transceivers from large lead-acid batteries of looking closely at the manufacturer's recommended minimum supply voltage: "At least one manufacturer lists a *lower* limit of 12.15V for a radio that has Field Day activities as a suggested use. Yet this voltage is reached with a 12V lead-acid battery at the 60% charge point. Such a lower limit thus significantly restricts operating time even when starting off with a fully charged battery with excellent low-loss power connections."

He notes that since an HF radio rated for 100W will draw a peak of about 20A, a total DC resistance of only 50 milliohms in the supply leads will impose a 1V drop. Thus with a maximum of13.8V and a lower limit of 12.15v there is only about 1.45V to play with, and when fully loaded the 50 milliohms will reduce this to about 0.45V between a fully-charged battery and the lower operating limit.

Food for thought when contemplating emergency or field operation. N1KHB believes that equipment manufacturers should address this problem in the design stage, and operators need to scrutinise all operating considerations when selecting equipment if you intend operating in an emergency, or a 100 miles offshore in a sailboat or in a car with the engine turned off.

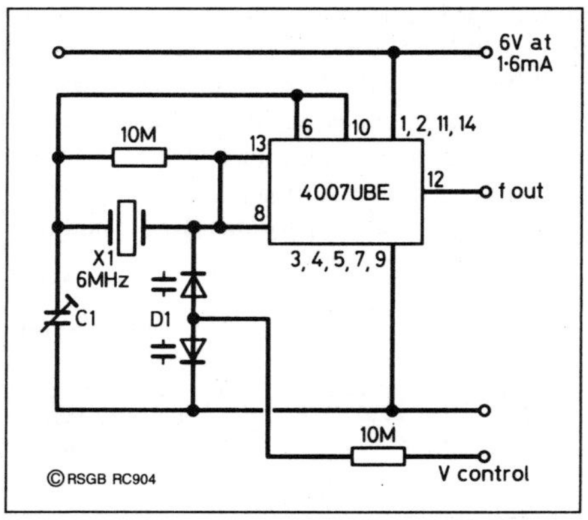

Fig 12: 6MHz voltage-controlled oscillator using ceramic resonator (*Electronics World*).

VOLTAGE-CONTROLLED CERAMIC OSCILLATOR

R G HARRISON IN THE 'Circuit Ideas' feature of *Electronics World* (January 1996, p63) - the magazine has have dropped 'Wireless World' from its title - provides details of a narrow-band VCO using a ceramic resonator: **Fig 12**. Although not intended primarily for amateur radio applications, it adds to the growing recognition that it is much easier to 'pull' a ceramic resonator than a quartz crystal. Based on a 6MHz resonator (RS type 656-215) with CMOS 4007UBE IC and BB102 varactor it has a sweep range of about 50kHz for a change of 33-220pF, the characteristic being logarithmic to give a falling frequency with increasing capacitance.

HERE & THERE

IN A *TT* (JANUARY 1996) item on elevated radials versus buried earths, I commented that it was regretful that none of the subsequent articles based on computer simulations had acknowledged the pioneering work of Arch Doty, K8CFU, *et al* who made thousands of field measurements. I made reference, amongst others, to Al Christian, KB8I, and his two papers in *IEEE Trans on Broadcasting*, in 1988 and in September 1991. R Dean Straw, N6BV, of ARRL points out that an article by KB8I 'Elevated Radial Antenna Systems' (*QST*, August 1988) did in fact cite Arch's work. [I believe this was the suggestion of QST editions - *G3VA*] KB8I has continued his work in this field, and N6BV hints that further intriguing developments will be reported later this year in the next volume of 'The ARRL Antenna Compendium' series.

What with the European Union EMC directives and the Health & Safety Codes, the amateur operator faces increasing difficulties. Recently Roy Clayton, G4SSH, chairman of the Scarborough Special Events Group reported that under the English Heritage Health and Safety Code of Safe Working Practice No 16, it is no longer permissible to operate electrical equipment at English Heritage Sites unless all equipment is submitted for testing, or a safety certificate produced (Para 8.5 of the Code). A formidable specification for temporary electrical installations at special events forms a six-page appendix to the Code. G4SSH was able to save the day by taking his FT100 into his own place of work (Composite Signals Organisation) and obtain the help of the head of the technical section (a licensed amateur) who was able to complete CSO's relatively straight forward 'Electrical Safety Acceptance Test for Single Equipment's'. This satisfied English Heritage - but it leaves open the problem for amateurs who are less fortunately placed. There is no standard IEE Electrical Safety Test covering radio equipment. Another barrier for portable and special event operation!

John Roscoe, G4QK, recalls that in 1939, the year he obtained his licence, his total income from all sources was just £5. Yet with this he managed to build a 6L6XG (ceramic base) tritet single-stage transmitter and an 0-v-1 (two valve) regenerative detector receiver. Taking inflation into account, I suppose this is equivalent to getting on the HF bands today with a total annual income of less than £100! ♦

Pat Hawker's Technical Topics

PAT HAWKER, G3VA
London 37/SE22 8SS

BATTERY LORE

TWO 1995 *TT* items - 'Advances in Lithium Rechargeable Batteries' (June p70) and 'Seeking Better Batteries' (September, pp64-65) - surveyed recent improvements in rechargeable cells, noting the growing rivalry between lead acid, nickel cadmium, nickel-metal hydride, lithium-ion, lithium-polymer and zinc-air. Basic characteristics of these in terms of nominal voltage, energy density, power density, self-discharge, cycle life were given in Table 3 of the September 1995 *TT*.

It was noted then that the energy density of the increasingly popular, on environmental grounds, nickel-metal hydride (NiMH) cells have improved energy density compared with nicad cells but greater self-discharge (about 30% per month). They have less energy density than lithium-ion and some under-development lithium-polymer cells, both of which have very low self-discharge.

Recently, Matshushita have announced that they are developing an NiMH battery with a claimed energy density almost double that of current NiMH cells: roughly 300Wh/1 compared with around 170Wh/1. It seems likely that when such batteries are available they will prove formidable rivals to lithium batteries for such applications as portable handhelds and laptop computers. NiMH batteries have a lower environmental impact than Nicad or lithium based cells.

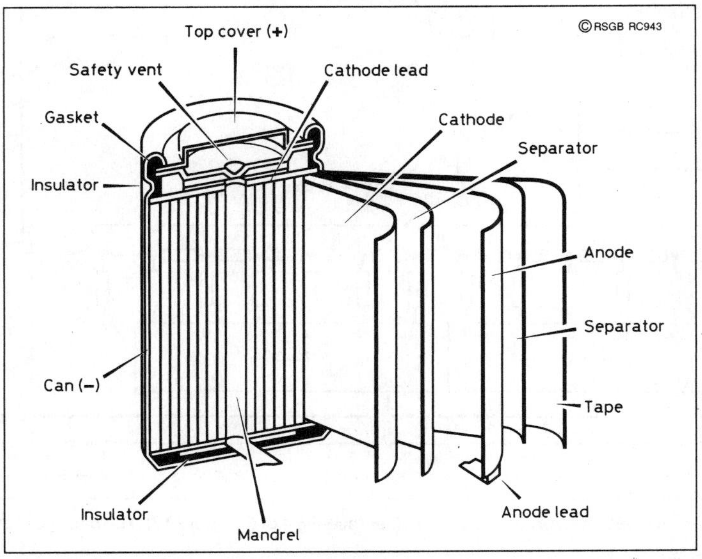

Fig 1: Structure of a lithium-ion rechargeable (secondary) cell (source: *Electronics Times*).

Progress is also being made in increasing the capacity of rechargeable zinc-air batteries. AER Energy Resources of Columbus, Georgia is claiming 50% increase in capacity and 78% output power for its rechargeable zinc-air batteries, claiming that when fitted in laptop computers they can extend the operating time to about 12 hours.

According to a report by Paul Jackson in *Electronics Times* (2 November 1995) on the Power 95 conference in California (before the Matshushita announcement) there was a sharp divergence of opinion between battery scientists from Matshushita and Sony Energytec. Sony is putting its faith in Lithium Ion despite admitting that some rechargeable Li ion cells have experienced the problem of overheating during recharging - a problem that seems not to have been detected during the design and development process.

Matshushita continues to be keen on NiMH rechargeable cells, pointing to the shortage of Li ion supplies and to the revival of interest in improving the NiMH cell. Sony believes that the demands of the battery-operated products now - and will continue to - outstrip the available performance of rechargeable cells, so forcing further battery development, with Li ion cells having the inherent advantage of 3V per cell and high-energy density. **Fig 1** shows the structure of a typical lithium ion rechargeable cell.

Marketing specialists pointed out that Japanese production figures show four times as many NiCd cells still being produced as NiMH and almost 60 times as many as Li ion.

Many 'maintenance free' secondary cells tend to have limited life due either to loss of electrolyte due to overcharging, sulphation, growth of dendrites which short circuit the plates. At various times, suggestions for renovating cells - at least for a time - have been given in *TT*. Recently, my two-year old Yuasa 6V, 2.6Ah gel-type lead acid battery which I use to power a small broadcast receiver, failed overnight. Several attempts to recharge it proved useless, the battery failing to power the radio for more than a few minutes. I soon came to the conclusion that at least one of the three cells must be short circuiting.

I tried the old dodge of discharging a large electrolytic capacitor across the battery in the hope of clearing the short circuit which I felt might be due to growth of dendrites. Despite several attempts this seemed to give only a very temporary respite.

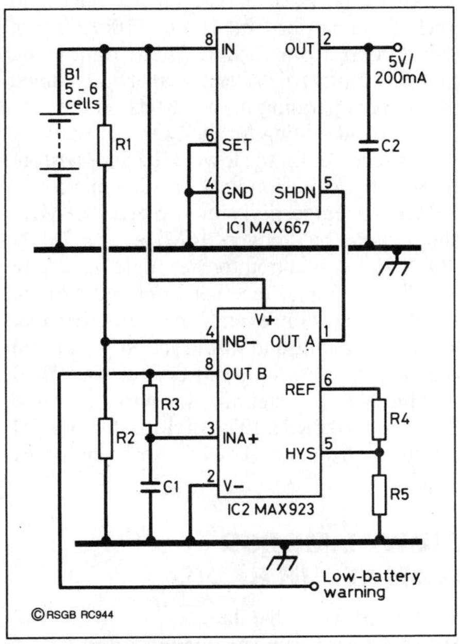

Fig 2: Low-voltage battery monitor with timed close-down (source: *Electronics World*).

LOW-VOLTAGE MONITOR SHUTS GRACEFULLY

CRAIG FALKENHAM and Larry Suppan of Maxim Integrated Products Ltd (Theale, Berkshire) in the 'Circuit Ideas' feature of *Electronics World* (February 1996, p154) describe a low-voltage battery monitor which on detecting the low voltage issues a warning signal and then shuts the power line down after a precise short time interval, rather than after the usual somewhat indefinite time taken for the battery voltage to decrease even further.

They point out that accurate voltage monitoring, achieved by the close tolerance of comparator threshold, allows the battery-low warning to be positioned exactly on the knee of a nicad discharge characteristic. While quiescent, only a few microamps are drawn from the battery.

In **Fig 2** a low-dropout linear regulator (IC1) supplies current to the output power line, dropping only 350mV at 200mA. IC2 is a dual-comparator +/- 1% voltage reference. When the fraction of total battery voltage at the junction of the R1/R2 voltage divider falls below the internal 1.182V reference voltage, OUT B (pin 8) goes high, serving as a warning and also charging C1 through R3. When the voltage at pin 3 rises to the reference voltage, OUT A (pin 1) shuts down IC1. For a hysteresis of +/- 25mV, R4 should be 49.9K and R5 2.4M.

The writers provide as an example: if R3 is 1M, and OUT B is 4.9V then C3 will give a time delay of one second with a value of 3.6uF (low leakage type). Note that during shutdown, C1 becomes charged and needs about 6s to discharge when operation resumes. A larger value electrolytic capacitor, or higher value R3, would give a longer warning time.

G3SBI DDS SYNTHESISER UPDATE

TT (FEBRUARY 1996, p73-74) included details of a hybrid DDS/PLL frequency synthesiser as implemented by Colin Horrabin, G3SBI, in late 1994. Since then a few modifications and new ideas have been incorporated providing improved performance. **Fig 3** shows those sections which have changed, together with some component values missing from Fig 4 in the February *TT*.

G3SBI explains: "(1) The EPLD phase detector (February, Fig 7) was designed so that the two charging pulses overlapped at balance so there was no phase uncertainty as found with a conventional 4046 phase detector.

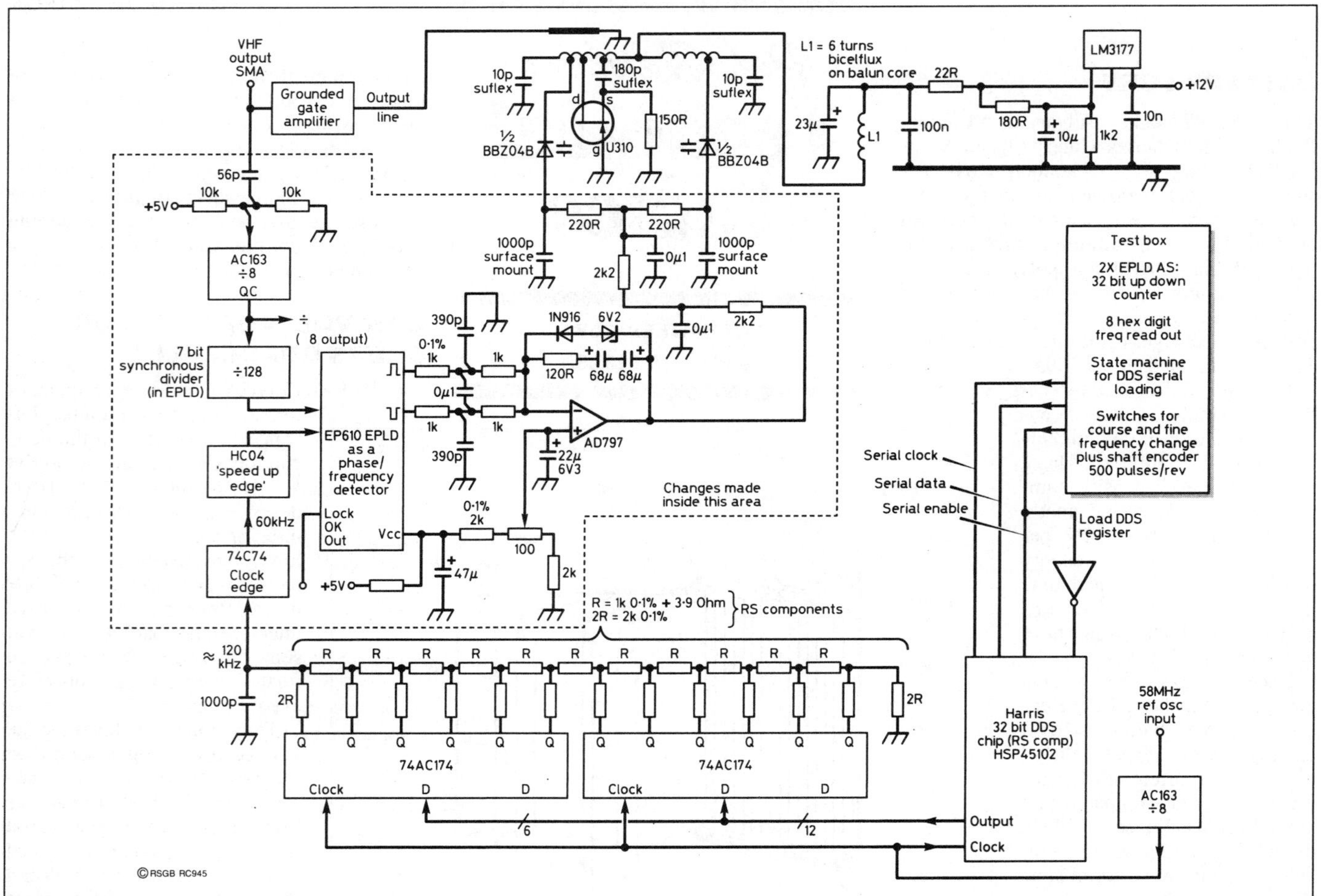

Fig 3: Update of part of the circuit diagram of G3SBI's hybrid-DDS synthesizer (see Fig 4 of February *TT*). The changes reduce the loop bandwidth and the noise plots of Fig 4 are with these changes.

Mullard (Philips) now also use this technique in one of their new phase detectors.

"(2) The cause of the reference frequency break through at the -100dBc level mentioned in February was found to be due to the phase difference pulses (5mA) which charged the two 10nF capacitors via the 1k resistors (a low value keeps thermal noise low). The return path for this current was via the ground pin of the EPLD and this modulated the output to the integrator (AD797). The solution proved to be the replacement of the two 10nF capacitors with a single capacitor between the two resistors so that the charging current went out and back into the chip - note the updated circuit in Fig 3. It then became possible to dispense with the AD797 active low-pass filter and rely on the following two passive sections.

"(3) The HC4024 divider should be replaced by a synchronous divider and I included this on the EPLD. Without this change the dividing of the VCO frequency by eight gave a flattening of the phase noise profile close to the carrier. It became possible to see this once the HP8640 was used as a reference oscillator.

"(4) It proved possible to reduce the closed loop bandwidth of the VCO to under 100Hz in order to further improve close-in phase noise, and reduce close-in spurs. The mechanical rigidity of the basic oscillator (Fig 3 in February *TT*) permits this without introducing microphony. But it should be noted that, in a transceiver, the reduced loop bandwidth may not be fast enough for break-in operation. Thus, electronic change of the loop bandwidth may be necessary to obtain a fast initial lock, with reduced bandwidth on the VCO once in lock. This approach was used in the VCO on the Icom 751A to give improved phase noise performance. It is believed that Racal Electronics have or had patents in this area.

"(5) The latest and hopefully final VCO design is a PCB zig zag with integral voltage regulator operating over the range of 60 - 66MHz. The intention is to divide by 12 to give a 5.0 - 5.5MHz VFO which will be mixed against a crystal oscillator for the required HF bands. To improve the LO purity it may be that on some bands the VCO is divided by 10 or 8 before crystal mixing. Typical noise performance when dividing by 8 is shown in **Fig 4 (a)** and **(b).** 150dBc/Hz is reached at about 1.5 to 3kHz from carrier, and 100 to 110dBc/Hz at only 100Hz from carrier. These figures are probably typical of the best permeability-tuned oscillators operating in the 5MHz region.

"To avoid mixing for the LO of a relatively simple receiver for the lower HF bands with an IF of 9MHz, division by 12 results in 3.5 and 14MHz coverage, division by 6 gives 1.8MHz and division by 4 covers 16MHz for a 7MHz LO. It is my intention to use single mixing in the H-mode mixer receiver front-end with an IF of 9MHz. However, one alternative that looked promising and would simplify the local oscillator design would be a crystal-controlled first LO per band and mix down to a tuneable IF at 10 to 10.5MHz. The divided down VCO would run from 4 to 4.5MHz for a final IF of 6MHz.

DIGITAL MORSE & AMATEUR RADIO

AS EXPECTED, the January *TT* item 'The Morse Requirement' resulted in a number of comments: a few dissenting but far more assenting views. I must admit that I still react with surprise when amateurs suggest that "Morse code is obsolete". It is true that manual Morse is gradually being phased out for professional communications and for the Armed Services (although there is still quite a lot of Morse to be heard outside the amateur bands).

But as several correspondents stressed, Morse is, above all, a valuable and unique system for experimental amateur radio both below and above 30MHz. Without it, there would be virtually no amateur moonbounce (EME), no meteor scatter, little QRP operation or 'mountain-type' simple portable rigs, few if any beacons, and far less chance of exploiting unusual propagation modes. Morse still remains a technically valuable operating mode for amateurs. Yet, without a mandatory test it would inevitably disappear in time. Newcomers need to be encouraged not only to pass the 12WPM test, but to persevere 'on air' until they achieve the ability to operate at, say, 18WPM, by which time Morse operating becomes a truly enjoyable as well as a uniquely useful mode. For those who consider Morse code "obsolete" perhaps we should refer to it as a "binary non-return-to-zero-digital code" - if that makes it sound more up-to-date!

Ray Soifer, W2RS, writes: "I wish to align myself with the comments of VK9NL, ZL2AZ and the others you quoted regarding the continuing importance of Morse. Although DSP systems have been shown to detect weaker signals than can the human ear, signal detection is not the same thing as transmitting information. No better substitute for Morse, copied by ear, has yet been devised for getting across essential information (eg, callsigns, signal reports and 'rogers' under very adverse conditions; indeed, none is likely to be, since slow

Morse channels approach fairly closely the theoretical limits for optimum signal-to-noise ratio in the bandwidths required for such applications.

"The EME work which I described in the cover article of the May 1992 issue of *RadCom* would not have been possible without the use of Morse, which is employed for virtually all the amateur EME activity going on today. The vast majority of intercontinental DX work on Top Band (1.8MHz) takes place on Morse, as does most QRP DX work on the HF bands. From what I have read, most of those who advocate the abolition of Morse have stronger-signal applications in mind, requiring data rates much higher than those mentioned above. Yet, weak signal CW cannot tolerate co-channel interference from wideband modes such as voice and packet, and if new HF amateurs are no longer to be required to have had at least some acquaintance with Morse, the survival of CW bands over time would seem rather unlikely."

One must remember that multi-tone telegraphy systems such as Kineplex and Piccolo which could work below the noise level required a 3kHz channel. For slow Morse, the bandwidth is much less (under 50Hz) so that the receiver noise floor on stable CW signals can be very much lower.

While most of the comments I received supported the views expressed in the December *TT*, it is clear that there are some readers who believe that the time is coming to abandon the mandatory Morse test, possibly substituting a more advanced technical barrier to the limited HF spectrum. The comments ranged from rational argument to personal abuse

For example, Walter Blanchard, G3JKV, believes that the real debate should not be the emotive issue of 'should knowledge of Morse continue to be a condition for a Class A licence' but the wider issue of 'incentive licensing'. In brief, he considers the big question as whether we should have more than one class of licence; whether Morse is the right delineator is a secondary question. His personal answer to the first is a "resounding yes" and the second an equally resounding no! "There is, to my mind, no doubt that Morse is obsolete, although that does not mean the RSGB should not encourage its use."

He challenges the statement that Morse will get through when all else fails. He points out that this is adhered to by those who do not know what up-to-date computer-controlled digital coding systems can do, with certain forms of spread-spectrum signals *well below the noise level*, and undetectable by narrow-band receivers, providing traffic reliabilities of 99.8% basis. He would like to see the RSGB campaigning for amateurs to be allowed to use advanced spread-spectrum modes.

I may be wrong but I believe that, in terms of transmitter power, slow Morse is still the winner and is environmentally more friendly. With typical spread spectrum systems, the signal is distributed over a wide band of frequencies so fast that it causes only a slight increase in the 'noise' heard by a narrow-band receiver - but if a large number of stations are using spread-spectrum then the background noise must inevitably rise. Thus, while it is true that a few spread spectrum signals may be undetectable on a narrow band receiver and cause no interference to others, a large number could seriously limit conventional weak signal communication on those bands where spectrum is severely limited.

At the recent IEE colloquium on HF frequency selection and management (see 'Automating HF' item), two Belgian engineers described the use of direct sequence spread spectrum on HF using a RAKE receiver. "In the direct sequence, receiver noise can be considered as a white gaussian process. It is clear that it would mean a very inefficient occupation of the limited HF spectrum if separate bands for the individual users were allocated. A single spread spectrum user can be superimposed on the existing narrowband traffic, or CDMA allows several spread spectrum users to occupy the same band at the same time... The noise will in general be negligible in comparison to the interference from other users in the same band. Each (spread spectrum) user contributes to the background noise affecting all other users."

They admit that there has been initial distrust towards bandspreading techniques on HF with its limited spectrum but "there are reasons to believe that it could be worth trying". But this is for professional traffic handling, not, surely, for the different nature of amateur weak-signal contacts where we are not primarily concerned with traffic throughput.

Des Shepherd, G3LCS, admits to being very pro-CW but recognises that, on a voting basis, he suspects that those against will ultimately carry the day. But one wonders whether licence regulations should be arrived at in this way. But G3LCS would like to see more attention paid on CW to better operating standards including correct procedure.

George Canning, G4GBN, believes that as a matter of humanity, pro-CW amateurs should recognise that many genuine and committed enthusiasts bound up with family or educational ties have their Class A licences delayed for many years by having to learn Morse. He believes that the confusion between using Morse and the mandatory Morse test should be strongly discouraged. Fair enough, but if there were no mandatory test how many newcomers would be persuaded to learn the Code? Certainly, most of us find this a rather lengthy process and a very few never succeed. On the other hand, if a more advanced technical examination were substituted as incentive licensing, it is likely that the number who would never make the grade would be significantly larger. As AA0QF has pointed out in *QST*, there are, apart from the technical value, many non-rational arguments for CW.

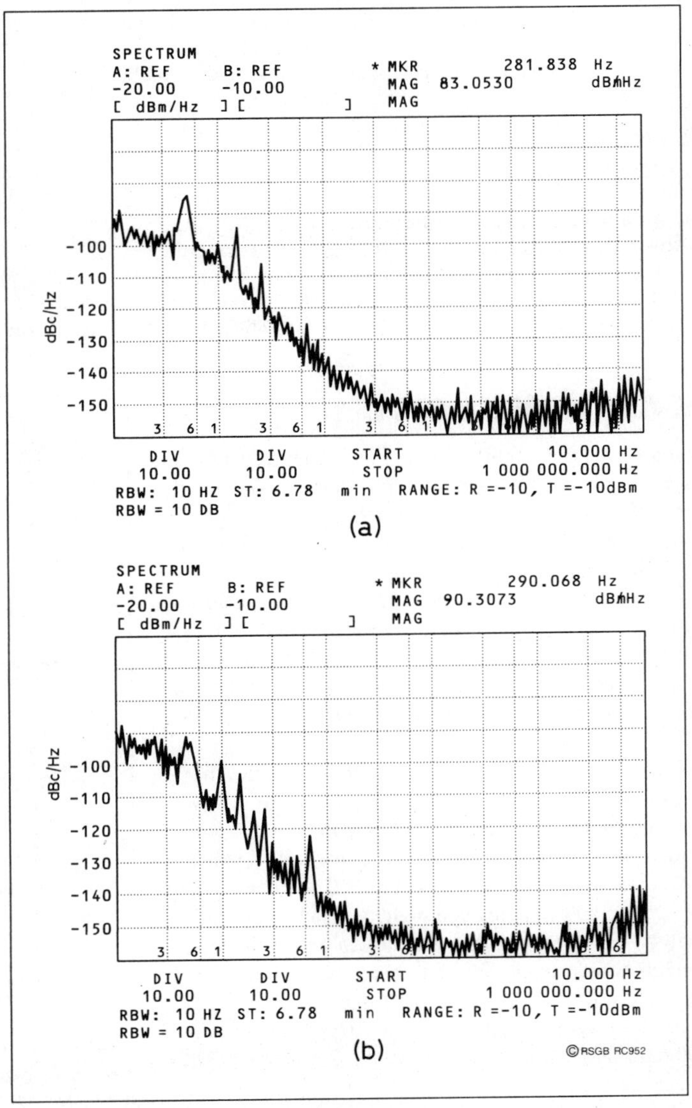

Fig 4: Noise plots of the up-dated G3SBI synthesiser (a) at 60MHz divided by eight, and (b) at 66MHz divided by eight.

AUTOMATING HF

A RECENT IEE two-day colloquium on 'Frequency selection and management techniques for HF communications', with virtually the status of an international conference, attracted more than 100 delegates from the UK, Belgium, France, Germany, Holland, Sweden, Australia, Canada and the USA. More than 25 papers were presented. The re-emergence of HF communications as an important aspect of long-haul traffic handling with particular reference to Defence communications was evident.

It underlined that "capricious" HF, with its ever-changing propagation conditions, not only in terms of signal-to-noise ratio, but also multipath effects including Doppler shift and spread even between fixed sites, provides channels of varying quality capable of supporting baud rates that may vary from 75 to a maximum of around 2400 bits per second (which can carry digitised speech). Even with skilled machine-telegraphy operators only about 50% of the potential data throughput is achieved, with less skilled operators only about 25%. With a fully automatic system the aim is to achieve near 100%.

Measurements show that Doppler shifts up to about 50Hz with spreads of 10 milliseconds or more may occur between fixed sites. This has a major effect on bit error rates particularly at higher data rates, and much effort is going into evaluating various waveforms (modulation modes etc) to provide the most robust signal. Clearly this is of importance to HF Packet users but the topic was not covered at this colloquium.

While automation and maximum data throughput are not generally of paramount importance to the still-human amateur operator, several of the university, Defence, and Industry research and standardisation projects described at the colloquium are not without interest. A major NATO project is currently aimed at standardising the elements of an automatic radio control system (ARCS) including ACS (automatic channel selection), ALE (automatic link establishment). A major participant in this work is the Defence Research Agency (DRA) at Malvern (the former Royal Radar & Signals Establishment) which is also tasked with ensuring commercial as well as Defence possibilities.

From an amateur viewpoint, much of the interest lies in the associated research into HF propagation, particularly in high-latitude zones, and the shortcomings of ionospheric predictions compared with real time measurements. There is a growing belief that the various computer prediction programs for mathematical modelling of the ionosphere, such as MINIMUF, VOACAP, IONCAP and the more advanced ICEPAC need to be supported by various real-time sounding and ionosphere monitoring techniques.

Dr Peter Bradley (long associated with propagation research at the Rutherford Appleton Laboratory) outlined the history of HF propagation predictions. The use of the optimum working frequency (OWF) at 85% of the median MUF compensates for day-to-day variability and is based on the assumption that the OWF is suitable for use on 90% of days. The need to update predictions to provide real time conditions over specific paths requires ionosphere sounding of monitoring, and suitable processing to give practical real time channel evaluation (RTCE).

He noted - as did other speakers - the strenuous efforts being made to standardise and co-ordinate the differing frequency management approach of the US Department of Defense (DOD) with that of the NATO committees: "Emphasis is being placed upon schemes embodying automatic adaption not only for frequency selection, but also of data rate. It remains to be determined what part propagation predictions will ultimately play."

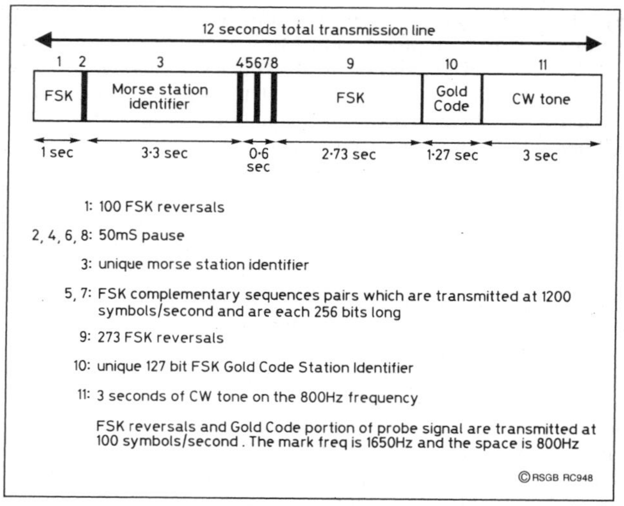

Fig 5: The signal format of the ITU beacons as used at LN2A, Oslo.

He concluded: "The classic approach of long-term prediction and forecasting is no longer valid as the sole aid to modern communications circuit frequency management, but better ionosphere variability statistics are needed in support of system designs. Predictions continue to be required and an ongoing need exists to develop best ways of updating these with real time vertical and oblique-sounding measurements in support of jamming and intercept avoidance and for the interpretation of measurement data from remote-sensing systems."

TT (January 1991, p33) drew attention to the ITU plan to set up a worldwide system of HF beacons in support of ITU planning of HF broadcasting frequencies. The 1kW beacons with broadband omni-directional antennas were to use time multiplexing along the lines of 14.1MHz amateur beacons, operating in sequence on five widely spaced frequencies from about 5.7 to 20.9MHz. It was then reported that one beacon, AUS1MLB, was already operating. But even then it was clear that the original hope of 15 beacons had been paired down to 5.

The signal (probe) format was carefully chosen to allow monitors to evaluate the signal not only in terms of signal strength but also the channel quality in terms of multipath etc.

TT (November 1993) noted that a beacon, LN2A, in Oslo, Norway had come into operation and that the Australian beacon had moved and was using another callsign. The Australian beacon has since closed down. At present only LN2A is operational, though there are still hopes that there will be five beacons operating by the end of this year (but with such a

REMOTE ANTENNA SWITCH

THE TECHNIQUE of powering mast-head pre-amplifiers over the coaxial feeder cable is a long-established technique for fringe-area TV reception. Phil Salas, (*CQ* January 1996, p23 & p28) shows how a similar technique can be used to provide a low-cost remote antenna switch, permitting a second antenna to be added without requiring another costly coaxial cable or the drilling of more holes in the house: see **Fig 6**.

He points out that you can run DC control power along the centre conductor and braid to a remote single pole double throw relay (a DPDT type may be easier to find) provided that this is decoupled by a simple filter from the RF power: Fig 6. For the outdoor switch, you need to use a weatherproof and waterproof container. AD5X uses "one of those aluminium electrical outdoor boxes that are meant for adding external switches and outlets to your home".

He writes: "The most common kind (in the USA) has three ½in threaded holes in it - one on the back and two on the sides. You will also need a blank aluminium cover for this box, such covers come with a weatherproof seal. Box and cover cost in the USA about $4. The box comes with two threaded hole plugs. In this application you need to plug all three ½in holes, so you need an extra hole plug. You first task is to drill a three-eighths-inch hole in the back of the box and secure a 1½in long three-eighths x 16 bolt protrudes from the back of the box. This will be used to attach the relay box to one of the tower (mast) legs with another three-eighths inch lockwasher and nut. Use stainless-steel hardware to prevent corrosion."

But any suitable enclosure could be used provided it is made fully weatherproof. Similarly the indoor unit could be built into any suitably sized box. The value of RFC chokes shown is for HF but this approach could be used on VHF or even UHF bands provided suitable chokes etc were used.

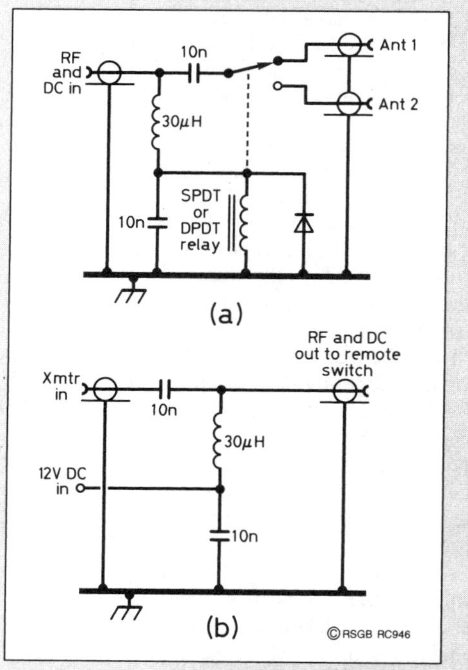

Fig 6: Remote antenna change-over switching using relay powered over the coaxial feeder. (a) Remote (outdoor) unit. (b) Indoor unit.

All stations	Minutes Past each hour (LN2A)
5.470MHz	8-12, 28-32, 48-52
7.870MHz	12-16, 32-36, 52-56
10.407MHz	16-20, 36-40, 56-60
14.405MHz	0-4, 20-24, 40-44
20.945MHz	4-8, 24-28, 44-48

Table 1: Frequency allocations and LN2A time slots.

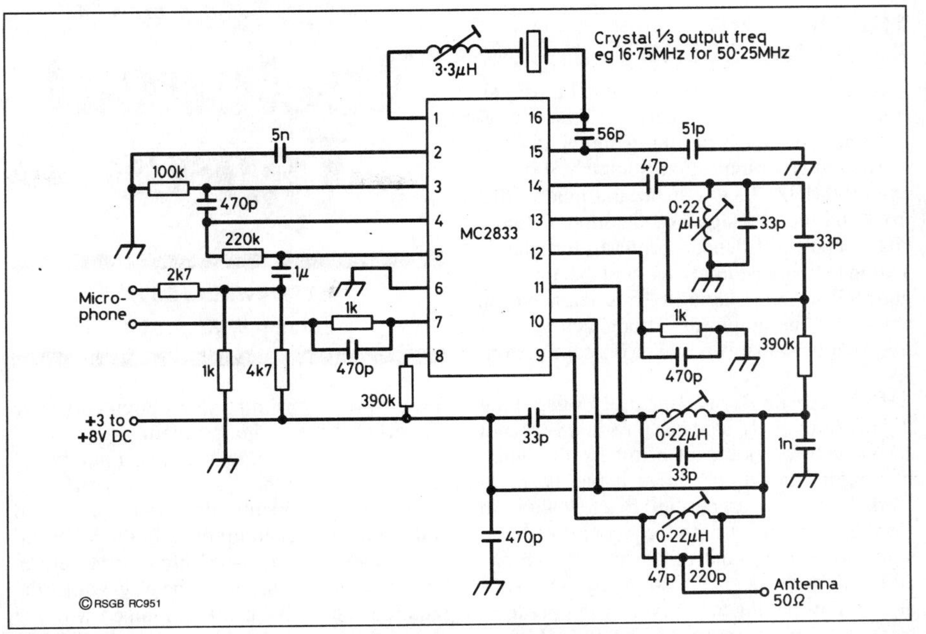

Fig 8: Circuit diagram of the versatile crystal oscillator as described in *RF design*.

frustratingly long gestation period there must remain doubts).

Meanwhile, as reported at the IEE colloquium, use is being made of LN2A at the University of Leeds where a PC-controlled, amateur grade receiver with a DSP board is being used. The signal format for these beacons is shown in **Fig 5** and the schedule for LN2A in Table 1.

As one American speaker emphasised, a number of buzz words and abbreviations have become current in HF data transmission technology including: ALOHA, TDMA, CDMA, PSK, FSK, TCM, asynchronous, synchronous, M-ary FSK, single tone, 16/39/52, FEC, CRC, Reed Soloman, Golay, Interleavers, narrow-band, direct-sequence, spread-spectrum, etc, etc. Full marks for any non-professional amateur who takes all these in their stride. The buzz words at this colloquium included ALE (automatic link establishment), LQA (link quality establishment), KBS (knowledge based system), DAMSON (Doppler and multipath sounding network), RTCE (real time channel evaluation), PFM (passive frequency monitoring), IONCAP (ionospheric communications analysis and prediction program).

ARCS is the NATO project for an Automatic Radio Control System including ACS (automatic channel selection), ALE (automatic link establishment) and ALM (automatic link measurement). The intention is that ACS will automatically select one or more channels to permit waveform and traffic requirement matching against propagation and interference conditions. ALE will automatically set up an appropriate two-way link using a selected channel or channels. ALM will automatically attempt to provide the continuing availability of an established communications link: **Fig 7** shows a block diagram of the modular testbed for ARCS being proposed by DRA, Malvern.

HERE & THERE

CIRCUIT DETAILS of 'A Versatile Wide-band Crystal Oscillator' are given by R Partha and Shanthi Krishnakumar in *RF Design* (December 1995, p50); see **Fig 8**. It is claimed that with suitable values of L and C2, the circuit will oscillate reliably from about 6 to 120MHz with an output of from -0.5dBm to +3dBm. The oscillator trimming capacitor, C8, will vary the frequency a few parts per million. With C5 at the value shown of 3.3pF with crystals between about 6 and 12MHz there may be a tendency for 2nd harmonic oscillation. This can be overcome by increasing the value of C5 to 68pF. The authors provide values of L (uH) and C2 (pF) for a number of frequencies: 6.4MHz 5.6uH, 4700pF; 12MHz 3.4uH, 4700pF; 20MHz 1.98uH, 470pF; 34.29MHz 0.78uH, 470pF; 45.45MHz 0.78uH, 47pF; 73MHz 0.39uH, 47pF; 104MHz 0.16uH, 47pF; 120MHz 0.1uH, 47pF.

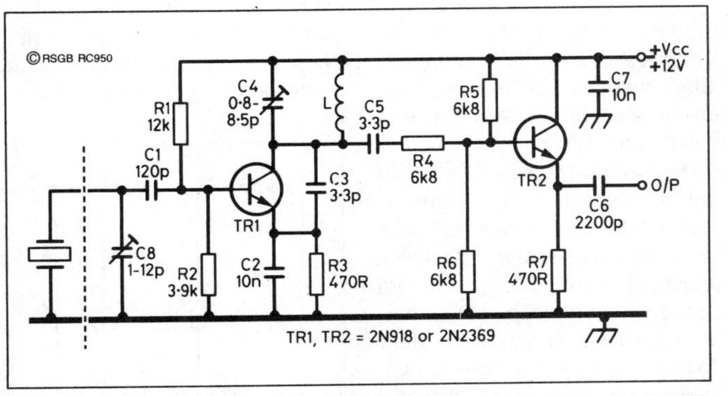

Fig 9: One-chip micro-power (10mW) 50MHz FM transmitter as described by WA2NDM.

Irwin Math, WA2NDM, in his Math's Notes column in *CQ* December 1995 provides details of "a 'one-chip' Micro-power 50MHz FM Transmitter" based around the Motorola MC-2833: **Fig 9**. This chip contains a microphone amplifier, FM reactance modulator, crystal oscillator, buffer, and two power transistors designed to be used as power amplifiers. The crystal is at one-third of the operating frequency. Power coupled to the antenna is about 10mW with 5kHz narrow-band FM. While 10mW is not much power it could be used to drive a Class C amplifier. Since this type of arrangement is widely used at 49MHz for cordless telephones, with a better antenna it should prove quite suitable 'barefoot' for local contacts."

WA2NDM writes: "To align the circuit, first apply power and check for RF at pin 14. With a high-frequency scope (at least 50MHz) connected to pin 11, tune the coil connected to pins 13 and 14 for maximum output. Finally, with a field strength meter loosely coupled to the actual antenna you plan to use, retouch all coils for maximum output and you are on the air." ♦

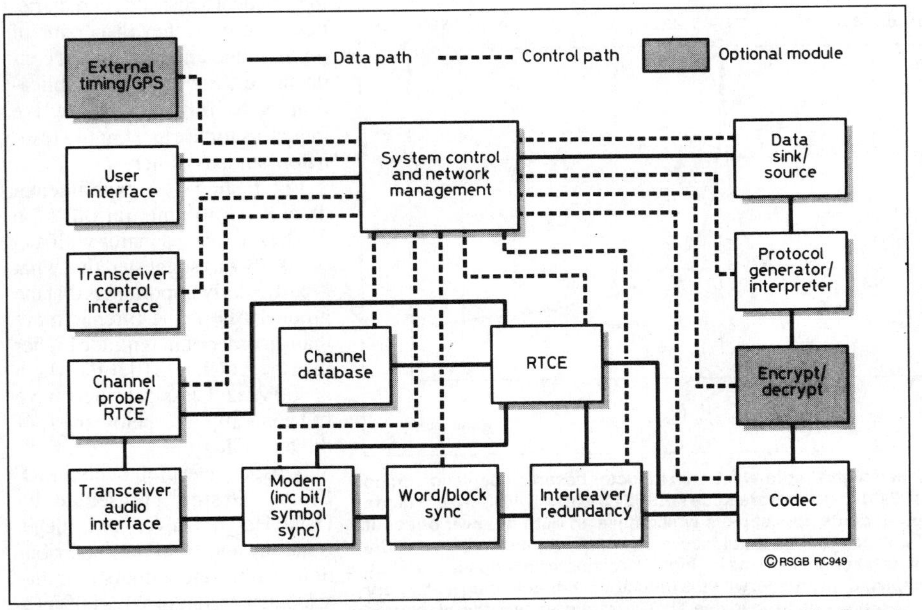

Fig 7: Modular block diagram of the Automatic Radio Control System under development at the Defence Research Agency (DRA) Malvern.

Pat Hawker's Technical Topics

PAT HAWKER, G3VA
London 37/SE22 8SS

THE VHF/UHF RF DESIGN SCENE

A RECENT IEE colloquium on 'The RF Design scene' focused largely on receiver requirements for the burgeoning digital personal and mobile communication systems at 800MHz and 1800MHz. But it soon became evident that front-end circuit design considerations for these UHF bands still depend primarily on classic analogue RF considerations - and that the art of good RF design is having to be re-learned in an era where the emphasis among young engineers has been firmly based on digital electronics.

As Dr Alison Payne (Imperial College) put it: the analogue RF side has for some years not seen much development effort by the whiz-kids sitting at workstations; it has been regarded as a black art best left to a few wise old men. The effects at high frequencies of parasitic capacitances and inductances make circuits start to behave very strangely. This situation is now having to be reversed as operating frequencies move upwards. The wise old men of RF design are again in demand.

This reminded me of the views expressed some years ago by Peter Baxandall (of tone-control and hi-fi amplifier fame) who believed there was a distinction too often drawn between academic studies and practical engineering.

He wrote: "It seems to be sometimes overlooked that engineering is logically concerned with how to make things and solve practical problems . . . the path along which the competent engineering designer travels in order to end up with a first-class design, whose final theory is properly understood and in which the effects of component tolerances are known, is often very different from that implied in many text books and published articles . . . more discussion of such practical matters as parasitic oscillation, earthing techniques, minimising hum etc . . . would help less experienced design engineers to become aware of some of the things that have been found out the hard way by others . . ."

While written before the almost universal adoption of computer-aided-design (CAD), the sentiments remain true today. Those very practical engineers, Peter Chadwick, G3RZP, and his colleague Chris Shepherd, G4OOA (both of GEC-Plessey Semiconductors) pointed out in a paper presented by Chris Shepherd that the increasing use of Personal Radio Services such as cordless and cellular telephones, pagers, Wireless Local Area Networks, security systems, etc means that small, portable receivers are having to work in an increasingly hostile RF environment. The additional requirements of the European EMC Directive mean that front end selectivity requirements have tightened, while the move to lower power consumption requirements have exacerbated the design problem.

The large and rapidly increasing number of base stations for these services similarly impose serious problems for mobile, hand-held and fixed amateur equipment for the VHF and UHF bands. The 'near-far' situations where an unwanted local station is nearby while the required station is far off, combined with the significantly lower noise floor on VHF/UHF means that the requirement for front-end linearity and capability to handle strong signals - in other words the dynamic range - should ideally be even greater than that long demanded of HF receivers.

Indeed, the Chadwick/Shepperd presentation followed lines which have become established on HF: front-end passive filtering requirements; the four main receiver failure mechanisms (gain compression and blocking intermodulation spurious responses); reciprocal mixing; together with reception failure mechanism due to wideband noise and unwanted emissions from transmitters.

The limitations imposed by oscillator phase noise (resulting in reciprocal mixing) and by the emission of wideband noise were codified.

On power consumption it was noted that "power consumption is a parameter which is proving of great interest in the personal radio service, especially since the strides made in battery technology never seem adequate to meet the demands of the users . . . probably the prime example of very low power design may be found in pagers, but it is noticeable that the strong signal handling performance of the pager is compromised to achieve this . . . the statistical probability of missing a call has to be considered when deciding on what is, or is not, satisfactory".

Other speakers at the colloquium, mostly representing university research, described a number of new RF circuit developments some of which may well have application to amateur radio communication on UHF/VHF and microwaves.

Professor Chris Toumazou (Imperial College) in association with Dr T Vanisri (Dept of Telecoms, Bangkok) described a new hybrid design approach for high-gain, high frequency optoelectronic transconductors. This exploits the high output resistance of PIN photodiodes to realise high resistance current sources. In conjunction with submicron MOSFET and GaAs MESFET devices, photodiodes can be used to overcome their limitations which include lack of high performance p-channel devices (due to low hole mobility and poor Schottky barrier properties), backgating (due to non-ideal substrate characteristics), and frequency dispersion of transconductance and output conductance. One proposed application is to form high-Q active inductors for use as a high-Q resonator/bandpass filter.

Fig 1 shows an experimental discrete implementation of a 20MHz active resonator with a Q of 300. It was suggested that what is particularly important is that the product f_oQ of the filter is more than an order of magnitude higher than the f_T of the U310 JFET which is 500MHz. Classical filters have f_oQ generally well below the f_T of the technology.

A paper 'Developments in RF Circuit Design', presented by David Haigh (University College, London) and based on contributions from nine authors, emphasised that recent progress of integrated circuit design and technology has made it possible to dimin-

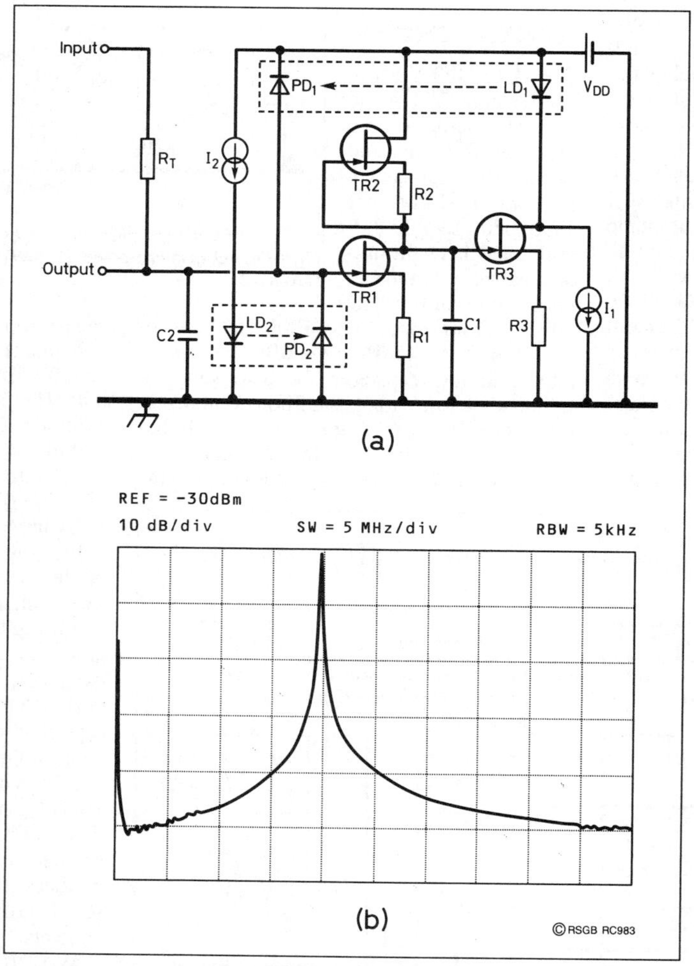

Fig 1: (a) Proposed high-Q opto-electronic inductor based on depletion mode silicon JFETs (U310) or gallium arsenide MESFETS with ML 4413 N laser diode (LD) and PIN photodiode optical pairs which have an output power of about 2mW, 850nm wavelength, threshold current 40mA. (b) In order to verify performance it was connected as a high-Q resonator/bandpass filter by introducing a capacitor in parallel with the inductor and terminating with a very large resistor in order not to load the Q. The curve shows the measured frequency response at 20MHz indicating a Q of about 300. (Source: Dr T Vanisri and Professor C Toumazou at IEE Colloquium on 'The RF Design Scene').

ish both the size and cost of communications systems. "The RF block especially has been at the focus of attempts at integration, and major advances have been seen in the fields of high performance amplifiers, mixers and oscillators."

Whereas filters tend to be realised as passive discrete components, leading to large bulk and high costs, "with the recent development of suitable circuit concepts, integration of highly selective narrowband filters is now becoming feasible".

David Haigh stressed that the enormous signal dynamic range in radio receiver systems provides a very demanding challenge in overcoming the distortion and noise of RF components - amplifiers, mixers and filters. Third-order non-linear distortion is of particular concern. The paper discussed device modelling and computer-aided design considered central to the tasks of RF circuit design in minimising power supply voltage and consumption, while still meeting performance specifications, including intermodulation distortion. Various methods of integrating active RF filters were also considered.

It was noted that while small signal modelling of FETs is already a mature subject, large signal modelling has been relatively poor.

Peter Saul (Elettronica (UK) Ltd) summarised aspects of IC design technologies relevant to the design of on-chip RF circuitry. While new processes continue to be introduced, and old ones improved, the driving force is rarely the RF performance requirements. He concentrated on silicon bipolar on insulator (SOI) processes which show performance improvements but at the cost of thermal problems. He concluded that "although the literally single chip radio, without off-chip tuning components, is still some way off, current and projected application requirements are such that it will form the basis for much research".

Another paper from Thailand proposed a new class of electronically-tuneable oscillators with low-distortion using several types of log-domain circuits as the frequency and loop-gain controlling elements. To achieve low-distortion (of the order of 1%) the practical highest frequency of operation is typically around either 40MHz or 100MHz, depending on the circuit approach, when using transistors with f_Ts in the GHz range. The paper also described temperature compensation and self-heating effects. It also stressed that, as a tuneable oscillator where the range of frequency tuning can be several orders of magnitude, some scheme of automatic oscillation level control is needed.

A paper 'Intermodulation distortion in Class F amplifiers' showed, on the basis of a computer-aided-design (CAD) investigation, that, for RF and microwave amplifiers, the use of Class F reduces the level of third order IMD as well as increasing efficiency. Improvement in the amplifier distortion performance is mainly due to the reduction in the 3rd order IM level resulting from terminating the even harmonics to a short-circuit impedance: **Fig 2**.

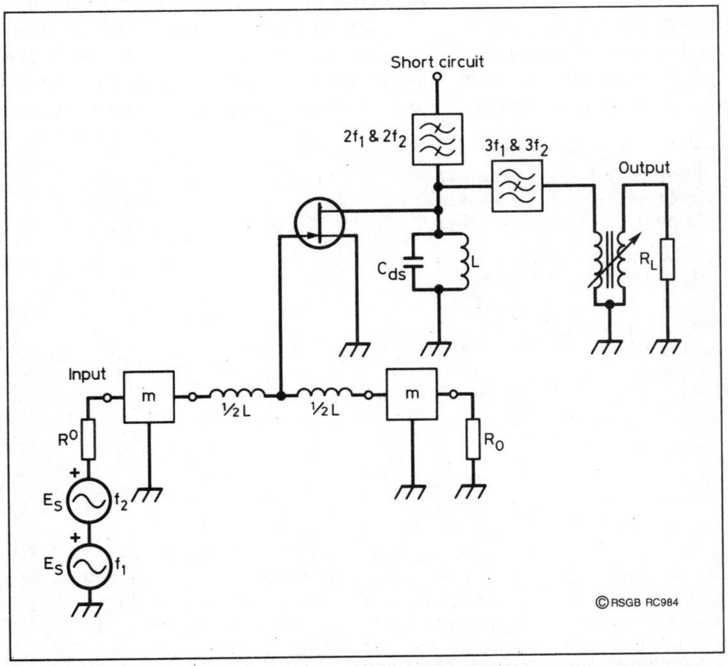

Fig 2: Circuit used for simulation of the effect of the class F mode of operation on intermodulation performance. (Source: M R Moazzum (University of Westminster and M Darvishzadeh (Rutherford Appleton Laboratory)).

MAST-HEAD PRE-AMP FOR 137 (OR 144) MHZ

AN ARTICLE, 'Low-noise Antenna Pre-amp' by Lt Cmdr D F Conway, RNZN, (*Electronics World*, March 1996, pp208-9) describes a low-noise, narrow band masthead amplifier in which the input from a balanced antenna element is shared between two MOSFETs. This eliminates the balun, enhances overload capability and reduces harmonic levels for strong, in-band signals. It is intended for reception of signals over a narrow band of about 2MHz, having originally been designed as a mast-head amplifier for receiving weather satellite transmissions between 137-138MHz. However, its centre frequency can be varied over a wide range by selecting suitable values for the tuned circuit components and should prove equally effective for the 144-146MHz band.

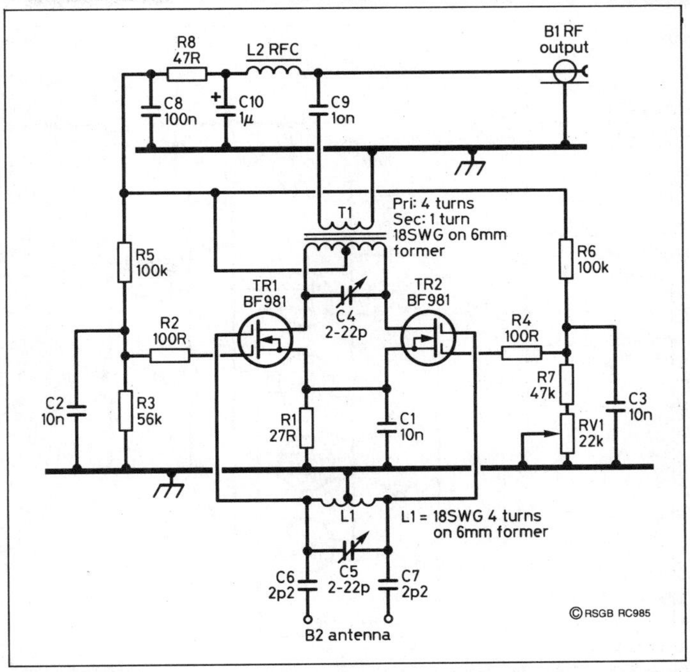

Fig 3: Narrow-band, low-noise masthead preamplifier optimised for 137.5MHz but readily adaptable to 144MHz etc. (Source: D F Conway, *Electronics World*).

The amplifier (**Fig 3**) has some 28dB gain and is powered from a 12-15V supply fed over the coaxial feeder cable. Mast head amplifiers are a particularly effective method of enhancing very weak signals since they provide gain directly at the antenna before the signal loss that occurs between antenna and the first stage of the receiver. The amplifier has two low-noise MOSFETS as a push-pull amplifier to remove the requirement for a balun at the input and improve the dynamic input power range compared with a single-device amplifier. At 137.5MHz the 1dB compression point occurs at about 5dB. Second harmonic is down about 30dB and the third down 50dB.

To ensure even power sharing between the devices, it is important that both signal paths have the same characteristics. This balance is achieved by a symmetrical circuit layout and by the input and output inductors. Correct operation of the high-gain amplifier depends on good circuit layout and proper shielding. DF Conway constructed his pre-amp on a double-sided PCB using surface mounted devices to minimise parasitic reactances.

The BF981 device is obsolescent and has been repackaged as the BF991, but the BF981 was used because the SOT-103 package can be mounted on its 'back' so keeping the RF signal paths symmetrical. Steel sheet shielding was used to isolate input and output signals to keep the amplifier stable. Air cored inductors, space wound using 18SWG enamelled wire are used. The centre tap is achieved using a short piece of wire soldered to the centre turn.

For setting up, solder a short piece of wire in place of C3, then measure the DC supply current which should be between 10 and 20mA. Replace C3 and adjust RV1 so that the supply current is now twice the initial reading. Connect the antenna to the amplifier and adjust C4 and C5 for maximum output at the required frequency .

Four BF981 MOSFETs in parallel to form a low-noise 50MHz pre-amplifier to a converted AM/SSB Cobra CB radio with a transistor linear power amplifier has been implemented

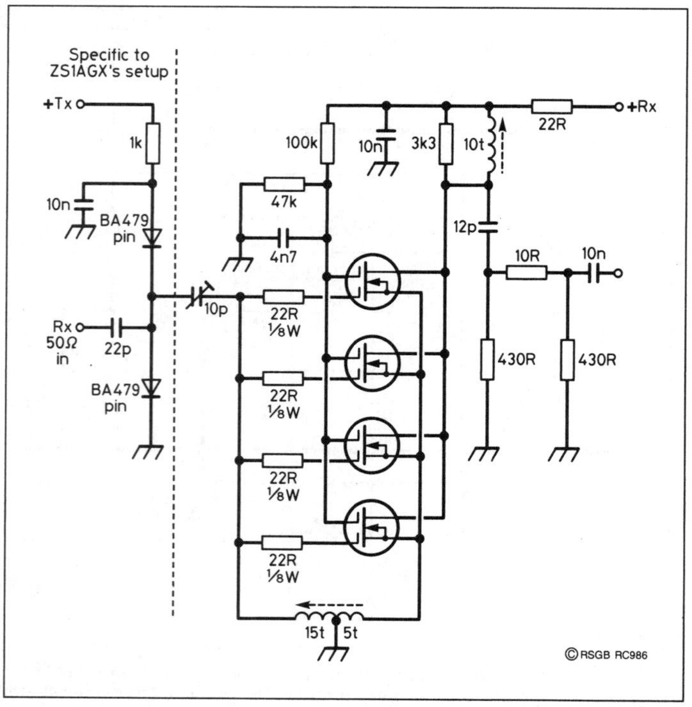

Fig 4: ZS1AGX's 50MHz low-noise MOSFET preamplifier using four BF981 (or BF981) devices.

by Robin, ZS1AGX (ex-G7HBL), who admits to being a confirmed "tinkerer". While I suspect that the use of multiple paralleled devices has little overall effect on the noise performance, it would significantly enhance the strong-signal performance and the forced feedback from drain to source should improve the input matching.

The circuit diagram is shown in **Fig 4**. ZS1AGX used BF981 devices but wonders if the later BF988 MOSFETs might not be more suitable, but were apparently not available in South Africa. The MOSFETs are assembled 'sandwich' style on top of each other. Coils are 5mm diameter 32SWG using slug (dust) tuning.

ELECTROMAGNETIC POLLUTION HAZARDS?

THE CONTROVERSY OVER the possible hazards of electromagnetic radiation has increasingly entered the public domain. Interest in this matter has been to a large extent stimulated by the concern of workers in the industry and by public concern at the possible effect on children living close to high-voltage electric supply lines. Another issue of concern is the possible biological effects on users of personal and mobile transceivers with antennas close to eyes and brain. Also in the public domain are the still equally unproven hazards from non-ionising RF radiation close to broadcast and amateur radio transmitting antennas at levels below recognised thermal effects.

It is now 21 years since the medical journal *The Lancet* (April 6, 1985 reported in *TT*, August 1985 and *Technical Topics Scrapbook*, 1985-90, p46) carried an editorial referring to an apparent increased risk of leukaemia among electrical workers in New Zealand and carrying a letter from Dr Samuel Milham who had collected "cause of death" information on 1691 radio amateurs formerly resident in California and Washington State. He had found 24 deaths ascribed to leukaemia compared to an expected 12.6 from statistical averages.

Since then there have been innumerable investigations into the possible links between low-level RF fields or electromagnetic fields near power supply lines and cancers. There is also great interest in possible hazards from low-power hand-held mobile phones.

One of the medical researchers prominent in the possible cancer link has been Australian born Dr Ross Adey, K6UIN, who is chairman of a committee commissioned by the US National Council on Radiation Protection and funded by the US Environmental Protection Agency.

After some nine years work, part of the committee's draft report was leaked to the press last autumn: see for example 'Leak Links Power Lines to Cancer' (*The New Scientist*, October 7, 1995). The draft is highly controversial in claiming: "There is strong epidemiological and laboratory evidence to suggest that children exposed to EMFs from power lines are at greater risk of developing leukaemia, and that adults exposed to EMFs at work run a higher risk of leukaemia and brain cancer. Other studies indicate that low exposure could help tumours to grow by causing genetic or chemical damage to cells. There is also evidence of effects on the immune and reproductive systems." It will be interesting to see if the final report (due this year) contains these assertions.

The draft recommends that future developments should be based on a safety limit as low as 0.2 microteslas (μT), and that offices should endeavour to keep exposure from computers, photocopiers and printers below this figure, which should also apply to new nurseries, schools and houses; it suggests that high-voltage power lines should be kept away from residential areas.

It should be emphasised that as a leaked draft, this US document has no official standing. The field immediately below a 400kV line would be some 40μT, or 7μT immediately beneath a 132kV line. Many domestic electric appliances, such as vacuum cleaners, drills, hair dryers etc would expose users to more than 2μT.

Over the years *TT* has reported that while some laboratory experiments have suggested that EM fields can affect biological cells, few if any such experiments have proved reproducible. Other scientists have pointed to possible contamination from sources other than RF fields. Recent work at the University of Bristol (sensationalised in the Channel 4 programme *Dispatches*) puts forward the possibility that there is a link between certain types of cancers and the inhalation of radon decay products (the natural gas that leaks into homes and is a known carcinogenic). The Bristol workers believe that radon gas becomes more concentrated in the presence of electromagnetic magnetic fields. But the report stresses that "at present, there still remains no persuasive biological evidence that power frequency electromagnetic fields can influence any of the accepted stages in carcinogenesis".

A long paper 'Mobile Communication Systems and Biological Effects on Their Users' by Professor Maria A Stuchly (University of Victoria, British Columbia, *The Radio Science Bulletin*, No 275, December 1995) notes that "influences of electricity and electromagnetic fields on biological systems were observed as early as the 18th century (Galvani) and the 19th century (d'Arsonval). A rigorous inquiry started after the Second World War . . . The main effort has concentrated in two frequency ranges: power line frequencies (50 - 60Hz) and radio frequencies (RF) including microwaves. Considerable progress has been made in understanding the interactions of RF fields with living systems . . . (resulting) . . . in very similar recommendations regarding safe exposure lev-

Standard	Frequency Range f (MHz)	Electric field strength E (V/m)	Magnetic field strength H A/m	Power Density S (μW/cm²)
IEEE C95.1 (1991)	0.003 - 0.1	614	163	-
	0.1 - 1.34	614	16.3/f	-
	1.34 - 3.0	823.8/f	16.3/f	-
	3.0 - 30	823.8/f	16.3/f	-
	30 - 100	27.5	$158.3/f^{1.668}$	200*
	100 - 300	27.5	0.0729	200
	300 - 3000	-	-	f/1.5
	3000 - 15000	-	-	f/1.5
	15000 - 300000	-	-	10000
CENELEC (1995)	0.01 - 0.042	400	16,8	-
	0.042 - 0.68	400	0.7/f	-
	0.68 - 10	275/f	0.7/f	-
	10 - 400	27.5	0.07	200
	400 - 2000	$1.37 \times f^{0.5}$	$3.64.10^{-3} \times f^{0.5}$	f/2
	2000 - 1500000	61.4	0.163	1000
	150000 - 300000	$0.158 \times f^{0.5}$	$4.21 \times 10^{-4} \times f^{0.5}$	$6.67 \times 10^{-3} \times f$

$*200$ or $9.4 \times 10^8/f^{3.336}$

Table 1: Exposure limits for 'general public' / 'uncontrolled environment'.

THE COHN 'MINIMUM LOSS' FILTER

WAY BACK IN 1970 (*QST*, July 1970, see also *TT*, July 1972 and subsequent editions of *ART*), William Sabin, W0IYH, of Collins Radio (since retired) in discussing solid-state receiver design, suggested that a Cohn three-pole, minimum-loss bandpass tuneable filter (**Fig 5**) was a good way of obtaining a high degree of signal frequency selectivity before the first mixer.

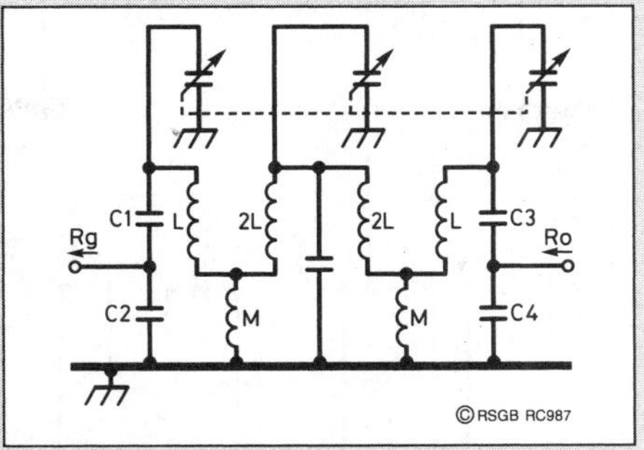

Fig 5: Tuneable three-pole minimum loss bandpass Cohn filter suggested by W0IYH in 1970 although noting that it is not an easy design and requires high-Q toroid inductors. Such filters can provide a high degree of signal frequency selectivity before the first mixer with minimum loss of signal (about 6dB). The ratio of C1/C2 should be adjusted for correct matching to the antenna, while the ratio C3/C4 provides the required output impedance.

W0IYH noted that this type of bandpass filter was unusual and relatively little known. It is a triple-tuned, bottom-coupled array using high-Q toroid inductors, based on a filter approach suggested by S B Cohn (*Proc IRE*, August 1959, pp1342-1348). It was pointed out that "when properly designed (and this may not be too easy) this type of filter has the property that for a specified number of tuned circuits, unloaded coil Q, and filter noise figure, it will provide the best possible rejection outside its passband: basically it is a bandpass filter derived from a low-pass prototype having equal elements".

Steve Dyke, G3ROZ, was reminded of the Cohn filter by the D11/R234 pre-selector (*TT*, March 1996). He writes: "The Cohn filter of the 1972 *TT* is shown also in the 'QRPers Bible' otherwise known as *Solid State Design for the Radio Amateur* (ARRL). Prospective builders beware! It is quite difficult to make these Cohn filters with the quoted 6dB insertion loss but it is incredibly easy to make them with insertion losses of 20dB or more.

"My recent attempt at one for 1.8MHz (part of my development programme for a Coherent CW receiver) resulted in an insertion loss of some 30dB. Placing the values in a simple computer 'network simulator' still resulted in a predicted loss of over 20dB! G3TPW of Cob Web fame assured me the design (**Fig 6**) was not flawed although he felt the use of the two centre coils, acting effectively as a single inductance, was really a waste of a coil." [I suspect S B Cohn or W0IYH would disagree - *G3VA*].

G3ROZ continues: "The problem is that these filters are based on coils with a Q of 175 at 1.8MHz. This is high. The usual Toko range won't make it. By comparison a typical home-wound slug-tuned Top Band coil on a ¼in slug-tuned former with about 1in length of close-wound enamelled copper wire will give a Q of about 60. G3TPW did suggest that 16 or 18SWG would make a Q of 175 no problem, but I wanted an input filter less big and heavy than an AR88!

"My final filter (Fig 6), which does make the 6dB loss specification, was made as follows: The coils were pile-wound with multistrand Litz wire on the usual slug-tuned ¼in forms. Pile winding actually gave a higher Q than making it 'look pretty'. A coat of the XYL's nail varnish to seal the coils completely fouled up the Q, so I ended up using unsealed coils. These were adjusted to give just over the required design inductance with the cores fully in. While on Top Band, virtually any old cores will work, this will not be true on other higher frequency bands. My coils have a Q of about 120, still far short of 175, but at least high enough to work. Ct was then made temporarily a three gang 500pF capacitor and the L/C ratio and the ratio of L2 to L3 was adjusted experimentally going for the minimum value of Ct to produce the required insertion loss of 6dB. In my case, it worked out at 100pF. The attenuation characteristics are quite impressive. With Ct peaked at 1900kHz, the band edges at 1.8 and 2.0MHz are over 30dB down.

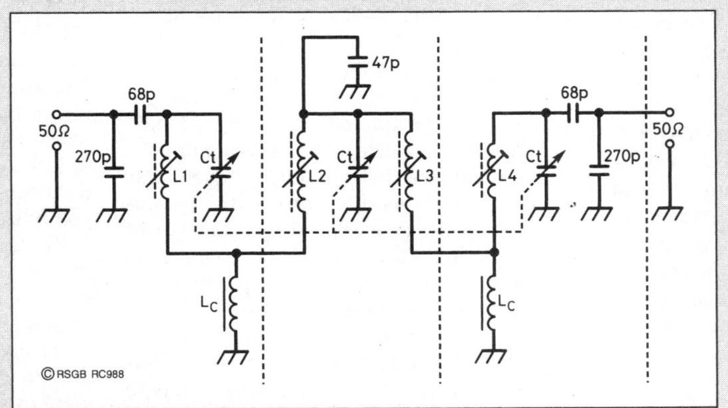

Fig 6: Final Cohn filter as implemented by G3ROZ as a front panel tuner for 1.8MHz. L1 and L4 70μH Q 175 (J W Miller 43A685C81 in S74 can). L2 and L3 140μH Q 175 (J W Miller 43A154C81 in S74 can). Lc 1.45μH on toroid core (about 20 turns on T-50-2 then set up on a bridge for exactly 1.45μH).

"My advice, if you want to make a Cohn front end filter and cannot measure Q and insertion loss, seek help from someone who can. Otherwise it will end up in tears!"

els in many national standards and international guidelines".

Her conclusions note that "development of RF dosimetry, a quantification tool, has provided means for extrapolation from animal experiments into assessment of risks to human health. By the late 1980s, health protection standards in various countries had converged to nearly identical limits . . . Current proliferation of devices and new services in mobile communications have raised new concerns. These mostly relate to the relatively large rates of energy deposition in some parts of the body, mostly the eye and the brain, and to the data transmission resulting in the amplitude modulation of RF signals at ELF . . .

"The question of ELF modulation is more difficult to answer. Independent of what is the final assessment of health impact of ELF fields, it does not directly translate to the same or even similar assessment for microwave fields modulated at ELF. There is no reason to suspect any specific harmful effects related to ELF amplitude modulation of radio or microwave fields if ELF studies indicate that the detrimental impact on human health is minimal or none. On the other hand, if ELF exposures result in risks to people, that would not translate into the same risks for exposures to microwaves modulated at ELF, but would provide motivation and rational research into the potential problem."

Another recent paper 'EM Field Measurements in the Vicinity of an Antenna Park for Radiation Hazard Purposes' by a number of engineers at the Aristotle University of Thessaloniki, Greece (*IEEE Trans on Broadcasting*, December 1995) provides a useful list of the American (ANSI) standard IEEE C95.1-1991 and the provisional European CENELEC (1995) reference levels for continuous exposure by the general public: **Table 1**.

A detailed list of 16 guidelines for amateurs were included in *TT* (January 1990, pp36-37). These were based on those drawn up by Ivan A Shulman, MD, WC2S (*QST*, October 1989). WC2S drew up a set of guidelines that were intended to cover all potential or suspected hazards affecting radio amateurs. As I noted at the time, these were considerably more restrictive than any previous guidelines and I confessed that several (based on the American higher legal power limits) were breached at G3VA.

THE G4DTC ULTIMATE SUPER-REGEN

RAY HOWEGEO, G4DTC, whose 1987 'ultimate' HF hybrid (PCC189 cascode RF stage, 7360 beam-tube mixer) HF receiver was described over several months in *TT* during 1987-88 (see also *Technical Topics Scrapbook* 1985-89) spotted the two simple VHF super-regen receivers in the March *TT*. He writes: "Readers may be interested in the circuit diagram (**Fig 7**) of the 'G4DTC Ultimate Super-Regenerative' receiver which I designed about six years ago and which was published with full details in *Short Wave Magazine*, June/April 1991. As far

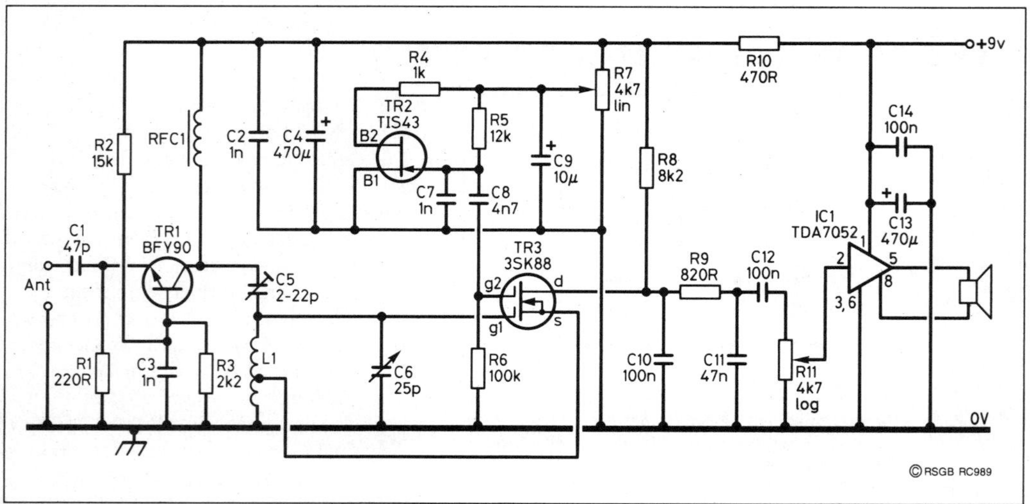

Fig 7: G4DTC's design of a super-regenerative VHF receiver with unijunction (TR2) quenching oscillator. For 90-170MHz range L1 is 4 turns, 10mm ID, stretched to 10mm length. Tapped at 1 turn. RFC1 5 turns of thin wire on a ferrite bead. Published originally in *Short Wave Magazine*, 1991.

as I am aware, this is an original circuit, unlike most super-regens which tend to be basically similar except for the use of different devices.

"My circuit has the advantages that: (a) By careful, once only, adjustment of R7 it achieves full sensitivity without the high level of inter-station noise associated with most super-regens. (b) It tolerates a degree of coupling from the RF stage (or antenna) which would normally stop most circuits regenerating. It is therefore extremely reliable and reproducible.

"TR3 operates as an oscillator tuned by L1/C6 to the received frequency. TR2 forms a unijunction sawtooth-wave oscillator at about 75kHz which provides the necessary quenching (or 'blocking') voltage to the second gate of TR3. R7 varies the amplitude of the quenching voltage and should be adjusted until inter-station noise all but disappears. TR1 functions as a buffer stage with a gain of about 10dB. The gain may be increased by reducing R2 but this may call for a more critical layout to prevent instability. C5 should be increased to the point where TR3 fails to oscillate reliably, then reduced a little. All components were carefully optimised but some experiment with C7, which controls the quenching frequency, may be justified. A low value (470pF) seems best suited to AM reception, while a high value (4.7nF) seems better for FM reception."

Although most recent super-regen designs have been based on self-quenching ('blocking') regenerative detectors, it was common in the inter-war years to use a separate quenching oscillator, and my junk-box still contains quenching coils as marketed by a number of firms including Eddystone. But I cannot recall having ever seen before a unijunction sawtooth oscillator for quenching as used by G4DTC.

G4DTC mentions that he has yet to see a convincing theoretical analysis of all aspects of super-regenerative receivers and admits to not fully understanding the operation of his super-regen receiver.

I recall that when discussing simple super-regen receivers in *TT* during 1980, I was reminded by Brian Bower, G3COJ, of a full-length book *Super-regenerative Receivers* by Dr J R Whitehead (Cambridge University Press, 1950). This was based on the intensive wartime investigations made into super-regeneration at TRE (later RSRE and now the Defence Research Establishment (DRA)) Malvern, when many thousands of pulse responders for radar identification of ships and aircraft (the IFF system) were based on super-regen valve circuitry. These required that the initial design should be stable and reproducible, and a lot of work was put into analysis of unreliability, and in solving some of the questions left unanswered by the early development of super-regeneration by Howard Armstrong and others.

In 1980 G3COJ found that the BBC library copy of this book had remained on the shelves ever since he had first borrowed it in 1961! The copy I obtained on loan from the National Reference Library also showed few signs of use! As I suggested in 1980, and would endorse in 1996, it seems a pity that relatively little deep interest has been shown for many years in super-regeneration in which the detector can provide a gain of up to 100dB!

Incidentally, for those readers who feel that *TT* is too inclined to resurrect ideas from the valve era, it is interesting to note that valve amplifiers are once again objects of desire for many high-fidelity enthusiasts. *Electronics World* has recently been running several long articles on valve amplifiers and pre-amplifiers. This provoked a letter from the former editor Frank Ogden who complained that "only a complete loony suffering from terminal nostalgia would seriously consider a valve for anything". This provoked Simon Yorke of Durham to reply that such a "flippant assertion betrays him as a devout philistine. The absurd notion that all past wisdom is rendered obsolescent in the face of technological advance is simply juvenile . . . the idea that 'new' is inherently better than 'old' is more in tune with adolescent truculence than with any considered intelligence.

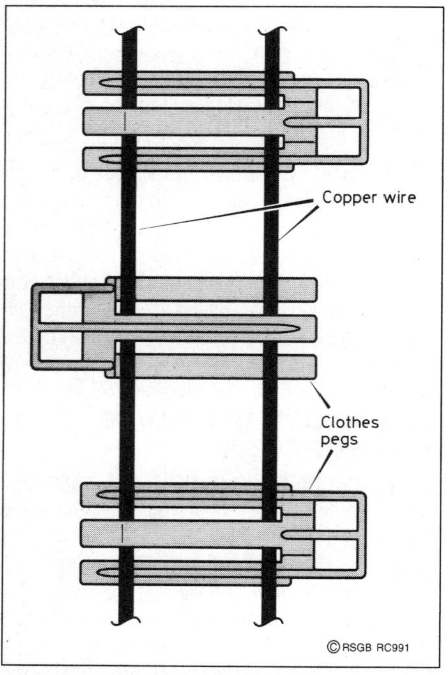

Fig 52: ZL3QQ uses plastic clothes pegs made in New Zealand by the Sunshine Peg Company as low-cost spreaders that can be slipped on pre-stretched transmission line from the side to give a 450Ω open wire feeder.

CLOTHES PEG 450Ω OPEN WIRE SPREADERS

FROM DR TOM SEED, ZL3QQ, comes a suggestion that possibly does not apply in the UK but might encourage some plastics manufacturer to introduce a form of clothes peg which is available in New Zealand and which sells there at 25 pegs for about the equivalent of 80 pence: **Fig 52**. Alternatively, it might be possible to order them from the manufacturers.

ZL3QQ writes: *TT* (December 1995, p73) had an item about the perennial topic of holding insulators on open-wire transmission lines. I think readers may be amused at the method I find convenient. It does not involve the need to drill holes and thread the insulators nor tie them as shown in the December Fig 6 - from which they always seem to slide away.

"Enclosed are some plastic clothes pegs made in New Zealand by the Sunshine Peg Company Ltd, Hamilton, New Zealand. These can simply be slipped on a pre-stretched transmission line from the side (alternate sides if preferred). They give a line spacing of one inch (25mm). I use them with 70lb/mile cadmium bronze wire which is sometimes available here as scrap metal from Telecom (formerly the Post & Telegraph Department). This wire is approximately 1.2mm in diameter, which makes, with 1in spacing, a convenient 450Ω line.

"They need replacing about every three years or so, but at three pence per spreader this is not unduly hard on the family budget." ♦

PAT HAWKER, G3VA
London 37/SE22 8SS

THE BITS OF SPEECH - DIGITAL RADIOPHONES

DIGITAL SPEECH over radio is rapidly becoming established in a wide range of telecommunications applications and seems likely to find its way soon into amateur radio. The key is the development of advanced coding techniques that enable communications-quality speech to be transmitted at progressively lower bit rates. However, there are already a considerable number of different ways in which the low bit rates can be achieved. It is highly desirable that the same coding technique should be adopted internationally for amateur radio since otherwise operators are likely to be faced with a variety of non-compatible coding systems that would seriously limit the value of digital speech.

The coding of analogue speech waveforms into the original pulse code modulation (PCM) system was devised by Alec Reeves, an Englishman then working for the ITT French subsidiary in Paris in 1936. Earlier he had devised various techniques, including pulse width modulation, for transmitting analogue speech waveforms in pulses. This formed the basis of the multiplexed system used in the British No 10 1944 microwave radio relay. But, from the beginning, Reeves had recognised that PCM, which is a true coded digital system, could become one of the most powerful tools ever invented to minimise the effects of interference, noise and distance on speech transmission.

The very high-speed switching needed to implement speech as a fast 'Morse-like code' requires a large number of electronic switches, which in the 1930s and 1940s could be achieved only with bulky and power-hungry thermionic valves. The first time that a crude form of PCM speech was achieved was for a wartime secure long-distance telephony system developed by Bell Laboratories, known as Siginy or 'Green Hornet' with its 'one-time' enciphering-deciphering key recorded on disc. One such circuit was established between Washington DC and London and was used by Prime Minister Winston Churchill and President Roosevelt. But the digital equipment occupied many racks of equipment with 30kW of electricity consumed at the terminal to produce a few milliwatts of deciphered audio!

As I noted in a 'Quick Guide to Pulse Modulation' (see *ART7*, pp233-234), the switching and digital logic operations involved in encoding, regenerating or decoding PCM need large numbers of switching-type operations at very high speeds: most unsuitable for valve circuits - but far, far simpler with semiconductors and integrated circuits. **Fig 1** shows the basic principles of PCM.

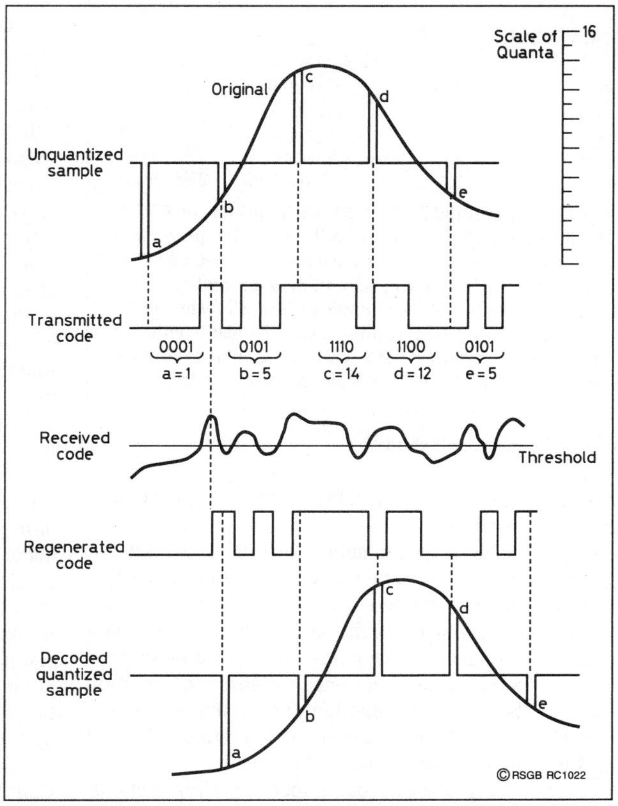

Fig 1: Pulse code modulation (PCM) forms the basis of digital transmission systems. The diagram illustrates how analogue waveforms are sampled and converted into coded pulses and transmitted in digital form. Distortion of the received pulses can be eliminated by regeneration of the original coded signal and then decoded back into analogue form. In the simplified example shown, a four digit binary code 'word' provides 16 quantized levels. Regenerating the code at the centre of the received pulse introduces a delay of half a digit. Decoding cannot be effected until all the digits of the code word have been received so the reconstructed samples are further delayed by a character period. Conventionally, speech analogue waveforms are sampled at 8000Hz into 128 levels requiring an 8 bit word.

It was not until the development of solid-state switches in the form of integrated circuits that PCM became really practicable. In the mid-1960s I recall watching a demonstration of the first experimental multiplexed PCM system, developed for the GPO by STL/STC, intended for use on urban junction routes in the public switched telephone network (PSTN). This was soon after Bell Telephones had begun using their basically similar T1 system in the USA.

The bit-rate required for conventional PCM is a function of the number of discrete levels and the basic sampling rate, and these in turn depend upon the bandwidth of the speech and the acceptable level of quantization noise. Good telephone 'toll-quality' speech (about 300 to 3400Hz) is typically sampled 8000 times a second and coded into 8-bit words (128 levels). This results in a 64Kbit/s bit-rate. In 1972 this rate was standardised internationally (ITU-G711) as the bit-rate for telephony channels.

A bit-rate of 64Kbit/s, however, would be most unsuitable for use for mobile or amateur radio communications. Today, the coding of speech and video, with their high degree of redundancy, has moved a long way from basic PCM, resulting in much lower bit rates with only minor degradation in quality. The result is that digital systems are already in widespread use for professional mobile and satellite systems, with bit rates, that could be

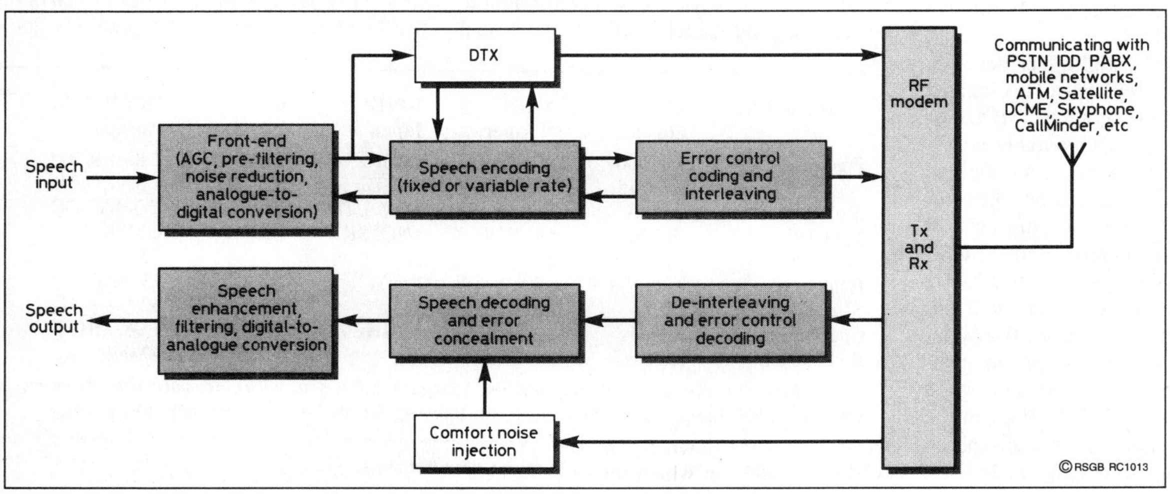

Fig 2: Block diagram of a typical mobile radio transceiver for digital speech or data systems.

However, a number of different, non-compatible bit rate reduction and error correction systems are used. If amateurs are to use digital speech systems it is desirable that an internationally agreed standard should be adopted.

Standard	Year	Coding Type	Bit Rate (Kbit/s)	MOS	Algor delay (ms)
ITU-G711	1972	PCM	64	4.3	0.125
ITU-G721	1984	ADPCM	32	4.0	0.125
ITU-G726	1991	VBR-ADPCM	16, 24, 32, 40	2.0-to-4.2	0.125
ITU-G727	1991	Embedded-ADPCM	ditto	-	0.125
ITU-G728	1992	LD-CELP	16	4.0	0.625
Inmarsat-B		APC	9.6/16	commun	20
GSM Full rate	1989	LTP-RPE	13	3.7	20
GSM-EFR	1995	ACELP	13	4.0	20
Skyphone	1989	BT-MPLPC	8.9	3.5	28
DAMPS Full rate	1991	VSELP	7.95	3.6	20
ITU-G729	1995	CSA-CELP	8	4.0	15
IS96	1991	Qualcomm CELP	1, 2, 4, 8	3.5	-
JDC Full rate		VSELP	6.7	commun	20
GSM half rate	1994	VSELP	5.6	3.5	24.375
ITU-G723	1995	A/MP-MLQ CELP	5.27/6.3	commun	37.5
US DOD FS1016	1990	CELP	4.8	3.0	45
TETRA	1994	ACELP	4.56	commun	35
ITU	(1998)	(to be defined)	4.0	(toll)	(25)
Inmarsat M	1990	IMBE	4.15	3.44	78.75
JDC half rate	1993	PSI-CELP	3.45	commun	40
US DOD FS1015	1984	LPC-10	2.4	synthetic	22.5 (min)

Notes: Coding abbreviations: ADPCM adaptive differential PCM; VBR variable bit rate; CELP codebook excited linear prediction; LPC linear predictive coding; APC adaptive predictive coding; ATC adaptive transform coding; IMBE improved multiband excitation; LD-CELP low delay code excited linear predictive coder; VSELP vector sum excited linear predictive; PWI prototype wave form interpolation; SBC sub-band coding; HC harmonic coding; STC sinusoidal transform coding; PSI pitch synchronous interpolation; RELP residual excited linear predictive (adaptive predictive LPC coding); ABS analysis by synthesis; MPE-LPC multipulse excitation linear prediction coding; VQ vector quantization; VECV voice excited channel vocoder.

Table 1: Comparison of speech coding standards for digital telephony.

implemented in amateur bands above 30MHz, and with the prospect of even lower rates that could be used at HF. On the market there are already systems offering relatively low quality speech over Internet modems operating at 14.2Kbit/s. Military digital vocoder systems can transmit intelligible but synthetic sounding speech at 2.3Kbit/s on HF.

A recent survey article 'Low Rate Speech Coding for Telecommunications' by W T K Wong, R M Mark, B M Cheetham and X Q Sun (*BT Technology Journal*, January 1996) provides a table listing over 20 different telephone band speech coding standards: **Table 1**. The authors note that the most recent techniques permit telephone quality speech at 8Kbit/s, but that there are still demands for even lower rates and more flexible, good quality coding techniques. They describe a new class of speech coding known as speech interpolation coding which has the potential to provide toll quality speech at or below 4Kbit/s, that is a bit-rate well below that now often used for VHF/UHF amateur PACKET radio.

Table 1 lists existing and under development standards. It lists: the type of coding; the bit-rate; the 'Mean opinion score (MOS)' as evaluated under the most favourable conditions but should not be taken as a definitive indication of codec performance; and the algorithm processing delay time in milliseconds. The 1-5 scale of mean opinion scores (MOS) for assessing speech quality in 200 to 3400Hz channels is worth noting: (1 bad, 2 poor, 3 fair, 4 good and 5 excellent). A score of 4.0 or better is needed for network (toll) telephone channels; a score of 3.5 to 4.0 represents good communications quality. MOS values in the region 2.5 to 3.0 implies 'synthetic' speech quality which although still highly intelligible, has reduced naturalness and poor speaker recognition.

From an amateur radio point of view, it would clearly be advisable to adopt an internationally defined system for which the necessary VSLI circuits are, or will be, in mass production. The delay time would be of relatively little importance for simplex operation but becomes important for full duplex speech.

Fig 2 shows the block diagram of a typical mobile digital radio transceiver.

Fig 3 shows basic speech coding methodologies based on a 1992 paper by Professor C Xydeas of Manchester University. It has to be admitted that for most of us Table 1 and Fig 3 are likely to be fairly indigestible, full of abbreviations and concepts unfamiliar to amateur operators. In general, decoded speech quality tends to be closely linked to output bit rate and codec complexity. As bit-rate decreases, codecs become ever more complicated, using complex algorithms having a direct impact on implementation costs and power consumption. To achieve good speech quality at the lowest bit rates requires processing that may introduce excessive time delay for real time, duplex speech. But where material can be stored, good quality speech can be recorded at very low bit rates by using repeated processing.

There is continued work on developing low bit rate codecs (2.4kbit/s) for such satellite applications as Inmarsat-P and Iridium. Currently, the American Department of Defence is developing a new 2.4kbit/s speech coding standard that is expected to have better performance than the 1990 FS1016 4.8kbit/s standard.

The BT article also notes: "An emerging packet speech transmission technology allows full-duplex conversation over the Internet. There are already products referred to as Internet phone or Iphone on the market... However, the Internet is not engineered to cope with the requirements of voice traffic. The transmission delay will be variable, and unacceptably high when the Internet link is heavily loaded. For current Internet voice products, the quality of service cannot be guaranteed. It will be interesting to see whether these systems ever become more than the equivalent of a '**CB radio**' channel for the Internet amateur."

The fact that BT would suffer a large drop in revenues from International telephone calls if Internet telephony catches on may be reflected in these sentiments!

With digital speech now becoming common for mobile and personal communications, it seems likely that new EMC problems may arise. *IEEE Spectrum* (January 1996) notes: "A number of observers have expressed concern about ubiquitous wireless telephony. Studies conducted in Australia, Italy, Switzerland and the United States have shown that digital cellular phones can interfere with heart pacemakers, hearing aids, automobile air bags, insulin pumps and other lifesaving equipment. Particularly risky seems to be the GSM technology, at least as implemented in Europe.

"To further investigate these concerns, the Cellular Telecommunications Industry Association (CTIA), Washington DC has presented $200,000 to the University of Oklahoma to conduct two separate studies on the interference of digital cellular phones with pacemakers and hearing aids. The US Food & Drug Administration also plans to conduct its own tests. The 'spillover' from cellular phones into the bands protected for study of the cosmos are of concern, too, to another group, the radio astronomers, led by a co-author of the *Handbook on Radio Astronomy* James Cohen of the Nuffield Radio Astronomy Laboratories (home

Characteristic	GSM Europe	NADC N America	J-PHP Japan	CT2 Eur/Asia	DECT Europe
Freq (MHz) down	935-960	869-894	1.9GHz	864-868	1.88-1.9GHz
up	890-915	824-849	1.9GHz	864-868	1.88-1.9GHz
Multiple access	TDMA	TDMA	TDMA/TDD	FDMA/TDD	TDMA/TDD
Modulation	GMSK	π/4-DQPSK	π/4QPSK	B-FSK	GMSK
Speech Kbit/s	13	8	32	32	32
Handset o/p	3.7mW-1W	2.2mW-6W	10mW	1mW-10mW	250mW
Modulation Kbit/s	271	49	384	72	32
Channel spacing	200kHz	30kHz	300kHz	100kHz	1.762MHz
Burst length	156bits	324bits			424bits

Notes: TDD time division duplexed. GMSK Minimum shift keying with Gaussian premodulation filter. DQPSK digital quadrature phase-shift keying. BFSK binary frequency shift keying.

Table 2: Selected Digital Radiophone System Characteristics.

of the Jodrell Bank Radio Observatory) south of Manchester, England."

It is perhaps relevant to note that many UK hospitals now require all cellular phones to be switched off on entry. Whether the digital radiophones are more prone to cause interference than analogue phones to vital medical equipment seems to be a moot point.

With TDMA (time-division multiple access) the available spectrum bandwidth is divided into channels and each channel is further divided into time slots. The speech waveforms are compressed as well as digitised, allowing three users to share each common channel. The GSM form of TDMA is already in use in over 85 countries. CDMA is a spread-spectrum technique that uses neither frequency channels nor time slots. Instead, it multiples the digitised version of each call by a unique pseudorandom sequence to spread it over the full available bandwidth. Calls sound like random noise to each other user. CDMA is claimed by proponents to carry more traffic in a given frequency band than TDMA.

It is clear that the digital revolution is sweeping into professional voice communications and is unlikely to leave amateur radio depending purely on analogue systems such as AM, FM and SSB. But it is equally clear that the present situation is chaotic with a multimode of coding and transmission standards. In 'Whither Speech - the Future of Telephony', A V Lewis and F A Westall in the same issue of the *BT Technology Journal* write: "Challenges to formerly immutable fundamentals should not really surprise us, because all revolutions inspire chaos. Science is concerned with the pursuit of knowledge, but engineering makes a business of accommodating uncertainty. That is why good engineering is an art as well as a science.... Telephony has an exciting and challenging future in which we will have to learn how best to exploit the transforming potential of technology - but more than that, we may need to re-examine, or even relearn, some of our fundamental ways of thinking."

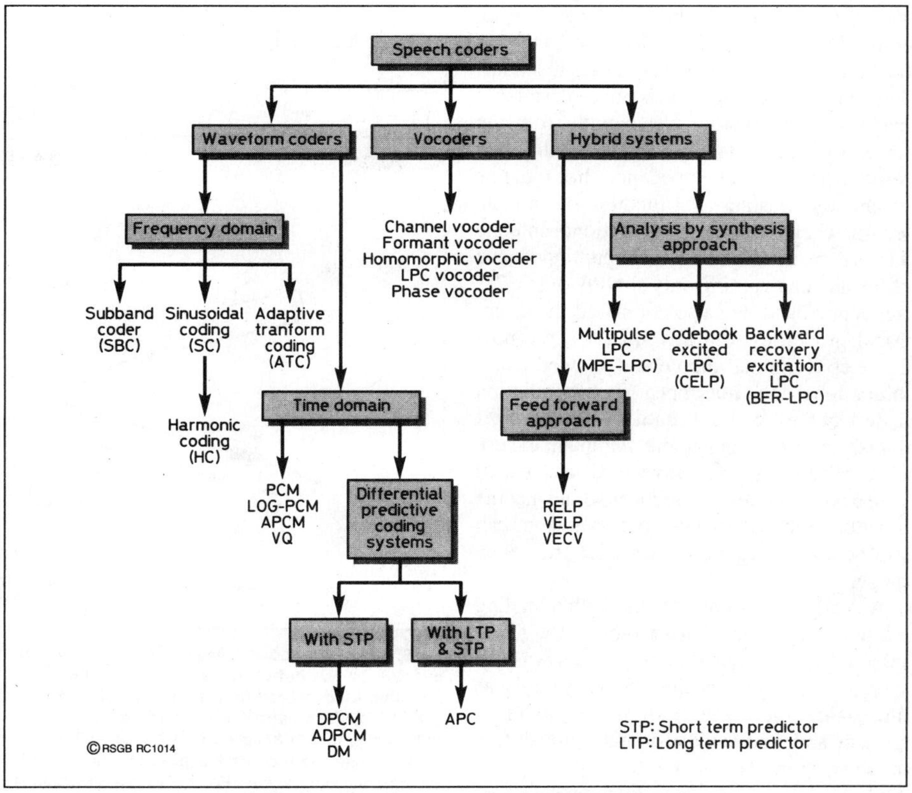

Fig 3: Speech coding methodologies as outlined in 1992 by Professor C Xydeas.

TRAVELLING WAVE BROADBAND ANTENNAS

THE TOPIC OF travelling wave antennas seems to have cropped up several times recently in *TT* but it is clear that there is some misunderstanding of the ways in which this concept is normally applied to long wire HF antennas.

As noted by Robert S Elliott in his book *Antenna Theory and Design* (Prentice-Hall, 1981), one of the simplest travelling wave antennas is the long horizontal wire at a distance *h* above the earth, fed at one end against ground and usually terminated at the other end in a matched load formed from a non-inductive high-wattage resistor. If *h* is not negligible compared to a wavelength, the wire and its ground image do not behave like a two-wire transmission line, but rather comprise an efficient radiating system.

Whereas with a matched low-loss transmission line the RF current (and voltage) remain near constant, the travelling wave of current proceeding outwards along the wire of a long-wire antenna is attenuated due to the radiation, and thus the power absorbed in the matched load may be reduced to an acceptable level by making the wire long enough. Indeed, the leakage (ie radiation) may be sufficient to obviate the need for a terminating load. The net current distribution on the wire is then not a standing wave, but is essentially an outward travelling damped wave.

The Bruce inverted vee antenna is a vertically polarised form of travelling wave antenna. Other TW antennas are the sloping vee, the vee-beam and the rhombic, regarded for many years as the Queen of all directional antennas for those with sufficient space. However it should be noted that the very long, very low, Beverage directional receiving antenna works on a different principle and cannot readily be used as an efficient transmitting antenna. It should not be forgotten that long-wire, vee and rhombic travelling wave antennas can also be used very effectively at VHF or even UHF where space requirements for such high-gain fixed beams are far more modest than on HF.

One of the major advantages of travelling wave antennas is that their impedance remains virtually the same over a wide frequency range, enabling them to be used as truly broadband antennas without the need for a variable ATU.

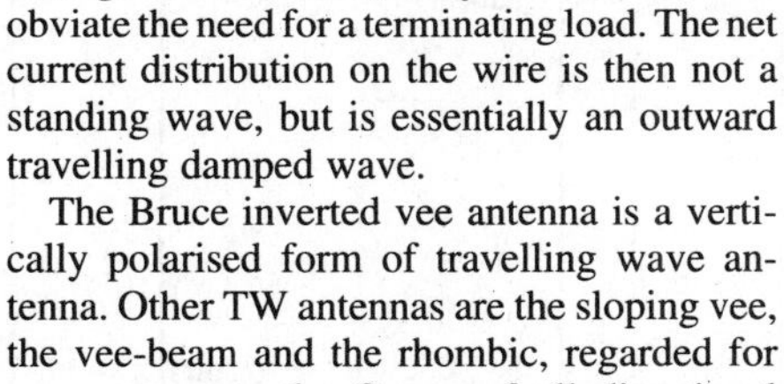

Fig 4: G3SRO's terminated long wire antenna with ferrite broadband matching transformer as describe in 1972.

With an impedance of somewhere between about 400-800Ω, they also lend themselves to the use of a fixed ferrite-toroid matching transformer fed from 50Ω coaxial cable. Of course, like all truly broadband antennas, care must be taken that the output from the transmitter does not contain significant spuriae which would be radiated along with the fundamental. The longer the wire, in terms of wavelengths, the sharper the main lobes, the lower the angle of elevation, and the less power wasted in the terminating resistor.

Many years ago, Richard White, G3SRO, described for *TT* a terminated long-wire travelling wave antenna suitable for use on all frequencies from 3.5 to 30MHz (this could today be extended to 50MHz) that does not require an ATU (at least for matching purposes) yet is capable of giving a satisfactorily low SWR across all HF bands (including WARC bands). He was fortunate in being able to run out some 400ft (122m) of wire at a height of 40ft (12.2m). At the far end this is terminated with a non-inductive high-wattage 500Ω resistor earthed to an 8ft aluminium stake via a vertical down lead (see **Fig 4** and also *ART7*, p305). The transmitter feeds the antenna via a ferrite toroidal step-up transformer using a toroid of about 1.5in or more diameter. The transmitter feed goes to a three turn winding and the antenna to seven turns bifiliar wound and spaced. This gave a worst SWR on any band of only about 1.8:1. In practice, the antenna provides the two main lobes to be expected from the theory of terminated long wire antennas, with the lobes showing up well on 14MHz and above. By directing the run of the antenna towards central America, he achieved good coverage of North and South America.

In 'The Travelling-wave Linear Antenna' (*IRE Trans on Antennas & Propagation*, July

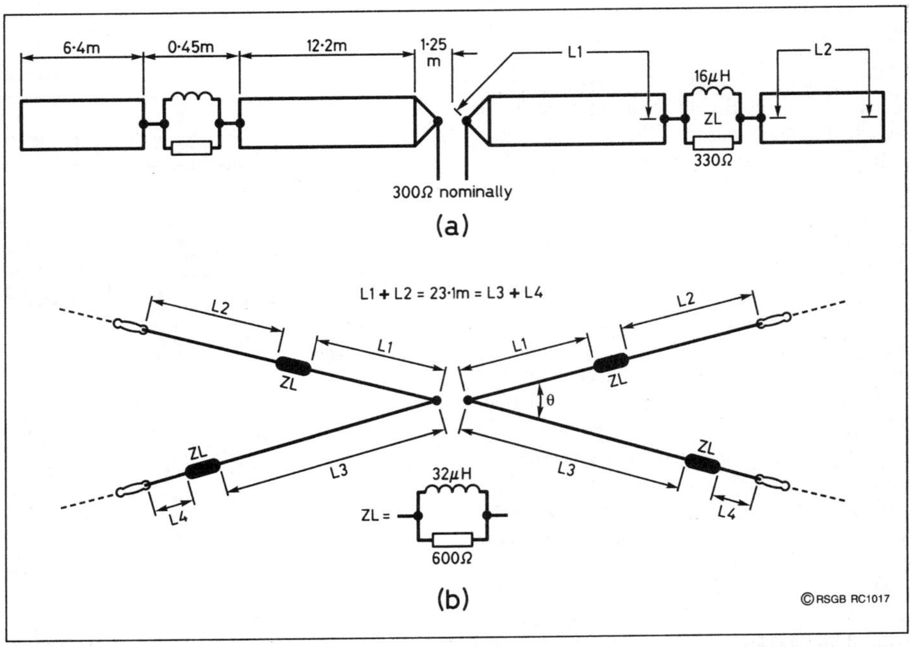

Fig 5: (a) The broadband 'Australian Dipole' antenna originally described by R J F Guertier and G E Collyer in 1973 for use between 2.3 to 30MHz (see *TT*, June 1974) but shortened for use from 3.5MHz. (b) The resistive loaded broadband inverted vee 'fat dipole' as developed in South Africa by Dr Brian Austin, G0GSF/ZS6BKW, and Andre Fourie for use over the entire HF spectrum 3 to 30MHz. The key dimensions (*TT*, September 1987) are L1 + L2 = L3 + L4 = 23.1m; L1 = 13.5m (hence L2 = 9.6m); L3 = 17m (hence L4 = 6.1m). While the included angle does not markedly affect the VSWR, the impedance of the feedline is affected to some extent. For a 5° angle optimum feedline impedance is 500Ω, but as this angle is reduced the impedance decreases to about 400Ω.

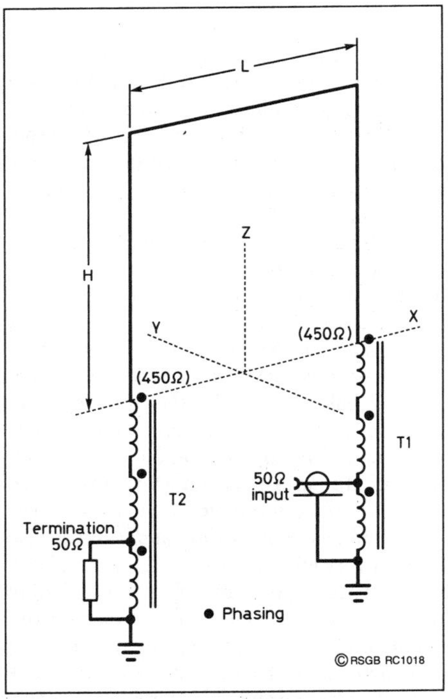

Fig 6: An end fed, end terminated travelling wave antenna as described in detail by WA2WVL in the November 1995 QST.

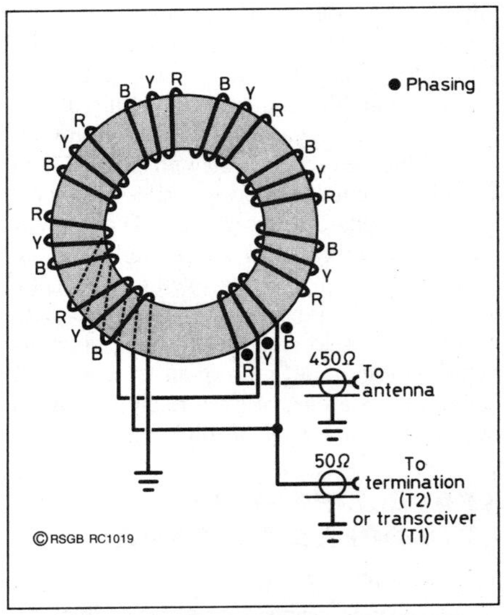

Fig 7: Winding details of the 50 to 450Ω (1:9) impedance broadband toroidal transformers as used by WA2WUL but which should also prove suitable for other TW antennas such as that used by G3SRO or the vertical half rhombic (VHR) design. It is advisable to use different coloured insulated wires to facilitate identification of the windings (eg Blue, Yellow, Red). Core is Amidon (FT-240-61) with each winding 9 turns of No 18 AWG Teflon insulated stranded wire.

1961), Dr Edward Altshuler showed that an essentially travelling-wave distribution of current can be produced on a linear (ie resonant) antenna by inserting a resistance of suitable magnitude one-quarter wavelength from the far end of the antenna. He showed that this resulted in an input impedance that remains essentially constant as a function of antenna length, whereas that of a conventional antenna varies considerably, with the input impedance of the antenna varying only slightly over a 2:1 frequency band. He also compared the directional properties of a travelling wave dipole and a conventional dipole and showed that a minor lobe does not appear in the radiation pattern of the TW dipole until it is much longer than the conventional dipole. Although the use of a resonant quarter-wave (artificial earth) wire beyond the resistor reduces somewhat the broadband characteristics, such an antenna can still be used effectively over a 2:1 frequency range.

An early form of an antenna with a loading resistance to increase bandwidth was the T2FD folded dipole antenna, although this was described originally (by the US Navy) only as increasing the bandwidth by a moderate amount, and involved considerable power loss in the resistor.

Dr Altshuler's resistance loading technique has since been used to provide effective broadband antennas covering the entire HF spectrum. Examples, which have been published in *TT,* include the so called 'Australian dipole' and an improved version developed by Dr Brian Austin, G0GSF: see **Fig 5**. These antennas were described in *TT*, June 1987 and September 1987 (see also *Technical Topics Scrapbook, 1985-90*). Earlier descriptions of the 'Australian dipole' appeared *TT*, June 1974 and September 1984.

A somewhat more conventional approach to 'Broadband Transmitting Wire Antennas for 160 through 10 metres' was described by Floyd A Koontz, WA2WVL (*QST*, November 1995, pp22-24). This noted that "a travelling wave antenna offers some attractive qualities: a 50Ω feedpoint at ground level, no tuner required, horizontal and vertical radiation components, no radials, nothing in the air but wire, and flexible size and height requirements. In his recommended two-mast configuration (**Fig 6**) he uses two broadband toroidal transformers to match to the 50Ω transmitter coax-line and to a 50Ω non-inductive termination to the 450Ω impedance of the TW antenna (high wattage non-inductive resistors are more likely to be available at 50Ω (dummy antenna resistors) than at 450Ω). The *QST* article provides tabulated information on the impedance and SWR for a typical TW antenna (50 x 50ft) together with calculated efficiency versus various antenna dimensions. Winding details of a 50 to 450Ω impedance transformer wound on an Amidon FT-240-61 core with a trifilar winding of nine turns of no 18 Teflon-insulated stranded wire are shown in **Fig 7**. (Such transformers would also be suitable for use with the G3SRO terminated long-wire antenna described above).

A commercially-available form of broadband travelling wave HF antenna that has proved popular among amateurs in North America is the 'vertical half rhombic (VHR)' antenna made by Barker & Williamson as Model AC-1.8-30 (for use between 1.8 and 30MHz). Dr Brian Austin, G0GSF, has drawn my attention to an article 'Practical Characterisation of a High Frequency Vertical Half Rhombic' by Mike Fanning (Motorola) and W Perry Wheless Jr (University of Alabama) in the *Newsletter of the Applied Computational Electromagnetics Society*, Vol 10, No 1, March 1995, pp41-50. This provides selected radiation pattern characteristics of the AC-1.8-30 antenna above finite ground investigated both experimentally and numerically using both NEC-81 with a PC and NEC3 running on an IBM 3090 mainframe computer.

While the prime aim of this paper is to compare and contrast the accuracy and reliability of NEC2 and NEC3 radiation pattern predictions for HF wire antennas in close proximity to finite ground, and with ground stakes extending into the earth, it serves also to draw attention to the VHR antenna which could be built rather than bought.

The authors note that the popularity of the HF rhombic is restricted in practice by the large area required: "A logical alternative to the individual who does not have the necessary room to construct a full horizontal rhombic antenna is to reduce the area required for installation by cutting the rhombic in half and rotating the assembly into the vertical plane. In this case, it is possible to reduce the overall surface area drastically, as the antenna can be placed so that the legs are terminated at ground level in a relatively small installation space. The remaining interior angle of the rhombus is mounted vertically away from the ground as high as practically possible. The distant ends of the antenna are grounded and tied together by a single counterpoise wire, and the whole structure is fed through a 9:1 balun.

"This configuration is commonly known as a vertical half rhombic (VHR), although an alternative description would be a 'balanced-feed, loaded pyramid' [It could also be considered a form of the original Bruce inverted vee antenna - *G3VA*], the VHR chosen for this study is the Barker & Williamson model AC-1.8-30 which is designed primarily for broadband amateur use in confined areas. The configuration is depicted in **Fig 8**. Antenna wire is stranded Nr 14 AWG, and the centre support for this study was a 10.7m fibreglass pole." [Non-conductive supports as high as possible are advisable for Bruce and VHR antennas. The terminating resistor is presumably again about 450Ω, non-inductive - *G3VA*].

50MHZ MICROPOWER RECEIVER

TT, APRIL 1996, p74 reproduced details of the micro-power 50MHz FM transmitter based around a Motorola MC2833 chip as described in *CQ* by Irwin Math, WA2NDM. Note that the captions to Figs 8 and 9 were accidentally transposed. In the March 1996 issue of *CQ*, WA2NDM provided a companion receiver using a single MC3367 for all functions other than a speaker driver which could be added if needed.

To quote WA2NDM: "The design (**Fig 9**) is single conversion for simplicity, but the technical specifications are quite impressive. Input threshold sensitivity is 0.2uV (for 3 dB of limiting) and 0.6uV for 20dB (S+N)/N. Operating voltage goes down to 1.1V and maximum operating current required is about 3mA, making an AAA penlight cell an ideal, long-life power source. The chip contains an on-board voltage regulator, low battery detection circuit, and an enable pin to reduce current drain even further (0.5uA) in a standby mode. The receiver can be made crystal controlled (single channel) or tunable."

Two standard 455kHz ceramic IF filters are used (they can be replaced by 0.1uF capacitors at a cost to selectivity) in addition to the Toko RMC 2A 6597HM 455kHz adjustable assembly (L1/C1). Crystal frequency for the single-channel version should be 455kHz lower than the required signal frequency. For a manually tuned receiver the crystal can be replaced by a tuned LC circuit. To use the Enable pin, connect pin 16 to ground instead of Vcc.

To set up the receiver, connect an antenna and adjust the 0.7μH coil and L1/C1 for maximum recovered audio with a weak input signal. It should be possible to achieve an output of 0.5 to 0.8V into 600Ω - enough to drive a small earphone. If a speaker driver is added it will greatly increase power consumption.

WA2NDM advises that construction is straightforward but good RF techniques should be used. All leads should be as short as possible and use good-quality ceramic bypass capacitors. The MC-3367 is supplied in a 28-pin surface mount package and those experienced in the use of tiny surface mounted components could use these to produce a truly tiny receiver.

Like the transmitter, the receiver chip was developed primarily for cordless telephones. When used as suggested by WA2NDM the image protection will undoubtedly be poor and dynamic range very limited. Nevertheless, it should provide a good 'fun' and sensitive companion to the micropower transmitter.

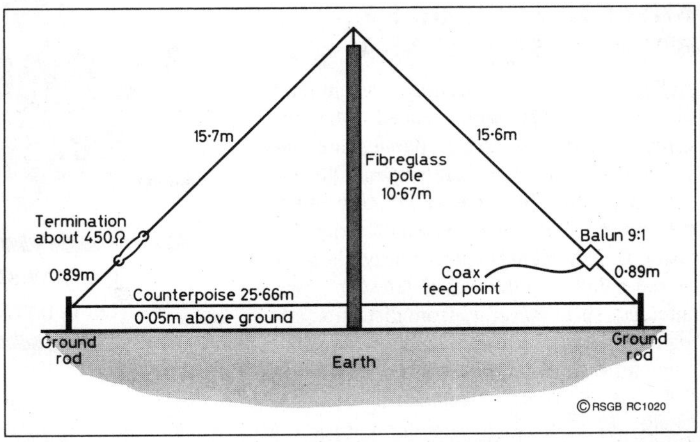

Fig 8: Configuration of the Barker & Williamson Model AC-1.8-30 vertical half rhombic antenna as used by Mike Fanning and W Perry Wheless to compare computed and experimentally determined radiation characteristics. The VHR forms a broadband TW antenna for use between 1.8 and 30MHz.

TRANSOCEANIC DUCTING

EUROPEAN VHF/UHF amateurs have long regarded with envy the California/Hawaii troposphere ducts that open every July to provide 4000km contacts across the Pacific. Over the years since July 1957 contacts have been made not only on 144MHz but also on the 220, 432, 902, 1296, 2304, 3456 and 5760MHz bands. Despite many organised attempts, it has so far proved impossible to span the Atlantic directly on VHF/UHF. Now the Brendan Trophies await the first 144MHz transatlantic contact.

A detailed and most informative article by Emil Pocock, W3EP, (*QST*, March 1996, pp41-46) advises on how, when and where transoceanic troposphere ducts are likely to exist. He believes that the potential for the Bermuda-Azores High to support long-distance VHF ducts has not been probed to its full extent adding that "there is ample evidence that the Atlantic can be spanned by troposphere ducting". He points to the 2500km-plus contacts made by English and Scottish stations with the Canaries. He believes that the Bermuda-Azores High suggests that VHF ducting paths from the Caribbean to Southern Europe and North Africa (6000 to 7000km) may well be possible. Among the paths he suggests as most likely are between Spain/Portugal and Puerto Rico or Florida and between the Canaries and Florida. ♦

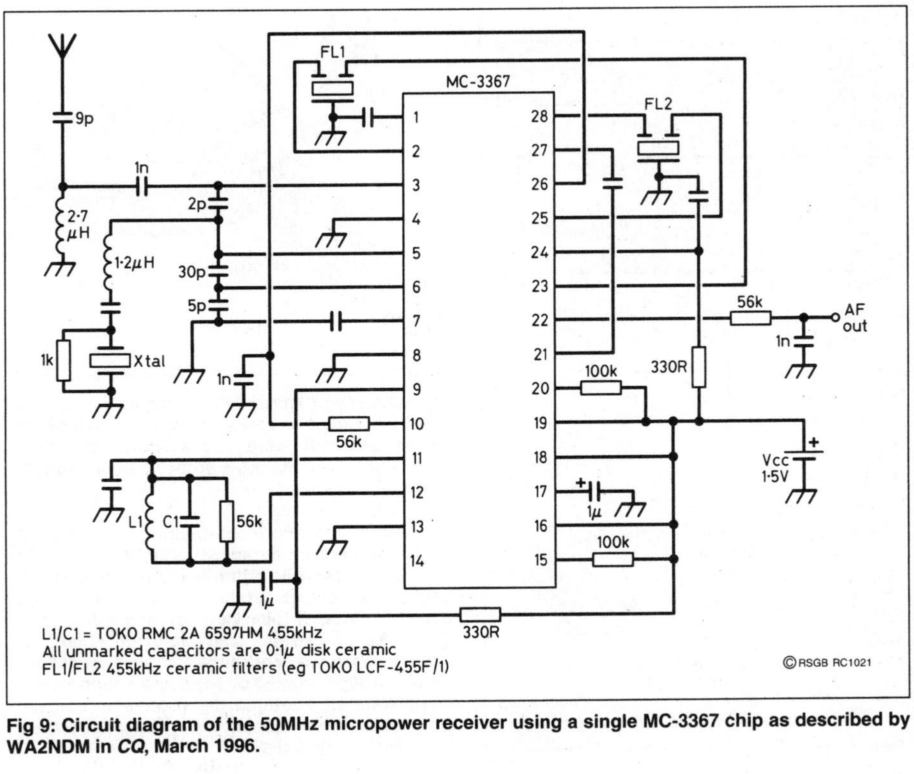

Fig 9: Circuit diagram of the 50MHz micropower receiver using a single MC-3367 chip as described by WA2NDM in *CQ*, March 1996.

Pat Hawker's

Technical Topics

PAT HAWKER, G3VA
London 37/SE22 8SS

ANTI-FADE INDOOR HF BROADCAST ANTENNA

ALTHOUGH HF broadcasting is due in future to change to SSB (with reduced rather than suppressed carrier) such changes are only gradually being made. At present, the vast majority of the high-power HF transmitters still transmit both sidebands and carrier. One result is that signals, particularly those received indoors on the typical telescopic whip antennas, suffer severely from distortion arising from frequency selective fading due to the multi-reflections indoors.

To quote from the article 'Improving the Quality of Short-wave Broadcast Reception Indoors by Using an Anti-fading Probe Antenna' by O G Villard, W6QYT, and colleagues G R Hagn and J M Lomasney (SRI International) in *IEEE Trans on Broadcasting*, June 1995, pp44-47: "When HF broadcasts are received on self-contained receivers indoors, there generally is a reduction in signal strength and an increase in the speed and depth of fading in comparison with reception in open spaces outdoors. Even when the received signal level is adequate, the quality of reception indoors is degraded. This comes about because the fields indoors are a superposition of components of radiated signal from the distant transmitter plus other components representing energy re-radiated or scattered by nearby conductors such as house wiring or plumbing. Since many of the near-field components originate very close to the receiver, it is possible to reduce their effect by deriving the receiver input from two or more probes - ie, pickup devices specially designed to respond to fields in their immediate vicinity without disturbing those fields."

It is shown that a straightforward means for achieving this goal is to deliver to a receiver at least an approximate average of the field over an appreciable fraction of the local multipath interaction field. Indoors this can be as little as a 1m diameter sphere. To obtain such an average, signals can be collected with two or more devices acting as field probes rather than as conventional antennas. Advantage can be taken of the high input impedance of portable, battery-powered receivers designed for use with small telescoping whip antennas. Then spacing of multiple probes only a few tens of centimetres apart suffices to provide an output significantly smoother and of higher quality than that obtainable with the receiver's whip antenna by itself. Although such an arrangement forms a relatively inefficient antenna, this is of little consequence for listening to high-power HF broadcast stations.

The article describes what is termed an "Anti-Fading Probe Antenna (APA)": see **Fig 1**. Although shown as used with a Sony 7600B receiver, it is claimed that it would be equally suitable for any of a number of similar small, modern, battery powered SW receivers. Three additional short telescopic whip antennas are used as 'probes' each connected to a different part of the receiver's own whip antenna via a 10pF capacitor which can be formed by wrapping a short length of hook-up wire around the passive whip. The receiver's whip (the passive whip) is laid flat above the metal 'ground plane', forming in effect a transmission line.

The exact number of 'probes' and their placement is not critical but for optimum results they should be spread as far apart from each other as practical. Three probes are better than two but a fourth probe will add only a little to results.

The authors show that in practice, the large extended ground plane (support plate) may be reduced by using one main strip plus two smaller strips of metal to two of the probes arranged so that they can be spread around the receiver mounted on the main ground plane: **Fig 2**. The width of the metal strips is not critical but the narrower the strip the closer the horizontal passive whip must be located with respect to it in order to cut down undesired pickup by the line itself. The strips can be hinged so that two of the strips can be folded back immediately adjacent to the receiver strip when not in use. For best results the receiver's chassis ground should be connected to the main metal strip on which the receiver is fastened (using rubber bands). In receivers not having an external ground terminal, the chassis ground can be accessed by connecting to the metal ring forming the outside of the earphone or tape recorder connecting plugs.

It will be appreciated that this form of 'diversity' reception is likely to prove of benefit only when used indoors where severe fading/distortion is normally experienced.

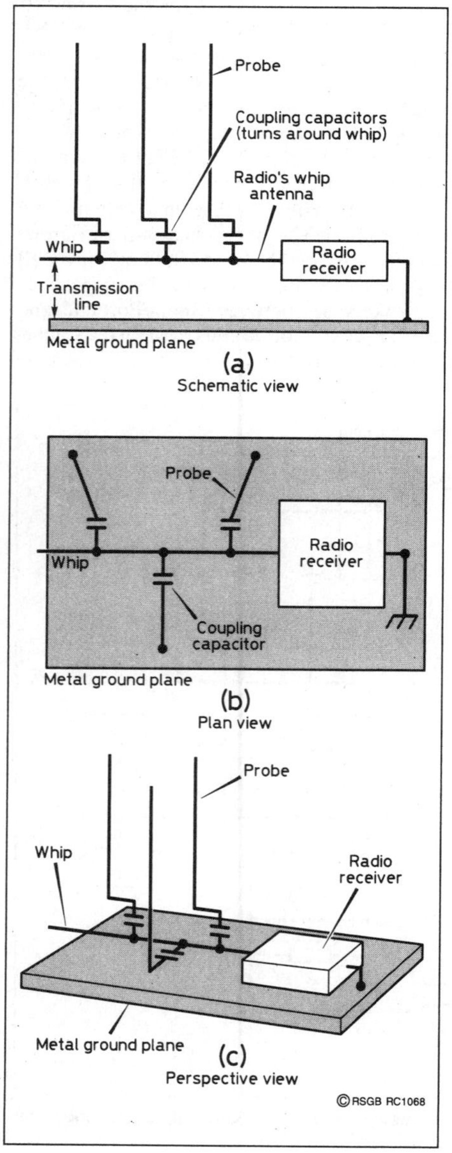

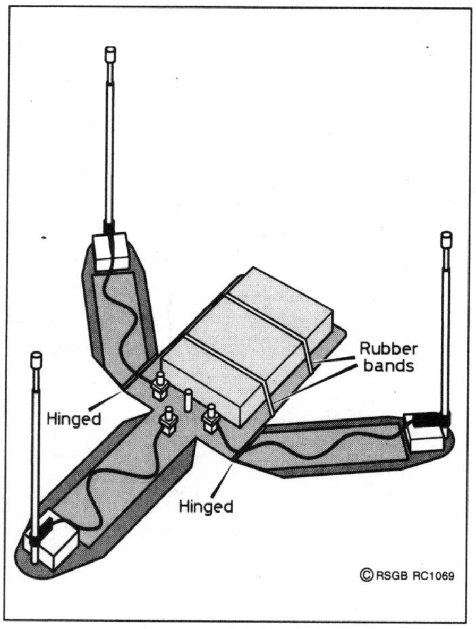

Fig 2: An experimental APA model as it would be used. Instead of one large metal plate, two of the 'active' telescopic whip probes (shown retracted). When not in use, the three probes can be brought together.

(Left) Fig 1: The indoor anti-fading-probe-antenna (APA) for use with broadcast-type HF receiver to reduce fading and distortion due to selective fading. Layout when a large metal support plate is used. (a) Schematic view; (b) plan view; (c) perspective view. The three probes are telescopic rod antennas, spread as far apart as possible. Coupling capacitors are about 10pF, formed by wrapping a short length of hook-up wire around the receiver's 'passive whip' which forms a transmission line when placed close to the metal plate.

WIDE RANGE FUNDAMENTAL SQUARER

IT HAS BEEN emphasised many times that optimum performance from a switching type mixer requires the injection of well formed square waves from the local oscillator. Colin Horrabin, G3SBI, in seeking new techniques in the development of a super linear receiver, has been collaborating with a number of American 'professional amateurs' including Bill Carver, K6OLG/7; Jacob Makhinson, N6NWP; Ed Oxley, formerly of Siliconix, etc. From this co-operation has emerged a number of high performance, wide dynamic range techniques that are suitable for use with G3SBI's H-Mode mixer, including an improved wide range funda-

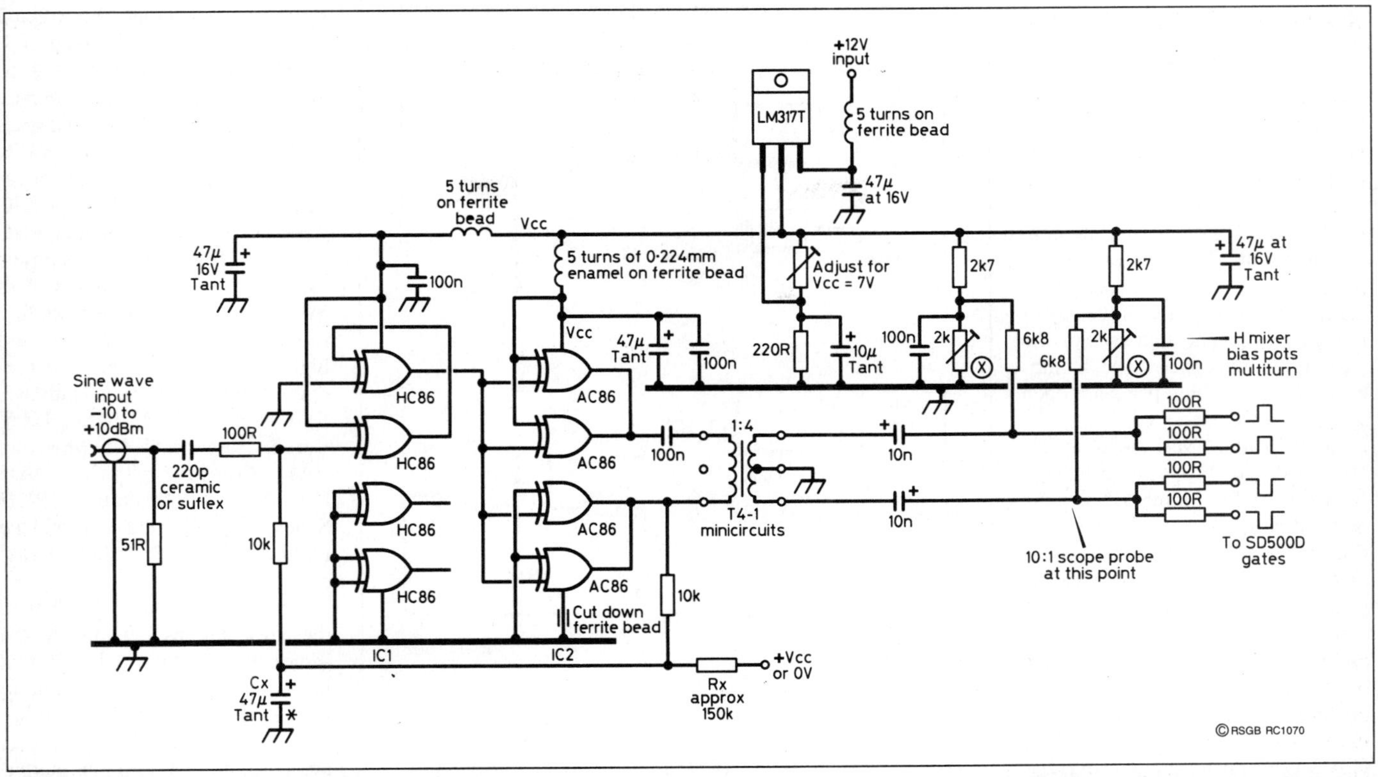

Fig 3: K6OLG and G3SBI's wide range fundamental frequency squarer for use with switching type mixers such as his H-mode mixer. Notes: Rx is a high value resistor about 150kΩ to connect to +Vcc and OV. Cx if such a high value of capacitance (47μF tantalum) is not fitted IC2 will self-destruct should the sine wave drive be removed. IC1 74HC86N (National Semiconductor) IC2 74AC86E (Harris). An earthplane printed circuit is used. G3SBI used a 400MHz bandwidth scope to check the squarer.

mental frequency squarer based on work by K6OLG.

G3SBI writes: "My friend Bill Carver, K6OLG/7, has built a squarer based upon frequency division-by-two using an AC74 bistable followed by paralleled advanced CMOS tristate buffers from each Q not-Q output. The tristate drives two minicircuits T4-1 transformers in parallel. Each transformer output is about 20V peak-to-peak. An advantage of this approach is that the tristate 'enable' can be used with a noise blanker. K6OLG is likely to report in detail on this system.

"In my particular case I want my proposed super-linear receiver to cover 50MHz plus HF with an IF of about 9MHz. This requires a local oscillator frequency of 41MHz. There are some advantages in a local oscillator scheme working only on the fundamental frequency. The circuit shown in **Fig 3**, developed from K6OLG's scheme, will do just that.

"Advanced CMOS and high-speed CMOS is specified to 7.5V Vcc. Use can be made of this together with a T4-1 transformer to deliver two square waves each 180° out of phase and having an amplitude of 15V peak-to-peak, double the Vcc of 7.5V. An adjustable voltage regulator is included so that the square wave amplitude can be set to 10V if desired.

"The basic principle uses the well-established dodge of connecting an odd number of inverters together with a feedback resistor from output to input. This biases the gates into the linear region with a large gain. If a small signal is AC coupled to the input, the output limits in square wave fashion which, to a certain extent, is self-regulating in the mark space ratio.

"The actual circuit uses AC86 and HC86 exclusive-or gates and the design is complicated by the output transformer. The AC86 is the transformer driver and each side of the transformer is driven by two gates in parallel. The other two gates have input high so that each inverts; the first two gates with input low do not invert. The transformer sees a 2 x Vcc swing at its input without time skew since the inversion/non-inversion is done by a one-gate delay. The HC76 gives two-gate 'signal gains' and an inversion. The input working point is DC-stabilised by the feedback from the two 10kΩ resistors, one of which connects to the transformer. The large value of CX is essential when the sinewave input is removed since it prevents destructive oscillation of the loop which kills off the AC86.

"The squarer operates with sine-wave input amplitudes of -10 to +10dBm over the frequency range 50 to 45MHz. An earthplane printed circuit must be used. Experience has shown that there are no problems in using up to 7V Vcc with AC and HC logic and is within specification."

G3SBI describes adjustment as follows: (1) If Vcc is adjusted initially to 7V the square wave amplitude to the mixer is about 13V peak-to-peak at each output. Adjust H-mode mixer operating bias to about 1.95V (see *TT*, October 1993). (2) With a 5MHz sinewave input 'Rx' is adjusted for a 50% duty cycle while checking on a high frequency oscilloscope. (3) If this has been done correctly, amplitude and frequency can be -10 to +10dBm and 5 to 45MHz with no observable change to the output square wave mark/space ratio.

LIFE OUTSIDE OUR BANDS - WORLD-WIDE BEACON NETWORK

IT HAS LONG been recognised that fashion moves in circles rather than straight lines. When communication receivers became established in the 1930s, they were invariably 'general coverage' receivers, often with electrical bandspread that made tuning on the amateur and HF broadcast bands easier. Only gradually were 'amateur bands only' receivers introduced, permitting accurate ganging etc within the amateur bands and, more importantly, reducing costs.

One of the first 'amateur bands' designs was the Tobe Deutschmann Model H tuner designed by Glen Browning for ready incorporation into home-built high performance superhet receivers. Catering for a single tuned RF amplifier stage, the tuner gave excellent bandspread coverage of the 1.8, 3.5, 7 and 14MHz bands only. This tuner kit was originally introduced about 1936, although it was some 10 years later that I acquired one from a Special Communications colleague, the late John Bowers, then G4NY. My much modified receiver, with cascaded crystal filters plus an external converter for 28MHz and for the post-war 21MHz band, remained in regular use until the early 1970s, and still works quite well. The problem, as with all amateur bands receivers, is that one missed out on all the interesting transmissions outside the amateur bands.

For many years, the majority of new HF receiver and then transceiver designs remained 'amateur bands only' (some covering only segments of the bands). But the introduction of frequency synthesis and up-conversion front ends have seen the resurrection of 'general coverage' at least for the receiver section. An increasing number of the equipments in the hands of SWLs, soon-to-be-licensed, and licensed amateurs now provide the ability to tune throughout the HF spectrum. Additionally, many of the better 'all-band' broadcast-type receivers now include a BFO permitting reception of SSB and CW signals.

One of the advantages bestowed by general coverage is the ability to make use of an extremely effective world-wide propagation beacon network. Although not designed for this purpose, the large number of maritime HF coast stations - sited throughout the world and transmitting continuously on frequencies spaced through most of the HF spectrum - can provide at any time of the day or night an excellent appraisal of the state of the ionosphere. The stations can also show which paths are open in which directions, etc.

Frank Wolle, NM7R, in his article 'A World-wide Propagation Beacon Network' (*QST*, March 1996, pp53 & 57) describes the use of coast stations as beacon transmitters to give a real time indication of propagation conditions. He also looks at how they can indicate which part of the HF spectrum is open, and to which part of the world at any given time of the day or night. Encouraged by this article, some checks have been made at G3VA using an old general coverage Hammarlund HQ129x receiver with an antenna wire draped around the room, independently of the amateur rig.

As NM7R puts it: "All you need to tap this font of propagation wisdom is a general coverage receiver and a basic ability to copy Morse Code." The beacons are marking signals transmitted by Maritime Coast Stations located all over the world and used by ships for commercial radiotelex traffic . . . the mode is Narrow Band Direct Printing (NBDP) radiotelegraphy. This mode is also called SITOR (Simplex Teleprinting Over Radio) and is the protocol from which AmTOR was developed by Peter Martinez, G3PLX.

When in traffic with a ship, the coast stations transmit the familiar 'chirp-chirp', but for most of the time the stations are idling, transmitting a marking signal that acts as a beacon for ship operators, but is equally suitable as a propagation beacon for amateurs and SWLs. The marking signal sounds like a 'purr-purr-purr' followed at intervals of a few seconds with the station's callsign, usually three letters, sent once in Morse. With the frequent repetition of just three letters, it should not require great Morse skill to decipher them (a beginner could even record them on a reel to reel tape recorder and play them back at slower speed. Morse still plays an important role in maritime radio).

The coast stations use powers of around 10 to 15kW, much lower than most HF broadcast stations and only a few S-points above the signal that might be achieved from amateur stations in the same part of the world.

While a few marker signals from British coast stations can be heard within the 1.8 and 3.5MHz shared bands, the most useful as propagation beacons are those in parts of the 4, 6, 8, 12, 16, 19, 22 and 26MHz maritime bands. NM7R points out these marker signals are located in the following narrow segments: 4.210 to 4.218MHz; 6.314 to 6.328MHz; 8.417 to 8.433MHz; 12.579 to 12.609MHz; 16.807 to 16.643MHz; 19.680 to 19.690MHz; 16.807 to 16.843MHz; 19.680 to 19.690MHz; 22.376 to 22.412MHz and 26.100 to 226.110MHz. The assignments at 19 and 26MHz are relatively new and used by relatively few coast stations. At the present low point of the sunspot cycle, the segments at 16.8MHz and above provide a useful indication of even weak sporadic E propagation, with strong signals from Europe.

The coast stations transmit simultaneously on their various bands and thus provide a most useful indication of current propagation.

While full lists of coast stations are given in the Klingenfuss Guide to Utility Stations, the following are among those that I found the most reliable and useful: A9M Manama, Bahrein (8 to 16MHz segments); 9VG Singapore (12 to 16MHz); XSG Shanghai; XSQ Guangzhou, China; WCC Cape Cod, USA; PPR Rio de Janeiro; WLO Mobile, Alabama; ZSC Cape Town; but there are hundreds of others. Most European countries, even those with no coasts such as Switzerland (HEC) come pounding in on all bands up to 12MHz and higher bands when open for medium distance: FFT near Toulouse, SVT Athens, IAR Rome, LZW Barna, Bulgaria, UAT Moscow, EAD Aranjuez, Spain, DAN Norddeich, Germany, PCH Scheveningen, Holland, etc and the multi-callsigned Portishead transmitters.

NM7R does note a problem in that a few stations transmit with the same callsign from widely different locations; eg, VIP may be coming from Perth or Sydney. But there are few such jokers. Callsign prefixes are in accordance with the International List so that the country is identifiable without the Utility Station list. Some coast stations outside the NBDP segments can be heard calling CQ de . . . with traffic lists etc in the traditional coast station manner. Listener DXing of coast stations can become a hobby itself!

Among the many important centenaries of radio that have recently, or will soon be celebrated, it is fitting that we should recall that it was on 20 August 1896 that Captain (later Admiral) Henry Jackson successfully transmitted and received a Morse transmission on the deck of *HMS Defiance*, the Naval torpedo

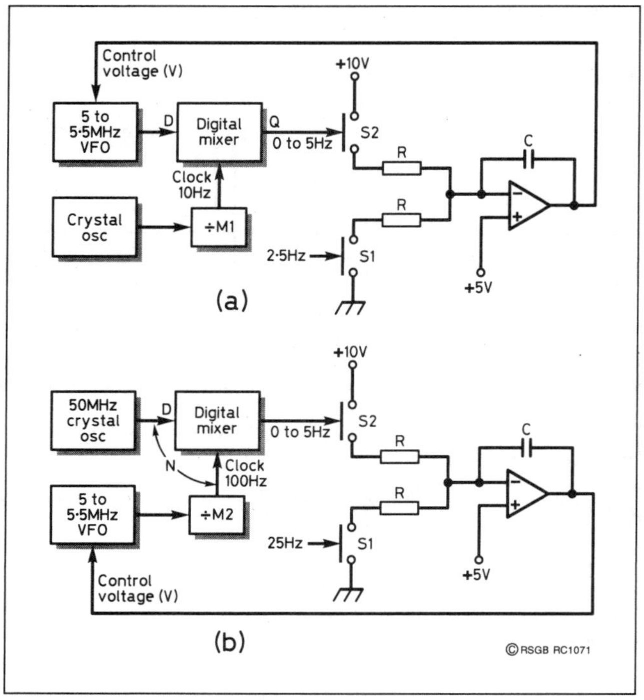

Fig 4: (a) Block diagram of PA0KSB's basic 'huff-and-puff' frequency stabilisation system. (b) Block diagram of the improved system.

GRANDFATHER OF ZERO-BIAS LINEARS

IN HIS article 'The 811A: Grandfather of the Zero-bias Revolution' (*QST*, April 1996, pp51-53), George Badger, W6TC, traces the long history of two of the most popular high power valves used in SSB linear amplifiers: the 811 (later 811A) and the 3-500 zero bias triodes. The 811 was introduced by RCA in 1939; as the 811A it is still in production in Russia for Svetlana. Originally intended as a Class B audio frequency tube, a pair of 811 valves could provide high level amplitude modulation of a pair of 812 medium-mu triodes as a half kilowatt class C RF power amplifier. The 811 then sold in the USA for $3.50 (then less than £1).

During the war, the 811 was used with the RCA813 by Collins in the airborne AN/ART-13, a famous auto-tune transmitter. After the war, George Senti, W0ROW, at Collins showed that, taking advantage of SSB's low duty cycle, four 811A valves could be used in a simple compact linear RF amplifier and this arrangement with RF feedback was used in the amateur field for the popular Collins 30L-1 amplifier, initiating a revolution in the form of the zero-bias grounded grid linear amplifiers that became the mainstay of amateur SSB for many years, and still remains popular. In 1960, Eimac developed the higher power 3-400Z and soon the 3-500Z and the even higher power 3-1000Z etc. The wartime ART-13 was supplied to the Russian air force, and with the Cold War, exact replicas of the RCA 811 and 813 valves began to be manufactured in the USSR, with currently an improved 811A made for Svetlana by Electron Pribor in Ryazan, near Moscow.

RCA was not the first to produce zero-bias (high-μ) triode valves. The late 1930s saw regular advertising in the *T&R Bulletin* of the Taylor TZ20, TZ40 etc as audio-modulator valves alongside the popular T20 and T40 medium-μ RF power triodes. But the 811 and 813 valves of 1939 and still in production must be among the valves which have seen the longest production timescale ever. But today the 811 costs a lot more than the $3.50 of 1939!

school, at that time situated in an old wooden ship-of-the-line in Devonport harbour. He had developed his equipment entirely independently of Marconi, drawing on the pioneering scientific work of Hertz, Lodge and the Indian scientist Jagadis Bose.

As R F Pocock and the late Geoffrey Garratt, G5CS, show in *The Origins of Maritime Radio* this was the first transmission to be made on board ship and, in this rather narrow sense, was the true start of maritime radio. Probably as early as 1891, Jackson, when still a young torpedo officer, had suggested using Hertzian waves as a means of identifying ships at night to distinguish between friend and foe. Although his own experiments continued in 1897, with ranges increased to some miles, in the autumn of that year he met Marconi and recognised that Marconi's system was more advanced than his own. Sportingly, he advised that the ships of the Royal Navy should be equipped with the Marconi system. As Admiral Jackson, he later became President of the Wireless Society of London and it was during his presidency that the change of name to the Radio Society of Great Britain was accepted. Long before then, maritime radio had become, for some years, the single most important application of Hertzian waves.

PA0KSB's IMPROVED 'HUFF AND PUFF' STABILISER

BACK IN 1973, Klaas Spaargaren, PA0KSB, sent to *TT* an ingenious method of stabilising L-C oscillators, capable of turning a reasonably good receiver VFO into the drift-free equivalent of a frequency synthesiser or crystal oscillator. Subsequently, acting on the late G3BY's comment that the system resembled the action of the classic gas engine, I took the liberty of naming it a 'huff and puff' stabiliser.

The stabiliser was intended to be fitted to older receivers and transceivers to overcome the problem of the frequency drift found in L-C oscillators and tending to mar even the best valved transceivers and non-synthesised solid state designs. Various minor improvements were introduced and featured in a number of overseas magazines as well as in *TT* during the 1970s and subsequently.

An improved form of the stabiliser has recently been published in ARRL's *QEX* (February 1996, pp19-23): 'Frequency Stabilisation of L-C Oscillators' by Klaas Spaargaren, PA0KSB, with the sub heading 'So you have a clean, pure L-C VFO, but it drifts. Here's a way of stabilising its frequency'.

PA0KSB writes: "Two simple principles exist for L-C oscillator frequency stabilisation. One method uses a digital frequency counter that measures a VFO frequency periodically. Any deviation between the measured values and a pre-set value results in automatic correction of the VFO frequency. The principle can be executed simply (the original huff & puff system: see for example the 1977 Ham Radio version in *Radio Communication Handbook*, 6th edition, p5. 12-13).

The *QEX* article describes the second principle, which uses a mixer to downconvert the VFO frequency to a low value for further treatment. It has significantly better performance than the first. **Fig 4 (a)** shows the original system; **Fig 4 (b)** shows the improved system. In effect, then processing of the VFO and the third overtone crystal oscillator is transposed. The main difference is the location of the divider M. The function of the digital mixer etc remains as before.

Fig 5 shows a practical circuit for the improved system (the VFO that is to be stabilised is not shown). Two cascaded binary dividers, IC1 and IC2, divide the VFO frequency by 32768 in 15 cascaded stages (7 in IC1 and 8 in IC2). This forms the clock signal for the 74HC74 D flip-flop. The crystal oscillator operates at 48MHz using a 16MHz crystal. Switches S1 and S2 of Fig 4 are formed by TR1 and TR2. Both are normally off with an on time determined by the differentiating R-C networks in their bases and is less than 1mS per pulse.

The output voltage of the integrator has a range of 0 to 10V. After turn-on it starts approximately in the middle of its range. In case the output should become saturated (0 or 10V) it can be brought to midrange by momentarily pressing S3, preferably panel mounted. Press it now and again when retuning the radio.

The tuning of the varicap in the VFO must be larger than the expected drift, and preferably with a sensitivity of about 1kHz/V. The amplitude of the VFO signal to the first digital circuit must be about 4V p-p. The unit needs a supply voltage of 12V, draws some 20mA and has its

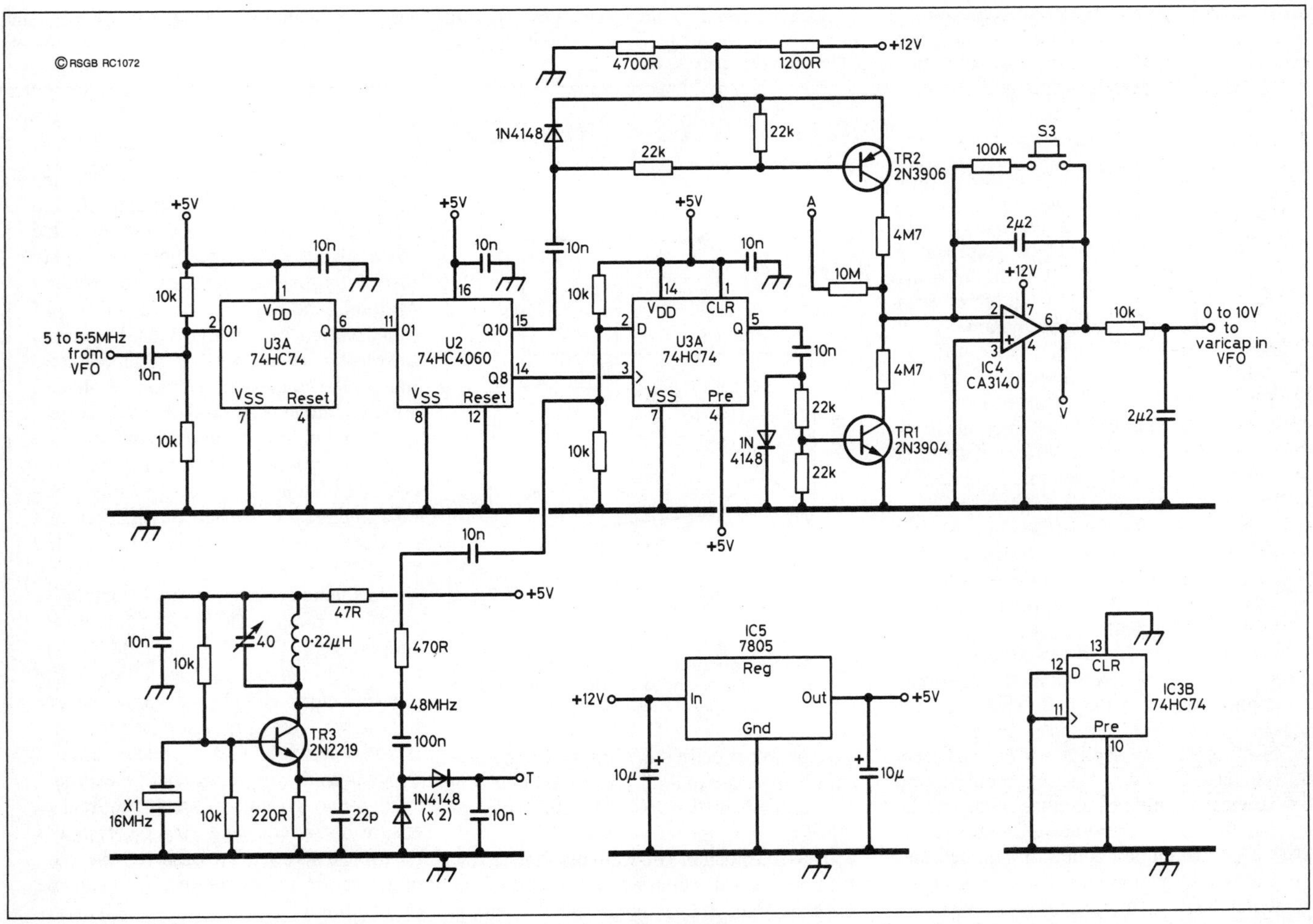

Fig 5: Circuit diagram of the improved frequency stabiliser.

own 5V regulator IC5 to supply the crystal oscillator and digital circuits.

PA0KSB used "ugly" construction, although he prefers the more elegant term "amateur type surface mounted technology". For the coil in the overtone oscillator he used a small 0.22uH RFC but reports that a ½in, 6 turn air coil worked fine. The 2.2µF capacitor in the integrator or polystyrene type. He recommends mounting the circuit in a small metal box to avoid RFI from the transmitter or to the receiver. The only adjustment is the 40pF trimmer in the overtone oscillator. Oscillation can be confirmed at test point T, which must be about 4V and can be adjusted by the trimmer. Precise frequency is unimportant.

With a VFO input at 5MHz the output of IC1, pin 6 should be 39kHz, and the outputs of IC2, pins 15 and 14 should be 152Hz and 38Hz respectively. Correct operation of IC1 and 2 can be checked with an oscilloscope, but a multimeter can also be used since, when operating correctly, the outputs will be square waves and a DC voltmeter should indicate a 2.5V. When not working the output will be either 0 or 5V.

PA0KSB stresses that the quality of the L-C oscillator must be very good; a T9 tone without appreciable hum, jitter, microphony, hand effects etc, and with long term drift the only deficiency. He admits there are two drawbacks to this form of stabilisation. Firstly, it is not easily possible to work split frequency or use an RIT function with the stabiliser in operation, since after a number of jumps from one frequency to the other, the VFO may lose its original lock point. Secondly, with a fixed value of M, the distance between the lock points varies with the square of the VFO frequency, making the system less suitable for a multiband VFO. In such a case, M should be selectable, which increases complexity. However, with this relatively simple, low cost circuit, the long-term stability of an HF L-C oscillator can achieve the stability of a crystal oscillator, while maintaining fine tuning steps.

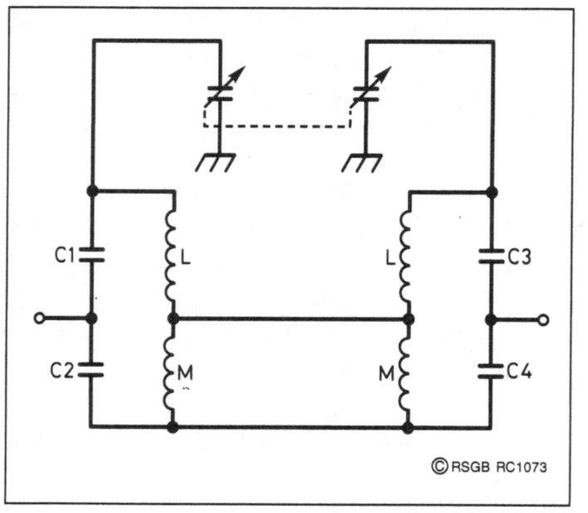

Fig 6: Two section bandpass filter as simplified form of Cohn filter. Vales of C1/C2 and C3/C4 determine the input and output impedances.

MORE ON COHN-TYPE FILTERS

GEOFF BAGLEY, G3FHL, noted with interest the *TT*, May item on "The Cohn 'Minimum Loss' Filter" as used by G3ROZ, despite the problems he encountered in achieving the specified low attenuation.

G3FHL writes: "I have been playing with this type of circuit using a General Purpose Ladder Analysis program (part of the software provided with the ARRL publication *Introduction to Radio Frequency Design* (ISBN 0-87259-492-0)). I find that the simpler two-section version is very easy to design and set up, whilst the three-section design (Fig 6 of May *TT*) is quite capable of large insertion loss as found by G3ROZ.

"The critical component seems to be the 'middle' tuning capacitor, and getting it to track the two end ones. In the two section version there is inherent symmetry, and the problem does not occur since both ends have similar terminations.

"The two inter-section coupling soils Lc do not seem to be particularly critical, presumably because they are both of the same value. I agree with your comment that the two centre coils of the May Fig 5 (2L) do not constitute a single inductance ("waste of a coil" comment by G3TPW) since they are associated with separate ports to the middle resonator.

"Incidentally, the software supplied with ARRL's IRFD contains some very useful items covering filters, amplifiers, oscillators, phase-locked loops, phase-shifters etc."

Personally, I feel that the two section filter (**Fig 6**) can no longer be ascribed to Cohn, since it becomes a version of the classic bottom-coupled bandpass filter which was quite popular for broadcast receivers in the 1930s to increase image rejection etc. In this case, I wonder if results would be affected by simply removing one of the Lc coils? Off-tune attenuation would be less than with the true Cohn filter but the performance should serve for most applications, with fewer ganging and high-Q problems. ♦

DANGERS OF GENERIC DEVICES

THE NOVEMBER *TT* contained a warning by DB1NV, in connection with switching diodes, that when using generic types (ie those bearing a standard type identification but possibly produced by different firms) "it cannot be calculated what effect variations in the manufacturing parameter will have (different production lines, different production methods)". This reminded Mike Clift, G3UNV, of a crisis that nearly lost his company an important customer due to the mistaken belief that all devices of the same reference number had the same characteristics.

He writes: "My expertise is in solenoid valve design... In order to ensure silent running of AC-mains powered solenoid valves one method is to apply rectified AC to the coil. In the interests of economy it is usual to use half-wave rectification with smoothing provided by a rectifier reverse-connected across the coil. This is a very compact arrangement and able to withstand much higher temperatures than an electrolytic capacitor. It is necessary only to ensure that the silicon junctions do not exceed their specified temperatures.

"We developed a new miniature solenoid valve in conjunction with a company which was to manufacture the solenoid operator. We specified BY133 rectifiers made by ITT. This was a well proven 1A device a little larger than the 1N4007 with a PIV of 1300V.

"The first batch of 100 were delivered to us, assembled to valves, tested and rushed to our customer. Catastrophe! When tested on appliances there was a 40% failure due to shorted diodes. I arrived on the customer's doorstep and was confronted by a dozen dissected solenoids all with failed diodes. I quickly discovered that the diodes were made by Motorola and not ITT. Our suppliers had found out that the ITT device cost 4p whereas the Motorola was only 2.5p. Only fast talking and the rapid supply of sample solenoid valves using the ITT diodes saved the relationship with our customer. Although the published specifications for the BY133 diodes for both makers were the same, practical application showed this marked difference! I agree wholeheartedly with the DB1NV comment.

"Incidentally the two-diode method is useful for running DC relays from an AC supply: **Fig 7**. The collapsing field through the coil during the off part of the cycle creates a current and hence magnetic force to hold the relay in the energised position. It will not work with all relays since it depends on the time constant of the circuit. A high-enough inductance is required to ensure that the current during the off part of the cycle does not fall to the level where the opposing spring load returns the relay to the off condition."

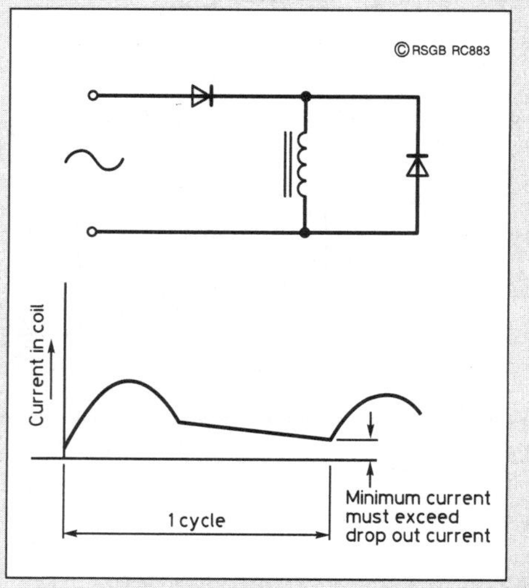

Fig 7: Tip for operating DC relays from AC power source. Effective in those cases where the time constant of the circuit is sufficient.

RHOMBIC LOOP & HULA HOOP ANTENNAS

SOME INTERESTING new concepts in antenna systems - which are capable of radiating circularly polarised waves and could well prove suitable for amateur UHF, VHF and possibly even HF applications - have been described by a team of Japanese antenna engineers at the Japanese National Defense Academy and the University of Tsukuba, Japan.

The original disclosure by H Morishita and K Hirasawa, 'Wideband Circularly-Polarised Loop Antenna', was at the 1994 IEEE Antennas & Propagation Conference (I have not seen this paper). A subsequent paper 'A Dual Rhombic Loop Antenna for Circular Polarisation', with additional authors T Hzuka and T Nageo, appears in the IEE Conference Publication No 407 (pp500 - 503); and the most recent development 'Circularly Polarised Rhombic Hula Hoop Antennas' by H Morishita, K Hirasawa and T Nageo is published in the IEE's *Electronic Letters*, (23 May 1996, pp946 - 947).

Basically, the concept uses rhombic loop element(s) with a perimeter of only about 1.3λ at the centre frequency (ie only a little larger than the usual 1.0 lambda quad element). The antenna, it is claimed, "can have a travelling-wave type current distribution, which has approximately a constant amplitude and a linearly changing phase. This current distribution is realised by making one point open (ie a small gap) or loading an infinitely large impedance on the proper position of the loop."

The experimental models described in the Japanese papers are based on 1.5GHz with the four sides of the rhombus only about 60 to 68mm long (ie perimeter of about 250mm) and with a bandwidth of roughly 20%.

The 1995 IEE conference paper extended the basic idea of a single rhombus loop in a plane parallel to and above a perfectly conducting ground plane to a dual rhombus system: see **Fig 1**. In this version, two types of single feed dual rhombic loop antennas for circular polarisation are proposed: **Fig 2**. The antenna consists of single or dual rhombic loops of the same size located above a theoretically infinite perfectly conducting ground plane. For the dual system, it is shown that by adjusting the perimeter, the rhombus vertex angle, and the gap position on the loop, the axial ratio 2dB bandwidth is more than 20% for both

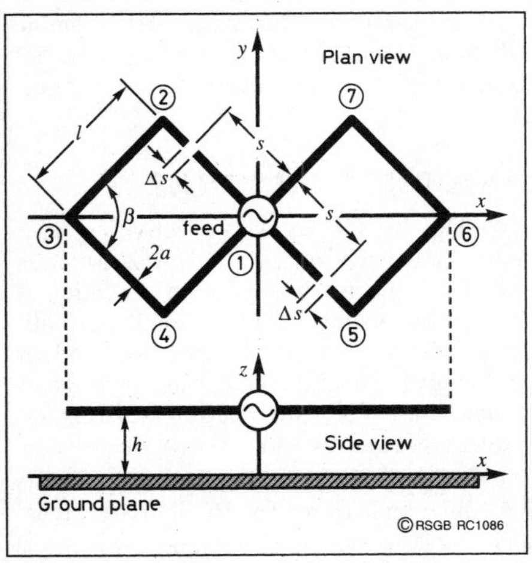

Fig 1: Dual rhombic loop antenna as described by Japanese engineers at a 1995 IEE conference.

Pat Hawker's Technical Topics

PAT HAWKER, G3VA
London 37/SE22 8SS

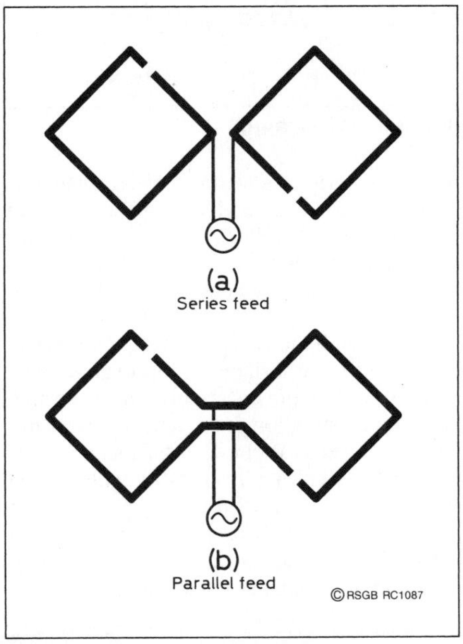

Fig 2: (a) Series feed and (b) parallel feed for the dual rhombic loop antenna.

types of feed and the gain of about 10dBi for the parallel feed and 11dBi for the series feed (about 3dB more than for the single rhombus). In addition, the possibility of controlling the input impedance is found by selecting the gap position, the rhombus vertex angle and the feed type.

I must admit that on first reading, I could not see how a horizontal rhombic loop and its ground plane image could radiate circular polarised waves, since there appeared to be no vertical component. The penny dropped (at least in my opinion) when I realised that there would inevitably be current flowing down the outer braid of the coaxial feedline between the ground plane and loop element.

This factor seems to have led the Japanese research engineers to develop an alternative form of the antenna utilising the 'hula hoop' principle (more correctly known as a Directional Discontinuities Ring Radiator or DDRR). The antenna was originally described for operation from 2 to 4MHz with the ring only 2.8 electrical degrees high by J M Boyer, W6UYH, of Northrop Ventura in *Electronics*, January 11, 1963 (see briefly in *ART7*, p272). This was later extended in 1965 to half-wave closed loops by Italian engineers (see *ART7*, p288).

These antennas (**Fig 3**) were designed to provide primarily vertical polarization from an electrically very short vertical section. Boyer pointed out that (with a perfect ground plane) a height of 2ft for the circular ring section could provide equivalent results to a quarter-wave vertical monopole, a saving of 30 times in height, but requiring real estate for the horizontal loop. The snags are that the system requires a highly conductive ground plane, for example when erected on the metal deck of a large vessel (or with the 120 radials of an MF broadcast station), with low resistive losses and a very high Q (comparable to a small transmitting loop) that requires returning for even small changes of frequency. It is possible that the 120 radials could be replaced by an elevated counterpoise system, or on UHF by a metal plate of reasonable dimensions.

The Italian half-wave closed loop version (although requiring more real estate) has a bandwidth some 10 times that of the original hula hoop DDRR when the ring has a vertical height/wavelength ratio of about 5%. Although there was interest among American amateurs of the original Boyer version, and a dipole version was also published in one of the amateur magazines, I cannot recall the *TT* or *ART* item on the Italian modified system (originally published in *Electronic Letters*) ever attracting amateur radio interest.

In their recent contribution to *Electronics Letters*, the Japanese team write in their introduction: "It is well known that hula hoop antenna is a variant of an inverted-L antenna and radiates a linearly polarised wave with vertical and horizontal components. Single feed rhombic loop antennas radiating a circularly polarised wave... can have a travelling wave type current distribution on the loop, which has approximately a constant amplitude and linearly changing phase... This current distribution is realised by making a gap at the proper position of the loop and by selecting the loop length."

They show that rhombic hula hoop antennas (RHHA), in either one or two rhombic loop configuration, radiate circularly polarised waves. As shown in **Fig 4**, the antennas are composed of a rhombic loop terminated in a gap. The main difference between the RHHA and their earlier rhombic loop antenna is in the feed configurations. The RHHAs are fed by a

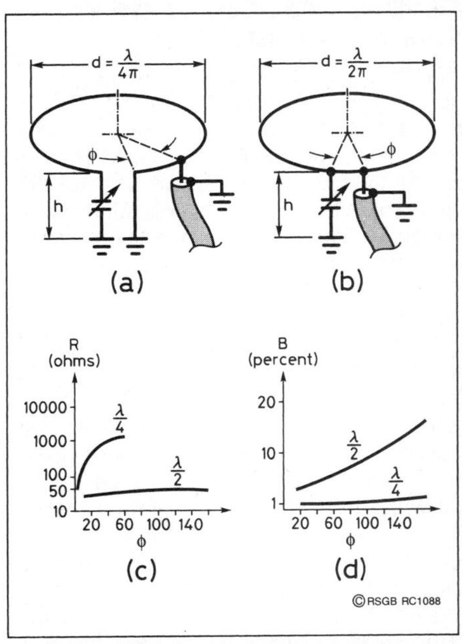

Fig 3: (a) Quarter-wave hula hoop DDRR antenna. (b) Half-wave closed loop DDRR antenna. (c) Resonance impedance versus feed-point angle. (d) Bandwidth versus feed-point angle.

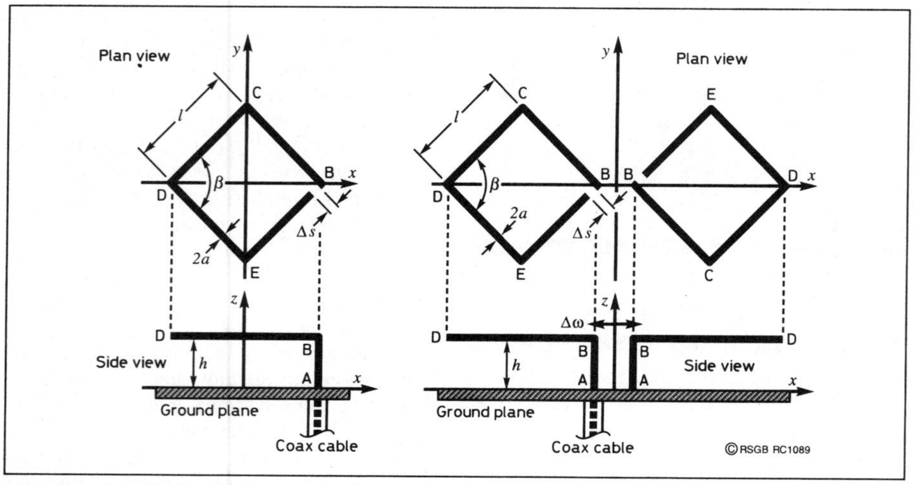

Fig 4: Rhombic hula hoop antennas (RHHAs). (a) Single RHHA; (b) dual-RHHA.

coaxial cable, of which the inner conductor is conductor is connected to the ground plane, whereas for the rhombic loop both the outer and inner conductor of the cable is connected to the horizontal loop(s) as shown in Fig 4.

Two RHHA antennas were investigated, a single and a dual rhombic with the loops having the same parameters: Beta = 90-°, Delta s = 0.016λ, a = 0.008λ and h = 0.21λ, where λ is the wavelength at the centre frequency of the antenna. The perimeter of the loop is chosen to be L = 1.29 λ for the single loop and 1.20 λ for the double-loop hula hoop antenna. No information is provided on the ground plane other than that indicated in Fig 4.

Calculated (Method of Moments) and measured data are presented on current distribution, frequency characteristics and radiation patterns, and the letter concludes: "Two types of rhombic hula hoop antennas are proposed to radiate a circularly polarised wave and their radiation characteristics are investigated. It is found that the axial ratio bandwidth (2dB) of approximately 10% and the gain of approximately 10dBi are obtained by selecting the loop configuration."

FEEDING F5LCI'S LOOP + 1.8MHZ ANTENNA

TT (DECEMBER 1991, p28) presented details of how Jean-Marc Bourdereau, F5LCI (then F1LCI), uses a large 84m horizontal loop for HF operation on 3.5MHz and above. He then utilises the fact that the far end is at zero RF voltage and can be earthed as protection against static electricity and also to form an inverted L-antenna for 1.8MHz with a large vertically polarised component, the loop providing top capacitance loading.

Fig 5 shows the antenna as given in December 1991. His loop is one full wavelength long for 3.5MHz, rectangular in shape (18 by 24m) to fit his garden. The mean height is 12m (varying between 8 and 14m), balanced, open-line fed from his attic shack for the HF bands, but with short-circuited feeders for 1.8MHz. Dr John Belrose, VE2CV, (*TT*, May 1992, pp38 - 39) subsequently commented at some length on this arrangement, showing that with both feeder arrangements, there would be some vertical and horizontal polarisation, and that the RF power absorbed from the long earth lead from the attic could not be ignored. However, he expressed the opinion that the system, which he termed a vertical electromagnetic ground-plane loop (EMGP) antenna, should prove a good 1.8MHz antenna, though he was less impressed by its calculated characteristics on the HF bands in the balanced feed configuration.

I must admit that, despite VE2CV's misgivings, I use a large horizontal loop, of about 80m length (exact length uncertain). This is formed out of an existing long wire antenna by bringing back a second wire to a balanced pi-network ATU, resulting in a most irregular shape and with the wires at differing heights. I find it provides a reasonably useful antenna on all the HF bands I use and is not dependent on the long earth wire to the upstairs shack or the counterpoise wire previously used with the standard long wire configuration.

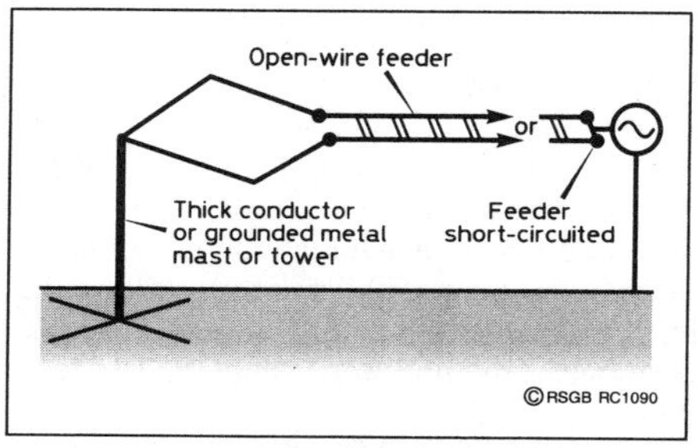

Fig 5: The F5LCI antenna as described in the December 1991 *TT* with the option of a short-circuit feeder for mainly vertical polarisation on 1.8MHz.

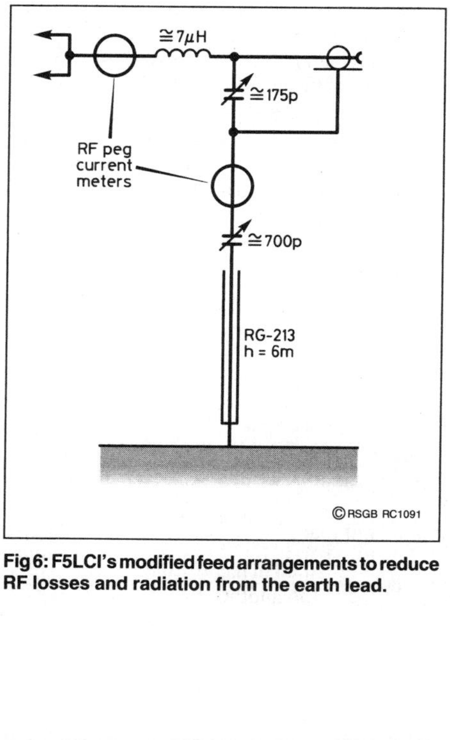

Fig 6: F5LCI's modified feed arrangements to reduce RF losses and radiation from the earth lead.

Recently, F5LCI had written further on using his 84m horizontal loop on 1.8MHz: "The 84m horizontal loop, although only a half-wave loop on 1.8MHz, was eventually found usable on 160m; its high input impedance is somewhat transformed down by the open wire line forming a linear transformer so matching is easier. Its radiation pattern, according to EZNEC, is almost spherical. Reception is very quiet. Short-skip (NVIS) contacts are more comfortable than with the shorted feeder configuration.

"Feeding of the vertical set-up ('mode') has been modified: **Fig 6**. The major drawback of the earlier system was radiation from the 'earth wire'. A Faraday screen (coaxial cable shorted at ground level) did not bring any improvement (noise pickup probably still coming from other parts of the system). I decided the reactance of the ground wire had to be cancelled. No direct measurement (with a GR bridge) proved consistent so a gross estimate was calculated from the straight-wire impedance formula. A magnitude-order of the neutralising series capacitor was then calculated. After many attempts to find a satisfactory way of tuning it, two methods were used: (a) finding a null on a portable receiver connected to this 'antenna'; and (2) the ever useful simple shunt flashlight bulb or, better, and RF current meter (peg type). I was delighted to find a marked increase in the wanted antenna current indicated on a second RF peg meter. This seemed proof of the validity of the setting of the 700pF capacitor.

"Next I measured, with the help of an Autek RF-1 meter, the impedance of the antenna. This led to my replacing the standard T-match ATU with a simple L network in order to minimise losses (TL program by N6BV or ARRL). I found that the SWR was still different on the shack side of the coax, and RF found its way to the PC keyboard. It seems likely that the only remaining cause is RF pickup from the antenna on the braid. A coil choke and a ferrite sleeve balun are used, although these are large on 1.8MHz!

"In short, radiation is now believed to be almost entirely confined to the vertical down wire from the middle of the loop, with the loop playing the role of a top-loading hat (its own radiation some 25dB lower)."

One cannot help feeling that F5LCI's 1.8MHz antenna has a striking similarity to some aspects of the Japanese UHF rhombic hula hoop antenna! But I am not sure whether he or VE2CV would agree with this suggestion.

THOUGHTS ON 73KHZ

FOR THOSE OF us whose activities have traditionally been concentrated on '200 metres and down', the availability of an allocation at 73kHz poses an interesting new challenge with a host of new problems in applying modern technology to the frequencies used in the dawn of radio communication and broadcasting.

Receivers for low and very low frequencies do not present many problems:

- A converter in front of an HF or MF receiver.
- A classic regenerative straight set using a couple of transistors, FETs or valves and tuning perhaps 15kHz to 150kHz.

It is now 56 years since Charles Bryant, GW3SB, and I described possibly the only VLF receiver ever presented in the Society's journal. This was a straight regenerative two valve (battery) receiver (**Fig 7**) in an article 'Ultra-long-wave Reception as an Aid to Morse Practice' (*The T&R Bulletin*, June 1940, p470).

The idea was to use the official wartime press bulletins from GBR Rugby on 18,750 metres (16kHz). For a tuning coil we used an Eddystone quench coil unit with the two coils in series and with 350 turns of fine wire wound between them to provide regenerative feedback. Two of the old 2nF compression pre-set capacitors were used for tuning. There was also a 200-turn plug-in coil in the antenna lead to prevent breakthrough from the local BBC transmitters at Washford Cross. In those days GBR transmitted A1 with auto-Morse at about 20WPM, ideal for getting Morse copying speed up during the period of the phoney war. It was not until well after the war that GBR changed to FSK RTTY.

I still have in my junk box some of the old two-pin honeycomb-wound Igranic coils ("What are the wild waves saying") one of which with 400 turns has a calibration chart that shows that with 500pF capacitor it can tune from about 1000 to 4000 metres and with 1000pF up to about 5500 metres.

Transmitters. In theory no problem, but there could be practical problems in coping with the large-value, high-voltage inductances and capacitors required. With the light loading of an electrically small antenna there will be very high RF voltages involved at these frequencies calling for well spaced, physically very large capacitors. Capacitor arc over could prove a limiting factor for high powers.

As good a way as any of generating low-level 73kHz RF would be to use a 7300kHz crystal followed by two decade divider ICs. For amplification broad-band ferrite transformers could be used. Again there are switched-mode power supply units operating at frequencies well above 73kHz, and there should be no problem in using relatively low cost power switching mosfets at this frequency. A switched mode system at 73kHz, if made to provide sine wave output, comes near to representing a modern form of high frequency alternator. The main challenge would be for high power units capable of the output of 1kW or more needed for an effective radiated power of 1 watt from the best (yet still hopelessly inefficient) antennas within the limitations of most amateurs.

Antennas are likely to prove the major problem for anyone who wishes to communicate over a distance and make use of the reasonably consistent day/night propagation on LF. Consider the following notes from 'Shipboard Antennas' by Preston E Law Jr, 2nd Edition, published by Artech in 1986: "It should come as no great surprise to learn that, since the initial use of radio was for long distance wireless communications at low frequencies, the first shipboard antennas were all made from haphazard lengths of wire strung as high as possible above the ship's topside. Evidently the thinking was that the longer and higher the wire, the better the results... By the early 1920s the most common shipboard antennas were the flat top, the T, the inverted L, and inclined wire types that remained in naval service with some modifications for another half century... Despite appearing primitive at first glance, one cannot help but admire upon closer inspection the careful and precisely detailed design evidenced by these old plans." **Fig 8** shows some classic LF/VLF antennas.

The US Navy antennas and equipment in use for many years were apparently designed to operate between 50 and 1500kHz. That is until close co-operation between the amateurs and the US Navy showed it was possible to use HF for long-distance communication with ships.

TT June 1993, p53, reported how Australian amateurs had succeeded in spanning the Tasman Sea (about 1200 miles) on 196kHz using about 100 watts to an umbrella-type, top-loaded 109ft (38m) steel lattice tower radiator and counterpoise: **Fig 9**. Effective radiated power was under 10 watts.

Although in terms of frequency the new allocation of 71.6 to 74.4kHz appears narrow it covers wavelengths between approximately 4190 and 4032 metres, a span of no less than 158 metres! Since the antenna will inevitably comprise an extremely short (electrically) radiator, it will have an extremely high Q with a bandwidth effective for only the narrowest modes. For the amateur with a single tall mast (or other support), probably the most practical approach would be to use an 'umbrella' antenna to provide a degree of top capacitance loading. For optimum results the spokes of the umbrella need to be open as widely as possible, requiring as much space as possible around the mast or tower. The old adage of as much wire as high as possible still applies. Fig 8 shows some antennas types on LF 90 years ago.

Historically, our new allocation is very approximately that of the very first 'broadcast' by Fessenden from Brant Rock on Christmas Eve 1906, using his alternator which was capable of generating power at up to 75kHz (the following year this was increased to some 100kHz). High frequency alternators were originally developed by Eliha Thomson and Nikola Tesla in the late 1889s but Fessenden was almost certainly the first person to apply this to radiate continuous wave electromagnetic waves. His first alternator, built for him in 1902 by Alexanderson at General Electric (US), had an output of 1kW and was built in 1902. By mid 1906, a number of practical difficulties had been overcome and a machine giving 50kHz was installed at the Brant Rock station. Various further improvements were made and by the autumn of 1906 the dynamo was working at 75kHz with an output of some 500 watts. With the aid of a water-cooled microphone (used to modulate directly the output from the

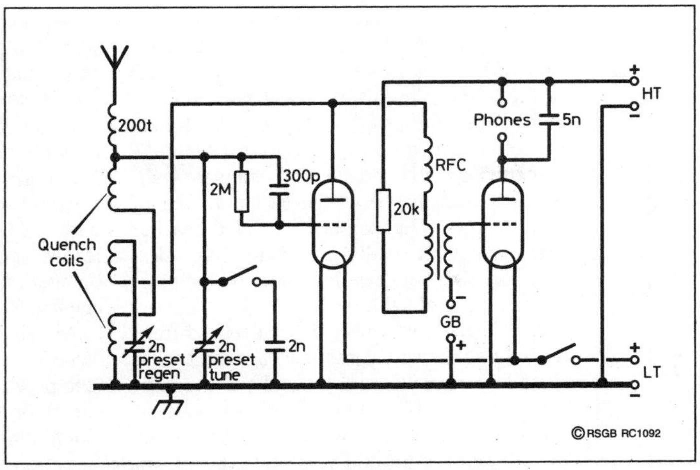

Fig 7: Simple VLF/LF two valve regenerative receiver as described by GW3SB and G3VA in June 1940 as an aid for improving Morse copying speeds by listening to the official wartime news bulletins transmitted from Rugby Radio, GBR on 16kHz. Today such a design could be easily implemented in solid-state.

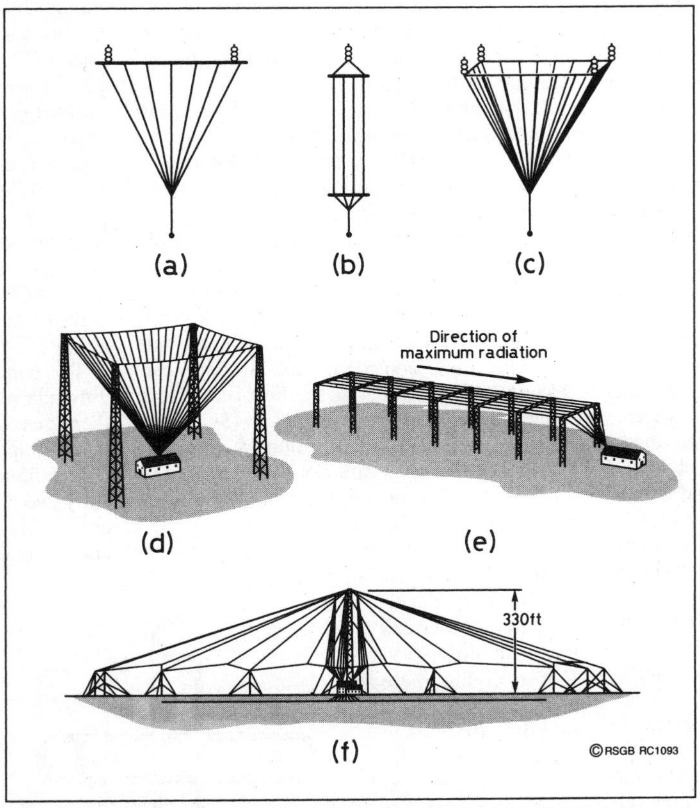

Fig 8: Classic LF antennas as used some 90 years ago. (a) Fan antenna. (b) Grid antenna. (c) Square cone antenna. (d) Square cone antenna as erected by Marconi at Poldhu, Cornwall in 1901 (but suffered gale damage). (e) An inverted L 'directional' antenna of 1905. (f) A large umbrella antenna as built at Nauen, Germany in 1910.

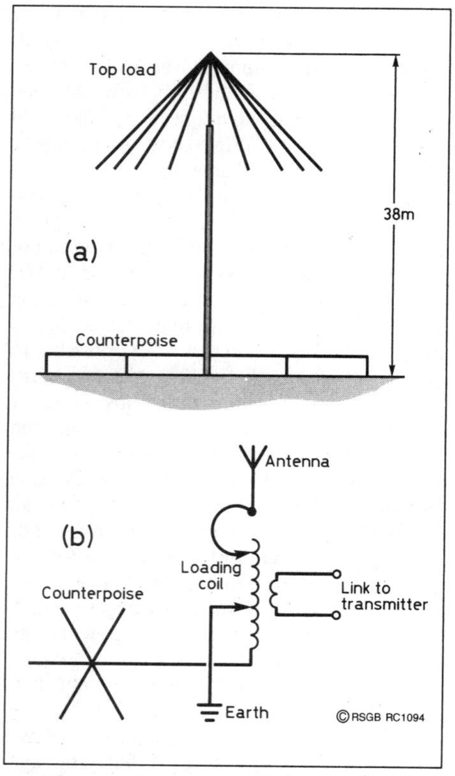

Fig 9: The amateur antenna that put signals across the Tasman Sea from Australia to New Zealand on 196kHz with transmitter power of under 100 watts (radiated power probably under 10 watts). (a) Outline of antenna (guy wires not shown). (b) Antenna loading arrangement.

dynamo fed to the antenna), Fessenden was regularly transmitting speech to a receiver some 11 miles away but capable of longer ranges.

In December 1906, he announced that a broadcast including music and speech would be transmitted on Christmas Eve and invited ships' operators to listen. Brant Rock by then had a 420ft high tower with an umbrella type antenna (originally developed by Muirhead-Lodge).

About 1929, the first duplex radio-telephony links across the Atlantic were opened using wavelengths of the order of 5000 metres (60kHz) and pioneering the use of SSB. But as amateurs, I think we will have to set a less ambitious target.

LF and VLF propagation permits fairly consistent 24 hour long distance potential (although at 73kHz there will be some day/night variation and some influence of the solar cycle) but requires high powers and massive antenna systems to be really effective. These frequencies are extremely vulnerable to lightning discharges in the form of severe atmospherics in some seasons. Nevertheless the ability of even low-power LF transmissions to penetrate underground and under the sea should provide some interesting challenges for amateurs, rather than simply a turning back to the pioneering days of radio.

RFI & NEW INDUCTION LAMPS

TT (AUGUST 1991, p32) was probably the first column to draw attention to the potential RFI problems that could stem from the introduction of RF-powered high-efficiency electric lamp bulbs, based on clippings from the trade press sent in by G3SWP and G1UKE. These induction lamps depend on RF generators working in the "industrial, scientific, medical (ISM) bands around 13.5MHz (American) or 2.65GHz (European). The basic arrangement of the Philips QL Lamp was illustrated. The topic turned up again in May 1993's *TT* and subsequently the EMC Committee published a comparison of the RFI potential of various types of lamps, showing that the earlier fears were justified although there is considerable variation between the induction lamps of different manufacturers.

Ray Nicholson, G4SQG, who retired from the lighting industry some years ago, still occasionally sees the trade press and has now sent some cuttings from the May 1996 issue of *Lighting Equipment News* in which the Editor Ray Malony reports on The World Light Fair in Hanover.

To quote a couple of paragraphs: "Philips has also gone a long way to solving the headaches of luminaire manufacturers trying to wrestle with the company's induction lamp, the longest life lamp on the market at 60,000 hours but a technology notorious for broadcasting electromagnetic noise.

"Up to now, the beast had to be contained in an interference suppressing Faraday cage to comply with EMC legislation, but by using a different array of windings in the coil, repositioning the components in the ballast and adding a triax instead of a coax cable, Philip's engineers have tamed it to such an extent that the lamp can even be used bare-mounted."

G4SQG comments: "This is all very well but I wonder if the 'el-cheapo' companies will show the same consideration." [I suppose it depends on how effective the European EMC Directive proves to be - *G3VA*]. G4SQG also draws attention to a description by Jonathan David of "two light sources vying to be the technology of the next millennium: the cluster lamp and the sulphur lamp".

Both the cluster lamp, developed by Philips in Holland, and the sulphur lamp, developed by an American research company, Fusion Lighting, are essentially high-power discharge lamps powered by microwave (2.4GHz) energy G4SQG points out that the example given of an experimental cluster lamp has a microwave input power of 5.9kW! The article illustrates how three 2.25m long light pipes, each powered by a single ultra-bright sulphur lamp has replaced 94 high-bay luminaires at the Smithsonian Institute's National Air and Space Museum in Washington, DC.

As G4SQG puts it: "As the lamp itself has to be screened from view it is to be hoped it is also RF proof. Apart from potential RFI problems I would not like to be too near a 6kW magnetron." One wonders whether a public which seems increasingly concerned with the milliwatts of UHF radiation from portable radiophones will be happy standing near kilowatts of microwave energy screened, it would seem, in both types of lamp only by wire mesh. One remembers the fuss about leaky seals on microwave ovens using magnetrons producing under 1kW of 2.3GHz energy.

Also in the May 1996 *Lighting Equipment News* is an article by Nils Borg stressing that "the poor power quality associated with compact fluorescent lamps (CFLs) has not attracted much attention - but the resulting harmonic distortion will eventually have to be addressed".

The CFL, in common with many electronic devices, has a poor power factor (50%, so that a 25W CFL consumes 50VA) and distorts the AC sine waveform increasing the total harmonic distortion (THD) as well as radiating harmonics that can cause electrical interference (RFI) problems. In *TT*, February 1990, p31, it was noted that, for example, when a number of high-power switch-mode PSUs are used, the load comprises short-duration, high peak-currents and that this can result in distorted AC duration pulses near the instants of peak supply tending to slice-off (clip) the peaks of what should be a near sine waveform. Other appliances that can contribute to the distortion of the AC sine wave include computer, microwave ovens, fax machines and answering machines.

The contribution of CFLs to THD is probably relatively minor. But, for radio amateurs, their generation of harmonics is recognised RFI problems. As Nils Borg puts it: "Electronics generate harmonic currents. There are basically two problems with harmonics. First, they are closely related to a reduced power factor, and second, harmonic currents can disturb

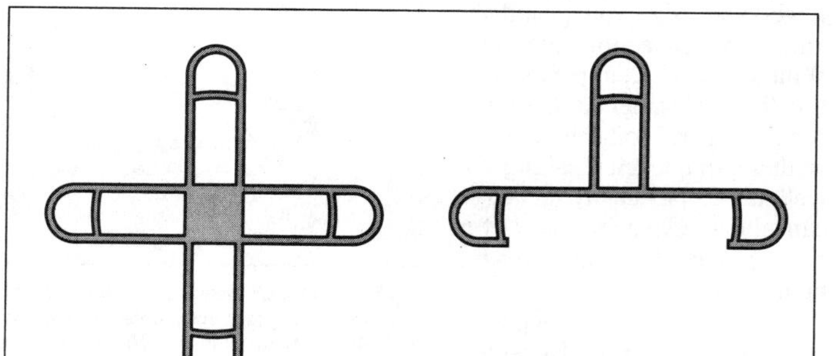

Fig 10: The Spanish-made 'spacers for wall and floor tiles' (Rubi Ref No 02905) as modified by EA1TH/G3URY to form spacers for twin-wire, open-line feeders. (a) Devices as supplied in packets of 50; (b) a few snips converts them into three-arm spacers held in place by a cord attached to the third arm; (c) Further snips result in a two-arm spacer; (d) Further plastic removed and wires held in place by 'thermal glue'.

other equipment. Very high harmonic distortion on a utility network will also reduce the efficacy of the power transformers and could ultimately damage them. An electronic CFL has a typical power factor of between 0.5 and 0.6 but the current cannot be compensated for with a capacitor. Instead, a filter has to be introduced, either in the lamp itself (in the ballast) or somewhere in the electricity networks.

"In countries where the government has adopted the IEC standards, lighting equipment must have a PF better than 0.96 and total harmonic distortion (THD) below 33 per cent. However, the IEC lighting standards make an exception for equipment with a rated power of less than 25W, such as screw-base CFLs. This is one reason why so much of the power quality debate has been focused on CFLs, although even worse harmonic hooligans, such the PSUs of TVs and computers, probably deserve much more attention... Experience gained in Swedish and US technology procurement programmes show that improvements in both the price and performance of products could be achieved if enough dedicated buyers get together to put pressure on manufacturers. Meanwhile, the poor power factor of CWLs does not seem to be a serious enough disadvantage to warrant that they should not be used."

It looks as though amateurs suffering interference from CFLs will have to put up with it for some time to come. The problem with international and national standards is that they always seem to have let-out clauses! Perhaps we should not complain since the European EMC directive has several useful get-outs for amateurs!

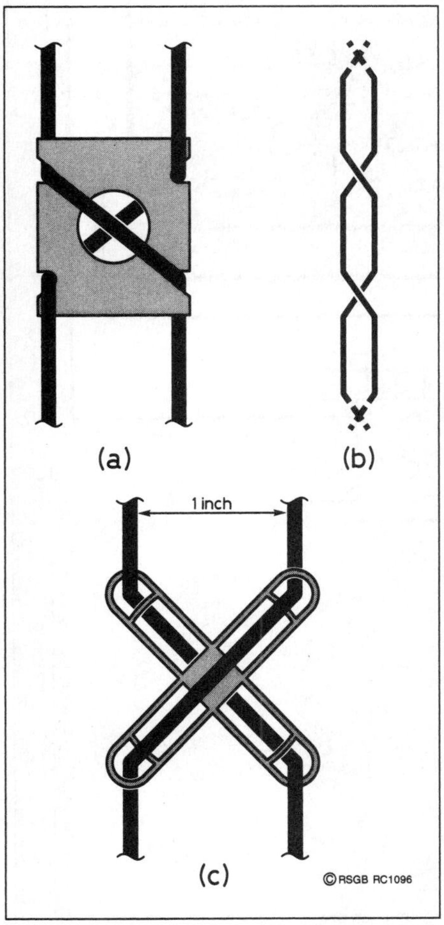

Fig 11: (a) Typical 'transposition blocks' marketed in the 1930s for improving the balance of twin-wire transmission lines, reducing pick-up or radiation from open-wire lines. (b) Length of twin-line with transposition blocks. (c) The Spanish spacers can readily be used as transposition blocks.

LOW-COST OPEN-WIRE SPREADERS/TRANSPOSERS

NOTICING THE New Zealand clothes-peg plastic spreaders for open-wire transmission lines in May 1996 *TT*, Bob S Wilde, EA1TH (ex-G3URY), was moved to send along what he feels is a noteworthy Spanish version. He writes: "Please find enclosed examples of this wonderfully shaped piece of plastic in various stages of creative destruction (**Fig 10**). The shape is ideal without modification for four-wire feeders [see also my comments below - *G3VA*].

"But it is also simple with a pair of snippers to remove one or two of the arms completely and also reduce the amount of plastic between the wires while still maintaining sufficient strength. A three-armed version allows a chord to be threaded through the 'top' arms thus supporting the feeder and controlling its potential unruliness.

"I use thermal 'glue' to hold the wires in place. Both materials, the plastic spacer and the 'glue', passed the microwave oven test with flying colours. The thermal 'glue' in question looks like soft polythene and is also excellent for 'ugly' construction... especially for such purposes as making A(S)TUs in plastic boxes, a favourite project here as I operate 'permanent portable' from a patch of ground up the hill from any neighbours.

"Having made two ZS6BKW/G0GSF multiband dipoles using the spreaders and threading the feed on to two lots of 12m long wires, I came to the conclusion that a few additional snips, together with the flexible nature of the plastic, allow the 'spreaders' to be clipped on the wires as with the ZL clothes pegs rather than threaded on. Then a squirt of the thermal 'glue' into the gap and in 30 seconds it is already hardening.

"The real test will be how long these materials will withstand the UV radiation from the strong Spanish sunshine. I can say with confidence 'at least two years'. Possibly a dab of black paint [white paint might give better protection against UV - *G3VA*] over the glued parts would help...

"So what are these plastic devices really intended for? They are used for laying tiles or paving stones, acting as spacers at the four adjacent corners, thus allowing accurate alignment before cementing or grouting. Maybe somebody will know what they are called in English and whether they can be found in UK DIY shops."

EA1TH enclosed the label of a pack of these devices made in Spain by 'Rubi'. As the label is printed in both Spanish and English it seemed likely that they could be found in the UK. The Spanish name is 'Crucetas 10mm' and in English: 'Spacers 10mm for wall and floor tiles' (the 10mm is presumably the tile thickness). The label bears a Reference Number 02905, with 50 units in a package.

Armed with this information, I popped along to the nearest local DIY store store but drew a blank. However, the helpful assistant suggested trying a specialist shop directly across the road: Tile Kingdom (124 Oakfield Road, Penge, London SE20, Tel: 0181 778 6447). The reference number was phoned through to their central stores, and it was confirmed that the 50 unit packages could be ordered through them at a retail price of £2.81. Tile Kingdom PLC have a number of branch outlets in the London area and possibly elsewhere. There should thus be little difficulty in obtaining these useful low-cost spacers.

Bob Wilde mentioned their unmodified use for four wire transmission lines (quite widely used in HF broadcasting) but looking at these devices, I was also reminded of the 'transposers' that were marketed by Eddystone in the 1930s. In those days, before coaxial cable became widely available, either twisted electric flex (poor HF insulation) or open-wire balanced line was widely used by amateurs and SWLs both for transmission and reception. A mid-1930s edition of 'A Guide to Amateur Radio' in a section on transmission lines noted: "Such lines are now very popular for conveying power from the transmitter to a radiating system, which, for practical reasons, must be erected at some distance from the transmitting apparatus. They may consist of a twisted pair of wires as is usual in ordinary 'flexibles' for electric lighting purposes; or a more robust feeder system of two insulated wires run close together may be used. In the latter case it is sometimes arranged that the wires are transposed regularly as shown in **Fig 11**. This ensures capacitance balance to earth, but the capacitance per unit length of the line in such cases must vary with position as the wires vary in their distance apart."

Where open-wire lines are used as resonant lines, the variations in characteristic impedance (**Fig 12**) caused by the transpositions would not be important, and the better balance to earth would, presumably, reduce radiation from and pick-up on the lines and might be a useful RFI precaution. Conventional open-wire lines are noted for some degree of radiation. Post-war with the availability of coaxial cable, transposers disappeared from the marketplace. It is a technique that might prove worth reviving, made easy by the Spanish crucetas! ♦

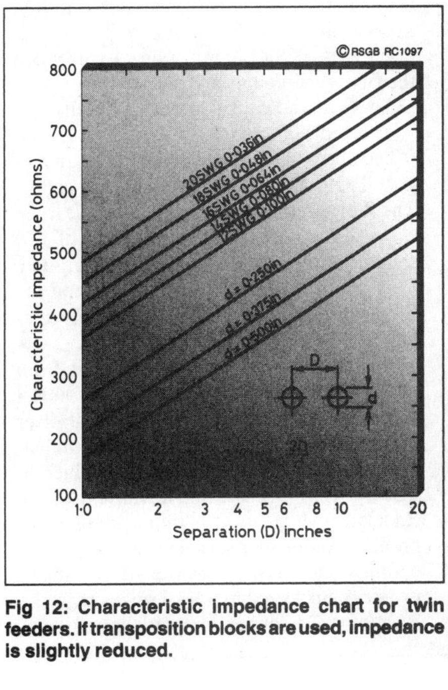

Fig 12: Characteristic impedance chart for twin feeders. If transposition blocks are used, impedance is slightly reduced.

Pat Hawker's Technical Topics

PAT HAWKER, G3VA
London 37/SE22 8SS

LONG LIVE HOME CONSTRUCTION

THE NOTES ACCOMPANYING the recent RSGB *Future of Amateur Radio Opinion Survey* stated boldly that "A high standard of technical knowledge, and interest in experimenting in radio communications, are essential to justify retaining our existing privileges. While there may be various classes of amateur licence, and many specialist interests in different aspects of radiocommunications, the nature of their operation must justify the privileges used. The future of the Amateur Service can only be ensured if high standards of technical knowledge and operating are maintained, and its activities are not seen to compete with other commercial radio services."

Nonetheless, it is also self-evident that genuine experimentation must include the learning of practical as well as theoretical knowledge of radio technology. This is surely an inherent part of the 'self-training' criterion of the ITU Radio Regulations. In other words, while we should not erect over-high barriers to obtaining a licence (Novice , class B, or full) we should use our licence, once obtained, to improve and extend both theoretical and practical knowledge of modern radio communications technology.

The Wireless Institute of Australia (WIA) has recently submitted a 40-page 'Exposure Draft' submission to all candidates standing for the federal election for their House of Representatives and half the Senate and to all sitting members of the Senate; some 500 copies in total were distributed. *Inter alia*, the submission (published as a supplement to *Amateur Radio,* April 1996) highlights key areas in which it was argued that the Amateur Radio Services are of value to the community:

(1) Education and self-training; (2) motivating young people to take up scientific or technological careers; (3) contribution to the advancement of scientific knowledge; (4) providing communications during emergencies; and (5) supporting community events.

Under self-training the submission stresses: "*Right from the outset, even before a licence can be obtained, amateur radio necessarily involves self-training.* Having obtained a licence, radio amateurs will *learn for themselves,* in a practical way, about the techniques and technologies. They have the comparative freedom and the flexibility to pursue an enormous range of radiocommunication techniques and technologies, ranging from: medium frequency transmission, propagation and recep-

HUFF & PUFF IN PRACTICE

CHARLES FLETCHER, G3DXZ, was interested in the June *TT* item on PA0KSB's improved huff and puff stabiliser, and decided to give one a try. He writes: "As a constructor over many years of L/C VFOs, like most other constructors I have been troubled by the variability in stability of virtually identical oscillators. PA0KSB's circuit looked like a miracle solution. So I built a purposely poor stability oscillator using ceramic plate capacitors in order to test the huff and puff stabiliser. I achieved the following results: over an hour's run of a 5MHz VFO with the stabiliser the measured drift was less than 2Hz. The drift without the stabiliser was 750Hz.

"Once an original design has been published, it is relatively easy to make minor improvements and/or simplifications. **Fig 1** shows my slightly simpler version, together with measured circuit parameters, of this super design. I feel that this form of stabiliser should become a routine addition to L/C oscillator designs which need high stability. I can't think why this outstanding contribution by PA0KSB did not appeal to me when the first version appeared in 1973!"

Several readers have pointed out that the "corrected" diagram of PA0KSB's circuit ("Technical Correspondence" RadCom August 1996, p80), still repeats some errors. The PIN numbers for IC1 should be the same as for IC2 and not as shown. PIN3 of IC4 should be connected to +5V and not to ground. Readers still having problems can send me (QTHR) a large SASE for further information.

Without in any way taking credit from PA0KSB, I recall that in the late 1960s I wrote up in *Electronics Weekly* a stabiliser intended for use with existing communications receivers developed (and explained to me) by Keith Thrower of Racal (called the Racaltor) which used a rather similar approach. As by then the new generation of communication receivers were beginning to use frequency synthesisers, I suspect that the Racalator did not become widely used.

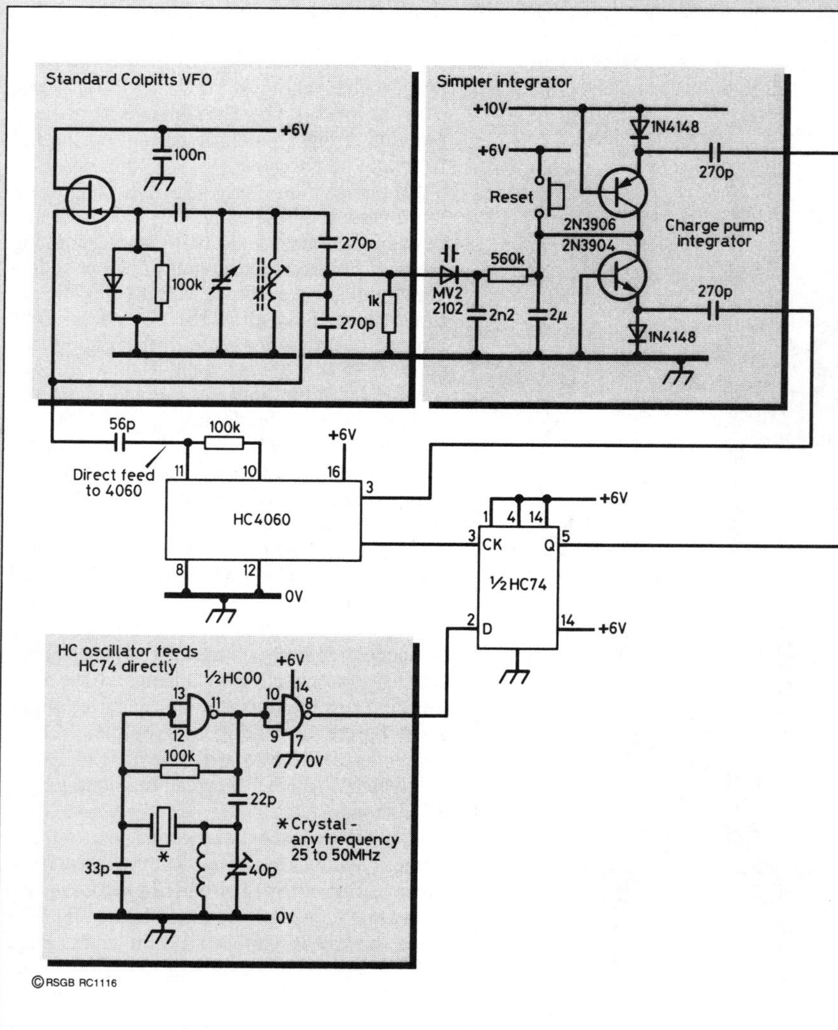

Fig 1: G3DXZ's simplified 'Huff and Puff' VFO stabiliser. Notes: (1) Using a 5MHz oscillator and 4060 outputs Q11 and Q13 gives a reference frequency of about 300Hz at B. (2) The integrator slew rate is about 0-5V/sec. (3) Varicap pulls oscillator frequency 2.5kHz/volt at 6V bias. (4) Integrator pump noise about 2mV p-p which is equivalent to 1.25Hz p-p ripple on VFO. These figures make good design centre values for experimentation.

tion, through high frequency (short wave) transmission, propagation and reception, earth satellite communications techniques and digital communications technologies, television transmission and reception to microwave technologies and communications techniques . . . People who have an interest in technology and gain an understanding of it through self-training are better equipped to adapt to and benefit from an increasingly technological world which is undergoing continuous and rapid change."

In the same issue of *Amateur Radio* Peter Parker, VK1PK, in his 'Novice Notes' presents some sound advice on homebrewing. He notes that "A longing to build a small transmitter, receiver or piece of test equipment is commonly expressed by many amateurs. All too often, however, the longing remains just that, due to perceived difficulties in obtaining components, a lack of test equipment, or not having a suitable circuit diagram."

VK1PH shows how these difficulties can be overcome, and stresses that satisfaction of successfully completing a project is immense. He suggests that most projects require only basic hand tools, listing a suggested priority order for tools and test equipment that will be found useful: (a) 15 - 25 watt soldering iron with fine tip; (b) benchtop power supply (to suit projects under construction); (c) multimeter (analogue or digital); (d) general coverage receiver; (e) various relative RF indicators such as field strength meters and absorption wavemeters [or just torch bulbs - *G3VA*]; (f) dip oscillator; (g) RF power meter and 50Ω dummy load; (h) frequency counter; (i) inductance and capacitance meter(s); (j) RF signal generator; (k) oscilloscope. All items except the oscilloscope and possibly the general coverage receiver can be built or assembled at home. The receiver can be an inexpensive short-wave broadcast receiver that can be used to give a rough indication of the frequency of VFO, verify the operation of crystal oscillator stages, or even act as a crude signal generator (local oscillator normally 455kHz below (or above) the dial frequency).

VK1PH refutes the perceived scarcity of components, suggesting that homebrewing has never been cheaper or easier than it is today, thanks to our throw-away society and the rapid obsolescence of much electronic equipment. Components not readily available from standard retail outlets can often be acquired at hamfests, rallies and junk sales. Dismantling old radios and TV sets may be time-consuming but the use of salvaged components can reduce construction costs: "Rather than remove all parts from printed circuit boards and placing them in jars, it may be preferable to keep the boards intact, and remove components as and when required. Valve radios are particularly desirable for the tuning capacitors and power transformers which can often be rewound as required or used as a source of wire for antennas. Second-hand shops and even rubbish dumps may be worth visiting. Items such as multimeters and tools can often be obtained second-hand. The homebrewer need not be bound by the old rule *you get what you pay for* - much can be obtained at zero or very low cost!

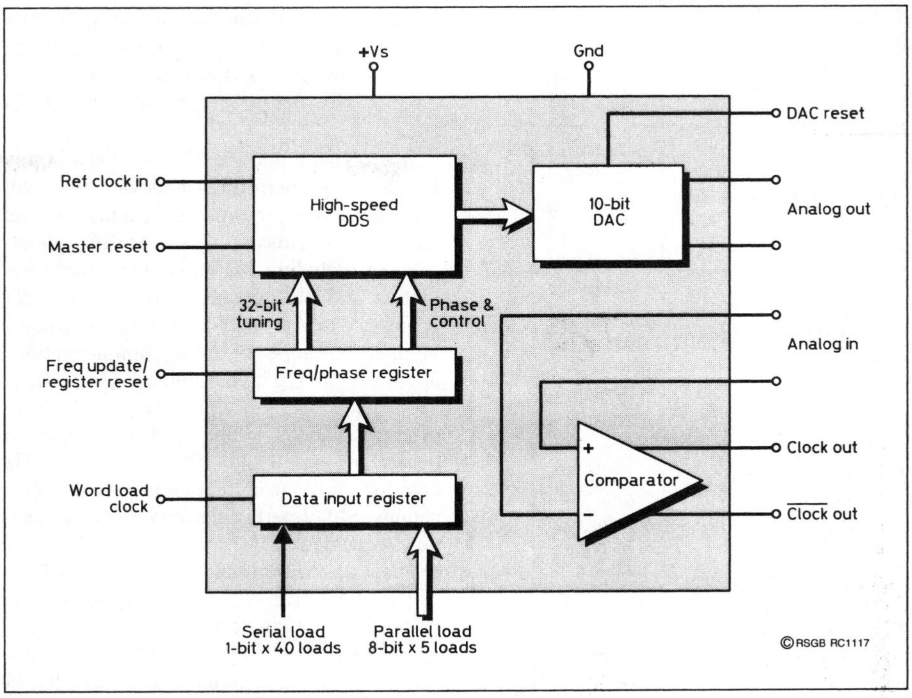

Fig 2: Functional diagram of the Analog Devices AD9850 chip.

It is surprising to what uses various (discarded) household items can be put!"

While factory-built equipment and kits normally use printed circuit board (PCB) or now, increasingly, surface mount technology, it has often been emphasised in *TT* and elsewhere that other forms of construction may be better and easier than PCBs for one-off projects. Among the methods noted by VK1PH is to use an etched PCB but solder components onto the copper side of the board, obviating the need to drill holes and making it easier to make changes: "For smaller projects that do not use ICs, the use of small pieces of adhesive tape placed on the parts of the board where you want the copper to remain is suggested as a simple alternative to conventional methods when etching a board."

Again, the 'paddyboard' form of construction is also suggested as suited to projects for which the ultimate in miniaturisation is not required. This uses PC board material but with no etching. Component leads not connected to the copper surface are soldered to small insulated pads, made from spare PC board material. The pads may be soldered to the main board. Here also, it is very easy to modify circuit layout and add extra components. Even ICs can be mounted on small pieces of Vero or matrix boards beforehand. The use of high-value resistors (several megohms) as standoff insulators, soldered to the main board, has worked well for VK1PK. Matrix boards also work well for RF projects well into the VHF region and are usually available from component suppliers.

A refinement is the Veroboard with a series of parallel copper strips which can be cut as required by using a drill bit held in the hand. But note that the stray capacitance between the parallel strips may impair performance for RF projects, and is more suited to power supply and audio projects. Veroboard, however, can be converted into a matrix board by immersing it in a bath of PCB etchant solution. A further increasingly well-known and very effective method is the so-called 'ugly' construction in which the devices are stuck upside down on PCB using wiring for interconnections and with the PCB copper used as a ground plane; this is capable of providing improved RF performance over a conventional etched PCB.

Relatively simple methods of removing surface mount devices were given in *TT*, February and November 1993 (which also gave an outline of surface mount technology). Alan Keech, G4PPW, has found that provided you have a fairly delicate touch, the following method of desoldering surface mount devices works well: Tools required: small soldering iron, small vice for holding the work; desolder braid; small-diameter Multicore solder; hot air gun (useful but not essential). Method: first place the solder braid over the device pins, then place the iron on top of the braid. This will remove the solder by capillary action from each pin. Then when this has been done to all the pins, place a small strand of wire under the pins in a loop and pull out from under each pin by placing the iron on each pin in turn. You should then be able to remove the device from the board without (hopefully) pulling out the track. This last part can best be done with the aid of a hot air gun if available.

A NEW LOOK IN DDS DEVICES

THE GROWING USE of direct digital synthesis (DDS) rather than the long-established PLL-type frequency synthesis seems to be growing apace, with new high-level-integration devices making it easier to implement high-performance DDS for receivers and transceivers. In the July 1996 issue of *Electron,* Rob van Dijck, PA0DCK, and Rob van Veen, PA3EPZ, present the first part of a two-part article 'DDS ook voor zelfbouw?' based around the established

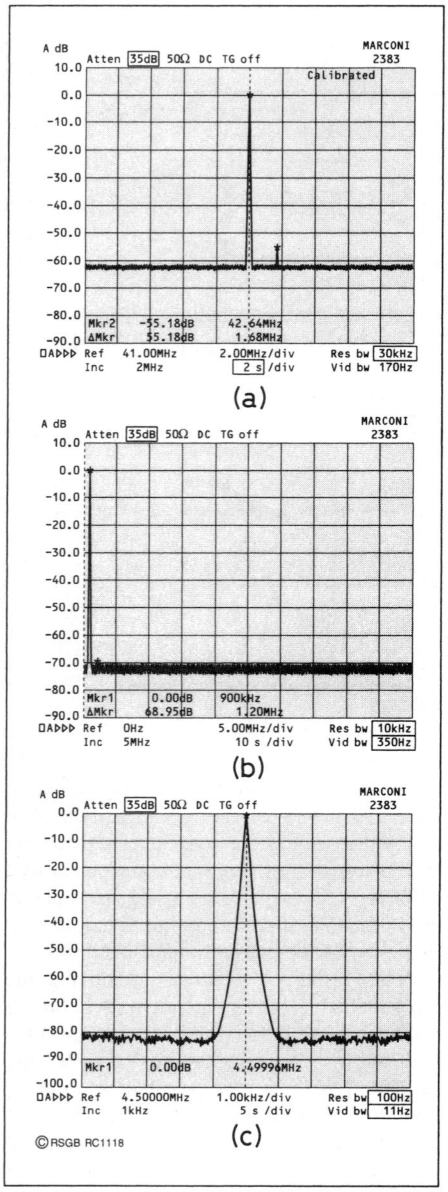

Fig 3: (a) Output spurious-free dynamic range (SFDR) plot of the AD9850 at 41MHz. (b) Output SFDR plot at 15MHz. (c) 4.5MHz output viewed narrowband (1KHz/Division).

Analog Devices AD7008JP50 which includes 32-bit serial and parallel registers and a 10-bit DAC.

A later but related DDS device - first of a new family of DDS devices -is described by Dave Crook and Jim Surber in *RF Design*, April 1996) both of Analog Devices in a feature article 'High level integration and performance comes to DDS'. This describes the AD9850 which is claimed to be "the world's first complete monolithic 125MHz high-performance DDS system. This innovative CMOS device integrates a proprietary high-speed 32-bit DDS core with a performance-optimised 10-bit digital-to-analog converter and high-speed, low jitter comparator." It is claimed that this device (**Fig 2**): "shatters the barriers associated with existing DDS and DAC synthesiser solutions by providing major advancements in virtually every feature and specification category: level of integration, power dissipation, size, speed, dynamic performance, ease of implementation and price." Even if these high-flown plaudits have to be slightly discounted as company hype, the authors do make a strong case for this new family of devices with the possibility of 12-bit DACs soon.

It is claimed that the "AD9850 is a highly flexible, high-performance device that will serve many of the traditional DDS markets and applications. However, its breakthrough attributes, including size, price, functional integration and performance makes it uniquely positioned to bring DDS technology into high-volume, consumer-oriented communications applications. The two foremost applications are: (1) low-distortion, frequency-agile synthesiser and (2) frequency/phase agile clock generator. Output bandwidth is given as 42MHz appropriate for many communications applications. SFDR performance is shown at various frequencies: **Fig 3** shows the output measurement the AD9850 at 41, 15 and 4.5MHz.

The alternative PLL form of synthesiser is described in 'Low-noise phase lock of high spectral purity 100MHz VXCO to a high stability 10MHz reference oscillator' by G Santarelli *et al* in work supported by the French Bureau National de Metrologie (*Electronics Letters*, 9 May, 1996, pp866 - 867). The work is aimed at the realisation of a microwave synthesiser suitable for use as a local oscillator for high performance atomic frequency standards.

However, a point of more general interest is that the report includes: "Recently, a new family of digital frequency-dividers has become commercially available (SP8400 GEC-Plessey). The manufacturer specifies a low phase noise which is potentially low enough to realise the new phase-lock circuit."

TOWARDS THE SUPER-LINEAR RECEIVER

THE DRIVE BY Colin Horabin, G3SBI, and his American collaborators continues, with a considerable amount of information being passed to *TT*, although space-limitations means that some of this has to be delayed and/or shortened.

TT, December 1995, pp70 - 72, provided detailed information on a high-gain, low-noise state of the art 4.43MHz IF amplifier using the Analog Devices AD600 dual, low-noise, wideband, variable-gain amplifier chips. This was developed last year in prototype form by Bill Carver, K6OLG/7. Subsequently, K6OLG has further developed this state-of-art design as a complete AGC/IF subsystem. A detailed write-up by K6OLG can be found in *QST*, May 1996, pp 39 - 44 including the circuit diagram occupying two full pages. Prior to publication, K6OLG passed his article to G3SBI, and I had hoped to include at least the circuit diagram in *TT*, but in view of its availability in *QST*, it would be difficult to justify using up so much of the limited space available for *TT*.

However, the specification of the subsystem provides a useful summary of what can now be achieved. It has a minimum discernible signal (MDS) level of 0.03μV in a 2.5kHz bandwidth, and its AGC can be set to have only a few decibels rise in audio with a range of signal

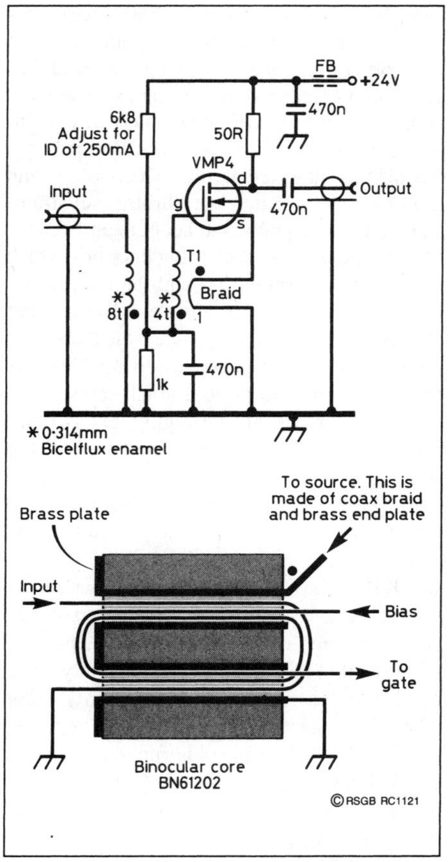

Fig 4: The basic low-noise, high-intercept-point amplifier with gate-source feedback and using the discontinued Siliconix VMP4 power FET as implemented by W4ZCB. To minimise leakage inductance, the single turn braid must use brass end plates so the braid can be opened up and pushed into contact with the sides of the hole in the binocular core. However, reasonable results can be obtained with just the braided turn with enamel inside and no end plates.

amplitudes of less than 0.1μV to over 0.2V. The input-intercept point is about +20dBm, so in-band intermodulation distortion (IMD) is 40dB down even for S9 + 70dB signals. The constant- gain time interval of the 'hang' AGC circuit is smoothly and continuously varied from 100ms to 2 seconds using a panel-mounted knob. The system is intended for use with a front end having a net gain of + 3dB and a 6dB noise figure (NF) and directly drives a +7dBm diode -mixer product detector. As the *QST* subtitle puts it: "If you're looking for exceptional receiver performance, you need an exceptional IF and AGC system - like this one."

G3SBI has been doing some important work on low-noise high-intercept amplifiers using RF power FETs with gate-source feedback, a type of feedback configuration analysed by Wes Hayward. W7ZOI, in his book *An Introduction to RF Design* available from the RSGB (see *Book Shop* on page 90 for details).

Some months ago, Harold Johnson, W4ZCB, a friend of and collaborator with K6OLG sent G3SBI a version of a high-performance amplifier originally developed a few years ago by K6OLG (**Fig 4**). This used the now obsolete (but excellent) siliconix VMP4 FET and a Fairite BN 61 202 binocular core. G3SBI tested this amplifier at the CCLRC (formerly SERC) Daresbury Laboratories but ran into difficul-

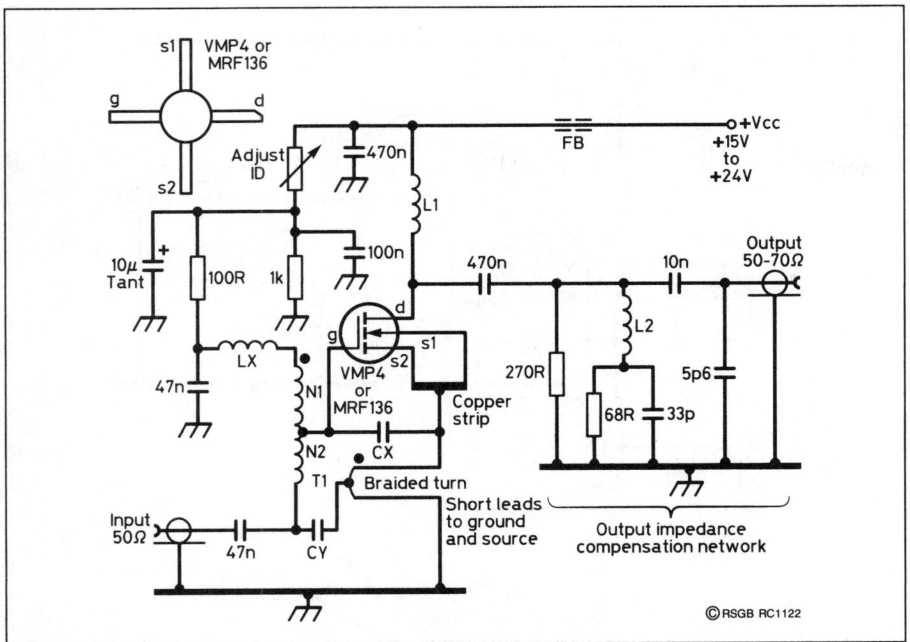

Fig 5: G3SIB's improved amplifier for use with VMP4 or MRF136. L1 55 turns 0.2mm enamel wire close wound on Micrometals T50-2 (toroid almost filled). L2 6 turns of 0.4mm enamel wire close wound (don't spread) on T37-6 toroid. Both these cores are available from Cirkit Components.

For the VMP4, Vcc is 22-24V, drain current 0.5A or 0.7A. Lx is not fitted and N1 start connects directly to the 0.047μF bypass capacitor. CY is not fitted. For N1=4T, N2=3T and BN61-202 core. Cx=56pF. With this arrangement gain is 11.4dB and the input and output impedances are 50Ω resistive ±2-deg to 60MHz (this has proved repeatable for two of the amplifiers built in the USA). IP3 out +59dBm.

For the MRF136,LX is 7t of 0.2mm enamelled copper wire wound using some 18SWG wire as a former and then removed. Cx is not fitted. Cy is 68pF.T1 is a BN43-202 core wound as in Fig 6. Gain 11.7dB with N1 2t and N2 5t, or 12.7dB with N1 3t and N2 5t. With Cy fitted, the input impedance is substantially flat to 100MHz. Drain current should be 0.3A if used as a post-mixer amplifier. IP3 out approximately +47dBm. For a pre-mixer amplifier drain current should be 0.5A to 0.7A for IP3 out of +56dBm at 14MHz. Unfortunately, Cy needs to be used to provide a purely resistive input impedance but it degrades IP3. Without Cy an IP3 out of +62dBm has been measured at 14MHz.

The VMP4 offers a much wider bandwidth with an IP3 of +50dBm measured on 50MHz compared with +40dBm for the MRF136 with reduced gain.

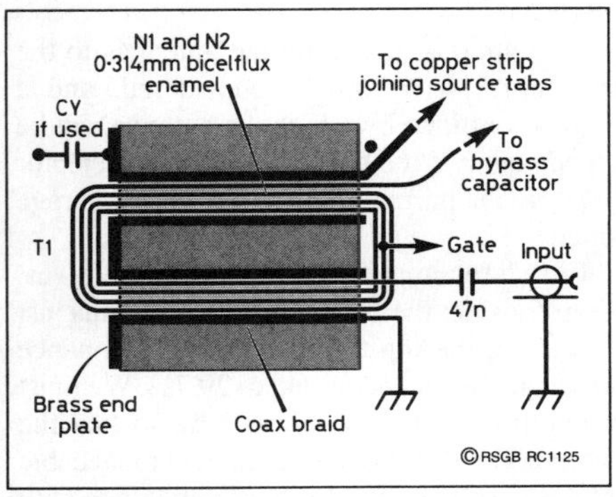

Fig 6: Detail of T1 winding.

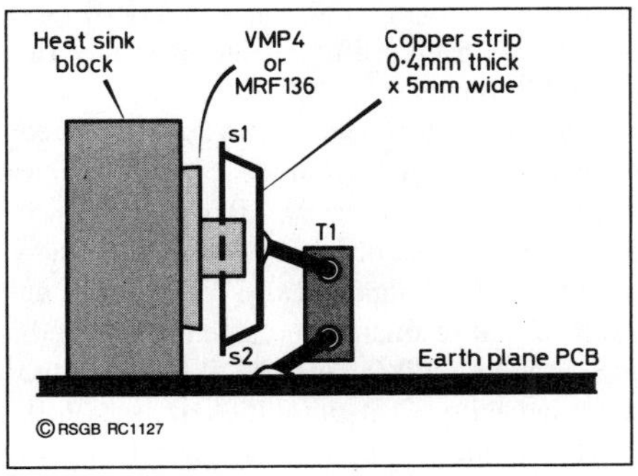

Fig 8: Basic construction: keep leads short around the transformer, particularly the 'source' winding.

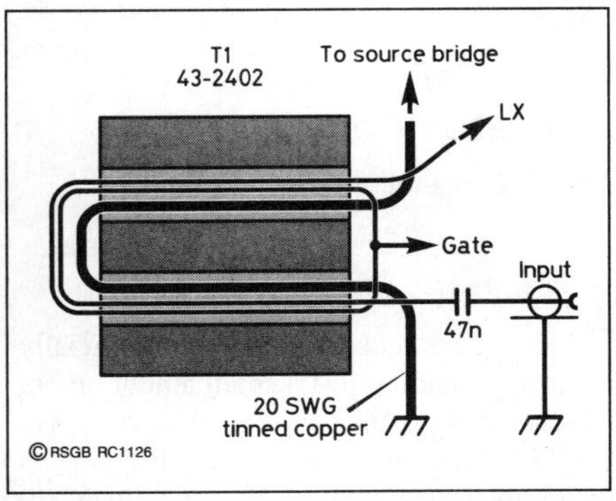

Fig 7: The smaller 43-2402 binocular core can be used for T1 with good results . A piece of 20SWG tinned wire is bent into a hairpin and pushed into the core. N1 and N2 are wound with 0.2mm Bicelflux enamel wire. This will be a very tight fit but provided that the 20SWG single turn has no kinks, it can be done.

ties, and temporarily dismissed this approach.

The following notes are based directly on notes supplied by G3SBI:

A super-linear receiver could possibly comprise a G3SBI H-mode Mixer; roofing filters; post-mixer amplifier; quality filters; followed by the K6OLG IF/AGC sub-system. The post-mixer amplifier would need to have a gain of about 12dB and a sufficiently low noise figure to handle the insertion loss of a high-quality 250Hz CW filter (9dB), as well as an input intercept of +35dBm which is rather higher than the in-band intercept of hybrid-connected SSB roofing filters.

So I looked again at the W4ZCB amplifier design but using an MRF136 RF power FET in my test circuit to compare its performance with the VMP4. It soon became clear that there must be important parametric differences between the two devices. The W4ZCB test circuit used just one FET source-tab to connect to the transformer primary. When this was done with the MRF136, the bandwidth performance was way down. *The test circuit was rebuilt with a copper strip bridge between the two source tabs and a printed circuit ground plane was stood at right angles to the device and used for everything earthy.* With this, the MRF136 would then perform reasonably satisfactorily.

At this stage, G3SBI blew up the VMP4 and for a time had to work solely with the MRF136. Later, however, as a result of an appeal on the Internet a number of VMP4 devices were kindly donated by David Stockton, Geoff Waters, GM4OSW and Paul Beastall, G1WBZ.

With the MRF136, various turns ratios and core materials were tried for the feedback transformers together with ways of making the input impedance look purely resistive. Performance figures with a BN43 202 core included: 14MHz, 50Ω input and output impedance 50Ω, gain 11.7dB, bandwidth 100MHz with no gate source capacitor, output intercept +56dBm at 0.5A drain current and +62dBm at 0.7A, noise figure 1.5dB, reverse isolation 45dB. from these figures one might suppose that this represented a super amplifier, but unfortunately there remain problems. The input impedance has a rising Jx inductive reactance term; this could be changed by the use of a series capacitor (or by other means) to a true input resistance at any one frequency or over 50MHz. However, if this is done, the output intercept is limited to about +51dBm the reason being that the feedback amplifier must see a broadband 50Ω source impedance for the feedback to work correctly.

At this stage, the donated FETs became available so that the comparison between the discontinued Siliconix VMP4 and MRF136 devices could be resumed. With the circuit arrangement shown in **Figs 5 - 8** with the VMP4 using four feedback turns and an additional three turns for the input match, with a single 56pF gate-to-source, the input impedance is, within a couple of ohms, purely resistance to 70MHz, the bandwidth over 100MHz and the IP3 out+58dBm. The output impedance, with the correction network fitted, is flat and purely resistive to 50MHz.

Unlike the MRF136, a 22V supply was necessary to depress fully the parametric capacitances, but it is clear that the discontinued Siliconix VMP4 device had an edge in a number of respects over the MRF136. It is a pity that the firm prematurely dropped out of RF power FETs. The VMP4 was a true V-MOS device whereas the MRF136 uses the planar process. No other firm now second sources the VMP4 which has about 12 source bonds from the die to the source tabs compared to only two on the current MRF136, considerably reducing the internal inductance, very important to the performance of a source-gate feedback amplifier.

While the MRF136 is usable in this application it is not ideal. G3SBI and his American friends, K6OLG and W4ZCB, now have enough VMP4 FETS for use in their own equipment, but they are anxious to develop designs using currently available devices. There is a faint hope that Macom (who took over siliconix) may be persuaded to produce the VMP4 again but they have discovered that the DU2810S FET from the Phi Division of Macom, despite having only two die source bonds but with a higher FT than the MRF136, provides a good compromise for this type of amplifier.

The DU2810S N-channel RF power mosfet uses U-groove technology developed by Dr Terry Heng for Westinghouse. In many ways the DU2810S is similar to the VMP4. Tests with a slightly modified feedback amplifier (**Fig 9**) have shown it can give an output IP3 of +53dBm up to 70MHz with 0.5A drain current, with gains of from 9 to 12.7dB (1,2,3 or 4 feedback turns). Unfortunately its gm (and hence IP3) peaks at about 0.55A drain current; if the device geometry had a gm peaking at 1A it would have outperformed the VMP4 in this application. From an input impedance point of view, the best amplifier with this device uses four feedback turns and is purely resistive to nearly 90MHz. It could be used effectively as either a pre-or post-mixer amplifier. As a pre-mixer amplifier the gain could usefully be reduced to about 9dB (1 feedback turn).

As a follow up to the work with the VMP4 amplifier, G3SBI has since driven it via a 3-pole ladder bandpass filter at 15MHz with an IP3 at 20kHz of +54dBm with no serious degradation when moving down the slope of the filter. It would thus seem that only the reactance present at the fundamental is important to the linearity of the feedback amplifier; the same is also true of G3SIB's H-mode mixer, considerably simplifying the input bandpass networks which will then also take out the 3dB image noise from the mixer output.

G3SBI recognises that some readers may have difficulty in obtaining the BN 202 binocular core. He therefore also tested the slightly smaller 43-2402 Fairite binocular core which is obtainable from Cirkit: see Fig 7. While this showed distinct signs of saturation on 1.8MHz with 0.5A drain current, nevertheless the IP3 out was about +48dBm. At 7MHz and above it was greater than +54dBm.

Finally, on a different topic, G3SBI reports finding that the use of ceramic trimmers in high performance oscillators (eg as a trimmer across a crystal) can significantly increase oscillator phase noise compared with air-dielectric or Sufflex-type capacitors.

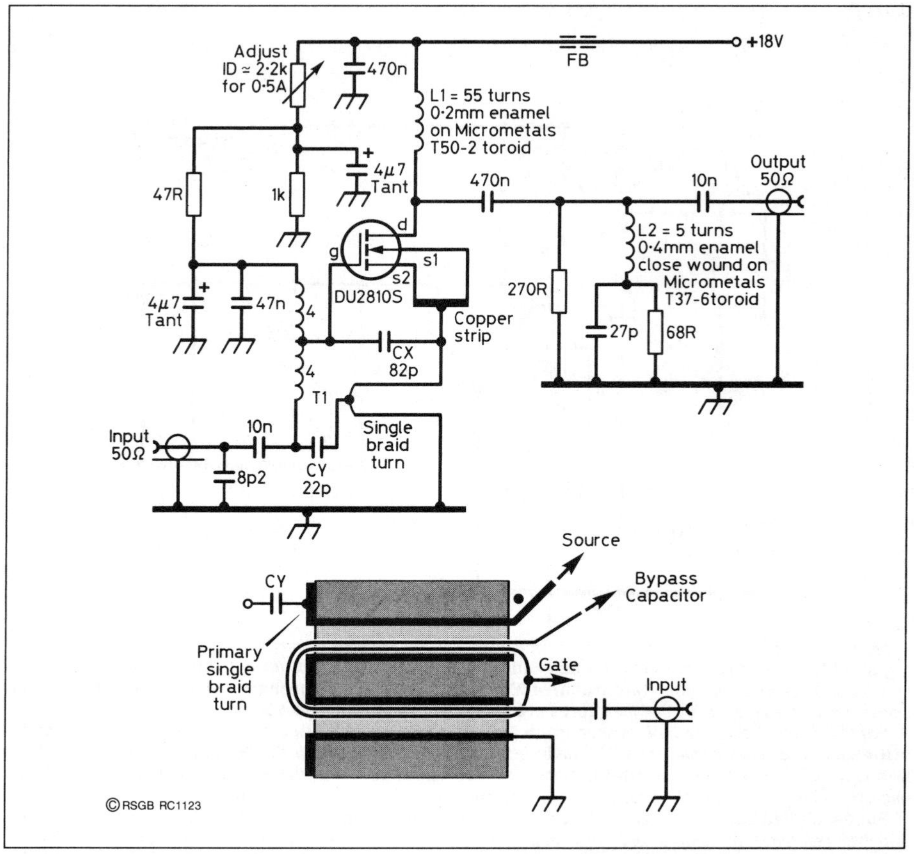

Fig 9: Low-noise, high-intercept amplifier using the PHI (Macom) DU2810S power FET. T1 uses Fairite BN61-202 binocular core with a primary winding one turn of braid. Each secondary winding is inside this using 4t feedback and 4t input match using 0.314mm Bicelflux enamel. Gain about 12.7dB, IP3 out about +53dBm, NF about 1.5dB, Z_{in} and Z_{out} approximately 50Ω. Construction is similar to the VMP4/MRF136 amplifier.

REDUCING SWR ON CENTRE-LOADED MOBILE WHIPS

DR JOHN (JACK) Belrose, VE2CV, whose experience of mobile HF antennas now covers some 50 years and whose first article on the subject appeared in a 1953 *QST,* recently purchased a Pro-Am 3.5MHz mobile antenna. He was reminded of the review of this series of antennas by RSGB HQ staff in the August 1995 *RadCom*, pages 44 and 80.

On re-reading the text he noted the comment "additional information on how to use shunt capacitor matching to obtain a lower SWR than shown in Fig 1 (PHF-20) and Fig 2 (PHF-80) would have been useful." [Figures in the August 1995 issue].

VE2CV writes: "That is easy. I have copied the pertinent page from my 1953 *QST* article [see below]. Fig 2 of the RSGB review gave an SWR at resonance of 1.9. I measured my pro-Am PHF-80 and obtained an SWR of 2. For an SWR of 2 at resonance the antenna's impedance is about 25Ω. To match 25Ω to 50Ω, the shunt capacitor C at say 3750kHz is 849pF. For the required series inductance lengthen the whip just a bit, looking for resonance. I put an 820pF silver mica capacitor (20% tolerance) across the base of my Pro Am and the SWR was as close to unity as I could measure."

Extract from *QST,* September 1953 p34, section on *'Tuning and Matching Centre-loaded Whips':*

A method of matching the antennas to the transmitter will now be considered, and a method outlined by which the antenna can be tuned using a GDO and an SWR detector (equipment normally owned by the average amateur).

First, let us consider what matching involves. If we consider the 16ft [3.81MHz] antenna just discussed, the input impedance at resonance was a pure resistance equal to 29.7Ω. We must transform this low value to 50Ω so that the antenna can be fed with standard coaxial cable (RG-8/U or RG-58/U). This can be done with an L-section matching unit as shown in **Fig 10.**

Here,

$$2\pi f L = \sqrt{(R_1(R_2 - R_1))}$$

$$1/(2\pi f C) = R_2\sqrt{(R_1/(R_2 - R_1))}$$

Substituting values,

$$2\pi f L = 24.5\Omega$$

$$1/(2\pi f C) = 60.6\Omega \text{ or } C = 690\text{pF}$$

Now the inductance, L, can be artificially obtained by adding just a small amount more inductance to Lo, the centre-loading coil, than needed for resonance, thus making the antenna input impedance inductive. An exact match to 50Ω can be obtained by adjusting the capacitor, C, and the centre-loading inductance, Lo.

The formula given by VE2CV is also applicable to the general design of L/C single-frequency impedance transformation as used for ASTUs etc. ♦

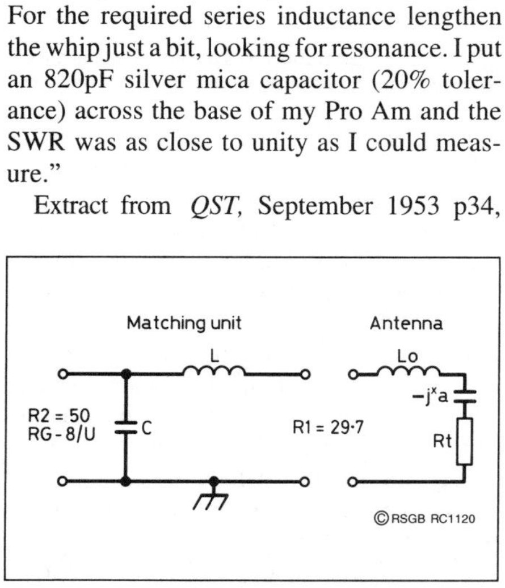

Fig 10: Example of the use of a L-section network to match a 50Ω line to the 29.7Ω impedance of an inductively-loaded mobile antenna. See text for the formulae for calculating C and L to reduce the SWR to near unity, as described by VE2CV in 1953.

Pat Hawker's Technical Topics

PAT HAWKER, G3VA
London 37/SE22 8SS

MORE 73KHZ ANTENNA BASICS

THE AUGUST *TT* included some thoughts on meeting the new challenges of 73kHz in respect of receivers, transmitters and transmitting antennas. But little was said about the lesser but still important problem of receiving antennas.

Comments on this subject have been made by G3NOQ who had a youthful interest in receiving ships and their coast stations on the 500kHz (600m) band but subsequently became a professional antenna engineer including work on VLF antennas. He stresses that on 73kHz the external noise from the atmosphere is so high that transmitting and receiving antennas may differ widely.

He writes: "External noise levels, on which a proper design of any receiving system depends, are published in CCIR Report 322 for a large range of conditions of frequency, location, time of day and season. Noise is quantified as the level in dB by which external noise power arising from a perfectly matched short monopole exceeds the kTB noise in the same bandwidth from a passive resistor. Remembering always that the signal power from a short monopole is independent of its height, provided that it can be perfectly matched - although, in practice, with a very (electrically) short monopole, it is impossible to achieve a perfect match.

"In the UK, external noise level at 73kHz is reported as varying between 95dB (winter afternoons) and 125dB (summer nights). The significance is that a receiving system with a noise figure of, say, 90dB will suffice. Systems with lower noise figures will not provide any improvement in signal-to-noise ratio. If we treat the noise figure of a conventional receiver with an FET preamplifier as negligible, a noise figure at 73kHz of 90dB arises with an antenna having 90dB of mismatch loss. It is for this reason that for 73kHz reception small antennas are entirely adequate.

"A distinct advantage arises from using small loop antennas. This is their seemingly magical rejection of locally generated noise when compared with end-fed wire antennas, attributed to the fact that in the near field region the noise is propagating with a wave impedance greater than in free space; in other words the E-field predominates over the H-field. The loop, responding to the H-field, thus discriminates well against local noise. This effect becomes increasingly marked as the distance between the noise source and the loop decreases below about one-tenth of a wavelength (at 73kHz this is about a quarter of a mile compared to about 120ft on 1.8MHz).

"An untuned loop with an area of 1 square metre has a minimum mismatch loss of around 95 - 100dB. If tuned with a Q of 50 - 100, the loss is reduced to 75 - 80dB and it will work very well if fed to an FET source follower and thence to a 50Ω receiver input. Such an arrangement represents an excellent receiving antenna system for 73kHz.

"The problems of transmitting are greater, as the article by Andy Talbot, G4JNT (*RadCom*, August 1996, pp21 & 56) shows very clearly. As he states, his loop (with an efficiency of -65dB) would require 3MW of power to produce 1W ERP. Small loops produce vanishingly low radiation resistance since this varies as the fourth power of the linear dimensions. The efficiency of any practical transmitting loop will thus inevitably be very poor at 73kHz.

"A few years ago, when high-temperature superconductors first appeared, I considered the effect of using such material for an LF loop antenna. The problem that surfaced is that while in theory high efficiency could be obtained, such an antenna, free of DC resistance, would have only a strikingly small bandwidth. G4JNT's loop, if the losses could be magically removed, would have a Q equal to the reactance divided by the radiation resistance. Using his figures the Q would be a thousand million with a bandwidth of 0.00001Hz, implying that it would take at least two hours to turn the carrier on or off!

"This would not be changed by using more or fewer turns on the loop or transformer. The bandwidth could be increased by adding resistance but this reduces the gain. The gain-bandwidth product can be improved only by using a larger loop.

"One application where loop transmitting antennas will work is for short-range underground communications. In this case, two loop antennas couple via their quasi-static magnetic fields. This technique is restricted to ranges of only a few hundred feet, since the coupling varies as the inverse sixth power of distance. Propagation will be better through dry rock than through wet ground.

"For transmitting beyond the near-field region I feel that monopole-type antennas will be more successful than loops. This is because the radiation resistance varies with only the square of the linear dimensions. It should thus be easier to achieve a viable radiation resistance with a practicable size of antenna.

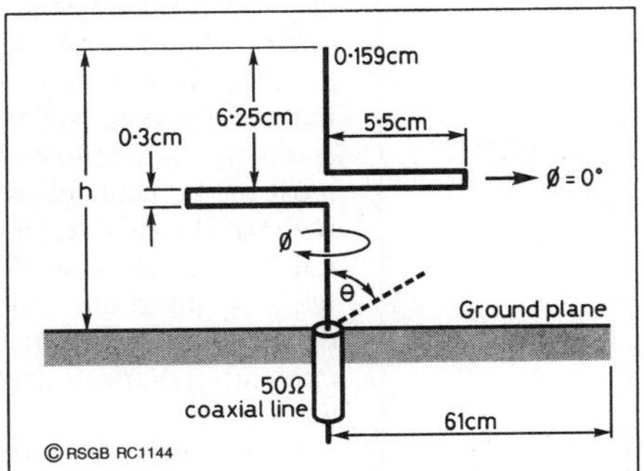

Fig 1: Altshuler 1.2GHz reasonably-broadband monopole loaded with a modified folded dipole a quarter-wave from the top. The total height, h, can be of any length since the lower segment has a travelling-wave distribution of current. An even broader bandwidth can be achieved if the folded-dipole is replaced by a 240Ω non-inductive resistor of suitable wattage but this reduces, by up to one-half, the radiated power. Dimensions are for the 1.2GHz antenna whose characteristics were measured in a model antenna range. Such antennas could be implemented as monopoles or dipoles.

"Approximate range may be found by calculating at what distance the field falls to the level of the external noise. Under the quietest expected UK conditions (external noise at 95dB above kTB) the equivalent noise field at this frequency, assuming a 3kHz bandwidth, is 3.8μV/m. The range for a 1W ERP transmission, with this bandwidth in the receiver, can be predicted from CCIR Report 368/4 to be around 1000km depending on the type of ground. Thus moderate DX is a possibility even with conventional equipment (provided that roughly the full permitted power of 1W ERP can be achieved). If the bandwidth is reduced to 30Hz, the noise drops by 20dB and the range rises to about 1800km. It is always possible that skywave propagation (not included in these ground-wave figures) could improve the range at night. The advantage of using narrow-band slow Morse is evident.

"With the upper level of external noise predicted in CCIR Report 322, the range falls to about 600km with a 30Hz receiver bandwidth, and an unimpressive 90km with a 3kHz bandwidth. This explains the need for radiated powers of 1kW or more when reliability is required under all conditions.

"It might just be possible to radiate 1W ERP by fully utilising the available power from a 13A mains socket (about 3kW) to generate as much RF as possible, aiming for an antenna efficiency of -30dB. With care, this might be achieved with an effective monopole height of 10m.

"The radiation resistance of a loaded monopole with an effective height of 10m will be 0.01Ω. This assumes a monopole a little over 10m high with plenty of top loading at the same height. The required antenna current in this case is 10A. The art of designing the antenna will lie in getting it to accept the required RF current when driven with the available RF power. The ground loss resistance will have to be less than about 10Ω.

"For someone with a bit more space, the technique which has been used for some very successful professional VLF antennas is to use several monopoles or umbrella antennas fed separately but in phase. If they are sufficiently close together the radiation resistance of each one increases by the square of the number of elements. Thus if four umbrella-type elements were used, each one would have a radiation resistance 16 times that of one antenna element on its own. This technique is also very useful in reducing ground losses, and has produced a radiation efficiency of 70 - 80% from an existing antenna using eight umbrellas a fortieth of a wavelength high.

"To sum up. I feel that 73kHz provides plenty of interest for those who like to experiment. A small tuned loop antenna with an FET preamplifier provides the basis of an adequate receiver. Short-range communication, possibly underground, should not be difficult. Longer distance working, using variants of the monopole as a transmitting antenna, will probably prove possible, but distances will be improved by a narrow bandwidth and a slow data rate: so the Morse key still has a role. It might be worth reserving a section of the band for DX (perhaps the lowest 400Hz)."

This interesting and informative letter from G3NOQ reminded me that many years ago I used to listen to maritime radio not only on the 500kHz band but also on the old LF maritime band of 100 to 160kHz. In the heyday of the transatlantic passenger liners, one could hear them sending loads of traffic to the LF coast stations from well out in the Atlantic. Large liners could have quite effective LF flat-top antennas. Calling frequency was 143kHz (2100m).

This band seems to have largely dropped out of use by ships sometime around the mid-1950s, but was still defined in the UK Merchant Shipping (Radio) Rules of 1952 and the following notes are taken from *Radio & Television Engineers' Reference Book* which I helped to edit in 1953-54.

Under a sub-heading 'Long-wave radio-telegraph sender', L P Learney of Redifon wrote: "Antenna power in this band (100 - 160kHz) is limited to 3kW and must be capable of progressive reduction in steps of 6dB or less, to 200 watts . . . At these frequencies efficient radiation is difficult to achieve, despite the low earth resistance [with a metal hull immersed in sea water - *G3VA*], on account of the relatively short antennas provided. In order to reach the rated power without excessive loss into the smallest specified antenna, a high-Q loading coil, capable of dissipating considerable power and of withstanding exceptionally high voltages, is necessary. Thus, at one extreme, the antenna constants are 3.6Ω + 300pF; to achieve even 200 watts into this load, a current of 7.5A is needed, with a resulting voltage of 37,500V RMS at 100kHz, and with some 500W or so dissipated in the antenna tuning inductance . . ."

This passage thus emphasises the importance at LF of the antenna loading inductance, which needs to have as high a Q as possible, be capable of carrying a large RF current, and withstanding very high RF voltages. For terrestrial installations, the need to reduce ground losses is equally important.

There seems a good case for the RSGB to institute some new awards - for example, for anyone whose signals are heard at a distance of over 1000km or perhaps even 1000 miles with a Supreme Award for a two-way contact over such a distance! Not easy but not impossible, particularly over mainly sea paths if you can succeed in radiating something like 0dBW without obtaining temporary use of a professional LF antenna.

MONOPOLE LOADED WITH A LOOP

ON SEVERAL OCCASIONS recently I have referred to the work of Dr Edward Altshuler, an eminent American antenna engineer currently working at the Rome Laboratory of the Hanscom Air Force Base. In 1961 he showed that a travelling-wave distribution of current can be produced on a linear antenna by inserting a resistance of about 240Ω one-quarter wavelength from its end ('The Travelling Wave Linear Antenna' *IEEE Trans Ant & Prop*, July 1961, pp324-329). Such an antenna can have the lower section virtually of any length and provides reasonably broadband characteristics. In 1993 he further showed that the main disadvantage of a relatively short resistance-loaded system - that such an antenna is only about 50% efficient because of the power dissipated in the resistor - can be overcome by replacing the resistor with a resonant antenna having a radiation resistance approximately equal to a 240Ω resistor (see *TT*, January 1994, p44). He showed that with a modified folded dipole replacing the 240Ω loading resistor, virtually the full power will be radiated with mixed horizontal and vertical polarisation. **Fig 1**, reprinted from the January 1994 item, shows his 1.2GHz antenna which could be constructed using stiff wire. The system would work also on HF although construction would be rather more demanding in support structures.

Dr Altshuler returns to this form of antenna in a recent paper 'A Monopole Loaded with a Loop Antenna' (*IEEE Trans Ant & Prop*, June 1996, pp787-791). The main objective is to obtain an antenna with hemispherical coverage which is circularly polarised. Both series and parallel-fed circular and square loops (**Fig 2**) were investigated both by NEC computer simulations and by actual measurements at 1.6GHz. He found that these very simple, low-cost antennas can operate effectively over the entire band 1.4 to 2.0GHz. It was shown that circular polarisation could be obtained over most of the hemisphere with both types of series-fed loops; the parallel-fed loop insertions produced more directional circularly polarised patterns.

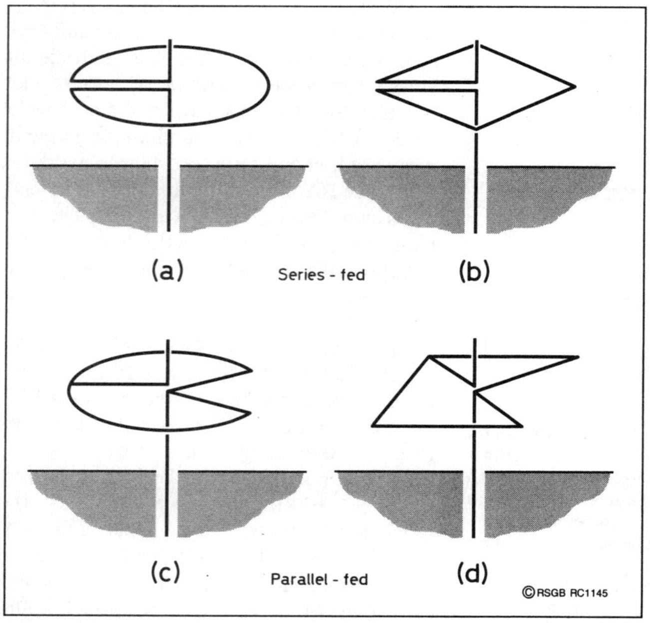

Fig 2: Altshuler-type monopoles loaded by circular or square loops, series- or parallel-fed.

Loop perimeters of approximately one wavelength produced the best results. Increasing the height of the monopole (and the corresponding distance of the loop over the ground plane) enhanced the low elevation angle radiation but nulls also appeared. When the monopole was 0.75λ (and the loop 0.5λ above the ground plane) there was, as expected, a deep null in the zenith direction. Optimum results were obtained when the loop was approximately 0.25λ above the ground plane . . . The antennas were constructed from 18AWG copper wire which has a diameter of about 1.0mm. However, for the wire to hold its desired shape, it had to be stretched; this process reduced the wire diameter slightly.

Dr Altshuler points out that as long as the receiving antenna has the same sense of polarisation as the transmitter, then a maximum polarisation loss of 3dB occurs when the receiver is linearly polarised. But if a circularly-polarized receiving antenna has the opposite sense of polarisation to a circularly polarised signal from the transmitter then the polarisation loss becomes very high.

His conclusions were that "all loaded monopoles operated satisfactorily over the frequency band from 1.4 to 2.0GHz. The agreement between computed and measured results was very good considering the computational model and the actual antenna were not exactly alike. Whereas this type of antenna can produce circular polarisation over most of the hemisphere and over a broad bandwidth without the use of phasing networks, it may prove to be a very low cost alternative for systems such as IRIDIUM [the planned low-earth-orbiting satellites for use with handheld radiophones - *G3VA*] and the Global Positioning System. Finally, the null that is usually present over a limited range of azimuth angles may be useful for some applications. For example, there is concern that radiation from a handset held to one's head may be dangerous. This antenna can easily be designed so that the null is in the direction of the head, thus minimising the radiation in that direction".

ELECTRICAL NOISE - UPSIDE-DOWN RADIO ASTRONOMY

THE LATE AND great 'Dud' Charman, G6CJ, who did so much to introduce many of us to the importance of antenna radiation patterns both by his many demonstrations of table-top models and by such classic articles as 'DX to Order' (*T&R Bulletin*, January 1937, pp297-303) once explained how, while living at Stoke Poges, he could detect the electrical noise coming from the industrial town of Slough some three miles away. For those of us living, even in the residential areas of large towns, the problem of the enhanced background noise stemming from a multitude of electrical machines and plants is a very real problem to which there appears to be no really satisfactory solution. Almost all the discussions on curing electrical interference are based on noise coming from a single source and direction. Moving to a remote location in the country is not a simple solution - and even there one is liable to encounter noisy farm machinery!

There seem to have been few attempts to quantify or monitor the noise stemming from urban environments. An exception is disclosed in a short article by a radio astronomer, George W Swenson, Jr of the University of Illinois entitled 'Radio Sociology - or should we have called it upside-down radio astronomy?' (*IEEE Antennas and Propagation Magazine*, April 1996, pp45-46).

He explains how in 1970 he was asked by a

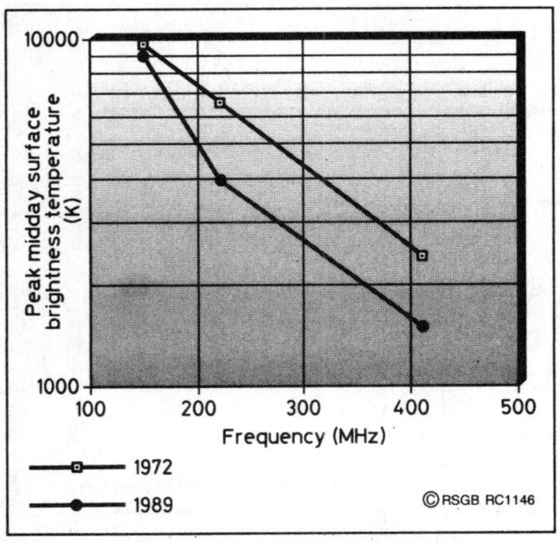

Fig 3: Peak midday surface brightness temperatures of the town Champaign-Urbana, Illinois as measured from aircraft in 1972 and 1989 at 74, 148, 222 and 440MHz. This composite diagram was originally published in 1990 in the Master's thesis of Mark C Cudak.

friend, Bill Cochran of the Illinois State Natural History Survey, to get a light aircraft and help him locate a migrating, radio-tagged hawk which had moved out of range of the associated D/F tracking vehicle. It might be possible to locate the missing hawk from an aircraft fitted with a 222MHz Yagi antenna looking sideways and downwards. The hawk was eventually located roosting in a tree.

But before this happened, while making gradual turns and listening intently on headphones, George Swenson and Bill Cochran noticed that when the antenna beam swept across a small town (population 8600) "there was a rush of noise-like sound in the headphones . . . We looked at a couple of other towns, all of which had similar noise signatures. It was apparent that concentrations of population correspond with areas of increased noise temperature. This was no real surprise as radio astronomers had long known they couldn't make sensitive observations from within or near cities or large towns, but I was intrigued by the very obvious effect we observed".

He continued: "On my way home I speculated on the sources of the radio noise. Over the years, we had experienced much trouble with man-made interference at our Vermilion River Observatory, usually being able to identify the source as local electric fences, automobile ignition, weather balloon-sondes, out-of-band spurious emissions from TV transmitters etc. But here was a phenomenon different in that the noise was broadband and featureless, like the 'signal' from a strong cosmic source as sensed by a radio telescope... and could similarly be characterised by its 'equivalent noise temperature'."

He felt it would be interesting to study the relationship between a town's radiation signature and its other attributes, such as economic type, population, geographic concentration, etc.

With the aid of a colleague who was as urban sociologist a proposal was put to various agencies in an effort to obtain funding with little success until a scaled down proposal was put to their own university's research board. The board granted a modest sum for airplane rental and some electronic components.

One of the university's training planes, a Cessna 150, was fitted with some downward-looking Yagis for 148, 222 and 440MHz and a trailing dipole for 74MHz (only one Yagi at a time could be carried). "We built low-noise converters for these frequencies, and modified a communication receiver to suppress its output during the airplane's ignition-noise impulses. Data was recorded on a Rustrak strip-chart recorder, along with noise-calibration signals from a semiconductor source. On the first test flight we confirmed that an open cornfield gave an equivalent noise temperature of 300K... and demonstrated that the ignition-noise blanker did not bias the measurement."

The aircraft was then flown over a large coal-fired power-generating station, high-voltage transmission lines, interstate highways and TV stations. None gave significant noise. Only towns with concentrations of infrastructure were observable. A standard route at 700m above ground was flown many times at all hours of the day and all days of the week. George Swenson reports: "I became so familiar with the noise signatures of the several towns that I could navigate under a hood with no other position reference, merely by watching the strip chart." He felt that this (in the days of the Cold War) could explain why US government agencies had not been enthusiastic to fund this work.

"Over many flights we saw consistent results. The temperatures of the larger cities showed strong diurnal variations, presumably related to human activity. Above 74MHz, the surface temperatures were inversely related to frequency. . . The city surface temperatures varied typically from 9600K at 148MHz to 2400K at 440MHz . . . The results of these first measurements were published in *Science* (Vol 181, 1973, pp543-545).

The smaller towns and hamlets proved harder to characterise. One small village (pp 532) showed an anomalously high noise spike, especially at 222MHz. It proved to be from the valved superregenerative receiver of an automatic garage-door-opener.

Sixteen years later, in 1988, again with Bill Cochran, the experiment was repeated in an effort to see if there had been any changes in the urban environment and to provide a graduate student with the topic for a thesis. Once again, he used a Cessna 150 but with more sophisticated equipment including dual receivers to enable two antennas to be carried at the same time; and with a downward-pointing broadband log-periodic-dipole array for 412 and 222MHz on the right wing, and a folded dipole on 148MHz under the left wing which served as a reflector.

The earlier campaign of measurements was repeated, this time with many more flights and more complete seasonal and diurnal coverage (see *Science* Vol 26, 1991, pp773-781). However, despite the increase in electrical apparatus, the spectrum of the cities still showed the same slope, varying from 9000K at 148MHz to 1600K at 412MHz. The midday noise temperatures at the town of Champaign were 2dB lower than in 1972, about the same within experimental error. A composite of the two spectra is shown in **Fig 3**.

George Swenson believes that the lack of increase in noise pollution, despite the increase in electrical apparatus, is a hopeful sign that regulation of spurious emissions from apparatus is having the desired effect. "In 1989 we did not see so many discrete noise spikes as before, suggesting that newer models of garage-door openers, in particular, have cleaner receivers."

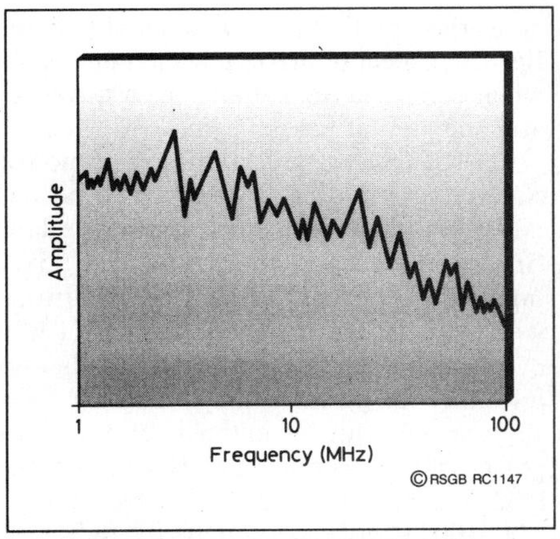

Fig 4: *QST* illustration of how the amplitude of electrical noise varies as a function of frequency between 1 - 100MHz.

However, he adds: "This somewhat optimistic result should not reduce our concern for spectrum pollution and conservation. The proliferation of transmitting services continues apace, and we should be cautious in our exploitation of this precious resource."

Guidance on troubleshooting discrete sources of electrical noise is given in the 'Lab Notes' feature of *QST* (June 1996, pp73-74) prepared by ARRL Laboratory staff. This discusses various sources such as electric clocks, aquarium heaters, food blenders, high-voltage lines etc and the use of portable receivers to trace sources, but admits that this only touches the fringe of the subject - source location, a few simple cures and working with the power company (on power-line noise which is less of a problem in the UK due to the use of more buried power lines). It also points out that the amplitude of electrical noise varies as a function of frequency (as noted above). Noise spikes are much more pronounced at LF, MF and HF (see **Fig 4**) than in the VHF-UHF range investigated by George Swenson; but since this is true also of atmospheric noise, the lower noise factors possible with VHF-UHF receivers means that a noisy urban environment is a significant problem on all bands below about 1GHz.

ZERO-BIAS TRIODES AND TETRODE POWER VALVES

TT (JULY 1996 p70) briefly traced the history of zero-bias linears with particular reference to the 811A power triode still available from Svetlana. This resulted in a letter from Derek Stanners, G3HEJ, from which the following extracts are taken:

"I was delighted to read your comments on the RCA 811A having just completed a quickly 'put together' linear amplifier for the July 70MHz VHF NFD in which I was a 'solo-competitor', the first time I have used a 811A.

"In the past, I have repaired/modified for friends a number of HF linears, including the Collins 30L-1 and Heathkit SB-200, and have noted the simplicity of zero-bias 811 amplifiers in respect of drive, power supplies etc when compared to tetrodes such as the 4-400 and similar beasts."

G3HEJ constructed his first high-power (Class C) amplifier in 1951 when he obtained his 150W-input licence (in those days first-year licences were restricted to 25W-input and CW only) using a 100TH triode with up to 3kV on the anode. This EHT was of much concern to parents and friends who felt 3kV was pushing things a bit far for a 17-year old! But his enthusiasm for high-power valves has remained. His garage still houses such classic power valves as the 811, 100TH, a brand-new 35T, HK54 and the better-known 813, 4-125, 4-400 etc.

His recent 70/50MHz classes B or C amplifier is fully contained, requiring only 240VAC, 12W drive, 50Ω antenna and a keyline to earth. The 1.6V/36μF EHT for the single 811A in uncooled, ground-grid, neutralised operation, with a tuned, link-coupled, push-pull anode circuit of unknown origin. There is switching is by two Jennings vacuum relay, interlocked. It brought him 100 contacts on 70MHz during the field day including a contact with the Mull of Kintyre (555km) using a 67ft vee-beam firing north-north-west from Surrey at a height of about 20ft.

A letter from George Badger, W6TC, of Svetlana notes that the Yaesu FL-2100 series of linears, using the 572B high-mu high-power triode, has proved very popular in the UK. He points out that the new Svetlana 572B has slightly higher gain than the original. When fitted in the FL-2100, FL-2100B, FL-2100B and FL-2100F linears they sometimes oscillate in the standby mode because, as Yaesu has confirmed, the cut-off bias is only slightly higher than cut-off. If the amplifier 'takes off' as an unloaded TPTG oscillator it can wreak havoc.

W6TC bought a used FL-2100B and modified the bias circuit: **Fig 5** shows the original arrangement and **Fig 6** the modification, incorporating a voltage-doubler configuration to increase the stand-by cut-off bias. After modification, the amplifier becomes unconditionally stable when fitted with the higher-gain Svetlana 572B.

The Heathkit SB-200 which uses a similar circuit has 100V cut-off bias and is stable when fitted with the Svetlana 572B. This applies also to the new Yaesu FL-2100Z.

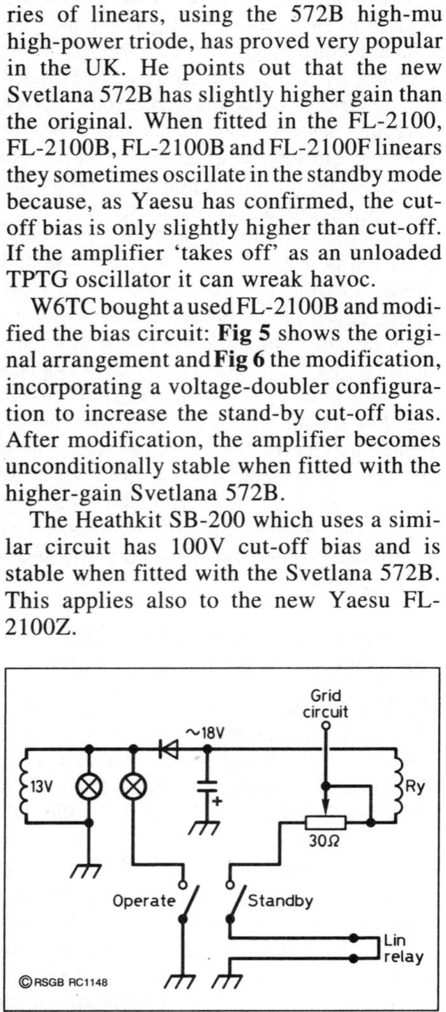

Fig 5: Stand-by bias arrangement as used in the Yaesu FL-2100B linear amplifier.

The Svetlana 572B features a massive graphite anode for high peak overload capability and an average anode dissipation of 160W. Typically, as a linear Class B grounded-grip amplifier with 2.4kV HT and 50W drive, it can provide a single-tone useful output power of about 300W. It can be used as a drop-in replacement for the 811A to provide up-graded performance.

Svetlana Electron Devices Inc (3000 Alpine Road, Portola Valley, CA 94028, USA) can also supply a 4CX400A tetrode for linear amplifier service and also a relatively inexpensive Svetlana SK2A ceramic socket for this valve. In the AB2 mode with 2.5kV EHT, it can produce over 600W PEP with low intermodulation distortion. A pair can produce 1.2kW PEP or CW at up to 500MHz.

David Kirby, G8WRB, has raised the question of air-flow requirements for power valves when there are high inlet air temperatures. He has recently been trying to determine the air flow requirements of various forced-air-cooled valves such as the 4CX250B at different inlet air temperatures. This arose when, to his surprise, he noted in the data sheet for the very high power (5kW) Eimac 3CX5000A7 triode that a larger flow rate (209 cubic feet per minute) was necessary for full 5kW dissipation if the inlet air temperature was at 25°C than when the inlet air temperature is 35°C, where it is stated that only 181 cfm is needed! [It seems likely that this must be a printing error since the figures given for 4kW dissipation seem more logical - *G3VA*].

The data sheet thus appears to contradict the accepted formula found in the Burle (formerly RCA) Application Note TP-118 and also in the book published by Eimac (Care and Feeding of Power Grid Tubes). These sources suggest that the air flow data at 25°C should be scaled by a factor Ft for any inlet air temperature T:

$$Ft = (273 + T)/ (273 + 25)$$

By this formula, air flow with an inlet temperature of 35°C, the volume of air flow at 25°C needs to be scaled up by a factor of 1.03; an increase of 3% more air.

G8WRB is convinced that the Eimac data sheet for the 3CX5000A7 is incorrect, but also believes that the formula given above still underestimates the air flow requirements as the inlet air temperature is increased. Also, if used to predict air flow requirements at high inlet temperatures, this could result in undercooling of these ceramic valves which have a maximum operating temperature of 250°C. He has been trying, not entirely successfully, to develop an improved scaling factor.

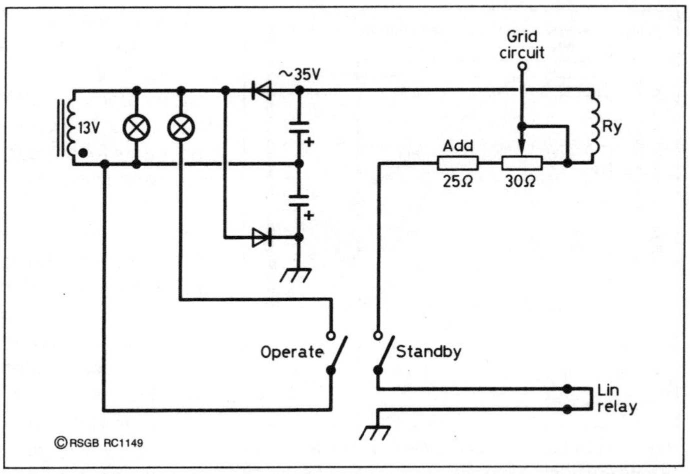

Fig 6: How W6TC modified the stand-by bias in his FL-2100B to increase the bias voltage by using a voltage-doubler arrangement in order to ensure unloaded stability when the higher-gain Svetlana 572B valve is fitted.

HERE & THERE

WHEN RECEIVING faint and often fading signals, most of us regard background noise as the principal enemy. Indeed, when Samuel Morse transmitted his first long-distance telegrams in 1844, his main concern was how to detect the signals above the noise. Yet a new hypothesis currently being advanced (see 'Noises On' by Kathryn S Brown, *New Scientist*, 1 June 1996, pp28-31) is that to pick up a faint message, background noise can be a help, not a hindrance. This extraordinary idea first emerged in 1981 in an entirely different context concerned with melting ice flows, but it appears to be increasingly recognised that in any non-linear system, background noise can actually make it easier to pick up faint signals on the very verge of detection. The problem remains how we can best take advantage of this revolutionary concept. Already one reads complaints in the hi-fi magazines that some digital signal processing for concert-hall recordings, which virtually eliminates the ambient noise, detracts from, rather than enhances, listening pleasure.

One of the saddest consequences of writing a column over a long period is to find increasingly often the names of former contributors in the Silent Key column. As recently as June 1995, *TT* included a long extract on coaxial connectors from a 1986 paper on 'Connectors Since the 20s' by the now late Robin Addie, G8LT. My personal regard for Robin dates back to 1941-42 when he was an Engineering Officer at the special RSS intercept station at Hanslope Park. At that time he was engaged on building the elaborate and very effective rhombic and vee beam antenna farm, with the expert advice of Canadian Professor Bayly and Dr Otto Baume, a German refugee scientist; and incorporating the excellent wide-band receiving amplifiers using 807 valves designed by the late 'Dud' Charman, G6CJ. The system involved a very large number of tall poles erected by a team of Non-Combatant Corps (NCC) soldiers. They were urged on with many vivid curses by 'Digger' Buick, G3XJ, a giant of an Australian who soon established himself as one of the unforgettable personalities of Hanslope Park, the wartime home of more than 100 pre-war amateurs plus many who obtained post-war licences.

FLEXIBLE ANTENNA WIRE - LESSONS LEARNED

ERIC THIRKELL, GM4FQE, feels that members contemplating the use of flexibly woven copper antenna wire may be interested in his experiences.

He writes: "I live in an exposed seaside site and during the past 15 years the winter gales have damaged and destroyed three 144MHz ground-plane antennas and countless G5RV feeder systems. A satisfactory, if limited, HF solution was found in the use of a ground-mounted 3.5MHz magnetic-loop antenna until I noticed an advertisement proclaiming the advantages of flexible antenna wire.

"A 3.5MHz inverted-vee dipole was duly made from this 'knottable' wire and hoisted to a height of 40ft on a single wall-mounted 10ft aluminium pole. Exactly two months later a clean break occurred in the wire, adjacent to the centre piece, during a 45-knot gust (confirmed by the Fife Ness Coastguards). Examination of the 5-thou diameter strands showed severe corrosion. The first lesson to be learned is that serious thought should be given to using this type of wire in an industrial environment or within five miles of the sea.

"A second lesson relates to the flexible nature of the wire. Perfect knots can be formed very easily, but as any sailor would confirm, a knotted line has half the tensile strength of an equivalent length unknotted. Admittedly my wire did not break at a knot but at the position of a straightened 'kink' formed in the process of measuring, cutting, assembling and erecting the dipole. In practice, the springy nature of this type of wire ensures scores of 'kinks' despite reasonable attempts to avoid them. Each 'kink' happily clutches at every twig and branch in the garden making it a three-man job for a trouble-free hoist. It is evident that in terms of resultant tensile strength, a 'kink' has the effect of a knot. Both kinks and knots should be avoided unless you enjoy repairing antennas!"

MOSFET CLASS F AMPLIFIERS

ON A NUMBER of occasions, *TT* has referred to the professional work of Dr Frederick Raab, WA1WLW, in developing high-efficiency power amplifiers using switching-type modes including Class D, Class E and most recently Class F. An article by him entitled 'An Introduction to Class-F power amplifiers' (*RF Design*, May 1996) makes interesting reading, although unfortunately space limitations seem to have brought about the regrettable absence of his Table 1 and possibly other important material. But WA1WLW makes it clear that

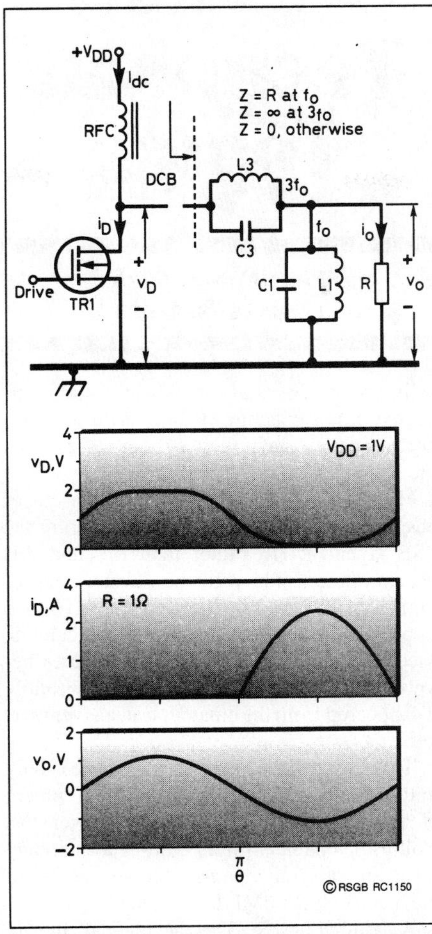

Fig 7: Third-harmonic-peaking class-F mosfet high-efficiency power amplifier - a modern form of the 1950s high-efficiency valve amplifier developed by V J Stokes of the Marconi Company as described by Dr Raab, WA1WLW. This approach is more suited to solid-state VHF/UHF power amplifiers than the true switching-mode class-D or class-E amplifiers.

Class-F is in effect the high-efficiency system developed for valve amplifiers by V J Stokes of the Marconi Company in the 1950s (see *TT*, April 1959).

WA1WLW writes: "High efficiency power amplifiers can be implemented by true switching-mode techniques (classes D and E) at frequencies from VLF through HF and sometimes even VHF. At UHF and higher frequencies, however, it is difficult to find transistors capable of switching fast enough for class-D and class-E operation. In addition, the drain/collector capacitance, lead-inductance, lead-length (including bond wires), and dispersion (frequency-dependent propagation velocity) make implementing of ideal tuned output circuits difficult. As a result, there is considerable recent interest in class F for improving the efficiency of PAs in cellular and personal-communication systems.

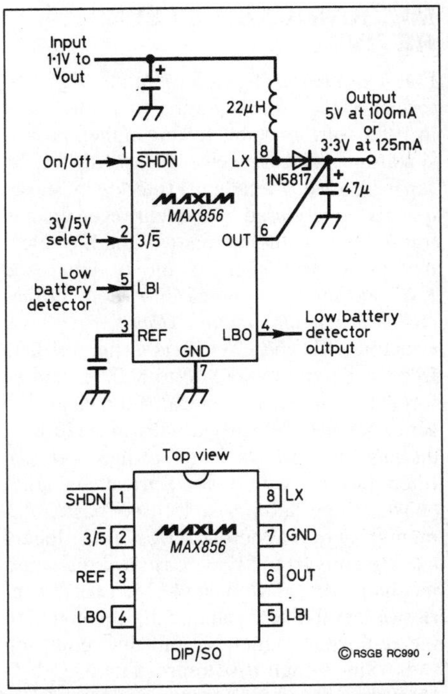

Fig 8: Corrected pin connections etc for the Maxim 856 step-up DC-DC-converter which can provide about 100mA at 3.3 or 5V output.

"A class F power amplifier uses a multiple-resonator output filter to control the harmonic content of its drain-voltage and/or drain-current waveforms, thereby shaping them to reduce dissipation and to increase efficiency. It is probably the oldest technique for improving the efficiency of an RF PA. This paper introduces class F through five basic circuits:

1. third-harmonic peaking
2. second-harmonic peaking
3. transmission line with parallel tuned output
4. transmission line with series-tuned output
5. arbitrary number of correct harmonic impedances"

Fig 7 shows basic details of a third-harmonic-peaking class F power amplifier.

FEEDBACK

DON WALKER, G7DNE, notes that in *TT* (January 1996, p65) the connections to the Maxim 856 DC-DC converter (Fig 2) are misleading. Connections as shown in the Maxim data sheet for their 3.3V/5V, 100mA step-up DC-DC converters are given in **Fig 8**. The 856 accepts a positive input voltage of 3.3V or 5V. The 857 is an adjustable version which accepts a 1.1V to 5.5V input voltage and generates a higher adjustable output voltage in the 2.7V to 5.5V range. Both devices have a typical efficiency greater than 85%. ♦

Pat Hawker's Technical Topics

PAT HAWKER, G3VA
London 37/SE22 8SS

MECHANICAL FILTERS REVIVED

FORTY YEARS ago, the Collins high-performance mechanical filters, first introduced in 1952, were a notable feature of their professional and amateur receivers. Basically similar Japanese-made mechanical filters were subsequently widely used in amateur receivers and transceivers including many of the KW Electronics models such as the widely used KW2000-series. As noted in several editions of *Radio Communication Handbook*: "Very effective SSB and CW filters at intermediate frequencies from about 60 to 600kHz can be formed from the mechanical resonances of a series of small elements, usually in the form of discs." **Fig 1** shows details of these original filters together with their electrical analogue.

The filters comprised three basic elements: two magnetostriction transducers used to convert the IF electrical signals into mechanical vibrations and vice versa; a series of metal disc mechanically resonated to the required frequency; and disc coupling rods. Since each disc represents a high-Q series resonance they form a bandpass filter with the bandwidth determined by the coupling rods. The original Collins 455kHz SSB filters could have 6-60dB shape factors as low as about 1.2, with low pass-band attenuation. There were also excellent narrow-band 500Hz CW mechanical filters.

A 1968 brochure on Collins filters listed over 250 symmetrical, single-sideband and voice multiplex sideband models at frequencies from 64kHz to 535kHz in about a dozen different sizes and shapes. When correctly installed they provided excellent adjacent channel rejection with steep-skirted selectivity while achieving a near flat-topped response and relatively low insertion loss. They were electrically and mechanically stable and could tolerate extreme frequency changes with only minor frequency shift; they did not suffer from the ageing effects of quartz crystals.

However, it was later found that transceiver design could be simplified by generating the SSB signal at HF rather than MF and this became practicable with the development of improved HF crystal filters. Higher IFs also improved image rejection in the receivers. Mechanical filters limited to frequencies below 600kHz fell out of favour and virtually disappeared from the amateur transceiver market.

This looks like now being reversed, if only in the receiver sections, with the development of a new generation of compact, lower-cost Collins mechanical filters. These have already made an appearance in the receiver section of the Yaesu FT1000MP transceiver.

A detailed article 'The mechanical filter in HF receiver design' by William E Sabin, W0IYH, was published originally in *RF Design* (March 1993) and more recently in ARRL's *QEX* (March 1996, pp14-20). The new Collins Low Cost Series (526-8634/5/6-010) are made by the Filter Products Division of Rockwell International (Collins Radio Company has for a number of years formed part of Rockwell International). These are centred on 455kHz and have 3dB/60dB bandwidths of 0.5/2.0, 2.5/5.2 and 5.5/11kHz with maximum insertion loss of 6dB. They use a recently perfected torsional-mode design which is claimed to lead to a performance/cost ratio and performance/size ratio "unprecedented in the long history of the IF mechanical filter". Input and output terminations should be 2000Ω, within 5%, and as purely resistive as possible across the 6dB bandwidth.

Dimensions are 1.25 by 0.5 by 0.25 inches; other miniature torsional-mode filters are also in the product line. With an SSB filter 6/60dB shape factor of just over two, the performance approaches but does not quite equal that of the original range of Collins mechanical filters.

In his article, W0IYH provides a useful review of HF communications receiver design considerations and the evolution of the modern HF receiver (**Fig 2**) as well as detailed information on the application of these new filters. He makes it clear that care must be taken in matching and lead-screening etc if full advantage is to be taken of the potential 100dB isolation. To quote briefly from his long article which deserves to be studied in full by anyone interested in the design of high-performance communications receivers: "A receiver which has perhaps 100dB or more of gain compression due to AGC has stringent stopband attenuation requirements. In order not to spoil the stopband rejection of strong adjacent channel signals, my experience has been that the inductors must be well shielded, both electrostatically and magnetically. Circuit layout, shielding and ground management are critical... I have measured better than 100dB of ultimate attenuation for these mechanical filters, therefore the stray paths around the filters must be at least that good.

"I have also noticed that small amounts of stray coupling can produce some anomalous and undesirable frequency-response distortions both inside and outside the passband. These leakage problems are a lot more manageable at the 455kHz frequency than at the much higher IF frequencies that are often used."

For those who believe that the future of IF/AF filtering is to be found in digital rather than analogue techniques, it is interesting to note that Brian Beezley, K6STI (well-known for his NEC-type antenna simulation software) is now offering a $100 ($105 outside USA) software package that provides DSP filtering without the need for add-on DSP hardware packages but instead uses an existing PC together with a 16-bit Creative Labs sound card. Such a set up can provide high and low-pass SSB filters, CW/Data bandpass filters, CW peaking filters, adaptive noise reduction and automatic gain control. Such a set up can display the received audio spectrum and this can then be used to select filter cut-off and to obtain an insight into the nature of the signals received. E-mail address K6STI @ n2. met (or by post Brian Beezley K6STI, 3532 Linda Vista Drive, San Marcos CA 92069).

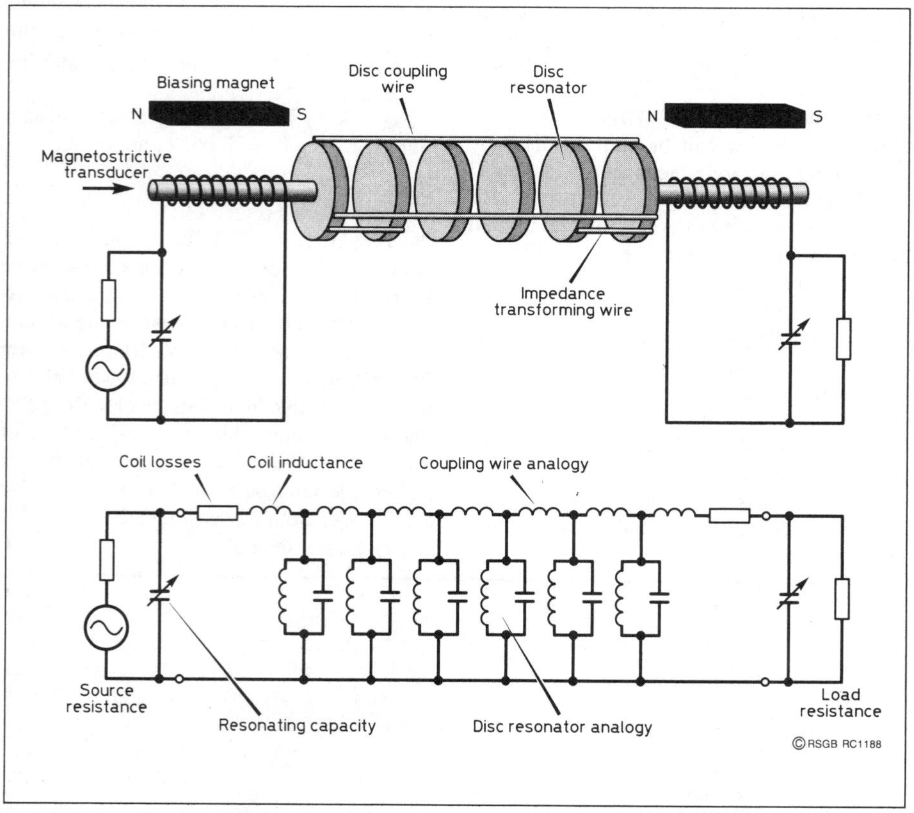

Fig 1: The Collins mechanical filter as introduced in 1952 together with its electrical analogue.

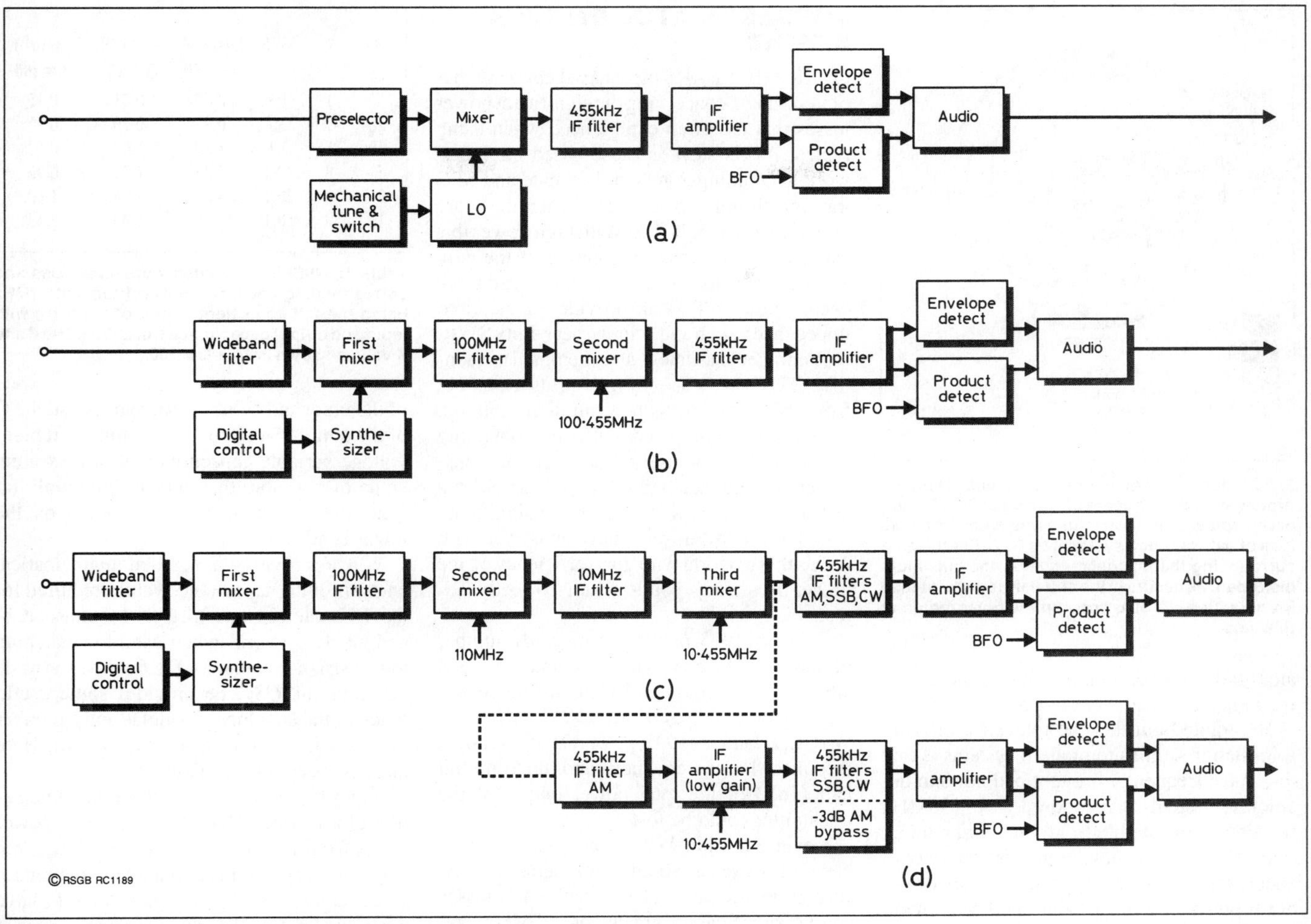

Fig 2: Evolution of the modern HF analogue receiver as described by W0IYH.

SUNSPOT CYCLE 23 & SATELLITE SCINTILLATION

IT SHOULD NOT be long now before a new sunspot cycle 23 begins to liven up the 21, 24 and then 28MHz bands after what seems a depressingly long period when these bands have been dormant for much of the time. Exceptions are mostly European signals during periods of Sporadic E, with just occasional openings in the evening to South America or to Japan around noon. Fortunately, the rise in solar activity at the beginning of a new cycle is customarily much steeper than the later decline.

Despite centuries of study, there are still many unpredictable features of the solar cycles, including (until they have occurred) the dates of maxima and minima, whether the cycle will have a higher or lower than average maximum, whether the cycles have any long term effect on the ozone layer or even global weather trends (see, for example, 'A weather eye on the sunspots' by E Norman Lawrence in *New Scientist*, 26 August 1996, p46. He believes the solar cycle could form a guide to long-range weather forecasting).

Fig 3 is an outline of recent cycles, reproduced from a short article by Alan Pickup in *The Guardian* (May 13, 1996), suggesting that solar activity should begin to rise steeply from about July, or at least Autumn, 1996.

This may have been a little premature, but certainly the new cycle should be having an effect by about Spring, 1997.

February and March 1996 were the first

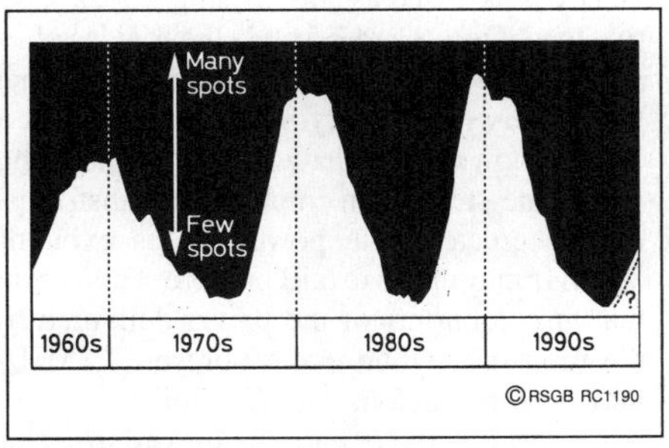

Fig 3: Outline of sunspot numbers over the past 30 years of solar cycles 20, 21 and 22 (and now or soon the start of 23).

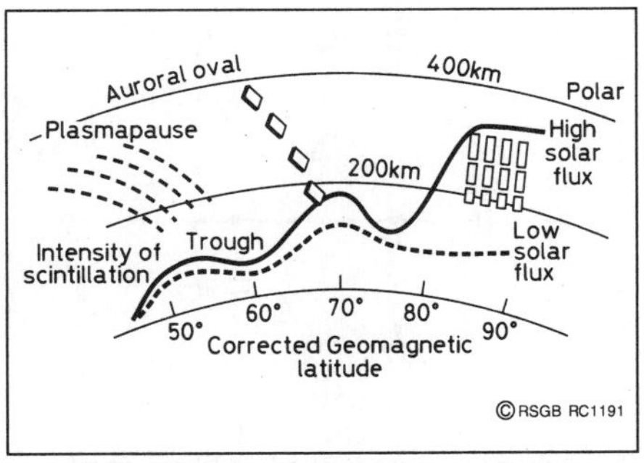

Fig 4: Sub-auroral and polar regions of the ionosphere with the lower latitudes showing irregularities during severe magnetic storms. The auroral region F layer irregularity intensity increases during magnetic activity. The primary parameter in the polar region is that of solar flux with intense scintillations of transionospheric signals during years of high solar flux. As described by Dr J Aarons at the '100 years of radio' conference (IEE Conference Publication No 411).

months this decade in which no sunspots at all were visible for more than half their days. On the basis of an average cycle period of 11 years (although recent cycles have tended to last only 10 years), the next maximum could occur as early as 2000 (11 years after the maximum of mid-1989).

It is interesting to note that where there has been a very low minimum, this tends to be followed by a very high maximum. Hence, there is a good possibility that the next few years will see unusually high maximum usable frequencies. There is a good outlook for F2 50MHz DX! **Fig 4** shows that the ionospheric layers and the effects of magnetic storms are far from uniform in all geometric latitudes.

A very high maximum could have effects well beyond our own concern with MUFs. It has been noted, for example, that the low-cost and hand-held receivers for GPS and the planned low-earth-orbit IRIDIUM systems, designed during the low-activity years, could be adversely affected by increased trans-ionospheric scintillation in the more vulnerable parts of the globe: see **Fig 5**. This point has been emphasised by Dr J Aarons of the Center for Space Physics at Boston University; for example in his paper '50 years of radio scintillation observations' at the 1995 IEE '100 years of Radio' conference. Fading, amplitude variations, scintillations - call it what you will - occurs under certain circumstances on all signals passing through the ionosphere although this is more pronounced at VHF than at UHF and above. It is now usually accepted that the

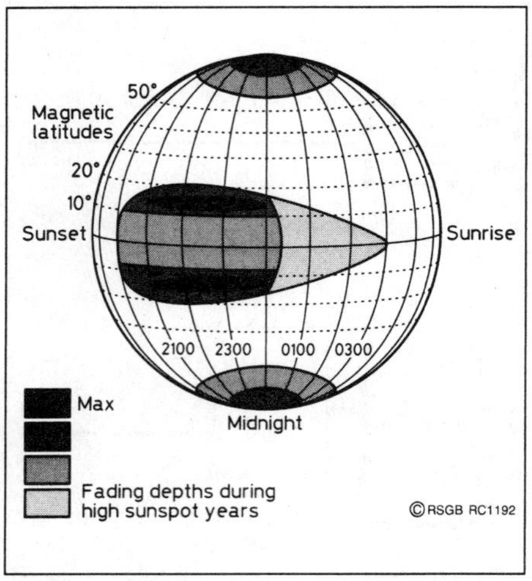

Fig 5: Intensity of fading due to transionospheric propagation during years of high solar flux. Fading depth could be sufficient in some areas to cause loss of data for such systems as GPS. For the area surrounding the magnetic equator, the maximum disturbances occurs in the anomaly region. Intense irregularities will also occur in the polar region at this time.

amplitude falls off with the ratio of increasing frequency to the power of -1.5.

He pointed out that the relevance of radio scintillation studies to modern systems is that they have frequently brought bad news to developing systems. "A proposed use of 136MHz for aircraft to satellite to ground communication showed that these transmissions would encounter the same problems that HF would during magnetic storms and the development was aborted. The use of 250MHz military satellite transmissions for both high and equatorial latitude communications meant that users had to be made aware of the strange fading on their transmissions at certain latitudes and under certain conditions.

"The signal statistics which have been measured allow digital transmission to circumvent some of the effects of scintillations. In many cases the needed redundancy or loss of information needed to minimise scintillation has not been accepted by system operators... Bits are lost when fades descend into the noise of the receiver.

"Today the users of Global Positioning System signals should recognise the conditions and areas where low SNR equipments might have difficulty; studies are underway cautioning users of GPS. New systems are being proposed and developed in the 1.2-1.6GHz range (IRIDIUM). Low signal-to-noise receivers will be affected by scintillation. Only recently a satellite system using the 135 - 140MHz band has been reproposed; the ionospheric problems should be critically evaluated.

"From the point of view of the ionospheric physicist, there is much to be learned about the phenomenon of the generation of F layer irregularities... It is hoped that understanding the forcing functions for the generation of these scatterers of radio transmission will yield techniques for forecasting and warning of scintillation."

All very revealing to those of us who think of the ionospheric layers as rather dirty HF mirrors in the sky rather than as misty microwave windows to the cosmos.

LOSSES IN ATUs ON 1.8 & 3.5MHZ

OVER THE YEARS there have been a number of items in *TT* concerning the significant power losses that are often experienced when using typical ATUs such as those based on T-networks, including Z-match, Ultimate and SPC transmatch units, particularly when these are used on 1.8 and 3.5MHz. With high power this can often be detected by heating of the coil. With QRP the losses may pass unnoticed since even a lossy ATU can provide the required impedance match and achieve near unity SWR.

A number of articles have attempted to quantify ATU power losses by measuring the output from the transmitter when run directly into a 50Ω resistor compared with that obtaining when the ATU is inserted between the transmitter and the matched resistive load. Such a test will often show that there are significant losses on the lower frequency bands even with ATUs that perform with good efficiency on the higher frequency bands when correctly adjusted.

TT, April 1995, pp62 - 63 reviewed a number of basic ATU configurations and provided advice on a recommended tuning procedure for T-networks, based on a *QST* article by W4ULD (January 1995, pp44-47). It was noted that on the lower frequency bands up to around half (3dB) of the total power output of the transmitter could be lost.

Kevin Schmidt, W9CF, in 'Estimating T-Network losses at 80 and 160 metres', *QEX*, July 1996 stresses that in practice the losses can be significantly greater on 3.5MHz, and particularly on 1.8MHz than when the efficiency is assessed as above using a 50Ω resistive load. On 1.8 and 3.5MHz the ATU is often required to match the transmitter into antennas with a very low or very high feed impedance. W9CF shows that under worst case conditions, where the step-down impedance transformation is greatest, the power losses even on 3.5MHz may climb to 6dB or more. This means that three-quarters of the power delivered by the transmitter (and even more on 1.8MHz) may be dissipated in the ATU coil.

The critical factors are the maximum value of the input and output capacitances and the Q of the coil. Most factory-made tuners fit high-voltage variable capacitors with a maximum value of not more than about 220-250pF, far from the capacitance really needed on the lower bands.

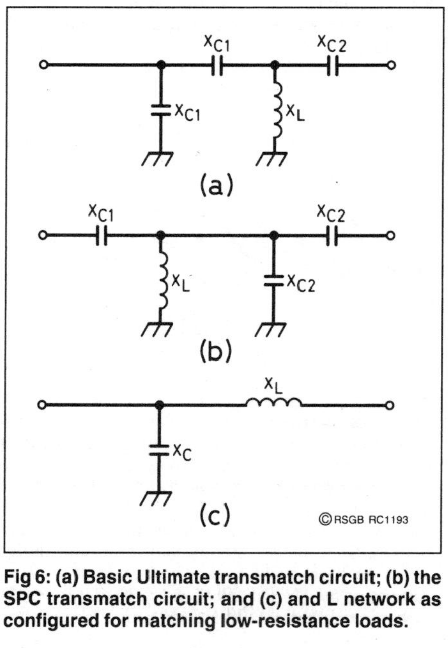

Fig 6: (a) Basic Ultimate transmatch circuit; (b) the SPC transmatch circuit; and (c) and L network as configured for matching low-resistance loads.

R	X	SWR	Loss (dB) Q = 50	Loss (dB) Q = 100	Loss (dB) Q = 200
50	0	1:1	0.62	0.31	0.15
26	-7	2:1	0.97	0.50	0.25
10	-8	5:1	1.89	1.02	0.53
5	-8	10:1	3.05	1.72	0.93
2.5	-8	20:1	4.69	2.84	1.60
1	-8	50:1	7.56	5.06	3.13

Table 1: W9CF's calculated worst-case loss and corresponding load for a T-network tuner at 3.7MHz using the full equivalent circuit of Fig 6 (b) with input and output capacitance's of 250pF. The R and X values shown are for Q = 100.

W9CF provides a detailed mathematical analysis of the capacitance values required for the 'Ultimate transmatch', the 'SPC transmatch' and an 'L network' when used for matching low-resistance antennas. **Fig 6** shows the basic networks. In *QST*, he provides some useful 'back of the envelope' formulae and a number of tables illustrating worst-case losses. Here there is space for just **Table 1**.

It may be worth noting that the implementation of a flexible ATU that offers low power losses from 1.8 right through to 30MHz is not easy since capacitors having a high maximum value suitable for 1.8MHz are likely to have too high a minimum value for effective use at 28MHz. It may be worth noting that in 1981, Professor Mike Underhill, G3LHZ, showed how a two-stage technique incorporating a prematch unit (PMU) in the form of a 50:200Ω coaxial wideband balun could be used with a switched pi-type network and roller-coaster inductance to cover 1.8 to 30MHz using a series capacitor of 1600pF maximum value provided that its minimum value was less than 40pF. To cover the full military requirement of 1.5 - 30MHz with a typical 10m whip antenna the maximum capacitances needed to be increased to 2200pF! The unit was designed to cope efficiently with up to a 20:1 SWR.

This technique was revisited in connection with PA0SE's 'comudipole' in *TT*, May 1993, pp54-55, which also showed how a simple L-network could be used by switching the one capacitor from input to output: **Fig 7**. This used a triple-gang broadcast-receiver-type 500pF variable capacitor to provide a maximum capacitance of 1500pF and was intended to be used in conjunction with a 4:1 PMU to permit open-wire feed to doublet antennas of any length elements (preferably not less than 40% of the lowest wavelength to be used). When correctly loaded and used in this configuration a triple-gang capacitor from an old valve receiver should not spark over (a 1500pF high-voltage transmitter-type variable capacitor would be massive, high-cost device!).

W9CF concludes: "There are some fundamental problems in using a 160 through 10m T-network tuner at 80 and especially at 160m. Even feeding a resistive 50Ω load with a tuner with 250pF capacitors and a coil Q of 100 gives

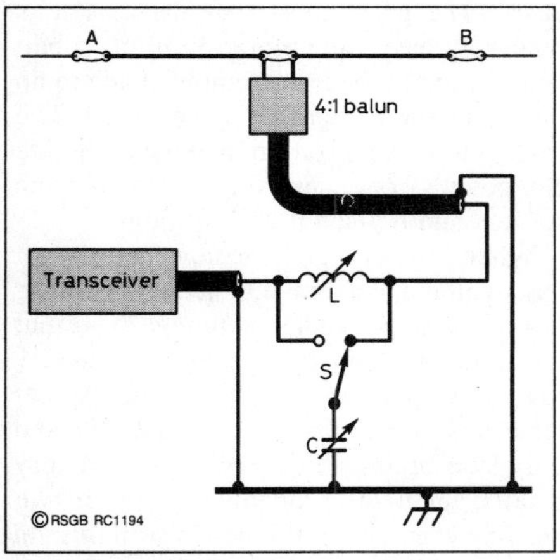

Fig 7: PA0SE's Comudipole (coaxial cable fed multiband dipole). The length AB can be of any value, but preferably not less than about 40% of the wavelength on the lowest frequency band to be used. L can be a roller-coaster or a coil with multiple switched taps. C can be one of the old style three-gang (3 x 500pF) receiver capacitors with the three sections in parallel (the PMU reduces the need for a special high voltage capacitor). If a proper match cannot be obtained with S in one position, the other switch position should be tried.

a loss of about 0.6dB. With a 1500W transmitter, the coil will have to dissipate about 180W. I doubt if many tuners can stand up to that. Increasing the SWR to three increases the worst-case loss to 1.3dB and the dissipation could increase to almost 400W. Clearly, a real 1500W 160m T-network tuner needs to have significantly larger capacitors and a high-quality coil in a large cabinet to minimise loss."

Most of us tend to think that any ATU that brings the SWR down to near unity must be satisfactory. What we forget is that it is not only a question of matching the antenna system, but also of doing so efficiently, without the loss of too much of the RF power.

TUNEABLE HF LOOP ANTENNA

CHRIS TOWNS, G8BKE, draws attention to an 'Ideas for Design' item in *Electronics Design* (July 22 1996, p112). M J Salvati describes a 'High-frequency Loop Antenna' pointing out that tuned loop active receiving antennas have tended to be confined to low and medium frequencies primarily due to the difficult requirements associated with the active circuitry: "Good performance in the 5 - 30MHz range requires an amplifier with extremely high input impedance and low noise that can drive 75Ω loads at high signal levels at frequencies to 30MHz. Combining dual FET source-followers and the new Maxim 436 wideband transconductance amplifier can produce just such as amplifier: **[see Fig 8 -*G3VA*]**."

He adds: "A balanced configuration is used for the tuned loop to preserve the symmetry of the figure-of-eight polar antenna pattern. As a result of using FET source followers, only the 1 Megohm gate resistors load the tuned circuit, so tuning is very s ρ and resistance to off-frequency interference is very high. The FETs drive the differential inputs of the MAX436 which amplifies the balanced signal and converts it to a single-end output."

Voltage gain is switch selectable (8dB or 20dB) into a 75Ω load. Maximum undistorted output is 1500mV across 75Ω. The 470Ω resistors prevent gate damage from accumulated static charge when handling the loops to change the frequency range. A three-turn, 15in diameter loop made from No 8 aluminium ground wire spaced ½in between turns) will cover from 4.4 to 16MHz with a dual 10 - 330pF variable capacitor. A single loop using a 48in long strip of 1¼in wide sheet aluminium should cover from 13 to about 55MHz although performance falls off beyond about 40MHz.

If the site has very strong AC fields, adding 100pF input capacitors can reduce hum pickup, otherwise they are not needed. The only adjustment is to zero (balance) the output by means of the 330Ω trimpot.

PACKET 'DISAPPOINTING'?

IT IS ALWAYS risky to include in *TT* adverse comments on particular modes. Experience shows that those who adopt a new mode henceforth hold steadfastly to the idea that by adopting it they should be praised for their initiative and contribution to 'progress' rather than criticised. But the views of Jon Tinsley, KC5VH (*QST*, August 1996), deserve consideration - if only for his courage in expressing them.

He writes: "I decided to give packet radio a try about a year ago... I expected CW-type QSOs where I would communicate with others via my computer keyboard. What I found was far from that! The packet world turned out to be little more than a proliferation of bulletin boards. They contained megabytes of messages, mostly classified ads and occasional pleas for real persons to respond to CQs.

"I tried the 'miracle' communication medium by exchanging messages with a friend in Germany. When and *if* we were successful, the trip was *five days one way*... What good is perfect noncommunications? Packet mail makes sense if stations 'A' and 'B' have something to send to each other frequently and live within a stone's throw. Of course, Ma Bell (US telephone system) would do just as well in that case.

"I am greatly disappointed with packet. Even with CW, I talk to real people in real time and get the corresponding satisfaction. With packet, I might as well hang my message on the doorstep and hope someone comes by and gets it!"

One cannot help wondering whether, with the rapid progress of Internet, there is any real justification for networked bulletin boards or electronic mail systems in amateur bands better suited to real-time person-to-person communications.

[Packet and Internet users have an enthusiastic following. If you come into this category perhaps you would care to comment, particularly on the latest innovations in packet, - *Ed*]

NEW TYPE OF RECHARGEABLE LITHIUM BATTERY

ITEMS IN *TT* of June 1995 (p70) and April 1996 (p69) described recent progress in lithium batteries, including the claims for a new Ultralife Batteries system (based on research in the UK by their subsidiary Dowty Batteries) lithiated manganese oxide cathode material.

New work on the synthesis of manganese oxide in the form $LiMnO_2$ as an alternative to cobalt as an electrode for rechargeable lithium batteries has been reported (*Nature,* 6 June 1996 and also on BBC Radio 4 in 'Science Now') by A Robert Armstrong and Peter Bruce (University of St Andrews). The *Nature* report is introduced as follows:

"Rechargeable lithium batteries can store more than twice as much energy per unit weight and volume as other rechargeable batteries. They contain lithium ions in an electrolyte, which shuttle back and forth between, and are intercalated by the electrode materials. The first commercially successful recharge

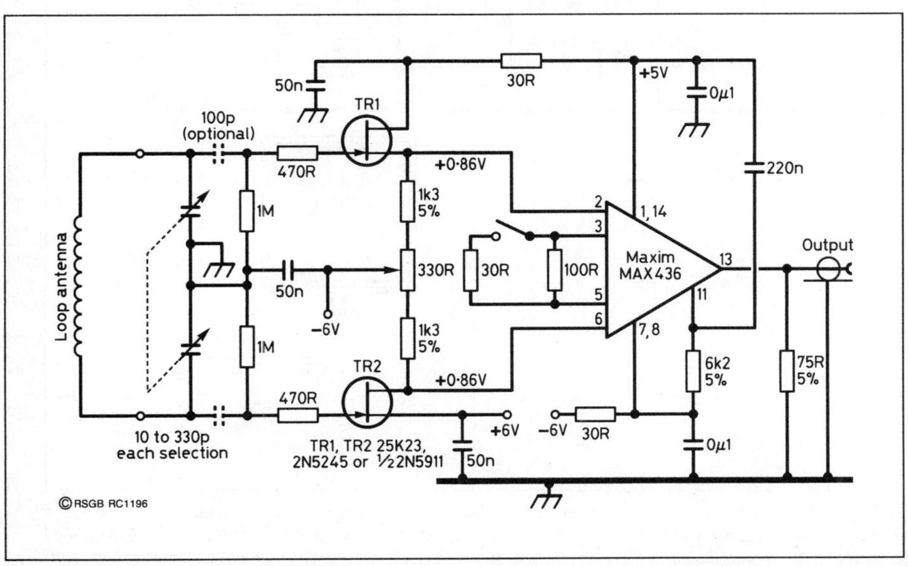

Fig 8: Balanced HF tuneable 'active' loop antenna using dual FET source-followers with Maxim 436 transconductance amplifier to create a high-impedance, low-noise amplifier that makes it possible to use effectively loop antennas throughout the HF spectrum.

lithium battery, introduced by the Sony Corporation in 1990, consists of a carbon-based electrode, layered $LiCoO_2$ as the positive electrode, and a non-aqueous liquid electrolyte.

"The high cost and toxicity of cobalt compounds, however, has prompted a search for alternative materials that intercalate lithium ions. One such is $LiMn_2O_2$, which has been much studied as a positive electrode material; the cost of manganese is less than 1% of that of cobalt, and it is less toxic. Here we report the synthesis and electrochemical performance of a new material, layered $LiMn0_2$, which is structurally analogous to $LiCoO_2$. The charge of $LiMnO_2$ (about 270mA/gram) compares well with that of both $LiCoO_2$ and $LiMn_2O_2$ and preliminary results indicate good stability over repeated charge-discharge cycles."

On the BBC, Peter Bruce admitted that the work was not yet at the stage where it was possible to confirm lifetime in terms of charge-discharge cycles: "The problem in a rechargeable battery is you have to be able to get the charge in and out many hundreds of times and this cycling is absolutely crucial and as yet we haven't been able to perfect that aspect . . . We have used a particular trick to make this compound. A number of groups around the world have been trying to prepare the manganese equivalent of the cobalt system for a number of years. The problem is that whereas it is relatively easy to combine lithium, cobalt and oxygen . . . many attempts made to do the same for manganese have failed. The trick is to make a sodium manganese oxide compound first... and then to extract the sodium and replace it with lithium."

When fully charged, the new lithium cell has a voltage of about 4.3V reducing during discharge to about 3.4V. Although at present the number of charge cycles is relatively low, it is pointed out that earlier spinal materials showed very poor cyclability, and only later optimisation yielded satisfactory results, and it will probably be between two and five years before this new type of lithium rechargeable cell appears on the market. ♦

GW0GHF'S VLF/LF 'TOBACCO-TIN' CONVERTER

BRIAN WILLIAMS, GW0GHF (formerly GW8CCA) provides details of his 'Tobacco-tin' VLF/LF converter which, in front of an HF communications receiver, provides reception on frequencies from about 10kHz up to about 650kHz. It is based on a crystal-controlled NE602 frequency converter chip with an optional FET pre-amplifier for use with an untuned loop antenna. He believes his design is more constructor-friendly and overcomes some possible snags than the PA3ACJ converter described in G4LQI's *Eurotek* (August 1996).

GW0GHF's converter - **Fig 9** - uses a 24.0MHz crystal with a receiver tuning the 24MHz band but it can equally well be used with 5.0, 10.0 or virtually any other exact MHz crystal into a general coverage receiver or 7.0 or 14.0MHz crystal into an amateur bands only receiver. In fact any crystal could be used if the preservation of direct receiver read-out (subtracting the crystal frequency) is not considered important.

His untuned loop (frame) antenna comprises some 5-7 turns round a frame as large as possible (minimum about 3ft or 1m square) fed by coax-line to the pre-amplifier. Random-length long wire antennas can be fed directly to the mixer via low-pass filtering. A 1000pF or 1500pF capacitor across the input of the mixer (pins 1,2) is important to prevent breakthrough of broadcast interference when tuning above 150kHz. The converter should be fully screened: a short length of coax to a screened receiver to avoid possible breakthrough of HF signals. If the converter is built in a plastic box, use screened leads to the panel-mounted VR1 to avoid HF pick-up.

GW0GHF admits to having become 'hooked' on listening to the many CW and MCW beacons between 280 - 420kHz. He has located CDF on about 385kHz as Cardiff Airport but is anxious to obtain a list giving details of these beacon stations which appear to use 'airport' call letters rather than international callsigns. He would welcome a detailed list of such beacons, or information on where one could be found (10 Pantycellyn Road, Llandough, Penarth, South Glamorgan CF64 2PG).

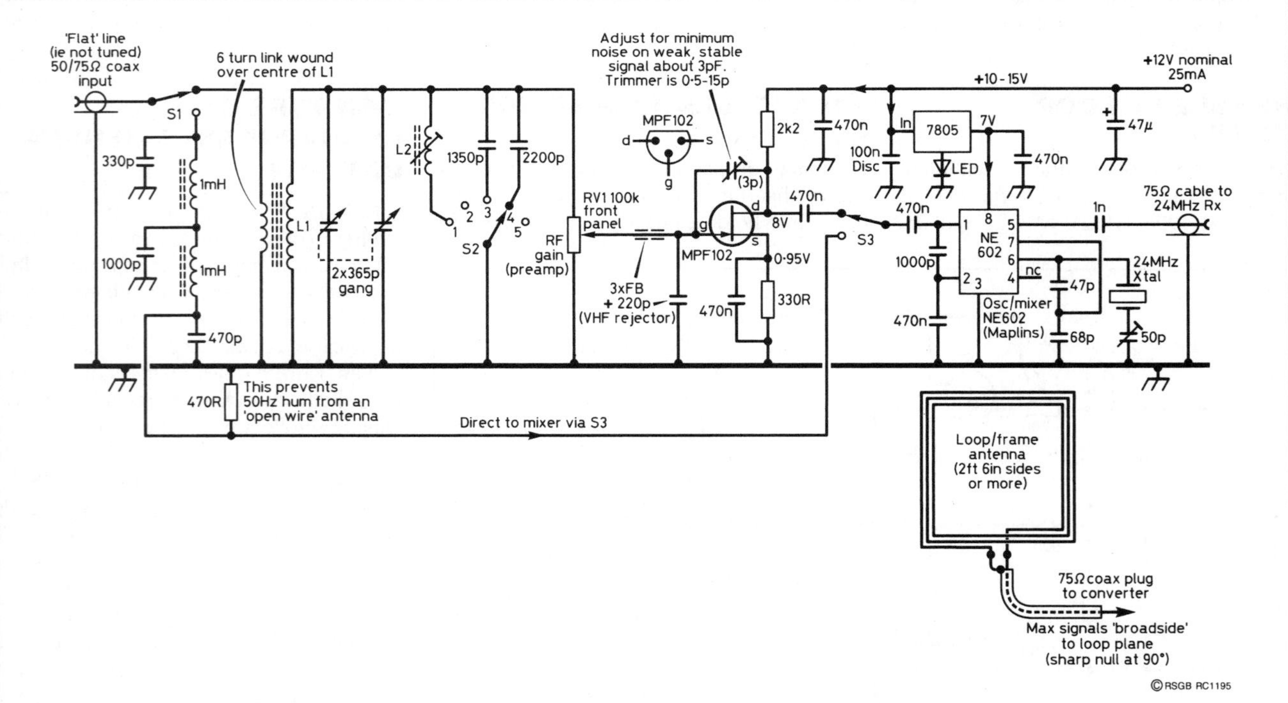

Fig 9: GW0GHF's tobacco-tin VLF/LF converter covering 10kHz to 650kHz with an untuned 'active' loop antenna or with a long wire antenna. S1 selects whether or not the pre-amplifier is in circuit; S2 frequency range: position 1 for coverage to 650kHz; 2, 150 - 450kHz; 3, 85 - 150kHz; 4, 50 - 85kHz. For lower frequencies, the antenna should be switched direct to the mixer via S3 (which can be ganged to S1). L1 is an old longwave coil fashioned from a broadcast radio ferrite rod, with 1½in rod inside (6 turn link). L2 is an old 465kHz IF transformer single winding less tuning capacitor.

Pat Hawker's Technical Topics

PAT HAWKER, G3VA

London 37/SE22 8SS

EXPERIMENTAL MICROWAVES

A REVIVED 'BACKYARD Range' feature, devoted to amateur radio and SWL projects, in the professional *IEEE Antennas & Propagation Magazine*, (February 1996) includes an outline of 'An experimental 5.7GHz SSB Communications System' by Richard Campbell, KK7B, of Michigan Technological University. The system has a transmitter output power of only -6dBm (0.25mW) but sufficient for line-of-sight communication between two similar units of over 16km with an antenna system that provides approximately 23dBi gain. Two 0.5m aluminium dishes are used with the feed shown in **Fig 1**. The feed support structure is UT-085 semi-rigid coax, and the feed is attached to the dish through an SMA male-male barrel connector, in the centre of the dish. It is stated: "This makes it very easy to remove the feed for repairs, after the tripod falls over in a small puff of wind. Fine adjustments in feed-focal position may be made by repositioning the SMA-barrel connector. The entire dish-and-feed structure weighs less than one pound. The 1.296GHz-to-5.7GHz conversion stage is mounted on the back of the dish to minimise transmission-line loss. The dish, feed and front-end electronics are all mounted on an inexpensive K-mart camera tripod."

KK7B (who is frequent contributor to *QST*) has also some pertinent remarks about our hobby. "In amateur radio, nothing counts unless a two-way contact is made... The amateur bands serve as a laboratory for a wide variety of communications and propagation experiments... amateurs (even those with PhDs) are fundamentally experimenters. Second, amateurs are passionate about building up impressive hardware capabilities at zero cost. Third, amateurs have a casual attitude towards publishing. And finally, by definition, amateur radio is pursued purely because it is interesting - there are no contract monitors or deadlines." The 5.7GHz equipment he outlines were evolved over a 10 year period, during which KK7B has been teaching senior-level microwave classes, advising microwave-graduate students, and actively participating in amateur-microwave-conference activities. He estimates that the amateur-microwave community has about 1000 active members around the world.

MORE SIMULATIONS OF ELEVATED VERSUS BURIED RADIALS

IN A DETAILED look at the question of elevated radials versus buried earths (*TT*, January 1996, pp69-71) I referred to work on this topic by Ralph Holland, VK1BRH, suggesting, *inter alia*, that there is an optimal height for an elevated groundplane, typically about 0.05λ and that large numbers of radials are not required with three or four being sufficient. In a follow-up note in the March *TT*, I mentioned that N6BV had reported that Al Christman, KB8I, had also uncovered further intriguing aspects of this work.

These findings are probably those disclosed in the article 'AM broadcast antennas with elevated radials: varying the number of radials and their height above ground' by Al Christman and Roger Radcliff (*IEEE Trans on Broadcasting*, March 1996, pp10-13). This covers much the same ground as the work of VK1BRH but with computer simulation for a quarter-wave monopole at a frequency of 1.0MHz. The American paper concludes:

"Computer-modelling exercises indicate that it is possible to construct an elevated-radial vertical monopole which is mounted too close to the ground to ever achieve performance equivalent to that of a conventional 120-buried-radial antenna.

"Generally, the necessary ground clearance increases as the soil conductivity decreases. However, once sufficient height over ground has been attained, it is also possible to use *too many* elevated radials, causing an actual diminution in the radiated field strength. A study was conducted in which three different soil types have been examined, along with numerous combinations of elevation height, feed methodology, and number of radials. Interestingly, it appears that using 8 radials in an isolated-feed configuration at a height of 10m above ground produces a competitive antenna for all of the soil types reviewed herein."

A height of 10m at a frequency of 1MHz (300m) is 0.033λ, quite close to VK1ABH's 0.05λ for one type of soil. The three sets of ground constants used by KB8I were: very poor soil (conductivity 0.0005 S/m, relative permittivity 4); average soil (conductivity 0.004S/m; relative permittivity 15); and very good soil (conductivity 0.03 S/m, relative permittivity 40).

It is noted that with 120 buried radials, the electrical character of the soil can cause the radiated field to vary by a factor of two or more. Improving the quality of the soil leads to an increase in the magnitude of both the field intensity and the input impedance.

DUAL POLARITY PSU

DAVID HEMINGWAY, G8EGG, has a very old signal generator - valved but good - and was dismayed when the mains transformer cooked. The original component supplied about 250V HT and also a negative 100V bias supply. A replacement transformer with two secondary windings was nowhere to be found.

He writes: "The only satisfactory way of providing both a positive and negative voltage from a single high-voltage secondary proved to be with two capacitors and a second bridge rectifier: **Fig 2**. I've never seen this idea used before but, as it works, it may prove useful to any other unfortunate person otherwise faced with ditching an old but useful piece of test gear."

Actually, the idea of using a capacitor and silicon diode to provide a low current negative bias supply was described many years ago in *TT* and still be found on page 247 of *ART7*. But good ideas deserve to live on.

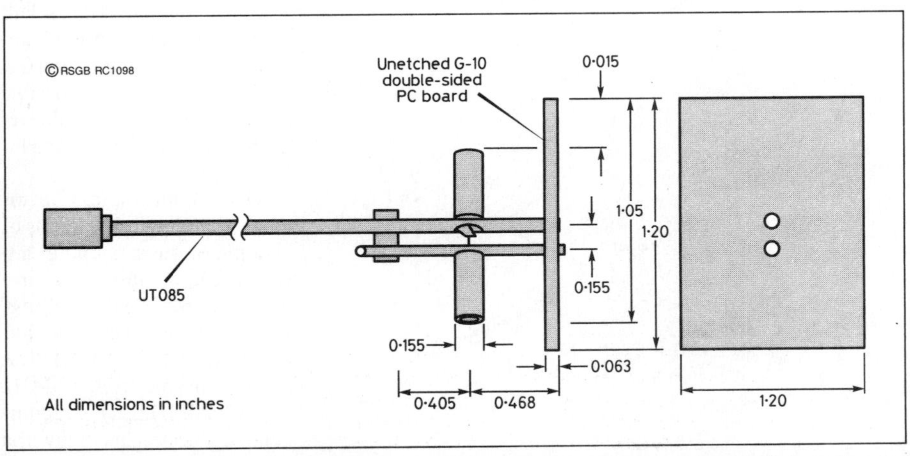

Fig 1: 5.7GHz dipole-splashplate feed as used by KK7B with 0.5m aluminium dish for his experimental milliwatt SSB system.

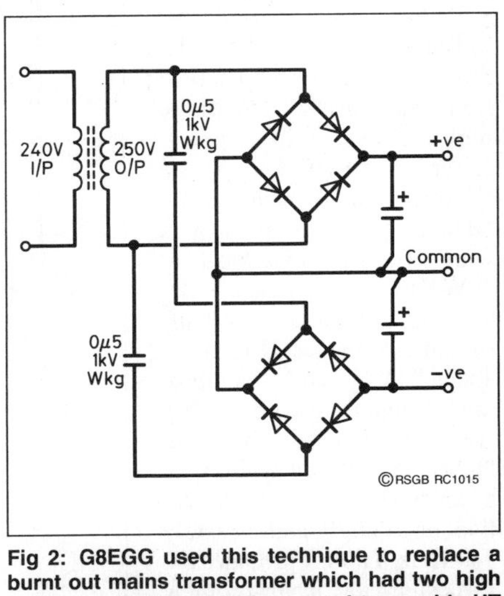

Fig 2: G8EGG used this technique to replace a burnt out mains transformer which had two high voltage secondary windings used to provide HT and a low current negative voltage. Using two capacitors as wattless resistors to isolate the two bridge rectifiers.

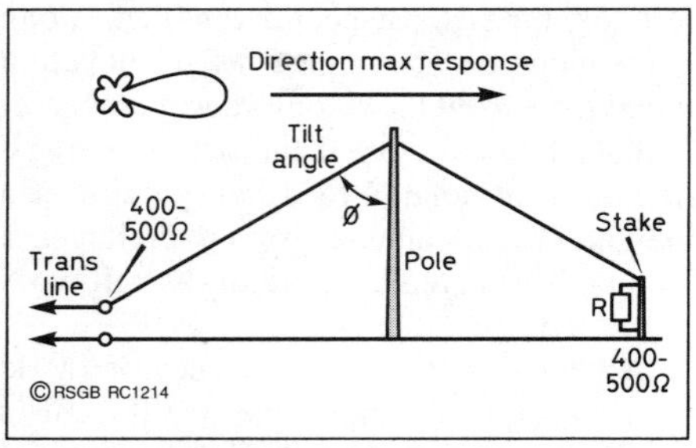

Fig 3: The inverted-rhombic (half-rhombic) antenna with counterpoise. Directivity is in the direction of the terminated end. Although originally developed (by Bruce) for HF, for which a very high (non-conductive) mast is required to obtain significant gain, this broadband travelling-wave antenna can provide a most effective fixed antenna for the 50 and 70MHz bands. The terminating resistor should be non-inductive and have a power rating up to half the RF power delivered to the antenna. For SSB an appreciably lower wattage rating (dependent on the duty cycle) can be used. For reception only a receiver-type resistor need be used if non-inductive.

HALF-RHOMBIC ANTENNAS

JUNE'S *TT* (p72) drew attention to the Barker & Williamson vertical half-rhombic attention with 450Ω resistor loading, a 10.7m Fibreglass pole and a low 25.66m counterpoise as a broadband travelling-wave antenna for use between 1.8 and 30MHz. I pointed out that this type of antenna is closely related to the original Bruce terminated inverted-vee antenna.

The Bruce antenna was for many years described in the amateur radio handbooks as an effective vertically polarised directive antenna. However, for optimum performance it was said to require a very high (non-conductive) mast and a very good earth connection in an area of good earth conductivity. The use of this antenna with a counterpoise to form a half-rhombic or 'balanced-feed, loaded pyramid' antenna, although less well-known, also dates back many years.

Recently, by chance, I stumbled on a detailed 50-year-old article by Capt John H Mullaney, W4HGU, entitled 'The half-rhombic antenna - a directive system for VHF' in *QST* January 1946, which stressed: "At lower frequencies the half-rhombic antenna is impractical for amateur use because of the required height of support. However, at VHF it provides a relatively simple means of obtaining high directivity and gain with vertical polarisation."

W4HGU discusses the design and performance of half-rhombic antennas (**Fig 3**) at about 41MHz and a compromise design for use on military frequencies between 27 to 38.9MHz. But clearly this type of antenna would prove effective on 50, 70 and possibly 144MHz. **Table 1** provides basic design data for half-rhombic antennas. He points out that for a design frequency of 41MHz, an optimum-performance half-

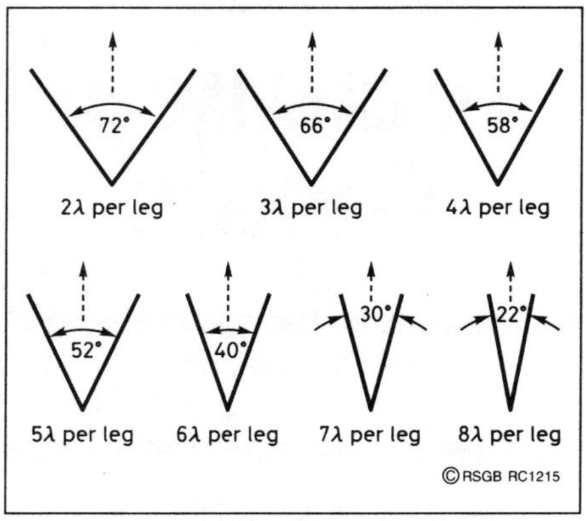

Fig 4: Approximate beam widths for the half power point for half-rhombic antennas of various leg dimensions.

rhombic with 2λ per leg, requires a tilt angle of about 50°, a pole height of 31.2ft and a counterpoise length of 72ft, and would have a half-power beamwidth of 72°: **Fig 4**. With 6λ per leg beam-width is cut to about 40°. At VHF, gains as high as 15dB can be achieved, although with compromise designs around 10dB is more likely. A half-rhombic designed for 50MHz should be usable from about 28MHz to over 70MHz.

W4HGU states: "The terminating resistor plays a dual function in the operation of a half-rhombic antenna. Upon it depends the

λ per leg	Tilt angle θ°	Pole height in λ	Counterpoise length in λ
1	30	0.87	1
2	49	1.3	3
3	57	1.6	5
4	62	1.9	7
5	65	2.1	9
6	67	2.3	11
7	68	2.6	13
8	70	2.7	15

Table 1: Half-rhombic design data.

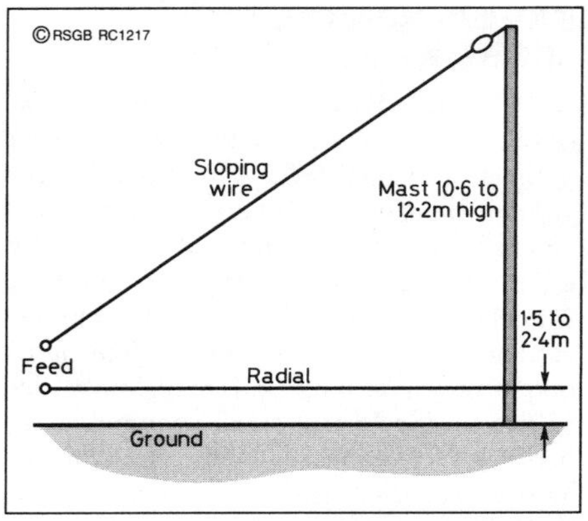

Fig 6: VK4FUQ's optimised sloper antenna for 3.5MHz. Radiator and counterpoise radial are both one-quarter-wave in length (length in metres = 71.37/(frequency in MHz) or in feet 234/MHz). The radial should be at least 5ft (preferably about 8ft) above ground and immediately underneath the radiator. Good quality insulators are need at the top of the preferably non-conductive mast and at the end of the radial.

unidirectivity of the antenna and the absence of any resonant effect. When a half-rhombic antenna is properly terminated it will offer a constant input impedance. This allows it to be operated over a wide band of frequencies without the necessity for readjusting the coupling at the transmitter. The resistor should be a non-inductive type and it may have a resistance between 400 and 700Ω without adversely affecting its terminating properties. It must be rated to handle approximately one-half the transmitter input power to the antenna. When using a half-rhombic for reception only, the resistor power rating is not important. The terminating resistor should be mechanically suitable for outdoor installation. Common practice dictates that it should be placed in a weatherproof box for protection from the elements.

"It should be noted that the terminating resistor affects neither the strength of the signal nor the field distribution in the forward direction. Its primary function is to decrease the radiation and reception from the back or reverse direction when it is connected. The power is wasted only in the sense that it is not radiated in the reverse direction... In other words, the terminating resistor is the factor responsible for making a half rhombic a unidirectional antenna. In cases where interference is negligible it is possible to remove or short-out the resistor and make the antenna bi-directional..."

While on this topic, it seems worth including brief details of a couple of other 'sloping' antennas. **Fig 5** shows one form of the unidirectional travelling-wave terminated vee-beam antenna intended for 14MHz (2λ legs) but usable from 3.5MHz to over 30MHz. The recommended pole height is shown as 50ft and should be at least 10m. The low tie wire between the

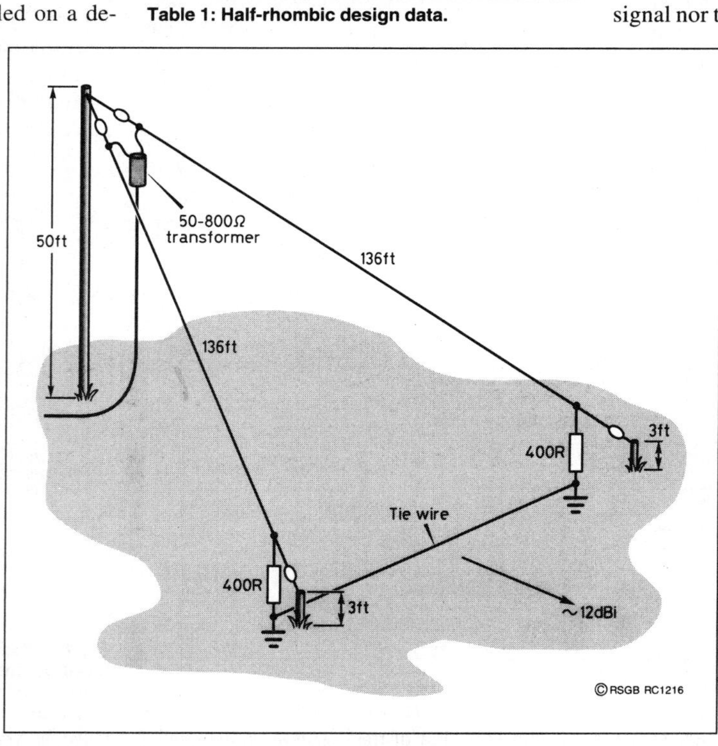

Fig 5: The terminated sloping vee beam antenna.

ends improves performance particularly in areas of relatively poor earth conductivity. This particular design comes from a survey of 'Some useful wire antennas for HF - Part 2' by Rob Gurr, VK5RG (*Amateur Radio*, April 1996) but originated in the unfortunately defunct *Ham Radio*, May 1990, p49). But note that the end wires are only a couple of feet above ground and thus represent both a physical and electrical hazard unless some form of protection is provided to prevent contact by humans or animals.

In 'An optimised sloper for 80 metres' (*Amateur Radio*, August 1996) Felix Scerri, VK4FUG, advocates the use of a sloper with an elevated radial: **Fig 6**. He claims that the use of an elevated radial is the key to success. He writes: "It was staggering to note during tests, the effect on both transmit and receive of varying the height of the radial. With the radial resting on the ground the results were very poor. When raised 1.5 to 2.4m (5 to 8ft) above ground, the difference was amazing." It would seem a wise precaution to clear heads etc with the radial at least 2.4m or 2.5m high.

He adds: "Direct coax feed can be used, but a tuner is probably necessary, as the feed resistance will probably be significantly less than 50Ω. Open wire line with a 4:1 balun is an option."

DOUBLE-FOLDED MONOPOLE

OCTOBER'S TT (pp73-74) provided some details of monopoles loaded by circular or square loops as described by Dr Edward Altshuler in *IEEE Trans Ant & Prop* (June 1996, pp787-791). In a second paper 'Hemispherical coverage using a double-folded monopole' (*IEEE Trans Ant & Prop*, August 1996, pp1112-1119), he describes a further variant of this approach, in which the 'resistive' loading is in the form of two crossed (orthogonal) folded dipoles, a technique which forms the subject of his US Patent No 5, 289, 198 of February 22, 1994.

The object is to produce circularly polarised coverage over most of the hemisphere using a double-folded monopole a quarter-wave above a ground wave, fed at the base from a coaxial line: **Fig 7**. Again, his calculations and measurements were made near 1.6GHz, a frequency close to that of the proposed IRIDIUM low-earth-orbiting satellite system and the L1 band of the global position system (GPS). For both these systems he suggests that the double-folded monopole could prove to be a very low-cost option as a ground antenna for such systems. To produce the circularly polarised wave it is necessary to tilt the folded elements up and down by about 40° as shown.

He points out that whereas many circular polarisation antennas that are used for hemispherical coverage require complex phasing networks, the loaded monopole is fed directly from a coaxial line and may provide a very low-cost option for achieving the desired coverage.

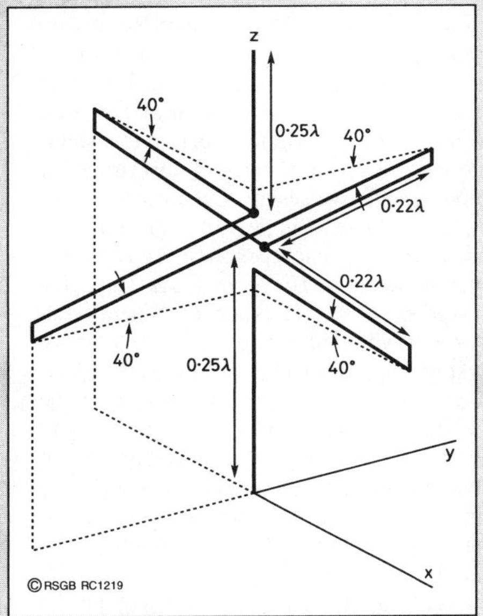

Fig 7: The double-folded monopole which provides circular polarisation over most of the hemisphere, as described by Dr Edward Altshuler.

HUFF & PUFF - PA0KSB COMMENTS

UNFORTUNATELY THERE were a few errors in the diagrams concerning PA0KSB's improved form of 'Huff & Puff' stabilisation of LC variable frequency oscillators. These errors blurred the very real advantages of the new approach system . . . For corrections to the main circuit diagram (July *TT* Fig 5) see *TT* August p80 and September p68.

Following the most recent item 'Huff & Puff in Practice' (*TT*, September, 1996, p68), which included G3DXZ's simplified circuit, I received a most useful letter from Klaas Spaargaren, PA0KSB, who originated the system in the 1970s and whose improved approach was fully described in *QEX* (February 1996), including an appendix showing the design calculations.

PA0KSB writes: "It is a pity errors were made in the conversion of the correct QEX schematics into *RadCom* drawings. A further important error concerns the quintessence of the improvement compared with the earlier versions. This is an error in Fig 4 (b) of the July *TT* which wrongly showed the output of the digital mixer as 0 to 5Hz, the same as for Fig 4 (a), instead of the correct figure of 0 to 50Hz.

"For a 5MHz VFO the distance between lockpoints is 10Hz in both Fig 114 (a) and (b), and in that respect both circuits are equivalent. However, in (b) the frequency of the correction pulses to the integrator is 10 times higher than in (a).

"This is, in fact, the core of the improvement offered in the new circuit: the frequency of the correction pulses is higher than the distance between the lock points. This means that either a higher drift rate of the VFO can be corrected, or that the remaining ripple on the VFO frequency, resulting from the correction pulses, is smaller.

"Another improvement not mentioned in the July *TT* outline is that the improved circuit is capable of operating with high VFO frequencies. I obtained reliable operation with a 38MHz VFO and a distance between lockpoints of some 17Hz. The original *QEX* article contains formulae and design guidelines for use at other frequencies."

PA0KSB and several other readers have commented on G3DXZ's simplified stabiliser (Fig 1 of the September *TT*). PA0KSB writes: "I found that a slight feedback exists from the internal dividers in the HC4060 to its input. This could frequency modulate the VFO slightly. The effect is larger with the biasing method used by G3DXZ than with my original circuit. So, if his direct coupling is used, make sure the VFO tone remains pure if only the divider is connected. If not, use an isolation amplifier in between, such as a dual-gate MOSFET.

"In G3DXZ's version with a single HC4060 divider the distance between lockpoints will be some 122Hz with a 50MHz crystal and twice this value with a 25MHz crystal oscillator. These values are fine for experimental purposes but I find them too large for general applications. I consider a distance between lockpoints of 40Hz to be the practical limit.

"His omission of the operational amplifier results in an extremely high impedance of the control signal to the varicap which means that a high impedance voltmeter is needed to verify the proper operation of the circuit. Also, after switch-on, the circuit is at the extreme end of its operating range (0V on the capacitor of 2uF) so the reset button must always be pressed to get it midrange. My circuit always starts automatically in the middle of its range.

"For those readers still confused by the earlier drawing errors I have good news: I understand that Hands Electronics are preparing a kit for the stabiliser. I have been promised a prototype for evaluation and will check the kit very critically."

For my part, I repeat the offer made in the September *TT* to those having problems to provide further information from the *QEX* article if they send me (QTHR) a large SASE. The Huff & Puff remains the most effective way of stabilising the frequency of good LC VFOs.

SINGLE-VALVE TRANSMITTERS

IN *TT*, JANUARY 1995, I included an item, 'Simple valve transmitter practice', with circuit details of a two-valve transmitter and a stabilised PSU (120-220V variable) with four valves for the EF50 oscillator valve and the screen-grid of the 807 or PE06-40 power amplifier from the design described by PA3ZFK in *Electron*, November 1994. I commented: "While this design is by no means the simplest possible, it is clearly intended that the performance as a 3.5MHz CW transmitter should meet modern standards."

John Roscoe, G4QK, was puzzled at why anyone should feel the need for such a relatively complex design when it is still possible to build a useful CW valve transmitter with far fewer components. He feels that in this solid-state era we are in danger of forgetting the simplicity of valve designs and drew my attention to an article, 'Single-valve transmitters - Rediscovering some Forgotten Knowledge', which he wrote in *The Short Wave Magazine* (December 1985). This began: "There is an obvious revival of interest in simple valve transmitters, in particular for low-power applications. It is also apparent that much of the experience gained in 50 years of living with valves has been forgotten. Developments in valve circuitry have now come to a complete stop, so it is an opportune time to look at some simple types of transmitter using a single triode-pentode audio valve."

In his 1985 *SWM* article, he noted that the ECF90 and ECF83 triode-pentode valve will run only about 2.5W input but the ECL82 will run up to 13.5W and the ECL86 (which has a maximum voltage rating for the pentode section of 500V and 9W anode dissipation) will run in Class C at up to 25W input.

Fig 8 from *TT* of September 1988 (see also *TT Scrapbook* 1985-89 p263) shows a simple 'Car Boot Special' crystal-controlled transmitter originally described by G4VAM in *Sprat* (Summer 1988) using a single ECL80, ECL82, ECL86 etc and which, as I noted at the time, reminded me of an earlier design by Al Rechner, VK5EK, who in 1981 had commented on designs for 5W CW solidstate transmitters containing almost 100 components as "solidstate technology gone berserk. If we are trying to overcome the black box syndrome by inducing people to build their own equipment, then we will maximise our chances of success by presenting simple, cheap designs". His design for a crystal-controlled triode-pentode transmitter could, he claimed, be built in "one hour".

From a safety point of view, G4VAM's design would be improved by connecting an RF choke directly across the output socket to prevent HT reaching the antenna in the event of the 1n coupling capacitor failing. This capacitor should be rated at least 1kV.

Crystal control can prove rather frustrating unless one has a good supply of crystals or is prepared to make CQ calls rather than replying to them. The triode section of such rigs can, however, be converted for VFO operation as in the transmitter shown in G4QK's 1985 *SWM* article(**Fig 9**)in which the Vackar-type oscillator was keyed using bias blocking. It may prove difficult to eliminate some degree of chirp with such a simple arrangement and in his article, G4QK also described the use of an ECC82 double triode as a bistable keyer. It should also be noted that the component values used in the Vacker circuit were those derived from the original articles by David Deacon, G3BCM, which as noted in *TT* (November 1994) unfortunately transposed certain component values and it seems likely that the stability could be further increased by increasing the value of the first 200pF capacitor and by reducing the value of the second (grid coupling) 200pF capacitor to about 20pF. As for the G4VAM design, an RF choke across the output will prevent HT appearing on the antenna in the event of the failure of the pi-network coupling capacitor. Unlike the crystal-controlled version, it would be advisable to use a voltage-regulated HT supply.

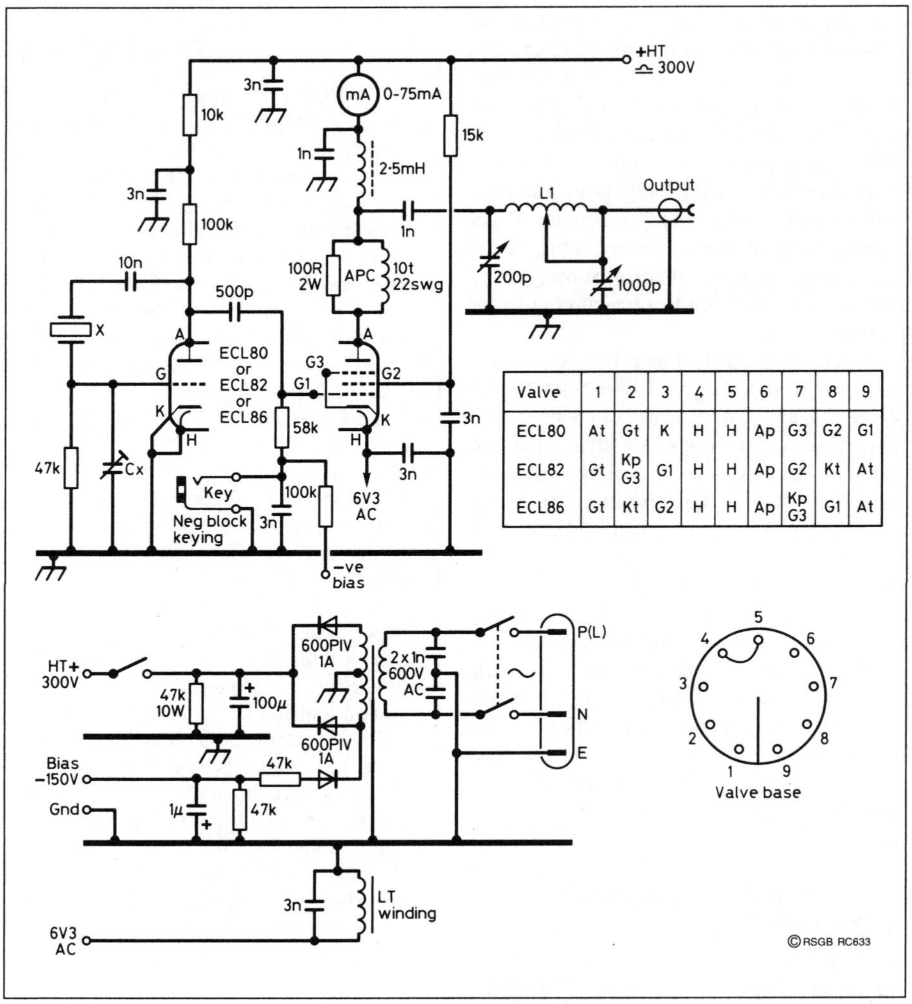

Valve	1	2	3	4	5	6	7	8	9
ECL80	At	Gt	K	H	H	Ap	G3	G2	G1
ECL82	Gt	Kp G3	G1	H	H	Ap	G2	Kt	At
ECL86	Gt	Kt	G2	H	H	Ap	Kp G3	G1	At

Fig 8: G4VAM's simple 'car boot sale' low-power crystal-controlled CW transmitter based on a single triode-pentode valve.

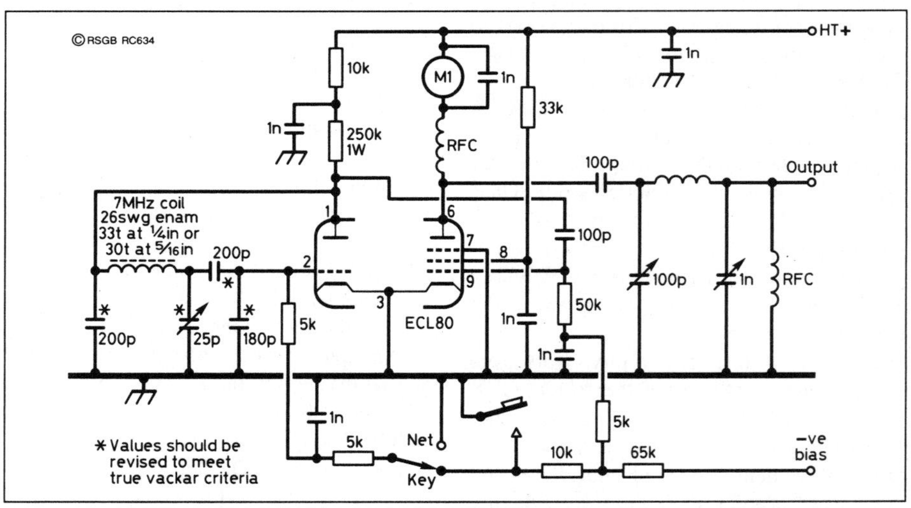

Fig 9: The Vacker VFO-PA transmitter based on a triode-pentode valve as described by G4QK in *Short Wave Magazine* in 1985. Note that the Vackar component values were based on the original description by G3BCM in which the ratios were reversed (see *TT*, November 1994). An RFC has been added across the RF output.

The May *TT* item referring to the WW2 clandestine Dutch Inland Radio Service included a photograph of one of the ill-fated stations using a converted Luxor 'Bestralings apparaat' diathermy massage unit (developed pre-war by Philips) but the single-valve 3.5MHz transmitter shown in Fig 1 was one of the lower-power units used by a sub-group in the Walcheren area which was fortunately liberated fairly soon after the Inland Service became operational in September 1944. **Fig 10** shows the higher-power push-pull oscillator derived from the Luxor units converted as crude transmitters for use between about 2850 to 3000kHz (100 to 105 metres) as used by the stations working to me at Eindhoven, although, as mentioned, the most successful station (Alkmaar) was modified as a MOPA.

CLANDESTINE RADIO IN AALESUND, NORWAY

THE RELEASE TO THE Public Record Office during 1995 of a large number of SOE files relating to wartime activities in Scandinavia, Poland, Australia and the Far East bring into the public domain some remarkable stories of wartime clandestine radio. I cannot resist a note on the highly successful 'Antrum' station near Aalesund which operated for a time early in 1942 and then almost continuously from March 1943 until the end of the war in Europe in May 1945. The operators were two Norwegian brothers - Karl Johan Aarsaether and his younger brother Knut Aarsaether.

Each brother landed in Norway several times from the secret 'Shetland bus' service of converted Norwegian fishing vessels and sub-chasers (operated jointly by SOE/SIS) which made some 200 trips to infiltrate well over 100 agents and several hundred tons of stores into occupied Norway. Knut operated a B-1 (Type 3 MkI) in 1942 and a B-2 (Type 3 Mk II) from February to September 1944 and from February 1945 until May 1945, with Karl taking over from March 1943 to January 1944 and again from September 1944 until January 1945. Each brother returned to the UK while the other took over. Unlike the usual SOE practice of sending in an Organiser accompanied by his own W/T operator, the brothers were combined Organiser/Operator for the Antrum project. They were clearly both skilful and resourceful.

The B-1 transmitter-receiver had a superhet receiver and a single stage 6L6 crystal-controlled tritet transmitter with pi-network output circuit. A circuit diagram of this transmitter was published in *TT*, November 1986. The better-known B-2 (of which some 7000 were produced by SOE) used loctal valves in the superhet receiver and had a two-stage transmitter (EL32-6L6). Both these equipments were designed by the late John Brown, G3EUR, for the SOE Signals Directorate. But many of the equipments put into Norway in 1941-2 were SIS Section VIII (Whaddon) models including the suitcase version of the Mk III and the simple Mk VIII (Paraset or Paracet).

The long periods of successful operation achieved by the Antrum operations called for extreme security precautions. The B-2 sets were used virtually daily for 18 months from the same two locations, one at a farm-house on the coast and the other on a nearby island. The suitcase sets, once installed, were never moved, despite being sought by German D/F teams using equipment installed on boats. When necessary, crystals, codes and signal plans were carried in a brief case with no attempt at further concealment.

At the farmhouse, the antenna (about 20m of wire) was actually built into the wall of the house, running straight up from basement to roof, with the B-2 in the basement: "On entering the basement there was, among other furniture stored there, a large heavy cupboard, which could be moved by pressing a button. When moved away from the wall a safe was disclosed; this was home-made but looked genuine. It opened with a key and contained genuine books, documents, money and private papers of the owner of the house. By crawling into the safe and opening the back of it, access was obtained to a strong room, which contained the B-2 and codes on a table, electric light, electric fire, electric van and ventilators, and all batteries and charging equipment in the corner."

Because of the practice of the German teams of switching off electric supplies to groups of houses etc to determine location of a transmitter, the B-2 was always run from car batteries.

Another B-2 with a different signal plan was installed on a farm on an island. Here the hideout was built on the upper part of a barn before the harvest was brought in, and was then concealed under hay, completely hidden. Access was from below.

All the many Antrum messages were passed during two-way contacts with no use made of the 'broadcast' system for receiving incoming messages without the risk of transmitting. An exception was the 'crack signal' iodoforms broadcast during BBC Norwegian-language news-bulletins. Among the stores sent out in December 1943 was an Avo Minor and a 'large' Morse key.

It would be interesting to learn from any Norwegian reader whether either of the remarkable Aarsaether brothers have ever, before or after the war, held amateur licences. They were trained in the UK at STS52 (Thame Park, Thame, Oxfordshire) and worked to the SOE Home Station, initially at Grendon Hall, Oxfordshire (STS53a) and then at Poundon, near Bicester (STS53b). The day frequencies were between 7 and 8MHz and night frequencies between 3 and 4MHz.

By comparison, many of the other SIS and SOE radio links with Norway had relatively short lives with many 'blown' by the work of the Police D/F teams and/or the V-men recruited from Quislings who infiltrated the Resistance and Milorg secret army. Incidentally, the Antrum group succeeded, on the orders of London, in liquidating one of the most notorious V-men informers, Ivar Grande. Antrum was one of several SIS and SOE stations that reported movements of the German battleship Von Tirpitz. A message to London from Antrum on March 7, 1942 reported: "Tirpitz lies at Langstein close in to the cliff wall. Camouflaged with trees. Condition re guard unknown. Seventeen other warships are said to lie in fjord beyond Trondheim. We try to start from here Sunday evening. Probable German aircraft patrol. Greetings from the harp-playing gang."

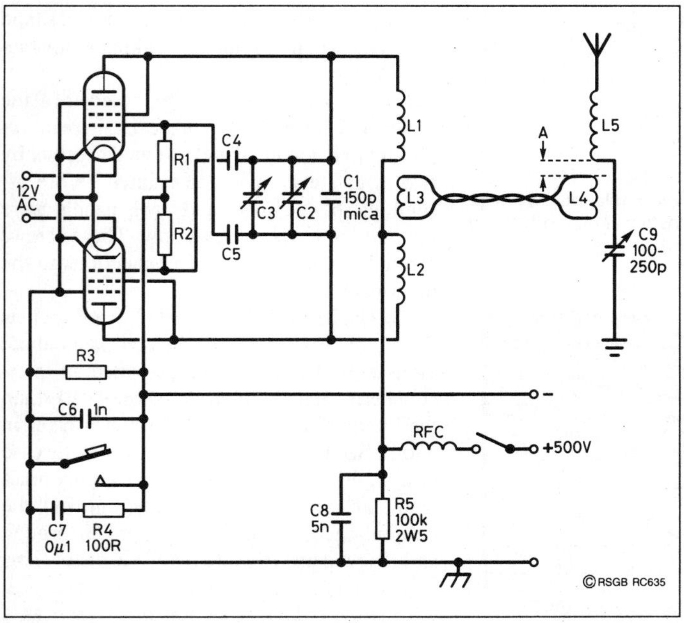

Fig 10: The crude, but relatively high-power, transmitter used by the main clandestine stations of the Dutch Inland Radio Service in 1944, based on the 'Luxor' home-diathermy equipment developed in the pre-war period and at the time claimed to benefit a number of painful conditions. See May 1995 *TT* for an illustration of one of these transmitters in use.

HERE & THERE

THE OCTOBER *TT* item on the finding by radio astronomer George Sewell that he could navigate an aircraft under a hood, by watching the strip chart recording VHF electrical noise from urban environments, rang a bell with Gus Taylor, G8PG, if only to prove that there is, as the man said, nothing new under the sun!

He writes: "In 1942 - 43 I was Radio Officer of the *Gypsum King* hauling bauxite between British Guyana and Trinidad. Radio publications were few and far between but in a junk shop in Port of Spain I came across some copies of an old American magazine on commercial radio. This magazine carried a monthly column on commercial aircraft radio which in those days largely comprised CW on frequencies around 330kHz. One piece described how, once the radio operator had got used to a route, he could help his pilot in thick weather by identifying towns being flown over by the type and intensity of the electrical noise they generated in his receiver."

On the topic of ATUs for low-power operation, Charles Michaels, W7XC, (*QST*, August 1996) writes: "Although at low-power levels a smaller - and hence less efficient inductor - may not become obviously hot or melt, the power loss in that inductor will be greater than if a large, properly designed inductor was used... the inductor in a QRP tuner should be just as large as one in a high-power tuner for the same efficiency." It could be argued that a low-power tuner requires an even higher Q inductor than one for high power since power losses are more significant. ♦

DIPOLES WITH COUNTERPOISES

ONE OF THE advantages of a dipole-type antenna over a monopole is that it does not require a loss-inducing earth connection. However, it should be remembered that all antennas that are situated close to the ground are affected by that ground to some extent. This point is stressed by Ralph Holland, VK1BRH, in his article 'Horizontal half-wave dipole above a counterpoise' (*Amateur Radio*, November 1996, pp12-13).

He points out: "The most obvious effect is that the ground forces the antenna's radiation pattern to appear in the half-space above the ground. This is illustrated by comparing the radiation around a monopole fed against ground to that of a dipole in free-space. The monopole has twice the power in the hemisphere of symmetry for the dipole in free-space. The nature of this reflection of energy above the ground is governed by the polarisation of the antenna and the ground effectiveness. The dielectric constant and conductivity of the ground determines how well or how badly the ground acts as a conductor and hence as a reflector.

"A less obvious effect is that the ground absorbs energy from the antenna; this energy is wasted in the intrinsic resistance of the ground."

VK1BRH goes on to show how - particularly on the lower frequency bands where a horizontal dipole is likely to be less than a quarter-wave above the earth - the placement of an artificial ground screen, or elevated counterpoise, can decrease the ground losses and enhance the performance of a horizontal antenna. He shows how the enhancement can be investigated either by NEC-type computer simulation or by comparative measurements of the feedpoint resistance before and after adding a counterpoise under the antenna.

He used NEC-2 computer specifications for antenna systems at 1.825MHz over average ground with a relative dielectric constant of 13 and a conductivity of 5 millisiemens per metre (13, 5). The simulations were performed for ideal (loss-less) antenna elements with a diameter of 1.2mm. He uses the term "displacement" for the distance between the counterpoise (which is the antenna system's lowest element) and ground. "Separation" is the term used for the distance between the driven-element and the counterpoise.

Fig 1 illustrates the efficiency of a horizontal half-wave dipole at various heights (displacements) above ground. In general, the efficiency improves as the dipole is raised higher. **Fig 2** illustrates the efficiency for various counterpoise system and a dipole above ground. Note that there is a knee in the counterpoise curves. This indicates that there is an optimum displacement. Be careful in interpreting the graphs because increasing the separation suffers from the law of diminishing returns. The efficiency of a counterpoise system must be compared with that of a dipole at the same equivalent height, ie the counterpoise displacement plus the driven element separation. Nevertheless, Fig 2 shows that under some conditions, a counterpoise improves the efficiency of a horizontal antenna system.

Pat Hawker's Technical Topics

PAT HAWKER, G3VA
London 37/SE22 8SS

VK1BRH concludes that "the placement of a counterpoise below a horizontal antenna can improve the antenna efficiency by a reasonable amount. The effect under a horizontal antenna is similar to the effect of elevated ground-planes for vertical antennas. You can measure this effect by observing an increase in field strength or by observing the lowering of the feedpoint resistance when adding the counterpoise. At 1.8MHz, with 13, 5 ground parameters, the effect peaks with counterpoise displacements of about 0.02-λ above ground. Increasing the antenna system efficiency with a counterpoise increases the gain at all angles, effectively lowering the effective vertical radiation angles."

RF ENGINEERING & ATUs

THE NOVEMBER *TT* emphasised that power losses in ATUs depend, particularly on 1.8 and 3.5MHz, on the maximum capacitance values and on the Q of the coil. But, as Ken Green, G1NAK, has pointed out, in practice even

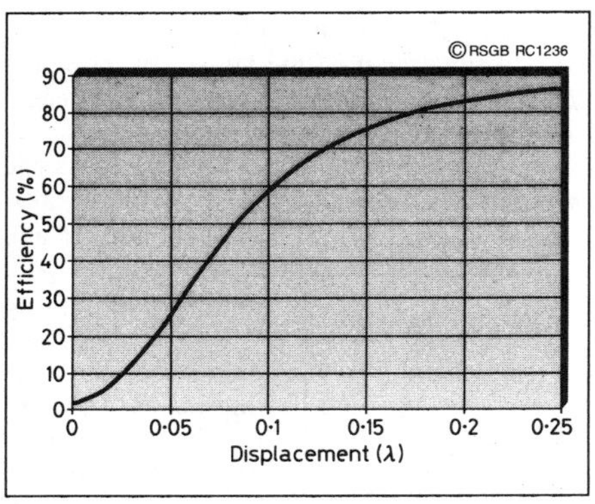

Fig 1: Efficiency of a half-wave dipole above average ground (13,5 = dielectric constant of 13, conductivity of 5 millisiemens per metre).

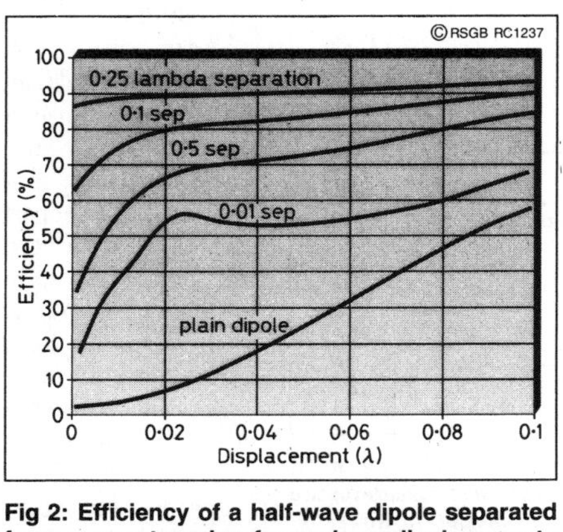

Fig 2: Efficiency of a half-wave dipole separated from a counterpoise for various displacements above real ground.

greater losses can result from poor RF engineering of the ATU.

He writes: "Your notes in *TT* November 1996 and April 1995 have overlooked one of the most important aspects of ATU design - namely the idiosyncrasies of RF currents and the need for good RF engineering.

"My wife is a keen amateur operator, G0ATS, and when she obtained her licence some 12 - 15 years ago, we set up a radio shack for the use of visitors in the hope, partly realised, that it might revitalise the failing Cornish holiday trade.

"With room to spare I was able to fulfil a long-held curiosity about the performance of long wire antennas and erected a 480ft wire held aloft by 21ft and 42ft masts. This had to be end-fed, requiring an efficient ATU. We took advantage of an offer by the now-defunct firm TAU to sell us an exhibition model of their 3kW ATU at half-price. It was claimed that this would match anything to anything and was regularly displayed feeding an amateur 150W transmitter into a metal hat-stand!

"Certainly, it produced a 1:1 SWR indication on all bands. But it did not take long to discover that when connected to the long-wire antenna, virtually no RF power was emerging from the ATU on about 98% of occasions!

"A first investigation was made into the power/SWR meter; this proved to be an Italian-made device which utilised a printed-circuit transmission-line directional-coupler. Since this was frequency-conscious and entirely devoid of any calibration, it was immediately rebuilt. The result was even 'curiouser'.

"The input and output capacitors were extremely fierce in operation; a slow-motion drive was fitted to the output capacitor. This made accurate adjustments possible and revealed that, for any frequency, there were between two to five distinct sets of control-settings that would produce a 'perfect match' without powering the antenna; sometimes a less-than-perfect match fired up the antenna.

"With the ATU dismantled I began to learn a great deal about these particular devices (and what needs to be avoided in making a satisfactory ATU):

(1) The case was made in four pieces: (a) the upper and lower halves were U shaped, approximately identical and fastened together by two screws each side and painted before assembly; (b) the front and back panels were mounted by corner screws only. These screws formed the only electrical connections to the main panels.

(2) The input and output SO239 RF sockets were mounted directly to the rear panel about 10in apart. The rear panel provided an interconnecting return which was shorter (3:1) than the internal path and several times greater in width. Because of the inadequacies of the case design, this may well have been the only practical path. The input socket was removed, the hole enlarged and the input cable taken through and plugged directly into the SWR meter. This removed many of the false resonances but still the device did not function other than as a dummy load!

(3) The variable-capacitors and the roller-coaster inductor were beautifully made (later proved capable of matching 5kW) *but* the

general layout and internal wiring of the unit showed supreme ignorance of good RF practice. I was unwilling (and lacked time) to alter the physical layout. However, I rewired the unit using braid in place of flexible wire and also contrived largely to eliminate the connections that formed common impedances in almost every mesh of the circuit. The earth returns via the casing were eliminated.

"The result was a complete success as far as the ATU was concerned. It loaded the long-wire antenna on all bands without even a grimace. A visitor who brought his own gear attempted to use his own ATU but (at around 80W) was presented with a magnificent display of blue corona which covered everything inside his box simulating a large neon light.

"With the modified ATU, the long-wire antenna proved enormously successful, providing contacts with most parts of the globe except one small area in Germany and a similar area surrounding London. The antenna ran alongside an 11kV SWEB feeder line (in Cornwall most distribution is overhead) and I began to suspect that the SWEB distribution network may have formed part of the antenna system. This had a major disadvantage in heavy rain when SWEB corona drowned out all radio signals and threatened eardrums."

I seem to recall that even at half price, the TAU unit was not cheap and incorporated high grade capacitors and a good roller-coaster. But, as G1NAK's story shows, there is an essential requirement also for good RF engineering practices in the design and implementation of such units. Unfortunately, it is not possible to detect poor practices from catalogues or even visual inspection of the outside of the 'box'. When buying expensive equipment it is a question of *caveat emptor* and/or buying from manufacturers with a proven track record for well engineered equipment.

'MINIMUM LOSS' FILTERS

THE MAY AND July 1996 *TT* included some notes on Cohn 'Minimum Loss' signal-frequency bandpass filters which showed that unless care is taken three-section filters can incur significant insertion losses, dependent largely upon the Q of the coils. Further, my notes, based on a bottom-coupled filter originally described briefly in the classic article 'The solid-state receiver - design problems and their solutions for high performance' by William Sabin, W0IYH, of Collins Radio (*QST*, July 1970, pp35-43), must have given a rather confused impression of the minimum-loss concept. In effect, this is a design approach which covers any bandpass filter comprising a series of resonant circuits, either top or bottom coupled.

For example, Dick Rollema, PA0SE, in 'Reflecties door PA0SE' (*Electron*, October 1996, pp415-418, Dutch text) describes a complete series of fixed-tuned (trimmer adjusted) four-section top-capacitance-coupled bandpass filters for all nine amateur HF bands based on the Cohn approach: **Fig 3**. Such filters - in front of a communications receiver employing only wide-band, octave or sub-octave filtering - can significantly improve intermodulation performance. PA0SE gives full details of the compo-

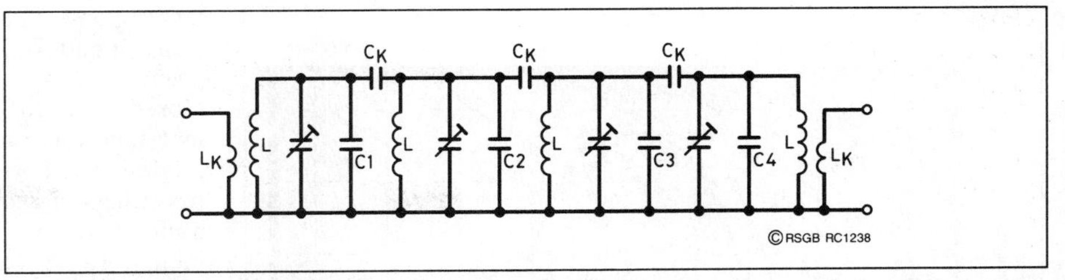

Fig 3: Circuit diagram of PA0SE's four-section bandpass filter based on the Cohn minimum-loss approach. Coils L have minimum's self-inductance and self-capacitance. Top coupling is through the small C_K capacitors.

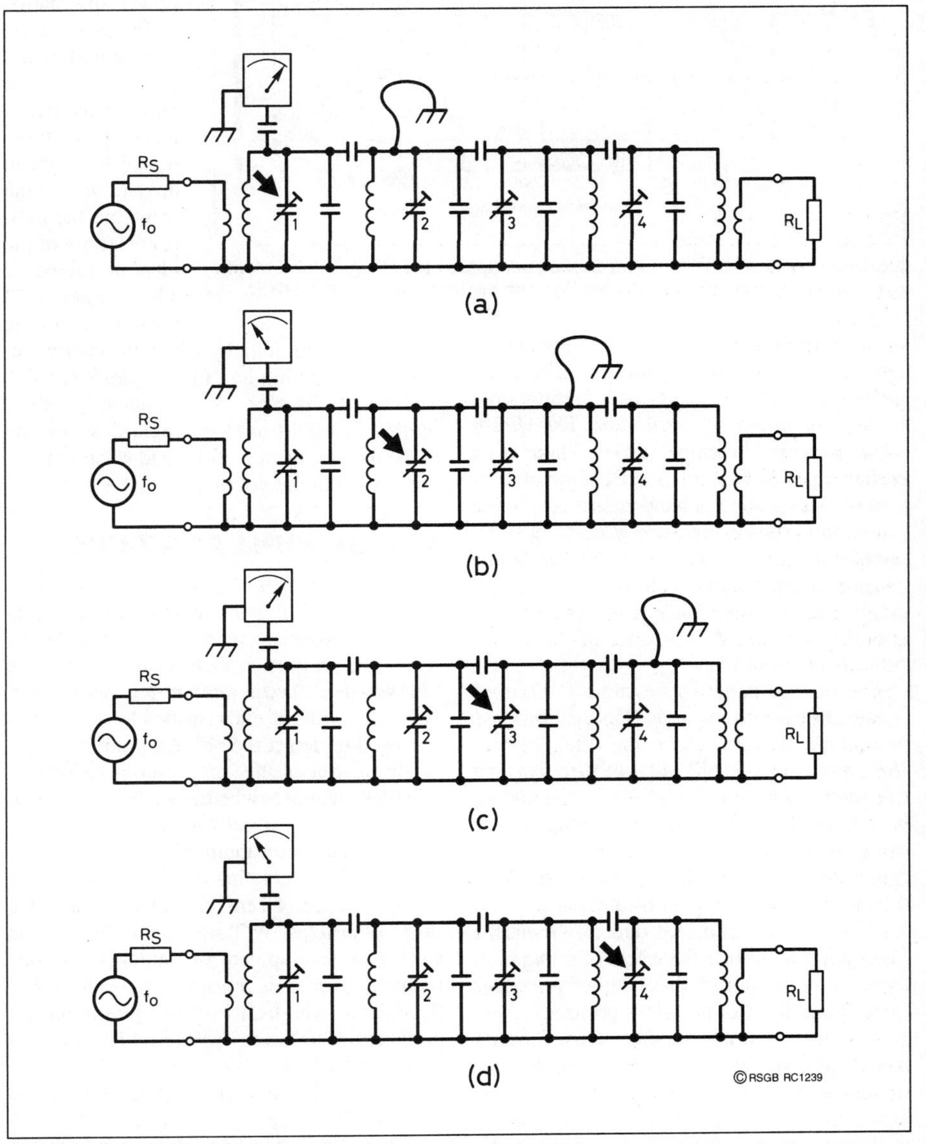

Fig 4: Alignment of a multiple-circuit bandpass filter using the 'maximum-minimum' method as described in 'Alignment and adjustment of synchronously tuned multiple-resonance-circuit filters' by Milton Dishal (*Electrical Communication*, June 1952).

Band m	Core (toroid)	Wire mm	Turns	L_K turns	L (µH)	Q	R (kΩ)	Cres (pF)	C1, C4 (pF)	C2, C3 (pF)	C_K (pF)	Loss (dB)	Bandwidth (kHz)
160	68-2	0.25 enam	100	5	54.5	178	25.8	131	129.1	125.8	3.3	4.0	66
80	68-2	0.355*	60	6	19.4	217	6.41	97-8	91	84.2	6.8	1.2	408
40	68-6	0.6*	34	2	5.72	282	15	89	87.5	86.0	1.5	4.0	184
30	68-6	0.6*	35	2	5.74	241	22	43	42.3	41.6	0.68	4.0	226
20	50-6	0.6*	25	2	2.67	240	12	47	46	45	1.0	3.0	510
17	50-10	0.6*	25	2	2.08	194	10.7	37	36	35.2	0.82	4.0	458
15	50-10	0.6*	20	2	1.43	193	7.49	39.2	38.2	37.2	1.0	3.5	653
12	50-10	0.6*	16	2	0.973	187	5.31	41.7	40.6	39.4	1.2	2.5	947
10	50-10	0.6*	16	3	0.973	170	3.68	28.5	30.6	29.1	1.5	2.2	2198

* Polyurethane-coated copper wire.

Table 1: Component details of PA0SE's four section minimum-loss bandpass filters.

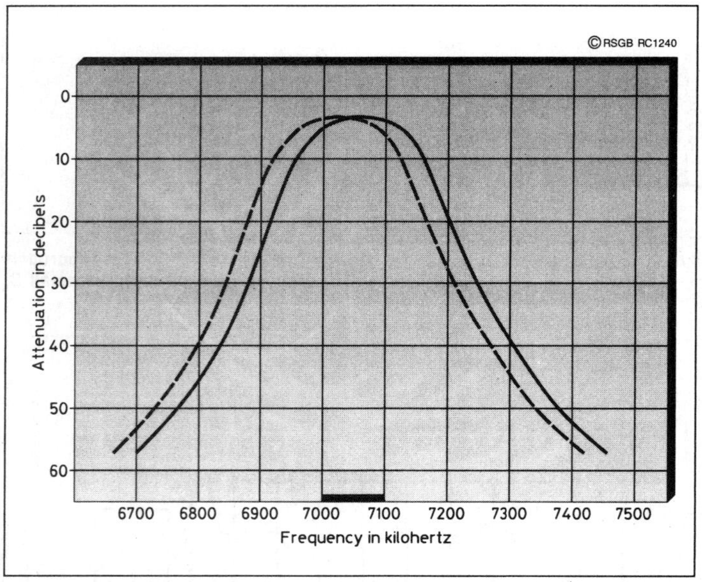

Fig 5: Response curves of the four-section bandpass filter of Fig 3: Full line with centre-frequency of 7050kHz. Broken line with centre-frequency of 7020kHz.

nent values, illustrations, bandwidths and an ingenious method of aligning the filters. He gives as a reference 'Design of Minimum Loss Band-Pass Filters' by Jesse Taub, *The Microwave Journal*, November 1963. [There is an earlier paper by JJ Taub and BF Bogner 'Design of three-resonator band-pass filter having minimum insertion loss' *Proc IRE*, May 1957, pp681-687 - *G3VA*]. **Table 1** shows the details and component values for the nine filters based on Fig 3. A useful method of aligning the filters is outlined in **Fig 4**. Response of the 7MHz filter is shown in **Fig 5**.

The original paper by Seymour B Cohn of Stanford Research Institute, 'Dissipation Loss in multiple coupled-resonator filters' (*Proc IRE*, August 1959, pp1342 - 48) from which this filter design approach has become known is, in effect, an analysis of how to design filters using the prototype low-pass-filter design concept to provide minimum insertion loss rather than for minimum ripple or flatness.

To quote some of his introductory remarks: "This paper examines the effect of dissipation on the response of multiple-coupled resonator filters designed from lossless prototype low-pass-filters. A simple approximate formula makes possible the computation of centre-frequency loss in terms of bandwidth, unloaded Q of the resonators, and the parameters of the prototype filter... It is found that symmetrical designs offer lower pass-band loss than unsymmetrical designs.

"Filters consisting of resonators coupled in cascade are widely used in all frequency ranges. At microwave frequencies the resonators are usually waveguide cavities or lengths of coaxial or strip line. At lower frequencies they may be resonant circuits of capacitors and inductors, or they may be quartz crystals or other mechanically-vibrating structures. Because of the importance of coupled-resonator filters, their design has received considerable attention. The most convenient design approach is to synthesise first a low-pass prototype filter having the desired insertion-loss response, and then to calculate the resonator and coupling parameters that will yield the same insertion-loss response on a transformed frequency scale . . . closed-form formulas are available for designing low-pass filters or any number of elements to have maximally-flat, or equal-ripple response. However, power dissipation is always present in a practical filter, and as a result the actual insertion-loss response function of the filter will deviate somewhat from the theoretical response of the ideal prototype. . . This paper will present a simple approximate formula by which the centre-frequency dissipation loss may be determined."

Thus a Cohn-type filter is one designed for minimum insertion loss regardless of the number of resonant sections and whether these are top or bottom coupled.

ELF ANTENNA FOR 73KHz?

GEORGE CRIPPS, G3DWW, while reading a book about ELF communications (by Burrows) was interested to find some details of the 'electrode pair' as an ELF receiving antenna. This was described as a centre-fed dipole about 10m long with the ends earthed by two spikes and used to detect the voltage gradient in the earth produced by the oncoming wave. G3DWW wonders whether such an 'antenna' would prove useful on 73kHz.

I am not too sure about this but certainly the technique has a long history. Dr Brian Austin, G0GSF, recently sent me photocopies of a series of articles 'W/T RE - an account of the work and development of field wireless sets with the armies in France' by Capt B F J Schonland (who later in WW2 played an important role in South African radar development), published in *Wireless World* over several issues. The following extract is from the August 1919 issue:

"No mention has so far been made of the Power-Buzzer, an instrument which we adopted from the French and later improved. In some respects hardly a 'wireless' set - for it was part of a method of signalling by conduction through the earth - this, too, was manned and run by the wireless sections (of the Royal Engineers). It may shortly be described as an induction-coil 'buzzing' a current into the earth by means of two earth-pins a hundred yards or less apart, connected to the coil by well-insulated cable. This buzzer current spreading out along the earth's surface was picked up by two similar earth-pins connected to a three-valve amplifier. As the Power-Buzzer required no aerial wires and could work over a range of from one to three thousand yards, it was much used for the very forward infantry communications over areas where the enemy's barrage fire was intense. Its working was, of course, affected by the nature of the country between it and the amplifier. Owing to its extreme simplicity, portability, and the ease with which the length of cable could be repaired if broken, the Power-Buzzer was very successful in trench-warfare."

Similar wires with earth pins were also used in the secret 'IT' apparatus in which a special three-valve amplifier placed in a dug-out close to the front line was used to pick up enemy telephone, buzzer and power-buzzer messages by means of earth-pins inserted in the ground at points as near as possible to the German trenches. The pins were placed in 'no man's land' at night with cable leads back from them to the amplifier. "Owing to leakage from the telephone cables, either actual or by induction, it was possible to hear German officers talking, sometimes louder than they could hear themselves. In this most important department of Intelligence the Germans were first in the field, hence their almost uncanny knowledge of such things as infantry reliefs. Our first listening sets were in use in the spring of 1916... manned by German-speaking operators attached to wireless sections... Information gained included strength of the enemy units holding the trenches, times and dates of reliefs, projected raids, mine-explosions and attacks, as well as the names of divisions, brigades and battalions in the line . . ."

ANTENNA-LOADING AND ATU COILS

ON 1.8 AND 3.5MHz (and of course for 73kHz), it is easy to lose a great deal of power in antenna loading or in ATUs by the use of inappropriate low-Q coils. We seem to have forgotten much of the work in the 1930s that went into the formulation of the rules governing the construction of large air-cored coils of optimum efficiency, as set out, for instance, in a series of charts and nomographs in 'Radio Data Charts' by Dr R T Beatty and published by *Wireless World* or as design formulae in such classic texts as the *Radio Engineer's Handbook* by F E Terman.

Matters such as the insulation and dielectric properties of the former, optimum length-to-diameter ratio, optimum wire diameter, skin effect, proximity effect, losses from surrounding objects, the use at LF and MF of Litz multi-strand wire etc were in those days all recognised as having a significant effect on coil efficiency.

One of the relatively few recent articles directed at radio amateurs on this topic was 'Loading coils for 160-metre antennas' by Charles J Michaels, W7XC (*QST*, April 1990, pp28-31). This is subtitled 'Why waste your transmitter power in lossy loading coils? Follow these guidelines for highest efficiency in coil-loaded antennas'.

W7XC stresses that losses in loading coils arise from several basic sources: AC resistance of the coil windings, loss induced by surrounding structures, the effects of shunting capacitance of the supporting and surrounding structures, the inherent distributed capacitance of the coil itself, and coil-form loss.

He provides a useful explanation of why the RF (AC) resistance of the wire in a coil is significantly higher than its DC resistance as a result of three basic phenomena: skin effect, proximity effect and shunting capacitance, including the distributed capacitance of the coil itself. At 2MHz, skin depth (the depth having a resistance equal to that which would exist if the current was uniformly distributed in a conductor of that thickness) is only about 0.002in (two thou) for a flat conductor, increasing with curvature and decreasing with frequency. A piece of hollow tubing with walls at least five times thicker than the skin depth will behave the same as a solid wire having the diameter of the outside-diameter of the tubing.

When a wire is wound into a coil, the electrons constituting the current in one turn are repelled by those in adjacent turns. This means that the current is forced to portions of the wire cross section farthest from the adjacent turns, incurring power loss. Proximity effect thus acts in opposition to skin effect, implying that when the wire is coiled there is an optimum wire diameter that balances skin and proximity effects against the resistance of the wire. Optimum wire diameter results in space-wound turns which also reduce the distributed capacitance below that of close-wound coils. Distributed capacitance cause circulating currents to flow through the coil, resulting in additional loss.

W7XC investigated the effects of varying the coil length, diameter and wire size for several inductance values (80, 150 and 190μH) in connection with the development of loading coils for use in top capacitance loaded antennas where the coil is placed immediately below the capacitance hat. He used coil form diameters of 2, 4 and 6 inches with winding length-to-diameter ratios of 0.5, 1, 2 and 4. **Fig 6** shows curves of AC (1.9MHz) resistance versus wire gauge numbers for all 12 coils with vertical marks at the minimum of each curve indicating the optimum B&S wire gauge for that coil. It will be noted that the AC resistance decreases significantly with large diameter formers - although for this type of antenna a 6in diameter coil could prove unwieldy; but 4in is better than 2in. Fig 7 also suggests that a coil with an axial ratio of roughly 1:1 is near optimum where this is practical.

W7XC points out that coil-form material for loading coils in a hostile environment should ideally have low-loss (dissipation factor), low dielectric constant, low water absorption, adequate mechanical strength, and the ability to tolerate heat, cold, rain, birds and sunlight. In practice, most amateurs settle for PVC drainpipe tubing which is readily available. However, the characteristics of this material can vary considerably, dependent of the type of filler used.

He bought a short length of 3in PVC (3.5in outer diameter) and subjected it to a number of tests, including boiling in water for 20 minutes and while still hot attempting to distort it with pressure. He then wound a coil inside which he could slip the PVC tubing and, using a receiver noise bridge, he resonated the coil at 6MHz with a mica capacitor. Then he checked that placing the PVC tubing inside the coil produced a small change in resonant frequency but little or no change in the coil's resistance at resonance. Next he soaked it in water for 24 hours and repeated the experiment, again without detectable change in Q. Finally, he placed the tube (together with the necessary minimum load of a cup of water) in his 700W 2.3GHz microwave oven. The oven was switched on until the water boiled. The tube was then found to be no warmer than the air exhausting from the oven.

These tests showed that at least some PVC is a suitable choice for a loading coil, although in view of the variation between different samples it would always be advisable to subject a PVC former to the microwave test.

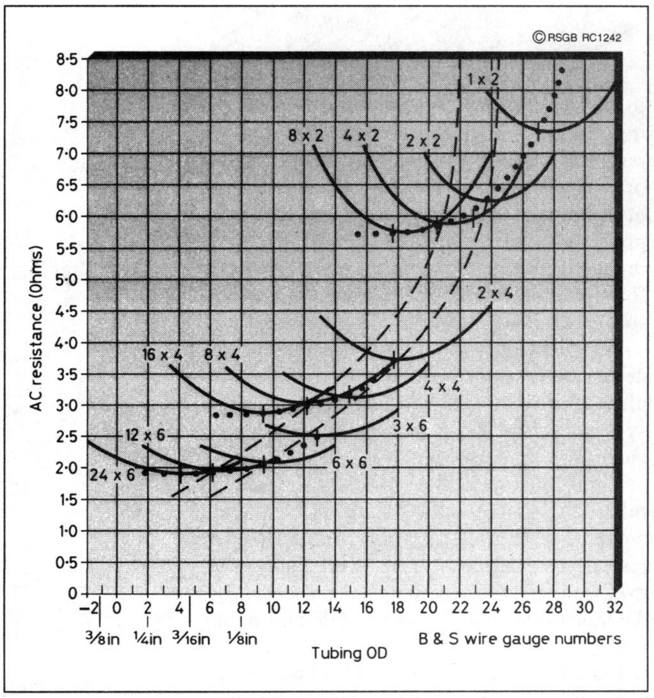

Fig 6: W7ZC's curves of 1.9MHz AC resistance versus American (B&S) wire gauge numbers for coil forms having diameters of 2, 4 and 6 inches. With each of the three formers, curves are shown for length/diameter ratios of 0.5, 1, 2 and 4. Numbers such as 8 x 4 indicate length by diameter in inches. Vertical marks at the minimum of each curve indicate the optimum wire gauge for that coil.

W7XC advises against grooving the PVC since this opens the surface for moisture entry. He points out that almost any varnish or coil dope will eventually crack away from the surface. He is also suspicious of hardware store ABS (plastic containing styrene) as an insulator feeling that its grey or black colour suggests it may have a carbon-black filler. ABS failed the microwave test.

To wind a tight coil on PVC he uses the following procedure: "Drill the form at the winding-end points and install brass machine screws with brass nuts and lock washers. Put the screw heads inside the form. Secure the far end of the wire around a screw on the coil form, in a tight U shape, over a brass flat washer. Do not cross the wire, and leave a generous pigtail. Secure the U with another brass flat washer, lock washer and nut. Tuck the pigtail inside the form. Check for clearance between the flat washers and the first turn. If necessary, add an extra nut to be sure the flat washers will not short to the first turn. Keeping the wire tensioned, turn the form as you walk towards the anchored end.

"When the required number of turns are on the form, wrap the wire around the other screw just enough to keep it from slipping off. Cut the wire, leaving a generous pigtail. Place the wound form in the freezer for a couple of hours... The coefficient of expansion/contraction of PVC exceeds that of copper, so the winding will be looser than it was at room temperature. Working quickly so as not to allow the PVC form to come up to room temperature, work the loose turns towards the loosely secured end of the coil, adjusting the spacing reasonably uniformly. The heat from your hands will warm the copper and help it expand.

"Do not try to finish this process in one pass. The PVC may warm up too much. Take up the slack at the end and again secure the wire to the screw with a partial turn around it. Put the coil back in the freezer for another couple of hours and then repeat the process, spacing the turns uniformly and taking up more slack. Bend the wire around the screw in a tight U-shaped loop over a brass washer, and secure it with another brass washer, lock washer and nut. When the form comes up room temperature, it will take a screwdriver to move the turns. They seem to squeeze so tightly that I suspect they dent the PVC.

"Although the coil is tight, after extremes of temperature cycling the soft-drawn copper may no longer be as tight. (Do not use hard-drawn copper; its conductivity is only 95% of that of the soft-drawn variety). I suggest a coat of Q Dope to ensure preservation of the spacing. This can be purchased or can be made by dissolving polystyrene shavings in carbon tetrachloride. Be careful - 'carbon tet' is a suspected carcinogen that effects the liver and kidneys when inhaled or absorbed through the skin!"

A useful source of polystyrene shavings is from the inner insulation of a discarded coaxial cable off-cut. A further reduction of AC resistance is possible at roughly 3MHz or below by the use of multi-strand Litz wire rather than a single copper wire, but that is a topic that will have to await another occasion.

RF POWER TRANSISTORS

AMONG EXPERIMENTERS the RF power transistor has long held the reputation of being the fastest fuse in existence; highly vulnerable to such conditions as over-voltages, including lightening discharges, wrong supply polarity, thermal runaway and the absence of a well-matched load. Modern factory-built designs take care of most of these problems (vulnerability to nearby lightening discharges can still take out devices) but for the experimenter they can still be the source of costly failures. To guard against some conditions, very fast-acting electronic devices are really needed to provide complete protection. Nevertheless, protection systems based on electromagnetic relays can still play a useful role.

Mirko Voznjak, YU1AD, has devised a relatively simple, relay-based system that provides good protection against: (1) wrong polarity connection of the power supply; (2) overheating of the heat sink (and consequently the transistor); and (3) excessive SWR on the load.

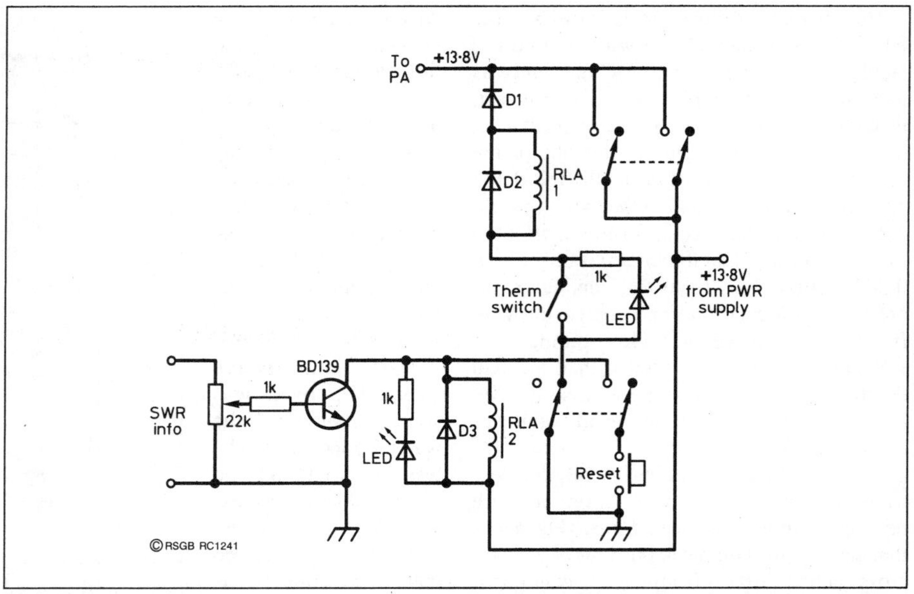

Fig 7: YU1AD's relay-based protection circuit for RF power transistors to guard against wrong polarity of supply, excessive temperature of heatsink, and excessive SWR of load.

Fig 7 shows that the heart of his system is a power relay, having contacts adequate to cope with the current supplied to the power transistor. Where this current is less than 20A, the plastic-enclosed type of relay with octal socket can be used with two sets of contacts in parallel. For 20 - 30A, three-sets of contacts should be used with a power relay having an 11-pin socket. 12V relays are suitable for use with 12V devices, 24V relays for 28V RF power transistors.

One end of the relay coil is connected to the power supply via a 4007 diode, connected so that the relay coil will be connected to the supply if the polarity is correct. The other end of the relay coil goes to a thermal switch mounted directly on the heatsink. The thermal switch should be of the type (available from RS or Electromail) that is normally closed but opens if the temperature of the heatsink reaches 70°C. With the relay RLA1 energised, current will be supplied to the power amplifier, interrupted by excessive temperature, or by the activation of RY2 (any low power DPDT relay) driven by the BD139 which is in turn driven by SWR information. The SWR level at which RLA operates can be adjusted with the 22k pot.

There remains the problem of providing the SWR information. One method would be to build an SWR line and use only the reflected-power half to drive the system. YU1AD, however, finds that low-cost 144MHz SWR meters can be used. He has found that if the potentiometer on such SWR meters is adjusted to read an SWR 2:1, the voltage on the reflection detector diode is approximately 1.5V, which is more than adequate to drive the BD139. The 22k potentiometer can then be adjusted so that RLA2 operates with an SWR of 2:1. A second pair of RLA2 contacts are used as latching contacts and can be disabled by a reset press-to-break button.

SWR voltage can be picked-up from the diode in the SWR meter via a 100μH choke with the other end blocked to ground by a 1nF capacitor. YU1AD notes that it is easy to mount close to the end SO239 antenna connector, a miniature 3.5mm socket, with a similar socket on the back of the power amplifier, and then to connect the two sockets via a piece of shielded wire having two 3.5mm plug connectors.

Note that while the optimum level for SWR cut-out is 2:1, most Motorola MRF series of RF power transistors have a permitted ratio of 30:1 and for these devices an SWR cut-out ratio of 3:1 is permissible. However, an exception is the Motorola MRF448 (which can provide 250W RF output at 30MHz) which has a permitted limit of 3:1 and the cut-out should be not more than 2:1.

The two LEDs are useful in showing immediately whether RLA1 cut-out is caused by SWR or thermal-level.

YU1AD considers that, provided your PSU already incorporates overvoltage and overcurrent protection, his system means that the output transistor is likely to be destroyed only by enormous over-drive or possibly by lightning. While this should generally prove to be the case, there must remain some doubt whether in all circumstances the relays will act fast enough. ♦

HERE AND THERE

IRWIN MATH, WA2NDM, in *CQ* (June 1996, pp50-51) draws attention to the new breed of low-dropout IC regulators stemming from National semiconductors. Whereas with the standard 7800-x series, the input-to-output differential is typically 2.5V, the recent 2940-x series has a differential of only 0.5V at 1A of output current. At 0.5A the differential is only about 0.3V, representing a dissipation of only about 0.15 watts. The output voltage in these regulators is indicated by the x. The LM2940 is available with output voltages of 5, 8, 9, 10, 12 or 15 volts. As an added bonus, the LM2940-x is reverse polarity protected up to an input of 15V which means that the protection diode shown in **Fig 8** is not necessary, saving an 0.7V drop across the diode. In a T0-220 outline it's a drop-in, pin-for-pin replacement for the 7800 series.

Another link with the pioneering days of radio has been broken with the death of Len Borrow, G3ZTK, who as a young man was an assistant at Kings College, London to Sir Edward Appleton, who discovered the 'Appleton Layer' and then modestly renamed the ionospheric layers as D, E and F. Len spent most of his working life in R&D for the Navy, right back to the spark era and later wartime work on submarine communications and radar. He took out an AA licence 74 years ago, remaining active with a daily contact with Bill Heggie, GM3NHM, for over 22 years until his death last May.

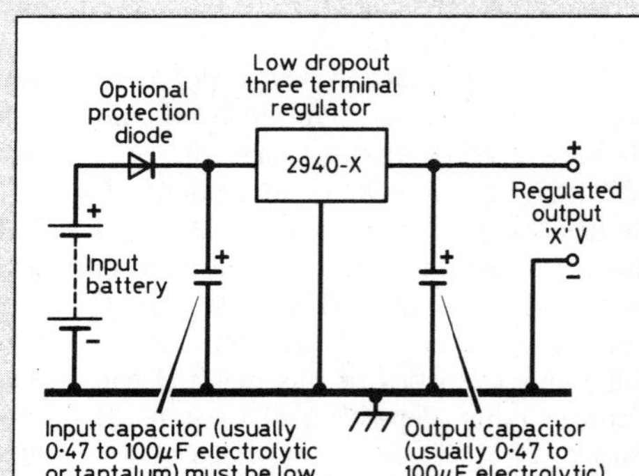

Fig 8: Low-dropout three-terminal IC regulators (2940-series) have much lower input-to output voltage differential compared with the standard 7800-series for which they are drop-in replacements.

Pat Hawker's Technical Topics

PAT HAWKER, G3VA
London 37/SE22 8SS

IMPROVED MULTIBAND TRAP DIPOLE

FOR SOME 40 or so years, the multiband two-trap dipole (as originally developed and popularised by W3DZZ and G8KW) has been recognised as one of the most useful and effective HF multiband wire antennas. Advantages include reduced span, instantaneous band switching and a standard dipole radiation pattern achieved at the cost of a fairly modest power loss in the traps, and having to support the weight of the traps. An alternative form of trapped dipole uses a series of resonant traps, one pair for each (except one) of the bands covered.

As noted, for example in *TT*, March 1984 pp220-221, the original and most popular version used only two 7.1MHz traps but when correctly dimensioned and implemented can provide a reasonably low-impedance centre feedpoint on 3.5 / 7 / 14 / 21 / 28MHz with an SWR of not much more than about 2.5:1 on at least some parts of each band. To achieve such results it is important that the traps are correctly positioned in the element and have the correct LC ratio. **Fig 1** shows an acceptable W3DZZ design requiring the use of very-high-voltage 50pF capacitors in the traps and excellent weatherproofing. The whole system should be symmetrical about the feedpoint, with the top span either horizontal or in inverted-V form but not as a continuously-sloping wire.

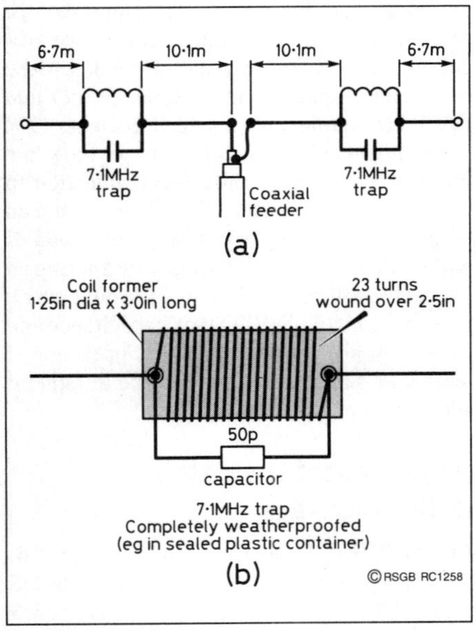

Fig 1: The original W3DZZ multiband trap dipole using only two 7.1MHz traps yet providing a low-impedance centre feedpoint on 3.5 / 7 / 14 / 21 / 28MHz with an SWR not greater than about 3:1. The 50pF capacitor must be a high-voltage type capable of withstanding high RF voltages. The whole system should be symmetrical about the feedpoint, with the top span either horizontal or in inverted-vee form but not as a continuously sloping wire. For use with solid-state transmitters an ATU is advisable on at least some bands where the SWR is greater than about 2:1.

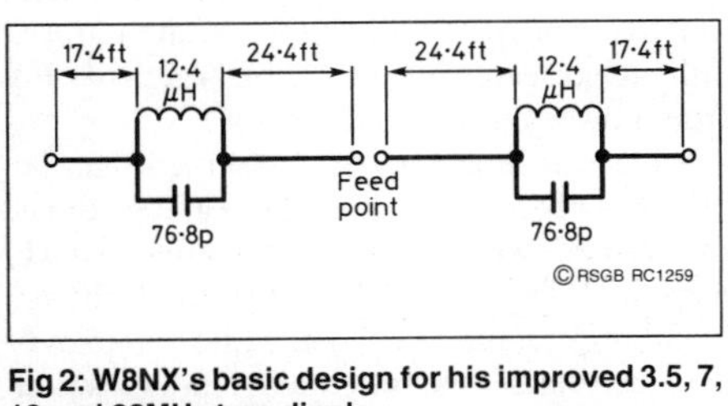

Fig 2: W8NX's basic design for his improved 3.5, 7, 18 and 28MHz trap dipole.

Some 12 years ago, Robert Sommer, N4UU, (*QST*, December 1984 - see also comments by Geoff Roberts, G3ENY in *TT*, May 1986, p338) showed that the use of coaxial cable to form the traps has the unique advantage of providing an in-built high-voltage capacitor as well as other useful features.

N4UU noted that there is an optimum way to use the cable to form an inductor, the whole acting as a parallel resonant circuit: "The trick is to use the shortest length of cable (minimum capacitance) which can be closewound (maximum inductance) on a former of such a diameter to make a coil of correct inductance having a length-to-diameter ratio to give a good value of Q."

G3ENY pointed out that this is not easy or simple to achieve using cut-and-try methods since it requires the use of a computer to solve, by an iterative loop, the cubic equation which ties together the LC ratio and diameter variables to obtain resonance. However, G3ENY provided two near optimum designs of trap for 7010kHz (probably more suitable for European bands than 7.1MHz), one using standard coaxial cable (RG58/U) and the other miniature coaxial cable: **Table 1**. He noted that since there is only a limited range of PVC pipe sizes available from builder's merchants, the constructor would be well advised to choose a diameter which gives a length-to-diameter ratio of about 0.6 certainly less than one (a square coil).

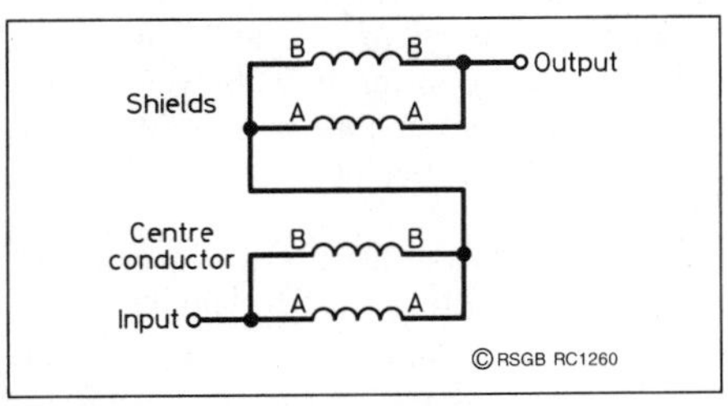

Fig 3: Diagram showing the connections of a double-coaxial-cable trap providing integral capacitance.

	RG-58/U	RG-174/U
Length of coaxial cable	61.6in	52.7in
Diameter of former	2.125in	1.6in
Number of turns	7.8	9.8
Length of former	2.5in	2in
Coil Inductance	3.35μH	3.78μH

Table 1: 7.01MHz traps based on coaxial cable.

Al Buxton, W8NX, in 'An improved multiband trap dipole antenna' (*QST*, July 1996, pp32-34) introduces a new trap design and a change in trap location: **Fig 2**. His design features double-coaxial-cable-wound traps (**Fig 3**) having a higher quality (Q) factor (**Fig 4**) than earlier coax-cable traps, and hence providing a useful reduction of power loss in the traps. The traps also provide a fourfold increase in power-handling capability over that of the earlier single wound coax traps.

The new design is for operation on the American 3.5 - 4.0MHz and 7.0 - 7.3MHz bands plus the 18 and 28MHz bands. At his location, with a 50-to-75Ω transformer and a 75Ω feed line, in an inverted vee configuration with the apex at 40ft above ground, the SWR performance achieved by W8NX is shown in **Fig 5**, showing that no ATU would be needed on 7MHz, on part of the 3.5MHz band and on part of the 28MHz band (SWR less than 2:1) but advisable with solid-state transmitters on 18MHz and for extending low SWR coverage of the 28MHz and 3.5MHz bands.

Fig 6 shows how the traps are made. Each trap requiring no external high-voltage capacitor consists of 6.88 double turns of seven-foot lengths of RG-59 (Belden 8241) coaxial cable wound on a 5.33in length of 3.5in outer diameter (schedule 50) PVC pipe (three-inch inner-diameter). Check that the form is actually 3.5in OD as the trap frequency (5.16MHz ±25kHz) is sensitive to the diameter of the former. Use a high-quality 1:1 balun; the apex should be at least 35ft high, the ends at least 15ft high. Feed the antenna with either 50 or 75Ω cable, the higher value being preferred to lower the feed-line SWR on 18 and 28MHz.

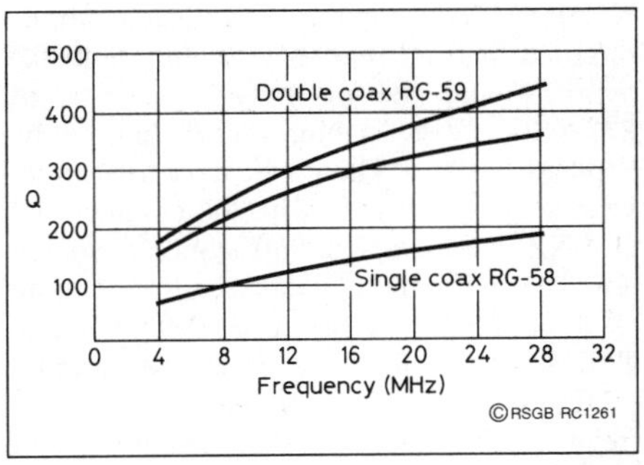

Fig 4: Quality factor (Q) versus frequency of coax traps.

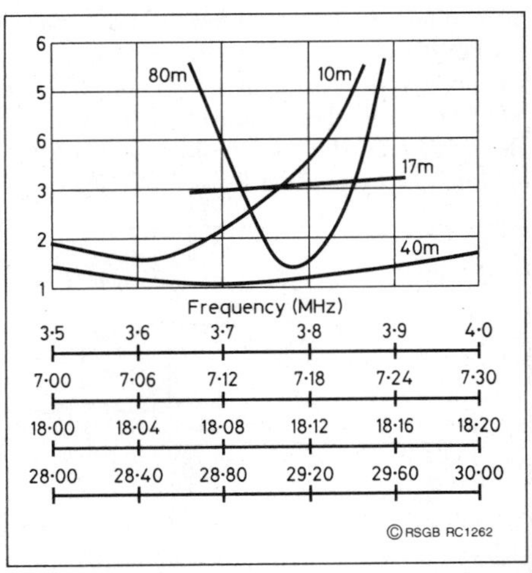

Fig 5: W8NX's SWR plot of his improved trap dipole for the 3.5/7/18/28MHz bands based on the design shown in Fig 1.

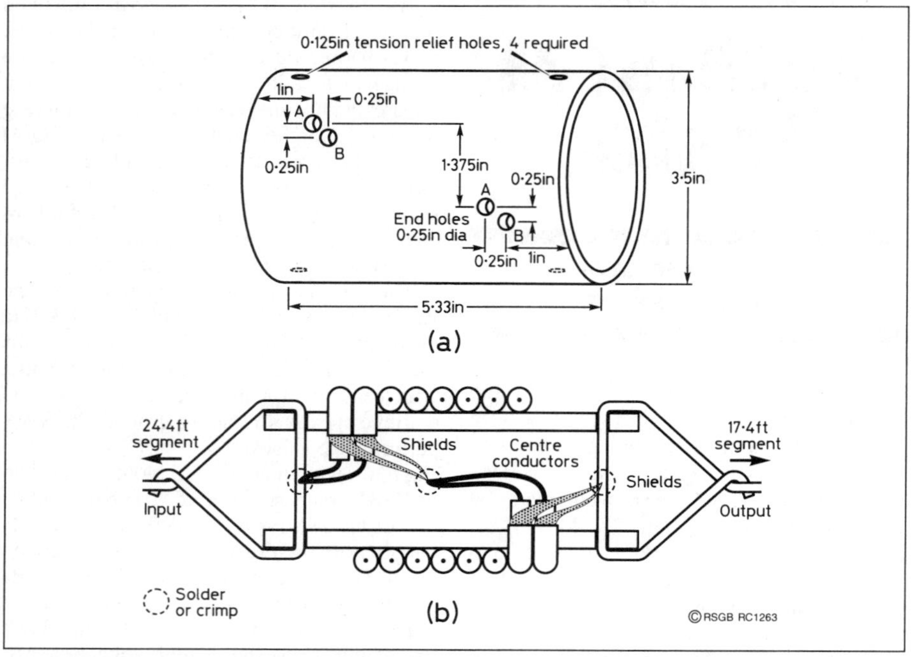

Fig 6: Constructional details of the W8NX double coaxial traps. (a) Position of holes in the trap coil former. (b) Cross-sectional view of a double-coax trap.

W8NX suggests that trap loss on 3.5MHz is 1.66dB but only 0.04dB on 7MHz, 0.001dB on 18MHz and 0dB on 28MHz. He provides further detailed constructional information on the traps but it is hoped that the brief outline given here will at least allow experimentation with his 'improved' system.

HUFF & PUFF OSCILLATOR

A NUMBER OF comments and ideas continue to come in from readers regarding both the original and the improved versions of PA0KSB's VFO stabiliser (*TT*, June, July, September 1996). Those who have tried either version seem to agree with the comment by G3DXZ that a Huff & Puff stabiliser should now be considered a standard addition to an LC-type VFO. One result, as Peter Lawton, G7IXH, points out, is that a stabilised oscillator forms an effective replacement for the low-frequency VFO plus crystal oscillator mixer configuration "with all their horrible unwanted frequencies".

G7IXH was one of those who have experimented and implemented the original type of PA0KSB stabiliser, before the appearance of the improved version in which the divided VFO clocks the crystal rather than the reverse. He found the earlier system most satisfactory, although he agrees that the new arrangement offers useful advantages (see PA0KSB's comments in *TT*, December 1996).

G7IXH writes: "G3ROZ's material on building a Huff & Puff stabiliser for the KW2000-series (note in *TT* December 1995, p73) enabled me to build stabilisers for a solid-state receiver as well as my KW2000A and these do an excellent job. The VFOs remain locked for indefinite periods after only a few seconds warm up, lock is regained immediately on retuning and no saturation of the stabilisers has been noted.

"**Fig 7** shows the circuit I developed for the input to the stabiliser from the VFO and this may be of interest to others. It combines high sensitivity with a high degree of isolation of the stabiliser/frequency-meter from the VFO. The input capacitance is only 1.8pF and at 5MHz it produces a 0 to 5V approximately square wave at the frequency of the VFO from an input as low as 60mV pk sine wave. It will work up to 30MHz with reduced sensitivity. A frequency meter can be connected to measure the oscillator frequency with very little effect on the frequency although in practice I include a further source-follower after the gate to improve isolation even further. At 5MHz, the change on connecting the meter is only a few Hertz.

"The output from the 2N3819 source follower is applied to one of the inputs of the 74HC132 Schmitt trigger NAND gate. Since the hysteresis of this gate is reduced by the feedback via RV1, then with careful adjustment of RV1 and RV2, a very small input change is sufficient to turn the gate on or off. To set up, adjust RV2 until the gate oscillates. Then adjust RV1 until oscillation ceases. Repeat this process until there is only one position of RV2 at which oscillation occurs. Then back off RV1 into the non-oscillating region and the circuit is ready."

G7IXH has also tried the original form of stabiliser at frequencies up to 20MHz and found it perfectly effective. However, he found that his transistor oscillators gave a rather rough note at these higher frequencies, possibly due to input capacitance variations during the course of a cycle. The cleanest note was from the Kallitron push-pull oscillator as described in the current RSGB Handbook (p6.45) and originally stemming from G4DTC's 'ultimate' receiver (*TT*, December 1987, and *Technical Topics Scrapbook, 1985-89*, p211, with subsequent notes on the origin of this type of oscillator for HF and VHF).

G7IXH continues: "One would expect input capacitance variations partially to cancel with a push-pull oscillator. The varicap diode(s) of the stabiliser also increase the roughness of the note at these frequencies. Back-to-back diodes are better, and this arrangement can also cope with higher oscillator voltages before rectification occurs. I built my stabilisers before the new version appeared in the June 1996 *TT* and PA0KSB's comments in the December *TT*. The 'universal' stabilisers I have built are modules which need only one connection to the VFO, a power connection, an earth and an output for the frequency meter. The modules will work with any VFO, simply by altering an internal resistor according to a formula which takes account of the VFO tank capacitance and frequency and ensures optimum operation. I feel that a stabiliser should now be a standard addition to any good VFO."

RADIATION HAZARDS QUESTIONED

ACCORDING TO a report in *Nature* (November 7, 1996) a panel of experts set up by the US National Research Council has advised the US Government that there is "no conclusive and consistent evidence" that exposure to electric and magnetic fields from overhead power lines causes cancer or other adverse health effects. The panel stated that the established correlation between the incidence of leukaemia and proximity to power lines was most likely due to other factors related to the vicinity of the power lines, such as poverty or high traffic density. It is suggested that the study could help to disabuse the widespread but unproven perception that overhead power lines are a health hazard, although research into a possible health hazard from power lines should continue.

American amateurs are concerned that the FCC has adopted new guidelines and methods for evaluating the environmental effects of RF emissions from FCC-regulated transmitters, including cellular telephone, radio and television broadcast and satellite communication transmitters. For the first time the guidelines

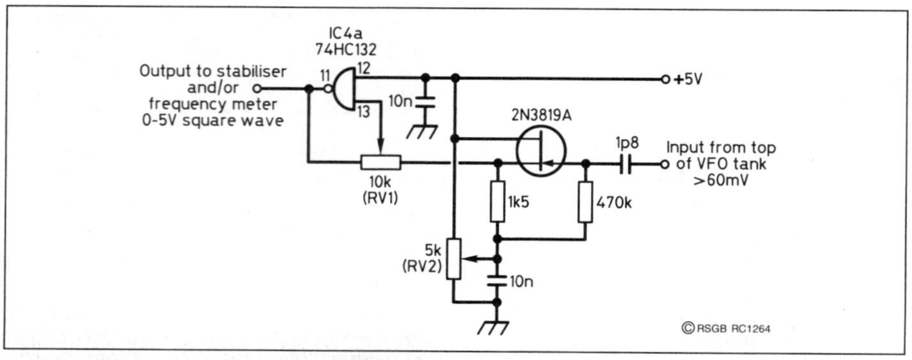

Fig 7: G7IXH's input amplifier for use with a Huff & Puff oscillator stabiliser.

requiring evaluation of possible hazards apply also to amateur radio transmitters. According to *CQ*, November 1996, earlier studies have suggested that while some field strengths from amateur HF transmitters can be relatively high, the potential hazard may be less than for lower field strengths measured at VHF where the highest specific absorption rates (SARs) occur in humans. Ground-level field strength readings at HF frequencies were relatively low; with significant radiation levels largely confined to main 'beam' paths. It seemed that vehicle-mounted amateur antennas closer to the ground create the greatest possibility for significant exposure to RF in publicly accessible areas, but generally, RF field strengths encountered inside amateur shacks were well below IEEE recommended exposure limits.

It would seem that unless the new guidelines are rescinded, amateurs will be legally bound to ensure (by evaluation) that their RF radiation levels are within the 1992 ANSI standards. ARRL are opposing the inclusion of amateur radio in the new FCC guidelines, but this has led to the resignation of their Biological Effects Committee of specialists.

A long article in *IEEE Spectrum* (September 1996, pp26-33) 'Do portable electronics endanger flight? - The evidence mounts' carries an editorial introduction: "According to a new study, the risk that RF emissions from carry-on electronic devices will affect avionics, although not high, is still high enough to warrant tougher government regulations." The article shows that a typical airliner carries some 20 communications, beacon, automatic direction finders, distance-measuring equipment, VHF omnidirectional range localiser, radio altimeters, glide slope (dual) and weather radar units etc positioned throughout the aircraft. Most of these are potentially vulnerable to interference from RF emissions from lap-top computers and other portable electronic devices (PEDs). An RCTA committee draft recommends that the current American policy of asking passengers to turn off PEDs during taxi, takeoff and landing, which is now voluntary on the part of the airline, should be made mandatory. Another is that the travelling public should be educated about the reason for concern over PEDs and which of these may cause problems. The report also recommends continued study of emissions from PEDs, especially the newer models that may contain built-in communications devices.

AMMETERS & THE UNIVERSAL SHUNT

BOB PRICE, GW3ECH/F5VDN, who divides his days between Brecon and Brittany, writes: "It is strange how old principles get forgotten and remain unknown to many of the current generation of amateurs. I mentioned building a multi-range ammeter to someone the other day. He replied that surely such an instrument required very heavy duty switches and was not a practical proposition.

"In practice, if you have a sensitive microammeter and can afford a small voltage drop in the circuit in which you wish to measure the current, the meter can be used as a multi-range voltmeter measuring the voltage drop across a low-value shunt unit. The range switch then only has to carry the full-scale deflection current of the microammeter.

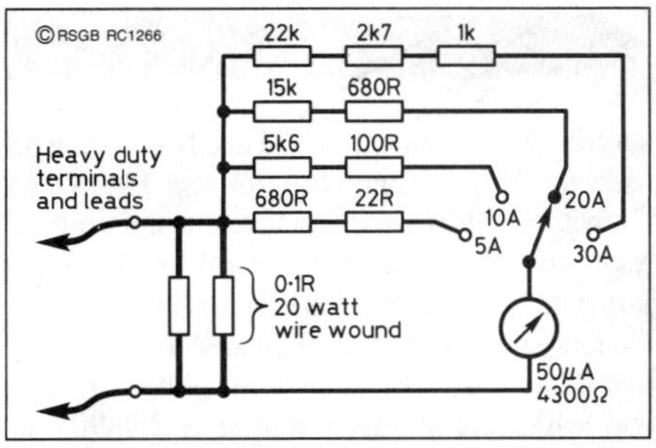

Fig 9: A multi-range ammeter capable of measuring currents up to 30A based on the universal shunt principle as described by GW3ECH.

"A typical microammeter currently offered in one of the large mail-order catalogues has the following specification: FSD 50 microamperes; internal resistance 4300Ω. (Similar or even better meters can often be found ex-equipment; my own multi-range ammeter uses a meter with an internal resistance of 1000Ω and FSD 100 microamps).

"Careful application of Ohm's Law (remembering the decimal points) will reveal that the voltage required across the microammeter for FSD current is:

V (volts) = I x R = FSD (A) x Internal Resistance (Ohms) = 0.00005 x 4300 = 0.215.

"If you make the shunt a resistance of 0.05Ω, made up of two 0.1Ω 20 watt wirewound resistors, the current through the shunt to produce 0.215V is I =V/R = 0.215/0.05 = 4.3A.

"With the meter scaled 0 - 50, it will be convenient if the meter measures 5A FSD; this requires a resistance in series with the meter of 700Ω in order to make a total resistance of 5000Ω. Similar calculations can be made to provide any other higher ampere ranges required, the limit being the allowable dissipation in the shunt. This could be increased if the shunt is metal clad and mounted on a heat sink and/or the measurements are taken quickly.

"**Fig 9** shows a design based on a 4300Ω 50µA microammeter with FSD ranges of 5, 10, 20 and 30 amperes. It uses preferred-value series resistors and a cheap low current switch. But remember, with the high currents involved, not to skimp on the size of the terminals or the diameter of the wires used to connect the completed meter into the circuit to be measured, or the wires or straps between the terminals and the universal shunt resistors."

TRANSISTORS AS VVC OR ZENER DIODES

DOUG DEMAW, W1FB, (*CQ*, November 1996, pp76 and 78) reminds us that small-signal NPN bipolar transistors, connected as diodes, can often be pressed into use to meet a requirement for voltage variable capacitor (varactor) diodes or for voltage regulating diodes. W1FB believes that this long-established trick, which can save delays in obtaining VVC or Zener diodes, is relatively unknown to many amateurs.

Fig 10 shows how common NPN transistors

OVERTONE CRYSTAL OSCILLATOR

TT, FEBRUARY 1996 and July 1994 pp53-55, described in detail a novel very-low-noise double-tank stripline Hartley VHF oscillator developed by Colin Horrabin, G3SBI, for use in a super-linear receiver. Since then, he has also tried a similar approach for a 50MHz overtone crystals oscillator: **Fig 8** (a 53.39MHz crystal was on hand).

He writes: "With the feedback point on the spiral resonator used, the output is nearly a square wave, rich in harmonics. Output after a 6dB pad is 2dBm and the 9th harmonic is -20dBm. Putting a finger on the drain output causes less than 100Hz shift of the 8th harmonic output confirming the high degree of isolation. The circuit could be constructed in a similar arrangement to the original using a -12V supply with the 56Ω drain resistor grounded. It seems possible to use such a crystal oscillator to obtain the 9th harmonic directly using a helical filter, and amplify it to simplify a local oscillator I am using at UHF. Note that the use of a ceramic trimmer in this circuit should add much more close-in noise and should be avoided."

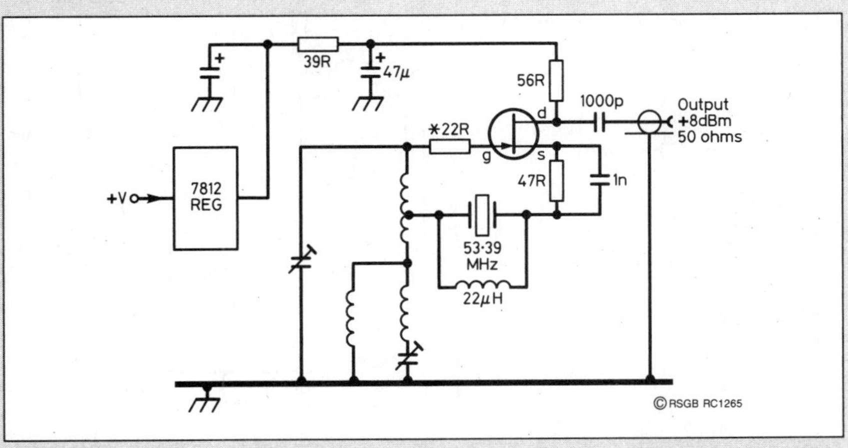

Fig 8: G3SBI's 50MHz overtone crystal oscillator using the low-noise split Hartley circuit fully described in *TT*, February 1996 (see also earlier version in *TT*, July 1994).

12 TO 24V DC-DC CONVERTER

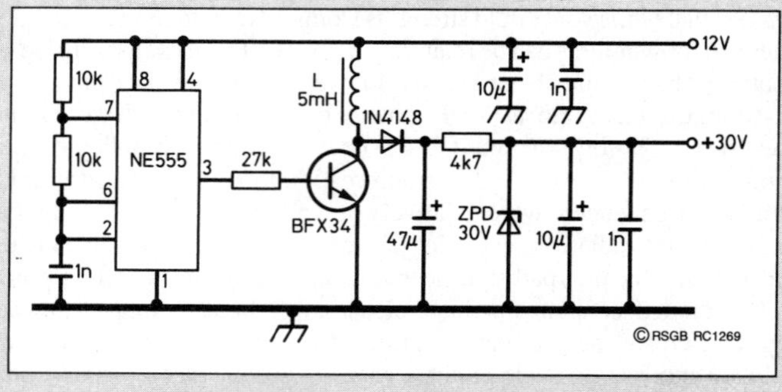

Fig 12: 12-to-30V DC-DC converter suitable for use with electronic tuning diodes etc, as described by DJ8ES.

IN ORDER TO obtain maximum tuning range with some VVC electronic tuning diodes, a voltage span of some 20V or so is required. With the usual 12V supply, this can be obtained by the use of a low-current DC-DC converter. In *VHF Communications* (Nr 2/1996), Wolfgang Schneider, DJ8ES, reviews a number of simple voltage converters including some based on a storage (flyback) choke. His preferred arrangement is shown in **Fig 12** in which an NE555 timer IC is used as a 40kHz oscillator (frequency determined by the two 10K resistors and the 1nF capacitor) to drive a BFX34 switching transistor. The 5mH storage choke (L) produces about 80V pulses at the collector which are then rectified and smoothed. The output voltage is then regulated to 30V by means of the ZPD30V zener diode. If other switching devices are used in place of the BFX34, the base resistor (47K) must be optimally determined for the transistor chosen.

DJ8ES describes the storage choke as having an AL of 250 and uses a Siemens shell core, in which case a high quality inductance of 5mH is obtained with 150 turns of enamelled copper wire of 0.1mm diameter. I note that in an earlier but basically similar design he described the storage inductor as using as air-gapped pot-core (AL = 250) wound with 80 turns of 0.2mm lacquered copper wire.

such as the 2N3904, 2N2222 or 2N4400 can be used to obtain a regulated voltage. Fig 10 (a) shows a 2N3904 used as a low-current voltage regulator. But note that not all small-signal transistors regulate at 5.4V since each transistor has its own characteristics. W1FB reports that the maximum 'zener' voltage he obtained was +8V. He points out that a 2N3904 is roughly on a par with a 400mW Zener diode in terms of the maximum power dissipation.

The regulated voltage can be raised by adding silicon diodes as shown in Fig 10 (b), with each diode adding 0.7V or, as shown in Fig 10 (c), a second bipolar transistor can be connected in series. Silicon power transistors and silicon power diodes can also be used to regulate huskier supplies, with several power diodes providing regulation at a rate of 0.7V per diode. When diodes (or diode-connected transistors) are used as regulators the positive unregulated voltage must be applied to the anode of the diode.

Although not mentioned in W1FB's article, it was noted in *TT* as long ago as January and April 1968 that a junction FET with gate and source connected together can form a useful "constant current diode" for such applications as noise or spike limiters, for biasing differential amplifiers or in conjunction with a Zener diode (or transistor equivalent) to provide pre-regulation making for better voltage regulation and temperature performance. Later, G3NOX pointed out that the resistor shown in **Fig 11 (a)** in the FET drain circuit is not necessary unless to protect the device or to form a voltage drop; but if a variable resistor is included in the source circuit the constant current can be adjusted to any value of current up to the device's I_{DDS}: Fig 11 (c).

Similarly, it has long been recognised that almost any RF diode, Zener diode or diode-connected bipolar transistor can be used as a voltage-variable-capacitance tuning diode, although normally it is preferable to use VVC (Varactor) diodes designed specifically for this purpose, since the limiting factor with the use of electronic tuning is often the relatively low Q of some diodes at HF and VHF. In *TT*, July 1967, a table was given showing the range of capacitance variation with then commonly available germanium, silicon and zener diodes for a negative voltage variation between -8V and 0V. Even today, apart from the question of Q, it must be stressed, as W1FB does, that VVC-tuned oscillators have greater short-term (temperature) drift than those which are tuned by well-designed air variable capacitors. However, for less critical applications, the thermal problems of both genuine VVC tuning diodes, and transistors used as varactors, can often be tolerated.

EDDYSTONE 75TH ANNIVERSARY

S A FAULKNER, GU3MLR, noted the reference in *TT*, November 1996, to the 1952 introduction by Collins Radio of the mechanical filter and its effective use in their high-performance professional and amateur-radio receiver. He felt, however, that space should be found in *RadCom* for some account of the contribution to our hobby of UK firms such as Eddystone, KW Electronics, Labgear (closely associated with Pye), Racal etc. I have to agree with GU3MLR's comments especially as I still use a number of excellent Eddystone components, a KW2000A transceiver made by KW Electronics, and a Labgear LG300 transmitter. Also, I have often told the story of how Racal came to introduce the Wadley loop into their RA17 receiver with the aid of the a single saw cut.

In practice, *TT*, June 1988, p425 told something of the interesting history of Eddystone. In 1922, 75 years ago this year, as an offshoot of the Birmingham costume-jewellery firm of Jarrett, Rainsford and Loughton - which had a factory turning out six tons of hairpins each

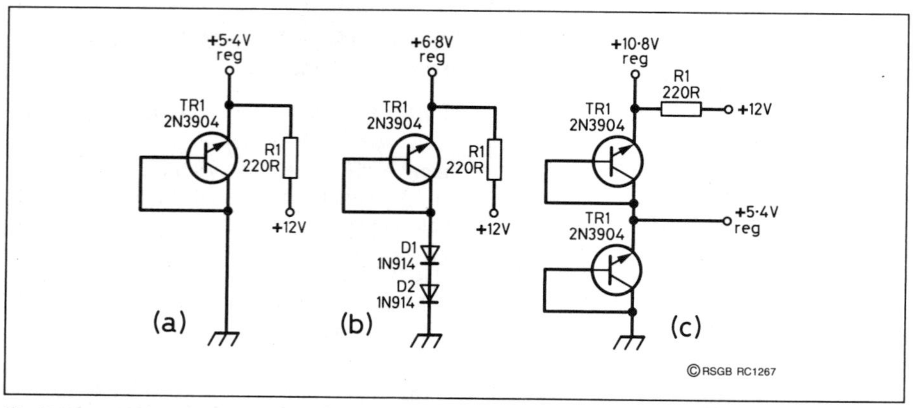

Fig 10: Ways in which common NPN transistors can be used as voltage regulating diodes. (a) 2N3904 used as a +5.4 V regulator; (b) regulated voltage may be raised by adding silicon diodes (extra 0.7V per diode); (c) two diode-connected transistors in series used to provide regulated 10.8V and 5.4V outputs.

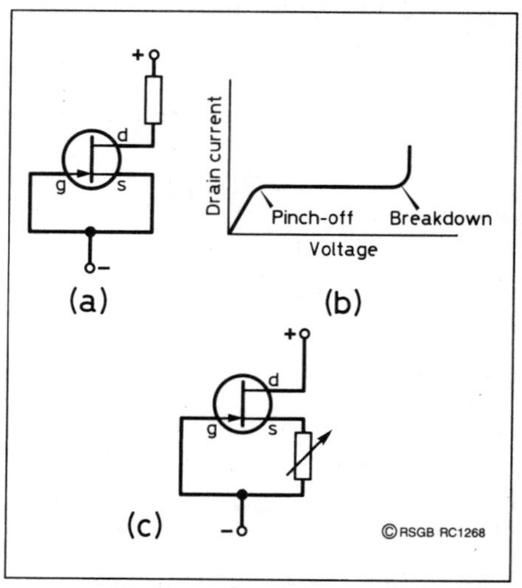

Fig 11: (a) FET used as a constant current diode; (b) typical characteristics; (c) variable constant-current arrangement.

week badly hit by the change of fashion when women began to cut their hair short in the Eton crop, the bob and the shingle - decided to enter the broadcast radio industry as Stratton & Company under G A Loughton, with production beginning in a modest way in 1923. The competition proved intense and four years later, in 1927, Harold Cox and Arthur Edwards, G6XJ, changed the emphasis from medium to high frequencies. This soon resulted in receivers such as the Eddystone 'All World Four' which in its 'ant-proof' aluminium die-cast box brought HF broadcast reception to many tea and rubber planters in the Far East, helping to pioneer the concept of 'tropicalised' radio equipment.

For amateurs and SWLs, the modest but effective 'All World Two' with its ingenious bandspread plus bandset tuning proved an extremely popular 'first model' in the late 1930s. I can still recall trying one borrowed from Charles Bryant, GW3SB, when he moved on to a Hallicrafters SX16 Super Skyrider.

Eddystone pioneered the use of VHF for police radio, providing an emergency communications network for the London police just a few weeks before the outbreak of WW2 in September 1939. The company suffered severely in the Blitz and were finally forced to move into the West Heath Lido swimming baths (known as 'The Bath Tub') where they stayed until last year, long after becoming a subsidiary of GEC-Marconi in the 1960s, and largely withdrawing from the amateur market. Their first superhet communications receiver was marketed in 1939 and during the early post-war period their communication receiver became virtually the mainstay of the British amateur radio market, together with an associated range of broadcast models designed for shipboard use.

Eddystone Radio, like the parent Marconi Company, has always been conscious of the importance of recording and preserving its history. A detailed hard-cover book on the company was published even before the Marconi take-over, though now long out-of-print. A booklet compiled by Pam Reynolds was published in 1973 marking 'The first 50 years, 1923-73' and a four-page article 'The Eddystone Radio Story' by Chris Pettit, G0EYO, managing director of Eddystone Radio Ltd, was published in *Practical Wireless*, February 1994.

During 1996, Eddystone Radio Ltd finally moved out of 'The Bath Tub' to a new location at Unit 8/9, Birkdale Avenue, Heeley Road, Selly Oak, Birmingham B29 6UB with a range of professional broadcast-transmitter and communications products that includes a high-performance VLF/HF Model 6200 communications receiver priced around £2500 plus VAT that would delight radio amateurs. The company's comprehensive museum of past Eddystone products has been retained at the new site.

G0EYO is also Patron of the 'Eddystone User Group' which (as mentioned in *TT*, February 1992, p38) provides an excellent newsletter, now approaching its 40th issue, compiled by Ted Moore, ex-G7AIR, but with the Group now administered by Graeme Wormald, G3GGL, (15 Sabrina Drive, Bewdley, Worcestershire DY12 2RJ. Tel: 01299 403 372).

SWITCHABLE RF MATCHING TRANSFORMER

PHIL SALAS, AD5X (*CQ*, April 1996, pp30-31) has described a switchable trifiliar-wound wideband auto transformer that provides a flexible method of matching to mobile or fixed electrically-short antennas. As the sub-heading puts it: "This ubiquitous device is easy to build and will add many pleasurable miles to your mobile operation." It should prove equally useful for such applications as 'tourist' operation using balcony-mounted whip antennas etc. By adding just one tap to a standard trifiliar-wound toroid transformer, with a two-pole, six position rotary switch the following ratios are provided: 1:1 (bypass); 1:1.44; 1:2.25; 1:4; 1:5.75; and 1:9 up or down.

The transformer is based on the well-established, high-power design by Jerry Sevik, W2FMI, who used a 1.5in diameter ferrite toroid with a permeability of 250. For use with RF outputs up to about 100 watts, AD5X chose an FT114-61 toroid (1.14in diameter) with a permeability of 125. With 10 trifiliar turns there is sufficient impedances for the lower-frequency (1.8 and 3.5MHz) bands, yet suitable for use with the low impedances of short whip antennas on any of the HF bands.

Details are shown in **Fig 13**. AD5X built his unit into a small aluminium bow with dimensions of 2.75 x 2.125 x 1.625 inches. He writes: "To wind the transformer, first take one wire and create a centre tap as shown in Fig 13(c). Then place three wires together (I held them together by placing heat-shrink tubing on either side of the tap) and carefully wrap the wires around the toroid starting from the centre tap - ie, place the centre tap on the toroid and wind turns on either side of it. Each non-tapped wire consists of 14 inches of 16AWG enamel-covered wire. The centre tapped winding consists of 116 inches of the same enamel-covered wire."

AD5X found that in practice the input impedance of mobile antennas tends to be significant greater than might be expected from the theory of electrically-short antennas due to the large ground losses in small vehicles.

AD5X concludes: "The W2FMI design, with appropriate switch wiring, gives you a very flexible impedance matching transformer for many HF antenna applications."

HERE & THERE

HILLAR RAAMAT, N6HR, (*CQ*, November 1996, p20) shows that a low-cost 'surplus' computer mouse can be readily converted into a piano-type iambic Morse keyer: "After you get used to the up-and-down motion of the 'mouse' instead of the side-to-side-and-squeeze motion of the iambic keyer, you can send as good CW as the microswitch movements allow.

David Long, G3PTU, writes: "I have been trying out some 'Avantek' integrated RF amplifier which visually can easily be confused with conventional transistors in a TO5 can, but are altogether more wondrous devices.

"Several of these devices, type GPD-403, were found recently on a surplus 'scrap board' bought cheaply at a rally. I believe that the GPD-403 is a broadband (1MHz to 2GHz) amplifier with 10dB gain. From examination of the board, it requires a resistor in series with the positive supply and heavy decoupling from that point to the case and another of the pins. The remaining two of the four-pins are for input and output, each at 50Ω impedance.

"To investigate the amateur-potential of these devices, I have built one into a small in-line box with BNC connectors as the ports. I find that the old absorption wavemeter is totally reborn when this unit is placed ahead of it, as is the detector unit on my RF sweeper, particularly when sweeping filters or other units with no integral 'gain'. As for my frequency counter . . .

"I believe that the devices I have represent an oldish design, and more modern equivalents are available such as the MAV series. Nevertheless, for a 'box' with gain, extremely wide bandwidth and docile temperament they appear to take a lot of beating." ♦

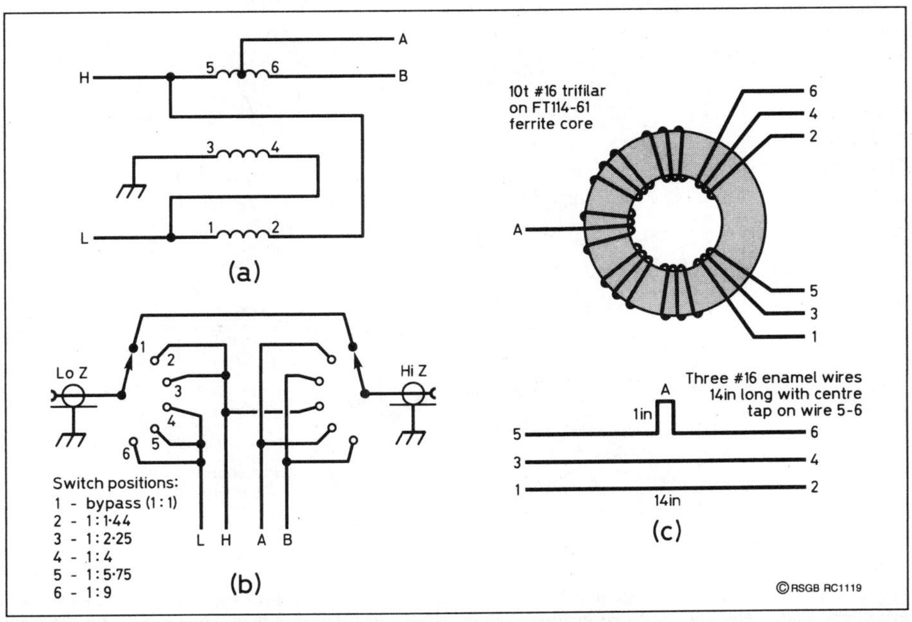

Fig 13: Broadband switchable impedance matching transformer as described by AD5X. It is based on the W2FMI design with the addition of one tapped winding. (a) Circuit diagram of the trifiliar-wound transformer. (b) Wiring of the switch-box. (c) Winding details of the transformer.

Pat Hawker's
Technical Topics

PAT HAWKER, G3VA
London 37/SE22 8SS

TAMING ELECTROSTATIC DISCHARGES

IN THE 1960s, with the introduction of MOSFET (IGFET) and CMOS devices, many warnings were given in *TT* and elsewhere of the destructive power of electrostatic discharges (ESD) and mains leakages which could easily puncture the thin insulated layers of semiconductor devices. *Amateur Radio Techniques 7th ed,* (p22), for example, includes a list of "special handling procedures" published originally by MCP Electronics some 30 years ago and intended to prevent the destruction of MOSFET devices not protected by built-in zener diodes before they were safely wired into equipment with bleeder resistors:

Shorting wire or foil should be kept wrapped around leads of devices whenever possible, and always during storage. When this shorting wire is to be removed, hold the leads between finger and thumb of one hand.

Never allow a gate lead to float since there is then no bleed path for electrons. Sufficient static charges can build up to cause permanent breakdown of the gate oxide.

Always use a soldering iron with an earthed tip. Beware of voltage transients in the test or power supply equipment (short duration voltage transients can damage MOSFET/CMOS devices even when only present on the source or drain).

Do not attempt to measure gate leakage current (the extremely high impedance of the gate would make this operation hazardous with conventional techniques).

Because of the extremely high input impedances, extreme care should be taken to avoid voltages in excess of the maximum rated breakdown voltage from being accidentally applied to the gate-to-source or gate-to-drain leads: "Do not connect, touch or handle improperly grounded power supplies or equipment, as excess voltage or static charges may cause permanent damage to the gate oxide."

As the years have passed, and devices often fitted with in-built zener diodes there has been a tendency to disregard such advice. However, modern chips with their mixture of bipolar and FET devices, and their ever thinner gates for low-voltage operation, remain highly vulnerable to ESD and other high-voltage transients.

This is emphasised in an article by Frank Harding 'Controlling Electrostatic Discharges' (*Television*, December 1996, pp94-95). He points out: "Static electricity is generated when two materials are rubbed together. With plastics and other nonconductive materials the charge remains, hence the name 'static'. We humans are the main source; the clothing we wear, the common plastics we use, the carpets we walk across - these all contribute to an electrostatic build . . . which often results in a tiny spark discharge to earth of over 3kV; far more than is required to puncture a delicate, and possibly expensive, microcontroller chip!"

Frank Harding notes that ESD can damage delicate semiconductor devices in all electronic equipment: "Sometimes the damage is minimal - perhaps nothing more than a disturbed clock pulse train that results in the microcontroller 'hanging up'. In this case all that is required is to press a reset button or switch off, wait a few seconds, then switch on again. At its worst, ESD can cause catastrophic failure of a device by 'punch through'.

"The strange fact is that in most cases the component appears to be undamaged but then degenerates in such a way that failure occurs at a later date . . . such 'latent failure' may happen quite quickly or after several weeks. The important thing is to avoid the initial static discharge."

Frank Harding advises that while static electricity cannot be eliminated it can be controlled by providing adequate paths to earth. **Fig 1**, redrawn from his *Television* article, shows ESD protection (but not the earthing of solder-iron tips and other test instruments etc) for a permanent professional repair work-station. It should provide some useful ideas for anyone who carries out a good deal of solid-state home-construction.

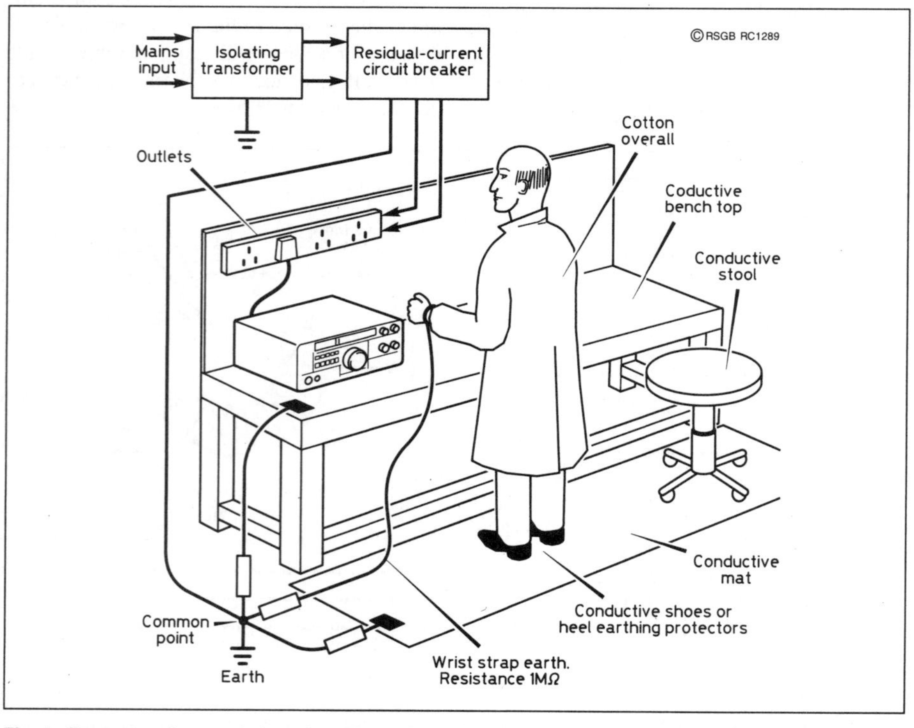

Fig 1: For extensive workshop handling of semiconductor devices based on FETs, CMOS and microprocessors etc, good protection against electrostatic discharges is highly desirable. This diagram illustrates the type of ESD protection recommended for professional workstations. Frank Harding suggests wearing a cotton overall and leather-soled shoes or heel earthing protectors; and using conductive compartment trays for all ESD-sensitive components.

144MHZ NOTCH FILTER

A NUMBER OF recent *TT* items have emphasised the increasing problem for 144MHz operators posed by the large number of commercial paging services operating on frequencies fairly close to the amateur band. The January 1994 *TT* recounted the experiences of P R Kemble, G3UYK, in dealing with bursts of 153MHz paging data which interfered with his two-metre mobile reception. Paging transmissions, particularly in urban areas, can represent a real problem in view of the relatively poor front-end selectivity of most VHF receivers.

G3UYK overcame the problem with the aid of a home-built, three-pole strip-line notch filter with the strip lines fashioned from some scrap double-sided copper-clad board (Fig 6 of *TT* January 1994, p40). Later (*TT*, February 1995, p69) Uwe Keonneker, DL8OBF, described the construction of a basically similar two-pole strip line filter with a rather sharper notch which he used to filter out a local 144MHz amateur digirepeater which was interfering with the reception of BREM-SAT signals on 137.800MHz.

I was reminded of these problem-solving filter designs by an article 'Intermod - A modern urban problem' by Ed Hare, KA1CV (*QST* Aug 96), sub-titled 'Getting squeaks and squawks from your H-T? Are pagers crunching your mobile rig's front end? Here's how intermod happens - and how to fight back!'

KA1CV contrasts the relatively empty VHF spectrum in earlier days with the "thousands upon thousands of transmitters operating today above 50MHz for business, public service, amateur, satellite and many other services". He stresses that it is likely that some of these systems will interact in undesirable ways as a result of intermod, harmonics, and the poor RF selectivity and limited dynamic range of many VHF receivers and transceivers.

The article describes in detail the intermod processes and how these are affected by the dynamic range of the receiver. He describes the use of external band-pass or band-reject (notch) filters illustrating a number of professionally-made filters available in the USA.

However to accompany KA1CV's article, Zack Lau, KH6CP/1, ARRL Senior Lab Engineer, describes a '2-Metre Notch Filter You Can Build' (**Fig 2**) which can reduce signals just above the 144MHz band without significantly degrading wanted 144MHz signals. He claims that "as a bonus it notches out some FM broadcast signals (88 - 108MHz) as well. The filter has a notch depth of 27dB, with 0.5dB of insertion loss."

He notes that, as shown, the design assumes that you want a notch *above* the 144MHz band (typically 152MHz) to which L1/C2 should be tuned. For a notch below the band, the unwanted reactance would be inductive, which would require only a capacitor to obtain parallel resonance, eliminating TRL1. The filter works both ways so that it can be attached directly to a BNC connector on a hand-held transceiver. C2 controls the notch frequency and should be adjusted first. Next tune C1 for best SWR. It should not affect the notch frequency although C1 may require retuning if you change the notch frequency. Construction should follow sound VHF practice. KH6CP does not give constructional details of L1.

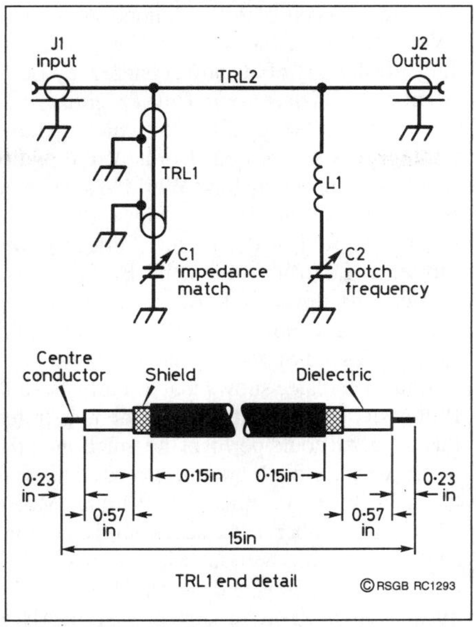

Fig 2: KH6CP's suggested notch filter suitable for permitting undisturbed 144MHz operation in the presence of strong 'pager' signals around 152MHz. C1, C2 1-14pF high-Q air-dielectric variable capacitors. J1 male BNC connector. J2 Female BNC connector. TRL1 15 inches RG-188 Teflon coax with ends prepared as shown. TRL2 1.75 inches, Nr 18 AWG enamelled wire between coaxial jacks (to hook the connectors together).

CW FOR 73KHZ & HF

MORE THAN 20 years ago ('World of Amateur Radio', *Wireless World*, March 1975, p142 and *TT*, June 1975, July 1976) I drew attention to the work of Raymond Petit, W7GHM, in developing for amateur applications the use of coherent (synchronous) CW systems in which start and finish of each symbol is accurately anticipated by the receiver. The system was first proposed by W7GHM in *The Miniwatt* (University of South Dakota) in an article later republished in *QST*.

Coherent CW depends, in effect, on the narrow-band receiving filter knowing in advance that if a dit or dah is to begin or end, it will do so at a precisely known time, allowing integration of the signal over a period of time. The technique permits some 20dB signal-to-noise ratio improvement over conventional CW at moderate signalling rates but requires a stability of the order of 2 - 3Hz using an electronic keyer based on an accurately controlled master clock. It was suggested that such schemes could provide the ultimate in micropower transmission.

But in the 1970s there were quite severe practical difficulties in achieving the required levels of stability and timing accuracy; the complex system also required special hardware at both ends of the contact. Not surprisingly, interest tended to be limited to a very small minority of dedicated micropower enthusiasts for whom it has proved of long-term interest. The system was based on a standard Morse rate of 12WPM.

One CCW enthusiast is Peter Lumb, G3IRM, who writes: "Coherent CW was proposed by Ray Petit, W7GHM, in 1975 and, despite its many advantages, never became popular. There were two reasons. Extreme stability to within 2Hz was needed and a somewhat complicated ±10Hz filter had to be built. All this has changed, and CCW could not be easier to implement.

"In 1993, Bill de Carle, VE2IQ, developed a computer program to replace the hardware

SCREEN REGULATOR FOR HIGH-POWER TETRODES

OVER THE YEARS a good deal of information has been given in *TT* and elsewhere on the requirement for good regulation of the screen voltage applied to modern tetrode tubes such as the 4X and 4CX series etc. Yet this continues to be an important topic for those building high-power linear amplifiers.

Chris Tran, GM3WOJ, writes: "**Fig 3** shows a design for a +325V DC regulator which I have been using regularly. Since 1992 I have constructed three HF linear amplifiers using 4CX1000A/4CX1500B valves and have experimented with a variety of screen regulator designs. All have proved to have some drawback - reliability, stability, insufficient current source/sink ability etc.

"I asked my friend Robin Kerr, GM4FDT, to design a regulator, and this is what he came up with (I happened to have S11E12 valves which are rather difficult to obtain and are expensive, but many other valves such as the EL34 are suitable). The design works extremely well and is suitable for any of the 4CX series. I find that 4CX1500B valves run with very little screen current, whereas 4CX1000A valves need a higher-capacity screen current supply."

***Notes*:**

(1) The +450 - 500V unregulated DC supply must be delayed for three minutes after switching on the amplifier valve.

(2) The V1, V2 heaters (63V pins 2, 7) must be fed from separate heater windings on the power transformer.

(3) Adjust RV2 'Set volts' for +325V DC output.

(4) Adjust RV1 'Set current' for 60mA through V2 by monitoring the voltage across R3 (100Ω) and adjusting RV1 for 6V across R3.

(5) V1 and V2 should be forced air-cooled by a fan.

(6) An additional VDR (360V) should be fitted directly from the screen pin on the valve-base to provide flashover protection.

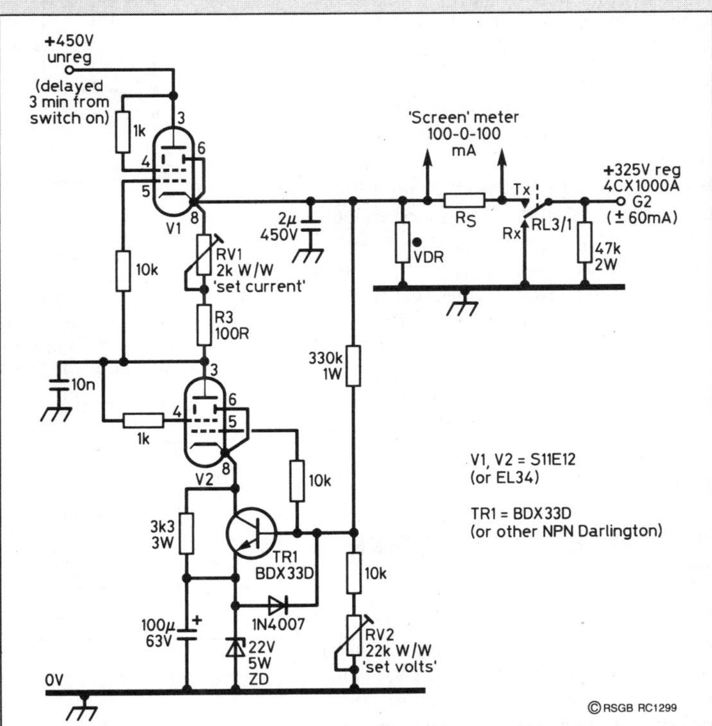

Fig 3: Design of a 325V screen regulator for use with high-power 4CX-etc valves as designed by GM4FDT for use in 4CX1500B/4CX1000A linear amplifiers built in GM3WOJ.

filter but had to add an analogue to digital converter in its place. This was certainly simpler to build than the original Petit filter, requiring only 10 fairly readily-available devices. Code was included in the program to compensate for slight frequency drift so that fairly modern transceivers could be used without the need for all the frequency compensation and stabilisation originally required. Keyboard sending was added and the received code was available from the computer loudspeaker.

"In 1995, Ernst Schroder, DJ5HS, wrote a new program based on the de Carle program, adding automatic decoding of the received code to the computer screen. Since the standard 12WPM speed was often considered too slow (although the slower the code the better the results), DJ7HS added two additional speeds, one slower, the other faster, than the standard rate: 8WPM with a bandwidth of ±6.7Hz; 24WPM with a bandwidth of ±20Hz. Think of how many stations using CCW could be fitted into the amateurs bands!

"However, despite the fact that the hardware and stability requirements have now been largely overcome, CCW is still not widely used. Version 2.0 of the DJ7HS program is now available and removes the last excuse for not trying CCW. Provided you have an IBM-compatible computer fitted with a Sound Blaster card, no further hardware is needed with the exception of an odd transistor or relay to convert the RS232 output from the computer to whatever is needed to key the transmitter . . . Most modern transceivers can be used and those which tune in steps of 1Hz are ideal as these can take advantage of the automatic frequency correction available in the program.

"Coherent CW is an ideal mode for groundwave propagation on the new 73kHz band [where the licence and antenna limitations impose low radiated power - *G3VA*] but also offers significant advantages over normal CW on the HF bands.

"The PCW program from DJ7HS is shareware and registration costs $30 or DM42 for which you receive the latest version with embedded callsign and the printed PCW manual. New versions will be priced about $5 to $10 to registered users.

"A copy of the shareware program can be obtained from Peter Lumb, G3IRM, 2 Brianwood Avenue, Bury Saint Edmunds, Suffolk IP33 3QF. A contribution of £1 to cover the cost of the disk, jiffy bag and first-class postage would be appreciated."

[For more on coherent CW, see 'Coherent CW Made Easy' by DSP, by G3IRM, *Radcom* Aug 95 - *Ed*]

ELECTRODE PAIRS, POWER BUZZERS & 73KHZ

THE *TT* ITEM 'ELF antennas for 73kHz?' (January 1997, p63) resulted in a most interesting letter from John Taylor, G0AKN, enclosing an article 'Earth Current Communications' which he has written for the journal of the Cave Radio & Electronics Group.

G0AKN has been experimenting in various forms of earth current communications for more than 30 years and has also researched the historical aspects of this technique. Among his references, he includes 'Earth Current Signalling - the history of the power buzzer' by Louis Meulstee (*Journal of the Royal Signals Institute*, Spring 1988, pp228-235) which includes photographs and circuit details of the equipment used during the First World War as mentioned in the January *TT*. **Fig 4** shows the circuit diagram of the French Parleur power buzzer No 2 from which the British power buzzer was developed. An earth speech set (Erdsprechgeraete) was still being built for the German Signal Service as late as 1944.

Louis Meulstee shows that as early as 1838 Professor Steinheil at Munich demonstrated that the earth could perform the function of the return wire for a telegraphic circuit. This was taken up by none other than Samuel Morse who in December 1842 successfully demonstrated signalling without wires across a 27m canal in Washington and later in 1845 over the Susquehanna River, a distance of nearly a mile. As shown in **Fig 5**, Morse simply keyed direct current to a pair of earth electrodes on one bank, and connected a sensitive galvanometer to the second pair on the other bank as the receiving instrument.

In his introduction, G0AKN explains: "More than 30 years have passed since a friend and I first used the earth current method to communicate between our houses, as an alternative to what seemed then to be an unattainable amateur radio licence . . . In our tests we used an army surplus Loudspeaker Amplifier No 9 as 'transmitter'. The high voltage output from the pair of 807s was connected directly to ground electrodes. The receiving electrodes were connected to the input of a record player amplifier. It was successful over the required 200 metres, but the 50Hz hum was enormous! Since then I have continued to come across references to the method, both ancient and modern, and have carried out further work of a simple but effective nature. It is interesting to note that the study of 'whistlers' at VLF, amateur radio developments such as DSP filters and the new frequency allocation of 73kHz may all have a part to play in modern earth current communications."

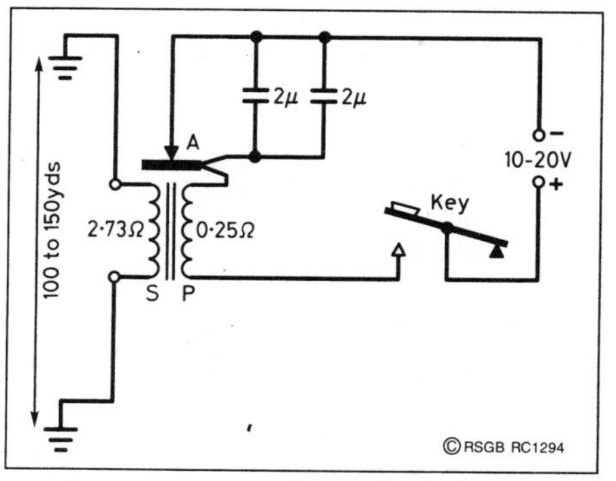

Fig 4: Circuit diagram of the French Parleur power buzzer No 2 of World War 1 vintage.

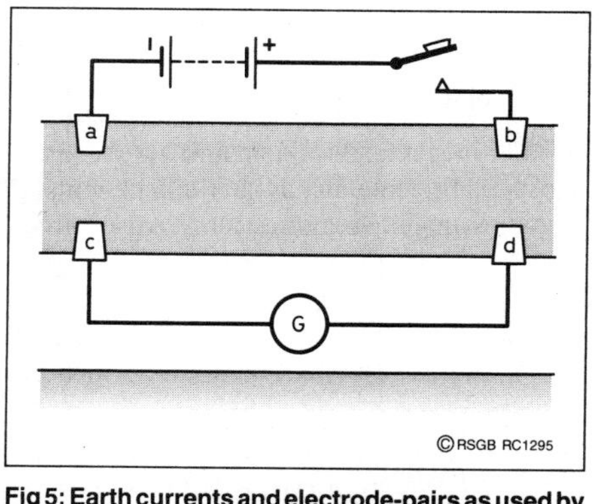

Fig 5: Earth currents and electrode-pairs as used by Samuel Morse in the 1840s for signalling across rivers. G is a sensitive galvanometer detecting the signals.

G0AKN's article includes a useful survey of available equipment suitable for this type of work, including reception of ELF earth current signals by conversion to VLF with a VLF converter in front of an HF communications receiver, or by direct reception of audio tones. He concludes:

"The use of amateur radio and 'natural radio' equipment has been found to be most effective in the improvement of reception of signals from the ground. It is not a cheap option but, as with much of today's electronics, construction of some of the equipment would not be feasible for the average experimenter."

In his accompanying letter, he writes: "Although I have been using audio frequencies, I intend to try 73kHz as well . . . I have Coherent CW now in operation and have a 1000 watt audio amplifier that handles up to 100kHz.

VALVE PRACTICE RECALLED

THE *TT* (DEC 96, p63) item 'Single-valve transmitters' - with G4QK expressing a fear that we are in danger of forgetting the simplicity of thermionic valve designs - prompted 'Jakey' Gould, G3JKY, to recall some aspects of valve practice applicable to simple transmitters. Evidently valves still have a useful role in amateur radio home construction.

G3JKY writes: "Whatever circuit is being used, it is sensible to keep the keying as far removed from the oscillator as possible. If the grid or cathode of the amplifier valve is keyed and grid current is drawn on key-down, this represents a varying load on the oscillator. Such variations are buffered by electron coupling in a pentode oscillator but it may still be enough to cause some chirp. Keying the screen electrode of the amplifier stage greatly reduces frequency shifts.

"I still use a VFO/PA rig built for the 1962 NFD in which the screen-grid of the PA is keyed by a relay. To improve the cut-off of the PA valve [and thus minimise the radiation of a spacer wave - *G3VA*] a high-resistance can be connected from the screen-grid to a negative voltage. In my rig the resistor is connected to the junction of the RF choke and the grid-leak as suggested by the late G3IIR: **Fig 6 (a)**." [I use a basically similar screen-grid keying system on my modified LG300 which originally used cathode-keying of the buffer stage with a screen clamper valve for the 813 PA. However the short life of clamper valves and the failure of the high-wattage screen voltage-dropping resistor persuaded me to change to keying the screen-grid of the 813 using a separate PSU supplying the screen-grid and deriving, for the key-up condition, a high negative voltage from the VFO PSU -*G3VA*].

G3JKY continues: "For simplicity the 'negative screen bias' could be omitted and the rig could be keyed by putting the normally closed contact found on most 'straight' keys to the PA screen as in **Fig 6 (b)** - although this is likely to result in some spacer radiation.

"With careful selection of resistors R1/R2 to form a potential divider feeding the screen it should prove possible to make the current consumption constant irrespective of key 'up' or 'down' eliminating the need for a voltage regulator. If R1 is the upper part of the divider and R2 the lower, then R1 = Vs/Ia and R2 = Vs x Vs/IaVs -IaVs - IsVs. R1 must dissipate as much power as the total HT power to the PA and for single-valve transmitters a 15 watt resistor would be appropriate. with typical values of Va 300V, Ia 30mA, Vs 200V and Is 5mA, R1 works out at 6.6k, 13.8W and R2 at 20k, 2W.

"On the topic of voltage regulation, those unfamiliar with valve practice may like to be reminded that a bleeder resistor across the HT supply not only discharges the electrolytic capacitors on switch-off but also draws sufficient current to bring the PSU voltage onto the flattest part of its output-voltage/load-current curve: **Fig 7** illustrates a curve drawn for a PSU that is still in use.

"The December *TT* mentioned the ECL82 as one of the possibilities for a single-valve, two-stage transmitter. I would steer well clear of this valve. There was one in my Sommerkamp FR1008 receiver with the pentode section used as a series voltage regulator, controlled by the triode section and a neon. Now and again the sidetone would become ear-splittingly loud due to the ECL82 losing its grip on the voltage. I eventually traced this intermittent problem to inter electrode leakage in the pentode, making the control grid go positive... the basically-similar PCL82, widely used in television sets in the 1960s, often showed similar inter-electrode leakage problems.

"On a different topic, one of the common problems encountered with old equipment is the 'dirty switch' effect, usually involving the RF wavechange switch. This often takes the form of the receiver going quiet or shifting frequency in a sudden jump. Normal operation can often be restored by rotating one or more of the switches but the fault is likely to re-occur. The switch cleaning fluids used by service engineers can be useful although less effective where the fault is due to deterioration and some loss of springiness in the switch contacts. I have had some success by treating each RF switching contact as though it were a switching diode, that is to say by arranging to pass a few microamps of DC through the switch contacts when these are closed. This tends to break down contact resistance where only small-signal RF voltages are present.

"**Fig 8** shows typical methods of providing a small direct current through a rotary wavechange switch. This can involve adding a blocking capacitor in the antenna lead with a high-resistance bleed resistor (say 4.7M) from the HT line to the pole of the switch. A similar arrangement may be needed for the grid coil, although in this case the DC may stem from the AGC line by connecting the bleed resistor to chassis (but remember that for CW reception older receivers may need to have their AGC switched off). Switches in the anode circuits will often already be carrying DC and unlikely to require modification.

"I used this DC technique also in the old VHF TV turret tuners with some success. I also used it on a UHF TV tuner of the type which still used mechanical tuning capacitors, connecting the bleed resistor via a pigtail to the spindle of the capacitor to ensure effective earthing. Personally, I have never found this dodge necessary for old HF receivers possibly because the switch contacts are larger."

The technique of 'wetting' RF-only switching contacts with DC to overcome oxidation is long established and has been discussed before in *TT*, although G3JKY remains uncertain of the chemical explanation.

On the topic of simple crystal-controlled transmitters, Gerald Stancey, G3MCK, does not entirely agree with my comment about these being limited in operation unless one has a large number of crystals. He came on the air in the mid-1970s with a classic 3.5MHz 35W CO/PA (807) and an HRO.

He writes: "I started out with three crystals but over time acquired a few more. However the extra crystals did not significantly increase the number of contacts resulting from my CQ calls. It seems that once you have about three or four crystals you enter the realm of diminishing returns. A little maths applied to this phenomenon led to a theory that supports it. However one proviso is that none of your crystal frequencies coincides with those of commercial stations. I still quite often fire up this old rig and have no trouble in getting contacts. If you are feeling a bit jaded then one good way of recapturing the thrill of amateur radio is to come on the air with a classic CO/PA rig and a vintage HRO receiver!"

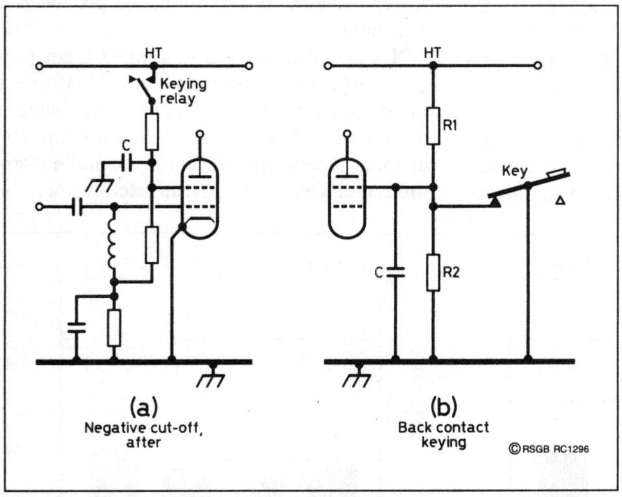

Fig 6: Screen-keying circuits recalled by G3JKY. (a) With negative screen cut-off voltage derived from the oscillator drive to eliminate radiation of a spacer wave (b) Use of back contact with potential divider (R1-R2). Values can be adjusted so that the total HT current drawn by the PA remains constant but there is likely to be some spacer. For both arrangements the value of C should be selected to provide a suitable trailing shape to avoid clicks.

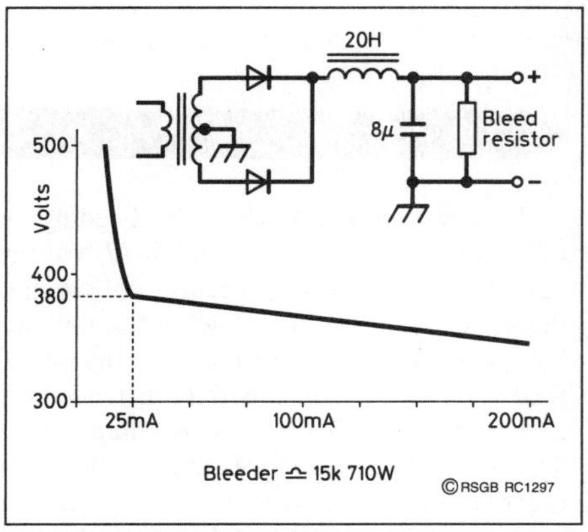

Fig 7: The use of a high-wattage bleeder resistor in a high-voltage PSU not only protects users by discharging the filter capacitors on switch-off but also improves voltage regulation. For a 300V PSU suitable for a simple transmitter, the bleed resistor should be about 15k and at least 10 watt rating.

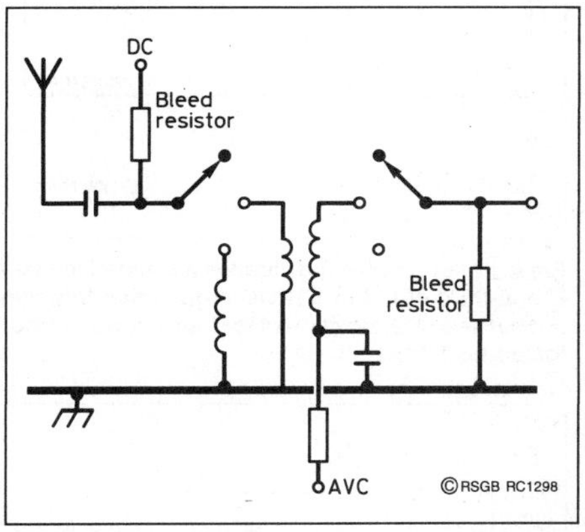

Fig 8: Provision of a low direct-current by means of bleed resistors to overcome oxidation of the contacts on rotary wavechange switches.

FEEDING THE OCFD

THE OFF-CENTRE-fed-dipole (OCFD) began life in the late 1920s as a single-wire-fed, half-wave dipole antenna developed by Bill Everitt and John Byrne at Ohio State University. First described in *QST* (September 1929) by L Windom, W9GZ, it soon became known world-wide as the Windom antenna.

Later, Jim MacIntosh, VS1AA (later GM3IAA), in 'Some experimental work with aerials' (*The T & R Bulletin*, November 1936) introduced the concept of the 'one-third' feed-point to form a compromise multi-band version. This still used a single-wire 'transmission-line' in which an essential feature was the use of a thinner-gauge wire for the feeder than for the element. In all cases where a single-wire feeder is used, even when correctly matched, there will inevitably be some vertically-polarised radiation from the feeder wire, which may or may not (TVI) be desirable. It should also be remembered that the VS1AA one-third tap antenna was developed before the allocation of the 21MHz band and cannot be expected to work as a matched antenna on that frequency.

From across the Atlantic in the 1950s came the concept of using 300Ω twin transmission line while still retaining the 'one-third' feed-point. Alternatively, it was possible to use coaxial cable feeder with a suitable balun transformer at the element feed-point.

TECHNICAL TOPICS

Richard Formato, K1POO, in 'Feeding the off-centre dipole' (*Electronics World*, November 1996, pp853-854) suggests - following a theoretical study - that while it is widely acknowledged that the best place for the OCFD feed point remains the one-third distance from one end (as originally recommended by VS1AA/GM3IAA) - **Fig 9** - this should not be regarded as a universal rule.

He points out that the OCFD is an attractive multi-band antenna "because it is simple, inexpensive, and requires no antenna tuner [at least on some of some bands - *G3VA*]. Improving its performance simply by moving the feed point makes the antenna even more attractive."

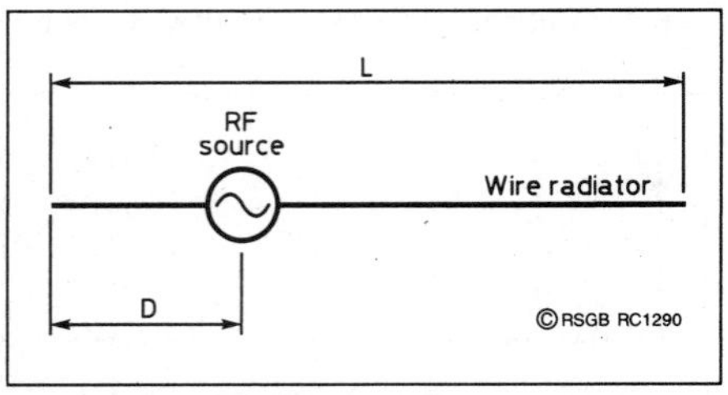

Fig 9: For off-centre-fed-dipoles it is usual to make D one-third of L, but K1POO argues that this may not be the best arrangement for four-band operation including 21MHz.

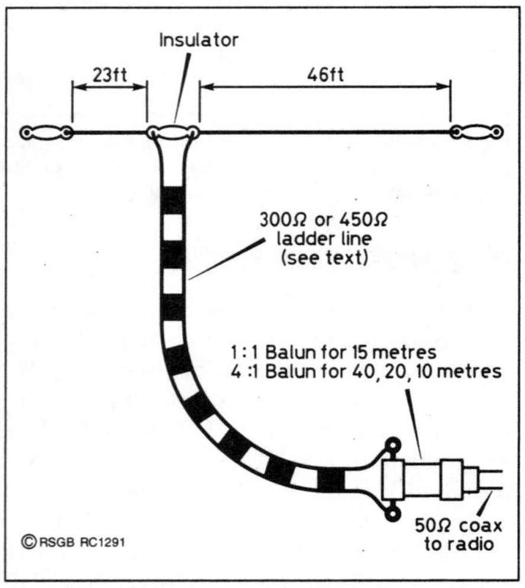

Fig 10: The OCFD four-band (7, 14, 21, 28MHz) antenna as described by G0FAH in *QST* using the ladder-line feeder as a quarter-wave impedance transformer by making this an odd multiple (electrical) of quarter waves. It is necessary to change baluns for 21MHz.

He based his computer studies on a wire element 21.03m (69ft) long, 0.2053cm diameter (No 12 AWG) - dimensions taken from a February 1996 *QST* article by Bill Wright, G0FAH, entitled 'Four bands, Off Centre' (7, 14, 21, 28MHz bands). The G0FAH design (**Fig 10**) uses the one-third point with (preferably) 300Ω ladder line which forms an impedance matching transformer on 21MHz by making the ladder line an electrical multiple of a quarter-wave. eg 55 or 111ft of 450Ω line. At the transmitter end, the ladder line is matched to 50Ω by the use of a 4:1 balun on 7, 14 and 28MHz and a 1:1 balun on 21MHz.

K1POO computed the band-centre travelling-wave ratios of the G0FAH design computed on the four bands at the antenna input terminals for a feed-point impedance of 200Ω located at 6.98m (one third) from one end: **Fig 11(a).** He then computed the SWR when the feed point is located 8.65m from one end: **Fig 11(b)**. While the SWRs on 7, 14 and 28MHz are slightly higher, that on 21MHz is very low with the result that all four bands can be fed from 50Ω coax with a single 4:1 balun at the element feed-point.

K1POO then shows that changing D to 3.65m provides excellent results on 7, 14 and 21MHz with the SWR on 7MHz only slightly above 2, and the SWR on the 14 and 21MHz bands below 2; the highest SWR is on 28MHz where it is approximately 2.4: **Fig 11(c)** showing the computed values. He points out that since balun and cable losses increase with frequency, the SWRs measured in practice at the transmitter end of 50Ω cable with a 4:1 balun at the element feed-point should be lower than those given in Fig 11(c), with the reduction greatest on 28MHz where it is needed most.

He concludes that feeding the antenna 3.65m from one end may well provide the best overall four-band performance. He notes however that in practice for any specific implementation the OCFD, like any other resonant antenna, must be "tweaked" for optimum SWR. "This is accomplished by adjusting the feed point location. Other antennas, nearby metallic objects, and the earth are typical factors that influence antenna performance. Since these factors are not included in the computer model, they must be dealt with empirically by adjusting the antenna on-site. The 8.65m and 3.65m points should provide a starting point... Depending on total antenna length L, height above ground, earth electrical parameters and feed system impedance, it should be possible to operate a single OCFD on four or more bands without an antenna tuner or special feed arrangement."

It will be noted that K1POO bases his results on computer simulations but his results - subject to the variations imposed by height, earth constants and nearby objects etc - are unlikely to be much in error. On the other hand, it is worth stressing once more that this is not true of all published computer-simulations.

Alan Taylor, G3JMO, for example, is rather dubious about the conclusions (based on computer simulations with NEC-2) reached by Ralph Holland, VK1BRH, (*TT*, January 1997, p61) that putting a counterpoise below a horizontal antenna (on the lower HF and 1.8MHz bands) is likely to improve the antenna efficiency by a reasonable amount and lower the effective vertical radiation angle.

G3JMO writes: "I have tried this out (mainly on 1.8MHz) but with inconclusive results. The counterpoise, like any parasitic element can be tuned to either accept or reject energy from the radiating element, in this case above it. One way to tune it is by placing a tuned circuit in the middle, or even one end of it.

"When the counterpoise accepts energy from the radiating element, the skyward is reduced and the local radiation about the counterpoise is increased. As this is nearer the ground, the ground absorbs more energy and the overall result is a lowering of signal strength reports received from my local stations.

"When the counterpoise is tuned to reject energy, giving very low radiation levels near the counterpoise, this increased the skywave radiation as it would if the antenna itself was lowered near the ground. However, I did find the local stations again reported lower signal strengths. One way to increase the high angle radiation from a horizontal dipole, to increase signal strength in the near field, is to lower it nearer the ground, providing the ground is a reasonable reflector.

"I was sufficiently discouraged from proceeding further. In summary I suggest that a counterpoise below an antenna does have a pronounced effect on the radiation from the antenna, dependent on the tuning of the counterpoise, and needs to be investigated further." ♦

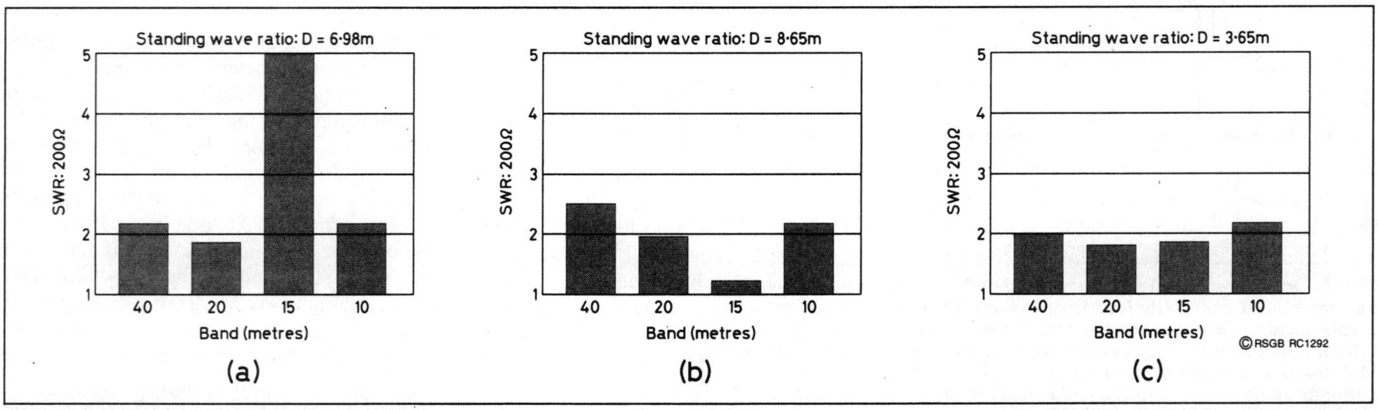

Fig 11: K1POO's computed standing wave ratios on 40, 20, 15 and 10 metres mid-band for a 21.03m (69ft) element (L). No 12AWG wire for a feed system impedance

Pat Hawker's Technical Topics

PAT HAWKER, G3VA
London 37/SE22 8SS

PA0SE COHN FILTERS

THE JANUARY 1997 *TT* (pp62-63) included details of the four-section 'minimum loss' bandpass filters for all nine amateur bands from 1.8MHz to 28MHz together with a 'maximum-minimum' method of aligning such filters as described (Dutch text) by Dick Rollema, PA0SE, in *Electron*, October 1996.

PA0SE has since provided translated information on the background to this project. He writes: "It was a pleasant surprise to see my Cohn receiver input filters in *TT*. These are to be used in my new transceiver for 10 to 160m although, given my slow rate of progress, may perhaps never be completed.

"Most of the dynamic range problems with modern Japanese transceivers stem from the fact that the receivers provide general coverage without tuning the signal frequency circuits... I have intentionally limited my transceiver to amateur-bands-only coverage, where possible making the input filters only just wide enough to pass those bands. On some bands they have to be rather wider than the ideal in order to keep the insertion loss acceptable, as noted later. But my philosophy is that unwanted strong signals outside the amateur bands are best stopped right at the entrance to the receiver.

"If this is done, the requirements for the mixers, etc are relaxed considerably. My transceiver will have a tuneable first IF of 2500 - 3000kHz, so the image (second) channel is only 5000 to 6000kHz away from the wanted channel. The fixed tuned input filters thus have to suppress the image-frequency bands dictating the use of a minimum of four tuned circuits for the 28.5 to 29.7MHz filters. I included four circuits on all the other bands since a number are close to broadcast bands with extremely strong signals that I want to attenuate as much as feasible.

"To add a few remarks on the design of the filters, which for an equal element filter is really very simple: Given a maximum allowable insertion loss in the passband, the minimum attainable bandwidth of the filter is dictated by the unloaded Q (Q_u) of the coils. For a Cohn minimum loss filter the insertion loss should not exceed N dB where N is the number of tuned circuits. For my four-circuit filters the insertion loss should not exceed 4dB. I found this to be true provided $Q_u > 5 \times f_o/B$ where f_o is the centre frequency and B the -3dB passband - width. For narrow amateur bands, such as 7, 10, 18 and 24MHz, it is therefore imperative to make the unloaded Q of the coils as high as feasible. For example, even with a high value such as $Q_u = 200$, the minimum bandwidth attainable on 24MHz is 5 x 24MHz/200 = 0.6MHz which is still much wider than the 100kHz of this band. In practice my filter for 24MHz is even wider (0.947MHz) since I wanted a lower insertion loss than 4dB on the 24 and 28MHz bands in order not to raise the noise factor of the receiver too much.

"I started the filter design by buying Amidon powdered-iron toroids of different sizes and permeability. I put on windings of different wire-sizes and then with the aid of a Q-meter found the combination of core and wire size that gave the highest feasible unloaded Q for each band. Within reason, of course. I did not want to use cores larger than 0.69in outer diameter.

"I could have saved myself the effort since the Amidon data sheet (provided with the cores) contained almost all the information required to achieve maximum Q. The maximum Qs I could achieve were shown in the January *TT*, Table 1, p62.

"Given the coils, the total capacitance C_{res} that resonates each circuit to f_o can be found. C_{res} consists of the parallel capacitor plus the one (end circuits) or two (middle circuits) coupling capacitors connected to it. This explains the difference between C_{res} and C1, C4 with respect to C2, C3 in Table 1.

"Next the value of the coupling capacitors C_k have to be determined. For a starting value the following equation can be used: $C_k = A \times B \times C_{res}/f_o$, in which A is a factor that varies between 0.60 for a relatively wide filter and 0.77 for a narrow filter.

"It then remains to find R, the resistance loading the input and output circuits of the filter:

'JUNK BOX' CONVERTER

ROD KREUTER, WA3ENK, in 'Junk Box Converters for 6 and 2 metres' (*QST*, January 1997, pp32-34) points out that four technologies have made simple VHF converters easier to build than ever: (1) General coverage HF receivers; (2) low-cost (computer) crystal-oscillator modules; (3) low-cost IC mixers; and (4) easy to make PC boards [even easier to use 'ugly' surface mounting by sticking ICs etc upside down on the copper side of PCB and then wiring components - *G3VA*].

WA3ENK describes two simple converters based on the NE602 mixer and oscillator modules (40MHz for 50MHz; 32MHz with quadrupling for 144MHz) with the 50 - 54MHz output between 10 - 14MHz and 144 - 148MHz output between 16 - 20MHz. Circuit diagram of his 50MHz converter is shown in **Fig 1**.

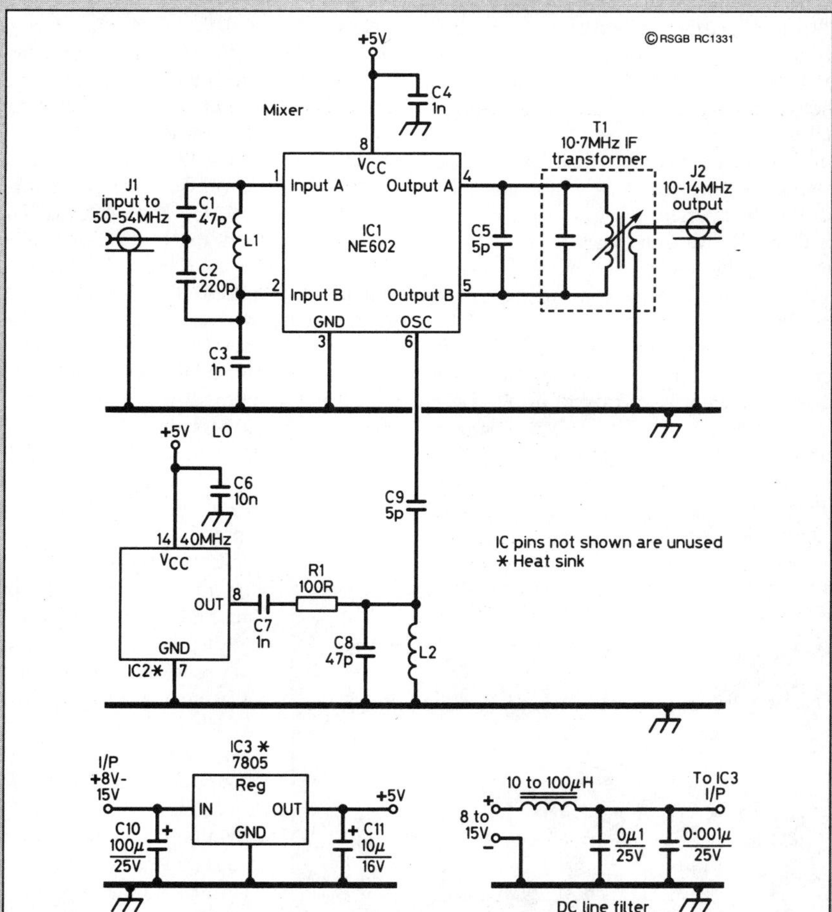

Fig 1: Simple 'junk box' 50MHz converter described by WA3ENK in *QST*. Component details: L1 8 turns, 0.25in diameter, Nr 20 AWG enam wire 0.4in long. L2 10 turns, 0.25in diameter, 20AWG enam wire 0.55in long. T1 10.7MHz IF transformer; 15KΩ primary, 300Ω secondary (ME42IFI22). IC1 NE602A mixer/oscillator. IC2 TTL oscillator module, 14-pin DIP, 40MHz. IC3 7805 IC regulator with TO-220 heat sink.

The nine-band Cohn input filter built by Dick Rollema, PA0SE, (circuit details in the January 1997, *TT*).

The Paraset Mk VII/7 'simplest' wartime clandestine transmitter-receiver as developed by Special Communications (MI-6/SIS) at Whaddon.

$R = 1/(2 \times \pi \times f_o \times C_k)$ or simply the reactance of C_k at f_o.

"Now the filter can be set up and aligned. If the passband is found too narrow or the insertion loss too high, C_k should be increased and the whole procedure repeated. Instead of actually constructing the filter, I used a near-obsolete computer network analysis program from 1989 called 'AC Analysis'. It was gratifying to find that the measured response after constructing the filter hardly differed from the computed curve. For example, in Fig 5 (January p63), the response of the filter aligned on 7050kHz was *computed* whereas the one centred on 7020kHz was measured. As can be seen the shape of the responses is almost identical! (I shifted f_o to 7020kHz to increase attenuation in the 40 metre broadcast band).

"I determined the number of turns for the input and output links experimentally by measurements on an impedance bridge. In some filters, exactly 50Ω could not be obtained because of the small number of link turns. One turn makes a significant difference when there are only 2 or 3 turns! On the other hand, there is little to be gained by aiming for a perfect match at f_o since in a Cohn filter the SWR always fluctuates over the passband, sometimes between 1 and 2 or more. But even an SWR > 2 should be of no concern in a receiver input filter.

"The maximum-minimum method of aligning a multi-circuit filter is most convenient. I found no improvement was possible by using a sweep generator after a filter had been aligned using this method. But it is important that the detector, which remains connected to the first tuned circuit all the time, is coupled very loosely via a small capacitance so that the circuit is not detuned by the detector. The circuit following the one being tuned must be either short-circuited (see Fig 4, p62, January *TT*) or detuned far from resonance.

The construction of the filters for nine bands is shown in the top left photograph.

REGENERATIVE 'STRAIGHT' RECEIVERS

THE COLLECTION, restoration and evaluation of wartime military and clandestine radios have become popular in many countries; but undoubtedly nowhere more so than in the Netherlands, home of such noted collectors and/or military historians as Arthur Bauer, PA0AOB; Cas Caspers, PA0CSC; Louis Meulstee and many others, with the active encouragement of Dick Rollema, PA0SE, who has over the years described in detail many of the British clandestine and German military equipment in the pages of *Electron* and elsewhere.

During 1996, PA0SE had the opportunity to evaluate three WW2 German transceivers made available by PA0AOB and PA0CSC. Two had superhet receivers, the third with a 'straight' (1-V-1) TRF receiver (Hagenuk Ha6K39b: see photograph, bottom left). He writes: "Although all the sets handled very well, I preferred the TRF. The Germans certainly knew how to realise a very smooth regeneration control, without any backlash and hardly any detuning. But the main factor for the success of the TRF set was, in my opinion, the differential capacitor-type gain control at the *input*, providing some 60dB of control!" The outline is shown in **Fig 2**.

I have noted previously PA0SE's belief that the Germans raised the art of the straight receiver to near perfection, resulting in receivers that were a joy to operate. I still recall while in 'occupied' Germany in 1945/46 using one of the German military transmitter-receivers with a first-rate turret tuner and very large, accurately calibrated, dial that allowed me to operate the set as a direct replacement for a crystal-controlled transmitter, coming up on the right frequency.

German WW2 military transmitter-receiver using an effective TRF renerative receiver.

At the other extreme, the Whaddon 'simplest possible' Mk VII (Paraset) transmitter-receiver using a 6V6 crystal power oscillator; and two 6SK7 valves as a regenerative receiver (circuit details in *TT*, November 1982, p961, etc) was widely used by secret radio-agents in Western Europe. It proved an effective simple rig, often putting in a relatively good signal into Weald from all over France etc. But by no stretch of the imagination could the receiver be regarded as good even by 1940s standards.

PA0SE recently had the opportunity of evaluating a Mk VII loaned by Cor Moerman, PA0VYL. It will be featured in 'Reflecties door PA0SE' in the May 1997 issue of *Electron*. He comments: "I used the set for a CW contact on 3.5MHz, which went well. But I used my station receiver. I can hardly understand how the 0-v-1 receiver of the Paraset could be used at all. It covers 3.15 to 8.15MHz with just 1.75 turns of the small tuning knob: no less than 2875kHz per revolution! Compare that with the commonly accepted maximum of about 20kHz per revolution."

I still have a Mk VII-type slow-motion drive on a VFO and I make it nearer two turns for a dial change from 0 to 100 but stiff and jerky to use. That would still represent over 2500kHz per turn of the knob. For many years I have advocated less than 5kHz per turn of the knob - achieved if necessary by fitting a second slow-motion drive to the tuning control (a practice I still use on my KW2000A).

The Paraset, was made in two versions, one in a compact metal box as shown in PA0SE's Mk VII/2 (see the top right photo); the other

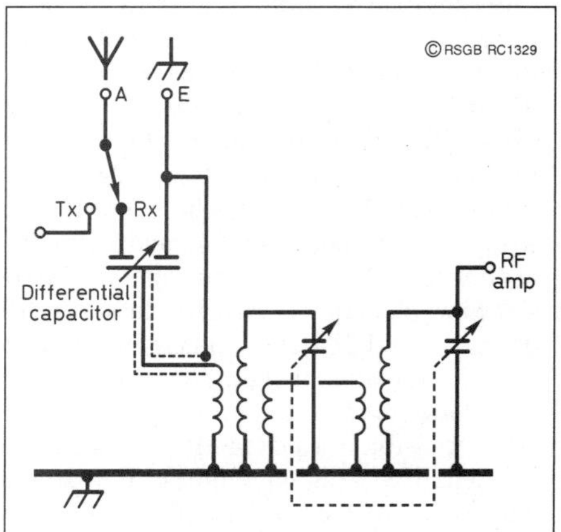

Fig 2: Simplified input circuit of the German military transmitter-receiver showing how a differential variable capacitor was used to provide a wide-range attenuator.

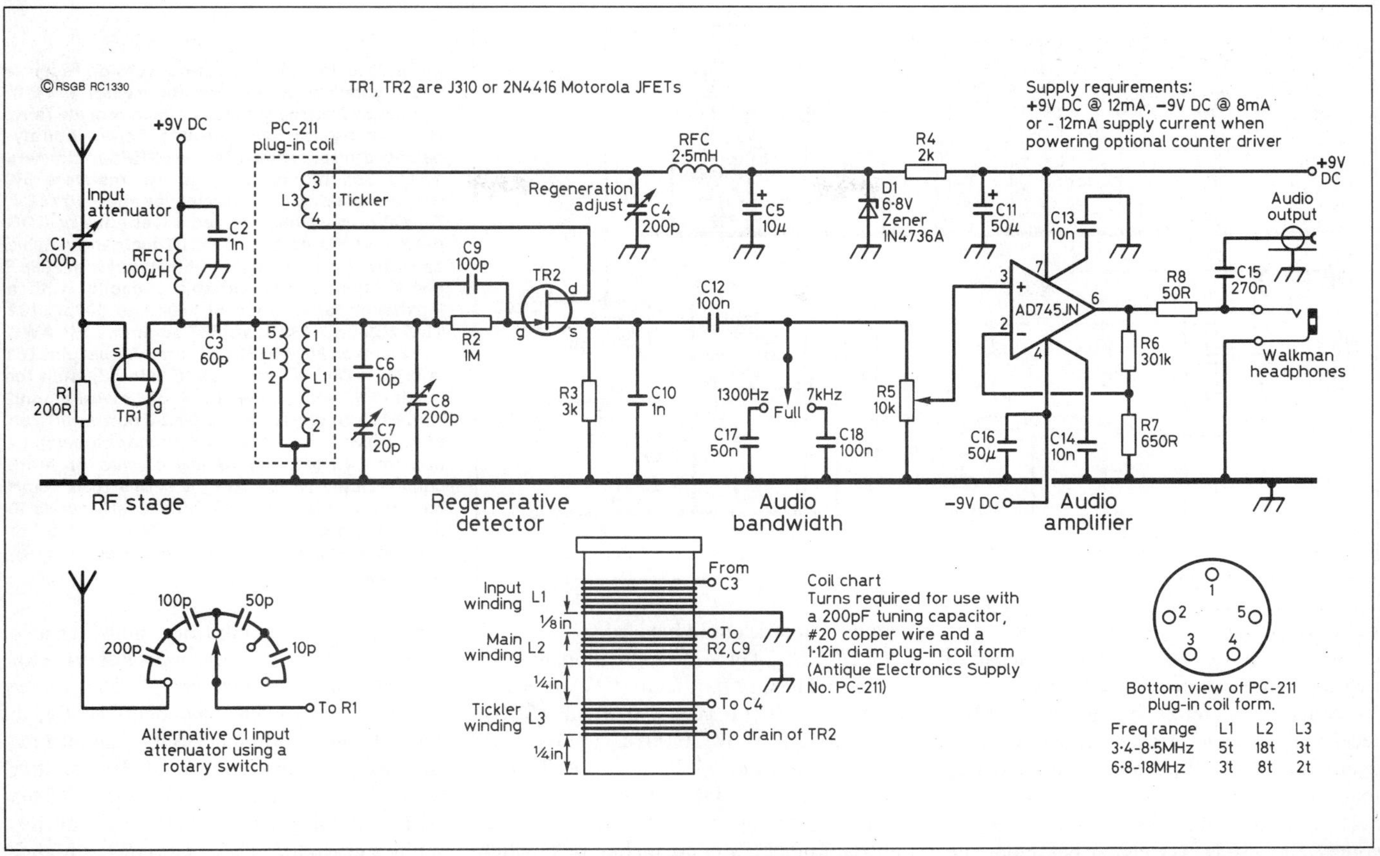

Fig 3: High-performance JFET regenerative receiver using modern active devices in a traditional circuit approach as built by Charles Kitchin, N1TEV, (*Communications Quarterly*, Fall 1995).

in a wooden box. A replica Mk VII built recently by Ian Haggart, G3JQL, is shown in the photograph below. Users had to rely on the type of small calibration curve shown in the photo. Some non-standard but similar Whaddon models did provide a 'fine tuning' knob, and this was usually provided on the 1-v-1 receivers for the suitcase versions of the Mk III, Mk V and Mk XV making them easier to use than the Mk VII, with better overall performance.

I agree with PA0SE that it must have been quite a feat for the radio-agents even to find the signals from our transmitters at Calverton, near Stony Stratford. Yet many did, and the Paraset remained in use from 1941 - 45.

A very long article 'Regenerative Receivers - past and present' by Charles Kitchin, N1TEV (*Communications Quarterly*, Fall 1995, pp7-26) covers the development of the genre from Armstrong's original circuit of 1913, the Reinartz 'short-wave' circuits of the 1920s through to the use of JFETs and IC audio amplifiers in a traditional TRF configuration (**Fig 3**) claimed to provide a 'high performance' SWL receiver and evaluated by John Hartman, NM1H, who reports: "Those who think that the days are gone when you could build your own simple communications receiver that rivals professional amateur gear in basic performance should consider this rig. I had the opportunity to evaluate it one night and was impressed by its sensitivity, selectivity, excellent audio, very low noise, and surprisingly good stability."

N1TEV concludes: "Since regenerative receiver circuit development basically ended many years ago - before homebrewers had low-cost semiconductors, op amps, or zener regulators at their disposal - it should now be possible to design some new features into this traditional circuit. These could include some type of regenerative AGC circuit for AM signal reception - using an op amp and rectifier, etc, to maintain a constant level of regeneration as the receiver is tuned. Crystal oscillator circuits could also be made regenerative to produce a very high gain, high Q, low drift, single frequency amplifier... Applying regeneration in a limited controlled way into several RF stages simultaneously (to raise their gain and Q) is also a possibility . . ."

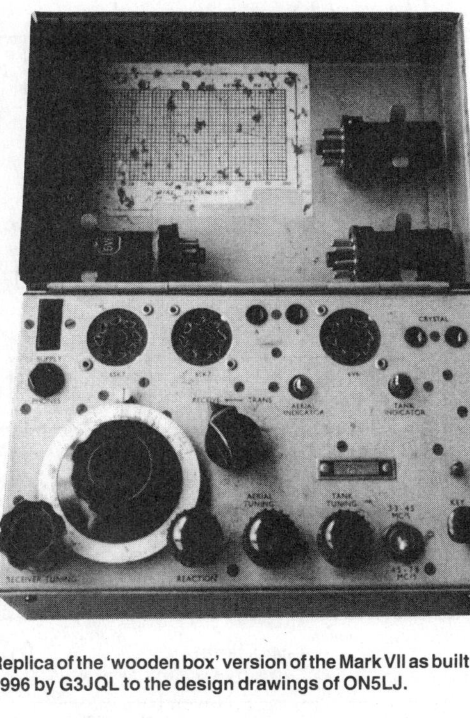

Replica of the 'wooden box' version of the Mark VII as built early 1996 by G3JQL to the design drawings of ON5LJ.

For my part, I would maintain a long-held view that a good 'straight' receiver (direct-conversion product detector or regenerative detector) can prove greatly superior (at least for CW reception) and cheaper to construct than simple superhet designs, although most unlikely to achieve the performance possible with a well-designed multi-stage superhet. Again, the use of a regenerative, crystal-controlled detector in a simple 'super-gainer' superhet can result in extremely high selectivity with high-gain, provided the regenerative control is smooth.

THE VXO IN QRP TRANSCEIVERS

DESPITE THEIR LIMITATIONS, the simple CW HF QRP transceiver based on direct-conversion receiver techniques continues to form one of the most popular home construction projects. Equally popular, particularly among newcomers, is the stand-alone single-band DC receiver. But, as emphasised by Randy Henderson, WI5W, in 'A Rock-bending Receiver for 7MHz' (*QST*, August 1995, pp22-25),

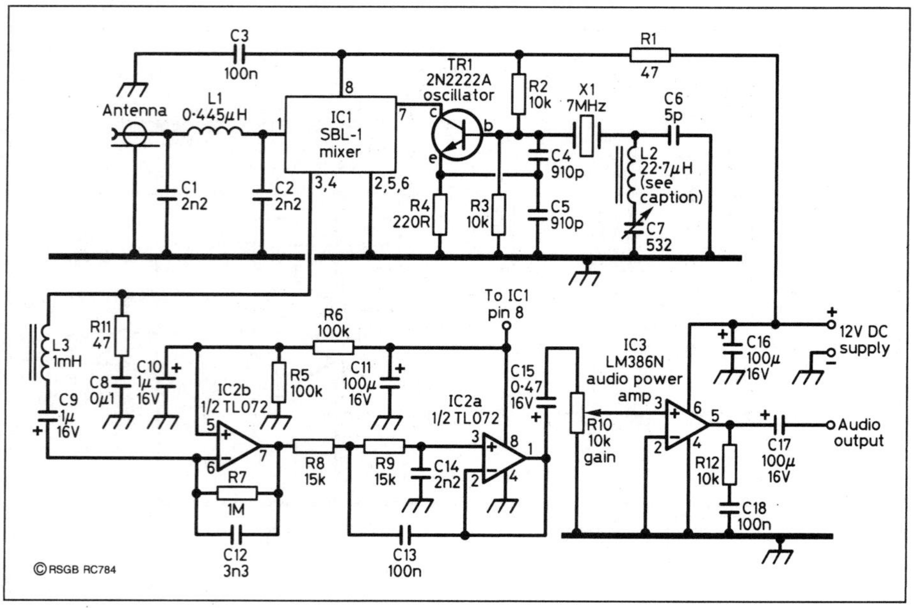

Fig 4: WI5W's 7MHz direct-conversion receiver with crystal VXO tuning over about 25kHz lower in frequency than the nominal frequency of the 7MHz (FT243) crystal. Mini-circuits SBL-1 doubly-balanced mixer contains two small RF transformers and a Schottky-diode quad. All resistors 5% tolerance ¼W. C1, C2 ceramic or mica, 10%. C4, C5, C6 polystyrene, dipped silver mica or COG (formerly NPO) ceramic, 10%. C7 dual gang variable capacitor (eg polyethylene-film type) with 268pF per section or any variable capacitor with a maximum capacitance of 350pF to 600pF. IC2 TL072CN or TL082 dual JFET op-amp. L1 4t, AWG Nr 16 wire on 3/4in PVC pipe form. Actual pipe OD is 0.85in. Coil length about 0.65in adjusted for maximum 7MHz signals (or 8t of 18 wire around 75% of circumference of a T-50-2 powdered iron core, expanded or squeezed for max signals). L2 approx 22.7mH, consisting of one or more encapsulated RF chokes in series (two 10mH chokes suitable). L3 1mH RF choke on 34t of No 30 enam wire around FT-37-72 ferrite core. R10 5 or 10k audio-taper pot. Typical crystal frequency for CW 7030kHz.

building and aligning a stable variable-frequency oscillator is often the trickiest part of getting your first home-made receiver going. He describes a simple but effective 7MHz receiver that uses a variable-frequency crystal oscillator (VXO) that overcomes most of these problems.

He finds that the older style FT243 and some HC-6 7MHz crystals can be 'pulled' up to 50kHz lower than their nominal frequency, provided that a relatively large value of inductance and capacitance is used. But for his receiver he limits the pulling to about 25kHz over which the stability remains very good. Note that plated crystals in HC-18 and HC-45 holders are unlikely to pull over more than a very few kHz unless stability is sacrificed by using large values of inductance and capacitance. As mentioned later, in connection with an Australian 80m QRP transmitter/receiver-oscillator, ceramic resonators are much easier to 'pull' than quartz crystals on 3.5MHz.

The WI5W design is shown in **Fig 4**. He readily admits that it represents no new earth-shattering concepts or technical advances: "All I've tried to do is put together a receiver design with few pitfalls for beginning builders. It's an easy construction project, and all of its parts are readily available from multiple sources." Yet he finds that when tested with a Hewlett-Packard HP606A signal generator, the receiver's minimum discernible signal (by ear) appears to be 0.3μV. He finds he can easily copy 1μV signals on his receiver.

The receiver apparently draws quite a hefty current (over about 100mA) from a 9-12V supply and it would be advisable to use during testing a supply with current limited to about 150mA. A mains PSU would be most economical although rechargeable lead-acid or nicad batteries could be used. The current could be reduced by using discrete rather than integrated active components for the high-gain audio amplifier/filter.

While WI5W recommends the use of PCB construction (PC boards are available from FAR Circuits, 18N640 Field Ct, Dundee, IL 60118-9629, USA price $1.50 for shipping in the USA of up to four boards), alternative construction such as the 'ugly' technique on unetched boards could be used.

C12/C13/C14 provide audio filtering: values should be C12 0.0033uF (CW), 0.001uF (SSB); C13 0.1uF (CW), 0,001uF (SSB); C14 0.0022uF (CW), 0,001uF (SSB).

In *Amateur Radio* (July 1995, pp4-7) Peter Parker, VK1PK, points out that "it has never been easier or cheaper to construct amateur radio equipment. Modern technology, which has given us black boxes with every conceivable feature, has also provided a host of components which can simplify home construction".

To prove his point, he describes a 3.5MHz CW QRP transmitter with an output of up to about 1.5 watts, which could form the driver for a higher-power amplifier and which also provides a stable variable oscillator drive for a direct-conversion receiver. The transmitter uses readily available components, costing in Australia not more than A$30 ie about £15. He stresses that most simple QRP transmitter projects are based on crystal control. "This assures stability but limits the rig's utility. While a crystal VXO is useful on the higher bands, the frequency pulling range with 3.5MHz crystals is small. The key to increased frequency agility on 3.5MHz without a conventional VFO, is the low cost 3.5MHz, ceramic resonator."

The use of ceramic resonators in VXO-type oscillators was first described by Albert D

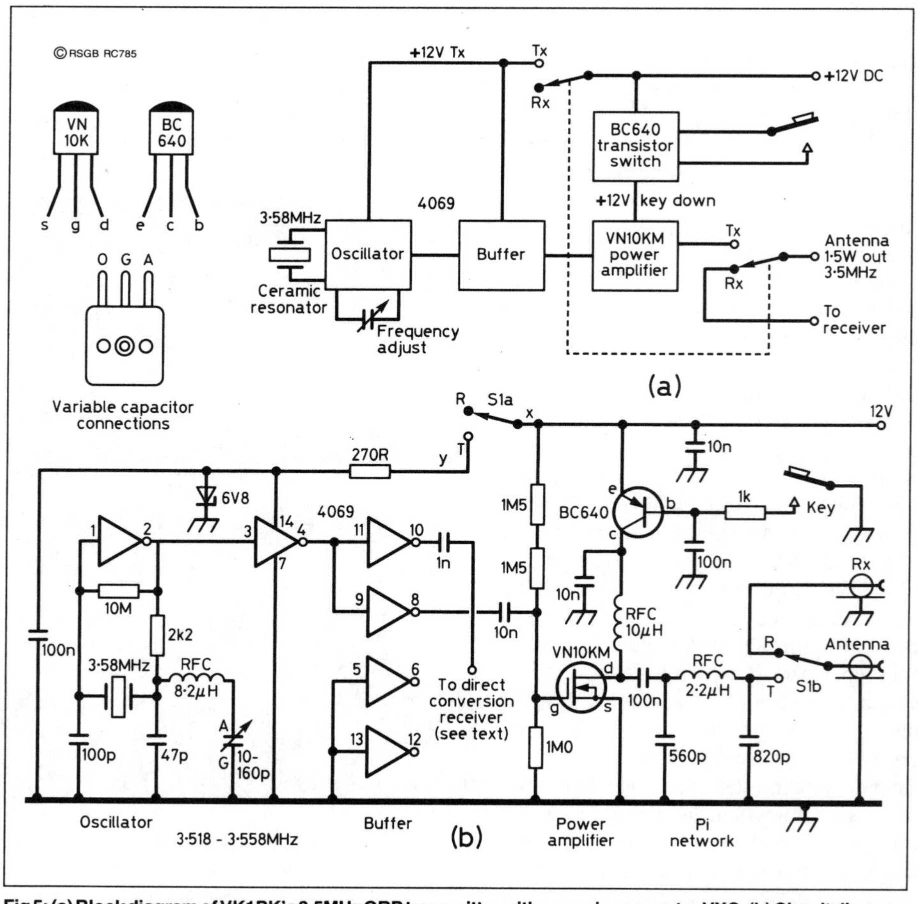

Fig 5: (a) Block diagram of VK1PK's 3.5MHz QRP transmitter with ceramic-resonator VXO. (b) Circuit diagram. Tuning range about 3518 to 3558kHz with values shown but see text for modified tuning ranges.

Helfrick, K2BLA, in *Ham Radio* (June 1985) and summarised in *TT*, December 1985, (see also *Technical Topics Scrapbook* 85-90, p69), with John Townsend, G3BBD, later exploring the use of ceramic VXOs in QRP transmitters (*TT*, February 1991 and see also *TT*, May 1994 which includes notes from LA8AK). A full-length article 'Using Ceramic Resonators in Oscillators' by Ian Braithwait, G4COL, appeared in *RadCom*, February 1994. The Q and temperature stability of ceramic resonators are significantly lower than for crystals, resulting in somewhat lower frequency stability than for a good crystal VXO, but significantly better than most LC oscillators - and much easier to construct.

An unusual feature of VK1PKs design is the use of a digital CMOS IC type 4069 for the oscillator and buffer stages. This IC contains six inverters of which four are used in this circuit: one as oscillator; two as buffers; and the fourth to provide an output for use in a direct-conversion receiver.

As shown in **Fig 5,** the VXO with a 3.58MHz ceramic resonator (a readily available frequency) in conjunction with the 47pF fixed capacitor, 8.2uF choke loading inductance and 10-i60pF variable capacitor, provides a tuning range of about 3518 to 3558kHz. By reducing the value of the 47pF capacitor or removing it, the tuning range can be moved up to cover the usual QRP calling frequency of 3560kHz and/or the main part of the UK Novice allocation (3560-3595kHz). The maximum frequency with this circuit is about 3590kHz. With a discrete Pierce transistor oscillator, rather than a section of the 4069 IC, it is possible to have a top frequency of about 3620kHz with a 3.58MHz ceramic resonator.

VK1PK notes some peculiarities of ceramic VXOs. Frequency tuning is non-linear with the spread more pronounced at the lower end of the tuning range. Although reasonably stable, every now and then the frequency may change abruptly (jump) by about 100Hz and then remain stable. While noticeable, such jumps should not interfere with CW contacts. Since a ceramic resonator is more susceptible to temperature changes, they should be mounted well away from heat producing components such as power amplifier devices.

At the lower end of the frequency range, there may be a tendency for the oscillator to be sluggish in starting due to excessive pulling capacitance. Either decrease this, for example by adjusting the usual trimmer mounted on most broadcast-type variable capacitors, or reduce the value of the 8.2μH choke inductance to 6.8 or 4.7μH.

The output should be about 1.5 watts with the key pressed and the transmitter drawing 200 to 300mA. If the VN1OKM becomes too hot to touch after a few seconds of transmitting then the 3MΩ gate resistor should be increased to limit device dissipation. If a suitable power meter is not available, connect a 6.3V torch bulb across the antenna socket. An orange to white glow is an indication that the rig is operating satisfactorily. As a final test, monitor keying quality with a 50Ω dummy load connected. The note should be free of chirps and clicks, although since the oscillator runs continuously during transmission, a backwave (spacer) may be heard.

The transmitter should preferably be enclosed in a metal rather than a plastics box; do not attempt to make the transmitter ultra miniature (box size about 5 x 15cm, with board size about 6 x 10cm). Use an IC socket, taking great care not to bridge the socket pins when soldering. The board is not etched but the copper side is underneath forming the ground plane, with pins and short lengths of wire used above the board: **Fig 6** shows a suggested layout.

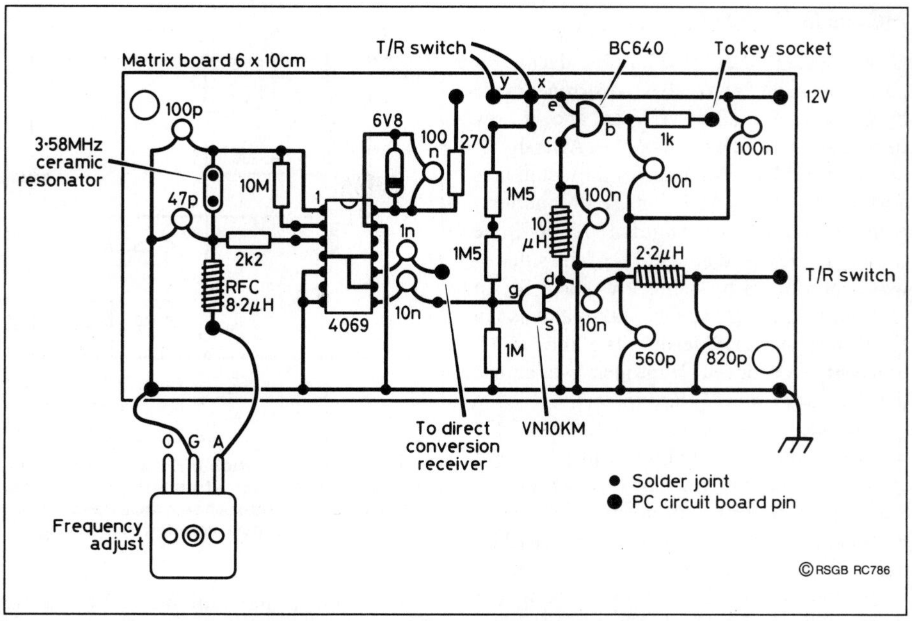

Fig 6: Suggested layout for the VK1PK QRP transmitter using pin and wire construction with an unetched PCB.

HERE & THERE

GERALD WROE, G4IUJ has found a use for an ancient audio inter-valve transformer made by Cossor some 60 or 70 years ago. It now forms part of a 73kHz transmitter. After dividing down 37MHz crystals using the techniques described by GW4ADL for use in toneburst generators (*RadCom*, August 1977, p603) he found that he could not get rid of the square-wave harmonics by using low-pass filters: "But putting 73kHz square waves across the 'grid bias' and 'grid' terminals of the old AF transformer produced clean sine waves across the 'HT' and 'Anode' terminals."

A simple time-out circuit useful in switching off battery-operated test equipment etc after about a minute or so and thus preventing draining of batteries is described by Rae Peraalaa in *Electronics World* (February 1997, p150): **Fig 7**. This uses a p-type enhancement-mode MOSFET (TR1 BS250) to switch loads of up to 50mA (for larger loads an IRF9530 could switch up to 2A) controlled by an n-type enhancement MOSFET (TR2 BA170). Switch-off time is determined by approximately twice the time constant of C-R2 with the values shown giving approximately one minute. Operating time can be lengthened, if required, by momentarily closing the switch; the operating time is then renewed to its starting value.

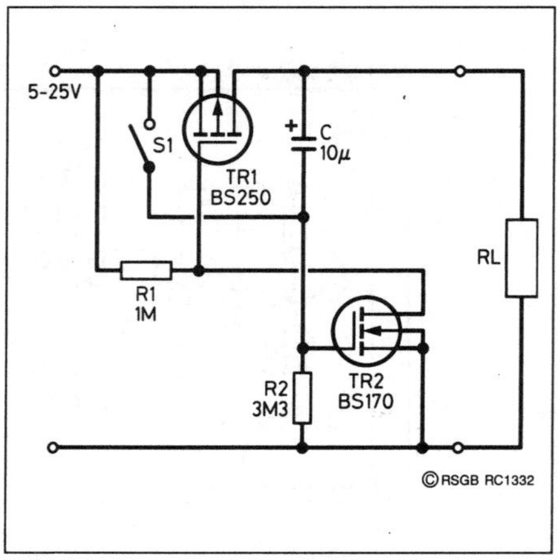

Fig 7: Automatic 'switch-off' timer to prevent battery drain from test instruments etc being left on after use. With values shown the delay is about one minute but his can be restarted by pressing the switch.

The reference (*TT*, February 1997, p77) to the 1930s Eddystone 'All World Two' with its ingenious bandspread plus bandset tuning, brought back memories to Doug Mepham, G4ERA, although in those days as an RAF Apprentice, his source of components tended to be the cheaper Raymart ones. But a few years ago he built a 'Nickey' direct-conversion rig, copying the Eddystone idea by using a 20-way switch recovered from a CB rig with a flexible coupler to a bandsetting 250pF variable capacitor having a 10pF bandspread tuning capacitor in parallel.

Readers have commented favourably on results achieved with the 'tuneable HF loop antenna' (*TT*, November 1996, p64) with its high-impedance balanced input permitting effective operation to over 30MHz. However, there was a fairly obvious omission in the circuit diagram (Fig 8) where the drain of TR1 should also be connected to the +6V supply line.

Jacob Makhinson, N6NWP, (whose work on high-performance HF receivers has been noted before in *TT*) has described 'A drift-free VFO' (*QST*, December 1996, pp32-36) as a complete design for a clean, quiet, stable VFO covering 6.0 - 6.4MHz and incorporating huff & puff type stabilisation.

He points out that the ability of receivers to receive weak signals in the presence of strong unwanted signals is limited either by the receiver IMD capabilities or local-oscillator phase noise: "The phase noise of a properly designed, free-running LC VFO need not limit dynamic range. Building a low-noise synthesiser is a serious undertaking, requiring considerable technical skill and possibly prohibitive cost and circuit complexity.

This article demonstrates that an LC VFO still offers excellent spectral purity, low cost, ease of construction and component availability. It shows that it is possible to combine in a VFO both very low phase noise and exceptionally good long-term stability." ♦

ELECTRON CENTENNIAL

SINCE 1995 WAS celebrated as the centennial of radio, 1997 should perhaps be considered the centennial of electronics. It was on 30 April 1897 that 40-year old Joseph John Thomson of the Cavendish Laboratory in Cambridge demonstrated his discovery of a body smaller than the atom, carrying a negative charge. He called this a corpuscle, later renamed the electron. The term 'electronics' came much later; it was not widely used until 1930 when an American journal launched a new technical journal using 'Electronics' as its title.

As noted in *Nature* (2 January 1997), 1897 was also the year that: Thomson's student Ernest Rutherford distinguished between alpha and beta rays; Marie Curie identified the uranium atom as the source of "uranium rays"; and Karl Ferdinand Braun quickly exploited the electron in designing a forerunner of the cathode-ray oscilloscope, which, in a modern form, still also forms the display-tube in countless millions of television sets and visual display units as a still-current thermionic device.

Thomson, who as a student in Manchester had taken a lively interest in the work of Clerk Maxwell and later took over Maxwell's chair in physics from Lord Rayleigh, later wrote: "At first there were very few who believed in the existence of these bodies smaller than atoms. I was even told by a distinguished physicist who had been present at my lecture at the Royal Institute that he thought I had been 'pulling their legs'. I was not surprised... it was only after I was convinced that my experiment left no escape from it that I published my belief in the existence of bodies smaller than atoms."

Thomson's experiment (taking advantage of earlier discoveries by Crookes, Perrin and Roentgen) sounded the death-knell of the earlier mechanistic and continuous approach to matter and was destined to provide the vital key to atomic physics, electronics, television and radio communications.

HF RECEIVER MEASUREMENTS & OPERABILITY

IN THE AUGUST 1984 *TT*, Peter Lonsdale, G3PVX, brought to our notice an unusually comprehensive tabulation of measured specifications of 38 HF receivers and transceivers. The list stemmed from Sherwood Engineering

Pat Hawker's Technical Topics

PAT HAWKER, G3VA
London 37/SE22 8SS

Inc, an American firm which then as now specialises in providing a modified IF filtering system for the Drake R4 receiver to provide a contest-grade CW receiver capable of coping well with pile-ups. The firm also markets an advanced synchronous demodulation system.

The table listed equipment in a pecking order based on their close-in dynamic range, as measured using two input signals only 2kHz apart (conventionally dynamic range is specified with signals 20kHz apart, or for close-in measurements 5kHz apart).

It was, and still is, clear that Sherwood Engineering has a commercial interest in producing tables which put their modified R4C/CF600/6 receiver in premier position; nevertheless, their table was based on laboratory measurements with high-grade test equipment, including the demanding measurements of oscillator phase noise.

In a further item ('Receivers of top performance', September 1984) I described how the Sherwood engineers had provided the much improved narrowband dynamic range on the R4C receiver not only by fitting a high-grade 5645kHz 600Hz-bandpass crystal filter as a first IF roofing filter but had also redistributed the signal gain through the receiver. The modifications were depicted in *TT*, August 1984, Fig 2, p858. The Sherwood approach was described by Robert Sherwood, WB0JGP, and George Heidelman, K8RRH, in 'Present-day receivers - some problems and cures' in *Ham Radio* (December 1977). The Drake R4C is a hybrid valve/semiconductor double-conversion (IFs 5645 and 50kHz) receiver manufactured from 1973 - 1980 using six valves in the front-end plus 15 transistors, 23 diodes and one IC.

Michael O'Beirne, G8MOB, has recently sent extracts from the latest Sherwood Engineering Inc catalogue including the updated **Table 1** now covering almost 60 fairly recent models but still with the R4C/CF-600/6 in top position. However, most recent models show significant improvement over those in the 1984 list.

G8MOB writes: "You have published earlier versions of the Sherwood table over the years together with timely warnings against its direct use for 'best buy' evaluation. The firm does not provide the information just for fun.

"Numbers alone, of course, scarcely tell half the story on the 'operability' of a receiver. The modified Drake R4C is no doubt excellent on CW but I wonder how good on SSB. My unmodified standard one (low in the table) is no great shakes. I find the tuning lumpy and too fast at 15kHz per revolution of the tuning knob. The Collins R390A is a superb example of mechanical engineering but is dreadful to tune and many professional engineers consider its AGC poor; indeed that Collins never got its AGC right in *any* of their receivers, whatever their merits elsewhere.

"Again, the new AOR 7030 has superb numbers, but I took an instant dislike to the display which I found difficult to read; moreover, the use of the tuning knob to control most functions is wholly alien to my tastes. Similarly, the Watkins-Johnson HF-1000 has a giant, uncomfortable-to-use, tuning knob, and the other knobs seem tacky, rubberised affairs. The meter looks like a cheap CB effort, a very far cry from the traditional gold-plated construction (and prices!) of their earlier models. I know one enthusiastic and well-heeled listener who returned his HF-1000 to Lowe's after a week, dismayed at the spurii. In the table, its oscillator phase noise figure is no better than that of the old Drake TR7!

"Surprisingly, the Racal RA6790 GM comes well down the table. This is the US version of the well-known Racal RA1792 and has been sold in large numbers to the USAF and other users. I have a RA1792 and it is one of my favourite rigs. Its dynamic range is a lot better than the table suggests, at least on SSB. I also have a Drake R7 and have measured the parameters with the same equipment and although the generators were old (HP 606Bs) the comparison was valid and gave better numbers for the 1792. A friend, G3YFK, measured the IP3 of his RA1792 as +31dBm using good modern professional test equipment.

"Whatever the numbers may say, the real test for me is a pair of decent headphones connected to trained ears. Here the 1792 wins hands down over most other rigs, though I can hear some computer-generated gunge low down. This can be reduced with an attenuator in series with the headphones (a 1kΩ wirewound pot).

"The only rigs I know to better it are the old Racal RA1779 which cost over £10,000 new in the late 1970s and the MEL/Philips PRC2000 HF transceiver (hardly on the average amateur's shopping list, assuming he has even heard of it!) The PRC2000 uses the Weaver Third Method of SSB generation and demodulation and has little of the coloration imparted by a conventional sharp crystal filter, yet gives excellent SSB selectivity on a par with a good filter. I currently use it as Q-fiver hanging on

SENSITIVE RF TEST PROBE

ALAN HOOPPELL, G4TKV writes: "Whilst experimenting with Superbright/Hyperbright LEDs, I have found that they will illuminate on low microamp levels. I tried one on the output of a crystal set and was amazed to see a dim glow. I have since made one into a useful test probe, following the usual arrangement of a diode probe but replacing the small meter with an LED: **Fig 1**.

"The probe glows quite bright when held a few inches from the output transistor/tank of a 0.5W QRP transmitting, and by touching can indicate oscillation in any circuit. Possibly not a new idea but not one that I have seen in print. I now have an idea that two LEDs and a small toroid might make the smallest of reflectometers".

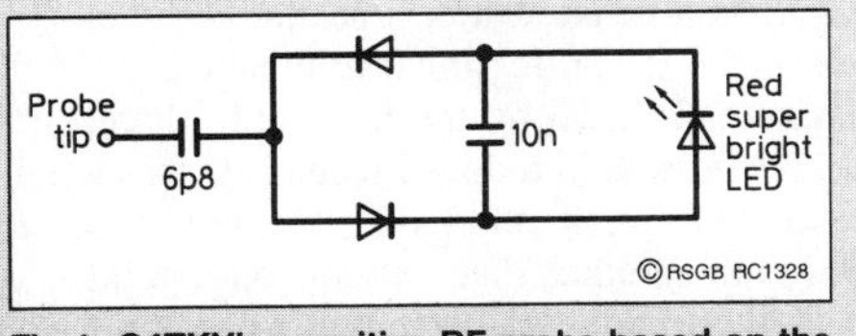

Fig 1: G4TKV's sensitive RF probe based on the use of a 'superbright' LED.

Device Under Test	Noise Floor	AGC Treshold		100kHz Blocking	Sensitivity	LO Noise/Spacing		Front End Selectivity	Filter Ultimate	Dynamic Range Wide Spaced		Dynamic Range Narrow Spaced	
	(dBm)	(µV)	(dB)	(dB)	(µV)	(dBc)	(kHz)		(dB)	(dB)	(kHz)	(dB)	(kHz)
Drake R-4C/CF-600/6	-139	0.7	3	130	0.15	135	10	A- Preselector	130	85	20	85	2
AOR AR-7030	-122/-128*	2.2	3	130	0.5/0.22	130	10	D Hi/Lo Pass	90	100	20	82	2
Atlas 350-XL	-131	1.0	11	117	0.2	125	4	C Band Pass	95	81	20	81	2
Kenwood TS-830/YK88	-129	1.5	3	122	0.1	114	2	C Preselector	85ø	84	20	81	2
Yaesu 901-DM	-135	1.6	3	124	0.15	109	2	C Preselector	85	87	20	80ø	3
Collins R-390A	-137	N.A.		130	0.2	130	2	A+ Trk Presel	85	81	20	79	2
Ten-Tec Corsair	-131#	0.1	14	130	0.2	132	5	C Band Pass	90	93	20	79	3
Icom IC-720A	-137	1.6	3	138	0.15	117	10	C 0.5 Octave	80	93	50	78	3
Kenwood TS-820S	-137	0.4	3	115	0.2	125	10	C Preselector	80	79	20	78	3
JRC NRD-515	-138	3.5	4	103	0.1	118	10	C 0.8 Octave	80	95	20	77ø	2
Ten-Tec Omni V	-134	1.2	6	135	0.18	134	10	C Band Pass	100	89	20	76	2
Atlas 210/215X	-120#	N.A.		123	0.5	N.A.		C Band Pass	95	76	20	76	2
Drake R-7	-135/140*	1.0/0.4*	3	145	0.28/0.15*	114	10	B 0.5 Octave	85	97	100	75	2
Drake TR-7	-134	1.3	3	146	0.5	116	10	B 0.5 Octave	90	99	100	75	2
Heath SB-104	-123	N.A.	92	0.5	N.A.			C Band Pass	75	79	20	75	4
WJ HF-1000	-129/136#*	0.11	3	123	0.23/0.13*	115	10	D Wideband	80	99	20	74	5
Ten-Tec Omni-B	-136	0.2	25	129	0.15	130	10	C Preselector	80	87	20	74	2
Icom IC-730	-140	1.5	3	135	0.1	118	10	C 0.5 Octave	80	92	50	74	3
Kenwood R-820S	-125	4.0	3	125	0.35	123	10	C Preselector	75	74	20	74	4
Collins 75-S3B	-146	1.1	15	122	0.1	120	4	B+ Preselector	85	88	20	74	2
Icom IC-781	-127/138*	2.0/0.5*	3	141	0.5/0.18*	129	10	B 0.5 Octave	90ø	94	20	73	2
Kenwood TS-930S	-135	2.0	3	143	0.15	115	10	B- 0.5 Octave	80ø	86	20	73	3
Icom IC-701	-129	5.5	6	130	0.3	125	10	C Band Pass	75	81	50	73	4
JRC NRD-525	-132#	0.9	3	123	0.2	120	10	B Trk Presel	65	95	50	72	5
Drake R-8	-128/131#*	0.6/0.3	3	130	0.25/0.18*	115	10	C 0.5 Octave	75ø	90/85*	20	71	5
Icom IC-R72	-127/135#	3.1/1.2*	3	129	0.28/0.11*	122	10	C 0.5 Octave	75	87*	20	71	5
JRC NRD-535	-135#	0.9	3	114	0.1	117	10	B Trk Presel	70	92	50	70	5
Kenwood TS-830S	-136#	0.9	3	122	0.1	113	2	C Preselector	80	84	20	70	3
Icom IC-761	-131/-139*	2.0/0.7*	3	145	0.4/0.17*	129	10	B- 0.5 Octave	90ø	87	20	70	2
Lowe HF-150	-126#	0.7	3	126	0.3	113	10	F No Bandpass	75	84	20	69	5
Kenwood TS-430S	-136#	0.6	3	134	0.1	102	10	C 0.5 Octave	70	78	20	69	5
Signal/One CX-11A	-122#	0.6	17	109	0.6	119	50	C 0.5 Octave	105	90	50	68	5ø
Kenwood TS-180S	-139	0.9	3	115	0.15	120	10	C Preselector	80	70	20	68	3
Drake TR-4C	-124#	1.2	3	105	0.4	130	10	C Preselector	80	74	20	68	2
Icom IC-735	-126/133*	1.5	12	135	0.35/0.18*	123	10	C 0.5 Octave	90ø	83	20	68	2
Drake SW8	-127#	0.9	3	125	0.32	113	10	N.A.	70	92	20	67	5
Racal 6790 GM	-128	0.3	1	145	0.7	130	10	D Broadband	85	95	20	66	2
Lowe HF-235	-126#	0.8	3	129	0.35	117	10	D Octave	80	71	20	66	5
AOR AR3030	-131#	2.0	10	130	0.16	117	10	C 0.5 Octave	85	90	20	66	5
Yaesu FRG-100	-133#	0.9	3	127	0.13	112	10	C 0.5 Octave	70	99	50	65	5
Kenwood R-5000	-131#	0.4	3	134	0.2	120	10	C 0.5 Octave	80ø	86	20	65	5
Yaesu FRG-7700	-130#	1.3	3	123	0.2	100	10	D Octave	65	83	50	64ø	5†
Kenwood R-1000	-130#	0.9	3	119	0.2	107	10	D Octave	70	76	20	64ø	3†
Heath SB-303	-134	N.A.		104	0.5	N.A.		C Preselector	70	66	20	64	4
Collins KWM-380	-127#	1.1	5	123	0.3	99	10	B 0.5 Octave	70ø	94	50	64ø	2
Icom IC-751	-127/133*	6.3	3	138	0.4/0.2*	127	10	B- 0.5 Octave	90ø	84	20	64	2
Kenwood TS-520	-139	N.A.		116	0.15	N.A.		C Preselector	70	63	20	63	3
Yaesu FT-One	-135	1.0	3	130	0.2	99	10	C 0.5 Octave	80ø	91	50	63ø	2
Collins 75-S3 Wing	-145	1.0	14	105	0.1	N.A.		B Preselector	75	75	20	63	3
JRC NRD-93	-141	1.6	3	128	0.15	133	10	A+ Trk Presel	80	94	20	63	2
Yaesu FT-980	-136	1.8	3	140	0.12	106	10	C 0.5 Octave	62ø	96	50	63	2
Icom IC-R70/R-71A	-129/135*	3.1/1.4*	3	132	0.4/0.2*	128	10	B- 0.5 Octave	90ø	86	20	62	3
Grundig Satellit 700	-127#	1.6	3	106	0.3	118	10	N.A.	85	76	20	62	5
Yaesu FT-101E	-141	N.A.		102	0.15	N.A.		C Preselector	70	60	20	59	3
Drake R-4C Stock	-139	0.7	3	0.15	130	10		A- Preselector	70	85	20	58	2
Yaesu FT-757	-120/134*	1.6	3	130	0.7/0.15*	109	10	C 0.5 Octave	70ø	86	20	56	3
Kenwood R-2000	-130#	1.4	3	115	0.15	105	10	D Octave	70	71	20	45	5
Kenwood R-600	-130#	0.8	3	109	0.2	99	10	D Octave	65	68	20	F.L.	5
Yaesu FRG-8800	-132#	0.6	3	122	0.18	N.A.		D Octave	70	87	20	F.L.	5

Notes

1 dB blocking (gain compression) test done at 100kHz to eliminate phase-noise interaction.

N.A. = Data not available; # = Measured with SSB filter; * = Built-in preamp actuated; ø = Measurement was phase-noise limited rather than intermodulation limited; † = Readings would have been lower if 2-kHz spacing had been possible; F.L = Filter limited, no measurement was possible due to signal leakage around filter.

Table 1: The updated receiver measurements as listed by Sherwood Engineering Inc.

the IF output of a receiver via a 20dB attenuator. The miniature keys make it hopeless for search tuning.

"Indeed, one of the really good rigs which is genuinely fun to use (almost to 'drive') is my ancient RA17L with a side-band adapter such as the RA121 or RA218. The audio quality is astonishing for a rig now 30 years old. The adapter is essential since the RA17 detector is poor for SSB, there is hum on the audio, and the selectivity is pedestrian. However, the SSB selectivity both in the RA17 and in the adapter is entirely determined by LC circuits at 100kHz and 18kHz which gives little of the phase distortion that can be recognised with crystal filters. It is also interesting that the F1-2 receiver, made for professional re-broadcasting by Phasetrack (used by the BBC until direct satellite feed was introduced recently), uses a patented phasing method of selectivity and demodulation rather than relying on a tight crystal filter. (Edward Forster of Phasetrack was one of the pioneers of effective low-cost synchronous demodulation of AM signals - G3VA). These receivers sound superb, albeit the selectivity figures are not as good as those of a conventional SSB rig, but who cares when the sound is right!"

In commenting on the 1977 *Ham Radio* article, I noted in September 1984 that WB0JGP/K8RRH pointed out that it is by no means as simple as sometimes suggested to improve factory designs, for example by fitting packaged ring mixers such as the MD108, unless detailed steps are taken to ensure that the correct injection-oscillator levels are provided. Altering just one part of a design may result in unexpected knock-on effects.

The R4C modifications included extra gain after the selectivity filters and an IF shunt. It was noted that while many receivers have just sufficient overall gain when used with an SSB filter they lacked the extra gain required to overcome the greater insertion loss of a narrowband CW filter. It was also pointed out that performance measurements on receivers are often made at 14MHz: "At this frequency (and above) the practical value of good tracked pre-mixer selectivity is unlikely to show up with a spacing of 20kHz or even 100kHz. Good pre-mixer selectivity really comes into its own below about 4MHz, and can be of real practical value to those seeking optimum performance on, say, 3.5MHz. As noted in 1984 it is important to remember that a receiver is only as good as its performance on your favourite bands in your shack, no matter how impressive its technical specification appears."

OVERVOLTAGE PROTECTION

A H RITCHIE, ZS5XK, in 'Tech Talk' (*Radio-ZS*, October, 1996, p8) describes a protection circuit to guard against failure of pass transistors in regulated power supplies. His system does not depend on 'crowbar' blowing of fuses but instead incorporates a relay that switches off the mains supply when the voltage exceeds a voltage that can be adjusted to a required voltage. His circuit (**Fig 2**) was adapted from one devised by W7RXV and originally published in *73 Magazine*, January 1980.

ZS5XK described construction as follows using the PCB layout shown in **Fig 3**.

1) Mount all components except the 47Ω resistor and TR3.
2) Connect a variable power supply to board and set for 14.8V.
3) Connect voltmeter to anode of ZD2.
4) Adjust the 10k multiturn trimpot until there is a sudden jump from 0V to two or more volts. If you cannot reduce the volts to zero then change the zener diode to one rated the minimum obtainable voltage plus 5.6V.
5) Mount TR3 and one side of the 47Ω resistor to the 12V latching relay (with suitably rated contacts for switching the mains supply to the PSU (ZS5XK gives Omrod GdRK-1)

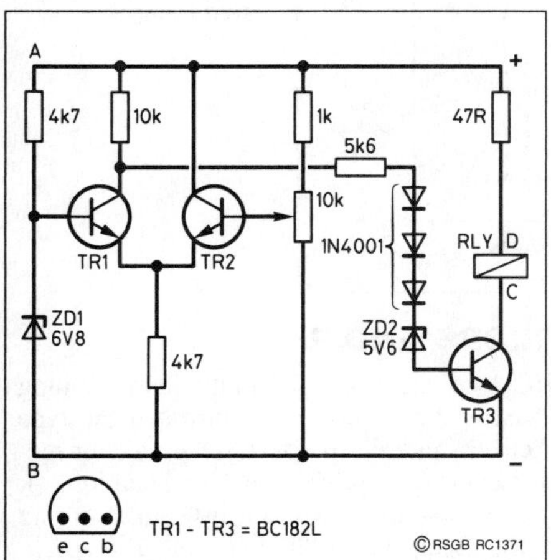

Fig 2: ZS5XK's over-voltage protection arrangement using a latching relay rather than a fuse plus thyristor to disconnect the mains supply to the regulated 13V PSU in the event of a failure of a pass transistor in the PSU. 10K multiturn trimpot (¾in rectangular style). Latching relay given as OMROD GdRK-1 12VDC (relay contacts should be suitably rated for 230V and current rating to suit the PSU rating).

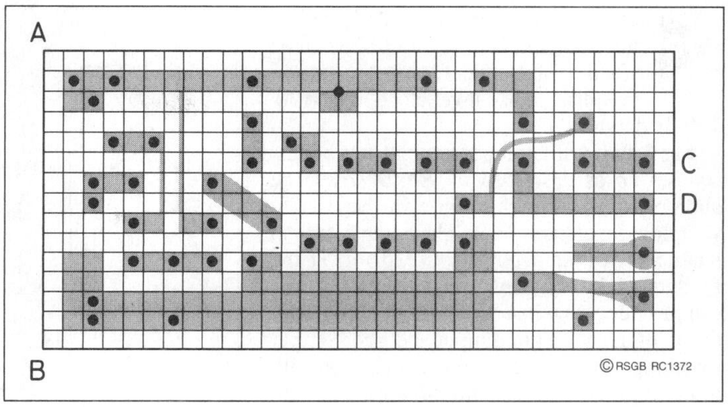

Fig 3: Layout of the PCB as used by ZS5XK, but other forms of wiring could be used.

6) Connect variable PSU as in (2).
7) With jumper leads connect 150mA FSD meter between free end of the 47Ω resistor and positive terminal of the power supply.
8) Then as the 1k trimpot is adjusted the current should jump suddenly from zero to about 70mA. If 70mA is exceeded then change 47Ω resistor to a higher value. Set the trimpot as the point where the current is at high point. Then reduce voltage from the variable PSU. The current through the meter should immediately drop to zero. Increase the voltage above 14.8V - the meter should immediately jump to 70mA. Disconnect power supply, mA meter and voltmeter and solder free end of 47Ω resistor to positive line of the unit.
9) Release the latch by applying 12V to points C(-) and D (+).
10) The complete unit can now be mounted near or in the main PSU connecting its output to points A+ and B-. Relay 250V contacts are inserted in the line mains lead. The latch release connections should be brought out to a convenient point on the PSU.

IMPROVING SIMPLE SPECTRUM ANALYSERS

UNTIL 1988-89, most amateurs regarded the spectrum analyser as a highly desirable but very high-cost instrument affordable only for well-equipped professional laboratories. Then Al Helfrick, K2BLA, published a prize-winning article in *RF Design* (January 1988) describing a unit suitable for use with a general-purpose oscilloscopes and providing a simple spectrum analyser that could display input signals from DC to near 100MHz. Suitable for home-construction it was based on a few low-cost ICs and a handful of active and passive components, with helical 145MHz and ceramic 10.7MHz filters.

When attention was drawn to this design in *TT* (April 1988) it immediately attracted considerable interest, and a series of comments and suggestions followed (*TT*, July, August, September and December 1988 including constructional hints from K2BLA, the original designer. All these items appear in *Technical Topics Scrapbook*, 1985-89).

Then Roger Blackwell, G4PMK, provided a detailed constructional article 'Simple Spectrum Analyser' (1 - 90MHz) based on the K2BLA design but with several modifications (*Radio Communication*, November 1989, pp44-49). This included full component details and layout diagrams and pointed out that PCBs for this project were to become available from Bonex. It was noted that, "for most people, a spectrum analyser is way out of reach - but this design makes a home-brew unit a realistic possibility". G4PMK's article has been republished in the current edition of the RSGB's *Radio Communication Handbook* (6th edition, pp15.23-29). It is believed that several hundred of the K2BLA/G4PMK simple spectrum analysers (SSA) have been built in the UK alone.

Recently, Brian Mitchell, G3HJK, brought to my attention an article entitled 'RF lumped-notch filter design for FCC compliance testing of wireless radio modems' by Lizhong Zhu *et al* (*RF Design*, November 1996, which carried an introductory note: "A notch filter circuit is useful for high-accuracy spectrum measurements because it can provide a 20-30dB improvement in a spectrum analyser's dynamic range. An improved dynamic range allows a less expensive analyser to be used . . . The filter improves analyser performance by notching the fundamental frequency, thus avoiding the need for adding external attenuation, which would limit the analyser's dynamic range. Notching allows the user to view low-level harmonics."

While the original authors designed their notch filter for 899MHz (mobile radio), G3HJK realised that a notch filter would greatly improve the dynamic range of his SSA, which with various earlier improvements, has a dynamic range of some 60dB. With a notch filter based on the *RF Design* approach this can now be extended by an extra 20-30dB, and with a further optional crystal series-resonant ladder filter he can add either 30dB or 60dB of dynamic range to his SSA. His SSA is now capable of showing harmonics

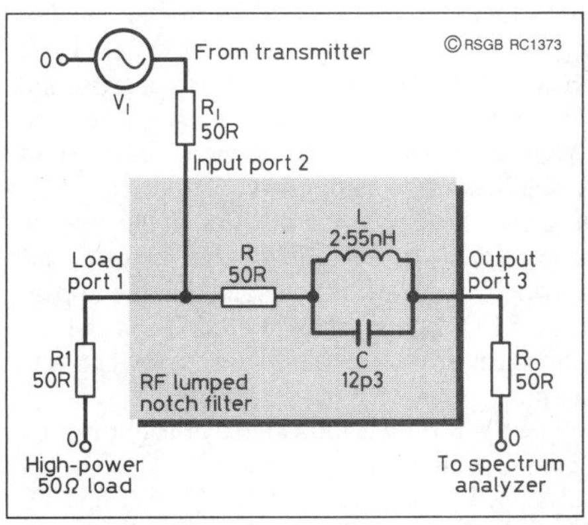

Fig 4: The 899MHz lumped-notch-filter and its test set-up as described in *RF Design* can improve the performance of simple, low-cost spectrum analysers.

72

etc down to near 90 or 120dB below the fundamental level. Using steep sided crystal ladder filters he can measure phase noise level close in at -130/140dB down.

Fig 4 shows the lumped-notch filter and its test set up as described in *RF Design*. The authors stress: "Notching allows the user to view low-level harmonics. The notch filter can handle high-output powers (>33dBm) at the fundamental frequency, around 899MHz, and can provide a low insertion loss (<-8dB) from DC to 6GHz with good input and good output matching (VSWR <1.5:1). It consists of only three lumped surface-mount elements: an inductor, a capacitor and a resistor . . ."

The SSA is limited to the display of signals below about 90MHz so that an HF rather than UHF notch filter is required - but the principle of using this type of notch filter holds good, as G3HJK has found.

On a related topic, an article 'See the signals you hear - PC-based audio spectrum analysers: a useful ham shack tool' by Donald Cox, AA3EK (*QST*, February 1997, pp28-32), underlines that "the introduction of personal computers (PCs) into many ham shacks and the advent of PC multimedia sound-card systems have placed the basic components of a digital signal processing (DSP) system at the fingertips of many hams and SWLs. This inexpensive, but powerful, technology can augment many Amateur Radio operations."

AA3EK notes that while AF spectrum analysers cover only audio frequencies, 0 to 20kHz, laboratory-quality AF spectrum analysers cost as much as their RF cousins. AF spectrum analysers are useful for measuring the fundamental-frequency components of narrow-bandwidth signals. They can also display received HF SWL and amateur radio signals, which seldom exceed a 12kHz bandwidth. AA3EK suggests: "The ability to see the jumbles of signals on frequently crowded HF bands is now within the price range of every ham with a PC... Identifying and tuning signals, or optimising and calibrating receivers are only two of the many applications this new technology brings."

A SMALL DIRECT-READING VHF SWR METER

IN A DETAILED constructional article entitled 'A Pocket-Size, Direct-Reading VHF SWR Meter' (*QST*, February 1997, pp33-36), Bill VanRemmen, KA2WFJ, describes an "easy-to-build LED bargraph SWR meter" which "goes with you anywhere". It uses an LED array as the readout, with its microstripline directional coupler feeding a sample of the feedline's forward and reflected waves to an LM3914 dot/bar display drive. The driver lights an LED whose physical position in the array is proportional to the ratio of the forward and reflected signals. It is arranged that the uppermost LED lights when the SWR is 3:1 or higher, improving resolution in the lower SWR range.

The SWR ratio is indicated by the number of LEDs that light in the array: 1, 1:1.105; 2, 1.222; 3, 1.353; 4, 1.500, 5, 1.667; 6, 1.857; 7, 2.077; 8, 2.333; 9, 2.636; 10, 3.0 or more.

Fig 5 shows the circuit diagram but for details of the construction and adjustments, it would be advisable to consult KA2WFJ's original *QST* article. However, some readers many find brief details of the basic construction of the microstripline directional coupler of interest. This is made of a sandwich of two PC-board sections (see **Fig 6**). When constructed on a double thickness of 1/16th inch glass-epoxy PC-board material, the 0.22in wide strip presents a 50Ω impedance. Two fine-wire pickup loops (4in pieces of AWG No30 or even finer bare wire locked in place with a drop of five-minute epoxy) run between the strip and the ground plane. The pick-up wire ends are terminated on the solder (foil) side of the board by 1N34A germanium (not silicon) diodes.

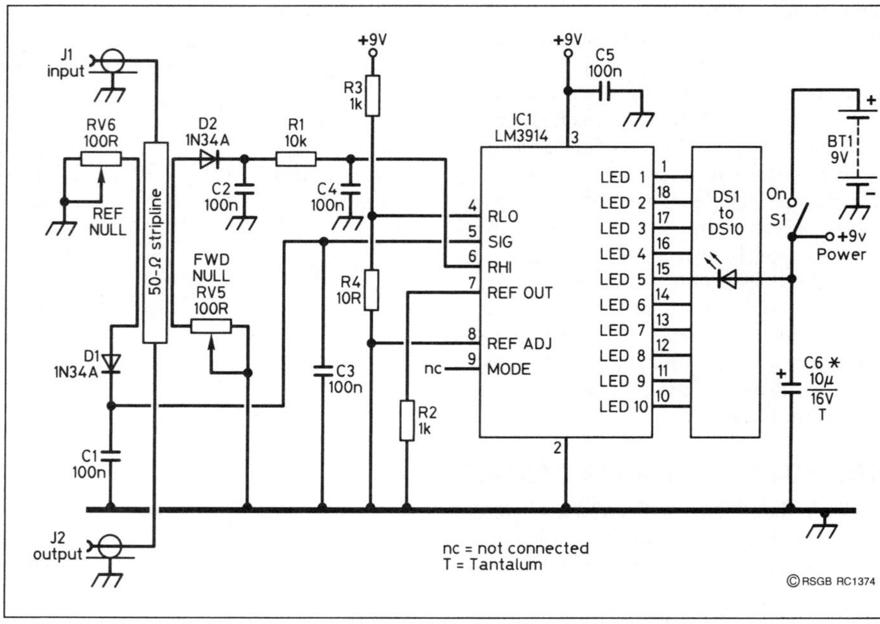

Fig 5 (above): Circuit diagram of KA2WFJ's pocket-sized. Direct-reading LED bargraph VHF SWR meter. Unless otherwise specified resistors are ¼W, 5% tolerance. J1, J2 BNC connectors. R5, R6 100Ω small trimmer pots. Use 1/16th inch glass-epoxy board which has a dielectric constant of 4.6 so that 0.22in wide strip with a separation of 1/8th inch yields an impedance of 50Ω. If readings above 3.0 are required replace R1 with a jumper wire and recalibrate the markings on the front panel. No LED should light with unity SWR.

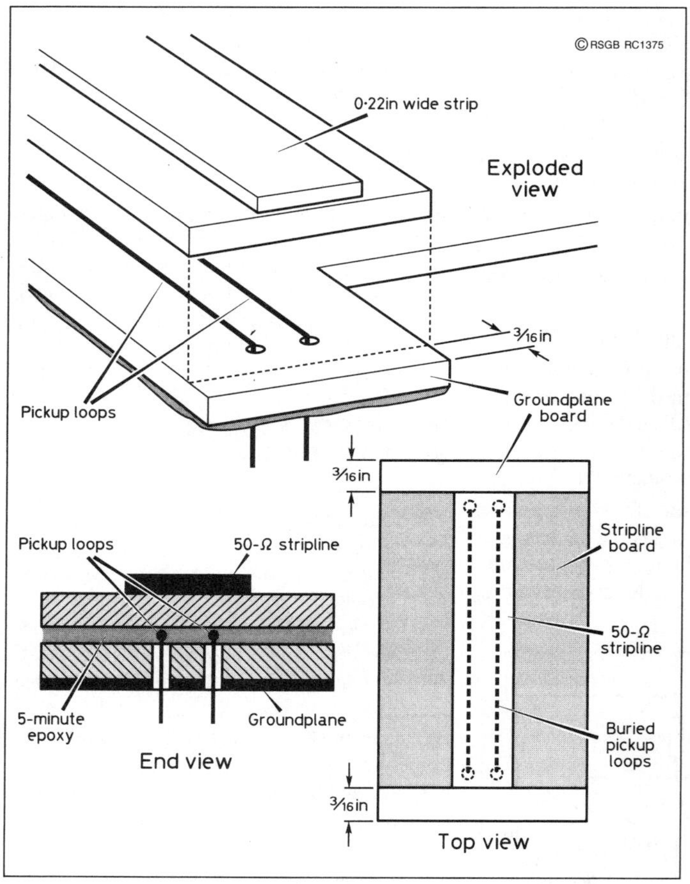

Fig 6 (left): Assembly of the directional coupler PC-board elements for the VHF SWR meter.

RCCBS AND PME

FIG 1 OF the March *TT* in the item 'Taming Electrostatic Discharges' illustrated the type of electrostatic discharge (ESD) protection recommended for professional workstations; it was redrawn from an article by Frank Harding in the December 1996 issue of *Television*. While the drawing effectively drew attention to methods of preventing electrostatic sparks being discharged from the repairer by providing adequate paths to earth, it unfortunately included misleading information on the use of residual-current circuit breakers (RCCBs, see **Fig 7**) when combined with isolating transformers. It also raised, once again, the continuing problem that many homes are now wired with protective multiple earthing (PME) with the need to avoid creating a common connection between the mains earth wire and true (RF) earths, or even the possibility of touching both simultaneously under fault conditions.

Letters commenting in detail on RCCB and PME practice have come from a number of readers, including R P Mew, G6HKF, Malcolm Perry, G8AKX, and David Long, G3PTU. It is clear that advice is still needed, although these topics have been discussed a number of times in *TT* and elsewhere; with the particular problems posed by PME well described by Peter Chadwick, G3RZP, in a full-length article 'The Killing Ground - earth your station safely' (*Radio Communication*, June 1987, pp404-5). Note especially that with PME all points at mains earth should be separated by at least 6ft from any point at real earth.

To quote briefly some of the comments from G6HJF and G8AKX. G6HKF writes: "In Fig 1 (March) the RCCB would not operate even in the event of a dead short across the mains output from the isolating transformer since there is no connection between the earths due to the double-wound transformer. It is imperative that the isolating transformer has part of its output, be it one end or a centre tapping, grounded out to real earth prior to the RCCB, otherwise, in the event of a 'case to live' fault, it will not detect any imbalance. The next time anything touches the 'other wire' a heavy fault current will flow, and even then the RCCB will not operate. Furthermore, the isolated output can, due to static charge, rise above ground to many thousands of volts, inflicting the damage to the solid-state 'three-legged fuses' that the workstation is intended to protect.

"If the property has a TT system or a cable sheath earth, then it is better to use the mains direct with an earth stake into the ground brought to a common bonding point. If the electricity supply is PME, do not bond down to an earth stake with anything less than 16mm earth wire, as used for the existing property earth, and do not attach yourself to it. With PME, the only effective solution is to use an insulated and isolated screened transformer, the screen connected to the PME earth and the live and neutral connected to the feed through an RCCB. The output should be centre tapped to real earth, and the 230V supply taken through an RCCB to the sockets. The real earth can then be used for the static return. It is essential to be aware that under no circumstances, must the two supplies come within working distance of each other since the earths could have 240V AC between them.

"Supplies from a PME socket should not be taken outside the building: think about that - no satellite dishes, antenna poles etc apart from metal outside lights, hedge trimmers etc. Personally, I will have nothing to do with PME . . . I had to get my area REC to put right five PME installations that I found had been installed incorrectly. Remember that if a PME system is involved, whilst it can be made usable, hazards exist that do not appear obvious. I despair for those living in flats with PME."

Malcolm Perry, G8AKX, points out that double-wound isolating transformers have mainly in the past been associated with the TV/radio trade due to domestic sets using "live chassis" techniques. Amateur equipment and surplus "ex-service" equipment normally comes with built-in double-wound transformers. However, with the increasing use of switch-mode PSUs, particularly in computer equipment and monitors, a large block of the circuit with associated components and heat sinks can be connected directly to the mains supply. There is also increasing use of capacitors in series with the mains to act as a voltage dropper on some equipment. There is normally no safety problem in using such equipment, but there can be a hazard when opened up for repair. There is thus increasing need to use isolating transformers when working on amateur radio and IT equipment. But, as with other correspondents, he notes that RCCDs which operate when there are unbalanced currents due to earth leakage are in effect invalidated by the use of an isolating transformer.

Clearly, the very real safety problems imposed by PME on workstations and amateur shacks where real (RF) earths are common deserve very careful consideration: for a start it is well-worth re-reading G3RZP's two-page 1987 *RadCom* article. ♦

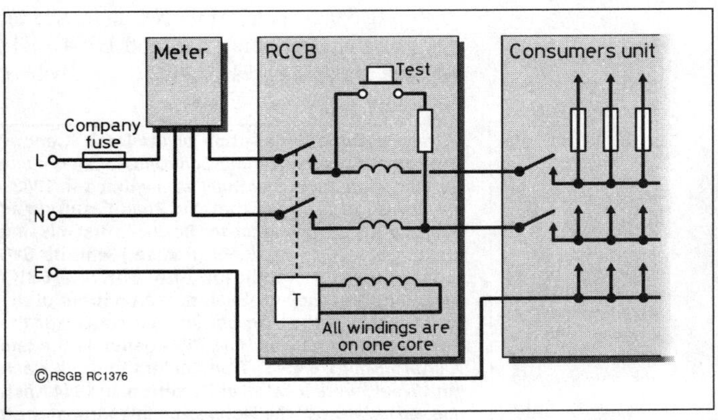

Fig 7: Showing an RCCB installed to detect excessive earth leakage currents throughout a domestic installation. Such units may have a trip sensitivity of 100mA or more. Higher sensitivity 30mA units should be used to protect the output from a single power socket or workstation. Even higher sensitivity units (to about 10mA) are available but may prove susceptible to false trippings.

BALL GATE ARRAYS

N D BONNETT, G0NNA/DL6EE, suggests that while amateurs are coming to terms with SMD (surface mounted device) technology, manufacturers of VLSI and ASICs seem set on defeating us all with their latest product using the Ball or Solder Gate Array (BGA). He writes:

"The production and use of ever more comprehensive and densely packed ICs has until recently been limited by the number of I/O pins available to the designer. Modern quad-flat packs with a pitch of between 0.3 and 0.4mm are notoriously hard both to place, paste and solder correctly in order to obtain a good board yield.

"Enter the BGA - **Fig 8**. On the underside of the chip is a matrix of solder balls, each one providing an I/O connection. The spacing between the balls is 1.27mm and the ball diameter 0.7mm. A matrix of 16 x 16 such balls gives 256 connections; the largest ICs currently produced have 25 x 25 arrays giving 625 connections! The multilayer PCBs have up to 12 layers. Connections to the lower layers are made by drilling through the board to the appropriate layer, inserting a connecting wire and then plating the hole to fix the connection. Initial soldering does not use a reflow bath but a hot air convection oven. It is claimed that placement, plating and soldering are easier.

"The only production problems are inspection (which obviously cannot be done visually) and board testing. Should a fault occur and a device has to be removed and replaced, this is done with hot air jets. Be wary if you come across scrap or surplus boards with BGA devices at rallies etc. This technology looks to be the one that will finally defeat home construction!

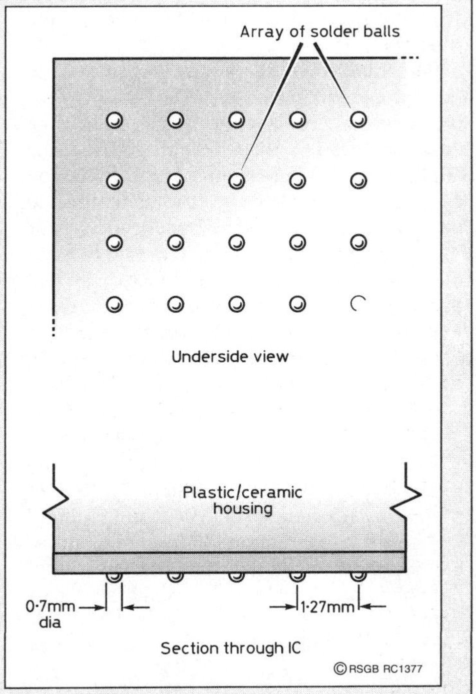

Fig 8: (a) Array of solder balls as now used on some BGA VLICS. The array may be as large as 25 x 25 solder balls (625 connections) and used with multilevel printed boards. (b) Section through a BGA IC.

Pat Hawker's Technical Topics

PAT HAWKER, G3VA*

REGENERATIVE RECEIVERS

THE APRIL *TT* item on 'Regenerative "Straight" Receivers' contained comments and illustrations (unfortunately the pictures of the genuine and replica sets were transposed) of the wartime Whaddon Mark VII (Paraset) clandestine receiver-transmitter. There was also a description of the modern high-performance JFET solid-state regenerative receiver built by Charles Kitchin, N1TEV. The item attracted the attention of a number of readers. It is clear that as an antidote to the modern highly complex 'black box' transceiver, there is a real desire among a significant number of readers to build simple equipment capable of being put to practical use. This seems to be coupled with a genuine interest in learning more about those who bravely used such equipment in the occupied countries between 1940 and 1945.

In this connection, the producers of the *Timewatch* television documentary 'Secret Memories' transmitted on BBC2 on 11 March, 1997 deserve congratulation for their admirable presentation of the authentic and vivid memories of several wartime radio-agents of both SIS and SOE. These included SOE radio-agents Brian Stonehouse and a Frenchwoman identified only as 'Rose' both of whom were caught but survived the concentration camps. Also featured were Oluf Reed-Olsen, a Norwegian SIS radio intelligence agent, plus a number of other non-radio agents and the redoubtable Vera Atkins who looked after the interests of SOE F-Section agents as assistant to Colonel Buckmaster. For those who missed this programme it is much to be hoped that, like most Timewatch documentaries, it will eventually be repeated.

One of the most remarkable features of the programme was that it included genuine film of Reed-Olsun ('Makiv/Aquila') using an attaché case Mark VII in the Kristiansand area of occupied Norway during 1943. Although shot as 'silent film' some appropriate Morse from the key of George Eddowes, G3NOH, was skilfully dubbed in.

In spite of the danger, quite a few still-photographs exist of radio-agents and/or their equipment in hazardous surroundings, but this must be one of the few, if not the only, genuine movie-film of a radio-agent 'in the field', to recall the wartime expression. Some further details of the two missions of Oluf Reed-Olsun together with his original Mark VII Paraset can currently be found in the 'Secret Warfare' section of the Imperial War Museum in London.

As a 21-year-old in 1940 he was an early member of the active Norwegian resistance. He then escaped in a small sailing boat, fortunately being picked up a fortnight later by a Royal Navy destroyer. In early 1943 he was recruited by MI6/SIS, and was almost certainly given instruction in clandestine operation at the SIS radio-agent training school in Hans Place, Knightsbridge, London ('Hans School'). In April 1943 he parachuted back into Norway with a Mark VII as part of the extensive SIS operations that kept watch on German shipping. For some six months he was in radio contact with the Nash 'Central' base station (see photograph). In October 1943, he escaped to Sweden but returned to Norway the following December, returning to England via Sweden in January 1944. As one of the surviving radio-agents he appeared on the *Timewatch* programme aged nearly 80 years, but still regularly climbing to the cliff-top mountain hide-out he used in 1943 - 44.

He parachuted back into Norway in May 1944, this time equipped with one of the more sophisticated SSTR-1 HF transmitter-receiver suitcase sets of the American Office of Strategic Services (OSS). This had an effective superhet receiver using loctal valves and including a tuned RF amplifier stage (7V7 RF amplifier - 7Q7 frequency changer - 2 x 7V7 IF amplifier - 7V7 BFO - 7F7 double-triode AF amplifier). The SST transmitter comprised a single 6L6G crystal power oscillator with a rotary antenna matching coil (roller coaster).

The photograph opposite shows a complete SSTR1 in its suitcase. **Fig 1**shows the circuit diagram of the 6L6 transmitter. According to information unearthed by Professor Gary P Cain, W8MFL, the cost to the US government of each SSTR-1 was $450. I recall trying to teach a young Dutchman how to use one of these sets at Eindhoven in early 1945 - he crossed into enemy-occupied territory carrying his suitcase set but was never heard of again, almost certainly caught by the Germans even before he could come on the air. Perhaps the rather distinctive green-fibre case did not help.

Coming up to date, C M Lindars (RS32010) built the 'high performance' solid-state regenerative receiver described by N1TEV in *Communications Quarterly* and included in the April *TT*. He writes:

"I find that it does indeed perform very well: the selectivity, sensitivity and tone are all one could expect from so simple a circuit.

"I have departed from the design shown in April Fig 3 in that an LM386N chip is used for the low-frequency amplifier and E304 FETs replace those suggested (J310 or 2N4416

A rare wartime photograph of the Nash 'Central' station of Special Communications, located close to Whaddon, near Bletchley as it existed in 1943 - 44. It was to this station that Oluf Reed-Olsun would have sent his Intelligence traffic on his first mission using the simple Mark VII (Paraset) with its 6V6 power crystal oscillator and 6SK7 + 6SK7 regenerative receiver. Seen at the controls of the HRO receiver, keeping continuous watch for the Norwegian coast-watching SIS agents, is the late Clifford Smith, G4SIY. Transmitters for both Nash and Weald were located at Calverton. In 1944 Nash closed down with the Norwegian links transferred to the large RSS (SCU3) intercept station at Forfar, Scotland and the Dutch links to Weald. Among the Nash operators was a well-known early blind amateur, W Krohn, G6KJ, and C S Francis, G3AVI, both of whom went 'silent key' many years ago.

*37 Dovercourt Road, London, SE22 8SS.

Motorola FETs). Although a coil was wound as described for the lower frequency range (with some difficulty), use is now made of the noval-valve-type Denco coils with the winding between pins 8 and 9 used for reaction, and the winding between pins 3 and 4 used for coupling.

"I experienced one or two problems, however. The 6.8V zener in the supply to TR2 is barely operating due to the voltage drop across R4. Also, the 'audio bandwidth' (top cut) control and audio gain potentiometer seriously affect the reaction setting. In an attempt to overcome these problems, one could play around with the value of R3 or use a lower zener diode voltage. The reaction interference has been greatly reduced by inserting a 2.2K resistor between C12 and the pole of SW (bandwidth). It is difficult to reconcile the figures given for the bandwidth. Possibly 7kHz should read 700Hz [No, but for 7kHz C18 should be 10nF *not* 100nF - *G3VA*]. A top cut is useful and I found that a 330nF capacitor offers a more useful cut than the suggested 100nF.

"The switched input attenuator is attractive. Apart from the more precise settings available, it obviated the need for insulated bushes. As a 6-way switch was available, I am using a small piece of wire on the final position to provide a low 'stray-capacitance' antenna coupling."

A short article 'Build a one-tube regenerative receiver' by Mark Starin, KB1KJ, (New Hampshire Office of Emergency Management) in *RF Design* (January 1996, p62) notes that "in today's world of one-chip receivers and transmitters, surface-mount components, and digitally-controlled tuning, it is hard for us to imagine the revolutionary impact Edwin Howard Armstrong's regenerative receiver design (with its 'tickler coil' providing positive feedback) had on the communications industry of 1915".

KB1KJ provides a circuit diagram **(Fig 2)** for an easily-duplicated electron-coupled regenerative receiver using a 6SK7 metal-octal valve (but virtually any other RF pentode could be used). He lists the number of turns required for the 3.5, 7 and 10MHz amateur bands (but gives no information on the formers he used). A typical valve / loudspeaker step-down transformer permits the use of modern low-impedance headphones.

I must admit that I still use 55-year-old Telefunken 4000Ω adjustable-diaphragm headphones which seem to be significantly more sensitive than modern 'phones in which emphasis is placed on high fidelity rather than on good sensitivity. The lightweight Telefunken 'phones have always appeared significantly more comfortable to wear than the once famous and equally sensitive Brown adjustable-diaphragm ones which after wearing them for some time clamped the ears painfully!

RESONANT FEEDLINE DIPOLES

THERE ARE MANY locations where an end-fed antenna can be more convenient than a centre or off-centre fed system. The end-fed 40m (132ft) long-wire antenna or even the 84ft W3EDP (see *RadCom*, August 1996 article by G3LCK) take a lot of beating as multiband

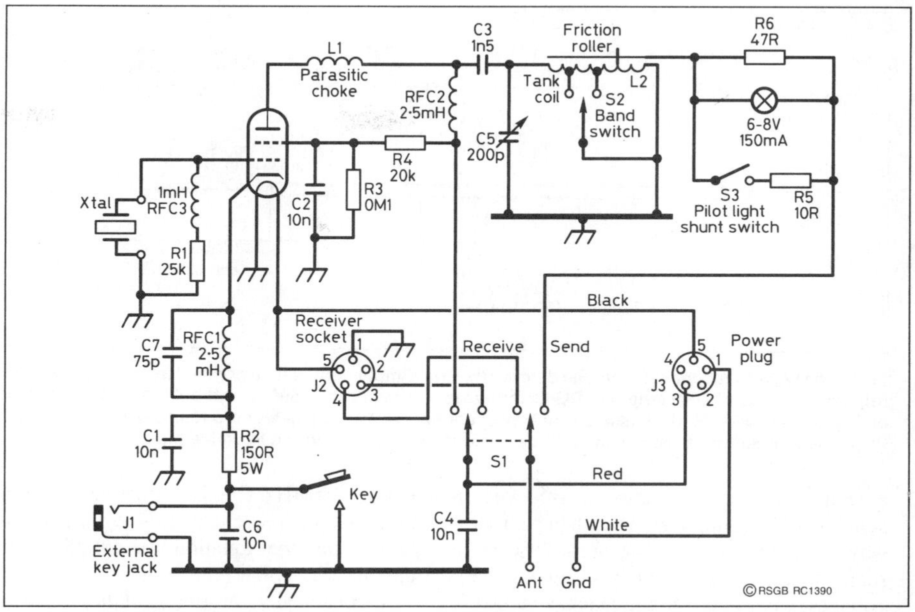

Fig 1: Circuit diagram of the SST-1-E transmitter, forming part of the American OSS's main WW2 clandestine 'suitcase set' type SSTR1 widely used in Europe during 1943 - 45. The crystal-controlled transmitter covered 3 to 15MHz in three bands (Band 1, 3 - 7MHz; Band 2, 6 - 12MHz, Band 3, 8 - 15MHz. (Information from Guido Roels, ON6PW).

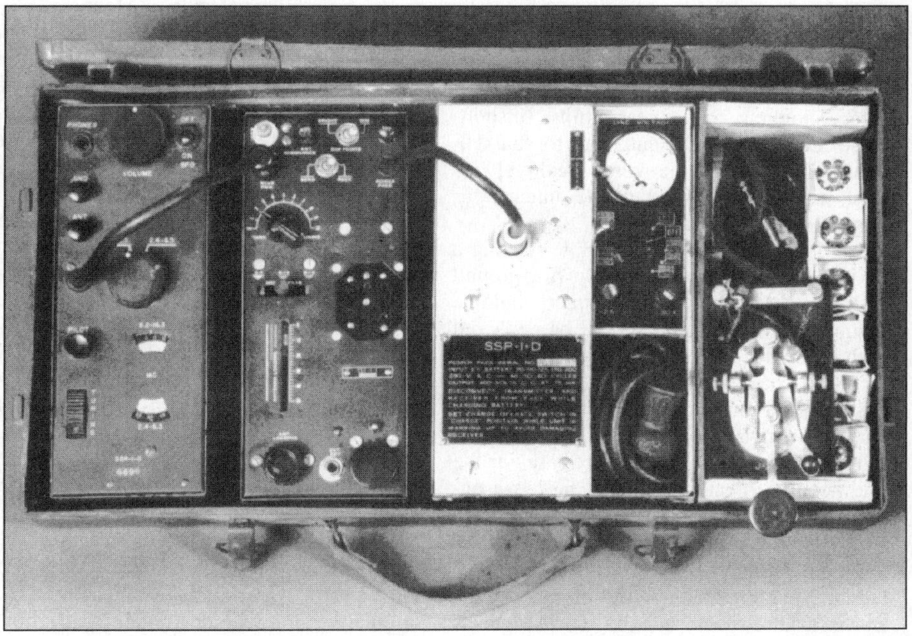

The American-built OSS suitcase transmitter-receiver type SSTR-1, comprising in three separate units, an SSR-1-G superhet receiver; SST-1-E 6L6 transmitter; and SSP-1 D power supply unit together with space for accessories and spares. This set was developed in the USA from information supplied by SOE on their A-2 suitcase set. Cost to the US government $450 compared to £292 for an RCA AR88 receiver. (Photograph courtesy ON6PW).

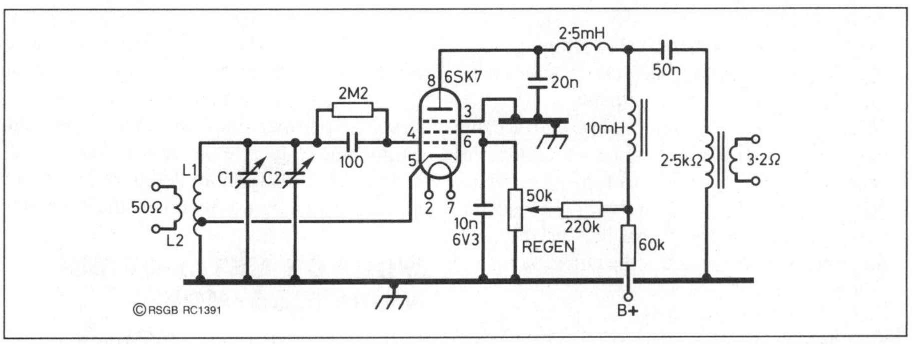

Fig 2: KB1KJ advises, "build a one-tube regenerative receiver" (*RF Design*) providing the circuit diagram of this electron-coupled regenerative design "which can be easily duplicated". He lists C1 as 50pF variable for band-spread tuning; C2 300pF (main tuning); L1 for 3.5MHz band 40 turns with tap at 5 turns from earthy end; 7MHz band 18 turns, tap 2 turns; 10MHz band 10 turns, tap 1 turn. L2 8t for 3.5MHz, 6t for 7MHz and 5t for 10MHz but does not specify the type or diameter of the former.

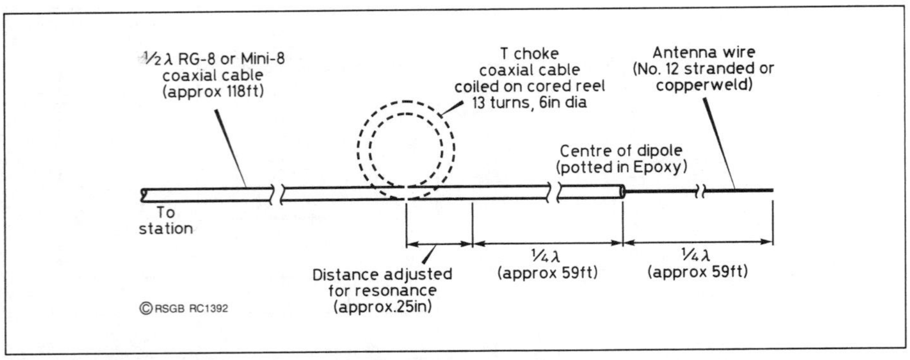

Fig 3: The W2OZH resonant Feed-Line dipole (RFD-1) as implemented for a nominal frequency of 3.95MHz requiring the use of 118ft length of RG-8x (minifoam) coaxial cable; 59ft length of No 12 (AWG) stranded or Copperweld wire; PL-259 male coaxial cable connector; 10in diameter cord reel; 3in length of ½in OD PVC pipe (for centre insulator) and sufficient epoxy potting compound to fill the pipe.

antennas provided that they are fed with a suitable ATU against a good earth or a quarter-wave or appropriate counterpoise. The main snag is that these antennas bring high RF voltages or currents into the shack, with possible health hazards and/or RFI problems when using high power.

James E Taylor, W2OZH, offers an alternative (single-band) solution in the form of the RFD (resonant feed-line dipole) antenna. His original presentation 'RFD-1 and RFD-2: Resonant Feed-Line Dipoles' (*QST*, August 1991, pp24-27) was sub-headed 'This unique design offers simplicity and general utility. In short, these are "reel" great antennas for the HF bands'. The play on the words "real/reel" reflects the ability to roll up the antenna for portability.

He developed his antenna while seeking a universal solution under the following ground rules. (1) No separate antenna tuner would be required; (2) the antenna could be deployed with no more difficulty than stringing up a length of coax; and (3) the antenna could be easily stored and unwound from a cord reel without a tangle of cable and wire.

In effect, his RFD antennas are based on the traditional vertical linear coaxial sleeve antenna in which a vertical dipole is constructed from a quarter-wave whip and a quarter wavelength of shielding braid with the coaxial feedline passing through the braid (sleeve). But instead of fitting an extra sleeve he takes advantage of the fact that the inner and outer surfaces of coaxial cable can carry separate RF currents. He uses the inner of a section of the coax braid as the feed-line but the outer of the braid as the sleeve. The problem then becomes letting the common-mode RF on the outer of the braid 'know' when to stop flowing down the coax and reflecting back towards the centre to form a radiating half-element.

One method is to use a coaxial choke balun (T) as shown in **Fig 3** comprising 13 turns of cable coil on a cord reel (13 turns, 6-in diameter) as shown. The remainder of the feed-line is then used as an electrical half-wavelength, resulting in a 50Ω match at the choke balun and at the transceiver output.

In his letter on 'Alternative end fed antennas', W2OZH writes: "Briefly, the basic RFD antenna offers a method of achieving an end-fed resonant dipole antenna which is well matched to a 50Ω feedline. Fig 3 shows the construction, schematically, for a frequency of about 3.95MHz (in the American 3.5MHz band). Other frequencies can be simply computed from the equation for simple dipole antennas.

"Mechanically, the joining of the two halves is accomplished by use of SO-239 coaxial sockets mounted in a suitable insulating cylindrical tube (I found the apothecary pill bottles 1¼in diameter and 2½in long, with the locking top to be very satisfactory). Mating PL-259 coaxial plugs are placed at the input ends of the two halves of the dipole radiator. The 13-turn T-choke coil was wound on a cord reel or other 6in diameter tube or reel. During adjustments the winding of this coil can be slightly re-spaced to achieve the 1:1 SWR at resonance."

RFD-2 was a two-band antenna covering 7 and 3.5MHz without severing the coaxial cable (but with a second T-choke when set up for 7MHz). The *QST* article also provided scaling of the T-choke for other (American) bands: 160m, 1.90MHz 19 turns; 40m 7.26MHz 9 turns; 30m 10.12MHz 8 turns; 20m 14.29MHz 7 turns; 17m 18.14MHz 6 turns; 15m 21.38MHz 6 turns; 12m 24.96MHz 5 turns; 10m 28.65MHz 5 turns.

Later, W2OZH authored 'The Improved Resonant Feed-Line Dipole' (*73 Amateur Radio Today*, July 1994, p44-ff) of which he has provided the following brief summary:

"The improved RFD retains essentially the same electrical and mechanical features as the original version with one notable exception - the somewhat cumbersome 6in diameter T-choke has been replaced by two compact toroidal coils which, in concert, act to provide the desired high impedance at the input-end of the dipole. Details and arrangement of these coils, which are wound on Amidon cores, are shown in **Fig 4**. They are mounted in a (Radio Shack) plastic box 6 by 3-3/16 by 1-7/8 inches.

"The above brief notes should be sufficient to start experimenters with either of these interesting and effective antennas. The original articles give many more details including calculated design figures for other amateur bands."

MORE ON METAL-OXIDE-VARISTORS (MOV)

THE OCTOBER 1994 *TT* (p63) drew attention to the views of Jim Sandox, N2MPT , on the use of metal oxide varistors (MOVs) to remove mains spikes and so safeguard solid-state equipment. N2MPT was critical of the use in the

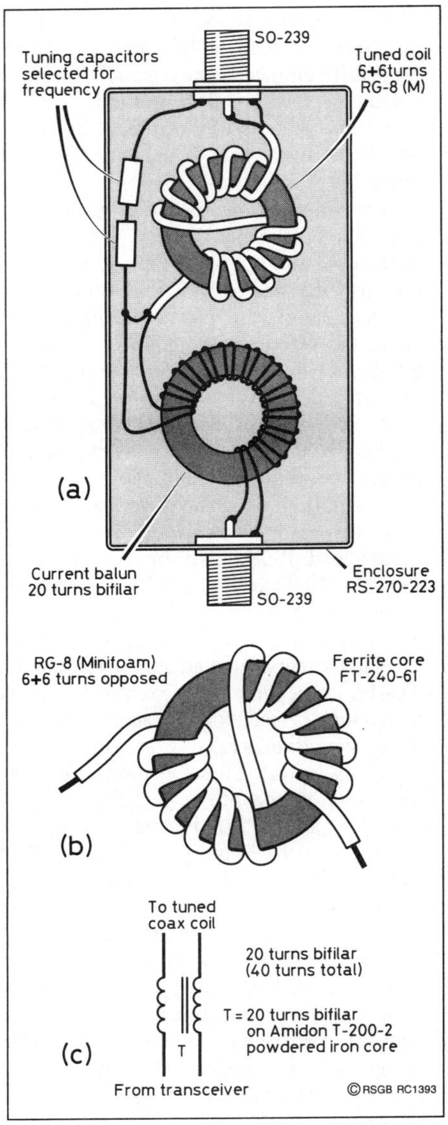

Fig 4: Improved version of the RFD-1 in which the T choke is replaced by two toroidal inductors (antenna coil and current balun. (a) complete unit in plastic enclosure; (b) antenna coil; and (c) current balun.

United States of 130V AC MOVs on 120V mains supplies and claimed that MOV voltage ratings should be at least 1.5 times the nominal mains voltage. For 120V mains supplies this would imply a minimum rating of 180V while for 240V a 380V AC rating. He stressed that MOV devices have a tolerance of up to 20% to which has to be added the variations experience in mains voltages.

I pointed out that the customary rating of MOVs in the UK was 275V and questioned whether any readers had experienced the type of problems outlined by N2MPT when using such MOVs. While 130V provides a tolerance of less than 10% with 120V mains - clearly giving the possibility that the device may turn on and stay on, 275V rating with 240V supplies gives a safety margin of around 15%.

Ian Braithwaite, G4COL, considers that N2MPT has raised some interesting points but, from experience, is convinced that for 240V supplies a 275-type device is satisfactory. He comments:

- Reputable makers do know what they're doing. The Farnell catalogue contains some useful data. Take the V275LA40

MOV, rated for 275V AC service. This needs at least 389V (the peak of a 275V sinusoid) across it before it passes 1mA. This is the lowest figure - most devices won't start to draw current and clip until the voltage is considerably higher, the maker's range being 389 to 473V, ie 430V ±10%.

- Varistors are 'aged' by transients - ie their clipping voltage drops after passing surge currents. Eventually, the clipping voltage falls to below the peak voltage of the mains supply and they undergo thermal runaway and can explode, spewing out their contents all over the equipment and creating a bad smell into the bargain. However, the above devices are actually very robust and will generally give many years service on the UK mains.
- If problems are experienced with transients on the mains caused by lightning or equipment switching, protection can be fitted externally in the mains lead (eg RS Components/Electromail 2390-949, which also incorporates RF filtering). This should follow the usual mains fuse, which for most amateur equipment can be 3A.

G4COL concludes: "In short, I would fit a varistor rated for the mains supply in use. Making a very generous allowance, as advocated by N2MPT, only means that more of the transient voltage appears across the equipment. Incidentally, varistors failing and blowing fuses is not always a bad thing. In one case I personally saw, a failed live-earth varistor led to the identification of a wrongly-wired standby generator. When this operated, the live-neutral voltage was correct, but there was about 200V between neutral and earth (hence 450V live-earth) at quite a sizeable installation".

Mike Cooper, BRS20751, writes: "Our trans-Atlantic cousins do seem to have trouble with their electricity supplies. Recently I wrote to the National and Tandy Users Group about an article in their April Newsletter, reprinted from the Silicon Valley Computer Society. This was on similar lines to the *QST* article quoted in *TT*.

"I have personal experience of using RS275V MOVs effectively to prevent spiky mains (in-house generated) from corrupting digitally generated data. These were in use without any failures for about five years.

"About 18 months ago, following retirement, I replaced an ancient portable typewriter with a PC system. Stirred on by an article by Stephen Waddington in Maplin's *Electronics* (No 67, July 1993), I constructed some suppressed mains socketed extensions (using 275V MOV) for the computer and domestic TV, etc.

"As there is always the possibility of some freak long-duration pulse occurring, I now take the simple precaution of putting a fuse in series with the MOV. Remember these are transient suppressors, and are not intended to cope with voltage surges.

"In a typical commercial mains distribution board with a built-in neon, it is fairly easy to install a 275V MOV, in series with a 20mm fuse and holder, wiring these in series across the L and N lines. If the inbuilt neon plus resistor (typical 100k) is disconnected at the (live) junction, and reconnected to the fuse/MOV junction, it will now serve as a 'mains connected' indicator showing that the MOV is intact and the fuse has not blown. Use a 50mA fuse (the lowest widely available value). The neon plus resistor typically takes about 3mA. If the fuse blows it shows that there have been really bad high-voltage surges and that possibly the MOV should also be replaced because it is not easy to functionally test it."

MULTIBAND OCFD ANTENNA

THE MARCH *TT* item 'Feeding the OCFD' stimulated Peter Pickering, G3ORP, chairman of the Maidstone YMCA Amateur Radio Society, to pass on details of an OCFD antenna which has been used by a number of amateurs in the Kent area for the past 30 years, developed from basic theory in the days before the emergence of NEC computer programs. He writes:

"The antenna (**Fig 5**) was developed as an 'idiot-proof' system to present a 50Ω load to a rig on most bands. It is currently used at the society station GX3TRF for operation from 1.8 to 28MHz (feeder strapped on the 1.8 and 10MHz bands as a Tee-antenna). A good outdoor earth system is used for lightning and EMC protection. We have called it 'a coax-fed Windom'. A smaller version (**Fig 6**) for 7MHz and non-WARC bands was also in use for a few years.

"The antenna could be fed with any length of 75Ω (RG213) or 93Ω (RG62) coax, but the system shown in Fig 50 gives a very low SWR on the none-WARC bands. The 93Ω line (Section E) is for improved matching on the 7MHz band (quarter-wave transformer). The 50Ω line (Section D) acts as a 1:1 transformer on most bands except 3.5MHz where it forms a quarter-wave transformer. Capacitor Cx is adjusted for resonance (lowest SWR) on your most-used frequency in the 3.5MHz band. Select-on-test within the range 200 to 470pF. I used a series/parallel combination of four heavy duty mica capacitors mounted on a glass fibre tag-board; it does not affect resonance on the other bands. On the CW-only end of 3.5MHz the capacitor can be 470pF or even replaced by a link.

"The 4:1 balun transformer T1 comprises 10 turns of 14AWG (16SWG) enamelled copper wire on an Amidon T200-2 core covered with a thin layer of PTFE plumbers' tape. Use an ATU on the WARC bands. On the non-WARC bands, the match can be near perfect with feeder balance and SWR particularly good on the 21 and 28MHz bands."

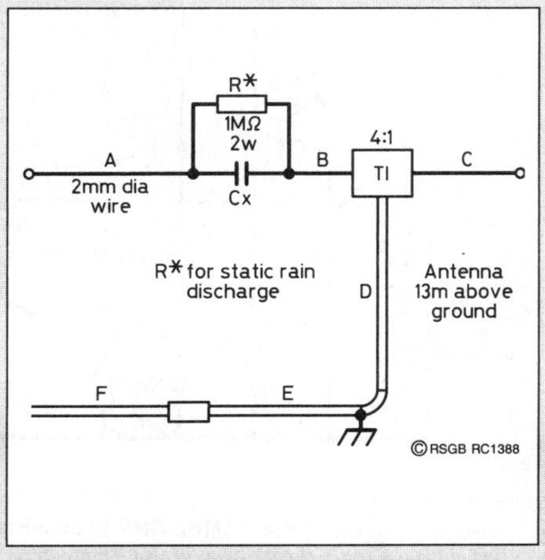

Fig 5: The 'Kent Windom' multiband off-centre-fed-dipole as used at GX3TRF and by amateurs in the Maidstone area of Kent. Dimensions as reported by G3ORP: Antenna formed from 12AWG aluminium wire: Section A = 20.8m; Section B = 3.048m; C = 17.754m. A + B + C = 41.6m. Coaxial cable D = 13.94m of RD213 (50Ω); E = 8.87m of RG62 (95Ω); F any length of 50Ω cable to rig. The antenna can be used as a Tee by strapping the feeder at the junction of coax sections D and E (useful on 1.8 and 10MHz bands).

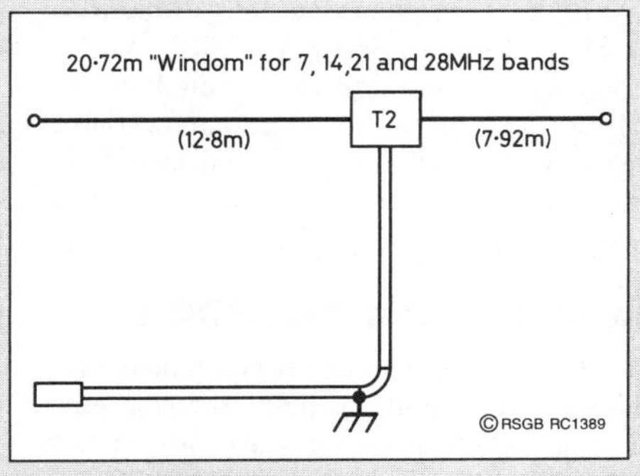

Fig 6: The 'Baby Kent Windom' for 7, 14, 21 and 28MHz bands. Do not run feeder parallel or close to antenna (drop it vertically at right angles to the element. T2 is 6:1 for 50Ω feeder and 4:1 for 75 or 93Ω feeder, wound on T200-2 core). Feeder balance and SWR are likely to be best on the 28MHz band.

THE UPDATED OPTIMIST

IT IS NOW A quarter of a century since Karl Kottenhoff, SM4BSN, first published in the Swedish journal *QTC* (see *TT*, August 1972 or *ART7*, pp213-5) an extremely simple 3.5MHz QRP CW transceiver using just one transistor, a couple of diodes and an IC AF amplifier. He called this ingenious little rig 'The Optimist'. It comprised a crystal-controlled transistor oscillator acting as a transmitter of up to about 1 watt. The oscillator also functioned, at reduced power, as the heterodyne oscillator of a fixed channel direct-conversion receiver, using a couple of germanium diodes as a voltage-doubler product detector. To the best of my knowledge this was the first time such a simple configuration had been shown to provide a complete and practical QRP transceiver - at least for the optimistic enthusiast.

Since then, this approach - often with variations - has become widely used by QRP enthusiasts, although its inception by SM4BSN 25 years ago is often forgotten. Five years ago, Oleg Borodin, RV3GM, published a 'Micro-80 Mini Transceiver' (*SPRAT* No 72, Autumn 1992, p20) which had a two-stage transmitter with the PA functioning also as the product

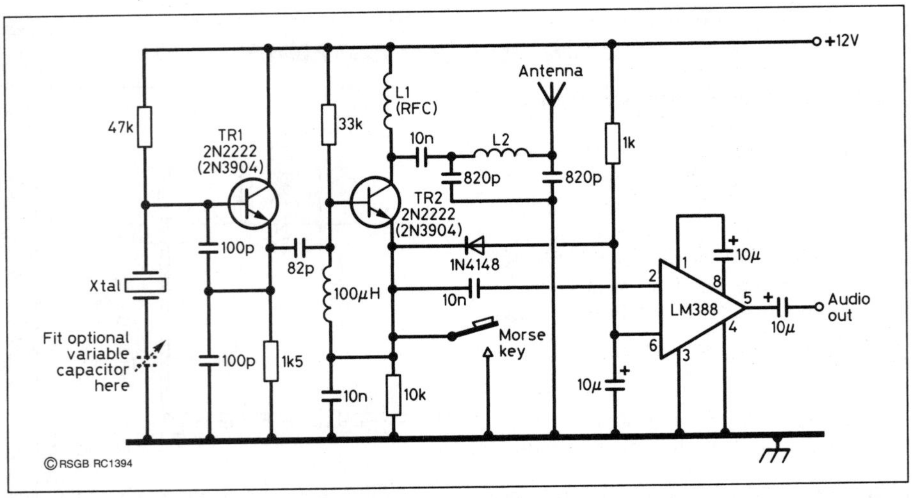

Fig 7: Circuit diagram of the 3.5MHz QRP Pixie-2/Pisce transceiver comprising two-stage transmitter combined with a simple direct-conversion receiver.

detector, and an AF amplifier with two transistors in a Darlington configuration, rather than an IC, but which involved many of the principles outlined by SM4BSB.

More recently, various basically similar designs have appeared under such names as 'Pixie' and 'Pisce' in various journals. **Fig 7** shows the Pisce/Pixie design attributed to David Joseph, WA6BOY, as reproduced recently in *Radio ZS* (October 1996) in ZS6AAW's QRP column. This design uses two transistors, a diode and an LM388 IC AF amplifier. An optional variable capacitor can shift the crystal frequency slightly (much larger shifts could be obtained using a ceramic resonator).

HF ANTENNA HANDBOOK

WITH SO MUCH amateur equipment now entirely factory built it is perhaps not surprising that the technical emphasis in amateur radio publications now centres on antennas, feed-lines and ATUs. One sometimes wonders whether anything new can be said on the basic designs and their variations. These have changed relatively little during the past quarter-century. Much of the current work of professional antenna engineers is directed at satellite and microwave systems and have yet to make a serious impact on amateur radio. For example, one finds papers on: dielectric rod antennas; quasi-planar leaky wave antennas; switchable ferrite micropstrip antennas; radial line planar monopole antennas; aperture coupled microstrip antennas; CPW-fed slot antennas; leaky-wire long-slot waveguide antennas - and all those in just a single issue of *IEEE Trans Ant & Prop* (August 1996)!

One author who manages to come up with an always interesting mixture of the old and the new in practical amateur radio antennas, in easily assimilated form, is William I Orr, known world-wide as Bill Orr, W6SAI.

Bill recently sent me a copy of his impressive new large-format (some 180 double column 8.5in by 11in pages), soft-covered book *The W6SAI HF Antenna Handbook*. It is packed with material derived from his monthly *CQ* magazine articles grouped into 10 chapters: A quick look at coax, ground loss, antenna height, SWR, radiation resistance (and all that jazz) (5pp); Feedlines, antenna hardware, and accessories ("Little things mean a lot") (5pp); Single-wire antennas that work (24pp); Multiband dipole antennas (30pp); Transmitting and receiving loops (15pp); 160-metre antennas - simple 'Top-Band' antennas you can build (22pp); Antenna tuners, baluns, and matching devices (21pp); Antenna analysis programs and how they work (10pp); Inexpensive beam antennas you can build (20pp); and Antenna instrumentation (10pp).

In his preface, Bill recounts his early experiences in ham radio - a lifelong hobby ever since his father gave him a crystal set and pair of headphones for his tenth birthday. He found he could hear several people talking back and forth; his dad told him they were radio hams and his first W2 call followed 1934.

Personally, I have always felt that one of the advantages of becoming intersted in radio while still an impecunious schoolboyis that it taught some of us to 'make do and mend ' searching around for odd parts and things that could be put in ways never intended by their makers.

Bill, however, seems to have had an unusual and possibly unique early experience while putting up his first outdoor antenna. A story he tells in his introduction to the new book: "Putting up a good antenna seemed a fine way to get things started. A nifty place to erect a transmitting wire was from my second-storey bedroom window out to an old tree at the back of the lot... Attaching the antenna to the house took but a moment one warm spring day. I reeled the wire out across the yard and climbed up about 25ft into the tree with a length of clothes-line to make a tie-off point... By chance I happened to glance into the next-door neighbour's yard. There was his daughter (about two years older than me) lying on a blanket taking a sunbathe! Starkers! I must have made a slight noise, as she saw me, grabbed a blanket and ran for her house, yelling she would 'fix me' when she caught me – next day on my way to school she planted a hard fist in my face..."

I must admit that nothing like this ever occurred to me while putting up antennas although I did once fall right through a low roof on to a heap of coke. And another time accidentally dropped a weight on my sister.

The book includes a timely warning: "CAUTION: Working on antennas and towers can be dangerous. All warnings on equipment and all operating and use in structions should be adhered to. Make sure that the antenna is disconnected from the station equipment before you begin to work on it. Make sure that your antenna is not close to power lines, and that it cannot drop onto a power line if wires or supports fail. Do not attempt to climb a tower without a safety belt. It is best to work on your antenna with someone who can assist you with tools and who will be able to help in the event of a problem of emergency."

HERE & THERE

ANDRE JAMET, F9HX, in *VHF Communications* (Spring 1/1997) describes a 10GHz super-regenerative receiver using a dielectric resonant oscillator (DRO) salvaged from a discarded satellite LNB often available for little or no cost. He shows how a dielectric resonator can be shifted in frequency to, typically, 10.368GHz.

He points out that a simple super-regenerative receiver can still provide surprisingly good sensitivity on VHF, UHF and microwave bands. He writes: "Nowadays when you hear people talking about super-regenerative reception, you think they are old-timers regretting the passing their youth! However, super-regeneration is still much used professionally: millions of units a year are produced for such applications as remote control, supervision, alarms, medical apparatus, etc with contributions from new techniques such as surface-wave resonators (SAW) and up-to-date semiconductors. Indeed, very simple and low consumption devices are valuable for those applications. We have shown that super-regeneration could still be useful for amateurs - from 144MHz to at least 10GHz."

Dr Dick Biddulph, G8DPS, comments on W7XC's suggestion (*TT*, January 1997) of making coil dope by dissolving scrap polystyrene in "carbon tetrachloride" which is no longer readily available in the UK because it is so toxic. In years past, we all used to splash it around to clean all sorts of things (including quartz crystals), but the Health & Safety Executive have now virtually banned it. However, you can make a good coil dope by dissolving polystyrene in cellulose thinners which are cheap, relatively non-toxic, readily available but flammable. [*this process should be done in a well ventilated environment with no naked flames - Ed*] Being a wet shaver, I use the packaging of my razor blades as a source of scrap polystyrene. I have tested this dope by measuring the 'Q' of a coil at HF before, during (wet) and after doping, using an old Marconi 'Q' Meter. There was no significant difference in any of the readings apart from an expected slight change in resonant frequency caused by the small increase in self capacitance.

"In regard to the suggestion of using shavings from scraps of discarded coaxial cable off-cuts, the insulation in coax is nearly always polyethylene which is a very good RF insulator but insoluble in any obtainable solvent. The only exception nowadays is the very high grade coax which has PTFE insulation - and that is even less soluble!" ♦

K1POO OCFD ANTENNA

TT (MARCH 1997, pp72 and 77) presented details of two variations on off-centre-fed-dipole (OCFD) antennas which Richard Formato, K1POO, has suggested as improved alternatives to the classical 'one-third' feed point for Windom-type multiband antennas fed by coaxial cable or twin feeder. The item included computed SWR information but emphasised that final adjustments were likely to be necessary to take into account variations imposed by height, earth constants, nearby objects etc.

Several readers, including Dale Gaudier, M0AOP/K4DG, and Don Elkington, G0PAN, have commented on K1POO's suggestions with particular respect to the difference between computer modelling and practical results in typical amateur locations, resulting from the limitations of some NEC-type programs.

Dale Gaudier, M0AOP/K4DG, writes: "The OCFD is regaining favour in amateur circles. With the advent of antenna modelling programs for personal computers, optimising an OCFD for four-band operation (40, 20, 15 and 10 metres) has become possible.

"The preferred design by K1POO, as reported in the March *TT*, comprised an overall length of 21.03m (68ft 5.5in) with the feedpoint located 17.4% (3.65m ie 11ft 10.5in) from one end: see March Fig 9 and Fig 11(c). The design has a nominal feed-point impedance of 200Ω so that it must be driven through a 4:1 balun to give a good match to a standard 50Ω coaxial feed-line.

"First, I modelled his design using the EZNEC Version 1.0 by Roy Lewallen, W7EL. Using K1POO's design criteria the free space SWR curves were a good match to those set forth in *TT* and Fig 11(c).

"I then built an antenna based on the 3.65m feed-point model. The dipole was constructed conventionally of 16SWG (1.628mm diameter) stranded copper wire mounted approximately 10m above ground. A 4:1 balun was inserted at the feed-point with RG-8X 50Ω coax used for the feed-line: **Fig 1**.

"The resonance and SWR of the antenna were then tested using an SWR analyser attached to the feed-line: see **Table 1**.

"Clearly the SWR minimum and the 2:1 SWR range for the 28MHz band were higher than desirable for an antenna whose purpose is to minimise the need for an ATU. However, the real puzzle was the lack of resonance anywhere near the 7MHz band! The SWR minimum around 8.5MHz was neither pronounced nor very deep, and certainly not what was predicted by computer modelling.

"I tried pruning the two legs of the dipole: this only caused the 8.5MHz minimum to shift somewhat. Pruning did not produce an acceptable (<2:1 SWR) minimum anywhere in the 7MHz band; it also caused the SWR to worsen for the 14, 21 and 28MHz bands minima.

Band (MHz)	Frequency of Minimum SWR (Resonance) (MHz)	2:1 SWR Range (MHz)
7	approx 8.500	(SWR >> 3)
14	14.170	13.250-15.500
21	21.580	20.420-22.350
28	29.690	28.550-30.480

Table 1: Resonant frequencies and 2:1 SWR bandwidth of the OCFD

Pat Hawker's

Technical Topics

PAT HAWKER, G3VA
37 Dovercourt Rd
London SE22 8SS

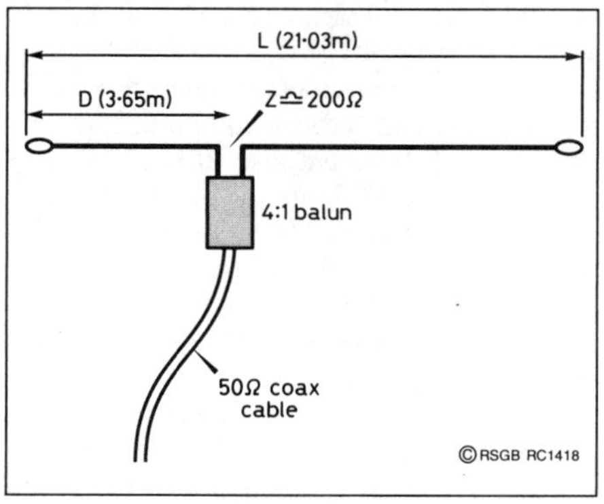

Fig 1: The K1POO-type Off Centre Fed Dipole design criteria for operation on 7, 14, 21 and 28MHz, as computer-simulated for free-space conditions. A practical version some 10m above ground failed to show a 7MHz resonance until an RF choke formed from five turns of the RG-8X coax with a diameter of 24cm was included close to the feed-point. With this addition, SWR ratios of not more than 1.5:1 were measured on all four bands.

"It then occurred to me that the effects I was observing might be due to unequal currents flowing at the feed-point (possibly due to the asymmetrical design of the OCFD). I then made a simple RF choke by coiling several turns of the RG-8X feed-line close to the feed-point and balun. After some initial adjustments of the choke diameter and number of turns, I had my 7MHz band resonance! The coil comprised 5½ turns of RG-8X with a diameter of 24cm (9.5 inches). **Table 2** shows the results with the RF choke fitted.

Band	Minima (MHz)	SWR at Minima	2:1 SWR Range (MHz)
7	7.000	1.1	6.850-7.220
14	14.100	1.2	13.520-14.420
21	21.220	1.1	20.420-22.340
28	28.340	1.4	27.460-29.050

Table 2: Measurements with the RF choke formed from coiling the RG-8X coaxial feed-line close to the feed-point and balun. Note: The 7MHz band minimum and SWR range can be raised by removing part of a turn of the RF choke.

Frequency (MHz)	R (ohms)	X (ohms)
7.05	314	+ 30
14.175	123	-23
21.225	99	-13
28.4	184	- 9

Table 3: Performance of the K1POO's improved OCFD antenna using NEC/NEC2D with true ground parameters and the dimensions given the impedance at the antenna input to the 450Ω twin feeder (21.262m electrical length) at 30ft height.

M0AOP/K4DG concludes: "As apparent from Table 2, the OCFD design of K1POO is surprisingly broadbanded and, with a little adjustment, capable of covering virtually all of the 7, 14, 21 and 28MHz bands at 2:1 SWR or less without involving the use of an ATU. However, it appears that an RF choke - the need for which is not apparent from computer modelling - is necessary for this design to work on its fundamental frequency".

Don Elkington, G0PAN, has a special interest in computer modelling of antennas. He is critical of the findings of Ralph Holland, VK1BRH, in modelling horizontal dipoles above counterpoises (*TT*, January 1997, p61). He comments: "This added nothing to what has been well known for some time, ie how antenna height and ground affect feed and radiation characteristics. Had the modelling looked at the extent of ground screen required to achieve particular results it would perhaps have added something.

"To do this VK1BRH would have needed to use NEC2D, not plain NEC2, in order to model a sufficiently fine ground mesh (Result = bigger the screen, more reflection, lower the feed impedance).

"You referenced some time ago in *TT*, the work of R F Proctor in the Ministry of Supply (Signals Research & Development Establishment) which was VHF related, but which relied upon the classic work of Sommerfield and Renner.

In connection with K1POO's improved OCFD antenna (*TT*, March) using NEC/NEC2D with true ground parameters and the dimensions, **Table 3** shows the given impedance at the antenna input to the 450Ω twin feeder (21.262m electrical length) at 30ft height.

"NEC2/NEC2D can model any nearby conductive structure, as well as feedlines, networks, loads and conductor specifications. The following are some general notes on NEC modelling that need to be taken into account.

"The NEC code employs the Sommerfield and Renner approach to evaluating the effect of near ground on antenna characteristics. For those who want it, their 28-page paper was published in three issues of *Wireless Engineer* (1942) including the detailed mathematics. The issues are available at the British Library.

"The Lawrence Livermore National Laboratories (LLNL) in the USA incorporated their findings into NEC2/3/4 which now also covers modelling of antennas in insulated/buried media rather than plain air. LLNL also did work on ground penetrating imaging radar and the permitivity/permeability of various materials such as buried reinforced concrete. Numerical modelling topics can be found on the Internet at http://www.llnl.gov with other leading edge material such as Planetary Defence.

Reference point	Distance from feed point to cross arm (inches)	Cross-arm length (inches)
A	27.5	30.5
B	50	42.5
C	77	12.0

Table 4: Dimensions of 1.3GHz dual-rhomboid antenna of Fig 3.

"NEC2D code allows larger models to be run than NEC with drawings of a variety of structures entered via digitisers.

"Nevertheless, I would raise a fundamental question about the modelling of amateur antennas at low heights: 'How do you know the nature of the ground below your antenna?' This will be frequency dependant and needs to be measured somehow. DC and 50Hz based techniques are probably wrong and errors will be significant."

UHF DUAL-RHOMBOID ANTENNA

ON MANY OCCASIONS, *TT* has drawn attention to the attractions of high-gain rhombic and long-wire vee antennas when scaled down to the more modest dimensions required on VHF/UHF. On HF, the fixed rhombic antenna long-ago earned the title of 'Queen of Antennas' but its size rules out its use by all but a tiny minority of amateurs. On UHF the dimensions are such that even a rotary rhombic-type beam becomes a practical proposition.

Recently, Terry Brown, G4TZB, was intrigued by a photograph published in the December 1996 *QST* of a dual-rhomboid antenna built by Dayton Johnson, W0OZI, for 1255MHz amateur television with a gain approaching 20dBi. G4IZB was one of a number of amateurs who got in touch with W0OZI to find out more about his antenna which it transpires was based on a 20-year-old article by Bill Parker, W8DMR, 'Dual Rhombic for VHF-UHF' (*73 Magazine*, August 1977).

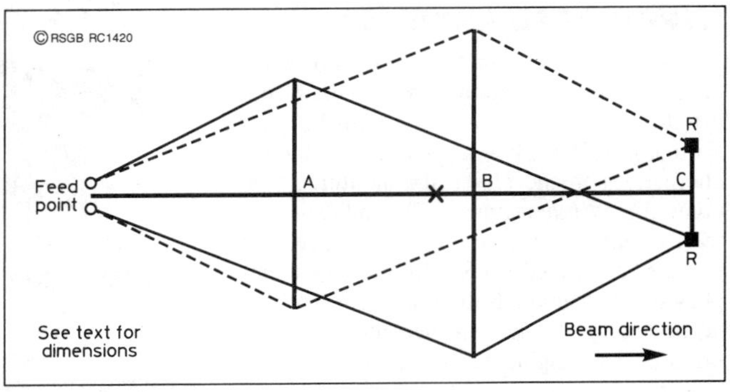

Fig 3: The 1.3GHz dual-rhomboid antenna which provides some 20dB gain with a boom about 8λ long. The longer sides of each rhomboid are 6λ and the shorter sides 3.5-λ. R marks the position of the 600Ω non-inductive termination resistors. See text for the non-critical dimensions of this broadband antenna.

In view of the considerable interest, W0OZI provided detailed information on the antenna not only to G4IZB but also to Emil Pocock, W3EP, for his *QST* (March 1997) column 'The world above 50MHz'. The following brief notes are based on both sources.

W3EP writes: "The rhomboid (a parallelogram with no right angles and adjacent sides of unequal length) is not new to antenna design. It is a variation of the rhombic, which has been used for many decades to achieve high-gain, primarily on the HF bands. The dual-rhomboid achieves even higher gain than a traditional rhombic by combining two antennas in such a way that useful sidelobes are reinforced, while those in unwanted directions are cancelled... The configuration of wires and supports that constitutes the dual-rhomboid looks confusing at first glance, but a close inspection shows that the antenna is simply two identical wire rhomboids in the same plane, with one flipped over. The rhomboids are fed in parallel and terminated at their opposite ends with 600 to 800Ω non-inductive resistors."

W3EP gives the essential dimensions for the 1296MHz dual-rhomboid shown in **Fig 3**. He states that this provides about 20dBd gain with a total boom length of just under 6ft 6in and other dimensions as listed in **Table 4**. The longer sides of each rhomboid are six wavelengths; the shorter sides 3.5 wavelengths. A major bonus for this type of antenna is its broadband characteristics making dimensions relatively non-critical and the possibility of it being used on more than one band.

For a 432MHz version all dimensions should be multiplied by three, requiring a boom length of just under 20ft, making it easier on this band to implement the antenna as a fixed-beam rather than a rotary system. This would certainly be the case on 144MHz. For 1.3GHz about 30ft of wire is required, preferably not thinner than No 12 AWG. The boom and cross-pieces must be non-metallic with W0OZI recommending clear fir, protected by paint or other proactive coating. Wires should be insulated from each other where they cross over. W0OZI used five plexiglass insulators with 6-32 bolt nut terminators at either end to interface wires to terminating loads and his input balun.

The two 600Ω non-inductive terminating resistors (R) should be capable of dissipating at least one-quarter of the power input to the antenna. For example, five 2W carbon-composition resistors in parallel would be sufficient for 40W continuous power supplied to the antenna, or some 80W SSB or CW. If the wires are left open, without terminating resistors, the antenna becomes bi-directional which may be preferable for fixed VHF/UHF antennas. The element can be mounted in either a horizontal or vertical plane depending on the required polarisation. The centre of gravity is between points A and B (closer to B) with the exact point governed by the constructional techniques but easily found by balancing the completed element.

Since the input impedance to each rhomboid is about 600Ω, with the two in parallel, the antenna can be fed directly from 300Ω balanced twin ladder-line with a suitable ATU or 4:1 balun into 50Ω coaxial line for connection to the transmitter. Alternatively, a common half-wave balun made from 75Ω coaxial line at the antenna feed-point can transform the 300Ω balanced load of the antenna to 75Ω unbalanced line

SOLID-STATE SCREEN REGULATOR

THE MARCH *TT* included an item 'Screen regulator for high-power tetrodes' providing details of a regulator suitable for meeting the requirement for good regulation of the screen voltage applied to modern tetrode valves such as the 4X and 4CX series. The item stemmed from Chris Tran, GM3WOJ, who had successfully used a circuit devised by Robin Kerr, GM4FDT.

Niels Rudberg, OZ8NJ, writes: "I found the circuit designed by GM4WOJ interesting. Clearly, this regulator configuration employing both a current source and a shunt-type voltage regulator is superior to any of the other screen-regulator circuits that I have previously seen or used.

"I tested the circuit, using two EL84 valves. As expected the voltage regulator part worked fine; however the current source, due to the low transconductance of the EL84, proved far from acceptable. A transistor in the grid circuit of V1 could improve the function of the current source but there would remain a major drawback. The circuit needs at least 200V for correct operation of the current source: in other words, the unregulated input voltage should be more than 200V above the required screen-grid voltage.

"A fairly simple circuit employing MOS power transistors rather than valves gives a significantly better performance in every respect: **Fig 2**. I used the Siemens type BUZ50A (1000V/2.5A). The 'drop-out' is then less than 30V.

"The shunt regulator loop gain is marginal; if, however, R2 were replaced by a 51kΩ resistor in series with a string of three small neon signal lamps, paralleled by an 0.1µF capacitor, regulation would improve.

"In use, the current source should be adjusted for a current only slightly above the maximum screen grid current for the high power tetrode used."

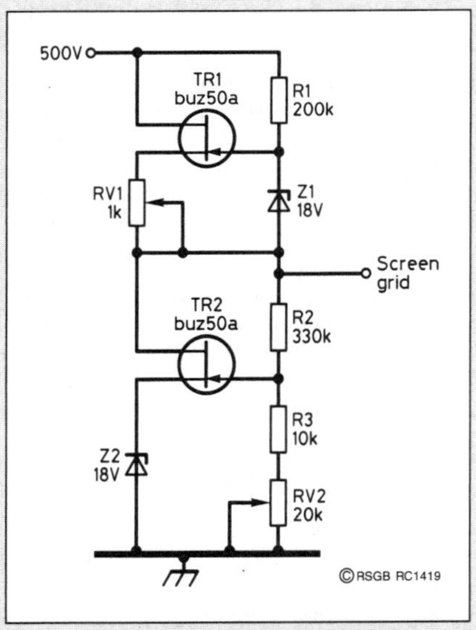

Fig 2: OZ8NJ's solid-state version of the GM4FDT screen-regulator for modern high-power tetrode valves.

to permit 75Ω coaxial feed-line to be used. W00ZI used an effective quarter-wave sleeve balun using 3/8th inch diameter brass tubing.

HANDSET ANTENNAS & BIOLOGICAL TISSUE

A RECENT IEE Savoy Place colloquium on 'Design of mobile handset antennas for optimal performance in the presence of biological tissue' included descriptions of some interesting antennas for use at 900 and 1800MHz designed to restrict radiation into the user. With the very short antennas at these frequencies held close to the user's ear 40 to 70% of the total radiated output can be absorbed into the head, hand and shoulder.

This has led to fears that the UHF radiation of these ubiquitous handsets could represent a potential health hazard. A number of current research programmes in British and other Universities etc, are researching possible hazards and developing antennas that direct the bulk of the limited (less than 1 watt) transmitter output away from the user.

Most speakers seemed slightly out-of-kilter with the views expressed by Daren Wood (BT Laboratories, Martlesham Heath) who appears more concerned with the very wide performance variations found in current handsets in which transmitter ERP can be from 10 or even 17dB down on those with the more efficient antennas. Although transmitter output is officially specified, antenna efficiency is not. He preferred radiation to be as omni-directional as possible.

Among the speakers was Prof F M Landstorfer (Institute fuer Hochfrequenztechnik, University of Stuttgart) whose work on HF antennas has been mentioned several times in *TT* over the past 30 years. He presented 'New conformal antenna concepts for handheld telephones with significantly reduced interaction with the user'.

His introduction stated: "The increasing use of mobile telephones, in particular of handheld sets, which have their transceiving antennas in the vicinity of biological tissue (head and hand) has caused public concern about possible radiation hazards. The majority of commercial mobile telephones use monopoles with omnidirectional radiation patterns. As a result of this unfavourable antenna concept, more than 40% of the radiated power is absorbed by the user when the antenna as customary is hold close to the head. Therefore, other concepts like the *planar inverted F antenna* (PIFA) or the *radiation-coupled dial-L antenna* (RCDLA) have been investigated and offer visible improvement of efficiency by mounting the antenna on that side of the transceiver which points away from the user's head. The disadvantage of these side- and back-mounted antennas consists in the increased influence of the user's hand on input impedance and radiation pattern. To overcome these drawbacks, we propose a new conformal antenna concept with significantly reduced interaction with both the human head and hand." However, it seems unlikely that the new antenna design could be readily adopted for amateur use on 2m or 70cm.

Other antenna designs for low absorption were presented by speakers from Aalborg University, Denmark and a screened printed monopole antenna developed at Sheffield University. A null-steering antenna for hand-

'NANO-L' INDUCTANCE BRIDGE

DREW DIAMOND, VK3XU, in *Amateur Radio* (May 1997, pp6-8), describes a useful piece of test equipment that overcomes the problem that many experimenters experience in measuring the small values of inductance needed for VHF equipment. He points out that this applies even when using a suitable dipper and the 'resonance dip' method, while many laboratory bridges will not go much below 5mH. In his case, he needed a unit that could measure inductances of less than about 500nH (0.5μH). As a result, he put together a handy little unit with a measuring range from less than 50nH up to about 500nH - which he feels should cover most tank and filter coil requirements encountered by those building VHF equipment.

Fig 4 shows the circuit diagram of the VK3XU Nano-L inductance bridge. A standard, reasonably priced, 50MHz computer crystal oscillator clock module generates the test signal, and a 50MHz resonant tank circuit removes most of the harmonic energy from the TTL output in order to provide a near sinusoidal signal to the 'unknown' terminal LX. When the coil under test is resonated by the 210pF variable capacitor (any swing of between about 200 to 400pF should prove suitable), the voltage fed to the detector diodes will be sufficient to turn on the 2N3904 transistor, causing the LED to glow. About 25mA is drawn from the 6V battery supply.

The capacitor dial can be directly calibrated, provided a set of test and calibration coils are constructed. VK3XU provides details (**Table 5**) of 50, 100, 200, 300, 400 and 500nH test and calibration coils made from No 18 B&S (1mm) tinned or enamelled copper wire, wound on the shank of a 6.5mm (0.25in) drill, spaced about one wire diameter between turns (or preferably, wound on an ordinary 5/16in Wh 18tpi bolt). Wind the turns on, then unscrew the bolt from the coil. The 400 and 500nH coils were wound on the shank of a 13mm (0.5in) drill, one wire diameter between turns. To avoid shorted turns, use enamelled rather than tinned wire.

To calibrate the unit, VK3XU recommends: "Start by fixing the 300nH coil in the LX terminals. Move the dial around the half capacitance point (ie roughly 100pF) so that the LED glows. Peak the 22pF trimmer across L1 for maximum glow. Now mark the dial for the remaining calibration points using the appropriate coil, peaking the LED glow for each one. The 50MHz clock also makes a handy signal source for receiver work."

Inductance	Turns	Winding details
50nH	2	No 18 B&S (1mm) tinned or enamel copper wire spaced 1mm between turns, 6.5mm inner dia.
100nH	4	As for 50nH
200nH	7	As for 50nH
300nH	10	As for 50nH
400nH	6	No 18 B&S (1mm) enam cop per wire, spaced 1mm between turns, 13mm inner diameter
500nH	7	As for 400nH

Table 5: Test and calibration coils for the Nano-L unit.

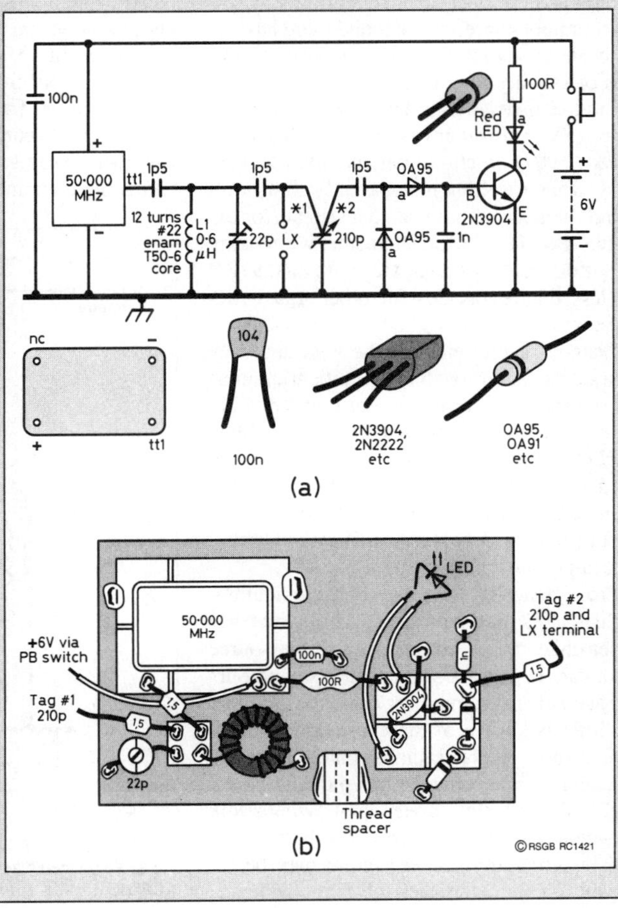

Fig 4: (a) Circuit diagram of VK3XU's 'Nano-L' inductance meter for small (VHF) coils within the range of about 50 to 500nH. (b) Component layout on the circuit board.

sets came from City University, London.

Papers representing work on biological tissue concern included one by Camelia Gabriel (Microwave Consultants Ltd) on the design of phantom models of the human head and hand for use in investigations into radiation absorption into human tissue. These models are being used at Bradford University by Dr Peter Excell in work on the 'Modelling of handsets, antennas, heads and hands in the FDTD method'.

While the lower frequencies of amateur-radio handsets involve rather less screening (and absorption) by the user, RF powers can be significantly higher than the 1W of the 900/1800MHz commercial units. Thus a paper 'Epidemiological studies of exposure to RF with particular note of mobile phones' by Professor R Cartwright (Leukaemia Research Fund, Leeds University) was of particular interest.

He surveyed briefly a number of recent studies including that of Rotham *et al* (1996) in which a cohort of some 25,000 phone users was used to contrast the causes of death etc between those using portable and long-established mobile phones. During 1994 there were 604 deaths but no statistical difference was found between the two groups in overall deaths. The study also investigated the billing records to find the correlation between the use of phones by individuals and charged time. Also, whether users always held the handset on the same side of the head, irrespective of whether they were right- or left-handed.

However, he emphasised that all studies so far published present problems: none directly address use of cellphones; they have produced inconsistent results; the measures of exposure to UHF are rudimentary or incorrect; the studies provide only a low statistical power that is unreliable in respect of rare conditions such as CNS tumours or leukaemia; and there have so far been no studies directed at acoustic neuromas or salivary tumours.

Studies investigating leukaemia near high-power TV transmitter towers (Hocking, 1996, Dock , 1997) on childhood and adult leukaemia), while resulting in apparently significant correlations, have not been consistent (for example, 1.83RR of adult leukaemia within 2km of Sutton Coldfield compares with an 0.97RR (ie less than expected) for other transmitter sites.

Lists were presented of the RRs and 95% confidence factors of some 25 epidemiological studies (based on profession or trade) of CNS cancer and leukaemia between 1978 and 1996 (including the Milham studies on the causes of death of American radio amateurs which have been referred to a number of times in *TT*). Results vary widely with 'military personnel' seemingly most at risk.

Prof Cartwright summarised the epidemiological bases for the public concern which has arisen on seemingly almost any artificial source of radiation. It does seem that CNS tumours and acoustic neuromas may be on the increase; but little is known of the causes of CNS tumours, salivary gland tumours and leukaemia in adults. While some of the epidemiological studies might be interpreted as showing a link between cancers and low levels of non-ionising irradiation other factors may be involved. Despite the lack of firm evidence, there remains considerable public disquiet about ELF from domestic sources and concern generally about living near masts and towers.

These concerns based on epidemiological (statistical) studies point to the continued lack of knowledge concerning: the biological bases for concern; lack of consistency in many areas or research; and lack of any robust programme of research which has led to numerous recommendations for further areas of research. It is possible (but not proven) that modulation of certain CNS enzyme systems may occur as a result of RF exposure; membrane conductance could be affected; and interactions with free radicals produced by biochemical reactions might occur.

Further studies of possible health effects related to the use of radiotelephones have been recommended (A F McKinlay, European Commission Expert Group, 1996). These would involve several studies of brain cancer in different countries; at least one study on risk for acoustic neuroma, salivary gland tumours and leukaemia in adults; personal data on exposure not based on billing; at least one cohort of sufficient size to examine range of deaths and cancers.

It seems clear that the queries hanging over possible, if only marginal, health hazards arising from exposure to low levels of non-ionising radiation are unlikely to be answered for a long time to come. There was one point which Prof Cartwright mentioned that struck a chord. He suggested that a *very small minority* of people did appear sometimes to suffer from bad headaches after use of 900 / 1800MHz handheld phones. This reminded me that some years ago, a reader reported that after standing beside an amateur VHF 144MHz mobile antenna while it was being used, he later experienced an unusual and very severe headache. I mentioned this to Prof Cartwright and he agreed that it is likely that such effects may extend downwards in frequency and include VHF. But there is the curious fact that this seems only to affect a very small number of individuals, and is not a common phenomenon.

MILITARY SURPLUS OPPORTUNITIES

THE NOTES FROM Michael O'Beirne, G8MOB, which accompanied the updated Sherwood Engineering receiver measurements (*TT*, May 1997) attracted attention, but note that for the unmodified (stock) Drake R4-C some columns slipped. The 100kHz blocking figure should be 133dB, sensitivity 0.15μV and LO noise 130dBc for 10kHz spacing.

Former professional radio-telegraphist Paul Cope, G4BNU, was stirred to add some comments on HF receiver 'operability' and also to deliver a broadside against the increasing use of 'off-the-shelf' data systems in the amateur bands. He writes:

"I agree with G8MOB's comments about the miniature keys on the MEL/Philips PRC2000 military-type HF transceiver being useless for search tuning despite, as a receiver using Weaver Third Method demodulation, it is hard to beat for SSB. However, I disagree with G8MOB that this rig is little known to amateurs and unlikely to be on their shopping lists.

"Quite a lot of amateurs have heard of it, have had it on their shopping lists or are now using one. Equipment of this class can still be found on what remains of the surplus market - at a price. You won't pick one up for £2.50 as you could a 19-Set some 30 years ago!

"The tuning problem with no rotary search tuning control, unfortunately applies to all military manpack/vehicle transceivers that I have come across.

"The PRC2000 will probably be considered old hat or semi-obsolete by now, at least by the 'computer-operated QSO and make-the-tea brigade'; it was first delivered for military service at the end of 1981. I recently obtained, as Gulf-war surplus, a Racal manpack transceiver of the same vintage with roughly the same specification as the PRC2000 but without the speech processing.

"But there is a further serious problem with such military-type gear - at least for those of us who are 100% CW operators - the lack of IF

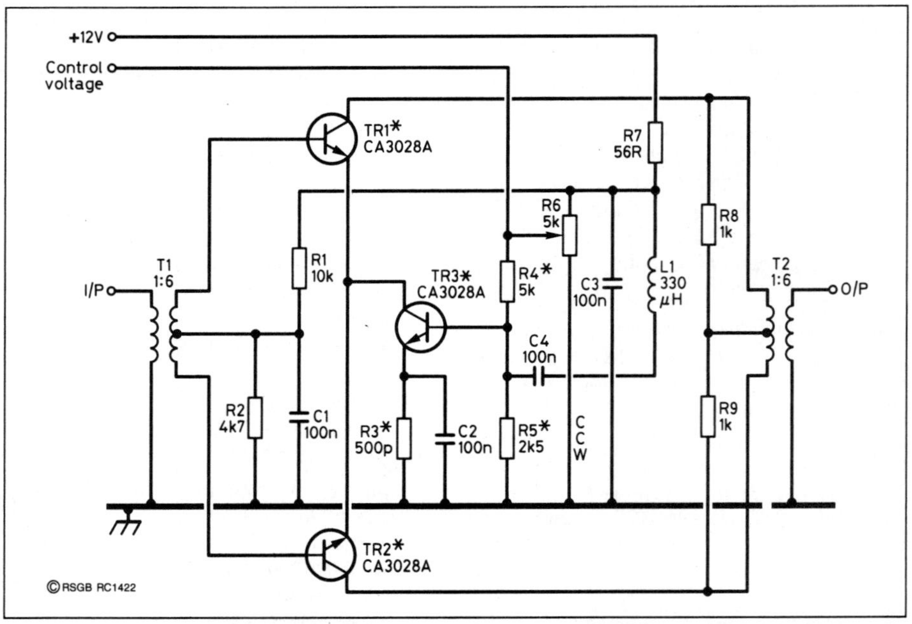

Fig 5: Test circuit using CA3028A chip showing use of feedback to improve harmonic performance of an AGC-controlled amplifier. Resistors marked with an asterisk (*) are integrated into the 3028A.

selectivity (no narrowband CW filter) and, possibly even worse, no facility for switching out the AGC and the absence of an RF gain control.

"The only manpack I have come across with any form of CW filtering is the Racal PRM4031 set (10W output), claimed to be the smallest in the world at the time of its introduction in the late 1970s. This covered 1.6 to 30MHz in 100Hz intervals.

"Also well worth looking out for is the Racal PRM4021 (2 - 16MHz) with a CW receive filter. For anyone interested in military communications equipment, it is well worth looking through the hefty *Jane's Military Communications* volumes. [Copies can be found in some good technical libraries - *G3VA*]. You may get a few surprises - even shocks.

"Even at the risk of appearing an 'Amateur Radio Luddite', I would like to ask what is the point of using off-the-shelf complex data systems including Clover and PacTOR, or even AmTOR, for amateur communications in the HF bands. Modern data modes are suitable for high-density commercial-military use but surely have no place on the HF bands for low-volume amateur-type traffic.

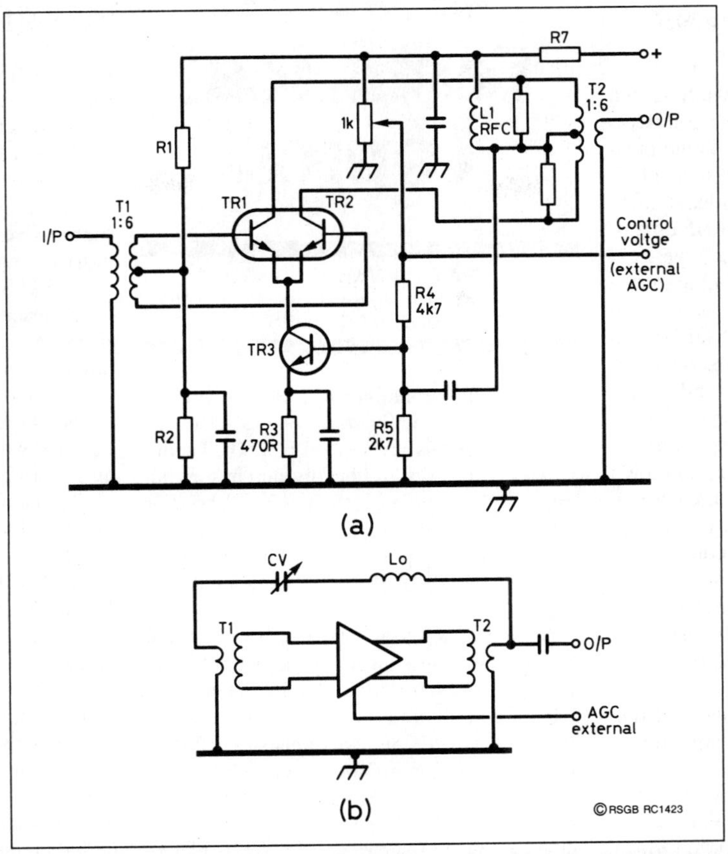

Fig 6: (a) I7SWX's suggestion for a self-balancing push-pull AGC controlled amplifier using discrete devices. (b) Possible arrangement of an oscillator with low harmonic output.

"I do, of course, fully understand the technical fascination of these modes for those who have worked hard to develop and implement them. But this surely does not apply to those who merely buy a modem/TNC across the counter and join it to a black box? Rather, it leads only to unwanted QRM and abuse - two things we can do without on the HF bands.

"Furthermore, none of these systems has the resilience or effectiveness over difficult paths of the Piccolo multitone frequency shift keying (MFSK) system first described in 1963 by Harold Robin, Don Bayley, T L Murray and J D Ralphs of the Engineering Department of the Foreign & Commonwealth Office (former Diplomatic Wireless Service). [Piccolo has been noted as a highly effective low-data rate system a number of times in *TT - G3VA*]. But nobody would want to see even this system with its near SSB bandwidth interference in the amateur CW sub-bands!"

G8MOB, on the other hand, has pointed out that the inevitable trend in military communications is towards flexible digital systems including bit-rate reduction for voice encryption and such techniques as automatic link establishment: "linear predictive coding"; sophisticated error detection systems including Golay 2; BCH and Reed-Solomon with in-band diversity using multiple tones. He gives, as an example of modern technology the flexible AN/PRC-138, an HF/VHF manpack transceiver with six data rates between 75 and 2400 bits/sec, made in the USA by the Harris Corporation. He notes that by the time 'Bowman' becomes fully operational in 2001 for the British Army, all of the military VHF nets will be data, either pure data or encrypted voice! He believes that such techniques will inevitably spread into amateur operation.

FEEDBACK IMPROVES AGC AMPLIFIER

GIAN MODA, I7SWX, draws attention to an article 'Feedback improves AGC amplifier harmonic performance' by Christopher Trask (*RF Design*, August 1996). This underlines that among the building blocks that system designers encounter, the AGC amplifier is right alongside mixers when it comes to intermodulation and harmonic generation. Both these blocks are essential to the design of receivers or transmitters.

A common form of gain controlled amplifier takes the form of a variable transconductance amplifier; in the valve era this often focused on vari-mu pentodes such as the 6AU6 and 6BA6. With such remote cut-off valves, any distortion was predominantly caused by non-linearity of the transconductance curve along the load line.

Solid-state technology introduced additional sources of distortion, far greater than experienced with vari-mu valves. This now included the exponential transfer characteristics of the diode (in the case of diode mixers) and of the base-emitter junction of transistors used either as amplifiers or mixer. When used as variable transconductance amplifiers, this non-linearity is additional to the non-linear transfer characteristics of the device.

The use of a matched pair of transistors in a push-pull (or differential) configuration offers certain advantages: (1) Even harmonics and even-order combination frequencies cancel at the output; (2) no fundamental frequency energy appears at the power source, thus minimising the need for bypassing; and (3) common-mode signals, such as power supply noise, are balanced out by the output transformers.

Christopher Trask stresses that all these advantages depend on the assumption that both transistors are perfectly matched, and that input and output transformers are perfectly constructed - rarely the case in the real-world.

He believes that the answer can be found in material published in the UK on self-balancing push-pull circuits during the valve era (EMI Laboratories, *Electronic Engineering*, June 1946, p189 and D R Birt, *Wireless World*, May and June, 1960). He writes:

"The explanation is quite simple. Regardless of its construction, the centre point of the output transformer is a summation node that has voltages relating to the imbalance of the two sides of the amplifier (in terms of odd-order content) as well as the sum of even-order signals (common-mode noise and even-order harmonics). By applying these voltages in a feedback path as an error signal, the imbalance of the amplifier can be corrected, thus enhancing the overall performance. This serves to reduce greatly even-ordered non-linearity products, such as intermodulation distortion."

Trask presents a test circuit based on the venerable RCA CA3028A as he has been familiar with this device for more than 25 years: **Fig 5**. His test data is based on an input frequency of 8MHz at an input power of -18dBm with control voltage varied between about 2-12V (measured at the wiper of potentiometer R6). It shows that the outputs of the second and fourth harmonics are significantly reduced, while the signal output power and the third and fifth harmonic are relatively unaffected.

He concludes: "The performance of a variable-transconductance AGC amplifier can be greatly improved with little increase in circuit complexity or cost. By reducing the generation of second harmonic energy in transmitter and signal source systems, the difficulty and complexity of removing these unwanted signals with subsequent filtering is avoided. We also reduce costs and improve system performance."

I7SWX comments: "I am sure that no problem would arise if the CA3028 is replaced by one double transistor (for TR1, TR2) and by a single one for TR3 (**Fig 6 (a)**) provided that replacement resistors are added to compensate for those integrated into the CA3028 (R4 4k7, R3 470, R5 2k7). It would be interesting to see if this circuit could be used as an oscillator by applying a feedback resonant circuit between input and output: Fig 6 (b)." ♦

Pat Hawker's
Technical Topics

PAT HAWKER, G3VA
37 Dovercourt Road
London SE22 8SS

G3HEJ'S SINGLE-CHANNEL CONTEST RIG

RECENT *TT* ITEMS on QRP, wartime and modern regenerative receivers have reflected a keen, if minority, interest in simple home-brew equipment among those who feel that 'black boxes' (no matter how effective) leave them feeling that there must be more to HF operating. The May 1997 *TT* recalled the 25 year old Swedish 'Optimist' and the subsequent Pixie and Pisce designs - all of which utilise a single crystal-controlled oscillator (or VXO) for both transmission and for a simple direct-conversion receiver. This is an ingenious and a 'fun' single-channel rig, but is neither intended nor suitable for serious HF contest operation.

Derek Stanners, G3HEJ, however, has shown that single-frequency operation, combined with relatively high-power, can still provide a high score in the 1997 RSGB 7MHz contest. His development of an unorthodox rig based on two (now obsolete) low-cost Howes kits came about as the result of an attempt to 'energise' the creative efforts of his local club. But let G3HEJ tell this part of the story:

"The equipment used in this year's 7MHz contest evolved from a frustrated Farnborough Radio Club home-construction project. Late in 1995, as HF representative on the club committee, I attempted to encourage members to build a valve (or transistor) crystal-controlled power oscillator similar to those put to good use in wartime and for some years afterwards, using such valves as a 6V6, 6L6 or 807. I could provide a quantity of cheaply purchased, but still excellent, ex-WD (1954) 10XAJ crystals capable of coping with the crystal currents involved in valve power oscillators, all closely matched on 7019kHz (series resonant frequency).

"Unfortunately, interest then was minimal (though there are signs of it picking up following my contest results). So, over the 1995 Christmas holidays, I modified an already-built Howes CTX.80 four-transistor 3.5MHz transmitter, following their specifications for 7MHz. I purchased a Howes direct-conversion RX-40 kit and assembled the pair with an antenna change-over relay in a small plastic chocolate box. No connecting mother boards were used, just bits of coax cable and 'back-to-back' sticky pads were used in the construction. The transmitter seemed fine with the 7019kHz crystal, giving about 3.5 watts RF output but keyed with a large (about -10dB) spacer wave! The DC receiver was very poor, due to the open (non-shielded) construction; the massive 40m (and 49m) signals from broadcast transmitters; hand-capacitance effects; and 50Hz FM hum on the local oscillator.

"A rescue plan emerged from a study of RSGB and ARRL publications; I also recalled having read that some Foreign Office embassy radio stations had utilised in-line signal-frequency crystal filters in the antenna inputs to their receivers to overcome cross-modulation and overload problems."

[The use of front-end signal-frequency crystal filters has been described a number of times in *TT*, including designs using 3.5MHz crystals by P W Haylett, G3IPV (*TT*, July 1985 and February 1986 - both items appear in *Technical Topics Scrapbook 1985-89*) and was advocated over 30 years ago by Stuart Meyer, W2GHK, following the trial of a series of Hammerlund bandpass filters at 4U1ITU, Geneva, covering small segments of the 14, 7 and 3.5MHz amateur bands - *G3VA*.]

G3HEJ continues: "Some lash-up filters, with from one to four 7019kHz crystals in series connected with crocodile clips, were

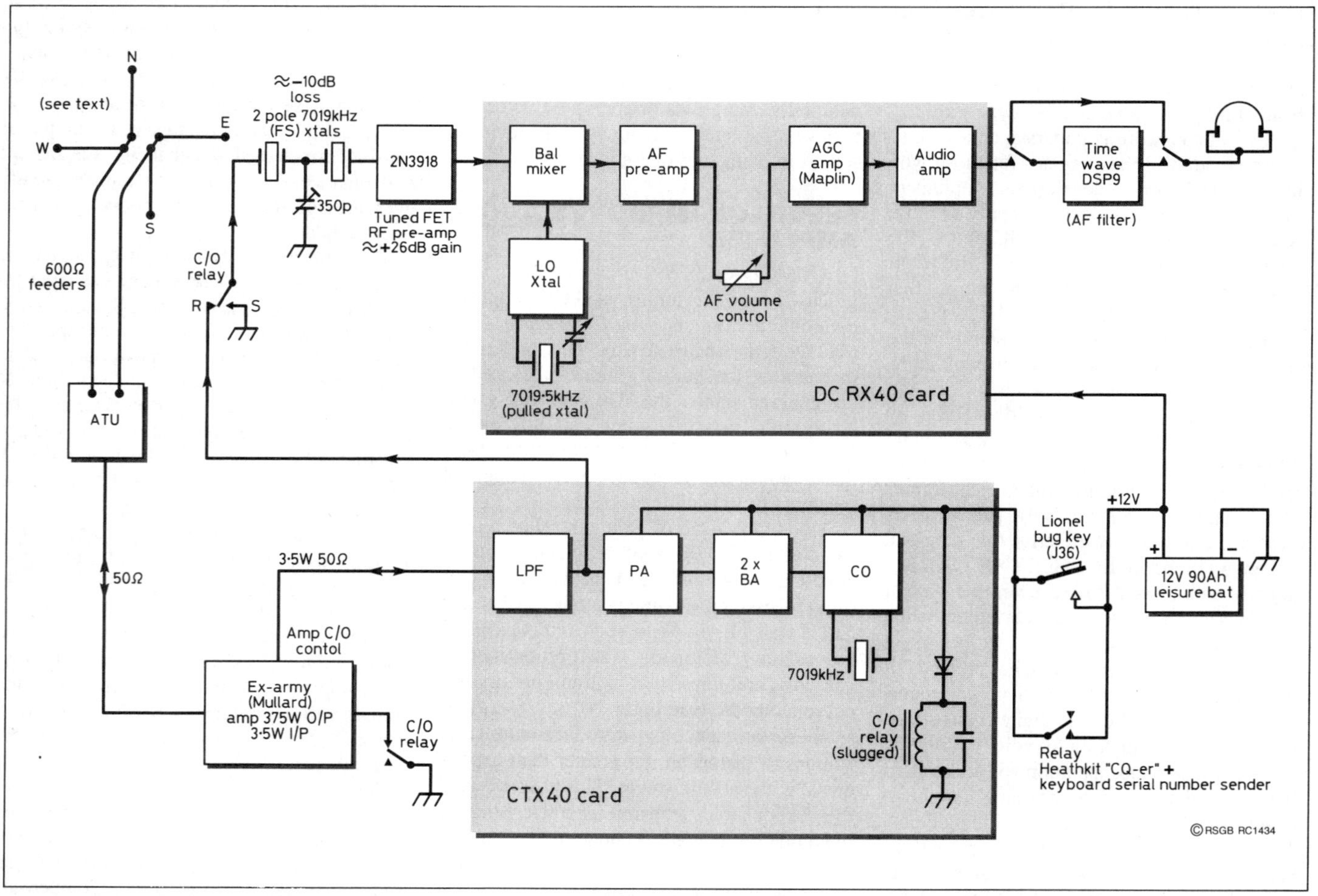

Fig 1: G3HEJ's 'fixed frequency' (7019kHz) station used during the RSGB 1997 7MHz DX Contest based around two low-cost Howes kits built into a small Ferrero-Rocher box. The simple direct-conversion receiver modified by having a crystal-controlled oscillator, tuned FET pre-amplifier to compensate for insertion loss of a two-crystal signal-frequency input filter, and added Maplin VOGAD amplifier. For the contest the 3.5 watt transmitter was used to drive an ex-Army (Mullard) 4-400 amplifier to provide 375 watts RF output. The only operator control was an AF volume control. The Timewave DSP9 filter was usually set to 500Hz bandwidth centre on the 500Hz beat note. All four 7019kHz crystals were circa 1954 10XAJ ex-WD crystals purchased from J Birkett of Lincoln at £1 each. Keyboard sender used while writing paper log and 'dupe sheet'.

tried on my Yaesu FT-One; while spurious capacitances and direct pick-up inhibited any deep notches, the idea seemed promising. I made up a noise-generator using a 9V zener diode and a PP3 battery and was able to measure roughly 60dB dynamic range over some 10 - 15kHz around 7020kHz. With just two crystals in series and no other components ahead of the DC RX40, the filter eliminated all blocking from the strong broadcast signals etc while providing strong amateur CW signals around 7020kHz. The downside was that tuning became even more delicate.

"The final plan became clear: to build a fixed single-channel station for 7019kHz capable of receiving CW stations over a range of about ±1kHz, either themselves calling CQ or replying to my CQ calls. It would be possible to contemplate adding more RF and AF sensitivity to the basic RX-40.

"A miniature FET RF tuned pre-amplifier was added between the antenna crystal filter and the RX-40 providing some 26dB gain, more than enough to overcome the crystal filter insertion loss of some 10dB; additionally, a Maplin microphone/VOGAD low noise AF amplifier (1 inch square PCB) was added to the middle of the existing AF chain, allowing the possibility of real, audio-derived AGC with loudspeaker level background noise at maximum volume level. The *only* control fitted is an AF volume control! The modified receiver now uses five transistors and three chips and runs from the same 12V battery supply as the four-transistor transmitter.

"The transmitter's 'spacer wave' problem was resolved by keying the complete transmitter board 12V battery-line without introducing any chirp or too sharp on/off edges at up to 35WPM (mechanical bug-key or external keyer relay). A second change-over relay was added in parallel to the existing antenna c/o inside the plastic chocolate box in order to provide for external c/o control of an outboard higher-power amplifier.

"Receiver measurements using a Marconi TF144G signal generator, digital frequency meter and additional RF stepped attenuation gave the following measurements (repeated several times): *Bandwidth* ±6dB 282Hz;± 60dB 4800Hz; ± >90dB 7500Hz. *Minimum detectable signal* -133dBm (0.033µV pd/50Ω), 10dB SNR 0.1µV pd/50Ω; *dynamic range* approximately 110dB.

"My FT-One with all filters in use (including audio narrow band filter) has a minimum detectable signal only 3dB better (ie -136dBm). The audio image beatnote normally present in any direct-conversion receiver (ie 500Hz LSB wanted, 500Hz USB unwanted) is suppressed by the antenna filter to -45dBc, making it a virtually single-signal receiver with hardly any audio-image signal. Adding two crystals between the FET amplifier and the Howes PCB mixer gave a ±6dB bandwidth of only 10Hz and a 500Hz beat-note USB/LSB ratio of -75dBc with the MDS improved by 6 to 7dB to -140dBm pd/50Ω but would not be practicable for amateur operation where experience shows that ±250Hz is about optimum."

G3HEJ was satisfied that he now had virtually a competition-grade receiver for a narrow frequency range of 7019kHz plus and minus a few hundred Hertz. The 3.5W transmitter could drive his ex-Army 4-400 valve amplifier to an RF output of 357 watts, as measured on a Bird wattmeter. His current 7MHz wire antenna comprises four 66ft wires fed from 600Ω feeder at the apex (60ft high), the wires drooping to 30ft and strapped N+W/E+S and virtually omnidirectional on 7MHz. His complete contest set-up is shown in **Fig 1**.

He used this single-channel rig over the 18 hours of the 1997 RSGB 7MHz contest, chalking up 546 valid contacts with 82 multipliers. This compares well with his 1995 results of 77 multipliers and 568 valid contacts using his FT-One, tuning the CW band. Many of his better DX contacts - KL7, XE, BV, HL, 9M2, JA, VK, ZL, all USA call areas etc - resulted from CQ calls. QRM, inevitably, was bad from time to time, but in general, stations moved away from his fixed receive/transmit frequency of 7019kHz. He feels that praise must be given to the Howes design engineers for the excellent balanced mixer of the RX40 based on two 2N3918 FETs.

He hopes that this unusual experimental project may encourage others, including the financially-stressed, to participate in amateur Grand-Prix-type contests without the support of big towers, beams and the latest £3000-plus transceivers. He admits, however, that contacts on 7019kHz using his barefoot 3.5W choc-box rig are "very, very few indeed" despite his using all the cunning that he can muster to attract "would be flies" into the web!

CLASS-E 300-500W MOSFET AMPLIFIERS

A TEAM OF amateurs associated with student projects at Caltech have produced a detailed two-part *QST* article (May & June 1997) showing how, with from 3 to 12W of drive, it is possible to obtain 300 to 400W RF output at 7MHz from a single switching MOSFET costing only $11 powered from a 120V DC bus line. **Fig 2** gives a block diagram of the 7MHz Caltech Class-E amplifier station.

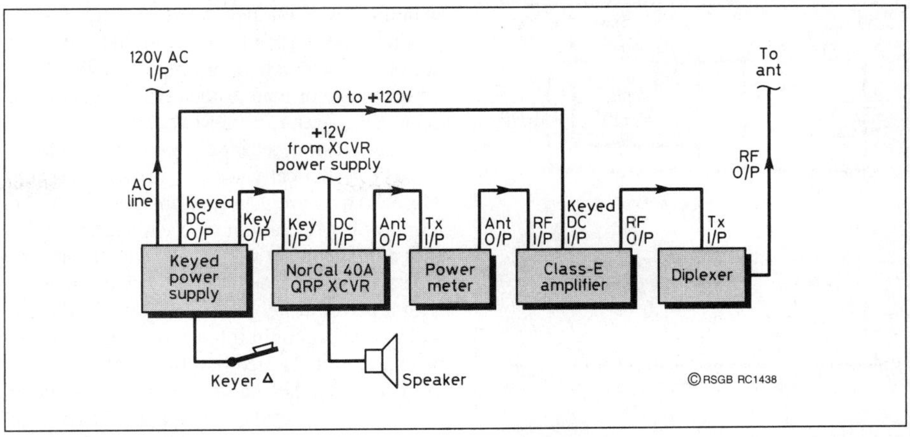

Fig 2 (top): Block diagram of the 7MHz Caltech Class-E amplifier station.

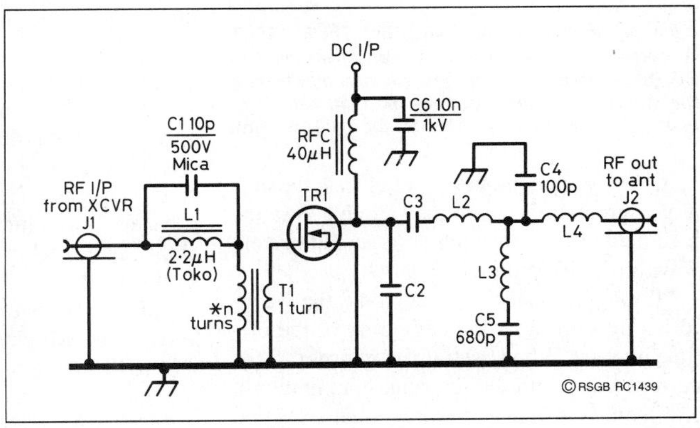

Fig 3 (right): Circuit diagram of both 300W and 500W Class-E high-efficiency amplifiers. MOSFETs, component values and coil details vary for the two powers. Shown only to provide an introduction to Class E. For constructional details etc see *QST*, May/ June 1997. For 300W, T1 primary 5t.

Details are presented for both 300W and 500W amplifiers, both based on an identical circuit diagram (**Fig 3**) and PC board, but with some different component values and coil details. The 300W version uses an International rectifier MOSFET type IRFP440 (cost in the USA only $8); the 500W version the IRFP450 costing $11.

High-efficiency Class D, E and F switching-type amplifiers have been discussed a number of times in *TT*, most recently in the form of a Class F amplifier as used by Dr Fred Raab, WA1WLW (*TT*, October 1996, p77). The Caltech team favour the Class E approach for home construction, as being less complex, requiring only a single transistor without gate bias or output transformer and capable of being driven from a 3-5W QRP transceiver.

With an efficiency of 90% and little heat loss, it can produce 21 times more power than a Class A amplifier with the same device dissipation working at an efficiency of 30 per cent. There is a downside. A Class E switching amplifier, like Class C, is not linear and is unsuitable for SSB operation; but it is fine for CW (Morse Code) or FSK. A further restriction is that a given design is suitable only for single-band operation, as mentioned below.

It is noted that, "The Class E amplifier was invented and patented by Nathan Sokal, WA1HQC, and Alan Sokal, WA1HQB, in 1975. It minimises heat loss by having as little overlap as possible between voltage and current. **Fig 4(a)** shows an idealised Class E cir-

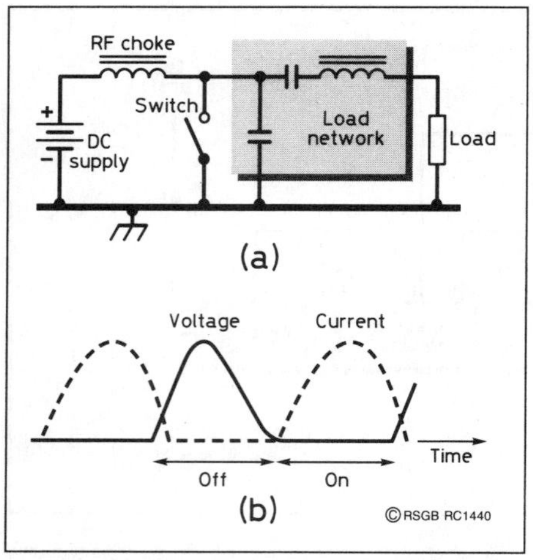

Fig 4: (a) Idealised Class-E amplifier. The MOSFET is represented by a switch that opens and closes at RF. (b) Class-E amplifiers reduce loss by keeping the overlap between voltage and current low, resulting in an efficiency of up to about 90 per cent.

cuit. As the switch opens and closes, the current alternately flows in the switch and in the load network. The switch voltage and current waveforms are shown in **Fig 4(b)**.

"It may be easiest to understand the waveforms by starting at the beginning of the off-time interval. When the transistor turns off, the current flows into the resonant load network, and there is a transient voltage that rises and falls. With a properly designed load network, the voltage returns to zero smoothly with zero slope. The transistor switches on when both the voltage and the current are small, keeping losses low even if the switching is slow or slightly mistimed. Once the transistor turns on, the current rises smoothly until it switches off again, and the cycle repeats. The resonant load network does limit a Class-E amplifier to single-band operation."

The information presented here is intended only as an introduction to Class E amplifiers for detailed practical information, the *QST* articles should be studied.

EXPERIMENTS WITH A HELICAL-WOUND VERTICAL

EVER SINCE THE early 1930s, it has been accepted that vertical HF antennas are likely to prove variable in performance, significantly affected by site factors. These include the local earth constants, the efficiency of the earthing system (buried earth mats, radials or counterpoise), and the proximity of surrounding objects to which there may be strong mutual coupling, resulting in absorption or re-radiation. All this implies that not only may the overall results vary greatly when a given type of vertical antenna is used at different sites, but also that virtually any resonant antenna, vertically or horizontally polarised, usually requires careful tweaking if it is to achieve its full potential.

One recalls the controversy that in the 1950s surrounded the short 'Joystick' helically wound antenna, made and sold in fairly large numbers in the UK by G3VFA. Some of those who purchased the antenna swore by it; many others swore at it.

Things have moved on since then. Antenna manufacturers, on the whole, strive to design, produce and market reliable products, even if at times they tend to over-emphasise the SWR performance or gain or the results likely to be achieved without further tweaking.

An interesting recent design is the Chelcom CA HFV1 helically wound vertical antenna. This has two fibreglass sections (each 2.74m in length) plus two alternative whip sections (to adjust the resonant frequency) which can provide a 3.5MHz resonant quarter-wave antenna with an overall height of just 7m (23 ft). No tuning unit is required for part of the 3.5MHz band and it is claimed that the antenna can be used on other HF bands provided a suitable ASTU is employed. A favourable two-page review of this antenna by John Heys, G3BDQ, appeared in *Practical Wireless* (April 1997, pp21-22).

John Brodzky, G3HQX, writes: "I have for some time been using a Chelcom 3.5MHz vertical antenna Model HFV1. This is helically wound and comes in two 10ft sections that screw together. There are also 23in and 51in rods, either of which is intended to be installed at the top as a whip for use at the HF or LF end of the band.

"My experiences have not worked out quite as expected. My findings agree neither with the *PW* review nor with the specification sheet provided by Lowe who market this antenna in the UK. Lowe show a response across the 3.5MHz band of around 3:1 with a minimum of about 1.25:1 at 3640kHz. G3BDQ suggests an SWR of 4:1 at 3500kHz reducing to 1:1 at 3800kHz, he also gives SWR figures for 10, 14, 21, 24 and 28MHz bands. Chelcom consider it a single-band (3.5MHz) antenna, although Lowe suggest it is good also for 14 and 28MHz.

"When I first had the antenna, I found that at my site both whips were too long. Instead I used an 11.5in rod and achieved good results using just four radials. I have since installed eight radials ranging from 68ft to 44ft and plan to install at least two more.

"I did not find it possible to obtain a good match on 14MHz and decided to try using a close-coupled resonator as described for wire dipoles by Gary Breed, K9AY, (*TT*, June 1995, pp67-69 and October 1995, p78). Rather to my surprise it worked, although the 14MHz resonator needed to be just under 12ft long, rather than the 16ft that I had expected. I used some aluminium rods which were on hand, varying from 0.75in to 0.5in diameter and spaced them from the HFV1 using empty 35mm film canisters, secured using plastic adhesive tape. There is no electrical connection to the 'resonator'.

"I have now been using this modified 3.5/14MHz antenna for about six months. I find it works well on both bands. On 3.5MHz SSB I worked VK, ZL, four VE districts, for W districts, JA, PY, LU, TI, VP2, YN; all W call areas except 7 and 3.5MHz CW. On 14MHz there is not much that I hear that I cannot work.

"But, for me, the bonus is that I achieve a low SWR right across both bands: see **Fig 5**. It appears that the introduction of the close-coupled 'resonator' has loaded the vertical in such a way that the system has become much lower-Q and hence has a markedly greater resonant bandwidth. On 14MHz, I have been using a 11ft 10.5in 'resonator', recently reduced by one inch to improve the SWR at the HF end of the European 3.5 to 3.8MHz band.

"I feed the antenna with a 50m length of 50Ω coaxial cable; this is slightly more than I need so I have coiled up the surplus at the base of the antenna, forming a coil of nine turns of approximately 14 - 15in diameter. This is presumably acting as a form of choke balun (or rather unun).

"The SWR readings were taken at the base of the antenna, using a simple L-section matching device at the transmitter to ensure maximum power transfer.

"It all seems too good to be true, but it certainly works. Whether it would be exactly reproducible at other sites must be left to others to find out. Meanwhile, I hope to continue experimenting with this rewarding system."

[The close-up rod appears to lower the Q of the HFV1 rather than form a resonator in the K9AY sense - *G3VA*.]

VE2CV ANALYSES THE DROOPING (INV-V) G5RV

FOR SOME 50 years a popular multiband antenna has been the G5RV system developed by Louis Varney, G5RV, in the immediate

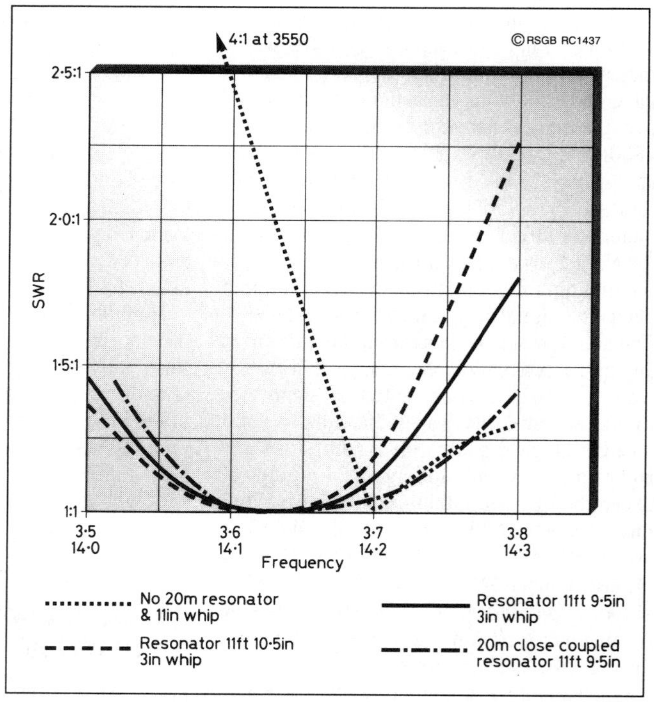

Fig 5: SWR measurements by G3HQX on his HFV1 helical wound vertical antenna showing effect of adding a close-coupled 14MHz resonator.

Freq (MHz)	A = 180°	A = 130°	A = 115°	A = 90°
3.75	5.97 (90)	5.13 (90) [5.13]	4.66 (90) [4.66]	3.65 (90) [3.65]
7.15	6.7 (39)	5.7 (48) [5.49]	5.2 (51) [4.88]	4.11 (55) [3.61]
10.15	9.74 (27)	7.4 (35) [6.98]	6.33 (38) [5.67]	4.23 (43) [3.1]
14.15	8.21 (19)	5.26 (21) [5.21]	7.11 (49) [3.16]	6.26 (64) [0.68]
21.15	10.11 (13)	7.97 (17) [7.37]	6.59 (18) [5.71]	2.44 (11) [2.26]
29.0	10.25 (9)	7.87 (11) [7.8]	6.83 (12) [6.15]	3.44 (12) [2.86]

Table 1: Gain dBi (and take-off angle) for a 31.1m multiband dipole versus drooping angle (A).

post-war era. After enjoying widespread use in Europe, in recent years it has also come to the fore as a simple wire antenna in North America. The original full-size 31.1m centre-fed dipole version was developed before the 21MHz band was released to amateurs and it has normally been essential to use an ASTU on that band, and preferably on other bands. A few years ago, Dr Brian Austin, G0GSF/ZS6AKW, presented an improved version - made possible by computer simulations - taking 21MHz into account (see, for example, *TT*, January 1993).

On the other side of the Atlantic, Dr John (Jack) Belrose, VE2CV, has, over several years, criticised some aspects of the G5RV particularly when used in the drooping dipole (inverted vee) configuration (see for example *ARRL Antenna Compendium Volume 4*). This has tended to upset some amateurs who have told VE2CV "the G5RV is a great antenna and you should not be critical of it". In reply, VE2CV admits that he has commented at times adversely on the method of feed and the difficulty of achieving resonance in several amateur bands. But his main criticism is concerned with its use in the drooping dipole configuration.

In response to a recent enquiry from Bob Ackman, VE6RI, Dr Belrose replied: "Yes, I have modelled this dipole in the drooping dipole configuration over a larger range of variable (droop and frequency) than in my *Antenna Compendium Vol 4* article . . . In **Table 1**, the reference angle A is the included angle between the arms of the dipole. The bracketed () number following the maximum gain figure (in dBi taking into account the ground image) is the elevation angle for maximum gain (take-off angle). The antenna is modelled for an apex at 15.24m (50ft) over average ground. The computer code is NEC-2, considered to be the industry standard for simple wire antennas. For 'grounded' antennas one has to use NEC-4.

"The dipole gain is a maximum for the horizontal configuration (A=180°) but maximum gain is not necessarily in the broadside direction for frequencies at or above 14MHz. The azimuthal pattern becomes lobed. In the drooping configuration, the gain is reduced, and for A=90° the performance is particularly poor on the 14MHz band, *the band on which G5RV wanted his antenna to work best*. This is because not only is the gain lower, but the elevation angle for maximum gain is increased, and increased significantly. On the higher bands, most amateurs want to work DX and seek a low take off angle. The take off angle for the horizontal configuration changes with frequency in about the right way. That is unless you (like me) wants to work coast-to-coast on the 3.5MHz band.

"If I calculate the gain as a function of frequency at the corresponding take off angle for the horizontal configuration, the gain in the drooping configuration is poor. The gain difference (drooping compared with horizontal) increases with increase in frequency and decrease in A as referenced in the square-bracketed numbers in Table 1.

"I fail to understand the popularity of the multi-band inverted-V. For years our Handbooks stated that the optimum angle A was between 90° and 110°, but now it is stated that A should be greater than 90° (optimum angle is 180°).

"But all this relates to horizontal and drooping dipoles supported by a tree limb or wooden mast, and fed with a balanced transmission line. The G5RV type of feed - balanced feeder with connecting coax usually without a balun since baluns do not perform very well when working into a mismatched load - can result in the conducting tower, or the coax itself (current on the outer surface of the shield) becoming at some frequencies an important part of the antenna system.

"In my files, I have a letter from an amateur reporting that when a balun was added to his drooping G5RV system, the ability to work DX on 14MHz went the way of the birds. Clearly, for this amateur, his metal support tower and/or coax was acting as part of the antenna system. I can include a tower, and the effect of an unbalanced coaxial feeder in my modelling, but this is all very geometry sensitive, and dependent also on how effectively the tower/and or coax is 'grounded'. The effects of re-radiation may be small, but only

THE BENT DIPOLE

THE HORIZONTAL RESONANT half-wave dipole centre-fed from coaxial cable remains probably the most popular HF wire antenna. It is capable of providing near 100 per cent radiation efficiency, is relatively easy to adjust and if erected in the clear and with the feeder descending vertically can be relied upon to work well without an ATU or balun. But in these days of small gardens, it may prove difficult to provide a half-wave span even on 14MHz, more often on 10 or 7MHz, and normally on 3.5 or 1.8MHz.

Nizar Mullani, K0NM, ('The Bent Dipole' *QST*, May 1997, pp56-57) uses NEC-2 computer simulation to show that surprisingly little radiation efficiency is lost, and matching to 50Ω coax may be improved, when the ends are bent down by up to about 40% of its length; even when half the dipole element is bent down, radiation may be reduced by only about 0.6dB providing that the horizontal portion is at a reasonable height above ground: see **Figs 6** and **7**. The radiation pattern is also relatively little affected, although the vertical take-off angle may be raised and the SWR bandwidth reduced a little.

As K0NM puts it: "Don't be afraid to bend your dipole antennas if you're cramped for space."

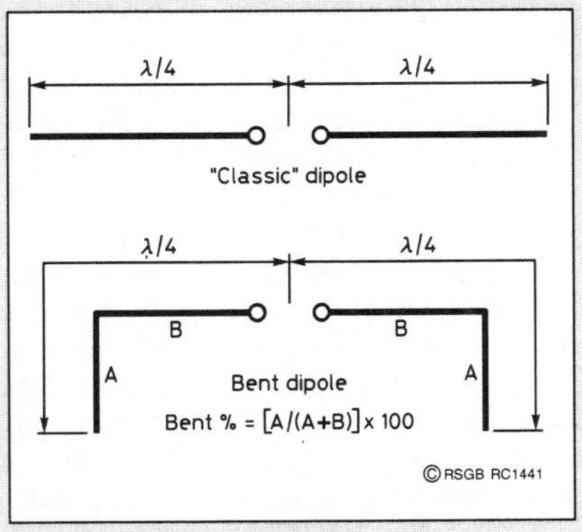

Fig 6: Comparison of a half-wave dipole with a level horizontal span and a bent dipole. The total length of the bent dipole is one-half wavelength and the percentage of bend is defined by the ratio of A divided by A+B.

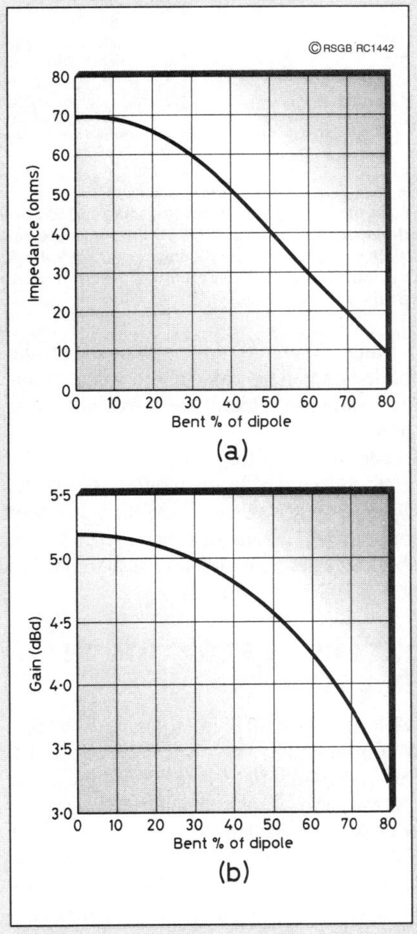

Fig 7: (a) Bent dipole feed-point impedance as a function of the percentage of bend, assuming a height of one-half wave above ground. There will be some variation of feed-point impedance at other heights. (b) Gain of the bent dipole (including ground image) as a function of the percentage of bending. 50 per cent bending reduces the span to 50 per cent of a dipole without bending.

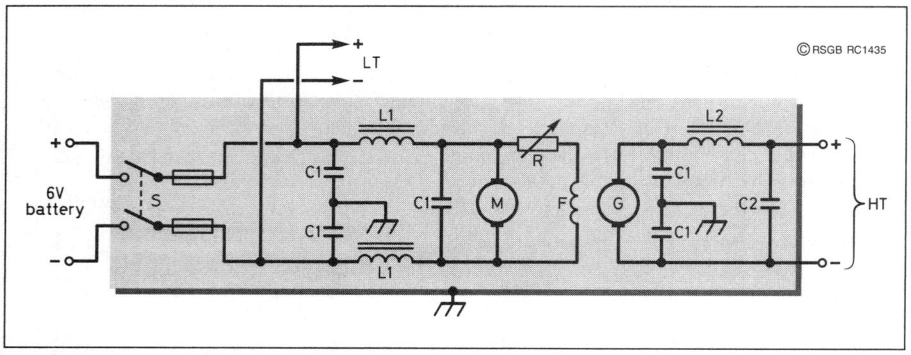

Fig 8: Dynamotor (rotary transformer) used to convert low-voltage DC to high-voltage DC during the pre-solid-state era. A starting switch S connects the battery to the motor armature and excites the field winding F. A field rheostat R regulates the generator voltage. Commutator ripple voltage in the output is smoothed out by ripple filter L2-C2. Sparking at the commutators could cause severe RFI unless suitable precautions were taken (L1-C1).

providing there are no resonances in the grounded conductors associated with the antenna system, in which case re-radiation can be dominant."

It seems possible that these criticisms of the inverted-vee configuration would apply to the G0GSF/ZS6AKW versions and indeed to other drooping dipole antennas, although clearly this configuration has the advantage of requiring only one high support mast.

PORTABLE OPERATION IN THE VALVE ERA

SEVERAL RECENT *TT* items have described how, throughout WW2, use was made by secret radio-agents of the Whaddon-built Mk VII receiver-transmitter comprising just a 6V6 crystal-controlled power oscillator and 6SK7 + 6SK7 regenerative receiver. This simple equipment was powered by either a compact mains PSU or from 6V vehicle batteries using a vibrator-type DC-DC converter. Incidentally, I gather that the Norwegian SIS radio agent Oluf Reed-Olsun (see *TT*, June 1947, p62) was trained at a Norwegian training school rather than the nearby Hans School in Knightsbridge; on his second mission taking the American SSTR-1 he decided the transmitter was too powerful and presented an easy target for the German D/F teams, and used instead another Whaddon set!

Mechanical vibrators were developed in the 1930s primarily for car-radios. They were electrically very efficient (up to 90 per cent) and had good frequency stability. Such units, however, were usually limited in total output power to under 30 watts and tended to deteriorate rather rapidly with the contact points becoming pitted and tending to stick, and the springs losing tension. They were, however, widely used by Whaddon, SOE and the London Poles for single-valve transmitter plus receiver sets and remained popular for some years after the end of the war. By the mid-1950s, mechanical vibrators were giving way to more reliable transistorised DC-DC converters and DC-AC inverters which soon became capable of outputs of well over 30 watts.

Before then, for higher powers, rotary transformers (dynamotors) or rotary converters were used. A rotary transformer (dynamotor) is primarily a machine which changes the voltage of a DC supply, and was commonly used to convert a 6- or 12-volt battery supply to HT for receivers or transmitters: **Fig 8**. Efficiencies were lower than for vibrators. Machines delivering 50 to 500 watts could have overall efficiencies ranging from 50 to 60 per cent. In those days most car-electrics were 6V but with heavier Army vehicles using 12V.

A rotary converter (often confused with a DC-DC rotary transformer) converts AC to DC and vice versa. Rotary converters were usually higher power machines than dynamotors. They were used, for example, in Whaddon signal vehicles in order to run from large vehicle batteries several HRO receivers and Mark III or even Mk X transmitters (with 813 PAs) having AC PSUs.

The Whaddon Mk XV set (successor to the Mark V) was built in three separate boxes: transmitter 6F6 to 6L6 CO + 6L6 PA; receiver three 6SK7 pentodes as a 'straight' 1-V-1 with regenerative detector; and three alternative PSUs: (1) Mains PSU with 5T4 rectifier; (2) Continental mains PSU with 6X5 rectifier; and (3) Power unit for 6V vehicle batteries using a Carter 417 XV dynamotor. Dynamotor units were heavier than vibrator supplies but could deliver more power. After the war many American-made dynamotors came on to the surplus market.

Photographs exist showing that even 350W Onan petrol-electric generators were delivered to SIS intelligence groups in Norway. Personal experience of these heavy devices, which were often reluctant to fire-up using their starting ropes, makes me wonder whether the local Funkabwehr were all considered deaf. One would have thought that in the still Norwegian countryside a noisy two-stroke petrol engine could have been heard miles away!

In the long-defunct *Short Wave News* (May 1945) an unsigned article (apparently written by the late Dr Arthur Gee, G2UK) described 'A portable transceiver for field days' (**Fig 9**) that has a marked similarity in concept to the Mark VII except that an 807 was used as the power oscillator. The two-valve receiver used a 6J7G as regenerative detector and a 6C5 triode AF amplifier, all in a wooden box with fold out front panel on which the Morse key was mounted. For portable operation on 1.8MHz CW a 'surplus' rotary transformer was used, acquired for the modest cost of 30 shillings (£1.50). ♦

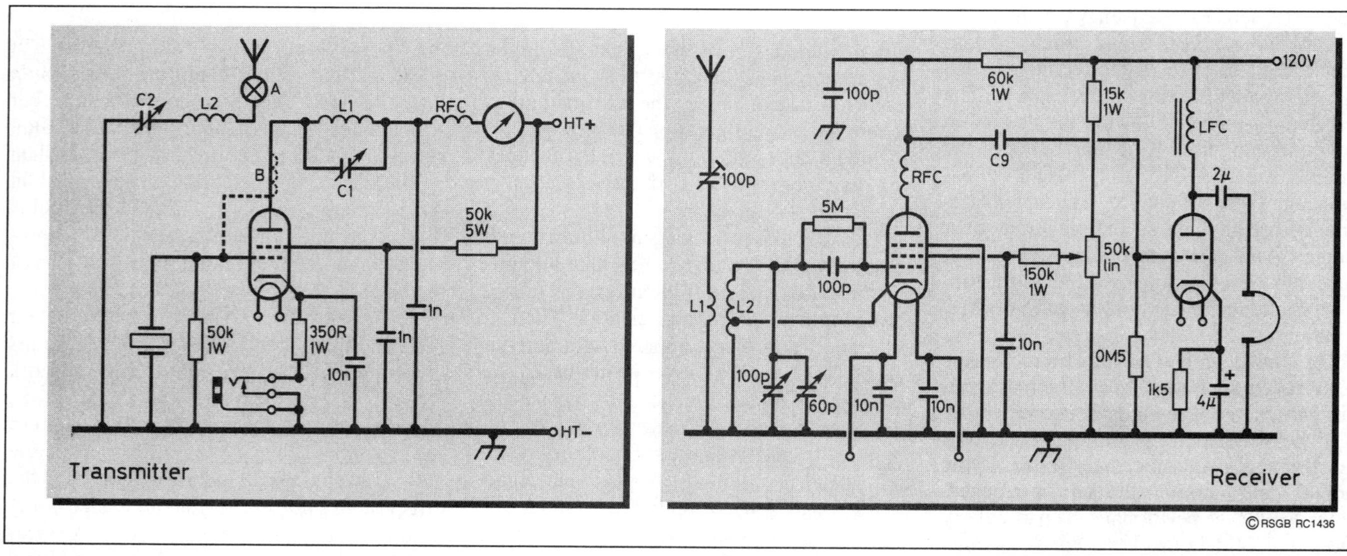

Fig 9: The 1.8MHz portable transmitter-receiver used by G2UK circa 1947. In portable operation the transmitter derives its HT from a 350V, 50mA dynamotor. C1 and C2 were ex-broadcast receiver of about 350pF maximum capacitance. Transmitter coils L1, L2 wound on 3-in diameter former, both 14 turns about ¾in apart; receiver coils on 1¾in outside diameter former, L1 9 turns spaced 1in from L2 40 turns with tap four turns from 'earthy' end. G2UK suggested it could be modified for 3.5 or 7MHz.

Pat Hawker's

Technical Topics

PAT HAWKER, G3VA
37 Dovercourt Road
London SE22 8SS

TETRAHEDRAL DIPOLE ANTENNA

THESE DAYS THE bulk of effort by the professional antenna engineers is directed towards satellite, UHF, microwave and gigahertz antennas. Conversely, relatively little attention is given to HF with the exception of improved near-vertical-incidence-systems for mobile units. Therefore, it was with something like relief that I came across a paper presented last April at the '10th International Conference on Antennas & Propagation' (ICAP97 - *IEE Conference Publication No 436*, pp 1.428 - 1.430) describing a novel HF antenna that could have promise for amateur as well as professional applications.

This is the 'Tetrahedral Antenna', invented by Alan Boswell and Barry Peters of the GEC-Marconi Research Centre (UK Patent GB 230.2990 granted February 1997). Basically, it is a development of the centre-fed biconical dipole, but using a spaceframe made from 12 metal tubes with good RF conductivity (copper or aluminium alloy etc) in the form of two regular tetrahedrons mounted one above the other with their apexes separated by an insulated platform into which three tubes are fixed on each side: see **Fig 1**. The self-supporting framework is held in place either by insulated rods or by insulated halyards (**Fig 2**) made from non-stretchable plastic rope etc.

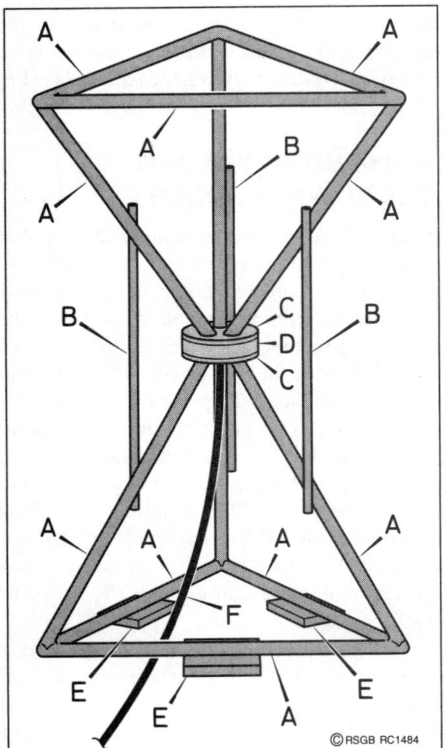

Fig 1: Initial implementation of the GEC-Marconi Tetrahedral vertical antenna - a novel form of the classic centre-fed biconical - to provide a self-supporting dipole antenna with good bandwidth characteristics, permitting operation on several amateur bands and not requiring an earth connection. The height of the 'dipole' is roughly the equivalent of a quarter-wave monopole antenna. A = 12 metal (eg alloy or copper) tubes of equal length (2.9m in the Marconi prototype) forming two skeletal regular tetrahedrons. B = 3 rigid insulated supports holding the structure in shape. C-D-C central sandwich of two metal plates (C) separated by insulated layer (D). E = 3 support blocks insulating the element from earth (about 0.1m height). F = coaxial feeder (preferably 75 ohms impedance).

A tetrahedron is a pyramid with four triangular faces; a regular tetrahedron has all four faces in the form of equilateral triangles. The tetrahedrons are substituted for the cones of the biconical antenna. By making all edges conductive, the dimensions of the dipole are further reduced. In the Marconi prototype all four tubes are 2.9m in length, and the resonant frequency is approximately 22.5MHz. The height of each tetrahedron is just over 2.1m, making the overall height approximately 4.2m, roughly similar to that of a 22.5MHz quarter-wave monopole. The structure can be stood on insulating spacers (E) raising the low base a few inches above ground. It was also tried raised about 0.5m above ground in which case the frequency of minimum return loss (ie resonant frequency) rose to 26.1MHz, an increase of 16%. Constructional methods can be varied to reflect whether the antenna is to be fixed or needs to be readily dismantled for easy transportation and erection on site.

The prototype antenna was made with two small conductive plates at the centre of which the six tubes are attached (for example, by means of a flattened end to the tubes as in **Fig 3**), with three to each plate with the plates separated by an insulated centre to form a 'sandwich' (C-D-C in Fig 1 or see the enlarged drawing of Fig 3). A coaxial connector may be mounted on the lower plate with its inner conductor connected via braiding to the upper plate as shown. The coax feeder is attached and should hang vertically downwards and then run horizontally along the ground. In this way, any currents induced in the outer surface of the coax will decay rapidly in the horizontal run.

The tetrahedral dipole antenna has a bandwidth of around 35% for an SWR of less than 2:1 (return loss more than 9.54dB) when fed with 75Ω coax-cable (less with 50Ω cable) compared with roughly 10% or less for typical vertical dipoles or monopoles of equivalent element diameter: see **Fig 4**. At 22.5MHz the measured input impedance of the Marconi prototype was 68-j14Ω with the antenna in the normal position, and 42-j32Ω when raised 0.5m above ground.

It should thus be possible to use this antenna on all bands from 14 to 24MHz with the simplest of ATUs or a broadband 50-to-75 **ohm** matching transformer at the transceiver end of the feeder. By adding a series inductance of 1μH, the resonant frequency of the prototype was lowered to 15.5MHz with a 9.5dB return-loss bandwidth of 2.7MHz: see **Fig 79**. The broad bandwidth of the antenna means that dimensions are far less critical than for a conventional dipole.

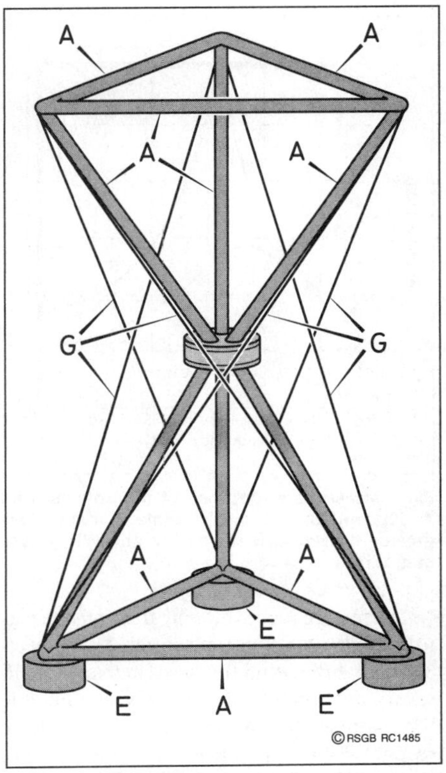

Fig 2: Second implementation (on which the IEE paper was based) in which the rigid supports are replaced with six halyards (G) of insulated, non-stretchable rope etc). 0.1m support blocks (E) are now at the corners.

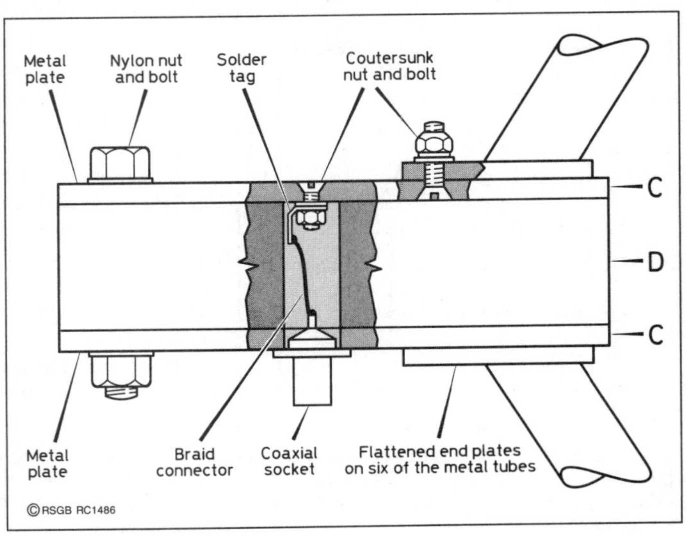

Fig 3: Detail of a section of the centre 'sandwich' showing how the antenna is fed and how six of the tubes are fastened to the metal plates (C) with flattened ends and countersunk nuts and bolts.

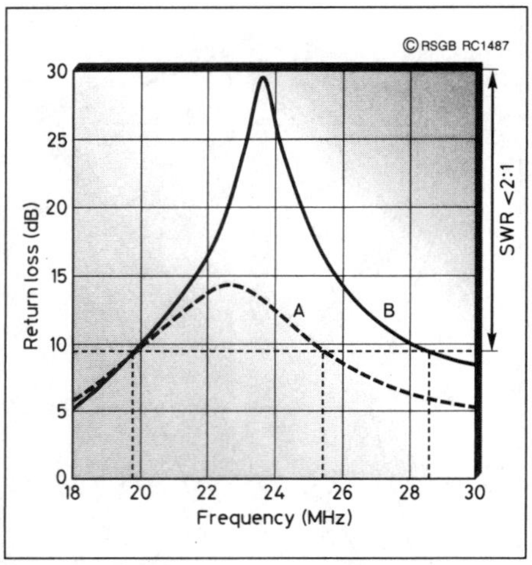

Fig 4: Measured return loss of the antenna with A 50Ω and B 75Ω coaxial cable. A return loss greater than 9.54dB represents an SWR of less than 2:1.

The motive for developing this antenna was to provide an easily transportable and erectable self-supporting (no mast) antenna for HF radars sited on beaches, but it is also suitable for communications. Some readers may have recognised that Alan Boswell holds the amateur callsign G3NOQ. In his IEE paper he reports that "the antenna was used for two-way contacts with several countries on all amateur bands from 7 to 24MHz".

It will be appreciated that, as with any non-expired patented design, it would not be possible for an amateur radio version of the tetrahedral antenna to be marketed commercially without a licence from GEC-Marconi. But this does not prevent any amateur from building and testing such antennas on an experimental basis.

In his paper, Alan Boswell provides an illuminating comparison between vertical dipoles and monopoles, applicable to all such antennas and not merely his tetrahedral design. Although the conventional vertical dipole does not depend on a low-loss earth connection or ground plane, it requires double the height of a ground-mounted monopole. The tetrahedral antenna provides the combined advantages of the monopole (small height) and the dipole (no wires in the ground).

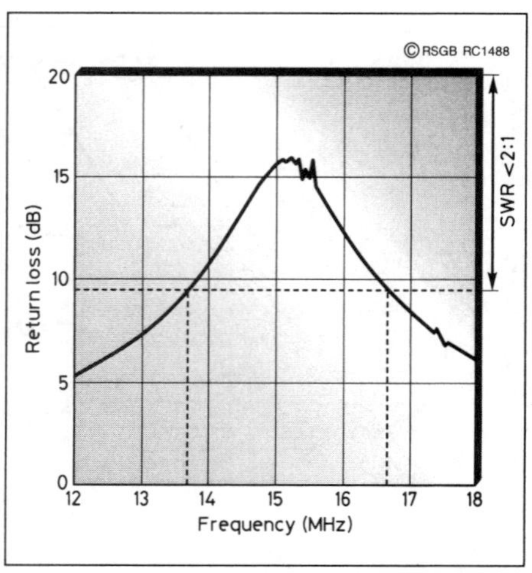

Fig 5: Measured return loss of the antenna with series coil (1uH) lowering resonant frequency but decreasing bandwidth.

Tubes at the base and at the top can be joined in various ways, for example by welding for a fixed structure, by flattened ends joined by nuts and bolts or by plumbers joints etc.

Ground losses occur for both dipoles and monopoles, but overall the dipole is superior to the extent that it is possible to dispense altogether with an earth mat although in the absence of good ground conductivity it may still be an advantage to use an earth mat or buried radials in order to improve the low-angle radiation. The absence of a direct earth connection may result in a small loss of gain, but a monopole would undoubtedly suffer a severe loss of gain without an effective earth mat. Radiation efficiency of the antenna itself has been shown to be approximately 100% although there will inevitably be some ground losses and possible absorption in nearby structures etc.

It should be noted that as with any vertical dipole close to ground, there will be high RF voltages at the base as well as potentially strong RF radiation (although this will be less than with a ground mounted monopole). Care should always be taken to ensure that there is no danger to children, adults or pets.

This antenna should function effectively when mounted on the ground via short insulating supports, and form an effective low-angle non-critical DX radiator for 18, 21 and 24MHz and give reasonable results on the 14 and 28MHz bands. However, results would be further improved if mounted at a height (possibly with slightly longer sides to the tetrahedrons to lower the resonant frequency). One way of achieving this would be to mount the antenna securely on a flat roof or similar high platform; the skeletal framework which could be made from lightweight alloy tubes should minimise wind resistance. It would be relatively easy to carry the tubes and the central 'sandwich' up to a flat roof or elevated platform in dismantled form and then assemble them.

The patent notes that the tetrahedral antenna can form part of an antenna array (eg phased array).

SAFETY WITH NICADS

R O BROOME, G1YKO, is concerned that when fitting nicad batteries to power radios, handlamps etc intended for use with conventional dry batteries, there is a possible danger of a fire or meltdown should a short-circuit occur. This is because of the very heavy current that can be delivered by a nicad in such circumstances.

He writes: "After purchasing four 4AH nicad D cells at the Elvaston Castle Rally for my recently purchased handlamp, I felt there was a need to fit a fuse. The handlamp is the type that will take a lantern battery or, with the adapter supplied with the lamp, will take four D-type cells.

"My simple and cheap solution was to cut a piece of 3mm *double-sided* copper board to slide between the top connection of one of the cells. I then soldered to one side of the board a piece of thin wire (about 40SWG); passed it over one edge of the board and then soldered it to the other side of the board. The copper

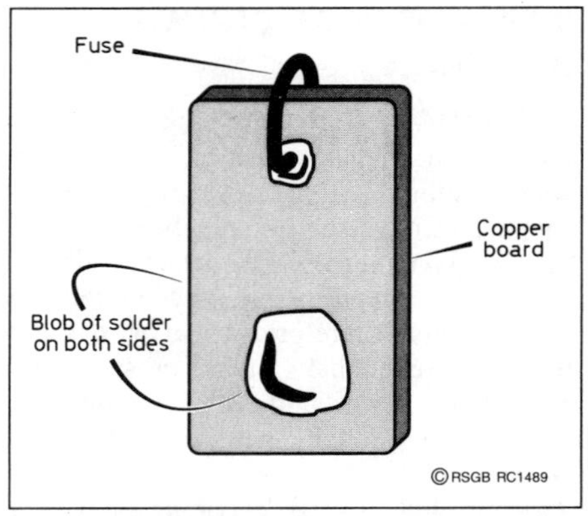

Fig 6: How G1YKO adds a fuse to a radio receiver or handlamp etc when using nicad batteries to prevent damage from the very heavy currents that can result from a short-circuit when using nicad cells. The blob of solder on both sides of the double-sided PCB is not likely to be needed for handlamps.

board was then inserted between the battery and the lamp connection. I tested the device by short-circuiting the spring terminals on the adapter in which the cells were fitted. The wire fused instantly.

"More recently I purchased a Roberts Radio, model R701 which runs off six 'C' cells when not connected to the mains adapter. Again, I am using nicad batteries and have provided another similar fuse this time also putting a blob of solder in the centre of the copper board on each side as it ensures good electrical contact between the battery (see **Fig 6**)".

END-FED RESONANT FEEDLINE DIPOLES

ONE OF THE rewarding features of *TT* is that some items stimulate further comments and suggestions from readers; at times it seems that the column is able to draw upon the collective memories as well as the practical experiences of the world-wide amateur-radio community. Two examples of such follow-up comments can be cited this month. Roberto Craighero, I1ARZ, exercises his excellent memory to add to the June 1977, p63-64, item on 'Resonant feedline dipoles'; and Chas Reynolds, GW3JPT, describes his experiences with double-wound coaxial traps (see 'Improved multiband trap dipole' *TT*, February 1997, pp70-71).

I1ARZ felt there was something vaguely familiar about the resonant feedline dipole based on the *QST* (July 1996) article by Al Burton, W8NX. A check of his collection of *73 Magazine* confirmed that an article 'New dipole feeder – tuned feeders, yet!' by A F Stahler, AA6AX, appeared in the June 1978 issue making use of his earlier article on what he termed the 'T2LT' (Tuned Transmission Line Trap): **Figs 7** and **8**.

In his case, the line trap was tuned to resonance (high Q) with the aid of a variable capacitor across the coaxial coil formed by coiling up a few turns in the feeder cable: see Fig 7. The number of turns should be kept low by using a large capacitance. The higher the Q the more efficient the dipole (but the narrower the useful bandwidth). Again, the technique is

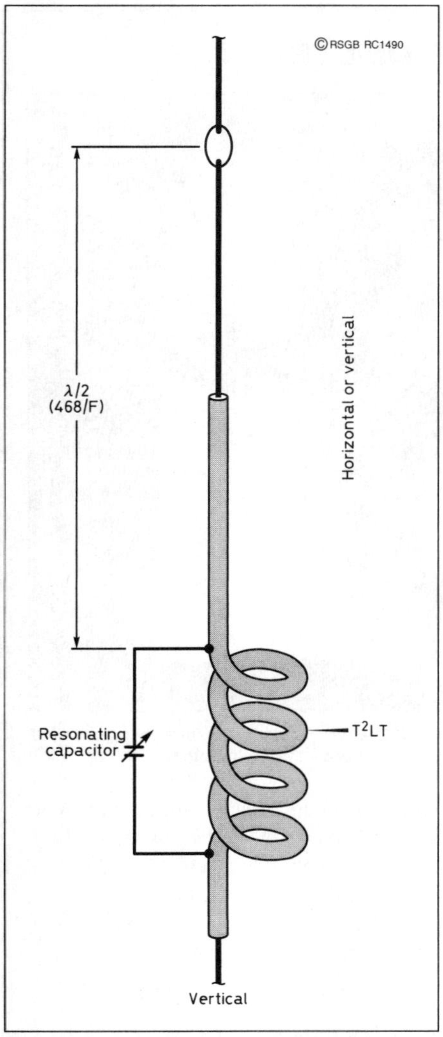

Fig 7: The end-fed dipole antenna in which a tuned transmission line trap (T2LT) is inserted in the coaxial cable a quarter-wave from the centre of the dipole.

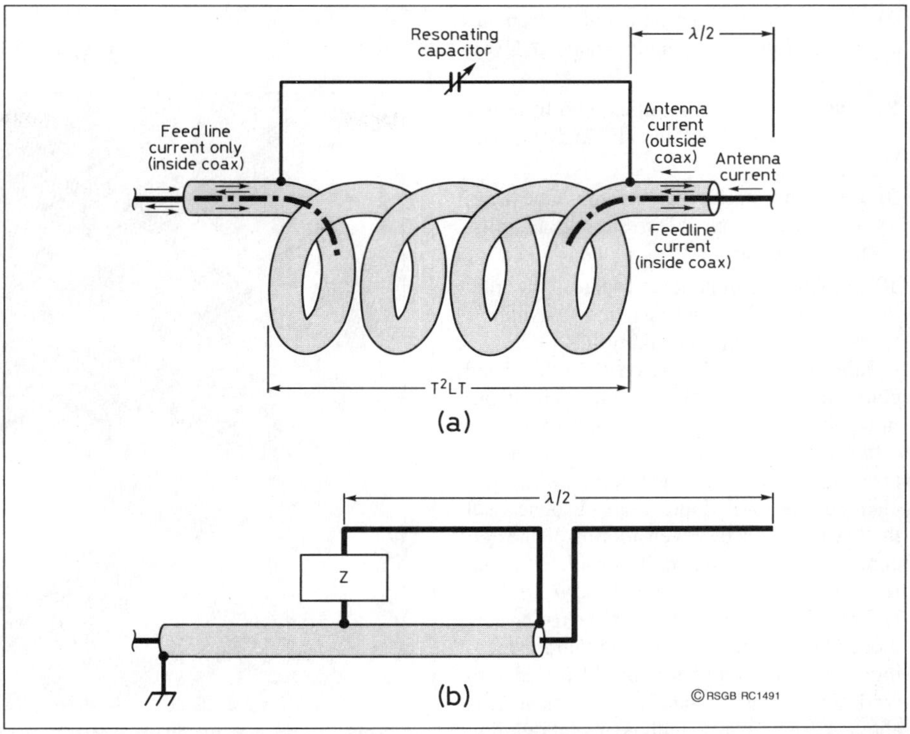

Fig 8: (A) Current distribution in the T2LT trap with a standing-wave formed on the outer-braid of the coax while the inner of the braid carries the low-impedance drive current. (B) Equivalent configuration.

to use the outer braid of the cable beyond the trap as a quarter-wave element while the inner conductor and the inner of the braid provides the current feed for the antenna. The length of the feeder before the trap should, at least theoretically, not be critical.

AA6AX pointed out that the use of such a high-Q resonant trap rather than an RF choke (as in the W8NX and the earlier designs of end-fed dipoles of this general type) provides superior performance since it copes with the considerable difference between the end impedance of half-wave dipoles fed against real rather than ideal earth.

In practice, AA6AX used his system as an end-fed 14MHz vertical dipole for QRP operation (2W HW-7 transceiver) and from the photographs appears to have used a fairly conventional variable capacitor. Presumably, for QRO operation this would need to have the vanes spaced more widely to avoid RF arcing. As a vertical dipole this antenna does not depend on radials or an earth mat, making it a good choice for portable operation.

GW3JPT as 'Beacon Traps' of Welshpool (see *Product News*, April 1996, p57) has been supplying amateurs with pairs of high-quality 7MHz traps together with a rain-proof dipole centre-piece to enable amateurs to build their own W3DZZ/G8KW multiband trap dipoles.

He writes: "I was very interested in the details given in *TT* of the double-wound coaxial traps. Over the years I have made various versions of dipole traps, and finally - a couple of years ago - settled on the form of construction illustrated in the 1996 New Products item. The coaxial coil is fully enclosed and the connections brought out via stainless steel bolts. There is also an extra hole for use with loopback and the shape was designed so that water did not collect in the end and gave extra thickness for fixings.

"Having studied the details given in *TT* I made up two double-wound coils with RG58 for 5.16MHz and put up the antenna as an inverted-vee. The results proved much as detailed by W8NX. The traps used a larger size pipe but were constructed in much the same way as that shown in the February Fig 6.

"I then decided to make up a pair of double-wound traps for 7.00MHz using RG58 and the larger pipe size. Again, as for the 5.16MHz traps, they showed a big improvement of 'Q' over single-wound traps. The 3:1 SWR readings were only 50kHz apart with an SWR of 1:1 in the centre. This compares with 150kHz apart with the single wound coils. **Table 1** shows the SWR readings taken on a horizontal trap dipole with an SWR Analyser."

IN DECLINE - OR TOO GLOOMY?

AMATEUR RADIO publications the world over have been engaging in what seems to be a fit of the minimum-sunspot blues. Is the hobby in terminal decline or is there just too much gloom and doom? The dichotomy is well represented by two divergent views expressed in the correspondence column of the June 1997 *QST*.

Nick Leggett, N3NL, ascribes the decline in activity to a set of specific factors, including:

1) A decline in the faith that individuals can invent new technology and develop commercially useful products. Many have used amateur radio as a research and development lab. Now people say inventing is for corporations only, encouraged by contracts that claim all inventions made by an employee including those made on his own time.
2) Equipment less operator dependent than it used to be. New gear does not make the same demand on operator skill.
3) New gear does not offer as much opportunity for modifications and design changes that made gear more personal and a point of pride.
4) Restrictive housing convenants and now RF exposure regulations (in the USA).
5) New housing without barns, outbuildings and sheds.
6) Smaller (building) lots reduce space available for antenna experiments.

Freq	SWR	Freq	SWR	Freq	SWR	Freq	SWR	Freq	SWR
3.5	3.1	7.0	2.1	14-15	3.1	20.5	1.2	28.0	2.5
3.55	1.7	7.05	1.2			21.0	1.7	29.0	1.7
3.6	1.1	7.1	1.2			21.45	2.5	30.0	2.1
3.65	1.5	7.3	3.1			21.9	3.1		
3.7	1.6								
3.75	2.5								
3.8	3.1								

Table 1: SWR readings on horizontal dipole using double-wound 7.0MHz coaxial traps. Frequency in MHz.

7) Constant media hype that cellular telephones and the Internet are modern and that CW is old-fashioned.
8) Two-wage-earner families with little free time after work, child care, and commuting.
9) Declining real family income. One wage earner used to be able to support a family, house and car(s).
10) Individual components replaced by integrated circuits resulting in reduced interest in homebrewing of linking ICs.

Ken Gilcrest, AB7KT, on the other hand claims there is too much written about our impending demise, a topic in which he has lost interest . . . The latest catastrophe on the horizon, so they say, is the Internet, which it is claimed has made ham radio obsolete. But then why didn't the even more reliable telephone make amateur radio obsolete? Amateur radio is the exploration of *radio itself.* "It is about building new antennas and trying new modes of propagation. It is about understanding electronics and putting that knowledge to work building and repairing equipment. It is about communicating with people at random, not knowing who might answer your CQ, or where they might be located. It is about providing communications in emergency situations. It is about sharing all of the above with people who share a common interest in radio.

A D-I-Y VARIOMETER

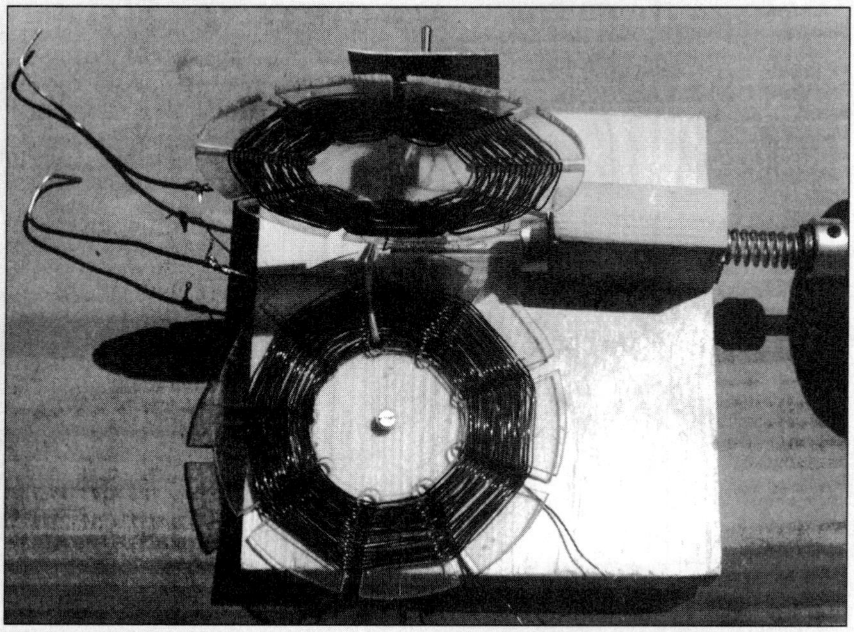

G8DPS's home constructed variometer based on the 1920s form of basket-weave coils, arranged so that the inductance of one coil can either reinforce or oppose that of the other.

IN THE PAST, there have been several references in *TT* to the advantages (and some disadvantages) of the classic variometer as an alternative to the roller coaster form of variable inductance. In particular, *TT* has noted the extremely fine mechanical construction of the German wartime variometers used in a number of their airborne sets.

Dr Dick Biddulph, G8DPS, recently sent some photographs of a very simple form of home-made variometer which he points out is based on the 1920s form of basket weave coils where the inductance of one coil can reinforce or oppose that of the other (see photograph).

He writes: "It was constructed from polystyrene sheet of the type used as a cheapo glass-substitute in horticulture although acrylic (Perspex) would probably do. You start by cutting out two discs, in my case 88mm diameter, then draw a circle inside on the same centre, in my case 337mm diameter. On this mark the sites for an odd number of holes, drill them (in my case 3mm diameter) and cut slots to the periphery. I cut my slots with a jewellers slitting saw which has a frame the size of a junior hacksaw but has a metal cutting fret-saw blade. A Junior hacksaw would do just as well, preferably lubricated with washing-up liquid. The slitting saw needs no lubrication. The wire is wound in and out of the slots, as can be seen in the illustration.

"Results were as follows: Number of slots 9; turns 34; wire diameter 0.5mm; inner winding diameter 37mm; outer winding diameter 72mm. This gave a reinforcing inductance of 235μH; coils at right angles 150μH; coils opposing 47μH. 'Q's respectively 90, 100 and 50. Exact inductance's and 'Q's cannot be guaranteed as they were measured on a very old Marconi 'Q' meter which has been re-calibrated since I bought it ten years ago, but should be good enough for comparison purposes."

HERE & THERE

L VANPROOYEN, K3KWD, in a contribution to the 'Hints & Kinks' column of *QST*, June 1997, pp72-73, points out that many classic valve communications receivers have sufficient stability and selectivity to cope with modern band conditions but lack a product detector, having been designed for AM/CW reception. SSB audio can be much improved by adding a simple product detector using a heptode frequency-changer valve such as the 6BY6, 6BE6 (EK90): **Fig 9**. He describes in detail fitting this arrangement in an Hallicrafters SX101 but has also used it to modify a Collins 75A2 and 51J3, various older Nationals and a Hammarlund HQ-140X.

Mike Gibbins, G3FDW, confesses to a dislike of using coaxial cable feeders at UHF/VHF where, unless expensive low-loss cable is used, losses can become excessive. To overcome this problem, he mounts his 25-watt power amplifier, with its in-built RF amplifier at the top of his mast. He then installs a 12V DC lead and a length of UR47 low-cost coaxial to provide drive for the PA box which is within 4 - 5ft of the antenna. The result is a low-cost low-loss system for 144MHz, and a similar arrangement for his 25-watt 70cm PA/RF amplifier. A strong plastic bag and some black tape keeps the amplifier dry, while the breeze keeps it cool. His 144MHz set up was illustrated in *RadCom*, July 1994, p13, with a 7-element log-periodic-Yagi (LPY) antenna on a 6ft boom (the bulge in the mast is the MM PA unit).

Chas Reynolds, GW3JPT, draws attention to 'Supertube' conduit which is a form of flexible conductive tubing intended for such applications as providing EMC screening etc for power, signal and data cables. He feels this might have application for constructing magnetic loop antennas for use in a roof space or for outside use, etc. From the adverts for Supertubing it would seem to be easy to bend by hand to fit the available space, light in weight, welded construction. It comprises an internal polyethylene layer, an aluminium layer and an outside polyethylene layer. Whether the aluminium layer would have sufficiently low RF resistance for use in small magnetic loop antennas is open to question; also, since it is apparently supplied in 100m coils to major cable-TV firms etc, it may not come cheaply if only a small length is required. The advertisement states, "for further information call Peter Teah on 01827 712744".

Stephen Bolt, writing from the Netherlands ('Letters', *Electronics World*, August 1997) in answer to a query about bio-galvanic batteries, recalls that a long time ago RCA developed a bacteria power source which was subsequently described in *Elektur*. The recipe was: "A mixture of brown powdered chaff of rice, together with single-celled fungi such as yeast will provide a few milliamps of current at some tenths of a volt. The mixture is put into a small plastic box. Add a little water, and collect the energy using a copper strip as the positive terminal and an aluminium strip as

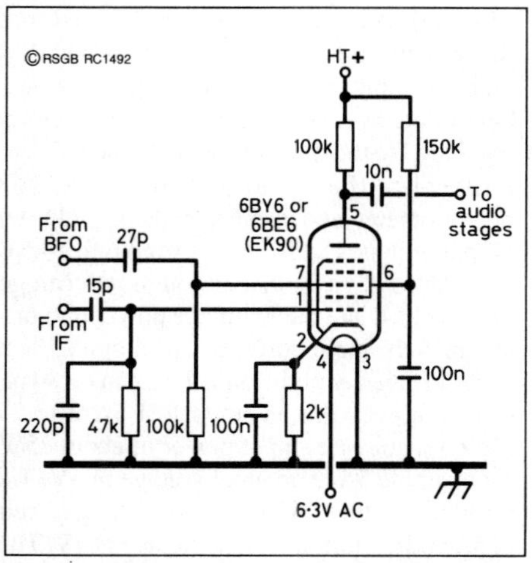

Fig 9: Classic valve-type product detector suitable for use in older communications receivers with AM diode detectors.

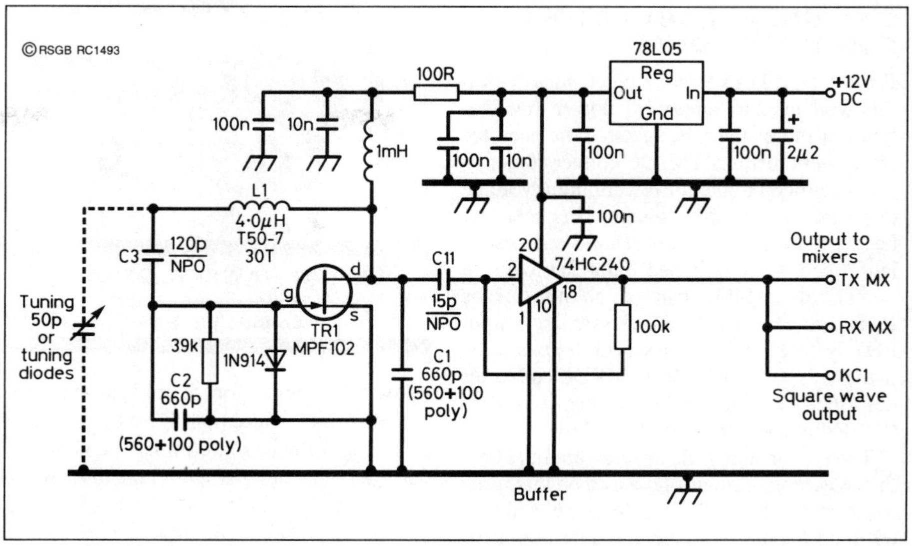

Fig 10: Basic FET Vackar oscillator used in the *QEX* design by WUOL simplified by the omission of the two voltage variable capacitance diodes he uses for main tuning and RIT. The use of a high-quality variable capacitor is significantly less susceptible to temperature changes, making it considerably easier to provide only a limited amount of temperature compensation.

the negative terminal. The process will keep itself going for several years. Twelve cells in series are sufficient to feed a small radio receiver, a little lamp or a tiny motor."

It is good to see that the ARRL Experimenter's Exchange journal *QEX* has resumed publication with the July 1997 issue after a break of six months. This issue includes 'Building a Direct Digital Synthesis VFO', 'Meet the Vacker: The simple, stable VFO you've been looking for' and 'Exploring the 1:1 current (choke) balun' by William E Sabin, W0IYH. The Vacker VFO by Mark L Meyer, WU0L, bases his VFO on a basic FET Vacker oscillator (**Fig 10**) driving an 74HC240 IC buffer providing square-wave outputs. His unit becomes rather more complex because of his requirement to use variable-capacitance diodes for tuning and RIT but admits that he has not found these to be stable without temperature compensation by use of capacitors having a negative temperature coefficient, adding: "If you wish to use a high-quality variable capacitor for the main tuning control you can eliminate this problem." Ideally, L1 should be wound on a ceramic former but WU0L used a T50-7 core wound with 30 turns of No 24 (AWG) enamelled wire to obtain 4.0mH for his VFO operating between 4.570 and 4.646MHz.

TT, June 1994, p53 showed how a 500 watt, 230V, 50Hz petrol-electric generator had been made by ZL2AXC using a discarded lawn-mower motor and a single-phase induction motor salvaged from an old pump motor or domestic appliance. Yaniko Palis, VE2NYP, in 'The 12V Pup; A DC generator you can build' (*QST*, June 1997, pp45-48) provides a detailed account of how a lawn-mower engine and an alternator can be used to provide a 12V 50A power supply for Field Day or emergency use etc. The Pup itself without a battery weighs about 45lb but with batteries, fuel, oil and cables can amount to somewhere between 70 and 100lb but can be disassembled into packages weighing less than 20lb each. Since even internally-regulated vehicle alternators are likely to put out ugly inductive spikes at a dangerous 20V or so a sizeable lead-acid battery is essential to stabilise the output to a ripple-free 14.5V. He used a modified Delco A-1238 alternator, but equivalent 14.5V, 50A units would be suitable. VE2NYP stresses the need to provide protective mechanical shielding of all moving parts. The four-page article includes full details of a suitable control box with the entire circuit protected by an internal, automotive, 50A automatic-reset circuit breaker.

In brief, Frank Trier, G8VH, found a 1949 readers ad in *SWM* offering a 'Special HRO Model MRB-3F' fitted with two Acorn valves (954, 955) in the RF stages and claimed to be factory-built rather than a home conversion. Yet in *QST* September 1936, James Millen the original HRO designer claimed that, despite enquiries, National had a variety of reasons for not using Acorns in HF receivers. G8VH wonders whether, despite James Millen's views, National Co later relented and built a few sets with acorn valves? . . . Christopher Henn-Collins (former GU5ZC) wonders whether anyone has already investigated, or is willing to do so, the use of the base of (discarded) compact fluorescent lamps as a small switched-mode power supply of up to 20 watt rating which also provides mains isolation. He writes: "The sample I have in front of me has a ferrite output transformer to drive the bulb which could easily be rewound on an amateur basis to provide the required voltages. Rectification would be no problem and, because of the high frequency the smoothing capacitors, would be quite small." He recognises that the proposal may seem heretical but worthy of consideration. ♦

Pat Hawker's Technical Topics

PAT HAWKER, G3VA
37 Dovercourt Road
London SE22 8SS

SWITCHING REGULATORS FOR MORE VOLTS

FROM THE DAYS when mechanical vibrators were used to supply HT power to valve-type car radios there has existed the need for convenient forms of DC-DC conversion, providing either a higher or lower output voltage compared to the input voltage. This need can be met basically by means of solid-state oscillators driving a switched amplifier with an output voltage built up across an inductor (as found, for example, in the classic line-output EHT system of television sets or, for example, in the low-power 12 to 24V DC-DC converter described by DJ8ES in *VHF Communications* (*TT*, February 1997, p76, Fig 12)).

Today, for many electronics applications, the switching regulator IC based on this technique is rapidly replacing the popular three-terminal IC linear regulators. The new generation of IC switching regulators overcomes the significant power loss arising from the significant input/output differential necessary with popular three-terminal linear IC regulators such as the 7805 etc. But unlike the 7800-series, switching regulators not only operate at efficiencies of the order of 80 or even 90 per cent but also can provide a DC output less than or greater than the input voltage.

An article 'My all-purpose Voltage Booster' by Sam Ulbing, N4UAU, (*QST*, July 1997, pp 40-43) carries the editorial comment, "this simple state-of-the-art switching regulator is a learning tool - and a handy gadget to have around". N4UAU introduces his booster as follows:

"Did you ever wish you could run your QRP rig using four nicads instead of a heavy duty 12V battery? Boating hams: did you ever want to recharge your laptop's or hand-held's batteries from your boat's system, without having to run the engine to get a full 13.5V? Do you need 24V for that nifty relay you bought at the hamfest? Do you want to use a 6V battery to get 12V? With the voltage booster described here, you can do all of this.

"My booster gives you a DC voltage greater than the input DC voltage. The booster is quite flexible, efficient and easy to build. DC-DC converters are usually built for a specific output voltage, and often for a limited output-current range. My voltage booster is different. It has an adjustable output voltage (up to 65V) and works with loads drawing from less than 20mA to over 3A."

With the component values specified in **Fig 1** the unit provides an output adjustable over the range 6 to 14.5V from an input voltage 4 to 13V. With some simple component changes other output voltage ranges are possible.

N4UAU emphasises the advantages offered by switching regulators over linear regulators but also draws attention to some problems, including the need to minimise RF-noise output. He points out that, until recently, switching regulators were unfortunately considered one of the most difficult linear circuits to design. "Mysterious modes, sudden seemingly inexplicable failures, peculiar regulation characteristics and just plain explosions are common occurrences" - so ran Linear Technologies Application Note 25 of September 1987.

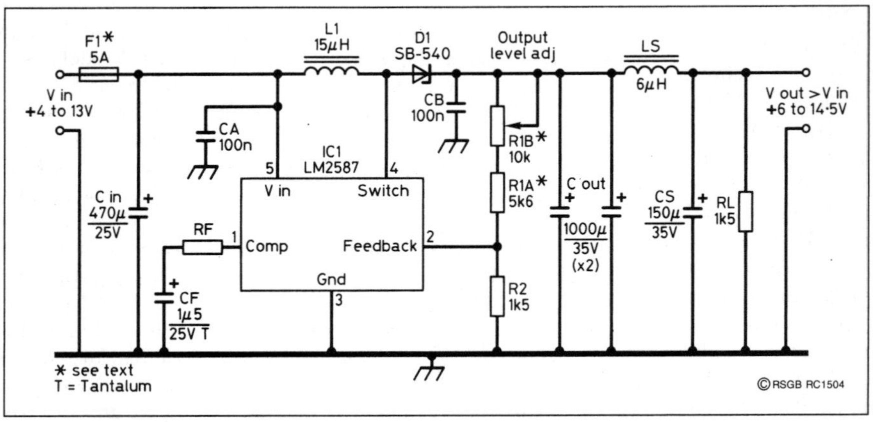

Fig 1: Circuit diagram of N4UAU's voltage-booster circuit. Unless specified otherwise, resistors are ¼W, 5% tolerance carbon-composition or film units. Use low-ESR capacitors. Component details: CA. CB 0.1uF, 35V monolithic, CF 1.5uF, 25V tantalum, COUT two 1200 or 1000uF, 35V electrolytics, CS 150uF electrolytic, D1 5A Schottky (SB540), F1 5A fast acting fuse, L1 15uH, 17 turns No 18 AWG on T-68-52A core, LS 6uH, 14 turns, no18 enam on T-50-18B core, R1B 10k trimmer pot, RF2K National Semiconductor LM2587.

Recently, however, National Semiconductor have introduced 'simple switcher modules' intended to remove a lot of the pain and aggravation for novice power-supply designers. N4UAU uses a National Semiconductor LM2587 for which free software and information on other of their Simple Switcher modules are available on the National Semiconductor Web page: http//www.national.com/sw/SimpleSwitcher/0,1043,0,00.html. However, the information provided by N4AUA should enable anyone to duplicate his unit.

The LM2587 operates like all 'boost mode' switching regulators. Pin 4 is connected to an internal switch - a high speed MOSFET that switches on and off at about 100kHz. When the switch is closed, supply current builds up through it and the inductor. When the switch opens, the voltage across the inductor rises above the input voltage because of the fact that inductors generate back EMFs to try to keep the current flowing when the connection is broken. Current is forced to flow through diode D1 to C_{out} with energy transferred from the inductor the capacitor at a higher voltage than the input voltage. The output voltage is controlled by the time the switch is closed compared to the time it is open. In the N4UAU unit with the combination of a 10K potentiometer and a 5.6K resistor (R1), the output voltage can be varied from 6 to 14.5V.

N4UAU stresses that although you can get more volts than you put in, the power output is less than the input power by an amount related to the efficiency of the unit. This can be 80 per cent or more so that an output of 1.5A at 13.5V (20W) requires an input power of about 25W, for example 5A at an input voltage of 5V. The limitation of the LM2587 is a switching current of 5A and a minimum input voltage of 4V. This means that with an input voltage of 5V the output at 13.5V must be limited to under 1.5A. However, if the input voltage is, say, 12V and you require 3A at 13.5V, then the input current will be about 4.2A, well within the LM2587 rating. For heavier loads, heat sinking of the LM2587 is necessary.

It is emphasised that the circuit components should be of high grade and carefully selected rather than drawn haphazardly from the junk box. Low effective-series-resistances (ESR), generous ripple current ratings and voltage ratings are required; similarly, the inductors must be selected for proper saturation current, inductance value, etc. The diode (D1) should be a fast recovery Schotty diode. For minimum RF noise output, National Semiconductors advises: "Keep the length of the leads and traces as short as possible. Use single-point grounding (earthing) or groundplate construction for best results. Separate the signal grounds from the power grounds. Keep the programming resistors as close to the IC as possible." N4UAU recommends the inclusion of a 5A fast-acting fuse in the input circuit.

By changing the value of R1 it is possible to obtain up to 17V output; additional component changes are needed to obtain up to 65V output.

A PC board, all board-mounted components and detailed instruction for this project are available as a kit from N4UAU for $21 plus S4 ($2 in USA/Canada) shipping costs. Foreign orders should include an international money order or US currency cheque payable at a US bank - charge cards not accepted. (Sam Ulbing, N4UAU, 5200 NW 43rd Street, Suite 102-177, Gainesville, FL 32606 USA, E-mail n4uau @afn.org.)

USING EX-MICROWAVE-OVEN EHT TRANSFORMERS

A FEW YEARS ago, *TT* ran a couple of items on salvaging transformers and other components from discarded microwave ovens: 'Cook-

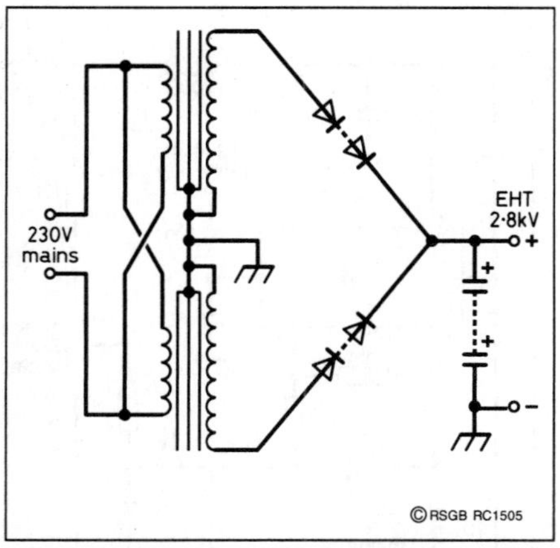

Fig 2: Using two ex-oven transformers for full-wave rectification as described by G3IZM.

ing up high power transformers' (*TT*, November 1993, p47) included notes from Dave Penny, G3PEN, reporting that a discarded oven can be the source of some very useful components including a sturdy EHT mains transformer suitable for high-power valve linear amplifiers. He showed that the EHT transformer from a faulty Toshiba Model ER-672 was capable of delivering up to about 300mA DC (half-wave rectification) at about 2000V DC.

G3PEN pointed out that microwave-oven transformers normally comprise an untapped primary winding, a well-insulated heater winding, and a single untapped HT connected to the substantial transformer core (earth). He stressed that it would be unwise to 'lift' the HT winding from the core in order to use bridge rectification since the winding would then be insufficiently well insulated from the core.

In a further item 'A 3kV 600mA Oven PSU' (*TT*, February, 1994, p47) John Harper-Bill, G3IZM, carried the debate further by using two identical 600W ex-oven transformers and four ex-oven high-voltage rectifiers: **Fig 2**. In his arrangement, the 'earthed' ends of the EHT windings formed the centre tap of the secondary windings, enabling full-wave (bi-phase) rectification to be used to provide easier and more effective ripple-filtering. In practice, G3IZM also used an input Variac, and ten 400V electrolytic filter capacitors to form the PSU for a 4CX1000 linear amplifier, variable from 0 to 3kV output at up to 600mA.

One must repeat the usual strong warning that extreme care must be taken with amplifiers and PSUs working with kV DC voltages!

It seems a good opportunity to remind readers that, as noted by Dr Ian White, G3SEK, (*In Practice*, January 1997, p79) and by Ken Stuart, W3VVN, ('Technical Correspondence', *QST*, July 1997, p68) the development of rectifier diodes with reverse-mode avalanche breakdown characteristics has brought about changes in practice. Originally, it was necessary to protect silicon diodes in series-connected strings (where unequal leakage resistance between diodes can exist) by connecting external resistor strings. The leakage resistances were swamped by the lower-value external resistors, forcing equal voltage sharing by swamping the internal leakages.

As W3VVN puts it: "For the last couple of decades, however, diode manufacturers commonly add protection against arcing by creating an avalanche reverse mode - similar to that seen in a Zener diode. With this characteristic, the diode conducts with a reverse voltage at a potential that is lower than the arc voltage. Unlike an arc, which once established, exhibits a high current and low voltage, the avalanche does not produce a low-voltage condition. Therefore, avalanche diodes connected in a series string will *not* allow a dangerous, internal junction arc to occur... resistors are *not* needed because each diode self-protects against excessive voltage. Equipment builders who want to shunt diodes with resistors should be aware that most resistors have only a 300V maximum rating. If the reverse voltage across each diode exceeds this, the resistors can fail and destroy the rectifier string!

"Shunting capacitors across diodes is not recommended. Momentary dielectric breakdown in capacitors is not uncommon, and this can overstress the other diode/capacitor pairs in the string causing catastrophic failure."

TWO FOR THE PRICE OF ONE

JOHN BRODZKY, G3HQX, has added some further notes to his "experiments with a helical-wound vertical" (*TT*, August 1997, p78) in which he showed how he had been able to obtain improved SWR bandwidth on both 3.5MHz and 14MHz for a Chelcom HFV1 helical wound vertical monopole antenna by adding a second 'resonator' 12ft rod element closely spaced (although not electrically connected). He points out that, "the main difference between the K9AY approach and my experience seems to be that the driven element in my case could be considered as a long inductance, and hence the amount of coupling between it and the added element could be expected to be considerably greater. I have measured the capacitance between the two elements as some 120pF.

"I would speculate that under these conditions, the resonant frequencies of both 'elements' are modified and could account for the very short length of the 14MHz 'resonator'.

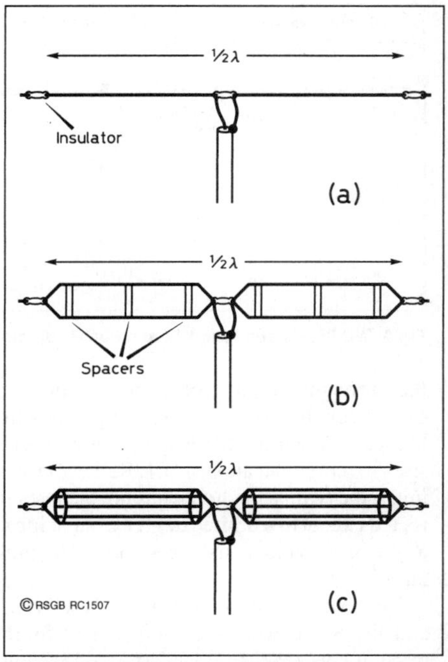

Fig 3: (a) Conventional single-wire half-wave dipole antenna. (b) Use of twin wires increases bandwidth. (c) Multiple-wire 'caged dipole' that can result in broadband operation over about an octave.

My guess is that both the resonant frequency and the Q of each element are altered by the mutual coupling.

"That the added element has a direct influence on the main helically-wound element was demonstrated when I reduced the 'resonator' length by one inch to 11ft 9½ in. The SWR at 3800kHz increased from 1.35:1 to 1.45:1, thus shifting its response upwards in frequency, when measured at the base of the antenna. Using a rather insensitive GDO, I have attempted to find the resonant frequency of the added element although the close coupling to the main element makes this difficult. I do obtain a shallow dip around 27.89MHz."

It would seem worth recalling that the bandwidth of a conventional half-wave is related to the diameter of the wire or rod and can be increased by using twin closely spaced wires. This is usually done in the form of the 'folded dipole' with the feed impedance of one wire increased to about 300Ω. But in years past, before the development of the folded dipole by Dr Kraus, W8JK, it was quite common to break and centre feed two or more wires: see **Fig 3**. In its multiple wire form, it became the 'caged dipole' or the 'biconical dipole' or 'conical monopole'.

With valve transmitters there was little advantage to be obtained by reducing the SWR much less than 5:1 (see 'Persistent SWR Myths' based on the writings of Walter Maxwell, W2DU, *TT*, December 1992, pp31-32). So, a single wire dipole would work satisfactorily across any of the European HF amateur bands, and the twin-wire dipole fell into disuse. It is important to remember that the usual formula for resonance length is intended for use with elements having length/diameter ratio of over 1000. The resonant length of 'fat dipoles' is significantly less but the low-SWR bandwidth greater. It was the coming of the solid-state power amplifier with the need to guard against excessive SWRs,

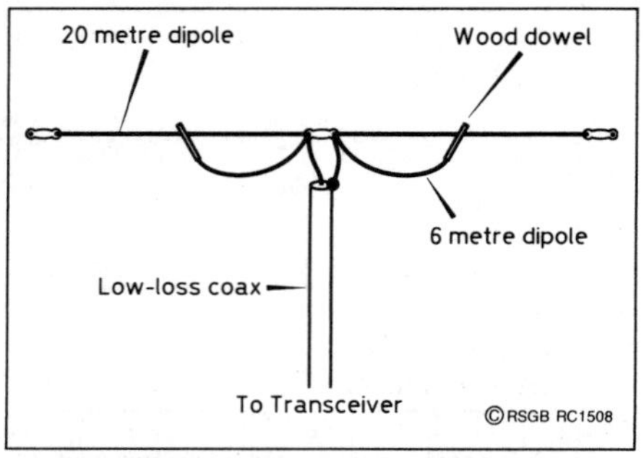

Fig 4: WB8IMY's combined 14 and 50MHz dipole.

that brought about protection circuits that often begin to reduce transmitter power automatically when the SWR at its output socket rises to more than about 1.5:1. Rather surprisingly, the twin-wire dipole has not been resurrected except to a limited degree in such forms as the broadband, resistance-loaded 'Australian dipole'.

However, as the experiences of G3HQX and also Steve Ford, WB8IMY, ('Two for the price of one' *QST*, July 1997, pp56-57) show, the presence of a second closely coupled non-resonant element, whether electrically connected or not, can result in improved bandwidth.

The prime aim of WB8IMY was to add a 50MHz capability to an existing 14MHz wire-dipole antenna by using the well-tested arrangement of simply adding a 50MHz element connected to the same feed-point as the 14MHz wire: **Fig 4**. Before adding the new element, WB8IMY plotted the SWR of his existing 14MHz wire: **Fig 5**. After adding and resonating the new element to 50.1MHz, the SWR plot on 14MHz was significantly improved: Fig 5(b). The SWR on the 50MHz was also satisfactory: Fig 5(c).

WB8IMY provides some pertinent constructional advice: "Never, ever, cut a wire dipole to the exact length the formula dictates. That's the first step on the road to Looneyville. The purpose of the formula is to get you in the ballpark... that adage should be carved in granite. Always cut your wires with a generous surplus of length to anticipate the trimming that must surely come later. With that in mind, I added about 12 inches to my calculation (6 additional inches for each wire)... The next challenge was keeping the 6 and 20 metre dipoles separated from each other. I needed something mechanically sturdy that could stand up to a stiff nor'easter. It also had to be aesthetically pleasing . . . I opted for thin wood dowels (each about one foot in length) treated with preservative. I passed the antenna wires twice through small holes drilled in each end. You might use strips of plastic or what is at hand or a multiple-spacer scheme to keep the 50MHz dipole parallel to the wire above it. Using extremely stiff wire would probably help."

VARIABLE CERAMIC RESONATOR OSCILLATORS

SOME 12 YEARS ago, Al Helfrick, K2LBA, described in the late lamented *Ham Radio* (June 1985, pp18-26, briefly summarised in *TT*, December 1985) the use of low-cost ceramic resonators as mechanically and varactor tuned oscillators on 10 and 14MHz. In due course this led John Townend, G3BBD/N3YZR, to emphasise (*TT*, February 1991) that a particularly useful feature of the ceramic resonator is the ability to swing (pull) its frequency over a significantly greater frequency range than a quartz crystal while retaining good stability. The downside is the greater susceptibility of ceramic resonators to temperature changes.

Now G3BBD writes: "My original *TT* item seems to have resulted in considerable interest and there have been a number of references to the use of variable ceramic resonator oscillators in *TT* and elsewhere since then. Readers may be interested in some further comments.

"I have found that 4MHz resonators are about the highest frequency that can be used without deterioration of thermal stability making them unsuitable for VFOs. As a result, use needs to be limited to the lower frequencies unless mixing with a stable crystal oscillator is employed.

"I have tried a number of ceramic resonator oscillator circuits using FETs and MOSFETs but still find my original 1991 design using a CMOS 4069 Hex Inverter chip hard to beat. With this device, the resonators will always oscillate and the additional inverters can be used as buffers and for tailoring the output waveform.

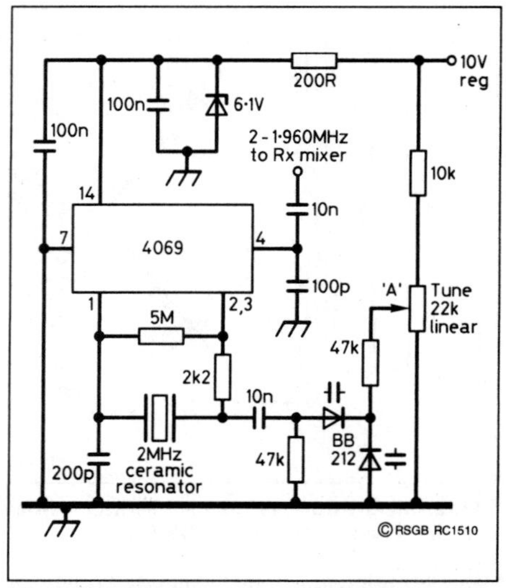

Fig 6: G3BBD's ceramic resonator VFO for 7MHz transceiver.

"**Fig 6** is such an oscillator using a 2MHz resonator which I use as the VFO for a 7MHz QRP CW transceiver having a 9MHz IF with a 4-pole ladder crystal filter. The resonator can be pulled from 2005kHz to 1958kHz which, with the 9MHz IF and subtractive mixing, covers the 7.000 to 7.040MHz CW section of the 40m band. The oscillator is very stable over long periods. Tuning, provided by two BB212 varactor diodes with potentiometer control, is smooth and has a reasonably linear characteristic.

"**Fig 7** shows the optional RIT control, a simple relay-controlled arrangement providing some 1kHz shift either side of the transmit frequency.

"It occurs to me that this ceramic-resonator circuit would provide an eminently suitable replacement for the free-running VFO in the 'Epiphyte' 3.5MHz low-power SSB transceiver designed by VE7QK and described by G3RJV in the July 1997 *RadCom* (pp17, 20-22). Using a 4.19MHz resonator and the 455kHz IF, a tuning range of at least 3.65 to 3.74MHz should be achieved. Although some earlier details of the 'Epiphyte' transceiver, in the *QRP* column of *RadCom* August 1995, showed that ceramic resonators were used as oscillators, the recent circuit (July Fig 2) shows a FET Vackar circuit for the VFO. I feel that the ceramic resonator circuit described above

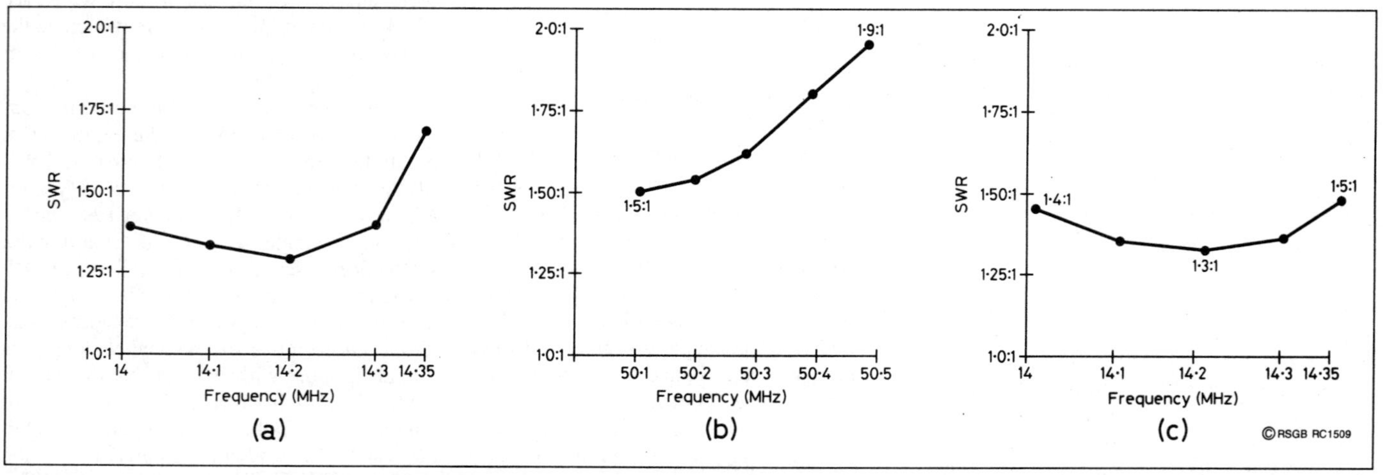

Fig 5: (a) SWR curve of the 14MHz dipole before the 50MHz addition. (b) Final SWR curve on 50MHz after trimming. (c) Improved 14MHz SWR curve after addition of the 50MHz element.

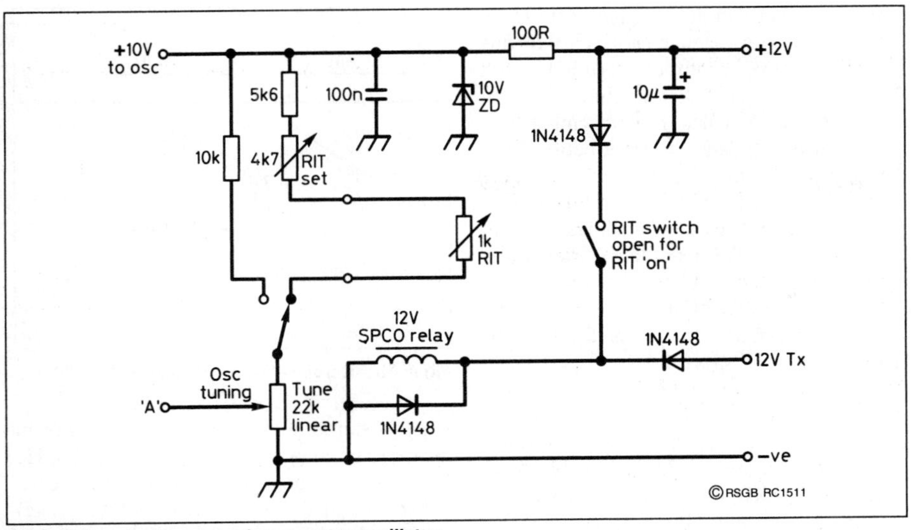

Fig 6: RIT control for ceramic resonator oscillator.

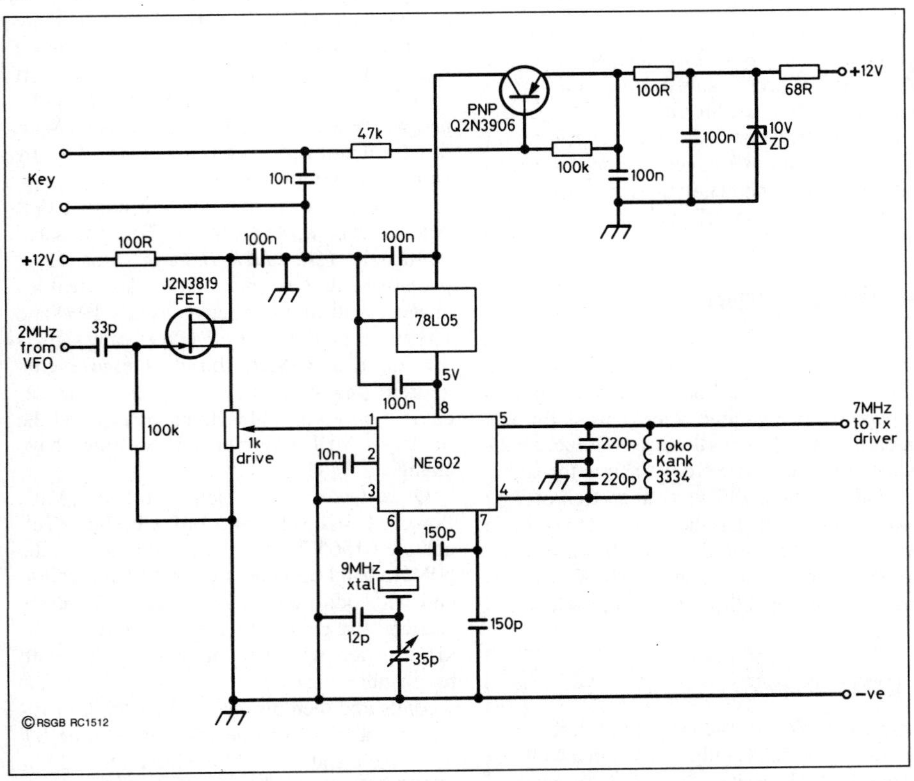

Fig 7: Keyed crystal-oscillator/mixer using NE602.

would probably be more stable than the on-board and unscreened free-running Vackar VFO. It might also be worth trying the use of other hex inverters on the 4069 chip for the 455kHz BFO although this might result in problems with signal breakthrough within the IC chip.

"Finally, another circuit arrangement which I have found very interesting and useful is a keyed mixer/crystal-oscillator using the NE602 chip: **Fig 7**. I came across this in an MFJ QRP CW transceiver and tried it in my 40m transceiver. Viewed on an oscilloscope it produces a near perfect keyed waveform with excellent rise and fall characteristics. When I first determined to try it I had doubts that the combination of a keyed on-chip crystal oscillator and a keyed voltage regulator would prove satisfactory. But looking at the waveform on an oscilloscope and listening to the signal on a nearby receiver underlined the quite remarkable absence of key clicks and splatter. I do not know if MFJ have any patent on this circuit but it would seem to be one that constructors would find very useful."

PLOTTING THE PROGRESS OF CYCLE 23

THE DOLDRUMS OF the solar minimum between Cycles 22 and 23 have lasted much longer than most people expected. In *TT*, November 1996, I noted several published predictions that solar activity should have begun to rise steeply from about July, or at least Autumn 1996 - adding the comment, "this may have been a little premature, but certainly the new cycle should be having an effect by about Spring, 1997". How wrong can one be!

But as these notes are written at the beginning of August 1997 there seems a good chance that we shall soon be seeing more consistent DX rather than Sporadic E openings on 21MHz.

In my view, one of the best - if not the best - ways of keeping track of solar conditions is by copying the transmissions from the German amateur beacon station DK0WCY on 10.144MHz which transmits continuously throughout the day and can usually be well received in the UK.

DK0WCY transmits on CW (Morse speed about 16WPM) with the propagation message sent at 5 minute intervals (05, 10, 15 etc minutes after each hour). Messages take the following form: "INFO 04 AUG 06 UTC KIEL K3 = FORECASTS 04 AUG SUNACT QUIET MAGFIELD QUIET = 03 AUGR 27 FLUX 72 BOULDER A9 + 03 AUG KIEL A24 + DK0WCY". K, R, FLUX and A numbers are sent twice.

The interpretation of this sample message is: "Solar information August 4. At 0600UTC the geomagnetic activity at Kiel, Germany for the period 0300 - 0600UTC was K3. Forecasts for 4 August is sun-activity quiet, magnetic-field activity quiet. Measurements for 3 August were Sunspot Number ® 27. Solar Flux 72. Average geomagnetic activity (A-index) at Boulder (Colorado, USA) was A9. Average geomagnetic activity (A-index) at Kiel, Germany on 3 August was A24."

The K rating at Kiel is updated every three hours, with time of origin of the message changed accordingly. I find that if these propagation messages are copied regularly, at least once per day, and then tabulated, the result gives an excellent overview of what is happening.

Details of the different scales used for the three-hourly K-index and the previous-day average A-index together with information of solar flares were given in *TT*, September 1995, pp63-64, but for convenience **Tables 1** and **2** are reproduced again this month.

Looking over tabulated details compiled daily from DK0WCY between July 12 and August 5, 1997 I note that 'sunact' was forecast as 'quiet' on all days except for July 26 and 27 when it was 'eruptive'. R numbers varied from 0 on several days to 65 on July 25. Solar flux reached a maximum of 80 on July 25 from a minimum of 67 on July 12/13. I found that HF band conditions closely followed the Kiel A index which varied between A3 and a maximum of A26 on 4 August (the only day on which is was difficult to copy DK0WCY for much of the morning on the 50-

Level	A-index	Potential for impact (solar flares)
Quiet	0-7	Low
Unsettled	8-15	Low
Active	16-29	Moderate
Minor Storm	30-49	Moderate
Major Storm	50-99	High
Severe Storm	>100	Very high

Table 1: Levels of Geomagnetic Activity.

k-index	0	1	2	3	4	5	6	7	8	9
A-index	0	3	7	15	27	48	80	140	240	400

Table 2: Conversion from K-index to the A-index

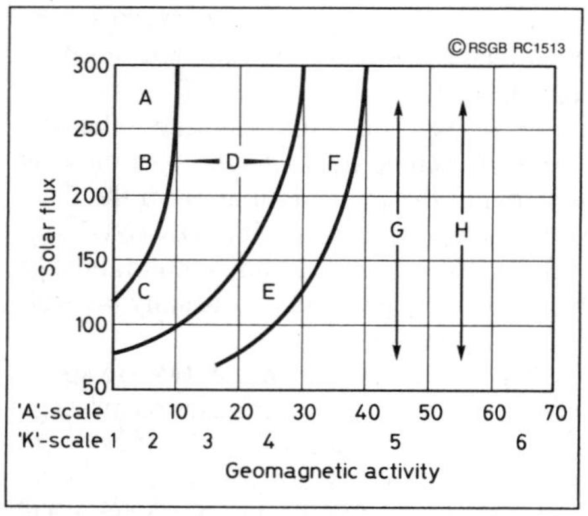

Fig 8: Chart relating solar flux and geomagnetic activity to likely band conditions as originally published by G3ENI in the TVARTS Newsletter. See text for explanatory notes.

year-old HQ129X I use for this purpose). Some British amateurs rely on Boulder data from WWV or WWVH but it is clear that the A-index from Kiel often differs significantly from that measured at Boulder. *Late extra*: By the end of August/early September Solar flux numbers of over 90 were common (96, Aug 31).

In the April 1992 *TT*, p40, John Pegler, G3ENI, provided a handy guide to translating basic propagation data into likely band conditions. Two main items are considered. First, the 'solar flux' which is noise measured on 2800MHz each day and which ranges from a minimum of about 66 units or less during sunspot minima to about 300 at the peak of a really high sunspot cycle. Good conditions on the higher HF bands require a high flux, preferably over 90-100. Second, 'geomagnetic activity' which acts against the benefits of a high flux; it may be presented in either A or K units. The K scale is based on readings taken eight times a day, each covering the previous three hours. The A scale is based on the eight K values for the previous day.

Fig 8 shows roughly the relationship between solar flux and geomagnetic activity in determining likely band conditions. On this chart, solar flux is plotted vertically with geomagnetic activity shown horizontally in both the K and A scales. But remember that, apart from the K data, the information from propagation stations is based on the preceding day, with significant variations occurring from one day to the next - as soon becomes apparent if one copies DK0WCY regularly. Remember that the higher the flux the higher the likely usable bands. But the higher the A or K measurements the worse the likely signal conditions throughout the HF spectrum.

In the diagram, the intersection of the flux and geomagnetic activity values determines likely band conditions, divided into eight zones: Zone A: Above normal, 50MHz probably open, 21, 24 and 28MHz open for up to 24 hours. Zone B: Above normal, 21 24 and 28MHz open for up to 24 hours. Zone C: Bands up to 28MHz open for part of the day. Zone D: Normal conditions. Zone E: Below normal, unsettled. Zone F: Below normal, disturbed. Zone G: Disturbed, sub-storm level, auroras may form. Zone H: Storm level, auroras. But remember that a sudden solar flare, etc may change conditions markedly. For a refresher on solar flares and magnetic storms see 'Solar Flares - predicting effects', *TT*, September 1995, pp63-64.

HERE & THERE

JOHN BRODZKY, G3HQX, draws attention to a novel form of 3.5MHz OCFD antenna, known as the 'Carolina Windom' which is apparently being quite widely used for DX operation. **Fig 9** shows the basic outline, a half-wave horizontal wire fed off-centre through a 4:1 balun from a 23ft vertical section of 75Ω coaxial cable, which is then isolated by a choke wound on a ferrite rod or ring from the 50Ω coaxial transmission line. G3HQX believes that the DX capability of this antenna arises

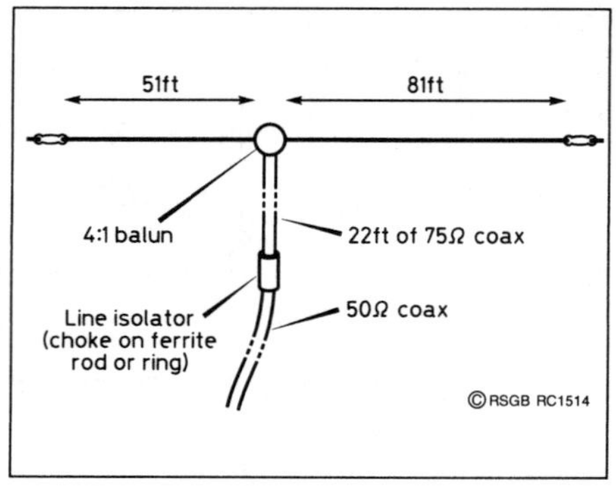

Fig 9: The 'Carolina Windom' antenna for 3.5MHz.

largely from vertically-polarised radiation from the outer braid of the 23ft vertical section. He comments that the system reminds him of the 14MHz vertical-Tee (inverted, ground-plane) antenna that I introduced in *TT* back in July 1970, pp450-451 following its successful use at G3VA and G3HQX, derived from the more elaborate 'Bobtail' array. I wonder if anyone can refer me to a published description of exactly how this antenna functions and who and where it was originated? 'Carolina Window' antennas are marketed in the UK by Haydon Communications.

Almost 60 years ago, the still-missed Bert Allen, G2UJ, wrote an article 'Four years and no signals' (The *T&R Bulletin*, June 1938, pp688-689 & 716) in which he described his unsuccessful attempts between early 1934 and 1938 to receive at Tunbridge Wells any signals on the old 56MHz band. Despite this unpromising start, Bert later went on to become a post-war 144MHz enthusiast and the Society's VHF columnist. How times have changed.

Quite the reverse problem is bothering Mike Powell, G3IJE, of Knutsford, Cheshire (Telephone 01565 733530) who complains, "the 50MHz band up here is polluted from end to end with odd carriers and noises. There are carriers that gently drift up and down a few kHz, carriers with data on them, carriers with rough notes, carriers that pop up for a few seconds and then disappear. You think of it - we've got it! I live line-of-sight of three BT microwave and several broadcast TV and FM VHF/UHF stations."

I wonder if the problem is intermod garbage from these stations? Presumably, C3IJE has checked carefully that the problem does not rise from intermodulation products (IMPs) in his receiver front-end. But certainly telecommunication and broadcast multi-transmitter sites do radiate IMPs arising from their transmitters and/or antenna systems. There could also be a problem of harmonics from the 27MHz ISM band, etc, etc. One wonders whether the new digital broadcasting systems with their multi-carrier COFDM could be radiating a mass of out-of-band IMPs - probably well down but possibly audible to those searching for weak VHF signals. He wonders if this problem is widespread. He would also be interested to know whether anybody knows of any commercial developments on how to eliminate the effects of 'static rain' which affects several stations in his local 50MHz FM net, even when using antennas that are shrouded inside a plastic tube. ♦

POLYSTYRENE COIL DOPE

ARISING OUT OF G8DPS's warning (*TT*, June 1997, p71) of the toxicity of carbon tetrachloride coupled with his suggestion that cellulose thinners could be used in a well ventilated area to dissolve polystyrene, J Wilshaw, G3MIX, writes: "Is polystyrene in short supply I wonder? I have a feeling that the white moulded packing containers used to transport Japanese transceivers is expanded polystyrene; similarly insulation/ceiling tiles, although this may have changed to meet fire regulations."

I should have reminded readers that the topic of D-I-Y coil dope was aired at some length in *TT*, October, 1993 when it was stressed by G2DYK that the polystyrene cement made by Humbrol and sold by model shops is ideal; while G2ARU added that the general-purpose polyurethane varnish similarly to be found in model shops is a useful replacement for coil dope, although taking some hours to harden. G0PRZ described how to make your own styrene in a small still, adding that genuine oil of turpentine (not turpentine-substitute or white spirit) is a suitable solvent for polystyrene.

The August 1993 *TT* included an item culled from *Amateur Radio* in which VK3IY noted that "there is an abundance of waste expanded-polystyrene in the form of 'Styrofoam' which dissolves readily in turpentine, although it takes quite a large amount of Styrofoam to make a little coil dope".

Jorge Dorvier, EA4EO, aware of the danger of asbestos widely used in the electrical industry for many years, wonders whether there is any danger from the fibre-glass used to fabricate PCBs. While the chemical composition is quite different from asbestos, he notes the fine powder generated when PCBs are cut and drilled. He recalls meeting a veteran dentist with a chronic lung disease brought about by the fine tooth powder generated from his dental lathe. Just in case, EA4EO when working on PCBs wears a protective mask of the type used today by dentists and industrial painters.

TUNED LOOP TRANSMITTING ANTENNAS

Pat Hawker's Technical Topics

PAT HAWKER, G3VA
37 Dovercourt Road
London SE22 8SS

IT IS EXACTLY 30 years since *TT* reported (November 1967, pp732-733) the development at the US Army Limited War Laboratory of an efficient 2-5MHz octagonally-shaped (eight 5ft tubes) transmitting loop antenna for use in the jungles of Vietnam. This was claimed as "usually doing as good a job as a full-length half-wave dipole 40ft above ground".

It was this development, soon taken up by amateurs, military and diplomatic HF users throughout the world, on which all the many subsequent compact transmitting loops have been based. It was emphasised that in order to overcome the extremely low radiation resistance of electrically small loops it was essential to reduce the RF ohmic resistance to a very low figure and to minimise losses in coupling RF into the loop. For HF diplomatic links, commercial models using copper tubing of up to about 6in diameter were used, with vacuum-type tuning capacitors. Amateur designs were soon increasingly based on the use of a small matching loop.

Most descriptions of loop antennas repeated the claim that, if well constructed, performance at the higher end of the frequency range could approach that of a half-wave dipole well above ground. It was not until 1991 that Tony Henk, G4XVF, FIEE in a clearly expressed two-part article 'Loop Antennas - Facts not Fiction' (*Radio Communication*, September/October, 1991) threw well-reasoned doubt on the basic efficiency of small loop antennas. Efficiency is difficult to measure accurately but G4XVF based his study on calculating Q from the measured bandwidth of a small loop whose inductance could be calculated. He concluded that radiation efficiency as a percentage was often in single figures (significantly below 10 per cent) whereas that of a resonant and well matched half-wave dipole approaches 100 per cent.

But there has continued to be difficulty in relating theoretical efficiency calculations, as exemplified by G4XVF, with the on-air results achieved by some amateurs using either home-brew loops or, more recently, the loops offered for amateur use by a number of manufacturers. Four models from three manufacturers were reviewed by Peter Hart, G3SJX in 'Loop Antennas for the HF bands' (*Radio Communication*, July 1994, pp41-44). He concluded: "I was really quite surprised at how effective these small loops could be. Being roughly equivalent in performance to a dipole or a multiband vertical, effective performance on the HF bands is still feasible when space for other antennas is not possible."

A most interesting new proposal about how small tuned loop transmitting antennas work is put forward in a paper 'Magnetic Loop or Small Folded Dipole' by Professor Mike Underhill, G3LHZ and M J Blewett, G4VRN of the University of Surrey (*IEE Conference Publication No 411*, pp216-225). The paper, presented at the 'HF Radio Systems and Techniques' IEE conference in July 1997, is based on theoretical and practical studies primarily centred on the Model AMA-3 compact loop antenna made by the UK firm Advanced Antennas and Ancillaries (one of the loops reviewed by G3SJX in 1994).

In their introduction the authors write: "The single turn tuned loop antenna is a remarkably effective small transmitting antenna . . . Effective performance can be obtained for loops with diameters down to as little as two per cent of the wavelength. For the HF band of 1.5 to 30MHz two per cent corresponds to loop diameters of 4m down to 20cm. For deployment in the confined space allowed to many, if not most, of the users of the HF band such a small size is of great importance . . . The main disadvantages are its narrow fractional bandwidth and, when used for transmitting, the high RF voltages which occur across the tuning capacitor. Both disadvantages are a consequence of the intrinsically high Q of this and any small antenna.

"A further important feature is that a tuned loop antenna can have a frequency tuning range of up to 3 to 1, depending primarily on the minimum-to-maximum capacitance ratio of the tuning capacitor. Furthermore, the on-tune match resistance is remarkably constant over the whole of the tuning range provided that the antenna is not used in an electromagnetically confined space . . .

"Most of the evidence for the 'effective' performance of the tuned loop is anecdotal and qualitative, not quantitative . . . Theoretical predictions remain pessimistic with respect to the performance actually being achieved by users, but it is not easy to find evidence adequate to be able to prove whether there is a real discrepancy between theory and practice . . .

"The main purpose of this paper is to pro-

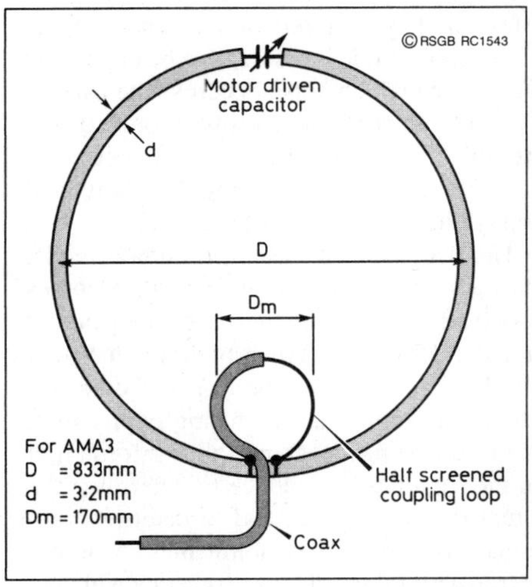

Fig 2: Representative tuned loop antenna with dimensions of the AMA3 model.

'UGLY' CONSTRUCTIONAL TIP

D M MACKENZIE, GM4HJQ, passes on a constructional tip that he uses to facilitate the 'ugly board' deployment of surface-mount components. He writes:

"The German company 'Wainwright' manufactures small pieces of self-adhesive PCB with land patterns suitable for a wide variety of components. These can be placed on a plain cooper board and inter-wired to build prototypes or 'one-off' units rapidly. The pads are available in the UK from Wessex Electronics but since they are intended for professional use they tend to be expensive by the standards of amateurs whose time is not costed!

"I manufacture my own Scottish (low-cost) equivalent by sawing Veroboard with a Junior Hacksaw as shown in **Fig 1** to form strips of component-pads or busses. These can be cut to size as required with snips and then superglued to a plain copper board ground-plane. Surface-mount resistors and capacitors etc fit easily between the pads or form a pad to ground. It does not take much imagination to see a number of possible variations.

"It would be convenient if some company such as Vero produced this type of product at a reasonable price geared to amateurs, but in practice it does not take long to make a small stock of parts. The next problem to solve is how to cope with fine pitch ICs with lead spacings of one-twentieth or one-fortieth-inch. For these one would need 0.025in Veroboard!"

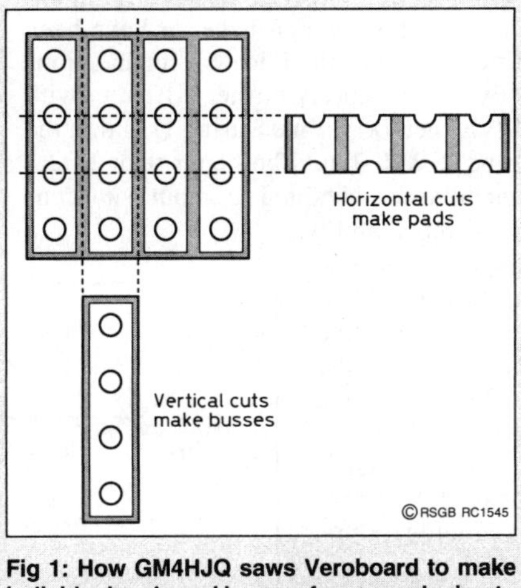

Fig 1: How GM4HJQ saws Veroboard to make individual pads and busses for supergleuing to a copper board for easy production of 'one off' prototypes as a form of 'ugly' construction.

…e an additional and different mode of operation for the loop antenna and to give some initial evidence that this 'small folded dipole' mode is usually the dominant mode. The dipole mode has a polarisation and radiation pattern which differs significantly from the normal loop mode. However, as will be seen, this difference can in practice be expected to be less apparent in the confined environment of typical deployment and operation."

The AMA3 antenna is intended for use over a range of 14 to 28MHz and has a diameter of 833mm and 16mm tube radius: **Fig 2**. At 14.0MHz the loop resonates with a capacitance of about 73pF and is shown by the authors to have an inductance of some 1.79µH. At 14MHz the radiation resistance is calculated to be only 0.0439Ω, increasing with the fourth power of the frequency, becoming 0.7024Ω at 28MHz. They conclude that for this antenna the loop-mode of radiation cannot contribute significantly to radiation at 14MHz, although at 28MHz some significant loop-mode radiation could be expected.

How, then, can the AMA3 loop antenna work quite well, as it undoubtedly does, on 14MHz? The authors propose that a dipole-mode rather than a loop-mode of radiation is usually dominant for small tuned-loop antennas. They believe that the antenna works, in effect, largely as a very small folded dipole. "The dipole mode is assumed to be excited by the voltage which appears across the capacitor when power is transformer-coupled into the tuned loop from the source connected to the subsidiary loop." See **Fig 3** for a typical voltage distribution on a tuned loop.

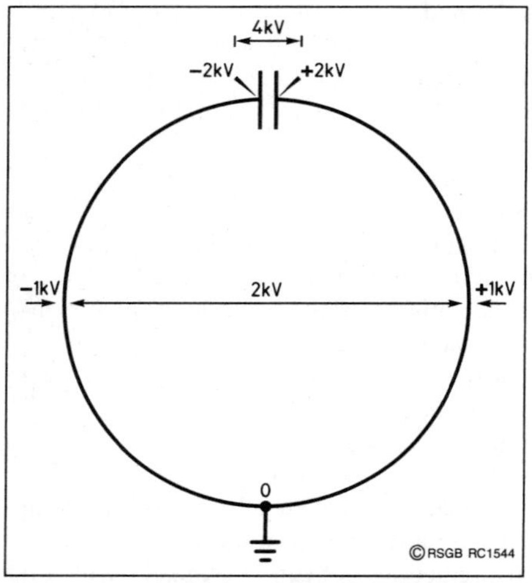

Fig 3: Voltage on the tuned loop antenna.

They calculate that in the loop-dipole mode, the radiation resistance at 14MHz becomes 0.484Ω (some 12 times that for the loop mode) and at 28MHz 1.94Ω (nearly three times that for the loop mode). The paper provides a detailed if somewhat confusing explanation of these findings, together with information on their attempt in practice to determine the actual radiation mode, and to determine antenna efficiency. They admit that, "it is not easy to determine antenna efficiency with any accuracy".

Using a network analyser with the AMA3 loop in a screened room, the authors obtained a Q value of 790 at 14MHz, giving a worst case dipole-mode efficiency of 68% (corresponding to a reduction of only 1.7dB on an antenna radiating with an efficiency of 100%). They add: "This predicted value is probably pessimistic. Some evidence is that the AMA3 antenna does not get perceptibly warm to the touch even after prolonged operation with 400 watts PEP of peak SSB speech power."

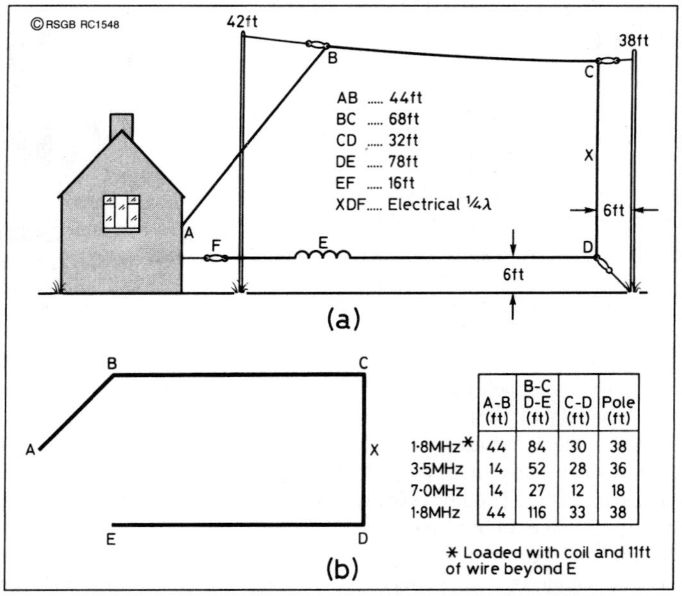

Fig 4: The original G8ON antenna providing some vertically polarised radiation from relatively low antennas, primarily for use on 1.8MHz.

The paper discusses in some detail the practical difficulties of using SWR/bandwidth measurements to determine the Q and/or efficiency of an antenna. "In principle this is a very good method for measuring the Q of a tuned loop antenna. In practice, because the antenna has a high Q, any matching unit losses markedly degrade the measurement accuracy. In fact, none of the commercially available AMUs were found to be sufficiently low loss. The main problem would appear to be contact resistance in the variable capacitors. On receive, or for the low powers used in network analysers, 'roller coaster' mechanically variable inductors are unusable. Contact resistances were found to be highly variable and a few watts of power obviously are needed to 'microweld' the contact points into a low resistance condition. The best variable capacitors are those which have, or have been modified to have, flexible wire connections directly to the rotor. A further difficulty with the SWR method for measuring Q is that the 'on tune' SWR should be better than 1.02:1 for results to be obtained to about 3% accuracy and repeatability."

It is pointed out that the L-network AMU originally described by G3LHZ in 1981 and which has been shown in *TT* several times (most recently *TT*, May 1993, pp54-55) is suitable for this purpose.

They conclude: "The tuned loop antenna does not appear to operate significantly in its loop mode. The small folded dipole mode proposed gives much better agreement between theory and measured results. It provides an explanation of why the loop antenna performs so well in practice . . . The radiation resistance of the loop antenna is sensitive to the environment but no more so than for a full size dipole or monopole. The tuning point is almost unaffected by environmental changes . . .

"Because the dipole mode in practice always is the dominant mode and never the loop mode, it is proposed that the term 'magnetic loop' antenna should not be used. The term 'small folded dipole' is a better term to use. However, the term 'tuned loop' antenna is more generic. It would remain correct if new modes of loop or dipole excitation prove to be possible with new, as yet, unexplored tuned loop arrangements."

It is possible here only to outline this interesting contribution (with its detailed mathematical support) to compact transmitting tuned-loop antenna theory and practice. In putting forward the suggestion that these antennas radiate primarily as small folded dipoles rather than in the magnetic-loop mode, G3LHZ and G4VRN appear to have shown for the first time how it is feasible for a well-

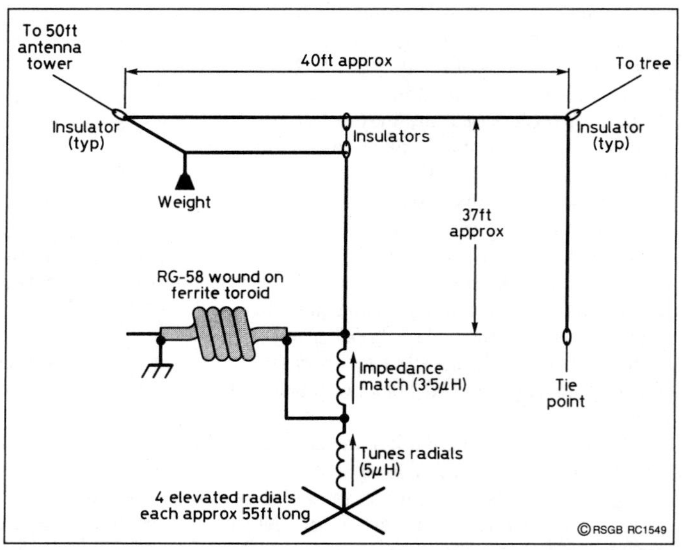

Fig 5: The 1.8MHz 'DX' antenna as used at S92SS. Inductance values are approximate.

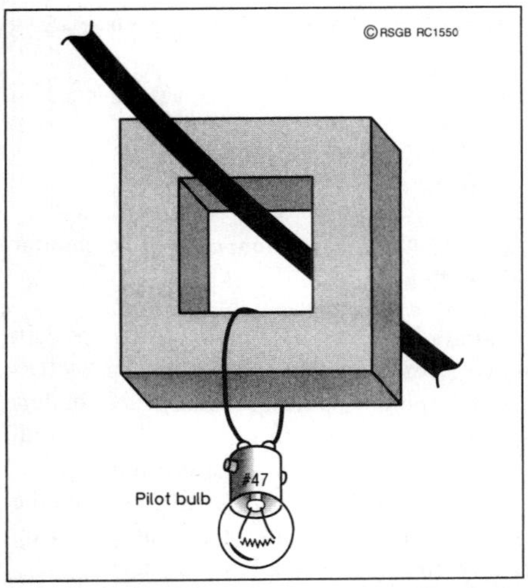

Fig 6: A low-cost RF current detector using ex-TV split ferrite core. Keep power low to avoid burning out the bulb.

constructed and well-matched loop really to compare well with a full-size half-wave dipole - a position from which it was to a marked extent toppled in 1991 by G4XVF.

I drew the paper to the attention of G4XVF who points out that it is not an easy paper to follow, contains many typographical and some fundamental errors and generally bears the hallmarks of a conference paper designed to impress rather than elucidate.

But to my mind, the important thing is whether or not the basic proposal of significant dipole-mode radiation from a tuned loop is fact or fiction. Hopefully that will become clear in time.

ANTENNAS - HERE & THERE

MANY YEARS ago, the late H S Chadwick, G8ON (father of RSGB Council member and past president Peter Chadwick, G3RZP) developed a 1.8MHz voltage-fed (end fed) half-wave antenna which could be fitted into a restricted space. It had an 84ft horizontal span at a height of some 40ft but with an all-important vertical 30ft section in the maximum current section. It did not require an earth connection but included a loading coil to bring the wire to half-wave resonance: see **Fig 4**. *ART7* (pp299-300) provides some further details. It tends to be most effective in sites having poor earth conductivity.

Another 1.8MHz limited-space antenna as used successfully by S92SS appears in the 'Radio Fundamentals' column by Bill Orr, W6SAI, in *CQ*, August 1997, pp85-86. A quarter-wave of wire includes a vertical, current-carrying section of about 37ft with the remainder of the wire folded around: see **Fig 5**. S92SS also found that with voltage feed the antenna works well on 7MHz. To break the ground return path to the transmitter, he wound as many turns as practical of the RG-58 feedline on a ferrite toroid salvaged from a TV set flyback transformer.

During the development of his limited space antenna, S92SS made use of a simple and inexpensive RF current detector: **Fig 6**. He used a rectangular split-toroid which could be clipped around the conductor under test and the loop with a small pilot bulb. While using low power to prevent burning out the bulb, he was able to check that current at the feed-point divided reasonably between four radials.

'A short, two-band vertical for 160 and 80 metres' is described by Paul Carr, N4PC, in *CQ* (April 1997 pp20, 22-23) including remote switching for the antenna tuner located at the base of the antenna: see **Figs 7** and **8**. N4PC used only an 8ft ground rod but is convinced that results would be further improved by adding a large number of buried radials [or a few elevated radials - *G3VA*].

Stewart Cameron, GM4UTP, was very interested in the GEC-Marconi 'Tetrahedral Antenna' (*TT*, September 1997, pp69-70). He writes: "So much so that I scaled it down and built it as a 144MHz antenna.

"Results were: Frequency coverage 142 to 161MHz to the 2.5:1 SWR points, with unity SWR at the centre.

"Then as an experiment I removed the six horizontal elements shown as A in the September *TT*, the antenna becoming more like a conventional VHF biconical dipole. Frequency coverage increased to 130 and 170MHz with the SWR reducing to 1.7 at these limits, making it a very useful antenna for both the 144MHz amateur band and the 156MHz maritime mobile band."

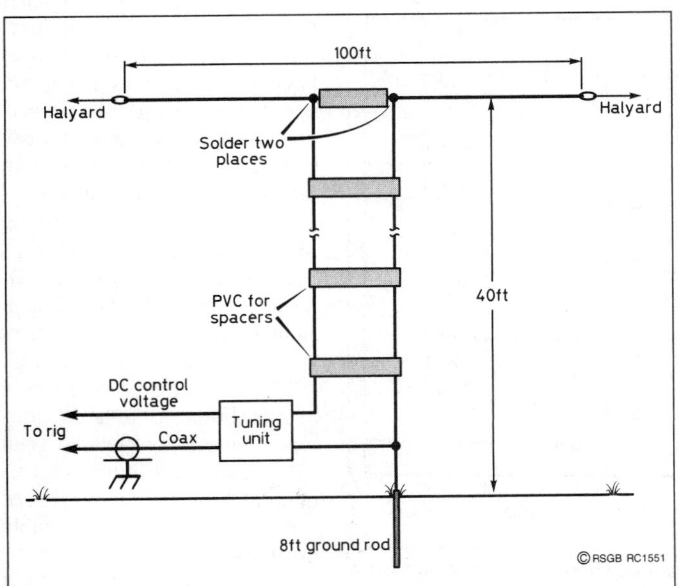

Fig 7: Details of the 1.8 and 3.5MHz vertical antenna as described by N4PC.

Richard Formato, K1POO, in 'Design Wideband Antennas' (*Electronics World*, October 1997, pp825-829) provides detailed guidelines on adding impedance loading - resistance and reactance - to antennas in order to create wideband antennas. He gives a design example for continuous coverage from 6 to 150MHz without resorting to tuning and matching. He points out that in 1961, E E Altshuler pioneered resistance loading of antennas by building a centre-fed dipole whose input impedance was, for practical purposes, flat over a 2:1 frequency range by inserting a single 240Ω resistor in each arm a quarter-wavelength from the end. The snag was that for transmission the radiation efficiency was reduced by about 50% with almost half the RF power absorbed in the resistor. [A much earlier use of resistive loading to increase bandwidth at some loss of radiation efficiency was the T2FD (Terminated Tilted Folded Dipole) antenna developed by the US Navy. Another impedance loaded antenna is the 'Australian Dipole' which has similarly been described in *TT*, most recently in June 1987, and in *TT Scrapbook*, p171 - *G3VA*]. K1POO has developed computer software to optimise the design of impedance loaded wideband antennas of reasonable efficiency and this is being offered through *Electronics World*.

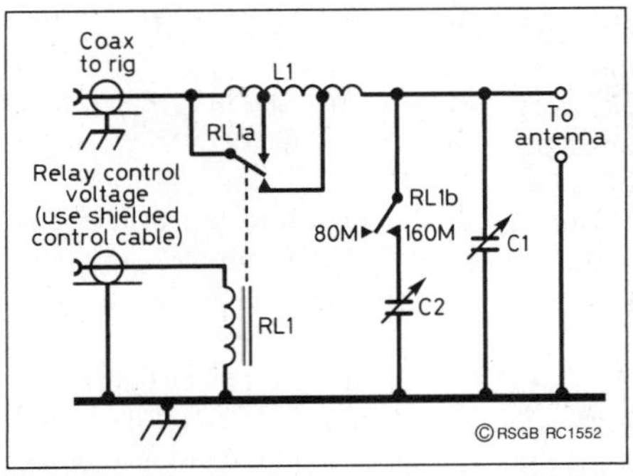

Fig 8: Remote two-band switchable antenna tuner as used by N4PC. L1 10-in long Nr 14 AWG wire, 2½-in diameter, 6 turns per inch (original used Air Dux 2006T). C1, C2 200pF, 4.5kV rating for high power, 2kV for low power. RL1 DPDT relay with 10A contacts.

ANTENNA EFFICIENCY

FROM TIME TO time, Lew McCoy, W1ICP, reminds us that antenna efficiency is not a question of gain, radiation patterns or directivity but is nevertheless an important factor that is too often overlooked. *In Antenna 'Efficiency' - What is it?* (*CQ*, April 1995, pp14, 17 and 18) and *Antenna Efficiency One More Time* (*CQ*, February 1997, pp18, 19), he stresses that the most efficient antenna, which can radiate about 97% of the power fed to it, is the basic half-wave resonant dipole. Compare this with some mobile and aircraft antennas which at about 2MHz may radiated only about 10mW for every 100 watts output of transmitter power.

W1ICP is primarily concerned with the RF ohmic losses in antenna elements, but remember that wasted power may also be a feature of any matching unit and the feeder.

As a simple example, he points out that a beam antenna with a 3dBd gain, but an efficiency of 50%, will radiate in the favoured direction only the same power as a dipole having a near 100% efficiency and consider-

less power in other directions [the directivity of the beam may, however, be advantageous for combating the reception of unwanted signals and in its low vertical angle of radiation - *G3VA*].

But why should a beam be less efficient as a radiator than a half-wave dipole? And tend to be even less efficient within a few months (or even weeks) of being erected?

W1ICP stresses that the power that is fed into an antenna element is spent or dissipated in two ways. The RF ohmic resistance of the wire or tubing (greater than the DC ohmic resistance due to skin effect) results in power being dissipated in heating up the element or any traps, earth losses, rusty connections etc. Only the remaining power is spent in the 'radiation resistance' of the antenna - in other words launched into space. Efficiency depends on how much greater is the radiation resistance than the loss resistance.

Whereas the radiation resistance of a half-wave dipole a half-wave above earth is about 73Ω, that of a close-spaced Yagi array may be as low as 4 or 5Ω (which is why some form of matching is required between a coax feeder and the driven element). The ratio between radiation resistance and RF ohmic resistance will almost always be much lower than for a dipole.

As W1ICP puts it: "A beam with 5 - 10Ω impedance is certainly going to have RF ohmic losses as an important factor . . . telescoping element connections, possible boom losses, the matching network itself, and so on. With a 10Ω impedance made up of, say, 5Ω radiation-resistance and 5Ω RF ohmic resistance, then we are looking at 50% efficiency. Although aluminium tubular elements are good conductors, depending on the atmosphere, can get corroded and 'scummy'. If this happens, ohmic losses can skyrocket, and performance and efficiency can go to pot: "If your beam has been up a long time, take it down and clean and polish all connections."

Salty and industrial atmospheres can result very quickly in deterioration of sliding joints in antenna elements; moisture ingress into coaxial feeder cables and baluns etc may significantly increase losses and lower overall efficiency. Traps and loading coils used with electrically short elements need to be of excellent high-Q construction if they are not to reduce efficiency seriously.

W1ICP notes that delta-loop or two-element quad-type (wire) antennas with one-wave or longer elements have relatively high impedance largely comprising radiation resistance, resulting in efficiencies of the order of 90% or more, a significant advantage over a typical tri-band Yagi.

For all types of antennas, VHF as well as HF, he stresses always solder any antenna connections, particularly wire antennas: "Don't be lazy and just twist leads of wire together. It only takes days for unsoldered connections to corrode and resistance to sky rocket."

Nearer home, Mike Gibbings, G3FDW, has been endeavouring to find effective ways of measuring the true power gain of beam antennas that take account of their efficiency - something which is seldom attempted by manufacturers or by those using NEC programs. He proposes that instead of dBd, a new unit is required - dBby or gain in decibels in your own back yard.

He noted the case of a local amateur with a three-element tribander of well-known make: "One day in the depth of winter the beam was seen to be covered with a thick coating of ice, after using the beam for an hour, the ice had all melted from the six traps, whereas the sub-zero weather still held the rest of the beam in its icy grip. As the transmitter was running only 100 watts he was surprised. As there had been no change in the SWR it was evident that traps could lose in heat a considerable amount of input power . . . The moral of this tale is that a prudent amateur does not spend hard cash to put a dummy load on the top of his mast. As G6XN has pointed out, a two-element beam may give more dBd (and dBby) gain, certainly more dBd per dollar, than a three element beam at twice the price."

MOV AS PROTECTION AGAINST SUPPLY FAULTS

PETER GRAVES, G0KSJ, has found that metal-oxide-varistors (see *TT*, June 1997, pp64.69) can do more than protect against mains transients (spikes). He fitted one to his burglar alarm in the hope of preventing false triggering. In the event the MOV saved the device from destruction when a faulty underground supply cable delivered 415V instead of 230V to his home for a period of more than one hour, wreaking havoc on a number of electronic appliances. These included VCRs and radios (in which the transformers remain powered even when is switched off) as well as several appliances using rechargeable batteries such as power screwdrivers etc. Several neighbours lost their central heating controllers and gas valves. But the MOV saved the burglar alarm by disintegrating and blowing the 3A fuse in the supply to the alarm.

All this happened one early December morning and it was evening before the supply was restored. It took three householders to advise the supply company that there were fault conditions before any notice was taken - presumably since supply companies are convinced that consumers know nothing about electricity. While compensation was paid by the supply company, this did not reflect the inconvenience caused.

All replacements have been fitted with MOVs, a move justified when it was discovered at the Chelmsford club that there are several hundred such supply faults in the UK every year [on the day this was written my own electricity supply was cut off during the entire afternoon while an emergency crew were busy digging up the pavement, but at least I do not think there were any prolonged over-voltages - *G3VA*].

Protocols Tested	Modem	Computer Interface
AmTOR	A&B	Serial port
Clover	C	PCbus(plug-in card)
G-TOR	D	Serial port
PacTOR	A&B	Serial port
PacTOR2	B	Serial port

Table 1: Protocols, modems and computer interfaces.

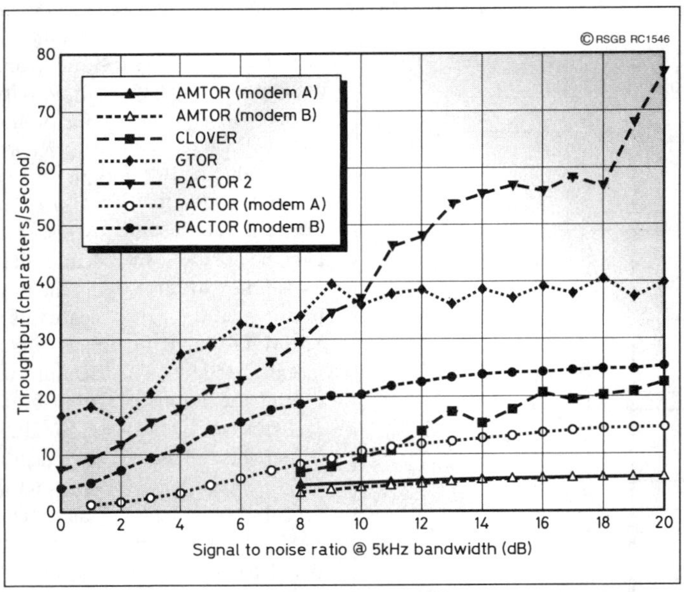

Fig 99: Comparison of the throughput versus signal-to-noise ratio of digital data protocols under simulated HF 'CCIR Good Conditions' as found at the US Institute of Telecommunications Laboratories, Boulder, Colorado.

PROFESSIONAL MORSE DECODER

FOR THOSE WHO believe that Morse is now virtually obsolete, it may be worth drawing attention to the fact that Rohde & Schwarz have recently introduced a new professional Morse Radio Decoder GM094 which serves for automatic detection of Morse messages under the particular conditions of short-wave propagation and manual keying: "It relieves personnel from tedious routine tasks. Its particular advantages are that it can work around the clock without fatigue and that it runs a classification procedure distinguishing Morse from other signals."

Introducing the decoder, Dr Klaus Rieskamp in 'New dimensions in Morse decoding', *News from Rohde & Schwarz*, No 155, 1997/III, writes: "Morse has a long tradition as a reliable short-wave transmission technique and still remains an important medium along with more advanced techniques. This is presumably due to two factors:

(1) Compared to other kinds of signal, Morse signals using ro-

bust amplitude modulation are affected relatively little by the interference present on the transmission path.

(2) Neither transmitter nor receiver is operated automatically but by trained operators. Routine and experience enable them to compensate for interference on the transmission path as well as for any shortcomings by the person they are communicating with.

He adds: "Because Morse continues to be widely used, its monitoring is an important focus. The nature of Morse, however, makes monitoring the signals highly labour-intensive . . . GM094 achieves excellent decoding results in real operating conditions."

But it is clear from the article, that to achieve these results requires a highly professional, relatively high-cost set-up. Thanks to G0GSF, G3KHZ and RS 92452 for drawing my attention to this item.

COMPARING DIGITAL PROTOCOLS

ALTHOUGH, LIKE many other CW operators, I am not a fan of HF packet - or at least those digital operators who insist on clobbering the amateur beacons or placing continuously transmitting packet stations increasingly far into the CW sectors of the HF bands - it must be conceded that the popularity of data modes continues to expand exponentially and has contributed to a resurgence in the use of the HF bands for professional as well as amateur operation. But the problems experienced under typical HF propagation conditions have in recent years led to the introduction of a number of ever-more complex, non-compatible protocols.

The IEE 'HF Radio Systems and Techniques' *Conference Publication No 411* includes 'A Comparison of HF Radio Digital Protocols' by T J Riley of the US Institute of Telecommunication Services (ITS). This covers an investigation into five amateur radio digital protocols. The intention is that the US Federal Emergency Management Agency (FEMA) should use these test results to determine the protocols that could serve as an interchange with the amateur community in the event of a national emergency. Tests at the ITS laboratories in Boulder, Colorado were carried out under a well-defined, repeatable set of simulated ionospheric-channel conditions.

Throughput was used to evaluate the performance characteristics of the protocols, modems and computer interfaces indicated in **Table 1**, covering AmTOR, Clover, G-TOR, PacTOR and PacTOR2. Earlier, in 1993, FEMA had run a series of on-air tests of some of these protocols but it did not prove possible to make a direct comparison between the laboratory and the over-the-air results.

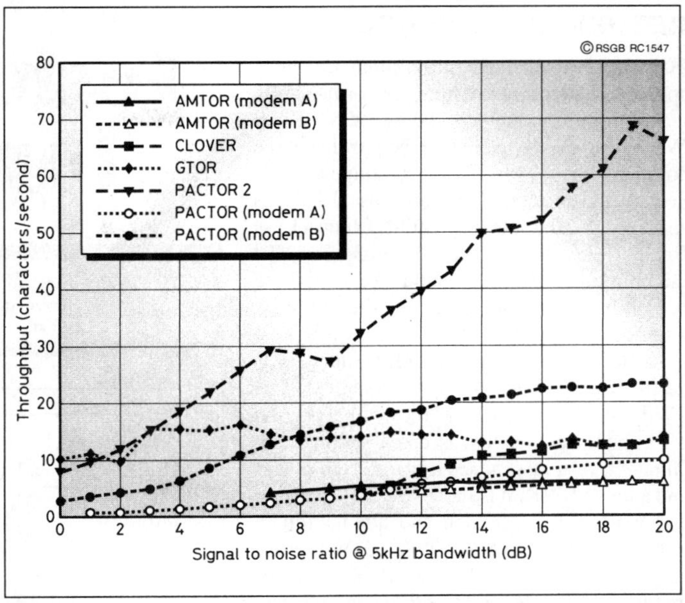

Fig 10: Throughput version signal-to-noise ratio of the protocols under simulated HF 'CCIR Poor Conditions'.

Among the conclusions presented by T J Riley was that, "by using adaptive techniques and more sophisticated modulation, error detection/correction and compression methods, the newer protocols exhibit significant throughput improvements over the 'first-generation' protocols. Compared to the baseline AmTOR protocol, the newer PacTOR2 protocol improves throughput by an order of magnitude under high SNR conditions, and continues to transfer data at a reasonable rate under conditions that render AmTOR inoperable. One drawback to such sophisticated protocols is the increased number of settings available to the operator. Depending on the propagation conditions and the type of data to be transmitted, the proper settings can significantly affect performance. This is especially true of Clover.

"Of the five protocols tested, PacTOR and AmTOR are the most widely implemented. Clover and G-TOR are proprietary protocols implemented by their respective developers only. PacTOR2 is beginning to appear in modems of several manufacturers but at the time of the tests was available from only one. To determine if the throughput of a particular protocol was specific to the manufacturer's implementation, throughput of PacTOR and AmTOR was measured on two modems from different manufacturers . . . in the case of PacTOR the difference is significant with the throughput of modem A nearly twice that of modem B even though the test conditions were identical for both modems.

Results are shown graphically in **Figs 9** and **10**. While it could be argued that in amateur operation high throughput is not often required, nevertheless throughput under poor conditions and at low SNR is a useful measure of the general reliability of these protocols. It is interesting to note that the pecking order differs to come extent under good and poor conditions. The CCIR conditions are based on the Watterson model taking into account SNR, fading frequency spread (Hz) and multipath delay (ms). For details of the equipment set-up used for performance testing, etc if would be necessary to consult the original paper in the IEE Conference Book which should be available in some good technical libraries. But the brief details given here should provide at least some illumination to data-mode enthusiasts. ♦

RETURN LOSS BRIDGE

FOR MANY YEARS most amateurs have specified antenna matching performance in terms of the Voltage Standing Wave Ratio (VSWR or SWR) and tend to be puzzled at the common use by professional antenna engineers of the concept of "Return Loss (RL)". As Jim Ford, N6JF, puts it in 'Build a Return Loss Bridge' (*QST*, September 1997, pp34-36): "To the uninitiated RL may sound like something bad! It may come as a surprise that with RL more loss is better."

In fact, an infinitely large RL is much to be desired, since it is equivalent to unity (1:1) SWR. N6JF defines RL as "the decibel portion of the input power that is taken by the load. Greater return loss means less reflected power and a lower SWR. It is much easier to measure and work with mismatch and attenuation in terms of return loss than SWR. RL will probably never replace traditional SWR measurements, but knowledge of both techniques can be very helpful. Despite some limitations, a Return Loss Bridge is a very good way to measure mismatch (especially for low-power applications) and coaxial cable loss."

In an introductory section, N6JF writes: "I've often thought it unfortunate that most amateurs are not familiar with Return Loss measurements. Let's look at a couple of examples. Suppose you know your coax loss and the SWR at your transmitter, but want to know the SWR at your antenna feed point. Factoring transmission-line loss into the SWR equation is difficult, but the problem becomes a piece of cake when you use Return Loss measurements:

$$RL_{ant} = RL_{TX} - 2 \times L_{coax}$$

where RL_{ant} and RL_{TX} are the return losses at the antenna and transmitter respectively, and L_{coax} is the coax loss. All three quantities must be in decibels. The reason we double the coax loss in the calculation is because the power returned by the mismatch travels twice through the feedline, once as forward power and then again as reflected power." [It is worth remembering that if we connect a long, lossy feedline to a transmitter, we can achieve near unity SWR no matter how great the mismatch between the feedline and the antenna element! - *G3VA*].

Table 1 shows the equivalent RL for a range of typical SWRs. It is possible to measure actual coax losses (rather than using a maker's optimistic specification for brand new cable) by making two measurements with a power meter and a dummy load. A Return Loss Bridge (RLB) can measure coax loss directly in decibels much more simply.

Another advantage of the RLB is that it can measure the mismatch at the input of a receiver

Pat Hawker's Technical Topics

PAT HAWKER, G3VA
37 Dovercourt Road
London SE22 8SS

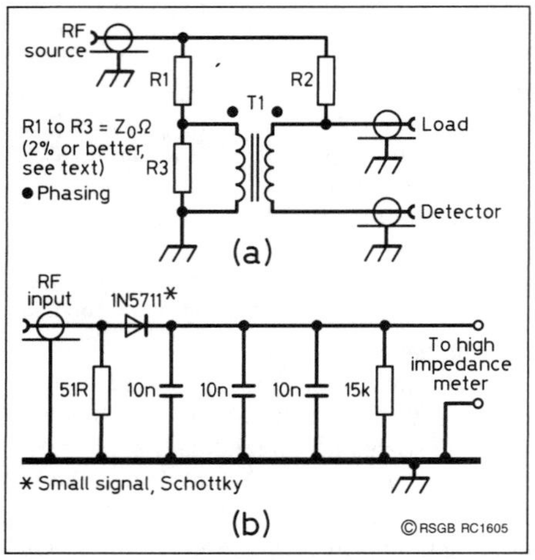

Fig 1: Return Loss Bridge for use with a digital voltmeter and diode detector. (a) Return Loss Bridge. T1 is 10 turns bifilar wound No 30 AWG enamelled wire on an Amidon FT-23-77 core. For good accuracy use 1% tolerance carbon-composition or metal film resistors, and keep the leads as short as possible to maximise bandwidth. For 50Ω systems R_1- R_3 are all 50Ω . (b) The diode detector used by N6JF. Unless otherwise specified, use ¼W, 5% tolerance carbon composition or metal film resistors.

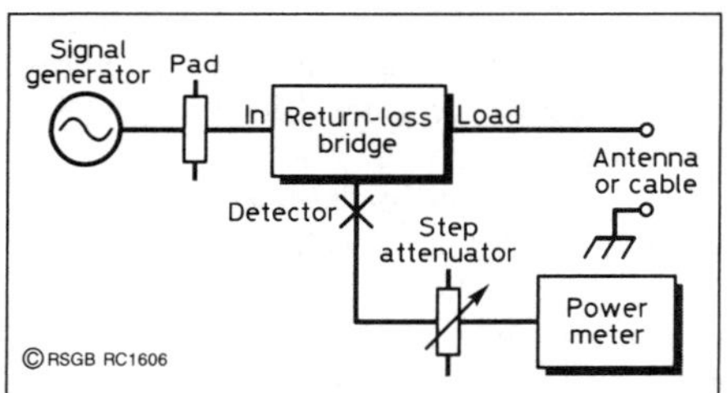

Fig 2: An RLB (Return Loss Bridge) set-up, to measure antenna system match or coax feeder.

or other low-power device without needing 100W of input power: "In fact, 10mW or sometimes less (depending on the sensitivity of your detector) will do the job. A signal generator will suffice for most purposes."

N6JF provides details of a simple RLB (hybrid combiner) using just four components: **Fig 1(a).** Note that for optimum accuracy close-tolerance resistors should be used and care should be taken with the detector: e.g. **Fig 1(b). Fig 2** shows an RLB test set-up to measure antenna-system match or coax loss. N6JF also provides details of the calibration of his diode detector at 30MHz, using a 1N5711 small signal Schottky diode.

For those wishing to use an RLB rather than just to understand the concept of return loss, N6JF's article provides a useful helping hand. It includes a straightforward discussion of the theory and operation of a RLB, and also warns of the importance of good isolation between the source and detector ports to avoid leakage at the detector port, even with a perfect load. Directivity serves as a floor for mismatch measurements. It is also pointed out that "probably the biggest disadvantage of this return loss bridge is that you cannot leave it in the transmission line during normal station operation".

The concept of Return Loss rather than SWR is increasingly found in amateur as well as professional literature. The 1997 *ARRL Handbook* has further information on RLBs (Chapter 26) and there is a detailed mathematical discussion of transmission lines and their associated losses in the *ARRL Antenna Book* (Chapter 24) These books are avaliable from the RSGB and for further details see pages 90 and 91.

RL (dB)	SWR	RL (dB)	SWR	RL (dB)	SWR	RL (dB)	SWR	RL (dB)	SWR
∞	1.00	15.6	1.40	10.5	1.86	6.0	3.00	1.9	9.00
26.4	1.10	14.0	1.50	10.2	1.90	5.2	3.44	1.4	12.33
20.8	1.20	12.7	1.60	9.5	2.00	4.4	4.00	0.9	19.00
20.0	1.22	12.0	1.67	9.1	2.08	3.7	4.71	0.4	39.00
17.7	1.30	11.7	1.70	8.0	2.33	3.1	5.67	0.0	∞
16.5	1.35	10.9	1.80	6.9	2.64	2.5	7.00		

Table 1: Return loss and equivalent SWR.

SUPER-LINEAR MIXERS AND FILTERS

ALTHOUGH HIS PROFESSIONAL WORK and family ties have delayed the completion of his super-linear receiver, Colin Horrabin, G3SBI, has been able to tie up a few loose ends on his H-Mode Mixer (*TT*, October 1993, corr. November 1993, and *Radio Communication Handbook*, 6th Edition, pp6.52-53) and has also checked the performance of a number of filter cores, etc. He writes:

"With an H-mode mixer, for some reason if transformer drive is used for the SD5000 DMOS FET array gate-drive, the input intercept is limited to about +25dBm. Only when driven by the original over-volted AC74 flip-flop arrangement (as described in 1993) will it give the superb +53dBm (although an over-volted AC86 would probably be entirely satisfactory from a non-divide-by-two drive, ie one gate with a feedback resistor for centre point and AC-coupled drive).

For some reason the broader-band low-output impedance presented by the AC74 seems to be very important in achieving the very high intercept levels possible with this mixer. I have recently tested a complete front-end built by Harold Johnson, W4ZCB, using hybrid-connected crystal filters; it measured +55dBm at 20kHz for a noise figure of 11dB.

I have also looked at the performance of three-section ladder filters used as RF bandpass filters with loaded Qs of 8, including their third-order intercept points (IP_3).

A filter using TOKO tuneable inductors measured +37dBm (coil Q of 70).

The Loadstone Pacific L45-6 former was +47dBm (coil Q of 125).

The Loadstone Pacific L41-6 former was +55dBm (coil Q of 70).

The Micrometals T50-6 toroid was +50dBm (coil Q of 125).

It is evident from these measurements that the Loadstone Pacific tuneable formers give excellent performance. It is a pity that the company does not market a range of ready-wound inductors, as is available from TOKO. The reason why the L41-6 provides the best performance is that this type has the largest diameter core and its design provides more air gap in the magnetic circuit.

My choice for a high-performance RF amplifier stage would be to use the L45 in the input filter (bandwidth $f_o/8$) and the L41 in the RF amplifier-to-mixer circuit (bandwidth $f_o/4$). With no RF amplifier my choice would be the L41 with $f_o/8$ bandwidth on the input to the mixer.

It is important to remember that the higher the filter-loaded Q (narrower bandwidth), the worse the IP_3 will become. This is because the coil carries a greater signal current.

To summarise: In a filter, tuneable inductors wound on Loadstone Pacific formers can outperform a toroidal core. This is useful to know since it makes printed circuit design much simpler when using Loadstone L45 and L41 adjustable formers rather than toroids with trimmer capacitors."

CAROLINA WINDOM & SWITCHED POLARISATION WITH IGP ANTENNA

THE NOTE IN THE OCTOBER *TT* querying the origins of the Carolina Windom brought me a plethora of information from Roy Price, GW3SYL; Geoff Dyson, M0AOG; Tony Morton, G0MDZ; and Ted Younge, G3IVH. M0AOG also provided a copy of an article by Mike Lee, G3VYF, from the *Short Wave Internal Frequency Handbook* covering Windoms generally as well as the Carolina type. This included a reference to a comprehensive article on the standard form of multiband windoms: 'The Off-Centre-Fed Dipole Revisited: a Broadband, Multiband Antenna' by Dr John Belrose, VE2CV, and Peter Bouliane, VE3KLO (*QST*, August 1990, pp28-34). Peter Waters, G3OJV, of Waters & Stanton PLC added further information and also pointed out his firm was the original importer of - and continues to market - this antenna, made in the States by Radio Works Inc. Several writers have pointed out that it is designed not only for 3.5MHz but, with a suitable ATU, can function effectively on all bands from 3.5 to 28MHz. The antenna was illustrated in the October *TT* but, for convenience, is shown again in **Fig 3**.

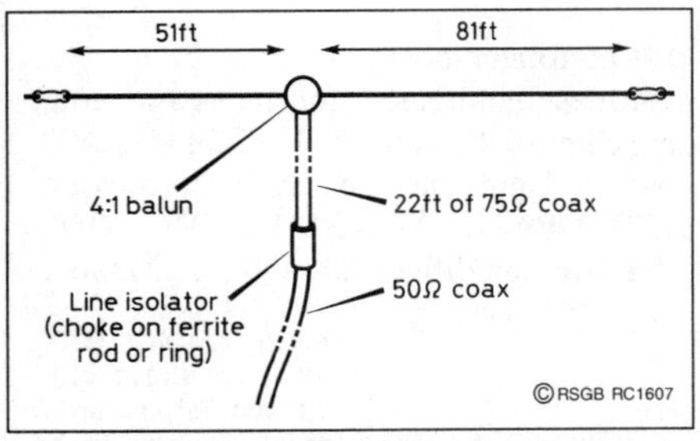

Fig 3: The multi-band (3.5-28MHz) Carolina Windom antenna, manufactured by Radio Works Inc of Virginia and marketed in the UK by Waters & Stanton PLC, and Haydon Communications.

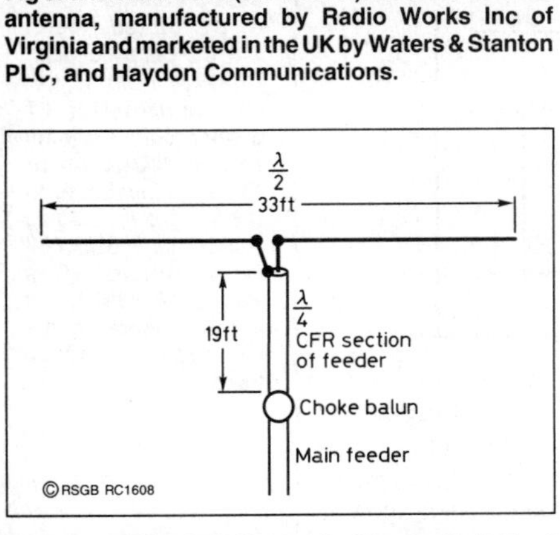

Fig 4: A 14.2MHz CFR (Controlled Feeder Radiation) half-wave dipole antenna, as described by G2HCG in the May 1990 *Radio Communication*.

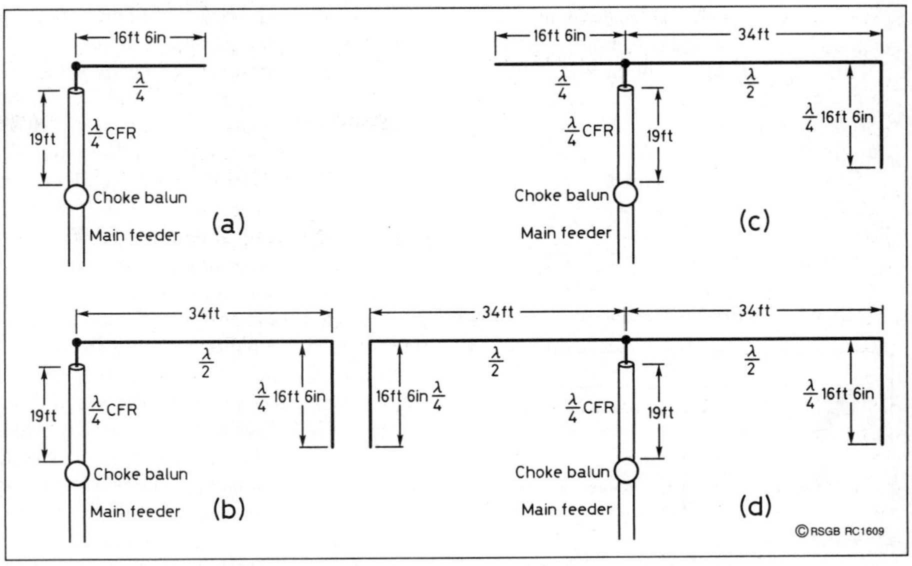

Fig 5: G2HCG's suggested variations for CFR. (a) isotropic, (b & c) half-square, (d) Bobtail.

The Carolina Windom was developed by Jim Wilkie, WY4R; Edgar Lambert, WA4LVB; and Joe Wright, W4UEB, then optimised by Jim Thompson, W4THU, and copyrighted. In 1989 a licence was granted to The Radio Works Inc of Portsmouth, Virginia, USA. It was developed as a multiband antenna, radiating both horizontally and vertically polarised signals. The 22ft vertical section provides vertical polarisation, part of the transmitter power being reflected down the outer braid of the feeder.

Although claimed to provide pragmatically a gain of up to 10dBd, this claim is clearly based primarily on the modification by the vertical element of the overall vertical radiation pattern rather than conventional forward HRP power gain, with low-angle radiation from the 22ft vertical radiator boosting the DX capabilities of the antenna. The sales literature makes what must appear extravagant claims, but users generally appear to have found the antenna a satisfactory performer with some suggesting that as an OCFD it is a world beater. Well, perhaps.

As pointed out by G3HQX, there is some resemblance to my 1970 Inverted Ground Plane

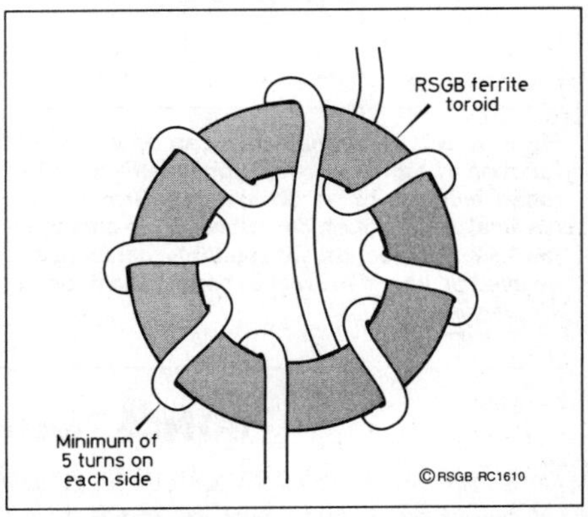

Fig 6: G2HCG's suggested method of winding a choke balun on a large toroid.

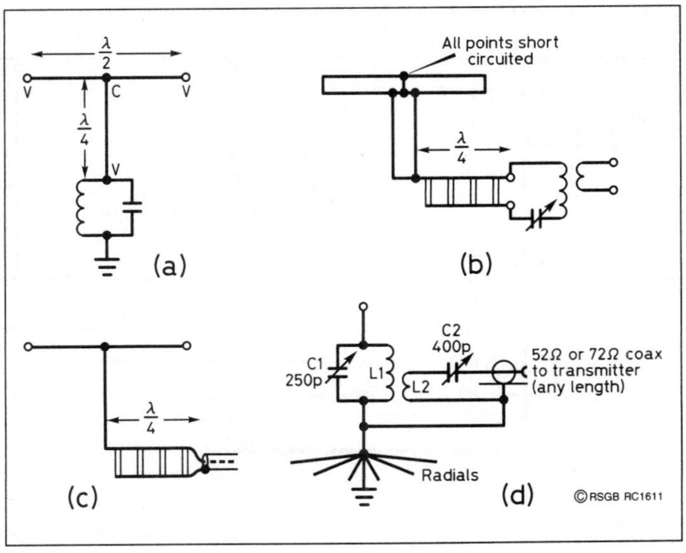

Fig 7: The basic single-band IGP (Inverted Ground Plane) or vertical T antenna, as first described by G3VA in 1970. Virtually all radiation is vertically polarised. Readily converted from a conventional dipole or folded dipole. Voltage fed at the base.

(Vertical-T) antenna, but whereas the IGP is intended to radiate almost entirely vertically polarised signals the Carolina Windom is designed to radiate primarily horizontally polarised signals but with a significant vertically-polarised component. As pointed out by G3OJV, and in the article by G3VYF, the American design is a good example of the 'Controlled Feeder Radiation (CFR)' technique described by B Sykes, G2HCG in *Radio Communication*, (May 1990, pp40-41).

G2HCG discovered that the polar diagram of an antenna can be modified and often improved by controlling the radiation from its feeder. He placed a choke balun some distance down the feeder line of a dipole element fed from an unbalanced coax feeder: **Fig 4**.

G2HCG provided several other examples of Controlled Feeder Radiation (CFR) antennas based on variations of the IGP, half-square and bobtail antennas etc: **Fig 5**. In these cases the outer braid of the final length of coax feeder was not connected to the element. **Fig6** shows how G2HCG recommended winding a choke balun on a large toroid.

As noted in *ART7* (p315). "In first describing the Vertical-T or Inverted Ground Plane antenna, **Fig 7**, I wrote "The ease with which a dipole antenna can be converted to a vertical-T suggests that it should be relatively easy to develop an antenna which could be switched from horizontal to vertical polarisation and vice versa." In *CQ* (June 1972), Ken Glanzer, K7GCO, followed-up this suggestion and described "The 10°/90° antenna for 75 to 40 metres": the two angles referring to the predominant vertical radiation angle - 90° with the fairly low horizontally polarised dipole and 10° with the antenna in its Vertical-T configuration: **Fig 8**.

"The twin open-wire feed is used as such for horizontal polarisation, but switched so that both wires go to the same side of the matching coil to form the voltage-fed quarter-wave top-loaded vertical. In his implementation of this arrangement, K7CGO incorporated remote switching and remote (selsyn) tuning of the matching unit, in his belief that it is very useful to be able to switch rapidly between the two polarisations to take advantage of propagation quirks, noting that there may even be occasions when it is better to switch to a different polarisation for reception to that used for transmission."

While the IGP and the K7GCO versions were designed as single band resonant antennas, either version can be used over a wide range of frequencies, provided that suitable impedance matching networks are located at the base of the vertical section. MF broadcast stations radiate efficiently with T-antennas significantly less than a quarter-wave high.

The manuals for the Carolina Windom stress that the vertical section *must* be kept well away from metallic objects such as towers, masts, gutters, metal roofing, etc. If a tower is used to support the antenna, a 10ft length of 2-inch PVC pipe may be used to hold it away from the metalwork. The vertical radiator section and the line isolator should both be more than 15ft from the tower. With the IGP-type arrangement and with the vertical section centred between two supports, it should be relatively easy to provide sufficient clearance from metal objects.

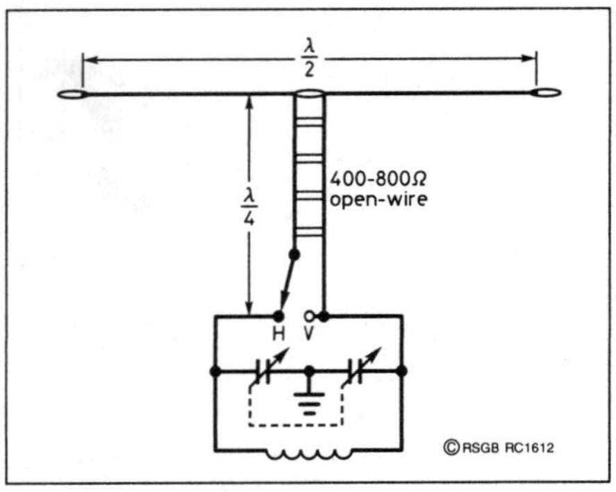

Fig 8: K7GCO antenna which can be switched to function as either a conventional half-wave dipole (open feeders) or as an Inverted Ground Plane (vertical T). Although described as an antenna for the 3.5 & 7MHz bands, with suitable matching it can be used on any of the higher bands as a doublet or T-antenna.

MINIATURE 7MHz DIRECT-CONVERSION RECEIVER

STEVE BERNSTEIN, K8IDN, in 'The MRX-40 Mini Receiver' (*QST*, September 1997) describes a miniature receiver simple enough to be built in a single evening. It was originally intended as a companion to a 7MHz Micronaut CW transmitter, a kit rig marketed by Dave Ingram, K4TWJ.

It is a crystal-controlled direct-conversion receiver comprising an NE-612AN oscillator/mixer chip, followed by an LM-380N audio amplifier: **Fig 9**. By using a 1N4004 diode as a varicap, the receive frequency can be shifted about ±1.5kHz the crystal frequency - a rather restricted tuning range, but eliminating the need for a bulky variable capacitor.

The voltage supply to the NE-612 and tuning circuit is regulated by a 78LO6 so that the 8V limit of this chip is not exceeded. The full battery voltage is applied to the LM-380, with the audio output stated to be more than ample for Walkman-style headphones. Current drain from the battery is about 16-17mA.

Definitely a 'fun' rather than a 'high performance' receiver, it can be built on a small piece of perforated board.

Paul Debono, 9H1FQ, believes that too many designers of direct-conversion receivers, with their requirement for high-gain audio amplification, overlook the inherent noise of the older IC amplifiers: "They keep on recommending chips such as the LM380, LM741, etc". He lists a number of examples of low or ultra-low noise operational amplifiers which can be substituted, provided the correct pin-outs etc are observed: Single package: OPA620KP, NE5534N, OP07N. Dual package: NE5532, ME5534, OPA2107AP, LM833N, MC33078. Quadruple pack: OP470GP, MC33079. An ultra low noise (professional) chip is the SSM2017. Power amplifiers: TDA7052, LM386.

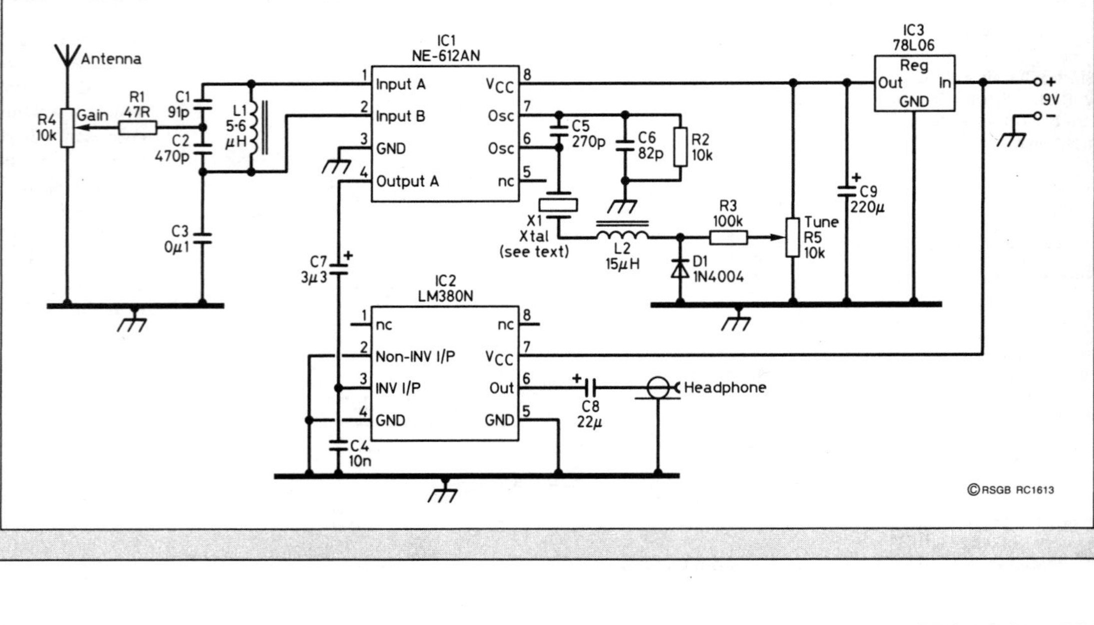

Fig 9: K8IDN's MRX-40 mini direct conversion 7MHz receiver, small enough to be built on a small piece of perforated board. C1, C2 ceramic disc; C3-C6 monolithic; C8, C9 electrolytic; L1 5.6uH moulded choke; X1 in HC49U holder for 7122kHz (available in USA from Doug Hendricks, KI6DS, 862 Frank Avenue, Dos Palos, CA 93620, but any frequency in the CW sector could be used).

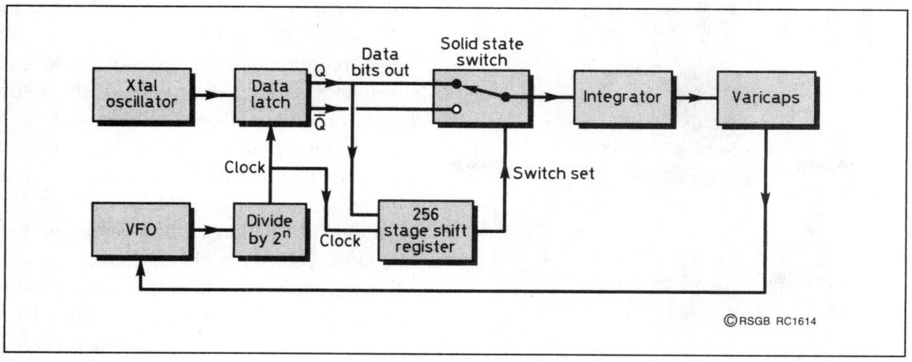

Fig 10: G7IXH's fast Huff & Puff stabiliser, suitable for frequencies up to 30MHz.

THE 'FAST' HUFF & PUFF STABILISER

PETER LAWTON, G7IXH, has recently developed what he has found to be a further improvement on the 'improved' version of the Huff & Puff LC oscillator stabiliser described by Klaas Spaargaren, PA0KSB, in ARRL's *QEX*, February 1996 (noted in *TT*, July 1996 with corrections in subsequent issues). G7IXH has written a detailed account of his 'fast' stabiliser in an article now with the *RadCom* editors but I feel that a brief outline of the new approach is likely to be of immediate interest to readers. He writes:

"Although PA0KSB's latest version is a great improvement in many respects over his original version, I found it inadequate at frequencies above 20MHz. Too much frequency ripple is put on to the stabilised oscillator.

"On examining the operation of the stabiliser, the limiting factor seemed to be the rate at which the oscillator frequency is sampled and the correction updated. If this sampling frequency could be increased, the excursions of the oscillator about its stabilised frequency would be diminished." Compare this with a thermostat; the more often it checks the temperature the closer it can keep it to a constant value. The improvement in PA0KSB's new version is primarily due to the increased sampling rate. [It is entirely satisfactory at frequencies below about 15-20MHz - *G3VA*].

"In all published designs to date, increasing the sampling rate increases the incremental frequency 'steps'. However, I have found it possible to increase the sampling rate without increasing the 'step'. **Fig 10** is a block diagram of my new 'fast stabiliser', so called because it samples the oscillator frequency and refreshes the corrections at 257 times the rate of the basic stabiliser, resulting in a diminution of ripple to about 1/60th and an increase in its drift-correction by about 20 times.

"The differences lie in fewer stages of VFO division, resulting in the faster sampling rate

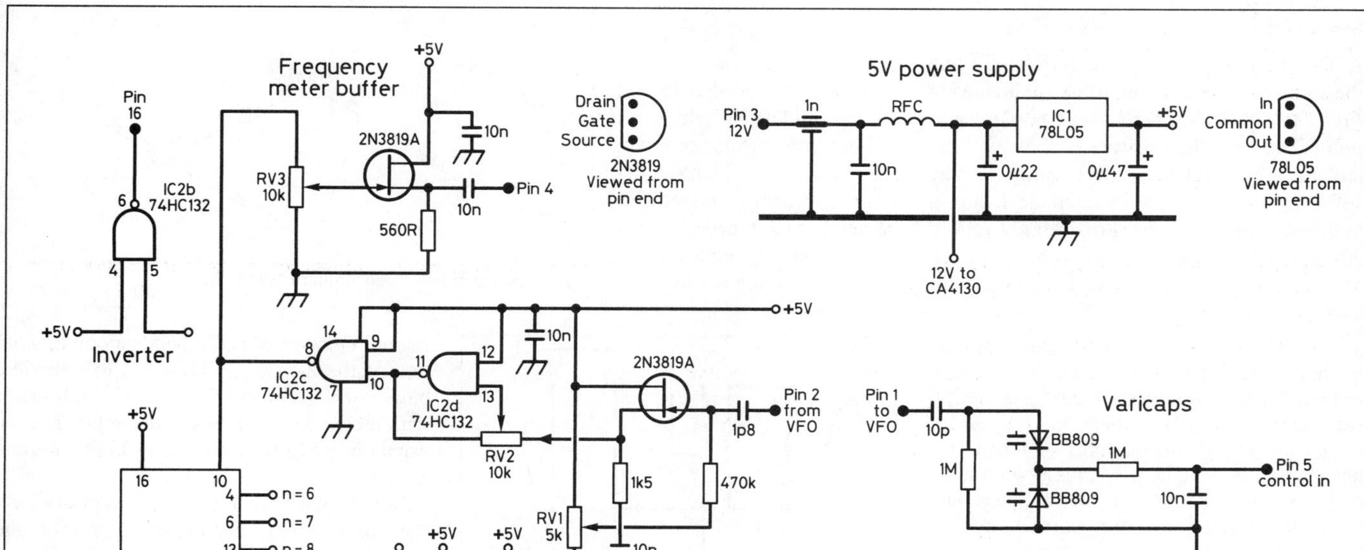

Fig 11: The fast stabiliser. Each IC is decoupled by a 10nF capacitor from Vcc to ground.

TECHNICAL TOPICS

RADCOM NOVEMBER 1997

'TUNED LOOP TRANSMITTING antennas': The IEE Conference Publication referred to should have been No 441.

'Two for the price of one': A dropping of part of a sentence near the top of page 78 column three may have misled readers. The sentence beginning "The SWR at 3800kHz", relating to G3HQXs adjustments to the Chelcom HFV1 antenna, should have continued: "was reduced from 2.3:1 to 1.8:1 and at 3500kHz increased from 1.35:1 to 1.45:1".

Pat Hawker, G3VA.

A/D CONVERTER

RADCOM OCTOBER 1997

Please note that the component layout shown in Fig 6 is as seen from the track (under) side of the PCB.

EXTREME NARROWBAND RECEPTION

RADCOM NOVEMBER 1997

IN HIS ARTICLE, G3PLX refers to modifying the software of my PIC interface, published in *RadCom* April 1997, to allow complex sampling. The new PIC software is now available and may be used with all existing display software, as well as in complex mode. It allows both real sampling at rates of 10,000 / 7,142 / 2,500Hz (the 1,000Hz real rate has been removed), plus complex sampling at 1,000Hz for an audio band which must be pre-filtered to the range 500 to 1,500Hz. Two DIP switches are used to select the sampling mode. Early versions of the PIC code only supported one DIP switch and two sampling rates, as shown in the original article. The second switch can be easily added. Later supplied versions of the code, SERADC06, included the four real sampling rates and the modifications needed to add the second DIP switch.

For those who can program their own PIC devices, the code including assembler source and HEX code is available either by requesting it via E-mail to me at drassew2@interalpha.co.uk, or via receipt of a disc and SASE to G4JNT, QTHR.

New programmed PICs are available from myself at £10.00 each. Alternatively, return your old programmed one along one with £2.00 and I will reprogram this device with a version of the code that overwrites the old code stored in the one time programmable devices with 'no operation' instructions and appends the new code at the end of the old data space.

Software to make use of this complex sampling mode is currently under development.

Andy, G4JNT

A PORTABLE 3-ELE 6m YAGI

RADCOM NOVEMBER 1997

AN E-MAIL FROM Dave Reid, PA3HBB/G0BZF, alerted us to the fact that on page 44 the content of **Fig 9** appeared as a second copy of Fig 8. Here is the correct Fig 9.

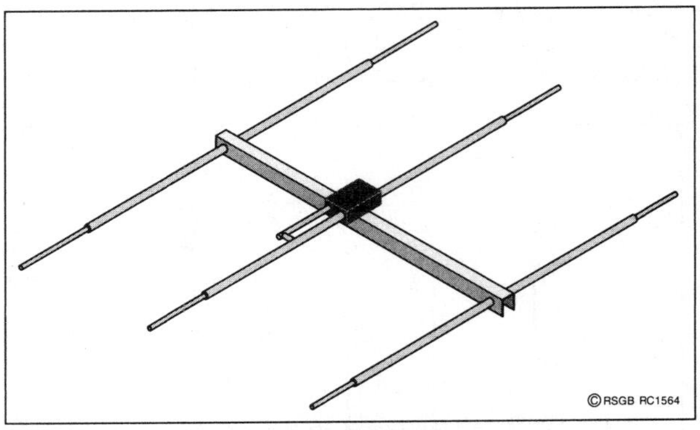

Fig 9: The complete three element 50MHz beam.

and in the decoding circuitry after the data latch. Data from the latch is stored temporarily in the 256-stage shift register and it takes 256 sample periods for it to emerge.

"By the time it does emerge, sufficient time has elapsed for it to be used along with the bit just emerging from the data latch at that instant to determine the direction of correction without increasing the step. So although the two pieces of data used for each correction are separated sufficiently in time to keep the 'step' as small as in the basic version, the corrections are refreshed at a much greater rate. The 'fast' Huff & Puff could be considered as 257 synchronised stabilisers working simultaneously on the same oscillator; although the circuit is no more complex than that of the basic stabiliser.

"The new stabiliser has proved entirely satisfactory at 30MHz. The crystal oscillator is a Maplins 66MHz unit, divided by two to give a unity mark-space ratio. Although the circuit diagram (**Fig 11**) may seem rather complex, much of the apparent complication represents refinements to ensure reliable operation and ease of setting up. The basic arrangement remains simple, as indicated in the block diagram."

In his detailed article, G7IXH reports on computer predictions, discusses the factors affecting the power of a stabiliser, provides a detailed explanation of the operation of the fast stabiliser, choice of components, and setting up. He also points out that the fast method can be applied to the less powerful crystal clock type of stabiliser in which the positions of the VFO and the clock are interchanged. It is to be hoped that it will prove possible for G7IXH's full article to be published before too long.

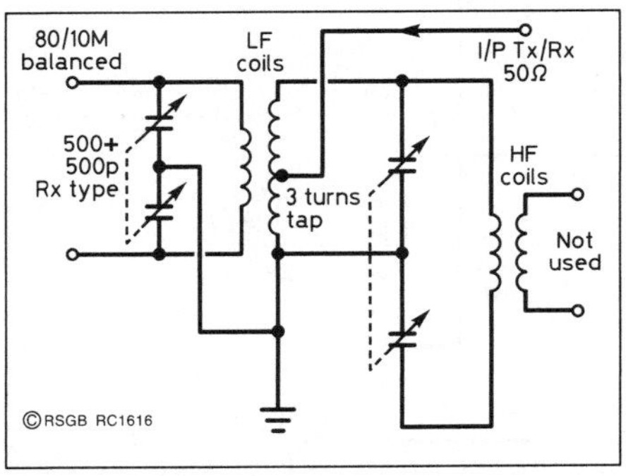

Fig 12: 9H1FQ's modified Z-match ASTU.

HERE & THERE

I AM NOT SURE whether someone, who shall be nameless, is pulling my leg or whether his leg has been pulled. He has sent me a copy of an article 'The Multi-Slot - a High Efficiency Compact Multi-band Aerial for Confined Spaces' by 'Professor Sebestian Vizz' which appeared in the *April* 1965 issue of the *RSGB Bulletin*. Not surprisingly, he reports that he has been unsuccessful in working out the dimensions for 40 metres as "the formula is not very clear to me". Since I was once caught out by a filter diagram that appeared in an *April* issue of *Radio-RE*, I have been rather jaundiced against spoof articles.

Paul Debono, 9H1FQ, sends along a variation of the Z-match ATU, used to provide a balanced 300Ω output on the 3.5-28MHz bands without band switching and using two ganged capacitors: **Fig 12**. Coils are as in the original Z-match.

Roberta Barmore, KB9GKX, has most kindly sent me a copy of her limited reprint of the 1935, 64-page 'The Radio Amateur Newcomer' by Frank Jones, W6AJF (original editor of *The Radio Handbook*), roughly equivalent to the RSGB's 1935 edition of 'A Guide to Amateur Radio'. I find this of particular interest as that year marked my entry into the wonderful world of short-wave radio. An era when the regenerative straight receiver and the simple supergainer superhet were the standard means of HF reception; although, for the well-heeled, high-performance communications receivers were beginning to appear. Nostalgic, yes, but the coil-winding data for plug-in coils is still useful.

I continue to find an interest - not entirely nostalgic - in valves and valve circuitry. Considering the LF noise of bipolar transistors, valves may still be a logical choice for LF (30-300kHz) receivers, ie for 73 and 136kHz. Terry George, GW3NJG, was particularly pleased to see the valve product detector in the September 1997 *TT* and writes, "Could we have more of this please? The terms HT+ and 6.3VAC bring back happy memories of amateur radio in the 1950s - perhaps it is time for the return of grid bias, reaction and the inter-valve transformer." ♦

BY PAT HAWKER, G3VA*

AN ALTERNATIVE TO NBVM

TWENTY YEARS AGO, *QST*, December 1977 carried what might now be considered an 'over the top' article "A new era in voice communication" by Dr R W Harris (University of the Pacific) and J C Gorski (*QST* editorial), which opened by claiming that "Amateur radio is about to undergo a fundamental change. After many months of investigation three researchers (R W Harris, J Cleveland, WB6CZX, and T Lott, VE2AGF/W6) have perfected a means of narrow-band voice transmission (NBVM) that promises to do for the Amateur Radio service what single sideband did 30 years ago."

They had successfully developed a practical means of reducing the bandwidth required for speech by a factor of two. To quote the introduction: "Called NBVM, it will radically alter radio and telephone communications the world over."

Basically the technique was to emphasise the most important information-bearing parts of speech (ie the consonants) but discard the mid-range vowel sounds, yet still retain after reconstruction the original timbre and voice identification characteristics. On transmission the consonants are emphasised and electronically folded, in real time, into the blank spaces no longer occupied by the dominant vowel sounds. "This folding action accomplishes the all-important spectrum savings by a factor of two . . . . On transmission the consonants are increased in energy so that there is a relatively constant speech energy output, as opposed to the case in normal speech where the vowels dominate the consonants by 20 to 25dB. On receiving the folded speech, the vowels and consonants are unfolded in a continuous manner and vowels amplified to restore the natural energy balance between vowels and consonants . . . . Preliminary experiments indicate that a rather high-quality speech signal can be achieved with a transmission bandwidth of 1500Hz. If intelligibility only is desired, however, a transmission bandwidth of 1200Hz or even 1000Hz is possible."

Although NBVM later found its way into the ARRL *Radio Amateur's Handbook*, it has never achieved its original promise, and today most professional work on bandwidth compression is based on all-digital rather than analogue techniques. However, Albert E Weller, WD8KBW believes that there are still benefits to be drawn from the 1970s NBVM experiments. He writes:

"The signal-to-noise ratio (SNR) for a received signal is maximised if the receiver bandwidth is just sufficient to accept the intelligence-bearing components of the signal. For the male human voice this bandwidth has been determined as extending from about 300Hz to roughly 2000-3000Hz. Consequently, SSB receivers designed to maximise the communication of intelligence by voice have a -6dB bandwidth of about 1700 to 2700Hz. (300 to approx 2700Hz): **Fig 1(a)**.

"Harris and Gorski have indicated that a portion of the normal voice band, about 600 to 1000-1500Hz, conveys very little intelligence. In NBVM they heterodyne the upper portion of the voice spectrum to fall in this intelligence gap, and thereby transmit a 'voice signal' within a bandwidth of about 1300Hz, but conveying virtually the same intelligence as an unmodified voice signal with a bandwidth of 2200Hz. This reduction of bandwidth provides a nominal SNR improvement of 2.3dB. The signal is processed within the receiver to restore the original relationships.

"NBVM has not become popular in amateur radio circles. It has the disadvantage that complex (and hence relatively expensive) signal processing equipment is required at both transmitter and receiver. Furthermore, NBVM is incompatible with stations not equipped for NBVM. These disadvantages have evidently been judged to outweigh the greater spectrum efficiency and the fairly modest improvement in SNR.

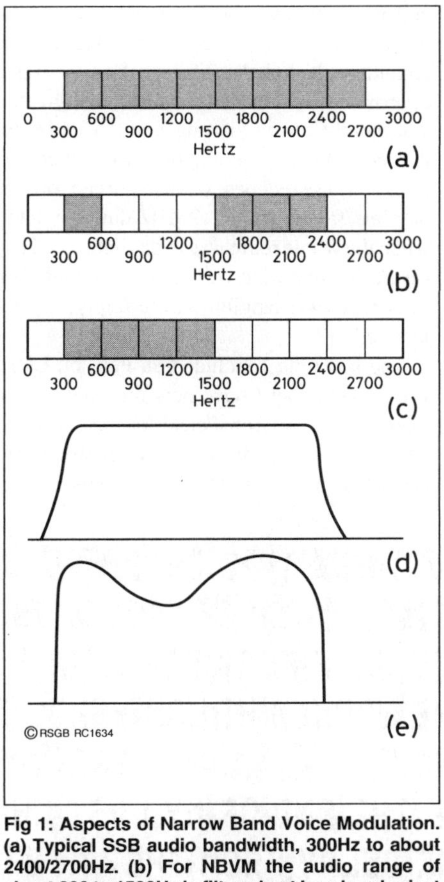

Fig 1: Aspects of Narrow Band Voice Modulation. (a) Typical SSB audio bandwidth, 300Hz to about 2400/2700Hz. (b) For NBVM the audio range of about 600 to 1500Hz is filtered out by a band-reject filter. This is also the transmitted audio spectrum suggested by WD8KBW. (c) In the original NBVM system, the range of about 1500 to 2400Hz is folded back to occupy the gap between 600 to 1500Hz, resulting in a continuous spectrum bandwidth of about 300 to 1500Hz. (d) Typical SSB filter tuned for maximum flatness (minimum ripple). (e) SSB filter with double hump provides sharper edge and slightly improved SNR.

"Furthermore, NBVM appears to have been devised with channelised operation in mind, whereas amateur operation on HF is not channelised. Interference (noise) is, to a reasonable approximation, uniform across the band. Under these conditions most of the advantages of NBVM can be obtained with much simpler processing, while retaining compatibility with unequipped SSB stations.

"At both transmitter and receiver the audio frequencies between 600 and 1000 to 1500Hz are filtered from the voice signal: **Fig 1(b)**. The major gain occurs at the receiver where interference and noise in the filtered band are rejected. Frequencies that do not contribute to intelligibility should not be received. The gain at the transmitter comes from conserving transmitter power for the transmission of intelligence-bearing frequencies. If a specific band of frequencies does not contribute to the intelligibility of the signal, those frequencies should not be transmitted.

"The only additional equipment required is a good audio band-reject filter at receiver and transmitter. The resulting signal occupies two narrow bands with a silent mid-range gap of the order of 500Hz or more, amounting to the same spectrum occupancy as true NBVM. It can be argued that in the absence of channelling, the continuous compact signal of NBVM offers little or no advantage except improved AGC operation (assuming the IF bandwidth of the receiver is reduced to just pass the compact NBVM signal).

"Other schemes for maximising voice signal intelligibility in the presence of noise have been reported. For example J D Griffiths in *Improved Analog Voice Communications* (AFCRL-68-0227, AD 671486, Air Force Cambridge Research Laboratories, April 1968) has demonstrated an improvement corresponding to about 2.7dB of transmitter power at a bandwidth of 3000Hz, by using a transmitting filter approximating a 3.5dB per octave rise above 500Hz. No additional receiver filter is required."

It seems to me that the design of an effective AF band-reject filter has been greatly facilitated by the use of Digital Signal Processing (DSP). However one would need to be convinced that the simple elimination of the mid-range audio frequencies does not result in unnatural sounding speech. Intelligibility is not the sole criterion in Amateur Radio. However, as a result of the original NBVM work, it has long been recognised that SSB crystal filters can be improved by not aligning them for maximal flat-top characteristics. A double-hump response, with greater attenuation in the mid-range (**Fig 1 (e)**) can have steeper sides than a maximal flat-top response (**Fig 1 (d)**) without significantly impairing speech quality.

*37 Dovercourt Road, London SE22 8SS

LONG DELAY ECHOES ON TAPE

ALMOST 30 YEARS AGO a *TT* item 'Long-Delay "Cosmic" Echoes' noted that a team at Stanford University was planning a new investigation into "the baffling, 40-year-old puzzle of long delayed echoes (delay times measured in seconds) that have been reported only extremely rarely".

It was noted that these strange Long Delay Echoes (LDEs) were originally reported in 1927 in a series of experiments by Professor Stormer at Oslo and Dr Balthazar Van der Pol at Eindhoven. The 1960 translation of *'Radio Wave Propagation and the Ionosphere'* by the Russian scientist Ya L Al'pert includes a summary of this work: "The first cosmic echo was observed at the end of the summer of 1927. Searches were organised on 31.4 metres from March 1928, but it was not until near noon on 11 October that a large number of echoes were received in Oslo with delay times between 3 and 15 seconds (most about 8 seconds) . . . . during the night 120 echoes were observed at Eindhoven . . . . in 1930 echoes were reported in Indochina . . . . in 1934 more than 70 echoes were observed between 30 May and 8 July."

Post-war investigations showed LDEs being heard only very rarely, and some researchers concluded that all reports must be mistakes, hoaxes, subjective effects originating within the listeners own brain, etc, etc.

Al'pert distinguished LDEs from the far more common HF Round-The-World echoes, on which a good deal of research was carried out by the Germans during WW2 in the course of devising a means of locating a distant transmitter by measuring the difference in time of arrival of the short and long path signals. It was shown that the average delay time of more than 750 RTW echoes was 137.78 milliseconds, with a spread in the individual measurement of 218 echoes less than 10^{-4} seconds.

In the following 11 years I included at least eight further items on LDEs, including several reports of such echoes being heard, or not heard despite long periods of listening, and also several hypothesises of how they might occur.

Two of the more convincing reports of LDEs in the 1930s - not only heard but also recorded - were included in *TT* (November 1975, p846). R L A Borrow, G3ZTK, who was one of those who carried out listening tests together with Professor Appleton at King's College, London, during the original transmissions by Van der Pol from The Hague in 1928-29 wrote: "In addition to hearing a number of echoes I did succeed in obtaining what was then the only known photographic record of an echo. This record was examined by Appleton and he was quite satisfied that it was undoubtedly genuine (see his letter to *Nature*, Vol 122, December 1928) . . . . On a very few occasions I distinctly heard the signal (an X) . . . . on one occasion an echo was heard by me in the laboratory at King's and also by Appleton at his home at Potters Bar."

L A Reeves, G4CEM also verified that even if LDEs were not being heard in the 1970s, they were heard and recorded by himself and many members of the commercial telegraph fraternity. He wrote: "The last time I heard this phenomenon was around 1938 on the Cable and Wireless London-to-Bangkok route . . . . The Bangkok station when 'idling' had the habit of putting on a call-band of spaced Vs and callsigns. The space between each V was about 5 seconds, and following each V there could sometimes be heard a perfect echo-type signal approximately 2 seconds behind the original. It was nearly always strong enough to be recorded on the pen-undulators used at that time . . . . These echoes were recorded on the daytime frequency of 19MHz during winter, when the whole route was in daylight . . . . I can assure you that what we heard and recorded on these tapes was tangible and certainly not subjective."

The last time I included an item on LDEs was in 1980, but I have now received a most interesting report from Peter Martinez, G3PLX, who has recently observed and recorded echoes while using his electronic Hellschreiber equipment. The delay time of 209 millisecond ±5 milliseconds is clearly too long for an RTW echo, although less than what has generally been regarded as the delay time of an LDE. He writes:

"The subject of 'long-delayed echoes' is a bit like the Loch Ness Monster or the Abominable Snowman. However implausible their existence may seem, reports still appear from time to time describing them. Here is another, but this time I captured it 'on tape', as it were, that I can show to those who may be interested.

"I was in contact with PA0OCD on 3.5MHz at 1930UTC on October 25, 1997, using the 'Hellschreiber' mode (described in detail by the late Stan Cook, G5XB, *'Hellschreiber - What it is and how it works'* (*Radio Communication*, April 1981, pp320-323). In this mode, letters are transmitted by sending patterns of dots in on/off keying mode, the patterns representing directly the letter shapes as they appear on the printed tape. The original German direct printing telegraph equipment used mechanical methods, but several amateurs are using DSP software and computers. My own set-up displays the incoming signal on the computer screen as black letters on a white background, stronger signals giving a blacker image. The sending software works so that whenever I hit a key, the transmitter immediately transmits the letter shape, then drops back to receive, in quick-break fashion.

"Imagine my surprise when, over a period of about one minute, I started to hear short 'pips' in the receiver speaker at the ends of my words. I then realised that I could also see the right-hand edges of the final letters of my words appearing on the receive screen, but shifted out-of-phase so that they did not line up with the transmitted letters.

"This appeared to be due an echo, so I did a quick test by sending a series of full stops in rapid sequence. Sure enough the dots showed up on the receive screen, appearing just after the transmitted dots but at the top of the text-line rather than the bottom. I quickly 'saved' the entire page. The effect faded and did not occur again.

"**Fig 2** shows a fragment of the screen-shot. At the top, the signal from PA0OCD; then my own test, with band-noise appearing between my words as a result of the quick-break operation. On the bottom few lines can be seen my remarks about the presence of this echo, followed by my tests with the string of dots. The echoes can be seen on some of the letters at the ends of words on the lines above, for example the 'K' at the beginning of line 5 and 'N' and 'CAN' in line 6. Where you see my text, this is 'forced' onto the screen by the software equivalent of 'sidetone', and are not signals received through the receiver which is muted while transmitting.

"I have spent some time analysing this screen-shot, picking out the actual sequence of received signals. The Hellschreiber signal scans

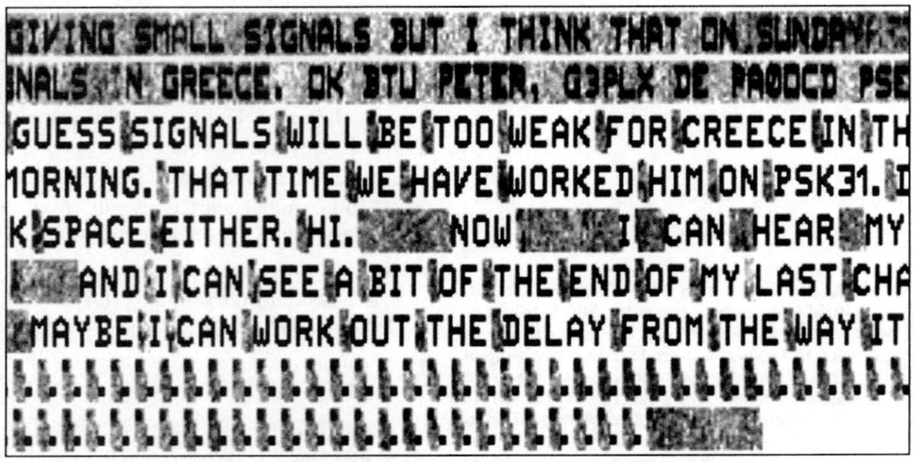

Fig 2: Screen-shot of a part of G3PLXs Hellschreiber-mode contact with PA0OCD.

in columns, bottom-to-top, the columns being transmitted at a precise rate of 17.5 columns per second. Each column is divided into 14 pixels or dots, each of 4.08mS duration. A full stop consists of 4 pixels in a square, and the waveform of a repeated string of dots is thus a pair of pulses 8.16mS long, separated by 57mS. They repeat at intervals of 457mS.

"I processed the screen-shot to unravel the rectangular characters into strips 457mS long, repeating down the page to show the repeated full stops. This is shown in **Fig 3**. I have added a line which is 49 pixels long, which represents exactly 200mS. As a separate exercise, by

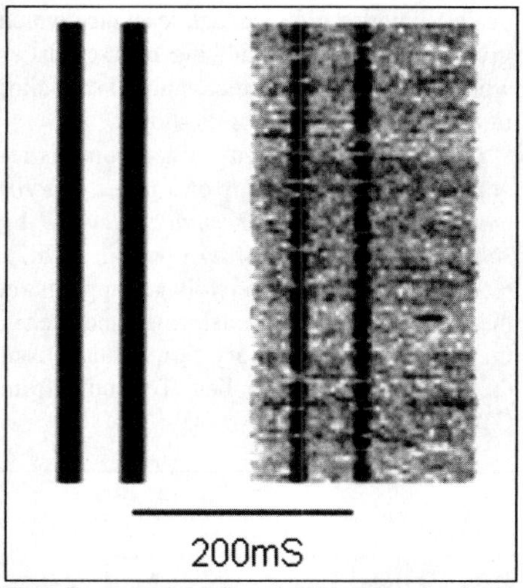

Fig 3: Processed version showing more clearly the echoes received 209 milliseconds after the transmission of the series of dots shown at the bottom of Fig 2.

listening to my own transmission on a separate receiver and displaying that on the screen while I was transmitting, I measured the delay from the 'sidetone' image on the screen-shot to the actual transmission. This was 55mS, formed mostly from a fixed software delay and a few milliseconds of delay through transmitter and receiver SSB filters. This enabled me to move the sidetone image from Fig 3 back by 55mS so that it represents the transmitted signal. The distance between the transmitted signal and the echo is seen to be about 209mS, with an uncertainty estimated at ±5mS.

"I have heard a similar echo once before on 3.5MHz, while transmitting AMTOR, but on this occasion I was fortunate in being able to record it for posterity. To put 209mS into perspective, it represents some 1.5 times the delay of a Round-The-World echo. Remember, this is 3.5MHz not 21MHz!"

[A check of DK0WCY beacon propagation forecasts (see *TT*, October 1997, pp80-81) shows that during October 25 there was a warning of a minor geomagnetic storm expected, and this must have been pronounced since, most unusually, I did not hear the beacon on October 26-27. There must be some suspicion that the echoes and the magnetic storm may have been connected, with the echo possibly arising from a storm-enhanced Van Allen belt - G3VA]

"BBC EXPERIENCE"

An opportunity arose to see a preview of the new permanent "BBC Experience" exhibition at Broadcasting House, London. This includes, in the reception area, the Marconi Collection which was so nearly dispersed by GEC-Marconi at public auction. To my mind, this was the most interesting part. The Experience proper, although extremely well presented and good fun, is directed squarely at the lay listener or viewer, rather than those genuinely interested in the history and technology of broadcasting. Those in charge cheerfully admit that they consider transmitters "boring", but then of course the BBC has now sold-off all its domestic transmitters to an American firm.

Even so one would have liked to have seen some mention of P P Eckersley, G2OO, responsible as both engineer and presenter of the original Writtle 2MT broadcasts, who became the BBC's first Chief Engineer (and Honorary Vice-President of the RSGB) and responsible for the original National/Regional twin-transmitter network. The BBC's historian, (Lord) Asa Briggs, has written of him: "Peter Eckersley had more ideas about broadcasting than any other man in the country. Some he put into effect, some were never realised, and some never could have been." Eckersley, unfortunately, was sacked by (Lord) John Reith for having been named in a divorce case, while still a young and highly talented engineer.

Broadcast engineering in the UK and elsewhere has always included many licensed radio amateurs, yet the only reference at the Experience to the amateur connection is a mock-up of the Tony Hancock's "The Radio Ham" television set - an extremely funny broadcast but one that by no stretch of the imagination could be construed as a credit to amateur radio!

TINY CMOS KEYER

GARY DIANA, N2JGU and Bradley Mitchell, WB8YCG in *'TiCK-2 - The Tiny CMOS Keyer 2'* (*QST*, October 1997, pp42-45) describe a keyer which vies for the title of being the smallest, lowest-cost, feature-packed keyer in the world! Based on one of the latest microcontroller technologies, it uses an 8-pin DIP microcontroller from Microchip Corporation, the PIC 12C509. The authors claim this IC is a perfect candidate for all sorts of Amateur Radio applications, because of its small size and high performance capabilities. It permits an iambic keyer at a cost that makes it accessible to virtually all amateur radio operators, and should encourage designers to include the keyer in their transceiver designs.

It features one memory message (eg "CQ CQ DE callsign callsign K"); mode A and B iambic keying; speed control; low cost; low current requirement for portable operation; small size; low parts count; simple interfaces both to the rig and the user; tune function to hold the keying line active while adjusting equipment; sidetone; paddle select allowing operator to swap the dot and dash paddles; manual keying to permit interfacing a straight key or external keyer; and weighting (3:1 default).

While the *QST* article provides detailed instructions, including component sources and programming, **Fig 4** at least provides the complete circuit diagram.

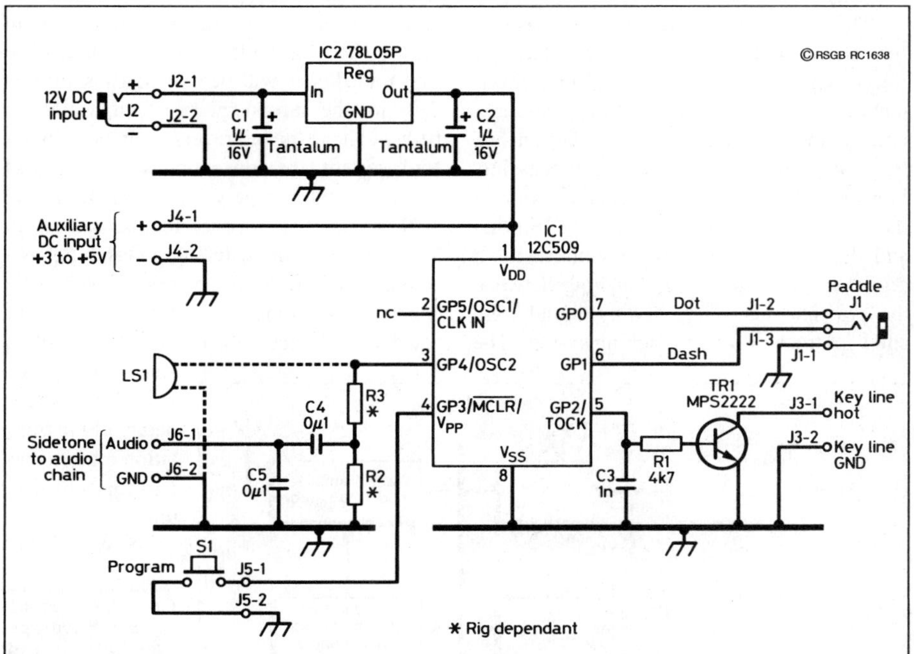

Fig 4: The low-cost TiCK-2 electronic keyer. The PIC12C509 IC must be programmed before use (see *QST*). C1, C2 tantalum; LS1 is an optional piezo element; TR1 NPN MPS222A or equivalent; S1 normally open push-button; IC2 5V, 100mA regulator.

SSB WITH CLASS C AND CLASS E AMPLIFIERS

THE AUGUST 1997 *TT* showed how low-cost switching-type MOSFETs could be used in high-power, high-efficiency, Class E power amplifiers at frequencies up to 7MHz. These 300W and 500W Class E switching amplifiers, like the better known Class C, are not linear and although fine for CW (Morse Code) or FSK, are unsuitable in this form for SSB. However, as pointed out by David Freeman, W2VLX (*QST*, October 1997, p82), there are a number of techniques that permit a Class C amplifier to be made suitable for SSB by the use of additional limiting and modulation circuits.

Several suitable approaches have been described in *TT*, most of which have been stimulated at least in part by the original work of Leonard Kahn. His approach was described in *'Single-sideband transmission by envelope elimination and restoration'* (*Proc IRE*, July 1952, pp 803-806) and *"Comparison of linear single-sideband transmitters with envelope elimination and restoration single-sideband transmitters"* (*Proc IRE*, December 1956, pp 1706-1712).

In the UK, a leading figure in the development of techniques that permit the use of high-efficiency non-linear amplifiers for SSB has been Dr V Petrovic. At the Universities of Bath and Bristol he worked for over a decade in the development of what he originally termed 'polar-loop feedback' and more recently 'Cartesian-loop feedback' systems (see for example, *TT*, April 1990, pp29-30).

W2VLX writes: "In view of modern technology, it may be time to take another look at the Envelope Elimination and Restoration (EER) approach. A simplified block diagram is shown in **Fig 5**.

"The upper channel contains the phase information for the RF amplifier, while the lower channel contains the amplitude information for the modulator. With present technology, the circuitry within the dashed box may be performed with digital processing. Starting with analog-to-digital conversion of the input (audio or other signal), the limiting and detection are obtainable with standard digital techniques. Phase matching between the two channels is achieved by digital logic and kept track of by the clocking system. The modulator would take on the appearance of a high-efficiency variable-voltage, switched-mode power supply. Typically, these supplies can operate easily at switching speeds of 100kHz, well above the sampling rate required for normal input bandwidths. The switching power supply will also satisfy the requirements of DC turn-off of the modulator and RF output during the normal quiet periods of signal input (suppressed carrier). Advantages of this approach include: light weight - because almost no iron or large transformers are required; small size - because of the integration of much of the circuitry, and high efficiency of the Class E RF amplifier and modulator."

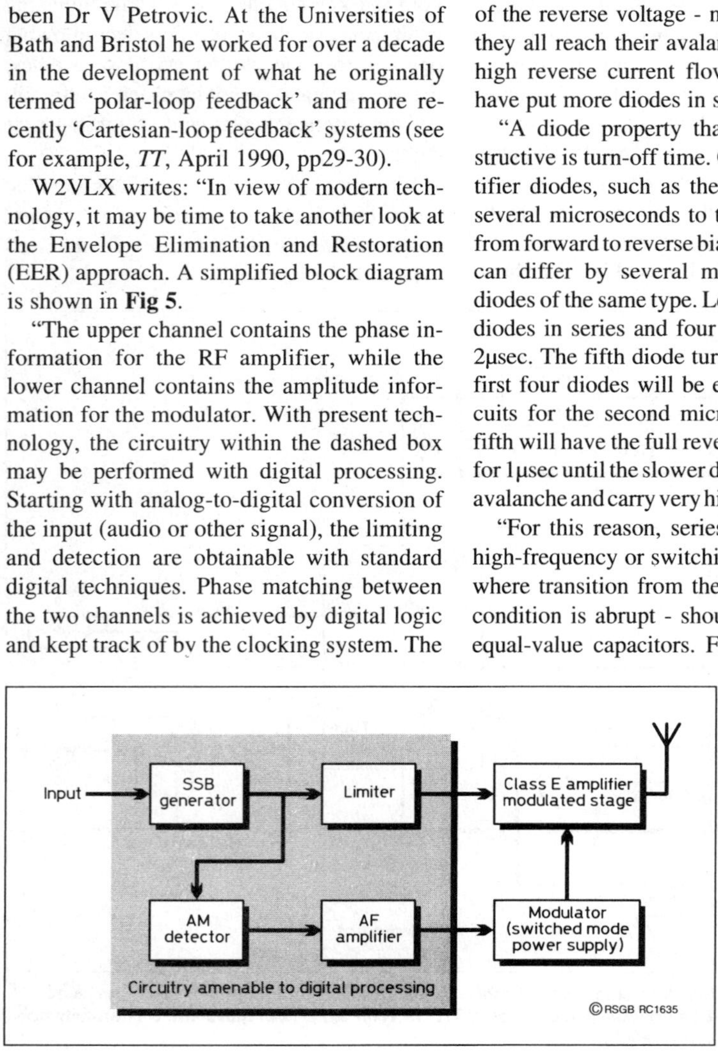

Fig 5: Block diagram of an Envelope Elimination and Restoration (EER) system, based on the work of Leonard Kahn in the 1950s.

DIODE PROTECTION AND OVEN TRANSFORMERS

WITH MODERN AVALANCHE power diodes there is no longer any need when using strings of diodes to connect 'equalising resistors' across each diode, as noted in *TT*, October 1997 p78. Another review of diode protection appears in the *'Technical Correspondence'* column of *QST*, October 1997, pp82-83 by Dan Metzger, K8JWR. He writes:

"Avalanche itself is not destructive. If one diode in a string reaches its avalanche voltage before the others, the others simply take more of the reverse voltage - no harm done unless they all reach their avalanche voltages. Then high reverse current flows, and you should have put more diodes in series.

"A diode property that is potentially destructive is turn-off time. Ordinary 1Amp rectifier diodes, such as the 1N4006, may take several microseconds to turn off when going from forward to reverse bias. This turn-off time can differ by several microseconds among diodes of the same type. Let's say we have five diodes in series and four of them turn off in 2µsec. The fifth diode turns off in 1µsec. The first four diodes will be essentially short circuits for the second microsecond, while the fifth will have the full reverse voltage across it for 1µsec until the slower diodes turn off. It will avalanche and carry very high current for 1µsec.

"For this reason, series rectifier diodes in high-frequency or switching power supplies - where transition from the forward to reverse condition is abrupt - should be shunted with equal-value capacitors. For 60Hz (or 50Hz) sine-wave rectifiers, the gradual nature of the transition makes capacitor addition unnecessary."

K8JWR shows how the value of the capacitors advisable in high-frequency or switching type power supplies can be calculated. However, it is not amateur practice with HT and EHT supplies to use other than 50/60Hz mains supplies, so equalising resistors and capacitors can normally be dispensed with.

John Lawrence, GW3JGA, was interested in recent items concerning the use of ex-microwave-oven EHT transformers in amateur high-voltage power supplies. He writes:

"Some years ago I carried out an investigation into the *'Characteristics of power supplies used in microwave ovens'* and feel it may be of interest to point out that these transformers are designed with a high leakage reactance which gives them special regulating characteristics when used with the correct value of capacitor in a microwave oven power supply."

GW3JGA enclosed an extract from his report, together with a copy of a paper *'Microwave Oven Power: A Technical Review'* by John E Gerling (*J Microwave Power*, E^2, 1987).

There are normally no voltage tappings on the primary of an oven transformer, and even in the UK voltages may vary significantly from the nominal 230V supplies. To quote from GW3JGA's report:

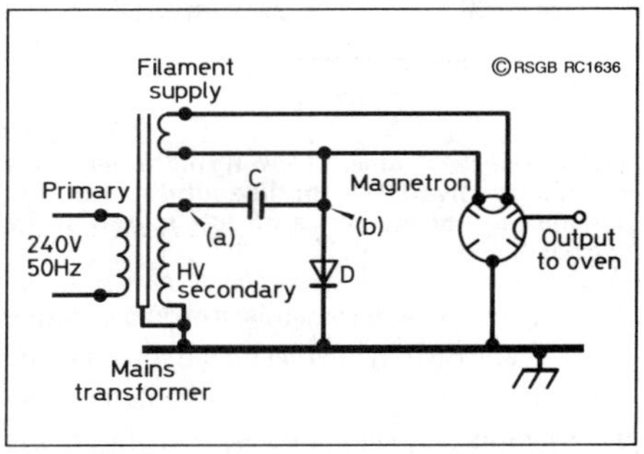

Fig 6: Circuit diagram of the usual microwave oven power supply, providing a pulsed negative supply of about -4kV for the magnetron.

"The RF output from a magnetron is reasonably proportional to anode current, but very dependent on anode voltage. A small change of 5% in the anode voltage produces a large change of around 50% in anode current and consequently in output power . . . . some form of regulation is essential in order to stabilise the output power from the magnetron.

"A practical solution which is used in all domestic microwave ovens is to employ a special type of mains transformer which has a high leakage reactance in the secondary circuit. When this is operated in a simple form of voltage doubling circuit (**Fig 6**) with a specific value of series capacitor (C), the effect is to provide a constant current feed to the magnetron over the normal range of supply voltage variations.

"The leakage reactance of the secondary winding and the series capacitor are chosen to resonate at a frequency about 15-20% higher than the incoming supply frequency. At resonance the effective impedance is at its minimum value. Any increase in the supply voltage will cause an increase in the secondary current and a reduction in the leakage inductance due to saturation effects. The resultant increase in resonant frequency raises the series impedance at the mains frequency, thus limiting the rise in current and stabilising the output power."

This simple, low-cost form of regulation dates from the mid 1960s and was developed by Feinberg of Advance Transformers (US patent 3,396,342). The oven EHT has no ripple filter and is essentially pulsed DC. As GW3JGA puts it: "In a practical circuit, due to the very non-linear voltage/current characteristic of the magnetron and the current stabilising effect of the transformer, the voltage waveform across the magnetron is substantially square in shape, as shown in **Fig 7**, curve (b). The positive half cycle is occupied by the recharging of C and no RF output power is generated. The voltage difference between the waveforms at (a) and (b) of Figs 6 and 7 represents the voltage present across the capacitor at various points in the mains cycle.

Such a pulsed EHT voltage would, of course, be unsuitable for a transmitter amplifier. As described in the October *TT*, oven transformers can be used in conventional half-wave or (with two transformers) full-wave circuits incorporating ripple filters, in which case the high-leakage reactance should not prove a major factor, although it may adversely affect rather than improve regulation on a widely fluctuating load.

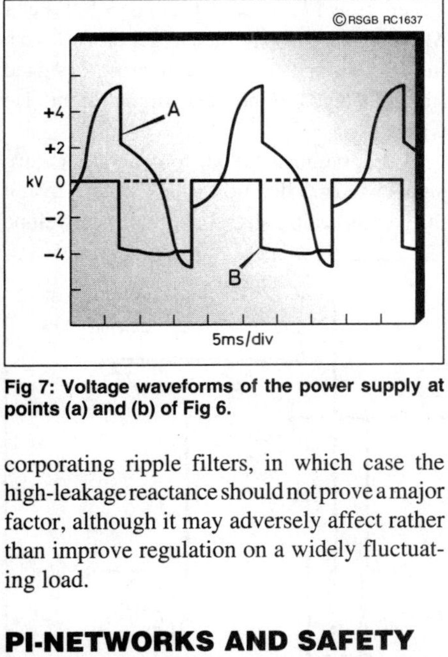

Fig 7: Voltage waveforms of the power supply at points (a) and (b) of Fig 6.

PI-NETWORKS AND SAFETY

WHATEVER THE PUNDITS may say, there is still an important role in amateur radio for high-power valve amplifiers with classic pi-output networks. But because many newcomers have grown up with solid-state and many old-timers may have all but forgotten good valve practice, it seems necessary from time to time to point out that unless precautions are taken it is all too easy to end up with kilovolts of lethal DC reaching the antenna. The use of suitable blocking capacitors and safety chokes is an essential requirement that can be easily overlooked.

Alex Allan, G3ZBE, who has recently completely rebuilt GM3VEY's Yaesu FL-2000 linear to use a single QY4-400 valve found this an interesting learning experience in understanding the grounded-grid mode properly, designing the output network correctly and the biasing and ALC. He writes:

"I have noticed on a few occasions the safety chokes being omitted from pi-networks in *RadCom* diagrams, although in *TT* this is usually picked up. However, the following story may be of interest.

"In the course of rebuilding the FL-2000, I used a 'new' 0.01uF 8kV disc capacitor, purchased at a rally, as the 2.4kV HT blocking capacitor. It blew up in short order. I replaced it with another capacitor from the same source although of a different make; this has proved entirely satisfactory. The point I would make is that many surplus/rally capacitors may not be intended for or suitable for high RF currents, or indeed for RF applications. Ideally, one should use a capacitor designed for RF with a known specification. This is likely to be expensive, but it is a safety critical component. A fact that is not always being stressed enough. Think of the consequences of a long wire antenna carrying 3kV DC!

"Of course, I fitted a safety choke across the output socket of the amplifier, but alerted by the experience of the blocking capacitor blowing, I began to look more closely at the requirements of this choke. If one is using an HT of say 3kV with a smoothing capacitor of the order of 25μF (not unusual these days) with a 50Ω current limiting resistor, at the instant the blocking capacitor blows there is a potential current of 60A, or even more if there is no current-limiting resistor. I took a selection of typical RF chokes from my junk box and subjected them to DC from a current-limited supply, set to supply 30A, for one second. None survived!

"'Ah!' you may say 'a fuse would blow in a lot less than one second'. Really? Are you prepared to stake your life or other people's on that? And did you use a proper HT fuse which won't continue arcing over for some time after it is supposed to have blown?

"Perhaps in connecting any old RF choke across the output of high-power/high-voltage amplifiers we are lulling ourselves into a dangerous sense of security. Few amateurs are in a position to be able to conduct a valid test to ensure that their safety choke would not go open circuit in the event of a blocking capacitor breakdown."

TEMPERATURE AND UHF PROPAGATION

WITH CLEAR EVIDENCE of the effects of the new Sunspot Cycle 23 coming from the DK0WCY beacon in early September (with solar flux of 119 on 8 September and a sunspot number of R101 on 9 September) it may seem perverse to turn to minor variations in UHF propagation. Perverse but interesting!

Ian Waters, G3KKD, of Cambridge writes: "I regularly monitor the transmissions of the Cambridge ATV (amateur television) repeater GB3PV. My computer path-prediction program shows this as a 14km path which, as far as the earth's profile is concerned, is unobstructed and has a path loss at 1316MHz of 121dB. Adding transmitter ERP and the effective gain of my antenna system gives a calculated input to my receiver of -59dBm, which is some 6dB more than that required for a P5 colour picture. There are, however, some large trees (wooden attenuators), mainly towards the ends of the path, which must be causing additional and variable loss (state of the foliage, moisture in trunks and branches).

"Over a period of time I noticed some small variations in picture quality, which led me to record signal level together with temperature and atmospheric pressure. My analysis of these show:

(a) A daily variation with temperature. Signals are strongest in the morning when the air is cool, and weaker in the late afternoon. This variation can be 2dB for a 10°C change.

(b) These are superimposed on a slower variation, due to changes of mean temperature from day to day and also on a still slower variation due to the change of mean temperature from summer to winter. The net effect can be a variation of 6dB from a hot summer afternoon of 28°C to a cool winter night of 5°C.

(c) There is also evidence of an additional 2-3dB rise of signal strength in the autumn, when the leaves fall off the trees (and presumably the reverse in the spring).

"No variation with air pressure could be established. There may be some additional variation between wet and dry conditions, but this could not be separated from the temperature effect.

"I have two antennas. The above results were while using my 'lookthru' antenna which has a gain of 18.5dBd and height of 7.5m. Generally similar results, but with somewhat smaller variations, were obtained using my main antenna with a gain of 22dBd at a height of 10m. The two masts are separated by 5m. I suspect the variations are due to changes in refractive index with temperature, which would explain why they are greater for the lower antenna which would be more affected by a slight 'lift'.

"It is also interesting that the lookthru antenna, with a lower gain at a lower height, gives an almost identical signal level as the main antenna. This is presumably due to it being in a slightly different position relative to obstructions along the path. This makes a good case for 'fishing' to find the best signal that one can on the premises, when wishing to receive a signal arriving from a particular direction."

This latter point is well established among TV antenna installers, who find that due to reflections etc it is advisable to move an array about within a metre cube while a colleague observes the picture or the readings on a signal strength meter. ♦

250W BROADBAND MOSFET AMPLIFIER FOR 136KHZ

THE DECEMBER 1989 *TT* included an item '50W RF from Low-cost FET at 7MHz' contributed by David Bowman (then G8PDW, now G0MRF), using a 2SK413 n-channel MOSFET costing under £5. Subsequently this was developed into a broadband 1.8 to 21MHz push-pull amplifier using two of these devices (fully described in *RadCom*, February-March 1993), capable of 100W output from 1.8MHz to 14MHz, and slightly less on 18MHz and 21MHz.

With the imminent introduction of the 136kHz allocation, he decided to re-examine the use of low-cost MOSFETs as a means of providing high-power amplification at LF. The amplifier shown in **Fig 1** provides some 250W output from 9W input. It is based on his 1993 design, but is optimised for LF operation. It uses four 2SK413 devices in a parallel push-pull configuration.

G0MRF writes: "Changes to the 1993 amplifier include reduction of the power supply voltage from 48 to 28 volts; paralleling of two MOSFET devices on each side of the push-pull amplifier to increase output power; and re-design of the output transformer and drain-feed choke using a more appropriate ferrite.

"The 28V power supply allows use to be made of the large number of surplus switched-mode power supplies (SMPSU) now appearing at rallies, etc. The 250W output is limited by my 28V, 12A SMPSU, rather than the amplifier itself.

"The technique of paralleling pairs of FETs has been used in the H Granberg designs for Motorola, but this is the first time I have tried it. In practice, simple paralleling works extremely well. The devices share current evenly if operated in class B (ie just on the point of conduction). Better RF linearity can be achieved in Class AB, with approximately 200mA per device idling current, but to set this accurately would really require a separate bias source per device.

"The inclusion of 1Ω resistors in series with the gates is important for stability, as is the 4700pF capacitor across the input transformer. The 22Ω resistors to earth on each device provide a passive gate technique, broadening the bandwidth and further enhancing stability. With the circuit as shown (50-Ω in/out) the input VSWR is less than 1.4:1 from 60kHz to 1MHz.

"For a second version, I have reduced the number of turns on the primary of the input transformer, to allow the amplifier to be driven from a TDA2030 audio-amplifier IC. The 'volume control' gives excellent control of RF output power.

"I had some difficulty in selecting core material for T2 and T3. At 136kHz the ferrite used in the 1993 HF design was on the edge of saturation at 100W, with its temperature rising to over 80° after five to ten minutes. It appeared that a core of higher permeability and greater mass was needed. Finally, 3C85 material was selected. This ferrite is extensively used in switched mode power supplies, but usually in the form of an E-type core. As I had experienced flash-over with some E-cored transformer designs on 73kHz, I searched for a toroid so that the windings could be separate. Finally I discovered that Philips Components manufacture several 3C85 toroid cores, and these are available from Farnell Electronic Services. These worked perfectly.

"The output core is very large and barely warms up at 250W. It would be suitable for use at powers up to close on a kilowatt. Another advantage of the toroid is the possibility of a series of taps for impedance matching. 50Ω output is the preferred value, since it facilitates power output measurements. My Bird Throughline with a 2-30MHz sensor reads zero, but the cheap SWR bridge from my shack reads just 30% low on 136kHz, a usable figure.

"Since the amplifier is likely to require matching to electrically very short antennas (eg Marconi-T), my original thought was to have a very low impedance output, perhaps less than 2Ω. However, based on 73kHz experience, the addition of losses in practical loading coils and ground systems raises the resonant feed impedance to a level higher than 50Ω; 80 to 200Ω would be a good estimate. For these impedance levels the toroid can be tapped to give a good match directly into the antenna, but don't forget to use a low-pass filter."

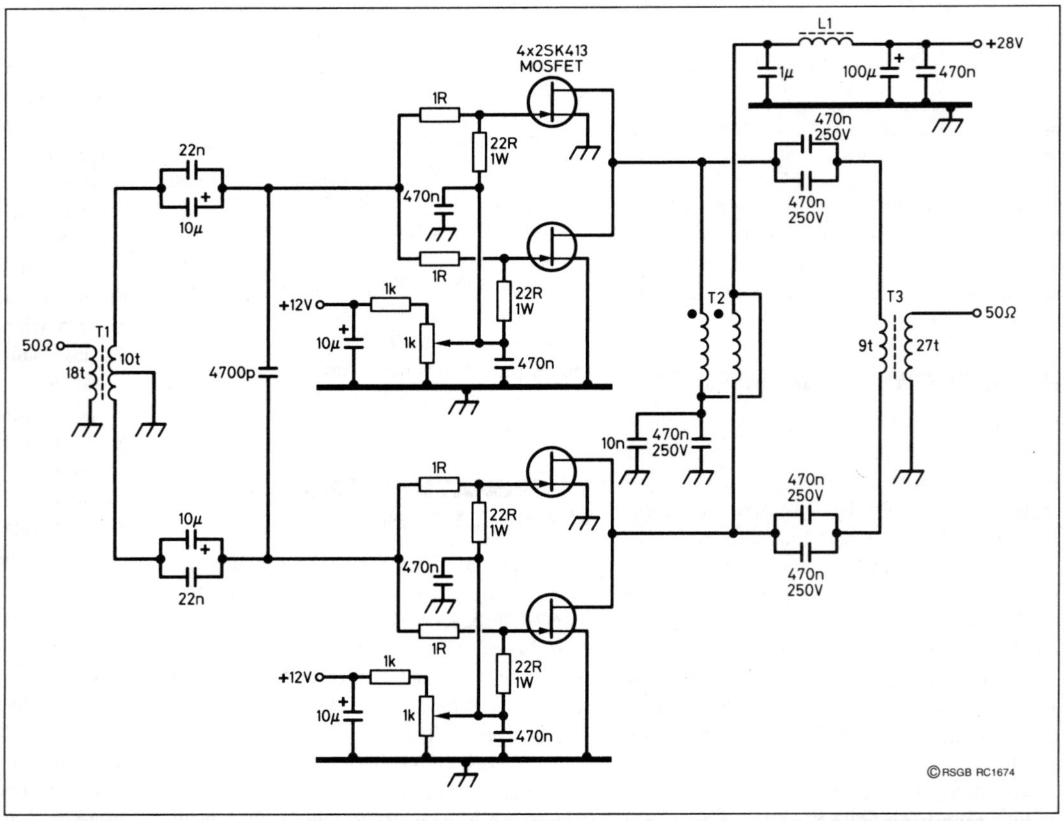

Fig 1: G0MRF's broadband LF amplifier, using four low-cost MOSFETs in parallel push-pull, providing 250 Watts or more at 136kHz. T1 Primary 18 turns 0.56mm wire, secondary 10 turns 0.8mm wire on Fair-Rite 59-61000401 toroid. T2 10 bifiliar turns, 0.8mm twisted pair on Philips 3C85 25/15/10 toroid. T3 Primary 9 turns, 1mm dia. Secondary 27 turns, 1mm dia enamelled copper wire on Philips 3C85 58/41/18 toroid. L1 RF choke rated at 12A.

G3PLX'S DSP ECHO-DETECTION SYSTEM

FOLLOWING THE DETECTION of 209-millisecond echoes on his electronic hellschreiber system (see *TT*, January 1998, p58-59), Peter Martinez, G3PLX decided to get to work with his DSP kit and establish a purpose-designed Long Delayed Echo detection system. He writes:

"I programmed the kit to generate a chirp signal of 32mS duration which sweeps over a 2kHz bandwidth, shaped so that its envelope

**37 Dovercourt Road, London SE22 8SS*

amplitude looks like a carrier that is 100% amplitude-modulated with one cycle of a 31.25Hz sinewave. This is a well-documented technique for obtaining enhanced radar resolution. My pulse, when processed by a matched filter in the DSP, gives a 1mS resolution, but with 12dB more energy per pulse than from a normal 1mS pulse. I arranged my TS930 transmitter so that it would send one of these chirps every two seconds, switching quickly back to receive, with the DSP processing the signal for about 400mS after each chirp, plotting the result (on the screen) in real time and saving it (in a file) for later study.

"I left this running on a number of days during November, trying different bands and times, but heard nothing. Eventually I began to lose interest. However, at about 2100GMT on the evening of December 4, while about to call CQ on 3.5MHz, I put out a single burst of carrier to check that the antenna was tuned. As the receiver came back on, I heard a distinct echo. I tried again and it was still there. My heart skipped a beat as I quickly loaded the echo-sounder program. The result was a clear, sharp image of an echo, which gradually faded out over a period of about 10-12 minutes. (G3PLX enclosed a copy of the image that fully confirms his claims. He could supply the image in electronic form to anyone seriously wishing to study it closely: Tel 01539 56870. Fax on request. Internet: peter.martinez @btinternet.com - G3VA).

"The image shows that the echo is very sharp in the time-domain, being at least as narrow as the resolution of my system, with a signal level about 25dB above receiver noise at its strongest. The delay was 218mS for most of the time, but in the later stages wandered about a bit. Microscopic inspection shows three distinct delay times, at 218, 220 and 223mS, which fade in and out.

"In order to eliminate various possibilities, I changed radio frequency several times during the 10 minutes or so that the echo was audible, but there was no change in the echo. I also changed the pulse interval to a random value to ensure that I was not listening to an echo at 2.218 seconds."

My personal guess remains that these echoes are in some way connected with reflection of vertically radiated signals from ionisation of a Van Allen belt, occurring for a brief period as the critical frequency of the much lower F layers drops below 3.5MHz, permitting near-vertical signals to pass through the lower F layers etc and reach a Van Allen belt sufficiently ionised to bounce some of the signal back down. There could well be other hypotheses!

As to the reason why the really long delay echoes of the 1920s and 1930s seem to have vanished, my guess is that this may in some way be connected with the "heating" of the electron-temperature of the lower ionospheric layers by the vastly greater number of high-power transmitters that came into operation from about 1940 onwards. The raising of the electron temperature of the plasma by RF was found to be the cause of non-linear inonospheric cross-modulation (The "Luxembourg Effect", first observed in the 1930s). If this is correct, then we are unlikely ever again to observe echoes lasting seconds rather than milliseconds (other than in the form of moonbounce), unless there is a mass switch-off of the world's high-power transmitters.

PA0KSB ENDORSES 'FAST' HUFF & PUFF

TT, DECEMBER 1997, pp71-2 presented a brief outline and circuit diagram (December Fig 11, p71) of a new generation of 'fast' huff and puff oscillator stabilisers, developed by Peter Lawton, G7IXH.

There were a couple of minor errors in the diagram, the corrections for which are given in **Fig 2(a)**. For the 2N3819A input amplifier, the slider of RV1 and the 470k resistor are connected to the junction of the 1k5 resistor and the 10nF capacitor (shown as a cross-over). Similarly for IC8 (CA4130), test pin 15 and the slider of RV5 are connected to pin 3 of IC8. G7IXH has also slightly modified the 5V power supply circuitry, the new arrangement being shown in **Fig 2(b)**. With the integrator given its own lead from the 12V input and two 4.7µF capacitors added across the 5V supply, it helps to ensure that no sampling frequency gets on to the VFO.

These minor errors and changes did not prevent Klaus Spaargaren, PA0KSB (who developed the original Huff & Puff stabiliser in 1973 and the improved version a couple of years ago) from trying out G7IXH's new 'fast' arrangement. He comments:

"With great interest I have read the contribution of Peter Lawton, G7IXH. It really is a great improvement, technically sound, and simple to implement. Please spare a few lines publicly to convey my congratulations to him. 'Why have not I thought of that myself?' was my first reaction when I understood the principle. It is so obvious and simple.

"The heart of the matter is indeed to increase the update frequency without increasing the small spacing between the stabilisation points. This is what he has done, although it took me some time to understand fully the way it worked. I am still puzzled by the statement 'refreshes the corrections at 257 times the rate of the basic stabiliser', which seems misleading, unless there is some further trick that I have failed to understand."

PA0KSB has used a specific example to derive his own explanation of how the system works, and feels that others may find this useful: "The shift registers operate as a delay line. Suppose that the clock frequency is 10kHz. Every 0.1 milliseconds, the contents of the register are shifted one position towards the output. There are 256 stages, so it would take 256 times 0.1 milliseconds, ie 25.6 milliseconds, before an input appears at the output.

"An input signal with a period of 25.6 milliseconds (about 39Hz) is shifted in phase by 360° and appears at the output in phase with the input signal. A similar situation exists for 2, 3, 4 . . . N times 39Hz, all of which get phase shifts of 2, 3, 4 . . . N times 360° and are thus in phase with their input signals. So there are a large number of frequencies where inputs/outputs of the shift registers are in phase.

"The switches in the HEF4053 IC operate as synchronous detectors, and produce average DC levels depending on the phase shift between the input signals and the control signal. The average DC output is a maximum (5V) when the signals are in phase, and a minimum (0V) when they are 180° out of phase.

"With the input frequency of the shift register swept slowly upwards, a DC voltmeter connected to the output of the phase-sensitive rectifier HEF4053 (Test Point 15) would show a voltage gradually increasing from 0 to 5V,

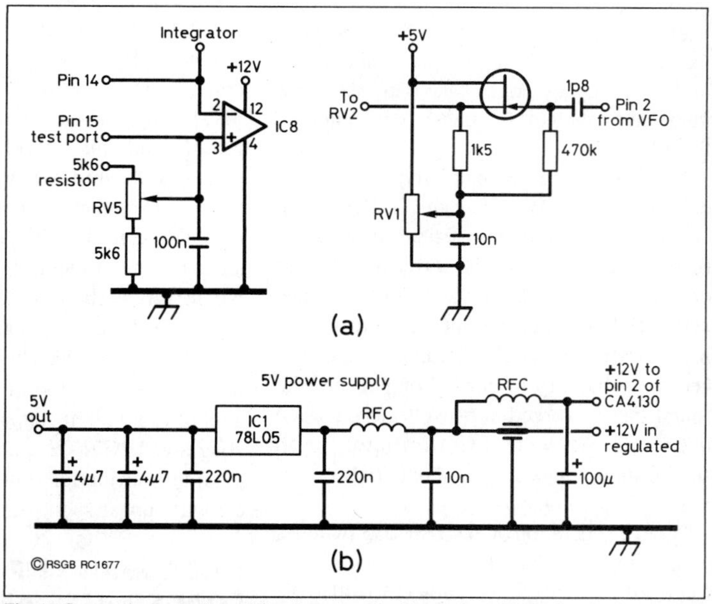

Fig 2: Corrections and modifications to the December 1997 circuit diagram of G7IXH's "next generation" Huff & Puff stabiliser.

then as the input frequency rises further it would gradually return to 0V and then again rise, and fall, and so on.

With a phase shift of 90° or 270°, the average DC output would be half way between the minimum and maximum values, ie approximately 2.5V. In this situation the integrator output remains stable, with its positive input connected to 2.5V. If the first frequency started at 39Hz, the first frequency where the 90° phase shift occurs is at about 49Hz, representing a stable operating point in a closed-loop situation.The next one would be 39Hz higher, and so on.

"Because of the presence of the digital mixer, the lock points of the VFO frequency will be spaced less than 39Hz apart, so long as the VFO frequency is lower than the 32MHz drive frequency of the digital mixer, which itself can vary between 0Hz and half the clock frequency. As in my original design, G7IXH exploits this effect.

"Now consider the update and the resulting ripple frequency. In the above example, where the closed loop would stabilise a 49Hz output frequency of the digital mixer, then owing to the configuration of the switches in the phase sensitive detector HF4053, the output ripple would be 98Hz, ie twice the input frequency. At the next possible lockpoint (49 + 39 = 88Hz), the ripple frequency is 176Hz."

As PA0KSB sees it, the update frequency is not constant. It varies at twice the value of the output frequency of the digital mixer, which itself can vary between 0Hz and half the clock frequency. In the example given above, the update frequency could vary between a low value and the clock frequency, depending on the VFO setting.

Even though the update frequency is not constant, PA0KSB agrees with G7IXH that it is much higher than in his original designs - and that the 'fast' stabiliser can correct much more drift with less ripple on the VFO frequency. PA0KSB built and experimented with a stabiliser based on the G7IXH circuit of the November *TT*, subject to available components. He found that it worked very well indeed when used on a 28MHz LC oscillator, with stabilisation points spaced about 40Hz.

PA0KSB comments:

"(1) The operation of the HEF4053 switches (apart from the initial action at switch-on, forcing the output of the analog CA3140 to the mid-range) resembles the operation of an exclusive OR gate. I used a digital HEF4030 EXOR instead of the HEF4053 analog switches. The results appear identical.

"(2) Shift registers type HEF4517 are seldom found in the lists of IC retailers, probably because of low demand. Instead of using two HEF4517 devices, I used four HEF4031 ICs, which are readily available in Holland. Each 4031 contains only one 64-bit shift register, but with four cascaded devices an overall register length of 256 bits is obtained.

"(3) Whilst I initially doubted whether the 'fast' stabiliser would behave well when the output from the digital mixer was at the extreme ends of its range - especially since the high output frequency looks irregular on an oscilloscope - I was very pleased to find that the stabiliser continued to operate perfectly normally at these extreme frequencies. I could not find any stabilisation point where the system loses control and where the VFO frequency would unintendedly jump to another lock point.

PA0KSB also draws attention to the drawing errors mentioned above. He also feels that the analog integrator should have been shown as a CA3140 rather than CA4130, but G7IXH has not confirmed this.

In retrospect, PA0KSB feels that the new 'fast' stabiliser has a conceptual affinity with the 1970s oscillators stabilised by the early mechanical 64-microsecond PAL colour TV delay lines (glass or steel), first described for professional/military VHF applications by Mike Underwood, G3LKZ (see *TT*, August 1972 and July 1975) and later as an HF "PALO" VFO for amateur use by Brian Rose, G3ULR (*TT*, September 1975). I recall that G3ULR sent me his VFO for evaluation. It was extremely stable on its 15.625kHz lock points, but had an unfortunate tendency to occasionally jump lock points, making it to my mind operationally impractical (apart from having 15.625kHz frequency steps).

Finally, PA0KSB believes there is considerable room for further "improvements". As the spacing between the stable VFO frequencies depends on the delay time (number of stages in the shift register and the clock frequency) a larger number of shift-register stages than 256 seems feasible. He wonders whether static computer memory ICs with some simple driver circuits could be used to make extremely long shift registers, working at a high frequency. That should upgrade the performance even further. He concludes: "I had a lot of fun with my experiments and want to congratulate G7IXH again, and thank him for his fine contribution. I hope his detailed article will appear soon."

BROADBAND HF IMPEDANCE-LOADED KITE ANTENNAS

"NEW DESIGNS OF ultra wide-band communication antennas using a genetic algorithm" by Zwi Altman, Raj Mittra and Alona Boag (*IEEE Trans Ant & Prop*, October 1997, pp1494-1501) describes a two-step design technique that can result in vertically polarised kite antennas capable of effective, low vertical angle operation over ultra wide low VSWR bandwidths of 4 to 30MHz (four-arm) or even 2 to 30MHz (eight-arm). The four-arm version (**Fig 3**) requires 12 impedance-loading 'tanks' (resistance loaded traps), and the eight-arm version 24.

Although basically a design paper, it provides worked examples of the impedance loading tanks and matching networks for both versions. The loads enhance the antenna characteristics to yield high gain and low VSWR by modifying the current distribution, and can

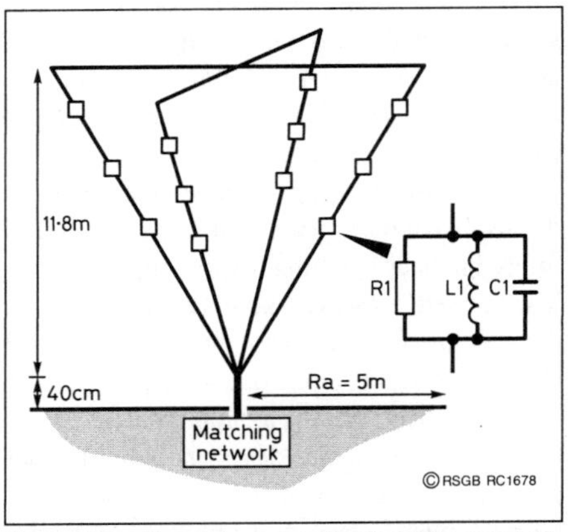

Fig 3: Broadband four-arm loaded kite antenna, mounted over an infinite, perfectly conducting ground plane.

force it to radiate nearer to the horizon in the elevation plane with an almost omnidirectional pattern in the azimuth plane.

Note that "kite" refers to the shape of these antennas. They are not intended to be carried aloft by kites.

VOLTAGE REFLECTION COEFFICIENT

BERNARD SPENCER, G3SMW writes: "I was very pleased that Jim Ford, N6NF is speaking out against the common practice of quoting the match in terms of VSWR (see *TT*, December 1997, p68). Standing Wave Ratio itself is not relevant, unless one is using a slotted line (or 3-Probe device) for measurement, which is seldom the case nowadays, or if the peak voltage on a cable is being calculated.

"In RF matching, the easiest to visualise and to manipulate is Voltage Reflection Coefficient. This has a magnitude which can be used instead of VSWR, and also a phase angle for those who want it to specify completely a complex impedance. The magnitude is simply the fraction of the forward voltage which is reflected, expressed as a decimal. Also, the magnitude is zero for a perfect match, and unity for the worst possible match (short- circuit or open-circuit). These are easy to visualise and to put into equations or into calculators. This is unlike VSWR, where a short- circuit or open-circuit is infinity, or Return Loss where a perfect match is an infinite number of decibels. Admittedly it is convenient to measure a mismatch in terms of Return Loss if your measuring system uses an attenuator or scale marked in decibels, but it is equally convenient to make and use a system reading Reflection Coefficient if you are constructing an instrument.

"Of course, it is not difficult to calculate Reflection Coefficient from VSWR or Return Loss, or vice versa, but it is an unnecessary step in most cases when the Reflection Coefficient is measured. The fixed resistive unbalanced bridge is called different things by different makers, such as 'Reflection Coefficient Bridge', 'SWR Bridge', 'Return Loss Bridge', or 'Autotester' etc. I prefer 'Reflection Coefficient Bridge' (or simply 'Reflection Bridge'), because the basic output (before any diode or other non-linear device) is proportional to the Reflection Coefficient."

TUNED TRANSMITTING LOOPS CONTROVERSY

THE NOVEMBER, 1997 *TT* item reporting the belief of Professor Mike Underhill, G3LHZ and M J Blewen, G4VRN that the compact tuned transmitting loops tend to function primarily as small folded dipoles rather than as magnetic loops has certainly attracted serious attention, indeed it would be possible to fill *TT* for several months with the many thoughtful comments received. On one point, virtually all agree: these compact antennas are surprisingly effective no matter whether their total radiation efficiency is or is not as low as was suggested by Tony Henk, G4VXF in 1991.

While the majority of those who wrote to me - including Jack Hollingworth, ZF1HJ; Dr John (Jack) Belrose, VE2CV; B Edginton, G0CWT; Jimmy Bolton, G3HBN - remain rather sceptical of the dipole-mode hypothesis, there has also been some disbelief in the validity of the 1991 article of Tony Henk, G4XVF. I would stress that G4XVF's findings as relating to the magnetic loop mode are not disputed in any way by G3LHZ. I understand that most of those who commented directly to G3LHZ supported his view that dipole-mode radiation from the loops may be the dominant mode.

But first a word of explanation about the "AMA-3" loop, used by G3LHZ/G4VRN, and also reviewed by G3SJX in 1994. These loops were, as stated, manufactured in the UK by Advanced Antennas and Ancilliaries (no longer trading), apparently based on a design by Dipl-Ing Christain Kaeferlein, DC5CZ, (Weingbergstrasse 5, Darmstadt D-64285, Germany), whose firm has, since 1983, manufactured a range of Abstimmbare Magnetische Antennen (AMA) loop antennas, including the AMA-3. The mathematical/electrical design of the German AMA-3 loops was based on the work of Hans Wuertz, DL2FA (summarised in *TT*, October 1986 and *TT Scrapbook 1985-89, P*127-8). Detailed information on the original AMA loops, still in production, has been published by DL5CZ in German and English. **Fig 4** shows the basic design of the AMA loops. **Fig 5** shows two AMA loops on a single pole.

ZF1HJ is representative of the sceptics. To quote briefly from a long, well reasoned letter: "There is no doubt that claims for the efficiency of magnetic loop antennas by both manufacturers and writers of construction articles are often greatly exaggerated. They appear to result from a calculation of radiation resistance derived from physical dimensions and an estimate of resistive losses only in the loop itself, based on its dimensions and material, perhaps with a fudge factor worked-in to allow for skin effect. As my friend G4XVF pointed out in his excellent article, there are other significant sources of loss which result in much lower than anticipated efficiencies.

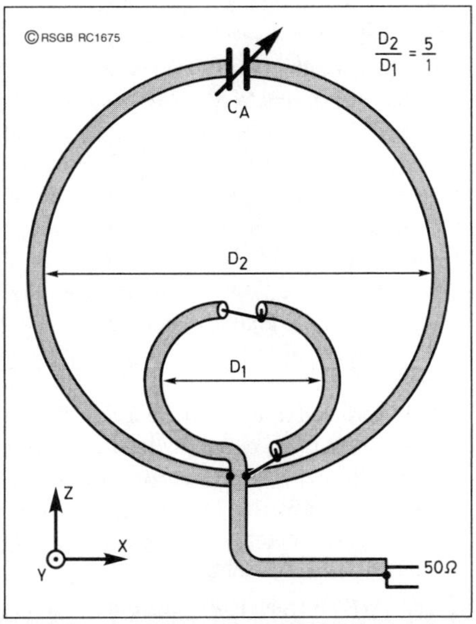

Fig 4: Basic design of the DL5CZ "Abstimmbare Magnetische Antennen" (AMA) tuned transmitting loop antennas with coupling loops, as manufactured in Germany since 1983. Currently there are some 24 models, grouped in four diameter sizes (0.6m, 1.3m, 1.7m and 3.4m).

"Nevertheless, I do not share Tony's pessimism. Like others, in practice I have found these antennas, even when mounted close to the ground, perform almost as well as a dipole at a much greater height. I have (at least until now) considered the explanation to lie in there being a substantial lobe in the vertical plane at a relatively low angle to the horizon, so that the loop, despite its low efficiency, radiates more power than a dipole at the low angles required for DX communication." (On the other hand, the attraction of loops for military applications is the strong vertical radiation required for short/medium range NVIS working - G3VA).

VE2CV has also found from personal experience that loops can be an excellent choice for HF camp/portable/mobile operation. Nevertheless he believes that "the proposed dipole-mode for compact loops is in fact fiction . . . It is surprising how well compact loops work, even on bands where the efficiency is very poor (HF can be very forgiving)." With access to genuine AMA loops of 0.8, 1.7 and 3.4 metre diameters, he has carried out both rigorous measurements and NEC-4 (EZNEC pro version with the NEC-4D machine) studies on the AMA-11 loop at 3.852MHz, showing efficiencies of 4.7% and 5.5% respectively."

VE2CV concludes: "A small compact loop behaves like a loop. The method of tuning and matching, resonating by a capacitor, and magnetically coupling power to the loop by means of a small coupling loop, is an efficient way to resonate and match the loop. To keep conductor losses low, the diameter of the loop's conductor must be large. The conductor diameter of the DK5CZ AMA loops is 32mm, the loop being made from a formed (almost a perfect circle) aluminium tube."

Due to space considerations I drastically condensed the information from ZF1HJ and VE2CV. My apologies to these - and the other correspondents - if I have misrepresented or omitted their views.

I passed the detailed (critical) comments made by G4XVF and VE2CV to G3LHZ, who, although very much bound up with his professional academic work, has provided additional notes in support of his dominant dipole-mode hypothesis. One of the contentions raised by G4VXF related to the space in the IEE Conference Paper given over to the calculation of inductance. He points out that this is well covered in NBS Technical Note C74. Using this, G4XVF calculated the loop inductance as 1.748µH without introducing a fiddle factor, as against the measured value of 1.77µH. G3LHZ has not yet had a chance to study C74, but he has checked the most quoted NBS inductance formula for a single-turn loop, which appears in some editions of the ITT *"Reference Data for Radio Engineers"*:

$$L = a/100\ (7.353 \log_{10} 16a/d - 6.386\ )\ \mu H$$

where 'a' = mean radius of ring in inches, 'd' = diameter of wire in inches, and $a/d > 2.5$

G3LHZ comments "There is no doubt that this is a very good formula, giving excellent predicted results at low frequencies. Out of curiosity I have manipulated and changed the units of the formula for direct comparison with the formulas given in my paper. Interestingly, all formulas can be encompassed within exactly the same expression. The only difference is a single constant k (a fiddle factor?) which differentiates between the formulas . . . Theory derived from direct solutions of Maxwell's equations for wires gives k=1. The NBS loop inductance formula corresponds to k = 2.166. Our paper calculated k as equal to 2.09, but suggested it could be rounded down to 2.0. Since 2.09 is nearer to 2.166, I am happy for others to accept a rounding up to 2.166 on the basis of the NBS measurements. The effect on the calculated inductance values is a little over

1%. The real mystery is why there is a two to one difference between theory and practice. It implies that the effective diameter of the wire or tube is doubled over its real diameter. Further investigation is clearly necessary!

"Also in our paper we suggested that at higher frequencies, approaching the first self-resonance of the loop where the circumference of the loop becomes a half-wave, transmission line effects should be taken into account. This can be achieved with all the above inductance formulas by multiplying by the factor (tan y)/y, where y equals $\pi^2 D/\lambda$. This give inductance values which are 27% higher at half the resonant frequency, +5.5% at a quarter wave, and +1.3% at one-eighth-wave. (At 14.3MHz for an 0.833m diameter loop, the increase of 5.5% should be applied)."

G3LHZ then turns to the $64k question of whether dipole-mode radiation is fact or fiction. He writes:

"Judging by letters and phone calls received or passed on, the dipole-mode proposal has raised a lot of interest. There would appear to be slightly more believers than sceptics. The case I would make for the dipole-mode now includes the following points, intended to counter the negative comments so far received.

1) Q values measured by SWR and antenna-bandwidth methods are much lower than predicted by the traditional loop radiation formulas.
2) There is a large increase in Q, typically by two to three times, if the loop antenna is placed in a large, reasonably low-loss tin box (three different screened rooms have been tried). This is Wheeler's Cap method for measuring the ratio of radiation resistance to antenna loss resistance.
3) The loop antenna does not get hot (enough), even with 400W supplied to the antenna at the lowest frequency of operation.
4) There is a strong, horizontally-polarised field, radiated from a vertical loop. 3ft long fluorescent tubes remain ignited to at least two loop diameters distant from the loop (with 100-400W) but only if deployed horizontally. This component is retained (dominantly) in the far field.
5) The SWR and match resistance at the minor loop input remains approximately constant for up to a three to one frequency range on a given loop antenna. The traditional loop formulas would predict a nine to one change in antenna input resistance over this range, provided antenna losses are kept small.
6) The patterns predicted by the NEC simulation code as given in VE2CV's 1993 *QST* article are exactly those given by the assumption that either the current around the loop is exactly uniform or that any out-of-balance currents do not radiate. These patterns do not show any dipole-mode radiation. B Edginton, G0CWT has supplied loop measurements showing a two to one ratio in currents (I think he used thermocouple-type RF ammeters, although this is for somewhat larger loop sizes). G4VRN is checking this for smaller loop sizes, and we also intend measuring the relative phases at the two sides of the loop. Using a simple, ferrite rod current transformer, first indications are that the currents are exactly as one would expect at the ends of a shorted-transmission line of length equal to half-loop circumference. At 14MHz a ratio of 0.95:1 and at 28MHz a ratio of 0.75:1 were measured."

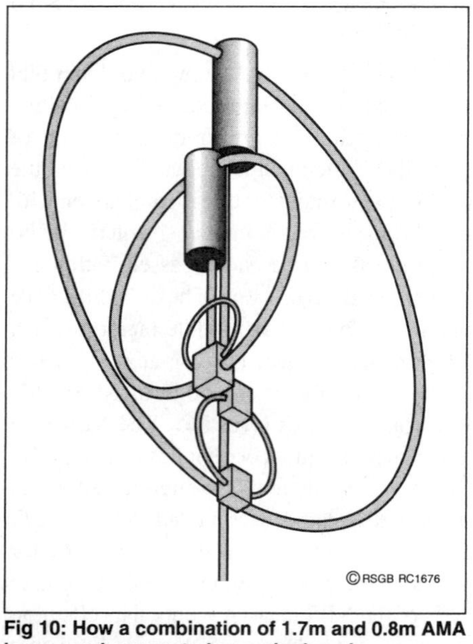

Fig 10: How a combination of 1.7m and 0.8m AMA loops can be mounted on a single pole.

"STEREO" RECEPTION OF CW

IT HAS BEEN NOTED several times in *TT* that pseudo-stereo reception of weak CW signals can benefit from the so-called 'cocktail party' effect. This was first investigated and patented by the late Dud Charman, G6CJ as long ago as 1958 and was first described in *TT* in October 1973. A practical analogue-type two-channel (I/Q) stereocode filter was described fully by G6CJ and G3OTK in *Radio Communication*, September 1975, pp74-81. Much simpler circuits providing phase shifts have also appeared in *TT*.

The *'Digital Dimension'* column by Stan Horzepa, WA1LOU in *QST*, October 1997, p97 notes that Brian Beezley, K6STI, well known for his NEC-type antenna software packages, has recently announced a new 2.0 version of his 'DSP Blaster' software. This now includes coherent stereo digital processing of CW signals, providing the listener with a flexible choice of: "listening to the I channel for monophonic SNR enhancement; monitoring the Q channel to check signal coherence and phase-lock quality; or you can select the incoherent bandpass filter output to obtain slow roll-off noise shaping without carefully tuning for phase lock; or - best of all - you can use stereo . . . . as a result the CW signal appears centred in your head while the noise becomes diffuse, non localised and non directional. This provides a striking spatial separation of signal and noise with off-frequency signals appearing off-centre but spatially distinct.

While it is probable that the digital processing in this software is superior to and more flexible than the 1975 analogue Stereocode filter, it does need a fairly sophisticated set up: a 486 (or better) PC, math co-processor, VGA monitor, 16-bit Creative Labs sound card and mouse. The software package costs $125 (K6STI, 3532 Linda Vista, San Marcos, CA 92069, USA. e-mail k6sti@n2.net)

G3LHZ concludes: "It is a pity that NEC fails to model the tuned loop correctly. This is because it fails to predict or take account of any imbalance in current around the loop. While for most applications NEC is enormously successful, like all simulations it is only as good as its internal assumptions, no matter what arithmetical precision it has.

"For the time being, I rest my case for the dipole mode of tuned loops on the above six points, but I am sure we have not heard the last word on this subject. For example, I intend to make measurements of the far field phase and will try to relate this to the loop current phase. This should give further information on the exact mechanism of radiation for this or any thick or thin wire antenna."

HERE AND THERE

PETER HALLS, G4CRY noted the 300W and 500W Class-E single MOSFET amplifiers mentioned in *TT*, August 1997, pp77-78. After searching the Internet he discovered that kits of parts are being offered by Caltech (Puff Distribution). Further details of these low-cost Class-E kits are given in *QST*, November 1997, p94. Component-only kits cost $60 (outside USA), including shipping by surface mail. For more information, call or write to Puff Distribution, MS136-03, California Institute of Technology, Pasadena, CA 91125, USA. Tel 818-395-2251; Fax 818-395-2137; http://www.systems.caltech.edu/EE/Faculty/rutledge/puff.html.

G4CRY mentions that he has purchased the SGC500 linear from Nevada. It covers all HF bands without retuning after a QSY and is operated by RF sensing, although PTT is an option. It can be remotely sited and is a lot cheaper than some competitive solid state amplifiers. ♦

DARC'S DK0WCY AURORA BEACON

THE OCTOBER 1997 *TT* item 'Plotting the progress of Cycle 23' remarked on the extremely useful solar data service provided by DK0WCY. Dr Ing Hartmut Buettig, DL1VDL, DARC's HF Manager, was glad to see my appreciative notes on this German "Aurora Beacon" near Kiel (JO44VQ) and has sent me detailed information on this excellent guide to HF band conditions and VHF aurora alert service.

Apart from continuous operation on 10,144kHz (30W into a triangular, horizontal loop antenna), DL0WCY operates also on 3.5MHz from 0700 to 0800 UTC and 1500 to 1800 UTC (an hour earlier during European summer time) on 3,579kHz (30W into a dipole). There is a first alternative of 3,567.5kHz and a second of 3,557.5kHz that may be used if there is commercial station interference. In London, the 3.5MHz signals are proving extremely reliable at times when the 10MHz skip during winter early mornings and evenings puts me into the zone of silence. Except during very disturbed conditions, I can usually copy the beacon on one or the other frequency, from 0700 to 1800 hours, despite the data interference on 10MHz.

The aurora alert takes the form of changing the idling loop, so that instead of the few seconds of continuous carrier there is a series of dots. During the solar data messages sent every five minutes, only the following expressions are used to forecast solar activity (**Sunact** forecast): **Quiet**, less than 50% probability of a C-class flare; **Active**, C-Class flares expected (probability greater than 50%); **Major flares expected**, X-class flares expected; **Proton flares expected** (probability greater than 50%); **Warning condition** (activity levels expected to increase); and **Nil** (end of alert period).

The **Magfield** forecast comprises one of the following:

- Quiet
- Active
  Conditions Expected
  k = 4, A ≥ 20
- Minor Storm Expected
  k = 5, A ≥ 30
- Major Storm Expected
  k = 6, A ≥ 50
- Severe Magstorm
  Expected
  k = 7, A ≥ 100
- Magstorm in Progress
  k = 4, A ≥ 4
- Warning Condition
  Activity levels expected to increase
- Nil, end of warning period.

Note also: 'NA' means 'not available'.

While the **R**, **Flux** and **Boulder A** readings are those supplied from Boulder, a flux-gate magnetometer (see photograph below) constructed by Ulrich Muller, DK4VW at the Kiel beacon, provides the Kiel k and A ratings. R is the relative support number (SSN), and is one of several items of information automatically retrieved by a computer at the University of Marburg via Internet from the Space Environment Center in Boulder. Once per day (in the morning) these details are forwarded by telephone to DK0WCY, again automatically. Further information can be added manually at Marburg or at the beacon. The range of **Kiel** k, 0-9 (with 9 greater than 600 nano Teslas). The range of **Boulder A** and **Kiel A**, 0 to 400nT.

As an aurora warning beacon, DK0WCY aids VHF operators by warning them of the more severe disturbances of the geomagnetic field, associated with visual and radio aurora. If the Kiel k figure is ≥ 5, it will normally denote that auroral contacts on 144MHz may be possible for stations in Northern Germany (and similarly in the northern part of the UK). Should the k figure reach or exceed 7, then 144MHz auroral contacts throughout Europe can be expected. On HF, when the magnetic field is much disturbed (high k or A numbers), radio conditions, particularly on paths crossing the polar regions, will be degraded in quality (see also November 1997 *TT*).

DARC publish a useful DK0WCY logsheet (one sheet per month), designed to enable amateurs to obtain a good 'feel' on how to interpret the data when compared with their own observations of conditions on the bands. Further information on this remarkably successful DARC project can be found (mostly German text) on http://members.aol.com/DARCM15/Bake.htm.

Just one word of warning. In the *Technical Correspondence* column of *QST*, December 1997, Carl Luetzelschwab, K9LA, points out that for those using computer propagation prediction programs, it is advisable to use the monthly smoothed solar flux (or smoothed sunspot number) rather than the daily numbers. He points out: "That's how the model of the ionosphere was developed. Using the daily solar flux will significantly compromise the prediction accuracy."

ANTENNAS VRPs, HRPs AND GAIN

THE NOVEMBER 1997 *TT* discussed the question of the lower radiation efficiency of most HF Yagi antennas, compared with simple half-wave dipoles. This was based on points made by Lew McCoy, W1ICP, who, as a simple example, pointed out that a beam antenna with a 3dBd gain but an efficiency of 50%, will radiate in the favoured direction only the same power as a dipole having a near 100% efficiency - and considerably less power in other directions. However, I added the comment: "The directivity of the beam may, however, be advantageous for combating the reception of unwanted signals and in its low vertical angle of radiation."

Gerald Stancey, G3MCK sought clarification of the comment that one of the advantages of a beam (Yagi) over a dipole is its lower angle of radiation: "This statement is often heard, but I would welcome a definitive confirmation. In the 14th edition of the *ARRL Antenna Handbook* there is a diagram which shows the calculated vertical radiation pattern (VRP) of a Yagi 1.25 wavelengths above ground. The lowest major lobe occurs at exactly the same angle as that given by a dipole at the same height. Could you provide a definitive statement on this?"

To answer this question adequately involves some appreciation of the fundamentals of antenna operation. In particular, it is essential to remember that the radiation is three-dimensional, rather than as shown in single-plane drawings. We need to think in terms of

The DL0WCY flux-gate magnetometer, which provides the Kiel k and A data. Constructed by DK4VW.

*37 Dovercourt Road, London SE22 8SS

solid geometry, with solid angles. Radiation from a short length of wire is in the form of a 'doughnut' or 'toroid' threaded on the wire, rather than only two-dimensional. Similarly, it is important to remember that a directional antenna, regardless of type, concentrates radiation in both the vertical and horizontal planes; eg a long-boom VHF/UHF Yagi, with many elements, radiates in a solid angle along the axis of the boom. For amateur satellites, moonbounce, etc, a long-boom Yagi needs to be directed in both azimuth and elevation.

The following notes draw in part upon the *HF Communications Data Book,* published in the 1970s by Rockwell International.

The power gain in decibels of an antenna in a specified direction is the ratio, expressed in decibels, of the radiation intensity of the antenna in that direction, compared to the intensity produced by some reference antenna, when equal powers are delivered to the two antennas. Radiation intensity is defined as the power radiated per unit solid angle from the antenna. Note that polar diagrams of directive, horizontally polarised antennas, show the radiation pattern at the optimum elevation, rather than parallel to the ground.

The presence and conductivity of the ground beneath, and for many wavelengths around the antenna, profoundly affects the performance of an antenna: (1) antenna efficiency is reduced due to the finite conductivity of the ground; and (2) radiation in the vertical plane is modified. As noted in many standard texts, the vertical radiation pattern of either horizontally or vertically polarised antennas is determined by the directional properties of the antenna and by the additive and subtractive interference between the direct waves and reflected waves. That is, the waves that are radiated from the antenna at angles below the horizon are reflected from the ground: **Fig 1(a)**.

At some radiation angles, the direct and reflected waves reach the same point in space in phase, and the resultant field strength is the sum of the two. Under perfect ground conditions (usually approached with horizontally polarised antennas at least a quarter-wave above ground), the ground-reflected wave from the 'image antenna', **Fig 1(b)**, is equal in amplitude to the direct wave, giving a gain of 6dB at these angles.

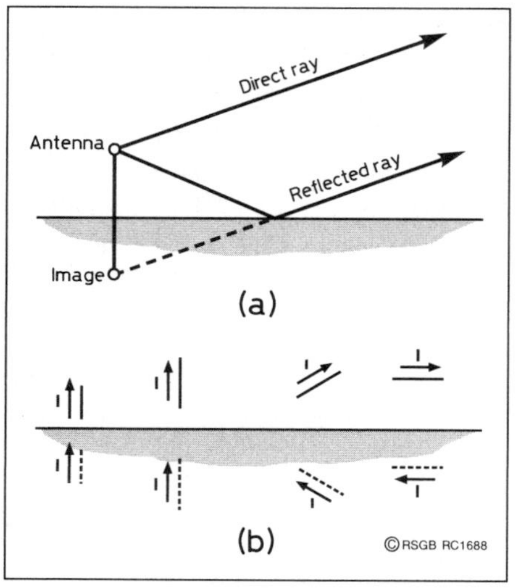

Fig 1 (a): Image antenna and ground effect, resulting in a direct and reflected ray. However, the effect of ground losses is more severe with vertical polarisation, where additional ground losses occur in the vicinity of the ground reflection. The relationship between the direct and reflected wave in any direction depends on the height of the antenna above ground, plus the electrical properties of the ground. The vertical plane pattern can be determined by substituting an image antenna for the earth. (b) To simulate perfect ground, the image antenna is fed with a current of the same amplitude as the actual antenna. The correct phase is indicated by the relative directions of the current arrows. The image concept is useful in antenna modelling for determining the patterns of antennas fed against ground. At some radiation angles, the direct and reflected waves reach the same point in space in phase, the resultant field strength then being the sum of the two. With horizontally polarised antennas at least a quarter-wave above ground, total radiation will closely approach that with perfect ground. (Source: *HF Communications Data Book*).

Table 1 was originally published in *TT*, July 1987. It shows the maximum gain of lossless elementary antennas with reference to an isotropic source and to a dipole. In practice, gains for horizontal antennas are normally specified relative to a half-wave dipole at the same height above ground as the antenna's phase centre. Gains of vertical antennas are normally specified with respect to a quarter-wave monopole with either 100% efficiency or with 60 quarter-wave radials, which is close to 100% efficiency.

In practice, the gain and direction of maximum gain of vertical antennas are affected very much more by ground conductivity than horizontal antennas. The elevated λ/4 ground plane antenna with sloping λ/4 radials is more akin to a λ/2 vertical dipole than a λ/4 vertical monopole. A power gain of about 5 to 6dBd is usually the maximum that can be achieved in practice for an HF three-element close-spaced Yagi beam (theoretical 7.5 to 8dBd). About 4 to 5dBd (theoretical maximum 5.2dBd) can, with care, be obtained from two close-spaced elements. Better overall results may often be achieved by raising a lightweight 2-element Yagi, rather than by increasing the number of elements at a lower height. Field strengths over a long, low-angle path, when compared to an antenna at λ/5, are roughly x2 at 2λ/5, x3 at 3λ/5, x4 at 4λ/5 and x5 at 1λ. 1λ loops give roughly a 1dB gain over dipole elements (ie the gain of a 2-element quad can be up to about 6dBd). At a VHF, height gain in the 3 to 10m range can amount to about 5.5 to 6.5dB for each doubling of the height in rural areas (8 to 9dB in urban areas) on paths up to 50km, and half these gains for distances exceeding 100km. For consistent HF DX, one needs to radiate and receive signals at vertical angles to ground of about 8 to 15°, although, in chordal hop (grey line) conditions, contacts may sometimes be made with antennas radiating at relatively high angles.

When it comes to a directional, horizontal antenna above an earth of finite conductivity, it becomes very difficult to determine the actual vertical radiation pattern, since - as noted above - this will depend both on its directivity and its height above ground. In most cases there will be a major lobe and one or more minor lobes at greater angles. In general terms, the greater the gain of the antenna, the greater will be the reduction in the angle of the major lobe, although the height above ground remains the dominant factor. Note that this does not apply only to Yagi antennas, but to all antennas exhibiting directivity. Long-wires, vees and rhombics will all have lower vertical angles than simple dipoles at equivalent heights. The greater the directivity, the greater the effect on the vertical angle of the major lobe.

With broadband antennas, such as rhombics, the vertical angle of maximum radiation will vary with frequency, since the height of the antenna varies in terms of wavelength. Even with the aid of computers, the calculation of vertical plane radiation patterns of antennas

Type	Location	Max gain (dBi)	Max gain (dBd)	Direction of max gain
λ/2 dipole	Free space	2.15	0	Plane perpendicular to axis
λ/4 vertical monopole	Over perfect ground	5.15	3	On horizon
λ/2 vertical dipole	Centre at λ/2 above perfect ground	8.2	6.05	On horizon
λ/2 horizontal dipole	Immediately above perfect ground	9.1	6.95	Straight up
λ/2 horizontal dipole	λ/4 above perfect ground	7.4	5.25	Straight up
λ/2 horizontal dipole	λ/2 above perfect ground	8.2	6.05	Plane perpendicular to axis, 30° above horizontal
λ/2 horizontal dipole	3λ/5 above perfect ground	9.2	7.05	Plane perpendicular to axis, 24.6° above horizontal

Table 1: Maximum (theoretical) gain of lossless basic antennas.

over finitely conducting earth is tedious, and such calculations are extremely difficult to validate. For professional HF antennas, one of the few practical methods is to fly an aircraft around the antenna at various heights. One must be very cautious in relying on NEC-type computer programs, since the amateur has no way of checking the results.

A 1940s book on rhombic antennas includes VRPs for a high-gain rhombic at a height of 16m above perfectly conducting ground. At 4.7MHz (64m), where the rhombic is a quarter-wave above ground, the main lobe peaks at about 40°, whereas, if it were a dipole a quarter-wave above ground, the main lobe would be directly upwards (90°). At 9.4MHz (32m), where the height is a half-wave above ground, the rhombic lobe peaks at about 25° compared with 30° for the corresponding dipole (see Table 1). At 18.8MHz (16m), with a height of one wavelength, the rhombic main lobe is down to about 11°. The directivity clearly influences the VRP at all heights, but the height is the main determining factor on how low an angle can be achieved. It is thus entirely possible for a rather low Yagi to have a greater vertical angle of radiation than a dipole at a good height. Incidentally, Table 1 shows that a lower angle of radiation is achieved by a dipole at 0.6λ, than the usually quoted 0.5λ.

Thus, any horizontally polarised antenna exhibiting directivity, will inherently have a lower vertical angle of optimum radiation than a dipole. However, since we are dealing with solid angles, this does not mean that the power gain of an antenna, as shown by conventional radiation patterns, can then be increased by the VRP directivity. The power gain of a two-element Yagi thus remains about 4dB, even though, over a specific DX path, it may provide, under some propagation condition, signals very many decibels above that provided by a low dipole. On the other hand, a dipole may out perform the Yagi over short, NVIS paths.

LEAD-ACID BATTERY PROTECTOR

DAVID GREENSPAN (*Electronics World*, December 1997, p1035) notes that when a lead-acid battery has been completely flattened, it often become useless. He provides the circuit diagram (**Fig 2**) of a protective circuit that will automatically disconnect a battery if its voltage falls below 10V, although permitting the occasional voltage drop caused by equipment drawing a high inrush current. It also automatically connects the battery to a correctly applied charging voltage.

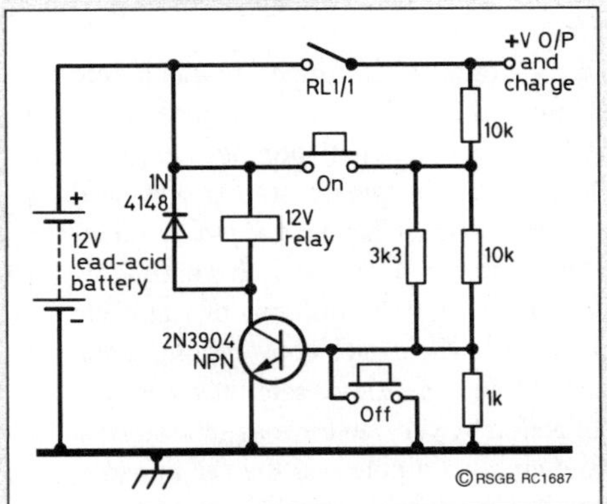

Fig 2: Protection circuit intended to prevent lead-acid batteries completely discharging, by automatically opening the relay contacts when the voltage falls below about 9.5-10V.

Assuming the relay is 'on', the potential divider applies 0.81V to the base of the transistor provided the battery voltage exceeds 10V. This is sufficient to turn the transistor 'on', keeping the relay closed. A falling battery voltage reduces the base current. At roughly 9.5V, the transistor turns 'off' and the relay drops out, disconnecting the battery. Similarly, when pressing the 'off' button. Pressing the 'on' button momentarily induces base current, turning the transistor 'on'.

THE ORIGINAL COSMIC ECHOES

FASCINATED BY THE 200 millisecond echoes heard by Peter Martinez, G3PLX (*TT*, December 1997 and February 1998), I am increasingly convinced that they must be connected with the stream of charged particles stemming from the Sun and their interaction with the Van Allen belts which can be some 20,000 miles above Earth. Recently I took the opportunity of checking on the original reports of long delay echoes, with delay times of up to 15 seconds (*Nature*, Vol. 122, No 3078, November 3, 1928, p681 and No 3084, December 8, 1928, pp978-9).

In the November 3 issue, under the heading *'Short Wave Echoes and the Aurora Borealis'*, Professor Carl Stormer wrote: "On February 28 (1928) I received a letter from Engineer Jorgon Hals, Bygde, Oslo, in which he says: 'I herewith have the honour to advise you that at the end of the summer 1927, I repeatedly heard signals from the Dutch short wave transmitter station PCJJ (Eindhoven). At the same time as I heard the telegraph signals I also heard echoes. I heard the usual echo, which goes round the earth with an interval of about 1/7th second, as well as a weaker echo about 3 seconds after the principal signal had gone . . . I suppose that the amplitude for the echo 3 seconds after lay between one-tenth and one-twentieth of the principal signal in strength. From where this echo comes I cannot say for the present. I will only herewith confirm that I really heard this echo."

Prof. Stormer wrote: "Immediately I heard of this remarkable observation, it struck me that the wireless waves were reflected from those streams and surfaces of electrons to which I was led by theoretical investigations on the aurora borealis, in my paper published in 1904 in *Videnskabsselskabets Skrifter*, Christiania [former name of Oslo - *G3VA*] ("Sur le mouvement d'un point matérial protant une charge d'électricité sous l'action d'un aimant élémentaire."). In reference to that paper. . . one of the most striking features of the theory was that streams of electrons coming from without, towards the earth, were deviated by the earth's magnetic field in such a way that an immense space was formed from electric particles, having the shape of a torus described by revolution of an oval tangent to the magnetic axis of the Earth at the centre. . . If now the wireless signals could penetrate the Heaviside layer, they would pass into this empty space and might be reflected by the walls of the electrons forming its outer boundary. The long time interval between the principal signal and the echo agrees well with the immense dimensions of these toroidal spaces . . ."

Today we know infinitely more about the streams of energetic particles erupted from the Sun during magnetic storms, and the manner in which this Solar Wind impinges on the Earth, the Van Allen belts, the magnetosphere, etc, etc. We know that radio waves above the critical frequencies of the E and F layers can penetrate these layers and pass into outer space. We also know how layer tilts can trap signals into orbital 'waveguides' formed between layers, etc.

We also know that high power transmitters can 'heat' the ionosphere, causing subtle changes. I suggested in the February *TT* it could be that these changes to the ionosphere's 'environment' may well mean that really long delay echoes may now very rarely, if ever again, be heard by terrestrial observers, although quite possibly feasible on spacecraft well above the F layer.

Balth van der Pol, of Philips, Eindhoven, in his letter of December 8 1928, provided a diagram illustrating the echoes heard at two sites in Eindhoven and on the Oslo receiver on October 11, 1928 (**Fig 3**). These show echoes between about 3 and 30 seconds. In his letter,

Prof. Appleton confirmed that R L A Borrow (later G3ZTK) had succeeded in getting photographic registration of the echoes from PCJJ, with a delay of one second.

The importance of further research into predicting which of the coronal mass ejections from the Sun will impinge upon the earth as severe magnetic storms was well described during a Channel 4 *Equinox* television programme a few months ago. This showed how, in 1989, millions of Canadian homes were blacked out by power failures and in 1994 Canadian Telesat satellites were put out of action by very severe magnetic storms. Storms are also a potential danger to astronauts, military satellite tracking operations, etc. Unfortunately, it is still difficult to determine the magnetic polarity of the solar clouds leading to the false alarm issued on April 6 1997, when widespread disturbances were forecast but did not materialise. An important satellite project (CLUSTER) was lost when last year the Ariane V rocket failed to go into orbit.

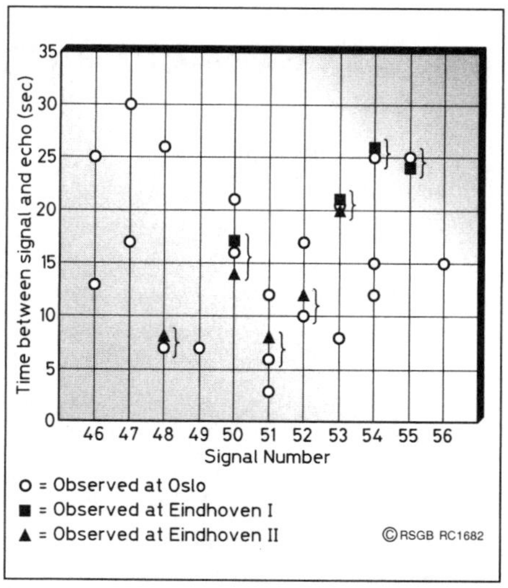

Fig 3: Reported echoes received at Oslo and Eindhoven on the transmissions from PCJJ (Eindhoven), during the night of October 11 1928. Transmissions consisted of three short dots in rapid succession every 30 seconds between 2100 and 2200 local time, a total of 120 signals, with echoes heard on transmissions number 46-56.

PASSIVE (LC) NARROW-BAND AF FILTERS

With the increasing use of active filters and Digital Signal Processing (DSP), less thought is being given to the continued value of passive LC filtering for CW, RTTY and HELL operation. Jan Smeets, ON4ASZ/EA3DPB, submits a number of circuits for narrow-band filters, together with useful comments.

He writes: "I cannot understand why Ten Tec use in their Corsair, Omni, etc, high-performance transceivers, active four-section filters which have the relatively poor performance of about -40dB at 400 and 1500Hz.

"My passive LC filters have been realised using the once popular 88mH (centre-tapped) toroid coils, but since these are no longer readily available it is possible that TOKO 719VXA9032 80mH coils, for which a Q of 45 is claimed, could be substituted. In the past, PA0SE has calculated that the Q of the standard 88mH toroid is about 40 at 450Hz, and more at higher frequencies.

"Few of the published LC filter circuits perform really well, with the exception of one by PA0KLS, which has a 640Hz bandwidth of -60dB. This has five coils, but the termination is at the odd impedance of about 130Ω. He added resistors to modify this to 600Ω, as is commonly used in professional (rather than amateur) equipment. The filter was designed for Hellschreiber operation, with a -6dB bandwidth of about 150Hz.

"**Fig 4** shows some of the LC filters that I have built, together with their characteristics. These are intended for a centre frequency of 780Hz or 800Hz. They are suitable for connection to the 8Ω audio output of modern amateur equipment, but require high impedance headphones (or other high-Z devices, such as old style RTTY or Hellschreiber impedances). The matching is perfect and it is possible (and preferable) to make these filters without resistors. In the parallel branch, 10μF may be replaced by 20Ω resistors and 20μF by 10Ω resistors, which results only in additional insertion loss."

ON4ASZ notes that these filters are suitable for RTTY as well as CW. Designs Fig 4 (a), (b) and (c) have been tested with a single tone (primitive) demodulator. Design (d), which eliminates the need for the AF transformer of (c), was completed later. The estimated -60dB bandwidth is about 200Hz, with an output voltage amplitude about 0.7 that of (c). For all practical purposes (c) and (d) are the same, and compare well with some excellent German telex filters, which have an attenuation notch 120Hz on either side of the centre frequency (the multiplexed RTTY channels were separated by 120Hz). Maximum attenuation of these German filters was between 5 and 6 Neper, or in decibels between 50 and 60 (5.76N = 50dB), a further indication that designs (c) and (d) are of high standard.

ON4ASZ adds some general comments on LC filters: "When building filters with more than one LC, in order to achieve the sharpest possible peak (for CW), it is advisable to add

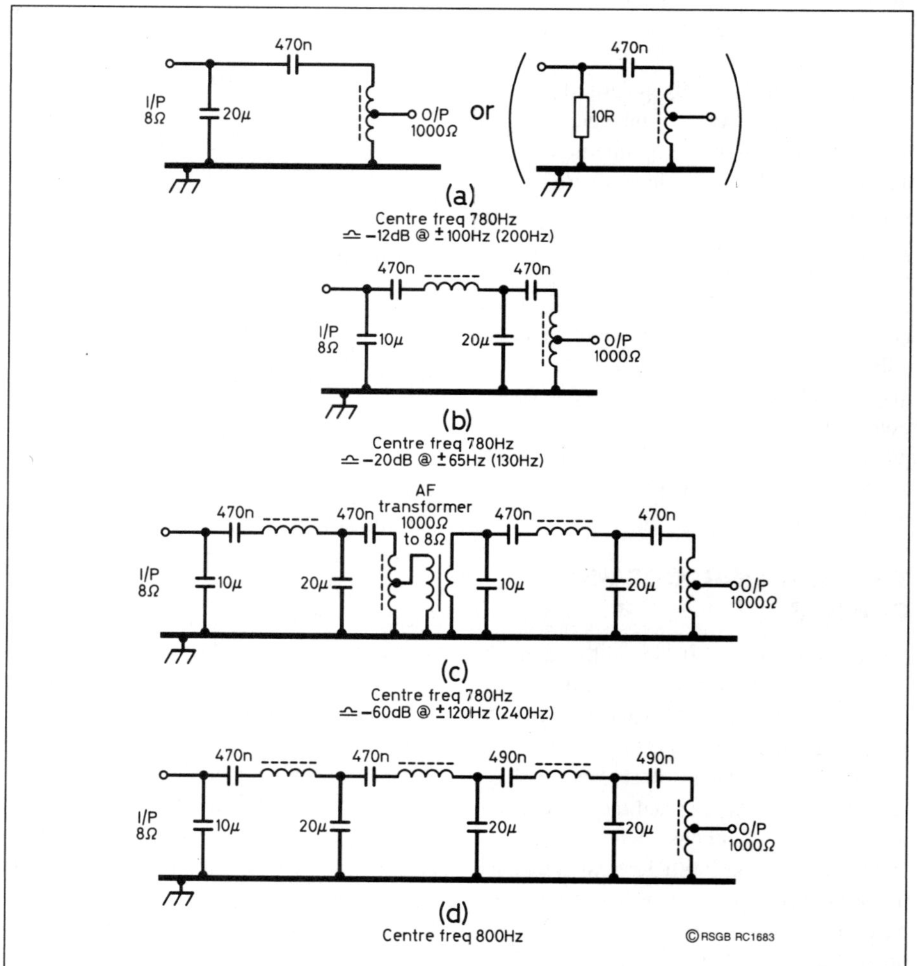

Fig 4: LC audio filters implemented by ON4ASZ/EA3DPB, using 88mH toroid coils and with Hi-Z (1000Ω) output impedance.

a small capacitance to the LC which resonates at the lowest frequency, probably due to the wide tolerance of the capacitors.

"I also once made a double filter for low impedances: **Fig 5**. I never measured its performance, although this seemed good, but possibly with greater insertion loss. I have always preferred a high-Z termination, which is more suited to driving my old Hellschreiber or an RTTY preamp. Moreover, high-Z headphones are so nice!" [A sentiment with which I fully agree! My old Telefunken 4000Ω adjustable-diaphragm 'phones remain my favourite 'cans' - *G3VA*].

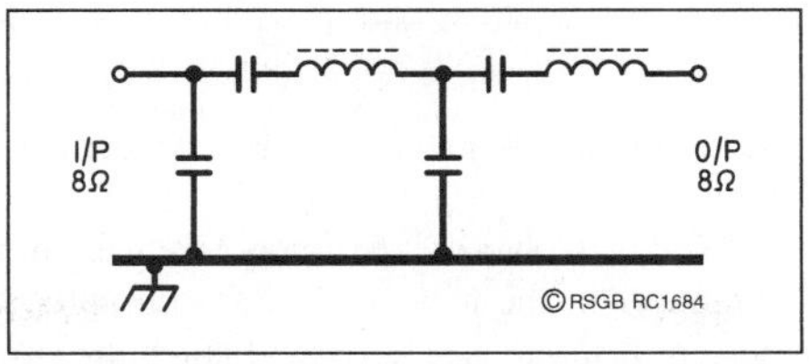

Fig 5: How a low-impedance termination can be implemented.

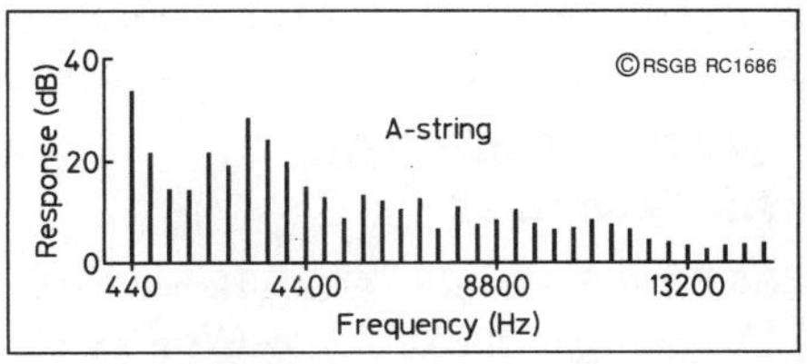

Fig 7: Musical sounding signals can be more pleasant than a pure sine-wave. The audio spectrum of an open A string of a violin, showing the amplitudes of the fundamental and harmonic frequencies composing the note. (Source: *High Fidelity Sound Reproduction*).

ON4ASZ/EA3DPB continues: "When listening to high speed CW through a very narrow filter, the sound becomes 'woolly' or 'soft', due to truncation. **Fig 6(a)** shows an arrangement that has been found to restore the hardness of the signals. When viewed on an oscilloscope, the output signal looks rather like the waveform shown in **Fig 6(b)**. Much the same effect can be found on the output of the Telefunken communications receiver type E127. I believe the distortion and harmonics induced on the signal by the diodes improves the readability of the signal. In fact, for my primitive RTTY single-tone demodulator, I have incorporated this 'threshold/limiter' arrangement and find it works well without a 'real' 471-type limiter, which was omitted in the interests of simplicity."

Fig 6 (below): A non-linear threshold/limiter network removes the woolly/soft characteristics of a narrow filter, by introducing harmonics, etc.

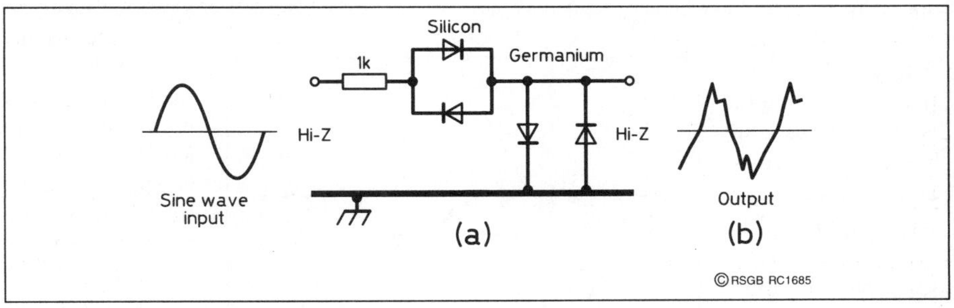

ON4ASZ's threshold/limiter reminds me strongly of a basically similar set-up for CW, which I used many years ago and described in an article *'The T5-er - a different approach to CW reception' RSGB Bulletin*, September 1962. I argued that listening to a sine-wave CW signal could represent a fundamentally wrong approach, leading to unnecessary fatigue and discomfort in CW operating: "Consider how seldom the sine wave occurs in nature . . . on the other hand, the notes of musical instruments consist of fundamental notes together with their related harmonic series', which musicians term 'overtones' or 'upper partials' (**Fig 7**) . . . For Morse, a roughly saw-tooth waveform, rich in harmonics, or commercial A2 ICW signals are much to be preferred . . . Why not simply change the normal audio output from a sine wave to a more acceptable waveform? . . . The basic process of turning a sine wave into a more complex waveform is to distort or shape it, by passing it through a non-linear network. . . So was born a simple device which we originally termed the 'SNAFU' (Signal Note All Fouled Up). Later, to avoid possible misinterpretation, this was renamed the 'T5-er', on the grounds that the intention was to produce a 'musically modulated note' - the old definition of a T5 signal."

I used various T5-er circuits, based on diodes and diode-connected germanium transistors - usually in conjunction with a narrow crystal filter or wartime FL8A audio filter - for a number of years in the 1960s. Then, for no good reason (other than the difficulty of providing genuine T reports) I discarded the technique. Perhaps it is time to try it again. ON4ASZ's Fig 6a should prove a good start. ♦

CLICKLESS SIDETONE

DENTON BRAMWELL, K7OWJ ('Hints & Kinks' *QST*, December 1997, p70) believes that good sidetone (or Morse practice) oscillators are hard to find. He writes: "One common approach is to build an audio multivibrator - and that does produce an acceptable sidetone. Unfortunately, it produces a square wave, and my attempts to eliminate clicks have all produced an unacceptable chirp. Other approaches seem to have one or more undesirable characteristics, chirp, square wave output, or no shaping of the keyed waveform. My oscillator (**Fig 8**) is different. It has shaped keying, no chirp and a sine wave output.

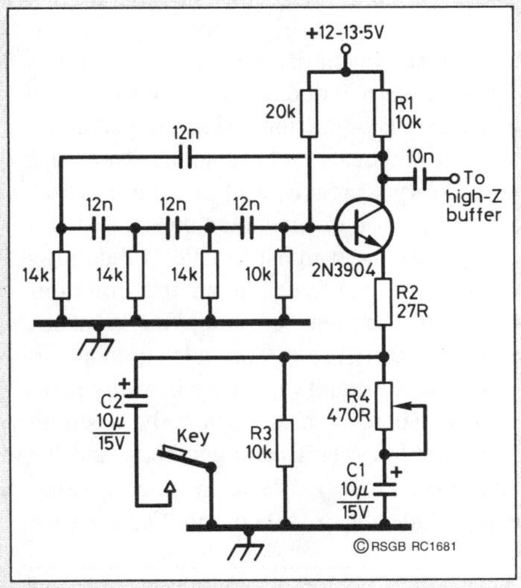

Fig 8: K7OWJ's sidetone oscillator. Use ¼W, 5% tolerance resistors, and ceramic disc capacitors (except for C1 and C2, which are electrolytics).

"The basic circuit is a phase shift oscillator with an RC network supplying the necessary 180° phase shift between collector and base. In order for the oscillator to work, it must have enough gain to more than overcome the losses in the phase shift network. We can, therefore, manipulate the gain to start and stop oscillation by adjusting R4. By shifting the gain just above and below the critical point, I can shape the envelope. Simply adjust R4 a bit past the point where oscillation ceases with the key up and it will be about right. Since this happens without keying any of the DC paths, there are no clicks. With a 12V supply the output is about 5V peak-to-peak, and keying is good over the 12 to 13.5V operating range.

"There are a couple of minor bugs: The output impedance is about 10kΩ, so the circuit requires a buffer with a high input impedance. Also, the circuit loses about three dots when you first switch it on. Since the oscillator draws so little power, I just always leave it on."

BY PAT HAWKER, G3VA*

'MAGIC' FRACTAL ANTENNAS?

BOTH JOE KESSEL, G7RWH, and I found 'Aerial Magic' by Mike May (*New Scientist*, 31 January 1998, pp28-30) of more than passing interest. It claims: "Your mobile phone, the gas meter under the stairs and the vending machine down the hall may soon sport the world's smallest, weirdest antennas." Professor Nathan ('Chip') Cohen is presented as the inventor of a new family of 'fractal antennas' which may come to be used in everything from mobile phones to huge receiving arrays. The idea emerged from an experience in 1988 when Cohen wanted to use his ham radio equipment at a flat in the centre of Boston, where the landlords stipulated no outside antennas.

Instead of using a difficult-to-disguise conventional antenna, Chip Cohen cut a sheet of aluminium foil into the shape of a mathematical pattern known as an inverse Koch curve, and stuck it onto a sheet of A4-sized paper: "Cohen connected the foil to his radio receiver, to see if it might serve as a covert antenna if he mounted it outdoors. To his surprise, the fractal foil pattern worked well and Cohen was able to continue his hobby without arousing suspicion [no indication is given of the frequency band(s) concerned - *G3VA*].

"Today, Cohen's experiment has made him a pioneer in the new field of fractal antenna design. It turns out that fractal antennas have many advantages over their conventional counterparts. For a start, they are smaller - a fractal antenna for a mobile phone can be made the size and shape of a 35mm photographic slide and can be built into the casing . . . fractal antennas are also better at picking up signals and can receive them over a wider range of frequencies. But . . . physicists are being left behind. Nobody is exactly sure why fractals make good antennas. Now the race is on to find out."

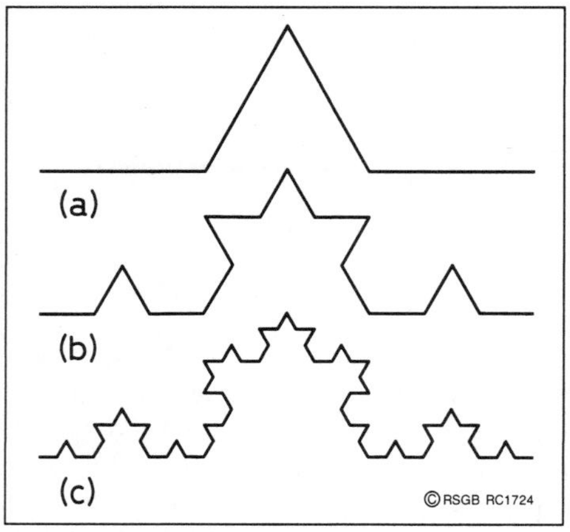

Fig 1: Construction of a von Koch 'snowflake' fractal, formed from the basic curve of (a), with successive iterations shown in (b) and (c). As stages are added, the total length of the curve tends to infinity, although the curve is confined to a finite space.

It seems, according to the article, that a fractal pattern antenna provides the 3-4dB gain of two half-wave elements in phase, but it is by no means clear how this relates to its useable bandwidth and what this amounts to.

Although the article describes in general terms what constitutes a fractal pattern, it does not attempt to provide precise details of the dimensions, impedances, etc, of a practical fractal antenna. Nor, in a quick library search, could I discover any further information on fractal antennas. My dictionary does not even list 'fractal'. I did, however, find out that fractal geometric structures are of particular interest to computer scientists, because of their use in computer graphics to depict landscape and vegetation in a very natural way. There is also apparently a strong connection between fractal geometry and the trendy chaos theory. Fractal geometric structures include the so-called snowflake curve, devised by Swedish mathematician Helge von Kohn in 1904 (**Fig 1**). Fractal patterns exhibit a 'self-similarity' when the distance at which they are viewed is changed. They consist of many repetitions (iterations) of a basic shape.

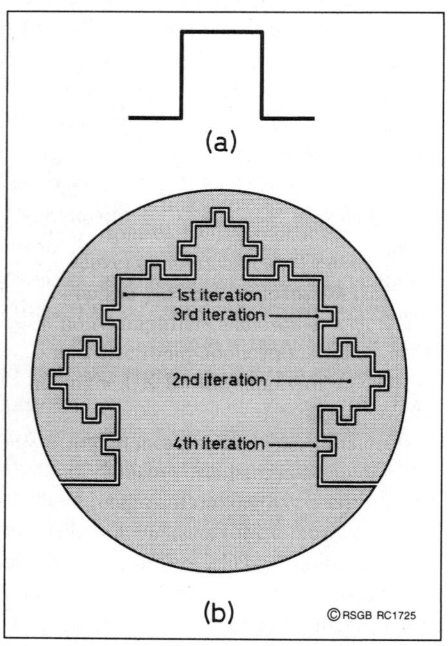

Fig 2: Illustration of a fractal antenna that could be built into a military helmet, as described in the *New Scientist* article, built up from the iteration of the basic curve (a).

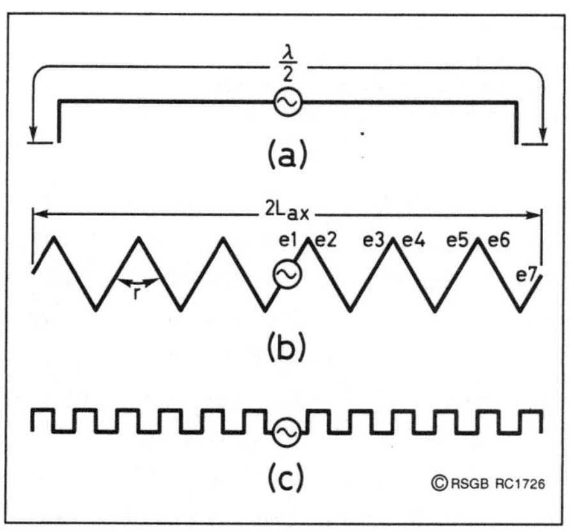

Fig 3: Significant shortening of the overall span of a conventional half-wave dipole element can be achieved without significant loss of efficiency or directivity. (a) Ends dropped downwards or bent inwards. (b) Zigzag element, where e1 = e7 = 0.0208λ and e2 = e3 . . . e6 = 0.0416λ. (c) Meander-line element as noted in *TT* of February 1992.

The inverse Koch curve used by 'Chip' Cohen for his covert amateur radio antenna was apparently a fractal, comprising a series of triangles stacked up on top of each other like a pagoda, but his UHF micropatch fractal antennas seem to be variations of the Koch snowflake curve: **Fig 2**.

There is of course nothing new about using elements bent into regular shapes in order to reduce overall length, as in the 'Meander' and 'zigzag' antennas (see for example *TT*, February 1992, from which **Fig 3** is taken).

MORE ON 'STEREO' CW

THE FEBRUARY *TT* included a note on the 'stereo' reception of CW, recalling the pioneering work of G6CJ, and featuring digital signal processing using a PC with 16-bit Creative Labs sound card and mouse, in conjunction with a software package developed by Brian Beezley, K6STI. This resulted in a letter from Oliver Tillett, G3TPJ, drawing attention to the KISS approach offered by his 'Soundfield Adaptor', which is marketed by Waters and Stanton Electronics, 22 Main Road, Hockley, Essex SS5 4QS. The circuit details are currently subject to a patent application and G3TPJ offers no explanation of the way in which the pseudo-stereo effects are obtained (although I recall that in the mid-1980s a number of ICs were introduced to permit stereo effects to be obtained from MW AM broadcast receivers, etc, and it is possible that such devices are used).

A sales leaflet prepared by G3TPJ sets out the disadvantage of relying on extremely narrow-band CW filters, particularly for contest operation. It continues: "With normal bandwidths we hear not just one signal but the accumulation of all the signals in one complex audio waveform. The operator at the receiving end has the task of mentally selecting the signal he wishes to copy, whilst trying to ignore all other sounds. It is a skill that can be developed, but nevertheless does result in fatigue and there is a personal limit that varies from person to person.

"The Soundfield Adapter sets out to tackle the problem in a completely new manner. Used in conjunction with a set of stereo headphones, it simplifies the waveform complexity and adds a spatial characteristic. This enables the operator to hear signals as if they are coming from different directions. More importantly, the wanted signal (800 - 1000Hz option) is modified so that its source sounds wider and thus closer, almost 'larger than life'. All other

**37 Dovercourt Road, London SE22 8SS*

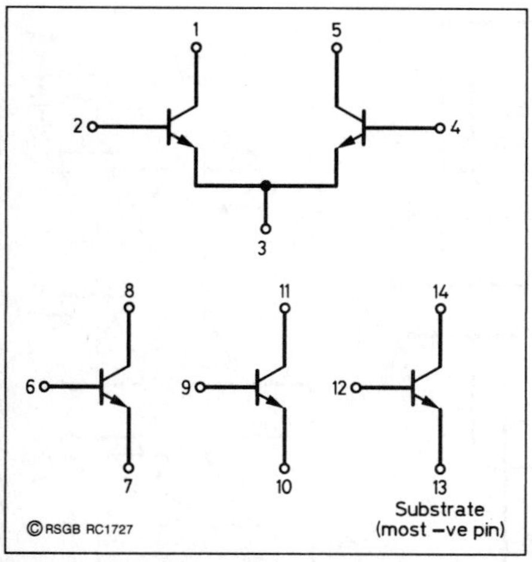

Fig 4: Internal circuitry of the 3046 IC transistor array.

tones retain their original small source size. The effect is very natural to listen to, dramatically reduces the problems associated with filters, and has the added bonus of reducing the effect of static crashes and noise bursts on the chosen signal."

While one needs to show a degree of scepticism in accepting claims not supported by full disclosure of the technology, the basic premise of the improvement possible with pseudo-stereo reception of CW was well proven by G6CJ's pioneering work, and there seems no reason to doubt that G3TPJ has come up with an effective add-on unit. He writes: "After 30 years of involvement in audio and HF line transmission, I decided to do my own thing and pursue a 'Soundfield' idea I had back in the sixties. After two years the final product came to fruition. The principle is quite different to those previously tried, and it works."

GILBERT CELL MIXERS - MAKING DO WITHOUT THE SL6440

FOR MANY YEARS, the Plessey SL6440 double-balanced Gilbert Cell IC mixer has proved a convenient and effective approach for receivers of high dynamic range (see for example *TT*, June 1980, etc). Unfortunately, with the decision to discontinue the entire range of SL linear IC devices, it is now vanishing from the market. This undoubtedly leaves a serious gap for those constructing high-performance equipment without the complexity of G3SBI's H-mode mixer.

Ian Briathwaite, G4COL, has been conducting experiments to find out how this gap can be bridged. He writes: "I was disappointed when I heard that the SL6440 high-level mixer IC was being discontinued. This was based on the Gilbert Cell arrangement of long-tailed pairs, and gave a high dynamic range by virtue of its low distortion, reflected in the high third order intercept point (IP_3), stated on the data sheet as typically +30dBm.

"Although this was achieved at the cost of a high supply current of 50mA, the Gilbert Cell mixer is handy to interface to, and the low local oscillator power requirement reduces the need for high supply current in other stages.

"Other Gilbert Cell mixer ICs are available. In recent years the NE/SA602 has been much used in home-brew receivers, on account of its external circuit simplicity. However, its low current drain (a few mA) results in a non-ideal IP_3 performance (typically -13dBm, rather better with the NE602A), far below that of the SL6440's +20-30dBm. Used with good input preselector filters there should be few problems on the higher HF bands, but in Europe on 7MHz on a winter's evening, an excessive

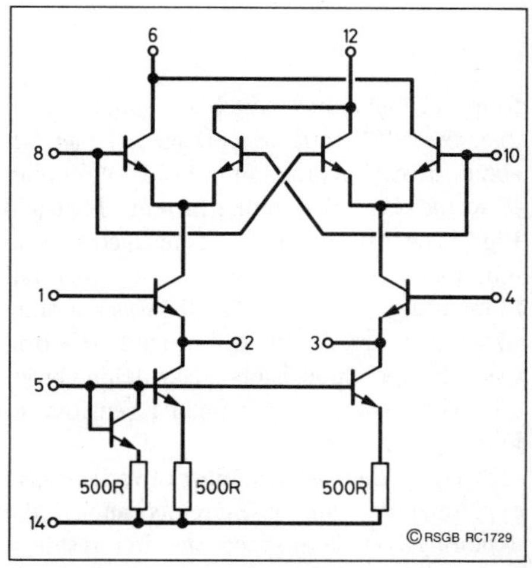

Fig 6: Internal circuitry of the 1496 IC.

number of signals may be evident! Intermodulation distortion can give the impression to the unwary of a receiver with 'lively' sensitivity. In reality this may reflect the extra, unwanted signals generated within the mixer circuitry.

"I decided to attempt to make a high-performance Gilbert Cell mixer, using low-cost matched transistor arrays, type 3046, which have been around many years: **Fig 4** and **Fig 5**. All experiments were made with an input frequency of 18.1MHz and local oscillator input at 3.4MHz, amplitude 2V peak-to-peak, examining the output at 21.5MHz. Distortion was found to reach a minimum with a 'tail' current of 17mA (supply current about 22mA, due to the bias chain). Third order intermodulation measurements are somewhat tedious and prone to error, but I found I was getting an intercept point in excess of +20dBm and up to +25dBm, depending on input level.

"This is somewhat short of the SL6440 per-

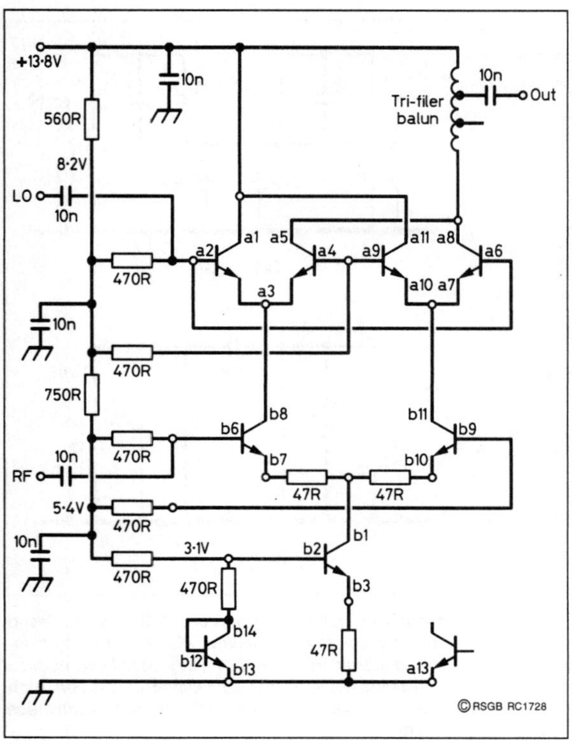

Fig 5 (left): Test circuit of G4COL's Gilbert Cell Mixer, based on two 3046 transistor arrays.

Fig 7 (right): G4COL's test circuit of the Gilbert Cell mixer, using a 1496 IC. Although this requires more external components than the NE602, these are mostly low-cost resistors, so the total cost may be less for a much improved third-order intermodulation distortion.

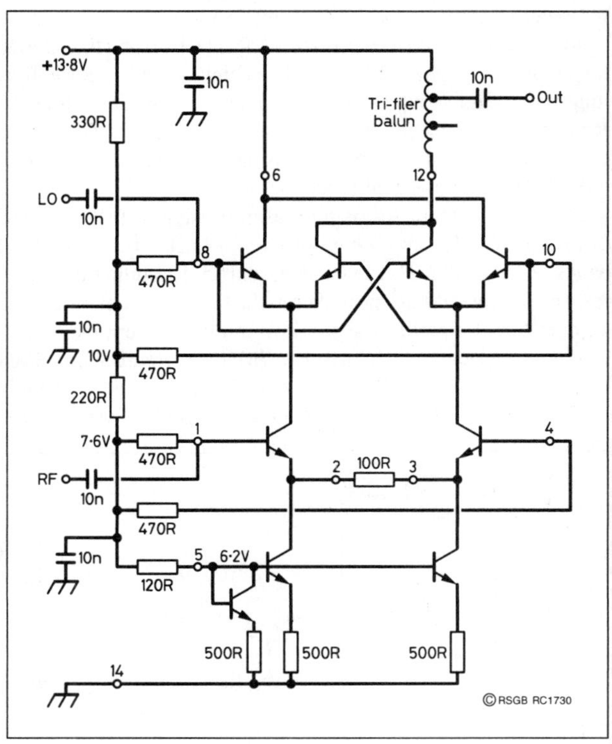

formance, but quite creditable and much better than the NE/SA602. Encouraged, I then set about seeing how the faithful old 1496 mixer IC would perform at similar currents: **Fig 6** and **Fig 7**. The distortion products reached a broad minimum with a current of 11mA per tail (33mA supply current). The IP_3 was not quite so good as before, but still around +20dBm. Once the upper transistors are switching properly, performance is not unduly sensitive to local oscillator level.

"Clearly, there are a number of variables, so experimenting with these circuits can feel like exploring a very large space. The circuits shown in the accompanying diagrams (Figs 5 - 7) are presented as the basis for experimentation. Although no guarantee of a specific performance can be given, I feel confident that they will give good performance when duplicated. I shall be basing future designs on them. The circuits show the full inner circuitry of the ICs, which makes them look formidable, but this will help interested readers to get a good idea of how the devices work. Remember that most of the components which would not be required for the NE602 are low-cost resistors, so the overall parts cost can even be less, depending on where they are sourced."

MULTI-BAND DC RECEIVER WITH IC HARMONIC MIXER

THE USE OF a 74HC4066 IC as a switching-type double-balanced mixer/product detector is a feature of a 'Multi-band, direct-conversion receiver' described by Hannes Coetzee, ZS6BZP (*Electronics World*, February 1998, pp138-143). He shows how low-cost components can be used in a reasonably high-performance receiver for the 7, 14, 21 and 28MHz bands, with a VFO covering 7.0 to 7.15MHz, by using the 4066 as a harmonic mixer on 14, 21 and 28MHz. He achieves performance figures that are impressive and compare well with that of many factory-built superhet receivers and transceivers, other than the drawback of all direct-conversion receivers not incorporating a phasing-type I/Q detector being unable to provide true single-signal reception.

Despite this, there seems no reason to suggest that this receiver, with a distortion-free dynamic range of over 100dB (105dB on 7MHz) on three bands (90dB on 28MHz) and 2.4kHz and 850Hz audio filters for SSB and CW respectively, would not provide satisfactory results at much lower cost than a single-signal superhet design. The minimum discernible signal with 2.7kHz bandwidth varies from -128dBm on 7MHz to -104dBm on 28MHz, but could be improved by the addition of a pre-amplifier ahead of the mixer.

There are several interesting features of ZS6BZP's design, some of which

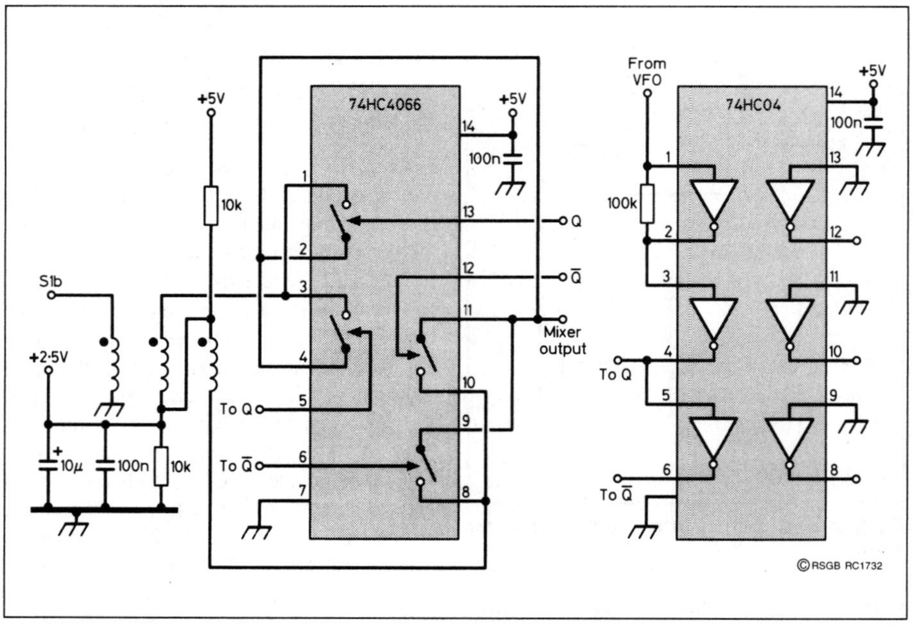

Fig 9: 74HC4066 mixer and 74HC04 hex inverter, used to generate a square-wave switching signal. On 14, 21 and 28MHz, harmonic mixing is used with some slight reduction in overall performance, but permitting the use of a local oscillator tuning only 7.0 - 7.15MHz. The input transformer is 7 turns, trifilar wound on 9mm diameter Philips 4C6 (purple) core.

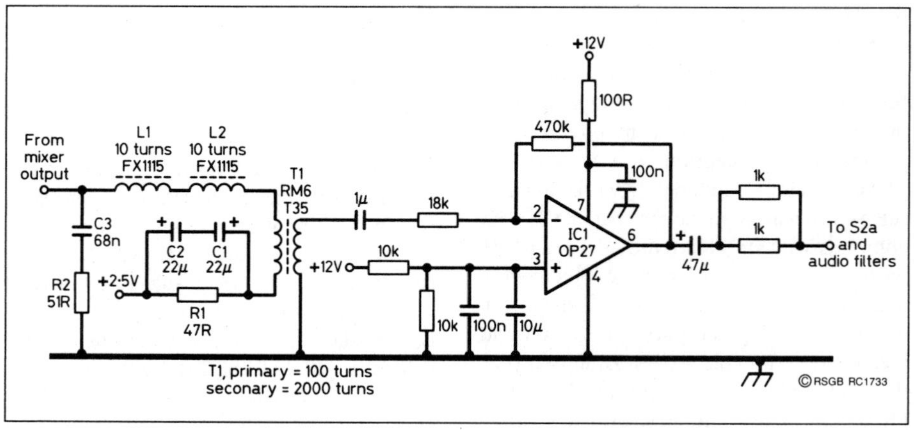

Fig 10: Diplexer used after the mixer, presenting a 50Ω load to the front-end, converting this to high-impedance for the op-amp.

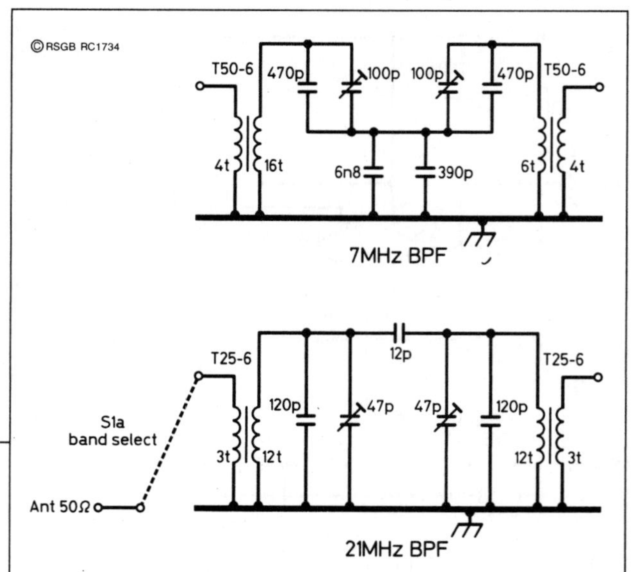

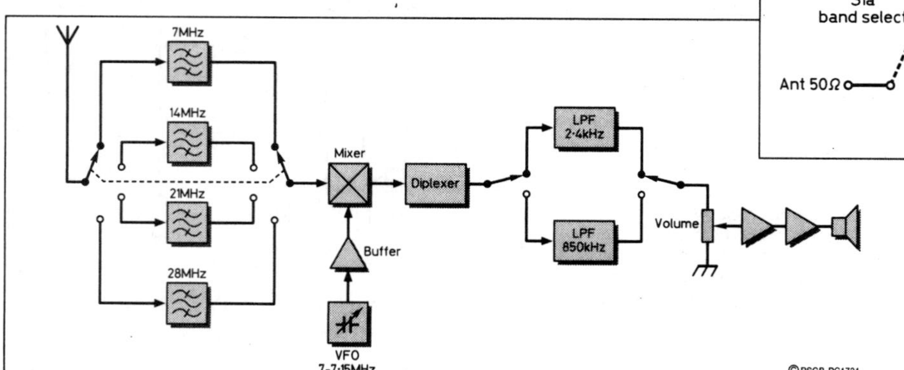

Fig 8 (left): Block diagram of ZS6BZP's four-band direct-conversion receiver, with good distortion-free dynamic range on 7, 14, 21 and 28MHz, using a switching-type IC product-detector (mixer) with square-wave injection, as described in *Electronics World*.

Fig 11 (above): Two of the four switch-selected RF band-pass filters. Good signal-frequency filtering is essential with a harmonic mixer.

are illustrated here, although for further details it would be necessary to consult the original article. **Fig 8** shows the block outline, and **Fig 9** the 74HC4066 mixer with the gates of a 74HC04 hex inverter used to spilt the local oscillator signal into two signals with a 180° phase difference to provide a balanced square-wave switching signal for the mixer. **Fig 10** shows the diplexer, which presents a 50Ω load to the mixer output and feeds a high-impedance signal to the op-amp. This is followed by the 2.4kHz and 850Hz AF filters.

With harmonic mixing, it is essential to provide good front-end RF bandpass filters: **Fig 11** shows the 7 and 21MHz filters used by ZS6BZP. The AF filters (see *Electronics World*) use hand-wound inductors on ferrite cores, with the author pointing out that: "Fortunately many modern multimeters can measure inductance, which makes confirming the inductance values at audio frequencies a piece of cake."

DSP FOR NBVM

IN THE JANUARY *TT* item on 'Narrow Band Voice Modulation' it was noted that this system, developed in 1977, permits communications-quality speech to be conveyed in a compact bandwidth of about 1200Hz. However, the system never achieved its potential due to complexity and non-compatibility with other SSB stations. In presenting a modified form of NBVM, I questioned whether it could today take advantage of modern DSP systems.

This resulted in the arrival of an audio cassette from Charles Brain, G4GUO, clearly demonstrating that indeed the original NBVM system, with a compact bandwidth of 1.2kHz, can be implemented with DSP.

G4GUO writes: "You may be interested in a tape I have just made up. It contains three passages, the first original speech which has frequencies up to 16kHz or so, the second the NBVM encoded signal contains only the frequencies between 300-600 and 1500-2400Hz 'folded', so that the total bandwidth is about 1200Hz. The third passage comprises the recovered speech. This implementation was done solely from the *TT* description. It runs on a Motorola 56002 Evaluation Module, which costs about £150 in the UK. It took only an afternoon to do and is by no means perfect, having slight distortion on speech peaks due to the A/D range being exceeded; this could probably be improved by adding a simple VOGAD circuit.

"I have also implemented a simple, direct-conversion transceiver using this module, and will add a suitable 300-1500Hz filter response to it for potential NBVM operation. One of the most difficult aspects of using DSP in receiver design is achieving a reasonable AGC response, but that is another story.

"If anyone wants to try NBVM and has one of these 56002 modules, they are welcome to a copy of the DSP code. The software runs in full duplex, so can be simply put in line with the microphone and headphones. I am not expecting a rush!"

The audio tape shows clearly that while the recovered speech is of very reasonable quality, the encoded passage is virtually unintelligible, so it would be necessary for both ends to be equipped for NBVM. Charles Brain (7 Elverlands Close, Ferring, West Sussex BN12 5PL) can be reached by telephone on 01903 245400 or e-mail chbrain@dircon.co.uk

SEMI-ELEVATED GROUND-PLANE ANTENNAS

I HAVE COMMENTED a number of times (most recently in *TT*, January 1996, pp69-71) on the professional interest of broadcast engineers in the revival of the 'counterpoise' elevated radial as an alternative to the standard 120-wire buried earth that stemmed from Dr George Brown's classic MF work in the 1930s.

It was Arch Doty, K8CFU, and his elderly friends who, in the early 1980s, challenged the buried radial concept on the basis of extensive measurements. This led to computer simulations by Al Christian, KN8I, prompting several American broadcast stations to try using four elevated radials (instead of 60-120 buried radials each ¼-½λ long), with mixed results. Some professionals now swear by elevated radials; others swear at them.

Two well-respected writers have faced up to this controversy, and both seem to have come to much the same conclusion: elevated radials *must* each be resonant electrical quarter-wavelengths if they are to equal or even improve on the extensive ground screens provided by large numbers of buried radials. This may come as no great surprise to the many users of elevated HF/VHF ground-plane antennas with sloping or horizontal radials, but was by no means clear from the earlier work, with even George Brown postulating radials of up to 0.5λ.

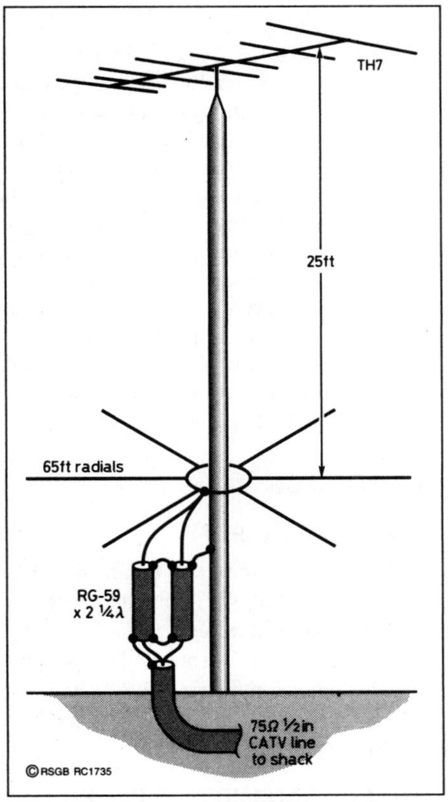

Fig 12: A 3.5MHz top-loaded 'reverse-fed' elevated ground plane type antenna ,with earthed tower and top-loading provided by the TH-7 Yagi antenna for higher frequency bands, as illustrated in VE2CV's ICAP paper, and based on the article by Russell in *QST* (June, 1994). Note that the tower at the feed point shown, carrying the TH7 triband Yagi antenna, is resonant on the band of interest.

Dr John S (Jack) Belrose, VE2CV, presented a paper 'Vertical Monopoles with Elevated Radials: An Update' at the 10th International Conference on Antennas and Propagation (ICAP) held in April 1997 (IEE Conference Publication No 436, pp1.190-195). His conclusions provide a useful summary of his findings, based on computer studies and practical experience:

"Elevated radial ground systems can be used effectively for many applications: (1) to realise achievable gain for ground plane type antennas over real ground with only a few radials; (2) to provide versatility, eg control of radiation pattern, and (3) for numerical modelling purposes, to simulate connection to ground. To realise achievable gain and not change the resonant frequency of the antenna system, the radials must be resonant. While this conclusion has been discovered during the course of our study, in retrospect it should have been anticipated at the outset. The magnitude of the difference (antenna gain and impedance), however, is particularly striking when there is a significant difference between the electrical and physical length of the radial wires, such as a radial wire system at a very low height."

VE2CV also concludes that phased array directional systems can employ elevated resonant radials, thus simplifying the installation of ground mounted antenna systems that employ great numbers of buried radial wires.

Inter alia he notes that radio amateurs have used elevated radials in a sort of reverse-feed arrangement as a method to feed a grounded tower (Russell, 'Simple, Effective, Elevated Ground-Plane Antennas' *QST*, June 1994, pp45-46). VE2CV wrote: "The radials are attached to a ring, centred on the tower at the point of feed, but insulated from the tower. The feed, using coaxial cable, is between the tower and the ring, with the shield of the coax connected to a tower leg: see **Fig 12**. The legs of the tower should be strapped together at this point. The radiator is the part of the tower above the radial system. Evaluation by impedance measurement and over-the-air QSOing support the view that this method of feeding a grounded tower works very well indeed. This method should be modelled (using NEC-4) and evaluated by full-scale field strength measurement. Elimination of the base insulator for guyed towers is a distinct advantage."

W A (Bill) McLeod, VK3MI points out that "Over many years the counterpoise has remained non-dimensional in most texts. So in a follow-up to my previous work on 'Near to

Earth Antennas' (*TT*, February 1995, based on VK3MI's 'Low Radiators and High Ground Planes', *Amateur Radio*, November 1994, pp10-14) I wrote 'The Semi-Elevated Ground Plane - GIRFU style' (*Amateur Radio*, July 1997, pp13-15): 'I believe that this corrects the traditional view that the length of a counterpoise is not important. This is based on full-size antenna tests, not modelled, and the computer simulation of similar systems by VK1BRH (see 'Dipoles with Counterpoise', *TT*, January 1997) generally confirms my results.'

"My new article presents a procedure for tuning the counterpoise to the electrical equivalent of the radiator, for a low-loss balanced system with respect to the feed point. The displacement height above ground is a matter of judgement of the affordable losses, based on the results presented in 'Near to Earth Antennas'. The unexpected result was the necessity to isolate the system from actual earth, but, in retrospect, it is quite logical to isolate the return circuit from the lossy earth capacitance."

In case you are wondering what 'GIRFU' in the title of his article stands for, the answer is 'Get-It-Right-First-Up'. VK3MI gives a nine-point procedure for ensuring that an unbalanced antenna system resonates in both dimensions. For general information on VK3MI's 'saggy' approach to elevated radials, in which the ends are raised higher above ground than the base of the vertical element, it may prove useful to refer to the January 1996 item. **Fig 13** shows a 10.1MHz antenna with two 'saggy' wire radials. Note the difference in physical length of the radials to the vertical element, a difference which would be even greater if the entire radial was at the lower level.

PROPANE GAS GENERATORS

STEVE EDWARDS, G8GEF, has sent along some useful notes and additional safety precautions for those considering the use of 'Gas Conversion for Petrol Generators', as described by A J (Jakey) Gould, G3JKY, and Martin Brown, G0DCG (*RadCom*, January 1998, pp27-29).

"The recommended maximum draw-off rates for propane cylinders in order to prevent serious icing-up and pressure-drop are:

Cylinder Size	kg/hr	lb/hr
4.5kg	0.528	1.16
13kg	1.054	2.32
19kg	1.319	2.90
47kg	2.373	5.22

"As with all liquefied compressed gases, drawing gas from cylinders results in cooling and occasionally condensation on the outside of the tank, even when operating below the above-quoted maxima (but without pressure loss). This information is taken from a catalogue issued by Bull Finch (Gas Equipment) Ltd, Diadem Works, Kings Road, Tyseley, Birmingham, B11 2AJ (Tel: 0121 706 6301).

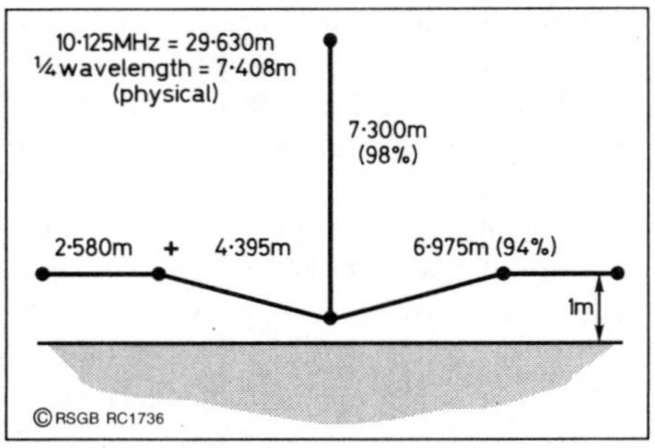

Fig 13: VK3MI's suggested antenna for 10.1MHz, with 'saggy' radials.

This firm provides a wide variety of LPG tools, fittings, adapters, etc. Their catalogue contains much safety information, conversion data and a list of LPG application and safety publications.

"For paralleling multiple cylinders, the firm's T-manifold and Flexible Hose Pigtail would appear to be the most useful. The Calor 'Wall Block Double' mentioned in the *RadCom* article is very similar to the Clesse unit described more fully by the authors. This is intended for installations needing higher flows, and provides automatic no-break changeover between pairs of cylinders operated in parallel (eg 2 x 2 x 47kg runs my domestic heating).

"Despite the euphoria of not having to refill a petrol tank adjacent to a hot engine, the safety aspects of using LPG in cylinders should not be taken too lightly. To the list of safety precautions given on p29 of the *RadCom* article should be added:

1) Propane gas is heavier than air. If leaks occur, gas will tend to accumulate unseen in low places, hollows, ditches, etc. Remember the gas explosions in small boats!
2) 'Empty' cylinders are still potentially hazardous, and should be treated much as full ones. Empty cylinders still contain Propane gas at atmospheric pressure. Keep valves shut and avoid sources of ignition.
3) Store spare cylinders securely, away from the generator and other potential sources of ignition. For portable events, this includes away from areas where 'live' antennas may conceivably drop out of the sky!"

FEEDBACK

'FAST' HUFF & PUFF

G7IXH has confirmed that PA0KSB was correct in suggesting (*TT*, February 1998, p61) that the analog integrator should have been shown as a CA3140 and not CA4130. G7IXH's detailed article on the fast Huff & Puff stabiliser is now expected to be published in the ARRL's *QEX*.

TUNED TRANSMITTING LOOPS CONTROVERSY

TT, February 1998, pp62-61, the German callsign of Dipl-Ing Christian Kaeferlien, manufacturer of the original AMA loops is DK5CZ (not DL5CZ or DC5CZ). G4VRN's surname should have been spelled Blewett, not Bluwen. Apologies to those concerned!

POSTSCRIPT

THIS MONTH IS the 40th anniversary of the start of *Technical Topics*, and in profoundly thanking all those who knowingly or unknowingly have contributed to its continued acceptance by readers, I hope that I may be excused for reflecting on the many changes (some for the better, some for the worse) in amateur radio during those 40 years. I have to confess that as one grows older, it is not always easy to summon-up the energy needed to seek out or to understand new developments and face yet another series of sessions with my ancient word-processor (the sole concession to Information Technology at G3VA), needed to produce a monthly column.

It seems appropriate to quote Rudy Stevens, N6LF, who has just taken over the editorship of the ARRL's *QEX*: "I, and I am sure most of you, have noticed (and lamented) the steady decline in technical content in amateur publications. Some of this decline is understandably due to changes in our hobby, but I have never accepted that we are, or should become, nothing more than appliance operators. That is not in keeping with the tradition I know . . ."

That just about says it all. My own view is that for amateur radio to survive and flourish, it is essential to preserve the sense of wonder at the 'magic' of communicating over long distances by radio with relatively simple equipment. It was this that brought many of us into the hobby and kept us there, together with curiosity about those who have made it all possible. As the late Dr R V Jones put it in 'Most Secret War': "My main hobby in my schooldays was, as with many other boys of my generation, the making of radio receiving sets. There has never been anything comparable in any other period of history to the impact of radio on the ordinary individual in the 1920s . . . it was as near magic as anyone could conceive, in that with a few mainly home-made components simply connected together, one could conjure speech and music out of the air." This remained true in the 1930s and beyond. Even more thrilling was the thought that one could use radio to communicate across countries and continents!

Of course, to be a long-time columnist one has to come to grips with not only the personal interests, idiosyncrasies and foibles of oneself, but also those of editors with whose views one may or may not always fully agree. So far *TT* has survived (just) under the encouragement of a succession of editors: the late John Clarricoats, G6CL; the late John Rouse, G2AHL; Trevor Preece, G3TRP (two spells); John Adey; A W Hutchinson (1969-89); David Bobbett, G4IRQ; Mike Dennison, G3XDV, and Technical Editor Peter Dodd, G3LDO; and now Steve White, G3ZVW. We have not always agreed unreservedly on editorial matters and content, but I acknowledge my debt to one and all! ♦

SSB PRODUCT DETECTOR FOR VINTAGE RECEIVERS (RA17 ETC)

MORRIS ODELL, VK3DOC, in 'An SSB Product Detector for "Boat Anchor" Receivers' (*Amateur Radio*, March 1998, pp7-8) notes that "despite advances in modern communications technology, and perhaps because of them, there is still a great deal of enthusiasm in some parts of the amateur community for older style vacuum tube gear. There are many pieces of equipment still around which were 'state of the art' in their day and which are capable of performing as well as, or better than, some modern solid-state rigs. Because of their heavy power transformers on solidly built chassis, these are affectionately known as 'boat anchors'. Many were produced before SSB became popular . . . the diode-detector/low level BFO combination, common before the 1960s, is not well suited to SSB reception and may result in distorted and noisy demodulation, requiring much fiddling with the RF gain control for a barely satisfactory result."

Like many UK amateurs, VK3DOC has a Racal RA17 receiver, which he considers a beautiful and very sophisticated design, but with a particularly unsatisfactory performance on SSB. While Racal produced an SSB adapter, these "are not easy to find". His solid-state product detector adapter (**Fig 1**) takes advantage of the output socket found on the RA17 and many other professional receivers. It provides IF at 100kHz, after the selectivity filters and IF amplifiers. The design can be modified for use at other final IF frequencies including 455kHz.

*37 Dovercourt Road, London SE22 8SS

VK3DOC writes: "As there is no BFO socket, I originally chose to generate an independent BFO, with TR1 in Fig 1 forming a Colpitts oscillator at 100kHz or, with some component changes, at 455kHz. For 100kHz operation, four varactor diodes are needed to obtain an adequate frequency swing, with the lower limit of the tuning voltage at about 1.8V (the text-book back-to-back varactor arrangement would need eight varactor diodes). A Motorola MC1496 forms the double balanced product detector. The output of TR1 is isolated by the emitter-follower TR2, which feeds the MC1496 oscillator port through an attenuator pad to ensure the correct injection level of about 300mV. **Fig 2** shows a suitable PSU for the adapter.

Once the adapter was built, it became apparent there were useful advantages in bringing out the BFO signal from the receiver. One snag with the independent BFO is that it does not work properly on AM: "Unless the BFO is in phase lock with the AM carrier, objectionable heterodyne effects occur. If the independent BFO is disabled, there is no output from the double-balanced mixer. VK3DOC points out that a simple (non-permanent) modification to the receiver gets round this problem . . . The anode of the set's BFO valve is coupled to the detector diode through a small capacitor. This means that when receiving AM with the BFO off, there will be a volt or two of the AM carrier on the BFO anode pin. When the BFO is on, there will be several volts of BFO signal at the same point. If this signal is extracted from the receiver and connected to the oscillator of the adapter through a suitable isolating network, three desirable results occur.

1. If the receiver BFO is on, it locks the adapter oscillator, allowing pitch to be controlled at the receiver front panel and overcoming any drift in the adapter oscillator.
2. When the BFO is turned off, the AM carrier from the BFO anode captures the adapter oscillator, permitting synchronous AM detection.
3. By unplugging the receiver BFO connection, the adapter functions as an independent SSB detector with its own BFO, and can be used with unmodified receivers.

VK3DOC brought out the BFO anode point by simply wrapping a piece of fine wire around the valve pin and passing it out through the top of the valve shield. This is connected in series with a 1nF capacitor to a coaxial connector mounted on a scrap of aluminium angle plate: **Fig 3**.

The TR1 oscillator coil can be any RF inductor with a value of around 3.9mH for 100 kHz or 300μH for 455kHz. A disused transistor IF transformer of the appropriate frequency would be ideal. Tuning range is a little over 3kHz at 100kHz with the values shown. The Colpitts divider capacitors should be reduced to about a quarter of their 100kHz values of 22nF and 10nF shown in Fig1.

VK3DOC concludes: "The difference in SSB reception when using the adapter is quite spec-

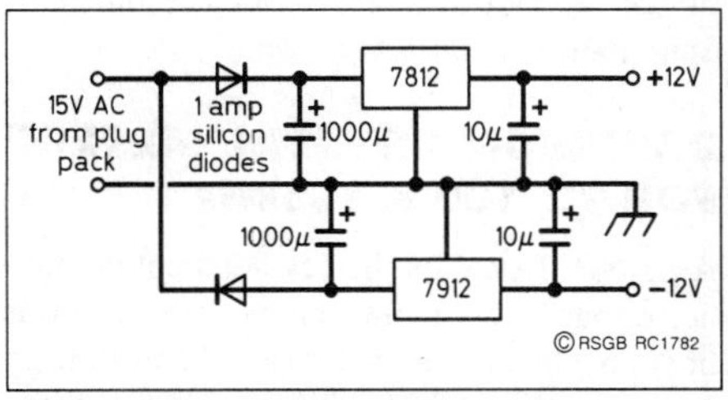

Fig 2: Dual-polarity PSU for the SSB adapter.

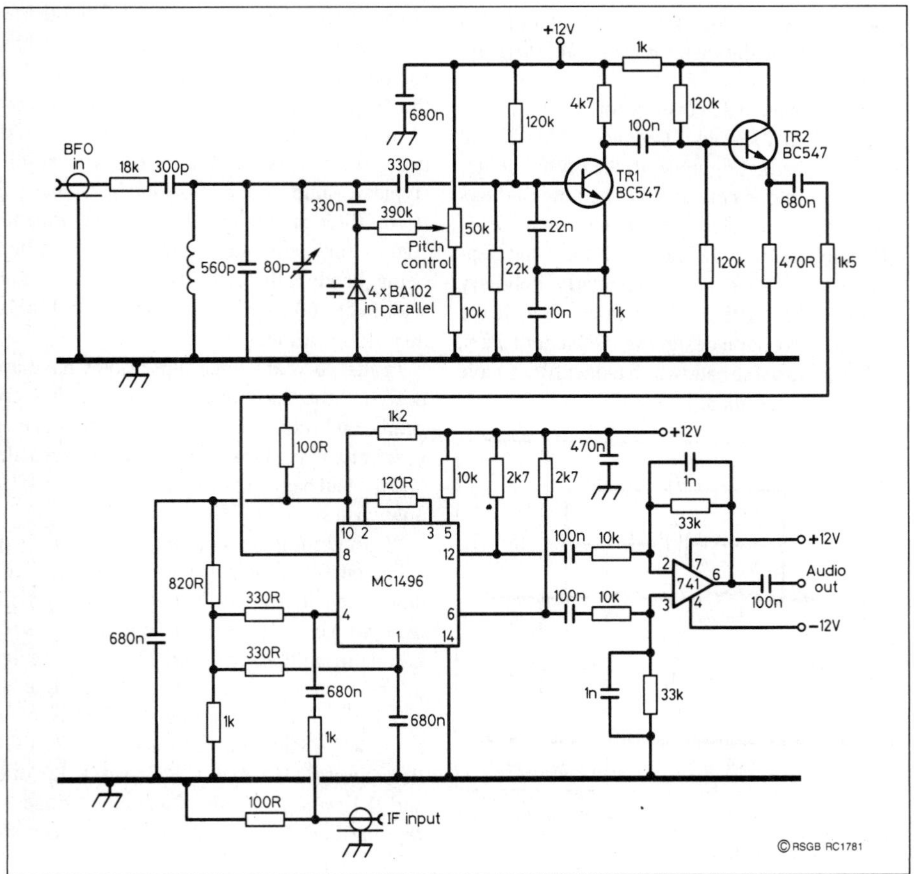

Fig 1: Circuit diagram of VK3DOC's SSB product detector adapter. Values shown are for 100kHz IFs, such as the RA17.

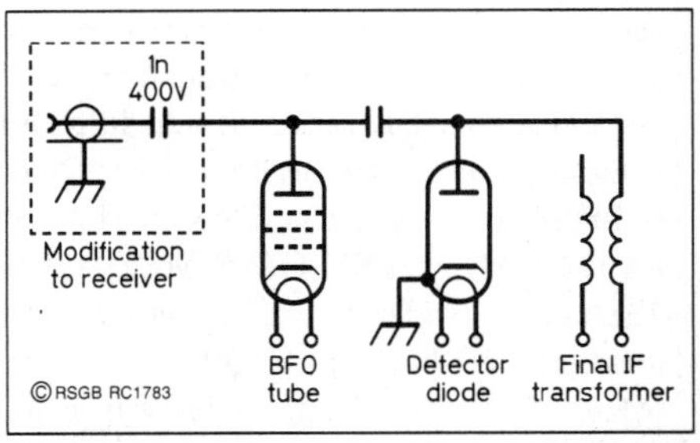

Fig 3: How the SSB product detector is attached to the receiver for use as the set's BFO, in addition to the adapter BFO.

tacular, and converts this magnificent receiver into a useful workhorse in the shack. By adding the BFO modification, the adapter functions 'automatically' on AM as well."

BRIDGED-T CRYSTAL FILTERS FOR 73, 100 & 136kHz

A LANDMARK IN the development of high-performance communications receivers was undoubtedly the description in *QST* in 1932 by James Lamb of ARRL of the crystal-gate filter, based on the earlier 'stenode' receiver work in the UK by Dr J Robinson and Earnest Gardiner, G6GR.

The Lamb crystal-gate filter used a single crystal, usually at the standard American IF of 455kHz, with a 'phasing control' to shift the deep notch of any asymmetrical characteristic. The effective bandwidth could be changed by varying the output impedance, either continuously or switched. Generally, the filter was unsuited to the ±3-6kHz bandwidth of amplitude-modulated phone, but was, and remains, excellent for single signal CW reception, provided always that the operator understands the procedure involved with the correct setting of the then customary variable BFO.

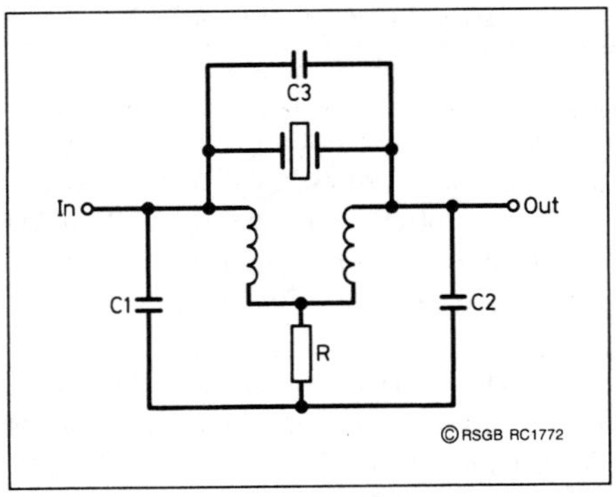

Fig 4: Basic bridged-T crystal filter as given in *TT* in June 1962. C1 and C2 are adjusted for maximum response at the series-resonant frequency of the crystal. C3 adjusts the positions of the maximum rejection points.

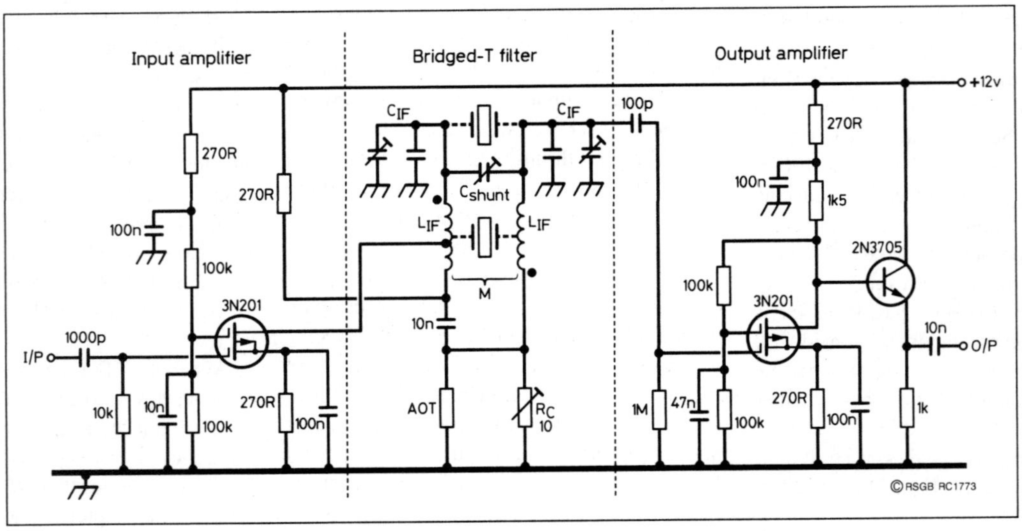

Fig 5: Bridged-T crystal filter and test circuit as used by G3JIR.

By the mid-1930s, crystal-gate filters were being incorporated in most up-market HF receivers such as the HRO and those made by Hammarlund and some by Hallicrafters. In wartime, it was found that many of the hastily-trained operators were unable to take advantage of crystal-gate filters and, for example, they were omitted from some wartime HRO receivers.

In the UK and Germany, considerable effort was put into developing both single-crystal filters having symmetrical characteristics and the half-lattice twin-crystal bandpass filters which for many years remained the popular, if expensive approach. The development of bandpass filters with centre-frequencies well up into the HF region was a feature of the late 1950s and 1960s. However, it was difficult to home-brew such filters, leading to the popularity of the HF crystal ladder filter using low-cost PAL (4.43MHz) or NTSC (3.58MHz) crystals. The crystal ladder filter was introduced to amateurs by J Pochet, F6BQP in *Radio-REF* (May, 1976). An outline of his article duly appeared in *TT* (September, 1976) and I also provided a translation for *Wireless World* (July, 1977).

It so happened that even before my notes had appeared in *TT*, Jack Hardcastle, G3JIR had begun investigating in depth the crystal ladder filter and in due course published his findings in a series of articles in *Radio Communication* (December, 1976, January, February and September 1977 and also particularly February 1979 and May 1983). These articles remain the definitive work on designing HF ladder filters, although variable bandwidth ladder filters have also been described.

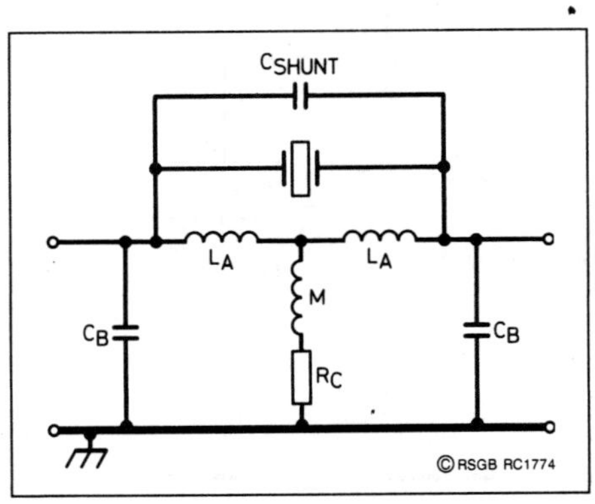

Fig 6: Theoretical equivalent circuit of the filter shown in Fig 5.

Essentially, the crystal ladder has asymmetrical characteristics and can provide an excellent filter for SSB. However, there can still be a requirement for symmetrical response curves at LF and MF provided by a single crystal.

In *TT* (June, 1962, summarised in all editions of *Amateur Radio Techniques*) it was noted: "Most of the professional texts on IF crystal filters devote considerable space to the bridged-T type of filter (sometimes called 'combined crystal and mutual inductance coupled circuits'). With a single crystal, two points of infinite rejection can be placed, one each side of the crystal frequency, thus providing a symmetrically shaped response curve, roughly similar to that of the more familiar two-crystal, half-lattice bandpass filter. The basic circuit is shown in **Fig 4**, though some variations can be noted between different authorities....

"Although this type of filter was successfully used in some Service receivers (R201 and R206) as a plug-in unit (see *Proc IEE* Vol 94, Part 3A, 1947), it never seems to have been widely adopted for amateur receivers. A possible drawback is that for optimum performance the inductance of the IF transformer coils and the mutual coupling have to be specified, and the series resistance (sometimes omitted) also affects results (typical value about 5000Ω). Yet clearly a useful filter can be made, quoted performance figures for one of the Service 465kHz filters being bandwidth (at -6dB) 2.5-3kHz, bandwidth (at -60dB infinity points) 4.5-11kHz; and -40dB at lobes."

In the 36 years since those notes have appeared, little additional information has appeared on bridged-T crystal filters. However, a year or two ago, I learned that Jack Hardcastle, G3JIR, had begun researching these filters. He now writes:

"I am writing to report progress on bridged-T crystal filters, which you may remember I have been investigating (intermittently!) for over two years. It has taken so long because of initial difficulties sorting out the correct design information and then overcoming component problems. The design side was finally overcome when you obtained for me the Post Office report (C F Floyd, E R Broad, R L Corke and H Lewis 'Analysis of a simple bridged-T crystal filter', Post Office Engineering Depart-

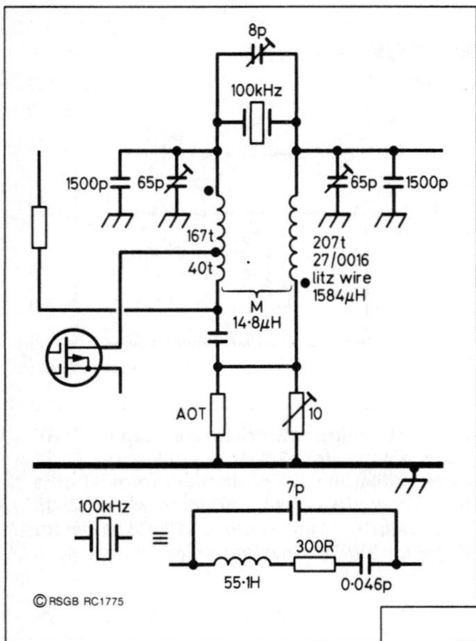

Fig 7 (left): Circuit details of the 100kHz filter developed by G3JIR.

Fig 8 (below): Constructional details showing mounting of the iron dust cores.

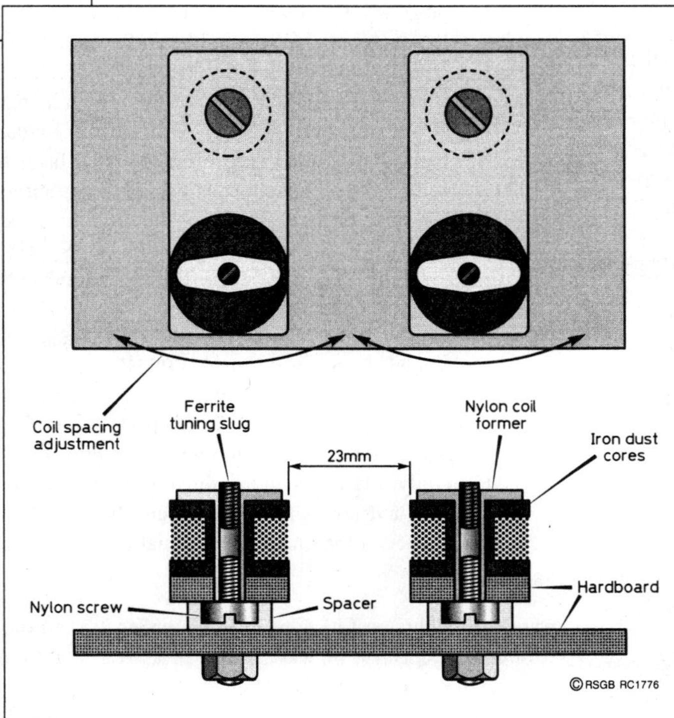

ment, Radio Report No. 1747 (37 pages) January 1948). The practical problems are outlined in my accompanying report. Although I have described only a 100kHz and a 455kHz filter, I have also tested filters up to 844kHz. I found that the lowest frequency filters were limited to CW bandwidths, while at 459kHz SSB could be accommodated.

"If supplies of components can be obtained I feel these filters could fill a useful niche at 73 and 136kHz, where they could be useful as an RF-stage roofing filter. Also, if a calling channel were established, a narrow-band filter and a TRF receiver could make a cheap monitor for continuous use, while the main station receiver was in use on other bands. Perhaps kit manufacturers could take note and persuade crystal suppliers that a potential market exists."

G3JIR has produced a detailed 12 page report 'The Bridged-T crystal filter' providing a wealth of design information and formulae, together with practical constructional information. With the aid of an his brother, G0JUA, he hopes to put all this information on the Internet (http://www.users.rapid.net.uk/enid/g3jir/crystal.htm). However, if the number of requests does not prove excessive, I am willing, with G3JIR's permission, to provide photocopies at a nominal cost, including postage, of £1 to those seriously interested. Only short extracts from his text and a few of his illustrations, including **Fig 5**, the circuit used to test the performance, are given here.

To quote briefly from G3JIR's report: "Although bridged-T crystal filters are mentioned in many of the earliest papers on crystal filters, the only known practical application is in the WW2 R206 army receiver. However, it is likely that these filters have been used by the GPO (as it then was) in carrier telephony systems, since the only available detailed design analysis was produced by a team working at the Post Office research laboratories at Dollis Hill (see above). The work of these authors as the originators of the equations used in my report and designs is fully acknowledged.

"There are several reasons why the bridged-T design could be attractive to radio amateurs. The principal one is that a symmetrical passband, with sharp rejection notches on either side, can be achieved using only a single crystal. Furthermore, if the design requires a multi-stage filter, all the crystals are on the same frequency, rather than being spaced apart by a critical amount (as for half-lattice filters). Also, the inherently unbalanced nature of the circuit avoids the difficulties of making the balanced (centre-tapped) transformers required for half-lattice filters . . . . It is at the lowest frequencies that it is considered that these filters may have their most useful application; that is as RF roofing filters for use in the 73 and 136kHz bands, prior to the main amplifier and any subsequent digital filtering.

"Fig 5 shows that the filter consists of two parallel-tuned inductively coupled circuits, resonated at the passband centre frequency, with a crystal shunted across the 'hot' end of the windings. The bridged-T description becomes apparent only when the equivalent circuit (**Fig 6**) is examined. Note that the phasing of the transformer windings is important. . . . Part of the alignment procedure is to adjust the mutual inductance to the value needed to produce the rejection notches on either side of the passband. **Fig 7** shows circuit details of a practical 100kHz filter, with **Fig 8** showing the coil mounting arrangement adopted to allow the coil spacing to be adjusted as required. The most satisfactory way to carry out this adjustment is to use a wobbulator . . . The design of the coils is most important, since the narrowest bandwidth achievable depends on their having a high Q . . . Litz wire and dust-iron or pot cores are needed . . . **Fig 9** shows the computed and measured frequency response of the 100kHz filter, compared with the response of the IF transformer without the crystal . . .

"Several factors limit the degree to which the basic design

Fig 9 (right): Frequency response of the 100kHz filter, compared with the response of the IF transformer without the crystal.

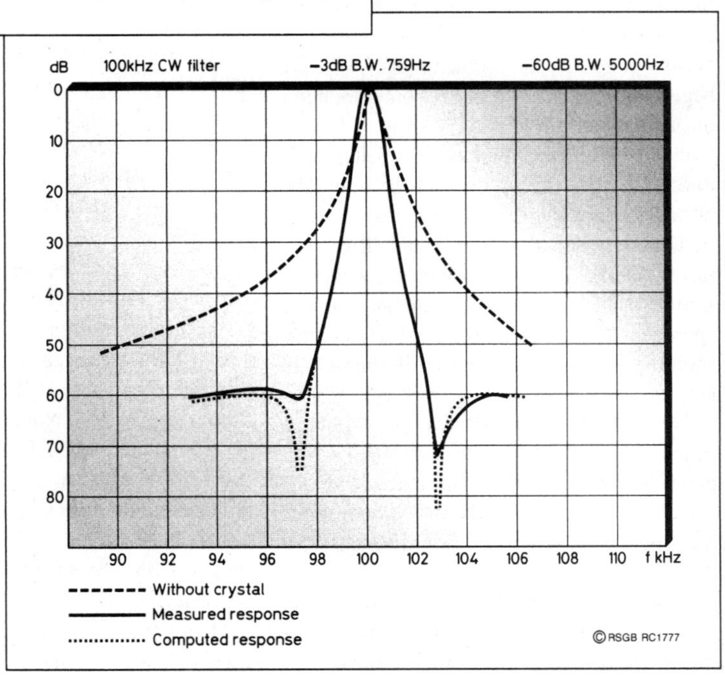

features can be changed. For example, to move the rejection notches closer to the passband requires a higher mutual inductance than when they are further apart . . . But as the notches are moved closer the response beyond the notch rapidly deteriorates . . . The width of the passband is increased with higher mutual inductance . . . and is subject to practical constraints. As the bandwidth is increased the design impedance also increases, and because the filter must be operated between source and load impedances several times greater, an impedance of 100kΩ, or so may be regarded as a practical limit . . . "

G3JIR concludes: "As may be thought appropriate for what has been an exercise in 'industrial archaeology', these tests were made possible only by using components which were accumulated over a period of many years, on the assumption that they 'might come in useful'. Unfortunately, this has meant that they are no longer in the catalogues of the major amateur radio suppliers. Hopefully, the potential usefulness of these filters for the new low-frequency bands will encourage kit suppliers to find sources of suitable components."

ADAPTING A G5RV FOR 136kHz

STEVE RAWLINGS, GW3ALG, writes: "When the new 136kHz VLF band was released at the end of January, I began building a 136kHz transverter for use with my FT707 operating at 10MHz. I then set about trying to devise a simple antenna for the band.

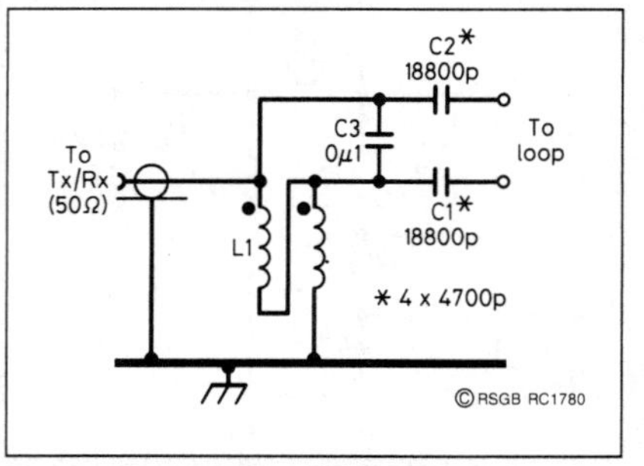

Fig 10: Matching network for the adapted G5RV 'loop' antenna for 136kHz. 4:1 balun with L1 22 turns, bifilar wound on surplus toroidal core (the one used was coloured yellow/white, 33mm OD, 20mm ID, 11mm wide). C1/C2/C3 adjust for minimum SWR (or maximum signal on receive).

"The (full-size) G5RV dipole antenna at GW4ALG had been installed in an inverted-vee configuration (apex about 35ft above ground) with about 40ft of 300Ω feeder, and was in regular use on 3.5MHz with my home-made 5W transceiver. To get this antenna working on 136kHz, I initially tried strapping the G5RV feeder connections and tuning the antenna against ground. However, I had great difficulty in matching the antenna, and on receive the noise level was very high. I decided to try configuring the G5RV as a loop by joining together the far ends of the dipole with about 100ft of multi-strand cable. The result was a form of top-fed delta loop, having an overall perimeter of 200ft, with its base about 5ft above ground.

"An idea for feeding the loop was found in *Amateur Radio Techniques*, 4th edition, p204, which describes how single-turn loops can be matched through the range 2 to 5MHz. [This was the low-loss matching unit based on capacitors designed at the US Army Limited War Laboratory for the original transmitting loop used in Vietnam, and appears in most editions of *ART* including the 7th edition, p279 - *G3VA*]. The application of this idea, together with the inclusion of a bifilar-wound ferrite balun, resulted in the choice of the feed arrangement shown in **Fig 10**.

"When used with my home-made 136kHz transverter (15 watts output), this arrangement provided an S5 report from GW0GHF over a distance of 42km. Note that, if used for transmitting, the capacitors in the matching network need to be low-loss, high-current types. On average, the local noise levels received on the loop were lower compared with the original G5RV configuration with a strapped feeder.

"Generally, I have found the single-turn loop configuration to be easier to adjust than the Marconi-fed arrangement; to be an effective, low-noise receiving antenna; and an adequate antenna for carrying out initial transmitting tests on 136kHz."

PACEMAKERS AND AMATEUR TRANSMITTERS

THE TOPIC OF pacemakers and their vulnerability to strong RF fields came up several times in *TT* during the 1970s and 1980s, the last time in an item 'EMC and pacemakers' (*TT*, February 1989 - see also *Technical Topics Scrapbook, 1985-89*, p209). In that item it was noted: "From time to time, *TT* has referred to the uncertainty that has long surrounded the use of implanted cardiac pacemakers in close proximity to medium and high-power transmitters and their associated antennas. There can be no doubt that early designs of these vital life-saving devices were often vulnerable to strong RF fields. It is only within the past decade or so that the problems of EMC have begun to be taken seriously in some segments of the electronics industry."

It was noted that attempts were being made to establish standards covering the immunity of pacemakers, but that German research carried out in 1988 still showed wide differences (up to 40dB) between the best and worst designs. It is clear that today while one needs to approach this problem with great care, the fitting of a pacemaker need not write *finis* to amateur operating.

Tony Tuite, GW0NSR, writes: "About a year ago, I was fitted with a pacemaker implant. It is the type which has a sensor in one chamber of the heart and a stimulation electrode in another. The two leads are each about 10-12 inches in length.

"I was given dire warnings about using mobile phones, radio transmitters, microwave ovens, etc. It seemed my amateur radio days were over, although I have been active under various calls since 1951. I was none too pleased to lose a life-time hobby, but delighted to be fit and active again.

"Over a period of weeks, I began with my FT101E, putting out just 2 watts RF on 3.5MHz, then tried other bands, steadily increasing power to 100 watts CW. No problems were experienced, despite my using a 250ft end-fed wire with the consequent amount of RF in the shack. I have an RF meter sitting on a shelf, and if the meter rises above one-tenth full-scale deflection, I re-adjust the Z-match ATU to reduce it.

"CW is normally my only mode. However, I tried SSB at 25 watts and had an immediate change in heart-rate. It jumped from its normal 70-ish to over 150. Not so funny!

"Then I tried 144MHz and 430MHz FM. No problems at up to my maximum power of 40 watts. Mobile phones seem to present an odd situation. No problems with analogue units but digital (GSM) ones have a slight effect, if held on the left side of the body. None if on the right.

"Our none-too-modern microwave oven has no effect, but a commercial microwave over in a cafe did cause the heart-rate to increase to 115 at a distance of some 12ft.

"I know one amateur who decided not to accept the offer of a pacemaker, as he had no wish to give up the hobby. I have tried to convince him that this may not be necessary but to no avail - it seems to me his priorities are rather confused.

"My overall view is that with good practice and attention to detail in setting up a station, particularly if good quality low-loss coaxial cable is used for patch-leads and feeders, no problems need arise, even with SSB. I cannot comment on the use of high-power linears."

If I might add a comment, it does seem that pulsed forms of RF, as found in SSB and regular pulse rate digital systems (GSM etc mobile phones are recognised as having an EMC problem), have appreciably more effect on most types of vulnerable devices than constant carrier FM, presumably AM and manual Morse. For those with pacemakers, the really important matter is always to carry out careful checks, initially at low power, along much the same lines indicated by GW0NSR.

UN-GLUING HOT-MELT GLUE

GILES READ, G1MFG, recently made a small electronic assembly, securing some components with hot melt glue. Unfortunately, one of the components failed and needed to be removed and replaced.

By Murphy's Law the faulty component was "at the bottom of the pile", well stuck down and largely obscured by other components. He writes:

"Having previously had little success in using a hot-air gun for removing hot-melt glue, I decided to immerse the entire assembly in very hot (just off the boil) water. I was surprised and delighted to see the glue becoming transparent . . . after a minute or so, when the warmth penetrated, the glued parts were easily separated.

"Needless to say precautions are necessary to avoid scalding oneself, and tools present problems. Metal tools tend to act as heatsinks and immediately become attached to the assembly one is trying to dismantle. I found a reasonable compromise between insulation and practicality was to use kitchen paper to hold the various items. This has the added advantage that the paper can be used to wipe off excess glue.

"Consideration needs to be given also to the effect of hot water on the components. Most electronic components are sealed to a greater or lesser extent and can be safely immersed temporarily in hot water, but some components (microphones, loudspeakers etc) can be immediately ruined by contact with water. Others, such as variable capacitors, may have their electrical characteristics altered subtly or severely. Mechanical components can suffer oxidisation-related problems."

SIMPLE 3.5/7MHz CW TRANSMITTER

THOSE OF US involved with wartime clandestine radio came to recognise the value of relatively simple CO-PA transmitters, providing some 5 to 15 Watts RF output from 6V6, 6L6, 7C5, etc. valves, on frequencies between 3 and 10MHz. Today this type of CW transmitter could easily be implemented using a single ECL86 triode-pentode, but many newcomers (and some old-timers) feel that it is infra dig to continue to use valve technology in low power transmitters.

The answer for those disbelieving (in my view wrongly) in the continued value of thermionics can take heart from a design for "A simple CW transmitter for 80 and 40 metres" by Charles Kitchen, N1TEV (*QST*, February 1998, pp40-42), sub-headed "If you've never used a MOSFET-powered transmitter before, give this one a try!"

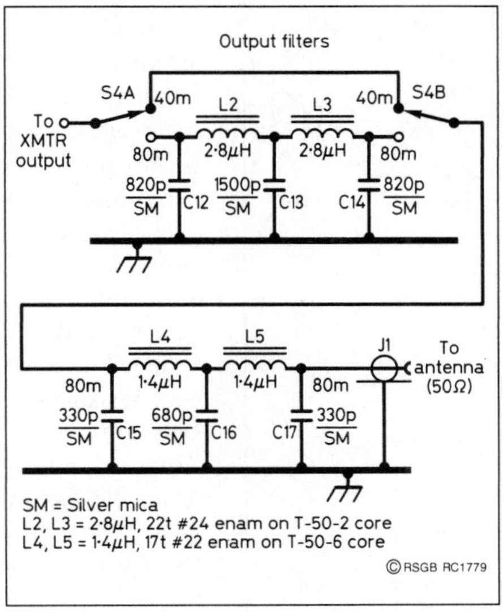

Fig 12: Output filters used to ensure the transmitter meets FCC spectral purity requirements. Both filters are in circuit for 3.5MHz operation. On 7MHz the final filter is bypassed. All capacitors are silver mica.

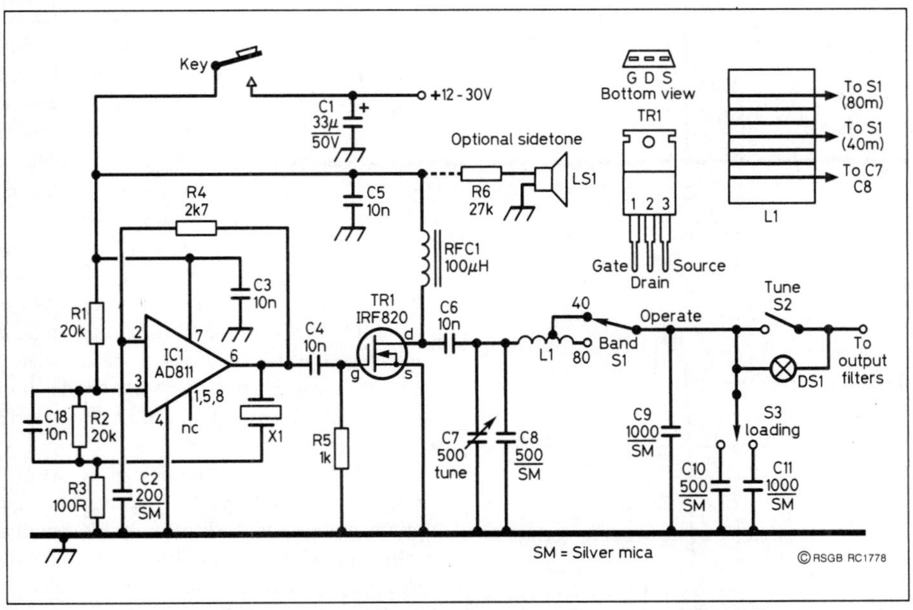

Fig 11: Circuit diagram of the simple 3.5KHz CO-PA MOSFET transmitter described by N1TEV. LS1 is a piezo buzzer for optional sidetone. C7 is air-dielectric variable capacitor, providing a total capacitance of about 500pF (discarded broadcast 350pF 'gang' with sections paralleled is suitable). N1TEV used (Y1) a 3.6864MHz microprocessor crystal, but this would not be a suitable frequency in the UK.

N1TEV points out that this transmitter (**Fig 11**) is "a modern-day equivalent of the classic two-valve CO-PA transmitter used by so many hams for decades - usually their first transmitter, and usually home-brewed. Building this equivalent is much simpler and probably less expensive. The two-stage transmitter uses a low-cost, high speed op-amp in the crystal oscillator circuit and a modern HEXFET device (an N-channel, enhancement-mode power MOSFET) for the output amplifier. Unlike some other solid-state designs, this little gem requires few components and can be built in an evening or two - ARRL Lab measurements indicate power outputs of 6W on 3.5MHz with a 30V power supply for a DC input of about double this. On 7MHz, 1W output with a 25V supply." With about a 24V supply it could make a very useful 4W Novice transmitter on 3.5MHz.

The use of a HEXFET PA provides a much higher load impedance for the oscillator than a bipolar device, eliminating the need for a step-down transformer generally required between bipolar transistor stages, and permitting the use of simple RC coupling. With a typical breakdown voltage of 500V and built-in Zener diode over-voltage protection, the HEXFET PA can operate without a load and under conditions of high SWR without damage. Suitable low-cost microprocessor crystals can be used.

An Analog Devices AD811AN op-amp is used as the oscillator stage. The AD811 was carefully chosen because it has a beefy 100mA output stage and very low output impedance. Other op-amps with these characteristics may work, but are likely to be very much more expensive. Similarly, the IRF820 was chosen because of its fairly low input capacitance (350pF). Other HEXFETs such as the IRF830 and 840 have very high input capacitances (about 800 or 1500pF), and would require the use of a step-down transformer.

N1TEV suggests that series-connected 6V lantern batteries are ideal for powering this transmitter. Apply no more than 30V, otherwise the AD811 may be destroyed. Use of a 'hefty' mains PSU, with low internal resistance, may cause the AD811 to overheat, resulting in a 'chirpy' note. Higher voltages could be applied to the IRF-820 only. 'Dead bug' or similar constructional techniques are suitable, provided normal HF RF practice is employed. It is advisable to wire and test the oscillator circuit before adding the output stage. To avoid static damage, keep the IRF820 inside its protective foam as long as possible, and install it as the final component.

To meet current FCC requirements on spectral purity, N1TEV provides circuits for a suitable low-pass filter: **Fig12**. ♦

BY PAT HAWKER, G3VA*

HIGH PERFORMANCE CRYSTAL LADDER FILTERS

JACK HARDCASTLE, G3JIR, recently sought my help in locating a copy of a 1992 German article on crystal filters. This was located through the National Science Reference Library but my visit there brought an unexpected meeting with Dr Tom Going, a historian of electronics, and Arthur O Bauer, PAOAUB, who is chairman of the 'Centre for German Communication and Related Technology, 1920-45' and author of at least two excellent, well illustrated, books (German text) *Deckname 'Wuerzburg'* (the secret development of the Wuerzburg radar between 1937-45) and (with the help of Ralph Erskine and Klaus Herold) '*Funkpeilung als allliierte Waffe gegen deutsche U-Boote 1939-45*' - the story of the Allied use of HF D/F (huff-duff) and Ultra intercepts against the German U-boats during WW2.

PAOAUB was at the library researching the history, development and application of quartz crystals for a new book, which it is intended will be published with a bilingual text (English/German). Our subsequent discussion revealed, amongst other things, that my long held belief that the original development of crystal filters was by Dr Robinson and then applied to HF communications receivers by Lamb of ARRL in 1932 overlooks earlier work.

PAOAUB has since provided me with a copy of British Patent Specification 293,446 'Improvements in or relating to Electrical Filter and the like Frequency Selective Circuits'. This has a United States convention date of 7 July, 1927, an application date in the UK of 18 May, 1928, and an acceptance date of 9 May, 1929. The British patent was applied for by Marconi's Wireless Telegraph Company Limited, as assignees of Clarence Weston Hansell of River-head, Rocky Point, Long Island, New York. PAOAUB has found that Hansell's name appears on many patents issued during the 1920s and 1930s.

The patent includes (see **Fig 1**) the familiar basic crystal-gate filter introduced to amateurs by Lamb and soon afterwards incorporated in many of the better factory-made communications receivers. Undoubtedly, credit for the crystal filter belongs to Clarence Hansell. His patent also included the use of three crystals in parallel to form a bandpass filter, although the later filters connected the second crystal in place of the phasing (neutralising) capacitor of the Hansell patent.

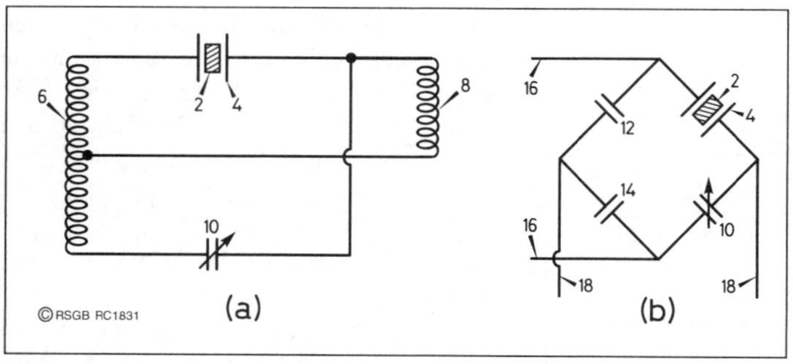

Fig 1 (above): Selective crystal filter with equivalent bridge illustrations taken from British Patent Specification 293,446, issued May 9, 1929, and assigned to Marconi's Wireless Telegraph Co Ltd by Clarence Hansell of Rocky Point, Long Island, and previously patented in the USA (July 7, 1927).

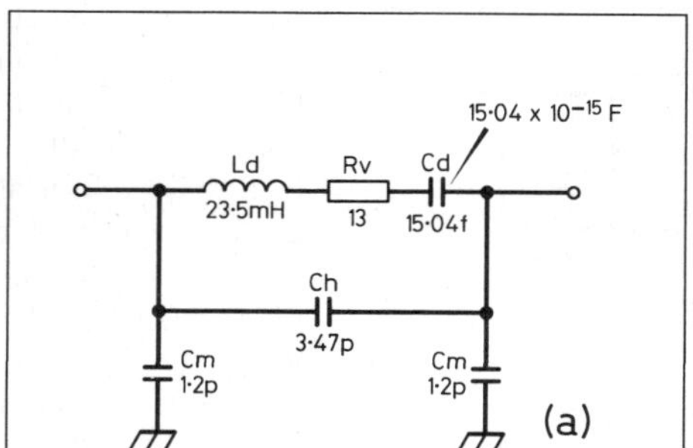

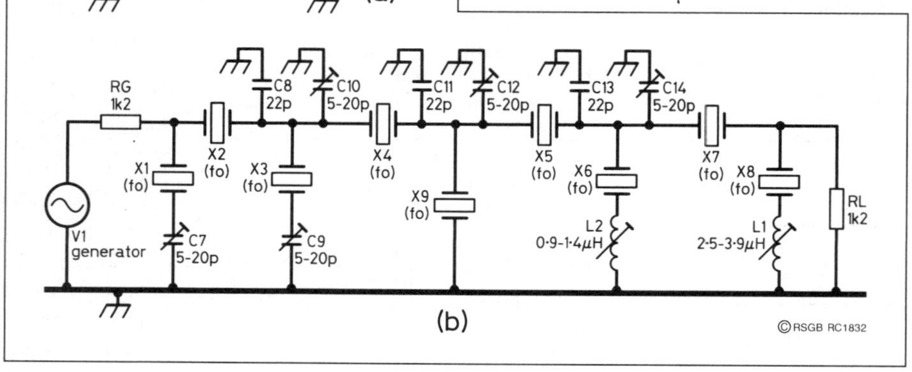

Fig 2 (right): (a) Equivalent circuit of an 8.44MHz crystal, as used in the nine-crystal ladder filter based on low-cost crystals (b).

But to return to the present decade, the German article sought by G3JIR was 'Quarzfilter aus Standardquarzen' (Quartz crystal filters from standard crystals - good correspondence between simulation and measured results) by Prof Dipl-Ing Christian Dirks and Dipl-Ing (FH) Michael Braun (*Elektronik*. 5/1992, pp74-75). Jim Leviston, G3NFB, has translated this article and G3JIR has kindly supplied me with a copy.

G3JIR notes that the Germans, in developing a high-performance ladder filter suitable for professional applications, have used the same technique using shunt crystals for improving the LF band edge, which he used in one of his early designs (Fig 19, February 1977, p124 and Fig 22, September 1977, p687). They have succeeded in producing a symmetrical filter response by using X6 and X8, which have had their series resonance lowered by inserting series inductance, but by using a more scientific approach they have produced a very desirable frequency response.

To quote from G3NFB's translation of the German article: "Many designs for building crystal filters already exist in the literature. . . and there are efficient (computer) programs. Most designs require crystals that have specific (different) frequencies. This article shows how readily available crystals of the same frequency can be used to build filters with acceptable performance characteristics.

"Very reasonably priced crystals are available. . . such as those manufactured in very large quantities for the television industry. The use of these crystals for filters is difficult; one must have expertise to use crystals of a single frequency throughout the whole filter. . . With the aid of a modern high frequency CAD program and an associated network analyser, it is possible to develop professional-quality filters from cheap quartz crystals."

The authors outline the design procedure, using SUPERCOMPACT program for optimisation, for an SSB receiver filter with a 2.5kHz bandwidth and better than 89dB sideband suppression, using 8.4667MHz crystals. **Fig 2(a)** shows the equivalent circuit of the 8.4667MHz crystals and **Fig 2(b)** the complete filter, although for a professional receiver two are combined with separate amplifiers and buffers. Measured response is shown in **Fig 3**, corresponding very closely to the simulated

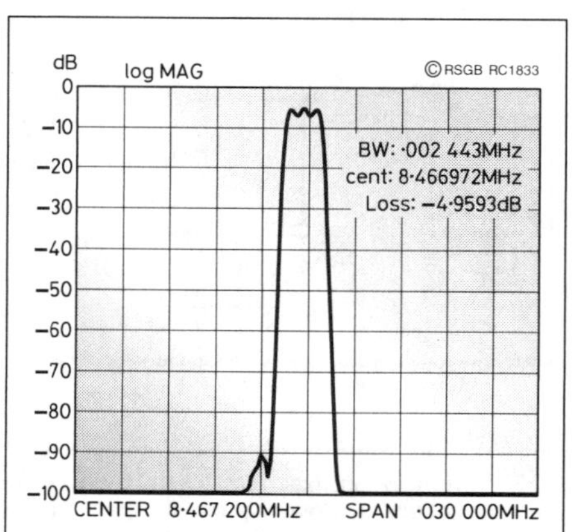

Fig 3: Measured symmetrical response of the ladder filter after optimisation by computer aided design (SUPERCOMPACT).

**37 Dovercourt Road, London SE22 8SS*

response after optimisation. The measured values are: centre frequency 8.467MHz; -6dB bandwidth 2.443kHz; sideband suppression better than -85dB down to -90dB; form factor (6/60dB) is 1.5. At -80dB the form factor is still only 1.6.

The authors conclude that it is possible with specialised, modern CAE tools, to develop high quality quartz crystal filters from economically priced, mass produced quartz crystals.

SHORT VERTICAL ANTENNAS AND THEIR RADIALS

ALTHOUGH AT PRESENT, with sunspot cycle 23 increasingly making good progress (the daily sunspot number reached 192 on 9 April, with a Solar Flux of 141 on 11 April, a Kiel A of 25 on 11 April, with the solar flux exceeding 100 throughout the first half of April, resulting in a number of 28MHz openings) it may seem perverse to discuss how a reasonably efficient 1.8MHz antenna can be fitted into a postage-stamp-sized garden or backyard. However, Thomas Kuehl, AC7A, in 'Build efficient, short vertical antennas' (*QST*, March 1998, pp39-44) provides useful advice on both inductively and capacitively loaded vertical antennas based on 44ft, 1.5in diameter tubular masts intended for a segment of the 1.8MHz band centred on 1825kHz. A capacitively top-loaded antenna is likely to provide a significantly wider low-SWR bandwidth than with inductive loading.

Of paramount importance for those with little real-estate, where quarter-wave resonant elevated radials are out of the question, is AC7A's advice on 'grounds'. His advice is that more short buried radials are better than a few long ones: "Set radial length and orientation to fit the available space. As a minimum, 20 radials is a good objective, while more than 80 is likely to be of little benefit. Don't depend on ground rods, they are an ineffective earth return in many locations."

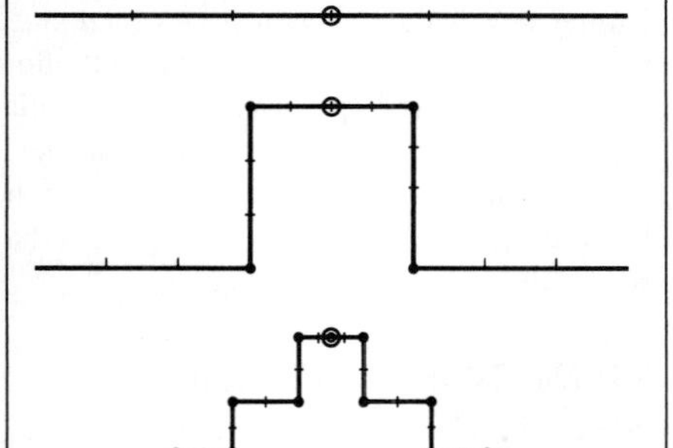

Fig 4: The Minkowski (box) fractal for iterations 0 (straight line), 1 and (bottom) 2.

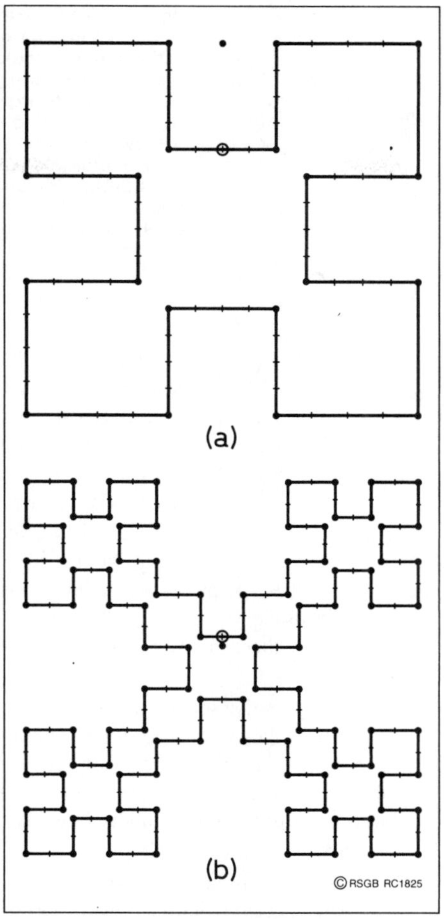

Fig 5: (a) First iteration Minkowski Island (MI1) fractal developed from the quad square element and used in several N1IR designs. (b) Second Iteration Minkowski Island (MI2) has also been used by N1IR.

MORE ON FRACTAL ANTENNAS

ALTHOUGH APPARENTLY some readers thought the item 'Magic Fractal Antennas?' in the April *TT* was an April Fool joke, I am grateful to Peter Chadwick, G3RZP, (and subsequently VP8AEM) for drawing my attention to the series of articles by (Professor) Nathan ('Chip') Cohen, N1IR, in *Communications Quarterly*. These articles, which I have located in the RSGB Library at Potters Bar, provide in considerable detail (totalling over 50 large pages) background information on antennas having their elements shaped in fractal geometric forms as dipoles, folded dipoles and loop antennas. Many of the looped designs are based on various iterations of the Minkowski Island (MI) fractal, starting (MI0) from a quad (square) element. The basic Minkowski fractal, with four iterations, forms the basis of the military helmet antenna shown in Fig 2 of the April *TT* (p59); see also **Fig 4** for M0, M1 and M2 geometry. Among the fractal antennas illustrated are an MI3 antenna for 144MHz, about 8in per side and a MI2 for 50MHz about 30in per side. **Fig 5(a)** shows the geometry of a first iteration Minkowski Island (MI1) and **Fig 5(b)** a second iteration, MI2.

In 'Fractal Antennas: Part 1 - Introduction and the Fractal Quad' (*Communications Quarterly,* Summer 1995, pp7-22), N1IR provides a good introduction to the geometry of antenna elements, pointing out that although antenna design has become a mature field, there is still room for a new approach that uses fractal geometry to produce very small antennas of high efficiency and with some other useful attributes. He presents a design for very small area single-element cubical quads, along with their comparative results. He makes it clear that although work on fractal antennas has blossomed recently, they are not entirely a new idea. One of the original forms of microwave log periodic microwave antennas, as built over 40 years ago by DuHamel and Isbell, was a toothed spiral fractal.

Then, in a decade of work some 15-20 years ago, Prof F M Landstorfer and R R Sacher at the Technical University of Munich, investigated the use of bent antenna elements in order to develop an optimum shape. This work was noted several times in *TT*. For example 'Optimum-shaped antennas' December 1982, pp1054-5, from which **Fig 6** and **Fig 7** are reproduced. See also 'Directivity of optimum-shaped antennas' (February 1983, pp131-2). Although the Landstorfer element was not a fractal, nevertheless his work showed that the conventional straight element was not necessarily the optimum approach.

'Fractal Antennas: Part 2 - A discussion of relevant, but disparate qualities' (*Communications Quarterly*, Summer 1996, pp53-71) came a year after Part 1. But before then a further article by N1IR on fractal antennas had appeared: 'Fractal and Shaped Dipoles - Some simple fractal dipoles, their benefits and limitations' (*Communications Quarterly*, Spring 1996, pp25-36). Then, together with Robert G Holfeld, N1IR, presented 'Fractal loops and the small loop approximation - Exploring fractal resonances' (*Communications Quarterly*, Winter 1996, pp77-81).

After attempting to digest, at least in outline, this mass of information, I remain rather uncer-

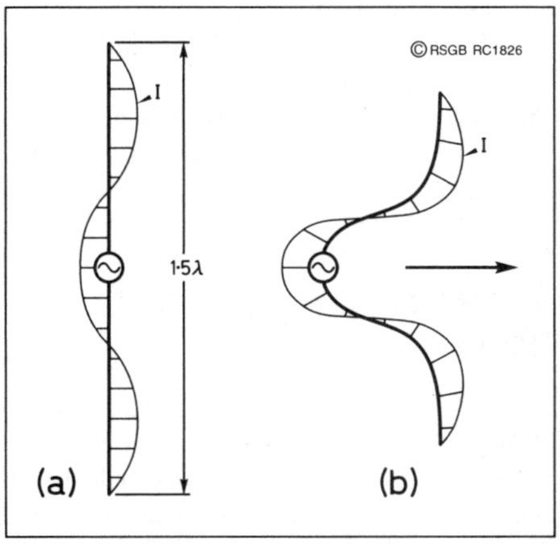

Fig 6: Showing the current distribution on (a) a straight 1.5λ dipole where the phase reversal reduces radiation normal to the axis of the dipole; and (b) with a Landstorfer gain-optimized shape, which causes radiation to increase significantly in the forward direction as a single-element 'beam'.

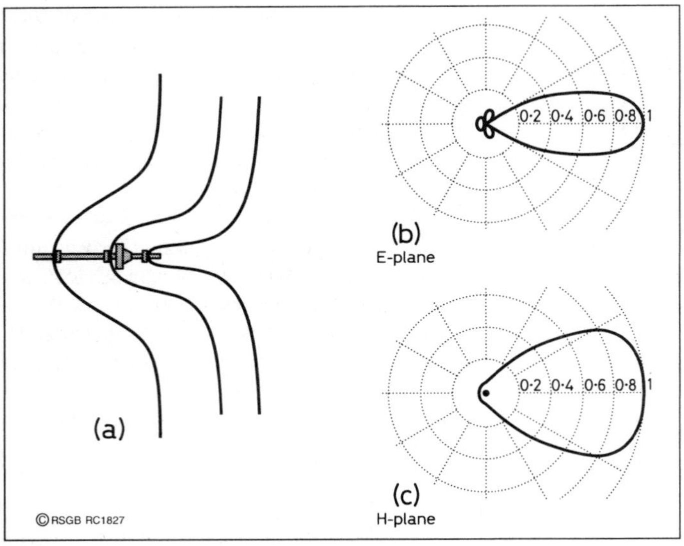

Fig 7: A Yagi array using three gain-optimised elements each approximately 1.5λ long. A VHF array of this type, reported in 1982 by Dr Landstorfer, provided a forward gain of 11.5dBi, sidelobe attenuation better than 20dB and a front-to-back ratio of 26dB.

tain as to its present practical value to amateur constructors. Clearly, as with the Landstorfer designs, it would not be easy to implement multi-iteration fractal elements on HF, where the space condensation would be most valuable. On 144MHz, fractal designs could be implemented using aluminium foil pasted on a dielectric, as in N1IR's original 144MHz antenna (which he reveals was eventually spotted and cut down by his landlord). N1IR has also implemented VHF fractal loop-type antennas using heavy gauge wire for 50 and 144MHz, and these should not prove too difficult to implement and mount.

N1IR shows clearly that this approach appears to be quite practical at VHF or UHF, offering several resonances and capable of providing resonant feedpoints with little reactance. I suspect that a practical difficulty would arise for an experimenter in dimensioning the more complex fractal antennas so that the major resonance fell in the desired amateur band without a fair amount of trial and error. Nevertheless it is clearly an important field for experimentation and may possibly come to represent the 'magic' breakthrough suggested in the *New Scientist* article on which the April *TT* item was based.

After writing the above item, further references to published articles on fractal antennas have been sent in by readers, including the use of fractal radials for vertical antennas (*73 Amateur Radio Today*, September 1996, pp34-35) from PA3CQQ.

MORE ON GILBERT CELL MIXERS

THE APRIL *TT* item stemming from Ian Braithwaite on making do without the SL6440 Gilbert Cell IC mixer attracted the attention of Peter Chadwick, G3RZP, who, as many readers may recall, was heavily involved in the original work almost 20 years ago on the design and application of the SL6440. He has maintained an interest in Gilbert Cell mixers ever since. He writes:

"Barrie Gilbert often pointed out that he never claimed that he originated this particular configuration, but was the first to discover its value for mixing. It had originally been introduced by RCA as a means of producing a four-quadrant multiplier.

As a mixer, there are two basic modes of operation:

1) The multiplicative mode, in which the signal level applied to the top switching transistors is of sufficiently low amplitude to ensure that these transistors are not switched, but remain in the linear mode. This gives the most gain.
2) The switching modem, in which the top transistors act as switches, steering the current produced by the bottom transistors. This mode tends to produce best linearity, and hence dynamic range. Switching the top transistors requires a few hundred (300) millivolts of base signal, but it does help if this signal is a square wave, minimising the period in which these transistors are in their 'linear' region. For this reason, better performance can be sometimes gained by overdriving, especially with sine-wave drive.

"At the 1986 IERE radio receivers conference at Bangor, a paper from the University of Delft examined in some detail the intermodulation produced in the switching, and showed that at quite low current levels - about 1mA or so - the switching produced little degradation of the IMD. The major source of IMD is the non-linearity of the voltage-to-current conversion in the base-emitter junction of the bottom transistors of the tree, ie those to which the signal is applied. The linearity can be enhanced by resistive degeneration (as in the SL6440) or by other feedback methods, such as transformers. I attempted to file a patent for this some years ago, but my boss at the time refused to sign it, on the basis that we could never incorporate transformers onto ICs! My 1981 paper for the Wescon Convention in Califormia plotted the improvement that can be obtained in linearity by increasing the current through the bottom transistors. In this respect, a high intermodulation intercept point is of little use if the gain compression (or incidentally the phase noise) is not adequate to allow the improved intercept point to be used.

"Another feedback method is to use PNP transistors around the bottom NPN devices; when this technique was first published in my 1981 paper, it was impossible to implement, but IC processes have developed, and it could today be a practical approach for use on HF.

"My personal belief is that the Gilbert Cell has an advantage with regard to the output matching because of the (hopefully) very high output impedance of the current sources the transistors represent, depending upon their *Early* voltages. This means that the reflection of the image frequency has very little effect on the commutation at the top of the tree, unlike the normal commutative (switching) mixer where the output wide-band match is of importance; in fact, the signal handling of the Gilbert Cell mixer can be improved by providing a low impedance load at the image frequency at the output. This reduces the voltage swing at the collectors by up to 6dB. When pushing for every last bit of performance, this helps by reducing the effects of any non-linearities in the collector circuits.

"Although I have not yet got around to trying it, I suspect that some 2N5109 FETs would perform well, especially with cross-couple transformer feedback between the collectors and emitters of the transistors in the bottom of the cell. Fed with sharp edges from a CMOS or AS TTL gate to drive the top transistors, tuned output of low enough impedance not to run out of signal handling on the wanted signal and to provide a very low impedance for the image. Running at about 25mA in each side, there is potential for very good performance. Matching, although desirable, only really shows up as poor oscillator balance, with possibly some effects on even-order IMD and perhaps on spurious responses at frequencies half the IF away from the wanted signal.

"The major advantages of the Gilbert cell - sometimes called the 'transistor tree' - mixer lie in its relative insensitivity to load match, especially at the image, its ability to provide gain, and its not-too-excessive noise figure. In the SL6440, this was determined partially by the transistors, but a major factor was the 18Ω emitter degeneration resistors. As mentioned above, transformer feedback may well allow this to be improved upon. Where local oscillator radiation from the antenna is a consideration (which with the EMC Directive it must be in commercial equipments) it certainly has some advantages over the diode ring."

TWIN COAX MATCHING SECTIONS

THE APRIL *TT* (p62, Fig 2) included an illustration of a 3.5MHz top-loaded 'reverse-fed' elevated ground-plane antenna, using an earthed tower with top-loading provided by a TH7 (7-element) Yagi antenna used on the higher-frequency bands. This illustration was taken from an article by a paper presented by

Dr John Belrose, VE2CV, at ICAP97, but came originally from the *QST* (June 1994, pp45-46) article 'Simple, effective, elevated ground-plane antennas' by Thomas Russell, N4KG. This included a means of calculating the equivalent loading in feet for a number of common Yagi antennas (**Table 1**), based on an empirical formula developed by John Devoldere, ON4UN, and given in his book *Antennas and Techniques for Low Band DXing*, published by ARRL in 1994.

N4KG also gave further information on the matching section formed from two quarter-wave (electrical) lengths of 75Ω RG-59 coax cable (**Fig 8**) used to match the 75Ω ½in CATV line feeder to the 17Ω radiation resistance of the antenna. With a velocity factor of 0.66, a 46ft length of RG-59 represents an electrical quarter-wave at 3.6MHz. By paralleling two pieces of RG-59 and connecting them between the feedpoint and a run of the rigid coax to the transmitter, the quarter-wave matching transformer formula can be used:

$$Z^2 = R1 \times R2$$

where Z is the line impedance, R1 the input impedance and R2 the output impedance. With R1=17Ω and Z=37Ω (half 75Ω), R2 will be 80Ω, an almost perfect match to the 75Ω CATV cable.

By coincidence, shortly before the publication of the April *TT*, Brian Clowes, GW4HBZ, a former BBC transmitter engineer, submitted an idea similarly based on using paralleled quarter wavelengths of coax for matching into low impedances, a technique he had not previously seen in print. He wrote:

"It occurs to me that if a 2 x ¼λ phasing harness with 50Ω cable could match 50Ω to 2 x 25Ω, these could be paralleled to match to a single 12.5Ω impedance. In effect the two paralleled 50Ω quarter wavelengths behave like a 25Ω quarter wave line. Three or more could be paralleled and 75Ω cable used, or even perhaps a mixture of 75 and 50Ω cables. Possible applications include matching into the low impedance of a Yagi, eliminating the need for a folded dipole or gamma match. The miniature RG-316 PTFE coax (only 3mm outside, diameter, yet capable of handling high power) could be used and fed through a ferrite ring or also form a choke balun. The idea could also be used for matching in and out of the low impedances of transistor power amplifiers."

Fig 9 illustrates three possible arrangements using paralleled coax cables. GW4HBZ notes that RG316 is capable of handling 300 watts

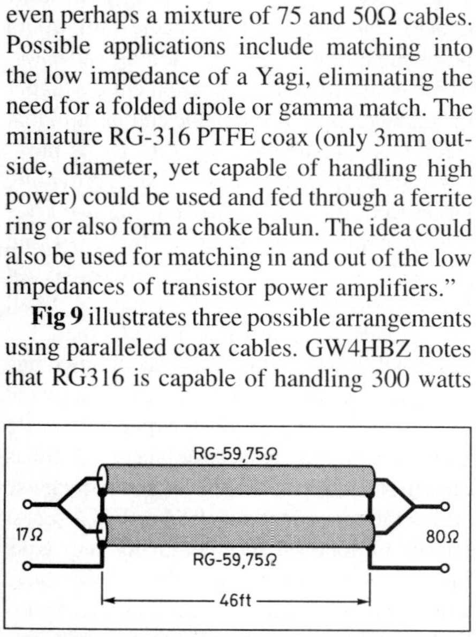

Fig 8: Dimensions of the 3.6MHz matching network, using paralleled RG58 coaxial cables in a quarter-wave transformer to match the 17Ω radiation resistance to 75Ω line.

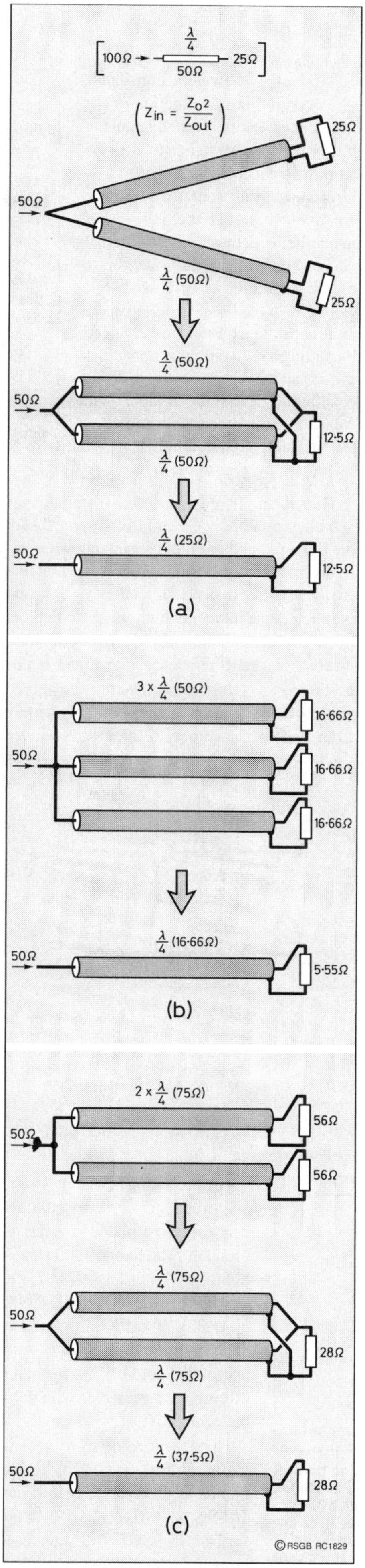

Fig 9: (a) How parallel lengths of 50Ω cable can match 12.5Ω. (b) Use of three parallel lengths of 50Ω cable to match 5.55Ω to 50Ω line. (c) Use of two parallel quarter-waves (electrical) of 75Ω cable can match 50Ω to 28Ω.

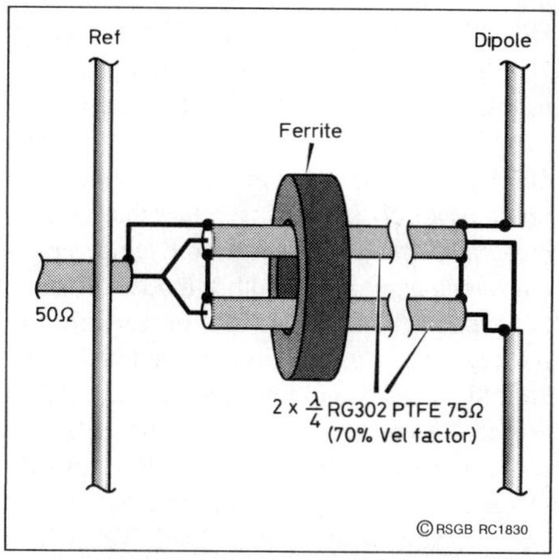

Fig 10: Use of two lengths of RG302 PTFE miniature cable as used by GW4HBZ in connection with a ferrite sleeve balun to feed a 144MHz three-element Yagi array.

CW (600 watts SSB) at 150MHz where the SWR is low. When used for a quarter wave transformer, power should be reduced to a little above 100 watts CW or 200 watts SSB. If two lengths are used in parallel, each cable takes only half the transmitter power. GW4HBZ points out that with PTFE dielectric cables such as the miniature RG316, the velocity factor is approximately 70% instead of the 66% of polyethylene cables.

In a later letter (in which GW4HBZ acknowledged that the April *TT* showed that similar techniques had been used by N4KG for his grounded tower ground plane antenna), he reported that he had just built a 3-element 144MHz beam using two quarter wavelengths of 75Ω coax in parallel (PTFE RG302 2 x 36.5 cm) fed through a largish ferrite cylinder to form a choke balun at the same time (**Fig 10**): "It was easy to get a 22dB return loss (better than 1.2 SWR) with 50Ω feeder. I had previously found it difficult to get a return loss much better than 10dB with a folded dipole."

THE SUPER VXO

THE TOPIC OF variable frequency crystal oscillators has long held a high degree of interest to those amateur operators interested in portable QRP operation, where the power requirements of frequency synthesisers are often difficult to meet. Recent interest has tended to be concentrated on the use of ceramic resonators, which are much easier to 'pull' than quartz crystals, but tend to have a higher

Antenna	Boom length (ft)	Area of Yagi (sq ft)	Equivalent loading (ft)
3-el 14MHz	24	768	39
5-el 21MHz	26	624	35
4-el 21MHz	20	480	31
3-el 21MHz	16	384	28
5-el 28MHz	24	384	28
4-el 28MHz	18	288	24
3-el 28MHz	12	192	20
TH7	24	-	Estimated 40
TH3	14	-	Estimated 27

Table 1: Effective Loading of Common Yagi Antennas.

temperature coefficient and are readily available only at limited frequencies. Unfortunately, the modern miniature plated crystal units in HC18/U and HC25/U, and even the larger HC6/U types, are much more limited in their usable frequency swing than the older, non-plated, FT243 and the large (wartime) 10X and 10XJ crystals. But for the largest shifts, one really needs obsolete X-cut rather than AT-cut crystals, although these are more temperature-conscious than AT and other modern cuts.

Much of the original work on variable frequency crystal oscillators was undertaken by Harold Stanesby and P W Fryer at the old Post Office Research Station at Dollis Hill, and published in *Proc IEE*, 94, IIIA, p368 onwards, 1947. But for amateurs, the first publication was probably that of J G Reed, VK2JR, in WIA's *Amateur Wireless,* now *Amateur Radio*, December 1948. I included a brief (unsigned) digest of VK2JR's article in the *RSGB Bulletin*, November 1949, p144, including his 6V6 oscillator circuit: **Fig 11**. VK2JR claimed that with X cut crystals, frequency shifts of about 1kHz per MHz (ie 7kHz with a 7MHz crystal) could be obtained without significantly impairing stability, but warning that with AT-cut crystals used in this circuit care was needed to avoid oscillation on spurious frequencies.

In the 1960s, 'A Guide to the specification and use of quartz crystal oscillators' was published in the UK by the Radio Communication and Electronic Engineering Association. This included, under 'Special Requirements', a note (6.3) on 'Frequency Pulling'. Although this was concerned only with the use of series or parallel variable capacitors, it stressed that the choice of putting the capacitors in parallel or series with the crystal depends on the type of oscillator circuit:

"Since the equivalent impedance of a crystal unit is a function of the frequency, it is possible by causing this impedance to change to alter the working frequency. This is most easily effected in an oscillator by putting a variable reactant either in series or in parallel with the crystal, the actual choice being decided by the nature of the oscillator circuit used. For example, in the Pierce oscillator, the load capacitance could be changed by putting a variable capacitor across the crystal socket . . . For series oscillators, with the variable load capacitance in series with the crystal, the same equation holds:

$$(fa-fb) / fb = C1/2 \times (Ca-Cb) / (Co-Ca)(Co+Cb)$$

Here fa and fb are the oscillator frequencies at load capacitances Ca and Cb. C1 and C2 are the dynamic and parallel capacitances respectively of the crystal unit. It is seen that the frequency shift depends upon the dynamic and parallel capacitance, which in turn depend upon the crystal unit design. Consequently, where consistent frequency adjustment is important, the crystal parameters must be agreed with the manufacturer, for they may vary widely from one manufacturer to another, even for crystals of the same frequency and in the same holder."

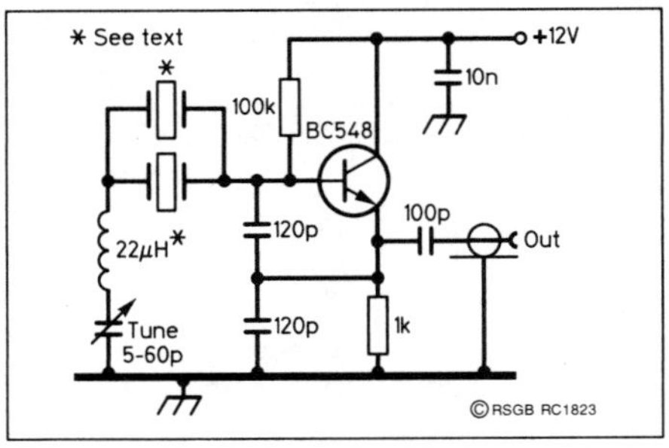

Fig 12 (above): Circuit diagram of the twin-crystal "super-VXO" investigated by VK1PK, based on a Japanese technique described in 1980 by JA0AS and JH1FCZ.

Crystal freq (kHz)	VXO induct (uH)	No of crystals used	Minimum freq (kHz)	Maximum freq (kHz)	Shift (kHz)	% change
3579	22	1	3579.26	3580.52	1.3	-
3579	22	2	3579.28	3581.33	2.0	54
4433	22	1	4430.78	4435.24	2.6	-
4433	22	2	4431.13	4436.93	5.8	120
4433	22	3	4431.93	4437.90	6.0	130
5366	22	1	5364.27	5368.83	4.6	-
5366	22	2	5364.28	5369.86	6.6	44
5366	44	1	5362.63	5368.13	5.5	-
5366	44	2	5360.07	5369.65	9.6	74
7159	22	1	7156.52	7162.65	6.1	-
7159	22	2	7154.55	7165.71	10.5	72
7159	44	1	7151.75	7160.80	9.1	-
7159	44	2	7135.93	7162.95	27	197

Table 2: VK1PK's Test Results

An interesting new look at variable-frequency crystal oscillators using modern plated crystals is found in 'The Super VXO Heads South' by Peter Parker, VK1PK (*Amateur Radio*, April 1998, pp31-32). VK1PK stresses that "maximum pulling range with acceptable stability" is the aim of every experimenter who dabbles with VXOs".

His article reports on investigations into a "super VXO" technique, originally used by the late JA0AS and JH1FCZ, and described in 1980 in a Japanese homebrew QRP magazine. It has more recently become available on Internet. Basically, the idea is simply to connect a second identical frequency crystal in parallel with the first crystal.

Internet states that 7N3WVM has reported achieving a 40kHz pulling range on 10MHz and 90kHz on 13MHz with his twin-crystal VXO, although VK1PK does not provide details of the crystals used by 7N3WVM. Instead, he provides an account of his own experiments using a similar circuit arrangement: **Fig 12**.

The frequency shifts (**Table 2**) he reports on 7MHz compare well with those claimed by 7N3WVM on 10MHz, and he concludes that adding an extra crystal substantially increases the pulling range of VXO circuits, suggesting that "builders of VXO-controlled equipment should always order two crystals instead of one - the extended shift obtainable will probably be worth the extra expense . . . but the expense of using three crystals is not justified".

VK1PK admits that he made no attempt to optimise component values to obtain larger frequency swings. Neither were detailed measurements made of the stability. He used one or two 22µH RF chokes as the series inductor and a 5-60pF variable capacitor. Clearly these would not be optimum at all four crystal frequencies investigated. He notes, for example, that larger shifts were obtained when using two 22µH chokes, and wonders what stable shift would be obtainable with a 100pF tuning capacitor. There is also room for improvement if a higher Q coil were used. It should be emphasised that VK1PC's investigation was directed at proving the concept of the "super-VXO", providing larger frequency ranges when using two crystals in parallel. Furthermore, his 3579 and 7159kHz crystals were the less suitable HC25/U types; the others in HC18/U holders. Thus all his crystals were modern, plated types, but nevertheless he obtained a 27kHz tuning range at 7MHz. As noted above, large, non-plated crystals should provide even larger shifts, with their greater inherent capacitances. There is clearly room for further investigation by those still possessing identical 10XJ or FT243 crystals. It is doubtful whether many will have matched X-cut crystals, and these would have a higher temperature coefficient, probably little better than ceramic resonators. ♦

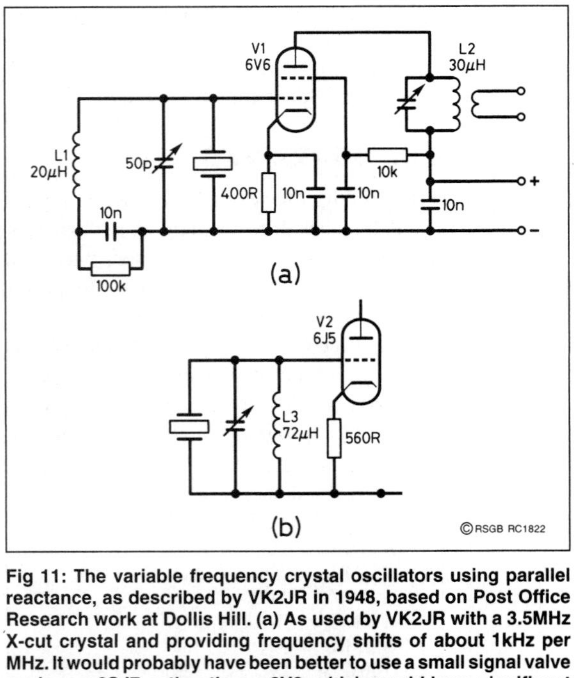

Fig 11: The variable frequency crystal oscillators using parallel reactance, as described by VK2JR in 1948, based on Post Office Research work at Dollis Hill. (a) As used by VK2JR with a 3.5MHz X-cut crystal and providing frequency shifts of about 1kHz per MHz. It would probably have been better to use a small signal valve such as a 6SJ7, rather than a 6V6, which would have significant crystal current, resulting in drift due to heating of the X-cut crystal. (b) An alternative input circuit for valve oscillators.

BY PAT HAWKER, G3VA*

SOLDERING AND DRY JOINTS

A COUPLE OF years ago, a *TT* item 'Is home brew and DIY servicing passing into history?' (February 1996, p74) drew attention to the continuing problem of 'dry' soldered joints developing after new equipment has been working well for some months. "Oxidation, heat cycles and the like, can affect any less than perfectly soldered joint, and result in an intermittent or continuous high resistance or open connection." It was noted that even for professionals the time taken to trace intermittent faults can quickly mount up and result in high costs, despite the fact that nothing has to be replaced.

Geoff Baxendale, G4IBS, who recently retired after spending 32 years in the electronics industry - much of it at the Mullard Blackburn valve manufacturing plant, which over the years has declined from 6000 to about 600 staff - contributes a number of relevant thoughts on the topic of dry joints:

"I think dry joint faults are less the fault of 60/40 solder than the way the joint is made. I recall my father telling me as a small boy to 'make a mechanical joint first, and then solder.' I still have a very old Lowry electronic organ, all valves and tag strip, and I don't think there is or ever has been a bad joint in it!

"The advent of printed circuit boards and wave soldering has much to answer for, although reducing manufacturing costs and hence prices. The problem with 'wave', I believe, is having to solder at the same time components of widely different thermal capacities. It is far too easy to leave the larger components with far too little solder. With some of these components mechanically supported by the solder, then subjected to thermal cycling, there is your dry joint.

"Another problem seems to be with power transistors and IC voltage regulators, with straight leads that are fastened firmly to a heat sink, which in turn is fastened to a PCB. With these components heating up and cooling down each time the equipment is turned on and off, the joints become badly stressed and eventually go dry.

"Some manufacturers remove as much copper as possible from their boards; this reduces the heat dissipation area, resulting in higher temperatures and more stress. It is difficult to understand the logic of this practice.

"Through-board connections are also a problem; this seems to be due to differential expansion. My IC-240 rig went on the occasional blink many years ago; it was ages before I could induce the fault on the service bench with a multimeter at hand. Eventually I found 9V on one side of the board with a variable voltage on the other side. Icom had used little pins as 'vias'. I replaced them all and the problem went away for a year or two, but then returned. My final solution was to use wire in a loop: see **Fig 1**. Fortunately, modern glass/epoxy boards with plated-through holes seem to be completely free of this problem, but I believe other makers besides Icom used the pin trick. I believe later IC-240s used rivets, although I am not convinced that these were much better than the pins.

"The current trend is towards surface mount technology. I have not yet had enough experience of these devices to know what they will be like long term; however, in the short term, joints without solder are not unknown. In a couple of equipments I have come across, they have worked for over a year (and so become out of warranty) and then gone intermittent. The 'J-pin' ICs (where the leg is bent under) seem most prone. In seeking to find poor joints, there is no substitute for a strong light and a 10x magnifier. With these aids such faults are easy to spot, but need a very fine soldering bit and a steady hand to fix!

"On the topic of good soldering practice, never be tempted to tin multistrand cable and then put it in a screw type connector block. The

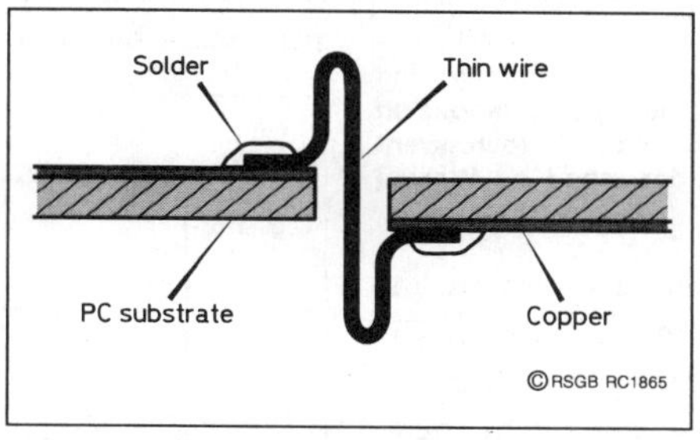

Fig 1: G4IBS's suggested form of 'via', for non-plated-through holes in printed circuit boards - preferred to the small pins or rivets used on some equipment.

solder will cold flow and result in an intermittent connection. This takes about a year or so and probably depends to some extent on the temperature. Similarly, when putting a mains plug on a new appliance, chop off the tinned bit of wire or the same thing will happen. If you don't believe me, take a light bulb out of its socket and note the dents in the solder pads! I think this effect led to the demise of lots of diodes in the Philips G11 series of television sets. This was brought to my attention by the appliance tester at work going intermittent, the fault traced to tinned leads!

"As to what causes equipment problems, I feel that faulty electrolytic capacitors must be a close second to dry joints. Time and temperature seem to be the problem. If you have some weird faults in a piece of equipment, it may be worth changing all the 'Elcos' and give it a new lease of life. I find the component tester described some time ago in *RadCom* that produces ellipses on an oscilloscope invaluable. Problems in switched mode PSUs can usually be traced to a faulty electrolytic, but sometimes you get the Christmas Tree light syndrome as a result! Several that I have met have just one elco that fails, nothing else (eg the old BBC computer). Sometimes they work off load, but fail on load.

"Finally, my main advice: If you want to keep electronic equipment working, *keep it cool*! Don't stack it in racks or cabinets or on top of other units, unless you use plenty of fans. There are some useful small 12V fans around these days (convenient and quiet running) stemming from the computer world. One of these pointed at your heat sink can result in a tremendous reduction in temperature."

SSB ON THE RA17 AND OTHER CLASSICS

THE MAY *TT* (pp57-58) included a design for an SSB product detector adapter unit, suitable for use with classic 'boat anchor' receivers, designed in the days when diode detectors were often coupled with low BFO injection. This item was a precis of an article by Morris Odell, VK3DOC, in *Amateur Radio*, March 1998, pp7-8. VK3DOC suggested that the Racal RA17 series, whilst a "beautiful and very sophisticated design" has a "particularly unsatisfactory performance on SSB".

This contention is strongly rebutted by John Truppen, VK6XJ (Technical Correspondence, *Amateur Radio*, May 1998, p46). He writes:

"I would not in any way criticise VK3DOC's quite delightful method of achieving SSB demodulation. His article is excellent, but I would question *why?* I have been involved with so-called 'boat anchors' on and off for some 40-odd years, and have repaired, restored, and operated many such receivers, RA17s in particular. In my opinion, the RA17 in its various models is a good candidate for being one of the very best valve HF communications receivers ever. When operated correctly, it is one of the best performers of all 'boat anchors' for receiving and resolving SSB signals.

"The correct procedure for SSB is: Set up the receiver for the frequency of interest. Set the IF bandwidth (selectivity) to 3kHz. Turn on the BFO. With the BFO TUNE (note, offset the BFO 1.5kHz *above* the IF centre frequency if copying *lower* sideband or *below* the IF centre frequency if copying *upper* sideband). Do *not* make further adjustment to the BFO tuning, unless going to the opposite sideband. All further adjustment to resolve the signal is done on the main receiver tuning.

"Check that the AGC response is set to *long*. Tune the receiver normally across the band, and you will find that it behaves just like a modern, solid state rig.

"The above assumes that the receiver is correctly aligned, particularly the IF, detector and BFO. It is possible that VK3DOC's receiver has an alignment problem.

"This technique works with all of the older valve receivers, although with varying suc-

*37 Dovercourt Road, London SE22 8SS

cess. As VK3DOC correctly states, many older types just do not have sufficient BFO injection, a prime example being the immortal RCA AR88. When this problem is encountered it is necessary to reduce the RF gain in order to cope with strong signals. This can also lead to the need to turn off the AGC. [In most of the older sets it is essential to turn off the AGC when using the BFO for SSB or CW - *G3VA*].

"A further complication with some receivers is that on some bands they change to opposite side local oscillator injection; in this case requiring a reversal to the BFO offset given above. Another problem with some of the older, less classic designs, is the lack of both short- and long-term frequency stability.

"But a correctly set-up RA17 does not suffer any of these problems. Yes, Racal made SSB and ISB adapters, but for the type of operation that we, as hobbyists, are involved in, it just is not needed. They are, and were, an unnecessary complication.

"A couple of other British-made 'boat anchors' are excellent SSB demodulators: the Murphy B40D and the GEC BRT402 series (on the lower HF frequencies). The well-known American Collins 51J-4 and R-390-A are not quite so tractable; but using the above technique they will do the job."

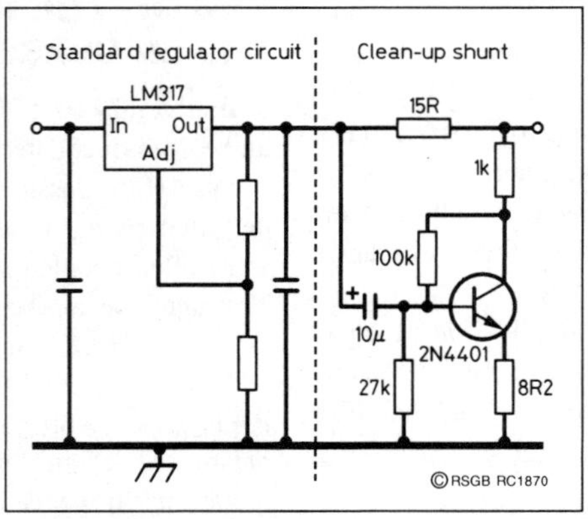

Fig 2: Clean-up circuit, to remove perturbations on the output of standard IC regulators. For low current loads (about 10mA). As described by Wenzel Associates, Inc.

FINESSE VOLTAGE REGULATOR NOISE

WITH THE HELP of Harold Johnson, W4ZCB, Colin Horrabin, G3SBI, has been discovering the enormous amount of technical information (including circuit diagrams) available from American firms on the World Wide Web. As just one example, he passes along a copy of a three page web article 'Finesse voltage regulator noise!' from Wenzel Associates, Inc (http://www.wenzel.com/documents/finesse.html). Wenzel appears to be a company specialising in low noise amplifiers and oscillators, etc, for radio and electronic applications. They have agents in the UK for the supply of their products.

The importance for high performance receivers of reducing oscillator noise in order to minimise reciprocal mixing has been stressed many times in *TT* (see, for example, 'Oscillators and synthesisers', February 1996, pp72-74).

To quote briefly from the web article: "System designers often find themselves battling power supply hum, noise transients, and various perturbations that wreak havoc with low noise amplifiers, oscillators, and other sensitive devices. Many voltage regulators have excessive levels of output noise, including voltage spikes from switching circuits and high flicker noise levels from unfiltered references. Ordinary three terminal regulators will have several hundred nanovolts per root-hertz of white noise, and some reference devices exceed one microvolt per root-hertz. . . many systems have offending devices that 'dirty up' otherwise clean supply rails.

"The traditional approach for reducing such noise products to acceptable levels could be called the 'brute force' approach - a large value inductor combined with a capacitor, or a clean-up regulator inserted between the noisy regulator and the load. In either case, the clean-up circuit is handling the entire load current in order to 'get at' the noise. The approach described in this paper uses a bit of finesse to remove the noise, without directly handling the supply's high current.

"The key to understanding the 'finesse' approach is to realise that the noise voltage is many orders of magnitude below the regulated voltage, even when integrated over a fairly wide bandwidth. For example, a 10V regulator might exhibit 20μV of noise in a 10kHz bandwidth - six orders of magnitude below the DC. . . by adding a tiny resistor, R, in series with the output of the regulator and assuming that a circuit somehow manages to reduce the noise voltage at the load to zero, the noise current from the regulator may be calculated as V_n/R. If, in this example, the resistor is 1Ω, the noise current will be 10μV/1Ω, that is 10μA. If a current sink can be designed to sink this amount of AC noise current to ground at the load, no noise current will flow in the load. By amplifying the noise with an inverting transconductance amplifier with the right amount of gain, the required current sink may be realised. The required transconductance is simply -1/R, where R is the series resistor."

The article then provides two examples of clean-up circuits, one for a low current supply of about 10mA using a 2N4401: **Fig 2** which is capable, without trimming, of reducing the noise by some 20dB or, after trimming the resistor values, over 40dB.

The simplicity of this configuration makes it attractive also for high current loads, although in this case it is desirable to have a much lower series resistor (1Ω rather than 15Ω). One approach is to replace the single transistor by a compound transistor arrangement, as shown in **Fig 3**, using an additional MPSD54 Darlington transistor. Further design information is provided in the web article.

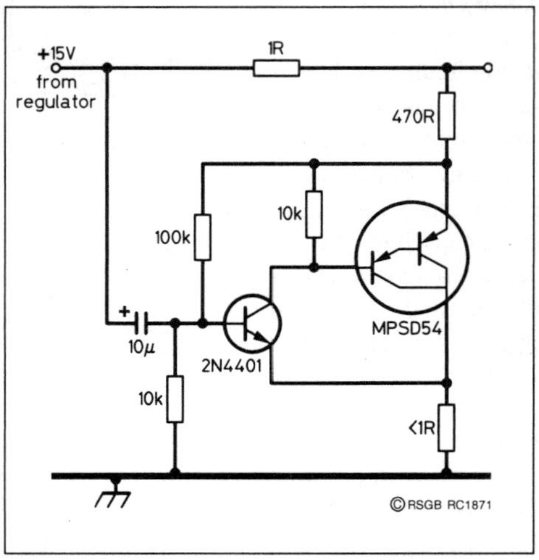

Fig 3: Clean-up circuit for higher current loads.

MORE ON HEAT SENSITIVE ADHESIVES

DR DICK BIDDULPH, G8DPS, comments on the method proposed by Giles Read, G1MFG, for removing components secured by hot-melt glue (*TT*, May 1998, p61) by immersing the entire assembly in "very hot (just off the boil) water". He writes: "The melting of hot-melt adhesive would be done better either with distilled or de-ionised water, since that would leave no residue. An alternative, for the same reason, would be steam from the spout of a kettle, which would also have the added advantage that it could be localised. One could heat the ends of tweezers in the steam, to avoid the 'heat sink' effect."

Ian Burns, G0AFH, points out: "Another method of removing hot melt adhesive is to freeze it with a quick blast of component freezer. This causes the adhesive to become very brittle, so that it can be broken away. I have not tried this on large areas of adhesive, but it works well on small areas."

This reminded me that Bruce Carter, GW8AAG, ('Tips and Topics', *TT*, December 1995, p73) noted that the labels on screw-top glass jars are stuck on with heat sensitive adhesive, and can be floated off by immersion in hot, soapy water, although some small areas of adhesive will usually remain: "This will spread but not dissolve. Complete removal of the adhesive can be achieved by placing the dry jar in the freezer or freezing compartment of a refrigerator, and leaving it there for three or four hours, then taking it out of doors and shaving off the adhesive with a knife using a 'pencil sharpening' motion. The (brittle) adhesive will shatter and drop off."

THE LUXEMBOURG EFFECT

FOLLOWING THE publication of the February *TT* item mentioning the effect of high power transmitters on the ionosphere as a possible reason for the disappearance of really long delay echoes, and referring to the 'Luxembourg Effect', I received an audio cassette from Trevor Lumb, G0ARJ, of Bury St Edmunds, Suffolk. He queried whether the interference audible on the ionospherically propagated RTE programmes from Athlone (567kHz), in the form of a ghostly accompaniment from BBC Radio 4 (Droitwich 198kHz), could be due to the Luxembourg Effect.

Of this I have no doubt whatsoever, as the circumstances fulfil all the criteria established in the 1930s for the occurrence of this type of ionospheric cross-modulation, *viz* the two stations are roughly on the same great circle bearing from the receiver, with the interfering station geographically between the receiver and the wanted transmitting station.

Furthermore, the R4 interference sounds woolly and muffled, due to loss of the higher audio frequencies, a characteristic of the Luxembourg Effect first reported in 1933, when a listener reported hearing the programme of Radio Paris (1725m, 75kW) as a faint background when his receiver was tuned to Radio Luxembourg (1190m, 200kW). The theory of this form of interference brought about by nonlinearity in the ionosphere was first explained by Professor V A Bailey (University of Sydney) in *Wireless World* in 1937.

Thirty years later, in his 1968 IEE presidential address, J A Ratcliffe pointed out: "We must remember that the electrons are not alone in the ionosphere. They are moving about in random directions and making collisions with heavy atoms and molecules. These collisions interrupt the oscillations which the electrons are executing under the orders of the oncoming (signal) wave, so that, after they have occurred, the electrons bounce off in random directions. The ordered oscillatory energy which the electrons acquire from the wave is converted into disordered energy of random movement. The electrons are then left moving in random directions a little more rapidly than before, as though they were at a slightly higher temperature. Energy has been abstracted from the (signal) wave and has gone into heating up the electrons. This increase can be quite appreciable: for example, when the long-wave broadcasting station at Droitwich is switched on, the temperature of the electrons at a height of about 100km increases by about 45°C."

HIGH LEVEL MIXERS

INTEREST CONTINUES in high level RF mixers for VHF as well as HF receivers, designed to provide excellent linearity and consequent good intermodulation performance in the presence of strong signals. IP_3 performance is every bit as important on VHF as on HF (some would argue even more important).

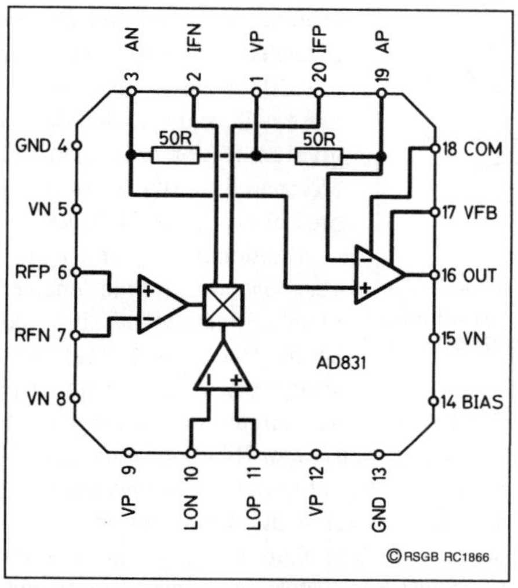

Fig 4: Example of a high level, double balanced mixer. For optimum third order intercept (IP_3) performance, the AD831 typically draws some 100mA quiescent current, although this can be reduced to about 45mA for minimum power consumption.

Darren Conway of the Royal New Zealand Navy in *Electronics World*, June 1998, pp464-460, provides a useful survey of mixer configurations and packages, with the emphasis on VHF but also applicable to HF. One suspects that the article was written before the announcement by GEC-Plessey of the discontinuance after almost 20 years of the SL6440, which still offers excellent performance up to about 100MHz and reasonable performance up to about 400MHz (noise figure roughly 11dB), although for optimum linearity requiring some 75mA or so current, and needing rather more external components than some of the other packages.

Lt Cmdr Conway examines the factors that make a good VHF high level mixer; listing in order of importance: frequency range, dynamic range, noise; gain, local oscillator drive, isolation, impedance matching, and simplicity ("overly complex design is a form of inefficiency that can have a significant effect on the through-life cost of equipment").

As a modern example of a high-level active double-balanced IC mixer, he gives pride of place to the AD831, which I cannot recall mentioning before in *TT*. This IC (**Fig 4**), housed in a PLCC package, includes a buffer amplifier with 50Ω input and output, the gain of which can be set by feedback resistors and configured as a balanced to unbalanced amplifier without the need for transformers. For the best third-order intercept performance, it draws typically 100mA quiescent current. Its high noise figure (20dB) means that an RF amplifier ahead of the mixer is mandatory. Performance generally is stated to be very good up to 100MHz ,with a two-tone intercept figure of some +44dBm, but performance varies significantly at higher frequencies, particularly around 200MHz. Darren Conway considers that "the ability (of the AD831) to combine a high performance mixer, low-pass filter and buffer amplifier into a very small space with no tuneable components means that this device would be a good choice for many applications".

In summary, he writes: "The selection of a mixer for any receiver application depends on many variables. The information in **Table 1** has been obtained from a number of sources. The values are representative of actual devices operating under typical

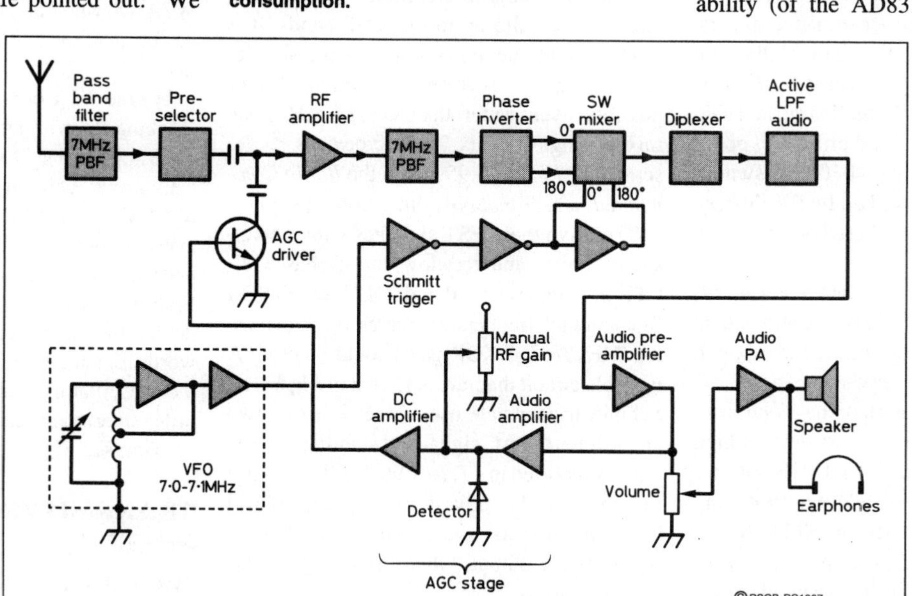

Fig 5: Block outline of the 7MHz direct conversion receiver constructed by I7SWX, using a 74HC4066 switching mixer and diplexer, similar to the receiver described by ZS6BZP in *Electronics World* and noted in the April *TT*.

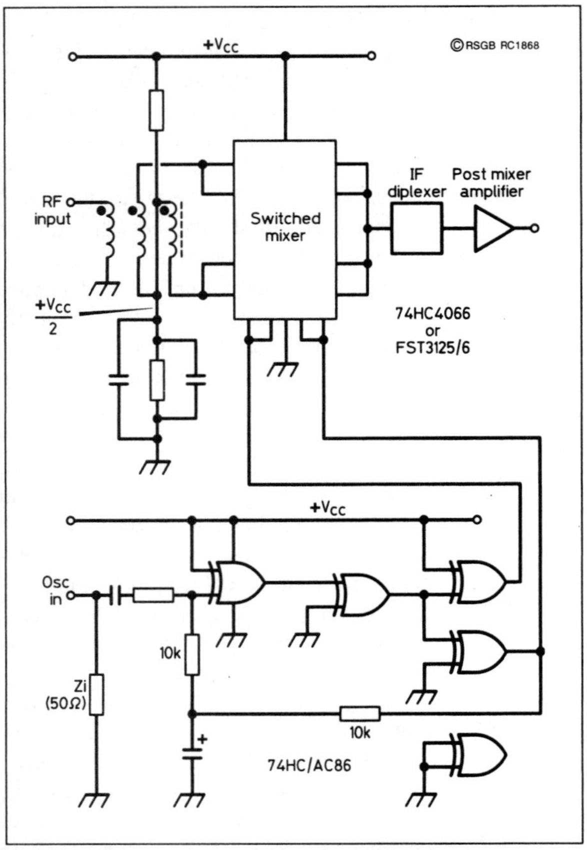

Fig 6: I7SWX's suggested high level mixer solution, based on the 74HC4066 configuration, but using Fairchild Fast Bus Switches and a 50/50 square wave 0-180° phase drive.

Characteristic	Active JFET (+)	Diode (SBL-1)*	Diode (RAYII)*	MOSFET (+)	Bipolar (AD831)	Bipolar (SL6440)
Freq range (MHz)	50-250	0.5-500	0.01-2500	0.2-500	0-400	0-90
Isolation LO-RF (dB)	35	45	32	30	65	-
Isolation LO-IF (dB)	60	40	20	25	28	29
Overall NF (SSB)	8	6	7	9	20	11
LO drive level (dBm)	+15	+7	+23	+16	-10	-5
2-tone intercept pt (dBm)	+34	+16	+25	+44	+24	+30
Conversion Gain	+4	-5.6	-6.2	-7 to +18	0	1
1dB compression	+13	+1	+15	+16	+10	+10

*Mini-circuits. +Active JFET Siliconix U430/U530. MOSFET Siliconix SD5000, Calogic SD8901.
Table based on Electronics World but with corrections to printing errors spotted by G3SBI.

Table 1: Comparison between high level mixers

conditions, and in most instances these values will be different from those specified in data sheets.

"As with most design choices, the selection of the right mixer for the right application is based on compromise. There is no single mixer that is suitable for all tasks... selecting the right mixer is a matter of determining the most important (required) characteristics and matching these to the appropriate mixer type."

On the topic of high performance mixers, Giancarlo Moda, I7SWX, notes that the multiband DC receiver described by Hannes Coetzee in *Electronics World*, February 1998, and reported briefly in *TT*, April 1998, pp61-62, uses a basically similar 74HC4066 switching mixer to the one described by P J Coetzee in *RF Design* (June 1995), based on an idea by J M Groundlingh.

I7SWX drew attention to this mixer in *TT*, January 1996, pp65-68, as a result of practical experience of using one as the second mixer in a modified FT29A transceiver.

Following the publication of the *Electronics World* article, I7SWX has constructed a high performance monoband 7.0-7.1MHz DC receiver using another 74HC4066 switching mixer, preceded by an RF amplifier (block diagram **Fig 5**). However, within a few days of completing this receiver (which gained a prize plaque at a local amateur radio exhibition for home construction), I7SWX received technical information from Fairchild Semiconductors that appears to open the way for further improvements in this form of switching mixer, potentially outperforming Schottky diode rings, FETs and active ICs. These are fast bus switches, with very low ON resistance, CMOS TTL compatible. He believes that two of these switches in parallel should give an R_{on} of only 2 to 4Ω (compared with the several hundred ohms of the 74HC4066).

Additionally, T_{on} and T_{off} are very fast, with a time 'enable' of only 5 nanoseconds. It would seem that these fast bus switches, used with the 74HC4066 mixer circuit bias arrangement, might well be used for RF applications, possibly up to VHF, providing a high performance switching mixer with an attenuation of only about 1dB. The Fairchild type numbers for these fast bus switches are Quad Bus Switch FST3125 and FST3126.

I7SWX writes: "I believe that these fast bus switches with the 74HC4066 bias circuit should give no problems in RF mixer applications, and may be valid up to the VHF bands. It is possible that the FSTs could, with simpler circuitry, give a performance comparable or possibly even better than G3SBI's H-mode mixer using Siliconix SD5000 devices, as described in *TT* (1994-95 and in the *Radio Communications Handbook*, 6th edition, p6.51).

"To drive these FST switches with a 50/50 square wave - and very low out-of-phase ON-OFF dis-alignment - the K6OLG and G3SBI fundamental frequency squarer principle using 74HC/AC86 XOR gates could be used. A possible circuit diagram is shown in **Fig 6**. The Schmitt trigger gate uses three resistors, two are to filter the AC signal and stabilise the DC bias, as reported in *TT*, July 96, Fig 3. The third resistor is used to polarise the input gate level better, in order to have a well-balanced 50/50 square wave output, when driven by an AC sine waveform."

I7SWX's letter was written in late April, at a time when he was waiting for some sample FSTs, to try out his ideas in practice. As he is due to start a new job outside Italy, this work may have to be delayed. I passed I7SWXs letter to G3SBI, who in turn alerted Bill Carter, W7AAZ, who quickly confirmed the excellent performance of the FST3125 as an H-mode HF mixer.

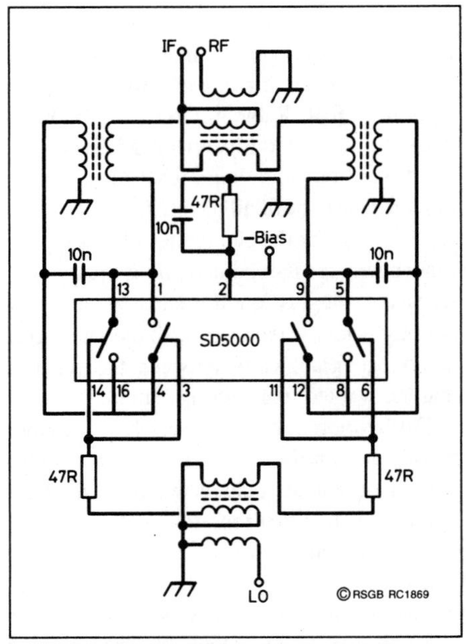

Fig 7: AG4R's suggested HF mixer replacement for the SL6440 chip, using the Siliconix SD5000. Transformers, 10 turns on an FT-37-43 core.

Wayne W Cooper, AG4R, comments on 'Making do without the SL6440' (*TT*, April 1998, pp60-61): "I used the SL6440 Gilbert Cell mixer chip in a receiver modification (*QST*, October, 1983) with good results. However, in considering replacements, there is no need to pass over the SD5000 quad DMOS FET array. The G3SBI H-mode mixer uses this device to achieve very good results, but it is worth remembering that even with sine wave local oscillator injection one can get a third order intermodulation product (IP_3) of about 25dBm: see **Fig 7**".

TRANSMITTING LOOPS - IN BRIEF

THE TOPIC OF whether the dominant radiation mode from small transmitting loops may be that of a small folded dipole rather than a magnetic loop continues to excite strong feelings, with Jack Belrose, VE2CV, Tony Henk, G4XVF, and Jack Hollingworth, ZF1HJ, and others still highly sceptical of the dipole-modehypothesis of Mike Underhill, G3LHZ. Mostly the writers are convinced that despite the undoubtedly good performance of small loops, their radiation efficiency is really very low. HF can be extremely tolerant of efficiencies as low as 5% or so. I have to admit that the case against dominant dipole-mode radiation seems to be strong.

VE2CV is particularly dogmatic, concluding: "If the compact loop looks like a loop, behaves like a loop, it must be a loop. There is no controversy - G3LHZ and G4VRN are wrong in their view that a compact loop does not radiate like a loop. The patterns that I have calculated and everything I have said about loops are correct."

G3LHZ remains firmly convinced that there is more to single-turn loops than presently-accepted theory, NEC calculations and field measurements suggest, although he points out that he has never denied the existence of one or more true loop modes: "All I have said is that the dipole mode is stronger." He recognises that with a loop of three or more turns, the dipole mode is suppressed below the true loop mode. He suggests that the true lope mode radiation resistance of a loop decreases as $(D/\lambda)^2$ and not $(D/\lambda)^4$ as the text books (and NEC) state.

There the matter must rest for this month! It still remains to be seen whether G3LHZs enthusiastic belief in the dominant dipole-mode is justified or whether his equally convinced critics have demolished his hypothesis! As pig-in-the-middle, I am expressing no opinion.

MODEL-SHOPS YIELD USEFUL ACCESSORIES

DR STEWART BRYANT, G3YSX, recently became interested in aeromodelling, and discovered once again that having multiple interests teaches you that what was a problem in your old interest often has a simple solution when you apply the common knowledge derived from your new interest.

He writes: "Aeromodelling has led me to discover a whole range of new materials and techniques that are readily applicable to amateur radio. For example, most amateurs equate glues to 5-minute epoxy or superglue from the local DIY superstore. A visit to good quality model shop will introduce many new adhesives: a wide range of epoxies of various viscosities and cure rates, super-glues (cyno-acrelates) in a range of consistencies from treacle to water, together with a whole host of other glues. Reading the model-making publications leads to the discovery that many white glues, previously thought only suitable for sticking wood together, can, in fact, form a very powerful general purpose adhesive (for

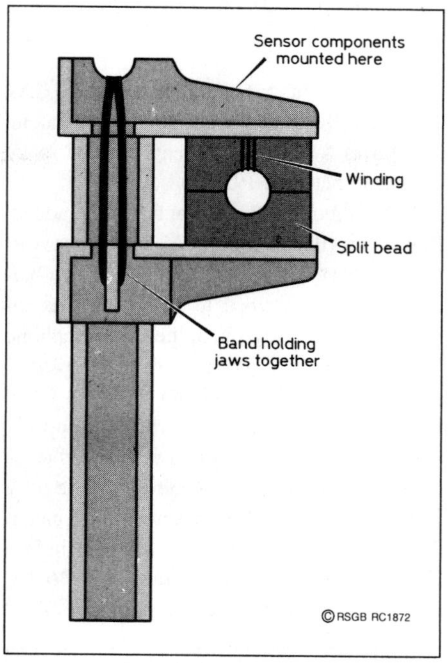

Fig 8: How G3YSX uses an inexpensive aeromodeller's plastic clamp as a replacement for spring clothes pegs in an RF ammeter.

example, I used a white PVA-like glue to bond fibreglass to wood in a very high joint, with good results).

"Recently, I needed to build an RF ammeter. On G3SEK's web site I found a design for a current sensor, in which a transformer based on a split ferrite core is clamped around the cable in which the current is flowing. The design was published in *RadCom* some time ago. In the original, the author relied on the plastic clamp supplied with the split head, but this was designed only to be clamped on a cable and left there permanently. It tended to fail relatively quickly. G3SEK built a clamp based on a couple of clothes pegs and fibreglass board. Whilst this appears more robust, it struck me as being a little unwieldy.

"Proops of Leicester (tel: 0115 2403 400) supply a range of tools and clamps to modellers. One particular clamp seemed ideally suited to the construction of the clamp needed to hold the split ferrite around the cable. First, I wound the transformer winding around the ferrite; then glued it to the fixed jaw of the clamp with epoxy, taking trouble to put a thin covering of epoxy over the winding to protect it. The bond was assisted by using the clamp itself to hold the ferrite in contact with the fixed jaw.

"When this was set, I used epoxy to glue the other half of the ferrite to the movable jaw, again using the clamp to hold the ferrite in place while the epoxy cured. **Fig 8** shows my arrangement. The plastic clamp costs only about 40p. The two halves of the clamp are kept in tension by a small elastic band (which could easily be replaced). In my case I drilled a number of small holes in the fixed plastic arm and mounted the sensor components within a few millimetres of the transformer."

CLOVER UNDER-VALUED?

JACK HOLLINGWORTH, ZF1HJ, was vastly surprised to note that the November 1997 *TT* item 'Comparing Digital Protocols' (based on a paper presented by T J Riley of the US Institute of Telecommunication Services in IEE Conference Book 441) showed (in the November Figs 9 and 10) Clover well down the Y-axis, along with Amtor and Pactor, rather than somewhere up around the plot for Pactor-II. He writes:

"My own experience is that, under 'good' propagation conditions, Clover delivers a slightly higher data rate than that shown for Pactor-II, even at very low SNR. Under truly appalling conditions, and low SNR, it continues to deliver data at 10-15 characters per second (I have copied signals which I could not even hear under the noise in a 3kHz bandwidth).

"I queried the ITS results with Bill Henry at HAL Communications. He believes that the ITS data are flawed, mainly because the computer used for the trials was a 286 machine running Windows, and was consequently unable to deliver data to the Clover board (which is a plug-in PC-bus device, rather than a stand-alone unit like Pactor-II) as fast as it could be sent over the HF link. Also, apparently all the tests with Pactor-II were conducted with data compression on, whilst the Clover tests were run without data compression. Understandably, Bill Henry is reluctant to publicly critise the ITS results . . ." [Some lines are apparently missing from the FAX as received from ZF1HJ via RSGB HQ - *G3VA*].

ZF1HJ concludes: "A couple of years ago I wrote an article intended for *RadCom* entitled 'The Case Against HF Packet', which attempted to explain the truly abysmal performance of (AX25) packet on HF. It was not published, perhaps because it came too soon after 'Stop the Packet Racket on HF' by Colin Richards, 9M2CR (September 1992) and the resultant outburst of indignation from the packet community. Compare the poor results of packet with the 'commercial' version of Clover (Clover-2000), which can send over 200 characters per second in a 2kHz bandwidth at high SNR, and still passes 5 characters per second at -10dB SNR in a 3kHz bandwidth.

"I am truly surprised that packet still continues in use on HF, when so many far more efficient alternatives are now available. My hypothesis is that whilst most amateurs understand antennas and transceivers and regard modes and computers as mysterious black boxes which one can connect to them in order to do interesting things, the packet community understands the working of computers and modems but regards radios and their attendant antennas as mysterious black boxes which can be connected to a computer and in some arcane way manages to get data from A to B." ♦

BY PAT HAWKER, G3VA*

WHO INVENTED THE CRYSTAL FILTER?

THE JUNE *TT* noted that Arthur O Bauer, PA0AOB, (not AUB) had sent me a copy of British Patent Specification 293,446 granted in July 1927 to CW Hansell and assigned to Marconi's Wireless Telegraph Company. This covered a bridge type crystal filter and preceded the patents of Dr J Robinson whom I had always considered responsible (in connection with his Stenode Radiostat broadcast receiver) for the development of the 'crystal gate' filter, as used by James Lamb of ARRL in his pioneering 1932 design of the first truly 'single signal' HF communications receiver.

It is now clear that the story was rather more complicated than I suggested, with several others working along similar lines, but with Robinson's work emerging as a vital contribution.

First I sought out a copy of Lamb's landmark article 'Short Wave Receiver Selectivity to Match Present Conditions - Constructional and Operating features of the Signal-Signal Superhet' (*QST*, August 1932). This included the following acknowledgement: "Dr Cady first suggested ('The Piezo-Electric Resonator', *Proc IRE*, April 1922) the use of the quartz resonator as a sharp electric wave filter over ten years ago, and recently its use as such has been greatly accelerated. We recall that the series resonator as a filter in a receiving circuit was tried several years ago by Paul Zotta in Dr Cady's laboratory at Wesleyan University. A wide variety of applications has been made by the Bell Telephone laboratories, Radio Corporation of America, and others. A quartz filter is also used in the stenode broadcast receiver, recently given considerable publicity in England and this country, its use being to secure high RF selectivity for modulated signals (particularly broadcasting). . . "

My belief is that Lamb either attended or read about one of the lectures given by Dr Robinson in the States and it was this that encouraged him to fit a similar crystal filter in his new receiver.

The notes in the June issue stimulated Lorin Knight, G2DXK, to send along a copy of 'Simple Quartz-crystal Filters of Variable Bandwidth' by Dr Geoffrey Builder and JE Benson (*Wireless Engineer*, April 1943, pp183-189, reprinted from the Australian *AWA Technical Review* 1941, Vol.5, No.3, with an addendum in Vol.5, No.5, 1941). The original AWA article included the comment "the filter circuits in general use all seem to have been developed from the selector or 'crystal gate' described by J Robinson, 1931, for the highly selective tone-corrected type of receiver exemplified by the Stenode Radiostat". But an added footnote and an addendum on the history of piezo-electric crystal filters credited a number of others, listing US patents by Cady, Espenschied (RCA), Marrison (Bell) and the Hansell British Patent, as well as four British patents granted to Dr Robinson between 1928 and 1930.

In the footnote, Builder and Benson added: "Since this (their 1941 AWA article) was written, we have discovered that the bridge-balanced circuit ascribed to Robinson was invented by WA Marrison of the Bell Telephone Laboratories (US patent 1,994,658, filed 7 June 1927, granted 19 March 1935).

Lorin Knight, G2DXK, writes: "It appears that the earliest disclosure of a balanced crystal-gate filter similar to Robinson's was on 7 June 1927, when WA Marrison filed US patent 1,994,658. This was exactly a month before Hansell filed his British patent 293,446 describing a similar balanced crystal-gate filter.

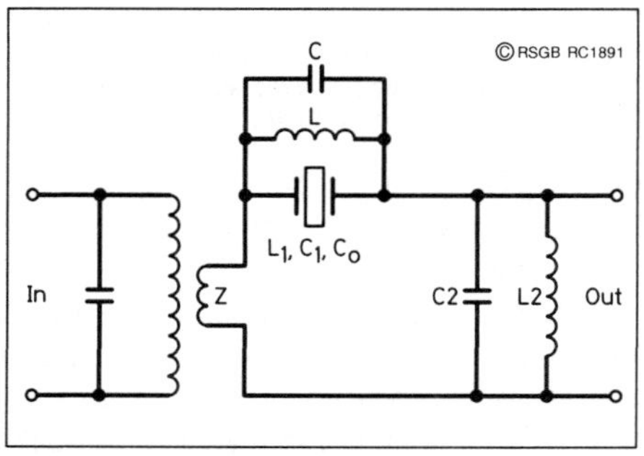

Fig 1: The m-derived crystal filter described by Builder and Benson in 1941. The effective parallel capacitance C_0 of the crystal is neutralised by LC. Effective variation of bandwidth can be attained by (a) drastic mismatching, obtained by lowering the impedance terminating the filter at mid-shunt or by detaining the parallel arm L2/C2, which may be regarded as loading the filter at mid-shunt with a reactance.

Even earlier, in January that year, L Espenschied of BTL, filed US patent 1,795,204 for a multi-crystal bandpass filter."

G2DXK adds: "Although Robinson was not the first to patent a balanced crystal-gate filter, he certainly gave his plenty of publicity. Further, he emphasised that by suitable adjustment of the balancing capacitor it was possible to put a rejection notch anywhere in the passband - a virtue which was exploited later in communications receivers. This very useful property appears to have been missed by Marrison and Hansell."

It thus appears that as amateurs we are largely indebted to Dr Robinson and James Lamb, even if Robinson was not the first to develop a balanced crystal-gate filter. One of those who played an important role in Dr Robinson's work on crystal filters was Earnest Gardiner, G6GR, later President of the RSGB (1944-46).

Incidentally, the Builder and Benson article points out that "a bridged-T single-crystal filter described by Masson (1934) does not seem to have been used in communication receivers, although the arrangement is simple and economical in components (in reality bridged-T filters were fitted in the British R206). Nor does there appear to have been any use of conventional ladder-type filter structures for the purpose." **Fig 1** shows the m-derived crystal filter described in detail in their 1941 article.

FRACTALS AGAIN

JOHN HARBOTTLE, RS96257, points out that further information on fractal antennas (see *TT*, April and June 1998) can be found on a number of web sites. A check by RSGB HQ shows the following were still to be found during June:

http://www.mbvlab.wpafb.af.mil/rkutter/thesis

http://www.mbvlab.wpafb.af.mil/rkuttere/thesis/thesis2.html

This provides the undergraduate Honours Thesis of Richard E Kutter at the University of Dayton, Department of Electrical Engineers. It includes some useful introductory notes: "Why fractals? Fractals are perhaps the most efficient way of increasing the perimeter of an area. It is well known that the perimeter of a loop antenna is the most important factor in determining its resonant frequency. A fractal antenna with a given perimeter takes up less area than a comparable square loop antenna (I realise this is a rather simplistic assumption, but is does work - to a point!).

Why a fractal antenna? (The following comparisons are made against a square loop antenna, assuming the user wants directive gain).

Advantages:

- Smaller cross-sectional area.
- No impedance matching network required.
- Multiple resonances.
- Higher gain in some cases.

Disadvantages:

- Lower resonant bandwidth than loop antenna.
- Creation is more difficult.
- Lower gain in some cases.

THE TERMINATED RECEIVING LOOP

GARY BREED, K9AY, in the September, 1997 *QST* (pp43-46) presented an article that offered a novel solution to the amateur with limited space seeking a compact, low-band directional receiving antenna. He described an easy to construct antenna having a polar diagram that could be remotely switched to the four quadrants of the compass, yet capable of being fitted into a 30ft diameter (10m) circle. He called this 'the K9AY terminated loop', and acknowledged that it had been developed from the much larger EWE antenna developed by Floyd Koontz, WA2WVL (*QST*, February 1995), which required an 80ft diameter circle.

*37 Dovercourt Road, London SE22 8SS

Subsequent comments in *QST*'s Technical Correspondence column (April 1998, pp77-78) by Bill Bridges, W6FA, revealed that the origin of the terminated loop can be traced back to HH Beverage (W2BML) of RCA and his US patent 2,247,743 of July 1941. W6FA emphasised that the terminated loop is not a wave antenna with which the name of Beverage is more usually associated and which was developed by him in the early 1920s. The only commercial application of his terminated loop appears to have been as a VHF television receiving antenna.

W6FA points out that the K9AY design is basically a half loop, requiring an earth connection, whereas the Beverage patent described a full loop which does not operate with a reflected earth image and requires no earth connection. In a full circular loop the feed line is connected across a break in the loop, with a terminating resistor connected across another break in the circle diametrically opposite the feed line.

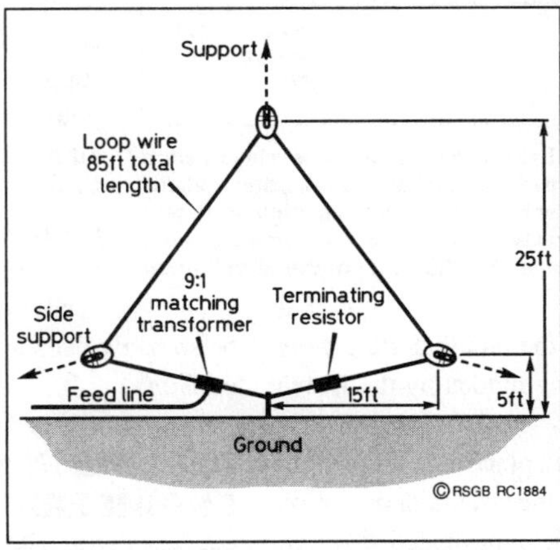

Fig 2: The basic design of the K9AY terminated (half) loop directional receiving antenna, suitable for use below 4MHz in an easy-to-construct quasi-delta configuration. Exchanging the feed point and termination reverses the pattern. A four direction system uses two similar loops installed at right angles to each other with a relay controlled switching system.

Fig 2 shows the basic design of a single K9AY type terminated half loop receiving antenna in an easy-to-construct quasi-delta loop configuration. In this form the antenna is unidirectional, in the sense that it receives best signals from the left hand side of the page, with a sharp null at an elevated angle in the opposite direction. If the feed point and termination resistor are exchanged the pattern is reversed, receiving signals better from the right hand side.

In practice, K9AY developed a four direction system by using two of these loops hung from a single support and installed at right angles to one another. A relay controlled switching system, located at the base of the loop (**Fig 3**) enables a single terminating load and matching transformer to be switched to each loop in each configuration. Later, K9AY provided modified circuits of his switching system (*QST*, May 1998, p73) to overcome some hum problems reported by users: **Fig 4** and **Fig 5**.

A single or double terminated loop system could clearly have useful applications from LF to VHF, either as a fixed directional antenna, switchable bi-directional antenna, or as the K9AY dual loop system switchable to any of four quadrants.

To quote K9AY: "The terminated loop is physically and electrically quite simple. It consists of a wire loop of any convenient shape (diamond, delta, etc), hung from a single support and with a ground rod at the bottom. A 9:1 impedance matching transformer connects from one end of the loop to ground, a terminating resistor connects the other end of the loop to ground. This antenna is directional, favouring signals arriving from the feed point end, rejecting by several S-points any signals arriving from the end connected to the terminating resistor. . . A very small terminated loop maintains its directional pattern. As the loop's size gets smaller, however, the desired signal strength is also reduced."

K9AY suggests that the value of the terminating resistor should lie between 390 and 560Ω, with 390Ω about optimum on 1.8MHz, 560Ω optimum for 3.5MHz and 470Ω "pretty good" on both bands. For amateur two-way operation with a separate transmitting antenna, the resistor should be rated at least 1 watt (K9AY uses a 2 watt carbon resistor), in case some of the transmitter power ends up on the loop. "The matching transformer is a 9:1 impedance, 3:1 turns-ratio type that should be familiar to many readers.

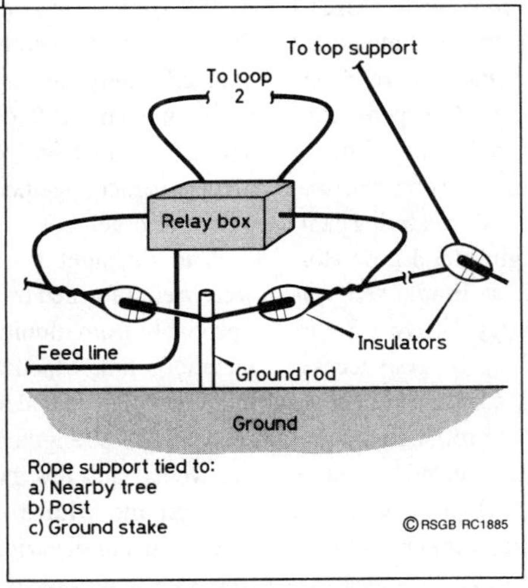

Fig 3: How the relay-control box, which also contains the matching transformer and terminating resistor, is located at the base of the K9AY terminated loop.

It is claimed that with an optimum value of terminating resistor, the rejection null can be as deep as 40dB at 20 to 55° elevation, depending on the shape of the loop and local ground conditions. Unless the loop is short and wide or tall and skinny, the null will be at 30 to 40° elevation and very useful for rejecting medium distance signals, although such a loop would not be suitable for DF applications on ground wave signals. The circumference of the loops (essentially half loops reflected in the ground) should not exceed a quarter wavelength at the highest frequency band on which they are to be used. With full loops, without the ground connection, circumference should be less than a half-wavelength.

K9AY points out that the terminated loop antenna is an excellent antenna for weak AM broadcast reception and, if scaled in size, can be used for HF broadcast or amateur reception. For reception from a fixed area, a single loop without remote switching could be used.

DIGITAL POWERLINES, RFI & A WW2 CARRIER SYSTEM

DR GARY HINSON, G4IFB, draws attention to a worrying aspect of plans by Nortel, in conjunction with Norweb, to send high data rate Internet traffic over electricity power lines (see also the July *RadCom* News, page 10 and *EMC* column in this edition of *RadCom*, pages 72 and 73). According to a front-page lead story in the trade newspaper *Network News* (UK edition) of 27 May 1998: "Nortel has been forced to change its technology for sending Internet traffic over power lines, as its transmissions interfere with emergency service frequency bands. One result is that later products have been changed to 2.5 - 8MHz. As G4IFB puts it: "Are they seriously talking about sending data over power lines at 2.5 - 8MHz? Oh Boy!"

In the story, a Radiocommunications Agency spokesperson is quoted as saying: "The early prototype of the equipment interferes with the 2 - 5MHz band, which is used for emergency communications." According to *Network News*, the later products had not then been seen by the RA.

Nortel's Digital PowerLine (DPL) technology enables data transmission at over 1Mb/s along electricity power lines. A joint venture involving Nortel and Norweb, DPL is planned to reach around 2000 homes this year.

According to Nick Long of the Low Power Radio Association: "DPL causes alarming levels of interference."

DPL is a digital HF version of the classic carrier systems used by electricity distribution companies for many years. I have a personal reason for questioning whether even at LF or VLF such carrier systems radiate and are - or at least were - far from secure.

At the end of October 1944, as an SCU9 radio operator with Special Communications. I travelled by jeep from Brussels to Eindhoven and then, a few days later, to Nijmegan to act as the signals operator for an officer of No.2 Intelligence (Underground) Section. The reason for going to Nijmegan was that there existed, at the local PGEM power station, a private carrier-type telephone link over which it was still possible to speak directly across no man's land between the Waal and Lower Rhine (Lek) rivers to a Dutch resistance group in enemy held Ede, near Arnhem.

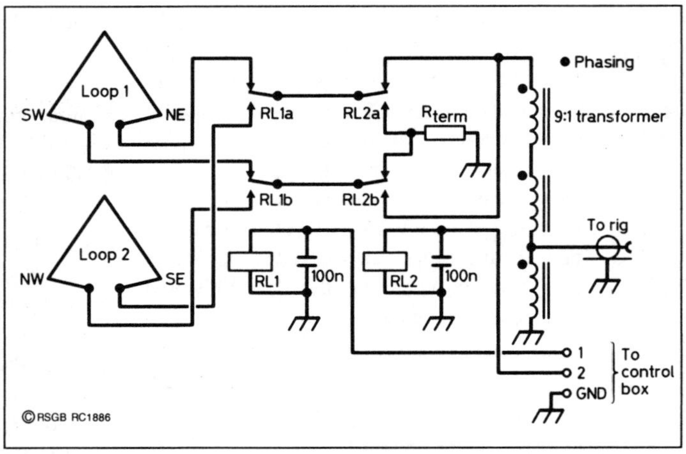

Fig 4: Circuit details of the relay box at the base of the antenna. The relays switch between the two loops and also reverse the lead point and termination resistor connections. This diagram is from the later K9AY note, with his original system modified to avoid hum pick up from the relay power being fed over the coaxial feeder. K1 and K2 are 12V DC, DPDT relays. For the value of the termination resistor R, see text.

Hardly had I set up my Whaddon Mark III transmitter, HRO receiver and 150 watt Onan petrol-electric generator in a house on the outskirts of the town when Captain Fraser (the late Hugh Fraser MP, one-time SAS Phantom officer), second in command of the small IS9 (WEA) unit, arrived next door with the Forward section including their own signals operator with a B2 suitcase set. IS9 was the free-wheeling Intelligence Unit (MI9) concerned with assisting 'evaders' and 'prisoners of war' escape from enemy held territory.

Concerned that the IS9 B2, working to Special Forces (SOE) at Poundon, and my MK III would create mutual interference, I attempted to argue prior rights. My claims were cheerfully swept aside by the simple expedient of IS9 moving into both houses and virtually taking me over in a manner, I soon realised, that had characterised the unit's roistering campaign across France, Belgium and now Holland. The main IS9 unit at Eindhoven was commanded by Major Airey Neave ('Saturday'), the first British officer to have successfully escaped from Colditz Castle and later, as an MP, killed by an Irish terrorist car bomb. The IS9 forward party had expanded to include sundry Canadians, Belgian SAS parachutists, Belgian and Dutch resistance workers. A lively bunch of tearaways!

IS9 was soon busy planning Operation Pegasus II, using the private powerline telephone to make arrangements with the Dutch Underground, who were still hiding many Arnhem evaders in homes and woods around Ede. I spent much of my time reading a copy of *Gone With The Wind*, belonging to the pleasant Dutch family whose home we were sharing. Nijmegan was still recovering from a terrifying American heavy bomber daylight raid some months before, when almost a thousand Dutch civilians were killed when a navigational error mistook the town for Kleve, just across the German-Dutch frontier.

An earlier IS9 operation, Pegasus I, had been successful in bringing out some 130 Arnhem survivors, but an unfortunate press leak must have alerted the Germans to the presence of the escape unit and the fact that there were still many evaders being hidden by the Dutch across the rivers.

Pegasus II finally took place during the night of 23 November 1944 - the last major operation of IS9 - but proved an unmitigated disaster. Only one evader and two of the Dutch guides reached safety across the rivers. The main party of some 120 evaders was ambushed, about 36 killed and most of the remainder taken prisoner, together with many of the Dutch guides and helpers.

With hindsight, it is clear that the Germans were well informed on the plans for Pegasus II - probably from monitoring the PGEM carrier telephone link which had been thought to be unknown to them and hence secure. This belief seems to be strengthened by the fact that the line went dead a few days before the operation.

It left me with a distrust of carrier-type systems using electricity power lines - a distrust that may turn to positive hatred if high speed data begins radiating from the local lamp standards at frequencies including the 3.5 and 7MHz amateur bands, not to mention the 6 and 7MHz HF broadcast bands.

My weeks in Nijmegan also showed me an unusual way of ruining an Onan generator. Electric power in the town was switched on for only about two hours a day, and it occurred to me that this could be extended in the house by switching off the mains supply when this was not available and plugging the output of my Onan into a domestic socket, which would then provide a limited supply of electricity throughout the house. This worked fine until one day I forgot to remove the Onan plug before reconnecting the house to the mains supply. The result, when the two hour supply was restored, was burned up field coils on the Onan. Luckily I was able to get a replacement quickly from Brussels, though my colleagues there could not understand how the field coils came to be so badly damaged. I thought it wiser not to explain!

DECLINE IN CLASSICAL ENGINEERING?

THERE IS CLEARLY widespread concern at the world-wide decline, particularly among the young, of interest in amateur radio. It seems to me that the real competition for youthful interest comes from mobile telephones, electronic mail and Internet, with the accompanying downplaying of interest in classical engineering, particularly electrical, electronic and radio frequency engineering. Listening to SSB on 3.5MHz, most of what technical conversation there is seems to concern the use of computers and, more rarely, antennas and ATUs. I can see the day coming when there will be a mandatory test in information technology, with little or nothing about the design of receivers or transmitters, which it will be assumed have all been bought across the counter.

This change of outlook is well reflected in one of the regular 'Reflections' columns written by Robert E Lucky in *IEEE Spectrum* (May 1998, p21): 'Electrical engineering diminishing role?' To quote briefly:

"I see a trend in the downward enrolment in electrical engineering and in the changing focus of the workplace around me . . . Attention in technology has moved increasingly to applications, irrespective of the mechanisms of the physical platform that supports them. . . Remember when computers were called 'electronic computers'? . . . Now nobody seems to care that they run on electricity. They could contain gears and pulleys, as far as most people are concerned. No one bothers about what is in-

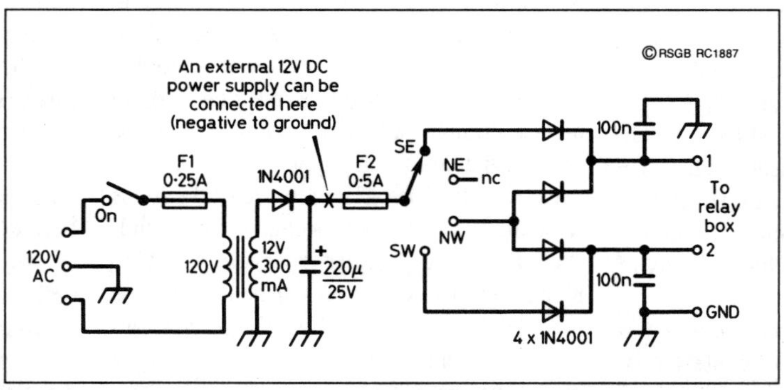

Fig 5: Circuit details of the control box located in the shack. This is the modified arrangement designed by K9AY to overcome hum pick-up from the relay power being fed over the coaxial feed line.

side today's computers, because for most of the world it's irrelevant.

"I blame all this on Heathkit. By going out of business, they took the fun out of electrical stuff. For a while after their demise it was possible to play with the hardware in home computers, but that soon ended... A lot of the fun things disappeared into the Lilliputian mazes of faceless black chips... Now the huge boom is in what is called information technology. People in the most nontechnical of occupations talk knowingly about their problems with the registry in Windows 95... the world is filled with users. In contrast, far fewer designers are needed. It's not that design isn't important. It is terribly important. The world rests on its shoulders... the sacred legacy of Ohm's Law is sinking into an abyss of complexity. Even circuit designers can seldom be concerned with the details of electronics... When a billion transistors sit on a chip, how many people will care about a single circuit? ... If current trends endure, future computers will consist of a single chip. No one will have the foggiest idea what is on it. Somewhere in the basement of Intel or its successor will be a huge computer file with the chip's listing. The last electrical engineer will sit nearby, handcuffed to the disk drive... That engineer will be the last keeper of the secret of the universe, E = IR (**Fig 6**)"

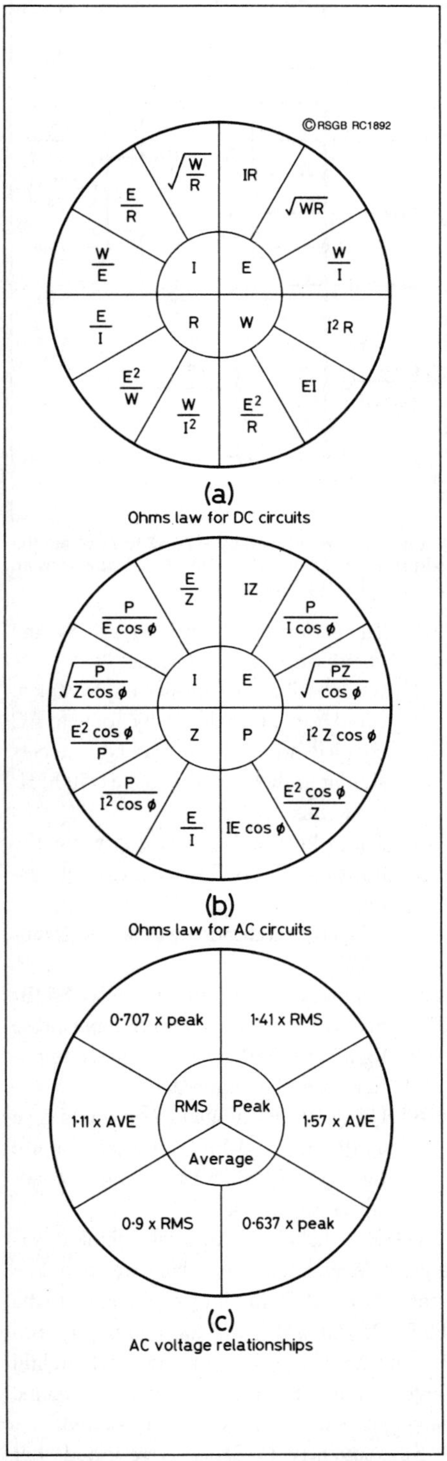

Fig 6: Robert Lucky believes that the secret of the universe rests in Ohm's Law, the fundamental building block of electrical and electronic engineering. These diagrams form an aide-memoir to Ohm's Law and AC voltage relationships.

VARIBLE BANDWIDTH CW FILTER

ALMOST THIRTY YEARS ago, I included in *TT* (and later editions of *Amateur Radio Techniques*) brief details of an AF filter capable of providing a 3dB bandwidth variable from about 250Hz right down to about 2Hz. This design (**Fig 7**), brought to my notice by Ron Shelton, 6Y5SR/G3IHP, was based on a design by RE Scheme published in *Wireless World*, August 1967.

It used two Q-multiplying (regenerative) stages, with the Vinkor pot core inductors tuned to slightly different frequencies (separation of 5 to 7Hz) to reduce 'ringing'.

An updated version of this filter is described by PM Bakker, PA3FZK, in *Electron*, June 1998, p245, **Fig 8**. (Dutch text): "Centre frequency with the component values indicated is about 1000Hz, but could be lowered by increasing the inductance of the pot cores or the value of the resonating capacitors. Many CW operators prefer a signal about 700Hz, or even 500Hz. PA3ZFK gives the 6dB bandwidth variable from 5Hz with RV1a, b set to 25kΩ, to 500Hz when this is reduced to 0Ω."

H-MODE MIXER WITH FST3125

GIAN MODA, I7SWX, recently drew attention (*TT*, July 1998) to the Fairchild Quad Bus Switch FST3125 device as a potential substitute for the SD5000 in G3SBI's H-mode superlinear mixer.

I7SWX emphasised the potential advantages in switching mixers offered by the maker's specified characteristics of the FST3125 and FST3126 devices, but reported that he had not yet been able to try out his ideas in practice but hoped that others would do so and report their results in *TT*.

I felt it was well worth alerting - in advance of the publication of the July *RadCom* - Colin Horrabin, G3SBI, of I7SWX's belief in the potential value of these new very fast switches. In turn, G3SBI alerted Bill Carver, W7AAZ, who has done a lot of work on super-linear receivers and who was able to obtain immediately some of these new surface mounted devices.

Within a few days, I received the following report from W7AAZ. This came via G3SBI, who has added some further comments on what is clearly an important development in the search for mixers of exceptional dynamic range. These devices clearly now make possible an excellent high level mixer that requires virtually no adjustments, has sufficiently low attenuation to allow its use without an RF preamplifier, and does not require the use of over voltages (as in the original G3SBI H-mode mixer).

William N Carver, W7AAZ, writes: "The Fairchild FST3125 provides four independent high frequency, low resistance switches and fast switch drives in one package, making it [as suggested by I7SWX - *G3VA*] very attractive for H-mode mixer service. My interest in the FST3125 was piqued by the very low 'on'

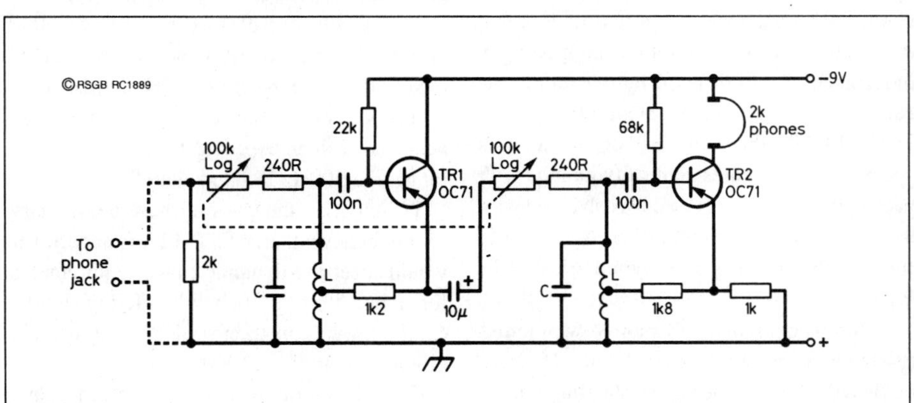

Fig 7: The variable bandwidth AF filter as used some 30 years ago by 6Y5SR to determine minimum usable bandwidths for CW operation. L is 150 turns of 24 SWG insulated wire on Mullard (Philips) Vinkor LA210. C is chosen to tune L to the desired frequency (about 1.25μF for 1kHz). Described in all later editions of *Amateur Radio Techniques*.

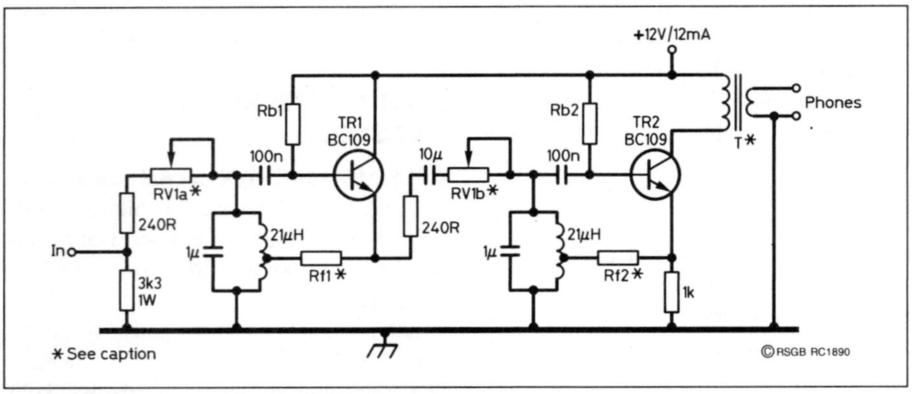

Fig 8: PA3FZK's updated version of the variable bandwidth AF CW Filter. RV1a, b ganged 2 x 25k log pots (bandwidth control). Rb1 and Rb2 are 1k trim-pots. Rf1 and Rf2 are 1M trim pots. T is an audio transformer to match the impedance of the headphones. PA3ZFK describes (Dutch text) how the filter can be accurately adjusted by observing Lissajous patterns on an oscilloscope.

resistance of the switches, much lower than the SD5000, and I expected lower conversion loss.

"The device is characterised for operation from a single nominal +5V supply, designed to switch digital signals with amplitudes between 0 and +5V. In my test arrangement (**Fig 9**) for RF use, one side of all switches, and the centre tap of the RF port (antenna) transformer were all held at Vcc/2 with a pair of bypassed 2k2 resistors. Incoming signals swing symmetrically about this Vcc/2 bias. Anticipating some difficulty in handling the surface mount chips, a PCB was designed and etched for the test mixer.

"The test equipment used to check results is surplus, but has been cross-checked with other instruments and considered to be reasonably accurate. The set-up comprised a Hewlett Packard 8640B local oscillator, set at four times the required 18.868MHz, a USM-323/8640 at 14.2MHz, and a lovely old General Radio 1003 at 14.0MHz. Amplitude measurements were made with an HP-3586B, essentially an HF receiver displaying signal amplitude with 0.01dB resolution.

"All three transformers were six trifilar turns of No.26 AWG wire, wound on BN302-61 'balun' cores to obtain high coupling coefficient. The series connected secondaries had about 6μH inductance and a loss of less that 0.25dB from 3 to 30MHz. Interestingly enough, inductance of two differed by 6% and I was curious to see how that might affect the mixer.

"At 14MHz the balance of the mixer was 45dB. This compares quite favourably with packaged diode mixers, and was obtained with the transformers as wound (there's nothing to 'tweak') on a reasonably symmetrical PCB layout.

"With pairs of switches shorted with jumpers, the transmission loss was 0.7dB. This was predicted from the measured coupling coefficient of 0.98 in the transformers. With switches operated by DC voltages the loss rose to 1.0dB, a low number indeed. This is not operating the device as a mixer, just confirming proper transformer construction and connections.

"Signals were applied and the conversion loss measured at 14MHz. It was 4.85dB, an excellent figure, almost exactly as predicted. Given a low-noise post-mixer amplifier (1.5dB can be readily obtained) and with a low loss RF filter, the overall receiver noise figure with this mixer can be as low as 9dB, thus producing quite a 'hot' HF receiver without an RF amplifier being necessary.

"To measure the intercept figure, two +21dBm signal generators were combined and their sum attenuated by 10dB to produce a pair of +5dBm signals at the mixer input. This resulted in the desired 4244kHz and 4434kHz IF signals being 0dBm. The two third-order products were -78dBm and -80dBm. Using the -78dBm number results in a calculated input intercept of +44dBm.

"A second set of measurements was made with signals 10dB lower, and the calculated intercept was +43.5dBm.

"To sum up, I expected and confirmed a low conversion loss, but I was surprised at the excellent intercept figures with such large signals; the peak amplitude was over 1V. Since the FST3125 could safely be operated at its rated 6V maximum, even better results are likely at that voltage. While the intercept performance falls short of that of G3SBI's SD5000 H-mode mixer, it is still good. If care is taken to keep the noise figure of the IF amplifier low, a receiver noise figure of 9dB should be achieved with an expected 118dB of dynamic range in a 500Hz IF bandwidth."

W7AAZ, on the basis of these preliminary results, believes that an FST3125 H-mode mixer would meet the dynamic range requirements for some 98% of users, while offering significantly fewer constructional and adjustment problems than the SD5000.

G3SBI, while waiting to obtain some FST3125 chips from UK sources, believes that the experimental results reported by W7AAZ underline a number of important points:

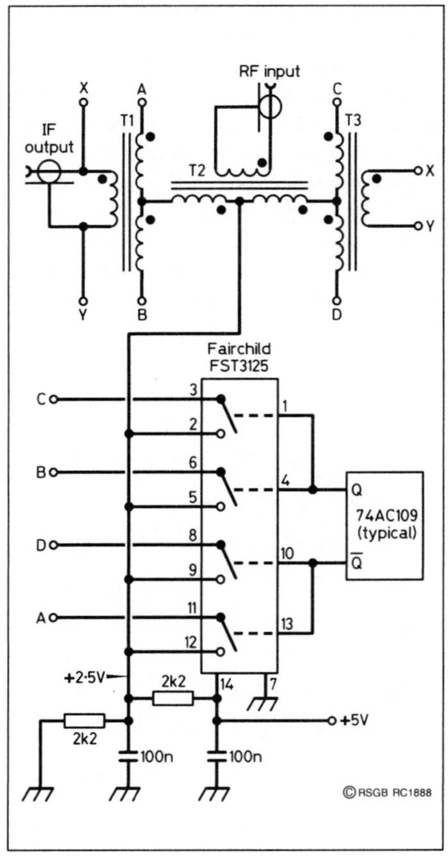

Fig 9: Test circuit used by W7AAZ to evaluate the performance of the Fairchild FST3125 device as an H-mode HF switching mixer.

1. The performance achieved is at 5V and no adjustments were needed.
2. It is possible that further improvement could be made to the IP_3 by going to AC coupled drive, with bias adjustments similar to those used in the SD5000 H-mode mixer.
3. The +44dBm IP_3 may represent the limit of W7AAZ's measurement system.
4. RF to IF feedthrough was an excellent -42dB.
5. The very low insertion loss (4.85dB) makes possible a receiver input noise figure of 8-9dB, so no pre-mixer amplifier would be required.
6. Using a 5dB RF attenuator would give an IP3 of + 49dBm for a receiver noise figure of 13dB, which would be adequate for use below 14MHz.

G3SBI concludes: "All-in-all I think this is a great development, for which we should be grateful to I7SWX for drawing attention to the FST3125 and to W7AAZ for so promptly confirming the value of this new, low cost Fairchild device as an HF mixer of excellent dynamic range and exceptionally low attenuation."

Subsequently G3SBI has reported that FST3125 devices are obtainable from Farnell Components at under £1 for single units and that he has obtained generally similar results to W7AAZ. ♦

BY PAT HAWKER, G3VA*

SOLDERING AND DRY JOINTS

A COUPLE OF years ago, a *TT* item 'Is home brew and DIY servicing passing into history?' (February 1996, p74) drew attention to the continuing problem of 'dry' soldered joints developing after new equipment has been working well for some months. "Oxidation, heat cycles and the like, can affect any less than perfectly soldered joint, and result in an intermittent or continuous high resistance or open connection." It was noted that even for professionals the time taken to trace intermittent faults can quickly mount up and result in high costs, despite the fact that nothing has to be replaced.

Geoff Baxendale, G4IBS, who recently retired after spending 32 years in the electronics industry - much of it at the Mullard Blackburn valve manufacturing plant, which over the years has declined from 6000 to about 600 staff - contributes a number of relevant thoughts on the topic of dry joints:

"I think dry joint faults are less the fault of 60/40 solder than the way the joint is made. I recall my father telling me as a small boy to 'make a mechanical joint first, and then solder.' I still have a very old Lowry electronic organ, all valves and tag strip, and I don't think there is or ever has been a bad joint in it!

"The advent of printed circuit boards and wave soldering has much to answer for, although reducing manufacturing costs and hence prices. The problem with 'wave', I believe, is having to solder at the same time components of widely different thermal capacities. It is far too easy to leave the larger components with far too little solder. With some of these components mechanically supported by the solder, then subjected to thermal cycling, there is your dry joint.

"Another problem seems to be with power transistors and IC voltage regulators, with straight leads that are fastened firmly to a heat sink, which in turn is fastened to a PCB. With these components heating up and cooling down each time the equipment is turned on and off, the joints become badly stressed and eventually go dry.

"Some manufacturers remove as much copper as possible from their boards; this reduces the heat dissipation area, resulting in higher temperatures and more stress. It is difficult to understand the logic of this practice.

"Through-board connections are also a problem; this seems to be due to differential expansion. My IC-240 rig went on the occasional blink many years ago; it was ages before I could induce the fault on the service bench with a multimeter at hand. Eventually I found 9V on one side of the board with a variable voltage on the other side. Icom had used little pins as 'vias'. I replaced them all and the problem went away for a year or two, but then returned. My final solution was to use wire in a loop: see **Fig 1**. Fortunately, modern glass/epoxy boards with plated-through holes seem to be completely free of this problem, but I believe other makers besides Icom used the pin trick. I believe later IC-240s used rivets, although I am not convinced that these were much better than the pins.

"The current trend is towards surface mount technology. I have not yet had enough experience of these devices to know what they will be like long term; however, in the short term, joints without solder are not unknown. In a couple of equipments I have come across, they have worked for over a year (and so become out of warranty) and then gone intermittent. The 'J-pin' ICs (where the leg is bent under) seem most prone. In seeking to find poor joints, there is no substitute for a strong light and a 10x magnifier. With these aids such faults are easy to spot, but need a very fine soldering bit and a steady hand to fix!

"On the topic of good soldering practice, never be tempted to tin multistrand cable and then put it in a screw type connector block. The solder will cold flow and result in an intermittent connection. This takes about a year or so and probably depends to some extent on the temperature. Similarly, when putting a mains plug on a new appliance, chop off the tinned bit of wire or the same thing will happen. If you don't believe me, take a light bulb out of its socket and note the dents in the solder pads! I think this effect led to the demise of lots of diodes in the Philips G11 series of television sets. This was brought to my attention by the appliance tester at work going intermittent, the fault traced to tinned leads!

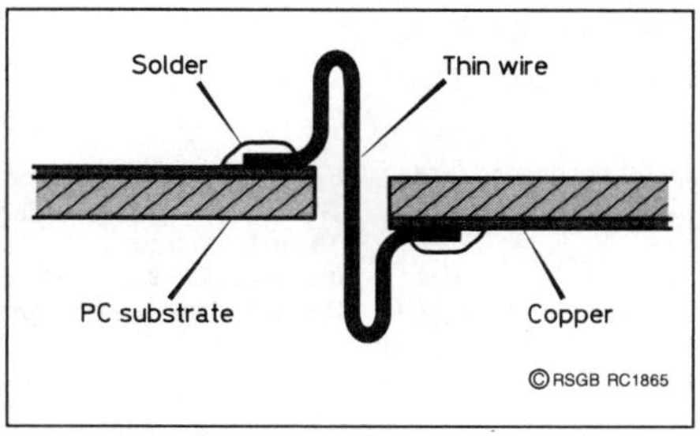

Fig 1: G4IBS's suggested form of 'via', for non-plated-through holes in printed circuit boards - preferred to the small pins or rivets used on some equipment.

"As to what causes equipment problems, I feel that faulty electrolytic capacitors must be a close second to dry joints. Time and temperature seem to be the problem. If you have some weird faults in a piece of equipment, it may be worth changing all the 'Elcos' and give it a new lease of life. I find the component tester described some time ago in *RadCom* that produces ellipses on an oscilloscope invaluable. Problems in switched mode PSUs can usually be traced to a faulty electrolytic, but sometimes you get the Christmas Tree light syndrome as a result! Several that I have met have just one elco that fails, nothing else (eg the old BBC computer). Sometimes they work off load, but fail on load.

"Finally, my main advice: If you want to keep electronic equipment working, *keep it cool*! Don't stack it in racks or cabinets or on top of other units, unless you use plenty of fans. There are some useful small 12V fans around these days (convenient and quiet running) stemming from the computer world. One of these pointed at your heat sink can result in a tremendous reduction in temperature."

SSB ON THE RA17 AND OTHER CLASSICS

THE MAY *TT* (pp57-58) included a design for an SSB product detector adapter unit, suitable for use with classic 'boat anchor' receivers, designed in the days when diode detectors were often coupled with low BFO injection. This item was a precis of an article by Morris Odell, VK3DOC, in *Amateur Radio*, March 1998, pp7-8. VK3DOC suggested that the Racal RA17 series, whilst a "beautiful and very sophisticated design" has a "particularly unsatisfactory performance on SSB".

This contention is strongly rebutted by John Truppen, VK6XJ (Technical Correspondence, *Amateur Radio*, May 1998, p46). He writes:

"I would not in any way criticise VK3DOC's quite delightful method of achieving SSB demodulation. His article is excellent, but I would question *why?* I have been involved with so-called 'boat anchors' on and off for some 40-odd years, and have repaired, restored, and operated many such receivers, RA17s in particular. In my opinion, the RA17 in its various models is a good candidate for being one of the very best valve HF communications receivers ever. When operated correctly, it is one of the best performers of all 'boat anchors' for receiving and resolving SSB signals.

"The correct procedure for SSB is: Set up the receiver for the frequency of interest. Set the IF bandwidth (selectivity) to 3kHz. Turn on the BFO. With the BFO TUNE (note, offset the BFO 1.5kHz *above* the IF centre frequency if copying *lower* sideband or *below* the IF centre frequency if copying *upper* sideband). Do *not* make further adjustment to the BFO tuning, unless going to the opposite sideband. All further adjustment to resolve the signal is done on the main receiver tuning.

"Check that the AGC response is set to *long*. Tune the receiver normally across the band, and you will find that it behaves just like a modern, solid state rig.

"The above assumes that the receiver is correctly aligned, particularly the IF, detector and BFO. It is possible that VK3DOC's receiver has an alignment problem.

"This technique works with all of the older valve receivers, although with varying suc-

*37 Dovercourt Road, London SE22 8SS

feel what it's like to bring a detector in an out of oscillation, to actually create the beat frequency oscillations that bring a whisper of a CW signal to life! This may be why - all technical reasons aside - the regen still survives after 85 years."

G3SBI'S H-MODE FST3125M MIXER - CONSTRUCTIONAL DETAILS

AS MENTIONED briefly at the end of the item on W7AAZ's validation of the value of the FST3125 device as a super-linear HF mixer (*TT*, August 1998, pp60-61), Colin Horrabin, G3SBI, was able later to acquire some of these low-cost surface mount chips (from Farnell Components). The FST3125M is a 14-lead moulded small outline package (SOIC JEDEC): **Fig 3**.

His results agree closely with those reported by W7AAZ, but he has now provided some further measurements, comments and constructional details which should help those taking advantage of this useful new development in high-performance switching mixers.

First, G3SBI finds that the commercial transformers made by Minicircuits - TT4-1A, style X65 (6-pin DIP) - give better results than the home-made variety using available toroid cores: **Fig 4**. This applies both to overall insertion loss and third-order intercept point (IP3) although adding significantly to cost (about £5 each). Apparently the core mix used by Minicircuits is superior to the standard mixes. The transformers are available from Minicircuits Europe (Tel: 01252 835094).

The circuit diagram, **Fig 5**, represents also the basic layout of the FST3125 used as an H-mode mixer. Note how balance and symmetry is preserved with short

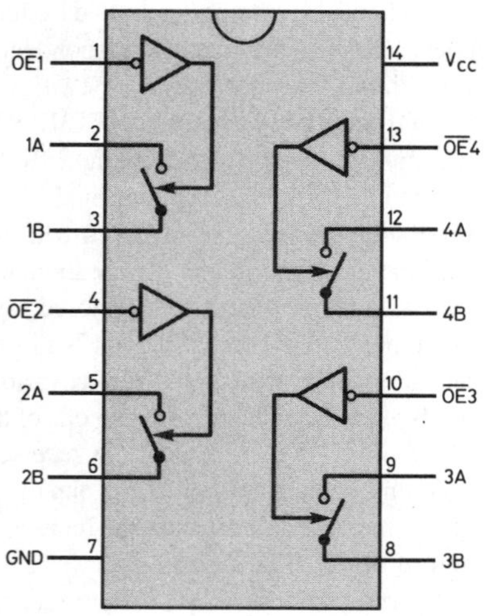

Pin name	Description
OE1, OE2, OE3, OE4	Bus switch enables
1A, 2A, 3A, 4A	Bus A
1B, 2B, 3B, 4B	Bus B

Pin descriptions

Inputs	Inputs/Outputs
OE	A, B
L	A = B
H	Z

Truth table

©RSGB RC1947

Fig 3: Details of the Fairchild quad fast bus switch surface-mounted chip, device type FST3125M.

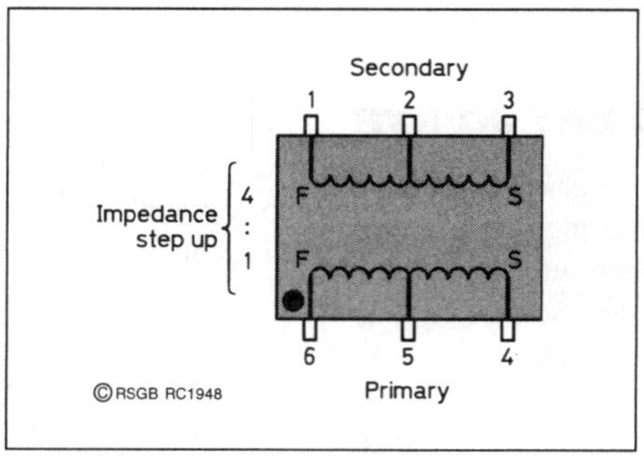

Fig 4: The commercial transformers made by Minicircuits (Type TT4-A). Although relatively costly, they give lower insertion loss and better IP3 results than are likely to be achieved with home made transformers using standard cores.

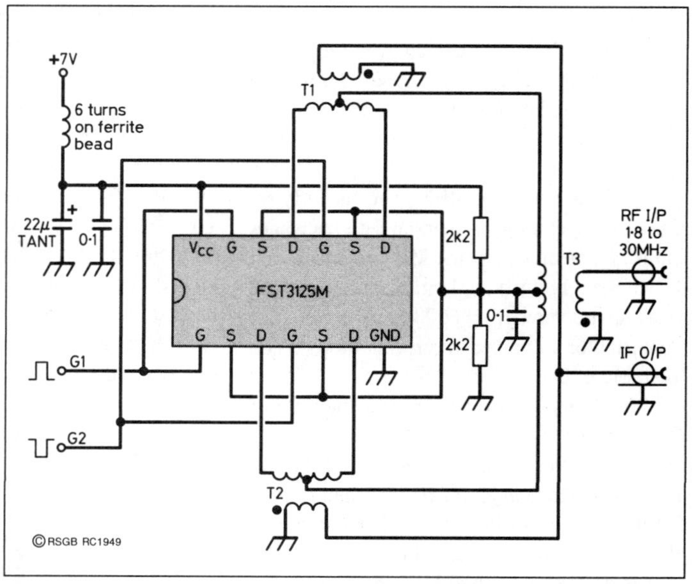

Fig 5: Circuit diagram of the basic FST3125 H-mode mixer, as implemented by G3SBI. Note that the diagram is drawn to show the symmetrical layout. T1, T2, T3 are Minicircuits TT4-1A style X65 transformers (6-pin DIP) and may be plugged into 6-pin DIP bases. G3SBI uses bases for prototypes. Note that the track joining the switch sources together goes underneath the chip and is 0.1in wide.

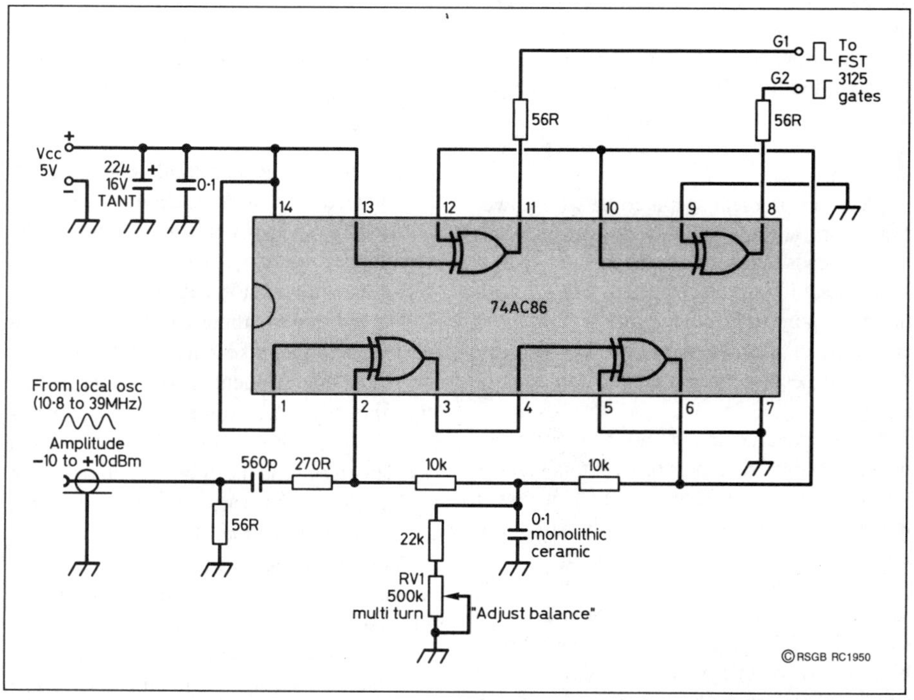

Fig 6: Fundamental frequency squarer, for use with FST3125 mixer. The unit should always have an input frequency. Use a ground plane PCB. The local oscillator input amplitude should preferably be +10dBm, as the squarer then takes less current. Balance adjustments for 9MHz IF using FST3125 mixer: (1) Set RF signal generator to 9MHz at 0dBm. (2) Set LO to 37MHz. (3) Adjust RV1 for minimum feed-through at 9MHz. The mixer should now be correctly balanced. Do not omit the 270Ω resistor, or the squarer will become unstable.

leads to the transformers from the switch; output tracks from T1 and T2 are the same length, before combined to form the IF output. Also 7 volts Vcc is used.

Basic performance on all bands (1.8 to 28MHz) is IP3 +45dBm, insertion loss 4.6dB, maximum input 7dBm (each tone) using 9MHz as the IF and high-side local oscillator. An IP3 of up to +50dBm is possible using over-voltages which are within the design maximum (Vcc) of the device, but in this case the long-term reliability may be degraded.

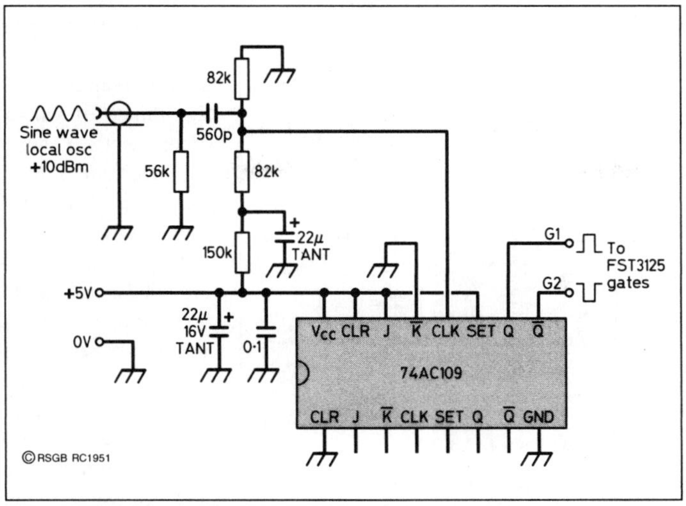

Fig 7: The ÷2 squarer, as used in the original H-mode mixer.

A squarer operating on the fundamental frequency is shown in **Fig 6**. This accepts a wide input amplitude (-10 to +10dBm), for square wave switching output. The original squarer circuit design operating on a local oscillator output at twice the required mixer switching frequency is shown in **Fig 7**. The track side of the PCB for the complete mixer is shown in **Fig 8**. Note the symmetry.

A significant feature of this design is that the mixer is almost as good with a fundamental-frequency local oscillator rather than one based on divide-by-two using a JK flip flop. The fundamental squarer is very easy to adjust for balance, using the IF amplifier to detect minimum feed through.

As G3SBI puts it: "What we need to see now is an army of constructors putting this exceptionally high-performance mixer to use." I would add that we should all be grateful to G3SBI and W7AAZ for investigating these new chips, whilst not forgetting the role of I7SWX in spotting their potential.

Finally, on the general topic of high-level mixers, I would like to correct the text statement in the July *TT* (p58) that the AD831 is capable of an IP3 of +44dBm. The correct figure is +24dBm, as shown in Table 1, with the +44dBm IP3 napplying to SD5000/SD8901 mixers.

HERE AND THERE

D A BUNDEY, G3JQQ, entirely agrees with John Truppen, VK6XJ, that the classic 'boat-anchor' Racal RA17 series of receivers when operated correctly is very good at receiving and resolving SSB signals (*TT*, July 1998, pp56-57). However, on one of his Racal receivers (Model RA17W) he has converted the noise limiter double-diode (V21) into a product detector: **Fig 9**. The BFO feed is taken to the strapped cathodes and the front panel limiter switch used to provide an SSB/AM changeover function. This simple modification has been in use for several years. Comparison with his unmodified RA17L receiver shows an improved performance on weak SSB signals, estimated at some 6dB.

Colin Horrabin, G3SBI, draws attention to a new complete DDS synthesiser chip which is expected to go into production in 1999 by Analog Devices and which will represent state-of-the-art for direct digital synthesis tuners. The AD9852 with specified applications including "agile local oscillator frequency synthesis in amateur radio tuners", features an on-chip 12-bit DAC and high speed comparator, a 300MHz internal clock rate and capable of generating a highly stable frequency/phase programmable output sinewave at up to 130MHz. It will be available in a 48-pin TQFP surface mount package. An AD9835, due later this year, is less advanced (10-bit DAC, 50MHz clock rate) but is small for a 16-pin package with serial control, and supports 12-bit phase modulation as well as two frequency registers.

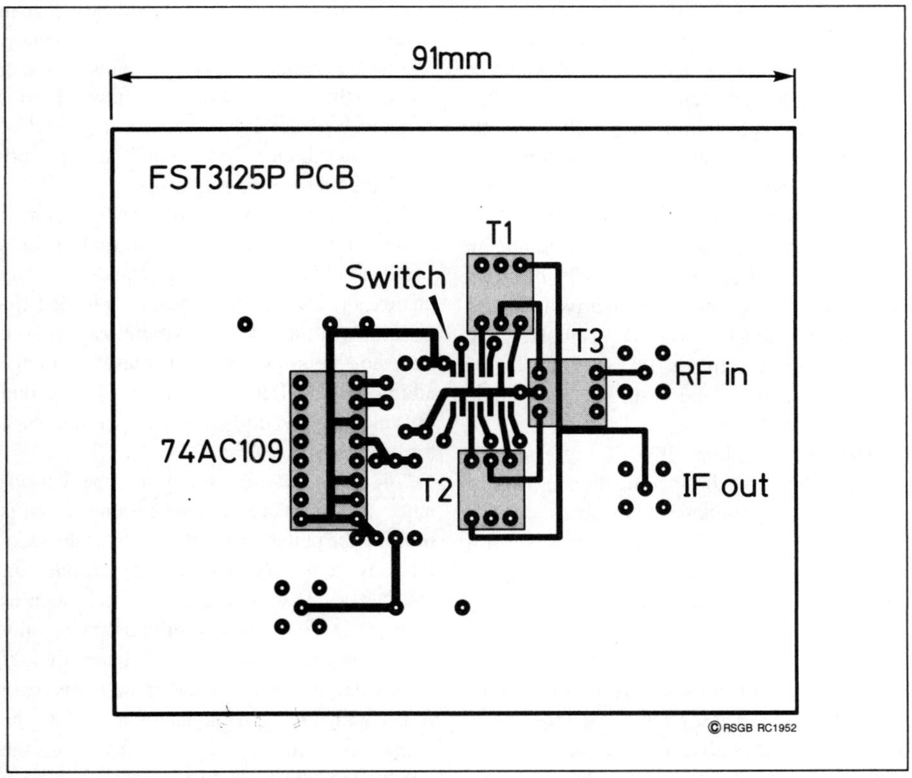

Fig 80: Track side of the PCB for the FST3125 H-mode mixer.

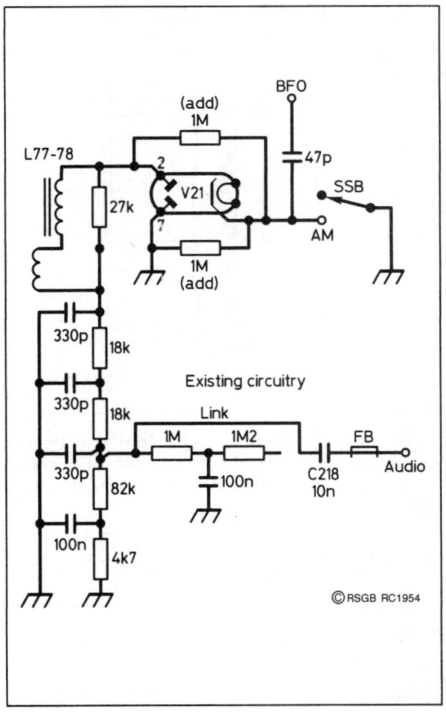

Fig 9: Simple modification to the Racal RA17W receiver converts the noise limiter into a product detector to improve performance on weak SSB signals.

FIGHTING COAX POLLUTION

A SHORT NOTE by Roger Gould-King, ZS6QL (*Radio-ZS*, February 1998) emphasises the importance of regular, preferably monthly, inspection of antennas and feeders to check the integrity of connections and to ensure that moisture is not entering any lengths of coaxial cable. Air-spaced and semi-air-spaced cables can quickly be ruined along long lengths if any moisture can enter through the termination.

ZS6QL writes: "Increasing levels of pollution covers corrosion of solder joints and copper persuaded me to find a better coax to antenna connection than the usual perspex strip and joints protected by rubber sealant or self amalgamating tape. **Fig 10** shows a simple solution which is mechanically very strong and provides excellent waterproofing and corrosion protection. The coaxial cable is inserted into the base of a plastic container, pigtails soldered, and the entire assembly encased in epoxy casting resin. The result can be left transparent so that one can inspect the connections, or coloured, using a pigment obtainable from the epoxy supplier.

"When the resin hardens, it shrinks and crimps the coax, providing a strong bond. The system can be used for dipoles, inverted vees, delta loops and the like, as well as connecting lengths of coax where mechanically strong connections are needed.

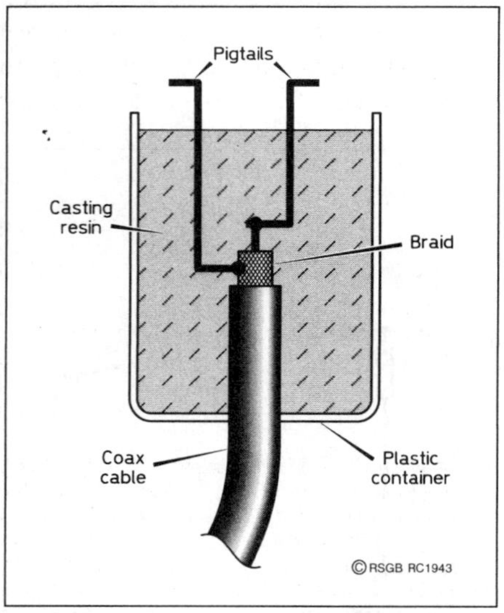

Fig 10: Coaxial cable/antenna joint protected against moisture and corrosion by the use of a plastic container and casting resin, as recommended by ZS6QL.

REPLACING THE SL6270

THE DISCONTINUANCE of the Plessey SL range of linear ICs is regretted by many amateurs. Quite widely used in home built equipment has been the 6270 gain-controlled microphone pre-amplifier (audio-derived AGC). John R Hey (presumably G3TDZ) in the 'circuit ideas' feature of *Electronics World* (August 1998, p660) needed a replacement for an audio

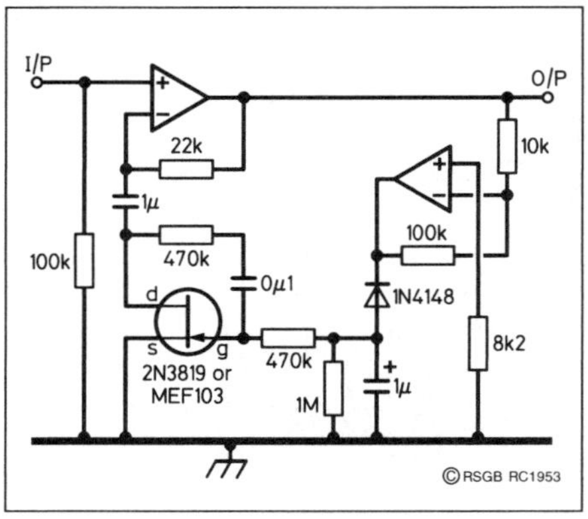

Fig 11: Audio frequency AGC system, to replace the discontinued Plessey SL6270.

AGC system, but found other suitable ICs somewhat expensive. He devised the circuit shown in **Fig 11** which he claims: "Works well and is cheap." Gain in the non-inverting op-amp is set by the 22kΩ feedback resistor and the return via the 1μF capacitor and FET, with local feedback preventing waveform distortion. Overall feedback is supplied by the other op-amp, with a gain of 10, with its output rectified and smoothed. Input signals in the range 4mV to 4V all emerge at a level of about 200mV.

CRYSTAL OSCILLATORS AND VXOS

MY INTRODUCTORY notes to the June *TT* (pp60-61) item on the Super VXO, in which two crystals are used in parallel to increase the frequency shift, attracted several long letters from Mike Hall, G3USC, who has had considerable professional experience as a supplier of quartz crystals. G3USC has previously contributed information on basic crystal oscillator technology (*TT*, October 1991, p29) where he underlined the problems that amateurs often experience due to failing to specify correctly the crystals they order, with the result that the crystals do not work as intended. He emphasised: "It is important that amateurs should recognise that crystals behave differently in different circuit arrangements, with different load impedances, with different drive levels, or in series-resonant or overtone oscillators. **Fig 12**, reproduced from the 1991 item, shows the family grouping of the basic forms of single transistor oscillators.

Then, as in the June 1998 *TT*, I referred to a booklet: 'Guide to the specification and use of quartz oscillator crystals' drawn up in the 1960s by 'a committee of experts' representing a number of crystal manufacturers and suppliers. This was published under the auspices of the Electronic Engineering Association. G3USC has managed to borrow a copy of this now ancient publication and is not impressed! He writes: "I have to say that if that booklet was ever used as a crystal bible, it is clear to me why there have been so many misconceptions about how crystal oscillators actually work. Really, it explains little in meaningful terms. . . "

He adds: "Please have another look at what you reproduced from section 6.3. It states that the choice of where to place the trimming component depends upon 'the nature of the oscillator circuit used', and there follows the formula reproduced in the June issue. I can say only that for the oscillator configurations in use today, that statement is misleading, and in most circumstances untrue, which both measurement or computer simulation and the formula when interpreted properly will show."

G3USC is critical of both the early (1947) circuit of (Fig 11, June) and the VK1PK super-VXO circuit (Fig 12, June) although not of the basic technique of using two crystals to increase possible frequency swing which he intends to investigate. While Fig 11 was intended only to illustrate the early history of the VXO as investigated by Post Office Research in the 1940s, the comments on the VK1PK circuit seem of significance since G3USC believes that it may not perform reliably from specimen to specimen, as would be required for a commercial product. He writes:

"(1) For 'modern', plated crystals (we stopped making anything else 40 years ago), the crystal drive level is far in excess of permissible figures for good stability, and it is a wonder that it/they have not shattered. Remedy: feed from a regulated 8V supply (a Zener diode will be better than nothing), and connect a 68k resistor from the transistor base to ground. It seems likely, anyway, that the 12V rail shown is the raw supply, which it usually is for amateur applications.

"(2) For 7MHz, change the values of Cbe and Cce from 120pF to 220pF and vary them in inverse proportion to the working frequency. More capacitance here should allow a greater swing in the negative direction below the series resonance of the crystal, before the onset of spurious oscillation. There should still be plenty of loop gain to spare as a result.

"(3) The chances are that the inductor in series with the crystal, if of normal Q for a small moulded type, 40 say, stands a strong chance of causing non-crystal-controlled oscillation via the crystal's parallel capacitance Co, the likelihood being increased with the additional parallel capacitance of a second crystal. Remedy: de-Q it, with a parallel resistor of the order of 15kΩ.

"(4) Apart from the small motional capacitance of AT-cut crystals, the parameter that is limiting the pullability of the circuit is their Co. Remedy: tune it out with a parallel inductor. On 7MHz this will be about 100μH in series with a 1kΩ resistor in an oscillator having only a single crystal. Halve the inductance value if using two parallel crystals. Being effectively in series with the 22μH inductor as far as the transistor is concerned, the purpose of the resistor is again to de-Q the new inductor to prevent the possibility of spurious oscillation

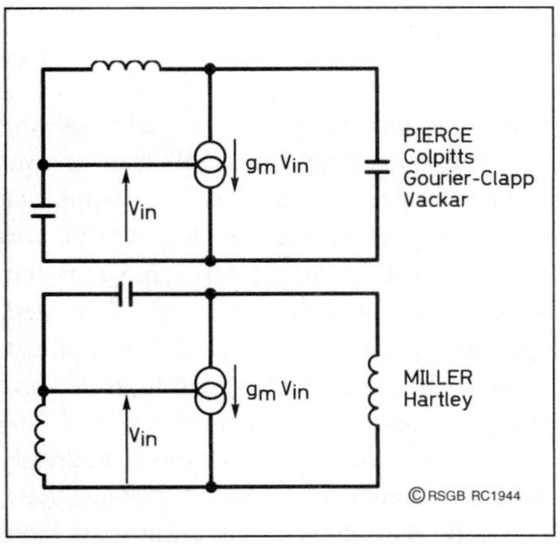

Fig 12: The two basic single active-device oscillator circuits, with some related derivatives.

at a non-crystal-controlled frequency.

"(5) Contrary to the remarks in the June *TT* item, high-Q inductors are a positive disadvantage in VXOs. We want the crystal to control the frequency, not the inductor. Any oscillator will function at a frequency where its loop gain is greater than unity and where the loop phase shift is zero. If there is more than one frequency where the phase shift is zero it will tend to favour the one having the greatest loop gain eventually. 'Eventually' because the circuit Q will probably be lower at another frequency where the conditions for oscillation are satisfied, causing a burst of oscillation at that frequency first, or even possibly both at the same time later.

"Small crystals in their fundamental mode have a permissible motional resistance (not ESR) of around 25Ω, so we should not use inductors having a lower figure unless steps are taken to roll off the gain at the inevitable spurious frequency. A 22 μH inductor having a Q of 40 will have an effective resistance of 24Ω at 7MHz! Due to a striction-like effect commonly called 'sleeping sickness' or 'drive-level dependency', crystals may exhibit an even higher resistance before oscillation has actually started.

"(6) In response to VK1PHK's query about the affect of increasing the tuning capacitor to 100pF, I predict that this would result in only a marginal increase in the pullability. The swing of the circuit of Fig 12 (June) as it stands, at least with a single crystal is pretty non-linear and most of it will occur at small values of the tuning capacitor."

G3USC adds that the concept of using two parallel crystals in an oscillator is new to him: "Amongst other things, it makes me wonder what its spectral output looks like. I must give it a try."

It is perhaps worth noting that his comments on the VK1PK design is based on the considerations that would be required to produce a commercially reliable and reproducible VXO. The design presented in *Amateur Radio* by VK1PK was clearly a test circuit used to investigate the parallel crystal technique, the results of which were given in the June *TT*.

The earlier 1947 circuit (June, Fig 11) design was presented for its historic interest rather than intended to be copied. In this connection, I have since had a chance of reading the original informative and well-written paper by Stanesby and Fryer ('Variable-Frequency Crystal Oscillators' *Proc IEE*, 94, IIIA, pp368-378). To the best of my knowledge, this remains the first description in a journal of using inductance to increase significantly the possible frequency shift of a crystal controlled oscillator. However, a footnote to their paper states: "Since writing this paper the authors have become aware of British Patent Specification No 537,167 which relates to the use of inductance with a crystal in a crystal-controlled oscillation, for increasing the range of frequency

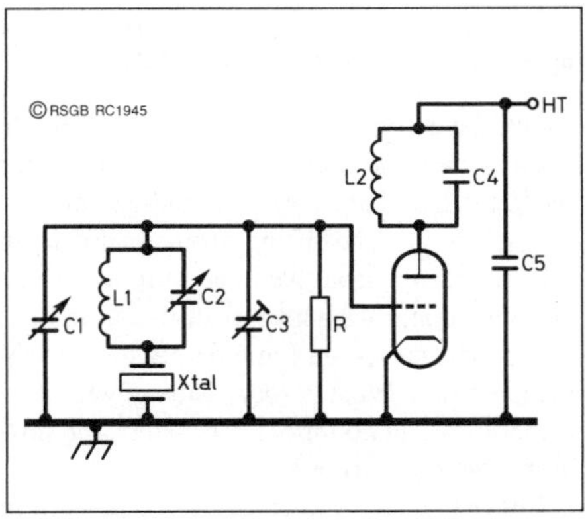

Fig 13: Koerner's original VXO circuit, as described in British Patent Specification No 537,167 of 1941.

adjustment and achieving a linear relationship between the variable element and the percentage change in frequency."

Following this up, I found that British Patent 537,167 (application date 30 April 1940, accepted 11 June 1941, with a US Convention Date of 3 May 1939) was taken out by STC as assignees of Lawrence Frederick Koerner, who presumably was an American employee of ITT which had a controlling interest in STC. This clearly preceded by several years the work at Dollis Hill. Koerner's VXO was developed to provide a frequency monitor for American FM broadcast transmitters which had then just started on about 50MHz. **Fig 13** based on the Patent illustration shows a basic VXO arrangement with the inductance-capacitance in series with the crystal, the usual practice with modern VXOs. It was claimed that the frequency could be varied by about ±1%, about 50 times the deviation possible with capacitance alone, much the same as that claimed later by Stanesby and Fryer.

Eric Elsley, G3YUG, raises the question of the low power Mizuho transceivers (see 'A Handful of Watts-QRP on 20m' by Rev George Dobbs, G3RJV, *RadCom*, February 1991, pp46-47, which presents a user view of the MX-14S). This model provided two 50kHz shifts on 14MHz with VXO control, which is on a par with the 27kHz on 7MHz measured by VK1PR on his twin crystal Super-VXO. But G3YUQ also notes that there are Mizuho versions for 3.5MHz and 7MHz with 25kHz VXO tuning ranges. To shift a 3.5MHz crystal by 25kHz seems extraordinary, although it would be possible with ceramic resonators. Alternatively, it would be possible with a heterodyne oscillator using two high-frequency crystals, one of which is pulled by 25kHz. G3YUG wonders whether any reader knows how Mizuho achieved their wide tuning range VXOs.

By coincidence, within a few days of receiving the letter from G3YUG, the Summer 1998 issue of *Sprat* Nr 95 arrived. This included a short article 'Wide Range Crystal Control on 80m', by Walter Farrar, G3ESP, in which he provides a basic outline of a KISS form of heterodyne VXO using two crystals spaced by 3582kHz, (in his example 17642 and 14068kHz) with both crystals oscillated and mixed by a single transistor: **Fig 14**.

G3ESP writes: "The crystals were on hand and gave a basic difference of 3582kHz. As wired up, I can tune from 3509 to 3587kHz [78kHz tuning range]. The output was rock steady and drew 7.5mA from a 9 volt supply. The transistor oscillates on two frequencies simultaneously, and the difference is taken from the emitter. This is only the bare bones of the system, which will need filtering to remove the undesirable mixtures [spurii] produced. Other crystals would give other outputs. For example, 8.8633 and 8.9985MHz (together available for about a Fiver) should prove ideal

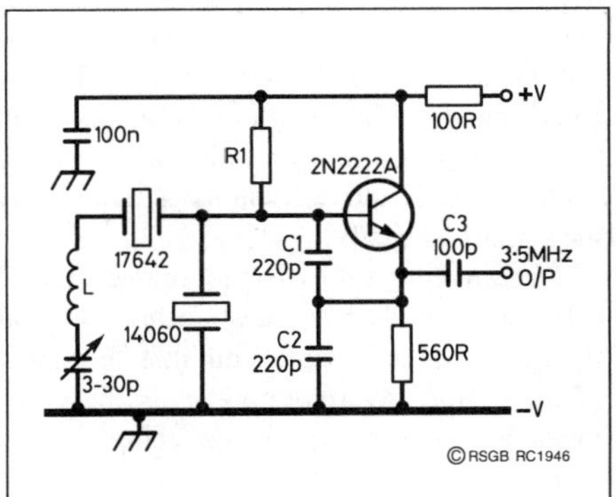

Fig 14: The experimental wide-range 3.5MHz heterodyne VXO described by G3ESP in *Sprat*. This is only the bare bones of the system, which needs filtering to remove the unwanted mixer products. V+ nominal 9V battery. C1, 2, 3 are polystyrene types. L is 44 turns of 0.4mm enamelled copper wire on a T68-6 core (about 9.1μH, but a 10μH RF choke would probably be suitable).

for the new LF band 135.7 to 137.8kHz. Any HF transistor should work if a 2N2222 is not available." The *Sprat* diagram does not show the value of R1, but presumably this should be 10kΩ or more.

I suspect that G3USC might not fully endorse this approach. Like all heterodyne VFOs, it would be essential to provide effective filtering-out of the unwanted mixer products, but it looks an interesting experimental approach for portable QRP rigs. ♦

TECHNICAL TOPICS

BY PAT HAWKER, G3VA*

UNDERSTANDABLE TECHNOLOGY

THE AUGUST *TT* noted Robert Lucky's belief that, with so much of the fun taken out of home-building, the emphasis has shifted to using electronic appliances without concern for whether they work from electricity or "could contain gears and pulleys, as far as most people are concerned". How many of us now really understand the appliances and equipment we use?

Robert Lucky was concerned with computers (no longer "electronic computers"), but his remarks apply far wider. One suspects that for most of us the inner workings of a modern, top-of-the-range transceiver is indeed a 'black box' that somehow, mysteriously lets us make radio contacts. Even if bought carefully on the basis of independent, detailed performance measurements or on-air recommendations, most owners would be hard put even to attempt to sketch out their basic block diagrams, while complete circuit diagrams are now most unlikely to appear in an owner's manual since home-servicing is considering all-but impossible without professional help from an authorised dealer holding the appropriate service manual.

As for the latest high-bit-rate digital techniques, how many of us, unless professional engineers, could claim other than the most rudimentary knowledge? But then, as Robert Lucky put it: "When a billion transistors sit on a single chip, how many people will care about a single circuit - no one will have the foggiest idea what is on it."

A recent editorial by David Sumner, K1ZZ (*QST*, June 1998, p9) 'New Technology and the Radio Amateur' points out that "For most of the century we [radio amateurs] have enjoyed a virtual monopoly on being able to communicate world-wide without an intermediary. . . however, in recent years, the cost of using intermediaries has dropped to the point where it is hardly a limiting factor, and no longer depends on the distance being bridged or the number of parties to the communication. . . As the new century approaches, telecommunications technology is advancing all around us at a dizzying pace - and the process of change has barely begun. What is the proper relationship between amateur radio and this new technology?

"Many amateurs aren't all that interested in using new technology in amateur radio. Dyed-in-the-wool Luddites are rare, more common are those who simply want to use their free time to escape from the technology at work, or who revere other ways of doing things, because of the values they represent. For others, the opportunity for hands-on experience with current technology is what keeps them interested. . . In the middle is the view that new technology

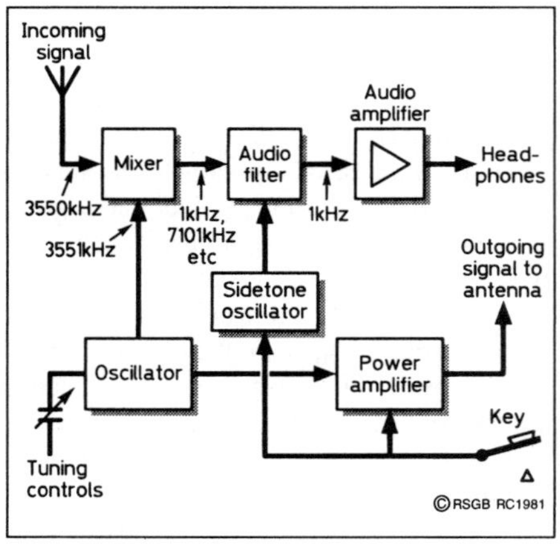

Fig 1: Block diagram of the ALIVO 3.5MHz battery-operated transceiver, designed by ZL2AMJ.

should be actively embraced whenever it can address and solve existing problems. . . By itself, the fact that every other radiocommunication service is shifting from analogue to digital doesn't mean we have to follow suit - but what could we accomplish if we did? . . . The opportunities are limited only by our collective vision. What is your vision of what new technology can do for amateur radio, and how do we get there from here?"

I have to confess to being personally still in the 'Luddite Analogue camp' other than in the sense that Morse is a binary non-return-to-zero digital code. My fear is that too rapid an advance into the high-bit-rate digital modes by non-professional amateurs will take us ever further along the road as 'black box' appliance operators. Remote from the days when it was expected that 'experimental' radio amateurs should understand the technology they used, and be capable of keeping that equipment, whether home or factory built, in service. 'Progress' does not always imply success. One cannot help observing the doubts that already surround Digital Audio Broadcasting (DAB), currently being hyped as the future of radio broadcasting. Unlike the set-top digital TV boxes, which are being heavily subsidised, these are burdened by very high receiver costs and are already being dubbed 'dead and buried' by some cynics.

It would be blinkered to suggest that amateur radio should return to the days of readily understandable technology and home construction, or even home servicing, but it is important that the flag of technical interest and home construction should be kept flying high and not lowered yet to half-mast.

What is described as the 'ALIVO Fun Transceiver' is reproduced in VK3AUI's 'Technical Abstracts' column in *Amateur Radio* (August 1998). The ALIVO is a simple 3.5MHz CW transceiver, originally presented by Fred Johnson, ZL2AMJ, in *Break In* (April 1998). The easy to follow block diagram (**Fig 1**)

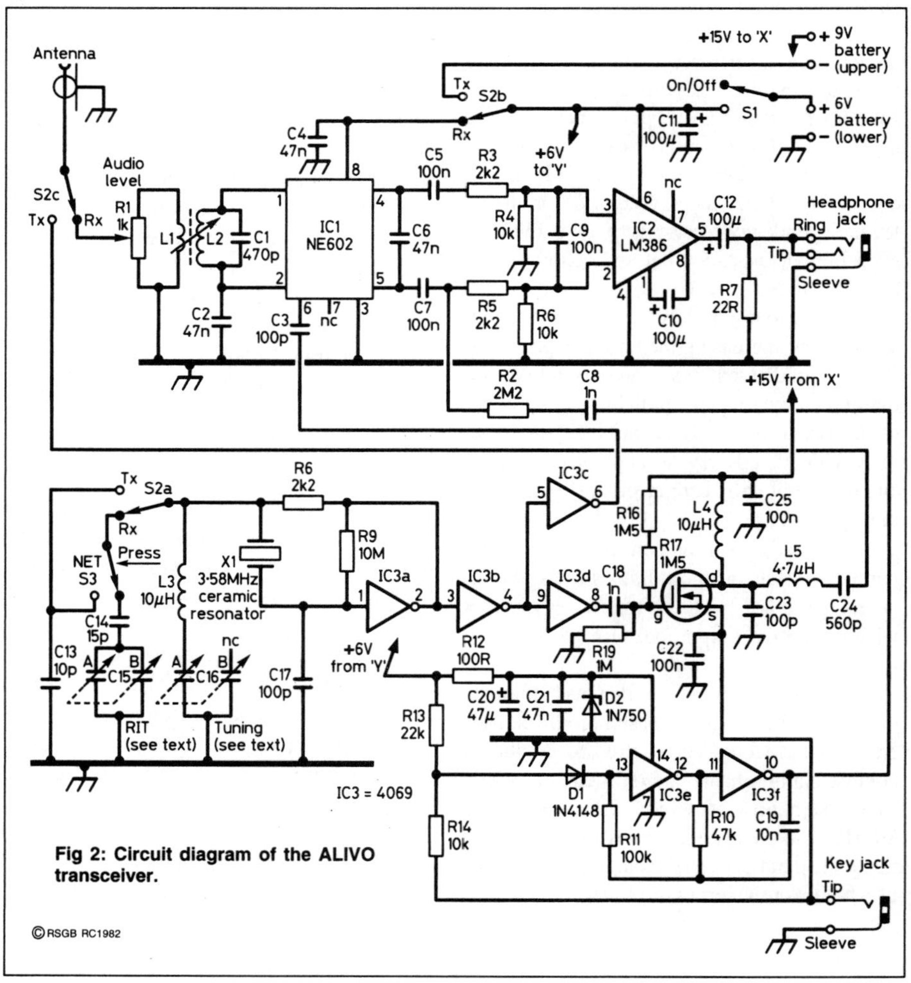

Fig 2: Circuit diagram of the ALIVO transceiver.

**37 Dovercourt Road, London SE22 8SS*

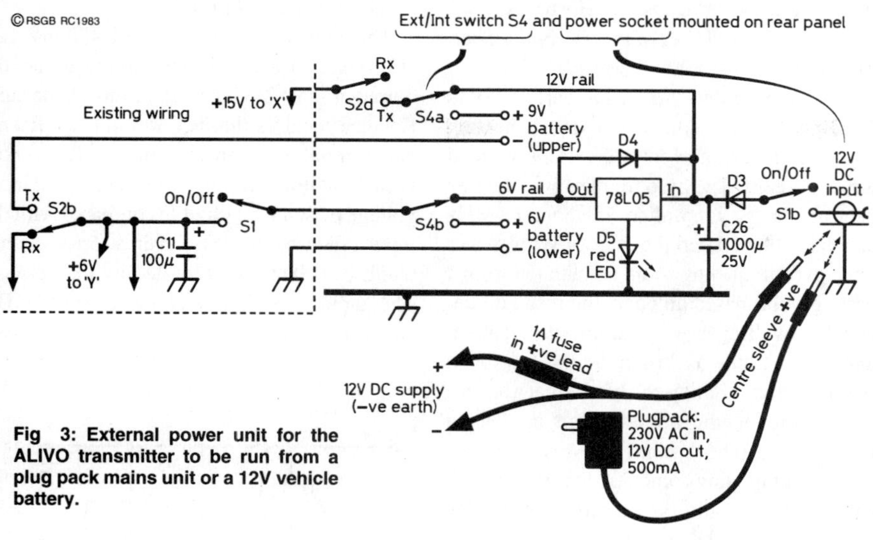

Fig 3: External power unit for the ALIVO transmitter to be run from a plug pack mains unit or a 12V vehicle battery.

shows that the ALIVO is based around a simple, direct-conversion receiver, with its oscillator used also to drive a low-cost VK10KM FET power amplifier, providing about 2-3 watts output from 9V plus 6V dry batteries: **Fig 2**. Later (*Break In*, May 1998), ZL2AMJ showed how the rig could be adapted so that it could be powered from a 12V car battery or AC mains plug-pack: **Fig 3**.

Although simple, this approach provides a flexible low-power transceiver with a ceramic-resonator VXO providing stable tuning over a range of some 50kHz, plus RIT and sidetone oscillator. It is all built around just three integrated circuits and the VN10KM power amplifier. The design could also form the basis of a higher performance rig by adding additional AF amplification and improved low-pass audio filtering to the receiver. It would also be advisable to add a low-pass filter at the output (suitable designs have been published many times in *TT*, for example most recently in the May 1998 *TT*, p 61, Fig 12).

The ALIVO can be built dead-bug style on a piece of PC laminate. The variable capacitors can be the plastic film variables used in transistor radios. The only inductor to be wound is L1/L2, and this uses a 6mm slug tuned former with L2 40 turns of 26 SWG (0.018in) enamelled copper wire. L1 is six turns of similar wire, wound over the earthy end of L2. The other inductors are the small moulded RF chokes available from most component firms. The output FET requires a small heat sink.

Originally the rig was intended to be run off batteries. A 6V battery runs most of the unit with, on transmit, a 9V battery connected in series to power the FET amplifier. For alternative power sources, as shown in Fig 3, a 12V DC supply (mains or battery) can be used with a 5V IC regulator plus LED used to provide the six volt supply, with the 12V (13.8V) supply used directly for the 15V rail.

LONG DELAY ECHOES AND THE LUXEMBOURG EFFECT

DESPITE SOME CRITICAL comments from readers, my belief remains that the 3.5MHz echoes delayed by about 210 milliseconds heard by Peter Martinez, G3PLX, on 25 October and 4 December 1997 (see *TT*, January, February and March, 1998) were due to his signals being reflected from the Van Allen belts in the magnetosphere, during short periods when the critical frequency was dropping below 3.5MHz and the electron density in the magnetosphere was enhanced as a result of minor magnetic storms. Similarly, the long delay echoes (>1s) heard in the 'thirties may quite likely also have been connected with magnetic storms, but are obscured by the ionospheric heating of many thousands of high power transmitters.

Some evidence in support of my views would seem to be found in a paper (URSI's *The Radio Science Bulletin*, No 285, June 1998, pp9-20) by a team of about a dozen scientists and engineers who are concerned with the Imager for Magnetopause-to-Aurora Global Exploration (IMAGE) satellite (due to be launched on 1 January 2000). The payload will include a radio plasma imager (RPI), comprising a swept-frequency digital radio sounder covering the frequency range from 3kHz to 3MHz, which will be capable of producing echo maps showing magnetospheric electron number density (N_e) structures.

It is pointed out in the introduction that while the lower portion of the ionosphere has been routinely investigated from the ground, using pulsed ionospheric radio sounders (ionosondes), "the shielding effect of the ionosphere has prevented the conduct of similar routine ground-based sounding of the magnetosphere". The ability of the magnetosphere to affect VLF radio waves in the form of 'whistlers' has been recognised since 1953, but it is interesting that the sounder will cover up to 3MHz, relatively close to the 3.5MHz frequency on which G3PLX heard his echoes. If I have understood this correctly, it would seem likely that for short periods the N_e of the magnetosphere might be sufficient to reflect vertical or near vertical 3.5MHz signals from a terrestrial transmitter, provided the ionospheric critical frequency is low enough to provide a temporary window.

A different view is expressed by Alan Goodacre, VE3HX. He writes: "The echoes are probably signals which travel along magnetic field lines to the southern hemisphere and are then reflected back again to the northern hemisphere. Such echoes were observed from the Canadian 'top-side' ionospheric sounding satellites at frequencies up to about 7MHz. Echoes interpreted to be of this type were observed and tape recorded on the 3.5MHz band several years ago in the vicinity of Seattle, Washington."

VE3HX enclosed a table of two-way travel times, calculated assuming that: (a) the earth's external magnetic field is a pure dipole; (b) the waves travel at the speed of light; and (c) the waves reflect off the earth's surface in the opposite hemisphere. The table shows two-way travel times (ie delay times) varying from 0.050 seconds for a geomagnetic latitude of 26° (geomagnetic latitude) to 0.571 at 65°. Interestingly, the delay time is 0.208s at 51°. I am not sure what the geomagnetic latitude of Cumbria (G3PLX's location) is, but the geographic latitude is around 54°. Nevertheless, I find it difficult to believe that the ground reflected back signals would be readily audible on 3.5MHz.

VE3HX adds: "In reality the picture is somewhat more complex than indicated above, but to a first approximation the earth's magnetic field is like that of a dipole with its axis tilted about 30° away from the geographic north pole towards Ottawa. So, although the geographic latitude of Ottawa (VE3HX's QTH) is about 45° N, the approximate geomagnetic latitude is about 55°. In Britain, the geomagnetic latitude would be approximately the same as the geographic latitude, so one would expect to observe delay times of the order of 200 milliseconds in southern England with delay times reaching double this in Scotland."

VE3HX makes the valuable suggestions that such echoes might be better referred to as "Medium Delay Echoes" (MDEs), leaving the term Long Delay Echoes for those with delays of 1s or more. The LDEs still require some other explanation.

Research into magnetic storms has suffered a setback by the announced failure of the SOHO satellite.

Prof Martin Harrison, G3USF, notes that 'How The Ionosphere Really Works' by Eric P Nichols, KL7AJ, (*QEX*, Mar/Apr 1998, p37) suggests that the mystery of LDEs has been solved. "No, these are not reflections from the

Van Allen belts or undiscovered planets in our solar system. Nor are they long-running hoaxes. They are purely ionospheric phenomena that can be consistently reproduced in plasma chambers, (with shorter time scales, of course). Yes, 20 second delays are possible in the Earth's ionosphere. These are elegant examples of *mode conversions*."

I seem to recall that this hypothesis has been around for some years, but personally I remain to be convinced that mode conversion is the explanation of the LDEs reported in the 1920-30 era. Though KL7AJ's article does provide useful explanations of some unusual propagation behaviour.

Charles Bennett, a long-retired chartered physicist with solar-terrestrial relations as a major interest over the past 20 years, adds some useful comments on the Luxembourg Effect (*TT*, July 1998). He writes:

"As I was 'radio-active' in 1933, a teenager cutting his teeth on medium-wave DXing, may I make some comments on the July item:

"(1) The first reports came from listeners in Eindhoven, who were tuned to Beromunster, but noted the recently built Radio Luxembourg programmes as background. Beromunster was the desired station, Luxembourg the interfering station. Hence the name given to the phenomenon.

"(2) Co-location of the two transmitters and the receiving station on roughly the same Great Circle is not a pre-requisite, it simply makes it more likely that the phenomenon will be detected by a particular listener.

"(3) The essential factor is that the modulating (ie 'interfering') transmitter shall be delivering a sufficiently high power signal into that part of the atmosphere which is the 'reflection' region for the signals from the desired station. Hence the reason why the phenomenon was detected only after the high-power Luxembourg transmitter came into action.

"I still interest myself in tracking present day examples of the L-effect. The BBC World Service transmitter at Orfordness is the culprit. On suitable winter afternoons, I have noted the L-effect on as many as 16 stations, with no significant colinearity! This suggests that the BBC engineers are not getting as good a ground wave signal into the Low Countries as they hoped, since obviously a large amount of energy is escaping into the upper atmosphere. It is rather amusing that Radio Luxembourg is now a 'victim', with the World Service frequently audible as a background on their 1440kHz signals!"

Dr Brian Austin, G0GSF, noted the July *TT* comment that the theory of the Luxembourg Effect was first described in 1937 by Professor V A Bailey of the University of Sydney. He writes: "As is so often the case, I came across Bailey's name for the first time just a day or two before *RadCom* arrived. . . [when] I received a copy of a paper written by my good friend Professor Russell Burns 'The Development of Methods of Detecting Hostile Aircraft at Night 1935-1941' (*Trans Newcomen Society*, Vol 69, No 1, 1997-98, pp1-21)."

This shows, *inter alia*, that as early as 1938 Prof Bailey suggested that ionospheric heating by power radio signals could create an artificial aurora borealis. The gyro frequency of the free electrons in the ionosphere is about 1.4MHz and Bailey postulated that if 500kW of power at the gyro frequency could be radiated from a large, ground-based antenna, the resultant celestial glow discharge would lie below 100km and be 50 times as bright as the sky on a moonless night. Bailey calculated that a square antenna array having a side of 2km, mounted a quarter-wave (about 50m) above the earth's surface, comprising some 800 horizontal half-wave dipoles would be needed. The dipoles would be arranged to produce circularly polarised radiation; the beam width 6°. Apparently this proposal was passed to TRE for detailed examination, but the Superintend (Rowe) wrote to Prof E V Appleton on 9 April 1941 that it would be necessary to radiate 100MW [presumably ERP] of power at the gyro frequency and thought that the development of the proposal would not be worth the enormous effort involved. In effect the problem of locating hostile night bombers was largely solved by both the Allies and the Germans by the use of co-ordinated airborne and ground-based radars.

A letter in *Nature* (16 August 1998, pp552-553) notes that "The temporal behaviour of solar activity (as manifested in sunspots) has long been debated. The 11-year periodicity in the total number of sunspots is well established observationally, as is a periodicity of 152-158 days in the occurrence of high-energy solar flares that was seen during cycle 21. . . Here we report a time-frequency analysis, using the wavelet technique, of sunspot areas between 1874 and 1993, which reveals a 158-day periodicity coincident with that of energetic solar flares. . .strongest in cycle 19, which was the most intense cycle of the century and disappearing after cycle 21. The analysis shows that the 158-day periodicity in both high-energy solar flares and sunspots is related to a periodic emergence of magnetic flux, which only appears near the maxima of some solar cycles.

Sunspot cycle 23 continues to make good progress, with the daily sunspot number and the solar flux both exceeding 150 on 9 August 1998, compared with 31 and 78 on 10 August 1997."

LOW NOISE, HIGH-Z-INPUT AMPLIFIER

PETER GOODSON of Bracknell (*Electronics World*, June 1998, p478) describes an amplifier developed for use with a guitar pickup, but which should prove suitable for other demanding applications where small voltages need to be amplified with very low noise and very high input impedance: **Fig 4**.

The input stage consists of a U430 matched JFET pair in a differential arrangement, the output op-amp having feedback to set the gain. Balance requires that R3, R4 and R5, R6 are equal. Input impedance is equal to R1, which, with input bias current at 2pA, causes an input voltage of 20uV, negligible compared with the input offset of the JFET pair at 5mV. Bandwidth is 1MHz and is limited by the op-amp. The input capacitor may be left out if DC coupling is needed.

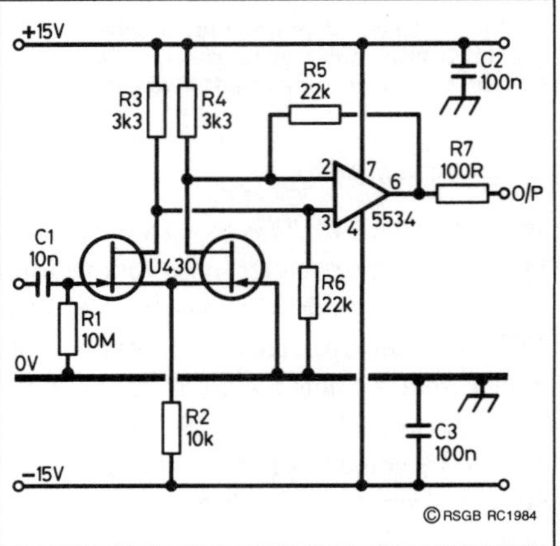

Fig 4: Simple, but effective amplifier, having very high input impedance and 2.5nV/√Hz noise voltage.

BATTERY MISCELLANY

LITHIUM BATTERIES have the highest energy density of all rechargeable batteries, and are favoured in applications where low weight or small volume are desired - for example, laptop computers, cellular telephones and electric vehicles. However, as pointed out by a team of MIT scientists (*Nature*, 16 April, 1998, p604-695), one of the limitations of present commercial lithium batteries is the high cost of the $LiCoO_2$ cathode material. Searches are continuing for lower-cost replacement materials. The MIT believes that such searches should be guided by first-principle calculations, and have predicted and verified experimentally that for $Li(Co,Al)O_2$ aluminium substitution raises the cell voltage while decreasing both the density of the material and its cost. However, it is clear that considerably more experimental work will be needed before this may contribute to reducing the cost of rechargeable lithium batteries.

There are still many applications where it is advantageous or even essential to use equipment powered from zinc-carbon or alkaline dry cells, despite the unfavourable economics compared with operation from mains supplies. It is by now well-established that provided suitable precautions are taken, it is possible to recharge several times modern 'throw-away' batteries. Care must always be taken not to overcharge or overheat sealed primary cells, which can explode or leak. A useful article

'Recharging the Unchargeable' by Michael Slifkin and Yaakov Levy (*Electronics World*, April 1998, p320-325) shows that batteries respond well to recharging not only by the well-established 'dirty-DC' or pulsed method (described several times in *TT* and *ART*), but also to relatively 'fast charging' from conventional chargers. But an editorial note warns: "Use at least eye and skin protection while experimenting, and don't risk building a primary cell charger into expensive equipment."

The authors believe that "recharging of zinc carbon and alkaline batteries is perfectly feasible using quite simple charging methods, without resorting to pulse-charge reversal. We consider that the dangers of these techniques have been greatly exaggerated. While the extension of the battery life is less than claimed in the advertisements we have seen for commercial ['smart'] chargers of this type, it is nevertheless useful." Overall, they estimate that you can only increase the life of the best of batteries by five times under optimal conditions, with three times the life more common. While under controlled conditions it is possible to recharge a cell up to about 40 times, each cycle gives successively less capacity.

A useful and still applicable guide to the charging of disposable batteries was given in *TT*, June 1990, p33.

A different approach to reducing the total ownership cost of battery-powered radio broadcast receivers is advocated by Duncan Grant of the University of Bristol (*Electronics Letters*, 11 June 1998 and UK patent application GB 2314474A, filed 21 June 1996 and published 24 December 1997). He is concerned primarily with remote regions, where mains power is not available but where, for example, radio broadcasting provides the main means of distributing information, including public health and safety messages, etc. for communities where the cost of batteries is often prohibitive.

"Very often the cost of batteries is the dominant element in total ownership costs... price of a medium quality radio is about £20, but if used for 10 hours a day, the yearly battery cost is over £60.

He believes the answer lies in radios that consume drastically less power. Whereas a typical small receiver consumes about 120mW, "this can easily be reduced to about 5mW, with any small increase in purchase cost likely to be a negligible proportion of the total ownership cost."

His patent outlines a power-saving arrangement for a receiver, using a class D output amplifier and a signal processing section powered from the amplifier: **Fig 5** shows the basic arrangement. While such a receiver would inevitably have poor dynamic range, it is a useful reminder that there must be some applications in amateur radio where Class D audio amplifiers could prove an economical and effective approach.

A low consumption receiver would not only be more economical on batteries, but would also be easier to run from a small solar panel or from a clockwork generator.

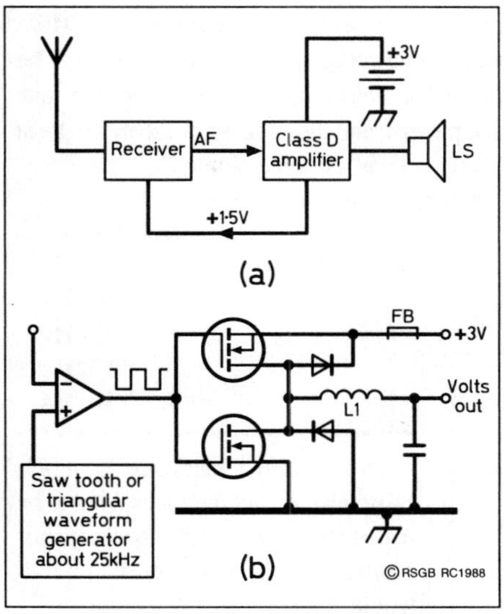

Fig 5: (a) Block diagram of the broadcast receiver consuming only about 5mW battery power instead of the customary 120mW or so, as patented by Duncan Grant (Bristol University). (b) Outline of the Class D audio amplifier.

BALANCED OUTPUT ASTU

BOB BASTOW, G3BAC, contributes a simple way of obtaining a balanced output ASTU without the complication of plug-in coils.

This is achieved by switching a parallel inductance. He writes: "In **Fig 6**, L1 is scramble-wound over the centre of L2 to avoid reflecting the unbalanced input into it. L3 is tapped to cover the HF bands above 3.5MHz. Meters can be switched between reading SWR and current in the feeders."

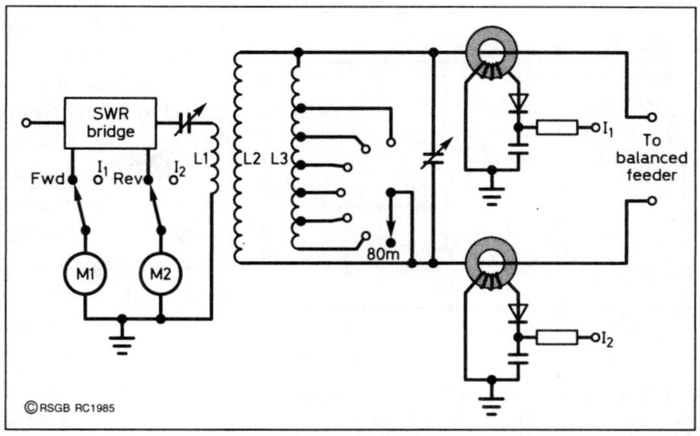

Fig 6: Balanced output ASTU by G3BAC, avoids the need for plug-in coils by switching a parallel inductance.

MORE ON CRYSTAL OSCILLATORS & VXOS

THE SEPTEMBER 1998 *TT* item 'Crystal Oscillators and VXOs' included an experimental wide-range 3.5 MHz heterodyne VXO. Further information on this simple approach to frequency synthesis appears in an article 'One Transistor Frequency Synthesiser' by Peter Parker, VK1PK (*Amateur Radio*, August 1998, pp8-9). This credits H Popov, LZ1BB (*The QRP Digest* Nr 6, 15 November 1997) as the original source. VK1PK wonders whether the frequency range could be further increased by adopting the parallel-crystal 'super-VFO' arrangement described in the June 1998 *TT*.

Fig 7 shows VK1PK's test circuit. VK1PK admits that he was sceptical whether LZ1BB's idea would work. He initially used available crystal frequencies close to those suggested by LZ1BB: X1 15,810kHz,; X2 12,250kHz; L1 4,7μH: and L2 14μH. Note that the higher frequency crystal is used as the VXO, to maximise frequency shift. VK1PK made no attempt to optimise the value of L1 and used an HC25-style crystal for X1. He achieved a respectable 22kHz frequency shift for the 3.5MHz output.

He writes: "After peaking the 200pF variable capacitor for maximum output on 3.5MHz, the prototype circuit covered 3.547 to 3.569MHz. . . The image output (sum rather than difference of the two crystal frequencies) turned out to be 28.039 to 28.061MHz - allowing covering of [CW] frequencies on two amateur bands simply by changing L2.

"My next experiment was an attempt to obtain increased frequency shift by using two crystals in parallel for X1, as in the Super-VXO (*TT*, June 1998, pp60-61): X1a, X1b 10.240MHz; X2 18.312MHz; L1 23μH; L2 1μH. This was the most successful circuit tried, with reliable coverage from 28.240 to 28.560MHz achieved. Coverage below 28.480MHz was possible, but the oscillator was then sluggish in starting. 10.240MHz crystals were used in the VXO section, because most constructors are likely to have some of these buried inside ex-commercial VHF equipment.

"Since twin-crystal VXO circuits are known to be less stable than with a single-crystal, stability measurements were made to assess whether the circuit would be acceptable for SSB. The oscillator was left running for a little under two hours. Drift was between 100 and 200Hz. This is an excellent result, probably better than many older commercially-built transceivers. The oscillator lends itself to use in a 10m direct-conversion receiver or double-sideband transmitter. However, care must be

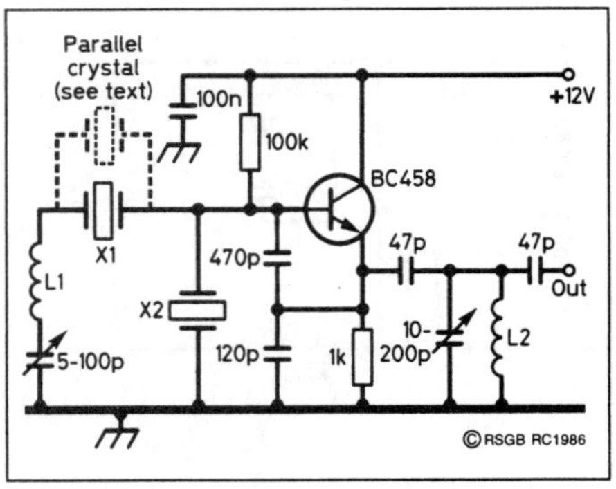

Fig 7: Test circuit used by VK1PK in investigating the one-transistor frequency synthesiser (Heterodyne VXO). See text for L1 and L2 values and for various crystal configurations.

taken to ensure suppression of spurious outputs, of which 30.72MHz (ie 3 x 10.240MHz) is likely to be the most troublesome.

"An experimental 80/40/20 metre exciter circuit, using a 3.58MHz ceramic resonator and a 10.7MHz crystal was tried, but did not work - the crystal would not oscillate while the ceramic resonator was connected. A similar circuit using a 3.58MHz crystal also failed to work. It seems that this technique is most reliable when the two crystals used are within about a 2:1 frequency range."

VK1PK concludes: "A novel one transistor circuit for generating a range of frequencies from a pair of crystals has been described. It is anticipated that the technique will find ready acceptance on the lower HF bands (where it is difficult to obtain large swings from conventional VXOs) and for equipment on ten and twelve metres, where fundamental crystals are difficult to come by but frequency agility is still required."

My own feeling is that as for all heterodyne-type VFOs rather more care must be taken to suppress unwanted frequency products than suggested by VK1PK and that this will be especially true where the sum rather than the difference frequency is used. Nevertheless, it does seem that this very simple form of frequency synthesis should prove a useful technique for low-power or portable 3.5/7MHz receivers/transmitters/transceivers.

SENSITIVE FIELD STRENGTH METER

MANY FIELD STRENGTH meters are passive devices, requiring relatively high levels of RF to provide a useful indication. Adding a DC amplifier not only increases sensitivity, but also enables 1mA FSD meters to be used. Such devices find a number of applications, not only in adjustment of antennas but also sniffing for unwanted RF energy in interference investigations, probing RF generating equipment, such as oscillators, and for the final 'closing-in' phase in fox-hunting activities, etc.

The untuned meter shown in **Fig 8**, described by Drew Diamond, VK3XU (*Amateur Radio*, August 1998, pp6-7) is claimed that as an E-field detector with a simple pick-up rod (fitted with a knob to avoid injuring your eyes), is capable of detecting at a considerable distance the signal radiated from a 5W transmitter from 1.8MHz to at least 148MHz. To detect (mainly) the magnetic (H) component, a suitable small pick-up loop can be used. A small loop with a BNC lead comprising two turns of insulated hook-up wire about 10mm diameter is useful for checking that an oscillator is working, or for checking that RF energy is not leaking into places where it should not be. A three-turn loop with a diameter of about 50mm is useful for detecting signals on transmission lines, radials, guy wires etc. A third loop made from 1.25mm gauge steel 'tie-wire' as a single-turn of about 350mm diameter provides extra sensitivity.

The indicator comprises a two-diode voltage-doubler detector with the positive DC voltage applied across a 100kΩ linear potentiometer, which sets the sensitivity. An LM386 is wired to form a DC or servo amplifier, establishing the no-signal quiescent DC output at pin 5 at half the supply voltage. The 1mA meter is connected in what is virtually a bridge arrangement. If it is required to be able to read the meter at a distance, while making adjustments, it may be advisable to use a meter with a relatively large, clear scale. The whole unit, including battery, can be built into a home-made or purchased aluminium box.

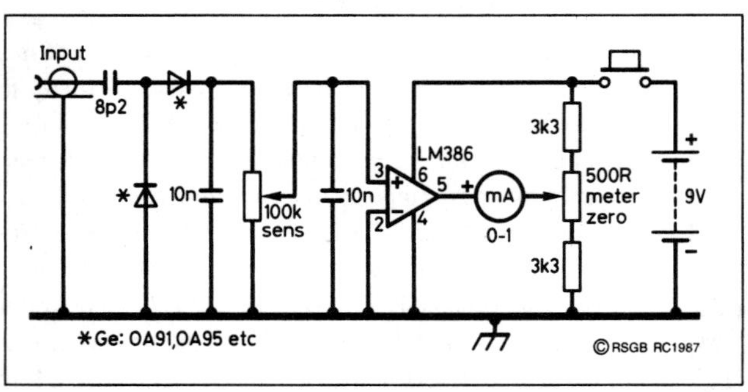

Fig 8: The sensitive field strength indicator described by VK3XU.

TWO-TONE TESTING WITH TELEPHONE TONE PADS

DR STEWART BRYANT, G3YSX, writes: "It is well-known that the ideal way to tune an SSB transceiver or linear amplifier is to use a two-tone signal. This allows the amplifier to operate at its designed maximum power without the excessive dissipation that would result from using a continuous audio signal. The two-tone signal also allows a rough assessment of amplifier linearity to be made, by looking at the amplitude envelope on an oscilloscope. Whilst a single tone generates a carrier with constant amplitude, two equal amplitude tones generate a pair of carriers that beat together to cause an envelope that varies between zero and twice amplitude of the tones. This in turn causes the power to vary between zero and four times the single amplitude power.

"Because amateur radio transceivers do not include an internal two tone generator, most amateurs tune up either by changing mode to a constant carrier mode such as FM, or by changing to CW and sending a string of dots, or by shouting 'wharlo' into the microphone. The first two approaches do not allow us to tune up and monitor the behaviour or the transceiver in a mode that simulates voice operation, whilst the latter approach does not yield a constant amplitude suitable for making measurements.

"*TT* has in the past noted that telephone dialling tone generators could form the basis of a two-tone generator, without the complexity of building a pair of oscillators and combining their outputs. Even if one built the two-tone generator it would then have to be wired into the transmitter, with all the issues of connector pin-out, microphone impedance, etc.

"During the past few years, a number of tone diallers have appeared on the market which are designed to be placed across the mouth piece of a pulse-dialling telephone, to allow it to access tone-based network services such as voice mail or computer integrated telephony applications, or to control remotely answer-phones etc. It occurred to me that these would make excellent two-tone generators for tune-up and testing. They are designed to generate a sound pressure at the microphone of about the same level as a person speaking. All one has to do is place the keypad over the transmitter microphone, key the transmitter (or switch to VOX) and press one of the numbers on the keypad, and a pair of audio tones will be generated. I have tried using one of these with my TS-940S, while monitoring the transmitter output. The transmit envelope shows the classic two-tone beat pattern, and the peak output seems to be at the same level as voice output.

"The tones generated are:

1209Hz	1336Hz	1477Hz	
1	2	3	**697Hz**
4	5	6	**770Hz**
7	8	9	**852Hz**
*	0	#	**941Hz**

"None of the pairs are harmonically related, and the tones need to be reasonably pure, or the tone detector normally used with these devices would register incorrect values.

"These keypads are available from BT, Radio Shack and a number of shops in Tottenham Court Road, priced about £10." ♦

VALVES - NOT DEAD YET!

EARLY AMATEUR and experimental radio depended on spark transmitters. It was not until about 1914-15 that the value of the thermionic valve as a generator of continuous wave RF began to be appreciated, and gradually crept into use, initially by American amateurs. The supremacy of CW signals over spark was firmly established in the early 1920s. Thereafter, spark quickly disappeared from the amateur bands, although I recall listening on 600 metres during WW2 to an Egyptian-registered ship using spark to send an SOS after striking a mine. It was not until the post-war period that spark as a maritime emergency system finally vanished, but the transition from spark to valve by amateurs was virtually complete in less than 10 years from the development of the valve oscillator.

It is now over 50 years since the transistor was invented at Bell Telephone Laboratories. The all-solid-state transceiver became established in the 1980s, following a decade of hybrid valve-transistor equipments. The manufacture of receiving valves in the UK and USA faded out during the 1970s, although much valved and hybrid valve/semiconductor remained (and remains) in use. Unlike the spark/valve situation, it proved difficult for solid-state to out-perform thermionics for some important applications.

Today, it is clear that thermionic devices will still be important, and in a few cases dominant, into the 21st century. All but a minority of TV sets still use cathode ray tubes; most really high power transmitters still depend on special power valves, including triodes and multi-cavity UHF klystrons; the majority of affordable amateur radio linear amplifiers still opt for valves, either air-cooled ceramic types or still available audio power valves such as the EL34 etc, although solid-state linears have been around for over a decade. The classic thermionic valve or tube continues also to be highly prized in a number of professional audio applications and, on less solid grounds, by a growing number of the 'golden ears' brigade.

*37 Dovercourt Road, London SE22 8SS

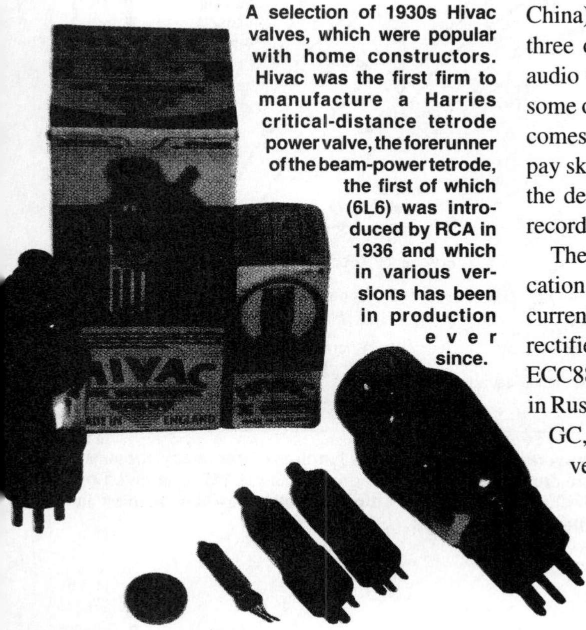

A selection of 1930s Hivac valves, which were popular with home constructors. Hivac was the first firm to manufacture a Harries critical-distance tetrode power valve, the forerunner of the beam-power tetrode, the first of which (6L6) was introduced by RCA in 1936 and which in various versions has been in production ever since.

A 12-page cover-story 'The Cool Sound of Tubes' by Eric Barbour of Svetlana Electron Devices Inc (*IEEE Spectrum*, August 1998, pp24-35) notes that "one of the last remaining tube domains is in music applications, but there the devices flourish and even innovate".

This long article begins: "Although solid-state technology overwhelmingly dominates today's world of electronics, vacuum tubes are holding out in two small but vibrant areas. They do so for entirely different reasons. Microwave technology relies on tubes for their power-handling capability at high frequencies ('Tubes Still Vital After All These Years', Robert S Symons, *IEEE Spectrum*, 1998, pp52-63, see later). The other area - the creation and reproduction of music - is a more complicated and controversial story.

"The complications and controversy stem from the fact that music is played to be heard by human beings, whose non-linear ear/brain hearing systems are far from fully understood. Since no one knows exactly what engineering measurements are appropriate to evaluating the performance of audio equipment, a smidgen of some kinds of distortion may sound worse to the ear than larger amounts of other kinds. Ultimately, the only way to judge audio equipment is by listening to it. Subjective human perception - especially when flanked by questions of artistic merit - is made to order for arguments and disputation. . ."

The author points out that a commercially viable number of people prefer the sound produced by tubed equipment in three areas: musical instrument amplifiers (mainly guitar amplifiers); some of the processing devices used in recording studios; and a small but growing percentage of top-of-the-range hi-fi equipment. A rough estimate suggests that the demand for valves in such equipments has been *growing* at 10% per year since the late 1980s, with much of the demand for audio valves coming from Asia (including Japan, Taiwan and mainland China). Electric-guitar amplifiers account for three out of four of the world production of audio valves. While Eric Barbour admits that some of the demand for valved hi-fi amplifiers comes from the same addicts who willingly pay sky-high prices for special speaker cables, the demand from the professional music and recording world is more solidly based.

The following main valve types find application in current audio equipment and are currently in production: 5AR4/GZ34 power rectifiers, still manufactured in China; 6DJ8/ECC88/6922 double-triodes, still manufactured in Russia and China; 6L6 (suffixes G, GA, GB, GC, TTY, etc) beam tetrodes, of which three versions are manufactured in Russia, many versions in China, and also in Slovakia; 12AT7/ECC81 double triodes, manufactured in Yugoslavia; 12AX7/ECC83 double triodes (Russia, Slovakia, Yugoslavia); EL34/6CA7 power pentodes (three versions in Russia and China, also manufactured is Slovakia and Yugoslavia); KT66 beam power valves, manufactured by M-O Valve Co from 1937 until 1988, now a collector's valve but with a version currently manufactured in China;

TRANSISTORS - ADVANTAGES

- Usually appreciably lower cost, especially for small-signal applications.
- Much smaller size and weight. More suitable for use with PCBs, etc.
- Many transistors can be combined in single integrated circuits, permitting the use of far more complex processing, including Digital Signal Processing, than is feasible with valves.
- Lower power consumption, especially for small-signal use, making transistors more suitable for operation from batteries.
- Less waste heat than equivalent valves. Reach thermal stability much more rapidly (an important factor for oscillator use), although long term stability more affected by changes in ambient temperature.
- Operate on low voltages, resulting in greater safety, lower component costs, smaller clearances.
- Audio matching transformers not required for low-impedance loads.
- Usually more rugged than valves, with less chance of accidental physical damage.

TRANSISTORS - DISADVANTAGES

- Tendency towards higher distortion than equivalent valves.
- Complex circuits and large negative feedback required to reduce distortion.
- Device capacitances vary with applied voltages (including signal voltages).
- Large unit-to-unit variations in key parameters, including gain and threshold voltages.
- Stored-charge effects add to signal delay, complicating high-frequency and feedback design.
- Device parameters vary significantly with temperature, complicating biasing and raising the possibility of thermal runaway.
- Cooling less efficient, since lower operating temperatures required for reliability.
- Power MOSFETs have high input capacitances, which vary with voltage.
- Less tolerant of overloads and voltage spikes than valves. Much more vulnerable to electrical faults, etc.
- Audio capacitive coupling usually requires high-value electrolytic capacitors, which give inferior performance at very-low and very-high audio frequencies, and tend to have limited life due to drying out.
- More vulnerable to RFI, owing to rectification by low-voltage diode junctions or slew-rate effects.
- Maintenance more difficult, since devices are less easily replaced by user.
- Older transistors and ICs often unavailable within about 20 years of their introduction. Replacement may be difficult or impossible.

VALVES - ADVANTAGES

- Highly linear without negative feedback, especially some small-signal valves.
- Tolerant of overloads and voltage spikes.
- Characteristics highly independent of temperature, greatly simplifying biasing.
- Wider dynamic range than typical transistor circuits, resulting from higher operating voltages.
- Inter-electrode capacitances hardly vary with signal voltages.
- Capacitance coupling for audio with low-value, high-quality (long-lasting) film, rather than electrolytic capacitors.
- Circuit design often simpler than for solid-state equivalent. Matching less critical with high impedances.
- Operation usually in class A or AB, minimising crossover distortion in push-pull configurations.
- Output transformer in AF power amplifiers protects high-cost loudspeakers from valve failures. Valves less vulnerable to no or mis-matched loads.
- Home servicing and maintenance much easier, since valves can be replaced by user without soldering, etc.

VALVES - DISADVANTAGES

- Bulky, less suitable for portable use.
- Require high operating voltages.
- High power consumption, including heater supply.
- Generate considerable waste heat.
- More prone to microphony, especially in low-level stages.
- Finite life of cathode emission, resulting (for power stages) in typical life spans of one to five years.

KT88 M-O Valve Co from 1957 to 1988 (exact replacement for the US 6550, except for higher anode voltage rating). Versions are currently manufactured in China and Slovakia, and a Russian version is due shortly. There is also a 'cult' audio power triode, the 300B, originated by Western Union in 1935 for cinema amplifiers and now commanding very high prices in the Far East and currently prompting copies in the USA, Russia, China and Slovakia.

Many of these valve types were widely used in home- and factory-made amateur radio equipment up to and into the 1980s, a considerable proportion still in current use. While in Western Europe and America some valve manufacturers built up reputations for highly reliable, long-life valves with close adherence to published characteristics, this has not always been the case for valves made during the past two or three decades in Eastern Europe and the Far East. The result is that NOS (New Original Stock) valves can command a premium price. However, with growing experience, some factories are now turning-out high grade products.

The following comment by an American engineer applies as much to amateur radio equipment as to audio equipment: "Properly designed tube equipment is extremely reliable. I have repaired many pieces of tube gear from the late 1950s or early '60s, where all of the tubes still worked. Usually, a filter capacitor dried out and failed. The poor reputation comes basically from two factors. One is that from about 1970 to about 1990, the quality of vacuum tubes went through the floor. With the resurgence of tube audio, very high quality tubes are now being manufactured. The second reason is that many contemporary tube designers don't know what they are doing. They are not engineers, they are salesmen. Tubes are forgiving and easy to work with, which allows the less sophisticated to easily enter the marketplace."

While this latter point is less applicable to factory-made amateur radio gear, it is clear that many amateurs have never learned or have forgotten how to set-up or tune a power amplifier requiring neutralisation or, in these days of universal 50Ω broadband output, with a pi-network output.

Again, many have no experience of or have forgotten how to safely handle DC voltages up to 1.5 or 2kV!

The IEEE article contains a detailed comparison of the advantages and disadvantages of valves compared with transistors. Although specifically directed at audio practice, many items are still relevant - with some additions - to amateur radio.

The *IEEE Spectrum* list is clearly rather biased against solid-state, and it would be pointless to suggest that small signal valves will ever make a significant comeback in amateur radio. One has only to think of the problems that used to be encountered in implementing valved PLL frequency synthesisers in professional communications equipment. Again, constructors benefit from the absence of the need for large chassis holes for the valve holders, indeed the absence of a requirement for metal chassis and heater wiring. Nevertheless the Svetlana engineer's article should encourage us not to think of valves as now 'obsolete' devices, or to under-rate the continued usefulness of some of the better 'surplus' or second-hand 'boat-anchor' equipment.

The April 1998 *IEEE Spectrum* 11-page article 'Tubes Still Vital After All These Years' by Robert S Symons (Litton Electron Devices Division) stresses: "For high power and frequency, electron tubes are not only smaller, lighter, and more efficient than solid-state devices, they are cheaper and more reliable as well." The author concentrates on the really high power devices such as the UHF multi-cavity klystrons still dominant for such applications as TV broadcasting, high-power radar, etc. As a frontispiece there is an illustration of a multi-cavity klystron captioned: "A technician stands next to a klystron of the type, circa 1958, that lasted 240,000 hours (roughly 30 years continuous operation) when used in the Ballistic Missile Early Warning System." The author highlights the readiness of politicians to blame 'old-fashioned' valves for the shortcomings of American air traffic control, etc, when the blame really belongs to old software, computers, displays, and problems due to unsuitable connectors used on 5V systems. A survey showed that more than 98% of FAA maintenance problems were not due to the remaining use of valves!

With high power (kW level) devices, the cost of valves increases only as the square root of power for single devices, but linearly with power when multiple devices have to be combined. At VHF/UHF, the maximum power output from a single solid-state device is still limited to about 100 watts, and about 150 watts at HF, so multiple units have to be combined. Symons claims: "Solid-state devices cannot replace tubes in high-power radar, countermeasures equipment, or communications equipment, without the user paying a very high price indeed. A good transmitter must be a good heat exchanger. Some high power solid-state equipment is therefore so heavy it could keep warplanes from flying and make warships capsize during sharp turns if located too

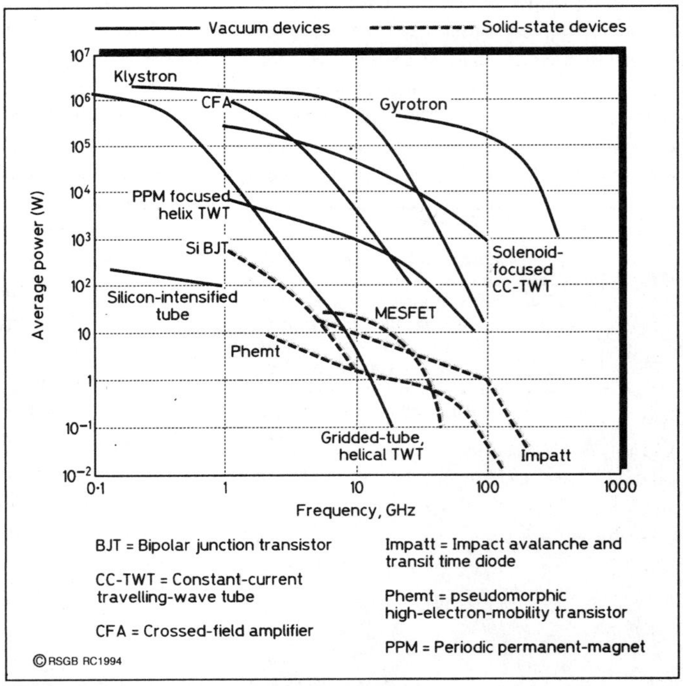

Fig 1: Single device average output power as a function of frequency, for state-of-the-art solid state and exotic thermionic devices, circa 1993, as given by Robert Symons, Technical Director of Litton Electron Devices Division in 'Tubes Still Vital After All These Years' (*IEEE Spectrum*).

far above the water line." In another section he points out: "Vacuum-tube technology is not standing still. In some ways and in some areas it is progressing more rapidly than is solid-state technology." **Fig 1** shows the power output versus frequency of current exotic thermionic and solid-state devices.

A recent press release from Svetlana Electron Devices of California draws attention to their new and improved web site: the Svetlana Tube Zone at www.svetlana.com The firm invites browsers to their site, which contains "a wealth of information on the tube industry", with access to *vast* amounts of specific technical information, including: data sheets on Svetlana products, downloadable in Adobe Acrobat format; and a tube search section, offering characteristics on virtually every popular tube type ever made, including which types are still in production and equivalents. A panel of experts can be reached directly by e-mail, covering applications including amateur radio, hi-fi audio and guitar: info@svetlana.com (by mail at 82000 South Memorial Parkway, Huntsville, Al, 35802; tel: 00 1 256 882 1344, fax: 00 1 256 880 8077).

SSB ON THE RA17

KEN GREEN, G1NAK is a long-time fan of the Racal RA17 receiver, and was dumbfounded by the view of Morris Odell, VK3DOC (*TT*, May and July, 1998) that it has a particularly unsatisfactory performance on SSB.

He comments *inter alia:* "I think that much of the troubles reported with SSB reception arise from failure to appreciate that the ratio of received RF to local oscillator injection levels is important and that it is necessary to jiggle with aerial attenuator and RF gain controls. I have not found that the carrier injection level is of great importance as such. The operative word is *skill*. . . What a splendid receiver the RA17 is. I met one in its first months of commercial launch [in the 1950s]. The BBC bought one to send out East and it fell to me to fit it with DC-amp, to drive a recording milliammeter. The job nigh broke my heart - it was sacrilege! When I removed the covers I found that every component was labelled and each was mounted so that its label was clear and easy to read, and each and every one could be accessed if necessary. It took a long while to do that job, as I was determined that it should look like a factory modification. . . I have had a couple in the workshop, because some twit had 'peaked' all the filters. The havoc that causes doesn't need much imagination if you understand how the Wadley Loop functions. Even with sweep gear it was a real headache getting them back in shape."

THE COMING STORM?

IT IS POINTED out in *Nature* that every 33 years or thereabouts the Leonid meteor shower is spectacular - so many meteors fall that it is called a storm. The last time was in 1996, so we could be in for another big show around November 17 this year or next.

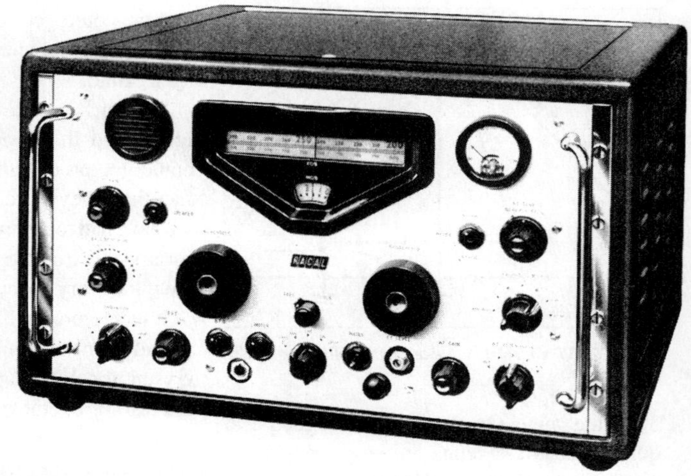

The original Racal RA17 valved receiver, with a drift correcting Wadley Loop triple mixing system. Introduced in the 1950s, it still remains an outstanding performer.

G3DWW'S 3.5/7MHz FOLDED UMBRELLA ANTENNA

WAY BACK IN July 1974, pp445-446, *TT* noted that while the use of guy wires to provide top loading for short vertical antennas in the form of 'umbrella' antennas was well known, an article by S U Nolan 'Developments in MF radiator systems' in the Marconi journal *Sound & Vision Broadcasting,* Vol 15, No 1, Spring 1974, described a useful looking variation in the form of a 'folded umbrella' antenna.

S U Nolan pointed out that many of the developments in MF antennas had been concerned with giving the users more for their money, both for broadcasting and maritime radio applications at 500kHz. He noted that typical low power umbrella antennas consist of a mast radiator with a number of wires (typically nine) attached to the top and radiating downwards at an angle of about 45° to the horizon.

The main limitations which arise from using any vertical radiator which is short in terms of wavelength are efficiency and bandwidth. Acceptable values of 90% efficiency and 10kHz bandwidth with an SWR less than 1.2 can be achieved with a mast height of only one-tenth of a wavelength at frequencies down to about 500kHz.

The July 1974 item continued: "It is claimed that a further simplification results from 'folding' the vertical element, to raise the impedance at the base to that of a coaxial feeder, so allowing the antenna to be fed directly without any matching network at the bottom of the mast, and also *allowing the mast itself to be grounded for lightning protection* (stress added by G3DWW).

"The way this is done is to form a cage of wires running parallel to the mast and with all wires directly connected to the top of the mast and connected together at the base, but insulated from ground and the mast: **Fig 2**. The antenna is made self-resonant, that is to say the inductance of the mast and cage of wires in parallel must tune with the 'guy wire' top capacitance. . . The input impedance at the base of the mast will be purely resistive and about 15Ω. This can be raised to the impedance of the coaxial feeder by the folding process and by adjusting the number of wires in the cage. . . The folded umbrella antenna looks quite a useful approach for 1.8 and 3.5MHz, although to be most effective it needs either good ground conductivity or an effective earthing mat of radials."

In the 24 years since the item was published, I cannot recall any reaction from readers until a letter arrived recently from George Cripps, G3DWW: "I have recently been using a form of 'Folded Umbrella' antenna based on the [1974] item in *TT.* In my case the mast is a 21ft, 1in-diameter ex-WD copper plated tubular mast and the triangular section resonated around 7MHz (**Fig 3**). It was fed from a TS520 transceiver via 50ft of 50Ω coaxial cable. Without an ATU, the SWR was 1.1 on 7MHz, 1.6:1 on 3.7MHz and 2.1:1 on 14MHz. With an ATU, all three bands show 1:1 SWR.

"I have not tried the antenna on 1.8MHz, but the performance on 7 and 3.5MHz has been very satisfactory, particularly at around the

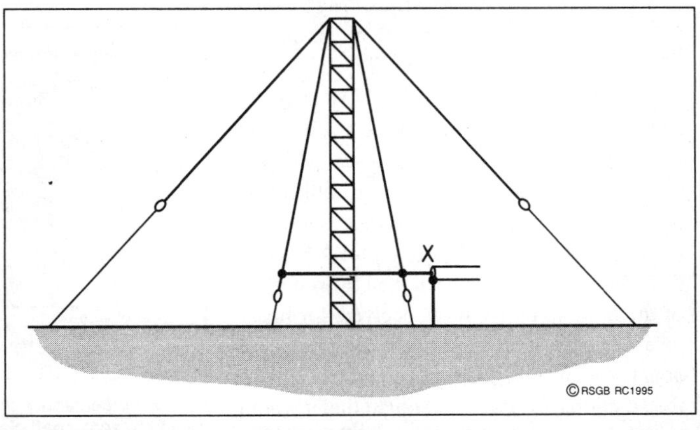

Fig 2: The folded umbrella antenna, which allows mast-radiators of as little as one-tenth-wavelength to be fed directly from low impedance coaxial feeder, as described in 1974.

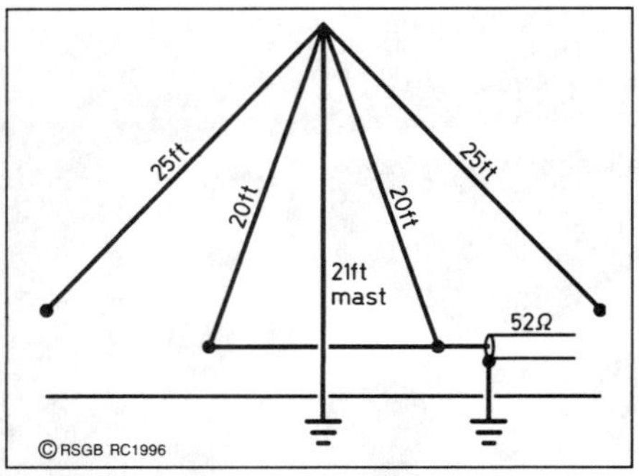

Fig 3: G3DWW's 7MHz version of the folded umbrella antenna.

300-mile range. On reception it is a good deal quieter than a T-antenna similarly used." I feel a little doubtful whether G3DWW's antenna is in practice a mast radiator radiating primarily vertically polarised signals, as with the Marconi design, but nevertheless it looks useful.

THE MIZUHO VXO

IN *TT* (September 1998, p61), Eric Elsley, G3YUQ (*not* G3YUG) raised the question of how the Mizuho range of low power transceivers could achieve 25kHz VXO tuning ranges on 3.5 and 7MHz, and 50kHz on 14MHz. This brought in some informative comments, though at least one element of the mystery remains.

MM0AYD wrote to G3YUQ, sending a circuit diagram and a spare crystal with which to experiment. He pointed out that "Mizuho claim that special crystals are needed, but are obtainable from Quartzslab (type X0-G, HC25 holder)". **Fig 4** shows the 7MHz VXO. This is a standard circuit, although used in a manner calculated to provide a wide frequency shift. The crystal frequency is 9.18996MHz, which is then frequency doubled to 18.37992MHz, which, minus the IF frequency of 11.2735MHz, gives 7.10642MHz. TC pulls the frequency more than 25kHz lower. For the 3.5MHz version a 14.8370MHz crystal is used. This pulls 25kHz, doubling to 50kHz. The system is thus a heterodyne type arrangement, with the pulling range of the crystals further increased by frequency doubling. The only mystery lies in the cut of the crystals used by Mizuho.

Trevor Brown, G3TNM writes: "My Mizuho MX-14S uses a VXO crystal at nominally 12.5MHz, doubled to a nominal 25MHz, which in conjunction with a nominal 11MHz IF results in the 14MHz. The 50kHz VXO swing comes from a 25kHz crystal swing, which is converted to an effective swing of 50kHz by the doubler stage. Very effective it is too!"

Peter Waters, G3OJV, of Waters & Stanton PLC writes: "We have been the UK importers of the Mizuho range of transceivers for many years. When we first brought them into the country we were concerned about the cost of the additional crystal and thought that it would be sensible to source them in the UK and save ourselves some money. However, we found that we could get a swing of only a few kHz, and certainly nowhere near the value achieved by Mizuho.

"When I was in Japan a few years ago, I questioned the designer on this point. All I could get out of him was a wry smile and the fact that 'they were a special cut'. Since then we have had to revert to paying the higher price for crystals from Japan, because they seem to have that very special elastic!"

It seems most unlikely that the Japanese firm would themselves have developed a special crystal for VXO operation, so it would be interesting to know what cut it is they use. While most crystals we use are AT-cut 'Zero temperature coefficient' types, which have a cubic form of frequency/temperature characteristic (**Fig 5**), AT-cut crystals are cut at an angle of about 35° 15' to the c-crystallographic axis, rotated about the X-axis. In the distant past there were X and Y cut crystals, which were easier to 'pull' but had a significant temperature coefficient. In *TT,* April 1980 (p381) I noted that a number of other 'cuts' were then under investigation, including TTC (Thermal Transient Compensated), SC (Stress Compensated), and FC (presumably Frequency Compensated), but that these cuts were more difficult to manufacture than the usual AT-types and were thus significantly more expensive. TTC types were then being manufactured in the UK by Cathodeon. It would be interesting to know the 'cut' of the 'rubber' crystals used by Mizuho, although in these days of 'professional amateur radio' I suppose they regard it as a trade secret!

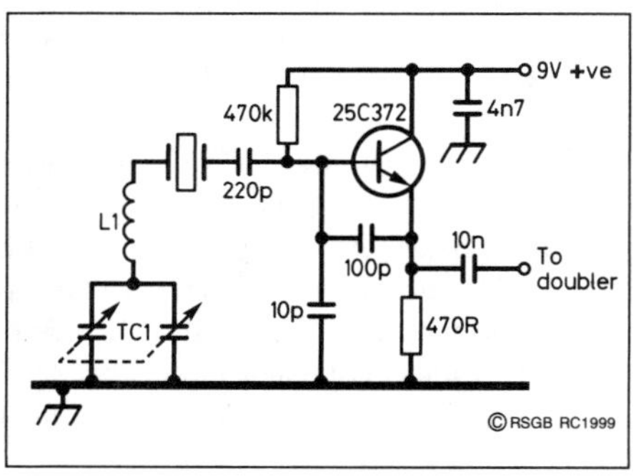

Fig 4: The Mizuho VXO uses a standard circuit approach, but it is claimed that a special cut of crystal is used to obtain a wide tuning range. Values of L1 and TC1 not given in manual.

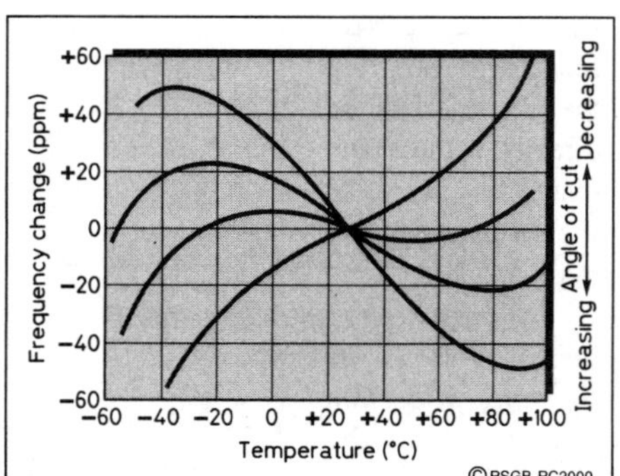

Fig 5: Typical family of a thin-plate AT-cut crystal with 27°C zero-coefficient crossing point, showing frequency/temperature variations with different angles of cut.

MORE ON DRY JOINT INTERMITTENTS

PAT McALISTER, G3YFK, agrees with Geoff Baxendale, G4IBS (*TT,* July 1998, p56), that intermittent faults are often the result of dry joints on pre-plated-through PCBs. To quote briefly from his long letter: "For some years I have been repairing various pieces of test equipment and a range of HF/VHF receivers, notably solid-state ones by Drake and Racal. Many faults on these are a direct result of dry joints. Drake equipment especially is prone to old crystallised joints, mostly on Amphenol-type interconnecting PCB pins. The R4 series are better than the DSR/MSR/TR7 line which go back to the mid 1970s, but suffer dry joints on the sub-boards. I still feel the solder itself should have stood the test of time better. I have never been a fan of indigenous American solder, Multicore seems superior.

"I constantly repair equipment with pre-plated-through hole technology causing intermittents. Some manufacturers did not perfect this technology until the late 1980s. Examples include the Racal 1771/72/1784 etc receivers that use through-PCB rivets, called (I think) 'Griplets'. These are soldered on both sides of the board, and it is worth re-soldering these as a matter of course when tracing an obscure fault. Possibly the prize for the worst offender goes to the Marconi Instruments TF2370/71 spectrum analyser, many of which are now available cheaply. They are repairable using components that can still be easily obtained, but suffer from intermittent faults that can drive users crazy. They could probably often be rejuvenated by re-soldering all the Griplets.

"Other offenders include the flat wire terminal posts found in Marconi and Racal equipment, which connect the wiring looms to the PCBs. Many of the faults do not show using aerosol freezer sprays. The later series of Racal equipment (ie the RA1792 receivers) used superior PCBs with plug/socket ribbon cable connectors, reducing the in-service fault rates considerably. I wonder whether this results from this receiver being a derivative of the Racal USA GM6790 series. It seems an unfortunate British habit to dispose of historically useful information – Racal at Bracknell seems to have no available history on the development of such items as the 1770 or 1790 series, with the original designers retired or gone elsewhere."

Ken Green, G1NAV, also heartily agrees with Geoff Baxendale in respect of dry joints: "Patience, lots of light, lots of magnification, and lots of tapping is the only cure - apart from good workmanship at the creation of the equipment." ♦

BY PAT HAWKER, G3VA*

HEADPHONE ADAPTER

I FEEL FORTUNATE in still having a couple of pairs of ancient Telefunken high-impedance headphones which are still in reasonably good trim and which I find excellent for CW reception on my old equipment, if necessary imposing a step-up audio-transformer (originally intended for valve output stages as a step-down unit) between the phones and the low-Z output sockets. In fact, it was a long time before I realised that such hi-Z headphones had vanished from the marketplace, superseded by low-impedance units including the high-fidelity stereo models and various lower-grade types, all low-impedance.

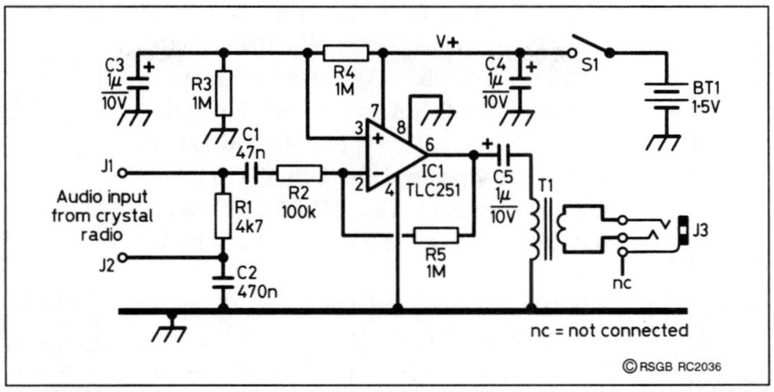

Fig 1: Circuit diagram of the HZX headphone adapter, to permit low-impedance headphones to be used on educational crystal sets, etc. C1, C2 = ceramic. C3, C4, C5 = 1µF, 10V electrolytics. T1 = audio transformer, 1kΩ primary, 8Ω secondary. IC1 = TLC251 op-amp.

An informative article 'The HZX headphone adapter' by Dan Wissell, N1BYT (*QST,* September 199, pp67-69) provides an interesting comparison of old hi-Z and modern low-Z phones. The main intention of his article is to point out the continued value of the simple crystal-set for AM broadcast reception as an educational tool, apart from being the only type of receiver that works entirely 'for free'.

But a classic crystal set will not work with modern low-Z phones, even the most sensitive, since they load the detector far too much. N1BYT presents an adapter that allows low-Z phones to be used, working economically from a 1.5V cell: **Fig 1**.

N1BYT has also brought to light some interesting general information on low-Z versus high-Z headphones. He writes:

"The most direct approach to using low-Z phones with crystal sets is to use an audio transformer with a high-Z primary winding (several thousands Ohms) and a lower-impedance secondary, to match the impedance of the phones. Unfortunately, this type of transformer is now rather difficult to find. Radio Shack offers a 1KΩ to 8Ω transformer (273-1380), but this loads the crystal detector severely. I find, as expected, only strong AM stations can be heard, and the excessive loading ruins the tuning.

"In experimenting, I also noticed that the type of low-impedance headphones had a significant impact on how well they worked with the transformer. I had assumed that 'low impedance' phones were 8Ω or 16Ω devices. After some measurements on an impedance analyser, however, I was amazed to find that the higher-quality (and better performing) headphones were actually 50 to 60Ω per phone at 2kHz, while the lower-quality, poorer-performing units, range from about 4 to 18Ω per phone.

"Because I had the phones in series, the better-performing units are actually presenting an impedance of 100 - 120Ω and, with the Radio Shack transformer, had a transformed load impedance of about 2.8kΩ, compared to the 1 - 1.5kΩ impedance presented by the lower-impedance phones."

N1BYT eventually located an (expensive) transformer with a primary impedance of 100kΩ and secondary of 100Ω, which provided encouraging results with Walkman headphones, but needed a 300ft-long antenna. The same set worked well with a 60ft antenna when using a 'no-name' brand of 1960-era high-Z magnetic headphones.

This encouraged a trip to an 'antique wireless fair' and the acquisition of a number of highly regarded 'Brandes Superior headphones' dating from the 1920s, and also various other high-Z phones, including magnetic and 'ceramic' (Piezo-electric) types, in order to gather test data on impedance, sensitivity and frequency response.

In summary, N1BYT found that there is not necessarily a direct relationship between impedance and sensitivity. Some with 1.6kΩ impedance at 2kHz had greater sensitivity than phones with an impedance of 25kΩ. But, in most cases, there was a direct relationship between the sensitivity and the frequency response, with the most sensitive models often having a lower limit of about 1.5 to 2kHz. Lower audio frequencies require more energy to drive the diaphragm. The newer Walkman phones, which have a response to below 50Hz, are, even with a correct matching transformer, some 12dB less sensitive than the best 1920s Brandes hi-Z headphones.

This is presumably why I am well-satisfied with my 50+- year-old Telefunken adjustable diaphragm phones for CW reception, but would not wish to use these for broadcast or tape-cassette listening. N1BYT found the answer with his HZX adapter, which allows him to use Walkman hi-fi phones on crystal sets.

If you have one of the old step-down audio transformers used in valved broadcast receivers, to match the output stage to a low-impedance moving-coil loudspeaker, an alternative arrangement for feeding Walkman-type phones from a set intended for use with high-Z phones, is shown in **Fig 2**. This comes from an item in *SPRAT,* Nr 96 Autumn 1998, p25, entitled 'Audio pre-amp for the Easy-Three receiver' by Rock Rockey, W9SCH, who wished to use Walkman phones with his 'Easy Three' simple regenerative straight set using three MPF102 FETs (*SPRAT,* Nr 92 Autumn 1997, p3) with the headphones connected in series for mono.

1/3 LAMBDA DIPOLE

THE CONCEPT OF a multi-band doublet antenna fed with a length of open wire, 450Ω ladder-line or 300Ω twin transmission line, with dimensions such that the end of the balanced feeder provides an impedance of roughly 50Ω (say 25-100Ω) and can thus be connected to a reasonably short length of 50Ω or 75Ω coaxial cable that is brought into the shack, has a long-established role in amateur radio. There was the Collins Radio three-band multi-band dipole of 1935 (**Fig 3** and see *TT,* April 1993) in which the open-wire feeder was brought into the shack. Then, in the late 1940s,

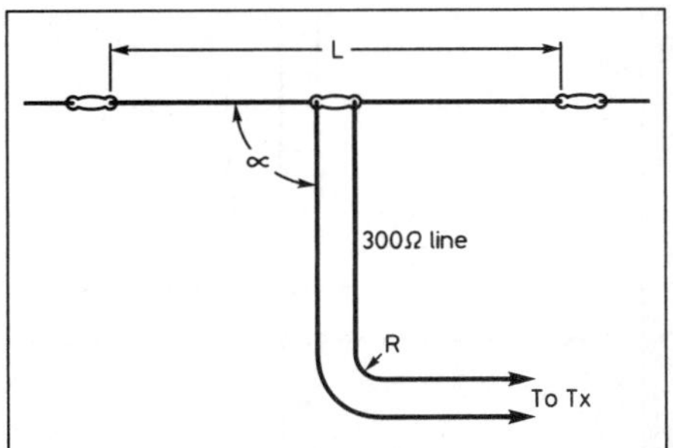

Antenna type	A	B	C
Length of antenna, Feet	99	133	266
Length of antenna, Feet	82·5	48·5	48·5
Frequency range, MHz	14·0 - 14·4 7·0 - 7·3 4·0 - 3·7	14·0 - 14·4 7·0 - 7·3 4·0 - 3·7	14·0 - 14·4 7·0 - 7·3 4·0 - 3·5 2·0 1·7

©RSGB RC2028

Fig 3: The Collins multi-band dipole antenna of 1935, in which the 300Ω line was brought into the shack and matched to the transmitter through a balanced pi-network tuner, to provide equal RF current in both feeder wires.

*37 Dovercourt Road, London SE22 8SS

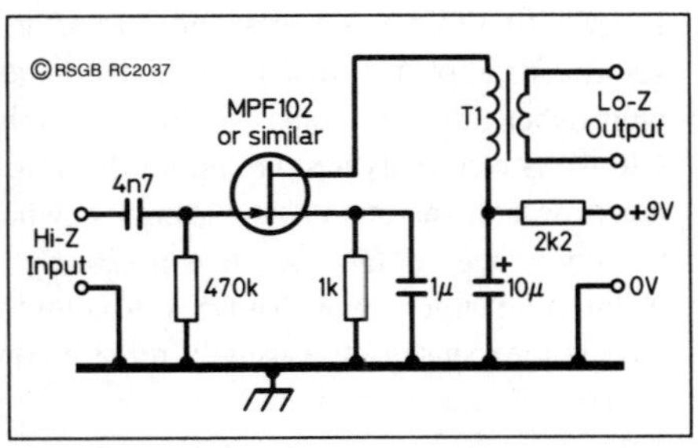

Fig 2: Alternative high- to low-impedance headphone adapter, as used by W9SCH.

Frequency (MHz)	Resistance (Ohms)	Reactance (Ohms)	SWR into 50ohm
3.56	6.7	-4	7.6
7.03	35	+28	2.1
14.06	40	-3	1.2
18.07	39	+69	4.3
24.90	98	+81	1.9
28.06	46	-31	3.5
With feeder length 8.8m			
10.1	27	+234	1.8*
21.06	24	-4	2.0
With feeder length 14.9m			
10.1	52	-422	1.0*
21.06	24	+3	2.0

Negative reactance is capacitive (positive is inductive) in series with the resistive part of the impedance.

*On 10.1MHz a 130pF capacitor in each leg of the feeder will tune out the reactance. With the 14.9m feeder, 3.3μH inductors should be used instead.

Table 1: G0FAH all-band HF antenna.

came the subsequently well-known 'G5RV', developed independently by Louis Varney, and incorporating the non-critical length of 75Ω coax feeder.

The 'G5RV' - **Fig 4(a)** - remains one of the most-popular multiband 'wire' antennas. This is despite it having been designed in the late 1940s, ie before the 21MHz band was available to amateurs, and long before the 10, 18 and 24MHz bands. It will work on all HF bands, provided that it is used with a flexible ATU, although on some bands the SWR on the coaxial cable will be very high.

In the early 1980s, as a result of an early use of computer-modelling, Dr Brian Austin, ZS6BKW/G0GSF, modified the dimensions of the G5RV to provide an antenna that could be used with 50Ω coax without an ATU on three bands - 7, 14 and 28MHz - and with a simple one also on 3.5, 10 and 21MHz (*TT*, May 1982, with a full length article by G0GSF in *RadCom*, August 1985): **Fig 4(b)**. Another modification to the G5RV was proposed by W5ANB in *QST*, November 1981: **Fig 4(c)**. A five-band version (7, 14, 18, 24, 21MHz) that is claimed to operate without an ATU has been described by Rick Hill, ZL1OK (*Break-in*, May 1992), although I do not have any further details.

In 1992, an 'all-band dipole' was described by Bill Wright, G0FAH, in *SPRAT*, Spring 1992, *TT*, June 1992, and later in *QST*. He drew on the computer-modelling work of G0GSF/ZS6BKW, overcoming the problem of covering 10 and 21MHz by arranging to change the length of the 300Ω matching section from 11.5m to 8.8m: see **Fig 5** and **Table 1**.

A basically similar arrangement turns up again in 'An improved one-third-wavelength dipole' by Andrew Griffith, W4ULD (*QST*, September 1998, pp41-45): **Fig 6**. This antenna represents a further development of an antenna he described in *QST*, September 1993, but now includes a remote matching network (**Fig 7**) for better performance on 3.5MHz. With the network remotely switched (from the shack), the antenna functions on 7, 14, 17.5, 24.5 and 28MHz with an SWR of less than 3:1, and less than 2:1 on most. The 1:1 balun used by W4ULD is a commercial unit capable of handling powers in excess of a kilowatt, but for his earlier implementation (powers up to 200 watts) his balun was made by stringing 10 FB-77-5621 beads close together on a length of RG-8X coax.

C1, C2, C3 and C4 matching network capacitors, which carry high RF currents or voltages when using the American legal power limits, are all home-made from readily available materials (construction details calling for some skill are given in the *QST* article). My own feeling is that if one wished to avoid the complications of the remote matching network, the 40ft of ladder line could be brought into the shack and then used with a flexible ATU providing a balanced output and manually adjusted for the band in use. With the dimensions shown, the ladder line should be current fed on most bands. The one-third lambda approach should prove an effective doublet antenna although, as for all the other G5RV variations and long-wire antennas, the horizontal radiation pattern will not be the same on all bands (not necessarily a disadvantage when chasing countries).

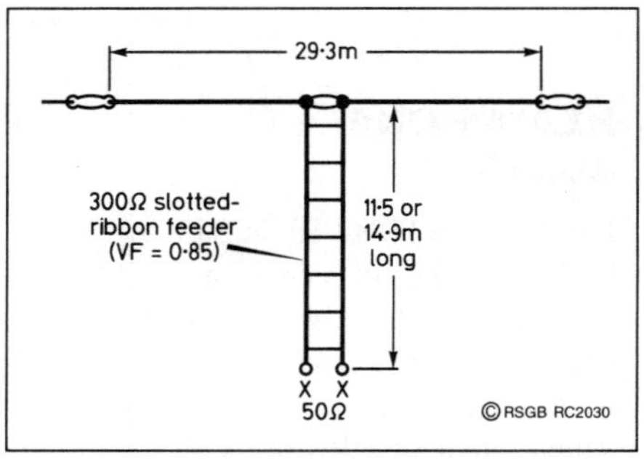

Fig 5: The G0FAH 'all-band' dipole, which can be used on all bands including the WARC bands from 1.8 to 30MHz, derived from the computer-modelling work of G0GSF in which the length of the balanced feeder section is changed on some bands. The SWR presented to the transmitter can be reduced by using the ATU shown in Fig 101(d).

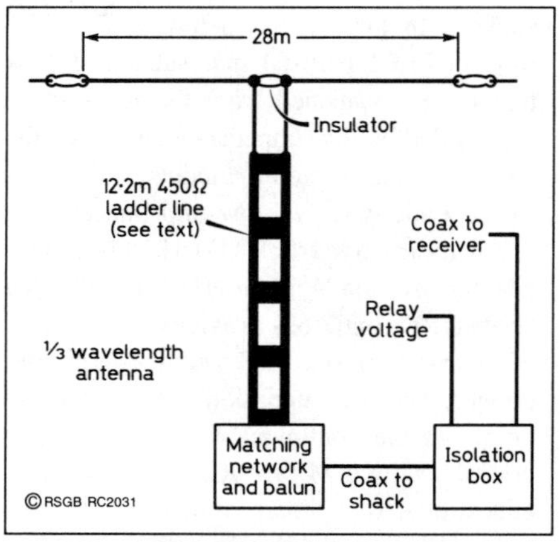

Fig 6: W4ULD's 'improved one-third-lambda dipole', with a remote switched matching network for improved performance on 3.5MHz.

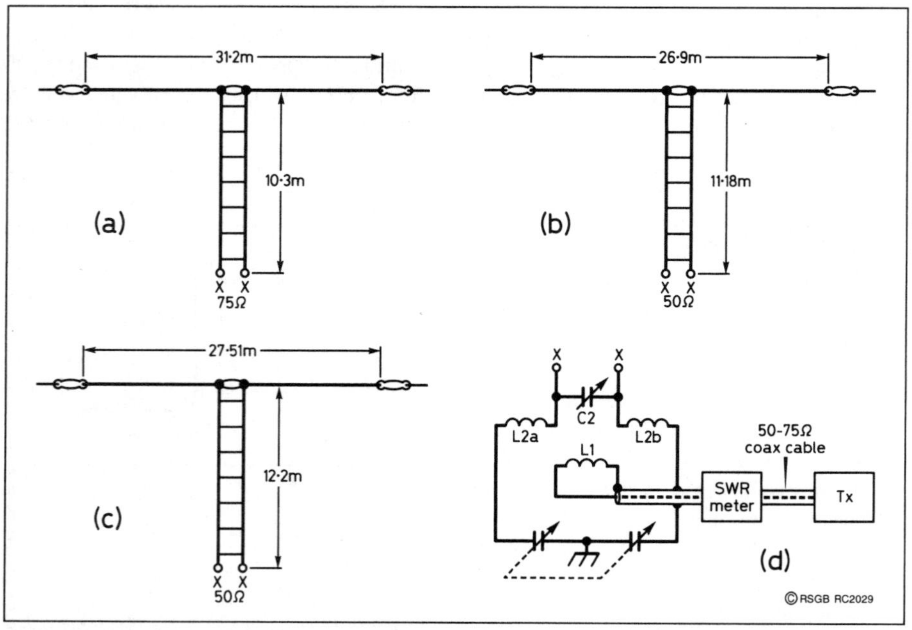

Fig 4: Three versions of multi-band dipole antennas designed to enable 50Ω coaxial cable to be connected to the end of the balanced line and brought into the shack. (a) The original 1940s 'G5RV', with 102ft (31.2m) top. (b) The computer-based 'ZS6BKW', developed in South Africa by Dr Brian Austin, G0GSF/ZS6BKW, which can provide five low-VSWR frequencies in the HF spectrum. (c) A version developed by W5ANB. (d) A suitable ATU for use on those bands where there is appreciable VSWR on the coax section.

POWER FROM HEAT

ELECTRICITY FROM solar cells has made a significant impact in many countries blessed with rather more consistent sunshine than the UK. Solar electricity has the serious drawback that in order to maintain power at night or when the sky is heavily overcast, it is necessary to incorporate large storage batteries. Apart from the extra expense, these usually require considerable maintenance.

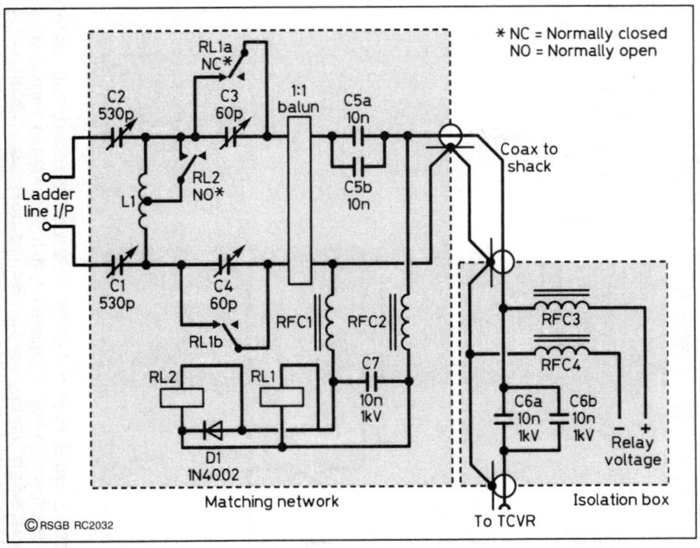

Fig 7: Matching network and isolation box used by W4ULD for powers up to the American legal limit. C1, C2 = 520pF howebrew variable capacitors. C3, C4 = 60pF homebrew variable capacitors. C5A, C5B, C6B = 0.01µF, 1kV disc ceramic (18mm diameter, minimum). K1 = modified double-pole double-throw 12V enclosed relay with 10A contacts. K2 = modified 4-pole double-throw 12V enclosed relay. L1 = 4.8µH (18 turns of No 6 bare copper wire with a 1.7in inside diameter), leads supported 1.625in above the mounting plate. RFC1-RFC4 = 220µH RF chokes (240 turns multilayer of Nr 38 or Nr 30 enamelled wire on 0.25 x 0.6875in long on non-metallic form). Note that to construct such a high power network it would be essential to consult the original *QST* article, but for lower powers less specialised components would probably be satisfactory.

However, it is often forgotten that photovoltaic cells can also provide electricity from infrared radiation, ie radiant heat. This can be generated by a source of energy such as oil or other fossil fuels. A six-page article 'Thermophotovoltaics' (TPV) by Timothy J Coutts and Mark C Fitzgerald (*Scientific American,* September 1998, pp90-95) is subtitled: 'Semiconductors that convert radiant heat to electricity may prove suitable for lighting remote villages or powering automobiles.'

An American company (JX Crystals of Issquah, Washington) is about to market a small 'Midnight Sun' 30W TPV generator, intended to run electrical equipment on sailboats. As a first launch, the price tag at US $3000 is significantly higher than for a conventional small diesel generator, but the unit has the advantage of running silently and is expected to be much more reliable, since there are no moving parts. Compared to a solar array it has the disadvantage of requiring a fuel supply, but has the major advantage of working at night without a storage battery.

More applicable to amateur radio are TPV generators currently under development, sponsored by the US DARPA (Defense Advanced Research Project Agency), to provide a power source for military communications equipment or for powering laptop computers in the field. These include 150 and 300W power modules, using arrays of heated ytterbium oxide fibres (providing infra-red radiation at a wavelength of 980 nanometres), matched with silicon photoconverters and with a multiple-layer insulating filter to recover unused heat energy.

In the long term it is forecast that thermophotovoltaics could generate megawatts of electricity using the large amounts of waste heat generated in the glass, aluminium and steel industries.

A thermophotovoltaic generator basically comprises a source of heat, coupled with a radiator that emits desired wavelengths of infrared radiation, which is then converted into electricity by an array of photovoltaic cells. The burners need to achieve temperatures above 1000°C. Finally, to perform efficiently, the system must rechannel unused energy back to the radiator or be used for other purposes, such as heating a room. Sources of heat could range from fossil fuel burners to sunlight to a nuclear fission reaction. Most systems currently under development use fossil fuels.

It is claimed that these systems can be designed to minimise pollutants, and can have substantially higher efficiencies than electric generators powered by natural gas or other fossil fuels. Because they contain no moving parts, they run silently and reliably, and should require little maintenance. Clearly a technology worth keeping an eye on for field days, DXpeditions, etc.

PCB TESTER

A SIMPLE BUT practical tester to detect the integrity of a 5V printed circuit board appears in the 'Circuit Ideas' feature of *Electronics World* (October, 1998), contributed by John Farbrother of Walsall: **Fig 8**.

The probe will detect 0V, 5V, and floating parts. On a broken (floating) track, both LEDs barely light up, in the absence of any current to drive the transistors, whereas if the probe touches 0V or 5V one or other LED lights up, provided the device is connected across the same 5V supply as the PCB under test. John Farbrother suggests that on a digital (switching) signal, the LEDs should light in proportion to the mark-space ratio.

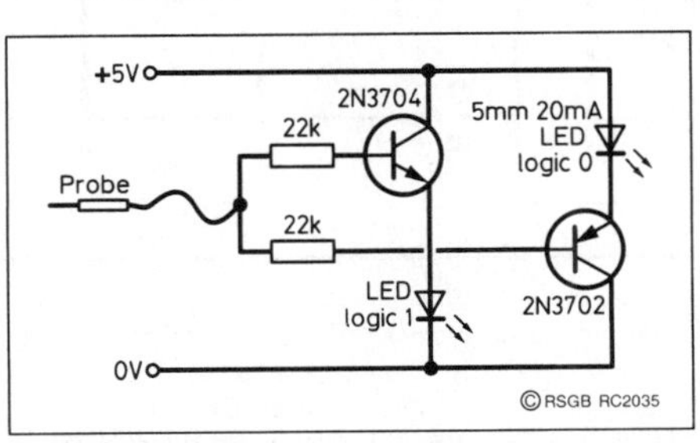

Fig 8: A simple tester to detect 0V, 5V, or a broken PCB track (*Electronics World*)

A REBUILT FL-1000 LINEAR

F H V GEERLINGS, PA0FRI, (several of whose 'Frinear' amplifiers have appeared in *TT*) was recently offered an old FL-1000 (4 x 6KD6 or 4 x 6JS6) that had changed hands and countries several times, in the course of which it had undergone various modifications. However, the HF transformer, tuning and loading capacitors and the main pi-network tank coil were in good condition. He duly purchased it with a view to rebuilding it with quality components collected during almost 25 years.

A circuit diagram of the rebuilt amplifier, using a single TB3/750 triode is shown in **Fig 9**. PA0FRI feels that the new features may be of interest and encourage other experimenters tackling a high-power linear:

(1) Nine amateur bands, one valve, 400W HF power output.

(2) Universal, all-bands, pi-input network. A 2-pole, 5-way switch, five small coils and two miniature air-spaced variable capacitors will match various types of valves needing 10W, 50W or 100W drive. J Birkett offers three-gang capacitors (365 + 365 + 365pF) for under £5.

(3) A simple tuning-indicator. This does not give scaled SWR readings but is an aid to zero tune-up of the input circuit. The sensitivity potentiometer can be omitted. The only critical component is the 47Ω, non-inductive resistor (two 100Ω, 0.25W resistors in parallel).

(4) A low-cost 1mA panel meter switched by a 1-pole, 4-way switch. Yes, the positive terminal of the meter is connected to chassis.

(5) Sequential change-over system. This ensures that the valve is 'on' only when the amplifier is connected to the antenna.

(6) Grid circuit earthing and pi-network choke decoupling. PA0FRI's experience is that parasitic oscillation and instability is usually the result of poor or inadequate earthing and decoupling. His circuit/layout is an example of how it is possible to overcome these problems without the use of parasitic chokes.

(7) The design of this amplifier has proved most reliable and permits various types (and makes) of valves to be used. It produces the same output on 1.8 and 29MHz. It was tested with the following valve types: RS686, TB 2.5/300, TB3/750, QB 3/200, QB 3/300, QB 3.5/750, QB 4/1100, 5D22, 3-500Z, 4-250A, 4-400PR, 6155 and 8802. With tetrodes the maximum output is lower, due to the lower maximum HT. It is possible to achieve the UK legal limit with an on-load HT of 1.7kV and 50W of

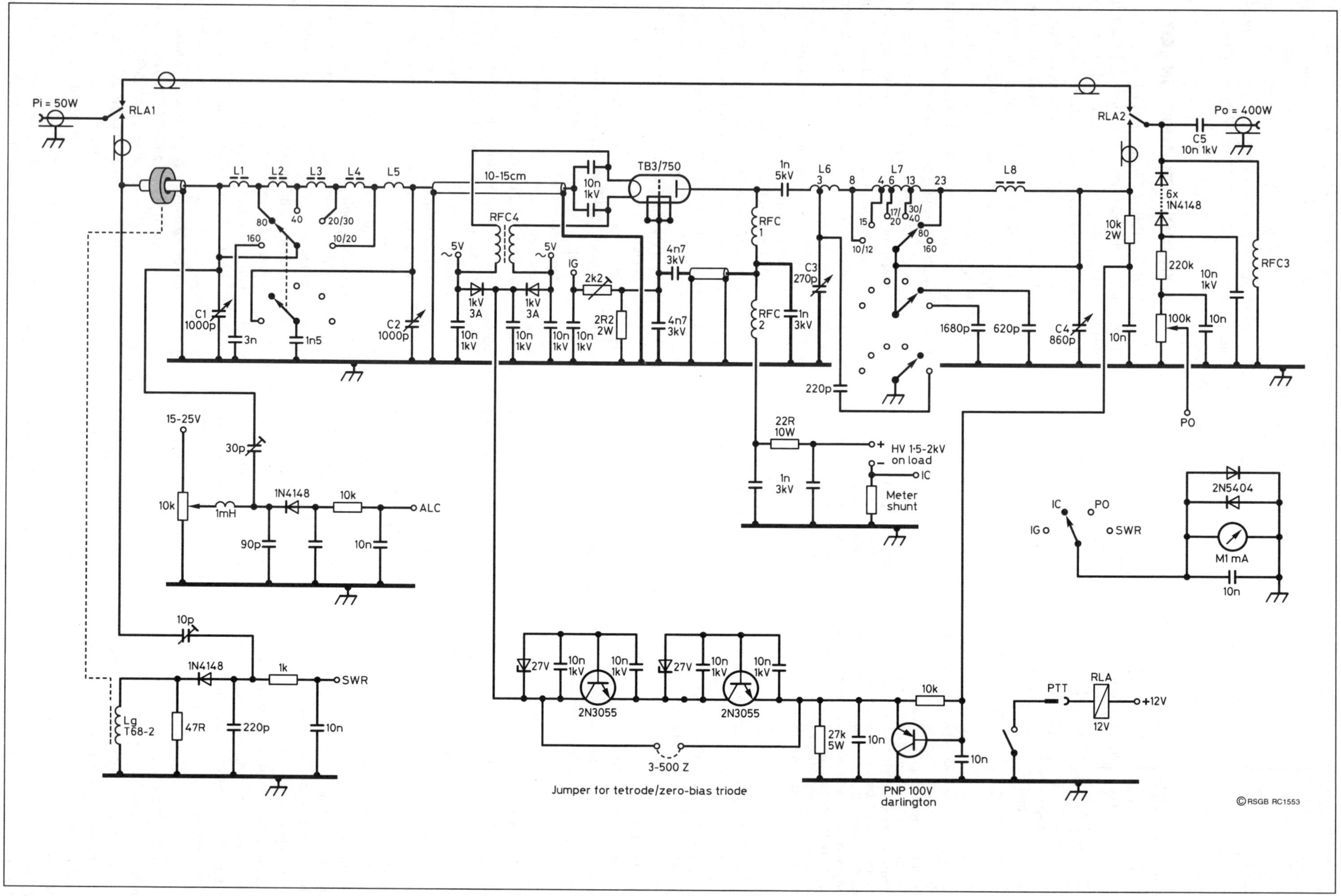

Fig 9: Circuit diagram of the high-power linear amplifier as rebuilt by PA0FRI. C1/C2 miniature air spaced variable, 57 x 32 x 30mm, 330 + 360 + 360pF. L1 17t, T68-2 toroid, 1mm 'trafo' wire. L2 21t, T68-2 toroid, 1mm 'trafo' wire. L3 11t, T68-2 toroid, 1mm 'trafo' wire. L4 7t, T68-2 toroid, 1mm 'trafo' wire. L5 7t, 1cm id, wire spaced, 2mm 'trafo' wire. L6 8t, 3cm id, wire spaced, 2.8 tinned wire, 4.5cm long, tap on 3t. L7 23t, 4cm id, wire spaced taps on 4t, 6t, 13t (FL-1000 pi-net coil). L8 28t, T200-2 toroid, 2mm 'trafo' wire. L9 60t, T68-2 toroid, 0.3mm 'trafo' wire. RFC1 2-2.2cm dia former, 0.3mm 'trafo' wire, single winding coil 5cm long approximately 180uH. RFC4 ferrite rod, d9.5mm dia, greater than or equal to 10cm long, bifiliar wound. C5 3 x 3.3nF, 1kV disk ceramic in parallel. RFC2/RFC3 non-critical, 200uH to 2nH. 'Trafo' wire is transformer wire.

drive to a TB/750. The key-down anode current with a 3-500Z was 420mA and standing current of 70mA, resulting in an RF output of 420W. The maximum output obtained with the TB 3/750 was about 560W, but this exceeds the maximum admissible cathode current. The QB 3.5/750, QB 4/1100 and RS 676 require about 85W of drive for 400W output. PA0FRI is not enthusiastic about the 4-250 and 4-400, with the gain on 10 - 20 metres falling below the European tetrodes. In general he does not recommend using American tetrodes.

(8) The tuning capacitor C3 is connected to a tap on the 28MHz coil in order to cope with the large minimum capacitance.

(9) The choke RFC4 has a bifiliar winding - 1.8mm enamelled wire on a 10cm, 9.5mm diameter ferrite rod from a broadcast AM radio. While the wire size might appear to be too thin for 14A filament current, it serves also to provide a resistance to limit switch-on in-rush current.

On the general question of the reliability of some high-power valves, PA0FRI notes that two Dutch PA3 amateurs sought his advice on their defective AEA LA-30 HF linear amplifiers fitted with EIMAC/VARIAN 3-500Z valves (serial numbers FYS 106 and FYS 132). It transpired that the anode spot-welding was poor, with the result that the structures had a 'leaning tower of Pisa' appearance. The sloping anode position caused an internal short-circuit and a burnt-out power transformer and rectifier circuit, all due to inadequate fusing and design. By knocking gently on the glass position, the valve performed as usual but then a slight movement of the desk and a big bang! He wonders whether poor anode spot-welding of the 3-500Z is common. Another Dutch amateur has told him that in his Heathkit SB-1000 amplifier the anode of the EIMAC 3-500Z dropped down on the internal base screen. PA0FRI points out that an excellent replacement for the 3-500Z is the Amperex 8802 with carbon anode.

This reminds me that Eric Parvin, G2ADR, last year questioned whether others had experienced the problem that later production of General Electric 6146B valves seem to have different characteristics to early production, requiring a change in bias, lower standing current and less RF output. He would be interested to know if others have experienced this problem and have found a solution or a source of early production.

WARC BANDS ON CUBICAL QUADS

BOB COX, G3PLP, has found a solution to the problems that can arise when attempting to add 24 and 28MHz to a two-element cubical quad antenna designed for the 18 and 21MHz bands. He writes:

"I built my first quad antenna in 1977 and have been using them ever since. My present two-element quad antenna uses the classical fibreglass spreaders and wire arrangement, with some metalwork to support and locate the gamma matches for each band. Recently, with the increasing sunspot activity, I decided to add 24 and 28MHz, taking the dimensions from the magnificent book on cubical quads by Bill Orr, W6SAI.

"As for the existing array, each driven element was provided with its own feed, switched at the mast, with its own gamma match. The result was a four-band quad for 18/21/24/28MHz. However, tests soon showed that something was very wrong. The front-to-back ratio on 21MHz, better than 20dB before the addition of the higher bands, had fallen to no better than 10dB, seemingly also with some loss of forward gain. Checking the antenna on all four bands showed that the new 28MHz section appeared to be working perfectly, but this was not the case on the other three bands, even though the gamma matches were adjusted close to unity SWR on each band.

"It then became obvious what the problem was. When operating on 28MHz, all four reflector loops were larger than the driven element - fine! But on 24MHz, the driven element could 'see' not only its own reflector, some 3.5% bigger, but also the 28MHz reflector loop, resonant in the region 26-27MHz and apparently acting as a director on the 24MHz band. Similar situations exist on 21 and 18MHz. Thus on all three-lower-frequency bands the array no longer comprised a driven element and an effective reflector, but instead a driven element plus opposing reflector and director loops.

"To verify that this really was the case, I decided to mount relays on the 21/24/28MHz reflectors (the 18MHz loop acts as a reflector on all bands and does not therefore require switching). The relays were in the normally closed position, with the relay coils RF bypassed. Switching was then arranged as follows:

On 28MHz	All loops closed
On 24MHz	28MHz reflector open-circuited
On 21MHz	28 and 24MHz reflectors open circuited
On 18MHz	28, 24 and 21MHz reflectors open-circuited.

"Eureka! All was now well again. Ground wave tests with a local station (G3AQM) about 7 miles distant confirmed this, with my 21MHz quad now having its front-to-back ratio of better than 20dB restored.

"I note that Bill Orr's book recognises some degradation in F/B ratio on tri-band quads (14/21/28MHz). Clearly the introduction of WARC bands, with the much closer frequency separation between bands, exacerbates this effect.

HERE & THERE

CHARLES KILPATRICK, WZ8A (*QST*, Technical Correspondence, September 1998, p80) experienced interference on 3510kHz, apparently from intermodulation products from three local AM broadcast stations (WLW, WCKY, WMOH), whose fundamental frequencies add up to 3.51MHz. He was about to register a complaint to the broadcasters, but first took some sandpaper to the alligator (crocodile) clip connected between his 80m vertical antenna and the centre conductor of the buried coaxial cable feeder. Result: an end to the cacophony from the AM stations. A corroded contact had acted as a mixer. Perhaps the moral should be that all outside antenna connections should always be carefully soldered!

How would you like to connect your HF receiver to a 3km-long linear array of almost 500 vertical dipole elements, forming several narrow 'finger' beams? Two such receiving sites are currently being built in Australia as part of a new over-the-horizon HF bistatic radar system (*The GEC Journal of Technology*, Vol 15, No 1, 1998, pp47-56): "When complete, the installations will probably represent the most powerful HF receiving system anywhere in the world."

Finally, I am seeking the circuit diagram and brief details of the 1940s Marconi CR150 receiver. Any offers? Tel: 0181 6934601. ♦

technical topics

by Pat Hawker, G3VA*

PA0SE ON 2197m

STILL MISSED ARE the 'Reflecties door PA0SE' articles which Dick Rollema contributed to the Dutch VERON journal *Electron* for 25 years until the end of 1997. However, PA0SE has not been idle. One of his subsequent long articles 'PA0SE op 2197 meter' (*Electron,* October 1998, pp425-429) describes his complete home-built 135.7 - 137.8kHz station, including a LF converter with tuned frame-antenna (**Fig 1**) and a crystal-controlled 200 watt transmitter using four IRF520 HEXFETs (**Fig 2**), in parallel-push-pull, powered from a 30V 10A supply.

PA0SE has worked a number of British and Belgian amateurs on this promising LF band, with an RF output of 60-80 watts. Keying can be either by hand or automatically from a cassette recorder, for which PA0GJH has computer-produced a test message: 'Test de PA0SE QTHR 40km SW Amsterdam tel 31715892734)'.

On a narrow band, a crystal-controlled transmitter is less of a disadvantages than it would be on the HF bands. A 136.567kHz signal can be generated directly from an HCT4060 IC-divider (which can divide by 2, 4, 8, 32, 64 or 128), using the divide-by-64 configuration from an 8739.89kHz crystal. An 8750kHz crystal would provide a 136.72kHz signal; an 8800kHz crystal 137.5kHz, etc.

*37 Dovercourt Road, London SE22 8SS

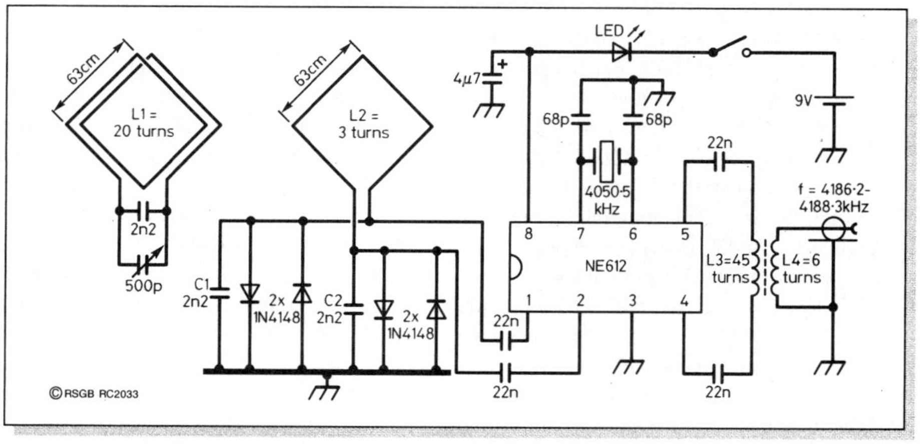

Fig 1: PA0SE's 136kHz crystal-controlled receiver converter. The large 20-turn tuned loop antenna is coupled to the mixer by means of a three-turn winding wound in the centre of L1.

The choice of crystal frequency for the converter is simply a function of the tuning range used on the main receiver, PA0SE's 4050.5kHz crystal placing 135.7 - 137.8kHz in the range 4186.2 -4188.3kHz, or alternatively 3914.8 to 3912.7kHz with L3 resonated to the middle of the chosen band. L2 is wound round the same large frame as L1, with L2 in the centre of the spaced turns of L1.

While PA0SE's article provides illustrations of his neatly constructed frame aerial and breadboarded transmitter, plus full details (Dutch text) of his LF station, it is hoped that reproduction of these few notes and circuit diagrams will at least provide some useful ideas for those seeking to make use of this band, which presents significantly fewer antenna problems than 73kHz and is attracting increasing activity.

THE BATTERY MISER

VINCENT EVANS, G4AVT, noted with great interest the *TT* (October, 1998, p41) item on a battery economy receiver for remote under-developed areas without mains supplies and where the cost of batteries becomes the major factor in cost of ownership. This was because the economical use of batteries in portable receivers has been a pet subject of his for many years. His article '*Battery Miser*' (*Radio Con-*

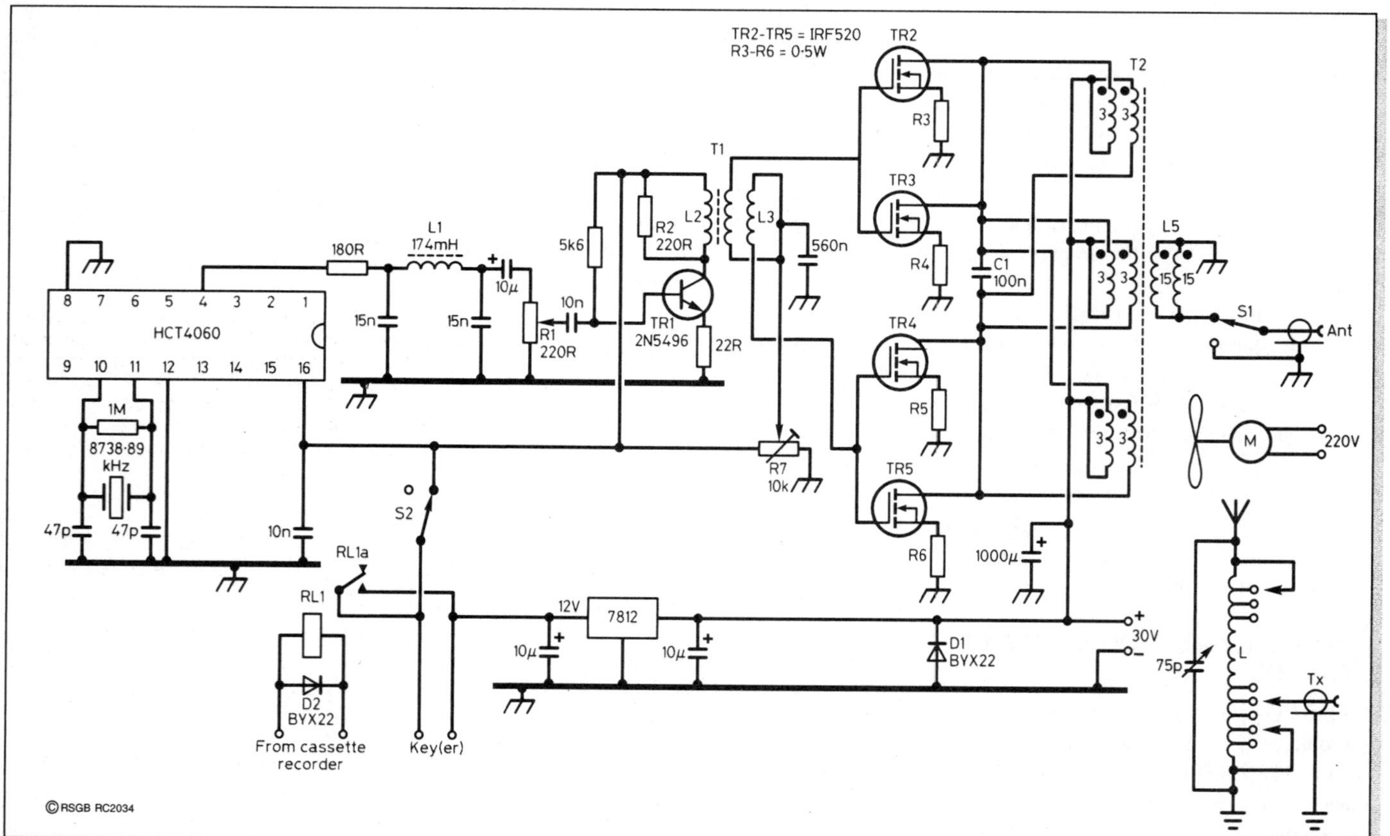

Fig 2: PA0SE's 135.55kHz transmitter, using an HCT4060 frequency divider and four IRF520 HEXFETs in parallel-push-pull as a power amplifier. T1 is wound on a Philips FXC362 ferrite ring core, with a diameter of 23mm. Construction of the output transformer and the coil-loaded antenna is described in the *Electron* article (Dutch text).

A selection of G4AVT's battery life extenders

structor, November 1974) described a receiver incorporating some of his battery-saving devices; he was later granted UK Patent No 2070294 'Battery Economiser Circuit', April 1984 for a more developed arrangement. He believes that with modern devices it could doubtless be developed further.

Like many patents, the concept is basically simple. G4AVT based his on the following considerations: A 9V battery when new delivers 9+V; the typical 9V transistor portable is designed to continue working satisfactorily at voltages appreciably below 9V; the quiescent current of a push-pull amplifier is related to the voltage on its power line; so that by reducing the voltage on this line to the minimum satisfactory working voltage the current drain will be reduced. His answer is to provide a voltage regulator at the lower operating voltage: **Fig 3**

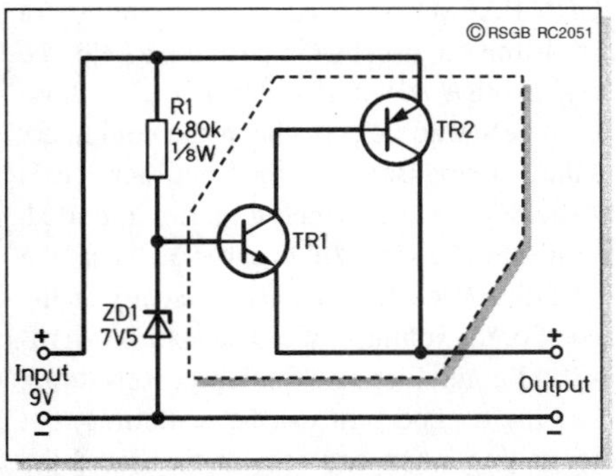

Fig 3: G4AVT's 'battery economiser', as patented in 1984.

is derived from one of his Patent illustrations.

But, as G4AVT stresses, the regulating device *must* itself draw minimal current and have minimal voltage drop. This means that the components used will be of *critical importance*, and should therefore be of high grade. His Patent illustrated various forms of enclosures for the regulator device. These include various shapes, fitted with connectors for batteries, but it would be possible to use other encapsulations such as heat-shrink sleeving etc, permitting the device to be tucked into the case of the receiver. His Patent also featured a unit built into a car cigar-lighter plug.

G4AVT did not pursue the project commercially with any vigour, but still has a limited number of economisers available in various enclosures, studded for PP9 or PP3 batteries, which he is willing to supply at cost (£5 d/d UK) to anyone interested in assessing performance and possible further development. He also still has a small stock of parts and working drawings: Vincent S Evans, Beacon View, Parbold Hill, Parbold, Lancs WN8 7TG.

In the letters column of *Electronics World* (November 1998), Reg Moores, G3GZT, notes that nickel-cadmium cells often fail well before the 500-1000 recharge cycles normally claimed for such cells. He believes that this is often due to chargers delivering much higher charging currents than the usually recommended one-tenth of the capacity of the cells. In some equipments very high charging rates are deliberately used to provide shorter charging times. To achieve the optimum number of recharging cycles, he recommends measuring the charging current and checking that this is not more than the one-tenth figure.

In the same issue, Frank Eliason of Plymouth raises the question of why, when powering equipment from two series-connected batteries, one battery virtually always becomes exhausted [ie high internal resistance] well before the other: "I have tried various pairs and the same thing always happens. I have even taken 'good' ones from each pair and paired those, but the same thing always happens. Can anyone suggest a solution to this problem?"

PHANTOM-COIL VXO

ALTHOUGH THE variable-frequency crystal oscillator (VXO) seems to have occupied a good deal of space recently in *TT*, there is little doubt that this topic is of significant interest to those remaining enthusiasts who still build simple equipment, particularly for low-power portable operation. It is still true that crystal-control can provide a frequency stability of up to 1000 times and a VXO some 100 times better than the average *LC* oscillator.

Gerald Stancey, G3MCK, in the course of some tidying up came across a five-page article ('*Phantom-coil VXO*' by Frank Noble, W3MT) in an old issue of the still-missed *Ham Radio*, January 1982, pp66-71, plus a corrected diagram in the July 1982 issue, p8. This not only introduced an interesting technique in showing how an artificial transmission line can create the illusion that a variable-capacitor is a large variable coil, as required by a low-frequency VXO, but also provides a useful survey of VXO basics for capacitor-only (CVXO) and inductor-only (LVXO) systems.

W3MT explains that the CVXO, using variable capacitance only, always operates above series resonance of the crystal and is nearly as stable as a fixed crystal oscillator. "But even with the greatest care in circuit design it is not possible to exceed a range of about 5kHz on 40 metres if standard crystals are used.

"On the other hand, the LVXO, using the equivalent of a variable inductance, can exceed a range of 50kHz on 7MHz, but the stability will decrease with range as shown in **Fig 4**. This figure underlines the enormous inductance required to shift the frequency. The LVXO always operates below the crystal's series resonance; typically this is 2-3kHz below the frequency of operation [parallel resonance] for 7MHz HC6/U crystals with a 32pF load capacitance.

"It is usual to obtain an effective variable inductance by connecting a large fixed inductor in series with a large variable capacitor. The minimum reactance of the capacitor must be small compared with the inductor's reactance, and its maximum must be equal to that of the inductor. This arrangement has a number of disadvantages: (1) A good coil must be inconveniently large mechanically on 7MHz and lower frequencies. (2) As a result of the large inductance, the self-resonance of the coil will

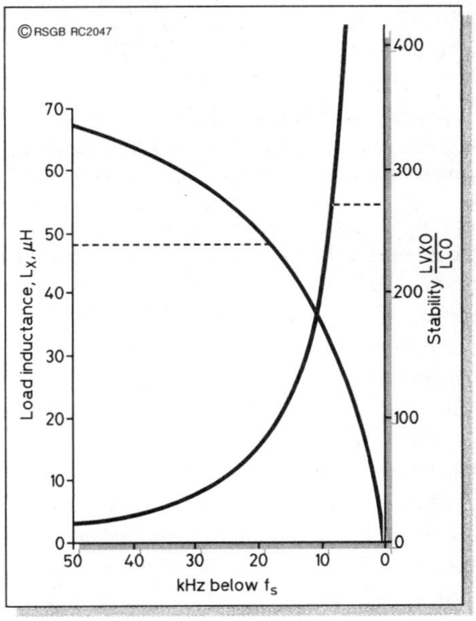

Fig 4: The LXVO can exceed a tuning range of 50kHz on 7MHz, but the stability will decrease with range. Left-hand ordinate, equivalent load inductance versus frequency shift below crystal series resonance. Right hand ordinate, stability ratio of the LVXO with respect to an LCO versus frequency shift below crystal series resonance (HC6/U crystal, 7MHz LVXO, Co 6pF, r = 250).

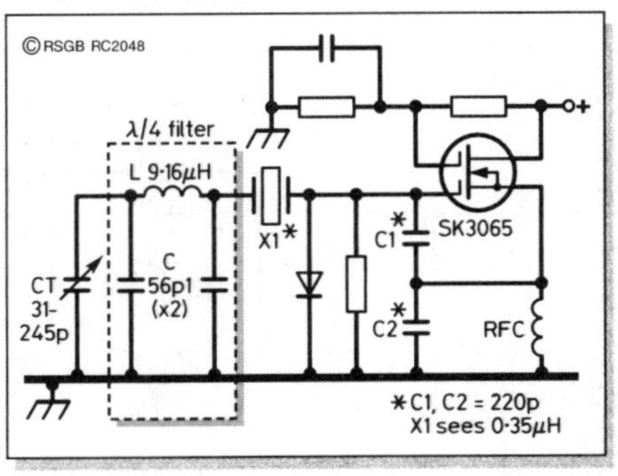

Fig 5: W3MT's tentative circuit for a VXO using an artificial quarter-wave line to transform a variable capacitor into a 'phantom' variable inductance. The crystal 'sees' an inductance that is variable continuously from zero to 35µH by means of CT.

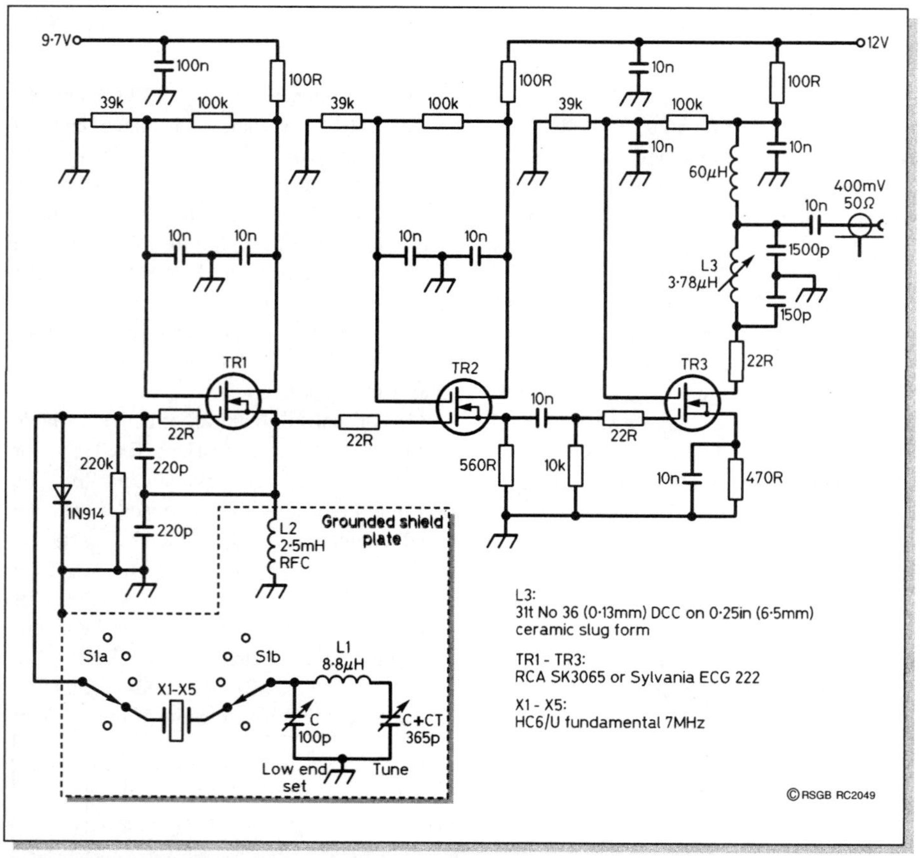

Fig 6: W3MT's prototype design for his 'phantom-coil' VXO, as a first model to test the theory. In his 1982 article he suggested various improvements. X1-X5 are 7MHz crystals, spaced 10kHz apart.

approach that of the operating frequency, reducing the stability of the coil. (3) Near series resonance, the net reactance is the difference between two large numbers, so the stability may suffer. (4) Bizarre capacitor-plate shapes are needed to obtain even reasonable frequency linearity."

W3MT argues that an improved tuning arrangement employs the reactance-inversion properties of transmission lines. A quarter-wavelength of coaxial cable terminated at the receiving end with a variable capacitor will appear as a 'phantom' inductance. While an actual transmission line could be used, its physical length on, say, 7MHz would be inconvenient in practice (although G3MCK hopes to try such an arrangement experimentally). Moreover, its characteristic impedance would be under 100Ω (typically 50 or 75Ω).

W3MT overcomes these problems by using an 'artificial' line based on lumped components; in his case the simplest possible - ie one-half of a 'half-wave' filter, comprising a single low-pass pi-section with all reactances numerically equal at the operating frequency. He provides the appropriate design formulae in his long article. **Fig 5** shows his tentative circuit in which the crystal 'sees' an inductance that is variable continuously from zero to 35μH by means of the trimmer capacitor CT which is 31-245pF or, omitting the left-hand C, should have the range of 87-301pF approximately.

W3MT stresses that by far the most important consideration in this design is the quality of the pi-section inductance, since the filter uses air capacitors, and the oscillator input capacitors can be excellent (air, polystyrene, silver mica, etc). The main requirements for the inductor L are: Low distributed capacitance (self-resonance far above 7MHz); High-Q, so that the oscillator coupling capacitors can be large; Small physical size, so that its field can be contained within shielding of reasonable size; Mechanical rigidity; Low temperature coefficient (less important if the heat dissipated within the enclosure is minimised and isolated from the coil). He stresses that any kind of ferrite or iron-core inductor should be avoided, since their temperature coefficient will be poor and the inductance may vary with signal level. He opted for the machine-wound B&W 'Miniductor' coils.

Fig 6 and **Fig 7** show how W3MT implemented the 'phantom coil' VXO as a first model, with the TR1 dual-gate MOSFET followed by two similar devices as buffer amplifiers, the complete unit providing an output of 400mV at 50Ω. He used five 7MHz crystals spaced 10kHz apart (7010-7050kHz) presumably limiting the VXO tuning range to 10kHz using partially plated fundamental AT cuts, type HC6-U specified for series-resonant frequency. The stability performance appears to have been on a par with an equivalent fixed frequency crystal oscillator. He concluded his article by listing a further six detailed improvements to this design, which he hoped other experimenters would develop further. I cannot recall reading of any later developments.

G3MCK suggests that the phantom-coil VXO appears to be a novel approach and would provide a good use for the Smith Chart as a design aid.

HERE & THERE

JOHN RABSON, G3PAI, draws attention to a note in the August 1998 issue of the *BT Retired Staff Section Newsletter,* from one of those volunteers who help maintain the equipment of the 'Talking Books for the Blind' service. H D Bickley writes: "When a fault is reported, the nearest Service Volunteer (SV) visits the subscriber to do the necessary - usually on the spot. For the volunteer it can be both a rewarding and a humiliating experience. Rewarding as doing a good turn can be rewarding, but humiliating when one sees at first hand the difficulties and hardships faced and cheerfully overcome by sightless people. As an example, one of my blind people built radio receivers for a hobby. He soldered all his joints. His joints might not have passed muster in BT workshops, but they did work. His method was to use a Weller Solder Gun (in which the 'bit' heats up almost instantaneously when the trigger is pressed and cools off as rapidly on release). He placed the wire on the tag, applied the business end of the gun, locating it with a finger of his left hand, replaced his finger with the end of the solder wire, and squeezed the trigger. He handled colour-coded resistors by having them threaded on a strip of paper which carried the values punched in Braille. . . Any-

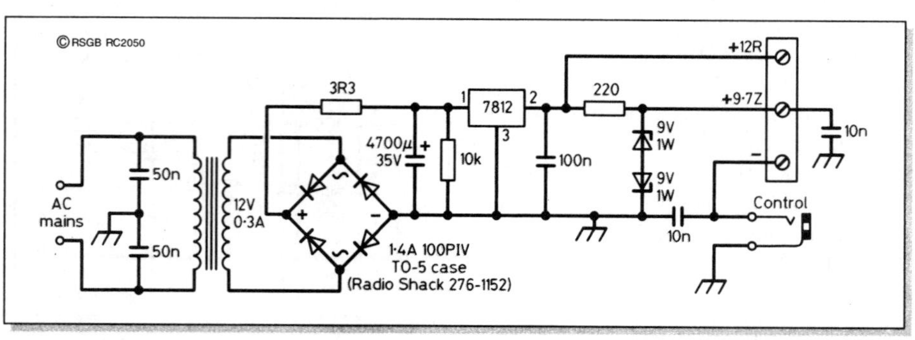

Fig 7: Power supply for the VXO. It was built in a separate enclosure, to isolate the VXO from vibration and heat.

one interested in joining the SV service can obtain details from the RNIB."

In respect of VXOs and many other aspects of quartz crystal practice, Etienne Nisolle, ON4KEN, draws attention to an article '*Crystal Testing*' by Detief Burchard, Dipl.-Ing. in *VHF Communications,* 3/93 pp162-177. This has nine sections: Introduction; Crystal resonances; More precise representation of resonances; Measuring the equivalent data; Measurement of temperature behaviour; Changing the resonance frequency; Applications; It is even simpler to measure the crystal; Literature.

DELAYED ECHOES - G3PLX's EXPLANATION

THE OCTOBER 1998 *TT* included some ideas on possible causes of the 210-millisecond delays recorded by Peter Martinez, G3PLX, in October and December 1997. Peter, himself, now writes:

"Geophysicists have given me more information about the Van Allen belts, and they no longer look very promising. All the indications are that although the energies of particles in the belts are high, the actual *number* of electrons is much too low, even in the strongest solar storms. There were, in any case, no significant solar events during my observed echoes. Also, my own studies of the ray paths involved show that when the F-layer just starts to show a 'hole' at the zenith with signals starting to escape into outer space (that is, a skip zone just starts to develop) the rays going through the hole are highly de-focused. In other words, rather than there being a narrow pencil beam shining through the hole, the signal fans out to cover a hemisphere. This means that the signal going straight up will be much weaker than the free-space value, although it will get stronger later on in the night. This is not consistent with my observed near-free-space signals in the evening.

"In addition, the 210mS echoes I hear correspond to a distance of 63,000km, which places the reflecting point 31,500km away. There are two Van Allen belts, but even the outer one is too low to be responsible - its inner edge is about 13,000km above the equator and is lower still at higher latitudes. The mean velocity factor of the intervening space would have to be less than 0.4 to account for this.

Before your piece in *TT* I had concluded that a topside field-aligned duct was the most likely cause, which I believe is what VE3HX is describing. The Canadian Alouette satellite in the 1960s did ionospheric sounding from the topside, and showed paths from the satellite at 500km above the equator which followed magnetic field lines towards each pole and reflected back from each end to the satellite. They even saw paths via both ends: see *Ionospheric Radio* by K Davies, p266.

"I believe that VE3HX assumed ground reflection and velocity of light to estimate the path length, but the signal is not actually reflected from the ground. As you point out, this would give the very high attenuation of a backscatter reflection. In practice, the reflection is a field-aligned critical reflection in the top side of the F-layer, in which the ray is brought to a standstill in the topside of the F-layer and then re-traces its path [**Fig 8** is my rough approximation (not to scale) of G3PLX's signal path - G3VA]. The signal velocity drops as it dips into the top of the F-layer, the effect of which is frequency-dependent, giving a longer delay at lower frequencies. But this part of the path is only a small fraction of the total, so the effect will only be slight. Last winter I tried to see if there was any change in delay time between 3.5MHz and 3.8MHz, but could not detect it, which means it was less than about 2mS. I could hear no echoes on 2 or 7MHz. Amateur bands at 2.5 and 5MHz would have been useful!

"To place a signal into this duct from the north end requires that the critical frequency is lower than the transmission frequency above the transmitter, so that the signal can bend around the top edge of the 'hole', until it lines up with the magnetic field and is then trapped along it. At the southern end, the critical frequency must be higher than the signal frequency, so that as the guiding field line gets deeper into the topside of the F-layer, the wave reaches a level where reflection occurs and the signal retraces its path. Such conditions would apply in northern hemisphere winter after sunset, and that is when I observed echoes.

"I am listening for echoes again this winter with the DSP set-up. Last year, in addition to the chirp sounder trace, I made a stereo tape recording of the echoes, with the transmitted chirp on one channel and the received echo on the other."

This all seems to be a logical solution to the problem of MDEs, although I remain a little puzzled at the apparent rareness of these propagation conditions.

G3PLX has also been investigating the Luxembourg Effect (see *TT*, February, July and October 1998), which he and other European amateurs have observed on a German 138.82kHz transmitter which radiates an unmodulated carrier between bursts of FSK data. Several amateurs noticed faint broadcast audio on the carrier after sunset, resulting in sidebands spreading into the amateur 136kHz band. It became apparent that the broadcast audio was (a) not from a local broadcast transmitters; and (b) was not the same for different observers. In Germany, the audio was from a German Long Wave transmitter in another part of Germany. In Sweden, it was from a Swedish Medium Wave transmitter (close to the gyro frequency). G3PLX in Cumbria found he was getting a mix of French LW transmissions and Radio 4 from Droitwich.

G3PLX continues: "Studying the books about the Luxembourg Effect shows that the thermal time-constant of the ionosphere is in the range 60-600μS, varying with reflection height. This means that the frequency response of the cross-modulating audio should be noticeably lacking in high frequencies [as noted in the earlier *TT* items], with progressively more 'top cut' later in the night. This characteristic should make it easy to eliminate the possibility that the cross-modulation effect is occurring in one's own receiver, which would have the same 'flat' response as the originating audio. I believe that I can detect this effect by ear, but have some ideas for using the DSP with its stereo inputs to correlate the crossmod with the original broadcast. I am hoping it will be possible to obtain a number of useful measurements from this technique."

While my linking MDEs directly with magnetic storms has to be retracted in view of G3PLX's studies, it seems germane to note that Joseph Borovsky in 'Still in the dark' (*Nature*, 28 May, 1998, pp312-313) wrote: "After decades of research, our understanding of the aurora is disturbingly incom-

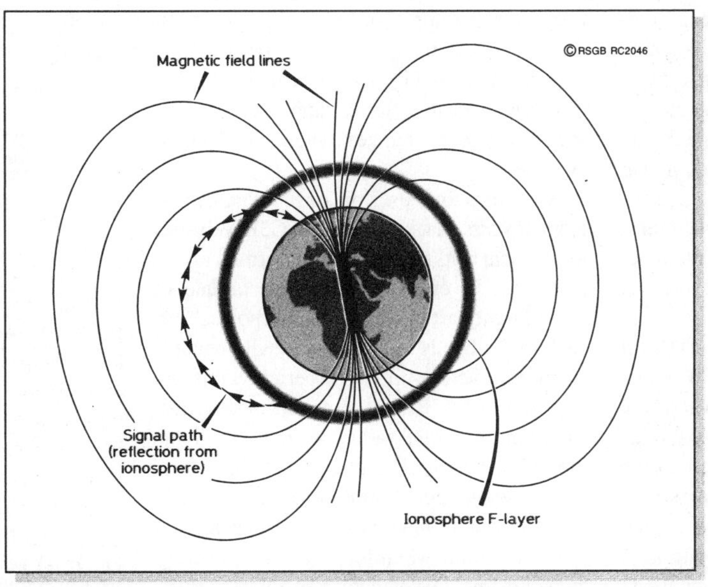

Fig 8: A rough approximation (not to scale) of the possible signal path that resulted in the 210mS echoes on G3PLX's 3.5MHz transmissions in October and December, 1997.

plete. One long-held belief, that aurorae are much more common following each peak in the 11-year solar cycle, has now been overturned by Newell and colleagues, reporting on page 342 of this issue."

Patrick T Newell *et al* introduced their report as follows: "The oldest documented relationship between the number of sunspots (the solar cycle) and terrestrial effects is the increased frequency of aurorae in the period immediately after the solar maximum (the peak of the number of sunspots). This correlation is, however, based only on observations of the relatively rare events of 'great aurorae' which are those that reach mid-latitudes or lower. The overwhelming majority of intense aurorae, and therefore most of the energy put into the ionosphere, occurs at high latitudes, where aurorae appear nightly. . . We find that, contrary to expectations, the total number of intense aurorae is uncorrelated with solar activity in darkness, and is negatively correlated with solar activity in sunlit conditions. These findings imply a causal relationship between aurorae and ionospheric conductivity (the latter is maximal at solar maximum) and therefore indicate that the occurrence of intense aurorae is a discharge phenomenon, similar to lightning."

N1IR, ON HIS FRACTALS

AN UNEXPECTED letter has arrived from Professor Nathan 'Chip' Cohen, N1IR, who has done so much pioneering work on 'fractal antennas' (see *TT* April and June, 1998), and whose attention had been drawn to the April item while in contact with G3BDQ. N1IR was using his 28MHz fractal quad-yagi antenna. He writes: "Given the sensational nature of the *New Scientist* ['Aerial Magic'] article, I thought *TT* did a very good job in presenting a sane perspective on this. . . There is, of course, no magic. But there are advantages when compared to other loading and/or band-broadening methods. The 4dBd dipole (Cohen dipole) mentioned, for example, is not for an electrically small antenna, but one that has a fractal stub at its centre, giving it echelon/colinear phasing. A Landstorfer dipole [see the June item] also shares this attribute, but is larger. I have found it convenient to look at meander and Landstorfer structures as the first 'pass' at a fractal.

"The main advantage of the fractal is that it provides ultra wide bandwidth (UWB) capabilities at the higher frequencies, where the fractally bent wire (for example) is very large electrically. This is of limited interest to the ham community - emphasis on shrinkage seems to be the issue. Here the fractal gives a small antenna with very high efficiency. Naturally, this is at the expense of bandwidth, and the Q increases for the lowest resonance. Upper resonances are usually *not* harmonic, and often merge from the 'fractal loading' effect. It is also noteworthy that the fractal generator and iteration allows one to 'tune' the antenna as desired, for multiband operation for instance.

"Readers may wish to look at the web site *www.fractenna.com* In addition to the ham page, a 'white paper' should be posted in the New Year."

EARTHING UPSTAIRS TRANSCEIVERS

FOR MANY YEARS a common problem when operating a rig from an upstairs shack was the tendency for the transmitter/ATU etc to become 'hot' to RF. This can lead to excessive coupling of RF into the mains wiring, or finger-burns when adjusting metal controls, etc, as well as acting as a potential source of RFI. While this most commonly occurs when using voltage-fed long-wire antennas, it can occur with most types of antennas. A 'real earth' wire to an upstairs shack is of little assistance in curing this condition, since the earth lead will inevitably be long in terms of wavelength.

However, in 1971, Dick Halls, G3EIW/ON8KM, reported (*TT*, December, 1971, p839) that he had often found himself operating from flats in various parts of the world where the only practical antenna was a long-wire strung out to a tree or other conveniently high point. In Singapore, as 9V1LK, he used a 500ft wire that proved highly effective, particularly on 28MHz. But he had found, as many others have, that in such circumstances the earth lead on the transmitter and/or matching unit often has standing waves with a voltage point where 'earth' should be.

His solution was that instead of trying to connect a real earth, he substituted a quarter-wavelength of wire for the band in use, causing the RF hot-point to disappear along with the other problems, while the radiation and radiation patterns improved. For multiband operation a series of wires can be left permanently connected as in **Fig 9(a)**; alternatively, it is possible to have a series-tapped or variable inductor which changes when the ATU is switched.

As a long-time user of long-wire antennas from upstairs shacks, I soon put G3EIW's idea to practical test, and found it performed just as predicted. In effect, G3EIW's technique of providing quarter-wave 'counterpoises' for each of the bands in use has virtually become standard practice. In my experience this not only eliminates 'RF hot spots' but also significantly increases RF radiation from multiband long-wire antennas, particularly when these are of random (not accurately resonant) length. Some amateurs have also found that RF hot spots can occur when feeding almost any antenna via unbalanced coax transmission lines, and the technique can be used to overcome all such problems.

Pault Debono, 9H1FQ writes: "My shack is situated on the third floor of my house, so I have some difficulty in providing an effective ground for my shack. I have experienced lots of RF flying about inside my shack. One way of taming this RF is to 'earth' my TS850S transceiver with quarter-wave radials, cut to the various operating bands. However, there is the problem of the limited space available inside my shack, because the radials have to run in straight lines, otherwise their frequency will change. Finally I came up with the idea of using quarter-wave coaxial stubs, connected as rejector [acceptor? - G3VA] tuned circuits. These stubs can be coiled, bent, twisted, laid on the floor, or in free air. Any configuration seems possible without detuning, unlike conventional wire radials. In my case, the stubs are hidden away behind the desk. The stubs should be left open and are tuned to the centre frequency of the band using (for RG58) the formula 234/MHz x 0.66, to provide the approximate length in feet, with final adjustments by GDO."

While I have no doubt that it is possible to use lengths of coaxial cable as the counterpoises with the braid providing an insulated 'fat' cable that can be coiled etc, I have strong reservations about the explanation and rough sketch (not shown) provided by 9H1FQ. I do not see how the lengths of cable connected his way could really act as coaxial stubs, but seem to be similar to G3EIW's wires, nor that the use of the velocity factor of RG58 is justified. It seems to me that the outer braid of the cables are simply acting as fat wires rather than as stubs. In my experience, insulated wire counterpoises can be bent around without affecting results, since accurate resonating of these wires is not usually critical. Nevertheless, 9H1FQ's letter is a useful reminder that this technique can be extremely useful in those situations where a truly effective RF earth cannot be provided. ♦

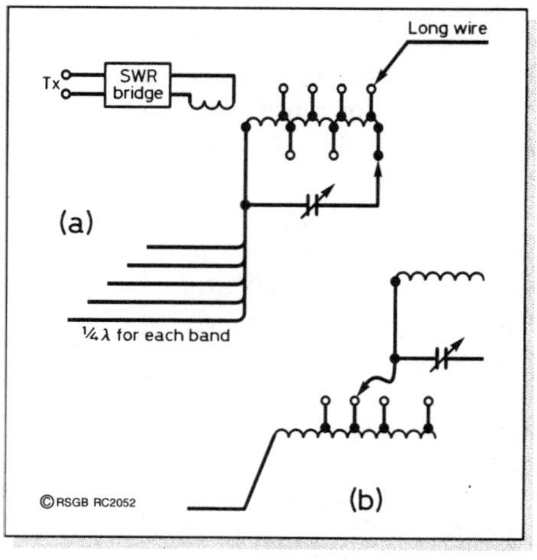

Fig 9: (a) Series of quarter-wavelengths of wire used to produce artificial earth for a multiband long wire antenna, for use in an upstairs shack. (b) It is possible to load the counterpoise wires using a tapped inductor with single wire.

technical topics

by Pat Hawker, G3VA*

SMALL LOOP TRANSMITTING 'MAGNETIC' ANTENNAS

IT IS NOW a year since *TT* reported the view expressed by Professor Mike Underhill, G3LHZ, and M J Blewitt, G4VRN, at the IEE 1997 conference on 'HF radio systems and techniques' that a "small folded dipole mode was usually the dominant mode". This represented a serious attempt to reconcile the anecdotal evidence that the performance of these small loop antennas appeared better than suggested by conventional theory (for example, as investigated by Tony Henk, G4XVF, in 1991). See **Fig 1** for various methods of matching small loops to 50Ω cable.

These views attracted mostly adverse comment from a number of readers, with ripostes from G3LHZ. I would like to report that this controversy has been settled one way or another; however, both sides continue to stick to their original opinions, presenting additional evidence. The matter is far from settled, though tending to move into exotic areas that have only limited bearing on the practical use of small mag-loop antennas by amateurs.

However, in some respects, both factions are coming to agree on certain aspects. There is convincing evidence that significant dipole-mode radiation does occur, increasing as the circumference of the loop approaches a quarter-wave, though it seems unlikely that in practice this is the dominant mode claimed by G3LHZ. There is also evidence that this dipole-mode radiation is taken into account in the NEC 'Method of Moments' software. The work at the University of Surrey also shows that Q factor made by measuring the bandwidth is significantly affected by coupling into the environment, ie by creating eddy currents in nearby structures.

From the viewpoint of the user, the eddy current losses account quite significantly for the differences in efficiency of the antenna calculated from RF-resistance losses in the loop and the overall radiation efficiency of the antenna. The eddy current losses contribute to the radiation resistance, thus, to some extent, they account for the relatively low heating of the loop as found by G4VRN and other users. Measurements made by G4VRN suggest at least 30% of the total radiated power may be absorbed by the environment and then either dissipated as losses, or re-radiated, causing distortion of the tuned loop antenna pattern. It would thus seem that the siting of a small loop antenna affects it performance and that it is wrong, as sometimes stated, that this is little affected by height and surrounding structures. It does not however invalidate G4XVF's finding that the effective overall radiation efficiency of a small loop, as found from far-field measurements, is relatively low (usually under 10%) compared with a dipole at a reasonable height.

The most significant new contribution to the debate comes from John Crabtree, writing from South Edina, MN, USA (e-mail: crabtreejr@aol.com). Here I quote briefly from his letter:

"I have read the item on the loop antenna debate with interest, and offer another perspective on the central issue, whether or not significant dipole radiation from a tuned loop is fact or fiction.

"The different current modes in loops have been extensively analysed in 'Antennas and Waves: A Modern Approach' by Prof Ronold W P King and Charles W Harrison jr, MIT Press, about 1969. Chapter 9 on 'The Loop Antenna: Currents, Impedances, Fields' includes a section 'The Small Loop: Electromagnetic Field' (pp577-581) which is particularly relevant. [Those seeking to consult this erudite book must be warned that it is highly mathematical, virtually double-Dutch to me, and I depend below on John Crabtree's interpretation - *G3VA*.]

"The section starts with the comment 'In numerous applications the loop antenna is electrically quite small... However, it is not always true - as often assumed - that all higher-mode currents may be neglected.' King goes on to develop a model of the parallel loop impedances for the loop and dipole modes (page 581).

"On page 578 King shows a graph of the ratio of the amplitude of the far field of a circular loop maintained by currents in the dipole mode to that maintained by currents in the circulating mode. The graph is approximately linear, with the maximum amplitude of the fields being the same when the loop circumferences/wavelength is approximately 0.4. I conclude that there can be significant dipole mode radiation from a small loop."

John Crabtree also shows that King makes it clear that the dipole mode plays a significant (but not dominant) role in the moderately small loop. He also draws attention to 'A note on the radiation resistance of loop antennas with short circumferences' by Peter Bertram, DJ2ZS (*Communications Quarterly,* Summer 1997, pp99-100). DJ2ZS modelled a loop antenna using EZNEC and obtained results in very good agreement with those of King... I conclude, rightly or wrongly, that NEC takes imbalanced currents into account. This is a point disputed by G3LHZ, who believes that NEC fails to predict the tuned loop correctly."

Alan Boswell, G3NOQ, a professional antenna engineer, has also confirmed by NEC that out-of-balance loop currents do exist when an AMA-3 loop is simulated and give dipole mode radiation from a tuned loop. His results indicate that the dipole mode becomes weaker than the loop mode by -6db per octave as the frequency is lowered below 40MHz. Even at 40MHz the dipole mode is not the dominant mode, despite the more favourable circumference/wavelength ratio. G3NOQ agrees with G3LHZ that the losses associated with loop antennas are largely the result of eddy currents

*37 Dovercourt Road, London SE22 8SS.

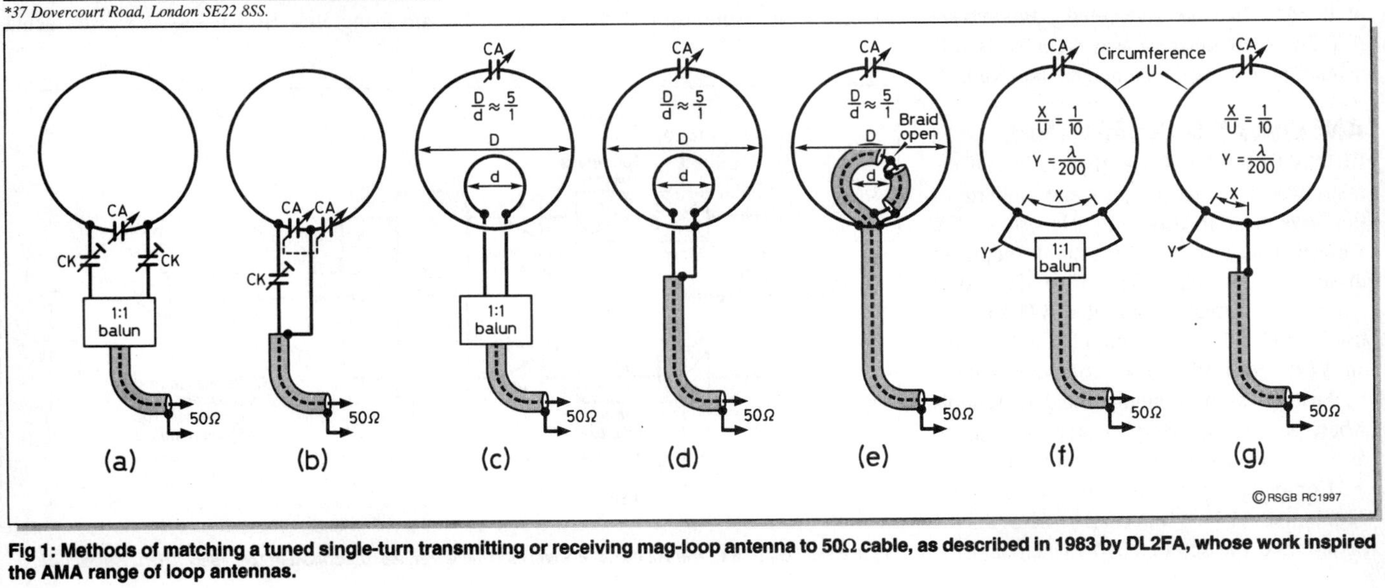

Fig 1: Methods of matching a tuned single-turn transmitting or receiving mag-loop antenna to 50Ω cable, as described in 1983 by DL2FA, whose work inspired the AMA range of loop antennas.

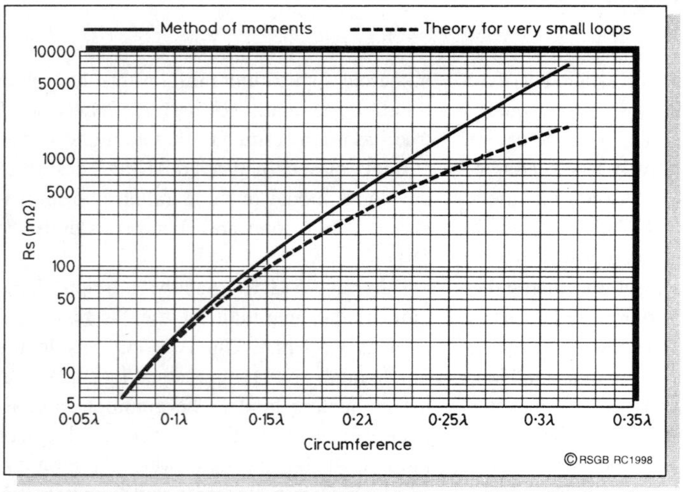

Fig 2: Radiation resistance of small loop antennas (assuming loss-less conductors) as calculated by DJ2ZS: Theory vs. Method of Moments.

in objects near the loop, and that the Q cannot be predicted by merely using the RF resistance of the conductor as the loss resistance in the equivalent circuit. DJ2ZS accepts that a mag-loop antenna works slightly better than predicted from the basic formulae that take no account of displacement currents: **Fig 2**.

John Crabtree concludes that there is significant dipole radiation from small loops, but is sceptical of G3LHZ's claim that on any tuneable frequency the dominant mode is that of a small folded dipole.

All this points to the validity of G4XVF's 1991 finding, supported by his far-field measurements, that the total effective radiation efficiency of small loops is very low (usually under 10%) and that the good results achieved by many users only underline the tolerance on HF of antennas with low overall efficiency. Tolerance to low efficiency is strained when using very low power. As many enthusiasts have found, QRP performs well with a good dipole, as can a mag-loop with reasonable transmitter power. A loop cannot be recommended for QRP unless made mandatory by space considerations, etc.

It should be appreciated that there is a limit to the space that can be devoted to this subject in *TT*, and that the above notes represents only a small portion of the comments received.

ANTENNAS & COMPUTERS

BRIAN CLOWES, GW4HBZ, is one of the many who feel that progress and modernisation have much to answer for. He writes: "We know that not many amateurs build equipment these days, but I feel things are getting worse when I see computer-controlled ATUs advertised. Surely an amateur should be able to tune an ATU? Over reliance on computers cannot be healthy, as now being found in schools where pupils often can't do simple arithmetic or even make rough estimates in their heads.

"Computers are fantastic tools if used with common sense, but I see articles where gains or losses are expressed to 0.01dB when 0.5dB is a barely perceptible change and difficult to measure, even with thousands of Pounds worth of test gear. There is a tendency to believe software is always correct. I asked an antenna manufacturer how they could claim a ridiculously high gain figure and the answer came back 'Our computer has calculated the figure'. They had made no attempt to actually measure the gain."

It is, however, necessary to recognise that *properly used* current antenna-modelling programs such as NEC-4 etc can provide realistic and acceptable performance figures, without the use of a professional antenna test range. ARRL, which, for 35 years, refused to allow antenna manufacturers to claim performance figures in *QST* and other ARRL advertisements, in 1998 reversed this policy and now permit these to be claimed, *provided* that the advertiser submits to the ARRL Laboratory their antenna computer models and the ad itself for checking *before* the claims can be published. The former ban covered forward gain, front-to-back ratio and radiation patterns. Claims made in catalogues, brochures and non-ARRL periodicals or by salesmen are, unfortunately, not subject to such rigorous validation and may need to be viewed with a degree of scepticism, depending on the reputation of the firm concerned!

N7VE's QUADRATURE MIXER/ PRODUCT DETECTOR

DURING 1998, the successive efforts of Gian Moda, I7SWX, Bill Carver, W7AAZ, and Colin Horrabin, G3SBI (*TT*, July, August and September, 1998) demonstrated the exceptional performance of the new generation of fast bus switches such as the FST3125 when used as H-mode front end mixers.

Now, from across the Atlantic, (via G3SBI) comes information on how these devices can be used as improved quadrature product detectors in high performance direct conversion receivers and the like.

In November 1998, Steven Weber, KD1JV, e-mailed G3SBI with information on a new switched quad-FET product detector (Tayloe mixer) developed by Dan Tayloe, N7VE, using the 74CBT3253 (similar to Fairchild FST3253 or Pericom PI5C3253 devices). KD1JV has tested the Tayloe mixer as a quadrature-type product detector in connection with the 'R2' image rejecting DC receiver designed originally by Rick Campbell, KK7B (*QST*, January 1993 and subsequently in various editions of *The ARRL Handbook* (1996 edition, pp17.72 -79)): block diagram **Fig 3**. G3SBI outlined how he felt the 3253 could be used in this application (**Fig 4**), but also got in touch with N7VE who was then currently putting the final touches to a 3.5/7MHz dual band CW transceiver using the new product detector. N7VE replied: "I am very excited, as the rig is working *very well.* I am currently measuring -136dBm MDS sensitivity (signal+noise 3dB greater than noise). I can just hear my generator at about -146dBm. This is *much better* than a stock R2 receiver and is even better than my NE602 type superhet receivers."

As KD1JV puts it: "The 3253 switch of the quad FET product detector is a high speed addressable one-of-four multiplexer (MUX) clocked by the Q0 and Q1 outputs of a binary counter (74ACT161 in my case) at a times-four receive frequency rate. RF input is through a 50Ω resistor to the switch common (the gate needs to be biased at one half Vcc also). The outputs are in the form of 0.22μF metal film capacitors. This results in a four phase audio output: 0, 90, 180 and 270°. The 0, 180 and the 90, 270° phase outputs are summed with a differential amplifier to give I and Q outputs. I am using Burr-Brown INA103 instrumenta-

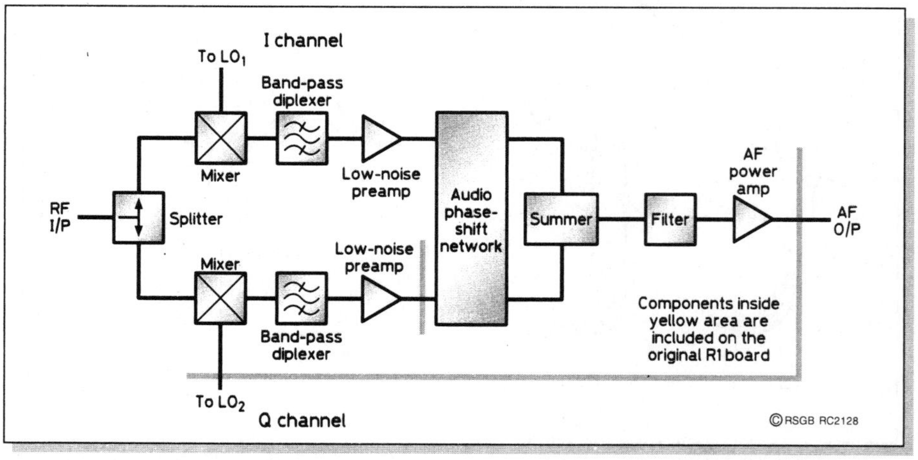

Fig 8: Block outline of the R2 direct conversion receiver, as described in the 1996 *ARRL Handbook*.

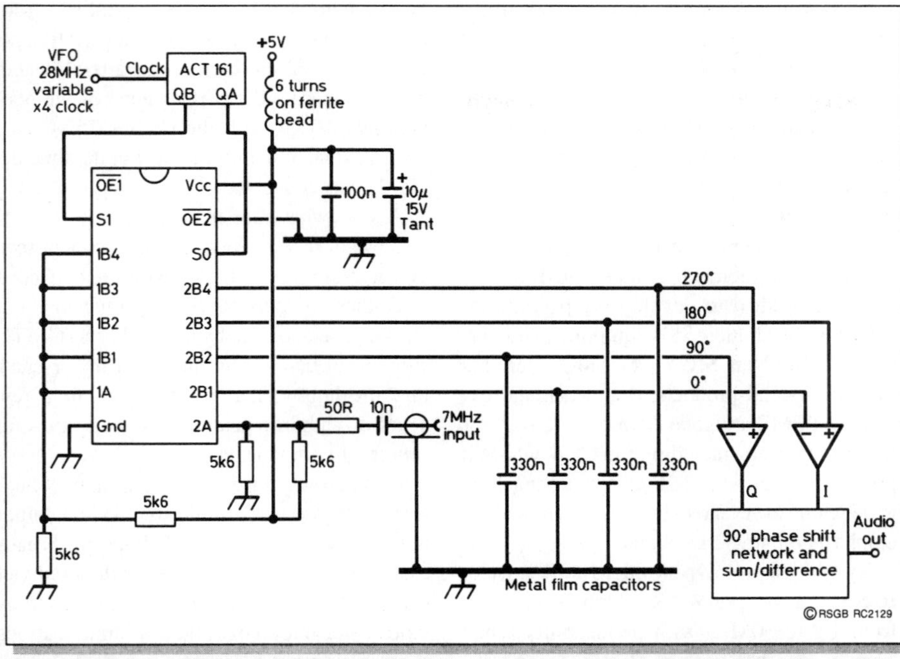

Fig 4: G3SBI's simplified form (not tested) of a 3253 used as a product detector. Note that the oscillator input is at four times the signal frequency.

tion amplifiers, which have very low noise and low third harmonic distortion (THD). This simplifies this part of the circuit significantly.

"The third order intermodulation product (IP3) is of the order of +30dBm and the minimum detectable signal is as good as the following audio amplifier permits. The maximum RF signal input to this product detector is limited mainly by the point at which the post mixer audio amplifier saturates.

"One interesting feature is that this detector is virtually immune to signals outside its bandwidth." G3SBI adds that as normal with an I, Q single sideband (image rejecting) receiver, a wideband audio phase network is used to provide upper or lower sideband reception. Opposite sideband rejection depends on the precise tolerances of the components in the phase shift network, but is typically of the order of 45dB suppression with the network described for the R2 receiver. A particularly useful characteristic of the system as implemented by Steve Weber with the Burr Brown Instrumentation op-amps type INA103 is the absence of the hum problem that is recognised as one of the main problems with high performance direct conversion receivers.

Fairchild describe their surface mounted FST3253 as follows: "The Fairchild Switch is a dual 4:1 high-speed CMOS TTL-compatible multiplexer/demultiplexer bus switch. The low on-resistance of the switch allows inputs to be connected to outputs without adding propagation delay or generating additional ground-bounce noise. It features: 4Ω switch connection between two ports; minimal propagation delay through the switch; low Icc; zero bounce in flow-through mode; control inputs compatible with TTL level; available in SOIUC and TSSOP packaging."

As noted above, G3SBI provides a simplified circuit diagram (Fig 4) of this detector (not at that time tested). The complete Tayloe detector as implemented by KD1JV is shown in **Fig 5**. Although the DC receiver tends to be thought of as a cheap and cheerful form of receiver for QRP rigs, it should always be remembered that the first published mention of a two-phase (synchronous) direct conversion receiver was that of the AN/FRR-48 (XW1) by John P Costas, W2CRR, in the classic SSB issue of *Proc IRE*, December 1956. The experimental AN/FRR-48 was a complex two-phase synchronous valve receiver with elaborate frequency synthesiser and phase lock loop demodulator, designed to more than complete with the best superhet designs of the era. As we have noted before, the idea of a high performance image-rejecting receiver is not a new one - it just happens that relatively few people have ever taken it seriously, leaving plenty of room for further experimental work.

An alternative approach to improving the R2 image-rejecting direct conversion receiver has been described by Rob Frohne, KL7NA, ('A High-Performance, Single-Signal, Direct Conversion Receiver with DSP Filtering' *QST*, April 1998, pp40-45). KL7NA added a DSP demodulator based on a Motorola DSP56002EVM evaluation board. While this

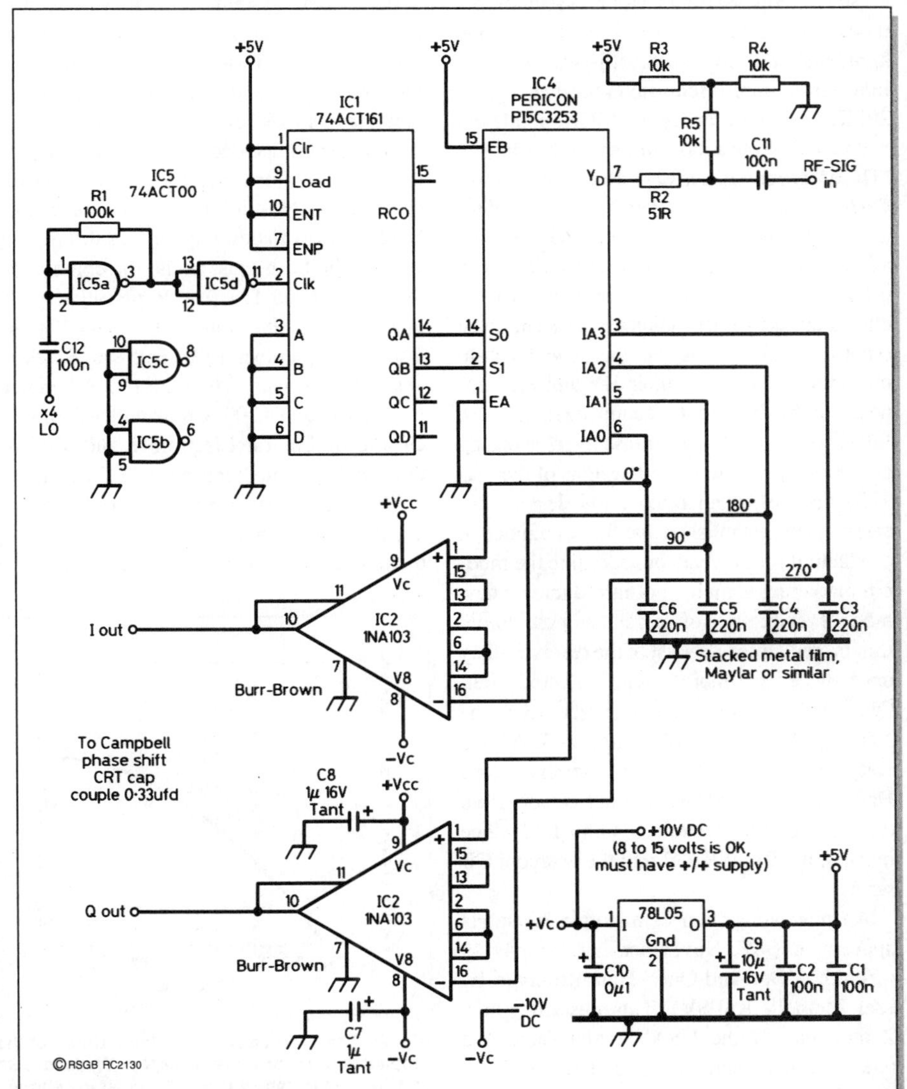

Fig 5: The Tayloe product detector, as devised by KD1JV.

approach is shown to be capable of providing much improved selectivity (1.1:1 shape factor, compared with 1.6:1 in the original R2 design), the blocking dynamic range and the IMD dynamic range are reduced, and the sideband rejection only much the same as can be achieved with the Tayloe detector/R2 approach.

CLASSIC VALVE RECEIVERS

I WOULD LIKE to express my profound thanks to the dozen or so readers who responded to my request (December *TT*) for information on the Marconi CR150 receiver. In particular to GMM Townsend, G3LSX, of Eltham, who not only phoned me on the day of publication but then promptly brought round and presented me with a copy of the original Marconi manual (covering the standard receiver plus the additional units when used in a triple diversity application) - an outstanding example of still existent Ham Spirit!

The CR150, although part of the family of designs (developed largely by Dr George Grisdale, G5GZ) that includes the well-known CR100 (Navy B28) series is of radically different design, covering 2 to 60MHz in five bands, as a double superhet with two stages of 465kHz crystal bandpass filters (four crystals). Although of limited sensitivity on the unusual high band of 30 to 60MHz, it is still an excellent HF communications receiver below 30MHz. Gerald Stancey, G3MCK, points out that G5GZ wrote a most informative article on 'The Marconi CR Series of Receivers' in *Radio Bygones* (No 31, October/November 1994).

G3LSX told me that when he had first tried a few years ago to dispose of his own CR150 by offering it to any young member of his local club, there had been no takers. This seems to be a not unusual response among newcomers to our hobby. Too often their HF ambition appears to be limited to acquiring a modern 'black box' solid-state transceiver, scorning all the many classic valved designs of the pre-1970 decades when often good designs became so well established that they continued in production for 20 years or so, unlike the modern marketing technique of introducing a new model every couple of years. I suppose all old timers could draw up a list of the receivers they once owned or, more often, hankered after. Often these included American sets such as the National HRO, RCA AR88, Collins 75A4 and some of their professional models, some Hallicrafters, some Hammarlund, etc, or - from this side of the Atlantic - some Eddystone models, the Racal RA17 and the Marconi CR series.

An interesting item of historical nostalgia appears in *QST*, November, 1998, pp47-49 ('Zenith's 'One-and-Only' Ham Receiver' by Joel Thurtell, K8PSV). This describes how Zenith, one of the USA's major radio and television manufacturers, entered but then withdrew from the amateur radio market. In 1952 the small independent firm of Central Electronics (Walter Schum, W9DYV, and Joe Bachelor, W4EGK, as chief engineer) introduced the low power Model 10A - one of the first SSSC (single sideband suppressed carrier, ie SSB) transmitters. By 1958 Central had developed the 100V and then the 200V models, but needed funds to expand to meet the growing demand. A merger with Zenith left the major firm in control, but keen to fund the production of amateur equipment. Schum and Bachelor had long wished to develop a top grade receiver to match their SSB equipment and persuaded Bill Van Slyck, W9EMB, Zenith's head of special products, that a demand existed. W9EMB assigned the project to some of his top engineers, including Jim Clark who had previously been with Hallicrafters, telling them to develop "the best receiver ever built, with an emphasis on single sideband".

By 1960, a prototype Model 100-R existed. It covered the pre-WARC ham bands only from 1.8 to 30MHz as a permeability tuned double superhet with a 50kHz second IF, achieving three degrees of selectivity for AM, two each for USB and LSB, and one for CW. It included a special bifilar compressor developed by Joe Bachelor, which was an RF-derived AGC system. This, it was claimed, made the front end virtually immune to strong signal overload. It was planned to produce five more prototypes and then begin marketing the receiver in 1962, but before then a new president at Zenith decided amateur radio was not right for his major corporation. Late in 1961 he ordered the closure of Central Electronics. Everything from design plans to sales receipts went to the trash bins, as did the components for the planned five 100-R prototypes. The lone 100-R receiver vanished for over 35 years. Then, in September 1997, Bill Van Slyck revealed that he had bought the 100-R, a matching speaker and a 100-V transmitter during the closure of Central Electronics and all three units had remained unused in his basement!

They were acquired by Joel Thurfell, K8PSV, but after playing with them and photographing them in his shack he has sent them on loan to Wes Schum, the founder of Central Electronics, who now has them on show at what he calls his 'Central Electronics headquarters', where he is comparing the performance of the one-and-only 100-R with the Collins 75A-4, the leading amateur bands receiver at the time the 100-R was conceived.

A somewhat similar happening in the UK was when, in the 1960s, the independent firm of Electroniques, which specialised in coils and other components for the amateur radio market, was acquired by the mighty STC. STC announced that it was making a major entry into the hobby market. Alas, within a few years, it withdrew and we lost a most useful source of components.

It sometimes puzzles me that many collectors seem far more interested in the rather mundane ranges of valved domestic broadcast receivers, notable more for their cabinet designs than their performance. The enormous contribution of valved communications receivers to radio seems often to be overlooked, yet many of the classic models still stand up well alongside the latest solid state designs. One firm that does cherish a sense of its history is Eddystone, both as a brand name of Stratton & Co for 40 years and then since the mid-1960s as Eddystone Radio Ltd (a member of the GEC-Marconi group). The firm has continued to support the *Eddystone User Group Newsletter,* a non-profit newsletter for Eddystone users, edited by Ted Moore (details from Eddystone User Group, c/o Graeme Wormald, G3GGL, 15 Sabrina Drive, Bewdley, Worcestershire, DY12 2RJ, Tel: 012399 403372). During 1998 this always informative and interesting newsletter passed its 50th issue.

Sadly, since writing these notes, it has been announced that Eddystone Radio Ltd is due to close shortly (but EUG will continue).

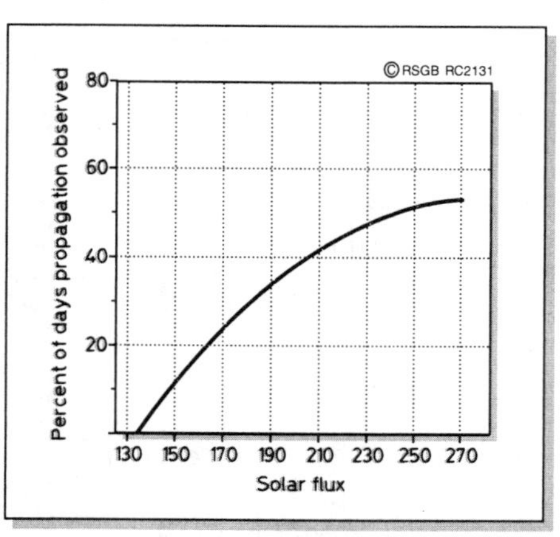

Fig 6: The percentage of days that 50MHz propagation was observed from New England, USA and Western Europe during November to February 1988-89, 89-90 and 90-91, plotted against the solar flux level.

50MHz TROPO

WHEN, IN 1985, the then UK licensing authority - the Department of Trade and Industry - accepted the recommendation of the Merriman independent advisory panel that the band 50 - 50.5MHz should be allocated to British radio amateurs, I noted in *TT* (September 1985 and subsequently *Technical Topics Scrapbook, 1985-89,* p50) that this band represents a most valuable and in many ways unique acquisition. I stressed the possibilities offered by the following 50MHz propagation modes:

- World-wide F2 ionospheric propagation in daylight at periods of very high sunspot activity (best determined from the daily solar flux number), with occasional, though possibly rare, openings towards the south in most phases of the cycle.
- Transequatorial and field-aligned F2 propagation (TE) in the evenings, even though the UK is often considered too far north for this to be at all common.

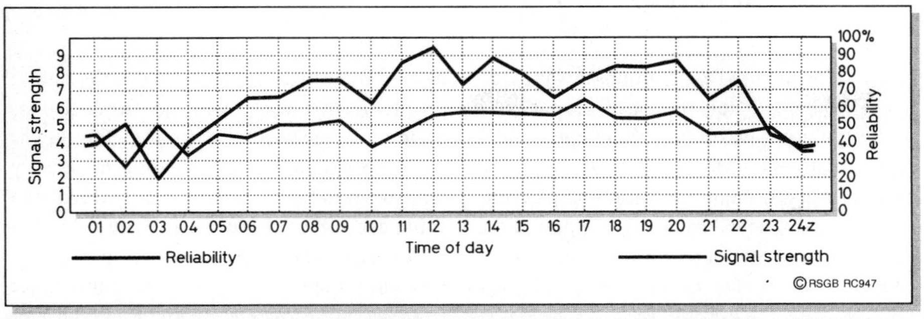

Fig 7: Results achieved during the initial 50MHz propagation tests between G4IFX and G2AHU during weekends and holiday periods between 17 November 1995 and 7 January 1996. The minimum number of tests on each hour was 14 and the maximum 21. The reliability shows the number of times a signal was received divided by the number of valid tests expressed as a percentage, and the signal strength as the average of the signal strengths when heard.

- Ionospheric scatter modes (with extremely high effective radiated power) which might provide reliable 24-hour communication over distances of from 500 to 2000km.
- Almost the optimum frequency for sporadic-E and meteor burst communications.
- Reasonable for auroral propagation.
- Some degree of tropospheric bending, ducting and tropospheric scatter modes, though less pronounced than at 144MHz and above.
- Significantly better local propagation over hills and down into valleys than at 144MHz.

In effect, 50MHz rather than 30MHz represents the real transition between HF and VHF propagation, supporting at times the modes of both, although this may not have been obvious during the recent minimum phase of the sunspot cycle, now safely behind us. I note from my daily logging of the DK0WCY beacon (see *TT*, March 1998) that the solar flux number has since 12 August 1998 been occasionally exceeding 150 and even reached 175 in early September. Emil Pocock, W3EP, in his regular *QST* 'The World Above 50MHz' feature noted in the November 1998 issue: "By early September the solar flux had exceeded 150 for more than a week, an encouraging sign that world-wide 50MHz propagation may be possible as early as this fall - perhaps as you read this column." W3EP reproduced **Fig 6** from his article 'Predicting Transatlantic 50MHz F-layer Propagation' (*QST*, March 1993, pp32-34) showing that with the solar flux exceeding about 170, transatlantic contacts were made on about 20 per cent of days during November to February in 1989-90 and 1990-91.

But it should be stressed that F2 DX is only one aspect of 50MHz propagation that deserves investigation and exploitation. In February 1996, Ray Cracknell, G2AHU, who as ZE2JV did so much to investigate and unravel the mysteries of transequatorial propagation (TEP), sent me some preliminary results he was achieving in co-operation with Chris Deacon, G4IFX. G2AHU had been regularly monitoring G4IFX's 50MHz signals over the 290km path between Darlington, Co Durham and G2AHU's QTH at Yarpole in northern Herefordshire during weekends and the Christmas holidays from 17 November and 7 January 1996. These initial tests preceded the issuing by the Radio Authority of a special beacon licence and had to be limited to times when G4IFX was at his home QTH. One result was an early publication on Internet pph@dc.dk of an item by P Preben-Hansen, OZ1RH 'Troposcatter at 50MHz: 700km QSOs at any time' in which it was asserted that "Troposcatter is always there, 365 days and nights a year". G2AHU and his collaborators felt that this was an over-estimation of the actual situation insofar as it related to amateur signals and were by no means convinced that the mode of propagation that they were investigating was due to troposcatter. An important finding of even these early tests was that the reliability of the path was significantly higher during daylight hours than at night, indicative of tropospheric refraction rather than scatter. **Fig 7** was prepared to illustrate the initial tests conducted between 12 November 1995 and 7 January 1996.

The issue of the 24-hour GB3SIX beacon licence in the summer of 1996 facilitated a more detailed investigation which commenced on 50.275MHz on 3 August 1996, and continued uninterrupted until 30 September 1997, this time including also 418km and 420km paths to Wimborne, Dorset (Ian Brotherton, G2BDV, at his home QTH, and also at the HQ of the Flight Refuelling Amateur Radio Society, G4RFR). The GB3SIX beacon comprised a Kenwood TS-690S transceiver sending automatic computer-generated Morse (A1B) and putting 30 watts into an antenna consisting of two stacked six-element Yagi antennas 20m above ground beaming due south. A detailed 30-page report 'An Investigation into Tropospheric Propagation at 50MHz', dated 11 September 1998 and prepared by G2AHU, G4IFX and G2BDV, has been submitted to the RSGB Propagation Studies Committee. The Report includes a hypothesis as a basis for further work into a propagation mode that has not previously been adequately explained. One of the many illustrations in this interesting Report, provides a comparison of daily average signal strength at G4SFR (420km path) and at G2AHU (264km path) during April 1997: **Fig 8**.

It is not intended here to forestall publication of this valuable Report, other than to note that their hypothesis is that the results do not point to troposcatter but rather to normal tropospheric propagation at 50MHz and that this takes place largely through the refractive index changes occasioned by rising moisture-vapour-laden thermals, adding "Since we postulate a moving target, the first task will be to measure any Doppler shift and the varying height from which refraction takes place. Techniques to perform these tasks are now available, and we hope to be able to do so ourselves or for others to undertake them."

I seem to recall that earlier work indicates tropospheric effects have sometimes been observed on 28MHz signals and even on 3.5MHz signals over sea paths.

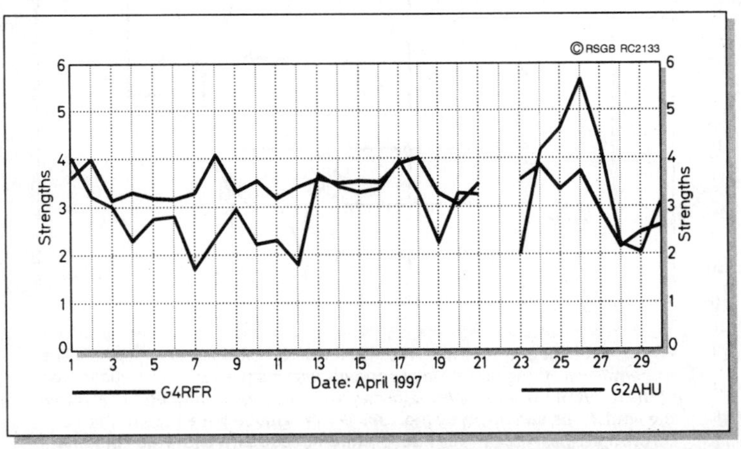

Fig 8: A comparison of daily average signal strengths at G4RFR and G2AHU during April 1997, from the detailed report prepared for the RSGB Propagation Studies Committee.

HERE & THERE

The October 1998 *TT* noted that research into magnetic storms had been set back by the failure of the SOHO satellite last June. Happily the cause has since been traced and rectified by the combined efforts of the NASA and ESA teams working in co-operation with the satellite manufacturers Matra Marconi Space. The problem arose when a string of management errors led to the satellite's solar panels turning away from the sun and draining nearly all the power from the batteries. The management team has been changed and now all 12 instruments are working again. ♦

technical topics

by Pat Hawker, G3VA*

THE SLOW DYING SPARK

IN INTRODUCING the November *TT* item on the continued importance of thermionic valves, I recalled hearing an SOS from an Egyptian vessel that struck a mine in 1940. I pointed out that spark transmission, at least for emergency purposes, continued in use in the maritime radio service until after the end of WW2. This reminded a former Radio Officer, Alan Wake, G3GIB that he may well have been the very last British operator to have used a spark transmitter to send a commercial radiotelegram. He writes:

"I may well be among the last, if not *the* last, British telegraphist to use a spark transmitter in commercial service. Towards the end of 1951, I was the Radio Officer in a ship anchored off Galveston, Texas with the ship's generators shut down to allow the Engineers to clean the main switchboard. The Master asked whether I could send a telegram to the agents ashore. I replied that I could by using the emergency spark transmitter which ran off 24V batteries, which also powered the receiver. I called the local coast station on 500kHz (600 metres) on spark. The coast station operator told me to change to the working frequency to send the QTC. I replied that the emergency transmitter could not change frequency, and sought permission to send the message on 500kHz, although this was against the rules. Somewhat reluctantly, the station operator agreed, and the telegram was duly sent and acknowledged. Only afterwards did it occur to me that I could have said that I would change to 480kHz and then carried on. The spread of the spark emergency transmitter was so broad that it would have been readily heard by the operator 20kHz off tune!

"The emergency transmitter was the standard pre-WW2 Marconi quarter-kilowatt quenched gap model, housed in a wooden box with the front held on by four clips: **Fig 1**. The operating procedure.was to remove the front of the case, mainly so that one could read the antenna current but also so that one could enjoy the amazing smell of ozone given off by the gap! The Galveston message was the only occasion that I ever used the spark transmitter other than regularly testing it."

G3GIB recalls that even in 1926 questions were being asked in the House of Commons about the continued use of spark by ships which were causing severe interference to broadcast listeners in coastal areas. The following year at the International Radiotelegraph Conference, held in Washington DC, a proposal was made to abolish immediately the use of all spark sets on ships but this was successfully opposed by representatives of Marconi Marine supported by the British Chamber of Shipping. A compromise was reached by which the period for the elimination of spark was extended to the end of 1939, a stay of 12 years, and then applied only to sets with a power of 300W and above. Even this was overtaken by the events of 1939: "hence your Egyptian ship sending SOS on spark in 1940; and my message in 1951, 24 years after the proposal to abolish 'spark sets' immediately!"

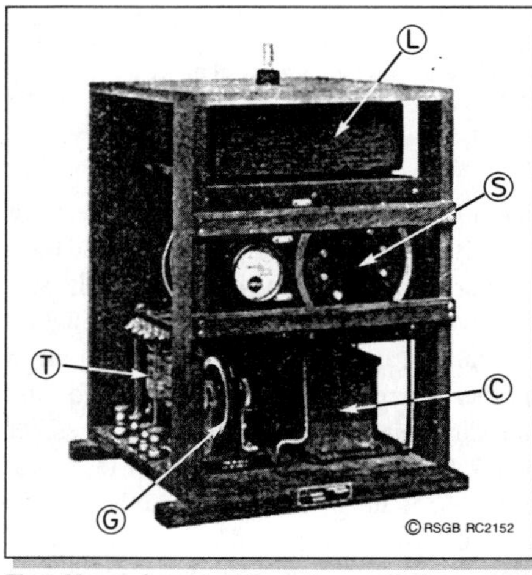

Fig 1. Marconi quarter-kilowatt quenched-gap spark marine transmitter probably similar to the one on which G3GIB may have been the last British operator to have sent the last commercial (non-emergency) telegram in 1951 on 600 metres. C capacitor, G spark gap, L inductance, S switch and T transformer.

Although G3GIB does not mention it, the failure to abolish spark in 1926 had one long-lasting and important outcome in the UK. The severe interference suffered by broadcast listeners in coastal areas led to the setting up of the first wired-wireless (Rediffusion) systems using twisted pairs in a few towns such as Southampton, so founding an industry that survives today in the form of broadband cable and optical fibre TV systems.

An earlier reference in 1994 to spark transmission brought a long and interesting letter from another former sea-going Radio Officer - Gordon Brown, G3FVW - who recalled that the first spark transmitter he ever saw was when as a young teenager he was taken to the North Eastern School of Wireless Telegraphy. He was to attend this a couple of years later when the imminence of war caused the Government to order the Training Schools to concentrate on the wartime 'Special' certificates, to meet the expected demand for many more Radio Officers and intercept operators (during early voyages G3FVW studied for and obtained his 2nd class PMG Certificate). The 1937 NESWT installation comprised a working mock-up of a ship's radio cabin with a 500W valve transmitter and a Marconi type 550 emergency quenched-spark set rated at 290-watts: **Fig 2**. Later he asked why such an odd power rating as 290 watts, and was told that this was to meet the 300-watt limit intended to be implemented in 1939!

G3FVW describes the equipment carried on the various ships on which he sailed during WW2 . Like G3GIB he came across the Marconi quarter-kilowatt quenched spark emergency transmitter. Illustrations of both the 250W and the 290W quenched-spark sets appeared in that one-time "Radio Officers' Bible" - H M Dowsett's *Handbook of Technical Instruction for Wireless Telegraphists.*

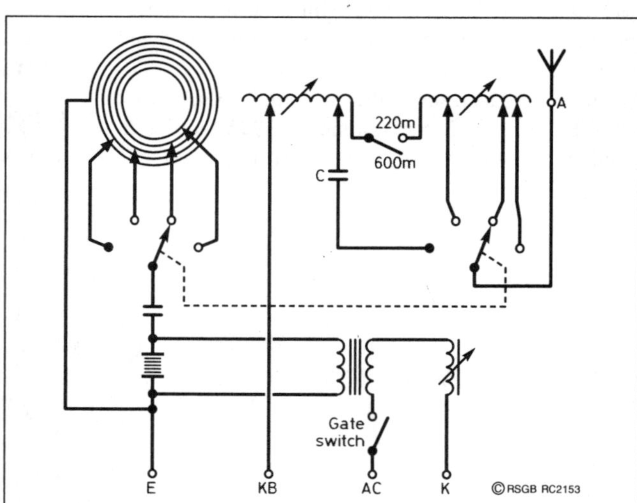

Fig 2. Circuit diagram of the Marconi 290-watt quench gap set, type 369. The transmitter working on full power consumes less than 300 watts measured at the input of the supply transformer and was thus designed to be within the limitations laid down by the 1927 Washington Convention intended to be brought into force in 1939 but postponed due to WW2. This equipment could work on 220, 600, 705 and 800 metres for use on trawlers and small craft. 600, 705 and 800 metres could be used on cargo vessels. Capacitor C was inverted when the antenna capacitance exceeded 500pF.

200 YEARS OF BATTERIES

IF 1998 MARKED "A Century of Amateur Radio" then without doubt we should be hailing 1999 as the bi-centennary of that key development: the first electrical battery. A historic invention that marked the entry into the world of current electricity as opposed to static electricity, leading inevitably to all the later discoveries of Faraday, Maxwell, Hertz, Lodge, Marconi and the radio and electronics age. As a discovery the battery was an outstanding example of scientific lateral thinking.

As noted by Hellbron and Bynum in *Nature* (7 January, 1999) it was in 1799 that Alessandro Volta first showed that dissimilar metals in salt water can generate an electric current, so making this scientific anniversary-of-the-year. He was following up the odd dis-

*37 Dovercourt Road, London SE22 8SS.

covery by Luigi Galvani that a freshly dissected frog could be made to kick by connecting a leg nerve to its corresponding muscle. The oddity was that the connector had to consist of two *different* sorts of metal joined together in series. Galvani wrongly deduced that the kick was caused by a special sort of 'animal electricity' stored in the frog's muscles much as static electricity could be generated in a machine and stored in Leyden jars. As a result, he began carefully to investigate the anatomy of the frog.

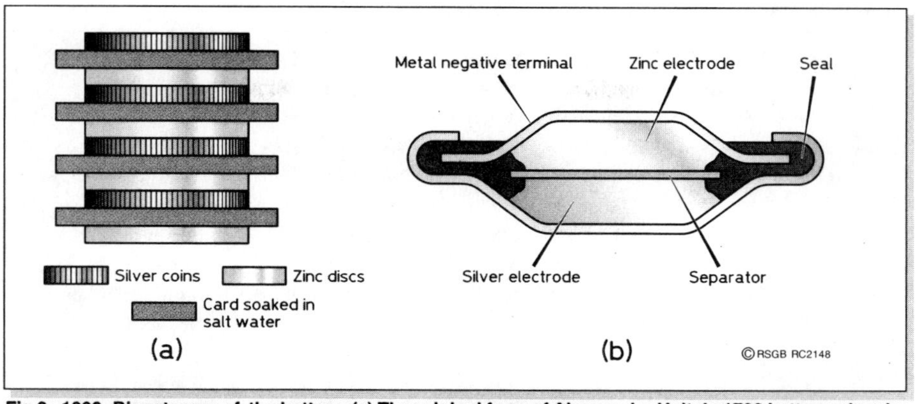

Fig 3. 1999 -Bicentenary of the battery. (a) The original form of Alessandro Volta's 1799 battery when he showed that dissimilar metals in salt water generate an electric current by taking silver coins and an equal number of zinc discs, putting them together in pairs with bits of card soaked in salt water between them. (b) Modern form of a silver-zinc button cell.

Volta however zeroed in on the bimetallic conductor, and came to realise that the key to the mysterious kick must lie in the need to use dissimilar conducting materials. This hypothesis was strongly refuted by Galvani with the result that to uphold his idea, Volta needed to eliminate the frog as the source and detector of 'animal electricity'. He refined the existing measuring devices (some of which he had developed). Then he multiplied the effect of the dissimilar metals by taking dozens of silver coins and an equal number of zinc discs, putting them together in pairs, and stacking the pairs with bits of card soaked in salt water between them: **Fig 3(a)**. Thus was the battery evolved, providing for tirst time a practical source of current electricity.. Volta's name is rightfully honoured in our designation of electrical potential in terms of "volts". Modern silver-zinc cells are still used in the form of button cells (**Fig 3(b)**) but with potassium-hydroxide rather than sea water as the electrolyte. Such cells have a potential of 1.5V, with double the capacity of equivalent-sized nickel-cadmium cells. They are however too costly for general use.

In the 200 years since Volta's first battery, many combinations of metals have been developed but the basic principles have remained essentially unchanged. Batteries, mostly of the rechargeable lead-acid and disposable carbon-zinc varieties, were for many decades considered a mature technology, capable of detailed refinements and improvements but offering few opportunities for radical innovation. The coming of personal mobile phones, laptop computers and above all the search for "electric cars" that could compete with petrol-fuelled vehicles have caused battery technologists to become once-again leading-edge scientists. Lithium-ion, nicklemetal hydride and the difficult "fuel-cell" are now firmly in the frame, with even the once-considered out-moded zinc-air cell being given a good going over. For amateurs the "12-volt car battery" (13.8V fully charged) has become a major source of power for the all-solid-state amateur transceiver whether used mobile, portable or in the shack.

A special report 'The great battery search' by Gary L Hunt of the Idaho National Engineering and Environmental Laboratory (*IEEE Spectrum*, November 1998) stresses that batteries are still the main stumbling block in the race to the widespread introduction of electric vehicles. Although storage batteries are getting better, they are still far from competing effectively with petrol as an energy source. The energy content of petrol is about 12kWh/kg compared with some 30Wh/kg for typical lead-acid batteries - a difference of about 400 times on a weight basis. Admittedly, the internal combustion engine has much lower efficiency than electric machines but the gap is still far too wide for general acceptance: "Advocates of the electric vehicle (EV) frequently find themselves explaining why no electric car can drive far without refuelling and as frequently promising that things will get better any year now, as soon as the latest advance in batteries reaches maturity. As the search goes on for new electrochemical couples, designers try to avoid the pitfalls of the old ones. Meanwhile the same few technologies gulp down most of the serious money in battery development, because new candidates do not seem to investors clearly superior to those on which hundreds of millions of dollars have already been spent."

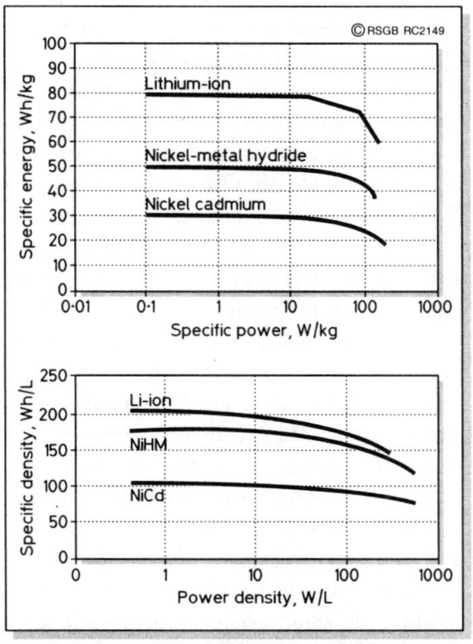

Fig 4. Plots of (a) specific energy versus specific power and (b) energy density versus power density Ragone plots of Lithium-ion, nickel-metal hydride and nickel cadmium cells. These illustrate one of the fundamental trade-offs in battery technology: boost a battery's ability to store energy and its power performance suffers. (source *IEEE Spectrum*).

It is stressed that the three main characteristics of a secondary (rechargeable) battery - energy performance, power performance and lifetime (both actual time and in charge-discharges cycles) - are inextricably linked: increase one and one or both of the others must decrease. For example, if the size of a current collector in the battery is increased to boost power density there will be less room for active electrode materials which will decrease energy density. A common way to improve the power performance is to use thinner electrodes, but this usually lowers both the energy density and the life expectancy. **Fig 4** shows Ragone plots of specific energy versus specific power and energy density versus power density illustrating one of the fundamental tradeoffs in battery technology: boost a battery's ability to store energy and its power performance suffers.

When using a car-battery to power an amateur transceiver, it is worth remembering that most vehicle batteries are designed to provide heavy currents for short periods but are not intended to be repeatably fully discharged. For operation from a fixed site, the available deep-discharge batteries will provide significantly better lifetimes, well justifying extra initial cost. An extreme example of a small battery designed to provide a terrific 1000A punch for a brief moment was reported in the January 1999 issue of *Electronics World* (p6):'New battery technology packs a 1.5kA punch'. Bolder Technologies of Colorado has re-spun lead acid technology to produce more instantaneous power weight-for-weight than any other battery. The cells are sub-C size just 23mm in diameter and 70mm long using very thin elecctrode materials. The 2V cells are designed to provide microsecond bursts of up to 1500A with complete discharge in 2 seconds. The capacity is just 1Ah giving a specific capacity of about 30Wh/kg much the same as a conventional lead-acid battery. I shudder to think of the effect of an accidental short-circuit!

In practice, the nickel-metal hydride (Ni-MH) batteries are now tending to supersede the nickel-cadmium cells, for long the usual choice for portable transceivers. The even-better lithium-ion batteries are becoming the preferred choice for laptop computers which require appreciable current-drain over extended periods, but are still costly. Much of the most interesting development work, from an amateur radio viewpoint, is in the efforts being made to bring down the cost of lithium-ion cells without jeopardising safety aspects or reducing the number of discharge-recharge cycles by development of lithium-polymer cells

The whole topic of "Seeking better batteries" - including comparisons between lead-acid, nickel-cadmium, nickel-metal hydride, lithium-ion, lithium-polymer and zinc-air cells - was covered in some detail in *TT* (September 1995, pp64-65) from which **Table 1** is reproduced:

Table 1: Comparison of some rechargeable batteries

Cell type	Nominal Voltage (V)	Energy dens. (Wh/L)	Power dens. (W/L)	Self-dis %/month	Cycle-life	Comments
Lead-acid	2.0	70	<400	4-8	250-500	Lowest cost
Ni-Cad	1.2	60-100	220-360	10-20	300-700	May show memory effect
Ni-MH	1.2	220	475	30	300-600	Possible slight memory
Lithium-ion	3.6	260	400-500	5-10	500-1000	Intrinsically safe
Li-polymer	3.0	150-350	>350	<1	200-1000	Not yet on market Contains metallic lithium
Zinc-air	1.2	204	190	<5	<200	Air manager needed to limit self-discharge

NOTES: Cycle lives are strongly dependent on how the battery is treated. Lithium-ion cells with petroleum-coke anodes drop from about 4.0V to about 3.0V fairly linearly as they go from fully charged to fully discharged. Figures for lithium-polymer cells are predicted (predating announcement of Ultralife cells). Power density figure for lithium-polymer cells apply at full charge and will drop as cell is discharged. Ni-MH data apply to 2.9Ah 4/3A cells of the kind made by Duracell and Toshiba. Zinc-air data are for a cell developed by Zinc-Air Power for a prototype battery for Electric Vehicles. It must be emphasised that these figures are at best only a rough guide. Data abstracted from a table in *IEEE Spectrum* (May 1995 p56)

The 1998 article points out that a consensus is evolving that lithium-based batteries offer the best long-term hope for commercially practical electric vehicles. The long development times for lithium EV batteries reflect in part the risks involved in using lithium in large batteries: "Lithium-ion consumer batteries, though, are fast ruling the roost in laptop computers and cellular applications, and the number of manufacturers worldwide is growing too fast to track . . . . Lithium-polymer batteries, using metallic lithium foil in their negative electrodes along with a solid polymer electrolyte have yet to find their way into consumer applications, but they are claimed to have the potential for even higher specific energies - up to 200 Wh/kg - and to cost less to manufacture than other types of lithium batteries because of their solid-state design. Their biggest disadvantages are twofold: they must be warmed to around 70°C for adequate performance and metallic lithium electrodes have historically been short-lived in rechargeable applications because dendritic growth of the metal conduces to internal self-shorting."

Over many years, there have been outbreaks of interest in 'fuel cells' which are electrochemical conversion devices that produce electricity directly by oxidizing hydrogen. In a fuel cell, the substances that react chemically at the electrodes in the cell are stored partially or wholly outside the reaction cell. When these substances are exhausted the current stops until more 'fuel' is added. The delays involved in 'recharging' batteries is thus overcome and, as with a petrol engine, the working capacity depends primarily on the amount of fuel that can be carried.

The first laboratory-model electric fuel cell was made by Sir William Grove over 150 years ago but it has never proved easy to develop practical fuel cells that can operate efficiently at low temperatures. Even today most fuel cells require inappropriately high temperatures, between about 300°C and 1200°C and use hard-to-manage electrolytes, such as phosphoric acid and molten salts.

However, in the November 1998 *IEEE Spectrum,* Tom Gilchrist (Ballard Power Systems Inc) suggests that proton exchange membrane (PEM) hydrogen fuel-cell systems have become more compact and could well solve the EV battery problem. He claims that: "The automotive industry is on the brink of a new era in alternative propulsion. Fuel cells are now being favoured for the title 'power source most likely to replace the internal combustion engine in the next generation of transportation vehicles'. In view of past experience this may seem an over-optimistic claim for a technology long regarded as at best a far-future hope.

Gilchrist explains that the PEM type of fuel cell, so called because a hydrogen ion is just a single proton, offers relatively benign characteristics as a power source for electric vehicles: "Operating at about 80°C and employing a thin plastic sheet as their electrolyte, they are easy and safe to handle in manufacturing and in later use. Unlike some other electrolytes, their solid plastic membrane can tolerate a modest pressure differential across the cell, making for easy presurization, which increaes power density, simplifies the rest of the system and reducs cost . . . other attractions include low emissions, high conversion efficiency, and fast start-up and fast response to transients." He admits that as a viable source of EV power they have been slow to win acceptance: their power density with respect to conventional car and truck enginess was too low. But in 1995, Ballard Power Systems of British Columbia reported the development of a stack design with a power density of 1000 watts per litre. The company developed a fuel cell stack that occupied just 30 litres yet converted hydrogen into electricy at 30kW rate, almost ten times the state-of-the-art in 1990". Could it be that the fuel cell concept that first saw the light in 1839 will at last come into its own soon after its 160th anniversary?

Most types of batteries pose environmental and safety problems - for example it has been claimed that the lead of lead-acid EV batteries might cause more pollution than petrol-engines though this is disputed as unrealistic. It is however conceded that almost any energy technology can be attacked as involving some risk: few, if any, are both inherently safe and enviromentally benign.

VALVES, DISTORTION & QUALITY

THE *IEEE SPECTRUM* article (August 1998) 'The Cool Sound of Tubes' which formed the basis of the opening item in *TT* (November 1998) has continued to provoke reader comments in that journal. A number of these revolve around whether there is any inherent difference in the audio quality of valve and solid-state high-fidelity amplifiers. The consensus seems to be that in most available amplifiers differences *can* be detected, at least by those with golden ears, even if the difference cannot be measured objectively. The key appears to be the increased amount of negative feedback used to achieve linearity with transistors. It is however claimed that it is possible, with extremely careful design, to build a solid-state amplifier that cannot be distinguished from the best valve amplifiers. But it is clearly much easier to obtain near perfection from valves than transistors. But the prime source of imperfections remains the loudspeakers.

On this side of the Atlantic, Ken Green, G1NAK writes: "It is my experience that few (if any) hi-fi enthusiasts know either why they prefer a valve amplifier, or, more importantly, what sort of amplifier it is that they prefer. Valves come in two type, basically triode or pentode. The triode produces predominantly second-harmonic distortion and the pentode predominantly third-harmonic distortion. Even-order harmonics (in reason- able quantities) produce a warm and rounded sound while odd-order harmonics produce at best an 'edgy' sound, and in any quantity, a harsh and extremely unpleasant sound.

"In my youth, the true earophile always demanded a triode power amplifier although few could say why, apart from what they had read or been told. The list of advantages and disadvantages in the November *TT* notes that it is necessary to use lots of negative feedback to obtain reasonable linearity from semiconductor active devices. I was taught that feedback can be used to improve a good amplifier but it cannot make a bad one into a good one. As seen on an oscilloscope, when feedback fails it does so spectacularly and, at the levels at which 'music amplifiers' are driven, this has to be a component in the final sound.

"Yet again, semiconductor hi-fi amplifiers are easily detected by their characteristic hiss which is usually ascribed to the foolishly-wide supersonic bandwidth which today's buffs insist they need. I have satisfied myself by tests and the three loudspeakers that I have lost, that the hiss is the result of supersonic and VHF/UHF parasitic oscillations which are close to mandatory in amplifiers exhibiting more feedback than is allowed by the learning of a true Designer"

Further confirmation of the manufacturing weaknesses of some high-power RF valves comes from Sean Lineman, EI7CV who writes: "Your mention of faulty 3-500 valves gave me consolation in knowing that I am not alone with my problems. I have a Heathkit SB1000 linear amplifier in which the 3-500 (serial number FTS 034) has developed an internal short-circuit causing high voltage to appear on the centre tap of the heater winding with consequent damage to other components."

At the time of writing he was hunting for a secondhand 3-500 since the price for a new one is over £160. I have to confess that during 1997 I ran into a comparable problem with the PA valve in my old LG300 transmitter. Although less than a year old, the Mullard CV26 which had replaced an RCA 813 that had given years of reliable service, suddenly blew its 50W filament. Luckily I was able to obtain an American 813 for a reasonable price, but I was left pondering why the European valve had had such a short life. The conclusion I came to was that in the LG300 the PA valve is mounted horizontally. The 1954 British Standard (CP 1005 etc) used to point out that valves should preferably be mounted vertically with the base downwards. Where mounted horizontally, the major axis of the first grid of indirectly heated valves of high mutual conductance and the plane of the filament of directly heated valves should be mounted vertically. With high-power RF valves it would seem that this advice is still relevant, if a hot filament can sag and cause a short-circuit then Murphy's Law suggests it will at sometime do so. But are European valves more prone to this than American ones or was it just my luck!

HERE & THERE

"TECHNISCHE NOTITIES VAN PE1PVB" (*Electron*, February 1999, pp3-5) presents a collection of HF antenna ideas drawn from many sources. One that caught my eye was a folded-dipole arrangement that overcomes the problem that a central 300Ω resistive feedpoint cannot cope with 2nd-harmonic frequencies: in other words while a half-wave dipole on 7MHz can be used on 21MHz it is not matched on 14MHz etc. **Fig 5** shows an arrangement, attributed to VA3ERY (original publication not given), of a 'Washing Line' arrangement of a folded dipole in which the position of the feedpoint can readily be moved to suit the band in use, and adjusted for minimum SWR.

The January *TT* provided details of the transmitter and receive-converter which has put Dick Rollema, PA0SE on the 136kHz (2137m) band. However in the February issue of *Electron* (p16), he provides an amended circuit for the LF converter, as a preferred arrangement of the crystal-controlled local oscillator for the NE612 particularly where overtone crystals are used: **Fig 6** shows the changes to the 612 circuitry, the frame antenna and static protection are unchanged. The crystal is now connected between pin 6 and earth, with as 68pF capacitor between pins 6 and 7. For overtone crystals an inductor L5 is connected in series with C3 which in combination with C4 is resonant on the fundamental of the overtone crystal. For example, with a low-cost 27MHz CB crystal, L5 is made resonant on 9MHz or for a 28MHz crystal on 9.33MHz (L5 2.58uH) with the main receiver tuned over the appropriate range. A standard 2.2mH inductor will resonate with 68pF to about 12.9MHz and should be satisfactory.

Since writing these notes, PA0SE has warned that parasitic oscillation can destroy the HEXFETs in the transmitter. He has provided a modified diagram and some additional information. I will send copies to anyone building or seriously interested in building this LF transmitter, if they send me an SAE plus a 26p stamp to cover photocopying.

In connection with the item 'Headphone Adapter' (*TT*, December 1998, p59), Rev George Dodds, G3RJV, draws attention to the 'High Impedance Headphone Adapter' available from Isoplethics (13 Greenway Close, North Walsham, Norfolk NR28 0DE, Tel: 01692 403230) which he points out is a firm offering a useful range of 'real radio' products, including plug-in coil formers, interstage coupling transformers, etc. Isoplethics describe their THA1 headphone adapter as having been developed to allow modern 32Ω stereo headphones to be used in high impedance applications. This unit is claimed as ideal for use with simple receivers, especially valve TRF sets and crystal sets, and giving good results with older valve receivers and test equipment such as the BC221 frequency meter. The input impedance is around 8000Ω. Modern stereo headphones, as noted in the December item, vary considerably in sensitivity but with examples having a sensitivity better than 100dBm the unit can provide a sensitivity approaching that of the 1950s Brown Type F 4000Ω headphones, The input transformer has been designed to simulate the peaky response of metal diaphragm headphones, in order to give good CW and communications grade telephony, with additional internal filtering to limit audio response to around 3kHz and with internal clipping to limit

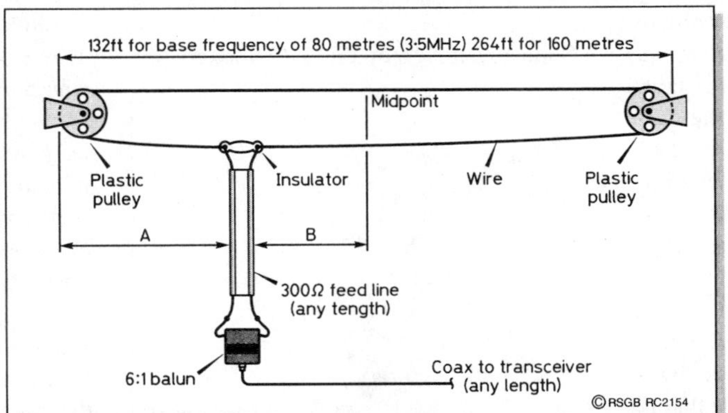

Fig 5. The washing-line folded dipole antenna credited to VA3ERY. For use on 7MHz and above the span could be reduced to 66ft.

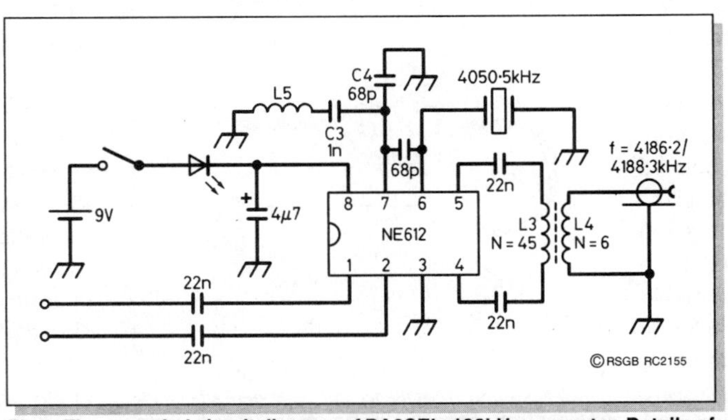

Fig 6. The amended circuit diagram of PA0SE's 136kHz converter. Details of the frame antenna etc remain the same as was shown in Fig 1 of the January 1999 *TT*.

noise spikes to a safe level. The input is shunted to eliminate 'threshold howl' with regenerative detectors. The unit is housed in a black, isophthalic resin-moulded box with output via a standard stereo socket; input is a flying lead terminated in a 6.35mm jack.

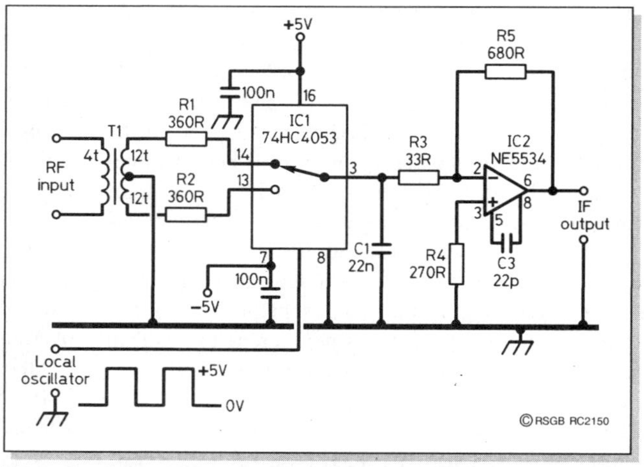

Fig 7. Bomb-proof RF mixer - as developed by Rod Green of Western Australia who calls it the Ned Kelly mixture in memory of the Australian bank-robber who wore bullet-proof armour.

ANOTHER "BULLET-PROOF" MIXER?

DURING 1998, *TT* featured a number of high-performance RF mixers including an SL6440-substitute by G4COL using the 74HC4066 and the exceptional H-mode mixer using the FST3125 surface-mounted device resulting from the combined efforts of I7SWX, W7AAZ and G3SBI.

Both Michael O'Beirne, G8MOB and I noted with interest an item 'Bullet-proof RF mixer' in the 'Circuit Ideas' feature of *Electronics World* (January 1999) contributed by Rod Green of Bedford, Western Australia, using a 74HC4053. It was designed for use in an image-cancelling superhet with a low IF but could equally well be used in a direct-conversion receiver: **Fig 7**. It is claimed to provide a dynamic range of 132dB and a noise figure of 12dB.

The RF signal input is applied via an input transformer which matches the source impedance to the mixer and provides two signals in antiphase to the 4053 multiplexer/demultiplexer chip. R1 and R2 terminate the mixer and provide current drive, no diplexer being needed in this type of circuit. The output current is amplified in the current-to-voltage converter IC2 with R3 isolating C1 from the op-amp input at high frequencies. The local oscillator should provide a square-wave switching signal.

CLEANING UP FAULTS

JAY CRASWELL, W0VNE, in the Technical Correspondence feature of *QST* (December 1998, p63) faced a quandary that is commonly encountered. One day his trusty Drake TR7 transceiver failed to provide any RF output. He had in the past always carried out his own repairs, but this time felt anxious about diving into this ageing but quite complex equipment. He pondered whether he would be able to fix it or only make it worse. Out of curiosity he phoned the Drake service-department technicians and explained his situation. They agreed they were still prepared to fix TR7s. Tnen he asked for a 'guesstimation' of the likely cost. The answer of something like $200 to $300 convinced him that he should risk a D-I-Y job. After all, he mused, we hams are supposed to be able to do it ourselves.

He disassembled the rig on a clean table using empty olive jars to save the *many* screws. Soon something caught his eye: Fuzz! The cooling fan over the years had sucked loads of the yacky stuff into the radio. Clearly this would have to come out despite resisting initial attempts to remove it. Then he recalled an article he had read on restoring antique radios. He borrowed a small paintbrush from his XYL, rolled out the vacuum cleaner and fitted its detachable hose. Then he held the hose near each patch of fuzz to be removed and used the brush to work it loose. The procedure is thus essentially the same as shown recently on Channel 4 for the six-monthly cleaning of dust from the delicate Grinling Gibbons carvings at Hampton Court, although apparently this does not have to cope with gooey fuzz!!

Soon the inside of the TR7 looked like

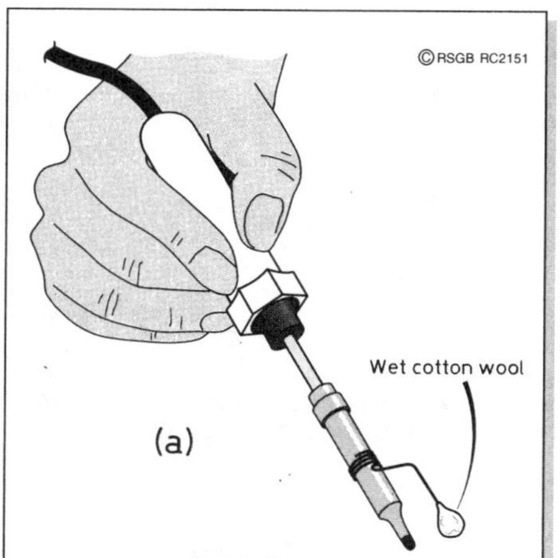

Fig 8. Two soldering hints, (a) A piece of wet cotton wool fastened directly to the iron removes heat rapidly from sensitive components. (b) A diode which can be switched in or out of the AC supply to the iron can reduce excess heat when dealing with small components or printed-circuit boards.

new. And wonder of wonders removing the fuzz also made the rig work like new - presumably some metallic fragments in the fuzz had finally caused a short-circuit. Success encouraged him to try, again successfully, the same procedure on a 144MHz rig that had also developed problems. Two rigs fixed in one week!

While the rigs are open, W0VNE advises, it doesn't hurt to give some of the dusty controls a blast of tuner cleaner or lubricant. He concludes: "So before you spend time and money to send your equipment to a repair shop, try cleaning it up! Even if it doesn't fix your problem it will make the job of the repair technician a little easier and maybe he will show some mercy when figuring out his bill!"

Unfortunately, simply removing the fuzz is not always sufficient to restore an ailing transceiver to good health and a soldering iron has to be used, if only to cure a dry joint or two. A letter from Scott Arnesen of Oslo in *Electronics World* (January 1999, p55) provides a tip on soldering and desoldering heat sensitive components which can be ruined if the heat is not removed quickly after using a hot iron. He writes: "A piece of wet cotton wool on a separate pin may be of great help. It is much quicker to have the cotton wool fastened directly to the soldering iron as shown in **Fig 8(a)**. Just twist the iron and apply."

Another soldering hint is given by Jerry Hemby, KE4TFE (*QST*, December 1998, p62) who points out that many inexpensive irons are too hot for printed-circuit work (particularly where surface-mounted and similiar miniature devices are involved), damaging or destroying heat-sensitive boards and components, or causing the foil to peel off the board. While large, high-wattage irons should be avoided at all cost, even medium / low-wattage irons may become too hot if left on for a time. KE4TFE advocates an old dodge: the use of a high-voltage 1A diode across an in-line on-off mains switch of a size that allows the diode to be put inside the switch. The new switch is then wired into the mains-lead to the iron: **Fig 8(b)**. When 'on' the iron is powered at its normal rating. When 'off', the diode acts as a wattless resistor passing only half the AC waveform, reducing the current through the iron to half and the power to a quarter of its normal rating. KE4TFE finds that with his Weller soldering pencil, the reduced rating copes with all but the largest components - for which the switch is closed. An editorial note warns that mains-transients can reach many times the nominal mains voltage so that in the UK the diode should be rated at least 1000V for long life and in some areas it might even be advisable to fit two such diodes in series. ♦

technical topics

by Pat Hawker, G3VA*

PA0SE's 136kHz TRANSMITTER

AS NOTED IN A brief addition at the proof stage to the March *TT*, Dick Rollema, PA0SE, has made several corrections and changes to the circuit diagram of his 136kHz transmitter (January *TT*, Fig 2). He writes: "At first the transmitter operated without any problem. But after returning from holiday one HEXFET after another failed. Fortunately they are quite cheap. After the demise of seven semiconductors I found the cause: parasitic oscillations on VHF or UHF. The remedy was clear from the valve era: stopper resistors (120Ω) in series with the gates.

"At first I used an HC4060 chip in the transmitter with 12V supply voltage. It worked perfectly, but someone pointed out that this device is rated for a maximum of only 6V, so I replaced it with an HCF4060 which can be used up to 18V (an HEF4060 could also be used). Because of the higher output resistance of the HCF family, the resistor setting the source resistance for the low-pass filter had to be reduced from 180 to 27Ω. Some corrections in the circuit diagram are due to mishaps at the printers: 'u' becoming m and omega becoming W: **Fig 1** shows the modified and corrected circuit diagram.

PA0SE adds: "I find operating on LF fascinating. Signals are usually very weak and some almost buried in noise and QRN partly caused by a LORAN-C transmitter in the North of

Breadboard construction of PA0SE's 136kHz transmitter.

France and which is even more troublesome in Southern England. It operates on 100kHz, but the steep pulses cause it to spill over the 136kHz band, despite complaints. I now use a special LF valve receiver made in the 1960s by Teletron in Germany, which was especially designed for FSK data transmission and has a small CRT showing the frequency shift. It covers 40-80kHz and 80-160kHz. Extremely good steep-sided filtering is essential on the LF-band, because at 138.8kHz there is a very strong station, used by the German electricity board for remote control of high-voltage switching stations, etc. It produces 6mV into 50Ω at my location, whereas amateur signals are in the microvolt range. I use an additional passive audio filter with three tuned circuits and about 200Hz bandwitdth between receiver and headphones.

"My antenna simulation computer program gave about 65mW as the actual radiated power . . . It is amazing that hundreds of kilometres can be covered with such small power. It helps to remember what the old *Radio Designer's Handbook* by F Langford Smith says: 'In the transfer of power between two aerials, the number of wavelengths, rather than the distance, is the significant parameter' . . . You will not be surprised that I am utterly against dropping the telegraphy test as part of the RAE. It is argued that the use of CW should be retained and only the test abandoned. But surely nobody is going to study for something not required for an examination and the result will be that in the course of time CW will disappear."

COILS, CORES & CHIPS

A USEFUL TIP COMES from Ray Loveland, G2ARU, who writes: "Although not presently active on air, I still do a fair amount of constructional work. I have recently been making some coils on the old Neosid 0.3in formers with ¼in (0BA) dust cores. Many readers will be aware of the problem of finding a way of locking these

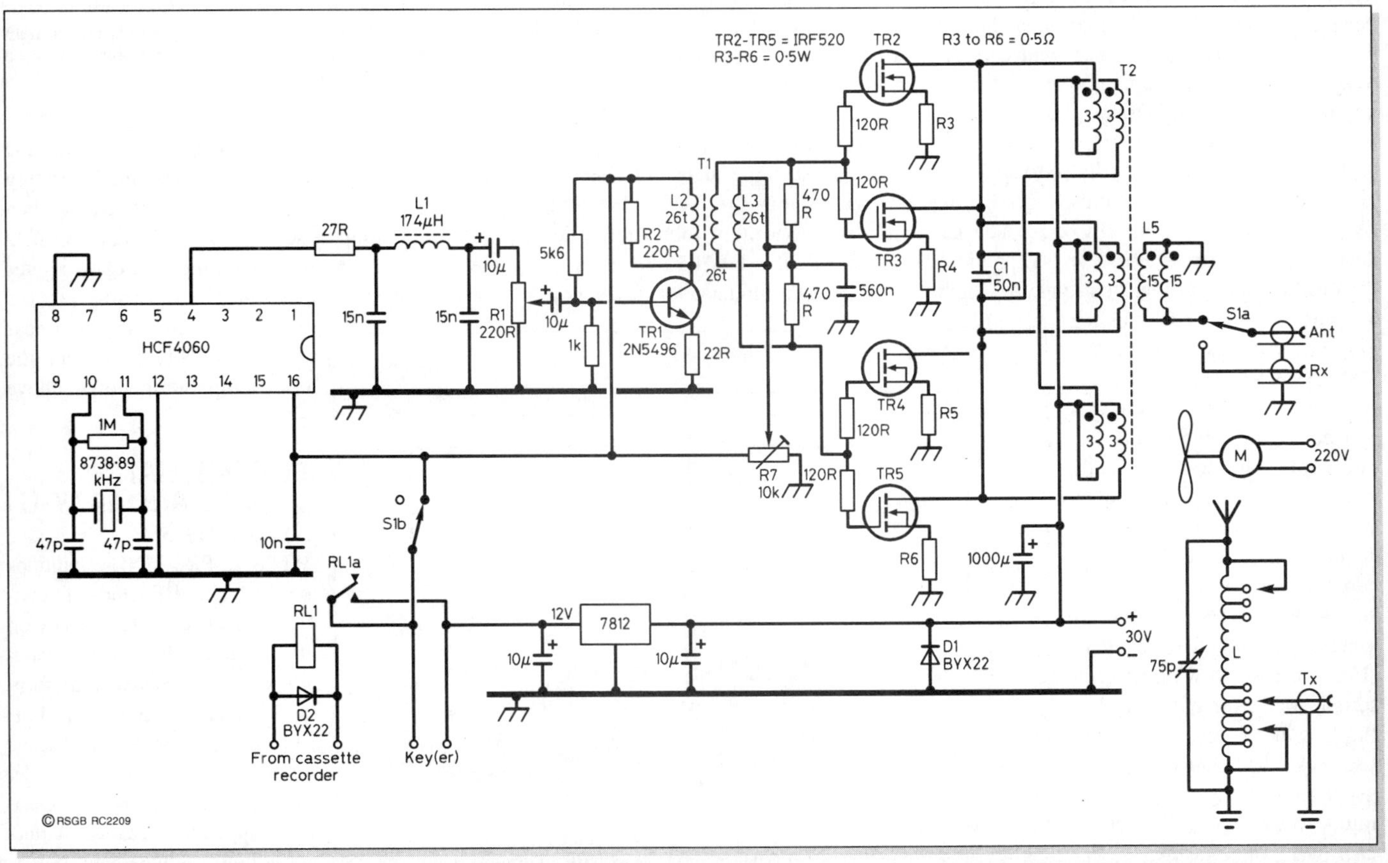

Fig 1: Modified and corrected circuit diagram of PA0SE's 136kHz transmitter published in the January 1999 *TT*. The 120Ω resistors in the gate circuits of the HEXFETs were found to be necessary in order to prevent destructive VHF/UHF parasitic oscillation. S1a and S1b is a DPDT switch, to prevent transmitting with the antenna switched to the receiver.

CORE	IPT	CORE	IPT
T-12	0.163	T-400	3.050
T-16	0.202	T-400A	4.350
T-20	0.252	T-520	3.720
T-25	0.327	FT-23	0.230
T-30	0.412	FT-37	0.438
T-37	0.426	FT-50	0.595
T-44	0.529	FT-50A	0.688
T-50	0.577	FT-50B	1.188
T-68	0.700	FT-82	0.809
T-80	0.800	FT-87	0.835
T-94	1.006	FT-87A	1.335
T-106	1.364	FT-114	1.045
T-130	1.394	FT-114A	1.070
T-157	1.760	FT-140	1.500
T-184	2.300	FT-140A	1.692
T-200	1.850	FT-150	1.250
T-225	1.950	FT-150A	1.750
T-225A	2.850	FT-193	1.930
T-300	2.080	FT-193A	2.180
T300A	3.080	FT-240	2.000

Table 1: Inches per turn factors for commonly used toroidal cores.

cores in a manner that does not prevent later re-adjustment. I have found that the core-locking compounds tend to harden with age, so that the cores cannot be adjusted later without damage.

"Some commercially made coils had a single strand of very thin elastic run into the former, parallel to the core. This works well on home-made coils, since supplies of very thin elastic can readily be obtained by carefully cutting lengthwise the wide elastic used on clothing. I have just recently tried another method: locking the cores by wrapping a few turns of plumber's PTFE tape around the core. This seems to work very well."

Recent practice in coil winding is to use powdered-iron or ferrite toroidal cores. Robert Olson, WD4OHO, in the 'Technical Correspondence' feature of the January 1999 issue of *QST* provides useful information on how to calculate the length of wire needed when winding popular toroids. He writes: "Many homebrew projects specify use of toroidal inductors wound on powdered-iron or ferrite cores. The articles usually provide the much-needed information on core types and the number of turns needed to approximate a given value of inductance. However, they don't generally tell you what *length of wire* you need to wind on each toroid. An easy way to calculate the lengths for most commonly used cores with single wire, single layer windings is simply to multiply the core's *inches per turn (IPT)* value [shown in **Table 1**] by the number of turns and add three inches to the result: ie (IPT x turns) + 3in. Then round-off the result to the next larger number. In this way you'll be able to conserve your supply of enamelled wire each time you need a toroidal inductor."

Colin Horrabin, G3SBI, provides some additional information on the Minicircuits TT4-1A Style X65 (6-pin DIP) transformer recommended in connection with the FST3125 H-mode mixer, described in the September 1998 *TT* (pp58-59). He writes: "This transformer has a centre-tapped primary (in addition to the centre-tapped secondary) and I wish to make it clear that the entire primary winding is used, leaving the centre-tap unconnected, so giving a 1:4 impedance step-up. The reason for choosing the T4-1A is that originally I used the T4-1 in my test set-up since I knew that this was capable of intercepts above 50dBm as I had used them in the SD5000 H-mode mixer. The insertion loss of the T4-1 is relatively high, so I tried the T4-6T since this has lower insertion loss, but the IP_3 bottomed out at about 40dBm. The Minicircuits handbook gave a low insertion-loss and is capable of an IP3 above 50dBm. These differences must be due to the grade of ferrite used in the transformer cores. It appears that the TT4-1A is the only available transformer providing both these desirable characteristics between 1MHz and 80MHz."

G3SBI adds: "While I was doing the IP_3 test on the FST3125 device, I also tested it as an up converter to 45MHz but using the normal IF port as RF input, with the normal RF input being the 45MHz IF output. I felt this might improve balance at 45MHz. With a Vcc of 7 volts (still driven by 5-volt logic on an AC109) I achieved an IP_3 of 50dBm up-converting a 14MHz signal with the local oscillator at 59MHz, using the divide by two squarer (74AC109). The fundamental frequency-squarer (74AC86) gave similar results, but the AC86 chip temperature increased noticeably. John Thorpe, designer of the excellent AOR7030 receiver, uses the 74AC86 as a fundamental frequency squarer

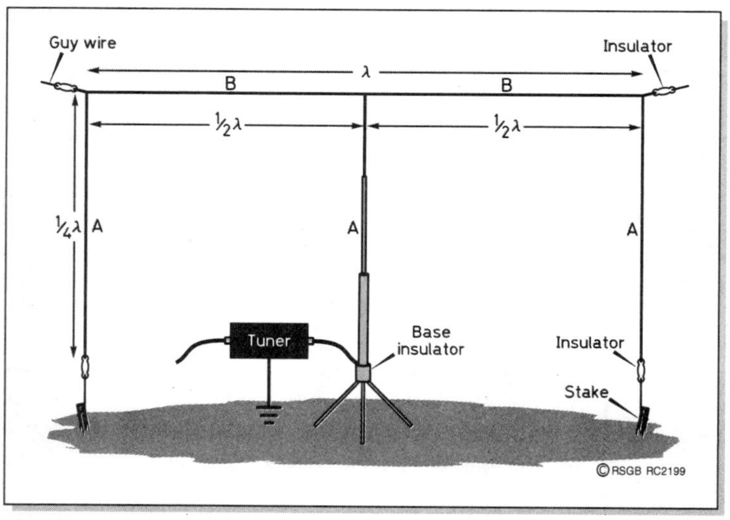

Fig 2: The classic Bobtail-curtain antenna with dimensions suggested by VE1TG.

Dimension	3.5MHz	7MHz	14MHz
A	66ft	33ft	16.5ft
B	132ft	66ft	33ft

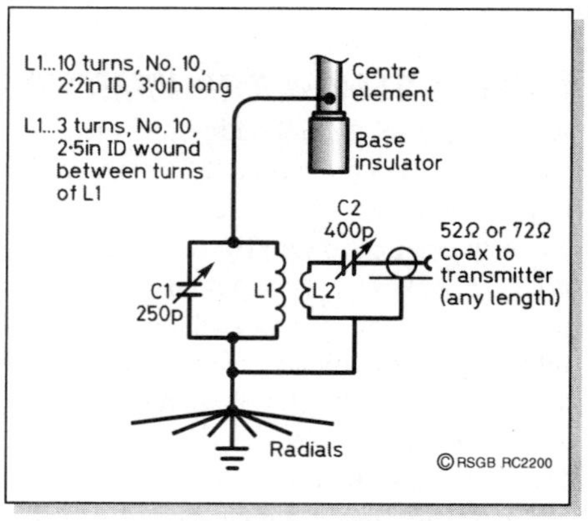

Fig 3: VE1TG's tuner for his 7MHz Bobtail-curtain array. For 14MHz L1 and C1 should be about half the values shown. C2 about the same. L1 10 turns of No 10 enamelled copper wire wound on former of inner diameter 2.5in over a length of about 3in. L2 3 turns wound between turns of L1 at the earthy end. Radiates predominantly vertically polarized signals.

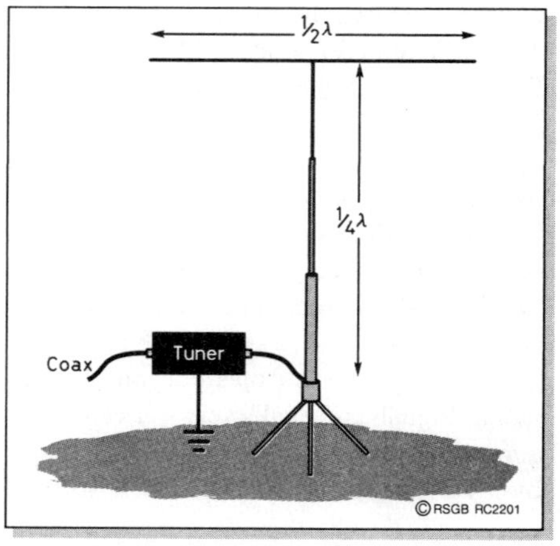

Fig 4: How the Bobtail can be reduced to form an omnidirectional inverted ground-plane antenna (Vertical-Tee)

77MHz). However, only one gate has the feedback resistor which sets the operating point. This would not be sufficient when designing the squarer to handle a 5 to 39MHz input range, as I found the gain of two gates were necessary at 5MHz. But using just the one gate biased in the linear region probably helps keep the chip temperature down above 45MHz.

BOBTAIL, HALF-SQUARE AND LAZY-U ANTENNAS

WITH IMPROVING conditions on the higher HF bands but continued interest in chasing DX on the lower HF bands, the search for wire antennas that perform well at relatively low vertical radiation angles continues to exercise many enthusiasts worldwide. For those with plenty of space and supports at least a quarter-wave high, a single-band broadside antenna that has been shown to be a particularly effective low

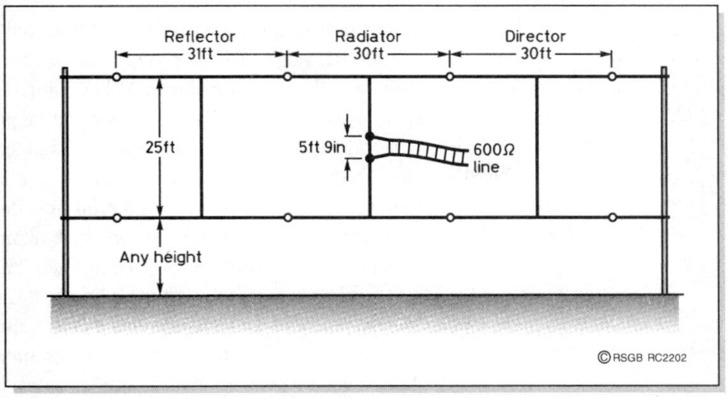

Fig 5: 7MHz three-element vertical Yagi array (wire) with element lengths much reduced by loading with the horizontal sections. Originally developed by Les Moxon, G6XN, in connection with experiments on extremely low-angle DX with low-power. It provided highly-reliable communication with Australia.

angle radiator (at least over good ground) is the Bobtail Curtain. This was included in many editions of *The Radio Handbook* (see also *ART* or *TT* July 1970). **Fig 2** and **Fig 3** show a version described by VE1TG with dimensions for the 3.5, 7 and 14MHz bands (dimensions for the 10, 18, 24 or 18MHz bands could be readily calculated from basic formulae). It should be appreciated that this antenna provides predominantly vertically polarized signals and over ground of good conductivity is capable of very low angle radiation. In its simplest single-element form, the Bobtail antenna reduces to the inverted ground plane (vertical-Tee) - a useful omnidirectional antenna: **Fig 4**.

Another loosely related configuration is the 7MHz low-angle directional parasitic array (a vertically-polarized wire squashed Yagi) developed by G6XN that provides good performance with relatively low supports: **Fig 5**. In this antenna the horizontal wires provide end-loading of the shortened vertical elements.

In all of the above configurations, the bulk of horizontally-polarized radiation is balanced out. This does not apply to the 'half-square' antenna, although this, in my view wrongly, is sometimes referred to as a bob-tail antenna. The half-square antenna (**Fig 6**) however deserves consideration in its own right. This is the basis of 'A visit to the Half-square Antenna' by Hannes Coetzee, ZS6BZP (*Communications Quarterly*, Spring 1998, pp83-90). In 1997, ZS6BZP began experimenting with a 14MHz version, based on a 7MHz half-square antenna described by W8HXR in *73* in May 1985. He used two 6m (later increased to 9m) masts. As a comparison antenna he used the modified form of G5RV developed by Dr Brian Austin, ZS6BKW/G0GSF, which has been described several times in *TT* and in a full-length article in *RadCom* (August, 1985), at a height of 12 metres. To the disappointment of ZS6BZP, on-air tests showed the G5RV dipole consistently out-performing the 14MHz half-square antenna: "After many comparative tests, I had to accept defeat. The half-square antenna had been thoroughly beaten by a multiband dipole. I had to find the reason why!"

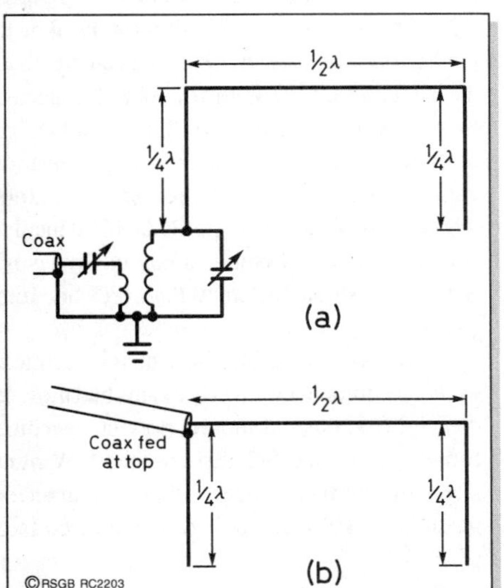

Fig 6: The basic Half-Square antenna. (a) voltage fed; (b) current fed. Mixed polarization.

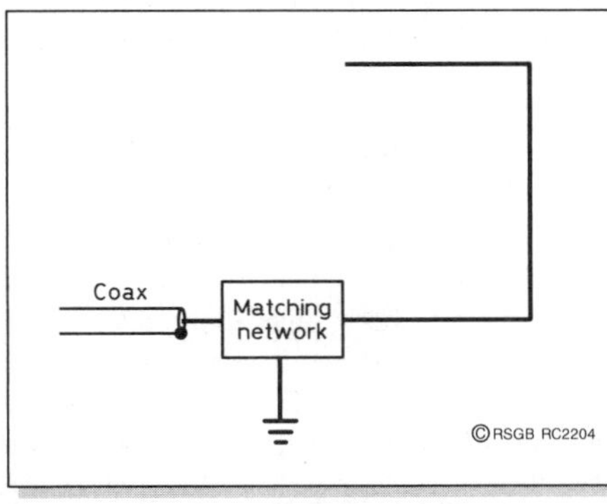

Fig 7: The basic Lazy-U antenna element. Produces predominantly vertically polarized signals.

Together with Danie Brynard, ZS6AWK, and his copy of EZNEC2 by W7EL, they simulated both antennas over average ground at a height of 12m. The bottom of the half-square verticals were 7m above ground, and the radiation at an elevation of 11° above horizon was calculated for both antennas. The simulated half-square gave 3dBi per main lobe, whereas the improved G5RV gain came top with 7dBi in the main lobe. ZS6BZP suggests that the reason for the higher gain of the G5RV antenna can, to an extent, be found in the fact that its legs are considerably longer than those of a 14MHz dipole, with results on higher frequencies producing even more respectable gains.

He notes that the half-square antenna was intended for use on 7MHz, delivering a bi-directional pattern with a simulated gain of 1.78dBi and nulls in the direction of the antenna nearly 13dB deep; a very useful performance. In comparison, the improved G5RV resulted in a figure-of-eight pattern with a maximum gain of 1 to 1.16dBi, a realistic figure for a 11° elevation angle for a dipole at close to a quarter-wavelength above ground, an improvement of some 0.5dB on the half-square antenna. In his article ZS6BZP describes methods of changing the direction of radiation from a half-square and also suggests that improved performance could result in changing it to a Phased Lazy-U antenna using Lazy-U elements. This idea might be worth pursuing, especially in the form of a four-square array that could provide selectable radiation direction on the lower bands.

The basic Lazy-U antenna resembles an up-turned half-square antenna, although with a height that can be less than a half-wave. This antenna was described in 1992 by Henry Elwell, N4UH, as a short vertically polarized antenna in which, provided the total element length is a half-wave or multiple thereof with equal horizontal sections, the radiation from the horizontal sections is largely cancelled out: **Fig 7.**

ZS6BZP concludes: "Despite all the claims in the amateur press over the years, the half-square antenna is not the ultimate solution to all your needs. At 14MHz and higher, a tri-bander (or even a G5RV dipole) at the same height, outperforms the half-square by a fair margin. At 7MHz and lower, the half-square antenna has distinct advantages over a small beam or dipole used at less than a third of a wavelength above ground. The directivity and off-axes nulls also count in its favour. The antenna is also simple to construct."

It should be noted that with all the above antennas there will be high RF voltages at the base of the elements and care should be taken to ensure the safety of people and animals.

CERAMIC RESONATORS IN LF/ MF LADDER FILTERS

JAN-MARTIN NOEDING, LA8AK, considers that too little has been written about the application of ceramic resonators to low-cost narrow bandpass filters. He has a number of Murata ceramic filters centred on frequencies which include 250, 440, 500 and 480kHz. He was planning an off-frequency standard that required some sharp filters. In order to check ceramic filters at these frequencies, he first set up a simple circuit to check the tuning ranges.

Murata CSB250D: Tuning range (with 0-140pF capacitor), 249.24 to 251 kHz. Single tuned circuit selectivity with 140pF capacitor, 248.0kHz -40dB; 249.0kHz -46dB; 249.5kHz -46dB; 249.5kHz -37.5dB; 249.8kHz -23dB; 250.0kHz 0dB; 250.2kHz -20dB; 250.5kHz -26dB; and 251.0kHz -28.5dB.

Murata CSB500E: Tuning range (with 0-140pF capacitor), 509.59kHz to 516.9kHz.

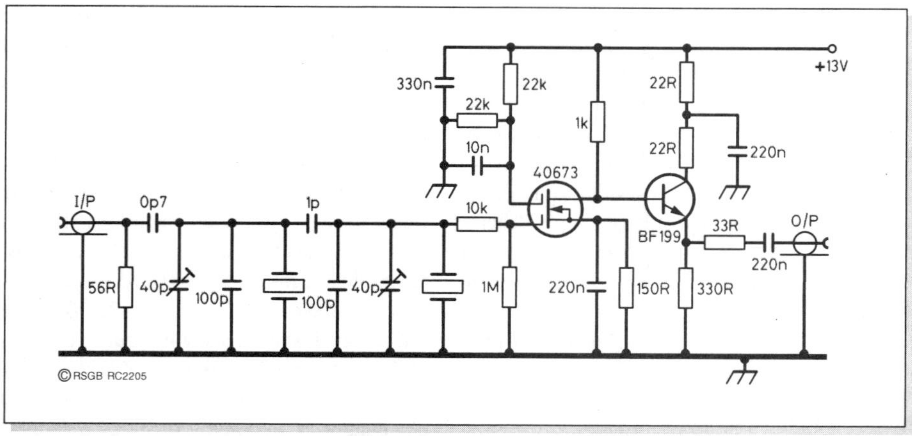

Fig 8: LA8AK's test circuit for evaluating CSB250D ceramic resonators as a narrow bandpass filter.

Murata CSB480E: Tuning range (with 0-140pF capacitor), 477.36 to 482.9MHz.

LA8AK writes: "These measurements indicated that the filters were sharper than expected, so I built a test circuit [**Fig 8**] on a piece of 80 x 40mm copper-laminate board using 'dead bug' construction. It is very important to use a screen between the two resonators. The filter using two CSB250D resonators was tuned to 250.0kHz. Selectivity proved to be extremely sharp, almost impossible to tune the signal generator (using a frequency counter). The -3dB bandwidth is 60Hz, with attenuation increasing rapidly to 70dB below 250kHz and to 60dB above 250kHz.

"Then two CSB440E resonators were tested. The filter could not be tuned higher than 443kHz, but perhaps could still form an alternative to the standard 455kHz intermediate frequency. With the two fixed 100pF capacitors removed (ie retaining only the two 40pF trimmers) the filter was trimmed to 442kHz, which seems to have optimum attenuation. The -3dB bandwidth is about 150Hz, ultimate attenuation was 70dB above and 80dB below more than 10kHz off the centre frequency. This test circuit was designed only for minimum loading on the tuned circuit, so it is difficult to know what the true insertion loss is, but with an improved filter circuit it seems possible to achieve some 6-10dB filter insertion loss.

"I also required an IF notch filter for incorporation in my receiver. Another simple test circuit (**Fig 9**) was built. I tried some different resistor values, and found that a deeper notch could be achieved with higher-value resistors. This indicates that a notch filter should be tapped on to a 5000Ω impedance point of a tuned circuit. The direct series resonance is at 244kHz and the notch is only 10dB deep. This is of no practical value, since it cannot be tuned. It is better to use the parallel resonance characteristics for the notch with the same capacitor values as for the bandpass filter. Notch depth is 31.5dB below the point of minimum attenuation. Measurements (with 150pF fixed capacitance): 240kHz -4.5dB; 245kHz -4dB; 243kHz -3.5dB; 246kHz -4.5dB; 248kHz -6.5dB; 249kHz -20dB; 249.5kHz -15dB; 249.61kHz -30dB; 249.69kHz -25dB; 249.73 -30dB; 249.77kHz -31.5dB; 249.79KHz -30dB; 249.99kHz -20dB, 250.14kHz -10dB, 250.38kHz -5dB; and 251.1kHz 0dB.

"These test circuit measurements indicate that ceramic resonators are fine business for building narrow bandpass filters, as well as notch filters. Perhaps the lower Q makes them easier to use than crystals for super-selective filters up to above about 500kHz."

THE CLASSIC SUPER-PRO

IN THE FEBRUARY *TT* item 'Classic Valve Receivers' I included a mention of "some Hammarlund models" but surely made an error in not picking out their "Super-Pro" models, which must have set a world record in being in production almost 40 years, from 1934 to 1972. Admittedly, the Super-Pro was a generic name applied to no less than (according to the 3rd Edition of Ray Moore's *Communications Receivers - The Valve Era 1932-81*) 30 variations of some five basic designs SP10, SP110, SP210X (military BC779, and the modified BC1004 version), SPC400X, SP600 (1950-72, military versions R274C/FRR, R320A). Even earlier, Hammarlund had introduced one of the very first true communications receivers with the 'Comet PRO' in 1932, although not of the same standard as their later Super Pro models. Hammarlund also manufactured the pre-war HQ120X and the basically similar immediate post-war HQ129X, using some of the design features of the Super Pro (including the six-position single-crystal filter, but with only a single RF stage) and to a lower cost specification. I still have a working HQ129X which is capable of reasonable performance, although it has always suffered a tendency to sudden small frequency jumps which appear to be due to the use of hygroscopic material for the oscillator coil formers that absorb and retain moisture. A *QST* contributor once advised that this could be overcome by fitting a low wattage (5-10W) electric lamp inside the cabinet and running this continuously, to provide a source of dry heat. Seems a good idea, although I have never got round to trying it.

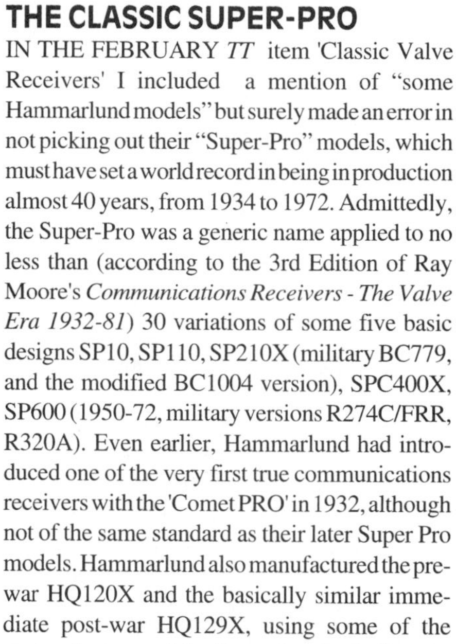

G3GGK's restored Hammarlund BC1009 Super Pro communications receiver, built around 1944 and still providing an excellent 16-valve plus rectifier receiver. A number of these were used on early long-haul US military radio-teleprinter circuits.

Peter Simpson, G3GGK, retains a strong interest in early valve receivers. He writes: "During my fifty-odd years in radio - both as a professional and as an amateur - I have restored countless AR88, HRO, CR100, several Eddystone and even a few Hallicrafters. In the 1950s I owned a Hammarlund HQ120X which gave me a lot of pleasure, but as a young amateur I needed to sell it to buy my first AR88.

"About six months ago, I decided that I would try to find and restore a Hammarlund receiver. A 'wanted ad' in *RadCom* produced some eight replies, varying from offers as free gifts, some at reasonable prices to others at prices best forgotten. One gift offer was of a BC1004D, the later US military version of the Hammarlund Super PRO SP210X, and seemingly closely akin to the SP400 produced in 1946 as a re-entry into the civilian market. With my wife, I collected this generous gift from Ron Lavis, G8DX, of Bath. [Ron is one of the diminishing number of wartime RSS/VI/SCU3 intercept operators whom I knew at Hanslope Park. Incidentally, another reunion is being organised by 'Nos' King, G3ASE, at Bletchley Park on May 16th - *G3VA*].

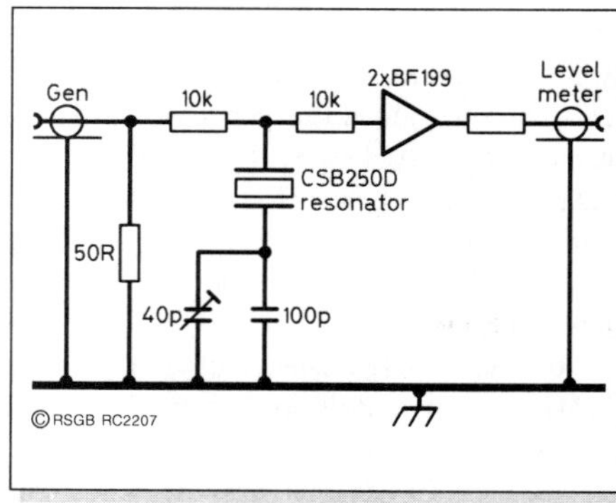

Fig 9: LA8AK's notch filter test circuit.

"I also obtained a BC794, which is identical except the frequency coverage. This had most of the original knobs and, more important, a genuine S-meter. Since the BC1004 covered MW right through to 20MHz, I made this the prime target for restoration. This has now been completed (see photo left).

"The construction and quality of this military receiver was absolutely outstanding. The plating on the chassis responded well to a little abrasive cleaning. Even the front panel was heavily copper then nickel-plated, so there was no rust. The wavechange switch is a plunger type operated by cams, and showed no sign of noise after some 54 years. The tuning heart is a magnificent piece of mechanical engineering and is easily removed from the chassis to permit work to be carried out. The tuning drive is by friction onto Ivorine scales and is much smoother than gears. Because each tuning range is limited to 2:1 in frequency, it came as no surprise to find that the scale accuracy was of a high order.

"So what is the result? I have ended up with: (1) a superb 16-valve plus rectifier communications receiver covering 540kHz to 20MHz; (2) 13 watts of very acceptable audio for listening to broadcast stations from push-pull 6F6 output valves; (3) a crystal-gate filter with six positions and phasing control; (4) an excellent pulse-type noise limiter; (5) smooth tuning controls through friction-type drives, including electrical bandspread above 2.5MHz; (6) A tuning heart built like a battleship, completely stable with superb tuning capacitors, antenna coils that have Faraday screens between primary and secondary (this unit alone would cost more to produce today than a complete Japanese transceiver!); (7) selectivity variable from 3 to 16kHz; and (8) last, but not least, real satisfaction every time I switch on the restored old-warhorse!"

CAMP/MOBILE ANTENNAS

FOR 50 YEARS, Dr John (Jack) Belrose, VE2CV, has been using HF antennas from remote sites, including camping parks and the like. Initially he used simple, easily erectable dipole antennas, including the single-wire fed Windom wire dipole. Later he devised a centre-fed dipole with the element wound up on bobbins so that the system could provide a resonant half-wave dipole on any frequency (or band) of interest. More recently he used a coax-fed, off-centre-fed dipole, again with the wire elements wound on bobbins and suitable for 3.5, 7 and 14MHz bands (41.5m long, fed 13.8m from one end with a home brew 4:1 ferrite bead over coax current balun).

While such dipoles could be supported from one tree with a sloping element, their length could be a nuisance, as they usually had to stretch over neighbouring sites or roads. In 1993, he evaluated a number of DK5CZ's AMA tunable loop antennas, and has used the AMA6 (6.7 to 25MHz) as a camp antenna for a number of years. Vehicle-mounted loops have been proposed for NVIS operation, although they are rather daunting for all but the keenest amateurs. The traditional HF antenna used with land vehicles is a vertical whip - relatively poor for NVIS propagation - although VE2CV considers that the coil-loaded monopole is more easily transported than a loop because it breaks down into sections and can be used as a camp or mobile antenna from 1.8 to 30MHz. ('A tunable all-bands HF camp/mobile antenna' by VE2CV and Larry Parker, VE3EDY, *Communications Quarterly,* Fall 1998, pp47-57). He points out that from a campsite it is better to go up, than up and out.

The article describes in detail a TABA (Tunable All-Bands Antenna) system for mobile/transportable applications. Sufficient details are provided to enable amateurs with the necessary tools and machine shop skills to construct it, together with measured data to determine efficiency of the antenna for the 3.5MHz band plus computer-calculated performance on this and several other bands.

It is stressed that for any vehicle-mounted, electrically short, base-loaded HF antenna, the loading coil is a key component, as high-Q and remote tuning ability are highly desirable requirements: "The inductance required for the lower bands (1.8/3.5MHz) can be near the self-resonant frequency for the coil, so a coil with a space between the turns must be used to reduce its self-capacitance and increase its self-resonant frequency. For typical mobile antennas, a separate coil is used for each band, with no ability to retune. An alternative system was originally devised by D K Johnson, W6AAQ, and a version described by Lodewijk Stuyt, PA3BTN, and Hans Spits, PD0NCF, (*Electron*, March 1997, pp98-103, Dutch text). A translation of the Dutch article 'The 'Screwdriver' Rapid QRV Antenna' translated and edited by Erwin David, G4LQI, appears in the February 1999 issue of *RadCom*, pp16-19. This describes the construction of a continuously tuneable W6AAQ-type 'DK3' 3.5-30MHz mobile or portable antenna with, for fixed operation, a radial earth mat.

In Canada, two versions based on the W6AAQ design, independently of the Dutch work, of this form of tunable whip antenna (one for 1.8MHz and above, the other for 3.5MHz and above) were fabricated by VE3EDY. Their radiation characteristics were determined by VE2CV, and their on-air performance extensively evaluated by both. Coil length is 55.9cm (22in) for 3.5MHz, 81.3cm (32in) for 1.8MHz. A 1.5m stainless steel whip is installed for mobile use and a sectionized 4.9m whip used on camp sites etc, although the full-length 4.9m antenna can be used on a stationary vehicle which then, in effect, forms the earth mat.

The W6AAQ ingenious remotely-tunable loading coil system clearly requires some care and expertise in mechanical construction and is well illustrated in the February *RadCom.* For the Canadian versions, the lower section of the antenna is a 5cm (2in) diameter stainless steel pipe. The Canadians do not use the toroidal impedance-matching transformer advocated by W6AAQ and the Dutch authors. They claim that no other antenna system tuning unit is needed.

Much of the Canadian article is devoted to a discussion of the radiation characteristics and efficiency, including comparisons with an AMA8 loop and with a low dipole, both for NVIS medium-distances and for long-haul working, showing the effect of the vehicle body on the radiation characteristics: **Fig 10**. Among their concluding remarks, they stress: "The important features of the TABA (and the compact loop) is the ability to tune for a minimum SWR on the frequency of choice. This is particularly important for 3.5 and 1.8MHz, where their bandwidth is narrow. The TABA is not confined to the amateur bands but can be used on any frequency in the HF spectrum." ♦

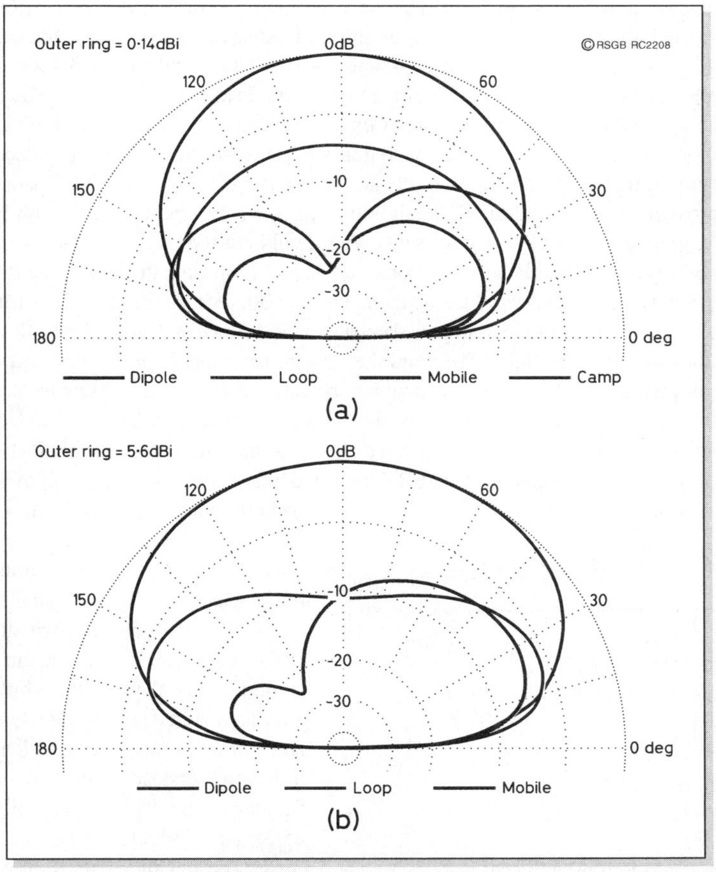

Fig 10: (a) Predicted vertical radiation pattern performance of VE2CV's 3.5MHz camp and mobile installations, compared with a dipole at low height and a vertical 1.6m compact loop (AMA8). (b) Predicted performance of the mobile antenna on 14MHz compared with the loop and dipole with an apex height and droop as for the 3.5MHz dipole.

technical topics

by Pat Hawker, G3VA*

CLOTHESLINE ANTENNAS

A TRADITIONAL FORM of covert antenna consists of a clothesline threaded with a thin, concealed wire, but in *TT*, March 1999, pp62, (shown again here as **Fig 1**) we reproduced a novel form of multi-band folded clothesline (washing-line) dipole in which the position of the 300Ω feedpoint could be adjusted by means of low-cost plastic pulleys normally intended for outdoor clotheslines of upstairs apartments etc. This came from a collection of HF antenna ideas in the Dutch VERON journal *Electron*, collated by PE1PVB and ascribed to VA3ERY, although the original source was not given by PE1PVB. This resulted in a letter from Roberto Craighero, I1ARZ, who provided the original *QST* article by VA3ERY and also revealed an earlier Italian endorsement of this ingenious approach. He wrote:

"The article on which PE1PVB drew was 'The Clothesline Antenna' subtitled 'Dry your laundry or work DX - Could this be the first dual-purpose antenna?' by Robert Victor, VA3ERY (*QST*, July 1998, pp56-58). I was impressed and decided to translate in into Italian for possible publication in the ARI journal *Radio Rivista*. It duly appeared in the February 1999 issue (pp43-44). Following publication I received a letter from Goffredo Navacci, I0TWA, who pointed out that a similar form of clothesline antenna had been developed and used by him since 1995, with publication in the Italian magazine *Radio Kit* (March 1995)."

I1ARZ enclosed copies of the relevant articles, which make it clear that while VA3ERY genuinely believed that he was the first to describe such an antenna, he recognised that: "Every once in a while, you run across an idea that seems so simple, so obvious, you can't believe it hasn't been done before. Surely (you say to yourself) you're not the first person in the universe to have thought of this... As far as I'm concerned, *I've* invented the Clothesline antenna. Still, I'm not going to be surprised if someone shows me that it's been done before! But even if this antenna design has been around since hydrogen, it may still be new to you.

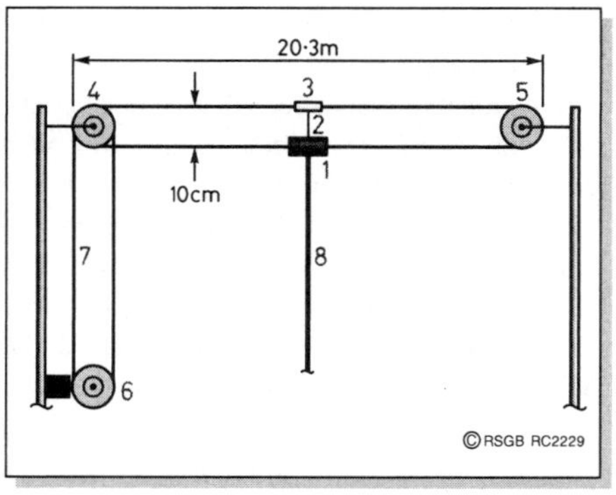

Fig 2: The multi-band clothesline folded dipole antenna as conceived and described in 1995 by I0TWA. (1) balun (see text). (2) Nylon cord. (3) PVC tube. (4), (5) and (6) plastic washing-line pulleys. (7) Nylon cord. (8) Coaxial cable feeder.

And even if you have seen something similar to the Clotheseline, this design may be worth a second look. For hams in a variety of situations, this could be just what the doctor ordered to cure your DX dilemma."

VE3ERY waxed enthusiastically on the usefulness of this approach for the many amateurs operating from flats, condominiums or apartments. He claimed that his 132ft span version (as reproduced again as Fig 1) "works all the HF bands, and it gets great results. It doesn't need a tuner because it's dead-on resonant on the 80, 40, 20, 15 and 10 metre bands. A tiny tweak will bring in 12 and 17 metres too. It's remarkably easy to build. There are no traps, no stubs, no loading coils, no variable or fixed capacitors, no screws, no clamps, and you don't have to drill anything. This antenna is so simple to put up, it hurts. It consists of little more than a piece of wire, a centre insulator, some feed-line and a couple of $2 hardware-store fittings."

VA3ERY advises that for covering the 17 and 12 metre bands, you should cut the original length for resonance just a little up from the bottom of the 80 metre band, at 3615MHz. The fifth harmonic is then smack on 17 metres. Cut it to 3.55MHz and the seventh harmonic is on 12 metres. In practice, because of the folded dipole form of element, exact resonance will be much less critical than with a single thin wire element. VA3ERY reported that "On every band I tuned for, the SWR was well below 1.5:1 without a tuner."

The 1995 I0TWA design, although basically similar, differed in some practical details: **Fig 2**. Instead of using a non-critical length of 300Ω feeder, he put the balun (6:1 for 50Ω coax, 4:1 for 75Ω coax) at the element feed point.

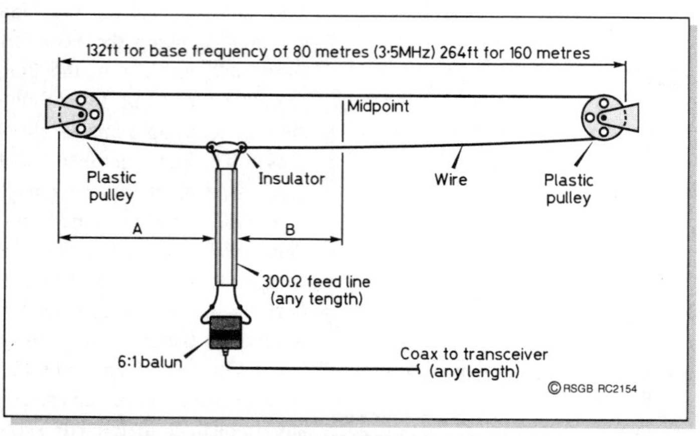

Fig 1: The multi-band clothesline folded dipole antenna as conceived and described in 1998 by VA3ERY.

He also used three plastic pulleys, with a nylon cord permitting the lower pulley to be mounted so that the feed point could be adjusted from ground level, rather than by leaning out of the apartment window. I0TWA suggested that a folded element with a span of 20.3m would resonate on approximately 7050kHz, 14100kHz, 21,150kHz and 28200kHz. With the longer span of 39.6m it will have a fundamental resonance at 3600kHz, second harmonic 7200kHz, third 10,800kHz, fourth 14,400kHz, fifth 18,000, sixth 21,600kHz, seventh 25,200kHz and eighth 28,800kHz - all reasonably close with the wide bandwidth of a folded element to the amateur HF bands. For each band the feed point should be adjusted by means of the pulleys to approximately a quarter wavelength from the near end and then 'fine tuned' for minimum SWR.

All-in-all, it seems that the clothesline antenna, whether you go for the Canadian or the Italian job, is a useful addition to the stable of multi-band HF antennas.

DX ON SPARK

THE ILLUSTRATION of the Marconi quarter-kilowatt quenched spark emergency transmitter in the March *TT* certainly brought back memories to a number of amateurs who were marine Radio Officers before they swallowed the anchor. Among these were Peter Bond, G3BEG; John B Tuke, G3BST; Alan Pink, G3RMZ; Adrian Ridley, G3ZLR; Keith Plumridge, G0MMA; and others, while Steve Richardson, G4JCC, sent along an interesting local paper cutting on the re-dedication of the grave on Hayling Island of Admiral Sir Henry Jackson, 1855-1929. As Captain Jackson, he was the first person to experiment (1896-1900) with maritime spark systems for use by the Royal Navy, developed independently of Marconi and based on the work of Hertz and Lodge. See *The Origins of Maritime Radio* by the late Gerald Garratt, G5CS, in collaboration with R F Pocock, published by HMSO on behalf of The Science Museum, 1972. Admiral Jackson was President of the Wireless Society of London in 1922, during the change of name to the Radio Society of Great Britain.

Spark transmitters were, of course, used by British amateurs in the pre-WW1 period, and again for a few years in the early 1920s. The photo (see page 58) shows Leslie McMichael, MXA (later 2FG), with his station as it existed in about 1912. The transmitter used a 6in spark coil, with a power of some 150 watts on 275 metres. The receiver consisted of a large inductance with a slider for tuning and a Perikon crystal detector, and could receive signals up to 10,000m. Leslie McMichael was a founder member of the Wireless Club of London (later RSGB) and

*37 Dovercourt Road, London SE22 8SS.

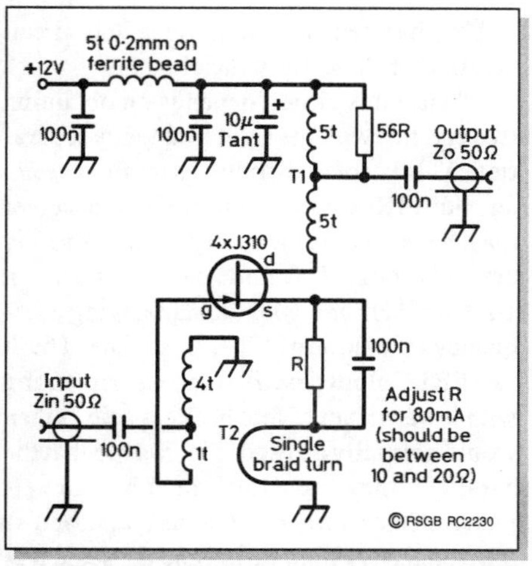

Fig 3: FET post-mixer amplifier as built by G3SBI from information supplied by W7AAZ and W4ZCB. Adjust R for 80mA current. Value should be between 10 and 20Ω. The 0.1µF capacitors are monolithic ceramic type. T1 five bifilar turns of 0.4mm Bicelflox enamel on Fairrite BN 61 302 core. T2 4 = 1 turns on Fairrite BN61 302 core using 0.4mm Bicelflox enamel wire. Primary single brand turn. See *TT*, September 1996, p72 for details of how to wire T2.

also founded the McMichael radio manufacturing firm. Some of the pre-WW1 amateurs used home-made rotary spark transmitters. A few achieved ranges of several hundred miles.

While most of the ex-RO reminiscences will have to wait another month, it is worth noting that it was apparently not until January 1956 that spark, even for emergency maritime use, was banned under ITU Regulations. Even then, Spain and China were at the time not members of the ITU and thus exempt. G3RMZ recalls hearing a Spanish vessel using spark as late as mid-1958. He also points out that many Radio Officers regretted the change to valve emergency transmitters such as the Marconi 'Salvare' (with two 807s), since a valve transmitter could lose a lot of its output power under damp conditions (not unusual at sea!) and required much careful attention in order to meet the UK requirement, regularly tested, that a lifeboat transmitter must deliver 15 metre-amps of notional power (antenna current times height of antenna). As G3RMZ puts it: "To meet the tests meant batteries on trickle charge all night, built-in drying heaters on for days, wire antennas erected at the last moment with the insulators washed in fresh water, polished dry and properly warmed. What a silly charade! But it did drive home the force of argument of those who considered that the spark set should still be in use in lifeboats. Self drying, so self-healing. Splutter and hiss a bit before they came up to full song, but probably capable of recovery from even full submersion. . . and audible whatever the listener's tuning or bandwidth! What more does one want from a lifeboat transmitter than reliability to make a broadband keyed noise, the very tonal quality of which simply screeches HELP!" And, one might add, far less liable to initiate false alarms than the current GMDSS automatic satellite system which is so often triggered off by mistake!

In the classic Transatlantic Tests of 1921, valve transmitters outperformed spark, but at least one amateur spark transmitter successfully spanned the Herring Pond. But this feat is outclassed by G3BST who writes: "I was intrigued by the illustration of the Marconi QG set, as I am sure were many others. Apart from reviving memories of long hours on watch on various 'tramps' with this transmitter and the MR4B receiver, it recalled my best DX of those days. I was at sea for three years before WW2, and one night steaming rapidly (four knots!) up Channel at about 0200z, I was on watch. We were 'navigating' by MF D/F (the Skipper did not believe in night-effect). While idly listening on 600m, I heard ZSC Capetown give his CQ and a QRU traffic list. I could not resist calling him with a TR [a TR comprises the name of the ship, location, next port of call and number, if any, of telegrams to be sent - *G3VA*] and we had a QSO which today would be called 449 - though probably more like 442 on my part! I was probably running a little more than the regulation 250 watts, as I had short-circuited the series inductance in the AC supply, as experience showed that while this may or may not have brought the primary circuit to resonance as it was supposed to do, in fact you got more output with it shorted! It would be interesting to know if any other ex-sparker can beat this DX contact. It was more than I ever achieved using a 5kW rotary spark set."

The 1912 station of Leslie McMichael (MXA), with a 6in spark coil, giving a power of some 150 watts on about 275m. The receiver consisted of the long inductance with slider and crystal detector. *Source* 'The Bright Sparks of Wireless' by GR Jessop, G6JP (RSGB, 1990).

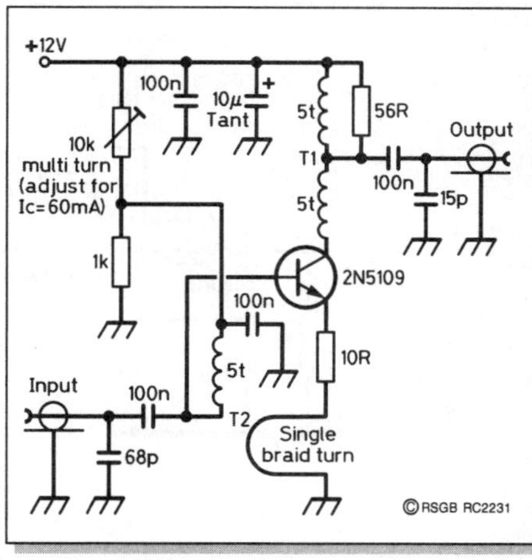

Fig 4: Bipolar post-mixer amplifier. The 0.1µF capacitors are monolithic ceramic type. T1 as for FET amplifier. T2 5 turns of 0.4mm diameter Bicelflox enamel wire in single turn of coax braid. See *TT* September 1996 for method. Fairrite BN 61 302 core.

A note from the former R/O Keith Plumridge, G0MMA, reports using spark to communicate with Adelaide (VIA) as late as July 1956, after gales had damaged the AWA transmitter during a voyage in Australian waters of *The Ironmaster* (3382 gross tonnage).

BIPOLAR & FET POST-MIXER AMPLIFIERS

COLIN HORRABIN, G3SBI, passes along circuit information on two post-mixer amplifiers that give reasonable results without requiring power-FETs with drain currents of some 0.3A to achieve third-order output intermodulation products (IP_3) above 40dBm. These designs have come from Bill Carver, W7AAZ, and Harold Johnson, W4ZCB, both of whom have previously contributed state-of-the-art information on topics relating to the development of receivers with super-linearity. G3SBI has confirmed their good performance.

G3SBI writes: "These amplifiers would be used as the load to the hybrid-connected roofing filters after an H-mode mixer to drive the high quality crystal filters for SSB and CW. The roofing filters normally govern SSB IP_3 performance and even more so on CW (my commercial IQD 250Hz CW filter has an input IP_3 of 45dBm). My roofing filters have an out-of-passband IP_3 when connected together of +43dBm. The passband IP_3 is 33dBm, so ideally the following amplifier needs an intercept figure above this. This can be achieved using

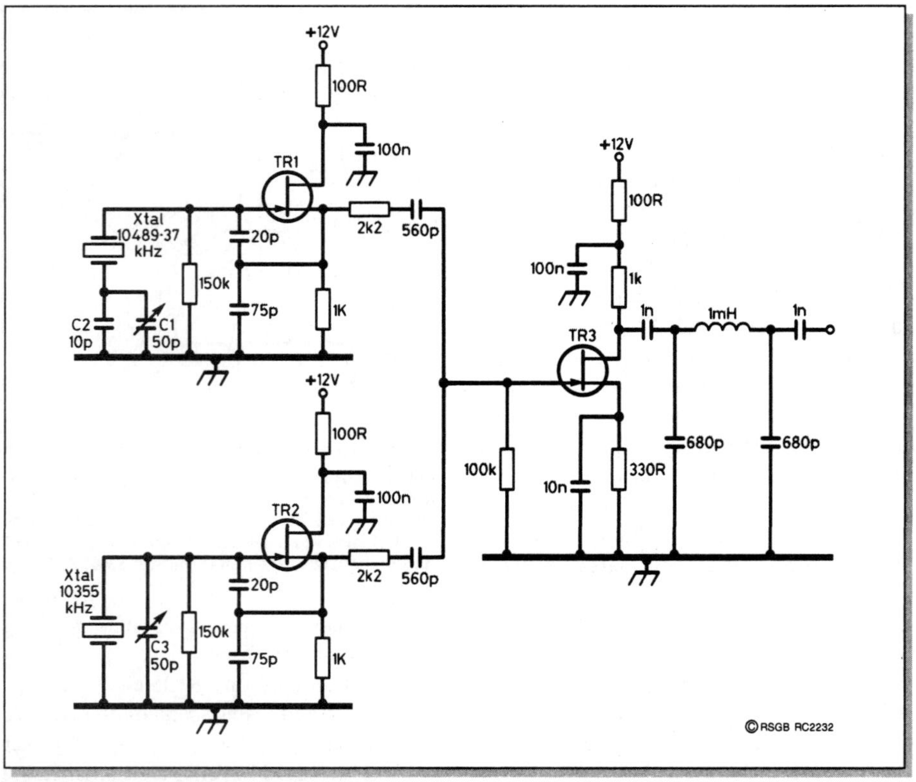

Fig 5: G3LDO's 136kHz mixer-VXO exciter.

power FETs, as described in *TT* September 1996, but at the cost of high drain current. The W7AAZ/W4ZCB amplifiers described here are a useful compromise.

"**Fig 3** shows the FET amplifier using four J310s in parallel (my own version has some slight changes). W7AAZ has measured this circuit as providing 12.4dB gain, 1.12dB noise factor, 25dBm input IP_3, 40dB reverse isolation at 9MHz. I constructed one of these amplifiers, but found I needed a different number of turns on the input transformer to obtain the match. The single source resistor is adjusted to give a total current of 80mA. This critically assumes that the J310s are closely matched. A better approach would be to have a resistor and bypass capacitor for each FET to set a current of 20mA in each device. The input and output impedances were slightly negative reactance, so small capacitors could not be added to obtain a 50+j0 match at one frequency.

"I found the bipolar design of **Fig 4,** using a single 2N5109 device, more interesting. Performance with 60mA collector current was: gain 14.5dB; output IP_3 40dBm (could be increased with more collector current). The input and output impedances were 50+jx with very small reactances, so that a small capacitor could make the impedances purely resistive at 9MHz. Without the capacitors, the impedance was flat to beyond 50MHz."

G3SBI also draws attention to and apologises for an error he made in his notes included in the September 1996 *TT*. The core type in the power FET amplifier should have been Fairrite BN 61 302 and not BN 61 202; and also BN 43 302 and not BN 43 202. The 202 core is actually bigger and should work in the circuit, but it may make the input compensation capacitors different for a resistive match.

VXO/EXCITER FOR 2300m

PETER DODD, G3LDO, was recently looking for a means of providing a variable but stable excitation for a 136kHz transmitter, after finding that most of the approaches used by the LF group involve synthesisers or digital dividers that seem not to work well for him. Then he remembered seeing a *TT* item on a mixer-type VXO.

This inspired him to develop the circuit described below. He writes:

"To use this circuit depends on obtaining two crystals with the right frequency separation. This is not as difficult as it might seem, an old VHF crystal-controlled transceiver may have two suitable crystals. From my crystals box, I found two crystals on 10489.37kHz and 10355.0kHz, giving a frequency separation of 134.37kHz. Then I built two FET Colpitts oscillators, experimenting with lower values of feedback capacitors than would normally be used. This was so that the parallel tuning capacitor would be more effective. This all worked, but the output across the frequency range varied and the scale was noticeably non-linear. I then tried the same 50pF variable capacitor in series with the higher frequency oscillator, which largely overcame the deficiencies of the parallel-tuned oscillator. The resulting arrangement is shown in **Fig 5**, where the tuning capacitor is designated C1 with the small fixed capacitor C2 to limit the frequency range. The parallel capacitance C3 (not used in my final circuit) is used to preset the lower frequency oscillator, should you not have crystals so conveniently spaced. C1 and C3 must be air spaced. C2 could be a Philips type air-spaced trimmer. Almost any type of FET seems to work. I used some 2152s, mainly because I had a lot of them. 2N3819s also performed well. The output is around 300mV p-p.

"In the course of deciding on a suitable mixer/filter, I read an e-mail from Peter, DF3LP, who described a basically similar VXO-exciter. He kindly sent me a copy of his circuit which uses 5MHz crystals (see **Fig 6**). This provided me with enough ideas to complete my exciter. Apparently the first circuit of this type, that I am aware of, was built by DJ1ZB. The main attraction of this approach is that if both oscillators are built in the

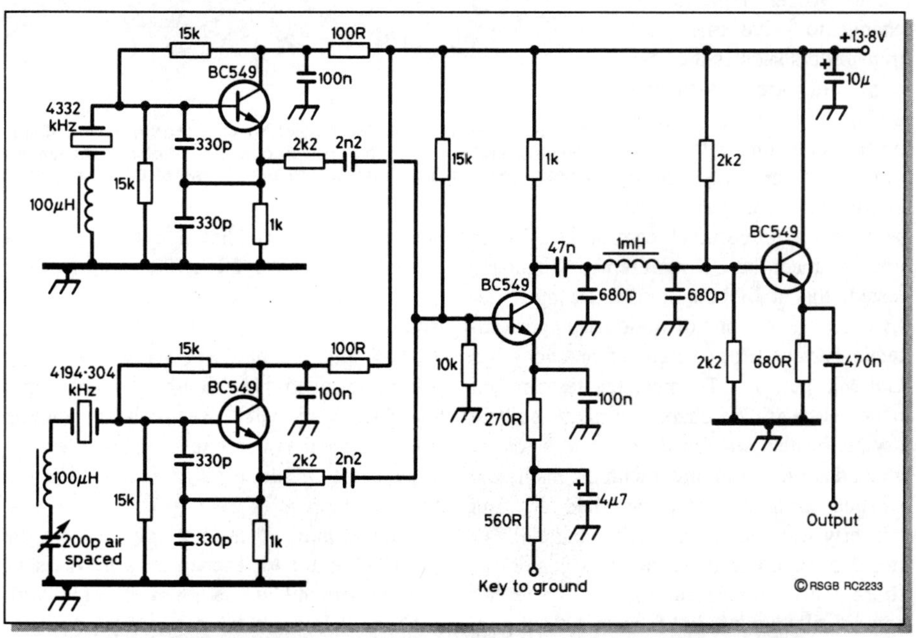

Fig 6: Mixer-VXO 136kHz exciter, as originally developed by DJ1ZB and modified by DF3LP.

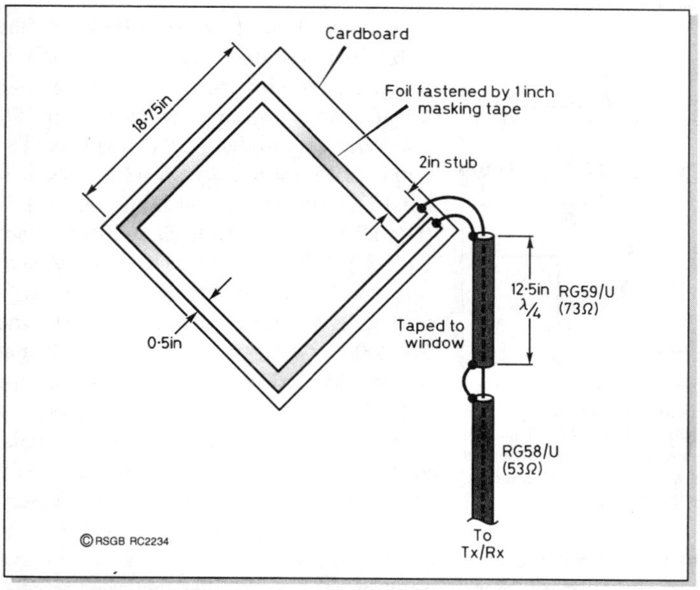

Fig 7: Window-pane single-element quad for 144MHz operation, using household aluminium foil mounted on cardboard. As originally described by W8AP in 1971.

same enclosure with identical components, the output frequency is held constant since any drift in one oscillator is compensated for by the equivalent drift in the other.

"My circuit is based on junk box components, yet over a three hour period the measured drift was only from 137.6530kHz to 137.6524kHz, less than 1Hz."

HI-FI & SOLID-STATE

THE RECENT *TT* items on the differences in audio quality between thermionic and solid-state hi-fi amplifiers has prompted E A Rule, G3FEW, to express his firm belief in the superiority of MOSFET amplifiers. G3FEW was for 25 years the chief design engineer for Armstrong Audio Ltd and later a consultant audio engineer. In 1979 he designed a stereo amplifier using the then new Power MOSFETs produced by Hitachi. Full details of this amplifier, called 'The Winton', were published in *Practical Wireless* (March, April and May, 1979) and many hundreds were produced from kits.

G3FEW writes: "The early transistor amplifiers used bipolar output devices. These required a wide area of safe operation to achieve reliability and a large gain-bandwidth product so that large amounts of negative feedback could be used at the higher audio frequencies to reduce distortion. They also had a positive temperature coefficient, which meant that thermal runaway was a possibility. As bipolar transistors are minority carrier devices, they suffer from storage effects at high frequencies. This can cause the most objectionable distortion. They were mainly operated in class B, and although various methods were used to overcome the problems this caused, nearly all transistor amplifiers suffered from 'cross-over' distortion to some degree. These various distortions led to what became known as 'transistor sound'. Even very small amounts of cross-over distortion caused objectionable sound quality, when compared with valve amplifiers.

This form of distortion could not be overcome by using large amounts of negative feedback. This, coupled with a poor 'slew rate', caused transient distortions as well, adding to the objectionable 'transistor sound'. At Armstrong we largely overcame this effect in the later Series 600 amplifiers, by using transistors with a wide bandwidth and by forward-biasing the output stages to the point where cross-over distortion was almost completely removed. However, it was still very difficult to equal the very best valve amplifiers of the time.

"My 1979 design using Hitachi Power MOSFETs completely overcame the distortion problems associated with bipolar transistors, and was widely praised for its purity of sound. . . The *TT* comment regarding supersonic and VHF/UHF parasitic oscillation in transistor amplifiers did not apply to the Winton amplifier, which proved completely stable under all conditions of use. The MOSFETs were almost indestructible, I have seen a piece of wire shorted across the output, glowing red with the full power applied, with no damage to the MOSFETs.

"Many hundreds of these amplifiers were produced from kits and no problems encountered in repeating the performance specification, provided they were built as instructed. Personally I was surprised that more hi-fi manufacturers did not adopt the power MOSFET. As the designer of the Winton amplifier, I suppose I may be biased in its favour, but I do not believe that many other amplifiers produced such a pure sound.

"Note the very low intermodulation products, even at full power with complex loads using a two tone test. The total intermodulation measured over a 50kHz bandwidth was still less than 0.1% with 30V peak output across a complex load. The worst case harmonic distortion was below 0.02% at all frequencies from 20Hz to 20kHz, at all power levels up to a maximum of 50 + 50 watts.

"Yes, there is a difference in the sound of valve versus transistor amplifiers, but I believe a MOSFET design can outperform them both!"

I cannot comment on the stability of G3FEW's Winton amplifier, but would not wish readers to assume that power MOSFETs, HEXFETs or plain FETs are always inherently free of destructive parasitic oscillations, unless precautions are taken. PA0SE discovered this with the destruction of a series of HEXFETs in the original version of his 136kHz transmitter, before fitting 120Ω gate resistors as parasitic suppressors.

TUNED LOOPS & VHF/UHF LOOPS

I HAD THOUGHT that the February *TT* item (pp77-78) on compact HF tuned loop transmitting antennas would put this topic to rest at least for a time. However, the item had been written before receiving a letter from Dr Jack Belrose, VE2CV. His letter, originally directed to John Crabtree, went a long way to explaining the reason why the measured bandwidth of a compact loop is significantly larger than would be expected from calculating the Q of the loop on the basis of the ohmic losses. He feels that the prime reason for this difference arises from the difference between the unloaded (theoretical) antenna bandwidth and the loaded operational bandwidth - and that discussion of eddy current losses in the February item is largely a red herring.

G8BTH's 50MHz (left) and 144MHz loop antennas, formed from copper strip mounted above ground planes.

To quote very briefly from

his long letter: "G3LHZ has noted that the Q values measured by SWR and antenna bandwidth methods are much lower than predicted by the traditional loop radiation formula. But this is to be expected. . . When we measure bandwidth by measuring SWR, we must remember to take into account the effect of the output impedance of the RF power amplifier, since for maximum power transfer the output or source impedance is the complex conjugate of the antenna's impedance (reference Belrose, VE2CV, Maxwell, W2DU and Rauch, W8JI, *Communications Quarterly,* Fall 1997, pp25-40) - and while not representing a loss, the source impedance (for narrowband antenna systems) increases the bandwidth of the antenna system. The operational bandwidth can theoretically be double the antenna bandwidth, but in practice the realized operational bandwidth is somewhat less (in my experience the increase in bandwidth is by a factor somewhat > 1.5 . . . For AMA-2 the measured operational bandwidth is greater than the antenna's bandwidth by a factor of about 1.7. The operational Q is lower than the antenna Q by this same factor."

It needs to be emphasised that the relatively low overall radiation efficiency of small transmitting 'magnetic' loop antennas arises from their electrically small dimensions, and that a large loop (for example a quad element) is a highly efficient radiator. On VHF/UHF a resonant or near-resonant loop antenna can provide a compact and attractive system. A useful form of indoor 144MHz antenna, comprising a quad element formed from household aluminium strip, stuck to a card and mounted on a window pane, has been described before in *TT*: **Fig 7**. Construction is roughly as follows: Cut a piece of aluminium foil to the outside dimensions, and a piece of cardboard of the same size. With 1in masking tape, fasten the foil and cardboard sheets together on the window. Cut the inside of the foil out with a razor, taking care not to go through the cardboard and scratch the window. Then tape the inside edge of the foil to the cardboard, leaving the 2in slit free of tape. The quarter-wave matching transformer reduces the feed-point of about 100Ω to match the 50Ω co-ax feeder. The antenna, when arranged as shown, results in vertically polarized signals; if it is moved around through 90° so that the feed point is at the bottom of the square, the signals will be horizontally polarized. Mount on a window broadly facing the most desired direction; the window glass will give little attenuation and results outwards should be much the same as an outdoor quad element antenna mounted at the same height. This form of construction could also be used to experiment with fractal designs. There is plenty of room for further exploitation of loops on VHF/UHF.

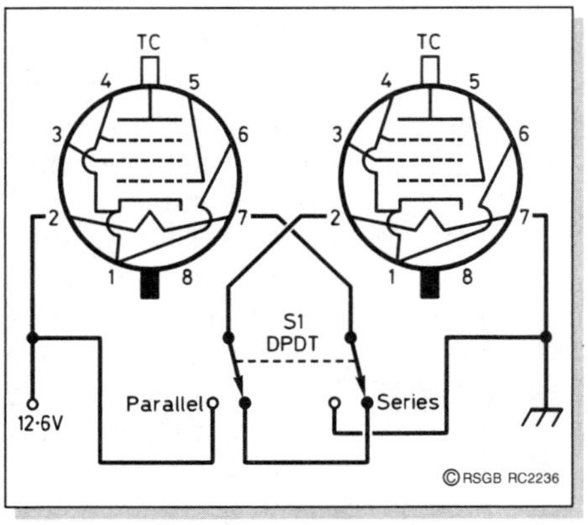

Fig 8: Switching arrangement for selecting series or parallel heater connection. S1 is a DPDT switch suitable for switching several amperes. For two 6146B valves and 12.6V supply, use series mode. For two 6883B valves, use parallel mode.

Colin Harlow, G8BTK, writes: "For many years I have experimented with various forms and types of antennas for VHF and UHF. This interest more than paid off, as some eight years ago we moved into a small flat with no possibility of an outdoor or loft antenna. Accordingly I have constructed, for various applications, many types, shapes and sizes of antennas suitable for use indoors.

"My main 50MHz antenna is a 20in loop of 9.5in wide cooper strip over a 48in square ground plane, mounted on a microphone stand. The loop is horizontally positioned over the ground plane. I have worked more than 30 counties and over 150 squares covering from North Africa to North Norway, Stornaway, etc, using a genuine 9 watts SSB from an FT680R.

"For 144MHz, a 6in diameter loop of 3.5in wide copper strip is used , with a tube 19.5in long and 3in diameter of wire mesh in addition. Also in use are some resonant 'delta-type' loops for 50 and 144MHz and flat resonant 'discs' for 144MHz and 70cm. These antennas may be mounted for either horizontal or vertical polarization. Some of these are shown in the photo opposite.

"I feel that small loops for the VHF/UHF bands could help many people who, like myself, have no hope of erecting antennas other than indoors - in my case at the end of my bedroom!"

HERE & THERE

BOB MAY, K4SE, in the 'Hints & Kinks' column of QST (February 1999, p70) notes that the popular RF power valve type 6146B/8298A, still widely used in many of the valve and hybrid transceivers designed in the 1960s and 1970s is, apart from the heater ratings, identical to type 6883B/8552 which, at least in the USA, is available at much lower cost. Whereas the 6146B heater requires 6.3V at 1.125A, the 6883 is rated at 12.6V, 0.625A. Since a number of hybrid transceivers fitted with the 6146B use the 12BY7A as the driver, the heater supply is often 12.6V with the heaters of the two 6146B PA valves connected in series. K4SE notes that this is true of the Kenwood TS-530S and TS-830S, Collins KWM-2, Heathkit SB-102, some Yaesu models such as the FT-901, FT-902, and FT-101ZD series (but not early FT-101 models, which used TV line-output valves). I would add that the fully-valved KW2000A (and presumably the KW2000B etc) has a 12V heater supply, but one would need to be careful, since all the valve heaters are arranged so that a current balance is achieved giving 6.3 + 6.3V rails, and this would be unbalanced by simply wiring the PA valves across the 12.6V supply. K4SE notes that with the switch arrangement shown in **Fig 8** it is possible to select between series or parallel connected heaters, to enable you to fit either type of valve (but be sure the switch is in the series mode when using 6146B valves!).

Steve L Sparks, N5SV (*QST,* February 1999, p34), describes an inexpensive RF current probe that can be built in a few hours: **Fig 9**. With the values shown, his unit has a half-scale-reading sensitivity of 0.2mA (10mW) into 50Ω at frequencies from 1.8 to 30MHz. By using a more sensitive meter (eg 50µA FSD) sensitivity could be improved. To use the probe, simply snap the core around the conductor you're checking and adjust the *sensitivity* control for a usable reading. The probe can be used to check RF current distribution in antenna elements, open-wire feed lines, guy wires and other conductors or, if bitten by RF in the shack, to help resonate the station's earth wire. For this, snap the probe around the earth wire, connect a variable capacitor (broadcast set type can be used) between the earth wire and your equipment connection and transmit using just enough power to obtain an indication on the meter. Then tune the variable capacitor for maximum meter deflection. It may take some trial and error to find the correct series capacitance, and it will be frequency conscious, but can cure help cure RFI problems, etc. ♦

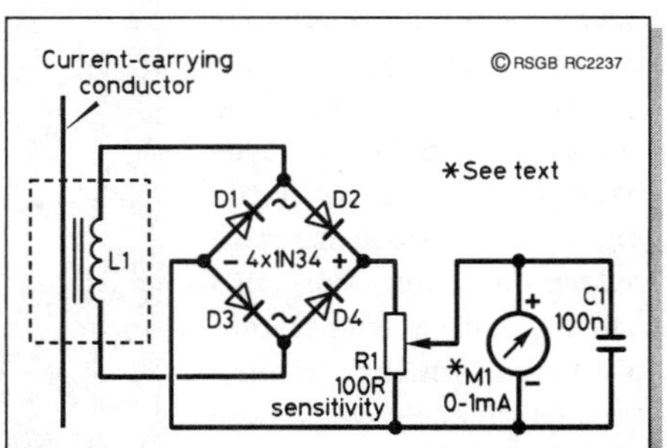

Fig 9: RF current probe as described by N5SV in *QST*. C1 0.1µF disc ceramic. D1-D4 1N34 germanium diodes (do *not* use silicon). L1 single turn of 14-gauge wire through snap-on ferrite choke (RS273-105). M1 0.1mA or greater sensitivity. R1 panel-mount pot, 100 to 500Ω (N5SV used a 10-turn pot).

technical **topics**

by **Pat Hawker**, G3VA*

HISTORICAL MISCELLANY

WITH THE PIONEERING work in practical radio communications now extending over a century, recent years have seen renewed efforts to identify and evaluate many of the significant steps along the way. While the history of the main active devices - thermionics and solid state - have received much attention, less regard has been given to the development of the many passive components on which all electronic equipment depends. An exception is 'Components: a Personal View' by GWA Dummer (*The Newcomen Society Transactions,* Vol 68, 1996-97, pp29-51, read at the Science Museum, 13 November 1996). This is introduced as follows: "Not one electronic invention - radio, radar, television, the telephone, the computer, etc - could have existed without the development of passive components. Sometimes known as the 'building bricks' of electronic equipment, they have been the cornerstone of electronics development over the last 200 years. During the last 50 years, until recently, they have been manufactured in huge quantities and in bewildering varieties. . . It is not possible to record a complete and detailed history because of, in some cases, the passage of time, and, in others, conflicting claims. . . this paper represents his personal views on the discovery, development and decrease of passive electronic components in the UK over the last 200 years." In a masterly survey, this recognised expert covers everything from resistors, capacitors, batteries, relays, transformers, iron dust cores, to standardisation, miniaturisation, reliability, maintainability, transport hazards, mechanical shocks, vibration, storage, low temperatures, high temperatures, humidity, high altitudes, resistance to nuclear explosions, the influence of assembly techniques on components, reliability evaluation and statistics, etc. **Fig 1** shows some early component layouts, illustrating the transition from the early breadboard to the completely screened metal chassis (roughly 1920 to 1940).

Although it was not until the development of the valve oscillator that an effective method of generating continuous waves became available to amateurs, it should not be forgotten that two earlier non-valve systems were developed and exploited for early telephony and commercial telegraphy. The 'HF' alternator in the early years of the Century made rapid progress but was limited to frequencies below at most 100kHz, with Fessenden using one at 80kHz to make the first 'broadcast' from Brant Rock in 1906. It was subsequently further developed by Alexanderson and then with a different approach by Goldschmidt, whose machines could provide up to 100kW for commercial telegraphy service.

But there was also the 'singing arc', originated in the UK by William Duddell but then much improved by the Danish engineer Valdemar Poulsen and often known as the 'Poulsen arc'. This, like the HF alternator, was widely used for high power commercial VLF/LF telegraphy, but was also the basis of a number of relatively low power telephony equipments, up to about 100 watts, used by Lee de Forrest and others in the period up to about 1912. A detailed article 'The Arc Method of producing Continuous Waves' by William J Byron, W7DHD appears in *Communications Quarterly*, Summer 1998, pp47-65.

In the UK the credit for being the first experimental 'amateur' is usually given to the young Lt MJC Dennis (later Colonel Dennis, EI2B), who set up a station at Woolwich Arsenal, London, in 1898. However, Berj J Ensanian, K3IU, has recently sent me the typescript of an article he has written 'The Grandfather of Amateur Radio', in which he notes how a Leslie Miller, AIEE, wrote an article 'The New Wireless Telegraphy - Some Interesting Experiments for Amateurs' published in the January 1898 issue of the London magazine *The Model Engineer and Amateur Electrician.* K3IU writes "This article persuasively invites the readership to begin wireless telegraphy experiments, and tells precisely how to get started . . . I believe it is hard not to consider Leslie Miller as the clearly identifiable 'Grandfather of Amateur Radio'."

Through a friend, K3IU obtained from the London offices of the *Model Engineer* a photocopy of the original article which begins: "When amateur electricians read in the daily papers that a special grant of money was asked for in Parliament for carrying out the experiments with Marconi's apparatus for signalling without line wires, and that a skilled staff had all the resources of the British Post Office at command, they probably thought trials on their own account were out of reach. The object of this paper is to prove this is not the fact; that the apparatus employed is of a character easy to construct, and experiment with, though it may be difficult to understand exactly the causes that produce visible results. . . Those readers who have induction coils with spark pillars and rods can screw round balls on to the ends of the rods and spark to a large central ball, or to the two balls, in case they are not frightened by Marconi's patent claims, valid or otherwise". Much of the article is devoted to a discussion of the claims in Marconi's patent which Leslie Miller suggests "had largely all been anticipated in various parts of the world without his knowledge, but he has, in sporting language, forced the pace, and done good work from which he deserves to benefit." Millar believed that "commercial interests were about to harvest financial gains from seeds they did not sow . . . and it will not be Professor Branly who will benefit financially by his discovery of the coherer."

K3IU (Berj N Ensanian, KI3N, 512 Grier Street, Willaimsport, PA 17701, USA) is anxious to discover more about Leslie Miller. Was he British or American? Was he the LB Miller who took out a British patents in 1897 on coherers? Is there a photograph of him in existence? K3IU hopes that further information may emerge about the man whom he considers the 'Grandfather of Amateur Radio' and whose article was not only clearly directed at encouraging the rise of a corps or community of amateur wireless experimenters but also gave so much detailed advice on how to get started, including advice on building transmitters and receivers. It seems entirely possible that it was his article that encouraged Lt Dennis to set up his station in London and there may well have been others in that fateful year of 1898.

A lightning visit to the old Patents vaults at British Library Science & Reference Section at Holborn (soon to move to the new British Library building in Euston) provided answers to most of K3IU's questions. Leslie Bradley Miller of 2 Gray's Inn Road, London, Electrical Engineer, applied for a British Patent on 16 December 1897 and this was granted on 15 October 1898. His Patent claims three improved forms of coherers: "My improvements relate to the receiving instruments used in Hertzian wave telegraphy, known as coherers, and especially to methods of decohering

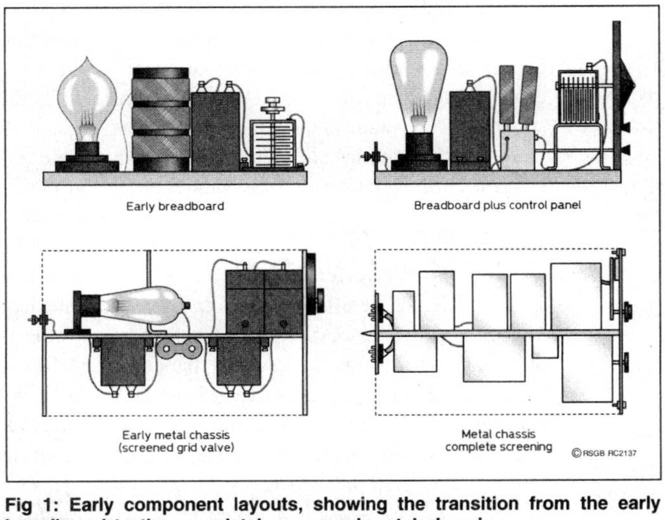

Fig 1: Early component layouts, showing the transition from the early breadbroad to the completely screened metal chassis.

37 Dovercourt Road, London SE22 8SS.

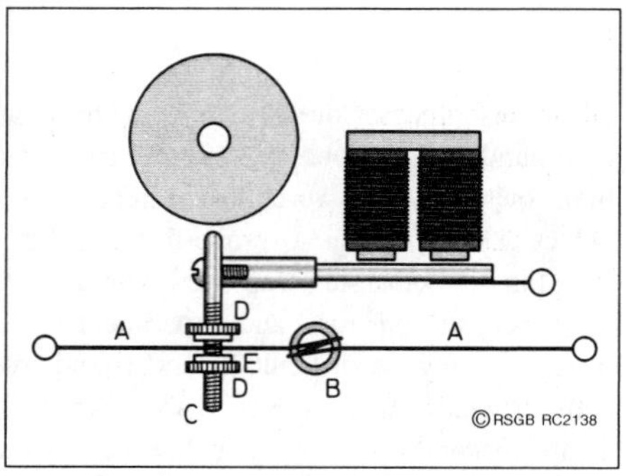

Fig 2: Trembler arrangement illustrated in the Patent.

the same. . . The coherer may be in the form of a metal tube with a central insulated wire passing through, or the coherer may be pivoted so as to be rotated by the motion of a solenoid core or by the expansion of a wire under the influence of an increase of temperature." Dr VJ Phillips in his book describes a large number of coherers and includes illustrations of both Miller's bell trembler and rotating systems for decohering the filings. **Fig 2** shows the trembler arrangement illustrated in the Patent.

FRACTAL & LOG PERIODICS UPDATE

FRACTAL ANTENNAS have figured prominently in recent *TT*s (April and June 1998, and January 1999) and I note that by the time these notes appear a poster paper 'Printed fractal antennas' will have been presented at this year's 'IEE Conference on Antennas & Propagation' (University of York, 31 March to 1 April) by R J Langley and R Breden of the University of Kent, showing that investigation into these antennas has spread across the Atlantic.

Meanwhile, Professor 'Chip' Cohen, N1IR, of Boston University, who has done so much to develop fractal antennas, writes: "Research continues to be heated at the fractal front. Although most hams have an especial interest in the shrinking ability of fractals as antennas, another salient aspect has been resolved which is most likely to affect much antenna use in the future.

"Dr Hohlfeld and I have just published an article in the journal *Fractals*, effectively revisiting the notion of what makes a log periodic frequency independent, that is, having the same gain, impedance, and pattern over a large wavelength range. I had previously shown that very broad bandwidths are ascribable to all fractals, but many were confused why there was changing gain over the very broad bandwidth.

"The new surprise is not that log periodics are fractal (a point I first published in the 1995 *Communications Quarterly* article) but that everyone missed, for over 42 years, the true motivator of this important antenna behavior. We find that Maxwell's equations require that an antenna has symmetry about a point of origin and self-similarity (of fractals), in order to be frequency independent. Log periodics become strictly a subclass of the overall family of frequency independent - fractal - antennas. Commensurately, not all fractal antennas will be frequency independent. The origin symmetry is just as vital, an 'A and B' situation.

"What this leaves us with is a huge variety of potentially useful frequency independent antennas - which are not log periodic. We have a new era of antenna development ahead of us for frequency independence - and a new independence from log periodic construction constraints."

It was the introduction of the non-harmonically related WARC bands at 18 and 24MHz that gave amateurs a special interest in frequency-independent antenna arrays and induced a number to tackle the by no means insignificant problems of constructing large periodic arrays, with their need for many elements in order to obtain forward gains comparable to those provided by far fewer Yagi elements. If it proves in practice that new forms of fractal antennas can provide similar frequency independence with or without significant forward gain but without so many constructional constraints, then indeed the new work by N1IR and his colleagues would seem to open up most interesting possibilities.

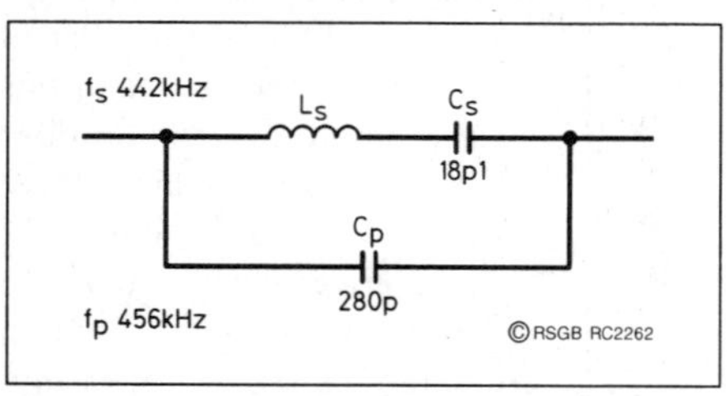

Fig 3: Equivalent circuit of a 455kHz ceramic resonator, as found by G3JIR.

LADDER FILTERS USING CERAMIC RESONATORS

THE APRIL *TT* item based on work by Jan-Martin Noeding, LA8AK, on the application of ceramic resonators to low-cost bandpass filters resulted in an interesting letter from Jack Hardcastle, G3JIR, (who did so much work on ladder filters using quartz crystals) and also a follow-up report from LA8AK on further work in this field.

G3JIR writes: "I have been watching with some interest the multiplying applications of ceramic resonators. A couple of years ago, following a *TT* note on their use in VXOs, it occurred to me that they could possibly be suitable for use in ladder filters at 455kHz. Measuring their parameters revealed equivalent capacitance values which appeared distinctly large when compared with quartz crystals, but which proved extremely useful when applied to a ladder filter. Because of the difference between the mechanical properties of quartz and the ceramic material used in these resonators, it is possible to produce filters having much wider bandwidths than is possible using quartz crystals. Some typical details and values for an eight-pole filter are shown are shown in **Fig 3**, **Fig 4** and **Fig 5**.

"When a batch of resonators was tested it was found that these devices are manufactured to have their *parallel* resonance at 455kHz [but see later - *G3VA*]. As a consequence the series resonance is not tightly specified. My batch had a spread over the range 13.2 and 15.2kHz below the parallel resonance. The Dishal approach to the design of filters, which I described in 1983 ('Computer-aided Ladder Crystal Filter Design' by J A Hardcastle, G3JIR, *RadCom,* May 1983, pp414-420), assumes a batch of crystals having essentially identical series resonance. This is then systematically stagger-tuned by inserting series capacitance. Because of the wide range of series resonance found in ceramic resonators it was not possible to achieve the required amount of offset by inserting precalculated values of capacitance. Instead each resonator needed to be tuned individually. Despite this, it was found possible, if a little tedious, to produce a quite respectable SSB filter. However, having satisfied myself of the feasibility of this approach, I decided that because of the extra work involved compared with designs using crystals, I would file this work away in my folder labelled 'it seemed like a good idea at the time!' - at least until I found a definite application.

"That time seems now to have arrived. Inspired by the 'Emma's Challenge' (see *RadCom*, November 1998, p8) the filter design has been dusted off and revived. Because of the severe cost constraints, the 'Challenge' seems to rule out using commercial

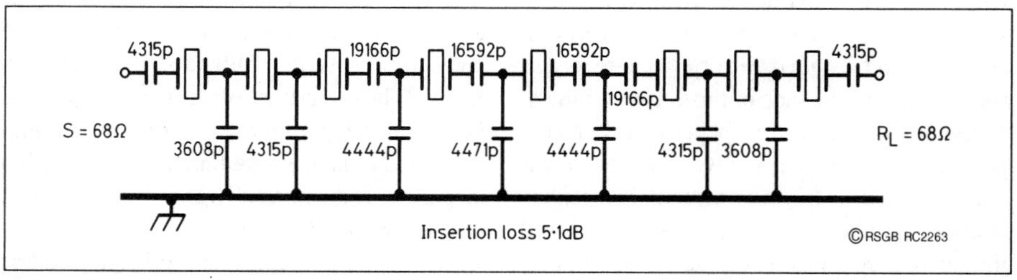

Fig 4: High-performance SSB ladder filter using eight ceramic resonators. All capacitors polystyrene. Values shown are as calculated. Actual values selected within 0.5%. A detailed selection procedure will be described in an article to be submitted by G3JIR to *RadCom*. Ceramic resonators are Maplin UL51R CSB455E.

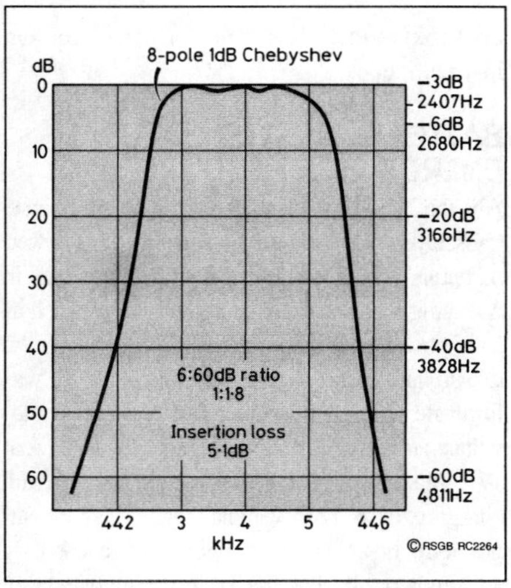

Fig 5: Measured performance of G3JIR's eight-resonator 455kHz SSB ladder filter.

crystal filters. I felt it was worth looking again at the possibility of using ceramic resonators in a filter sufficiently wideband to be suitable for NBFM. Fortunately their series resonance characteristics made this possible and I hope shortly to submit to the *RadCom* Editor details of an IF amplifier incorporating a ceramic ladder filter suitable for 12.5kHz channel-spacing receivers, which should find application in the modification of older VHF receivers to make them suitable for the new channel spacing.

"While seeking to uncover previous work in this field, the only paper I could locate was 'Design data for band-pass ladder filters employing ceramic resonators' by RCV Macario, *Electronic Engineering*, March 1961, pp121-127 [Dr Macario, GW8SRW, at the University of Swansea, has long been active in MF/HF/VHF projects and is a past contributor to *TT - G3VA*]. His design uses both shunt and series resonators, and so achieves a symmetrical frequency response. Unfortunately, ceramic resonators are currently only available at very few frequencies, so I am unable to reproduce his designs."

LA8AK, writing without being aware of G3JIR's work in this field, notes that while some ceramic resonators are specified for their parallel resonance some others are specified for their series resonance. He also draws attention to the problem of a spurious response at between 1.5-2 times the filter frequency. He writes: "It does not seem a good idea to build your own 455kHz ladder filters using ceramic resonators, but rather for those frequencies where suitable filters are not commonly available or too expensive. One such application is a 500kHz USB filter for use in some Collins receivers. I have now found a better configuration for the bandpass filter shown in the April *TT:* **Fig 6**.

"It appears that the CSB250D is specified for its parallel resonance, but the CSB500E3 for its series resonance."

Fig 7 shows a 500kHz ladder filter using four Murata CSB500E3 ceramic resonators with a -3dB bandwidth of 2.5kHz (with 2000pF capacitors, 5kHz with 1000pF capacitors). At -40dB the 2000pF filter has a bandwidth of about 14kHz). However there is a spurious response at 724.42kHz of -40dBr. Similar work by LA7MI also found a spurious response, leading LA8AK to feel that such filters should not provide the only selectivity in a receiver.

However, using a 243kHz ladder filter in his LF receiver enabled him to build "an impossible" design with 243kHz RF input, 250kHz IF and only 7kHz local oscillator frequency, resulting in an "incredible" 80dB IF rejection, but with a low impedance, since a 22nF capacitor has a reactance of -j35.

This month there is space to provide only the bare bones of LA8AK's latest report, but it is hoped that the combined notes on work by G3JIR, LA8AK and LA7MI, will encourage others to develop ladder filters using ceramic resonators, despite some of the problems outlined above.

GETTING GOING ON 136kHz

LAST YEAR Ron Kaye, G6RO, in Shipley, West Yorkshire, began to wonder what could be heard on 136kHz. He switched on his only receiver covering this band, an ancient Marconi B28 (CR100-series). He had to switch the IF passband to 300Hz to eliminate the strong signals on either side of this new amateur band. He was then thrilled to receive weakly EI0CF on Malin Island in the beacon mode and then a few days later G3KEV in Scarborough. The bug had bitten. He decided to have a go at

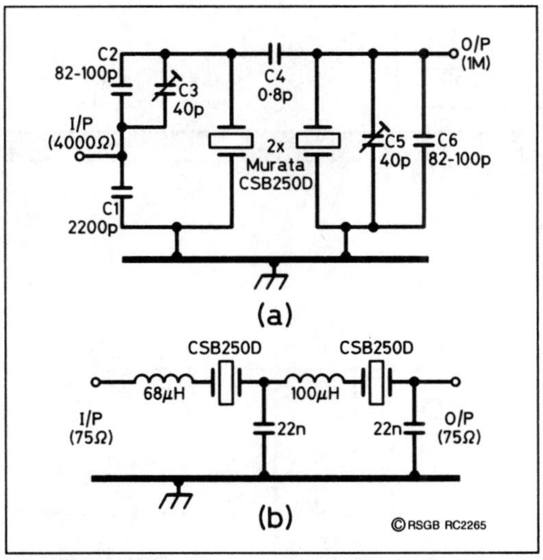

Fig 6: Ceramic-resonator filters built by LA8AK. (a) 250kHz CW filter with a bandwidth of 60Hz. (b) With the CSB500, C1=1000pF, centre frequency 513kHz, C2=47pF, F=512KHz, Z-input 300Ω, 4dB loss in bandpass, C4=0.5 or 5pF. (b) 243kHz filter with 75Ω input and output using CSB250D.

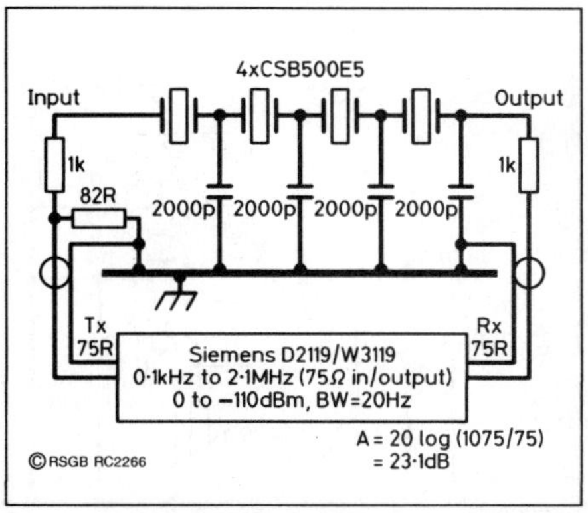

Fig 7: LA8AK's 500kHz filter using four CSB500E5 resonators with 75ohms input and output. Bandwidth about 20Hz, attenuation 23.1dB.

136kHz operating. He writes: "Having enjoyed 15 years of borrowed time, I knew it would have to be something quick and easy. The antenna system would be the main difficulty. My days of scrambling about on roofs, climbing trees and shinning up masts were over. I would have to use an existing antenna. I also recognised that I could not cope with gigantic inductors in my tiny shack or enormous loading coils at the foot of the mast. I had a 40m long-wire antenna which I can convert to a form of vertical by dropping a wire from the top of the 12m mast and connecting it at the base to an earthing system of four copper earth rods, each connected to a buried radial wire. This system, end fed at the shack end, has always served me well on the lower frequency bands.

"It occurred to me that it might be possible to make the system work as an earthed loop, which might then form an integral part of the PA tank inductance of a valve transmitter, avoiding the use of large inductors, ATUs and matching systems. From my junk box I took several 807 beam tetrodes, a three-gang variable capacitor in solid brass made by GEC in the 'twenties, some high-voltage 2000pF mica capacitors made by Muirhead, and an excellent RF choke and its associated blocking capacitor taken from the PA stage of an Air Ministry wartime T1154 transmitter. Finally, a very ancient 'Advance' signal generator." **Fig 8** shows the early form of his transmitter using an 807 PA.

G6RO chose to use valves rather than MOSFETs, since he had no PSU suitable for them, but did have quite a lot of RF power valves and could take the power supply from a 26-year-old FT101 via the accessories socket. Incidentally, the FT101 was the only commercial rig he had ever owned. Now, on 136kHz, he feels he is back in 'real' amateur radio. More recently he has replaced the 807 PA valve with two QQV06-40s (double beam tetrodes) in parallel, driven by a QQV03-10, generating

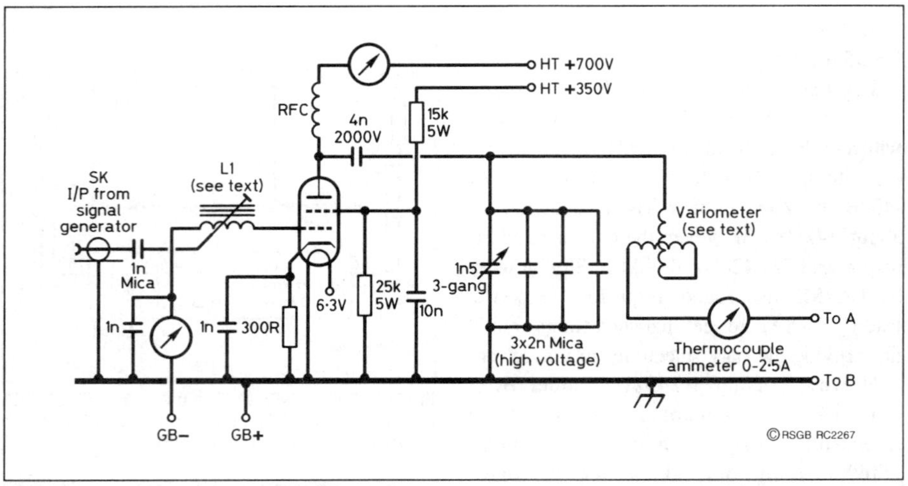

Fig 8: G6RO's 'getting-started' transmitter for 136kHz. Input from signal generator. The slug-tuned coil L1 came from an old TV. Feeding the input half-way down the coil gave a better impedance match. It is absolutely essential that the anode blocking capacitor should be of first class quality, as in case of failure the HT would short-circuit across the antenna.

about 110watts RF. With a current flowing in the antenna of some 5A now - only modest power for 136kHz - he has made many contacts on the band over the past months.

Originally G6RO built a large experimental inductance, comprising some 100 turns of 22 gauge plastic-covered wire, wound on a plastic liner from an old kitchen waste bin, tapped every 10 turns. The grounded loop antenna was connected in series with this temporary PA tank coil, initially with the full inductance in series with the antenna. There was no difficulty in dipping the PA to resonance. The current flowing through the external loop was measured with an RF ammeter. The temporary inductance was then progressively reduced, retuning to resonance and noting the current at each stage. As the external loop antenna made a larger contribution to the total inductance, so the current flowing through it increased. The anode current at resonance also increased, showing that the PA was being more effectively loaded. Eventually more capacitance had to be added, by connecting the 2000pF capacitors across the three-gang variable capacitor. By trial and error the optimum values of capacitance and inductance were found. At this point only ten turns of the inductor were needed. It was therefore dispensed with and replaced by an old variometer from the ATU of a No19 Army set. With this arrangement it was possible to tune very exactly for maximum current in the loop, at that time 2.2A with the 807 PA (anode current 85mA at 700V, giving a DC input of 60W). In the succeeding months, G6RO has added a lightly buried earth return to minimise the losses due to earth resistance: see **Fig 9** which also shows the modified output tank circuit. This resulted in a significant increase in the antenna current with the changed PA valves, and is now about 5A..

G6RO believes that the only matter of general interest is his antenna and the method of feeding it. However, I feel that readers will also be interested in his way of setting-up his valve-type transmitter and his use of a B28 receiver on 136kHz, aided by a well-stocked junk box of an earlier era. I rather doubt that his set-up meets current harmonic suppression standards, but with the low ERP on 136kHz that should not prove a problem in practice!

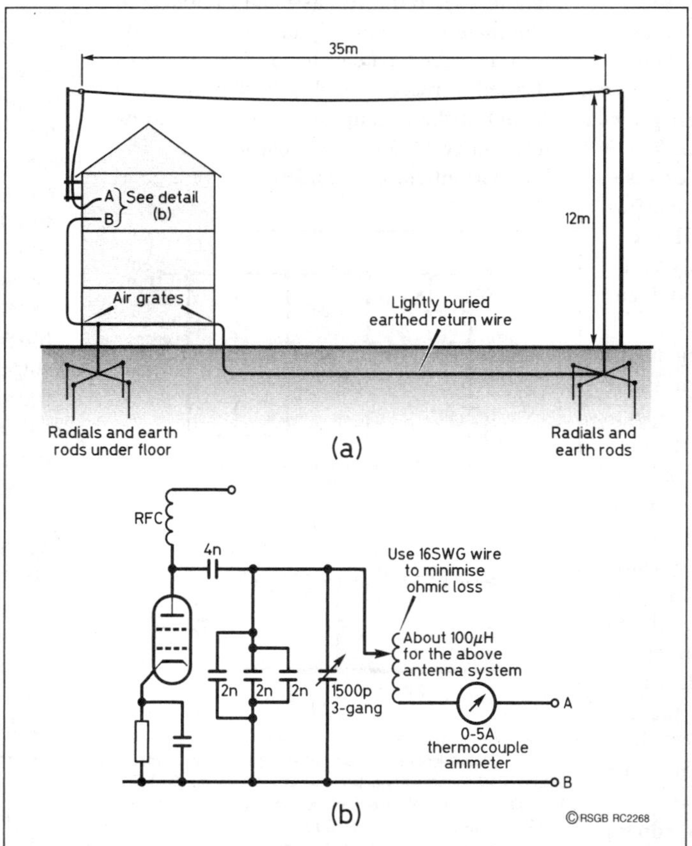

Fig 9: G6RO's 136kHz antenna system and current transmitter output.

BATTERY-ASSISTED NATURAL-ENERGY POWER SOURCES

AN INCREASING number of amateurs operate transceivers in locations remote from mains power, particularly boat owners and those interested in operating with naturally generated energy such as solar power or wind power, etc. Ten years ago, at a club junk sale, Roger Wilmot, G4PEY, was fortunate to obtain for just £13 a Solex photovoltaic array that can supply up to 17 watts (1A at 17V maximum). This was well suited to the charging of a 12V vehicle battery. The main problem, however, even with a large capacity accumulator, is that the battery cannot supply 13.6V for very long with loads of between 1.5A (receive) and 20A (transmitting).

For several years, G4PEY overcame the voltage shortfall of a single battery by connecting two 65Ah accumulators in series, delivering more than 24V, using a series pass design with an LM317K to regulate the output to a nominal 13.6V. This unit was well suited to contest use and capable of supporting a transceiver for between 2 to 4 hours depending on the receive/transmit duty cycle. A limitation was that he could not use his rig while the batteries (connected in parallel) were being charged. There was also significant heat dissipation and power losses through the series regulator to contend with.

Recently he decided to rationalise his arrangement, in part to use a step-up switching regulator IC, an idea he had first seen in *TT* (October, 1997, p77). He also decided to change the battery configuration to a 12V battery in series with a 6V battery, thus reducing the input from around 24V to about 18V and significantly reducing heat and power losses. At the same time he increased the capacity of both batteries to 100Ah.

The outcome was the system shown in **Fig 10**, in which the first part performs solar input control and battery regulation functions based on the LM2577T-Adj device. This IC uses few external components and achieves better than 80% efficiency as a DC-to-DC inverter. It is capable of supporting output currents up to 3A, has a soft start mode, internal protection against overcurrent, thermal limiting and undervoltage lock-out - characteristics well-suited to power generated by photovoltaic solar energy [or wind generators, etc - *G3VA*].

This part of the unit steps-up any input voltage between 3.5

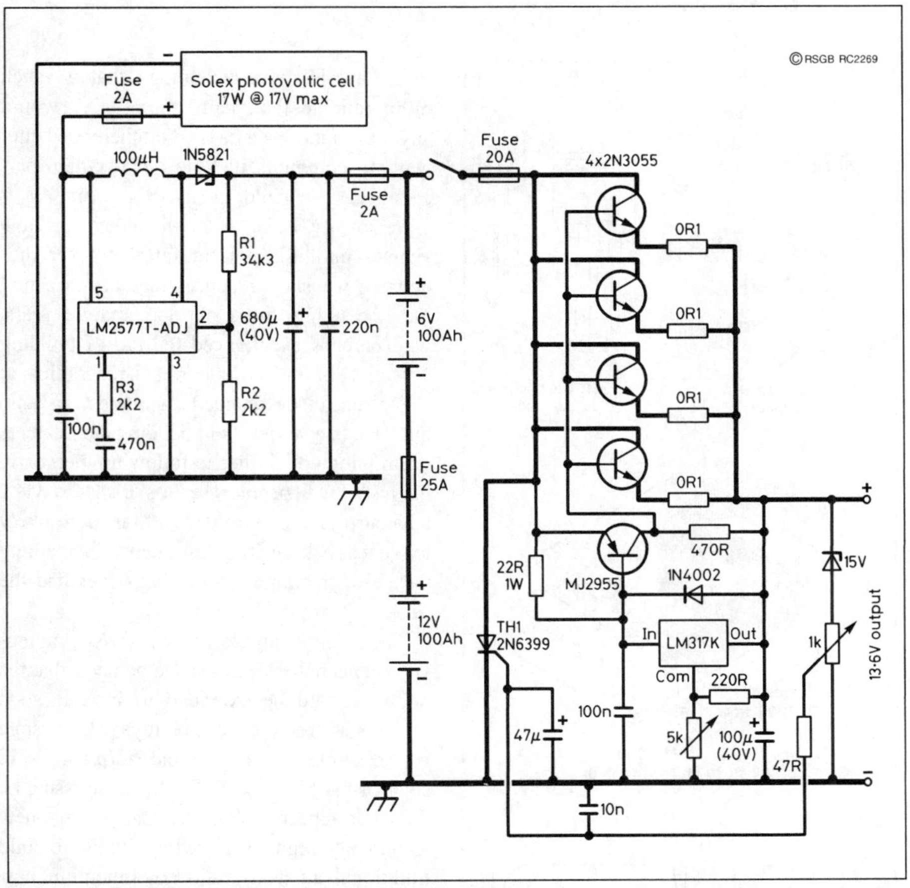

Fig 10: G4PEY's solar array step-up inverter and battery charge regulator, together with regulator for the transceiver. The circuit uses a LM2577T-ADJ chip to step-up a solar-generated input voltage in the range 3.5 to 17V, with the output regulated to 20.3V to charge 12V + 6V batteries in series. The second part of the circuit regulates the output from the batteries to 13.6V with overvoltage protection.

and the 17V maximum from the solar array, producing a regulated output voltage at up to 1A. G4PEY settled on 20.25V as the optimum charging voltage, representing 2.25V for each of the nine accumulator cells. These were of the non-spill, gel-filled type, which are easily damaged by bubbling or overcharge. Another benefit of this regulator is that at night the batteries cannot discharge back through the photovoltaic cells. The IC is supported by a Schottky diode. There is an insignificant discharge of about 4μA lost via the feed-back resistors R1 and R2, which is no problem. A major benefit of the new arrangement is that with both batteries in series, they can be charged and used at the same time, increasing operating duration and much reducing the complexity of the original switching.

The second part of the circuit, providing output regulation and overvoltage protection for the transceiver, is partly based on the heavy-duty PSU used by ZS2U (*TT*, February 1994, or *Technical Topics Scrapbook, 1990-1994*, p253) but with some modifications and the inclusion of overvoltage protection for the output devices.

G4PEY stresses that it is important to ensure that the batteries are well fused and treated with a great deal of respect, in view of the very high current potentially available from them. In practice he has found regulation to be better than on some commercially-available PSUs, with an efficiency better than 80%. The circuit diagram shown is of G4PEY's final version, which is proving entirely satisfactory. He hopes that the design may be of interest to amateurs remote from mains power such as boat owners or anyone interested in operating from naturally-generated energy sources.

HERE & THERE

HARRY LEEMING, G3LLL, adds to the hints on handling the fragile locking cores of old-style coil formers (*TT*, April 1999, pp 57-58). He writes: "My tool kit would not be complete without a candle. I do not know the dielectric properties of candle wax, but I have never found a drop of candle wax to do any harm when using it to lock cores. Its big virtue is that it only needs warming with a soldering iron to become a lubricant, and it then makes cores easier to turn, instead of jamming them up. Rubber bands are not to be advised. Some twenty years ago, Yaesu used rubber bands to lock the driver cores in the FT401, and these cores are now usually impossible to move. Fortunately, in these sets, the antenna and RF cores are locked only with wax which can easily be melted. The trick then is to peak the driver coil with the preselector, and then to tune the latter two coils to the same frequency."

In these days of multiplying satellite systems for digital mobile phones and paging, it is perhaps reassuring to hear at the IEE Colloquium on 'Frequency Selection and Management Techniques for HF Communications' (March 1999) of an ingenious short-data message system for mobile units (truckers, etc) currently established by Terion (also known as Flash Comm Inc) throughout the United States. It is based on an ingenious combination of low-power HF and high-power FM broadcasters' SCA sub-carriers. The mobile units transmit on HF channels that have been checked at the nine HF receiving sites distributed throughout the USA, using low-power (10-watt transmitters with an ERP limited to 1 watt). Magnetic loop arrays at receive sites optimise reception from vehicles in the 300km to 3000km range. Terion has found that a grid of stations not only increases the probability of receiving the weak signal at any particular frequency, but also stretches the range of frequencies over which the system can operate. The site receiving the clearest signal from the mobile passes this by line telecommunications to the Network operations centre: see **Fig 11**. Confirmation of reception is sent to the mobile via VHF broadcast stations. Mobile unit frequencies virtually span the HF spectrum and are automatically selected under the control of the Propagation Analysis Centre. On average, Terion has transmitted well over 2500 HF transmissions per day for the two years it has been operating the system. During these activities, no complaints or incidence of interference were recorded. The system operates under a special FCC licence with secondary status. Frequencies within the primary amateur bands may not be used by Terion. It would be interesting to know how the economics compare with satellite or cellular systems. ♦

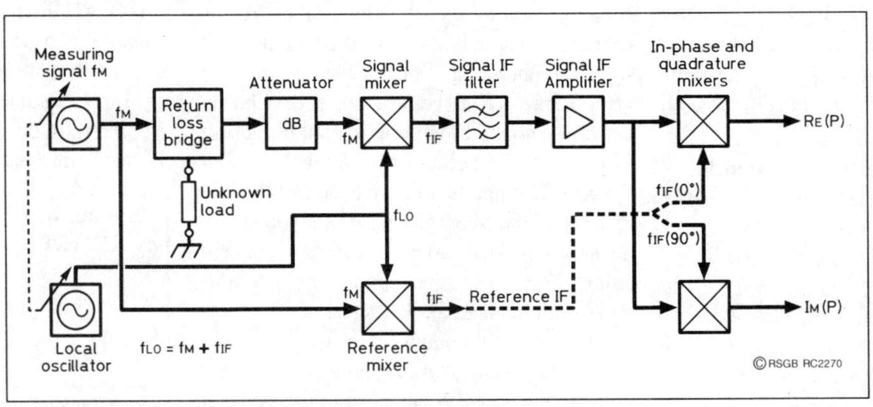

Fig 11: Block diagram of the superheterodyne reflection coefficient meter, capable of measuring real and imaginary reflection coefficients of an HF antenna.

by Pat Hawker, G3VA*

WORKING WITH SURFACE MOUNT TECHNOLOGY

SAM ULBING, N4UAU, in 'Surface Mount Technology - You *Can* Work with It!' (Part 1 in *QST*, April 1998*)* presents a convincing case why we should all start building projects using surface-mount devices, although I must admit that I feel that this will have to be left to younger eyes than mine. I seem to spend far too long searching for small parts that have fallen off the bench, without getting down on my hands and knees to look for those tiny chip capacitors. But, as N4UAU reports: "As I look through the various product data sheets, three things strike me. First, the large number of available ICs that perform functions formerly requiring several ICs. Second, the continuing shift to lower power requirements, smaller size and usability at higher operating frequencies. Finally, the increasing number of new products available *only* in surface-mount packages. New products today are smaller and more energy efficient. . . On the other hand, when I look at amateur radio constructional projects, I see continued use of many discrete components and bulky DIP ICs that perform limited functions. Recently I saw a voltage-controller project based on the use of transistors and relays! Frankly, it bothers me that there seems to be a growing divergence between the technology used by industry and that used by hams. . . I think hams are being left behind because they feel that SMT (surface-mount technology) is something they can't handle."

I must confess that at G3VA I still have a penchant for large, discrete junk-box components that belong to a much earlier age - the valve era rather than solid-state, let alone SMT. But then I like large bulky equipment and still make use of such hernia-inducing heavyweights as an AR88 receiver and LG300 transmitter. But I recognise that there is only limited appeal of these aged items to the youngsters that need to be attracted to amateur radio if the hobby is to survive well into the new millennium. All over the world the problem of attracting a generation increasingly hooked on computer games, e-mail and above all on Internet is looming ever larger on the agendas of national societies and local clubs.

Sam Ulbing, N4UAU, in the correspondence columns of the April *QST* has a view worth considering on this burning topic. He writes: "For several years now the ARRL has been concerned by the lack of interest in ham radio by younger people. I wonder, however, if we are doing the wrong thing. . . Perhaps we need to address how to make ham radio less boring [to the younger generation] and not how to make it easier." He quotes Seymour Papert, Professor of Educational Research at MIT: "Every maker of video games knows something that the makers of curriculum don't seem to understand. You'll never see a video game advertised as being easy. Kids who do not like school will tell you it's not because it is too hard. It's because it's boring."

EIA size code	Length (inches)	Width (inches)
0402	0.04	0.02
0803	0.08	0.03
0805	0.08	0.05
1206	0.12	0.06
1210	0.12	0.10
2512	0.25	0.12

Table 1: Dimensions and sizes of SM resistors and ceramic capacitors.

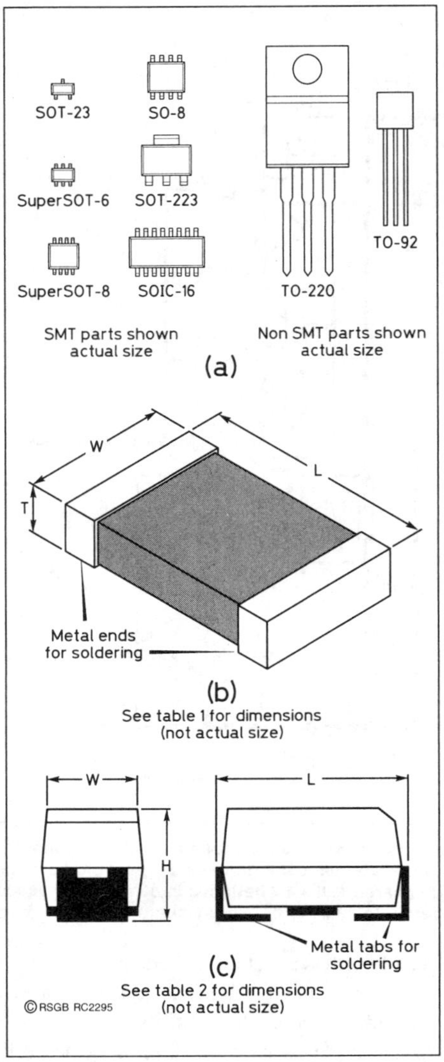

Fig 1: Size comparisons of some surface-mount devices against standard 'TO' outlines, from Part 1 of the *QST* articles on working with surface-mount technology by Sam Ulbing, N4UAU. (a) Size comparisons. (b) SMT resistors and ceramic capacitors. (c) Solid tantalum capacitors.

N4UAU suggests we need to address how to make ham radio less boring to the young and not how to make it easier. . . In the past, amateur radio offered a unique opportunity - worldwide communication. He stresses: "The reward of global communication was so attractive [and exciting] that people were more than willing to master CW and electronics. Today it is easy to communicate with people anywhere in the world, with technology that is much more reliable than ham radio. Why would anyone want to be a ham when there are better ways to communicate? No wonder ham parents hear their children say 'It's boring'."

Personally, I tend to the view that for a few years - until the present Internet fever dies down - we should concentrate on attracting into the hobby those of more mature years, who perhaps experienced the fun of building crystal sets and simple transistor radios at school and have retained an interest in radio but have been too busy in building their careers or bringing up families to follow up their early interest. Such people are less likely to view amateur radio as 'boring', and far more likely to see it as a fascinating fulfilment of something they always wanted to do, but never had the time.

The solution suggested by N4UAU, moreover, seems rather defeatist. He believes that the hobby should be extended to include such projects as remote control toys, short-range wireless links to hi-fi units, etc. Surprisingly, in spite of his SMT article in the same issue, he does not appear to consider that perhaps new technology might offer the opportunity to build traditional amateur radio equipment in new ways. But he clearly has a point in believing that youngsters need the challenge of new ideas and new problems rather than encouraging them, in effect, to make everything easy and become appliance operators of factory-built black-box radios. Social chatter on the established nets, interesting to the over-50s, cannot be expected to appeal to the under-20s, no matter how easy the licensing authorities make it to obtain an amateur licence. The excitement of chasing DX may be dimmed for the young by the realisation that obtaining QSLs via the bureau may take years. In contests, youngsters find themselves up against experienced operators with top grade equipment and elaborate, costly antenna arrays. On the other hand, they are not burdened by the weight of past practices and have an unrivalled ability to absorb and master new microprocessor-based technology. And the 20/20 vision and manual dexterity to handle SMT.

N4UAU lists some of the advantages of building with SM devices as including:

- Smaller, more energy-efficient projects. He built a time-out switch that fits on a PCB one-sixth the size of a postage stamp, able to fit into the battery compartment of his

Case size used by KEMET	EIA code	Length (mm)	Width (mm)
A	3216	3.2	1.6
B	3528	3.5	2.8
C	6032	6.0	3.2
D	7343	7.3	4.3

Table 2: Dimensions and sizes of SM solid tantalum capacitors.

voltmeter.

- SM devices may outperform earlier DIP versions, with lower operating voltages and microampere quiescent currents [for example, recall the remarkable performance of the FST3125 SMT devices as HF mixers, as revealed in *TT* last year - *G3VA*].
- VHF projects can benefit from the short signal leads, with SM capacitors used as bypass capacitors placed very close to an IC, etc.
- Experience gained from building SM projects gives confidence in tackling repairs to factory-built gear.
- With no need for through-hole parts, PCBs are easier to make for SM devices.
- New SM devices often have entire modules built into them, making it much easier to build a complex circuit than with older ICs and far easier than with discrete components.

N4UAU lists in some detail the equipment needed to work with SM, although emphasising that lots of expensive equipment is unnecessary and that you don't really need an eagle's eyesight if you use hands-free optical magnification.

He gives as a fundamental requirement an illuminated magnifying glass (eg 5in diameter lens giving 3x magnification with built-in circular light); also a low-power soldering iron with an earthed, screwdriver tip; used with *thin* (0.020in diameter) resin-core solder since 0.031in solder tends to flood the pads, resulting in bridging. Also useful is a wet sponge for cleaning the tip of the soldering-iron. A flux pen can come in handy if a clean flux is used; as can good desoldering braid to remove excess solder on a pad. ESD protective devices such as wrist straps may be necessary in dry, static-prone areas. A pair of non-magnetic, stainless steel drafting dividers with two very sharp needle points can be a great help.

Fig 1 provides size comparisons of some surface-mount devices and their dimensions as given in Part 1 of N4UAU's *QST* article.

DC CIRCUIT BREAKER (ELECTRONIC FUSE)

RUPERT CHANDRA (Madras, India) provides in the 'Circuit Ideas' feature of *Electronics World* (January 1999, p66) a useful arrangement

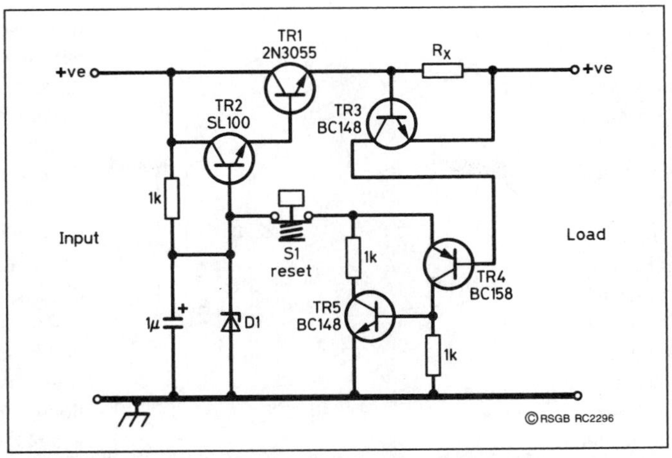

Fig 2: Circuit diagram of the electronic DC circuit breaker which can replace wire fuses and which includes a push-to-reset facility. For value of Rx see Table 1. *From Electronics World*.

that can replace an ordinary fuse in a DC circuit with the advantage of having a 'reset' switch. The 'fusing' current is set by the value of a single resistor (Rx), typically between 100mA and 5A: **Fig 2**.

TR1 is a series-pass power transistor, typically a 2N3055 for high currents, forming a Darlington pair with TR2, biased by the Zener diode ZD1. When the voltage drop across Rx exceeds 0.6V, TR3 turns on and triggers the thyristor-forming pair TR4-TR5, turning off the series-pass transistor pair. The reset switch SW1, when pushed, restores the *status quo*.

Rx should be selected to drop 0.6V at the required circuit breaking current. Rupert Chandra provides typical values, shown here in **Table 3**.

Fusing Current (mA)	Rx (ohms)	Rx rating (watts)
100	6	0.25
500	1.2	0.5
1000	0.6	1.0
5000	0.12	3.0

Table 3: Typical values of resistor required for Rupert Chandra's 'electronic fuse'.

AUDIO GAIN & THE IC-706 Mk2

NORMALLY, I TRY TO avoid including modifications to and comments on recent factory-built equipment in *TT*, but John Everingham, G4TRN, has found a solution (which has been received in the Bristol area as an answer to many prayers) to a problem which is commonly encountered by users of the popular, compact Icom IC-706 Mk2 when this is set up strictly in accordance with the Icom user instructions. The following is adapted from G4TRN's notes on 'How to Stop Shouting and Stay Happy'.

As a new owner, he soon found that these extremely compact rigs have minds of their own. In particular, the difficulty when following the recommended setting-up procedure of obtaining, and then holding, the full power output on SSB without acquiring a special, rather costly matching desk microphone. He writes:

"When first used, the rig seemed to be very deficient in audio gain on SSB. Using the settings recommended in the instruction manual makes it necessary to shout in order to achieve full SSB output. Historically, Icom have favoured the unusual but technically sound approach of putting an amplifier in their hand microphones. This is not the case with the IC-706 Mk2, although the rig gives the impression that this is needed. This is not the case on AM or FM modes.

"The service manual shows that all non-limiting audio functions are specified for 10mV at 1.5kHz at the microphone socket, and the Q2 menu setting of 5 for microphone gain. . . yet 10mV is an unrealistically high voltage output to expect from ordinary microphones.

"The first step is to ignore the book and set the Q2 mic gain to 10. The next step is to engage the compressor, and set up the external pot to keep the ALC in the acceptable range. For good results the ALC must be 'on scale'.

"The compressor does not do much in the way of speech processing, but is more of an audio AGC system [effective SSB speech processing requires RF rather than AF processing - *G3VA*]. The effect is as though the Q2 mic gain was turned up to 50. The down side is a high degree of bass cut, so that the signal sounds harder. It is often possible to recognise the '706 by this bass cut, which is fine for real mobile operation.

"It is rather frustrating to find the compressor cannot be accessed when in FM mode. However, this is of little importance since it is not functional on FM. It might have been more logical for the 'COM' flag to vanish on FM, but one can't have everything. Note that the Q2 mic gain should be set individually for each phone mode. Remember too that for duplex (ie SPLIT), the wide or narrow option is individually assigned to each VFO, so that you do not automatically send with the selected receive setting (the same applies also for CTCSS tone burst).

The above notes all apply to SSB operation at the Q2 power setting 'h' (high). As the output power is reduced, so the need for more microphone gain diminishes. The power setting is an ALC threshold setting, and the lower the power is set the easier it is to achieve the threshold value. The manual does not say, nor the rig do, but for tidy results the mic gain should be set to the value appropriate to the power setting. It is helpful that the ALC scale comes on automatically when setting the Q2 microphone gain.

"For those not confident of tweaking the engineering parameters of a complex, factory rig, this is about as far as it will be wise to go. A 'power mic' or the high gain desk mic approach suggested by Icom is then the next step. Those confident enough to undertake some tweaking should read on:

"The quick menu Q4 enables the carrier frequency to be altered (only on SSB transmit). Since all functions of the '706 are controlled by the in-built microprocessors, and, unlike older rigs, this does not change the output frequency. It is like moving the shift control, but on transmit. You will probably discover that your audio is improved, ie becomes more natural, if the carrier is set on the negative side of the default. This may be due to the difference between Japanese and European speech.

I have set my '706 to '-50' (rather unfortunately this setting has no effect on the receive function, as it does on my TS50, which now transmits the way I prefer to receive. I am a great believer in reciprocity).

"The final trick to obtain more output RF power is to turn up the 'SSB mic gain' control, if only there was one. However, there is 'Transmit Gain' which works like the RF gain control of a receiver but on the through RF transmit signal. Too little 'transmit gain' and the output varies with speech in an unstable way, and this signal lacks 'punch'. Too much and the result is overload and distortion. When set correctly, everything is clean and firm. Think of the ALC scale as being the transmit equivalent of an S-meter with delayed AGC. If you can't get enough through amplification - microphone to antenna with the Q2 setting (AF gain) then turn up the 'transmit gain' and do the job at RF.

"There is a variable resistor provided. The service manual says set it for 50W output at. . . yes. . . 10mV, 1.5kHz, Q2 setting 5! No wonder as '706 owners we have all been shouting our heads off! My suggestion is that the resistor is set for a decent ALC reading at Q1 power setting 'h', Q2 mic gain '6', compression off and a spoken 'FORE' into the hand microphone held the way you prefer, but fairly close. Afterwards, fine tune the Q2 value and the compression level.

"If you are happy with the above reasoning you can complete the adjustment in a few minutes. The 'transmit gain' potentiometer is on the main board (the one under the top cover), marked R511. It is at the front edge of the board, hidden by the printed ribbon lead, about one-third-way along from the left as you look at the front of the rig. You will need a small cross-point screwdriver, a magnifying glass, and a very small straight screwdriver. Connect your dummy load if you are going to monitor progress. Turn R511 about 20° clockwise.

"Success! Reports of 'nice audio' and 'sounds like you'. My PEP meter stays at 100W. I use the compressor now only when the going gets hard. What a relief to the vocal chords."

HERE & THERE

IN WHAT I HAD hoped would be a wind-up item on compact tuned transmitting loops [*TT*, May 1999, pp 60-61], I quoted Dr Belrose, VE2CV, as writing "we must remember to take into account the effect of the output impedance of the RF power amplifier, since for maximum power transfer the output or source impedance is the complex conjugate of the antenna's impedance (reference Belrose, VE2CV; Maxwell, W2DU; and Rauch, W8JI; *Communications Quarterly*, Fall 1997, pp25-48)." At the time, I had a feeling that this statement would be disputed and sure enough Bob Pearson, G4FHU, and Dr Andrew Smith, G4OEP, have expressed the opinion that VE2CV has mis-applied the maximum power transfer theorem, thus perhaps unknowingly re-opening one of the long-lasting disputes that have persisted for years among the leading technical boffins of our hobby as to the validity of the conjugate matching theory when applied to power amplifiers. The contrary view was first brought to my notice by one of my former IBA colleagues, Tony Harwood, G4HHZ. It eventually led to the ARRL removing references to conjugate load impedance matching from the *ARRL Handbook* and declining to publish articles by Dr Belrose and others protesting at their action. All I will do here is refer readers to the long (23-page) article in *Communications Quarterly* sited above, and in which such eminent engineer-authors as VE2CV, W2DU and W8JI argue strongly and in great detail for their belief in the validity of conjugate matching as applied to matching antennas to RF power amplifiers. This is a controversy into which I have no intention of entering at this time!

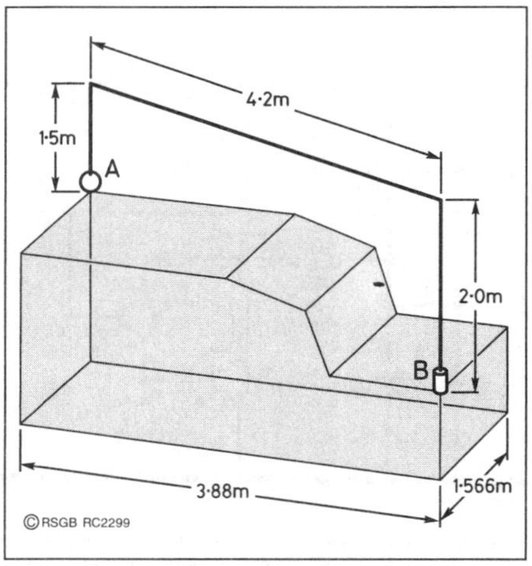

Fig 3: Configuration of the vehicle-mounted loop antenna for NVIS operation, developed and evaluated at Liverpool University by G0GSF *et al.* Generic Algorithm/NEC-2 simulation shows that optimum results were achieved when a capacitor is positioned at point B with a single RF source at point A.

FOR THOSE interested in the use of military-type vehicle transmitting loops as developed at Liverpool University under Dr Brian Austin, G0GSF (see for example: 'The Application of Characteristic-Mode Techniques to Vehicle-Mounted NVIS Antennas' by Brian A Austin and Kevin P Murray, *IEEE Antennas and Propagation Magazine*, Vol 40, No 1, February 1998), attention is drawn to a further contribution 'Genetic algorithm optimisation of vehicle-mounted loop antenna for NVIS' by B A Austin and Wen-Chung Liu (*Electronics Letters*, Vol 35, No4, 18th February 1999) in which a series capacitor with one rather than the two feedpoints in the earlier model is used to restore the uniform current distribution as the frequency approaches first resonance, as is likely for day-time operation and typical loop dimensions. This ensures the radiation of as much energy as possible towards the zenith, highly desirable for NVIS operation: **Figs 3, 4**.

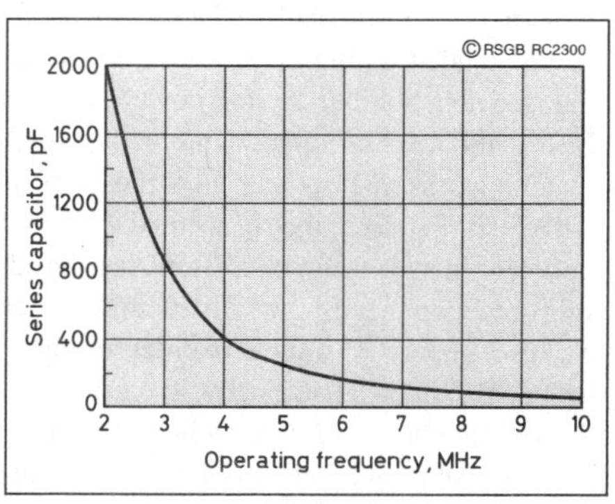

Fig 4: Capacitor values for optimum radiation toward zenith at frequencies from 2 to 10MHz. At 10MHz the percentage of power radiated towards the zenith increased from under 20% with a single source to about 35% with either dual, in-phase sources or with a single source plus series capacitor.

An overview of 'Short-haul Communications Using NVIS HF Radio' by S J Burgess and N E Evans (*Electronics & Communication Engineering Journal*, April 1999, pp95-104) includes an assessment of results using a 6793kHz NVIS simplex link over a 42km path to a monitoring receiver at the University of Ulster's Jordanstown campus. The paper provides advice on the use of typical horizontal wire antennas at low heights, as widely used by amateurs on 3.5 and 7MHz, and various amateur radio publications are listed in an extensive list of references, perhaps not surprising since Dr Noel Evans is presumably GI4BDR.

DANNY BEGG, GM3YXJ, notes the interest aroused by recent *TT* items on spark transmitters. He has sent some photographs (see opposite) of a straight Morse key which he still sometimes uses, but which has been variously identified as (a) an RAF key of the 1920s, or (b) a key intended for use with a spark transmitter. The Bakelite base is cracked and GM3YXJ has reinforced it with a block of aluminium, but otherwise the key is in its original condition. The base carries the legend 'No 1969'.

He writes: "Although difficult to see from the photograph, there are six terminal sockets on the base, and I wonder if any of your spark-operator correspondents could provide a circuit showing the use of all these terminals, and the adjustable spark gap which can be observed through the window in the cover. It makes a change from the standard two-terminal key!"

I skimmed through a number of back issues of the excellent *Morsum Magnificat* (now edited and published by Zyg Nilski, G3OKD, The Poplars, Wintanswick, Market Drayton, Shropshire TF9 2BA) and soon discovered an illustration of the 'S G Brown Ref No 1969' key (*MM* Nr 49, Christmas 1996, p48) from the collection of Tony Wilkes, ZL3SLH, together with earlier references and suggestions that this design was made for the Services by various makers. In Issue 25 there is a diagram stemming from G0FJT of the terminal connections - **Fig 5** - though this does not entirely solve the manner in which the key worked with the transmitter and possibly was used to mute a receiver. In Issue 26 (January 1993) Gus Taylor, G8PG, dimly recalled one of these keys at the old Liverpool Wireless College and known as a Siemens key: "My gut feeling is that they were made for the Air Ministry, probably for airship use, perhaps originally by the RNAS and then, after 1 April 1918, by the same people, now RAF." He believes that similar keys were sold pre-war in the surplus market, possibly by Electradix Radios as ex-Air Ministry keys. Tony Timme, G3CWW, recalled using such a key to instruct fellow cadets at Christ's College, Finchley, in 1940-42.

The "REF No 1969" Morse key believed to date from the WW1 or a little later and still in use by GM3YXJ, incorporating a spark gap. Possibly developed for use with spark transmitters in military airships.

To check G8PG's suggestion I found an Elextradix advert in a 1935 *T&R Bulletin*, showing an illustration of a key with a basically similar construction (but no spark gap) described as 'Brown's HT Model, price 8/6 [42.5p]'. The firm claimed "We have the best range of Wireless Keys in the country. Illustration shows Brown's H.T. model. Twelve models from 4/6 [22.5p] to 30/- [150p]".

DVB-T & TVI - PROBLEMS TO COME?

LAST YEAR, shortly before the introduction of both satellite and terrestrial Digital Video Broadcasting (DVB-S and DVB-T), I wrote an article for *Electronics World* (July 1998, pp570-574) in which I warned of several potential problems that seemed likely to affect the reception of DVB-T, particularly if dealers were to sell DVB-T sets or set-top digital units to viewers outside the recommended 'service area' of the local DVB-T (DTT) transmitter. I wrote: "Whereas the long-established UHF PAL service areas have required that the signal-to-noise ratio (SNR) at the receiver should be at least 40dB to achieve a substantially noise-free picture, digital transmissions can provide a noise-free picture with an SNR of 10dB.

"Theoretically, this feature of digital transmissions allows the transmitter power to be 1000 times lower. In practice, DVB-T service areas are being based on an SNR of about 15dB, to provide a margin for the inevitable diurnal and seasonal changes in UHF propagation. These are due to things like foliage and temperature, tropospheric lift, and seasonal changes.

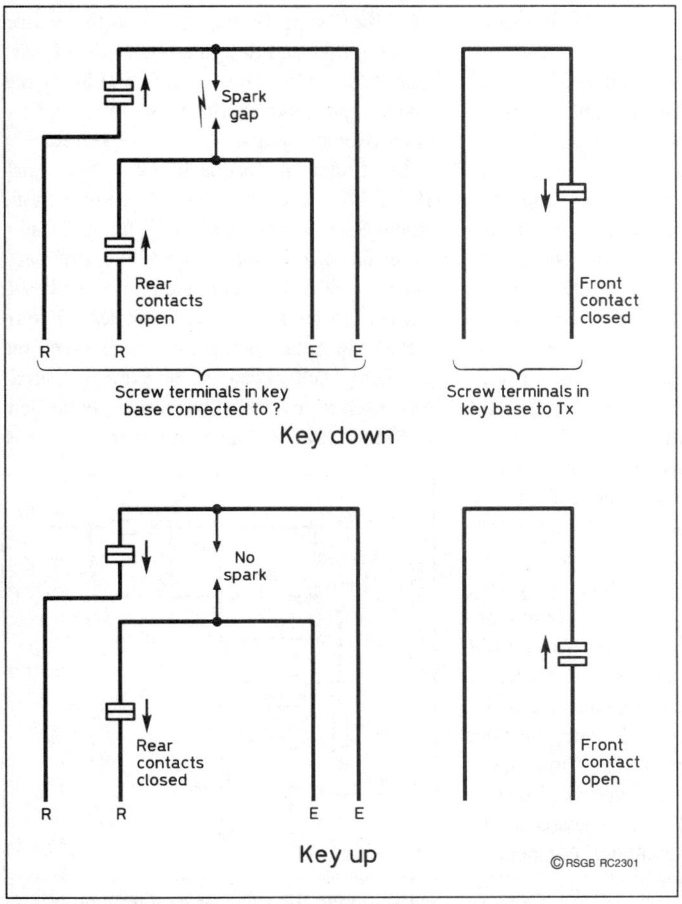

Fig 5: Circuit arrangements of the 'No 1969' key shown in the photo above. From G0FJT in a 1992 issue of *Morsum Magnificat*.

"Past experience however suggests that retailers competing for sales will result in DVB-T units often being installed outside the intended service areas, provided that on installation they achieve the bare 10dB or so SNR.

"Whereas with analogue PAL out-of-area reception usually results with propagation variations only in an increasingly noisy picture or co-channel interference patterning, with DVB it will lead on some days at some times in some seasons to complete loss of synchronisation and total absence of picture. This could result in DVB-T soon acquiring an unfortunate reputation for unreliability. . . Potentially even more serious is the likelihood that DVB-T transmissions in the so-called 'taboo' channels will cause interference to the reception of existing PAL transmissions [and vice versa]. . . High-speed digital data transmission in telecommunications has frequently resulted in unforeseen problems. It is an open secret that digital [satellite] distribution links for television have not always achieved 100% reliability [despite the relatively clean and consistent satellite channels]."

In that article I overlooked the fact that the much lower transmitter powers and greater susceptibility to impulsive noise of DVB-T than for PAL might lead to a significant increase in TVI from amateurs. Whereas slight patterning on analogue pictures can often pass unnoticed, or at least can be tolerated, interference to digital signals, even at low levels, can result in temporary loss of synchronisation and a silent 'red dot' on the screen. . . Furthermore, a weak interfering signal in a PAL channel is likely to affect only one channel, whereas with DVB it will inevitably affect all the channels carried on the particular multiplex. While I have not heard of many cases of digital TVI from amateur transmitters, a strong warning of related EMC problems is outlined in a letter from Dennis Glover of Stansted TTS, published in *Television* (May 1999, p469). To quote briefly from his long letter:

"Digital terrestrial TV has now been in operation long enough for us to be able to assess reception problems. . . Local EM conditions create the main reception problems. Those familiar with the [old] 405-line system will recall its susceptibility to impulse interference caused by electrical equipment, particularly vehicle ignition systems. The disadvantage of positive [video] modulation was not appreciated by those who designed the system [in the 1930s]. It seems that history is repeating itself. . . To avoid mutual interference with existing services, DTT transmissions are typically 20dB below the accompanying analogue ones. This has an unfortunate consequence - local interference within a DTT multiplex can be of sufficient amplitude to corrupt the modulation. The decoder's error correction circuitry may be unable to cope with this. Now here's the rub. The effects of impulse interference go largely unnoticed with an analogue transmission, mainly because of the use of negative modulation [noise driven into the blacker-than-black region] with any residual interference tending to appear in short-duration line flashes, which are generally not noticed by viewers. But the same level of interference has a devastating effect with a DTT channel.

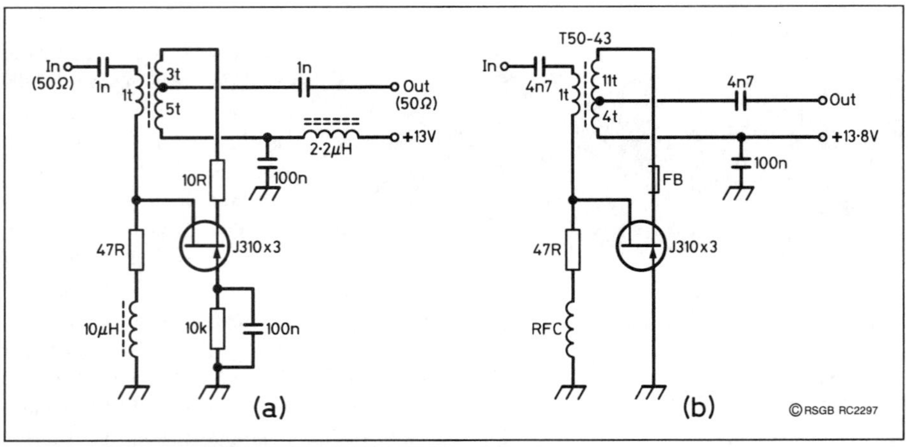

Fig 6: (a) 75MHz IF amplifier as used in the Kachina 505DSP transceiver, described in *QEX*. (b) As implemented by AG4R, with 0dBm input at 14MHz it has an IP3 of 40dBm with 10dB gain.

"The results vary from MPEG-2 mosaic blocking that typically lasts for one to two seconds to momentary total corruption, with the dreaded 'red dot' being displayed and no sound or picture. Other effects may include the sound and picture becoming unsynchronised for as much as two to three seconds, and lock-up with a frozen still picture (the sound may continue normally). The remote control unit then has no effect and the box must be switched to standby and back in order to recover. On the audio side you can get pops and cracks that are at much higher amplitude than the programme material, with the risk - if the volume is set much above average - of loudspeaker cone damage."

Dennis Glover reported that almost all the above effects were due to vehicle ignition interference, although the vehicle ignition systems had all been suppressed within the regulatory limits. He adds:

"Analogue terrestrial CTV signals degrade 'gracefully' and are far more robust than the 'all-or-nothing' digital alternative in the modern, hostile EM world. In terms of the crucially important EMC performance, the current DTT offerings are, overall, hopelessly outclassed by the existing, if ancient, analogue services."

You have been warned! Make sure your UHF harmonic suppression is of a high order.

HF ANTENNA IMPEDANCE MEASUREMENT

AT THE RECENT two-day IEE Colloquium on 'Frequency Selection and Management Techniques for HF Communications' Mel Maundrell of the Defence Evaluation and Research Agency, Malvern, emphasised that although HF research and development has been drastically contracted due to work on satellite communications, satellites are not a universal panacea and cannot answer all military and civilian needs. Recent developments in HF are providing traffic capabilities that in the past would have been thought impossible: "HF lives for ever". Many of the papers were concerned with such current techniques as automatic link establishment (ALE) and maintenance (ALM), improved HF data protocols (eg STANAG 5066), automatic frequency management, etc.

But a paper that potentially could be of great interest to amateurs was 'Measurement of HF Antenna Impedances', presented by Dr Y Sun with co-authors James Moritz and David Lauder [G4WTB] all at the University of Hertfordshire, Hatfield. This described the design of an antenna reflection coefficient measurement system, capable of ascertaining the real and imaginary components of the reflection coefficient of antennas between 1.8 and 30MHz. The system is able to operate with high level ambient incoming signals whilst the maximum measuring signal power delivered to the antenna is limited to just 1μW, providing 'quiet tuning'. The errors are less than 5% in magnitude and 4° in phase.

In their introduction it is pointed out: "An important parameter of an HF antenna is the impedance at its feed point. For practical antennas this is a function of the antenna geometry (and that of nearby conductors) and operating frequency, and is often variable and unpredictable. HF equipment is usually designed to operate with a fixed load impedance, typically 50Ω. An ATU, an adjustable reactive matching network, is used to optimise power transfer. The network components must be adjusted for a conjugate match between the antenna and the transmitter. An SWR meter is most often used to measure the amplitude of the signal reflected by the load due to mismatch. . . A disadvantage of this is that significant power must be radiated for the SWR meter to produce an indication, potentially giving rise to interference to other users or, for military users, the possibility of detection. Also, SWR measurement merely indicates the degree of mismatch, with the actual impedance remaining unknown, requiring the matching network to be capable of a wide range of impedance transformation to guarantee successful matching. Since voltage and current levels in the matching network are unknown, components must be very conservatively rated. . . This paper describes a system which provides true antenna impedance measurement capability."

A block diagram of a superhet form of this reflection coefficient was shown as Fig 11 last month and this is discussed in the paper, although circuit details were not given. During question time, I pointed out to Dr Sun that this antenna analyser would be of considerable value to radio amateurs and wondered whether it was intended to publish full details in one of the hobby magazines. Dr Sun replied that this was being considered, but warned that some 20 ICs were used, leading to complex circuit diagrams, although it might be possible to market such a device in kit form. A compact, portable instrument of this device would certainly be of interest to those experimenting with antennas, as well as hopefully leading to a reduction in 'tune-up interference'. An earlier, simple form of 'quiet tuning' described by Prof Mike Underwood, G3LHZ, (*RadCom*, May 1981, pp420-422) seems to have been largely forgotten, and did not provide the measurement capabilities of this new approach.

MORE POST-MIXER AMPLIFIERS

THE ITEM ON 'Bipolar and FET Post-Mixer Amplifiers' (*TT*, May 1999, pp58-59) attracted the attention of Wayne Cooper, AG4R, who provides some alternative circuit arrangements. **Fig 6(a)** shows the 75MHz IF amplifier of the Kachina 505DSP transceiver, described in *QEX* May/June 1998, p26. To try out this amplifier, AG4R put together a similar amplifier (**Fig 6(b)**). As a 14MHz amplifier with 0dBm input, the IP_3 was 40dBm with 10dB gain. He has also used the Norton RF amplifier and described this in *QST*, August 1984, p46. This amplifier (**Fig 7**) has given very good results in various applications over the years.

In 'Technical Feedback' (*RadCom*, April 1999) AG4R drew attention to 'A Logarithmic Audio Speech Processor' by William E Sabin, *Communications Quarterly*, Winter 1997, and mentioned that he was in the process of building one. In his latest letter he writes: "I completed the Sabin processor which gives me excellent results. I can use the extra 6dB, an S-unit boost to my 100W transceiver, equivalent to adding a 400W linear amplifier!" ♦

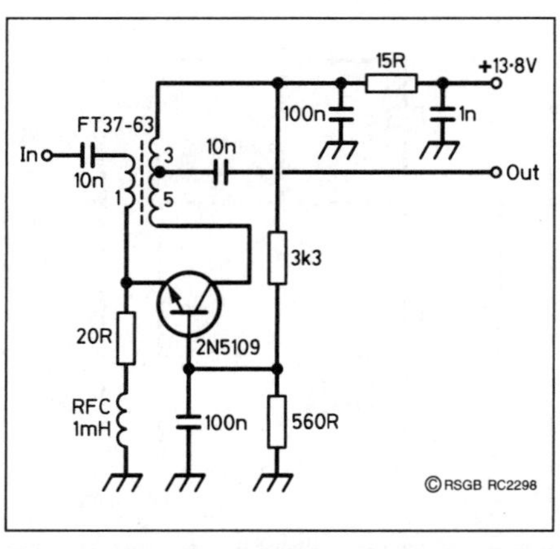

Fig 7: Norton RF amplifier as described in *QST* (Aug 1984), which has given AG4R very good results in various applications.

technical topics

by **Pat Hawker, G3VA***

EFFICIENT SHORT MEANDER ANTENNAS

IF YOU WERE asked to put up a 14MHz resonant dipole antenna with an overall span of just 3ft (1m) with a 2:1 SWR bandwidth of over 5 per cent of the centre frequency and a measured radiation efficiency of some 80 per cent, or a 21MHz monopole some 17in high over a 3ft by 4ft ground plane with a measured impedance of some 22Ω, you would probably regard such a request as asking for the moon. Such performance would seem to fly in the face of the classical theory of electrically short antenna elements. The trick, it seems, is to make the elements physically long but folded repeatedly into a short span, with for example a construction resembling the old caged dipole but with the short sections in series rather than parallel. This technique is described by Thomas J Warnagiris and Thomas J Minardo of the Southwest Research Institute, San Antonio, Texas in 'Performance of a Meandered Line as an Electrically Small Transmitting Antenna' (*IEEE Trans on Antennas and Propagation,* December 1998, pp 1797-1801). The techniques they describe would seem to open the way for the erection of LF/MF/HF antennas at sites where space will not support a conventional resonant antenna without incurring the restricted bandwidth and low efficiency normally associated with compact high-Q antennas such as magnetic loops.

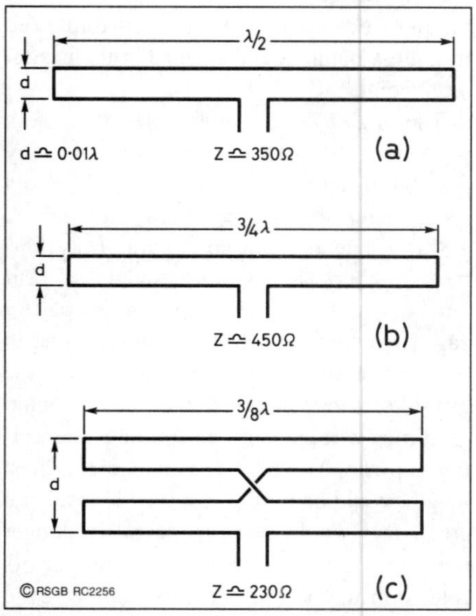

Fig 1: Three of the many folded-dipole antennas developed by Dr Kraus, W8JK, in the 1940s. (a) The conventional half-wave folded dipole with a resistive input impedance of about 300-350Ω. (b) The longer three-quarter-wave dipole; and (c) the shorter three-eighth-wave folded dipole.

The idea of multiple folding of elements is not new. Dr Kraus, W8JK, long ago described a three-eighth wave folded dipole and monopole, and these were included in *TT* back in May, 1971: see **Fig 1** and **Fig 2**.

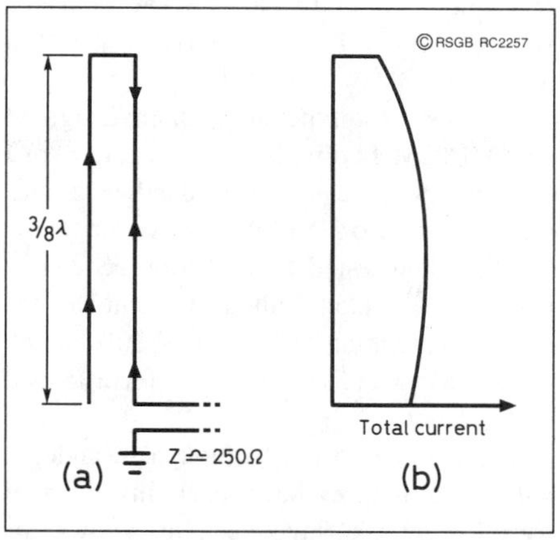

Fig 2: The three-eighth-wave folded monopole fed against ground. The resistive input impedance of about 250Ω is much higher than the conventional quarter-wave monopole.

There was also an ingenious *Radio Handbook* system based on the 3/8-wave folded element that could be switched to form an antenna resonant at half the frequency, eg for 7MHz and 3.5MHz, fed with 600Ω open line: **Fig 3**. But, even earlier, there had stuck in my memory an article I had read a few months after receiving my G3VA licence: 'Improving the Performance of Short Aerials' by M J Heavyside, G2QM (*The T&R Bulletin*, January 1939, p4). This described an antenna for the (then) 1.7MHz band, comprising 15ft of wire fed from a Collins pi-network tuner and going up to an attic. This, as might be expected, gave a relatively poor performance. But G2QM then added at the end of the 15ft wire some 120ft of 18SWG cotton-covered wire "wound in a zig-zag fashion on nails spaced six inches apart, driven into two beams situated five feet apart". This form of non-inductive 'loading' greatly improved results with his 7 watts (DC input) of AM phone. Reports increased from 3 to 5 S points, compared with the 15ft wire alone. Two other local amateurs, G6KU and G8CB, carried out tests with similar antennas and both reported gains in signal strength, particularly on reception. G2QM wrote: "When the 100ft outside aerial was compared with the loaded roof-space aerial, the signal-to-noise ratio of the indoor aerial was found to be so much better that it is now being used in preference to the outdoor one." He also tried the antenna on 3.7MHz, reducing the length of the zig-zag wire to 60ft: "Results on this band were found to be just as good, especially as regards reception."

Unhappily, all British amateur licences were withdrawn on 1 September 1939, before I and possibly others had much opportunity of trying out G2QM's ideas, and by 1946 they had been put on the back-burner. It is only in recent years that there has been increasing interest in the use of such meander line elements to reduce the overall span of resonant dipole antennas, reflected several times in *TT*.

The recent *IEEE Trans Ant & Prop* paper takes meandered elements a step further. The introduction reads: "At its operating frequency, an electrically small antenna is much shorter than a quarter wavelength. It has been occasionally pointed out that such an antenna, free of dissipation, could take from a radio wave and deliver to a load an amount of power independent of the size of the antenna. But the characteristics expected of an electrically small antenna are low-radiation resistance and large reactance and, therefore, very small instantaneous bandwidth with respect to the impedance of normal radio equipment. It's the matching of the antenna impedance to the radio equipment that can significantly affect the overall system performance. This is especially important for transmitting antennas, since a good impedance match is essential for maximum power transfer and hence efficient radiation."

The abstract of this interesting paper, apparently with amateur radio involvement, emphasises: "For antennas to radiate at maximum efficiency, their dimensions must be of the same order as the radiated wavelength. At frequencies below 30MHz, antennas with efficient radiation are often too large for mobile and portable applications. Smaller antennas can be made to radiate efficiently by use of matching networks. For convenience and ease of adjustment, these networks are usually placed between the transmitter and the antenna input, but it has been found that for best radiation efficiency, matching network elements should be placed at points on the antenna structure. Unfortunately, such matching networks must be tuned for each transmitting frequency, and when mounted on the antenna, they cannot easily be tuned. A meander element antenna was found to present some electrical and mechanical properties allowing convenient placement of tuning elements when configured as an electrically small transmit-

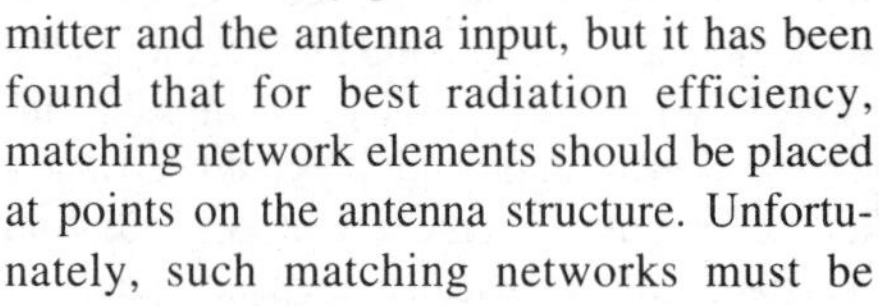

Fig 3: An adaptation of the Kraus three-quarter-wave dipole, which can be switched to operate on twice the frequency. For example with L = 98ft, the antenna is resonant on 14MHz with the switch closed and on 7MHz with the switch open.

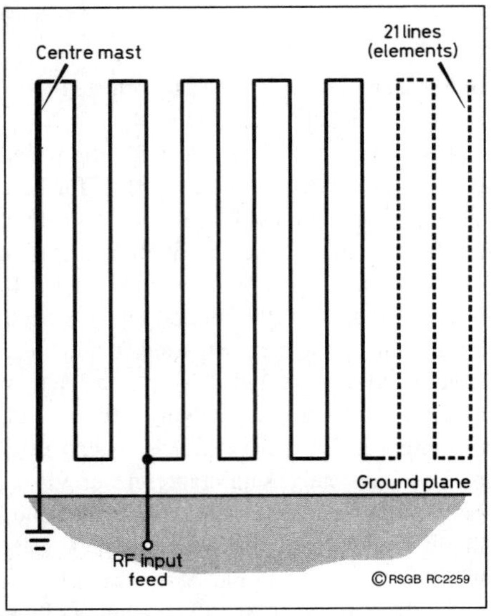

Fig 4: Electrical configuration of a very short multi-element meander monopole antenna with a height of only 17.25in with the 21 elements connected as a continuous single wire.

ting antenna. Some simplified design guidelines were derived from experimental data."

One technique for increasing the terminal resistance (maximum current point) and, therefore radiation efficiency, is antenna folding. This technique is beneficial to radiation resistance without decreasing antenna bandwidth. It was shown in 1958 that capacitive top loading and folding may be combined to produce a small broad-band antenna with high-radiation resistance.

The authors emphasise that the important factor provided by a folded structure is that the cancellation of opposite polarity currents from the various elements generally results in a higher average impedance along the folded element as the number of folds increases: "What apparently has not been heavily investigated is the effect on radiation resistance of electrically small antennas of many folds. A meander antenna is an extension of the basic folded antenna to include a large number of folded elements in various linear patterns." They point out that an interesting property of meander antennas is the number of resonant modes beyond the lowest resonant frequency [It could be argued that meander antennas are closely related to fractal antennas - *G3VA*].

Fig 4 shows the electrical configuration of a prototype 20.1MHz meander monopole antenna much shorter than a quarter-wave high investigated at the Southwest Research Institute. Mechanical implementation of this antenna is outlined in **Fig 5**. The input impedance of this 17.5in high antenna with 21 elements of No26 AWG wire was tested over a 3ft by 4ft ground plane. Measured impedance at first resonance (20.1MHz) was 21.9Ω, much higher than would be expected from an 0.03λ monopole. The bottom fold feed points were usually at points 10-20% of the total antenna length and produced a 2:1 VSWR bandwidth of about 2-4% of centre frequency (about twice as wide as bandwidths reported for reactively loaded monopoles 0.06λ long).

Another prototype antenna investigated was a 14.2MHz dipole with an overall span of only 3ft. A 2:1 VSWR bandwidth was measured over 6.5% of the centre frequency: **Fig 6**. When tested for radiation efficiency using the Wheeler method, this antenna exhibited a radiation efficiency of 80%, much higher than would be expected for a loaded 0.05λ dipole.

The authors write: "Although meandered antenna structures have been investigated by others, there have been no known attempts to construct practical meandered antennas for tuneable applications. SwRI has developed several basic meander structures that lend themselves to switch tuning by the low-loss connection or disconnection of elements at various points on the antenna to optimise performance. Practical RF switches can be solid-state devices such as PIN diodes, FETs, or even mechanical switches. The grounded mast of a meander monopole antenna can be a high-strength conductive mast that can be directly connected to the ground plane, offering also some protection from lightning strikes. Wide spacing between the elements and high RF voltage points allows transmission of high-power without arcing. . . Dipole configurations can be constructed consisting of two monopole meandered structures end-to-end, fed through a balun or unbalanced at their common point."

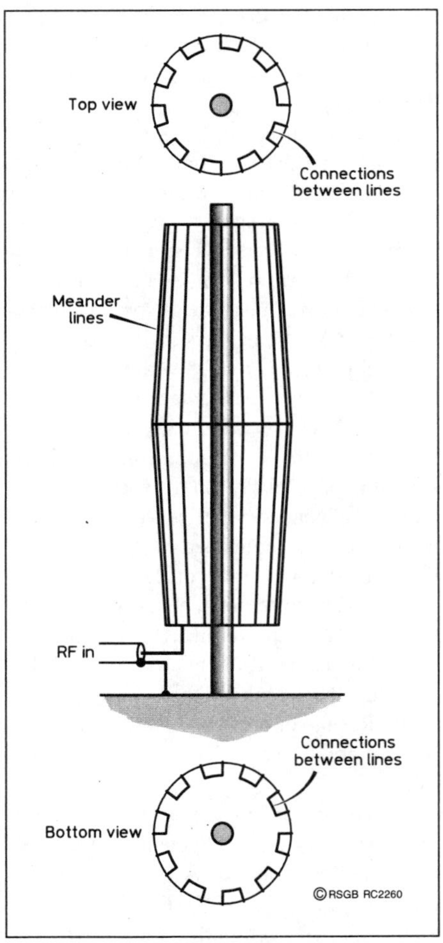

Fig 5: Typical mechanical structure of a meander monopole. The discs and central spacer ring or disc should be made from material offering good RF insulation.

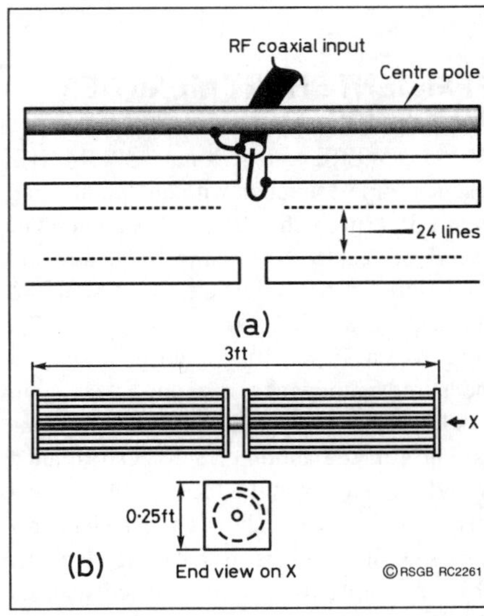

Fig 6: A 14.2MHz meander dipole with a span of only 3ft.

The *IEEE Trans* paper notes that more work is needed to characterise the many possible meander antenna variants. It would seem to me the one possibility could be a 1.8MHz antenna with possibly a 30ft (10m) metal support pole with a circular RF insulated disc at the top and then a series of zig-zag wires down to insulated stakes, rather in the form of an umbrella antenna but with the guy-like wires forming a single resonant mast plus wire, rather than forming a capacitive top-cap. The total wire-plus-metal mast length would be best determined by trial and error, and would need to be rather longer than an electrical quarter-wave due to its folded form. There would also seem to be the possibility of using this approach for a 136kHz Marconi-type antenna (a resonant antenna would presumably require over 500m of wire!) In this case, the higher the support pole and the more folded elements the better - but it could prove more efficient than most of the antennas now in use on this interesting and antenna-challenging band. At the other extreme, the meander antenna has possibilities for mobile or transportable operation on 1.8MHz, 3.5MHz or on the HF bands in restricted spaces.

NEW LIFE FOR THE FT-101

TT, MAY 1999 (p61) included the *QST* 'Hints & Kinks' suggestion that the RF power valve type 6146B/8298A, still widely used in many of the valve and hybrid transceivers designed in the 1960s and 1970s, is apart from the heater ratings, identical to type 6883B/8552. These, at least in the USA, are available at much lower cost. But some of the most popular HF transceivers of that era, including the Yaesu FT-101 and FT-101B, used valves originally intended for television line-output (sweep) stages, such as the 6JS6A and later the 6JS6C. It should be noted that the much-modified FT-101ZD (with WARC bands) used the 6146B, but the FT-200 also used 6JS6 valves.

'Hints & Kinks' (*QST*, May 1999, pp68-69) reprints from *Radio ZS* an article by Roger Davis, ZS1J, on providing a new life for the FT101. This points out that 6JS6A and 6JS6C valves have become rare and - even if you can find them - are often prohibitively expensive. Many FT-101 transceivers have been prematurely 'retired' because of the difficulty of finding replacement matched 6JS6 valves.

He writes: "Far-superior 6146B valves are available at a moderate cost, if not in most junk boxes. In the October 1996 issue of *Radio ZS* my article on the modification of the FT-200 to take 6146s was published. In it, I mentioned that it should also be possible to modify the FT101 to take the same tubes, but I did anticipate a slight drop in output power due to the lower plate voltage in the FT101, which is around 650V.

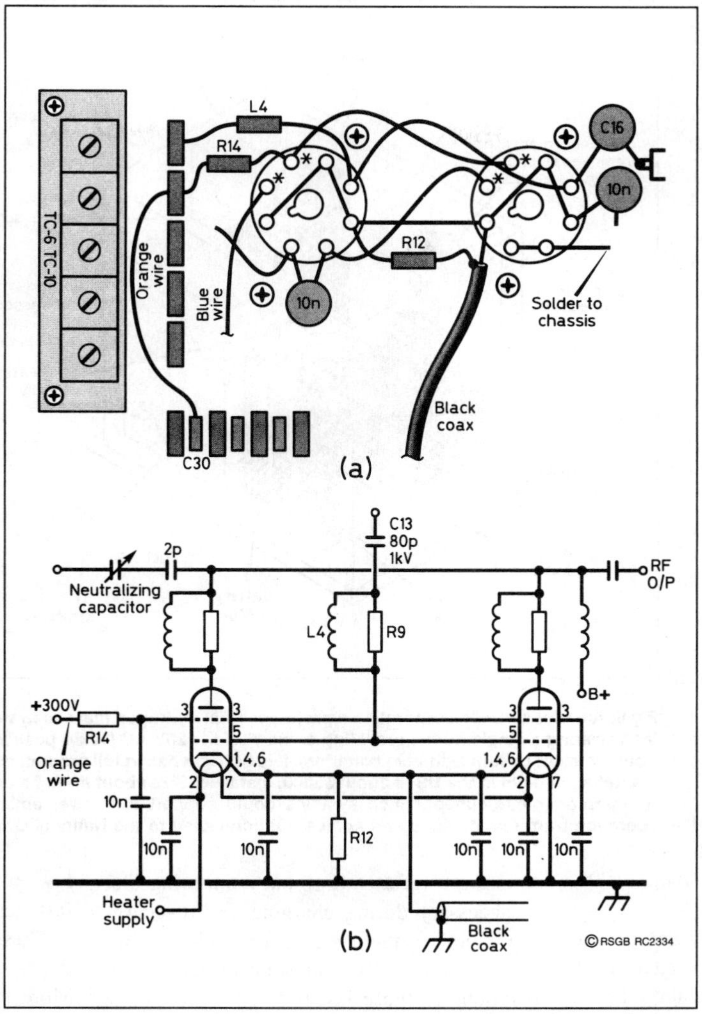

Fig 7: ZS1J's modifications to the FT-101 to use 6146B valves in the power amplifier, in place of the hard-to-come-by 6JS6 TV line timebase sweep valves. (a) pictorial view; (b) circuit diagram.

"Recently I was given an FT101B to repair. . . and decided this was a prime candidate for the modification. Before tackling this task I fitted a 0.01µF capacitor in series with C13 as a precaution. This 80pF often becomes leaky, destroying the final tubes."

ZS1J provides a blow-by-blow description of the modification. Although this is too long to be reproduced here, **Fig 7** gives an overview of the operation. The changes are listed briefly here.

The main points are the replacement of the two 12-pin valve sockets with octals (four Phillips screws). Use the decoupling capacitors from the old sockets between where indicated with an asterisk and the closest ground points. To increase screen voltage, trace the orange wire from R14 to the C35 and connect instead to C35, which is the 300V line for the 12BY7 driver. Due to the lower inter-electrode capacitances of the 6146B, two circuit modifications are needed. Since the permeability-tuned coil in the anode circuit of the 12BY7A (T103) on 28MHz uses the input capacitance of the PA valves, now reduced by some 25pF, this should be compensated for by fitting a small variable capacitance across T103 (this involves removing the two screws securing the circuit board carrying the trimmer capacitors, TC6 to TC10). The final modification is to reduce the neutralising capacitance by removing the 100pF C125 and replacing it with a 2pF capacitor (indicated on diagram) and adjust the neutralizing capacitor to minimum capacitance from below the chassis: "The set is now ready for testing, switching and final alignment." Adjust the bias control fully counter-clockwise then adjust for 60mA in the SSB mode with no audio. With the trimmer board unscrewed, set to 28.0MHz with preselector control at start of the 10-metre mark. With an insulated trimmer tool adjust the newly installed trimmer across T103 for maximum drive, being very careful not to touch the loose trimmer board, since there is high voltage on the components (see also the instruction manual page 25 section 3, para 2). Check and, if necessary, adjust the neutralisation as on page 24 of the manual. When completed the FT-101 should deliver in excess of 100W output on all bands.

Clearly it would also be possible to rewire the heaters in series to use 6883B 12.6V valves, as described in the May *TT*.

MORE ON BATTERIES AND FUEL CELLS

THE MARCH 1999 *TT* item '200 Years of Batteries' emphasised the renewal of research and development interest in batteries and fuel cells, resulting from the current interest in developing environmentally-friendly electric vehicles (EVs) and at the other end of the power spectrum for such portable items as lap-top computers.

With so many different types of rechargeable batteries now or soon to be on the market, selecting the best battery for your particular amateur radio application can be a compromise between cost and length of service. This is stressed in 'Which Battery Should You Use in Your Equipment?' by Don A Gagnon, WB8HQS (*QST*, April 1999, pp40-42). WB8HQS writes: "With several new battery types coming onto the scene in the last 10 years or so, it's hard to keep up with what's available today! . .

The tried and true standby batteries for hand-held transceivers, video cameras and portable computers have been nickel cadmium (NiCd). For less-portable applications, sealed lead acid (SLA or Gel) cells are often used. Today we have nickel-metal hydride (NiMH), Lithium ion (Li-Ion), lithium polymer (Li-Polymer) and reusable alkaline. Each has good and bad features, and selecting the battery for your needs can be a real headache unless you understand those features."

He provides a useful summary of the pluses and minuses of these batteries:

- Nickel Cadmium (NiCd) - Mature technology, used where long use life, high discharge rate and reasonable prices are important. Not good where long shelf life is essential.
- Nickel-Metal Hydride (NiMH) - Has been around for about 10 years, offers a somewhat better power density than NiCd, but also has a reduced total number of charge/discharge cycles and a lower peak output current (because of higher internal resistance).
- Sealed Lead Acid (SLA) or Gel Cells - Best suited for higher-current power applications, or those requiring low current for extended periods of time. The downside of these batteries is their large physical size and weight.
- Lithium-Ion (Li-Ion) - A somewhat new technology, really emerging in the last five to

seven years, although it has been experimented with for some time. Li-Ion cells provide a much higher power density (power available per unit volume) than NiCd. The downside is that they require very tight control of charge current, charge voltage and discharge, to defend against plating metallic lithium inside the battery.

- Lithium Polymer (Li-Polymer) - Not commercially available yet. Li-Polymer is a modification of Li-Ion technology and promises to be less expensive to manufacture, but will have a low maximum output current. Best suited for slow drain, low-current applications.
- Re-usable Alkaline - Cousin to disposable alkaline cells, the reusable cells are good for low-cost and low-power applications. They're known for having a very low self-discharge rate [good shelf-life].

Technical characteristics of most of these rechargeable battery types were given in *TT*, March 1999, p61, Table 1.

WB8HQS stresses that "for safety and battery-life reasons, each Li-Ion battery pack must be equipped with control circuitry to limit the peak voltage of each cell during the charge cycle, and to prevent the cell voltage from dropping too low during discharge. Most of these circuits also properly limit the charging and discharging current and monitor the cell temperature. There are many battery-monitoring microcircuits available that perform all these functions with a minimal number of additional components, components that are intended to be incorporated into the battery pack itself. By carefully monitoring and controlling all of these parameters, we can virtually eliminate the chance of plating metallic lithium within the battery. With such protection, charging the batteries involves simply attaching a proper voltage source to two terminals and the battery pack does the rest! A typical microcircuit is the Benchmarq bq2058.

"There are today many sources of Li-Ion cells including NEC, Energizer, Sanyo and GP Batteries. Although Sony is probably the world's largest manufacturer of Li-Ion batteries, at this time they have chosen to keep that manufacturing captive and not sell them to outside users.

"All-in-all, the hobbyist has several choices today in battery power, and Lithium-Ion appears to offer the best overall performance for use in handheld transceivers and similar electronic equipment. The one 'gotcha' you must pay very close attention to is the careful control and monitoring that is necessary during charging and discharging. Luckily, there are several microcircuit manufacturers that can supply ready-to-go solutions to those issues."

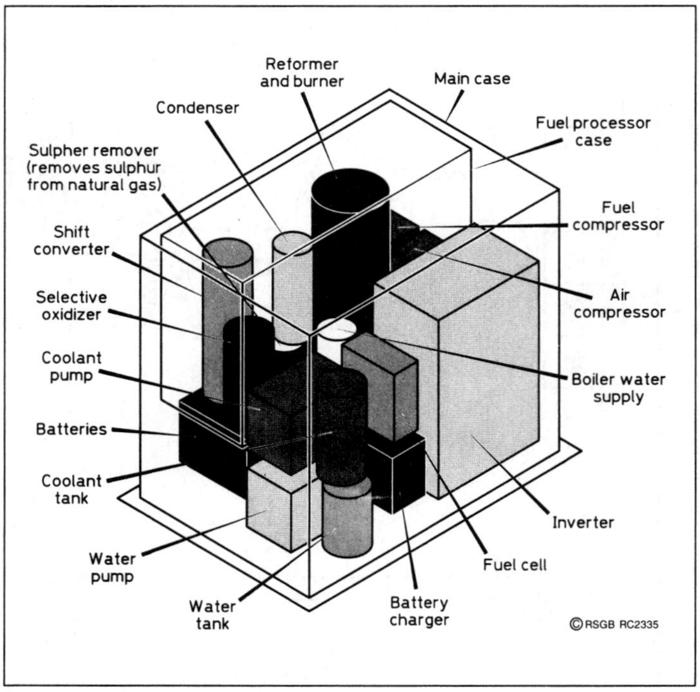

Fig 8: American Power Corporation's prototype PEM hydrogen fuel cell system, for powering a single home, providing some 3kW at 120V AC (10kW peak load from the eight built-in lead-acid batteries). It is hoped when in full production to market such units in the US at about $5000, installed. Size about that of a large refrigerator (metre-cubed). Such systems could support full-power amateur operation from remote sites and homes not connected to the National Grid.

Professor Colin Vincent in 'Lithium batteries' (*IEE Review*, March 1999, pp65-68) stresses that developments in lithium battery technology are currently being driven by the demands of consumer electronics products, although the emphasis will soon change to larger battery systems for electric vehicles and dispersed energy storage. He notes how the early lithium batteries were disposable primary button cells suitable for powering loads of a few microwatts over periods of several years, but stresses that energy from even the cheapest Leclanche consumer primary cells cost over a thousand times that of domestic mains electricity. A button cell provides energy at a cost of over £10,000/kWh!

He writes: "The first commercial lithium primary cells were introduced by SAFT 25 years ago, and the technology is now mature with over 50 companies in the market. . . The first volume production of AA-sized rechargeable lithium cells was by Moli Energy in Canada in the 1980s. The safety record of early lithium secondary cells was not good, and it was not until the introduction of the first commercial lithium ion cell by Sony in 1990 that the market became established with world-wide production of this type of cell now approaching 500-million units per year ,and with the first UK production due to commence in Thurso later this year."

While large rechargeable lithium batteries for EV applications have reached pilot production, the March *TT* drew attention to the potential role for electric vehicles of new forms of the 'fuel cell' that was originally invented some 160 years ago by the British physicist William Grove. Until recently their cost and their inappropriately high operating temperatures have restricted their use to military and space applications; for instance, aboard submarines and the space shuttle.

It was pointed out in March that the new form of proton exchange membrane (PEM) hydrogen fuel cell offers a relatively benign power source for electric vehicles: "Operating at about 80°C and employing a thin plastic sheet as their electrolyte, they are easy and safe to handle in manufacture and in later use. . . other attractions include easy pressurization, low emissions, high conversion efficiency, fast start up, and fast response to transients."

These remarks stemmed from an article in the November 1998 issue of *IEEE Spectrum*. More recently, a further article in the same journal 'Fuel Cells For Homes and Hospitals' by Ronald H Wolk (May 1999, pp45-52) shows that spurred-on by promising car and bus applications, designers are also developing fuel cells to power single homes and even large building complexes such as hospitals. It surveys a number of different fuel cell approaches and notes that: "Several designs are under way for power plants for the home employing PEM fuel cells. These are fuelled with hydrogen manufactured from natural gas in reforming systems that are part and parcel of a package that also contains battery packs to supply peak power, and a DC-AC inverter to supply power at 120V AC."

Ronald Wolk writes: "Although tremendous research progress has reduced the costs of these fuel cell stacks by a factor of 10 in the last decade, further reduction is required to meet the nominal economic goals set by the utilities industry of about $800-$1200/kW installed, whether at home or in a central power station. American Power Corp of Boston has a package containing two PEM fuel cells plus eight lead acid batteries. The cells produce 3kW of base load power, and together with the batteries support a peak output of 10kW. Each package is about the size of a large refrigerator and is targeted to sell for less than $5000, installation included, when mass-produced in moderate quantities. . . A prototype is in operation. Several dozen units have been ordered by the utility companies for a demonstration programme to gain experience." **Fig 8** gives an idea of the prototype one-metre-cubed Residential Power Generator from Analytic Power, with two fuel cells each producing 1.75kW and with battery support to provide peak loads of 10kW, 120V AC. The fuel cell itself is nearly dwarfed by the other components.

Clearly, if the cost targets can be met in practice, it will become feasible to consider seriously the use of PEM fuel cell packages to provide power at remote sites and houses not served by the National Grid. A simplified package without batteries would supply some 3kW of power and could presumably form the basis of a transportable or mobile power source for complete radio stations without the noise, fumes, etc of petrol or diesel electric generators. The PEM fuel cell may well come to have a role in amateur radio.

Much of the *IEEE Spectrum* article is concerned with much larger power units, up to about 250kW. It is warned that many long-term observers of fuel cell R&D programmes have repeatedly heard developers project that complete systems are within five years of attaining low enough prices for the competitive production of electricity: "Those claims have been belied by the problems normal to the development of any sophisticated technology. Difficulties with scale-up, materials that needed additional development, under-performing components and improperly assembled system have all arisen. But progress overall has been healthy, despite some discontinuities and delays. More vendors than ever are involved, more markets have emerged, and optimism is high.

"By now, commercially attractive fuel cell products are close to emerging in the 200-2500kW capacity range. That is enough to provide base-load power to a small neighbourhood. Recently, moreover, various systems using storage batteries for peak power requirements to serve as economically viable residential power plants are emerging. . . Industry observers are confident that these kinds of products will go on sale within five years. Success could lay a foundation for a future in which fuel cell systems could be scaled up to hundreds of megawatts and used in central station power plants."

ATUs FOR BALANCED TUNED LINES

AS A KEEN QRP operator, Glen Johnson, G4DUC, is convinced that the 'punch' achieved with his 5W signals is due to achieving maximum power transfer into balanced resonant antenna feeders. Originally he was well pleased with the results achieved using a conventional Z-match ATU with switched (tapped) coil on the balanced output link coils when using a G5RV-type antenna, but with the 300Ω line brought all the way into the shack. However, more recently he has found that significantly greater efficiency is obtained when particular attention is paid to the resistive/reactive combinations found at the transmitter end of a balanced line. He believes that the conventional transmatch type of ATU followed by a 4:1 balun at its output is inefficient when used with tuned feeder systems.

The result is that he has experimented with a number of arrangements, all placing the unbalanced-to-balanced transformation at their input rather than output of the ATU. It has proved possible to achieve excellent current balance in the feeders using either a simple unbalanced matching technique or a purely balanced arrangement. The balun never has to cope with a reactive load [an important consideration when using any type of balun - *G3VA*].

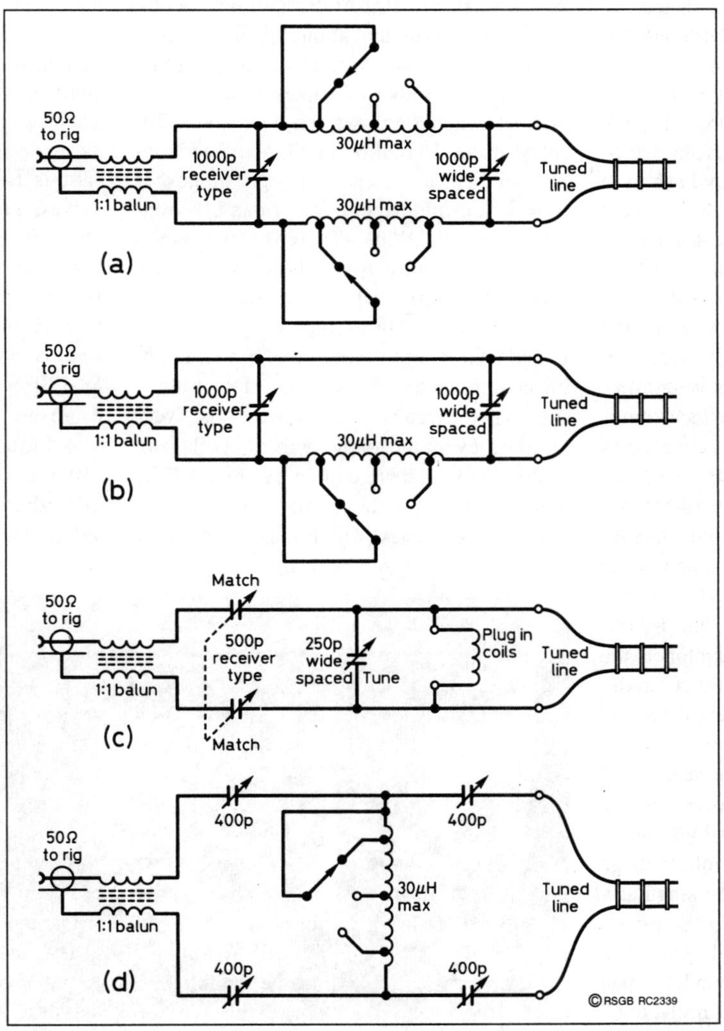

Fig 9: Four ATU configurations for feeding balanced resonant feeders recommended by QRP enthusiast G4DUC, in which the unbalanced-to-balanced transformation precedes rather than follows the matching network.

Fig 9 shows four of his arrangements: (a) is used extensively by G4DUC and is based on the classic Collins coupler, widely used in the days when transmitters provided balanced output. The only modification is the balanced drive derived from a 1:1 current balun. With switched or plug-in coils it will match virtually anything likely to be presented to it by a tuned line. (b) uses an unbalanced but floating pi-network design, but G4DUC finds the current balance still most acceptable (within 5% with careful layout) and with good power transfer. (c) is his favourite experimental arrangement, and he has so far had no problems matching anything with it (including small loops). The use of plug-in inductances of various values from 1µH to 30µH will cope with any amount of capacitive reactance, and the 250pF 'tune' capacitor with any likely inductive reactance. Where the line presents an inductive reactance, the inductor may be required. Matching to 50Ω is accomplished by the conventional Z-match arrangement of capacitive tapping, but in this case balanced. In operation the antenna system, including feeder, is brought to resonance with the 250pF 'tune' and/or 250pF plus inductance. The 50Ω match being obtained by the 500pF matching capacitors forming a low impedance resonance condition with the whole system in an apparent series-tuned condition. For very low impedances (below about 20Ω) it may be better to experiment with feeder length, although (d) could cope with low-impedances but is less suitable for higher impedance/reactance combinations.

G4DUC has found that the most efficient balun for those in all these circuits is the 'current' type. Typically for 160-10 metres this can comprise 14 turns of 50Ω coax wound on type 61 ferrite (not dust iron).

While receiver-type variable capacitors are suitable for low-power operation, the problem of arcing can arise when higher powers are used. Unfortunately, high-voltage 250pF, 500pF and 1000pF variable capacitors do not come cheaply these days for those without well-filled junk boxes. ♦

technical topics

by **Pat Hawker**, G3VA*

THE NOT-INVENTED-HERE SYNDROME

IN BRITAIN WE have of course every reason to be proud of the record of our scientists and engineers in the sphere of radio. We greatly respect the role of Marconi, even if he had an Italian father and Irish mother and was basically more of an entrepreneur than an academic scientist ("You know I have always considered myself an Amateur" - *Guglielmo Marconi*). But we are less mindful of the role played by those professional Marconi engineers H J Round (1881-1966) and C S Franklin (1879-1964). Then there is the Scot John Logie Baird who 'invented' mechanical television (downplaying the roles of von Ardenne, Karolus, Mihaly, C F Jenkins and other pioneers in this field). Then there was Robert Watson-Watt who 'invented' radar - better not mention the German engineer Christian Huelsmeyer who, in 1904, patented and later demonstrated his 'Telemobiloscope' as a crude but practical form of marine radar, or the collision-avoidance system installed on the French liner *Normandie* in 1935.

I was reminded how little we all tend to know of engineering work in other countries when I received an enquiry from an American amateur who was seeking information on Franklin antennas. He wrote: "One reference calls this antenna a Marconi-Franklin, which makes me wonder if Franklin may have been British since Marconi operated in Britain when Italy wouldn't support his work." I hurriedly replied that Charles Samuel Franklin was not only British but was responsible for much more than the Uniform Antenna and the Franklin Beam Array. He was the radio operator on board the *Philiadelphia* when Marconi (who was also aboard), provided irrefutable evidence that wireless waves could cross the Atlantic. He was granted 64 patents including the variable capacitor (1902) and ganged tuning (1907). He patented the regenerative receiver at much the same time as Armstrong's work in the USA. His greatest achievement was the Marconi short-wave beam system, for which he developed the coaxial transmission line and the Franklin Master Oscillator (still one of the most stable forms of VFO).

But before we think that Americans tend to be ignorant of all that happens outside God's own country, we should look at the motes in our own eyes. Because of language difficulties, most of us tend to be singularly unaware of what was achieved in Continental Europe. While British and American amateurs and professionsls did much to open the 'short waves' in the 1920s, it needs to be accepted that German scientists and engi-

*37 Dovercourt Road, London SE22 8SS.

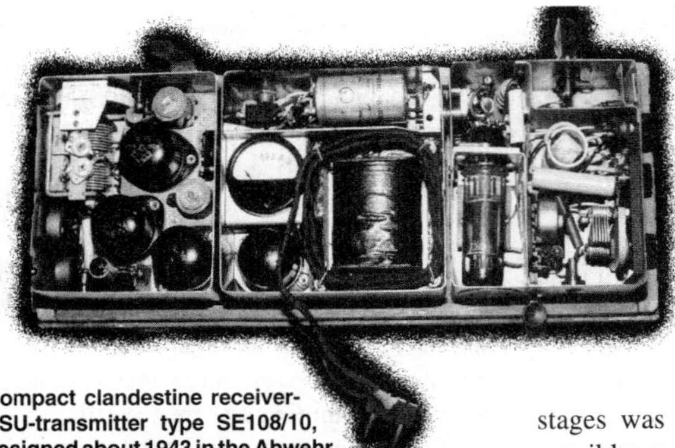

Compact clandestine receiver-PSU-transmitter type SE108/10, designed about 1943 in the Abwehr workshops and used operationally in 1944-45 in Italy, etc, after the Abwehr had been taken over by the RSHA. Transmitter output 10 watts, three-valve straight receiver and built-in Morse key. Receiver 3.8-5.9MHz. Transmitter 5-14MHz. Size, 380 x 140 x 60mm. Weight, 4.6kg (courtesy Rudi Staritz, DL3CS).

neers made unique advances in the 1930s and early 1940s - we tend for historic reasons to downplay the many important radio-engineering advances made in Germany during the Hitler regime (1933 - 45). One of those who have been attempting to put the record straight is the Dutch radio-historian Arthur O Bauer, PA0AOB, with his 'Centre for German Communication and related Technology, 1920-1945' who has in recent years presented a number of papers in the UK as well as conducting research here. Last February at the 'Evolution of Modern Electronics' seminar at Imperial College, he gave a most interesting paper on 'German Electronics between the Wars'. There is space here to quote only briefly from his paper, etc. He makes it clear that while, for example, most early communications receivers developed outside Germany were, more or less, enhanced broadcast receivers in a metal housing "Their circuit concepts were also very similar, even the same valves, coils and tuning capacitors were often employed. [The National HRO with its mechanical excellence due to James Millen is an exception - *G3VA]*.

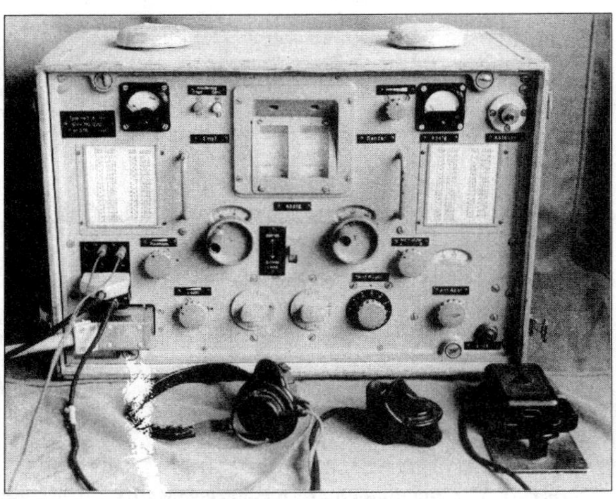

A wartime German Army field transmitter-receiver with superhet receiver, turret coils and temperature compensated master oscillators.

In the second half of the 1930s, the German Air Forces engendered wireless equipment whose technology could well be described as 'state of the art' at that time. Lorenz introduced three-dimensional construction technology, whereby optimal screening between stages was made possible and the optimal possible space-to-volume ratio was achieved. He makes it clear that until the end of the 1920s, general devopments in Germany were comparable to those in other developed countries. But then the Lorenz Company introduced magnesium-aluminium die-cast techniques ('elektron alloys') for chassis construction. At the same time, the Hescho Company developed ceramic substrates with stable dielectric properties, offering a very low-loss factor at HF and VHF. A third important factor was the invention of low-loss iron dust core material by Hans Vogt: "In Europe, both Switzerland and Germany had a penchant for the extensive use of complicated and fine mechanical construction, and these three developments were a contribution to that style of technology. . . The opportunity to create a new standard in valve design persuaded the military organisations to reduce the number of types of valves, greatly easing the supply of spare parts. For instance, the entire aircraft FuG 10 installation was equipped with only two different type of valves. The receivers employed either eight or eleven RV12P2000 valves and the transmitters only three RL12P35 valves with an anode dissipation of 35 watts (comparable to the 807)."

A further transmitter valve was the LS50 (50 watt dissipation), widely used in the Wuerzburg radar installations. Both power valves were designed for modular technology and needed only access from the top to be removed. The RV12P2000 was plugged upside down into its socket and could be easily mounted between two adjacent compartments. Both types could be used up to about 200MHz, unusual for large-production valves of the late 1930s. The RV12P2000 became the backbone of nearly all German commercial and military equipment, and more than 16 million were produced during WW2. It was really a major step forward to supply a maintenance organisation with only two different valve types for an entire radio installation. As Walter Farrar, G3ESP, wrote in 1947 "For use as triodes and diodes, the anode and grids were strapped together; and in one MF superhet a pentode was employed quite successfuly as a double-diode-triode, the suppressor grid acting as the diode anodes, and the screen-grid as the triode anode. This simplified both valve manufacture and provision of spares, and appeared to have no adverse effect on the performance of the receiver."

The German military equipment achieved a high degree of frequency stability without the use of quartz crystal oscillators, incorporating voltage regulator valves and both static and dynamic temperature compensation by its use after 1933-34 of low-loss titanium-dioxide ceramic capacitors. The problem of variable capacitor arcing, particularly at high altitudes was overcome by the use of improved iron-dust cores (developed by Hans Vogt) in the excellent German range of variometers which have been previously featured in *TT.* For the FuG 10 transmitter, both stages used variometers as the tuning device, with both tuned by only one knob. Initially new techniques had to be explored to counter the saturation problems of the iron dust cores, and later rather bulky blocks of sintered iron dust-cores were used inside the variometers.

PA0AOB also described the effective variable-bandwidth crystal filter used in the remarkable Koeln [Cologne] E52 HF communications receiver, covering 1.5 to 25MHz with a six-gang tuning capacitor. Some 2500 were produced between 1942/3 and 1945; and it was almost certainly the most advanced HF receiver of the decade, with a performance that is seldom bettered even today. It had a servo-tuning system which enabled control remote from the main receiver (a technique adopted post-war in the UK, etc), motor controlled tuning of four memories for each band. The high IF for this period of 1MHz gave an image rejection at 20MHz of better than 96dB! The *symmetrical* crystal filter response was variable between 200Hz and 10kHz, based on a filter developed by Kautter of Telefunken, with the crystal input and output LC circuits detuned in opposite directions, using ganged tuning: "Donald Prins in 1990 plotted the filter response of the IF module and measured 130dB down without any trace of side lobes! The two crystal filters were aided by six-resonant LC filter sections between the mixer stage and the first IF stage, followed by two further IF stages, each with its ganged variable-bandwidth crystal frequency. (**Fig 1**). A truly remarkable receiver for 1942/3!

THE ELUSIVE CONJUGATE MATCH

IN *TT* JULY 1999, p54 and *TT* May 1999, p61, I referred briefly to the belief that "for maximum power transfer the output or source impedance [of a tuned RF power amplifier] is the complex conjugate of the antenna's impedance", quoting as a recent endorsement of this view the 23-page article by Dr John Belrose, VE2CV; Walter Maxwell, W2DU; and Tom Rauch, W8JI (*Communications Quarterly,* Fall 1997, pp25-48) VE2CV coupled conjugate matching with the statement that the operational bandwidth of a tuned high-Q loop transmitting antenna system can theoretically be double that of the antenna alone, although admitting that he had found the increase to be somewhat less in practice.

In July I noted that there has been a long-lasting controversy among some of the leading (professional) boffins of our hobby as to the validity of applying the concept of conjugate matching to RF power amplifiers, pointing out that my former IBA colleague, Tony Harwood, G4HHZ (ex-BBC and Racal engineer), was one of those firmly resisting the validity of applying conjugate matching theory in this application, but adding: "This is a controversy into which I have no intention of entering at this time."

If only life were that simple. Now, having once raised the topic, it is clear that readers are not prepared to let it rest there.

Furthermore, having referred to the long pro-conjugate-match VE2CV, W2DU and W8JI article, it seems beholden on me to draw attention to a detailed rebuttal by Warren Bruene, W5OLY: 'The Elusive Conjugate Match in HF Tuned Power Amplifiers' (*Communications Quarterly,* Spring 1998, pp23-31). W5OLY was the senior Collins transmitter engineer who persuaded the ARRL to remove references to conjugate matching of HF power amplifiers from the *Radio Amateurs Handbook* and other ARRL publications.He writes: "Ever since I became a transmitter design engineer at Collins, I knew that a 'conjugate match' did not exist at the plate of a tuned HF power amplifier. But the author of a series of articles in *QST* in 1973 and 1976 and later writings claimed otherwise." [The articles were 'Another Look at Reflections' by W2DU, which ran to at least seven parts - *G3VA*]

W5OLY continues: "I did not challenge him until I noticed this material in an amateur handbook and learned he was the author. In personal correspondence in 1989, I thought I had changed his mind, but I have found that is not the case." This was rather embarrassing, since I recalled having quoted extensively from W2DU's excellent series not only in *TT* but also in the now out-of-print *Amateur Radio Techniques* and *A Guide to Amateur Radio* under the heading 'SWR Facts & Fallacies'. This material remains valid and does not directly mention conjugate matching. Nevertheless, references to conjugate matching in connection with RF amplifiers still exist in some current RSGB publications and the concept has been mentioned from time to time in *TT* and other *RadCom* articles. There can however no longer be any real doubt that W5OLY and

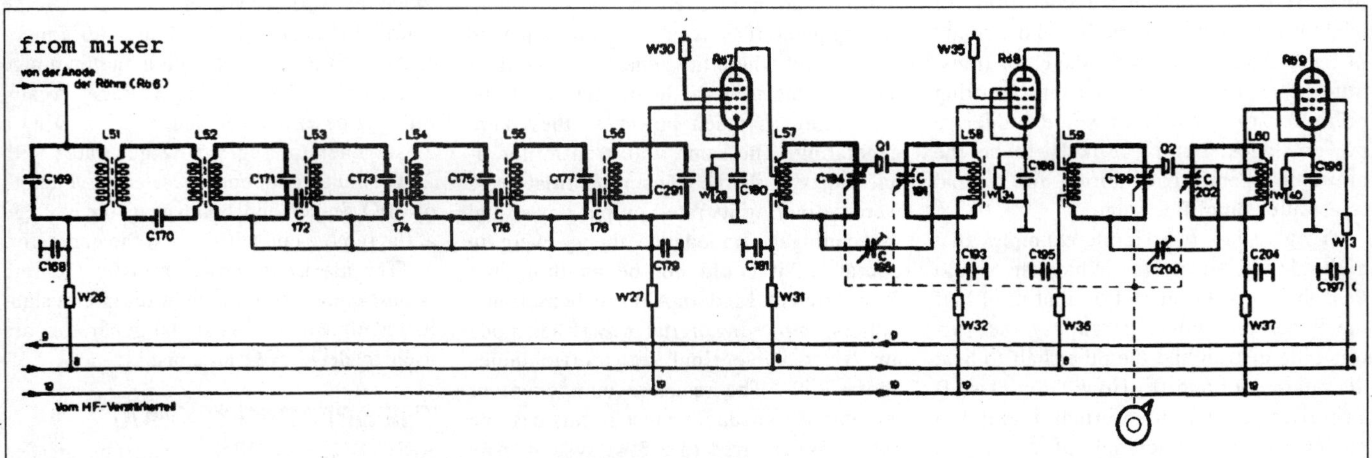

Fig 1: Original German circuit diagram of the 1MHz IF stages of the 1942 Koeln E52 HF communications receiver, providing a symmetrical response with bandwidth continuously variable from 200Hz to 10kHz and with six-gang preselection tuning, providing an image rejection at 20MHz better than 96dB (from PA0AOB's paper).

SINE WAVE AUDIO OSCILLATOR

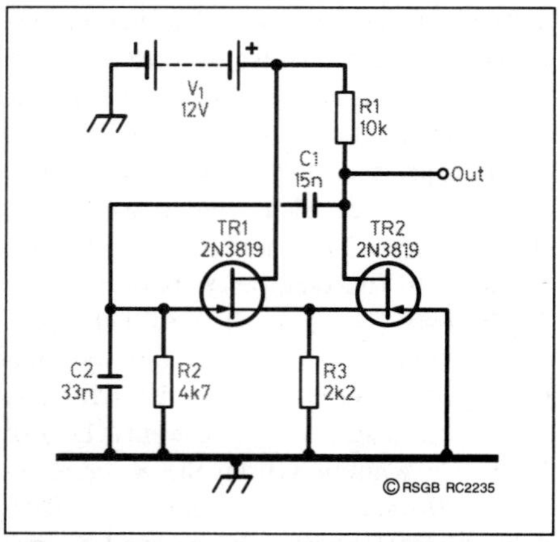

Fig 2: Simple sine-wave audio oscillator providing an output of about 8V peak-to-peak at about 1kHz, (Source *Electronics World*).

FROM THE 'Circuit Ideas' columns of *Electronics World* (April 1999) comes the very simple audio sine-oscillator, capable of providing a useful modulator drive, etc. shown in **Fig 2**. Using two small-signal FETs, type 2N3819, and just five other components it provides an 8V peak-to-peak output at about 1kHz. Davor Virkes of Zagreb, Croatia, claims that total harmonic distortion is around 10% and may be lessened by increasing the value of R3, although this will reduce the output swing and may make starting more uncertain. Similarly, he suggests that any resistance in the drain of TR1 will probably reduce performance. I recall including in *TT* many years ago, a basically similar arrangement as an HF oscillator using a tuned circuit, rather than an RC time constant.

those other engineers who believe that it is technically incorrect to apply the 'Maximum Power Transfer Theorem' in the form '. . . for maximum power transfer the output or source impedance (of the Power Amplifier) is the complex conjugate of the load (antenna system) impedance' have a strong case.

Bob Pearson, G4FHU, writes: "Alas, though often repeated in radio textbooks, for power amplifiers it is usually a misapplication of a theorem that applies only to ideal linear circuits. Its proof depends upon there being no limit to any signal level, and it may therefore only be applied safely to semiconductor amplifier circuits when they are operated at signal levels way below those limited by various considerations of supply, breakdown and dissipation, and where non-linear device behaviour is not significant.

"In circuits that do obey the ideal conditions of the theorem, the power dissipated in the signal source is equal to that in the load. It is an appropriate condition for many low level circuits such as those found in receivers and the early stages of transmitters. It may also be a useful starting point for the design of some moderate power stages, but is unlikely to be the optimum operating condition for a final transmitter amplifier stage.

"To give just two simple examples that highlight the error: (a) What use would there be for a power station that used half the generated power to supply the surrounding district and the other half to heat the alternator? and (b) How can a class B amplifier have a theoretical maximum power transfer efficiency of $\pi/4$ times 100%, ie 78.54%, if half the signal power is in the load and half in the source?

"Practical power sources have to be used with load resistances larger than the source resistance to permit practical efficiencies. In the case of an RF power amplifier it is not even easy to specify what the output impedance actually is, since the device parameters, such as junction capacitance, vary during the signal cycle. However, in any well designed amplifier, running near full output, the average output resistance should certainly be less than the chosen optimum load resistance (often wrongly called the output resistance). This would account for the overall bandwidth of an amplifier and antenna system being rather less than twice the bandwidth of the antenna system alone. However, in measuring the bandwidth of an antenna system, one could perhaps do it more precisely and meaningfully in terms of its receiving response with a pure resistance load."

Many of these points are made also by Tony Preedy, G3LNP: "Anyone who still believes that the output stage of a transmitter should have an impedance equal to its load should visit a modern broadcast transmitting station, or even an electricity generating plant. If a power station were loaded to the point where the generators saw their own resistance as the load, they would be dissipating as much power as they were generating! The same with transmitter final stages. . . If the transmitter final stage were delivering its rated output power and the conjugate impedance theory were in operation, it could not be anything but 50% efficient. Modern AM broadcast transmitters, employing digital or switched mode amplfiers, have final stage efficiencies above 95%. The resistive component of the output impedance must in this case be only 2.5Ω referred to a 50Ω system. Any attempt to present the conjugate impedance would overload the associated power supply by a factor of 20 and tend to reduce the efficiency to 50%.

"Linear amplifiers have by definition an efficiency that is proportional to output power. If they don't, they are not linear! A theoretical class B linear amplifier achieving 71% efficiency at PEP has a source resistance of 20Ω, referred to a 50Ω system. At half PEP it will have an efficiency of 35.5% and a source resistance of 90Ω. The source resistance appears as additional damping when connected to the antenna (analogous to damping of loudspeakers by low efficiency audio amplifiers). If one used this amplifier with a high-Q antenna, matched to 50Ω, the system bandwidth would be lowest at PEP and would increase at lower output power. There will in this case be an intermediate power level where the source resistance is 50Ω, and only at this power level will the system bandwidth be twice that of the antenna. This can be easily verified by checking the 3dB bandwidth, ie the frequency separation between points where the load current falls to 0.707 of maximum, at different power levels for a transmitter driving a high Q load. This phenomenon was known to plague broadcast transmitter engineers in the 1920s, when they used class B linear amplifiers to drive high Q LF antennas. The bandwidth at high modulation levels was significantly less than at lower levels.

"The reason that antenna engineers like to stick with the maximum power transfer theorem and the conjugate impedance theory is, I believe, that without it they will have to specify feeder cables of much higher voltage ratings than they can economically justify. Given their own way, all transmitters would be interfaced to antennas via attenuators!

"For example, if, in a 50Ω system, the transmitter impedance were 50Ω, as dictated by the conjugate impedance theory, the maximum voltage that could appear on the feeder under fault condtions would be twice the working voltage. However, if the source impedance is 2.5Ω, as with a modern amplifier, the maximum fault voltage is 20 times the working voltage. Again, this can be easily verified by connecting a 2.5Ω generator (eg a 50Ω generator with 2.7Ω shunt) to an open quarter-wavelength of 50Ω feeder and comparing the voltage at the remote end with that at the generator.

"The maximum power transfer theorem is best confined to antenna research establishments where 50Ω signal generators are used to drive 50Ω antennas!"

THE DREADED Y2K BUG

ARE YOU WORRYING whether the dreaded 'Millennium Bug' will affect your amateur radio, computer and/or microprocessor systems? An article in *QST* (April 1999, pp56-58) by Karl F Anderson, W6JUA, with a major contribution by ARRL staffer Joe Bottiglieri, AA1GW, seems generally reassuring in regard to transceivers, although satellite tracking software and widely used logging programs are likely to need patches. The main problem for the home arises from appliances with imbedded controllers that involve 'knowing' what date it is. W6JUA counsels: "The best advice is to expect the least, but prepare for the worst. Don't become swept up in the Y2K hysteria, but make some calm, prudent preparations so that you'll be ready to cope with whatever may happen." Personally I find my PC isn't waiting for 1 January 2000 to kick me in the teeth. I lost the entire text of the first three-quarters (over 3000 words) of one month's *TT* and had to start all over again from scratch - the very real worry is that I still don't know just how or why it happened (yes, I always make good use of the 'save' facility), so it could all happen again at any time! Such a personal catastrophe never happened when I used pen and paper or a typewriter, or even with my obsolescent but reliable Amstrad WP!

YAGI ANTENNAS USING GENETIC ALGORITHMS

FOR OVER at least a decade it has been widely recognised by both professional and amateur antenna designers that a major design-aid has resulted from widespread use of the Numerical Electromagnetics Code (NEC) based on Harrison's Method of Moments (MoM), permitting computer analysis of projected designs while still in the planning stage. NEC determines the current distributions on each wire of a selected antenna design and from these, the radiation characteristics. NEC has been progressively refined with recent software packages, usually based on NEC2 or NEC-2D (Numerical Electro-magnetics Code, Version 2, Double Precision). Many antenna designs appearing these days in the amateur radio journals include radiation patterns etc, based on the use of NEC software on PCs, preferably verified in practice.

Since about 1997, a highly significant new development has been the use of Genetic Algorithms (GAs) in conjunction with NEC. This provides a revolutionary and extremely powerful new approach to antenna design, which seems certain to play an important role in the future, for both professional and amateur designs, including, particularly, multi-element Yagi arrays. The following notes are based largely on three recent papers: Two articles by Dr Richard Formato [WW1RF, former K1POO] in *Electronics World* 'Genes and Yagis' (August 1998, pp646-648) and 'Improving VHF Yagis' (June 1999, pp505-509); and a conference paper 'An Optimized Shaped Yagi-Uda Array Using The Genetic Algorithm' by Dr Brian A Austin [G0GSF/ZS6BKW] and Wen-Chung Liu, both of the University of Liverpool (National Conference on Antennas and Propagation, 30 March - 1 April 1999, IEE Conference Publication No 461, pp245-247).

The aim here is primarily to show how GAs can improve antennas, even for those of us who are computer illiterates, rather than to attempt to explain how or why GAs work and how to use them. WW1RF looks at this question in the first article by comparing four 12-element Yagi antennas designed using genetic algorithms. In effect, this is done by the computer searching a 'decision space', which is defined by specifying minimum and maximum values for each antenna parameter. He stresses that only antennas falling within these limits are allowable solutions to the optimum solution, a characteristic of GAs that gives an antenna designer exceptional flexibility.

The process is introduced in the Liverpool paper as follows: "Until recently, traditional techniques used the Method of Moments (MoM) to determine the current distributions on each wire of a selected antenna configuration and, from these, the radiation characteristics were calculated and compared with a target specification. The procedure is iterative and both experience and some appropriate candidate configurations are usually required to ensure success. By contrast, the GA is based on the processes of natural selection and the 'survival of the fittest' in an evolutionary system, which leads iteratively to the desired specification with minimal foresight or pre-conditioning on the part of the designer. In using the GA, all design parameters are encoded into a string of binary bits known, unsurprisingly, as a chromosome, while each parameter is represented by several bits - the genes. The genes are the basic building blocks of the GA and map directly to the associated real, physical properties of the antenna. The immense power of the technique is its ability to satisfy a performance criterion without any a prior knowledge of candidate configurations, and the facility for finding the global optimum result."

The paper shows how a simple three-

BATTERY VOLTAGE INDICATORS

A USEFUL LITTLE device for keeping a wary eye on the condition of a storage battery up to the 13.8V range is described by Donald G Varner, WB3CEH in *QST* (October, 1998, pp50-51): **Fig 3**. It requires only a single Maxim 8211CPA-2 programmable voltage detector, a multi-turn trimmer pot, a 100k fixed resistor and a red LED as a visual indicator. Optionally it can be fitted also with a Sonalert or similar audio alarm having a working range of the order of 6 to 28V DC and current drain of 3 to 14mA or thereabouts, with a toggle switch to select which indicator. Quiescent current is only about 20μA, rising to 14mA with the LED activated or about 5mA with the sound alarm. R2 adjusts the threshold voltage at which the alarm operates. WB3CEH concludes: "This approach provides flexibility and reliability when monitoring battery voltage levels. It can be used in a variety of applications where low voltage detection and indication are desired. It's an inexpensive project, too."

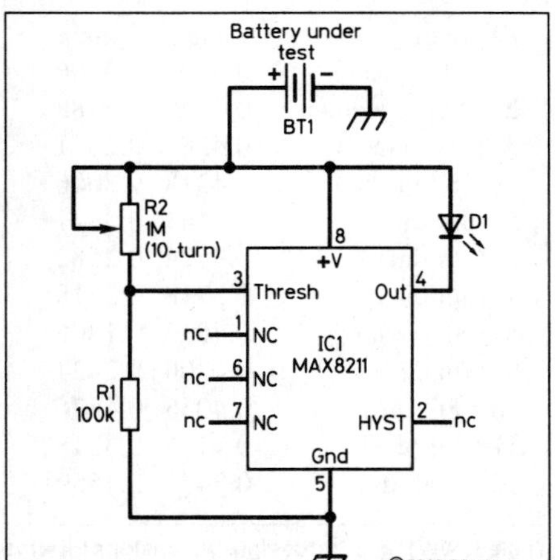

Fig 3: Circuit diagram of the low voltage indicator.

A low battery voltage detector suitable for use with 3V lithium backup batteries that draws a mere 0.3μA is described in the 'Circuit Ideas' feature of *Electronics World* (October, 1998) by Yongping Xia: **Fig 4**. This uses a Panasonic MN1381G in connection with a 74HC132. The MN1318G output is normally high, forcing IC1b to stay high. If the battery voltage falls below the threshold of 2.5V, the 1381 output goes low and allows IC1b to oscillate with a 5s period. Each negative-going edge is differentiated in C2/R2 and the output of IC1d drives the LED.

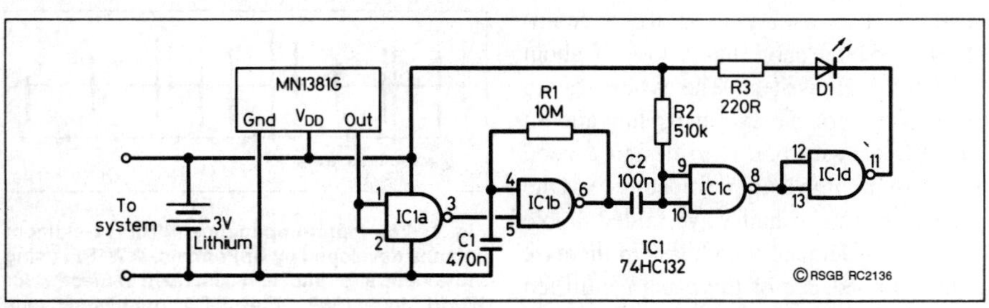

Fig 4: A low voltage indicator, suitable for 3V lithium batteries etc, normally taking under 0.3μA.

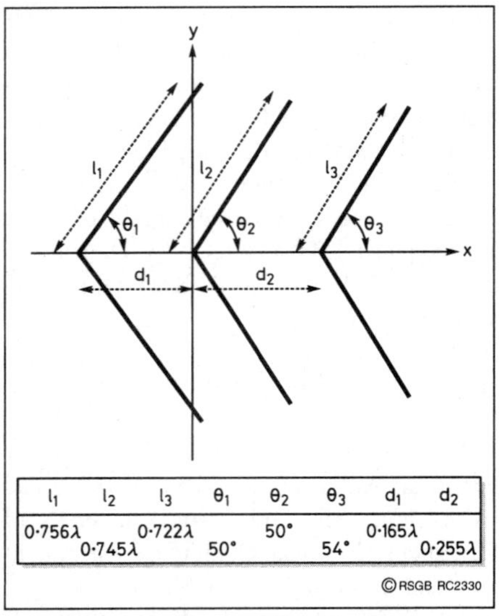

Fig 5: The optimised Vee-shaped Yagi-Uda antenna (L2 driven element) designed at the University of Liverpool by Brian Austin, G0GSF and Wen-Chung Liu, using the Genetic Algorithm. Computed gain of 12.1dBi. The linear elements of about 1.5λ do not present the constructional problems presented by curved elements of earlier 'optimum' designs, yet provide higher gain and directivity.

element Vee-shaped antenna with the elements roughly 1.5λ can achieve markedly better directivity than the elaborate element-curvature used for the Liang/Cheng (1983) or the Landstorfer (1977) antennas. The Landstorfer design with its curved 1.5λ elements has been described a number of times in *TT* (most recently June 1998, p59, Fig 7) in which it has been noted that a VHF antenna of this type, reported in 1982 by Dr Landstorfer, provided a forward gain of 11.5dBi, sidelobe attenuation better than 20dB and a front-to-back ratio of 26dB.

The Liverpool work with the GA produced an optimized array which would be relatively simple and straightforward to construct, at least for 50MHz and above (or possibly below 50MHz in the form of a fixed array). As shown in **Fig 5**, the GA-optimized array comprised three, linear elements with included angles of the order of 100°, in marked contrast to the elaborate element curvature that was found necessary in the Liang (11.8dBi) and Landstorfer (11.5dBi) arrays.

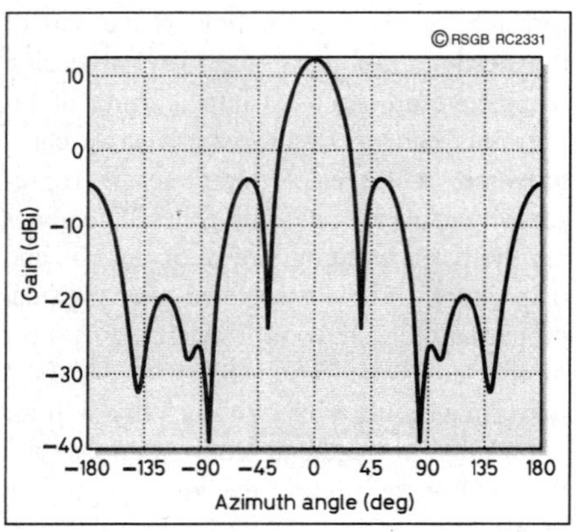

Fig 6: Radiation pattern of the optimized Vee-shaped Yagi-Uda using linear elements.

Fig 6 shows a plot of the gain against the azimuth angle of the Liverpool array, showing a maximum gain of 12.1dBi - some 0.6dB better than the Landstorfer design and 0.3dB better than that of Liang. The half-power beamwidth meets the specified 30°; the front-to-back ratio 16.6dB and front-to-side ratio 15.9dB. The computed input impedance is given as 20.4+j19.9Ω.

WW1RF's 'Genes and Yagis' article compares four 12-element Yagis designed using GA. He points out that each element in a Yagi-Uda array has three design parameters: length, spacing, and radius. The minimum/maximum range of each parameter can be set on an element-by-element basis, although the usual approach is to restrict all elements in the same way. For the four arrays described the range of elements (in terms of wavelength) were reflector 0.35 - 0.65; driven element 0.35-0.60; and directors 0.3-0.6; element spacing 0.05-0.5, limiting the longest boom to 5.5λ. For the first three designs, element radius was constant at 0.003369λ. His fourth design allowed the driven-element radii to range from 0.001 to 0. 0075λ. His first array was a sub-optimum design to illustrate that a GA does not produce a single 'best' design, but instead produces a group of designs, with each design ranked from best to worst. This feature, he claims, may be very useful, since even sub-optimal designs may be attractive in some instances.

His second design (**Fig 7** and **Table 1**) provided a maximum gain of 15.26dBi with an F/B ratio of 20.5dB and an input impedance of 19.8+j0.4Ω. Incidentally, his fourth design with slightly lower gain (14.53dBi) had an input impedance of 48.8-j0.1Ω, resulting in an SWR of only 1.02 when fed with 50Ω cable. Number 3 was gain maximised (15.86dBi) but had an input impedance of 5.3+j179.3Ω, making it unsuitable for use with a coaxial feeder.

In his later article 'Improving VHF Yagis', WW1RF uses GA to design a 12-element VHF Yagi antenna suitable for use with a 300Ω open-wire transmission line. He emphasises that high-impedance Yagis can improve VHF antenna system performance, because of the much lower attenuation of open-wire line (typically 0.25dB/100ft at 144MHz). The widely used RG-8 foam-dielectric coaxial cable has a loss of about 2.1dB/100ft). He writes: "The easiest way to take advantage of the extremely low attenuation of open wire line is to design a good Yagi with a high input impedance. One approach is to use a half-wave folded dipole as the driven element. Another is to increase the input impedance of the usual centre-fed linear dipole driven element by proper placement of the array's parasitic elements. His

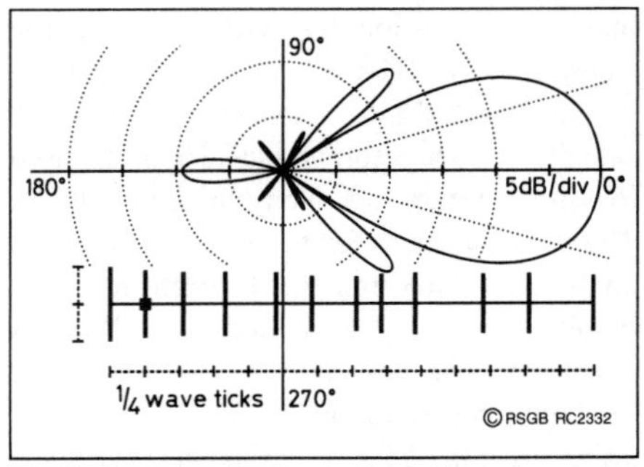

Fig 7: One of the four VHF 12-element antennas developed by Dr Formato, WW1RF, using the genetic algorithm. See text and Table 1.

design matches 300Ω line. The positions, lengths and spacings of the 'directors' are certainly unconventional: **Fig 8**. For full details refer to *EW*, June 1999.

Yagi Genetic Optimizer version 2 used by WW1RF is a freeware program, available on the web (Ray Anderson, WB6TPU's 'NEC Archive' web site, URL: http://www.qsl.net/wb6tpu). This uses NEC-2D, also available on the web or directly from the Applied Computional Electromagnetics Society (ACES) Attn; Dr Richard W Adler, ACES Executive Officer, ECDE Dept., Code ECAB, Naval Postgraduate School, 833 Dyer Road, Room 437, Monterey, CA 93943-5121, USA.

One thing seems certain, genetic algorithms are going to play an important role in future antenna designs and may result in some unexpected variations on traditional designs. And at least genetically-modified antennas present no new health or environmental hazards! ♦

Element	Length	Spacing
1 (reflector)	0.5006	0.0000
2 (driven element)	0.4549	0.2388
3 (first director)	0.4388	0.2671
4 (2nd dir.)	0.4318	0.3006
5 (3rd dir.)	0.4141	0.3641
6 (4th dir.)	0.3624	0.2582
7 (5th dir.)	0.3718	0.3218
8 (6th dir.)	0.4012	0.1806
9 (7th dir.)	0.4200	0.2424
10 (8th dir.)	0.4035	0.4876
11 (9th dir.)	0.4176	0.3218
12 (10th dir.)	0.3471	0.4559

Table 1: WW1RF's 2nd design. Dimensions in terms of wavelength.

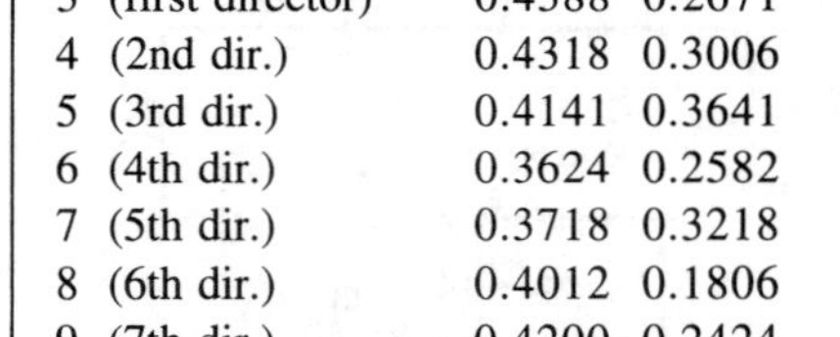

Fig 8: The 'high-impedance' VHF 12-element antenna developed by Dr Formato, WW1RF, using the genetic algorithm and described in *Electronics World*, June 1999. The 300Ω input impedance enables the array to be fed from low-loss open-wire line.

by Pat Hawker, G3VA*

xDSL AND RFI

THE POTENTIALLY serious problems presented by the development of Power Line Telecommunications [now cancelled - *Ed*] have been well covered in *RadCom*'s News and EMC columns, but there is less awareness of the equally serious and perhaps more imminent difficulties likely to arise from the introduction by the telecommunications industry of the family of high speed data services sent over the unshielded twisted copper wire pairs that are used for the 'plain ordinary telephone service' (POTS). The announcement at the end of July that British Telecom is to compete head on with the cable companies, by introducing technology capable of carrying 40 times the digital information conventionally carried on standard telephone lines, is a clear indication that we are on the verge of meeting the sort of RFI problems associated with PLT but which may build up more widely and more quickly. Many amateurs will have already experienced the problem of interfering with their own or their neighbours POTS, but there seems reason to believe that high speed digital signals carried on standard telephone lines will represent the additional problem of HF noise radiation affecting weak signal reception.

David Williams, G3CCO, recently drew my attention to a long article 'High-speed copper access: a tutorial overview', by I K Czajkowski, in the IEE's *Electronics & Communication Engineering Journal* (June 1999, pp125-148). This points out that the massive growth in the use of Internet has produced a requirement for fast and faster access to residential customers. To download a file of 10Mbytes (eg a 1.5 minute video clip encoded in MPEG-1) would take 46 minutes with a 28.8 kbit/s modem; 23 minutes with a 56.6 kbit/s modem; about 10 minutes on an ISDN line at 128 kbit/s; but under one minute with 'ADSL-lite' at about 1.5 Mbit/s; about 13 seconds with 'ADSL' at 6Mbit/s; and only 3 seconds with 'VDSL' at, say, 26 Mbit/s.

ADSL (Asymmetric Digital Subscriber Line) is a new service already standardised internationally and coming into use in North America and Europe. It can provide an upstream data rate of up to 1Mbit/s and a downstream data rate of up to 8Mbit/s, carrying the data in the spectral range of 25kHz to 1.1MHz.

VDSL (Very high speed Digital Subscriber Line) is due to be standardised this year or next [see also *EMC* column, October 1996 - *Ed*]. It will be capable of being operated in either the asymmetric or symmetric modes. For example, a downstream mode of up to 50Mbit/s and upstream of up to 2Mbit/s (asymmetric) or downstream and upstream each up to 26Mbit/s (symmetric). In this case the data signals occupy the spectrum 300kHz up to more than 10MHz (maximum 30MHz).

ADSL and VDSL are part of the family of high-speed telecommunication data services known collectively as xDSL. Both can be delivered to the home over an existing pair of telephone wires, with the first 4kHz of spectrum continuing to carry the normal telephony (POTS). Clearly the distance over which MF or HF signals can be carried over untwisted telephone pairs is limited, but the systems are designed to overlay the existing broadband optical fibre networks without incurring the high cost of running fibres all the way into homes. The xDSL technologies can deliver high-speed data over reaches of 4-5km, around 1.5km and less than 300m, depending on the architecture selected. The important point from the viewpoint of the radio amateur is that the entry into the homes is simply on untwisted copper wire pairs which can both pick-up and radiate signals.

In a section devoted to 'RFI issues', I K Czajkowski considers the problems both of *ingress* (ie signals from a local transmitter affecting the data transmission) and *egress* (ie RF noise radiation from the system affecting local weak-signal reception). Strong MW broadcast (AM) transmissions are seen the main ingress problem in the USA, where transmitters are often sited in town centres. This is uncommon in the UK, although I well remember when the IBA opened its London MW transmitters at Saffron Green, near Barnet, there were a number of complaints from north-west London that the programmes could be heard when using the telephone. It is claimed that the xDSL systems have been designed to operate in the presence of strong AM broadcast transmissions, while RF leakage from xDSL should not interfere with the reception of strong broadcast systems. It is stressed that xDSL systems must conform to EMC requirements and "cause no interference to other licensed users of the RF spectrum", though I doubt whether this would be held to include interference to weak DX-signal reception!

The article continues: "The consideration of radio amateur transmissions requires very different treatment from the AM broadcast situation. The radio amateur community is a large one, whose members are very experienced users of the spectrum allocated to them, comprising a number of separate bands, many of which fall in the HF spectrum under 30MHz. Radio amateur transmissions are not a relevant consideration for ADSL [this seems an unduly optimistic claim, since amateur transmissions can affect POTS and some modems! - G3VA] but must be considered for VDSL... Unlike AM broadcast transmitters, they are distributed among the population. . . The interference sources (the amateur antennas) are close to telephone network drop-wires. Because radio amateurs often listen to very weak signals, the issue of egress must be taken as seriously as ingress. Radio amateurs move house... the interference environment is neither a predictable nor a constant problem in any particular area....[Amateurs] use SSB and there is no carrier to lock on to, and the presence of an amateur

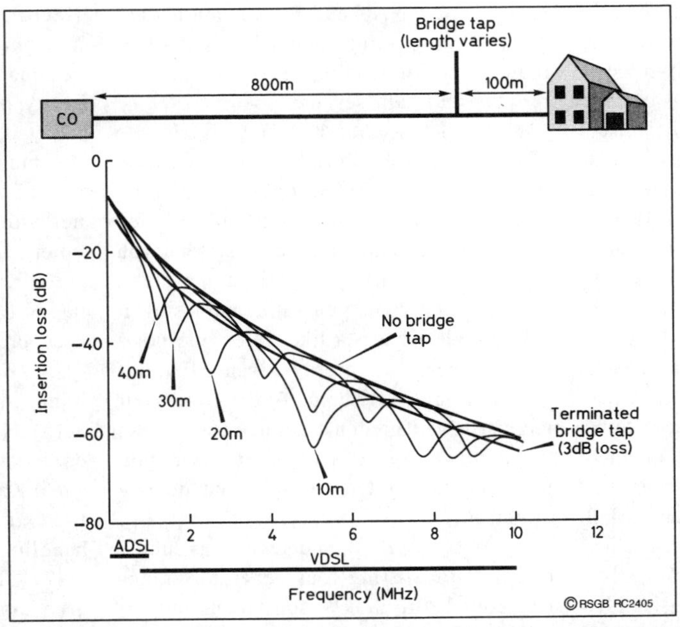

Fig 1: Simulated insertion loss as a function of frequency for an American National Standards Institute (ANSI) test subscriber loop incorporating a bridge tap to provide an extra 20dB protection on 7MHz, 10MHz, 14MHz and 28MHz with appropriate bridge taps as filters. Also shown is the insertion loss, assuming termination of the bridge taps (3dB loss). Diagram of the test-loop is shown at the top of the figure ('CO' is central office, ie local exchange). The spectral allocation to ADSL and VDSL is indicated at the bottom of the figure. (From *Electronics & Communication Engineering Journal*)

*37 Dovercourt Road, London SE22 8SS.

cannot be detected except while transmitting. Additionally, amateurs tend to have unusual and somewhat unique usage patterns. They are very frequency agile, often jumping between bands, and to complicate matters are often listening rather than transmitting.

"ETSI (European Telecommunications Standards Institute) is proposing a noise model for VDSL including a radio amateur interferer at 0dBm differential mode signal. Within ANSI (American National Standards Institute) signals as high as -5dBm have been discussed.... It is the careful consideration of the possible risks of egress interference in the amateur bands that has led to an inclusion of notching of the transmitted power spectral density in the draft standards requirements. VDSL systems will be required not to transmit powers greater than -80dBm/Hz in the amateur bands, whilst transmitting at up to -60dBm/Hz - which is still about 20dB less than PLT, and better balanced - in other parts of the spectrum. [These 20dB notches are thus similar to those proposed for PLT and while welcome are hardly likely to placate the DX community - G3VA]. **Fig 1** shows the simulated insertion loss (as a function of frequency for the more demanding ANSI test loop incorporating a bridge tap, showing the result for 10m, 20m, 30m and 40m bridge taps together with the insertion loss assuming termination of the bridge taps. The diagram also shows the result of omitting the bridge tap filters.

On the question of ingress RFI, the article reports the results of modelling the risk from ingress interference from radio amateurs in Raleigh, North Carolina. This showed the proportion of homes at risk of -5dBm worst case interference, assuming a 30dB balance in the amateur bands (or -25dBm interference assuming 50dB cable balance). Whilst the proportion of homes at risk in the inner city was found to be small (less than 1%), in the suburban area the risk is much more significant (nearly 10%). While these results should not be directly transferred to a UK situation, it does show "the possibility of amateur transmissions resulting in a sizeable interference level on an access line is a real one, and that VDSL systems will need to be completely robust in the presence of amateur radio signals." With different manufacturers and different levels of installation skill there must remain serious doubts whether "complete robustness" will be achieved!

The potential problem is emphasised by the comments: "By chance, the interference threat from/to CT1 cordless telephones was identified as an issue for VDSL. CT1 base stations in the UK transmit to the handset in the 1.6 to 1.8MHz band, and the base stations and the antennas are typically sited very near the entry point of telephone cables into the house.

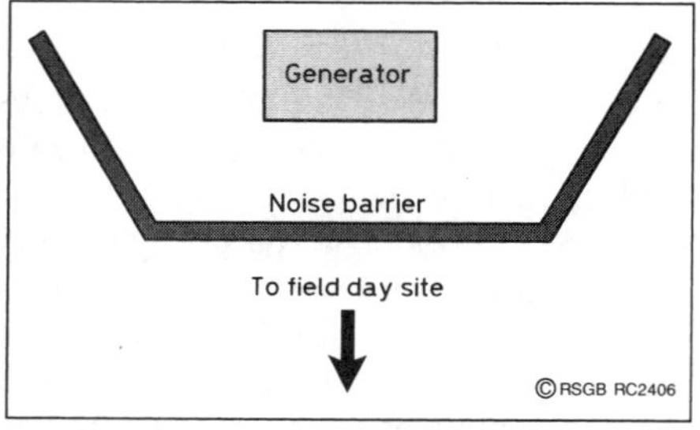

Fig 2: NT0Z suggests that a two- or three-sided "room-divider" noise shield is a useful adjunct when using a portable or emergency power generator. Such noise shields can be constructed from heavy cardboard or carpet-covered plywood, preferably with hinges. Never attempt to "box-in" a noisy generator in a covered or four-sided "box"! (*QST*).

We have measured interference levels in the laboratory from CT1 greater than those from AM broadcast stations. This example highlights the requirement for the VDSL line code to be flexible in dynamically notching out regions of spectrum with unforeseen RFI."

FIELD, EMERGENCY & STANDBY POWER GENERATORS

DESPITE THE STEADY progress in solar arrays, wind generators and the potential of fuel cells, the majority of field operation of amateur stations still depends, as it has done for over 50 years, on the use of an engine-driven portable power generator (PPG) fuelled by petrol or diesel (although paraffin [*US* kerosene], propane and natural gas PPGs are also available in the USA at higher cost). My own wartime experiences using ONAN and Tiny Tim units left me with an abiding realisation that PPGs require careful handling if they are to give reliable and safe service.

Kirk A Kleinschmidt, NT0Z, in 'How to Choose and Use a Portable Power Generator' (*QST*, June 1999, pp59-61) emphasises that if you're operating "off the grid" a portable generator may be just what you need. He provides practical advice about choosing the right generator and using it safely: "Generators - like all engine-powered devices - can injure or even kill you if you don't respect them. And unlike your engine-driven garden tractor, mower, etc, these powerhouses can *electrocute* you (or others). Don't be afraid - but do pay attention."

NT0Z considers such questions as choice of generator, listing four generators ranging from 1.4 to 3.5kW continuous output, calculating the capacity (power required), size and weight, engines and fuel, run time, noise, voltage and frequency regulation, DC output for charging batteries, etc. Under the heading of 'Using your generator' he includes the following advice: "Regardless of the earthing method you choose, a few electrical safety rules remain the same. Your extension cords *must* have intact, waterproof insulation, three 'prongs' and three wires, and must be sized according to loads and cable runs. Use 14-16 gauge, three-wire extension cords for low-wattage runs of 100ft or less. For high-wattage loads, use heavier 12-gauge, three-wire cords designed for air compressors, air conditioners or RV service feeds. If you use long extension cords to power heavy loads, you may damage your generator and/or your radio gear. When it comes to power cords, think *big*. Try to position extension cords so they won't be tripped over or run over by vehicles. And don't run electrical cords through standing water or over wet, sloppy terrain."

Among his handy tips are to keep the generator reasonably stable with a varying load (keying a transmitter for example) by running continuously a small load such as two light bulbs as a constant load to reduce power and voltage swings. After use, consider adding a small amount of petrol stabiliser to keep the fuel from oxidising and gumming-up the carburettor.

He admits that engine noise can be disturbing: "Set up the PPG in an out-of-the-way area and make a two- or three-sided sound shield from carpet-covered plywood or stiff cardboard (these look like small, folding room dividers): **Fig 2**. Keep the sound absorber/reflector between you and the generator. Do *not* make a four-sided shield or put the generator into any type of box. Generators need airflow to keep cool.

Some 12 years ago, Wendell Tulencik, K8OIP, (*TT*, November 1987 or see *TT Scrapbook, 1985-89* p200) provided ten commandments for operating an emergency or field-day installation:

(1) Use only the fuel recommended by the manufacturer.

(2) Pour fuel through a large-mouth funnel with a fine screen to filter out dirt and other contaminants.

(3) Keep waste cloth or paper towelling handy for blotting spills. Store properly in a covered metal waste bin.

(4) Keep a supply of lubricating oil handy.

(5) Have some 50 or 100ft extension cords available.

(6) Keep at least one CO2 or dry-powder fire extinguisher ready for instant use when handling fuel.

(7) A torch with good batteries is a must. Two torches are better than one!

(8) Check fluid levels and start the alternator at least once a week.

(9) If you have trees close to the house, keep a small chain-saw handy.

(10) Use a small trickle charge to keep the (starting) battery charged.

K8OIP stressed the need to provide safe storage of fuel and warned of the dangers

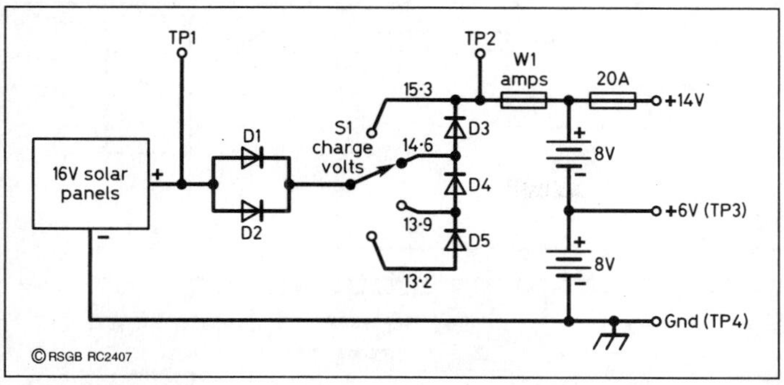

Fig 3: K5CN proposes this "KISS" form of an emergency or field solar-charging system. D1 to D5 are 6A, 50 PIV diodes. By using one 8V and one 6V vehicle battery, and power diodes as voltage-dropping 'resistors' there need be little of the solar power wasted. Charging may be done simultaneously with use of the radio equipment. W1 is a 20A switching circuit breaker and separately fused two-pin polarised sockets and a cigar lighter socket complete the output power access. (*QST*)

of fume and engine heat, pointing out that the exhaust fumes of petrol, diesel, natural gas or propane engines are all lethal and must be properly vented out of any enclosure without relying on natural ventilation.

A month or two later, C J Chapman added as 11th and 12th Commandments: "When you *have* spilled fuel over the generator, allow it to dry thoroughly *before* starting it up! Petrol/oil mixtures take quite a while to dry thoroughly. The twelfth commandment is to check the output voltage *before* connecting your radio equipment. Governors have been known to go wrong."

CHARGING BATTERIES FROM SOLAR CELLS

BOB WILKINSON, G3VVT, while on leave from the Middle East read with some interest the June *TT* item from G4PEY on the use of a step-up inverter and battery-voltage regulator in conjunction with a 17W solar array.

He writes: "In the Middle East I work as a communication technician at remote sites, few of which have public power. So the generation and provision of electrical power has always been a crucial item. Solar power has long been used for smaller power loads but, of late, the trend has been to provide even quite high power requirements in this way. 15kW solar arrays are now found being used to run completely a microwave site, including emergency air conditioning, etc.

"With daily exposure to the problems of providing power from solar sources, I found G4PEY's solution, while very elegant in providing a regulated source of 13.6V DC, still flawed as an efficient means of providing power from solar arrays. The major drawback is that his panel provides only 17W *maximum* output. During an average day, there are normally only 8 hours (or less in the UK) of usable sunlight. Thus his array would provide only 8 x 17 = 136W per day. Of this some 20% is being wasted in the voltage upconverter, reducing the supply to some 108W per day available to charge the batteries. However, of even greater significance is the voltage regulator between the batteries and load. At the worst case of the batteries being floated at 20.25V DC (the correct figure for gel/maintenance-free batteries at 2.25V per cell) the power that has to be dissipated in dropping that to 13.6V DC at 20A is a staggering *133 watts!*

"In practice, these days, all solar panels are fitted with blocking diodes to prevent the battery discharging back through the panel. With simpler systems, the solar array is formed from a group of panels in parallel to raise the power capability, and this is fed to a regulator(s) to control the battery charging. The regulator normally has a relay or contactor to feed the array current straight through on full charge to avoid power loss, then switching to a regulated feed at the nominal float voltage once the initial daily equalise period of the battery has been passed. With the more complex systems with multiple arrays the optimum method is to balance the power requirements after equalising the battery by reducing the number of arrays on line to provide the required float voltage. During the equalise period, series diodes are automatically added to the load feed to drop the voltage to safe levels (CEMF diodes). In the long term this method is more reliable than a series regulator, as it does not waste power or add unwanted heat.

"I feel that G4PEY is missing the point of efficient power conservation. The system might also be compared to the average mobile installation, where voltage variations of 12.0 to 14.0V DC are handled without any serious problems arising."

By coincidence, some of the same points are made in an item 'The Solar KISS', a contribution by Bill Edwards, K5CN, to the 'Hints & Kinks' column of the June *QST*. K5CN points out that one of the few negative considerations about using solar power is the loss inherent in dropping the typical 16V panel output to a safe value for equipment intended to operate at 13.8V. (KU7G, the column's editor, points out that most mobile equipment allows a +/-10% tolerance, ie operating from 12.4 to 15.2V). Charging some (nominal) 12V batteries at the full panel voltage output can be injurious, just as applying 16V to some equipment may damage it."

To avoid the hassle and cost of a voltage regulator, K5CN devised a simple system for his two solar panels producing 16V at up to 4A. He made a battery of 14V by series connecting 8V and 6V units, pointing out that suppliers now make 8V batteries to replace the 6V batteries used in older cars and using an isolating diode to protect the panels from reverse current that could flow if they were connected directly to the batteries and the panel voltage fell below that of the battery - a common occurrence in view of the dependence of solar cells on the strength of the sun. Using two 6A 50PIV power diodes in parallel as the isolator there will be a voltage drop of 0.7V, reducing the output to 15.3V which is only just above the recommended maximum charging rate for a 14V battery. As shown in **Fig 3** he has added three more diodes and a rotary switch to provide outputs at 14.6, 13.9 and 13.2V. W1 is a circuit breaker or fuse to protect the diodes and the 14V output from the batteries is fused at 20A or as required. Test points TP1 and TP2 allow the voltages to be monitored. This KISS arrangement may appear crude, but for typical amateur applications the maximum power loss (at 16V 4A, ie 64W from the solar panels) is only 2.8W.

ANTENNA CURRENT MONITORING

FOR MANY YEARS I have found it useful to monitor the RF current flowing to the antenna. Originally I used ex Air Ministry

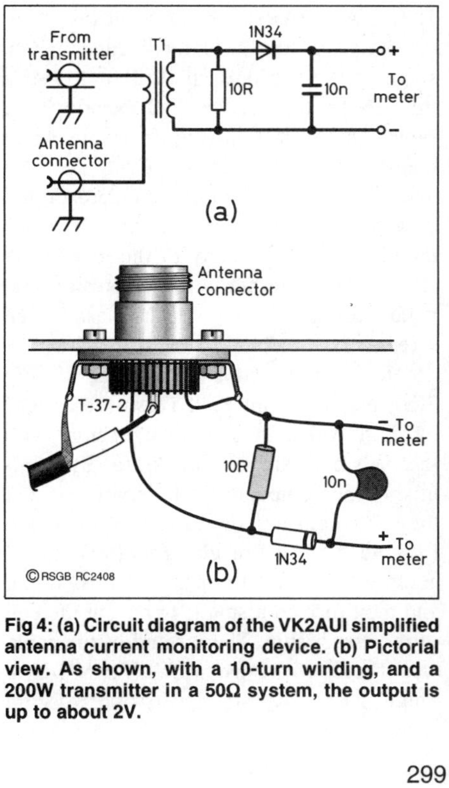

Fig 4: (a) Circuit diagram of the VK2AUI simplified antenna current monitoring device. (b) Pictorial view. As shown, with a 10-turn winding, and a 200W transmitter in a 50Ω system, the output is up to about 2V.

thermocouple ammeters, but soon found these very vulnerable to overloads. Since then I have tended to make use of small pilot lamps (torch bulbs) shunted by short lengths of wire. This works fine, but has the problem that the impedance of the shunt is frequency conscious and it is seldom satisfactory to use the same shunt on all HF bands. In the 'Hints & Kinks' column of the June 1999 *QST,* John S Innes, VK2AUI, revises his earlier system for simplified antenna current monitoring: **Fig 4**. This uses the established technique of a toroid used as a current transformer, utilising the fact that a wire passing through the toroid acts as a one-turn 'winding'.

VK2AUL writes: "The value and power rating of the load resistor depend on the current monitored and the voltage needed for the diode rectifier and meter. Let us assume the antenna current is 2A for a 200W transmitter in a 50Ω system. Practical values might be ten turns for the transformer secondary and a 10Ω load resistor across it. Because of the turns ratio, one-tenth of 2A will flow in the secondary circuit (in this example 0.2A through the 10Ω resistor produces 2V, a convenient value for the meter circuit). More turns produce a *lower* voltage because they divide the current by a larger ratio from the 'one-turn' primary. For these values the resistor dissipate 0.2 x 2 watts, ie 0.4W, a higher value of resistor gives more voltage for the meter but dissipates more power. . . the circuit is a very practical, useful aid to mobile work [in my opinion antenna current monitoring should prove equally useful with fixed, mobile or portable installations - *G3VA*].

RF VOLTAGE PROBE

JOHN A SHARE, G3OKA writes: "Recently, I had the need to discover which stage in the transmit chain of a 144MHz multimode transceiver had developed a fault. The logical procedure would be to measure the RF voltages along the path using an RF probe with a standard digital or analogue multimeter. A suitable probe circuit is given in many of the handbooks (**Fig 5(a)**) together with constructional details using a valve holder and associated screen can.

"However, my construction method (**Fig 5(b)**) may be of interest. This used a piece of single-sided PCB 7mm by 32mm with the pads cut using a craft knife. By minimising the component lead lengths so that they fitted close to the surface of the board I found that the resulting unit could be enclosed by a length of heat shrink tubing and then formed a small 'lump' on the end of the coax cable. The far end of the cable was fitted with two 4mm plugs which plugged directly into my low-cost digital multimeter (DVM).

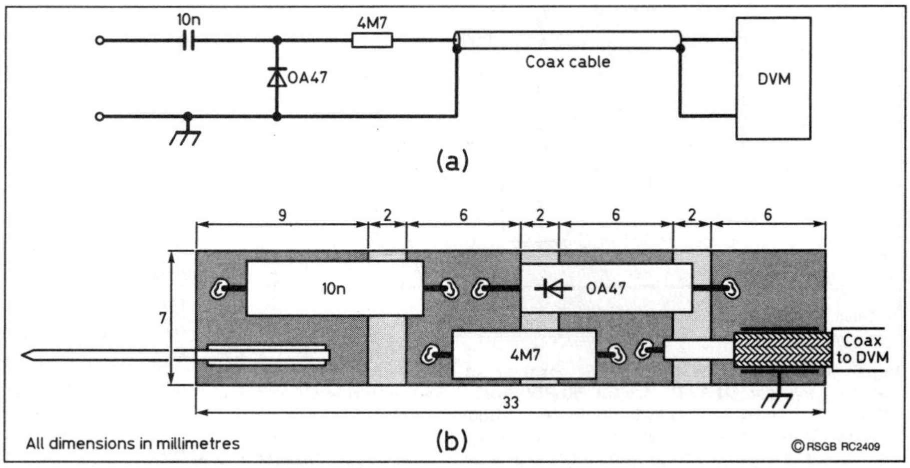

Fig 5: G3OKA's RF voltmeter probe for use up to about 150MHz. It can be enclosed in a length of heat shrink tubing. (a) Circuit diagram. (b) Component layout.

"I had anticipated that the detector diode would need to be an exotic VHF device, but I found that the unit functioned quite satisfactorily up to 150MHz using a general purpose OA47 detector diode. The little add-on RF probe now resides in the multimeter box."

HERE & THERE

GERRY WHITEHEAD, G2ACZ, reports an interesting outcome to the item on the Air Ministry 'No 1969' Morse key incorporating a spark gap, described and illustrated in the July *TT* (pp 54-55). He writes: "I was intrigued to see the reference to this key as I have had one for many years and never been quite sure of its true origins and age. After making an initial enquiry via RAF Cosford I was referred to the RAF Museum at Hendon, and subsequently received the following information which may be of interest to others:

"The key first appears in lists of Wireless Stores in June 1919 and it is assumed that it replaced the wartime '876' key sometime after July 1918 and saw service during the latter part of WW1 and the immediate post-war period. The key would seem to have been associated with the type '52' series of Spark transmitters which were used on large flying boats, small airships, and DH6 aeroplanes, and additionally for artillery co-operation. By AP1086 (1925 edition), the key is no longer listed."

G2ACZ continues: "The RAF Museum at Hendon advised me that they did not hold a type '1969' key. Therefore I have been quite happy to donate mine to the Museum."

Frank Cooper, G2QT, bought one of these keys in 1932 from Elextradix for about 2s 6d (12.5p) and used it for many years (it is still in his shack). He writes: "To me the big attraction was the back contacts. I used an HRO and separate finals for some years. Coupled to each tank coil was a coil in series, via a high-speed relay, with the antenna coil of the HRO. The back contacts keyed the 24V supply to the relay. This gave me full break-in. . . On my last visit to the Science Museum the key was on display as an early Morse key. . . I've had a paddle for some 10 years but make too many mistakes and still use the straight key! I envy the fellows bashing away at 25-30WPM on their paddles."

IAN WHITE, G3SEK, passes along the news that the August *TT* item on short meander dipoles and monopoles has attracted the attention of Rick Littlefield, K1BQT, a frequent contributor to *Communications Quarterly,* who has started an Internet discussion on the topic, has already built an experimental model of the 3ft 14MHz dipole and has contacted Tom Warnagiris, one of the two authors of the *IEEE Trans* paper. There is space this month for only a few comments: "Tom Warnagiris tried the 14MHz dipole described in the paper on the ham band. . . made some contacts. . . got good signal reports. . . people didn't believe it was a 3ft dipole." The work has been patented, but SWRI have no objection whatever to ham experimentation (US tax-payer money paid for the project), but due to other commitments would not welcome further enquiries. Tom sticks by the efficiency and bandwidth claims as published in *TT*, but feels that the primary application for this type of antenna might be for mobile or airborne communication. "It's complicated to make and easily detuned by surrounding objects in the near field. . . if you can't get it high and in the clear it would be better to use a conventional dipole." Unfortunately, practical trials by K1BGT and others, supported by NEC simulations, does not confirm the efficiency and bandwidth claims made by the SWRI authors in *IEE Trans*. The 1m dipole, constructed according to SWRI paper, appears about 15dB down on a reference full size dipole. Even so, me-

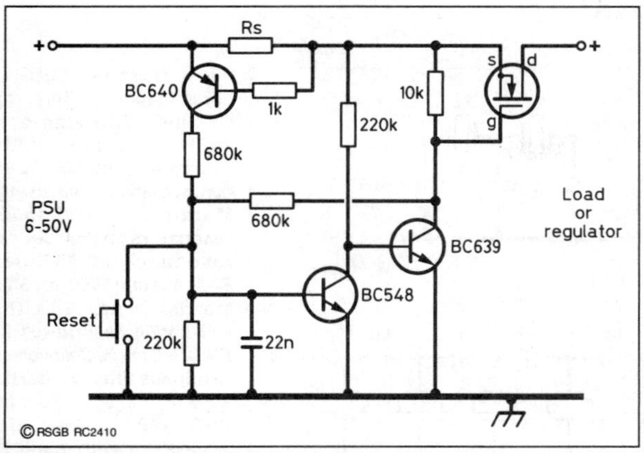

Fig 6: G3DXZ's electronic fuse.

ander antennas seem an interesting approach.

A SHORT NOTE from Ray Herbert, G2KU, comments: "Apropos the item in the August *TT* on the meander-type 1.7MHz attic antenna described by G2QM in 1939, I can confirm that the non-inductive zigzag 'loading-coil' really works. I have one in the loft." Ray is editor of the *Baird Television Newsletter* and as a pre-war staff-member of the company has done much to keep alive interest in John Logie Baird as a pioneer of mechanical television, his later work on airborne TV and his wartime work on electronic colour TV. He was also instrumental in ensuring that at least some recognition was given to the role of the Voluntary Interceptors (Box 25) and SCU3 operators in the recent IEE 'Secret Machines - British Code-Breaking in World War Two' exhibition at Savoy Place, London.

JEFF HARRIS, G3LWM, as a "proud" owner of an IC-706 Mk2 transceiver was interested in the July *TT* item 'Audio Gain & the IC-706 Mk2' in which G4TRN showed how the SSB audio gain could be increased in order to avoid the need to shout when using a microphone providing less than the specified 10mV output. However, he felt it advisable to obtain the views of the Icom Service Support department before re-adjusting his rig.

Chris Ridley of Icom replied: "The adjustment described will have the effect of increasing the microphone gain but care should be taken as too much gain will cause overloading and distortion. The levels used by Icom to set the 'Tx Gain' control are chosen to ensure that the radio will meet specification under all input conditions. After adjustment great care should be taken if other microphones or pre-amplifiers are used. *Also note that surface-mounted potentiometers will only take a limited number of adjustments before they 'self-destruct'* (emphasis added by G3LWM)."

In the outcome, G3LWM decided not to make the G4TRN adjustments and feels that the Icom cautionary comments should be passed to others using rigs with surface-mounted potentiometers. He adds: "Give me a nice old pot any day." The old pots may have had the tendency to become 'noisy' and need an occasional squirt of switch cleaner fluid, but you could go on adjusting them for years!

ELECTRONIC FUSE Mk2

CHAS FLETCHER, G3DXZ, feels that it was unwise to recommend the electronic fuse shown in the July *TT* to readers, who may have the impression that it is a surefire arrangement. He fears that it could prove unsatisfactory in practical applications than would appear to those who have not had professional experience of zener-controlled systems which can prove tricky to use.

He writes: "The circuit (Fig 2 of July *TT*, page 53) is basically a zener regulated supply, with TR1 and TR2 increasing the current supply capability. Zener regulators are not the easiest things to design and the value of the zener series resistor (1k in this example) needs to be calculated for each application since it is dependent upon the supply voltage/output voltage relationship and the current gain of the power transistors. To get it right is not at all easy. Second, the TR4/TR5 pair are not connected as a thyristor, as stated in the *Electronics World* description. If they were, an overload would reduce the output to zero volts. As it is, the output voltage will collapse to the point where the output current stabilises at the overload limit, or else the output volts at roughly half the input voltage. I am also puzzled why a 1μF capacitor is needed across the zener diode (perhaps the original application was prone to spikes of overload voltage?)."

G3DXZ adds: "IC regulators, although generally proof against short circuits and overheating, unfortunately draw nearly full load current from their supply when overloaded; so, if one wishes to guard against this problem and still wants a fast DC circuit breaker, then a MOSFET probably offers the best series element currently available. **Fig 6** shows a sample circuit arrangement which is intended for inclusion in a supply line liable to overload. If regulation is required, this circuit should be inserted between the PSU and the regulator. Its advantage is that under overload conditions it will reduce the drain on the PSU to virtually zero until the RESET button is pressed. Interestingly, if the highest speed of disconnection is not needed, you still cannot beat an assisted relay!"

In this DC interrupter arrangement the switch will work efficiently from 6 to 50V without component changes in the overload detector circuitry. Rs is set in just the same way as for the July arrangement. The MOSFET is selected according to the maximum supply voltage and load current: for example, a BUZ271 on a small heatsink will handle 20A at up to 50V. It will start in the 'on' state and is reasonably proof against transients.

END OF THE SILICON ROAD?

THE SHRINKAGE OF silicon-based chips continues to take place, with ever more transistors on a single chip.

In 1970, when the first LSI (Large Scale Integrated) devices appeared, some 100 to 5000 components were squeezed on a chip. Today the 64 Mbit memory chip contains some 64,000 times as many components as the first 1 kbit devices. This steady increase in components with corresponding decrease in the dimensions of the oxide gates are said to follow 'Moore's Law'. Gordon Moore in 1965 predicted that for each new generation of memory chip and microprocessor unit on the market, the device size would reduce by 33%, the chip size increase by 50%, and the number of components on a chip would quadruple every three years. This has proved an accurate prediction and still shows no sign of ending.

But, as noted by Max Schulktz in 'The end of the road for silicon?' (*Nature,* 24 June 1999, pp729-730) the discovery that a layer of silicon dioxide must be at least four to five atoms thick to function as an insulator suggests that silicon-based microchips will reach the physical limits of miniaturisation by about 2012. In 1997, a gate oxide of a microchip MOSFET had thinned down to about 25 silicon atoms thick, and it is now possible in the laboratory to produce gate oxides only ten silicon atoms thick. By about 2012 Moores Law predicts that gate oxides may be about five silicon atoms thick (1.3 nanometres), and thus reach the physical limits of micro-miniaturisation of silicon. He concludes "For the next 12 years, microelectronic circuits based on silicon technology will remain the basis for further 'electronisation' of our world . . . Silicon technology is still the most reliable and cost-efficient way to fabricate large microelectronic circuits. It is almost unthinkable that we are nearing the end of the silicon era, but in the meantime silicon chips will be further enhanced. Eventually science and industry will have to find new ways to build faster and larger computers." ♦

technical topics

by Pat Hawker, G3VA*

CRYSTAL & CERAMIC FILTER MISCELLANY

HOME-BREWING of IF and signal-frequency crystal or ceramic ladder filters is still an area of amateur radio which can bring substantial economic savings as well as high performance. While even simple filters can be better than nothing, a good filter requires a significant degree of expertise in order to achieve a specified response curve. A useful new guide to the design and construction of crystal ladder filters has been written by John Pivnichny, N2DCH, and published by MFJ Publishing (listed in the 1999 MFJ Group products catalogue distributed by Waters & Stanton, although not found in the associated UK price list).

N2DCH has kindly sent me a copy of his book, in which he has attempted to bring together for the first time almost everything that has been published in American and European journals on ladder crystal filters, including the early work in France, where the first notes for amateurs was published in *Radio-REF* by J Pochet, F6BQP, reported in *TT,* September 1976, followed shortly after by the first of the comprehensive articles by J A Hardcastle, G3JIR, *Radio Communication,* December 1976 (republished in *QST,* December 1978). N2DCH's book provides a list of 40 references to relevant material in professional and amateur journals and books. His own interest was stimulated by Wes Hayward's 'Designing and Building Simple Crystal Filters', *QST,* July 1987, and he has subsequently written a number of articles on ladder filters in various American periodicals (some noted in *TT*).

Certainly the book fulfils the publisher's claim that "anyone interested in the development and construction of Ladder Crystal Filters will find this a handy reference guide . . . [These filters] have had an appeal because crystals of just one frequency are required . . ."

With a wealth of practical information at a readable level packed into its 134 pages, reflecting European as well as American work, it may seem cavil to note the absence of any reference to the article by J A Hardcastle, G3JIR 'Computer-aided ladder crystal filter design', *Radio Communication,* May 1983, pp414-420. Also missing is any reference to *TT* (December 1980, pp1294-5), which described the first switched-bandwidth ladder filter as developed by Dave Gordon-Smith, G3UUR. I was also embarrassed to find a filter design (*TT,* January 1994) by Colin Horrabin, G3SBI, attributed to me, although I acted only as the reporter. Undoubtedly a useful book for anyone contemplating building ladder crystal filters, but not wanting to get too deeply into advanced design complexities.

Interest of late has turned also to the design of high performance ladder filters using low cost ceramic resonators, rather than quartz crystals (see for example the notes from LA8AK and G3JIR in *TT,* April and June 1999). After many years absence from *TT,* Dave Gordon-Smith, G3UUR, returns with a long letter covering *inter alia* his work on ceramic resonator and crystal ladder filters. Since 1983 he has produced an extensive range of high performance ceramic filters on 455, 500 and 560kHz and a few odd filters on such frequencies as 1, 2 and 7.075MHz for special purposes: see **Fig 1.**

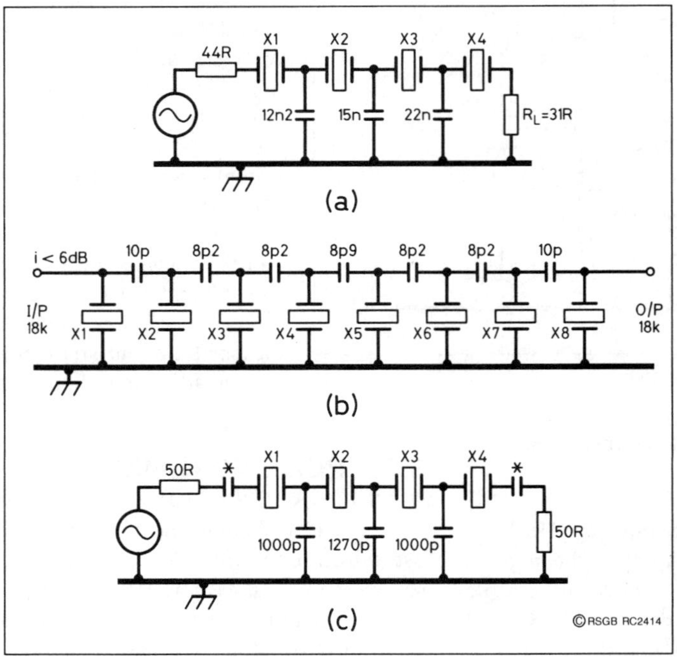

Fig 1: Three of G3UUR's ceramic ladder filters. (a) Original 1983 design for a 700Hz bandwidth Gaussian-transitional CW ceramic ladder filter, using Murata CSB455E resonators on their series resonance at 442kHz. Shape factor 6/60dB, 5.2. Insertion loss, 12dB. (b) USB configuration for 5kHz BW 0.3dB ripple Chebyshev bandpass filter on 1MHz. X1-X8 are Murata CSB1000J ceramic resonators with parallel capacitors to tune them to the correct design frequency. Shape factor 6/60dB, 2.0. (c) Chebyshev 0.1dB ripple design using ceramic resonators. Centre frequency approximately 7.075MHz, BW 50kHz. X1-X4 Murata CSA7.37 used on series resonance. Capacitors marked * selected to tune X1 and X4 to correct design frequency.

He writes: "I started with SSB and CW filters using the LSB configuration and got quite satisfactory results. However, remembering a 'mathematical doodle' that I did back in 1977, when I was deriving a way of calculating the effective inductance of quartz crystals due to the presence of the static capacitance, I decided to try the USB configuration. The doodle predicted that the apparent Q at the parallel resonance would be greater than that at series resonance. Looking at the motional parameters of the 455kHz resonators, it became apparent that they were ideally suited to this configuration, since they produced such convenient values of coupling and tuning capacitors when compared to the LSB configuration. I have always avoided the USB configuration for quartz filters because the termination resistances are usually in the megohms and the coupling capacitors are a fraction of a picofarad - pretty inconvenient values!

"My range of 455, 500 and 560kHz filters have termination resistances between 10k and 100k. The coupling capacitors are a few pF to tens of pF, depending on the bandwidth. They can replace Collins 455 and 500kHz mechanical filters. In all cases I can beat Collins on shape factor and insertion loss. What I have not been able to check is the temperature performance. During the last few years I have begun producing 560kHz filters with bandwidths of 2.5 and 4kHz, for use in R1155 receivers. These have been 8- and 10-pole designs, although I have made up to 14-pole designs on 455kHz and 12-pole designs on 500kHz.

"Earlier contributors have been concerned with the spurious responses of ceramic ladder filters. A look at the spurious responses of commercial ceramic filters show that these certainly require the use of additional LC tuned circuits to reduce spurious responses to acceptable levels. We should not expect otherwise from our discrete ceramic ladder filters. Remember that the anharmonic modes in quartz crystals can produce unacceptable spurious response levels in both ladder and lattice configurations. No designs using quartz or ceramic filters in the IF stages of a receiver should be totally devoid of additional LC filtering. [Consider the use of the six extra LC circuits used in front of the crystal filters in the 1MHz IF stages of the German E52 wartime receiver, shown as Fig 1 in the Septem-

Device & Frequency	Q High	Q Median	Q Low	Motional capacitance	Static capacitance	Motional inductance	Loss (Ω)
Murata CSB455E (455kHz)	3200	2400	1100	19.5pF	275pF	6.6mH	7.7
Murata CSB500E (500kHz)	2910	2150	1250	15.3pF	235pF	7mH	10
Murata CSB560J (560kHz)	1550	1300	950	12.1pF	182pF	7mH	18
Murata CSB1000J (1MHz)	2150	1730	1180	5.9pF	80pF	4.5mH	16

Table 1: Ceramic resonator Q ranges and typical parameter values.

*37 Dovercourt Road, London SE22 8SS.

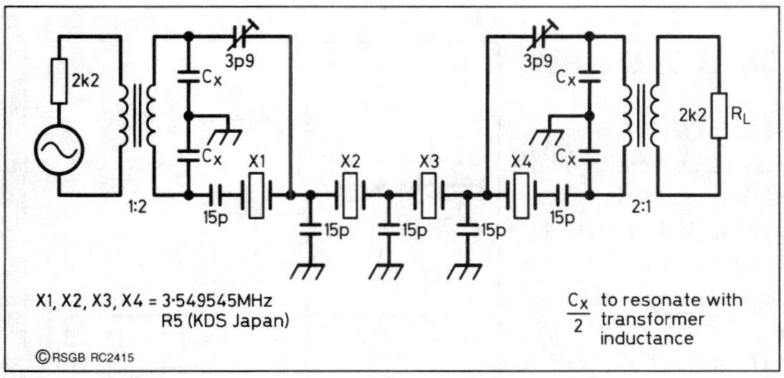

Fig 2: Simple 4-pole ladder crystal filter with four transmission zeroes giving symmetrical response. The adjustable 3.9pF capacitors set the two end section nulls on the low frequency side of the passband. 2.4kHz BW, 70dB ultimate attenuation, 1dB Chebyshev design.

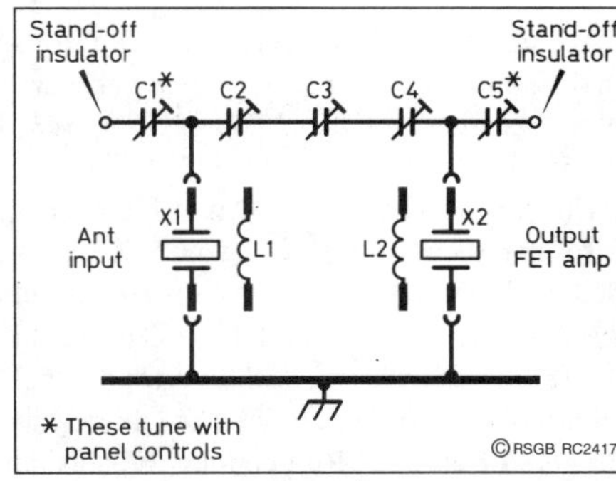

Fig 4: G3IPV's latest single-frequency crystal filter. With two similar-frequency crystals it is suitable for CW, but when coils are used it provides additional pre-mixer selectivity for SSB or AM.

ber *TT* - though this may be overkill! - *G3VA*].

"Ceramic resonators have a role to fill between quartz resonators and LC tuned circuits. I have found the Q of some 455kHz ceramic resonators to be as high as 3000. Some motional parameters and Q of a few ceramic resonators that I have measured are shown in **Table 1**. Incidentally, using one of the very high-Q 455kHz units in a Bridged-T circuit some 15 years ago I achieved the best tuneable passive notch filter that I have ever seen. It has a tuning range of 5kHz centred around 455kHz. The notch is less than 350Hz wide at the -6dB points and over 60dB deep. This, of course, required careful selection of the ceramic resonator. Similarly, for optimum results with the narrower ladder filters, the resonators must be carefully selected. You get out only what you put in!

"With careful measurement and selection, excellent results can be achieved. Variable bandwidth designs in the USB configuration can be easily produced and are more versatile than with quartz, since wider bandwidths can be accommodated. Filters covering 500Hz to 6kHz are quite easy to construct using switched fixed capacitors. This could be done with varicap tuning diodes or air-spaced variable capacitors, but the lower Q of the varicaps compromises the narrow bandwidths and the tracking problem degrades the performance, whichever type of variable capacitor is used, though the performance may still be acceptable for some requirements.

"I feel that the computer-designed symmetrical filter by G3JIR (*TT*, June 1999) departs from the KISS principle and appears extremely sensitive to circuit values. Even if symmetrical at room temperature, I wonder what it would be like 10 or 20° higher or lower?

"I cannot understand the desire of some people for perfect symmetry. I have measured many commercial mechanical and crystal-lattice filter responses and have yet to find one that is perfectly symmetrical. Production tolerances in the component values appear to cause some loss of symmetry and I imagine this varies also with temperature. There are several techniques for improving symmetry of crystal filters that are much less complex than some computer-design programs. The easiest is to make the bandwidth a smaller proportion of the frequency interval between the series and parallel resonance points. This requires the centre frequency to be raised for a given bandwidth and may not be convenient in practice.

"The next approach is to increase the number of resonators. The slope is increased and the asymmetry is masked to a greater extent than with fewer resonators. I have design tables for 1dB ripple Chebyshev filters produced using my own correction technique back in 1977/78 and these extend to 16-pole filters. Filters with 14 and 16 crystals show responses down to 100dB that are for all intents and purposes symmetrical.

"If neither of these approaches is feasible, then circuit modifications may be required to achieve better symmetry. There are a number of ways to do this:

- Use a design that has twice as many crystals as the number required to give the desired stopband and add compensating inductors across half of them to produce nulls on the low-frequency side of the filter in order to balance the effect of the nulls on the high side caused by the static capacitance of the crystals without parallel inductors.
- Use 4-pole hybrid ladder designs with half lattice end sections which overcompensate for the effect of the static capacitance of the first and last crystals; thereby creating two nulls on the high side and two on the low (**Fig 2**). If better selectivity is required, this basic filtering block can be used at two or more points in the IF chain.
- Use a conventional bandpass design with inductor compensation for the static capacitance of every single crystal, and coupling provided by parallel tuned circuits: see **Fig 3**. This method was inspired by Zobel, who was the first to realise the value of compensating the static capacitance of quartz crystals, but also recognised the weakness of this approach in that, away from the passband, the configuration degenerates into a poor Cauer low pass filter with little stopband rejection. He solved this problem by using another parallel tuned circuit at the load. The ultimate attenuation is then determined by the ratio of the two inductors . . . "

I have had to abridge G3UUR's very long letter and omit a number of his diagrams, although hopefully providing enough of his information to arouse further interest in crystal and ceramic filters. G3UUR has a lot more information on filters that he has never published. He writes: "As an academic material scientist, one of my jobs is to publish my research for the benefit of mankind, but I do the amateur radio work for my own amusement and not publishing it is hardly going to do any lasting damage to the development of mankind! Incidentally, during my wanderings around the globe doing X-ray crystallography experiments on large synchrotrons, I have used a number of times the facilities at Brookhaven National Laboratory (BNL) on Long Island, built on the site of the former army camp (Camp Upton) where John Reinartz did his basic army training in 1917.

"A radio club was set up at BNL in 1947 and was very active until the mid 60s. Then interest waned and nothing much changed in the shack until 1992. I discovered the shack in 1986 or 1987 and it was like walking into a time warp. It brought back happy memories of my youth, spending happy hours in that old radio shack 'playing' with the old gear until there was a resurgence of activity and new members decided it was time to re-equip. The old gear was still there at Easter 1992, but when I returned that September it had all gone and the only reminder was a pile of *QST*s that went back to 1947."

There is just space to mention the continuing work of PW Haylett, G3PIV, on signal frequency crystal filters for use in front of a receiver to provide very high degrees of pre-

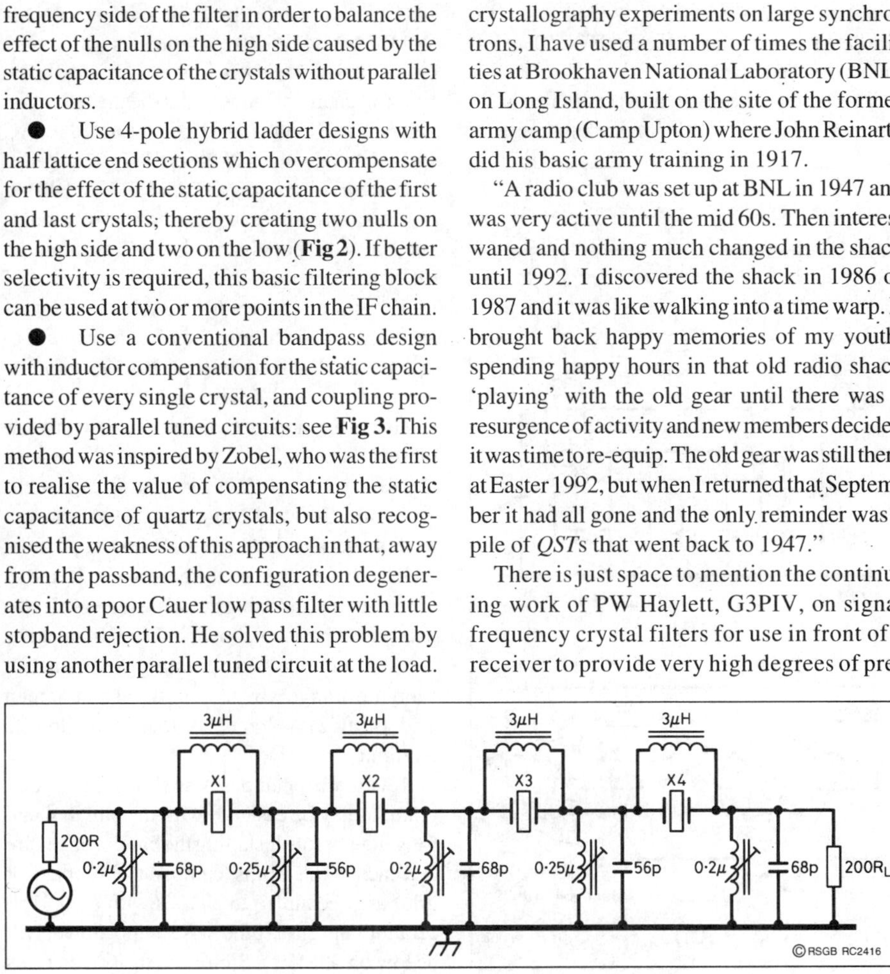

Fig 3: Symmetrical bandpass crystal filter centred on 42.5MHz, with 5kHz bandwidth and >80dB stopband performance. The 6/60dB shape factor is approximately 4. X1-X4 42.5MHz overtone crystals. All adjustable inductors were from Toko 3µH inductors wound on iron dust cores.

mixer selectivity. His previous contributions on this topic include *TT* July 1985, February 1986 (see also *Technical Topics Scrapbook 1985-89).*

He writes: "The filter shown in **Fig 4** comprises two sections: C1, X1, and C5, X2, coupled by three trimmer capacitors normally set near minimum values. The unit is built in a metal box (8 x 6 x 3in deep), with two front panel controls (C1, C5). The output is fed directly to the gate of an FET R-C coupled broadband amplifier and then to the receiver mixer. Crystals may be replaced by inductors (L1, L2), where extreme pre-mixer selectivity is not required. All components are preferably mounted on stand-off insulators. C1 and C5 150pF vane type trimmers. C2, C3, C4 vane-type 0.5 - 50pF. Intended primarily for CW reception, the bandwidth is roughly 1kHz on 3.5MHz, 5kHz on 7MHz and 9kHz on 14MHz. X1, X2 quartz crystals (30pF parallel resonance)."

BLOCKING NON-LOCKED PLL SIGNALS

DAVE PORTER, G4OYX/G3WOF, spotted an ingenious way of preventing non-locked PLL signals from appearing at the antenna terminals of some VHF/FM broadcast transmitters at switch-on. A 555 timer is connected in the emitter circuit of a low-power stage of the 20W amplifier board: **Fig 6** shows how an initial 5-6 seconds delay is obtained, to avoid accidental out-of-band emissions in the initial PLL lock-ing period.

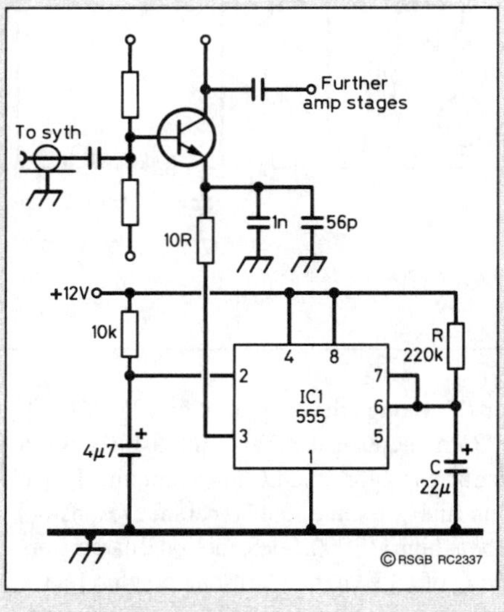

Fig 6: G4OYX draws attention to the use of a 555 timer IC to prevent out-of-band emission from a VHF broadcast transmitter during the initial lock-up period of the PLL frequency synthesizer.

NEW LIFE FOR THE FT-200/250

THE AUGUST *TT* item on restoring the FT-101 by fitting 6146 valves in place of the rarer (and less effective) original 6JS6A or 6JS6C power amplifier 'sweep tube' valves attracted considerable interest. But it also drew attention to an earlier *Radio-ZS* article by Roger Davis, ZS1J/ZS5L, on a similar modification to the once-popular Yaesu FT-200/250 which was also marketed as the Sommerkamp FT-250 and as the Henry Tempo One.

This has had the unintended result that ZS1J has been inundated with letters seeking copies of the original 1995 article. He writes: "I am sending the FT-200 modification for publication, as the cost of sending photocopies to the many amateurs worldwide who have written to me is rather prohibitive." [I know the problem all too well! - *G3VA*]

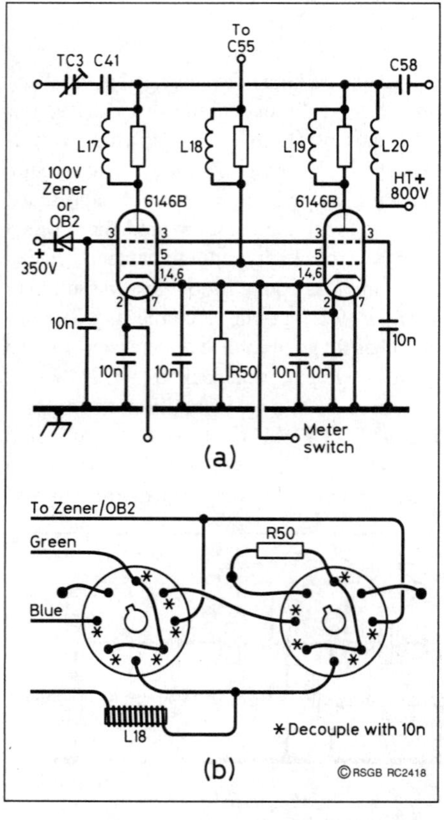

Fig 5: Circuit and wiring diagram of the FT200/250 power amplifier modified to use two 6146B valves. Note that the heaters must be wired in series.

The FT-200/250 was a low-cost SSB/CW transceiver marketed in the early 1970s as a basically valve design (but solid state VFO covering 5 to 5.5MHz). With a 9MHz crystal filter (shape factor 1.62), it covers the pre-WARC bands in 500kHz segments (often only one of the four 28MHz segments was fitted). The transmit chain includes a 7360 beam deflection valve. The FT-250 (£160) incorporated an AC PSU, whereas the FT-200 (£132) required a separate power unit (AC £35, DC 12V £40). It is not surprising that these low cost models became very popular in the 1970s, although often now unused owing to the problem of replacing the 6JS6A or 6JS6C valves. Many are still around and can be successfully revamped to provide a still effective HF SSB/CW rig for the 3.5/7/14/21 and 28MHz bands.

To replace the 6JS6C output valves with a pair of 6146B valves, ZS1J recommends the following: Remove the 6JS6C valves and disconnect all wires from the two 12-pin bases, taking care not to damage any of the decoupling capacitors which are then used on the new bases. Remove the two bases and using the same screws, fit two octal valve bases as shown in **Fig 5(b)**, making sure that the spigots are pointing as shown. Cut short the orange wire used to supply the screen grid (grid 2) voltage, as this is no longer required.

Rewire the octal bases as shown in the diagram. It may be easier to wire the pins 1, 4 and 5 together before installing the bases. Make sure that each of these pins (cathode and suppressor grid) are decoupled to chassis with an 0.01μF ceramic capacitor (otherwise output power will be low on 28MHz). Similarly, the screen grids (pins 2) on each valve must be decoupled, as well as the heater pins.

The new 250V g2 voltage can be obtained by wiring either a 100V 5W Zener diode or an OB2 voltage regulator tube in series with the 350V line which can be taken from the 'cold' end of the RF choke feeding the 12BY7A anode tuned circuits. To modify the neutralising circuit to cope with the lower inter-electrode capacitance of the 6146B valves, connect an additional 200pF capacitor rated at 1kV across the 200pF capacitor marked C40 on the original FT-200 circuit diagram. If this is not available, C40 is the capacitor connected to the cold end of the 12BY7A anode tuned circuits".

ZS1J continues: "In the 1970s, I diagnosed an inherent fault affecting all FT-200, FT-101 and FT-400/500 equipments: the 100pF coupling capacitor from the anode of the 12BY7A driver valve to the grids of the power amplifier becomes 'leaky' or short-circuited, resulting in up to 350V on the grids of the PA. Unfortunately, this normally occurs while switched to the 'receive' mode, with the meter reading 'S' units rather than PA current. As a result, the leaky capacitor immediately destroys the output valves, and in some cases the anode choke and power transformer. If on your unit, this has not yet happened - beware, it will! It is vital that this capacitor is changed to one with a voltage rating of at least 1kV DC - if a suitable capacitor is not available, fit a 1000pF 1kV capacitor in series with the original 100pF capacitor as a DC blocking safety measure.

"The final modification is to the Yaesu power supply. Open it up to expose the printed circuit board carrying the silicon HT rectifier diodes. Remove the four diodes with their parallel resistors and you will then find that the board is marked and drilled to take eight diodes and their resistors. Fit eight 1N4007 diodes with eight 470k resistors and then move the tap on the transformer from the 460V position to the 600V tap; this should increase the HT to 850V. As always, treat such high voltages with caution!

"When ready to test and align the modified set, switch it on to SSB, meter switch to IP, mic gain control to zero. Then, as soon as the set is warmed up, key the microphone and quickly set the anode idle current to 50mA using the bias control on the rear panel. Align the transceiver and neutralise the final amplifier stage as detailed in the original handbook. RF output should exceed 100W on all bands if modified and aligned correctly.

"Should you wish to carry out the modification but have available only the older 6146 valves, you will find that the hole size in the main upper chassis is too small for the valve to pass through. To make them fit, take a file and carefully remove the excess phenol from around the base of the valves until they just pass through the hole. The 6146 will give an identical RF output in this equipment to the higher rated 6146B.

"In use, the modified FT-200 runs much cooler than with the original TV-sweep valves, which were never intended for HF operation.

"Since writing the original article for *Radio-ZS* in 1995, I have modified many FT-200s and have heard of many others undertaking this task - all with entirely satisfactory results. These modifications will provide a new lease of life for equipment which can still hold its own against many modern solid state HF transceivers."

LAP-TOP FUEL-CELLS?

FUEL CELLS SEEM suddenly to have become the topic of the year for many of the professional scientific and electronic journals, rather as they did in the 1950s following the emergence of the Bacon fuel cell, though that never seems to have fulfilled its early promise. *TT* (March and August 1999) has reported on a number of articles in *IEEE Spectrum* covering the prospects of fuel cells for electric vehicles (EV) and for portable and static generators ranging from some 30 watts to 250kW. Another set of three articles on fuel cells has appeared in the prestigious *Scientific America* (July 1999); two along much the same lines as the earlier *IEEE Spectrum:* 'The Electrochemical Engine for Vehicles' by A John Appleby, sub-headed 'Fuel cells can power cleaner buses and cars, but key engineering and economic obstacles will delay widespread adoption of the technology', and 'The Power Plant in Your Basement' by Alan C Lloyd ("In the past, stationary fuel cells were megawatt behemoths, designed for the electric utilities. Now they are being shrunk for homes and other modest applications". But the third article breaks new ground and is of interest to those using hand-held transceivers, etc. 'Replacing the Battery in Portable Electronics' by Christopher K Dyer of Motorola ("Batteries are cumbersome and expensive. Miniature fuel cells could supplant them in cellular phones, laptop computers, camcorders and other consumer products"), introduced as follows:

"Despite major advances in portable electronics, the battery has changed little. Even so, small batteries remain the only choice for consumer products that need up to about 20 watts of power, for everything from toys to laptop computers. But batteries can be heavy and expensive, and they can expire without warning, requiring either replacements (presenting a disposal problem) or recharging (taking hours of precious time). Is there no better alternative? ... the answer may lie with fuel cells ... Theoretically the technology has the same consumer-friendliness as batteries: quiet and clean conversion of a material's chemical energy into electricity. But the real advantage lies in their amazing ability to liberate electrical energy from the hydrogen atom. A fuel cell running on methanol could provide power for up to 20 times longer than traditional nickel-cadmium batteries in a comparably sized package, but at a lower price and for a small fraction of the weight . . . fuel cells do not require lengthy recharging; they can instead be replenished quickly, simply by adding more fuel ... Recent attempts to miniaturise fuel cells could lead to substantial improvements, possibly cellular phones that can run continuously for months on standby power and laptop computers that allow more than 100 hours of operation from a single, compact device. But, as fuel cells shrink, various engineering problems become increasingly acute. Indeed, miniaturisation requires a tricky balance of several factors, including power, size, convenience and cost."

It is noted that hydrogen-based fuels do not develop much more than one volt per cell, compared with the 3 volts or so of lithium batteries, but that because of their lower weight and volume fuel cells nonetheless contain substantially more electric energy in a similar package. Both fuel cells and batteries extract electricity from chemicals at high efficiency, but the fuel cells have a natural advantage in using energy-rich fuels containing hydrogen. "For a given weight, the pure liquid form of hydrogen contains about 800 times the electrochemical energy of nickel-cadmium: See **Table 2**."

Unfortunately, liquid hydrogen has to be kept at a temperature below about -250°C, making it clearly unsuitable for consumer products. However, compounds containing hydrogen can provide an impressive performance. Theoretically, one litre of methanol could supply about 5000 watt-hours, enough to run a laptop computer continuously for more than a week. A comparable volume of lithium-ion, the most energy-dense rechargeable battery, has less than a tenth of that energy. The oxygen required for the fuel cell can usually be taken from the surrounding air.

A miniaturised fuel cell could be replenished quickly, possibly by inserting a new ampoule of methanol or by replacing a cartridge of solid hydride. With the reduction in costs "fuel cells may one day replace batteries for many consumer applications ... the price tag of any fuel cell must also include its manufacturing cost, and therein lies the difficulty. Indeed, designing a miniature fuel cell that can be mass-produced cheaply has been a formidable task. The engineering challenges are enormous, requiring no

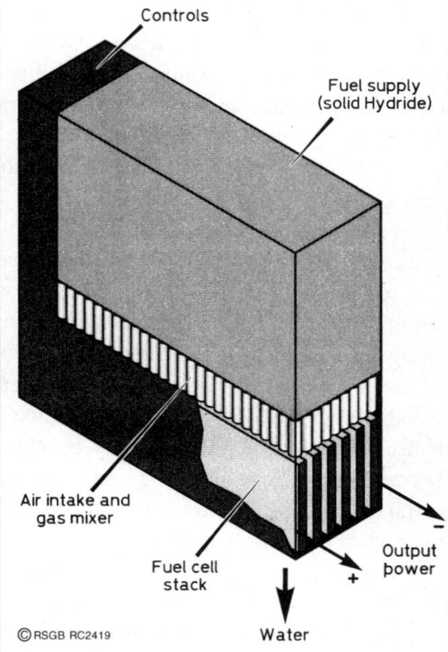

Fig 7: Thin film design opens the way to miniature fuel cell in which power is generated by drawing both hydrogen and oxygen from the same gas mixture, as opposed to traditional designs which require separate supplies. This experimental design comprises a stack of six miniature fuel cells, providing a total of about six volts output. Each cell contains an extremely thin (less than one micron) membrane, forming the electrolyte that separates the cathode and anode. At the anode of each cell hydrogen loses its electrons and travels through the membrane towards the cathode. Oxygen also diffuses through the thin electrolyte, while the freed electrons pass through the external load circuit. At the cathode, the hydrogen ions, electrons and oxygen combine to form water. (Source: *Scientific American*, July 1999)

Fuel Cells	By Mass (per kg)	By Volume (per litre)
Decalin	2400	2100
Liquid hydrogen	33,000	2500
Lithium borohydride	2800	2500
Solid metal hydride	370	3300
Methanol	6200	4900
Hydrogen in graphite nanofibres	Abt 16,000	Abt 32,000
Rechargeable Batteries		
Lead acid	30	80
Nickel-cadmium	40	130
Nickel-metal hydride	60	200
Lithium-ion	130	300

Table 2: Stored electrochemical energy of fuel cells and rechargeable batteries in watt-hours. Notes: Values are theoretical (thermodynamic maximum) except for metal hydride and graphite nanofibres which are for the extractable hydrogen. For hydrogen and hydrocarbon fuels, complete oxidation is assumed. Graphite nanofibres is a recent development currently being investigated, a density of two grams per cc is assumed. For batteries, values are actual available electrochemical energies. (Source: *Scientific American*, July 1999)

small amount of innovation, although, basically, a fuel cell is a very simple device. All that is needed is hydrogen fuel, a supply of oxygen, two catalytic electrodes and an electrolyte. From here, though, complexity creeps in.

"Various miniature fuel-cells are still in the very earliest stages of development, eg **Fig 7**. Although initial prototypes have been built, no one has yet definitively demonstrated a compact device that could be mass produced at a cost significantly lower than that of comparable rechargeable batteries . . . The commercialisation of small fuel cells would accelerate tremendously if scientists could find a better source of hydrogen. Methanol is rich in energy, but its use is complicated by engineering problems, and metal hydrides provide merely an adequate, if unspectacular, storage medium.

"Recently, Terry Baker of Northewestern University announced a form of carbon called graphite nanofibres that boasts almost unbelievable capacity, one gram reportedly able to deliver 10 litres of hydrogen. Such incredible energy density (equivalent to about 16,000 watt-hours per kilogram) would make other materials pale in comparison. In theory, just half a litre could power a 20 watt laptop computer continuously for more than a month!"

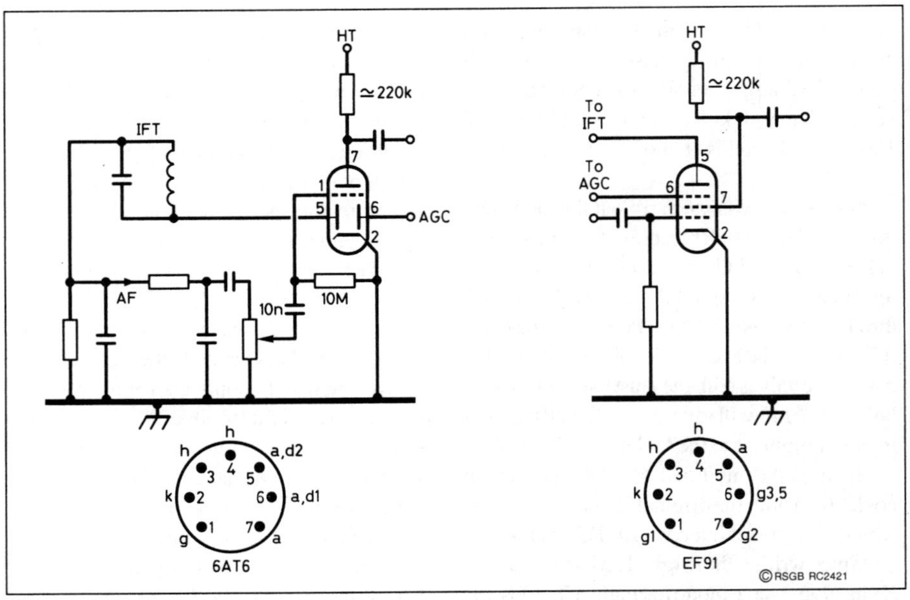

Fig 9: How the EF91/6AM6 pentode valve can be used as a plug-in replacement for the EBC90/6AT6 double-diode-triode. Both valves use B7G bases. Other pentode valves can be used as double-diode-triodes, provided they have the suppressor grid brought out to an external pin rather than connected to cathode, but will require some changes to the wiring of the socket.

PENTODES AS DOUBLE-DIODE-TRIODES

THE SEPTEMBER *TT* item on German technical developments in the 1930/40 era appears to have interested a number of readers. But first I must admit to an error in captioning the illustration of the transmitter-receiver as German Army field equipment. Richard Walker, G4PRI, spotted the error and identifies it as German Naval equipment, as indeed was the impressive Koeln E52 receiver.

RB Kerr, GM4FDT, noted the German military use of the RV12P2000 pentode valve in all stages of many receivers such as the widely used FuG10 aircraft equipment, etc. Its use as a double-diode-triode, for example, greatly simplified the manufacture of valves and the provision of spares. He writes: "I can confirm that a pentode works well in this application, provided the triode is always conducting, ie the valve is used in the small-signal class A mode.

"I first saw the idea in a 1933 *Wireless World*, wherein a screen-grid tetrode valve provided one diode and one triode: **Fig 8**. I have successfully replaced the valve type 6AT6/EBC90 with an EF91/6AM6 in various vintage radios. I have an old radiogram circa 1960 in which some 22 years ago I replaced a faulty 6AT6 with an EF91; and it still works! I also used an EF91 to replace the 6AT6 in an old KB HV40 Band 1 405-line TV. This set used many EF91 valves and by accident I plugged one into the socket intended for a 6AT6 valve. The set worked fine for another 12 years!

"An EF91 can be plugged directly into a 6AT6 socket without wiring changes: **Fig 9**. Other pentode valves such as the 6K7 and 6SK7 etc will also work well, but require some slight rewiring of the valve sockets. I suspect that if the 'triode' were made to oscillate using a crystal etc, an excellent product detector in which the diodes are switched on and off at the crystal frequency would result. Then, stopping oscillation by shorting-out the crystal, would provide AM detection."

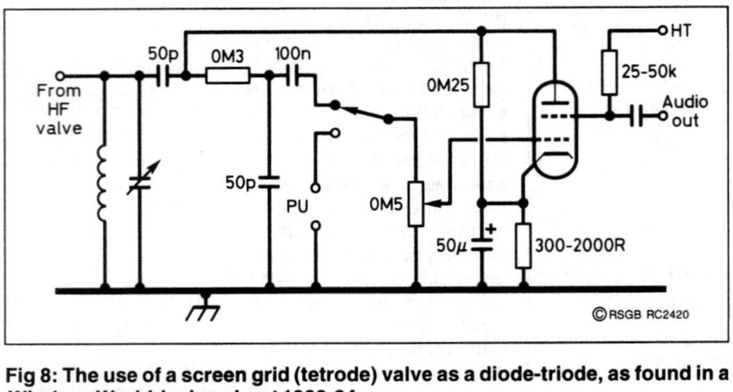

Fig 8: The use of a screen grid (tetrode) valve as a diode-triode, as found in a *Wireless World* design about 1933-34.

HERE & THERE

W G DURHAM, G3DNE, is one of a number of amateurs who have suffered considerable loss of hearing. He has devised a useful way of improving headphone reception, which could be of value to readers who wear the National Health type hearing aids that have a switch facility for use with induction-type systems. He writes: "My hearing loss is of the order of 80dB relative to normal hearing, and I use an aid for each ear. I find using headphones awkward, having to leave the aids in place to maintain a certain amount of built-in equalisation. I have therefore been carrying out some very simple experiments using the induction mode. I now use a simple loop hung around my neck consisting of 10 turns of insulated stranded wire with a series resistor equal to the output impedance of my communications receiver. I have been so pleased with the result that I now use the same system for listening to the sound on my television. For the minimal time it takes to set up, I thoroughly recommend other amateurs with hearing problems to give it a trial. The loop diameter I use is 10 inches, just big enough to slip over my head."

DAVID BUNDAY, G3JQQ, has been using a telecommunications tone coder in conjunction with a hand microphone as a convenient means of quickly tuning up on SSB (see *TT*, October 1998, p42 for an item on two-tone testing with telephone tone pads). However, he has found that the level set for telecom lines a little on the low side. On his 'Audioline' he has found that this can be increased substantially by shunting R2 with 1000Ω. Battery drain of around 3.5mA is not affected. He suggests examining other models if greater output is required.

FRANK VAN VLOTEN, ZR5CG notes that the 'Phantom-coil VXO' described in *TT*, January 1999, pp63-64 appears to use much the same system as the Marconi FMQ system for the direct frequency-modulation of quartz-crystal oscillators for wide-band FM broadcasting. FMQ was described by WS Morley in *Wireless World,* October,1951. A model developed for the original 1951 BBC VHF/FM transmitter at Wrotham used a susceptance modulator to deviate a 3.8033MHz crystal ±1 part in 1000. In other words, to swing a 3.8MHz crystal oscillator over almost 8kHz, followed by a string of frequency multipliers. The patented Marconi FMQ system became widely used in FM broadcast transmitters. ♦

technical topics

by Pat Hawker, G3VA*

FAREWELL PA0KSB, SILENT KEY

THIS IS THE FINAL *TT* of the 20th Century and inevitably ones thoughts turn to what the 21st Century may hold for the hobby which has occupied so many of our thoughts and brought so much interest and pleasure for so much of the present century. As someone born in 1922, I never really contemplated being in a position to be preparing, *Deo volente,* to greet the New Millennium. Still less to be still compiling *Technical Topics*! Indeed, some readers may suspect that my ideas and practices have not advanced much beyond the 1950s when *TT* was launched.

One of the sadnesses of writing a column over a long period is that so many of those on whose contributions it has depended are no longer with us. This has been brought home to me recently by the sad news of the death during September at the early age of 62 years of Klaus Spaargarten, PA0KSB. Regular readers will connect him immediately with the 'Huff & Puff' frequency stabiliser, the original version of which was published first in *TT,* July 1973 and soon attracted attention. The name 'huff and puff' was devised a few months later as a result of a comment from the late Joe Cropper, G3BY (*TT,* October 1973) who likened its action to an "old 'hit and miss' gas engine governor where this lifts one of the valves to reduce speed resulting in a series of thump, thump, gasp, thump, thump, gasp gasp" sounds. So I felt the action of the stabiliser could be compared to such a gas engine huffing and puffing. The name stuck and although this form of frequency stabiliser has since appeared in several modified versions, it is widely still known (even in Holland) as the 'Huff and Puff' stabiliser.

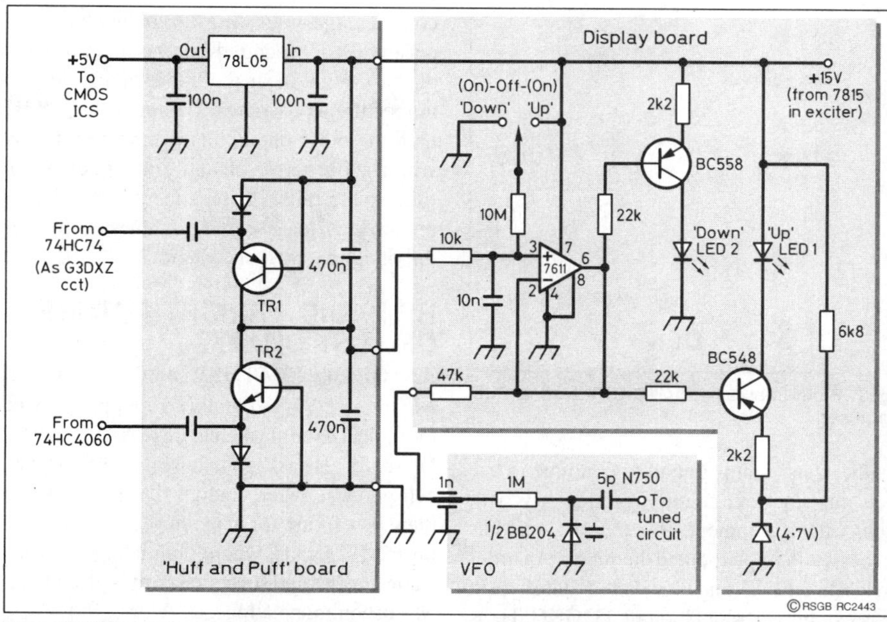

Fig 2: G3GKG's version of the Huff & Puff frequency stabiliser (circuit left of the dotted line) is similar to G3DXZ's (VFO not shown), but note that the 1μF capacitor has been replaced by two 0.47μF (Mylar film or tantalum) and the 'top end' of the integrator circuit goes to +15V. The buffer/display circuitry is built on a small PCB mounted behind the front panel, with legends as shown in Fig 3.

PA0KSB was a truly innovative member of the hobby. Dick Rollema, PA0SE, points out that Klaus wrote over 50 articles for the Dutch VERON journal *Electron*, not to mention his contributions to *TT* and *QEX.* He first attracted my attention with a short article in *Electron* (January 1967) which I picked up and included the circuit diagram in 'A future for the synchrodyne?' (*TT,* March 1967). His five-transistor 3.5MHz SSB/CW receiver was, to the best of my knowledge, the very first published details of a simple solid-state 'direct-conversion' receiver to appear anywhere in the world. A basically similar form of *valve* receiver had been described by James R White, W2WBI, (*QST,* May 1961, pp29-33) as a "Balanced Detector in a TRF Receiver - a novel tuner for 40- and 80-metre CW and SSB" using a 6SK7 RF stage followed by two 6SB7Y as a balanced product detector with a 6C4 oscillator. But it was not until November 1968 that Wes Howard, W7ZOL, and Dick Bingham, W7WKR, in *QST* published full constructional details of a solid-state receiver under the title 'Direct Conversion - a neglected technique' often given the credit for introducing this form of receiver in solid-state form. The advantage over the traditional regenerative detector receiver is that a balanced 'product detector' is far more linear and provides superior detection of SSB signals.

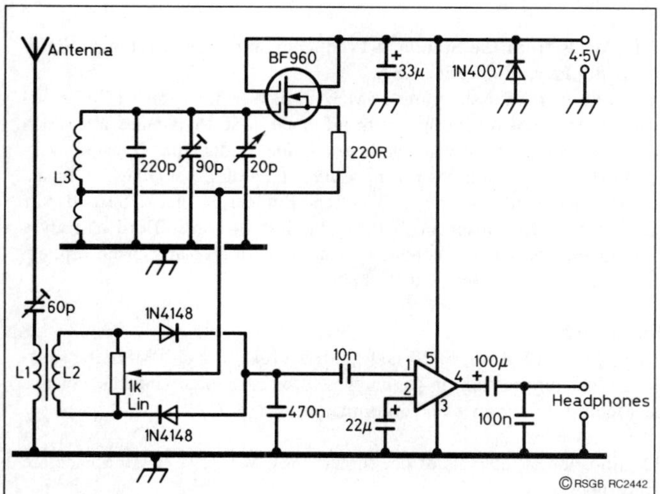

Fig 1: F5LVG's RX-20 simple 7MHz direct-conversion receiver from the 100th issue of *Sprat* dedicated to designs requiring less than 20 components. The antenna trimmer is adjusted for maximum signal. The 1K linear pot is adjusted to minimise detection of AM signals. L1 and L2 are moulded small chokes glued together. L1 22uH, L2 100uH, L3 15t, 8mm diameter, tapped 3t from ground, wound on a section of Bic-type ballpoint pen barrel. IC is a TDA2003 driving Walkman type phones. Oscillator BF960 with 220Ω resistor and NPO 220pF capacitor for increased stability.

Historically, of course, the homodyne or synchrodyne or 'direct conversion' form of receiver dates back to the early years of the century, but by the 1960s had virtually disappeared from the amateur scene. It is perhaps not too much of an exaggeration to suggest that PA0KSB's simple transistorised direct conversion receiver restored to amateur radio the possibility of a low-cost home-constructed receiver capable of reasonably good SSB and CW performance when built by relatively inexperienced enthusiasts. In a sense it proved even more of a landmark development than the Huff and Puff stabiliser. Many versions have appeared over the years, and although the NE602 IC device etc later brought an era of simple-to-build superhet designs, the simple direct conversion receiver lives on: **Fig 1** is the circuit of the RX-20 receiver described by Oliver Ernst, F5LVG, as a contribution to the recent special 100th issue of *Sprat*, devoted to items requiring less than 20 components. Offered as a fun receiver rather than achieving the high performance that can be possible, with care, with DC receivers.

To make one pioneering landmark contribution to amateur radio is worthy of high praise. To make two is beyond praise. Klaus's contributions must be rated very high indeed! RIP, and our sympathies to his wife, son and daughter.

The Huff & Puff stabiliser, despite the widespread use of low-cost fully fledged frequency

*37 Dovercourt Road, London SE22 8SS.

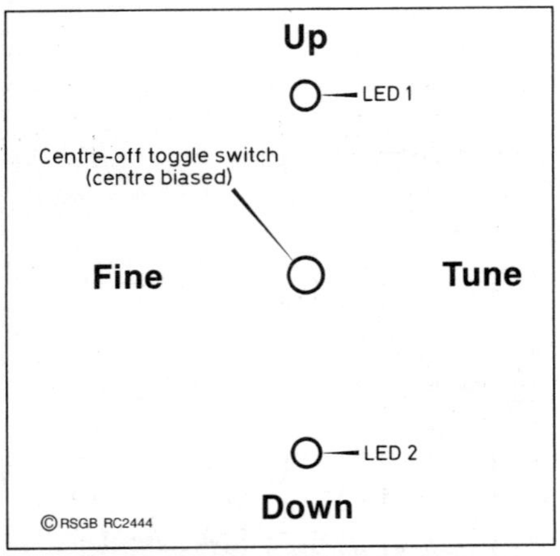

Fig 3: Panel display for G3GKG's Huff & Puff stabiliser.

synthesizers, seems bound to continue in use well into the next century, particularly with older valve equipment. One of those who has successfully implemented the scheme in order to stabilise an existing valve/solid-state hybrid SSB exciter is Brian Horsfall, G3GKG. He is using the system to control an EF91 electron-coupled Hartley type VFO which, as always recommended with this technique, was already pretty stable, but which now, once set, stays literally within a Hz or two over periods of up to four hours.

His circuit (**Fig 2**) combines the simple charge pump integrator developed by Charles Fletcher, G3DXZ (*TT*, September, 1996) with an op-amp integrator buffer which, apart from feeding the varactor diode, also drives a couple of transistors with LEDs to indicate the voltage level. G3GKG writes: "Dividing the integrator capacitance between two half-value capacitors ensures that the output now settles in approximately the middle of the 15-volt range at switch-on. It can actually go down to near zero, but the display guards against letting the control voltage drop below 5 volts, giving a margin which presents the single varactor diode being driven into conduction by the RF in the tuned circuit.

"My VFO tunes 3900 to 4460kHz so, rather than adjust the number of divider stages as suggested, I opted to reduce the clock frequency in the 'H&P' circuit to 13.9MHz, using a surplus crystal already to hand. This maintains the approximate 10Hz steps between lock-up points and provides increments of the same size for fine tuning - a useful adjunct which has gone missing in the later 'H&P' designs but which I have revived (I have incorporated G3DXZ's mathematics into a spreadsheet file which allows for easy manipulation of the various factors).

"The usable control voltage range provides for a total deviation of ± relative to the frequency set by the main VFO tuning knob. By holding or 'flicking' the centre-off, biased-to-centre switch (**Fig 3**) in the required direction it can then be adjusted to within 5Hz of the nominal 'net' frequency - enough to satisfy even the extremely frequency-conscious people I work. The balance of brightness between the two LEDs varies smoothly throughout most of the control range, indicating the voltage at the output and also the required direction of movement of the switch. Near the upper and lower extremes, the appropriate LED is at full brightness with the other one extinguishing completely within about half a volt of the integrator output 'hitting the rails'. In practice, the two LEDs remain at virtually equal brightness as long as the rig is switched on. Magic!"

HF RADIO - BRIGHT FUTURE OR LOST GLORY?

AN OP-ED FEATURE by Randy Pirtle, N6GN, in *QST* (September 1999, p91) raised the topical issue of whether there is "A future for HF radio". He asks: "In these days of Internet, cell phones, pagers and all the rest, how can there be a future for ham radio . . . particularly on the HF bands, where undulating F layers, static crashes and solar storms make for inconsistent communication?"

His conclusion is that: "Long after the Iridium satellite network has brought hand-held cell phones to everyone from the Amazon basin to the polar ice caps, there will still be hams on the HF bands. Technology will undoubtedly improve, but the vagaries of HF communications will remain a welcome challenge. What makes DXing on the short-waves fun is, paradoxically, the very inconsistency that many call its downfall. The sunspot cycle, solar spikes, the solar wind, temperature and weather all make hamming an enduring challenge. Let's stop worrying about the future of HF communication. I, for one, plan to enjoy that crackling music until my last breath."

I believe that many, although not all, would agree that it is the constant challenge of ever changing 'conditions' that makes HF an enduring enigma, constantly bringing surprises both pleasant and otherwise. If it is consistent and reliable world-wide *communication of information* to a specific DX destination that you are seeking, then clearly HF leaves a lot to be desired and is probably not for you. Satellites and Internet are the routes you need. But for anyone who enjoys the unexpected and the sheer fun of experimenting and operating and, yes, collecting QSL cards, there can be few media that beat the amateur HF bands - or for that matter anomalous propagation on VHF, UHF, or achieving good results on LF with under one watt effective radiated power.

Consider the present sunspot cycle. As I write these notes in early October, there seems to be a distinct and unexpected falling short in

DK0WCY	**aurora beacon (DARC) Kiel, Germany**
10144kHz	24-hour continuous service, 30W CW (about 16WPM). Geophysical information every 5 minutes.
3579kHz	0700-0800UTC and 1500-1800UTC (both -1 hour during European summer time). 30W CW. Similar information package to the 10MHz transmissions.
Info format	Kiel K measurement at end of last 3-hour period (updated every three hours). Forecasts of the day's solar activity and state of geomagnetic field. Relative sunspot number R. Solar flux at 10cm, measured at Penticton, Canada. Boulder A index, measured at Boulder, Colorado. Kiel A index, measured at DK0WCY, Kiel. All sections are dated. For explanations see *TT* March 1998, p59 and October 1997, pp80-81 (includes sample transmission). The long dash during each cycle is changed to dots during Aurora warnings. DK0WCY transmissions on either 3.5 or 10MHz are usually well received in the UK.
WWV, WWVH	**Geophysical Alerts from the Standard Frequency and Time Stations (NIST)**
WWV	**Fort Collins, Boulder, Colorado, USA** Continuous 24-hour transmissions on 2.5MHz (2.5kW); 5, 10 and 15MHz (all 10kW); and 20MHz (2.5kW). Geophysical information at 18 minutes after each hour (less than 45s long). Male voice with Amplitude Modulation. Updated every three hours. First part gives solar flux as measured at 1700UTC at Ottawa, Canada. The estimated A-index for Boulder and current Boulder K index. Second part gives solar activity and geomagnetic field for the last 24 hours. Third part gives optional information on current conditions (major flares, polar cap absorption, etc). Final part is a forecast for next 24 hours.
WWVH	**Kauai, Hawaii, USA** Continuoous 2.5MHz (2.5kW); 5, 10 and 15MHz (10kW). No 20MHz transmissions. Geophysical information at 45 minutes after each hour. Female voice, Amplitude Modulation. Format and information as for WWV.
Both stations make voice identification at intervals of one minute (WWV 15s, WWVH 7.5s before the minute). Signals relatively poor in UK.	
NIST	**National Institute of Standards and Technology.** Detailed information in NIST Special Publication 432 (Revised 1990) from Superintendent of Documents, Government Printing Information Service, Springfield, VA22161, USA (32pp large format).

Table 1: Geophysical Alerts DARC and NIST

the anticipated autumn solar flux values over the past few weeks. Thanks to the German beacon station DL0WCY, my daily records show that whereas June saw the flux peaking above 200 during June 25-27 and 30; and then between July 29 - August 3 and again August 24 - 29; September saw nothing above 160 after the first day, with the mid-September peak under 160, and not until mid October did the solar frux briefly reach 200. This was accompanied by an unusually large number of days of geomagnetic disturbances (including a Kiel K rating of 7 on September 27). Does this indicate that Cycle 23 reached its maximum around July/August, or more probably does it mean that Cycle 23, like the earlier Cycle 21, will prove to have two distinct peaks with the second one the greater and then decaying very slowly. Or will this all have been proved wrong even before these comments reach you? Surely, the beauty of it is that after so many years of ionospheric studies, nobody can say for sure what will happen over the next year or two. It is that sense of uncertainty even among the experts that makes amateur radio so interesting - not as a sure-fire means of communication but with a giant ionospheric puzzle in the sky!

One of the interesting features of the DL0WCY beacon transmissions (see *TT*, March 1998, p59) is that they provide the daily Ap ratings for both Kiel, Germany, and Boulder, Colorado. When I first began copying this beacon in 1998, I was puzzled at the very clear difference between the values recorded at each observatory, with the Kiel figure often considerably higher than Boulder. Puzzled, I eventually had a good look at my Atlas, and then the penny dropped. Kiel is at a significantly higher latitude than Boulder, and thus much closer to the Polar Cap Absorption zone. But, since the latitude of Kiel corresponds more closely with that of Southern England, while that of Boulder with Madrid, I now have no doubt that the Kiel Ap and K ratings are more applicable to UK amateurs (particularly those seeking North American contacts) than the widely-used Boulder values. And with DL0WCY transmitting good signals to the UK through daylight hours with information every five minutes, compared to Boulder alerts at one hour intervals and often not receivable in the UK off-air, my choice remains firmly with Kiel!

Brian Bower, G3COJ, has kindly provided a copy of NIST Special Publication 432 (Revised 1990) from which brief details of the WWV, WWVH geophysical service given in **Table 1** have been drawn. The 32-page *NIST Time and Frequency Services* by Roger E Beehler and Michael A Lombardi is available from: Superintendent of Documents, Government Printing Office, Washington DC 20402, USA. It covers much more than the Geophysical Alerts.

'An Introduction to Solar Indices' by Gwyn Williams, G4FKH, appeared in *RadCom*, September 1999, pp44-45. This emphasised that for the best chance of HF DX, look for high levels of SFU (say, 150+) and low levels of K (say, 0-2) staying around these levels for some days. It also seems worthwhile reproducing **Fig4** which last appeared in *TT* in October 1997 and was devised some years ago by John Pegler, G3ENI. In the diagram, the intersection of the flux and geomagnetic activity values determines likely band conditions. Zone A, above normal, 50MHz probably open for some hours; 21, 24 and 28MHz may be open for up to 24 hours. Zone B, above normal, 21, 24 and 28MHz may be open for up to 24 hours. Zone C, bands up to 28MHz open for part of day. Zone D, normal conditions. Zone E, below normal, unsettled. Zone F, below normal, disturbed. Zone G, disturbed, sub-storm level, auroras may form. Zone H, storm level, auroras may form. This chart has proved reasonably reliable over a number of years, but remember that a sudden solar flare, etc may change conditions markedly.

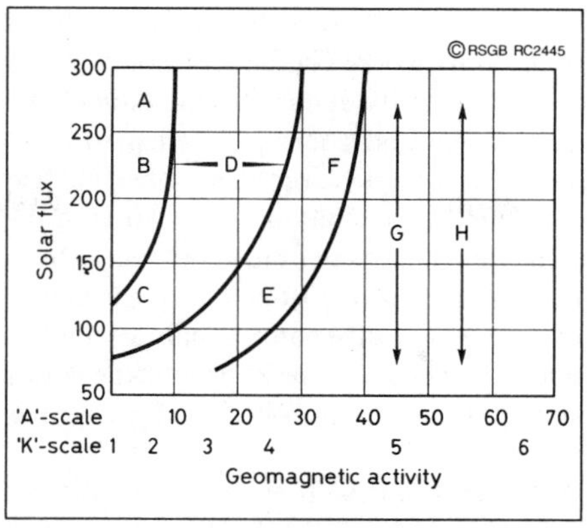

Fig 4: Chart relating solar flux units and geomagnetic activity to likely band conditions, as originally devised by G3ENI. See text for explanatory notes.

TWIN-TEE OSCILLATOR FOR RTTY/PACKET, ETC

RON GOULDSTONE, G3TAG, writes: "I wonder why so many people insist on using the 555 timer chip in their RTTY/PACKET terminals? These ICs are ideal for use in timers etc in purely digital equipment, but not when you want to end up with a clean sine wave. Their output is an asymmetrical square wave that is virtually impossible to clean up with the usual two-section R-C filter shown in most published circuits. The second harmonic of the 1275Hz and 1445Hz tones will almost certainly fall within the passband of the filters of most transceivers, resulting in 'false' signals being transmitted higher up the band. Admittedly, many people now use sound card-generated AFSK and this can, provided the transmitter is not over-driven, give perfect results. However, for those who are still using Versaterm-type units, myself included, I would like to suggest a simple modification. Replace the 555 circuit with a Twin-T oscillator, as shown in **Fig 5.** This gives a slightly clipped sine wave and is very easy to filter. It is no more complex than the dreaded 555 circuit, using only about three more passive components. Take my advice, modify your unit now and win back all those friends on the air you thought you had lost."

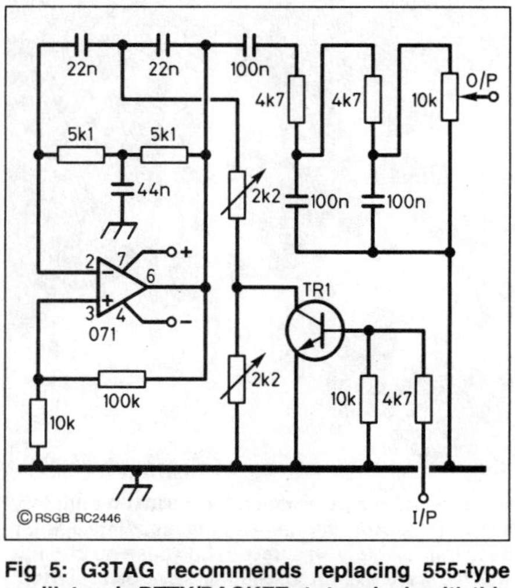

Fig 5: G3TAG recommends replacing 555-type oscillators in RTTY/PACKET etc terminals with this Twin-T oscillator, which provides a much cleaner sine-wave output.

MATTERS ARISING

SEVERAL USEFUL COMMENTS have been received on items in the October *TT*. Denzil Roden, G3KXF, writes: "The Antenna Current Monitoring device (October Fig 4, p62) caught my eye, as I have just retrieved one from my loft junk-box and am checking it prior to presenting it to a local amateur and note that this type of device seems to have disappeared from the amateur handbooks. But there is a discrepancy in Fig 4. The circuit diagram Fig 4(a) is correct but in the pictorial view, Fig 4(b), the connection from the meter minus lead to the earth-tag should be disconnected. The earth connection combined with the capacitance between the toroid and the co-ax inner introduces a voltage parameter. For genuine current measurement, the device needs to be isolated from earth and its stray capacitance to earth kept at minimum. My device uses two thumbscrew terminals, as current measurments/ indications are more valuable in antenna wires and earth leads."

Roy Mander, GW4DYY, noted the item on 'Charging batteries from Solar Cells' including the KISS system used by K5CN (Fig 3, p62) and writes: "My solar panel about 3ft long and 2ft wide produces just over 4A at 12V in very strong sunlight. I use it to power the caravan and to charge batteries as needed. Originally there was no form of regulation and I found that at times the batteries were getting quite a hammering. The published regulators seemed too complicated and I too looked for a KISS solution. Having a Triumph motorcycle, 1960 vintage, with an alternator and rectifier, I knew that the voltage was regulated by a high-current Lucas Zener diode which effectively is strapped across the battery when the ignition is on. This Zener commences to conduct at 14V and progressively passes more current until at about 15V it is passing about 5A, bypassing the battery and effectively limiting the charge to the battery. I have used this method of regulation for a few years with total success. The bike in fact has had a dryfit battery for the last four years and is still going strong, so the Zener does do its work. The only problem is that these diodes cost about £30 (including VAT and postage) and require a good heat sink, but totally outweigh the problems when you get into sophisticated circuitry. I also used the same Zener diode to protect my radios from overvoltage, having fitted one to a heat sink in a box between the power supply and the set, with a fuse between the Zener and the power supply. In this case only a small heatsink is required."

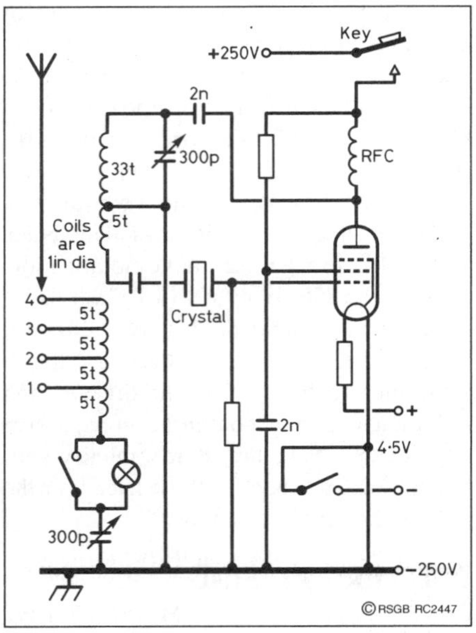

Fig 6: Circuit diagram of the German agent-radio (Afu) transmitter shown in the accompanying photograph and designed to be carried in a leather case slung over the shoulder, with the batteries in another. Antenna coupling coil taps allowed a rough match for different antenna lengths, the pilot bulb indicator lamp showing when tuned. Frequency range was 4 to 8.6MHz.

In respect of the hints on using portable power generators, GW4DYY wonders what is the 'petrol stabiliser' that NT0Z recommends. He suggested that after use of a PPG, a small amount should be added to keep the fuel from oxidising and gumming-up the carburettor. Is this something available only in the USA, or is it known here under some other, possibly proprietary, name?

Dr Dick Biddulph (formerly G8DPS but now with the brand new call M0CGN) adds to the 'Twelve Commandments' of PPG operation a thirteenth: "There is another rule which is regretful but necessary, viz: if using more than one generator, and one stops, investigate at once. A member of the Wimbledon Club waited a few minutes, then found the generator had gone walkabout! The insurance company was not best pleased!"

On the October topic of 'RF Voltage Probe', Dick writes: "Since only a qualitative reading is necessary, I use an LED and a single-transistor amplifier as an indicator. The LED is attached to my spectacles by means of a small paper clip, so I don't have to take my eyes off the work to see if there is RF about. I use the same system for a continuity tester."

1940 photograph of a British NCO examining Afu equipment taken from a German radio-agent on a military reconnaissance mission. The three-man team landed on the south coast during 1940 to 'broadcast' intelligence back to the Abwehr control station near Hamburg, but were immediately arrested. In this case no attempt was made to 'turn' the team into double agents and all three were executed. Similar low-power battery transmitters were also used with 'straight' regenerative-detector receivers for two-way contact with Hamburg. The photograph was originally published in *Wireless World* in 1941.

HERE & THERE

THE BBC2 TIMEWATCH programme 'The Spies Who Fooled Hitler' broadcast on 2 October 1999 was certainly watchable, though it suffered from the usual TV history-documentary drawback of using archival clips and stills that had no accurate relevance to the story, and some were misleading. But it did include a previously unseen clip from a 'home video' made by the late Ronnie Reed, G2RX. Ronnie was one of the first RSS/MI5 operators to supervise or take over the sending of cipher messages supposedly from German radio-agent transmitters (photo left and **Fig 6**) in the UK, but who in reality had been turned and were under the control of what became the XX ('Double Cross') Committee. I wonder if anyone knows where the full tape by G2RX can be seen? Two other amateurs, also no longer with us, shared this task: Russell Lee, G6GL, and Stan Riesen, G5SR. But the most annoying feature of the programme - at least to surviving former members of the wartime Radio Security Service - was the absence of any mention of that organisation which not only supplied the supervising operators but was responsible for intercepting the radio messages between the Abwehr outstations in France, Spain, Portugal etc and Germany. The Abwehr and the SD did their best to make this as tricky and difficult as possible. These intercepts were vital in monitoring the acceptance or otherwise of the deception plans and the 'chicken feed' disinformation transmitted from the UK, as well as providing a detailed study of the inner workings of the Abwehr. As usual, the programme assumed that Bletchley Park was responsible not only for decrypting the Abwehr traffic but apparently also plucking it out of the ether! Digging deeper into the true history might have shown that in 1939-40, BP were most reluctant to have anything to do with RSS Abwehr intercepts, that the original decryption of agent-traffic was made by RSS, and that Hugh Trevor Roper (Lord Dacre) was threatened with court-martial for distributing some of it!

New Scientist reports that NASA lost its $125-million Mars Climate Orbiter spacecraft as a result of a mistake that would shame a first-year physics student - failing to convert Imperial units to metric. The problem arose because propulsion people talk in pound-seconds of thrust and navigators talk in newton-seconds. No wonder one still hears that old saying: "Amateurs built the Ark - professionals the Titanic".

CONJUGATE MATCHING: QUESTION OF DEFINITION?

WHILE I FEEL that the topic of applying conjugate matching theory to obtaining maximum power transfer from a tuned power amplifier has by now had more than sufficient airing in *TT* (May, July and September 1999) it continues to generate correspondence that cannot be ignored. In a letter to the Editor of *Communications Quarterly*, Jack Belrose, VE2CV, stresses "The Conjugate Match Theorem is simple and absolute - when the energy being transferred across any linearly behaving connection cannot be increased simply by changing the impedance of either the source or load a conjugate match exists. There is no wiggle room."

To justify this statement he writes: "It is positively uncanny how easy it is for some people simply to ignore experimental evidence staring them right in the eye when one of their pet understandings is in jeopardy. Without exception, all concede that amateur (Class B and C) amplifiers typically operate at greater than 50% efficiency. They will also agree that it is common to tune for a power peak. They then wiggle and squirm to avoid agreeing that the tuning process is simply matching two impedances to a common conjugate. Their reason is that their 'resistance' (*IEEE Definition 1* for dissipative loss resistance) precludes the higher than 50% efficiency. However, the *IEEE Definition 2* for 'resistance' defines it as the 'real part of impedance'. Impedance is the ratio of voltage to current and so this impedance can be a non-dissipative parameter. To source impedance of Class B and C amplifiers at the input of the pi-network is the ratio of the sinusoidal plate voltage to the sinusoidal plate current appearing at the input of the network, and since power is not dissipated there, this impedance is a non-dissipative source impedance. The fact that there are these two separate and mutually exclusive definitions of the word 'resistance' doesn't seem to matter... The transmitter design engineer does not need to worry about conjugate match if that is his wish. He designs the amplifier to work into an optimum load impedance, and, having accomplished this he can then design an impedance matching circuit to match this to 50Ω - a dummy load for purposes of tuning which dissipates all of the power. The manufacturer sells the transmitter, which could be pre-tuned (commercial transmitters), and the radio operator tunes his antenna with an SWR bridge, so that it looks like 50Ω. He connects the transmitter and it works.

"But this narrow-minded view does not tell us anything about the principles of matching; it does not reveal that in fact we achieved what we achieved through a conjugate match; it tells us nothing about maximum power transfer; and this limited view tells us nothing about how an ASTU works, achieving zero mismatch loss by the reflection gain obtained with the ASTU matching network - all of which are important for our understanding of this complex subject... This simply amplifies what we said at the outset, that the IEEE Definition 2 of resistance defined as the real part of an impedance associated with transfer of power through networks is almost totally unrecognised and misunderstood by a majority of RF engineers. An understanding of this concept by them provides the basis for understanding why an RF power amplifier can deliver maximum available power into a conjugately matched load while achieving efficiencies significantly greater than 50%."

So it would all seem to boil down to which of the 'two mutually exclusive' *IEEE* definitions you use. In a later letter to *TT*, VE2CV writes: "Your correspondents all seem to be under the impression that the output impedance [of a power amplifier] is a loss resistance. The effective output impedance of an amplifier, the impedance associated with the generation of power, is a dissipationless resistance. We never said that this impedance was invariable, independent of output power. Dissipationless impedance is a concept that is misunderstood, or never even thought about by most radio engineers. Impedance is a very important concept which applies to all sources of power generation."

Tony Harwood, G4HHZ, sticks by the view that the output *resistance* of the PA must be low compared to the load seen by the transmitter [though this statement need no longer contradict VE2CV's views - *G3VA*]. He writes: "I would like to make an observation and raise a query, both of which are related. The observation is that the active device in a PA does not itself generate power, it switches power from the power supply (which should itself have an extremely low output resistance) at the required frequency, and hence produces an RF output. In this process the active device dissipates power and contributes to the output resistance of the transmitter. However, it is only that power dissipated at the operating frequency which affects the output resistance. Other losses at DC and harmonics are simply losses to be taken into account when considering the overall efficiency of the PA, which is different to its efficiency as an RF generator at the operating frequency.

"My question is: How is the RF output impedance calculated? I have consulted a number of the standard textbooks, but none show how this is done. My own gut feeling is that in the case of an amplifier where the angle of flow is small, such as a class C CW transmitter, then the output impedance becomes more and more dependent on the ratio of the loaded to unloaded Q of the tank circuit as the angle of flow decreases; but I cannot see how the characteristics of the active device or the PSU affect it. Perhaps someone with a better understanding of the topic would care to comment."

Since in an ideal Class E or F amplifier, the output device acts purely as a switch, then the output *impedance* must clearly be determined by the Q of the tank circuit, and this will vary with the degree of loading, so that we should consider this as a *dynamic impedance* rather than a fixed value. I seem to recall that at one time we were advised to aim at a loaded Q of 12.

My real hope is that we can now give the topic of conjugate matching a rest!

BALANCED SWITCHING MIXER

P GOODSON IN the 'Circuit Ideas' feature of *Electronics World* (July 1998) points out that the conventional dual-gate MOSFET mixer can be improved by changing to the balanced switching mixer arrangement shown in **Fig 7**. He writes: "The BF981 MOSFET is now used as an amplifier, allowing it to be biased for best noise figure. TR2, 3 provide the switching mixer. L1 with C4 resonate at the IF and R4 represents the IF load. Local oscillator input must be about 20mV pk-pk from a low impedance such as 50W and L1 is bifilar wound on a ferrite core for optimum balance. This arrangement provides higher gain, lower noise figure and improved third-order dynamic range than the conventional single dual-gate mixer. ♦

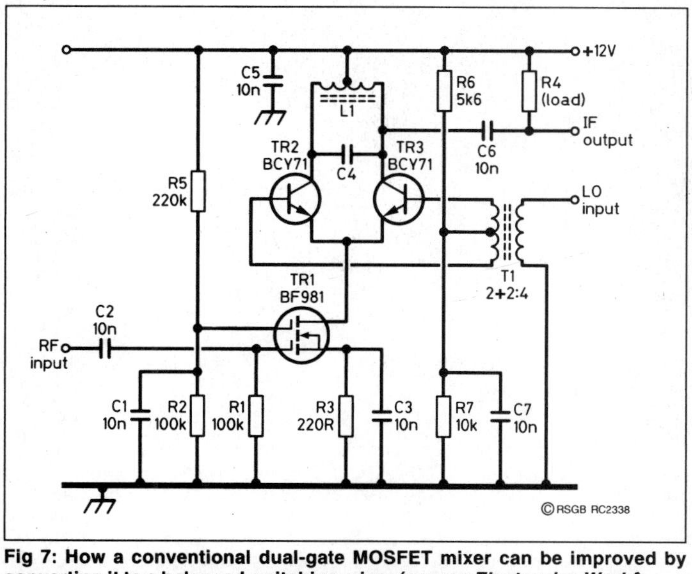

Fig 7: How a conventional dual-gate MOSFET mixer can be improved by converting it to a balanced switching mixer (source *Electronics World*).

Corrections

IN ADDITION TO corrections and updates appearing within the pages of *Technical Topics* (indicated by corr after the page number in the index), the following appeared in RadCom under the heading *Technical Correspondence*.

Two for the Price of One (page 179)

Dropping part of a sentence near the top of the page may have misled readers. The sentence beginning: "The SWR at 3800kHz", relating to G3HQX's adjustments to the Chelcom HFV1 antenna, should have continued: "was reduced from 2.3:1 to 1.8:1 and at 3500kHz increased from 1.35:1 to 1.45:1."

Tuned Loop Transmitting Antennas (page 183)

The IEE Conference Publication referred to should have been No 441.

Tuned Transmitting Loops (page 201)

Denis Walker, G3OLM, commented that: "The diagram shown as Fig 4 has been extensively reproduced, but it appears odd that the coupling loop is only half electrostatically screened and I suggest that the screen should extend over the full circumference of the coupling loop to provide a proper Faraday screened coupling loop, along the lines of that which is commonly employed as a 'braid breaker' to counter TVI. The consequence of failing to take this precaution will be to render the coupling loop itself capable of radiating E/M, whereas I feel it should be contributing solely magnetic coupling to the main loop by transformer action. "

Variable Bandwidth CW Filter (pages 231-232)

In the caption accompanying Fig 8, Rf1 and Rf2 should be 5k (pre-set to 2k) and Rb1 and Rb2 should be 1M.

Replacing the SL6270 (page 236)

David Lauder, G3SNO, wrote: "The circuit won't work because the electrolytic capacitor on the gate of the FET is the wrong way round. I have checked the original *Electronics World* article, and it was wrong there, too. The gate of the FET is driven *negative* by the diode, so the positive side of the capacitor must be grounded. The gain of the gain-controlled amplifier stage cannot decrease below 1, even when the FET is turned completely off, so an input of 4V would give an output of 4V, not 200mV. The 2N3819 FET has a very wide rang of Gate-Source cut-off voltage. This parameter can vary from 0.8V to 8.0V, and it directly affects the level at which the AGC circuit establishes. I suggest the J309 or J310 FETs as an alternative, as they have a much smaller spread of characteristics."

Cleaning up Faults (page 266)

In the soldering hint by KE4TFE (column 3, half way down) the addition of the diode will reduce the power to a half, not a quarter.

Camp/Mobile Antennas (page 271)

Larry Parker, VE3EDY, added: "There was a slight error in the description of the antenna I have made which tunes all frequencies. I *do* incorporate a feed-point matching device, in the form of a toroid to step up the impedance ration of the feed point to 50 ohms."

Here and There - Terion (page 281)

David Lauder, G0SNO wrote: "A reference was made to a paper presented by representatives of Flash Comm / Terion at the IEE Colloquium on Frequency Selection and Management Techniques for HF Communications, 29-30 March 1999. The *TT* text refers to Fig 11. This figure was in fact rfedrawn from a paper by J Moritz, D Lauder and Y Sun titled 'Measurement of HF Antenna Impedance', presented at the same IEE conference. Our paper relates to an instrument for measuring the impedance and phase angle of an HF antenna from 1.8 to 30MHz using very low power (less than 1microwatt)." The Editor added: "The 'Measurement of HF Antenna Impedance' text is included in this month's TT [page 280 in this book] and the diagram below shows what should have been Fig 11 on page 275."

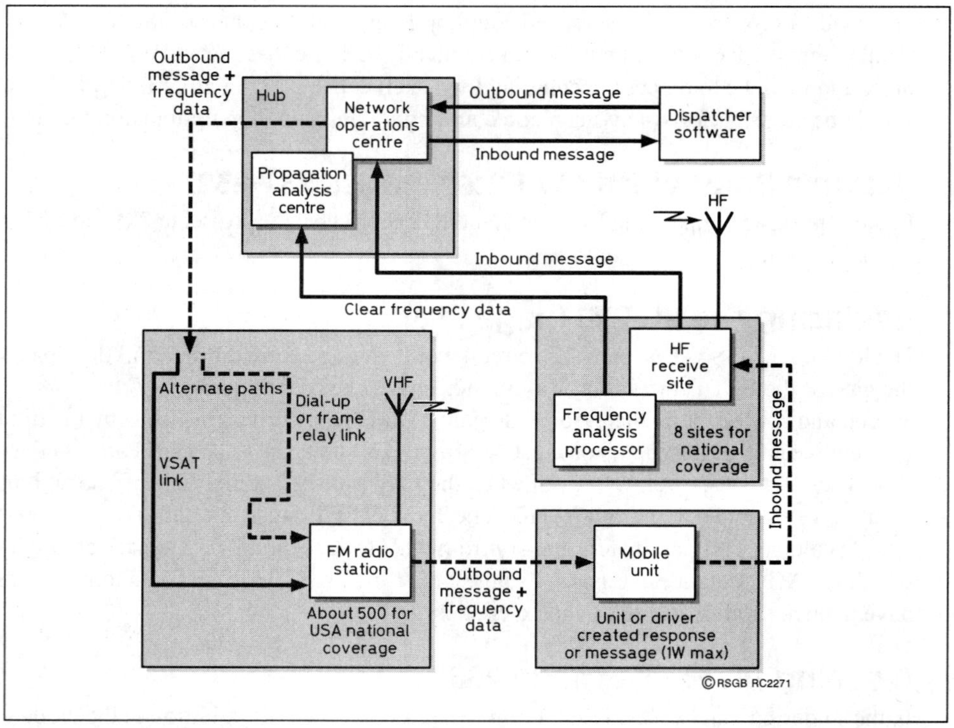

The Terion HF/VHF system for short data messages to and from mobile units throughout the USA, using 1W ERP HF on the mobile units and receiving data from FM broadcast transmitter SCA sub-carriers.

More on Post-Mixer Amplifiers (page 286)

The number of turns on the AG4R amplifier shown in Fig 7 is incorrect. There should be 11 turns nearest to the transistor, and the 4t close to the positive rail.

Index

The suffix 'n' denotes that the item will be found under the 'Pot Pourri', Here and there', 'feedback', etc headings. 'Corr' denotes that a correction to the item will be found at the page indicated.

About the author

PAT HAWKER was born in Minehead, Somerset in 1922. As a schoolboy he began a lifelong interest in radio, obtaining an "Artificial Aerial" licence (2BUH) in 1936 and a transmitting licence (G3VA) in 1938. Early in 1940 he was recruited as a "Voluntary Interceptor" (VI) by the Radio Security Service, leading in 1941 to enlistment for Special Duties in the Special Communication Units, nominally part of the Royal Corps of Signals, in reality part of Section VIII of the Secret Intelligence Service (MI6). From 1941 to 1943 he worked as a radio intercept operator for SCU3 at Hanslope Park covering German Military Intelligence (Abwehr) networks. He then transferred to SCU1 at the Weald control station for two-way clandestine radio links with France and Belgium, until the formation of SCU9 in preparation for the Normandy campaign, serving in France, Belgium and Holland and later in Germany, including a spell on loan to the Dutch Bureau of Intelligence as the war in Europe ended.

After the war he became a book editor and technical journalist, including periods with the RSGB and then Gerorge Newnes Technical Books where he edited and compiled the *Radio & Television Engineers Reference Book* and the early volumes of *Radio & Television Servicing* and many other books on electronics. From 1963 to 1968 he was Communications Editor of *Electronics Weekly.* covering the early development of satellite communications, broadband ocean cables, broadcasting including the controversy over the choice of the PAL colour television system etc. He broadcast frequently on BBC World Service from Bush House and in 1968 he joined the Engineering Information Service of the Independent Television Authority (ITA) - later Independent Broadcasting Authority (IBA) - based in Knightsbridge, London where he remained until his 'retirement' in 1987. He has contributed widely to the electronics and broadcasting press including *Wireless World (Electronics World), Everyday Electronics, Radio Bygones, Morsum Magnificat* etc and edited for the RSGB some 15 editions of *A Guide to Amateur Radio,* six editions of *Amateur Radio Techniques* and the HF Receivers section of the 3rd to 6th editions of *Radio Communication Handbook.* He is active on the HF bands on CW and SSB, working over 100 countries in most years in the 1990s.

FURTHER READING FROM THE RSGB

Technical Topics Scrapbook 1985-89
Technical Topics Scrapbook 1990-94

Two collections of all the very popular Technical Topics articles by Pat Hawker, G3VA, published in the RSGB members' magazine *Radio Communication* (RadCom). Pat has produced a wealth of ideas, modifications, and tips for the radio amateur for many years, and these books contain all the words, pictures and line drawings from this most popular column. The blend of clippings from other publications and contributed material is linked by his own unique commentary, enriched by a lifetime of being interested in the technical aspects of radio, both professionally and as a rado amateur. The pages are presented exactly as they appear in *RadCom,* but an overall index has been added. These invaluable collections of experimental antennas, circuit ideas and radio lore are a must for anyone keen on radio and electronics. (each book 320 pages)

RadCom on CD-ROM

Each year since 1996, the RSGB's members' magazine has been reproduced on a fully searchable CD-ROM. Every editorial page can be searched for any word or phrase, and any page can be viewed on screen or printed out exactly as it appeared in the magazine. A copy of Acrobat Reader for Windows 95 and 98 is supplied. (CD-ROM, please specify year required)

RSGB Journal Archive CD-ROMs

All of the RSGB's magazines from 1925 to 1996 are being archived on CD as Adobe Acrobat files (Reader for Win95 and 98 supplied) in sets of two or three CD-ROMs. The sets are:
Set1 Vols 67 - 71 RadCom Jan91 - Dec 95
Set 2 Vols 62 - 66 RadCom Jan 86 - Dec 90
Set 3 Vols 57 - 61 RadCom Jan 81 - Dec 85
Set 4 Vols 52 - 56 RSGB Bulletin Jan 76 - Dec 80
Set 5 Vols 46 - 51 RSGB Bulletin Jan 70 - Dec 75
Set 6 Vols 40 - 45 RSGB Bulletin Jan 64 - Dec 69
Set 7 Vols 29 - 39 T & R Bulletin Jul 53 - Dec 63
Set 8 Vols 15 - 28 T & R Bulletin Jul 39 - Jun 53
Set 9 Vols 1 - 14 T & R Bulletin Jul 25 - Jun 39
Each set has a composite searchable index covering all of the CDs in the set. Every page can be viewed on screen or printed out. The sets are being published as they are completed, and the entire project will be complete by the end of 2001. (CD-ROM. Call or e-mail to check for availability)

Backyard Antennas

Antenna guru Peter Dodd explains how, by using a variety of simple techniques, it is possible to achieve very high performance from a compact antenna. Also detailed is how to make an antenna efficient on several bands at once.

The book covers end-fed and centre-fed antennas, rotary beams, loops, tuning units, VHF/UHF antennas, antenna and mast construction, transmission lines, and how to estimate and measure the performance of your antenna.

Whether you have a house, bungalow or apartment, *Backyard Antennas* will help you find the solution to radiating a good signal on your favourite band. (208 pages)